THE MOLECULAR AND CELLULAR BIOLOGY OF THE YEAST *SACCHAROMYCES*

CELL CYCLE AND CELL BIOLOGY

COLD SPRING HARBOR MONOGRAPH SERIES

The Lactose Operon
The Bacteriophage Lambda
The Molecular Biology of Tumour Viruses
Ribosomes
RNA Phages
RNA Polymerase
The Operon
The Single-Stranded DNA Phages
Transfer RNA:
  Structure, Properties, and Recognition
  Biological Aspects
Molecular Biology of Tumor Viruses, Second Edition:
  DNA Tumor Viruses
  RNA Tumor Viruses
The Molecular Biology of the Yeast *Saccharomyces:*
  Life Cycle and Inheritance
  Metabolism and Gene Expression
Mitochondrial Genes
Lambda II
Nucleases
Gene Function in Prokaryotes
Microbial Development
The Nematode *Caenorhabditis elegans*
Oncogenes and the Molecular Origins of Cancer
Stress Proteins in Biology and Medicine
DNA Topology and Its Biological Implications
The Molecular and Cellular Biology of the Yeast *Saccharomyces:*
  Genome Dynamics, Protein Synthesis, and Energetics
  Gene Expression
  Cell Cycle and Cell Biology
Transcriptional Regulation
Reverse Transcriptase
The RNA World
Nucleases, Second Edition
The Biology of Heat Shock Proteins and Molecular Chaperones
Arabidopsis
Cellular Receptors for Animal Viruses
Telomeres
Translational Control
DNA Replication in Eukaryotic Cells
Epigenetic Mechanisms of Gene Regulation
C. elegans II
Oxidative Stress and the Molecular Biology of Antioxidant Defenses

THE MOLECULAR AND CELLULAR BIOLOGY OF THE YEAST *SACCHAROMYCES*

CELL CYCLE AND CELL BIOLOGY

Edited by

John R. Pringle
University of North Carolina, Chapel Hill

James R. Broach
Princeton University

Elizabeth W. Jones
Carnegie Mellon University

Cold Spring Harbor Laboratory Press
1997

THE MOLECULAR AND CELLULAR BIOLOGY OF THE YEAST *SACCHAROMYCES*

CELL CYCLE AND CELL BIOLOGY

Monograph 21

Printed in the United States of America
Book design by Emily Harste

ISBN 0-87969-356-8 (cloth : v. 3)
ISBN 0-87969-364-9 (paper : v. 3)
ISSN 0270-1847
LC 97-68084

ISBN 0-87969-363-0 (pbk. : v. 1)
ISBN 0-87969-355-X (v. 1)
ISBN 0-87969-365-7 (paper : v. 2)
ISBN 0-87969-357-6 (cloth : v. 2)

Paperback Cover: Organization of subcompartments of the yeast nucleus as visualized by indirect immunofluorescence. (For details, see Wente et al., p. 471, this volume.)

All Cold Spring Harbor Laboratory Press publications may be ordered directly from Cold Spring Harbor Laboratory Press, 10 Skyline Drive, Plainview, New York 11803-2500. Phone: 1-800-843-4388 in Continental U.S. and Canada. All other locations: (516) 349-1930. FAX: (516) 349-1946. E-mail: cshpress@cshl.org. For a complete catalog of all Cold Spring Harbor Laboratory Press publications, visit our World Wide Web Site http://www.cshl.org/

Contents

Preface

In 1981 and 1982, Cold Spring Harbor Laboratory published the two-volume monograph *The Molecular Biology of the Yeast* Saccharomyces. These volumes have been very successful, in part because they were produced just at the cusp between two eras. At the beginning of the 1980s, we were not ignorant of yeast biology—decades of work by physiologists, biochemists, microscopists, and classical geneticists (not to mention brewers and bakers before them!) had described many phenomena, outlined many pathways, defined many problems, and even begun to elucidate some molecular mechanisms—and the reviews in the 1981/1982 volumes captured this accumulated information in a way that makes them useful references even today. However, it was also clear that we were then in the early stages of a revolution in which the application of rapidly developing molecular genetic techniques would dramatically increase our understanding of yeast biology, and, in particular, of the molecular mechanisms of the phenomena and pathways that the 1981/1982 reviews so ably described.

In the event, this revolution moved faster and farther than even the most optimistic could have anticipated, and, by the end of the 1980s, it seemed clear that there was a need for a new monograph to supplement (not replace) the earlier one by summarizing the progress that had been made and redefining the problems that lay before us. With this goal, we undertook to produce such a monograph, in three volumes, with the help of a cohort of expert authors. We were not so young and foolish as to believe that the goal would be easy to achieve, but we did not fully appreciate how difficult it would be to hit the rapidly moving target that is yeast molecular and cellular biology. After some nontrivial agonies,

Volumes 1 and 2 were duly published in 1991 and 1992 and have, we hope, been useful. However, the present volume, covering most of the cell biology, has really been a killer. During its long and frustrating gestation, some of the authors originally scheduled have dropped out and others have been recruited (sometimes more than once). Chapters have been written, updated, and updated again. For a time, the project was so bogged down that it began to appear that it might never actually be completed. Thus, if this volume indeed proves useful to the community, we will all owe our thanks to John Inglis and Nancy Ford at Cold Spring Harbor Laboratory Press for sticking with it for so long.

Why was the production of this volume such a long and arduous struggle? Of course the human frailties of (some) authors and editors have played a role. However, we think that the deeper and more cogent reason relates to the nature of the field itself. Yeast cell biology is progressing at a truly dizzying pace and refuses to stop, or even slow down, long enough to make it easy to write a comprehensive and thoughtful review. This exhilarating but exhausting pace has been driven in part by the almost unbelievable improvements in the power of our tools during the past 15 years. Think for a moment of where we were in 1982: We did not have synthetic-lethal or dosage-suppression screens, much less two-hybrid screens; no immunofluorescence had been done on yeast, much less immunoelectron microscopy or observations of GFP-tagged proteins in living cells by computer-enhanced video microscopy; DNA sequencing was in its infancy, and Maynard Olson's proposal to produce a physical map of the genome had barely been made, much less ridiculed for its lack of realism (or, as it now has been, completed and then some); PCR was not yet even a gleam in the eyes of its inventors; we were happy to do even the crudest of reverse genetics experiments, much less demanding a perfect deletion from ATG to stop codon (and within a week, if you please!); and (except in the field of mitochondrial import) we had almost nothing in the way of techniques for in vitro reconstitution of cellular processes. In addition, it is important to recognize that the pace of yeast cell biology has also been driven by the progressive realization of the still-astonishing degree to which organelle and cytoskeletal systems, molecular mechanisms, and even the structures of individual protein species have been conserved throughout the entire evolutionary history of the eukaryotes. This conservation has both helped to attract talented cell biologists to work on yeast and allowed yeast cell biologists to exploit ideas and discoveries made in other systems. (Well, of course, usually *they* borrow from *us*, but *sometimes* it's the other way around!)

Is the frenetic pace of yeast cell biology likely to slow down any time soon? This hardly seems likely; indeed, with the completion of the genome sequence, the pace seems to have been ratcheted up even another notch. Given the awesome power of its experimental advantages and the remarkable conservation of cell biological mechanisms, yeast should continue to be both a major source of fresh ideas for cell biology generally and a crucial testing ground for ideas first suggested by studies in other systems. In this context, what are our hopes for this volume? We are aware of its imperfections: Although it covers a lot, it does not cover everything that it might have, and it will inevitably soon be out of date in many respects. Nonetheless, we are hopeful that its surveys of the field until about the end of 1996 will prove useful to several constituencies. For long-time yeast workers, these surveys should provide a measure of how far we have come and of how many problems remain unanswered. For those new to yeast, and for interested outsiders, they should provide a valuable starting point for understanding the rapid advances that will surely continue. For all three groups, they should (at least for a few years!) provide a valuable source for checking points of fact that we have not managed to learn or have forgotten in the sometimes overwhelming flood of new information.

In closing, we need to thank our authors, without whom, of course, there would be no volume. Some were quicker than others, but all have labored mightily to produce reviews that are as accurate, informative, and up-to-date as possible. We owe special thanks to the authors of our three *Schizosaccharomyces pombe* chapters. There have been many illustrations already, and there will be many more, of the special power that derives from comparing results obtained with the phylogenetically distant budding and fission yeasts. Thus, although this was always intended to be a volume mainly about *S. cerevisiae*, we were anxious to include chapters on those areas of *S. pombe* research where the comparisons have been particularly fruitful, and we think that the outstanding chapters included in this volume will serve this purpose very well. In addition, we take great pleasure in thanking the book-production staff at Cold Spring Harbor Laboratory Press, especially Joan Ebert, Dorothy Brown, and Nancy Ford, for their patience, prodding, and unbending commitment to the high-quality editorial work that has always been a hallmark of publications from Cold Spring Harbor.

John R. Pringle
James R. Broach
Elizabeth W. Jones

THE MOLECULAR AND CELLULAR BIOLOGY OF THE YEAST *SACCHAROMYCES*

CELL CYCLE AND CELL BIOLOGY

1

The Yeast Cytoskeleton

David Botstein, David Amberg, and Jon Mulholland
Department of Genetics
Stanford University School of Medicine
Stanford, California 94305

Tim Huffaker
Department of Biochemistry, Molecular, and Cellular Biology
Cornell University
Ithaca, New York 14853

Alison Adams
Department of Molecular and Cellular Biology
University of Arizona
Tucson, Arizona 85721

David Drubin
Department of Molecular and Cellular Biology
University of California
Berkeley, California 94720

Tim Stearns
Department of Biological Sciences
Stanford University
Stanford, California 94305

INTRODUCTION

An internal cytoskeleton consisting of a network of protein polymer filaments is a common feature of all eukaryotic cells. The cytoskeleton serves to organize the cytoplasm, to provide the means for generating force within the cell, and to determine and maintain the shape of the cell and its structural integrity. Three different types of cytoskeletal filaments are found. One type is based on long flexible polymers of actin called microfilaments (5–7 nm in diameter), another is based on more rigid polymers of tubulin called microtubules (hollow tubes 25 nm in diameter), and the third type is based on polymers of any of a number of related fibrous coiled-coil proteins called intermediate filaments (~10 nm in diameter). The microfilaments and microtubules of *Saccharomyces cerevisiae* closely resemble those of animal cells, and the structure and functions of these cytoskeletal elements have been extensively studied in this yeast (Botstein 1986; Huffaker et al. 1987; Drubin 1989; Stearns

1990; Welch et al. 1994; Cid et al. 1995). Potential nuclear lamins and 10-nm filament structures have also been reported in *Saccharomyces*, but relatively little is known about the structure and functions of intermediate filaments in yeast (Haarer and Pringle 1987; Kim et al. 1991; Chant 1996).

The cytoskeleton is a complex architecture of polymers and associated proteins that plays a part in many aspects of cellular physiology. Understanding the elaborate system of interactions upon which cytoskeletal function is based requires tools that can be used on the intact organism. Indeed, much of what has been learned about the yeast cytoskeleton has come from the application genetic methods, many of which were developed with application to the cytoskeleton in mind. These methods are discussed in considerable depth following a brief introduction to the actin and tubulin cytoskeletons. Since the actin and microtubule cytoskeletons in yeast (in marked contrast to most other eukaryotes) function largely independently of each other, we consider them separately. The few situations in which the two systems are thought to interact are pointed out specifically. For a number of useful previous reviews, see Huffaker et al. (1987), Winey (1992), Page and Snyder (1993), Bretscher et al. (1994), Hoyt (1994), and Welch et al. (1994).

II. THE YEAST CYTOSKELETON

Figure 1 shows a cartoon of the appearance of the yeast actin cytoskeleton through the cell cycle. The structures illustrated are polymers of a single protein, actin, which in yeast is the product of a single gene, *ACT1*. The two major types of actin structures are (1) patches at the periphery, or cortex, of the cell and (2) cables, also aligned along the cortex of the cell (see below), that tend to be aligned with the long axis of budded cells. The distribution of these two actin structures changes through the cell cycle, as illustrated in Figure 1. In terms of function, the actin cytoskeleton has been implicated in maintenance of cell polarity (especially with respect to membrane growth and protein secretion), in changes in cell shape (as, e.g., during mating), in resistance to osmotic forces, and in responses to environmental signals.

Figure 2 shows a cartoon of the appearance of the yeast microtubule cytoskeleton through the cell cycle. Microtubules are polymers of tubulin, which is a 1:1 heterodimer of α and β tubulins. All the β-tubulin in yeast is the product of a single gene, *TUB2*, whereas the α-tubulin derives from two nearly identical genes, *TUB1* and *TUB3*. All microtubules in yeast have one end at the spindle pole body (SPB), a

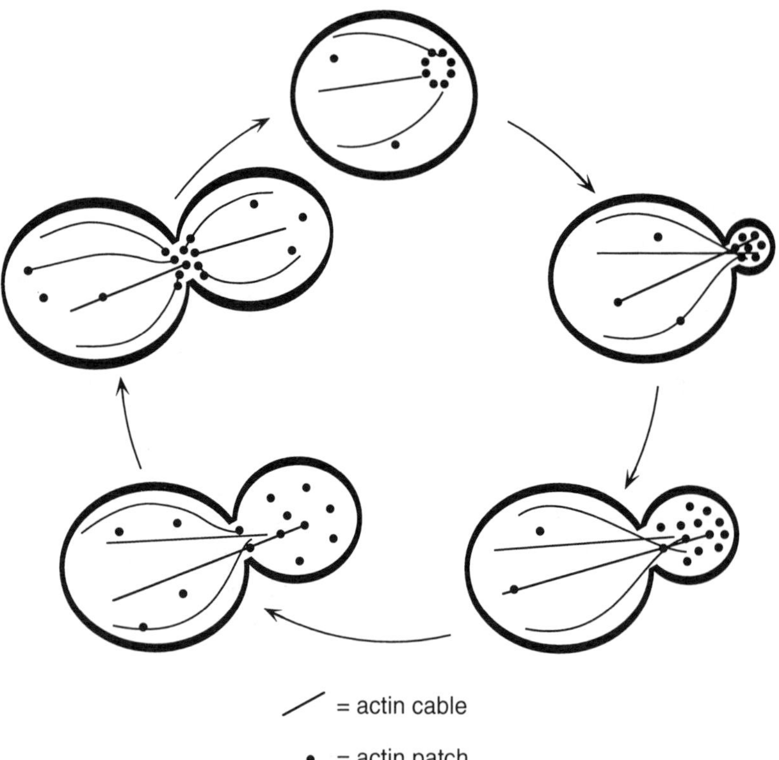

Figure 1 Cartoon of the yeast actin cytoskeleton in the mitotic cell cycle.

microtubule-organizing center that is embedded in the nuclear envelope, which does not break down during mitosis in fungi. Microtubules extend from both faces of the SPB, leading to an unambiguous distinction between two classes: intranuclear microtubules and extranuclear microtubules. The distribution and length of these two types of microtubules change through the cell cycle, as illustrated in Figure 2. The functions of the microtubule cytoskeleton appear to be limited to processes involving movement of the chromosomes (mitosis and meiosis and spindle orientation) and movements of the nucleus during the cell cycle (nuclear migration) and after mating (karyogamy).

The diversity of functions dependent on the actin and tubulin cytoskeletons, even in the simple yeast cell, suggests that the structures of microfilaments and microtubules under different circumstances must somehow be different. Many of these differences may be due to the presence of various accessory or associated proteins and structures whose function is to allow specialization of the basic actin and tubulin

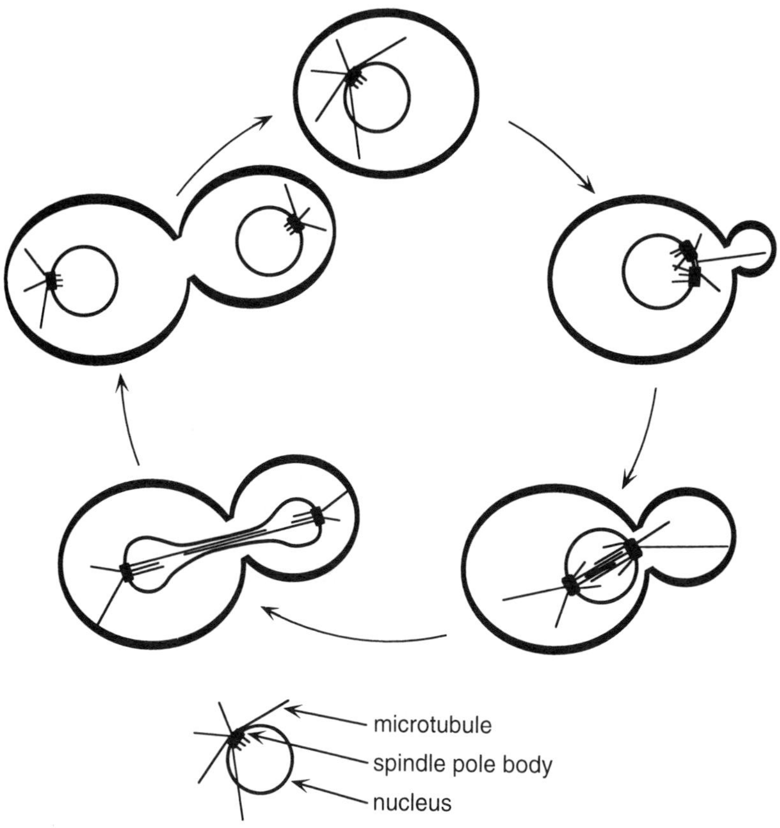

Figure 2 Cartoon of the yeast microtubule cytoskeleton in the mitotic cell cycle.

polymers. A great variety of proteins have been shown by one means or another to associate with the actin or tubulin cytoskeletons.

S. cerevisiae differs from most other eukaryotes in having only one gene for actin (*ACT1*), two genes for α-tubulin (*TUB1* and *TUB3*), and one gene for β-tubulin (*TUB2*). When the yeast actin and tubulin genes were cloned, and the predicted amino acid sequences were compared to those of more complex eukaryotes, it became clear that the degree of conservation is extraordinary. Yeast actin is 89% identical in amino acid sequence to mammalian actins (Gallwitz and Sures 1980; Ng and Abelson 1980), and yeast tubulins show about 75% identity to mammalian tubulins (Neff et al. 1983; Schatz et al. 1986a). Remarkably, many of the proteins found to be associated with the yeast actin or tubulin cytoskeleton turn out to be substantially similar in sequence to homologs independently found to be associated with the cytoskeletons of other eukaryotes. Thus, a substantial body of evidence has accumulated that the actin and tubulin cytoskeletons of yeast and other eukaryotes are highly conserved with respect to both structure and function.

The level of functional conservation must be, however, limited to the basic mechanisms (e.g., binding among components, force generation, orientation, and assembly and disassembly) as opposed to the ultimate biological functions attributed to the mechanisms. Yeast have no muscles and no flagellae, yet muscles and flagellae use actin and tubulin and a subset of associated proteins substantially similar to those of yeast. The diversity of uses to which microfilaments and microtubules are put in yeast is nevertheless very limited compared to the rest of the eukaryotes.

III. TOOLS FOR STUDYING THE YEAST CYTOSKELETON

The range of genetic approaches that can be applied to study of the cytoskeleton in yeast is peculiarly rich because of the well-known advantages of yeast as a genetic organism and, now, the availability of the entire genomic DNA sequence. For these reasons, it may be useful first to describe the tools used in yeast. The data and conclusions about the yeast cytoskeleton drawn from these largely genetic tools can then be seen and evaluated in their full context.

The actin and tubulin cytoskeletons are first and foremost extremely large assemblies of many diverse proteins organized around a filamentous polymer of actin or tubulin. Some of these proteins are always present, others are firmly bound only to parts of the cytoskeleton, and yet others may be bound only loosely or transiently. Some proteins may participate in cytoskeletal function by selectively binding components in their unassembled state. A central issue in cytoskeleton research, from the beginning until the present, has been to determine the "parts list," i.e., to determine the identity and role of all the proteins that participate as either structural elements, members of functional subassemblies, or functionally significant binders to polymers or monomers.

In most organisms, this task has been approached by a combination of biochemical and cytological methods. In yeast, we can supplement these methods with fundamentally genetic ones. For genetic methods to confer advantage, two conditions must obtain. First, it must be easy to isolate or construct, identify, recombine, or otherwise manipulate mutations at will. This is no longer an issue in yeast. Second, it must be easy to assess the phenotypic consequences of a mutation or combination of mutations. In the case of the cytoskeleton, this is not always straightforward in yeast. Despite this, research with *Saccharomyces* has contributed substantially to the cytoskeleton parts list, because one can apply genetic methods well-suited to finding interacting genes and proteins.

As a prelude to this survey of genetic methods for studying interactions among cytoskeletal genes and proteins in yeast, it should be emphasized that these methods are complementary to the more standard approaches. Biochemical methods that have been used successfully to identify and characterize yeast cytoskeletal proteins include direct purification of proteins based on assays for activity, affinity chromatography, and assay of assembly and motor functions in vitro. Finally, like the actin and tubulin genes, many yeast cytoskeletal genes continue to be found by taking advantage of sequence conservation, which makes it possible to find genes encoding yeast homologs of proteins whose cytoskeletal function has already been defined, at least in part, in other organisms.

By what criteria should a protein be accepted as a bona fide component of the cytoskeleton? The only criterion that has found general acceptance among cell biologists is cytological. Since the cytoskeleton has a characteristic morphology within most cell types, a newly discovered protein is accepted as part of the cytoskeleton if it is found to "colocalize" with one of the cytoskeletal polymers. Most commonly, demonstration of colocalization is achieved by immunofluorescence microscopy; an alternative is immunoelectron microscopy. Colocalization need not extend to all parts of the cytoskeleton: Indeed, some of the most interesting actin-associated proteins in yeast bind to only a subset of actin-containing structures. Of course, there may be proteins that affect the cytoskeleton by binding to monomers (e.g., the chaperonins or profilin); in this case, colocalization with polymer need not ever occur.

Other criteria are also useful, including coprecipitation with antibodies, coassembly with actin or tubulin filaments polymerized in vitro, modulation of assembly in vitro, affinity chromatography using monomer or polymerized actin or tubulin, and the panoply of genetic indicators of gene interaction, but none of these alone can be regarded as definitive. All of these methods are useful mainly in providing leads to interactions that can be assessed ultimately by some kind of colocalization experiment. The observation that a mutation (in vivo) or a protein (in vitro) has a profound effect on the cytoskeleton cannot be taken by itself as evidence for direct binding to the cytoskeleton. The diversity and complexity of cytoskeletal functions provide many avenues for indirect effects, and many such effects have already been observed. One of the advantages of studying the cytoskeleton in yeast is that such study can identify indirect effects susceptible to biochemical analysis that otherwise would never have been suspected.

Study of the cytoskeleton in yeast has concentrated on the identification of the genes that specify cytoskeletal proteins and the interactions among them. The parts list has been generated by a variety of methods ranging from the purely genetic to the purely biochemical and, increasingly, by methods that take advantage of the knowledge of the complete genomic sequence of *S. cerevisiae*. No matter how a gene or protein has first been associated with the cytoskeleton, in the course of these studies, a generic set of materials is generated: clones of the gene, a variety of mutations (deletions, point mutations, protein fusions to epitope tags or fluorophores, promoter fusions), specific antibodies, and often purified protein. With these materials, it is generally possible to acquire considerable information about the role of a gene and its product in cytoskeletal functions.

A. Constructing Mutations in Cytoskeletal Genes

The actin and tubulin proteins are essential for the viability of yeast cells, but surprisingly, knowledge of these genes did not begin with mutations. Despite their importance in cytoskeletal function, the actin and tubulin genes were not recovered from the classical collections of conditional-lethal mutations. This was particularly surprising for the tubulins, since conditional-lethal *tub1* and *tub2* mutations derived by directed mutagenesis schemes display a straightforward cell division cycle (cdc) defect with the expected phenotype for failure at mitosis. Instead, the genes encoding actin and the α and β tubulins were obtained through their sequence similarity with the cognate mammalian proteins. The cloned genes were then used to obtain mutations by directed mutagenesis. In this situation, several kinds of mutations can easily be made using yeast genetic technology. Each of these has different and often complementary advantages.

1. Null Mutations

Determination of the null phenotype is the starting point for virtually any investigation of a gene whose biological function is not known. Null mutations can be obtained in a variety of ways. The most reliable is deletion of the entire coding sequence of a gene, with or without replacement with a marker gene (Rothstein 1983).

For cytoskeletal mutants, the first phenotype of interest is generally the viability of haploid cells lacking the gene under study. This information is readily obtained from sporulation and tetrad dissection of a strain heterozygous for the constructed null mutation. In the case of many cytoskeletal genes, the result of such analysis is that the null phenotype is lethality. This is found, not surprisingly, to be the case of *ACT1* and *TUB2*, which are the only genes in *Saccharomyces* encoding authentic actin and β-tubulin, respectively. The results for *TUB1* and *TUB3*, both of which encode α-tubulin, were not expected in advance: *TUB1* is essential for viability in ordinary laboratory media and *TUB3* is apparently completely inessential. Some cytoskeletal genes are found not to be essential for viability. This result should be assessed in the context of the general observation that null mutations cause impaired viability or growth in less than 20% of all yeast genes tested (Goebl and Petes 1986; Oliver et al. 1992). Later in this chapter, we return to the issue of genetic redundancy and other possible explanations for the apparent dispensability of conserved proteins involved in important cellular structures.

Partial deletions turn out frequently to be null in phenotype, especially when the deletion removes all potential translation initiation codons. However, it should be recognized that there have been more than a few reports of significant residual function in partially deleted genes. In addition, the use of deletions that are larger than a single coding sequence has obvious hazards in a genome where intergene distance can be only a few nucleotide pairs; in such cases, considerable care must be taken to ensure that the phenotype being assessed is indeed due to the loss of the gene of interest. This can often be accomplished by testing such a deletion strain's phenotype when provided with a plasmid expressing only the gene of interest.

Transposon insertion mutations and frameshift mutations are usually a good source of information about the null phenotype when perfect deletions are unavailable or impractical. For example, the availability of the *Saccharomyces* genomic DNA sequence has stimulated interest in schemes aimed at determining the null phenotype of all of the yeast genes. One of these involves assessing the phenotype of several transposon insertions scattered throughout each open reading frame (Smith et al. 1995).

2. *Conditional Mutations*

Null mutations, although important, are not usually the most useful mutations for understanding the function of essential genes. The isolation of point mutations, especially those that have temperature-conditional phenotypes (i.e., heat-sensitive or cold-sensitive), is therefore the logical next step in trying to discern the functions of a cytoskeletal gene and its product. A variety of schemes have been successful in deriving new conditional mutations, including chemical mutagenesis of isolated DNA, mutagenesis of plasmids in mutator *Escherichia coli* (Echols et al. 1983), and deliberately inaccurate polymerase chain reaction (PCR) (Leung et al. 1989).

Conditional alleles of essential genes permit characterization of the resulting phenotypes using all of the methods already developed for mutations obtained by classical means. It is important to emphasize that not all conditional alleles affecting a gene or protein affect all of the functions of that gene or protein. For this reason, it is essential to examine a number of conditional alleles in order to be confident that all of the important functions are being assessed. This is particularly true for cytoskeletal genes, where multifunctionality is the rule, not the exception. Indeed, for actin, α-tubulin, and β-tubulin, multifunctionality of the protein has been inferred directly from the diverse phenotypes of different conditional mutations in the genes that encode them (Shortle et al. 1984; Novick and Botstein 1985; Huffaker et al. 1988; Wertman et al. 1992). Furthermore, one must be cautious in the assessment of which defects observed in a conditional mutant are primary and which are secondary; after a short period at the nonpermissive temperature, a culture consists of a mixture of still-growing, arrested, dying, and dead cells, and these can be difficult to distinguish. It is therefore necessary to follow the fate of cells as a function of time after shift to the nonpermissive temperature and to recognize that subtle early events may well be more significant than later obvious changes. Finally, it should be emphasized that shifting the growth temperature is not a trivial change in the environment even of wild-type cells; simple shifts of temperature have readily visible effects on the cytoskeleton that must rigorously be controlled for.

Alternative methods of producing conditional expression of proteins are available. Construction of strains in which the gene of interest is deleted and a copy of the gene is under control of the repressible *GAL1* or *GAL10* promoter allows the propagation of strains in media containing galactose as the only carbon source. When shifted to media containing glucose, such strains will eventually deplete their stores of the protein of interest, and during this process, some information about the consequences of the loss of this protein can be obtained. This method, although appealing in its simplicity, has significant disadvantages, among which are that the depletion may require many generations, depending on the stability of the protein and message and the fraction of the normal level of the protein that is minimally required. Typical experiments of this kind involve assessments of phenotype many generation times after the depletion was initiated. The extreme duration and asynchrony of the depletion process in these experiments compounds the normal difficulty of distinguishing primary and secondary defects and invites instead attention to cells that have been dead for hours instead of those that have just begun to die. Results from such depletion experiments are therefore to be regarded skeptically until confirmed with conditional mutations or other means.

Another alternative to isolation of conditional alleles in cloned genes has recently been proposed but has not yet been used extensively (Dohmen et al. 1994). The idea is to arrange for conditional degradation of the protein of interest, as opposed to conditional function or synthesis of the protein. The method involves constructing a fusion of the protein of interest to a mutant gene encoding a highly temperature-labile dihydrofolate reductase that is rapidly degraded at nonpermissive temperature. The fusion proteins turn out also to be highly temperature-labile, even when the original protein is very stable. In principle, this appealing idea provides conditional null mutations, thereby avoiding the complications in interpretation arising from the many different possible reasons a temperature-sensitive point mutation might cause loss of function at the nonpermissive temperature.

3. *Domain Deletions*

The multifunctionality of many cytoskeletal proteins is often reflected in the finding that separable functional protein domains exist. The presence of functional protein domains can sometimes be inferred from structural studies or evolutionary conservation but often is realized only by mutagenesis. There are numerous examples of multifunctional cytoskeletal proteins that carry two or more domains associated with similar functions in other proteins. A potent approach in studying such proteins is the use of intragenic deletions to remove selectively one or more domains. When these constructions succeed in producing a partially functional protein, the inferences obtained about which parts of the protein are responsible for which functions can be very strong indeed. A striking example is Kar1p, a protein involved in microtubule function, which was elegantly shown to have two functions, each of which could be provided by a domain deletion construct (Vallen et al. 1992). However, especially in proteins essential for growth, it may sometimes be difficult to produce deleted proteins with enough residual stability and/or activity to support even partial functions. This is particularly so when the boundaries of domains are not obvious or when different functions use the same domain(s) in different ways.

4. *Scanning Mutagenesis*

For the major structural proteins in the cytoskeleton (e.g., actin and tubulin), intensive efforts were undertaken to obtain diverse conditional mutations with diverse phenotypes with some success (Shortle et al. 1984; Huffaker et al. 1988; Schatz et al. 1988; Johannes and Gallwitz 1991). In the case of actin, however, it was clear that the existing mutations did not begin to cover the surface of the protein where interactions with other proteins must take place. Clustered charged-to-alanine scanning mutagenesis, which aims to cover the entire surface of a protein with mutations, has emerged as a particularly useful method in this context. The method involves the construction a complete set of mutations, each of which changes all members of a single cluster of charged residues to alanine. The idea, thus far borne out in practice, is that clusters of charged residues are generally found at the surface of proteins, and changes to alanine are minimally disruptive of the folding. For the case of actin, the set of mutations produced in this way has found many uses. Similar sets have more recently been constructed for *TUB2* (Reijo et al. 1994) and *TUB1* (K. Richards and D. Botstein, unpubl.)

It is worth noting that charged-to-alanine mutants are not likely to reveal functions mediated through hydrophobic interactions with other proteins. This point was strikingly clear in the case of calmodulin, where none of the charged-to-alanine mutants had any phenotype, and virtually all mutations of (conserved) phenylalanines to alanine contributed to one or more of the several distinguishable calmodulin phenotypes (Ohya and Botstein 1994b).

5. *Fusions*

Protein fusion technology has found many applications in the study of the yeast cytoskeleton and many are now quite standard, including the fusion of a marker protein (e.g., β-galactosidase) to study regulation and the provision of epitope tags to facilitate protein purification, localization, and quantitation without the necessity of raising specific antibodies. In many cases, the fusion proteins are fully functional, and under these circumstances, it can be inferred that the protein is localized correctly. Any such inference is greatly weakened when the fusion protein is overproduced, because of the possibility that excess protein is localized aberrantly. It is worth noting that fusion proteins that fail to provide normal function in vivo may nevertheless be very useful, for example, in providing material for raising specific antibodies. Noncomplementing fusions might still localize normally in vivo; however, unless good complementation is observed, other evidence supporting normal localization is required.

The recent discovery that the green fluorescent protein (GFP) from the jellyfish *Aequorea victoria* can be readily visualized when fused to a variety of proteins in different organisms (for review, see Stearns 1995) has opened the possibility of studying the localization of yeast proteins in real time in living cells. This technique has already been used to follow yeast SPB proteins, tubulin, actin, and a number of different actin-associated proteins (Kahana et al. 1995; Stearns 1995; Doyle and Botstein 1996; Marschall et al. 1996; Waddle et al. 1996).

B. Finding New Cytoskeletal Genes and Proteins

Obtaining mutations in genes encoding cytoskeletal proteins and assessing their phenotypes is only the first step in cytoskeletal genetics. The logical next step is to find other genes using the mutants (and the inferences made from them) already in hand. The yeast cytoskeleton has become a fertile testing ground for the various methods, especially the genetic ones, that allow the identification of interacting genes and proteins. Soon after conditional-lethal actin and tubulin mutations in yeast were isolated and characterized, searches began for related genes and proteins using genetic (Stearns and Botstein 1988; Adams and Botstein 1989; Novick et al. 1989) and biochemical (Drubin et al. 1988; Barnes et al. 1992) means.

The several methods discussed below each have advantages and drawbacks. All of the methods have been used to find new interacting genes and proteins. However, none of them have yet been accepted as entirely sufficient to demonstrate a functional interaction, as there are both "false-positive" and "false-negative" results for each of them. Convincing evidence for interactions have sometimes arisen through concordant observation using several of these methods; more often, follow-up studies that provide more direct evidence of interaction, such as colocalization and biochemical association, are required.

1. Common Characteristic Phenotype

The most straightforward way of finding different genes that together act to perform a common function is through a common mutant phenotype.

This approach works best when the phenotype is a relatively specific one. This approach has been extraordinarily successful for the tubulin cytoskeleton, where hypersensitivity or resistance to a specific microtubule-destabilizing drug, benomyl, was noted among many mutations in the tubulin subunits. This phenotype has been a rich source of mutations in many genes affecting the tubulin cytoskeleton (Thomas et al. 1985; Schatz et al. 1988; Stearns et al. 1990). Screens for mutations showing increased chromosome loss (without any concomitant increase in mitotic recombination) have yielded mutations in many of the same genes (Hoyt et al. 1990, 1992).

The common phenotypes need not appear to be quite so obviously related to the cytoskeleton or even very specific and yet still be useful. For example, many actin mutants show hypersensitivity to osmotic pressure in the medium (Shortle et al. 1984) as do mutations in many proteins known to bind or associate with actin by other means. Screening for randomized bud site selection in diploids yields mutations in actin-associated genes and is itself a phenotype of many of the known actin mutants (Drubin et al. 1993).

Finally, the diversity of phenotypes exhibited by some cytoskeletal genes raises the interesting possibility that this functional diversity might be detectable genetically, by intragenic complementation, for instance. Indeed, among the *CMD1* (encoding calmodulin) scanning mutations, four complementation groups corresponding to four different lethal phenotypes were found (Ohya and Botstein 1994a,b). One of these phenotypes suggests a function in nuclear division, another a function in bud emergence, and a third a function in forming or maintaining a normal actin cytoskeleton. More limited instances of complementation have also been observed among scanning mutations in *ACT1* and *TUB2* (K. Schwartz et al., unpubl.). It should to be pointed out that calmodulin is one of the points of proven interaction between the actin and tubulin cytoskeletons.

2. *Suppression*

The oldest and most direct purely genetic path to finding interactions among genes or their protein products is to find alleles of one gene that suppress defects in the other. Suppressor genetics is particularly easy in yeast, because after selection for reversal of a phenotype, crosses to wild type followed by tetrad dissection are immediately diagnostic of suppression, except in the rare cases of tight linkage between the suppressor and the mutation being suppressed. Simple suppressors of a cold-sensitive *tub2* allele were used, for instance, to identify *STU1*, a gene that encodes a novel protein that localizes to the mitotic spindle (Pasqualone and Huffaker 1994).

A variation on the suppressor technique (Jarvik and Botstein 1975; Moir et al. 1982) is to begin with a conditional-lethal mutation (e.g., heat-sensitive), select for survival under the nonpermissive condition, and then screen the survivors for a new conditional phenotype (e.g., cold sensitivity). The candidate suppressor mutations, when segregated away from the original mutation, often retain their conditional phenotype and can then be studied directly. This approach has been used to find new genes beginning with *act1* mutations (Novick et al. 1989).

Suppressors found in these ways can be either recessive or dominant as suppressors, although the new conditional phenotype is most often recessive. The recessiveness of suppression can be exploited in cloning the suppressor gene using plasmid libraries made from DNA of wild-type strains, as in the case of *STU1* (Pasqualone and Huffaker 1994). It is, of course, possible to select directly for dominant suppressors simply by carrying out the selection in diploid strains homozygous for the mutation to be suppressed. Sometimes the resulting suppressor mutations are themselves conditional-lethal (as above). A particularly well-studied case is the isolation of mutations in the gene encoding yeast fimbrin (*SAC6*) that suppress *act1* mutations (Adams and Botstein 1989). In this case, the suppression was reciprocal, i.e., the conditional phenotype of *sac6-ts* was suppressed by the *act1-ts* allele it suppresses.

There are very many useful variations of the suppressor idea. A particularly important one is suppression through overproduction. In this case, suppression is sought by providing a library of yeast genes either on a high-copy plasmid or under the control of a very strong promoter (e.g., *GAL1*). For example, this method was used in finding interactions among genes specifying SPB components (Vallen et al. 1994). The simplest interpretations of suppression by overproduction are based on the notion that the original mutation is defective in interaction or stability, which can be overcome by raising the concentration of the binding partner. A variant of this idea is to seek suppressors of the deleterious effect of overproduction of a protein (Meeks-Wagner et al. 1986), such as actin (Magdolen et al. 1993) or β-tubulin (Kirkpatrick and Solomon 1994; Archer et al. 1995). Suppression in this case can be effected by ligands that either sequester monomers or otherwise rectify the stoichiometry of the interaction. High-copy suppression, although clearly useful and readily interpretable in the context of stoichiometry, must be viewed with great caution under some circumstances, particularly when the protein being overproduced is an enzyme such as a kinase or phosphatase. In this context, overproduction may well result in loss of normal specificity, as opposed to revealing a normal interaction.

There is no universal mechanistic explanation of suppression even when it is clearly the result of the interaction of two proteins involved in a common function. Nevertheless, in some cases, good models for mechanisms can be proposed. Two particularly interesting cases are the suppression of loss-of-function kinesin mutations by similar mutations in other kinesins (Hoyt et al. 1993) and the restoration of binding between actin and fimbrin (Brower et al. 1995). Even though interpretation of suppression relationships can become complicated, the record suggests that where suppression of the defect in the original mutation is robust, and there is allele specificity, the chances that one is studying a real interaction are very good, and several of the most important interactions cataloged below were first discovered in this way. Frequently, tests of allele specificity result in the finding of other genetic indicators of interaction (see below), often synthetic lethality (e.g., *ACT1* and *SAC1*) (Novick et al. 1989; Pasqualone and Huffaker 1994) and sometimes also unlinked noncomplementation (e.g., *TUB2* and *STU1*) (Pasqualone and Huffaker 1994).

3. *Synthetic Lethality*

Synthetic lethality is said to occur when mutations that are themselves viable are found to be lethal in combination. Synthetic lethality can be observed among and between mutations of all kinds, including especially null mutations and conditional-lethal mutations. In the latter case, of course, the novel lethality is observed under conditions permissive for the component mutations. As mentioned above, synthetic lethality among genes encoding yeast cytoskeletal proteins was first noticed in studying the allele specificity of suppression of *act1* alleles by suppressor alleles of *sac1* (Novick et al. 1989), and indeed synthetic lethal interactions often affect the same genes that show suppressor interactions. Unlike suppression, however, synthetic lethality appears to be a very common phenomenon among genes affecting similar processes in the cell. Among the prominent cytoskeletal gene groups exhibiting synthetic lethality are some associated directly with microtubules (Stearns and Botstein 1988; Hoyt et al. 1990; Stearns et al. 1990; Pasqualone and Huffaker 1994) and others associated with the actin cytoskeleton (Adams et al. 1993) and the kinesins (for review, see Hoyt 1994).

Synthetic lethality can be exploited as a means of identifying new interacting genes by isolating mutations, although this is not quite as simple as finding extragenic suppressors. The usual strategy is based on the "plasmid shuffle": One first constructs a strain carrying a deletion of a gene of interest and also a low-copy plasmid carrying the intact gene and a marker gene (e.g., *URA3* or *ADE2*) that can be used to select or detect loss of the plasmid. This strain is mutagenized, and the phenotype sought is loss of the ability to lose the plasmid. Mutations that arise are frequently (although not exclusively) in genes that exhibit some kind of synthetic lethality with the gene of interest. This approach has been used to find new interacting genes among those affecting the actin cytoskeleton (Liu and Bretscher 1992; Holtzman et al. 1993; Karpova et al. 1993; Peterson et al. 1994; Li et al. 1995; Wang and Bretscher 1995) and among the kinesins (Roof et al. 1992; Saunders and Hoyt 1992).

As with suppression, there is no universal mechanistic explanation of synthetic lethality. One class of explanations invoke some kind of genetic redundancy. These range from the simple case where two genes are each sufficient to perform the same essential function (Basson et al. 1987) to the idea that the networks of gene interactions are so complex as to constitute an analog of a highly parallel processing system in a computer or computer network (Bray and Vasiliev 1989). Thomas (1993) has provided a useful categorization of the redundancy explanations: In addition to simple repetition of function, he envisions redundancies whose short-range advantage is modest (and thus would not impact on growth under ordinary conditions) but which have a long-term advantage that provided for their retention and for the fact that many of these genes encode proteins which have been highly conserved in evolution. There might be redundancy that ensures high precision in a process by providing two alternate mechanisms, or redundancy because it is convenient to do the same function in quite different contexts, resulting in two systems of overlapping capabilities or because two ways of doing the same operation may together have an "emergent" or "synthetic" effect that confers a subtle advantage other than fidelity.

There may of course be explanations for synthetic lethality that do not rely on genetic redundancy. In some cases, the particular partial defect produced by one mutation may make the organism peculiarly sensitive to the defect caused by another. To give a theoretical example, one mutation may cause an adventitious increase in the internal concentration of a metabolite and the other may confer sensitivity to the same metabolite. There would indeed be a significant connection or interaction but not one that can reasonably be termed as a duplicate or redundant function. A cytoskeletal example is the synthetic lethality seen between mutations in *MYO2* (encoding a type I myosin) and *TPM1* (encoding a tropomyosin);

on biochemical grounds, the tropomyosin was already implicated in the interaction of myosin I with actin (Liu and Bretscher 1992). Finally, care must of course be taken not to accept as significant synthetic lethality the observation that a combination of two mutations, each of which makes a strain sick, makes the strain so sick as to be inviable.

4. *Unlinked Noncomplementation*

When several different genes contribute subunits to the same polymeric protein structure, it is sometimes possible to find particular mutations affecting one subunit that would fail to be complemented by mutations affecting another, even in the presence of wild-type subunits of both kinds. Screening for unlinked mutations that fail to complement a *tub2* mutation yielded a mutation in *tub1;* subsequent screening of *tub1* mutations yielded mutations in *tub3* (Stearns and Botstein 1988). This scheme has also worked beginning with mutations in *act1*, where screening for unlinked noncomplementation yielded mutations in four new genes (called *ANC1–ANC4*), some of which also failed to complement or showed synthetic lethality with mutations in other actin-associated genes (Vinh et al. 1993).

5. *The Two-hybrid System*

The two-hybrid system (Chien et al. 1991) was designed to detect protein-protein interactions. It has the advantageous property that it can be used to scan the genome for fragments of genomic DNA or cDNA that encode polypeptides which interact with a given polypeptide. Any positive clones can then be quickly identified with a particular gene in yeast because of the availability of the genomic DNA sequence. The two-hybrid system has been successfully applied to actin: Sequences encoding several known actin-binding proteins were recovered along with three novel ones. In addition, as described below, use of the two-hybrid system in combination with scanning mutagenesis allows the inference of where on the three-dimensional surface of actin some of its protein ligands bind (Amberg et al. 1995).

6. *Biochemical Methods*

Although the unique advantages of yeast are in the genetic arena, it should not be forgotten that the basic biochemical methods for studying the actin and tubulin cytoskeletons have been applied to yeast with great effect. These methods can be divided into three broad categories: affinity methods, organelle isolation, and functional assays. When these methods are feasible, they may still represent the most direct and unambiguous approaches to studying cytoskeletal organization.

Affinity chromatography using polymerized actin or tubulin has yielded a number of associated proteins that are central to our understanding of the actin cytoskeleton. For example, Abp1p (an SH3 protein of yeast), a yeast tropomyosin, and yeast fimbrin were found in this way (although fimbrin also was found as a suppressor of *act1* mutations) (Drubin et al. 1988; Adams and Botstein 1989; Liu and Bretscher 1989; Moon et al. 1993). The most impressive and fertile organelle isolation in yeast has been the purification of the yeast SPB by Kilmartin (Rout and Kilmartin 1990; Kilmartin et al. 1993). This endeavor identified many of the known components of this organelle.

Functional assays have yielded many yeast proteins. There are two ways in which functional assays have been applied. The most direct is to apply a known assay, but surprisingly, this has been applied relatively little. One prominent example is the identification of yeast cofilin (Moon et al. 1993). A more indirect method is through the use of antibodies to heterologous proteins, with the expectation that yeast will have a similar protein. An example of this was the isolation of yeast actin capping proteins (Amatruda and Cooper 1992). A more serendipitous case might be the isolation of Spa1p, where an autoantibody directed against the human centrosome was used to find a protein that appears to be involved in yeast cell polarity (Snyder and Davis 1988).

Finally, it is worth mentioning that functional conservation can sometimes be demonstrated by functional assays. Central examples of this are the ATP-dependent ability of purified yeast actin filaments to move on purified rabbit muscle myosin (Kron et al. 1992) and the ability of mammalian microtubules to move on purified yeast Kar3p, again only when ATP is supplied (Endow et al. 1994; Middleton and Carbon 1994).

7. *Homology*

Probably the most important functional approach to finding yeast cytoskeletal proteins has been through homology with proteins of other organisms in which the relevant biochemistry and cell biology already were in place. Yeast actin and tubulin were identified in this way, as well as a whole host of other cytoskeletal proteins, including most yeast myosins (Sweeney et al. 1991; Haarer et al. 1994; Goodson and Spudich 1995; and see below) and most of the yeast kinesins (Roof et al. 1992; Saunders and Hoyt 1992).

In concluding this survey of methods for studying interactions among cytoskeletal genes and proteins in yeast, it should be reemphasized that the genetic methods, which are so well adapted to yeast, supplement more biochemical approaches. We can already see that the parts list and interaction tables are generating hypotheses that are beginning to be tested not only with more genetics, but also with biochemical studies of a quite definitive kind.

IV. THE ACTIN CYTOSKELETON

The distribution of actin within the cell and the changes as the cell cycle progresses are central to understanding the several functions of the actin cytoskeleton. The intracellular distribution of polymerized actin has been visualized with fluorochrome-labeled phallotoxins and immunofluorescence microscopy (Adams and Pringle 1984; Kilmartin and Adams 1984; Pringle et al. 1989), as well as by immunoelectron microscopy (Mulholland et al. 1994). With minor exceptions, the two very different fluorescent probes (anti-yeast actin antibodies and phallotoxins) have yielded very similar results, and double labeling reveals that the staining patterns are essentially coincident. Thus, polymerized actin appears to be localized in a set of cortical patches and in fibers that are often oriented parallel to the longitudinal axis of the cell (Fig. 3). Little is known about the distribution of actin monomers, although monomer-sequestering proteins such as profilin and Srv2p have been seen associated with some of the filamentous actin cortical patches described in detail below.

Immunoelectron microscopy (Mulholland et al. 1994) has provided a much more detailed view of the structure of the actin cortical patches. The patches are seen as electron-dense regions almost always associated with invaginations of the plasma membrane. The distribution of patches in cells seen by immunoelectron microscopy is entirely consistent with what is found by fluorescence microscopy and described in detail below. Images obtained in the electron microscope suggest strongly a model in which longitudinal actin fiber bundles begin in the cortical patch as windings around the plasma membrane invagination (Fig. 4). The plasma membrane invaginations are not seen with all ultrastructural techniques.

Fusions of GFP to actin and actin-binding proteins (Doyle and Botstein 1996; Waddle et al. 1996) provide fluorescent images, made in living cells, of the cortical patches that are entirely consistent with those obtained with antibodies and phallotoxins on fixed specimens. In addition, however, these studies indicate that the cortical patches, at least, are highly mobile, achieving rates of movement on the order of 1 μm per

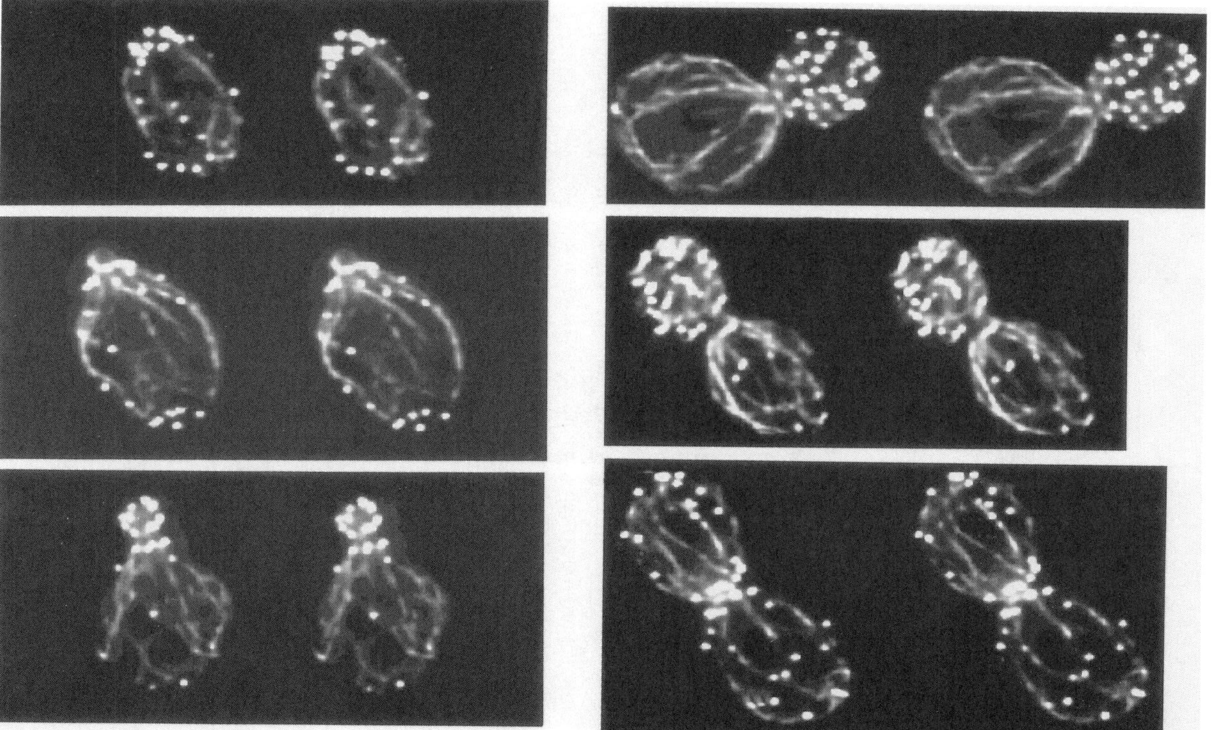

Figure 3 Stereo images of the actin cytoskeleton in yeast cells. Cells in different stages of the cell cycle were fixed and stained with rhodamine-conjugated phalloidin. The stereo images were created from a series of optical sections generated with a Delta Vision microscope system (Applied Precision, Inc.).

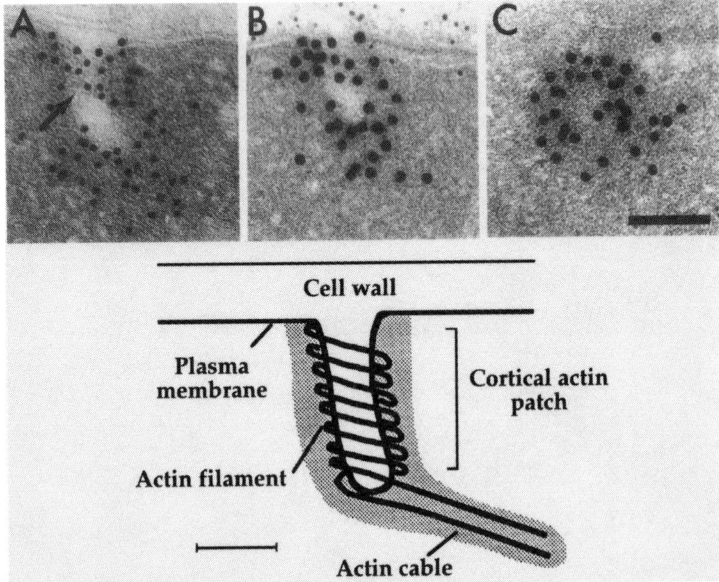

Figure 4 Ultrastructure of the cortical actin patch. In longitudinal sections through actin patches (*A* and *B*) ordered arrays of immunogold-localized actin are positioned almost perpendicular to the axis of the plasma membrane invagination. At least two labeled filaments (~7 nm), which appear negatively stained, can be resolved in *A* (arrow). On cross-sections through the actin patch (*C*), immunogold localization of actin produces ring-like patterns. The cartoon illustrates a model of the actin cortical patch as a finger-like invagination of plasma membrane around which the actin cytoskeleton is organized. The organization of actin in the actin patch as shown in speculative, although consistent with existing data; the actual organization, and the relationship between actin patches and cables, is not known. Bars, 0.1 μm; gold particles in *A*, 10 nm; gold particles in *B* and *C*, 15 nm.

second. Cortical patches, although mobile, appear to be largely limited in their movement to the regions in which they are found, using conventional fluorescence imaging.

As the cell cycle progresses, the distribution of actin (Fig. 3) changes in a manner which suggests that it may be involved in polarized growth (Adams and Pringle 1984; Kilmartin and Adams 1984). Thus, in unbudded cells (or at least cells in which no bud can be detected), actin patches are often found concentrated at one pole, where they are frequently arranged in a ring. Spatially, these rings appear to coincide with rings of chitin and are presumably located subjacent to them. When the bud first emerges, the ring of patches remains as a collar at its base (see also Novick and Botstein 1985), disappearing as the bud enlarges. This is again consistent with the hypothesis that actin has a role in the localized deposition of chitin, as it has been suggested that the chitin ring may continue to develop after bud emergence (Sloat et al. 1981; Kilmartin and Adams 1984).

In cells with small buds, actin patches are found concentrated in the developing bud. As the bud enlarges, at least in cells with more elongated growth forms (such as cells of *S. uvarum*, and hexaploid cells of *S. cerevisiae*), the patches become displaced toward the tip of the bud. This distribution again suggests an involvement of actin in the polarization of growth, as during the cell cycle, growth is confined almost entirely to the bud and, for much of the period of bud enlargement, to its tip (Tkacz and Lampen 1972; Farkas et al. 1974). A particularly striking correlation is seen in morphogenetic mutant cells that form multiple, abnormally elongated buds. In these cells, actin patches are found concentrated at the tips of some, but not all, of the buds of each cell. Staining of these cells to reveal sites of new cell-wall expansion shows that growth occurs primarily at the tips of the buds and that the cells are often able to sustain growth of more than one bud at a time. Moreover, double staining of the cells to reveal regions of active growth and concentrations of actin shows that essentially all growing bud tips possess high concentrations of actin, in contrast to nongrowing tips. Very similar conclusions about the relationship between tip growth and actin organization have been reached from studies on haploid yeast engaged in the formation of mating projections.

At about the time of nuclear division, when bud expansion has essentially ceased, the patches are again found uniformly distributed throughout the mother and bud (Adams and Pringle 1984; Kilmartin and Adams 1984). Toward the end of nuclear division, actin cortical patches can be seen in single or double rings at the neck between mother and bud. The ring is most apparent when viewed from the appropriate angle and is similar in location to the septin rings found at the neck. It is not clear whether any real differences exist between the single and double rings or whether the single rings are actually unresolved double rings. Both single and double ring structures are seen more frequently with phalloidin than with antibodies, but this may be due to differences in the abilities of the two probes to penetrate this region. The role of these ring structures is unknown, but their location in the neck region, prior to cytokinesis and septal-wall growth, suggests that they may be involved in preparing the cells for one or both of these events, presumably by directing secretion to these regions of the surface. Finally, at about the time of cytokinesis and septation, concentrations of actin patches (distinctly different in appearance from the ring structures) appear in the neck regions. Thus, throughout the cell cycle, regions in which actin patches are concentrated show remarkable coincidence with regions in which localized growth occurs.

The discussion above has focused on the changing distribution of actin patches. As mentioned earlier, however, fibrous structures are also seen in the cells. These vary in staining intensity and, as the fibers are usually stained more faintly than the patches, they are often difficult to follow in regions where patches are concentrated. For these reasons, it is not entirely clear whether they undergo similar changes in distribution as the cell cycle proceeds. In cells with more elongated growth forms, however (such as morphogenetic mutant cells of *S. cerevisiae* or wild-type cells of *S. uvarum*), very long fibers are often seen extending from the mother cells to the tips of the elongated buds, where the actin patches are concentrated. This, taken together with the observations in the electron microscope, makes it seem likely that the actin fibers are connected, at least some of the time, to the actin patches. Recent advances in fluorescence microscopy have made it clear that most of the fibers run along the inner surface of cells and that very few, if any, extend any distance into the cytoplasm (see Fig. 3, stereo).

The fibers seen by fluorescence microscopy appear to be similar to those seen in many other cell types. They are presumably bundles of individual actin filaments. Cytoplasmic bundles of filaments have been seen in yeast by electron microscopy (Matile 1969; Adams and Pringle 1984), but there is no evidence that these contain actin. The morphology of the filament bundles observed by Adams and Pringle (1984) does resemble somewhat the morphology of actin filaments arranged in "paracrystalline arrays" (Spudich and Spudich 1979; DeRosier and Tilney 1982; Tilney et al. 1983). A problem with this interpretation is that neither the diameter of the yeast filaments (apparently closer to 10 nm than the 7 nm expected for actin) nor the periodicity of the striations in the yeast bundles (apparently closer to 120 nm than 35 nm) corresponds well to what has been observed in other systems. More definite identification of these filaments must await their successful decoration with myosin subfragments or appropriate antibodies. Conditions have not yet been reported that allow both immunolabeling and visualization of actin fibers.

The cortical actin patches observed by fluorescence microscopy are a more unusual staining pattern for actin, although not unprecedented in other organisms (see Adams and Pringle 1984 and references cited therein). These patches also appear in *Schizosaccharomyces pombe* (Marks et al. 1986). The patches seen by fluorescence may correspond to sites of anchorage of the cytoplasmic network of actin filaments to the cell membrane or possibly the cell wall. Good evidence both for such anchorage of the actin cytoskeleton to the plasma membrane and for transmembrane attachments linking the cytoskeleton to the basement membranes or other substrata has been obtained in many other systems, using a variety of techniques.

Finally, the ring structures described above may correspond to the filamentous septal belt seen in other fungi by electron microscopy (Streiblova and Girbardt 1980). They are different from the rings of 10-nm-diameter filaments (also seen in the neck regions of the budded cells) reported by Byers and Goetsch (1976), as the actin structures are seen only late in the cell cycle, whereas the 10-nm filaments are present throughout most of the budded phase. The localization of actin to the neck region is consistent with its association with regions of growth, as the new cell wall must be deposited at the site of cytokinesis. It is also possible that the actin at the neck might correspond to the animal cell cleavage furrow, but there is as yet no evidence for contraction of the yeast actin ring during cytokinesis.

A. Actin Cytoskeleton Function

It was suggested above that the actin cytoskeleton has many diverse functions and that these functions are likely to be mediated by the many diverse proteins which have been shown to be associated with actin. The distribution of actin in cells, as we have seen, is consistent with a great variety of roles for the cytoskeleton. Virtually all the more substantial inferences about the function of actin derive from mutant phenotypes: (1) the phenotypes of mutations in *ACT1*, the actin structural gene itself, and (2) the phenotypes of the many genes that have been associated, by one or more of the methods discussed above, with the actin cytoskeleton.

1. Phenotypes of Actin Mutants

Early approaches to recovering mutations in the cloned yeast actin gene resulted in three conditional-lethal mutations (Shortle et al. 1984; Dunn and Shortle 1990). Much was learned about the primary actin mutant phenotypes from these mutations, as described below. Subsequently, two complementary site-directed mutagenesis strategies were employed to create a larger collection of actin mutations. One strategy was to alter residues implicated, for example, in interactions with myosin and to determine the effects of these mutations in vitro and in vivo. These studies (see, e.g., Johannes and Gallwitz 1991; Chen et al. 1993; Miller and Reisler 1995; Miller et al. 1995, 1996) contribute mainly to understanding the issues of mechanisms of interaction between actin and myosin and issues related to actin filament stability, as opposed to discerning the roles of the actin filaments in the living cell. The other site-directed strategy, charged-to-alanine scanning mutagenesis, was designed to target for mutation residues located on the surface of actin, without regard to prior biochemical information or bias. Studies of these mutations, especially in combination with the two-hybrid system (see below), have yielded considerable information about the in vivo functions of the actin cytoskeleton.

The charged-to-alanine mutagenesis resulted in the isolation of 11 recessive-lethal mutants and 16 conditional mutants and 7 alleles that had no discernible phenotype (Wertman et al. 1992). Two mutations were not recovered and may be dominant alleles. The design of the study resulted in substitutions of the mutations for the normal actin gene, so there are no issues of copy number or promoter strength. When the phenotypes of the mutations are considered with respect to the positions of the mutations on a three-dimensional model of the monomer based on the coordinates of Kabsch and Holmes (Fig. 5) (Amberg et al. 1995), a consistent picture emerges in which mutations in the regions implicated in intersubunit interactions along the filament, or with the nucleotide and divalent cation, are most likely to have severe phenotypes (Wertman and Drubin 1992; Wertman et al. 1992). On the other hand, mutations in regions that have diverged in the most distantly related actins had milder phenotypes (Wertman et al. 1992). These data thus provided valuable tests for the structural models of actin. However, the collection of mutations proved even more valuable because of the diversity of phenotypes found for the various alleles (Drubin et al. 1993) and the many uses that subsequently were found for these mutants in genetic and biochemical studies described below.

Assessments, at both permissive and nonpermissive temperatures, of the phenotypes of the original conditional-lethal and the temperature-sensitive subset of the charged-to-alanine actin mutants have shown great variety: Phenotypes observed for some, but not all, of these mutations include varying degrees of temperature sensitivity, osmotic sensitivity, sensitivity to ion concentrations in the medium, delocalized chitin deposition, a variety of gross morphological defects, abnormal nuclear segregation and/or cytokinesis, randomized bud site selection defects in diploids, alterations in organization and distribution of intracellular organelles, accumulation of intracellular secretory vesicles, endocytosis failure, and defects in the intracellular distribution of actin, a property shared by virtually all the mutants. It is important to note that none of the conditional actin mutants completely eliminate polymerized actin structures.

To make sense of this profusion of phenotypes, it is useful to consider not only the phenotypes of actin mutations, but also the phenotypes of mutations in the genes that have been associated with actin. If indeed the diversity of functions is managed through the diversity of associated proteins that bind or otherwise interact with actin, mutations in the genes that specify these proteins should have a more restricted range of phenotypes. As we describe in detail below, this turns out to be the case, although there is generally no simple one-to-one correspondence. Furthermore, individual actin alleles may have more restricted phenotypes because the primary defect is an aberrant interaction with some (but not all) of the binding proteins. There might therefore well be regions of the actin monomer surface where mutations have effects on some ligands (and the functions they control) and not on others. This also turns out sometimes to be the case, and we turn to that issue toward the end of the discussion of actin function.

Figure 6 and Table 1 provide two views of the number and complexity of the interactions of the genes and proteins associated with actin. The various lines of evidence linking each of the genes with actin are given by a color code in the figure, and the various genes are grouped spatially according to their apparent functions. We briefly discuss here the several different major functions of the actin cytoskeleton, as inferred from the phenotypes of mutations in the genes specifying actin and the proteins associated directly (and sometimes indirectly) with actin.

2. *Polarized Cell Growth and Secretion*

The first line of evidence that the actin cytoskeleton functions to provide polarized cell growth and protein secretion derives from the phenotypes of *act1* mutants. Virtually all of these mutants have some level of defect in actin staining with phalloidin or anti-actin antibodies, and virtually all show a reduction or elimination of the asymmetry in actin staining between mother and bud, with cortical patches abundant in mother as well as bud. Generally, the most defective in actin cytoskeleton asymmetry produce very large budded cells, indicating growth of both mother and bud. Several actin mutants also appear to accumulate invertase in intra-

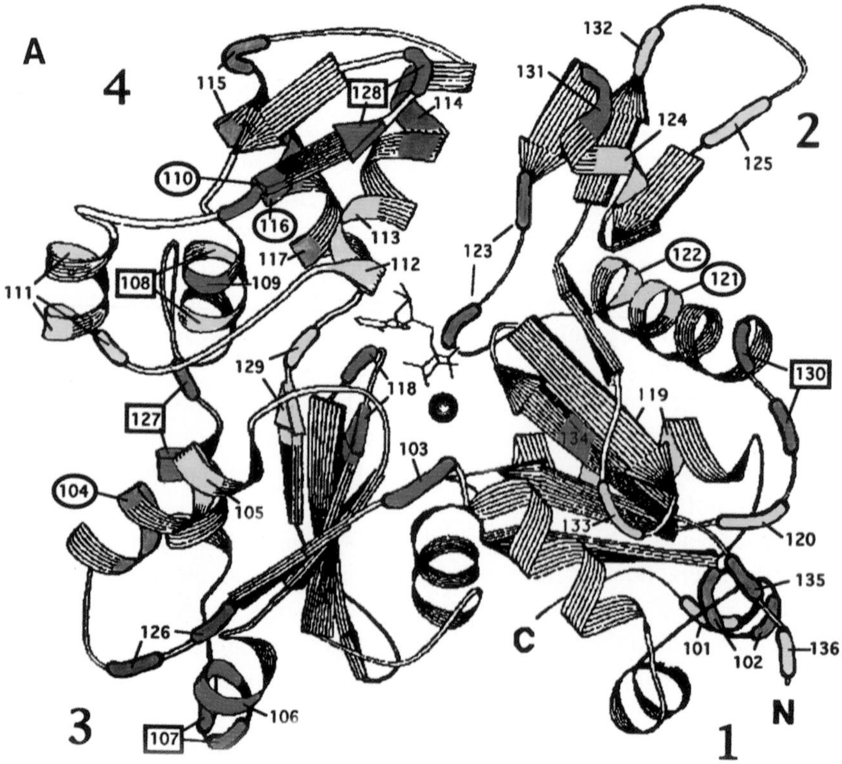

Figure 5 The alanine scan alleles of actin: correlation of phenotype with location of mutated charged-to-alanine clusters on the three-dimensional structure of actin. (*A*) The alanine scan alleles are shown on a ribbon diagram of rabbit muscle actin structure. This view is designated as the "front view" of actin. Actin subdomain numbers are indicated at the corners. Allele numbers are indicated; circled alleles were tested in two-hybrid interaction assays with known actin-binding proteins and found to disrupt at least one binding interaction; boxed alleles were similarly tested and had no effect. (*Blue*) Dominant lethal phenotype; (*red*) recessive lethal; (*yellow*) temperature-sensitive; (*green*) no phenotype. (*B*) The solvent-exposed surface of the rabbit muscle actin monomer structure, shown in front view, with positions of alanine scan alleles indicated with colors, as in *A*. (*C*) Same as for *B*, except back view of the actin monomer. *See facing page for B and C parts.*

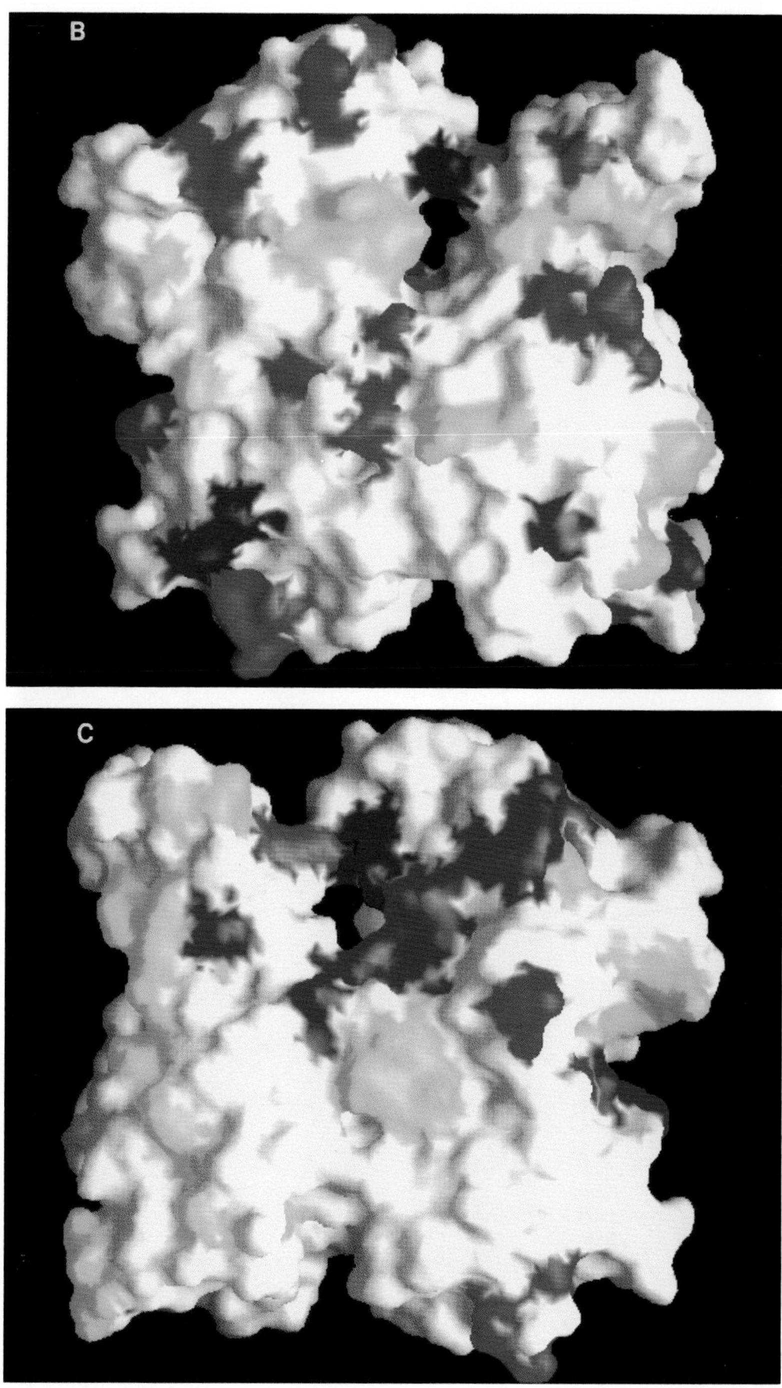

Figure 5 (*See facing page for part A and legend.*)

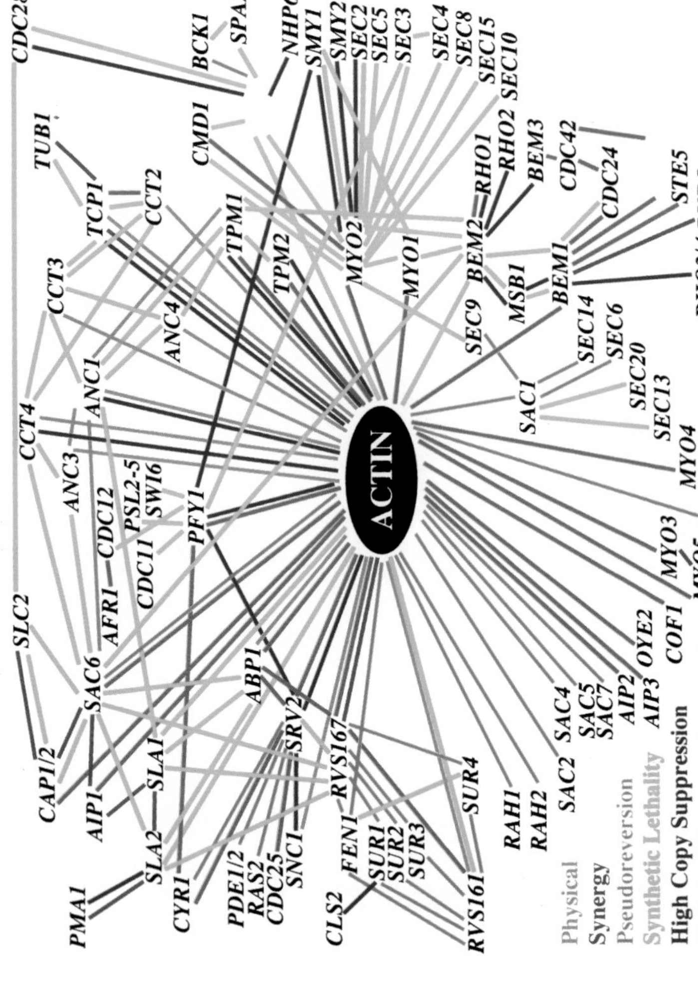

Figure 6 An interaction map for actin, and proteins involved in actin cytoskeleton structure. Interactions between genes or gene products are indicated by lines; the colors of the lines indicate the type of interaction.

cellular vesicles, a hallmark of a defect in secretion. Finally, many actin mutants show aberrant (usually delocalized) deposition of chitin, which normally appears only at bud necks and in bud scars.

A second connection of actin to polarized cell growth and secretion is the *SAC1* gene (Fig. 6, ~5 o'clock), which was associated with actin through suppression of *act1-1* and synthetic lethality with *act1-2* (Novick et al. 1989). The null *sac1* phenotype is cold sensitivity for growth, depolarized actin patches, and delocalized chitin deposition, but not suppression of *act1-1* (Novick et al. 1989). *SAC1* also turns out to be mutable to suppress mutations in *SEC14* (which encodes a phosphatidyl inositol transfer protein) including, remarkably, otherwise lethal null alleles (Cleves et al. 1989). Sac1p is an integral membrane protein that localizes to the Golgi and endoplasmic reticulum membranes (DeRosier and Tilney 1982). Some *sac1* alleles suppress alleles of *SEC9* and *SEC6*, each of which is involved in post-Golgi steps in the secretion pathway. Synthetic lethality is also observed between *sac1-6* (an actin-suppressing allele) and alleles of *SEC13* and *SEC20*, which have roles in early steps of secretion (Cleves et al. 1989). Thus, the actin cytoskeleton is implicated, through *SAC1*, in at least the latter part of the secretion pathway in yeast.

A third connection to the secretion pathway is through *MYO2* (encoding a type I unconventional myosin; Fig. 6, ~3 o'clock). *MYO2* (whose product presumably binds actin directly) is associated directly, by synthetic lethality among mutants, or indirectly, via the high-copy suppressor genes *SMY1* and *SMY2*, with a large number of genes in the secretory pathway (Burshell et al. 1984; Lillie and Brown 1992, 1994). Myo2p also directly binds calmodulin (the product of *CMD1*), and these two proteins are colocalized throughout the cell cycle (Brockerhoff et al. 1994). Antibody to either Myo2p or calmodulin stains cells at the presumptive bud site, the surface of small buds, the neck in large-budded cells, and the shmoo tip in mating cells. In short, these proteins stain the same regions in which one finds a high concentration of actin cortical patches, although it is clear that these proteins do not colocalize with the actin cortical patches (Brockerhoff et al. 1994). It is tempting to associate these localizations with secretion and membrane growth, since the bud site, small buds, the bud neck (especially during the formation of the septum), and the shmoo tip are the regions in which maximal rates of membrane growth and protein secretion occur.

In summary, the evidence that the actin cytoskeleton functions in polarized cell growth and protein secretion derives from the phenotypes of actin mutants, and the genetic connections between genes specifying proteins undoubtedly associated with actin (Sac1p and Myo2p) and

Table 1 Genes and proteins of yeast actin cytoskeleton

GENE	ESSENTIAL	SEQUENCE/MOTIFS	LOCATION	OSMOLARITY
ABP1	NO	SH3, cofilin	patches	
AIP1	NO	4xWD40	patches, cables	
AIP2	NO	FLAVIN		
AIP3	NO	Coiled-coil	pres. bud, small bud tip, neck	
ANC1	NO	Is a Transcription factor	Nucleus	Hypersensitive
ANC3	?	4xWD40		
ANC4	?			
BEM1	NO			
BEM2	NO			
CAP1	NO	Is CAP/alpha subunit	bud cortical patches	
CAP2	NO	Is CAP/beta subunit	bud cortical patches	
CCT2	YES	Is a TCP-1 Chaperonin		
CCT3	YES	Is a TCP-1 chaperonin		
CCT4	YES	Is a TCP-1 chaperonin		
CMD1	YES	Is Calmodulin	pres. bud, small bud tip, neck, spb	
COF1	YES	Is Cofilin	cytoplasmic, cortical patches	
END3	NO	EF-hand, PIP2 binding, coiled-coils		
MPK1	NO	Is a MAP kinase		Remedial
MYO1	NO	Is a Type II Myosin		Hyposensitive
MYO2	YES	Is a Type I Myosin	pres. bud, small bud tip, neck	
MYO3	NO	Is a Type I Myosin		
MYO4	NO	Is a Type V Myosin		
MYO5	NO	Is a Type I Myosin		
NHP6A	NO	HMG-1		Remedial
NHP6B	NO	HMG-1		Remedial
OYE2	NO	Is a NADPH Oxido-Reductase	cytoplasmic	
PFY1	YES/NO	Is Profilin	plasma membrane, cytoplasmic	Hypersensitive
PSL1/SEC3	NO	Coiled-coil		
PSL2	?			
PSL3	?			
PSL4	?			
PSL5	?			
RAH1	?			
RAH2	?			
RAH3	?			
RVS161	NO	Amphiphysin	plasma membrane (GFP)	Hypersensitive
RVS167	NO	Amphiphysin, SH3, Pro rich		Hypersensitive
SAC1	NO	HIV-1 Vpr	Golgi, ER	
SAC2	NO	unique		
SAC3	?			
SAC4	?			
SAC5	?			
SAC6	NO	Is Fimbrin	cables and patches	Hypersensitive
SAC7	NO	Bik1, Sla1 (both weak)		
SLA1	NO	3xSH3, Bindin Repeats	cortical patches	Hypersensitive
SLA2	NO	Talin	cortical patches (subset)	Hypersensitive
SLC1	?			Hypersensitive
SLC2	?			Hypersensitive
SMY1	NO	Kinesins, Coiled-coil	pres. bud, small bud tip, neck	
SMY2	YES		cortical patches	
SRV2	?			
SUR1	?			
SUR2	?			
SUR3	?			
SUR4	?	Transmembrane Domains		
TCP1	YES	Is a TCP-1 Chaperonin	cytosolic, cortical	
TPM1	NO	Is a Tropomyosin	cables	
TPM2	NO	Is a Tropomyosin		

ACTIN CABLES	CORTICAL PATCHES	ENDOCYTOSIS	BIOCHEMICAL INTERACTION	2-HYBRID INTERACTION
	Deloc'd if Overexpressed	WT	F-actin, Srv2	Rvs167
	Delocalized			Act1
				Act1, Bem1
None	Delocalized	WT		Act1, Aip3
Disoriented	Delocalized		RNA PolII, TFIIF	
	Delocalized		Act1, Cdc24, Ste5, Ste20	
	Delocalized		Rho1	
None	Delocalized		Cap2, F-actin	
None	Delocalized		Cap2, F-actin	
Disorganized	Delocalized		Tcp1	
Disorganized	Delocalized			
	Delocalized			
None	Delocalized	Internalization	Myo2, Nuf1	Nuf1
			F-actin	
None, Aggregates	Fewer in the Bud	Internalization		
?	Delocalized			
			F-actin?	
None	Delocalized	WT	Cmd1, F-actin?	
None, Chunks	Delocalized			
None, Chunks	Delocalized			
More	More		Oye2, Oye3	Act1
None, Bars	Delocalized		G-actin, Cyr1, PIP2	
None	Delocalized			
None	Delocalized	Internalization		Rvs167
None	Delocalized	Internalization		Act1, Abp1, Rvs161
None	Delocalized			
None, Bars	Delocalized			
None, Bars	Delocalized			
Reduced	Delocalized	Internalization	Act1	Act1
None	Delocalized			
None, Bars	Delocalized			
Disorganized	Delocalized	Internalization		
None	Delocalized			
None	Delocalized			
WT	WT			
			G-actin, Abp1, Cyr1, Cdc25	Act1
Disorganized	Delocalized		Bin3	
None	WT		F-actin	
WT	WT		F-actin	

Table 1 *(Continued on following pages.)*

Table 1 (Continued.)

GENE	CAN SUPPRESS BY MUTATION:	IS SUPPRESSED BY MUTATION IN:	CAN SUPPRESS BY DOSAGE
ABP1			
AIP1			
AIP2			
AIP3			
ANC1			*act1-1*
ANC3			
ANC4			
BEM1			*rho3+rho4*
BEM2		*GRR1, SRK1*	
CAP1			
CAP2			
CCT2			
CCT3			
CCT4			*ACT1 Overexpressed*
CMD1		*NUF1/SPC110*	
COF1			
END3			
MPK1	*slk1*		*cdc28*
MYO1			
MYO2			
MYO3			
MYO4			
MYO5			
NHP6A			*spa2Δ+mpk1Δ*
NHP6B			*spa2Δ+mpk1Δ*
OYE2			
PFY1			*ACT1 overexpression, srv2*
PSL1/SEC3			
PSL2			
PSL3			
PSL4			
PSL5			
RAH1	*act1-1*		
RAH2	*act1-1*		
RAH3	*act1-1*		
RVS161		*FEN1, SUR1-4*	
RVS167		*FEN1, SUR1-4*	
SAC1	*act1-1, SEC6,9,14*		
SAC2	*act1-1*		
SAC3	*act1-1*		
SAC4	*act1-1*		
SAC5	*act1-1*		
SAC6	*act1-1*		
SAC7	*act1-1, act1-4*		
SLA1			
SLA2			*PMA1 overexpression*
SLC1			
SLC2			
SMY1			*myo2-66, pfy1*
SMY2			*bet1, myo2-66, sec16-3, sec22, spt15*
SRV2	*cyr1, pde1/2, ras2*		
SUR1	*RVS161, RVS167*		*CLS2*
SUR2	*RVS161, RVS167*		
SUR3	*RVS161, RVS167*		
SUR4	*RVS161, RVS167*		
TCP1			*act1-1, act1-4*
TPM1			
TPM2			

SUPPRESSED BY DOSAGE OF:	SYNTHETIC LETHAL
	sla1, sla2, sac6
	cct3, anc4, tpm1, sla1
	cct4, sac6
	anc1, cct3, tpm1
MSB1	*bem2, cdc24, msb1*
BEM3, MSB1, RHO1, RHO2	*act1, bem1, msb1, myo1, myo2, sac6, sit4, tpm1*
	slc2, sac6
	slc2, sac6
	cct4, cct3, tcp1
	anc1, cct4, anc4, cct2, tcp1
	cct3, cct2, sac6, anc3
	myo2, mpk1
NHP6	*cdc28, spa2*
	bem2, myo2, smy1
SMY1, SMY2	*bem2, cmd1, myo1, sec2,4,5,8,15,10,9, mpk1, tpm1*
SMY1	*cdc11, cdc12, psl1-5, swi6*
	bet2, bet3, myo2, pfy1, sec4, gdi1, ypt1
	pfy1
	pfy1
	pfy1
	pfy1
	act1-2
	act1-1, sac6, sla1, sla2, srv2
	sec13, sec20
	abp1, cct4, anc3, bem2, cap1, cap2, rvs167, sla2, slc2
	act1-108,133, abp1, anc1, rvs167, sla2
	act1-3,101,108,115,116,120,124,125,129,133, abp1, rvs167, sac6, sla1
	cap2
	cap2, cdc28, sac6
	myo1, myo2
PFY1, SNC1	*rvs167*
	fen1
	cct3, cct2, tub1
	anc1, anc4, bem2, myo2, tpm2
	tpm1

Table 1 *(Continued on following page.)*

Table 1 (Continued.)

GENE	NON-COMPLEMENTATION	SYNERGY
ABP1	*act1-4*	
AIP1		*act1-119,125,133, SLA1, SAC6*
AIP2		
AIP3		
ANC1	*act1, anc3, sac6, tpm1*	
ANC3	*act1, anc1*	
ANC4	*act1*	
BEM1		*bud5*
BEM2		
CAP1		*sac6, slc2*
CAP2		*sac6, slc2*
CCT2	*act1-4*	
CCT3	*act1-4*	
CCT4	*act1*	
CMD1		
COF1		
END3		
MPK1		
MYO1		
MYO2		
MYO3		*myo5*
MYO4		
MYO5		*myo3*
NHP6A		
NHP6B		
OYE2		
PFY1		
PSL1/SEC3		
PSL2		
PSL3		
PSL4		
PSL5		
RAH1		
RAH2		
RAH3	*act1-1*	
RVS161		
RVS167	*act1-1*	*act1-4*
SAC1		
SAC2		
SAC3		
SAC4		
SAC5		
SAC6	*anc1*	*aip1Δ, cap1, cap2*
SAC7		
SLA1		*aip1Δ*
SLA2		*pma1*
SLC1		*sac6*
SLC2		
SMY1		
SMY2		
SRV2		
SUR1		
SUR2		
SUR3		
SUR4		
TCP1	*act1*	*act1-1, act1-4, tub1-1*
TPM1	*anc1*	
TPM2		

numerous *SEC* genes whose primary function is in polarized cell growth and secretion.

3. Osmoregulation

The second most striking actin mutant phenotype (after the defects in intracellular actin distribution described above) is hypersensitivity to high osmolarity in the growth medium. This intolerance to osmotic stress is a characteristic phenotype for some, but not all, actin mutants. In wild-type cells, shifts in osmolarity of the medium result in delocalization of the actin cortical patches and apparent loss of the cables. Hypersensitivity to osmotic pressure is a phenotype of *sac6* (fimbrin) and *sla2* mutations.

Mutations in *RAH1*, *RAH2*, and *RAH3* (~8 o'clock in Fig. 6) specifically suppress the osmotic sensitivity of *act1-1* without relieving the temperature-sensitive growth defect. Temperature-sensitive alleles of *RAH3* show delocalization of actin cortical patches only at the nonpermissive temperature. The temperature-sensitive *rah3-2* allele shows synthetic lethality with *act1-2* and unlinked noncomplementation with *act1-1* (Chowdhury et al. 1992).

Null mutations in the gene encoding profilin (*PFY1*), although viable, are temperature-sensitive for growth, have disordered actin cytoskeletons, and are osmosensitive (Haarer et al. 1990). Mutants in *RVS167* and *RVS161* (9 o'clock in Fig. 6) were first identified by their sensitivity to starvation for carbon, nitrogen, or sulfur. Mutations in each of these genes and double mutations in both genes are hypersensitive to salt in the medium (Crouzet et al. 1991; Bauer et al. 1993). Rvs167p binds actin in the two-hybrid system and this binding is affected by the same actin mutations as is profilin binding (Amberg et al. 1995). Mutations in both *RVS* genes exhibit synthetic lethality and unlinked noncomplementation with *act1-1* (M. Aigle, pers. comm.). Suppressor mutations of these defects have been found in four genes (*SUR1, SUR2, SUR3*, and *SUR4*); the mutations suppress the salt sensitivity of *act1-1* as well (Desfarges et al. 1993). In ordinary media, the actin cytoskeletons of *RVS* gene mutants are only slightly disordered, but in the presence of high salt or under starvation conditions, the cortical patches are delocalized and the cables disappear, but polarized cell growth and invertase secretion continue (Sivadon et al. 1995).

These data, taken together, strongly suggest that one of the functions of the actin cytoskeleton is to deal with osmotic stress in the medium. The ultrastructure of actin cortical patches, notably the intimate association with characteristic invaginations of the plasma membrane, has suggested that the cortical actin cytoskeleton may function to stabilize the cell during cell wall and membrane growth (Mulholland et al. 1994).

4. Bud Emergence and Bud Site Selection

One of the striking features of the normal intracellular distribution of actin is the ring of actin cortical patches that forms in cells just before bud emergence. Support for a role in bud emergence and bud site selection (as well as bud growth) comes from phenotypes of some actin alleles. Several alleles show large cells with multiple nuclei. A number of the charged-to-alanine alleles show a "wide-neck" phenotype in which the diameter of the bud neck is abnormal (Drubin et al. 1993); abnormally large bud scars was a phenotype noted in some of the original actin temperature-sensitive mutants as well (Novick and Botstein 1985). Many actin mutant alleles display a "large-cell" phenotype suggestive of difficulty in budding, and some display a multibud phenotype suggestive of poor control of budding.

Bud site selection in diploids (although not in haploids) is also affected by some actin mutations. Strains heterozygous for a deletion of actin show a marked increase in random budding instead of the bipolar pattern characteristic of diploids, and several actin mutant alleles, when homozygous, appear to show completely random budding (Drubin et al. 1993).

Mutations in *BEM1* and *BEM2* are defective primarily in bud emergence, as inferred from their phenotype, which includes large multinucleate cells even at permissive temperature (Breitfeld et al. 1990). Recently, Bem1p has been shown to exist in a complex with actin (Leeuw et al. 1995). Mutations in *BEM2* (which encodes the GTPase activating protein for the *RHO1* GTPase of yeast) (Peterson et al. 1994) display synthetic lethality with mutations in *ACT1*, *TPM1* (which encodes tropomyosin I, a major component of actin cables), *MYO1, MYO2*, and *SAC6* (which encodes yeast fimbrin) (Wang and Bretscher 1995). *bem2* mutants can also affect the microtubule cytoskeleton (see below) (Chan and Botstein 1993). Relatively little has been published on the *rho1* mutant phenotype, except that a temperature-sensitive phenotype caused by the partial complementation of a *rho1* deletion by a human *rho* cDNA results in lysis of cells at the nonpermissive temperature (Qadota et al. 1994). Recently, it has been shown that Rho1p is a regulatory subunit of glucan synthetase, which controls an early step in cell wall biosynthesis (Qadota et al. 1996).

AIP3 was discovered because it encodes a protein that interacts with actin in the two-hybrid system. Null mutants are viable, but they grow slowly and show aberrations in the actin cytoskeleton. Large cells with multiple nuclei are common, as are wide-neck and large chitin rings. Homozygous *aip3Δ* diploids have a defect in the bipolar budding pattern very reminiscent of the actin mutants described above; as before, haploid budding pattern is unaffected (D.C. Amberg et al., in prep.).

These phenotypes suggest that the actin cytoskeleton has a central role in bud emergence and bud growth and participates in bud site selection in diploids but not in haploids.

5. *Endocytosis*

At least some actin mutants (*act1-1* and *act1-2*) are defective in the internalization step of receptor-mediated endocytosis. Deletions of *SAC6* (but not *ABP1*, which is synthetic-lethal with loss of *SAC6*) are also defective (Kubler and Riezman 1993). A screen for additional defective mutants yielded four genes, two of which appear to involve the vacuole and two of which (*END3* and *END4*) are related to the actin cytoskeleton (Raths et al. 1993). *END4* is the same as *SLA2*, which had been isolated in a synthetic-lethal screen with *ABP1* (Holtzman et al. 1993). Deletions of *END3* are temperature-sensitive for growth and have disrupted actin cytoskeletons at the nonpermissive temperature. At the permissive temperature, they bud randomly as diploids, show delocalized chitin deposition, and sometimes fail to complete their septa, producing chains of cells (Benedetti et al. 1994). It is clear from these results that the actin cytoskeleton participates in receptor-mediated endocytosis as well as fluid-phase endocytosis.

6. *Nuclear Division and Cytokinesis*

Many actin mutant alleles produce large cells with multiple nuclei; several alleles also have obvious difficulty producing normal septa. Several of the associated proteins have similar mutant phenotypes. In the process of nuclear division (see Section V.D), the nucleus migrates to the neck between mother and bud, and the spindle is oriented with the long axis of the cell. Cytoplasmic microtubules are required for both of these processes. Disruption of the actin cytoskeleton in an actin mutant results in frequent misorientation of the spindle, indicating that the actin cytoskeleton might be interacting with the cytoplasmic microtubules to effect nuclear and spindle positioning (Palmer et al. 1992). Thus, nuclear positioning is another point at which the actin and microtubule cytoskeletons functionally intersect.

Mutations in *AIP3* produce large cells with several nuclei, fail to make complete septa, and display an aberrant number of long spindles, suggesting a defect in cytokinesis. Aip3p localizes to the developing septum well in advance of the appearance of large arrays of actin cortical patches (D.C. Amberg et al., in prep.).

Mutants defective in *BEM2* accumulate multinucleate cells and show aberrant spindles, but they are also hypersensitive to benomyl (see below). Mutations in this gene were recovered in a screen for mutants that increase their ploidy (Chan and Bostein 1993). It may be that the effect of this gene on nuclear division is indirect.

7. *Sporulation*

A common phenotype of actin mutants is failure to sporulate (D. Drubin, pers. comm.). This phenotype has been little studied in these mutants, but the phenotype is shared by recessive mutations in many actin-associated genes, including *sac1*, *tpm1*, and *sac6* (Novick et al. 1989; Adams et al. 1991; Liu and Bretscher 1992).

8. *Cytoskeletal Assembly*

Assembly of the actin cytoskeleton can be considered at two levels: the self association of actin into filaments and the superorganization of the filamentous network.

Yeast and higher organisms contain a complex of proteins called the TCPs that have a chaperonin role for both actin and tubulin (Lewis et al. 1996). Yeast contains at least four genes that encode proteins of this class, including *TCP1, CCT2, CCT3*, and *CCT4* (Ursic and Culbertson 1991; Chen et al. 1994; Miklos et al. 1994; Vinh and Drubin 1994). Deletion or mutation of these genes results in diverse and familiar defects in the organization of the actin cytoskeleton, as well defects in the function and organization of microtubules (Ursic and Culbertson 1991; Chen et al. 1994; Miklos et al. 1994; Vinh and Drubin 1994). In addition, mutations in these genes show diverse genetic interactions with themselves, mutations in *ACT1*, and the tubulin genes, as well as mutations in actin cytoskeleton-associated proteins (Vinh et al. 1993; Vinh and Drubin 1994; Welch et al. 1993; Ursic et al. 1994). It is not yet clear whether these mutant phenotypes, which seem to be specific to the actin and tubulin cytoskeletons, are the result of failure to fold the actin and tubulin monomers or to a defect in the association of the monomers to form polymer. Because pure actin and tubulin subunits are able to assemble into their respective polymers in vitro, the former seems more likely.

Yeast (like more complex organisms) show a higher-order organization of actin filaments, including the formation of the actin cortical patches and actin cables. Little is understood about the genesis of the actin cortical patch; none of the identified mutants result in the complete absence of this structure, and a drug that eliminates them has only just been reported (D.G. Drubin, unpubl.). It is clear that a large number of proteins colocalize with actin in the cortical patches, including Srv2p, Cof1p, Abp1p, Aip1p, Cap1p/Cap2p, Sla1p, Sla2p, and Sac6p (which is also on actin cables) (Drubin et al. 1988; Amatruda and Cooper 1992; Moon et al. 1993; Freeman et al. 1996), and that mutations in actin (*act1-1*) and *SLA2* cause a loss in the plasma membrane invaginations at actin cortical patches (J. Mulholland, pers. comm.). Sla2p is homologous in its carboxyl terminus to talin, a protein found in higher eukaryotes (Holtzman et al. 1993). Since talins mediate integrin-actin interactions at the cortex of more complex eukaryotes (Drubin and Nelson 1996), it is tempting to speculate that Sla2p might mediate membrane attachment of the actin cortical patch in yeast.

An in vitro system has been developed in which it is possible to observe assembly of actin cortical patches that measures the incorporation of fluorescently labeled actin into the cortical patches of permeabilized cells. Actin incorporation into the cortical patches is lost in *SLA1* and *SLA2* deletion strains and is stimulated by GTP-γS. The stimulatory effect of GTP-γS is believed to be mediated by the small Rho-like GTPase Cdc42p, which is known to be required for bud formation (Li et al. 1995).

Most mutations in *ACT1* and mutations in many of the actin-associated proteins result in loss of the actin cables; however, many of these mutations may cause this phenotype by indirect means. More likely candidates for proteins involved in organizing actin filaments are those that bind to F-actin and/or localize to actin cables in vivo. Sac6p does both and has been shown to be the yeast equivalent of fimbrin actin filament bundling proteins (Adams et al. 1991). Yeast has two tropomyosin-encoding genes *TPM1* and *TPM2*, and the products of these genes have been confirmed to bind and stabilize yeast actin filaments (Liu and Bretscher 1989; Drees et al. 1995). Certainly, the five yeast myosins are expected to bind F-actin, but only Myo2p has been localized in yeast cells and does not colocalize with actin cables (Brockerhoff et al. 1994). Yeast has one gene encoding a cofilin filament severing protein, *COF1*. Cof1p binds to actin filaments in vitro and severs actin filaments, but it is localized to the actin cortical patches and not the actin cables (Moon et al. 1993). The only other protein currently known to localize to actin cables is Aip1p; however, its biochemical function has yet to be determined (D.C. Amberg and D. Botstein, in prep.).

B. Regulation of Filament Assembly

Actin filament assembly is controlled spatially and temporally; presumably, the regulation of this process is coupled to signaling pathways via actin-binding proteins. Some proteins may regulate filament assembly by affecting exchange of ATP for ADP by G-actin. This has been shown for profilin in vitro but has yet to be tested adequately in vivo (for review, see Theriot and Mitchison 1993). Mutations in profilin have been shown to be synthetic-lethal with mutations in two of the genes encoding yeast septins *CDC11* and *CDC12* (B. Haarer, pers. comm.). The septins are clearly involved in cytokinesis, and genetic interactions with profilin indicate that profilin may be involved in cytokinesis in budding yeast. Profilin's role in cytokinesis in the fission yeast *S. pombe* is well established (Balasubramanian et al. 1994).

Another method to regulate filament assembly is through monomer sequestering proteins. Such an activity has also been ascribed to profilin. Other candidates for this class of actin-associated proteins include Srv2p, Oye2p, and Aip2p (Freeman et al. 1995; D.C. Amberg and D. Botstein, unpubl.).

Null mutations in *SRV2* are lethal (Fedor-Chaiken et al. 1990; Field et al. 1990). Domain deletions indicate that Srv2p binds actin monomer in vitro, and its binding is blocked by DNase I, a known monomer-binding protein (Freeman et al. 1995). Srv2p is a bifunctional protein; the amino terminus binds to adenylate cyclase and is not essential, whereas the carboxyl terminus which is essential binds to actin (Gerst et al. 1991; Freeman et al. 1995). Since the amino-terminal half of Srv2p is dispensable for in vivo function, it seems unlikely that the actin binding of Srv2p is regulated by the Ras pathway via its physical association with adenylate cyclase. Overexpression of profilin will suppress a deletion in the carboxyl terminus of Srv2p, supporting the idea that these proteins are functionally homologous for actin monomer sequestration (Vojtek et al. 1991).

Oye2p and Aip2p are predicted to be obligate monomer-binding proteins on the basis of two-hybrid data with actin mutants (see below), but this has yet to be confirmed biochemically. Oye2p is a known flavin-containing NADPH oxidoreductase (for review, see Schopfer and Massey 1991), suggesting that its actin-binding activity could be coupled to a redox cycle. Oye2p is also known to bind estradiols, raising the possibility that an estradiol compound could regulate actin binding (Burshell et al. 1984; Stott et al. 1993). Interestingly, Aip2p is homologous to flavin-binding proteins (although not to Oye2p) and therefore may contain a flavin group, suggesting (like for Oye2p) possible coupling to a redox cycle with respect to regulation of its activity in the actin cytoskeleton (D.C. Amberg and D. Botstein, unpubl.).

C. Organelle Distribution and Inheritance

Covisualization of actin cables and mitochondria in wild-type yeast cells has shown that mitochondria are organized along actin cables and that this organization is disrupted in certain actin mutants (Drubin et al. 1993). In addition, purified mitochondria can be observed to move on actin filaments. One of the five presently identified yeast myosins could be expected to have a role in this movement; however, mitochondria isolated from strains lacking any single myosin gene still support this movement (Simon et al. 1995). Presumably, two or more of the myosins are redundant for this function or a nonmyosin motor is involved.

As previously noted, mutations in actin as well as actin-associated proteins frequently result in multinuclear cells. In some cases, this is clearly due to a failure to complete septation, whereas in others (that do complete septation), it is probably due to a failure to position the dividing nucleus correctly, resulting in poor segregation of the daughter nucleus. Correct positioning of the dividing nucleus may be mediated through associations between the cytoplasmic microtubules and the cortex of the bud. Positioning of the nucleus in the mitotic cycle requires actin function (Palmer et al. 1992). Migration of the nucleus to the tip of the shmoo during treatment with α-factor is also dependent on actin (Read et al. 1992).

D. Delineation of Binding Sites on the Actin Surface

Yeast actin can be readily modeled in three dimensions because of the existence of a high-resolution structure of a highly homologous mammalian actin (Kabsch et al. 1990). The availability of a large number of yeast actin mutations with diverse phenotypes has provided an opportunity to correlate location of mutations with ligand binding and function. Four major methods have proven useful: (1) assignment by relationships of actin mutant phenotypes to the phenotypes of mutations (often null) affecting the binding proteins, (2) direct in vitro measurement of the ability of mutant actin proteins to interact with actin-binding proteins, (3) asking which actin mutants phenocopy genetic interactions between genes encoding components of the actin cytoskeleton, and (4) the differential effects of actin mutants on ligand binding in the two-hybrid system.

SAC6 (yeast fimbrin) was identified as a suppressor of *act1-1* (Adams and Botstein 1989), and subsequent analysis revealed that the original allele of *SAC6* could also suppress *act1-2, 3, 7, 8, 9, 10*, and *120*. These mutant forms of actin were purified and found to be defective in binding wild-type Sac6p but not Sac6p encoded by the suppressor allele (Honts et al. 1994). These actin mutations all map to subdomains 1 and 2 of actin, to that region of the actin surface known to interact with related proteins in mammalian cells (Levine et al. 1990; Mejean et al. 1992; Fabbrizio et al. 1993; McGough et al. 1994). The data indicate that Sac6p bundles actin filaments in much the same manner as fimbrins from more complex organisms.

A completely genetic approach was also used to identify actin mutants defective in Sac6p binding. A deletion of *SAC6* is synthetic-lethal with deletions in *ABP1* or *SLA2* (Holtzman et al. 1993). Mutations in actin that disrupt Sac6p binding would be expected to have a property in common with a *SAC6* deletion; i.e., we would expect these mutations to show synthetic lethality with deletions in *ABP1* and *SLA2*. This is true of *act1-3* and *act1-120*, in complete agreement with results obtained by the methods outlined in the previous paragraph (Holtzman et al. 1994).

Recently, another method for finding binding sites on actin was developed that employs the two-hybrid system. This method presumes that biologically relevant interactions detectable by the two-hybrid system are fundamentally as they occur in living cells. Therefore, altering the amino acids in the binding site for an actin ligand would be expected to disrupt the interaction in the cell as well as in the two-hybrid system.

Wild-type actin in the two-hybrid system was found to interact with several known components of the cytoskeleton including actin itself, profilin, Srv2p, and Rvs167p. Actin also interacted with old yellow enzyme (Oye2p), a previously identified protein not known to be a component of the actin cytoskeleton, as well as three new proteins called Aip1p, Aip2p, and Aip3p (*a*ctin *i*nteracting *p*roteins) (Amberg et al. 1995; D.C. Amberg and D. Botstein, unpubl.).

These ligands were tested for their ability to interact in the two-hybrid system with all 36 alanine scan-derived mutants of actin. Some of the actin mutants did not bind any of the ligands, presumably because they encode unfolded and/or unstable proteins. Most of this class of mutants are encoded by recessive-lethal alleles. At the other end of the spectrum, some of the mutants bound all ligands well. Most of these mutants are encoded by actin alleles with no deleterious phenotype. These two classes of mutants are uninformative with respect to describing the actin-ligand interactions, but they are informative with respect to understanding the effects that these mutations have on the structure of actin. Most of the mutants fell into a third class that bound some ligands well and others not at all and therefore show differential interactions in the two-hybrid system (Amberg et al. 1995).

The third class of actin mutants (those that show differential interactions with the actin ligands) are very useful for describing the interactions between particular ligands and actin. Modeling of disruptive mutations on the structure for rabbit muscle actin (a reasonable approximation of the structure of yeast actin) has delineated likely sites of interaction for these ligands on the surface of actin. One example is displayed in Figure 7 which shows in red those mutations that disrupt the actin-Aip1p interaction in the two-hybrid system and in green those mutations that have no effect. The data indicate that Aip1p binds to the front surface of subdomains 3 and 4 of actin in a manner consistent with Aip1p being an F-actin-binding protein (Amberg et al. 1995). This interpretation is consistent with the localization of Aip1p to F-actin structures in yeast cells (D.C. Amberg and D. Botstein, in prep.). Clearly, understanding where a ligand binds on the surface of the actin molecule can be very informative with respect to the functional consequences of the interaction.

V. THE TUBULIN CYTOSKELETON

The intracellular distribution of tubulin defines the tubulin cytoskeleton, and this distribution changes in characteristic ways during the cell division cycle. Understanding these changes, and how mutations in either the tubulin genes or genes encoding microtubule-associated proteins alter the distribution, is central to understanding the function of the tubulin cytoskeleton.

Microtubules, unlike actin microfilaments, are made of more than one subunit. Yeast β-tubulin is encoded by a single essential gene, *TUB2* (Neff et al. 1983), and yeast α-tubulin is encoded by two genes, *TUB1* and *TUB3*. *TUB1* is essential for growth and *TUB3* is not (Schatz et al. 1986a,b), and at least one of the differences between the two genes is that *TUB1* produces most of the α-tubulin in mitotically growing cells. A fourth tubulin gene in yeast, *TUB4*, encodes a protein that is similar to γ-tubulin from other organisms (Sobel and Snyder 1995; Marschall et al. 1996; Spang et al. 1996). Tub4p is localized to the yeast microtubule-

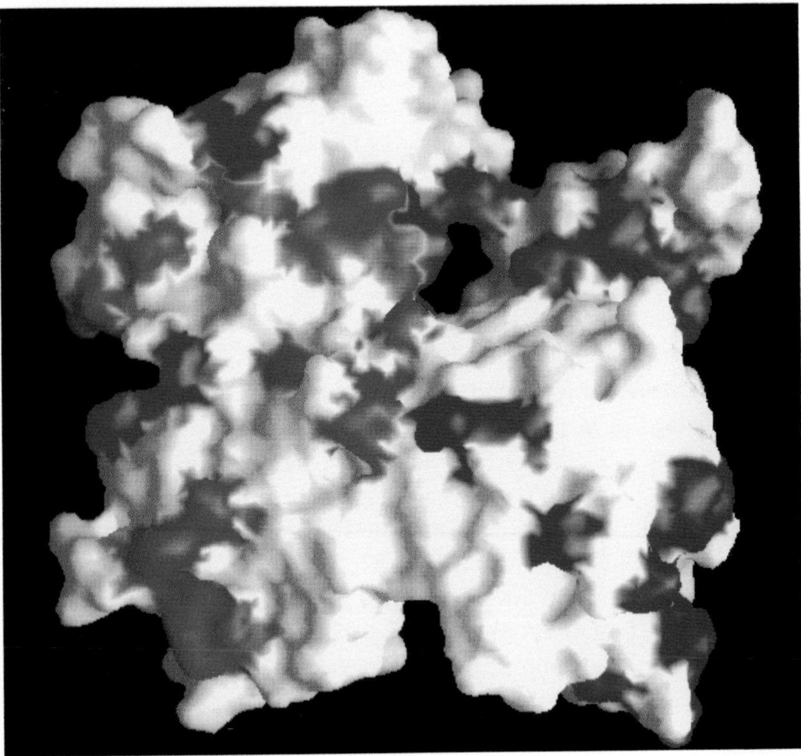

Figure 7 The solvent-exposed surface of the rabbit muscle actin structure, shown in front view. (*Red*) Locations of mutations in yeast actin that disrupt the interaction of Aip1p and yeast actin in the two-hybrid system; (*green*) locations of mutations that had no effect on the interaction.

organizing center and is not incorporated into microtubules (for a detailed discussion, see Section V.E).

A. Yeast Microtubule Cytoskeleton Morphology

The intracellular distribution of the microtubule cytoskeleton has been visualized in four ways: (1) electron microscopy, in which microtubules are revealed by their characteristic shape and dimensions; (2) immunofluorescence with antibodies against the α- and β-tubulin subunits; (3) differential interference contrast (DIC) microscopy; and (4) fusions of the tubulin genes to the gene for GFP, which allows observation of individual microtubules in living cells by fluorescence microscopy. The fundamental findings of each of these methods are described below with examples, followed by an analysis of the microtubule cytoskeleton during the cell cycle.

Landmark investigations by Byers and Goetsch (1975; Byers et al. 1978) used electron microscopy to establish the morphology of the yeast microtubule cytoskeleton. This work was preceded by that of Robinow (Robinow and Marak 1966), who first identified the mitotic spindle in yeast. Byers and Goetsch found two classes of microtubules in yeast: those that extend into the nucleus, which we refer to as intranuclear microtubules, and those that extend outward from the nuclear membrane, which we refer to as cytoplasmic microtubules. It is helpful to keep in mind that in yeast, as in all fungi, the nuclear envelope remains intact during mitosis.

Both intranuclear and cytoplasmic microtubules originate at a structure embedded in the nuclear membrane called the spindle pole body, which is analogous to the centrosome in animal species. Although it has not been proven in yeast, it is likely that the SPB is similar to other microtubule-organizing centers in that it nucleates microtubule growth with a particular polarity (Brinkley 1985), such that all microtubules have their minus ends at the SPB. The location of the SPB and the fact that it is present in only one or two copies per cell, depending on the stage of the cell cycle, thus define the location and orientation of the microtubule cytoskeleton (Fig. 8).

The arrangement of microtubules in fixed yeast cells as determined by electron microscopy was confirmed and extended by immunofluorescence microscopy (Adams and Pringle 1984; Kilmartin and Adams 1984). Although the sequence conservation of the yeast tubulins relative to those of mammals is not quite as striking as that of actin (~75% identity for the tubulins compared to nearly 90% for actin), antibodies that recognize both yeast and mammalian microtubules are easily obtained (Kilmartin et al. 1982). Staining yeast cells with antibodies against either α- or β-tubulin yields the same results; as in other organisms, there appear to be no structures containing either α-tubulin or β-tubulin alone. Examples of antitubulin immunofluorescence at different stages of the cell cycle are shown in Figure 9.

More recently, microtubules have been observed in living yeast cells by both light microscopy (Yeh et al. 1995) and fluorescence microscopy (Stearns 1995). The mitotic spindle can be observed by DIC microscopy (or Nomarski optics). Under optimal conditions, the spindle is visible as a bar bisecting the nucleus (Fig. 10) (Yeh et al. 1995). Remarkably, Robinow (1975) had seen a similar structure using phase microscopy more than 20 years earlier. The spindle can be seen by DIC, but the cytoplasmic microtubules cannot, presumably because of the greater number of microtubules in the spindle and the different refractive indices

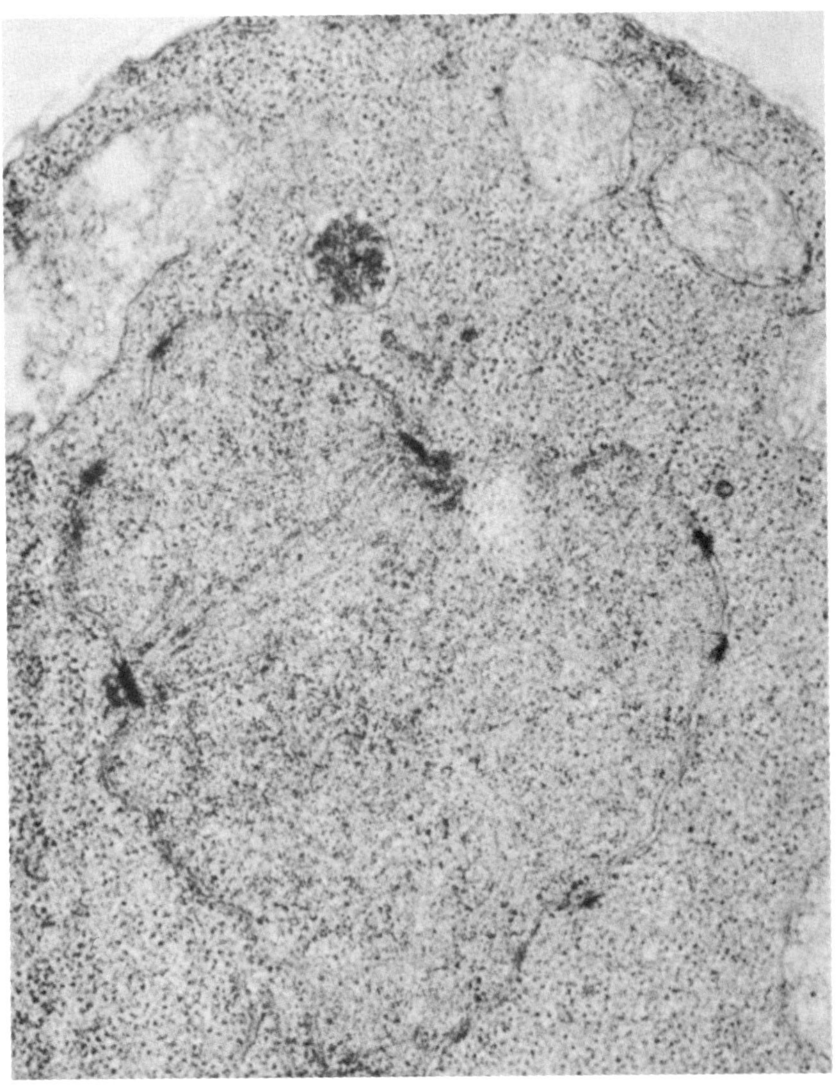

Figure 8 A complete spindle in a budded, mitotically growing yeast cell. The nuclear envelope and several nuclear pores are visible. The two spindle pole bodies are opposed to each other, embedded in the nuclear envelope with spindle microtubules running between them. Some of the microtubules are splayed out, presumably toward the kinetochores. Magnification, 50,000x (Reprinted, with permission, from Byers 1981.)

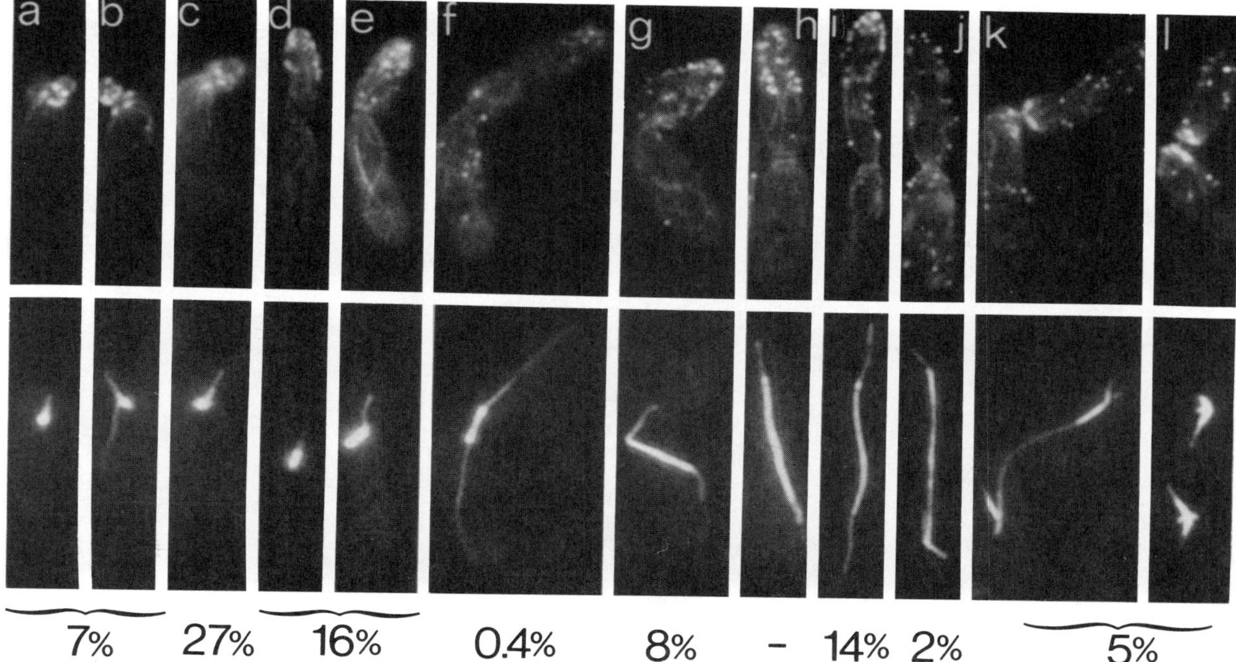

Figure 9 *Saccharomyces uvarum* cells double-labeled with rhodamine-phalloidin to visualize filamentous actin (*top*), and anti-tubulin antibody to visualize microtubules (*bottom*). The cells have been ordered from left to right according to bud size and spindle length. Note that *S. uvarum* has a more elongated cell morphology than *S. cerevisiae*. Magnification, 1800x. (Reprinted, with permission, from Kilmartin and Adams 1984.)

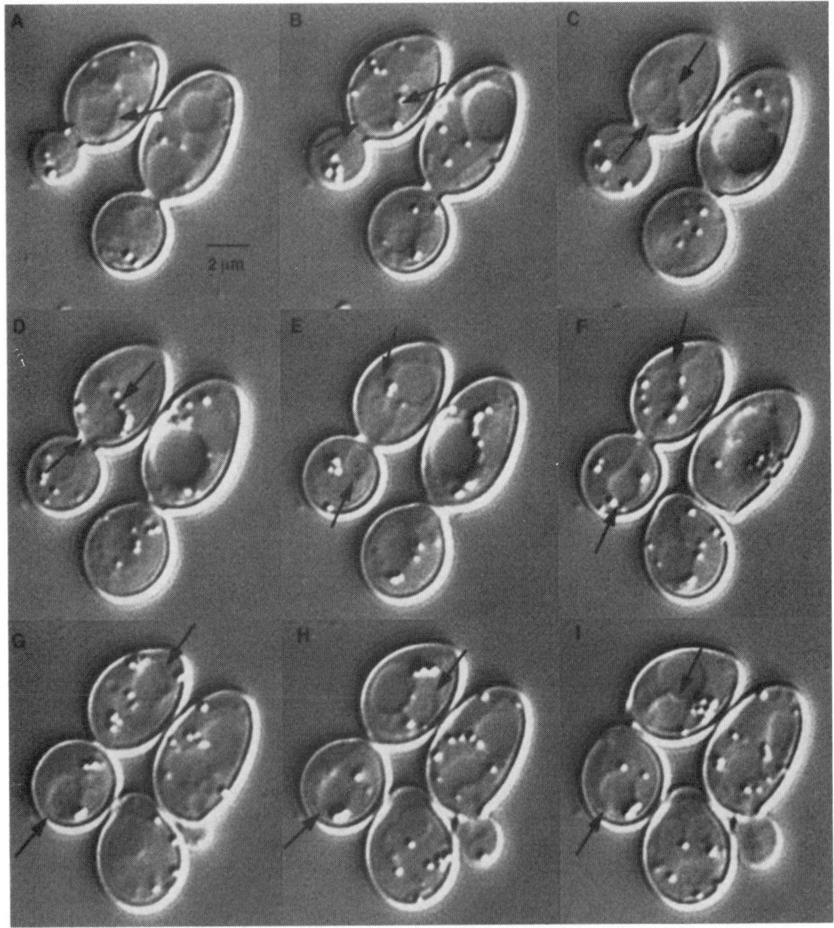

Figure 10 High-resolution DIC images of mitosis in yeast cells. A pair of arrows on each frame indicate the ends of the mitotic spindle in the upper left cell. (*A*,*B*) Spindle assembly. A short spindle, not fully spanning the nucleus, is visible. (*C*) Short spindle bissects the nucleus. (*D*,*E*) Spindle elongation through the neck. (*F*,*G*) Spindle elongation. (*H*) Nucleus moves from distal site in mother to the cell center. (*I*) Cytokinesis. (Reprinted, with permission, from Yeh et al. 1995.)

of the nucleus and cytoplasm. It has been possible to observe individual cytoplasmic microtubules using fluorescent tubulins created by fusion of yeast tubulins to GFP (Fig. 11) (Stearns 1995; Carminati and Stearns 1997). These fluorescent fusion proteins are incorporated into the microtubule cytoskeleton, allowing microtubules to be visualized by fluorescence microscopy. The fluorescence microscopy studies on living cells show that the yeast microtubule cytoskeleton is a very dynamic network of polymers, much like that of animal cells.

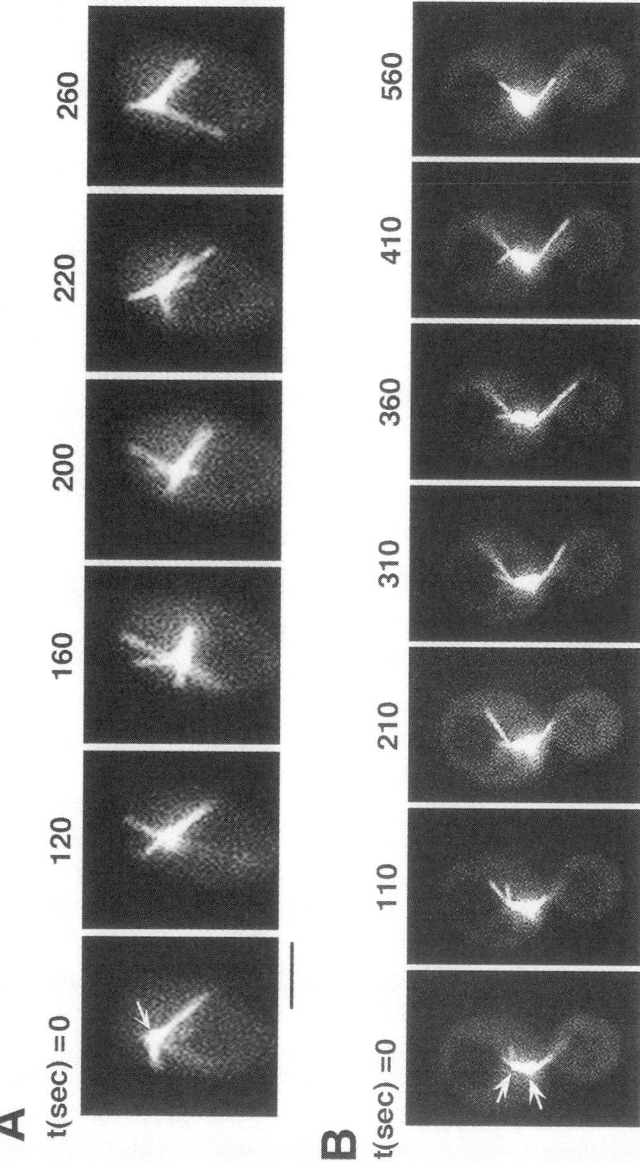

Figure 11 Microtubules in living yeast cells visualized with GFP-α-tubulin. (*A*) Time-lapse images of an unbudded cell. The spindle pole body, at the center of the microtubule array, is marked by an arrow. The time, in seconds, is noted above each frame. (*B*) Time-lapse images of a budded cell. The two spindle pole bodies at each end of the spindle are marked by arrows. Bars, 2 microns. (Reprinted, with permission, from Carminati and Stearns 1997.)

1. *Microtubules in the Mitotic Cell Cycle*

In the unbudded yeast cell, microtubules grow from the single SPB and extend into both the nucleus and cytoplasm. At this early stage in the cell cycle, the number of visually separable cytoplasmic microtubule strands (it is likely, but has not been proven, that each strand is a single microtubule) is greater than later in the cycle; as many as six have been observed in live cells (Carminati and Stearns 1997). This situation should be contrasted with the mammalian centrosome, from which hundreds of microtubles grow into the cytoplasm. The cytoplasmic microtubules are very dynamic at this stage of the cycle, sampling the entire cytoplasm by constantly growing and shrinking (Carminati and Stearns 1997). Nuclear microtubules have been observed by electron microscopy, but it is not known whether the chromosomes are attached to these microtubules prior to the formation of a spindle.

Near the time of bud emergence (and DNA replication), the SPB duplicates. Shortly thereafter, the new SPB begins to grow microtubules, after which the two SPBs separate. A "short" intranuclear spindle (<1 μm long) is formed between the two SPBs. In painstaking analysis of the ultrastructure of the yeast spindle, Winey et al. (1995) showed that the spindle consists of microtubules that go from each of the two SPBs to the presumptive kinetochores of the chromosomes (kinetochore microtubules) and microtubules that go from one SPB to interact with microtubules from the other SPB (interpolar microtubules). It is not clear exactly when the spindle attains this fully assembled form, as the timing of centromere duplication relative to spindle formation has not been well established. The electron microscopy studies provide information only on the structure of various spindles observed. Temporal information about spindle elongation has been obtained in living cells by DIC microscopy (Yeh et al. 1995) and by labeling of the SPBs with a Nuf2p-GFP fusion protein (Kahana et al. 1995).

In the short spindle, there are sufficient microtubules to allow one per kinetochore (32 total in a haploid cell), plus a few interpolar microtubules (Winey et al. 1995). The short spindle state persists through a substantial part of the cell cycle. It has often been assumed that the short spindle stage represents the equivalent of metaphase in animal and plant cells, but there is no indication that yeast has a true metaphase in terms of the morphology of the spindle. For example, the kinetochores, as defined by the ends of the kinetochore microtubules, do not line up in the center of the spindle as they do in animal and plant metaphase (Winey et al. 1995). It is difficult to make direct comparisons between the yeast and metazoan spindles because of the lack of nuclear membrane breakdown and the lack of a well-defined metaphase. For this reason, we only use the metazoan terms for the stages of mitosis when the analogy between other eukaryotes and yeast is unambiguous.

The short spindle elongates to form a "medial" spindle (~1.5 μm) with about eight interpolar microtubules and approximately enough kinetochore microtubules for there to be one per kinetochore (Winey et al. 1995). Finally, the medial spindle elongates, ultimately reaching a length of about 10 μm. During this elongation, the number of interpolar microtubules decreases, and those remaining increase greatly in length. The number of presumptive kinetochore microtubules also decreases as the spindles elongate. The transition to a "long" spindle, resulting in increased distance between the poles, is the equivalent of anaphase B in other eukaryotes. The experimental limitations described above make it difficult to determine whether there is an equivalent in *S. cerevisiae* of anaphase A, in which the kinetochores move to the poles. It is quite possible that the microtubules simply keep the kinetochores a fixed short distance from the SPBs, which are then separated by anaphase B. Alternatively, there may be a very rapid anaphase A that cannot be resolved by the methods used.

The rates of spindle elongation have been determined in studies of living yeast cells (Palmer et al. 1989; Kahana et al. 1995; Yeh et al. 1995), using fluorescence and light microscopy. These studies show that an initial period of rapid, continuous spindle elongation (1.0–1.5 μm/sec) is followed by a period of slower, discontinuous elongation (~0.7 μm/sec). About half of the total increase of spindle length occurs during the slower growth phase. Elongation of the spindle often begins in the mother cell, continuing as the budward SPB moves into the bud. The spindle ultimately extends almost the entire length of the cell.

Positioning of the mitotic spindle within the cell is critical to mitosis in yeast because the point of cytokinesis is necessarily fixed at the neck of the cell. The spindle must separate the chromosomes through the narrow opening of the neck. It appears that a set of movements of the nucleus within the cytoplasm of the mother function to position the spindle properly. After SPB duplication and before the onset of spindle elongation, the nucleus migrates to the neck, typically remaining on the mother side of the junction between mother and bud. The spindle is, by this time, oriented with its long axis parallel to the long axis of the cell, with one SPB close to the opening of the neck. It is important to note that there is great variability from cell to cell in the exact progression of nuclear migration and time of spindle orientation, and yet the end result is almost always a successful mitosis.

Throughout mitosis, cytoplasmic microtubules extend from both SPBs. In most cases, the cytoplasmic microtubules from the budward SPB enter the bud, often extending all the way to the tip of the bud, and the microtubules from the other SPB tend to stay in the mother cell. The number of cytoplasmic microtubules coming from the SPBs is lower during mitosis, typically only one to three microtubules. In addition, these microtubules exhibit slower dynamics than the cytoplasmic microtubules in unbudded cells. In many immunofluorescence images, it appears that the cytoplasmic microtubules interact with the cortex of the cell, particularly in the bud (Snyder et al. 1991). This apparent interaction has been confirmed in living cells with tubulin-GFP fusion proteins (Carminati and Stearns 1997). The cytoplasmic microtubules in the bud spend a significant amount of time with their distal end close to the cortex of the cell. This distal end seems to "sweep" the cortex, maintaining the association while moving laterally. Most interestingly, movements of the nucleus and spindle are usually only seen when at least one cytoplasmic microtubule is in contact with the cortex. The location and behavior of the cytoplasmic microtubules thus suggest that they are involved in nuclear migration and spindle orientation; this has been borne out by functional studies (see below).

At the end of mitosis, the spindle breaks down, starting from the middle. Spindle breakdown, nuclear division, and cytokinesis complete the cell cycle. Although it has not been examined in detail, it is likely that the meiotic spindle shares many of the properties of the mitotic spindle. It is possible that differences at the ultrastructural level will reveal something of the mechanism behind the functional differences between the meiosis I spindle and the mitotic spindle.

2. *Mating*

When two yeast cells mate, the cell walls and membranes fuse to create a common cytoplasm. The two haploid nuclei in this cytoplasm then fuse to form a diploid nucleus, a process called karyogamy. Observation of mating by DIC and fluorescence microscopy indicates that the nuclei become oriented with respect to each other even prior to fusion of the two mating cells. Mating cells grow toward each other, forming a projection. This projection is typically small in mating cells, but it can be quite large in cells treated with high levels of exogenous mating pheromone (the "shmoo" morphology). Nuclei migrate to the tips of shmoos, and this migration is microtubule-dependent (Read et al. 1992). After cytoplasmic fusion, the two nuclei are adjacent and joined by cytoplasmic microtubules extending from their SPBs. In the electron microscope, the nuclei appear to fuse at their SPBs (Byers and Goetsch 1975). The events of orientation and membrane fusion are genetically separable (Kurihara et al. 1994). Functional studies have shown that the cytoplasmic microtubules are essential for karyogamy (see below).

B. Yeast Tubulin Properties

Yeast tubulin has been purified and characterized (Kilmartin 1981; Barnes et al. 1992; Bellocq et al. 1992; Davis et al. 1993) with respect to polymerization properties, drug sensitivity, and stability of the polymer to low temperature. Although fundamentally the same as mammalian tubulin, there are some interesting differences. Yeast tubulin has a lower critical concentration for polymerization than mammalian tubulin, and both the growth and shrinkage rates for yeast tubulin are slower than those for mammalian tubulin (Davis et al. 1993). Yeast and mammalian tubulins will coassemble, and the mixed polymer can be stabilized with the drug taxol, whereas microtubules made from only yeast tubulin are not affected by taxol (Barnes et al. 1992). Yeast microtubules are sensitive to benzimidazole drugs such as benomyl and nocodazole, and these drugs have been used to great effect both in assessing microtubule function and in the isolation of mutations affecting microtubules.

The biosynthesis of tubulin in animal cells has been shown to be quite complex (Lewis et al. 1996), with the monomeric α- and β-tubulin proteins associating with several multimeric complexes before forming a functional tubulin heterodimer. Although the biochemistry of the pathway has not been established in yeast, components of several of the complexes have been identified genetically. These include the genes of the *TCP1*-containing complex (Ursic and Culbertson 1991; Chen et al. 1994; Vinh and Drubin 1994) and *CIN1* and *PAC2*, the homologs of which are found in a 300-kD complex involved in the folding of β-tubulin (Tian et al. 1996).

C. Tubulin: Genetics and Phenotypes

Microtubules, like actin microfilaments, have been implicated in a great variety of intracellular functions in higher organisms. In yeast, however, microtubules appear to be involved exclusively in the movements of the chromosomes and the nucleus. Some functions were obvious from the location of microtubules in the cell; for example, microtubules form the mitotic spindle that separates the chromosomes, as in every other eukaryote. However, virtually all other inferences about what microtubules do (or do *not* do) have depended on the same kind of examination of mutant phenotypes (or the application of microtubule-specific drugs) employed to infer the functions of the actin cytoskeleton.

1. Mutations in Tubulin Genes

The genes for the yeast α and β tubulins were cloned on the basis of their strong homology with their counterparts in other eukaryotes (Neff et al. 1983; Schatz et al. 1986b). α-tubulin is encoded by two genes, *TUB1* and *TUB3*, and β-tubulin is encoded by a single gene, *TUB2*. Disruption of *TUB2* results in lethality (Neff et al. 1983), as does disruption of *TUB1* (Schatz et al. 1986a). Disruption of *TUB3* results in cells that grow quite normally under most conditions (Schatz et al. 1986a). There is currently no evidence for any functional differences between the α tubulins encoded by *TUB1* and *TUB3*. The phenotypic difference in the genes might be explained by the difference in their levels of expression; *TUB1* makes most of the α-tubulin protein. Accordingly, overexpression of *TUB3* will suppress the lethality of a *tub1* null mutation (Schatz et al. 1986a). Although it is not clear why *S. cerevisiae* has two genes for α-tubulin, one essential and one not, this arrangement is conserved in the fission yeast *S. pombe* (Adachi et al. 1986; Yanagida 1987), suggesting that it may have some as yet undiscovered functional significance.

Conditional mutants in the tubulin genes have come from several sources, including classical methods such as the identification of mutants displaying increased rates of chromosome loss, sensitivity, or resistance to antimicrotubule drugs and suppressor analysis, as well as by the more direct method of mutagenizing the genes in vitro. Because the identification of tubulin mutants by the classical methods depended on their phenotypes, these will be introduced briefly here where necessary and discussed in greater detail below. Almost all of the conditional tubulin mutants that have been isolated by the methods described below are cold-sensitive. This presumably reflects the intrinsic cold sensitivity of the microtubule polymer (for review, see Dustin 1984). This tendency toward cold sensitivity has also been observed in genes for other multisubunit structures (Guthrie et al. 1969; Jarvik and Botstein 1975; Moir et al. 1982).

In the case of the actin cytoskeleton, there was no useful alternative to the use of conditional mutations to alter or eliminate function in vivo. This is not the case with tubulins, because antimicrotubule drugs, notably the benzimidazoles benomyl and nocodazole, work well in yeast. Benomyl acts preferentially on fungal microtubules (Kilmartin 1981) and is produced commercially as fungicide on a very large scale. Nocodazole acts on both animal and fungal microtubules (Davidse and Flach 1977). Application of high concentrations of nocodazole or benomyl to wild-type yeast cells results in the nearly complete and largely reversible depolymerization of microtubules (Jacobs et al. 1988; Stearns et al. 1990). These drugs have therefore been extensively used both to study microtubule function and to impose cell cycle blocks at mitosis, where cells without functional microtubules arrest.

Conditional yeast α- and β-tubulin mutations were identified in mutants with altered sensitivity to benomyl. Selection for resistance to benomyl yielded mutations primarily in *TUB2* (Thomas et al. 1985). Many of these mutants were temperature-conditional for growth as well, with the great majority being cold-sensitive. More recently, benomyl-resistant mutations have been recovered in *TUB1* as well (K. Richards and D. Botstein, unpubl.). One benomyl-resistant allele of *TUB2*, *tub2-150*, has the interesting property of requiring benomyl for growth at high temperatures (Thomas et al. 1985; Machin et al. 1995). This phenotype has been interpreted to mean that the microtubules in this mutant are more stable than those of wild type and require the destabilizing effect of benomyl to remain functional. This is supported by the observation that a mutation in one of the *CIN* gene products (see below) suppresses the benomyl requirement of the strain, presumably also by destabilizing the microtubules (Stearns et al. 1990), but microtubule stability has not been examined directly in this mutant.

tub3 null mutants are more sensitive to benomyl than wild-type cells (Schatz et al. 1986a). This phenotype is found in many of the genes involved in microtubule function and will be referred to here as the benomyl supersensitive, or ben^{ss}, phenotype. Screening specifically for benomyl supersensitive mutants yielded mutations in *TUB1*, *TUB2*, and *TUB3* (Stearns et al. 1990), as well as other genes (Stearns et al. 1990; Hoyt et al. 1991; Li and Murray 1991). The sensitivity of yeast cells to benomyl and nocodazole is strongly temperature-dependent; cells become less sensitive as the growth temperature is increased. It is therefore standard practice to assess benomyl sensitivity at 22–25°C and to report the temperature used.

Mutations in both α-tubulin and β-tubulin have also been isolated as mutants displaying increased rates of chromosome loss (Hoyt et al. 1990). This is not surprising given the central role of microtubules in the mitotic apparatus. It is remarkable though that neither *TUB1* nor *TUB2* was identified strictly on the basis of their cell-cycle-arrest phenotypes. Mutations in either gene can result in complete cell cycle arrest at mitosis (see below), a terminal phenotype that has been studied intensively.

Many more conditional mutations were isolated in both *TUB2* (Huffaker et al. 1988) and *TUB1* (Schatz et al. 1988) by in vitro mutagenesis methods. In both cases, mapping of the mutations showed that they occur throughout the coding sequence. As described below, when studied for phenotype, these mutants provided considerable diversity that allowed some conclusions about the functions of microtubules. For example, certain of the *tub2* mutations result in the loss of all microtubules, whereas others result in the preferential loss of the cytoplasmic or intranuclear microtubules (Huffaker et al. 1988). Some of the *tub1* alleles also result in the complete loss of microtubules (in a *tub3* null background), and others result in cytoplasmic microtubules that are longer than normal (Schatz et al. 1988). More recently, both *TUB2* (Reijo et al. 1994) and *TUB1* (K. Richards and D. Botstein, unpubl.) have been subjected to clustered charged-to-alanine scanning mutagenesis by methods analogous to those used for actin and described above. These studies have provided a wealth of information, but the lack of a three-dimensional structure of tubulin has prevented the sort of detailed correlation of structure with function that has been possible with actin.

Although the detailed structure of tubulin is not known, biochemical experiments have yielded some information on functional domains of the protein. Tubulin is a GTP-binding and hydrolyzing protein, and this activity is known to be responsible for some of the basic properties of the microtubule polymer. Tubulin does not have significant homology with members of the well-studied Ras superfamily of GTP-binding proteins, and the mechanism of GTP binding by tubulin is not known. However, cross-linking experiments with animal cell tubulin have identified some of the regions of the protein involved in GTP binding. Mutation of certain of these residues results in proteins that have properties distinct from those of wild type both in vivo and in vitro (Davis et al. 1994).

The phenotype of overexpression of α and β tubulins has also been studied (Burke et al. 1989; Bollag et al. 1990; Weinstein and Solomon 1990), revealing interesting differences in the two polypeptides. Overexpression of *TUB2* is lethal and cells ultimately accumulate aberrant structures that contain β-tubulin but little or no α-tubulin. Overexpression of α-tubulin is not lethal, but overexpression of both α-tubulin and β-tubulin is lethal but results in the accumulation of an excess of structures that appear to be normal microtubules by immunofluorescence microscopy.

2. *Phenotypes of the Lack of Microtubules*

The functions of microtubules in yeast cells have been assessed in two ways: (1) use of nocodazole to depolymerize microtubules and observe the phenotypic consequences (Jacobs et al. 1988) and (2) use of tubulin mutants to examine the consequences of losing all or only certain classes of the complement of microtubules (Huffaker et al. 1988). These two approaches gave largely similar results and defined a limited set of functions for microtubules in yeast cells: formation of the mitotic spindle, nuclear migration and orientation, and karyogamy.

Shift of the most extreme *tub2* mutations or application of adequate concentrations of nocodazole or benomyl results in the relatively prompt (~10 minutes) disappearance of the microtubules, as judged from immunofluorescence microscopy, although in most cases, a very small amount of fluorescence remains at or near the SPB (Huffaker et al. 1988; Jacobs et al. 1988). These phenotypes allow reasonably strong inferences concerning the role of microtubules in the cell. Treatment of growing yeast cells with adequate concentrations of benomyl or nocodazole results in a characteristic uniform cell cycle arrest with a large bud and an undivided nucleus (Jacobs et al. 1988). Shift of many *tub1* or *tub2* mutants to the nonpermissive temperature has the same effect (Neff et al. 1983; Thomas et al. 1985; Huffaker et al. 1988; Schatz et al. 1988; Stearns and Botstein 1988). Cells arrested at this point have duplicated their DNA, but they have not separated their chromosomes. Thus, the arrest is at or before mitosis, showing that in yeast, as in other eukaryotes, microtubules are required for mitosis. During normal yeast mitosis, the nucleus migrates from a random position in the mother cell to a position at the mother-bud neck, with the spindle aligned with the long axis of the cell. In the absence of microtubules, this nuclear migration does not place. Nuclear movements are also important in karyogamy, and in the absence of microtubules, karyogamy does not take place. Interestingly, the complete loss of microtubules does not block SPB duplication but does inhibit SPB separation.

The arrangement of microtubules in yeast, with the cytoplasmic microtubules on one side of the nucleus always associated with the bud, led to the hypothesis that microtubules would be responsible at least in part for the organization of polarized secretion and bud growth. It was surprising then to find that polarized secretion, bud emergence, and bud growth appear to occur normally in the absence of microtubules (Huffaker et al. 1988; Jacobs et al. 1988). The kinetics of secretion of invertase in an extreme β-tubulin allele (*tub2-401*) are normal at the nonpermissive temperature (G. Barnes, unpubl.). Segregation of mitochondria to the bud also occurs normally in the absence of microtubules (Stearns 1988). Thus, unlike in animal cells, there is no evidence that yeast microtubules have any role in vesicular transport or the movement of organelles other than the nucleus.

D. Yeast Microtubule Cytoskeleton Function

The analysis of tubulin function described above indicated that microtubules are involved only in nuclear functions in yeast. Although these functions are more easily defined than those of the actin cytoskeleton, they are nevertheless very complex: Mitosis itself is a fantastically complicated process involving the function of a great many proteins. As is the case in other systems, the function of microtubules in these processes is a combination of the properties of microtubules themselves and of the activities of other proteins, some associated directly with microtubules and others functioning indirectly in the process. Identification of these other proteins has been achieved through application of all of the methods described earlier in this chapter. The relationship between these genes, the microtubules, the SPB, and the kinetochore is depicted in Figure 12 and listed in Table 2. Our intent in the figure is to supply information about how a given gene product is involved in microtubule cytoskeleton and at what level that involvement has been determined. Table 2 includes this information and details of phenotype. The functions of the yeast microtubule cytoskeleton are considered in more detail below.

1. Mitotic Spindle

The most obvious phenotype of the complete loss of microtubules in yeast is the failure of mitosis and arrest of the cell cycle. As is the case for animal cells, yeast cells progress through the cell cycle in the absence of microtubules until they reach mitosis, where they arrest. This terminal phenotype is homogeneous, indicating that there are no other essential functions for yeast microtubules in the mitotic cell cycle. Cells accumulate with a large bud and an undivided nucleus located randomly in the mother cell. Cell cycle arrest in the absence of microtubules is mediated by a checkpoint that requires the function of *MAD* (Li and Murray 1991) and *BUB* genes (Hoyt et al. 1991) (see Lew et al., this volume). Mutations in these genes result in progression of the cell cycle despite the absence of a mitotic spindle. The cell cycle arrest point associated with the lack of microtubules is often used as a standard for mitotic arrest. Al-

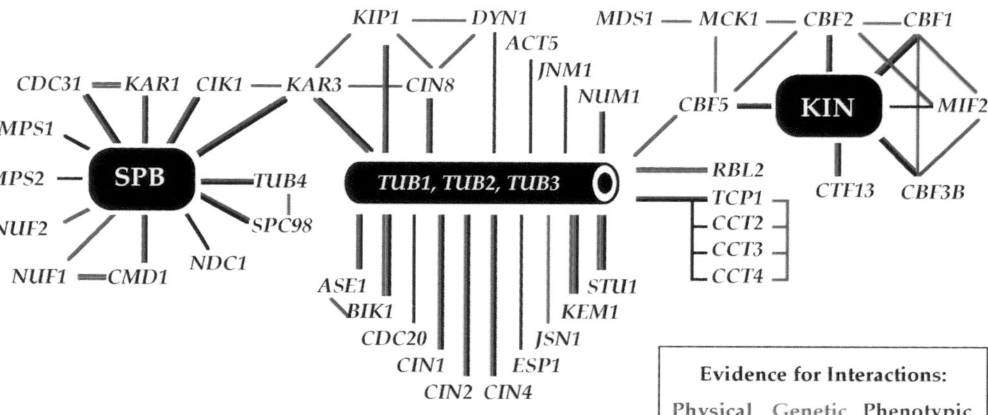

Figure 12 An interaction map for the microtubule cytoskeleton. Interactions between genes or gene products are indicated by lines; the colors of the lines indicate the type of interaction. Specific structures are also indicated: (SPB) Spindle pole body; (KIN) kinetochore. *Physical interactions:* Evidence for a physical association between a gene product and an organelle (SPB, kinetochore, or MTs) includes immunofluorescence and immunoelectron microscopy, cell fractionation, and MT-binding assays. Evidence for physical associations between two gene products includes coimmunoprecipitation, in vitro binding assays, and the interactions delayed by the two-hybrid system. *Genetic interactions:* Evidence for genetic interactions between two genes includes pseudo-reversion, dosage suppression, noncomplementation, and synthetic lethality. Genetic interactions between a gene and *TUB1*, *TUB2*, or *TUB3* are indicated by green lines extending to the MT. For clarity, this diagram does not indicate which of these *TUB* genes displays the genetic interaction; indeed, several genes display genetic interactions with both *TUB1* and *TUB2*. A number of genetic interactions have been observed between these *TUB* genes but are also not indicated in this diagram. *Phenotypic interactions:* Phenotypes include increased rates of chromosome loss, failure to duplicate or properly assemble SPBs, failure to form a functional spindle, failure to align the spindle in the bud neck prior to chromosome segregation, failure to attach chromosomes to the spindle, changes in sensitivity to benomyl, and alterations in levels of MT assembly. Unless some secondary criterion exists, genes that display ambiguous phenotypes are diagrammed as phenotypically interacting directly with MTs.

Table 2 Genes and proteins of the yeast microtubule cytoskeleton

	Genetics				Phenotype							Physical		
Gene	SUP	OX	SL	NC	KO	Ben	MTs	Kar	NM	SPB	Spn	Loc	Bioc	2-H
Tubulins														
TUB1	*TUB2*	*TUB3*	*TUB2* *TUB3*	*TUB2* *TUB3*	L	SS,R	X	X	X		X	MT		
TUB2	*TUB1*		*TUB1*	*TUB1*	L	SS,R	X	X	X		X	MT		
TUB3		*TUB1*			V	SS	X	X	X			MT		
TUB4		*SPC98*	*SPC98*		L					X	X	SPB		SPC98
MAPs														
ASE1			*BIK1*		V						X	Spn		
BIK1			*ASE* *TUB1* *TUB2*		V			X	X			Spn		
CDC20					L			X			X			
CIN1	*TUB2*		*TUB1*		V	SS		X	X		X			
CIN2	*TUB2*		*TUB1*		V	SS		X	X		X			
CIN4	*TUB2*		*TUB1*		V	SS		X	X		X			
ESP1					L						X			
JSN1		*TUB2*			V							Cortex		
KEM1				*TUB2*	V	SS		X	X		X		MT	
MIF2					L	SS					X			
STU1	*TUB2*		*TUB2*	*TUB1*	L						X	Spn	MT	
Kinesins														
CIN8			*KIP1* *DHC1*		V						X	Spn		
KAR3	*CIN8/* *KIP1*				V			X			X	SPB Kin cMTs	CIK1 MT Kin	CIK1
KIP1			*CIN8*		V						X	Spn		
KIP2					V									

Dynein												
DYN1			*CIN8*	V			X			SPB		
ACT3				V			X					
ACT5				V			X					
JNM1				V			X					
NUM1			*TUB2*	V			X			Cortex		
Kinetochores												
CTF13				L							Kin	
CBF5		*NDC10*							X		Kin	
CBF3b				L							Kin	
CEP3												
MCK1		*NDC10*		V	SS							
		CBF5										
MDS1		*MCK1*		V								
SPBs												
CTF14				L					X	Kin	Kin	
CDC31	*KAR1*	*KAR1*		L				X		SPB	KAR1	
CIK1				V		X		X		SPB	KAR3	KAR3
										cMTs		
CMD1			*MYO2*	L						SPB	NUF1	NUF1
											MYO2	
KAR1				L		X		X		SPB	CDC31	
MPS1				L				X				
MPS2				?				X				
NDC1				L				X		NcEnv		
NUF1	*CMD1*			L						SPB	CMD1	CMD1
NUF2				L				X?		SPB		
SPA1				V			X					
SPC42				L				X		SPB	SPB	
SPC98		*TUB4*	*TUB4*	L				X		SPB	SPB	TUB4

(*Continued on following page.*)

Table 2 (*Continued.*)

Gene	Genetics				Phenotype							Physical		
	SUP	OX	SL	NC	KO	Ben	MTs	Kar	NM	SPB	Spn	Loc	Bioc	2-H
Chaperones														
CCT1		*ACT1*	*TUB1*		L	X	X		X					
TCP1			*TUB2*											
			ACT											
			ANC2											
			BIN2											
			BIN3											
CCT2			*ANC2*		L	X	X		X					
BIN3			*BIN2*											
			TCP1											
CCT3			*ANC2*		L	X	X		X					
BIN2			*BIN3*											
			TCP1											
CCT4			*BIN2*		L	X	X		X					
ANC2			*BIN3*											
			TCP1											
CCT5					?									
CCT6					?									
CCT7					?									
CCT8					?									
RBL2		*TUB2*			V								TUB2	

Genetics: (SUP) Mutations in genes listed under SUP can be suppressed by gene in question; (OX) mutations in genes listed under OX can be suppressed by gene in question; (SL) mutations in genes listed under SL are synthetically lethal with mutations in gene in question; (NC) mutations in genes under NC fail to complement mutations in gene in question.

Phenotype: (KO) Gene knockout phenotype (L = lethal, V - viable); (Ben) benomyl sensitivity (SS = supersensitive, R = resistant); (MTs) defect in microtubules; (Kar) karyogamy; (NM) nuclear migration; (SPB) defect in spindle pole body; (Spn) defect in mitotic spindle.

Physical: (Loc) Location in the cell, usually as determined by immunofluorescence; (Bioc) biochemical determination of interaction with other proteins; (2-H) two-hybrid interactions.

though Cdc28p kinase activity is high at the arrest point, it should be noted that the terminal phenotype of a tubulin mutant is a state not found in normal cells and may not be representative of normal mitosis. Indeed, Cdc28p activity reaches higher than normal levels in nocodazole-arrested cells.

The roles of the intranuclear and cytoplasmic microtubules in mitosis have been tested using specific mutants that lack one or the other class of microtubules. At the restrictive temperature, *tub2-406* cells contain microtubules within the nucleus, but these microtubules fail to organize into a typical bipolar spindle; they also contain extensive cytoplasmic microtubules that are functional for nuclear migration (the undivided nucleus is invariably located at the bud neck) (Huffaker et al. 1988). The *tub2-406* mutant cells arrest uniformly with large buds and an undivided nucleus, indicating that functional cytoplasmic microtubules are not sufficient for nuclear division. A caveat of this experiment is that the lack of a spindle likely activates a checkpoint mechanism that might prevent any form of chromosome segregation, even if it were possible.

At a semipermissive temperature, cells bearing the *tub2-401* mutation assemble spindle microtubules but lack detectable cytoplasmic microtubules. Under these conditions, nuclear migration is completely blocked. Spindle elongation and chromosome separation are, however, unimpeded and occur entirely within the mother cells (Palmer et al. 1992; Sullivan and Huffaker 1992). This indicates that cytoplasmic microtubules are not essential for nuclear division.

2. *Microtubule Motors in Mitosis*

Mitosis requires that chromosomes be moved within the cell. Underlying this movement are molecular motor proteins that use microtubules as the tracks upon which to move. Two known classes of microtubule motor proteins, dyneins and kinesins, are found in yeast. Dynein is a very large multisubunit protein that moves toward the minus ends of microtubules (presumed to be the SPB end). The heavy chain of dynein, as characterized in other organisms, has microtubule motor activity, and there are several intermediate and light chains (for dynein review, see Holzbaur and Vallee 1994). There is a single gene for the dynein heavy chain in yeast, *DYN1* (Eshel et al. 1993; Li et al. 1993). Kinesin is a family of smaller proteins, usually consisting of a heavy chain with motor activity and light chains (for kinesin review, see Goldstein 1993). Both plus-end-directed and minus-end-directed kinesins have been identified. There are at least five characterized kinesin-related genes in yeast: *KAR3* (Meluh and Rose 1990), *CIN8* (Hoyt et al. 1992; Roof et al. 1992), *KIP1* (Hoyt et al. 1992; Roof et al. 1992), *KIP2* (Roof et al. 1992), and *SMY1* (Lillie and Botstein 1992) (a sixth kinesin-like gene has been identified by the yeast genome project). Of these, only Kar3p has been shown to have motor activity in vitro: It is a minus-end-directed motor (Endow et al. 1994).

Kip1p and Cin8p appear to be functionally redundant in mitotic function (Hoyt et al. 1992; Roof et al. 1992); single mutants are viable, but the double mutant is inviable. Cin8p and Kip1p have been localized to the spindle microtubules. A $cin8^{ts}$ *kip1*-Δ strain is temperature-sensitive for mitotic spindle assembly. Kar3p was first identified as being required for karyogamy (Polaina and Conde, 1982), but it was also found to have defects in mitosis and meiosis. Kar3p has been localized to microtubules in mating cells (see Section V.D.6) and the SPB in mitotic cells (Meluh and Rose 1990; Page et al. 1994). This localization requires the function of the *CIK1* gene, and Cik1p physically interacts with Kar3p (Page et al. 1994), suggesting that it might be a light chain for the Kar3p kinesin.

Interestingly, mutations in *KAR3* are able to suppress some of the phenotypes of the $cin8^{ts}$ *kip1*-Δ strain. If Kip1p/Cin8p activity is removed from cells that already have bipolar spindles, those spindles collapse inward. The suppressing alleles of *KAR3* are all mutations in the motor domain of the protein and are probably not null for function, as a true null mutant does not suppress (Saunders and Hoyt 1992; Hoyt et al. 1993). One interpretation of this genetic interaction is that Kar3p acts antagonistically to Kip1p/Cin8p in the function of the mitotic spindle. This is consistent with the known polarity of the Kar3p motor (minus-end-directed) and the presumed polarities of the Kip1p/Cin8p motors (plus-end-directed, by analogy to similar motors in animal cells). Kar3p is required for microtubule motility at the yeast kinetochore (Middleton and Carbon 1994) and this may be its principal site of action in mitosis.

How is the dramatic anaphase elongation of the mitotic spindle achieved? In a $cin8^{ts}$ *kip1*-Δ strain, some chromosome segregation occurs at the restrictive temperature but only very slowly. Deletion of the dynein heavy-chain gene *DYN1* results in the complete elimination of chromosome movement (Saunders et al. 1995). This indicates that although the most obvious phenotype of a mutation in the dynein heavy chain is a defect in nuclear migration prior to mitosis (see below), dynein has a more direct role in mitosis as well. It is not clear whether this mitotic role of dynein is carried out in association with the cytoplasmic microtubules, as is thought to occur during nuclear migration, or in association with some component of the intranuclear spindle. In addition to microtubule motor activity, some of the microtubules in the spindle must lengthen to account for the change in spindle length.

3. *Kinetochore Proteins*

The function of the microtubules in mitosis is to segregate the duplicated chromosomes. Therefore, there must a mechanism for attaching the chromosomes to the mitotic spindle. This attachment is achieved through the kinetochore, a specialized region of the chromosome that corresponds to the genetically defined centromere. The yeast centromere is the smallest and most well-characterized centromere: The DNA sequences required for function have been extensively analyzed, and proteins that bind to the centromere to form the kinetochore have been identified (for review, see Hyman and Sorger 1995). The centromeric DNA contains three sequence elements important for function, CDE I, II, and III. CDE III is responsible for attachment to microtubules and is the binding site for a complex of proteins called Cbf3 (Lechner and Carbon 1991). The four known components of the Cbf3 complex are all essential for viability. The genes for these components have been isolated both by purification of the proteins and by various genetic screens. *CEP3* encodes a zinc finger protein (Lechner 1994; Strunnikov et al. 1995), possibly linking the complex to the centromere DNA. *CTF13* and *CTF14* were identified in genetic screens for mutants with defective kinetochores (Doheny et al. 1993; Lechner 1994). *CTF14* was also isolated as *NDC10* and *CBF2* (Goh and Kilmartin 1993; Jiang et al. 1993). A fourth component, *SKP1*, is also essential for complex formation and is conserved in animal cells (Connelley and Hieter 1996).

The microtubule-binding activity of the kinetochore has been investigated in several different assays (Kingsbury and Koshland 1991; Hyman et al. 1992; Middleton and Carbon 1994). The Cbf3 complex is essential for binding of kinetochores to microtubules but is not itself sufficient (Sorger et al. 1994). The yeast kinetochore has been reported to have microtubule motor activity (Hyman et al. 1992) and to have associated Kar3p (Middleton and Carbon 1994). It remains to be seen whether these associations and activities are physiologically relevant.

4. *Other Mutants Defective in Mitotic Microtubule Function*

A temperature-sensitive mutation in *ESP1* causes a defect in nuclear division and produces cells that accumulate extra SPBs (Baum et al. 1988; McGrew et al. 1992). Although these cells assemble short spindles of normal appearance, spindle structure becomes abnormal when anaphase elongation would normally occur. In these abnormal spindles, microtubules radiating from the two poles fail to retain their interdigitation, leading to the generation of two independent half spindles. Cytokinesis occurs despite failure to complete nuclear division, causing massive chromosome nondisjunction. Generally, the daughter cell acquires most of the DNA and both spindle poles. At later times, the missegregated SPBs enter a new duplication cycle leading to the accumulation of extra SPBs in a single cell.

Yeast cells doubly mutant for *BIK1* and *ASE1* also display an extra SPB phenotype. Mutations in *BIK1* were originally shown to cause a bilateral karyogamy defect. Bik1p, which contains a region similar to the microtubule-binding motif of *Tau* and another region similar to the microtubule-binding motif of CLIP-170, colocalizes with microtubules in vivo. Synthetic lethality is observed between a *BIK1* allele and alleles of *TUB1* and *TUB2*, consistent with a physical interaction between these gene products. *bik1* null mutants are viable but display defects in both spindle orientation and elongation (Berlin et al. 1990). *ASE1* was identified in a screen for mutations that are lethal in combination with a *bik1* mutation (Pellman et al. 1995). *ASE1* is also not an essential gene, but a particular *ase1 bik1* double-mutant combination is temperature-sensitive for growth. At its restrictive temperature, this strain is able to assemble a short bipolar spindle, but anaphase spindle elongation is blocked. Like the *esp1* mutation, the *ase1 bik1* mutations do not cause a cell cycle arrest and produce cells with multiple SPBs and increased DNA content. Ase1p localizes to the spindle midzone, consistent with a role in the antiparallel sliding of polar microtubles during anaphase B.

Alleles of *STU1* were identified as suppressors of a cold-sensitive mutation in *TUB2* (Pasqualone and Huffaker 1994). In addition, alleles of *STU1* display synthetic lethality with other alleles of *TUB2* and noncomplementation with alleles of *TUB1*. *STU1* is an essential gene, and temperature-sensitive *stu1* mutations block spindle assembly and nuclear division; neither short nor long spindles are observed in *stu1* cells at the restrictive temperature (T.C. Huffaker, unpubl.). Similar to the mutations discussed above, mutations in *STU1* do not cause a cell cycle arrest, and cytokinesis leads to severe chromosome nondisjunction. Stu1p is localized to spindle microtubules in vivo (Pasqualone and Huffaker 1994) and binds to purified microtubules in vitro (T.C. Huffaker, unpubl.), indicating that it is a microtubule-binding protein.

Cdc20p appears to have a role in modulating microtubule assembly. Following a shift to the restrictive temperature, *cdc20* mutant cells rapidly accumulate anomalous microtubule structures as detected by immunofluorescence microscopy (Sethi et al. 1991). These cells arrest in the cell cycle as large-budded cells with an undivided nucleus located at the bud neck. The mitotic spindle in these cells is about the same length as a normal short spindle, but it appears to have many more microtubules in it than normal. Cdc20p contains five WD-40 repeats. Although these repeats are found in a variety of proteins that do not share any obvious functional properties, one WD-40 protein, EMAP from sea urchin (Li and Suprenant 1994), is a known microtubule-binding protein. Cdc20p shares the greatest sequence similarity with the *Drosophila* protein Fizzy (Fzy), which is required for progression through the metaphase-anaphase transition (Dawson et al. 1995). This raises the possibility that the microtubule defect in *cdc20* mutants is a consequence of arrest in the cell cycle at a point where the microtubule nucleation capacity of the SPB is increased.

CIN1, CIN2, and *CIN4* were isolated in two independent screens; one screen identified mutants that were supersensitive to benomyl (Stearns et al. 1990) and the other identified mutants with increased rates of chromosome loss (Hoyt et al. 1990). Null mutants of each of the *CIN* genes are viable but cold-sensitive for growth. At their restrictive temperature, *cin* null mutants display a cell cycle arrest at mitosis. Both nuclear migration and spindle formation are defective in these cells. The *cin* mutants are also defective in karyogamy. Given this pleiotropy, it seems likely that the *CIN* genes affect microtubule function in general, rather than mitosis specifically. The *cin* mutants are synthetically lethal with many tubulin mutations, but they suppress the benomyl requirement of *tub2-150.* That the *cin* mutant phenotype closely resembles the phenotype caused by moderate concentrations of antimicrotubule drugs suggests that the role of the wild-type Cin proteins is either to stabilize microtubules or to increase the amount of functional tubulin, both of which would be antagonistic to the effects of depolymerizing drugs. The latter possibility is supported by the finding that a protein showing homology with Cin1p is part of a complex involved in the biosynthesis of functional β-tubulin (Tian et al. 1996).

JSN1 was identified as a gene that when overexpressed could suppress the temperature-sensitive benomyl dependence phenotype of the *tub2-150* mutants (Machin et al. 1995). The benomyl dependence phenotype of *tub2-150* suggests that this allele causes an increase in microtubule stability. Overexpressing *JSN1* in wild-type cells produces only a small increase in benomyl sensitivity. Jsn1p is located at the cell surface where it could interact with the ends of cytoplasmic microtubules. Alternatively, Jsn1p may have a less direct role in microtubule stability.

5. *Nuclear Migration and Spindle Orientation*

Prior to nuclear division, the nucleus migrates to the neck and is most often oriented with the spindle "pointing" through the neck. Evidence that the cytoplasmic microtubules are required for this migration and orientation comes from studying the effect of the *tub2-401* mutation at 18°C. At this temperature, *tub2-401* cells assemble spindle microtubules but lack detectable cytoplasmic microtubules. A population of *tub2-401* cells was synchronized before the onset of anaphase such that cells contained short spindles correctly oriented through the bud neck. When these cells were shifted to 18°C, cytoplasmic microtubules disappeared and spindles fell out of the bud neck. Subsequent spindle elongation often failed to extend through the bud neck, resulting in a high frequency of binucleate cells.

Null mutations in the gene for the dynein heavy chain, *DYN1*, result in a partial defect in nuclear migration/orientation. Strains bearing such mutations are viable, but they undergo a mitosis in which separation of the DNA often takes place in the mother cell, followed by segregation of one of the DNA masses to the bud in the majority of cells, but failure in a significant fraction, resulting in binucleate mother cells and anucleate daughter cells (Eshel et al. 1993; Li et al. 1993). Careful analysis of the defects in *dyn1* null cells has shown that the back and forth oscillations of the spindle observed in wild-type cells do not occur in the mutant, and spindle elongation itself remains unaffected (Yeh et al. 1995). A Dyn1p-LacZ fusion protein is localized to the SPB in a microtubule-dependent manner and to the cell cortex. The wild-type protein has not yet been localized. Because *dyn1* null mutants are viable, there is likely to be a second mechanism for ensuring spindle orientation.

In animal cells, functional cytoplasmic dynein consists of the dynein heavy chain and several other associated intermediate and light chains (for a review of dynein structure, see Schroer 1994). In addition, the attachment of cytoplasmic dynein to "cargo" is mediated by the dynactin complex of proteins (Gill et al. 1991). One of the proteins in the dynactin complex is an actin-related protein, Arp1p. Mutations in *ACT5*, the yeast homolog of Arp1p, have defects in spindle orientation and nuclear migration that are very similar to those in dynein heavy-chain mutants (Gill et al. 1991; Clark and Meyer 1994; Muhua et al. 1994). Thus, it is likely that there is a dynactin complex in yeast mediating the action of dynein. Nuclear movement and positioning are also mediated by dynein/dynactin in *Neurospora crassa* (Plamann et al. 1994).

Orientation of the nucleus and SPB takes place quite early in the cell cycle (Snyder et al. 1991). The presumptive bud site is marked by the appearance of Spa2p early in G_1. Soon after appearance of Spa2p, a cytoplasmic microtubule(s) interacts with this part of the cell. Soon after, the nucleus is oriented with the SPB proximal to the Spa2p patch at the bud site, presumably through the action of cytoplasmic microtubules. As the bud grows, Spa2p remains localized to the tip of the bud. Spa2p is not required for viability and *spa2* mutants do not have a nuclear migration defect, so it is likely that Spa2p is only one of several proteins localized to this region of the cell. Given the above results, a simple model for nuclear migration/spindle orientation is that the cytoplasmic microtubules interact with some asymmetrically distributed component of the cell and that this interaction results in the application of force to the nucleus and spindle through the SPBs. Disruption of the actin cytoskeleton results in defects in nuclear migration and orientation (Palmer et al. 1992), suggesting that microtubules interact either with actin directly or with a component that depends on the actin cytoskeleton for its localization. Aspects of this model have been confirmed by video fluorescence microscopy using GFP–α-tubulin fusions to visualize microtubule behavior in living cells (Carminati and Stearns 1996). Throughout most of the cell cycle, cytoplasmic microtubules interact with the bud cortex, and movements of the nucleus are associated with contact of microtubules with the cortex.

6. *Karyogamy*

Successful mating of two yeast cells requires that the cells first fuse cytoplasms and then fuse nuclei. The fusion of nuclei is termed "karyogamy" and depends on microtubule function. Observation of mating cells reveals that the two nuclei appear to move toward each other led by their SPBs and that fusion ultimately takes place at the SPBs, resulting in one larger SPB in the newly formed diploid nucleus. Eliminating microtubule polymer either by treatment with antimicrotubule drugs (Delgado and Conde, 1984) or by mutation of the tubulin subunits (Huffaker et al. 1988) results in mating pairs in which the cytoplasms fuse, but the nuclei do not, creating a binucleate zygote. Fusion of nuclei requires activation of the cells by mating pheromone; spheroplasts that have been forced to fuse cytoplasms do not undergo karyogamy unless treated with mating pheromone (Rose et al. 1986).

Several nontubulin mutations have been isolated on the basis of their karyogamy defect. Some of these *kar* mutants require that both parents be mutant to express the defect (termed "bilateral *kar*"); in others, karyogamy fails even if only one of the parents is mutant (unilateral *kar*). The first *kar* mutants were isolated by Conde (Conde and Fink 1976; Polaina and Conde 1982) and have proven to be very useful in understanding the role of microtubules in the process. *KAR3* encodes a kinesin-type microtubule motor protein (see above). Its expression is induced by α-factor, and Kar3p is associated with the cytoplasmic microtubules in shmooing cells. *KAR1* encodes an SPB protein that is present in the half-bridge, is required for SPB duplication, and associates with Cdc31p, at least in mitotic cells (see below). *KAR2* encodes an Hsp70-type chaperone protein (homologous to BiP in animal cells) (Rose et al. 1989) that is localized to the lumen of the endoplasmic reticulum and is required for transport of some proteins across the endoplasmic reticulum (Vogel et al. 1990). It is not clear whether the defect in *kar2* mutants is in microtubule or SPB function or in more general nuclear membrane functions.

KEM1 was isolated as a mutation that enhances the karyogamy defect of strains carrying a *kar1* mutation (Kim et al. 1990). Strains lacking *KEM1* grow slowly, are hypersensitive to benomyl, display increased rates of chromosome loss, and show defects in nuclear migration and SPB separation (Kim et al. 1990; Interthal et al. 1995). The *kem1* null allele also shows genetic interactions with alleles of *TUB1* and *TUB2*. Biochemical experiments demonstrate that Kem1p binds microtubules and promotes tubulin assembly into microtubules in vitro (Interthal et al. 1995). This evidence suggests that Kem1p is a microtubule-associated protein that influences microtubule assembly, possibly affecting karyogamy only indirectly.

More recent genetic screens have isolated new karyogamy mutants, and the phenotypes of these mutants show that there are at least two separable steps to the process (Kurihara et al. 1994). Class I mutants isolated by these authors fail in nuclear congression and have defects in cytoplasmic microtubule function. Class II mutants congress normally but fail to fuse nuclei; these mutants have a defect in membrane fusion and not microtubules.

E. The Spindle Pole Body

Microtubules in animal and fungal cells grow from a microtubule-organizing center (MTOC); this organelle defines the spatial organization, and thus the function, of the microtubule cytoskeleton. The SPB is the MTOC in fungi and is functionally homologous to the animal cell centrosome. Like the centrosome, it occupies a perinuclear location in the cell, it nucleates microtubule growth, it anchors one end of the nucleated microtubules, and it duplicates once per cell cycle.

The morphology of the *S. cerevisiae* SPB has been defined by thin-section electron microscopy (Byers and Goetsch 1975) and is distinct from that of the centrosome. The SPB is a trilaminate structure with a central plaque embedded in the nuclear envelope and inner and outer plaques attached to the central plaque. The inner plaque is within the nucleus and the outer plaque is in the cytoplasm (see Fig. 13). Microtubules appear to be attached to amorphous material in the inner and outer plaques. All microtubules in the cell appear to be attached to the SPB throughout the cell cycle. The SPB end of the microtubules is closed and cup-shaped; this feature has not been observed in other MTOCs and might be unique to *S. cerevisiae*. MTOCs nucleate microtubule growth with a specific polarity; in all that have been examined, the minus end of the microtubule is proximal to the MTOC. The polarity of the microtubules attached to the SPB has not been established in yeast or any other fungi but is likely to be the same as in other organisms. The ability of SPBs to nucleate the growth of microtubules has been demonstrated both in vivo and in vitro. There is some question, however, about the validity of these experiments because there appears to always be some tubulin protein at the SPB, even with high concentrations of depolymerizing drugs or in tubulin mutants that cause depolymerization. This remaining tubulin could be polymeric and could be acting as a seed for microtubule assembly; this would be different from true nucleation in which a microtubule is created de novo from free subunits (see discussion of *TUB4* below).

1. SPB Proteins and Genes

Components of the SPB have been identified by producing antibodies against fractions enriched for SPBs (Rout and Kilmartin 1990). These antibodies recognize four SPB components of 42, 80, 90, and 110 kD. Immunoelectron microscopy was used to localize these proteins within the SPB structure: The 110-kD protein was localized between the central plaque and the inner plaque, the 90-kD protein to both the inner and outer plaques, the 42-kD protein to the central plaque, and the 80-kD protein to the inner plaque and to spindle microtubules. It is possible that the 80-kD protein is a spindle component and not a true SPB component. This

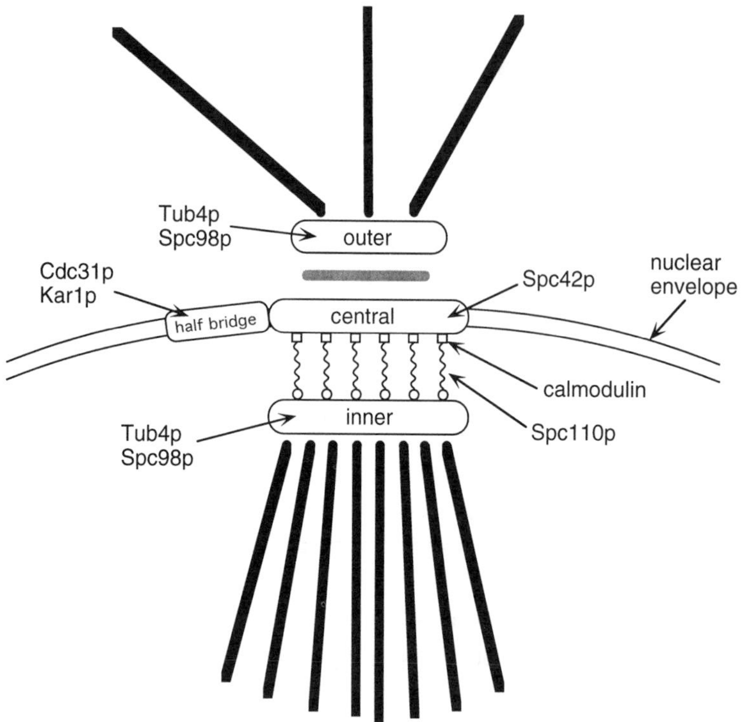

Figure 13 Cartoon of the yeast spindle pole body. The inner, central, and outer plaques are indicated along with known protein components and their locations.

raises a point about localizing and defining the components of an organelle such as the SPB. In an animal cell, many proteins have been found to be associated with the centrosome, but only a subset retain that localization if the microtubules emanating from the organizing center are depolymerized (see, e.g., Stearns et al. 1991); these are considered to be centrosomal proteins. We propose that the same standard be adopted for the yeast SPB.

The gene for the 110-kD protein, *SPC110,* was cloned (Kilmartin et al. 1993) and was shown to be identical to the previously isolated *NUF1* (Mirzayan et al. 1992). Spc110p is predicted to have a long central coiled-coil domain, and the molecule has a rod-like structure in the electron microscope. Remarkably, deleting portions of the rod domain of the protein resulted in a decreased distance between the central plaque and the inner plaque, providing proof that Spc110p is the spacer between these two parts of the SPB (Kilmartin et al. 1993). Spc110p was also identified as a dominant suppressor of a calmodulin mutation. Spc110p binds calmodulin (Geiser et al. 1993; Stirling et al. 1994), and the calmodulin-binding site is a small region of the carboxyl terminus of the protein. Immunofluorescence with calmodulin antibodies shows that calmodulin is also localized to the SPB (Geiser et al. 1993; Stirling et al. 1994). A temperature-sensitive *spc110* mutation in the calmodulin-binding region can be rescued by overexpression of calmodulin (Sundberg et al. 1996). This mutation causes the accumulation of an aberrant structure in the nucleus which contains both the 110-kD and 90-kD proteins and has associated microtubules. This structure has been named the intranuclear microtubule organizer (IMO). Spindle assembly and function are defective in this mutant, presumably because of interference of the normal spindle by the IMO (Sundberg et al. 1996).

The gene for the 90-kD protein, *SPC98* (calculated molecular mass of 98 kD), was isolated as a high-copy suppressor of a mutation in *TUB4* (Geissler et al. 1996). *SPC98* is essential for viability, and overexpression of *SPC98* is lethal but can be rescued by overexpression of *TUB4*. The gene for the 42-kD protein, *SPC42*, is also essential, and overproduction of *SPC42* results in formation of a polymer extending from the central plaque in the plane of the nuclear envelope (Donaldson and Kilmartin 1996). Like Spc110p, Spc42p has a potential coiled-coil region. Temperature-sensitive mutations in this region of the protein have a defect in SPB duplication, suggesting that the protein normally forms a polymeric structure which makes up part of the central plaque and either has a function in duplication or must be functional for duplication to take place.

Another component of the SPB was identified by its homology with γ-tubulin, a highly conserved component of MTOCs found in other fungi, animal cells, and plant cells (Oakley and Oakley 1989; Stearns et al. 1991; Zheng et al. 1991). An open reading frame on chromosome XII was identified as having homology with γ-tubulin and was named *TUB4* (Sobel and Snyder 1995; Marschall et al. 1996; Spang et al. 1996). Interestingly, Tub4p is considerably more divergent than other fungal γ tubulins, exhibiting only about 40% identity with animal γ tubulins, whereas the other fungal γ tubulins are all at least 70% identical with animal γ tubulins. This divergence is striking because the *S. cerevisiae* α- and β-tubulin proteins are as homologous to animal tubulins as are other fungal tubulins. The properties of *TUB4* indicate that it is likely to be the *S. cerevisiae* γ-tubulin, despite the sequence divergence: *TUB4* is essential for viability, Tub4p localizes to the SPB, and mutations in *TUB4* result in defects in SPB function (Sobel and Snyder 1995; Marschall et al. 1996; Spang et al. 1996). The mutant phenotype of *TUB4* is considered in more detail below.

The localization of Tub4p at the SPB is independent of the presence of microtubules (Marschall et al. 1996), as is the localization of γ-tubulin to the centrosome (Stearns et al. 1991). Using a Tub4p-GFP fusion protein, it was shown that Tub4p is able to associate with a preexisting SPB, suggesting that there is exchange between a cytoplasmic pool of Tub4p and the SPB (Marschall et al. 1996). Localization of the Spc98p to the SPB depends on Tub4p function (Marschall et al. 1996), and the two proteins are associated in the cell (Geissler et al. 1996).

Several genetic approaches have been used to identify genes important for SPB function. In some cases, the proteins encoded by these genes have been shown to localize to the SPB, and in others, the phenotype of mutants is the only link to the SPB. For many of the mutants, the primary defect is in SPB duplication rather than in the microtubule-nucleating and anchoring functions of the SPB. The first of these genes was *CDC31*, which encodes a small calcium-binding protein. Cdc31p is related to centrin, a protein found in the basal bodies of *Chlamydomonas* (Salisbury et al. 1984, 1988; Huang et al. 1988a,b) and in the animal cell centrosome (Lee and Huang 1993; Errabolu et al. 1994). Conditional mutations in *CDC31* result in a monopolar spindle (Byers 1981) that is larger than normal, indicating that the mutant is defective specifically in duplication and not simply defective in the assembly of SPB components. Failure to duplicate the SPB results in a monopolar spindle. Although the lack of a bipolar spindle causes a transient arrest at mitosis, many cells can complete a monopolar mitosis, resulting in all chromosomes segregating to one cell and yielding cells of twice the original ploidy (Schild et al. 1981). Immunoelectron microscopy shows that Cdc31p is localized to the half-bridge of the SPB, which lies adjacent to the SPB plaques and presumably has some role in the duplication process.

Mutations in *KAR1* lead to an SPB duplication defect similar to that for *CDC31*, resulting in a monopolar spindle. *KAR1* was originally isolated as being required for karyogamy (Conde and Fink 1976) and was subsequently shown to be essential for mitotic growth as well (Rose and Fink, 1987). Kar1p is localized to the half-bridge of the SPB like Cdc31p (Spang et al. 1995), and the SPB localization domain of Kar1p is separable from the domain required for karyogamy (Vallen et al. 1992). Kar1p is required for the localization of Cdc31p, and *KAR1* and *CDC31* interact genetically (Vallen et al. 1994) and physically (Biggins and Rose 1994; Spang et al. 1995). This suggests that the essential role for Kar1p in SPB function is to localize Cdc31p to the SPB.

Winey et al. (1991) used the characteristic monopolar spindle phenotype of SPB duplication mutants to screen for other genes involved in the duplication process. This resulted in the isolation of *MPS1* and *MPS2*. *MPS1* is essential for viability and encodes a dual specificity protein kinase (can phosphorylate serine/threonine and tyrosine). The *mps1-1* mutant has a defect in the satellite structure of the SPB, which occurs on the cytoplasmic face of the SPB prior to duplication. SPB duplication occurs in the *mps2-1* mutant, but one of the SPBs, presumably the new one, has a defective inner plaque and is unable to nucleate intranuclear microtubules. A similar phenotype has been described for mutations in *NDC1* (Winey et al. 1993). Ndc1p has potential membrane-spanning domains and is localized to the nuclear envelope. In *ndc1* mutants, as for *mps2*, one of the SPBs is defective, resulting in a monopolar spindle and subsequent increase in ploidy as the cells proceed through mitosis.

2. *Microtubule Nucleation at the SPB*

The main function of the SPB is to nucleate microtubule assembly and to organize those microtubules to perform functions. Remarkably, most of the mutations in the SPB proteins described above still have microtubules attached to the SPB, and the arrangement of those microtubules approximates wild type. Indeed, the most common phenotype of mutations in SPB components is a failure in SPB duplication, with the subsequent formation of a monopolar mitotic spindle. Work on the animal cell centrosome has identified γ-tubulin as being essential for microtubule nucleation from the centrosome (Joshi et al. 1992; Stearns and Kirschner 1994; Zheng et al. 1995), and mutations in the yeast homolog *TUB4* have been somewhat more informative regarding nucleation at the SPB (Marschall et al. 1996; Spang et al. 1996). The predominant phenotype of temperature-sensitive mutations in *TUB4* is also an accumulation of cells with monopolar spindles. However, examination by electron microscopy of serial sections through the SPB of such mutants shows that the SPB duplicates but does not separate, that one of the SPBs has a normal complement of microtubules, but that the other often has no, or only a few, associated microtubules (Marschall et al. 1996).

Interestingly, the same *tub4* mutant that is unable to assemble an SPB capable of microtubule nucleation is able to regrow microtubules from the SPB at the restrictive temperature after depolymerization with nocodazole (Marschall et al. 1996). The *tub4* mutant used in this experiment loses most Tub4p and Spc90p from the SPB at the restrictive temperature, so this regrowth is taking place with greatly reduced levels of the protein thought to be required for nucleation. These results suggest that microtubule nucleation in the true sense, assembly of a microtubule de novo, without a preexisting microtubule template, only occurs once in the yeast cell cycle, at the time of SPB duplication. Subsequent regrowth, either natural, through microtubule dynamics, or after drug-induced depolymerization, is likely to occur from the ends of short microtubules that remain embedded in the SPB.

REFERENCES

Adachi, Y., T. Toda, O. Niwa, and M. Yanagida 1986. Differential expression of essential and nonessential α-tubulin genes in *Schizosaccharomyces pombe. Mol. Cell. Biol.* **6:** 2168.

Adams, A.E. and D. Botstein. 1989. Dominant suppressors of yeast actin mutations that are reciprocally suppressed. *Genetics* **121:** 675.

Adams, A.E.M. and J.R. Pringle. 1984. Relationship of actin and tubulin distribution to bud growth in wild-type and morphogenetic-mutant *Saccharomyces cerevisiae. J. Cell Biol.* **98:** 934.

Adams, A.E., D. Botstein, and D.G. Drubin. 1991. Requirement of yeast fimbrin for actin organization and morphogenesis in vivo. *Nature* **354:** 404.

Adams, A.E.M., J.A. Cooper, and D.G. Drubin. 1993. Unexpected combinations of null mutations in genes encoding the actin cytoskeleton are lethal in yeast. *Mol. Biol. Cell* **4:** 459.

Amatruda, J.F. and J.A. Cooper. 1992. Purification, characterization, and immunofluorescence localization of *Saccharomyces cerevisiae* capping protein. *J. Cell Biol.* **117:** 1067.

Amberg, D.C., E. Basart, and D. Botstein. 1995. Defining protein interactions with yeast actin in vivo. *Nat. Struct. Biol.* **2:** 28.

Archer, J.E., L.R. Vega, and F. Solomon. 1995. Rbl2p, a yeast protein that binds to β-tubulin and participates in microtubule function in vivo. *Cell* **82:** 425.

Balasubramanian, M.K., B.R. Hirani, J.D. Burke, and K.L. Gould. 1994. The *Schizosaccharomyces pombe cdc3*$^+$ gene encodes a profilin essential for cytokinesis. *J. Cell Biol.* **125:** 1289.

Barnes, G., K.A. Louie, and D. Botstein. 1992. Yeast proteins associated with microtubules in vitro and in vivo. *Mol. Biol. Cell* **3:** 29.

Basson, M.E., R.L. Moore, J. O'Rear, and J. Rine. 1987. Identifying mutations in duplicated functions in *Saccharomyces cerevisiae:* Recessive mutations in 3-hydroxy-3-methylglutaryl coenzyme A reductase genes. *Genetics* **117:** 645.

Bauer, F., M. Urdaci, M. Aigle, and M. Crouzet. 1993. Alteration of a yeast SH3 protein leads to conditional viability with defects in cytoskeletal and budding patterns. *Mol. Cell. Biol.* **13:** 5070.

Baum, P., C. Yip, L. Goetsch, and B. Byers. 1988. A yeast gene essential for regulation of spindle pole duplication. *Mol. Cell. Biol.* **8:** 5386.

Bellocq, C., I. Andrey-Tornare, D. Paunier, B. Maeder, L. Paturle, D. Job, J. Haiech, and S.J. Edelstein. 1992. Purification of assembly-competent tubulin from *Saccharomyces cerevisiae. Eur. J. Biochem.* **210:** 343.

Benedetti, H., S. Raths, F. Crausaz, and H. Riezman. 1994. The *END3* gene encodes a protein that is required for the internalization step of endocytosis and for actin cytoskeleton organization in yeast. *Mol. Biol. Cell* **5:** 1023.

Berlin, V., C.A. Styles, and G.R. Fink. 1990. *BIK1:* A protein required for microtubule function during mating and mitosis in *Saccharomyces cerevisiae:* colocalizes with tubulin. *J Cell Biol.* **111:** 2573.

Biggins, S. and M.D. Rose. 1994. Direct interaction between yeast spindle pole body components: Kar1p is required for Cdc31p localization to the spindle pole body. *J. Cell Biol.* **125:** 843.

Bollag, D.M., I. Tornare, R. Stalder, D. Paunier, M.D. Rozycki, and S.J. Edelstein. 1990. Overexpression of tubulin in yeast: Differences in subunit association. *Eur. J. Cell Biol.* **51:** 295.

Botstein, D. 1986. Why study the cytoskeleton in yeast? *Harvey Lect.* **82:** 157.

Bray, D. and J. Vasiliev. 1989. Cell motility. Networks from mutants (news). *Nature* **338:** 203.

Breitfeld, P.P., W.C. Mckinnon, and K.E. Mostov. 1990. Effect of nocodazole on vesicular traffic to the apical and basolateral surfaces of polarized MDCK cells. *J. Cell Biol.* **111:** 2365.

Bretscher, A., B. Drees, E. Harsay, D. Schott, and T. Wang. 1994. What are the basic functions of microfilaments? Insights from studies in budding yeast. *J. Cell Biol.* **126:** 821.

Brinkley, R.R. 1985. Microtubule organizing centers. *Annu. Rev. Cell Biol.* **1:** 145.

Brockerhoff, S.E., R.C. Stevens, and T.N. Davis. 1994. The unconventional myosin, Myo2p, is a calmodulin target at sites of cell growth in *Saccharomyces cerevisiae. J. Cell Biol.* **124:** 315.

Brower, S.M., J.E. Honts, and A.E. Adams. 1995. Genetic analysis of the fimbrin-actin binding interaction in *Saccharomyces cerevisiae. Genetics* **140:** 91.

Burke, D., P. Gasdaska, and L. Hartwell. 1989. Dominant effects of tubulin overexpression in *Saccharomyces cerevisiae. Mol. Cell. Biol.* **9:** 1049.

Burshell, A., P.A. Stathis, Y. Do, S.C. Miller, and D. Feldman. 1984. Characterization of an estrogen-binding protein in the yeast *Saccharomyces cerevisiae. J. Biol. Chem.* **259:** 3450.

Byers, B. 1981. Cytology of the yeast life cycle. In *The molecular biology of the yeast* Saccharomyces: *Life cycle and inheritance* (ed. J.N. Strathern et al.), pp. 59. Cold Spring Harbor Laboratory Press, Cold Spring Harbor, New York.

Byers, B. and L. Goetsch. 1975. Behavior of spindles and spindle plaques in the cell cycle and conjugation of *Saccharomyces cerevisiae. J. Bacteriol.* **124:** 511.

———. 1976. A highly ordered ring of membrane-associated filaments in budding yeast. *J. Cell Biol.* **69:** 717.

Byers, B., K. Shriver, and L. Goetsch. 1978. The role of spindle pole bodies and modified microtubule ends in the initiation of microtubule assembly in *Saccharomyces cerevisiae. J. Cell Sci.* **30:** 331.

Carminati, J. and T. Stearns. 1997. Microtubules orient the mitotic spindle in yeast through dynein-dependent interactions with the cell cortex. *J. Cell Biol.* (in press).

Chan, C.S. and D. Botstein. 1993. Isolation and characterization of chromosome-gain and increase-in-ploidy mutants in yeast. *Genetics* **135:** 677.

Chant, J. 1996. Septin scaffolds and cleavage planes in *Saccharomyces. Cell* **84:** 187.

Chen, X., R.K. Cook, and P.A. Rubenstein. 1993. Yeast actin with a mutation in the "hydrophobic plug" between subdomains 3 and 4 (L266D) displays a cold-sensitive polymerization defect. *J. Cell Biol.* **123:** 1185.

Chen, X., D.S. Sullivan, and T.C. Huffaker. 1994. Two yeast genes with similarity to

TCP-1 are required for microtubule and actin function in vivo. *Proc. Natl. Acad. Sci.* **91:** 9111.

Chien, C., P.L. Bartel, R. Sternglanz, and S. Fields. 1991. The two hybrid system: A method to identify and clone genes for proteins that interact with a protein of interest. *Proc. Natl. Acad. Sci.* **88:** 9578.

Chowdhury, S., K.W. Smith, and M.C. Gustin. 1992. Osmotic stress and the yeast cytoskeleton: Phenotype-specific suppression of an actin mutation. *J. Cell Biol.* **118:** 561.

Cid, V.J., A. Duran, F. del Rey, M.P. Snyder, C. Nombela, and M. Sanchez. 1995. Molecular basis of cell integrity and morphogenesis in *Saccharomyces cerevisiae. Microbiol. Rev.* **59:** 345.

Clark, S.W. and D.I. Meyer. 1994. *ACT3:* A putative centractin homologue in *S. cerevisiae* is required for proper orientation of the mitotic spindle. *J. Cell Biol* **127:** 129.

Cleves, A.E., P.J. Novick, and V.A. Bankaitis. 1989. Mutations in the SAC1 gene suppresses defects in yeast Golgi and yeast actin function. *J. Cell Biol.* **109:** 2939.

Conde, J. and G.R. Fink. 1976. A mutant of *Saccharomyces cerevisiae* defective for nuclear fusion. *Proc. Natl. Acad. Sci.* **73:** 3651.

Connelley, C. and P. Hieter. 1996. Budding yeast SKP1 encodes an evolutionarily conserved kinetochore protein required for cell cycle progression. *Cell* **86:** 275.

Crouzet, M., M. Urdaci, L. Dulau, and M. Aigle. 1991. Yeast mutant affected for viability upon nutrient starvation: Characterization and cloning of the *RVS161* gene. *Yeast* **7:** 727.

Davidse, L. and W. Flach. 1977. Differential binding of methyl-benzimidazol-2-yl-carbamate to fungal tubulin as a mechanism of resistance to this antimitotic agent in strains of *Aspergillus nidulans. J. Cell Biol.* **72:** 174.

Davis, A., C.R. Sage, C.A. Dougherty, and K.W. Farrell. 1994. Microtubule dynamics modulated by guanosine triphosphate hydrolysis activity of β-tubulin. *Science* **264:** 839.

Davis, A., C.R. Sage, L. Wilson, and K.W. Farrell. 1993. Purification and biochemical characterization of tubulin from the budding yeast *Saccharomyces cerevisiae. Biochemistry* **32:** 8823.

Dawson, I.A., S. Roth, and T.S. Artavanis. 1995. The *Drosophila* cell cycle gene fizzy is required for normal degradation of cyclins A and B during mitosis and has homology to the *CDC20* gene of *Saccharomyces cerevisiae. J. Cell Biol.* **129:** 725.

Delgado, M.A. and J. Conde. 1984. Benomyl prevents nuclear fusion in *Saccharomyces cerevisiae. Mol. Gen. Genet.* **193:** 188.

DeRosier, D.J. and L.G. Tilney. 1982. How actin filaments pack into bundles. *Cold Spring Harbor Symp. Quant. Biol.* **46:** 525.

Desfarges, L., P. Durrens, H. Juguelin, C. Cassagne, M. Bonneu, and M. Aigle. 1993. Yeast mutants affected in viability upon starvation have a modified phospholipid composition. *Yeast* **9:** 267.

Doheny, K.F., P.K. Sorger, A.A. Hyman, S. Tugendreich, F. Spencer, and P. Hieter. 1993. Identification of essential components of the *Saccharomyces cerevisae* kinetochore. *Cell* **73:** 761.

Dohmen, R.J., P. Wu, and A. Varshavsky. 1994. Heat-inducible degron: A method for constructing temperature-sensitive mutants. *Science* **263:** 1273.

Donaldson, A.D. and J.V. Kilmartin. 1996. Spc42p: A phosphorylated component of the *S. cerevisiae* spindle pole body (SPB) with an essential function during SPB duplication. *J. Cell Biol.* **132:** 887.

Doyle, T. and D. Botstein. 1996. Movement of yeast cortical actin cytoskeleton visualized *in vivo*. *Proc. Natl. Acad. Sci.* **93:** 3886.

Drees, B., C. Brown, B.G. Barrell, and A. Bretscher. 1995. Tropomyosin is essential in yeast, yet the *TPM1* and *TPM2* products perform distinct functions. *J. Cell Biol.* **128:** 383.

Drubin, D. 1989. The yeast *Saccharomyces cerevisiae* as a model organism for the cytoskeleton and cell biology. *Cell Motil. Cytoskeleton* **14:** 42.

Drubin, D.G. and W.J. Nelson. 1996. Origins of cell polarity. *Cell* **84:** 335.

Drubin, D.G., H.D. Jones, and K.F. Wertman. 1993. Actin structure and function: Roles in mitochondrial organization and morphogenesis in budding yeast and identification of the phalloidin-binding site. *Mol. Biol. Cell.* **4:** 1277.

Drubin, D.G., K.G. Miller, and D. Botstein. 1988. Yeast actin-binding proteins: Evidence for a role in morphogenesis. *J. Cell Biol.* **107:** 2551.

Dunn, T.M. and D. Shortle. 1990. Null alleles of SAC7 suppress temperature-sensitive actin mutations in *Saccharomyces cerevisiae. Mol. Cell. Biol.* **10:** 2308.

Dustin, P. 1984. *Microtubules*. Springer-Verlag, New York.

Echols, H., C. Lu, and P.M. Burgers. 1983. Mutator strains of *Escherichia coli, mutD* and *dnaQ* with defective exonucleolytic editing by DNA polymerase III holoenzyme. *Proc. Natl. Acad. Sci.* **80:** 2189.

Endow, S.A., S.J. Kang, L.L. Satterwhite, M.D. Rose, V.P. Skeen, and E.D. Salmon. 1994. Yeast Kar3 is a minus-end microtubule motor protein that destabilizes microtubules preferentially at the minus ends. *EMBO J.* **13:** 2708.

Errabolu, R., M.A. Sanders, and J.L. Salisbury. 1994. Cloning of a cDNA encoding human centrin, an EF-hand protein of centrosomes and mitotic spindle poles. *J. Cell Sci.* **107:** 9.

Eshel, D., L.A. Urrestarazu, S. Vissers, J.C. Jauniaux, J.C. Van Vliet-Reedijk, R.J. Planta, and I.R. Gibbons. 1993. Cytoplasmic dynein is required for normal nuclear segregation in yeast. *Proc. Natl. Acad. Sci.* **90:** 11172.

Fabbrizio, E., A. Bonet-Kerrache, J. Leger, and D. Mornet. 1993. Actin-dystrophin interface. *Biochemistry* **32:** 10457.

Farkas, V., J. Kovarik, A. Kosinova, and S. Bauer. 1974. Autoradiographic study of mannan incorporation into the growing walls of *Saccharomyces cerevisiae. J. Bacteriol.* **117:** 265.

Fedor-Chaiken, M., R.J. Deschenes, and J.R. Broach. 1990. SRV2, a gene required for RAS activation of adenylate cyclase in yeast. *Cell* **61:** 329.

Field, J., A. Vojtek, R. Ballester, G. Bolger, J. Colicelli, K. Ferguson, J. Gerst, T. Kataoka, T. Michaeli, S. Powers, M. Riggs, L. Rodgers, I. Wieland, B. Wheland, and M. Wigler. 1990. Cloning and characterization of CAP, the *S. cerevisiae* gene encoding the 70 kd adenyl cyclase-associated protein. *Cell* **61:** 319.

Freeman, N.L., Z. Chen, J. Horenstein, A. Weber, and J. Field. 1995. An actin monomer binding activity localizes to the carboxyl-terminal half of the *Saccharomyces cerevisiae* cyclase-associated protein. *J. Biol. Chem.* **270:** 5680.

Freeman, N.L., T. Lila, K.A. Mintzer, Z. Chen, A.J. Pahk, R. Ren, D.G. Drubin, and J. Field. 1996. A conserved proline-rich region of the *Saccharomyces cerevisiae* cyclase-associated protein binds SH3 domains and modulate cytoskeletal localization. *Mol. Cell. Biol.* **16:** 548.

Gallwitz, D. and I. Sures. 1980. Structure of a split yeast gene: Complete nucleotide sequence of the actin gene in *Saccharomyces cerevisiae*. *Proc. Natl. Acad. Sci.* **77:** 2546.

Geiser, J.R., H.A. Sundberg, B.H. Chang, E.G.D. Muller, and T.N. Davis. 1993. The essential mitotic target of calmodulin is the 110-kilodalton component of the spindle pole body in *Saccharomyces cerevisiae*. *Mol. Cell. Biol.* **13:** 7913.

Geissler, S., G. Pereira, A. Spang, M. Knop, S. Soues, J. Kilmartin, and E. Schiebel. 1996. The spindle pole body component Spc98p interacts with the γ-tubulin-like Tub4p of *Saccharomyces cerevisiae* at the sites of microtubule attachment. *EMBO J.* **15:** 3899.

Gerst, J.E., K. Ferguson, A. Vojtek, M. Wigler, and J. Field. 1991. CAP is a bifunctional component of the *Saccharomyces cerevisiae* adenyl cyclase complex. *Mol. Cell. Biol.* **11:** 1248.

Gill, S.R., T.A. Schroer, I. Szilak, E.R. Steuer, M.P. Sheetz, and D.W. Cleveland. 1991. Dynactin, a conserved, ubiquitously expressed component of an activator of vesicle motility mediated by cytoplasmic dynein. *J. Cell Biol.* **115:** 1639.

Goebl, M.G. and T.D. Petes. 1986. Most of the yeast genomic sequences are not essential for cell growth and division. Cell **46:** 983.

Goh, P.-Y. and J.V. Kilmartin. 1993. *NDC10:* a gene involved in chromosome segregation in *Saccharomyces cerevisiae*. *J. Cell Biol.* **121:** 503.

Goldstein, L.S.B. 1993. With apologies to Scheherezade: Tails of 1001 kinesin motors. *Annu. Rev. Genet.* **27:** 319.

Goodson, H.V. and J.A. Spudich. 1995. Identification and molecular characterization of a yeast myosin I. *Cell Motil. Cytoskeleton* **30:** 73.

Guthrie, C., H. Nashimoto, and M. Nomura. 1969. Structure and function of *Escherichia coli* ribosomes VII. Cold sensitive mutations defective in ribosome assembly. *Proc. Natl. Acad. Sci.* **63:** 384.

Haarer, B.K. and J.R. Pringle. 1987. Immunofluorescence localization of the *Saccharomyces cerevisiae CDC12* gene product to the vicinity of the 10-nm filaments in the mother-bud neck. *Mol. Cell. Biol.* **7:** 3678.

Haarer, B.K., A. Petzold, S.H. Lillie, and S.S. Brown. 1994. Identification of MYO4, a second class V myosin gene in yeast. *J. Cell Sci.* **107:** 1055.

Haarer, B.K., S.H. Lillie, A.E.M. Adams, V. Magdolen, W. Bandlow, and S.S. Brown. 1990. Purification of profilin from *Saccharomyces cerevisiae* and analysis of profilin-deficient cells. *J. Cell Biol.* **110:** 105.

Holtzman, D.A., K.F. Wertman, and D.G. Drubin. 1994. Mapping actin surfaces required for functional interactions in vivo. *J. Cell Biol.* **126:** 423.

Holtzman, D.A., S. Yang, and D.G. Drubin. 1993. Synthetic-lethal interactions identify two novel genes, SLA1 and SLA2, that control membrane cytoskeleton assembly in *Saccharomyces cerevisiae*. *J. Cell Biol.* **122:** 635.

Holzbaur, E.L.F. and R.B. Vallee. 1994. Dyneins: Molecular structure and cellular function. *Annu. Rev. Cell Biol.* **10:** 339.

Honts, J.E., T.S. Sandrock, S.M. Brower, J.L. O'Dell, and A.E. Adams. 1994. Actin mutations that show suppression with fimbrin mutations identify a likely fimbrin-binding site on actin. *J. Cell Biol.* **126:** 413.

Hoyt, M.A. 1994. Cellular roles of kinesin and related proteins. *Curr. Opin. Cell Biol.* **6:** 63.

Hoyt, M.A., T. Stearns, and D. Botstein. 1990. Chromosome instability mutants of *Saccharomyces cerevisiae* that are defective in microtubule-mediated processes. *Mol. Cell Biol.* **10:** 223.

Hoyt, M.A., L. Totis, and B.T. Roberts. 1991. *S. cerevisiae* genes required for cell cycle arrest in response to loss of microtubule function. *Cell* **66:** 507.

Hoyt, M.A., L. He, K.K. Loo, and W.S. Saunders. 1992. Two *Saccharomyces cerevisiae* kinesin-related gene products required for mitotic spindle assembly. *J. Cell Biol.* **118:** 109.

Hoyt, M.A., L. He, L. Totis, and W.S. Saunders. 1993. Loss of function of *Saccharomyces cerevisiae* kinesin-related *CIN8* and *KIP1* is suppressed by *KAR3* motor domain mutations. *Genetics* **135:** 35.

Huang, B., A. Mengersen, and V.D. Lee. 1988a. Molecular cloning of cDNA for caltractin, a basal body-associated Ca^{2+}-binding protein: Homology in its protein sequence with calmodulin and the yeast *CDC31* gene product. *J. Cell Biol.* **197:** 133.

Huang, B., D.M. Watterson, V.D. Lee, and M.J. Schibler. 1988b. Purification and characterization of a basal body-associated calcium-binding protein. *J. Cell Biol.* **107:** 121.

Huffaker, T.C., M.A. Hoyt, and D. Botstein. 1987. Genetic analysis of the yeast cytoskeleton. *Annu. Rev. Genet.* **21:** 259.

Huffaker, T.C., J.H. Thomas, and D. Botstein. 1988. Diverse effects of β-tubulin mutations on microtubule formation and function. *J. Cell Biol.* **106:** 1997.

Hyman, A.A. and P.K. Sorger. 1995. Structure and function of kinetochores in budding yeast. *Annu. Rev. Cell Dev. Biol.* **11:** 471.

Hyman, A.A., K. Middleton, M. Centola, T.J. Mitchison, and J. Carbon. 1992. Microtubule-motor activity of a yeast centromere-binding protein complex. *Nature* **359:** 533.

Interthal, H., C. Bellocq, J. Bahler, V.I. Bashkirov, S. Edelstein, and W.D. Heyer. 1995. A role of Sep1 (= Kem1, Xrn1) as a microtubule-associated protein in *Saccharomyces cerevisiae. EMBO J.* **14:** 1057.

Jacobs, C.W., A.E.M. Adams, P.J. Staniszlo, and J.R. Pringle. 1988. Functions of microtubules in the *Saccharomyces cerevisiae* cell cycle. *J. Cell Biol.* **107:** 1409.

Jarvik, J. and D. Botstein. 1975. Conditional-lethal mutations that suppress genetic defects in morphogenesis by altering structural proteins. *Proc. Natl. Acad. Sci.* **72:** 2738.

Jiang, W., K. Middleton, H.-J. Yoon, C. Fouquet, and J. Carbon. 1993. An essential yeast protein, Cbf5p, binds in vitro to centromeres and microtubules. *Mol. Cell. Biol.* **13:** 4884.

Johannes, F.-J. and D. Gallwitz. 1991. Site-directed mutagenesis of the yeast actin gene: A test for actin function *in vivo*. *EMBO J.* **10:** 3951.

Joshi, H.C., J.P. Monica, L. McNamara, and D.W. Cleveland. 1992. γ-Tubulin is a centrosomal protein required for cell cycle-dependent microtubule nucleation. *Nature* **356:** 80.

Kabsch, W., H.G. Mannherz, D. Suck, E.F. Pai, and K.C. Holmes. 1990. Atomic structure of the actin:DNase I complex. *Nature* **347:** 37.

Kahana, J.A., B.J. Schnapp, and P.A. Silver. 1995. Kinetics of spindle pole body separation in budding yeast. *Proc. Natl. Acad. Sci.* **92:** 9707.

Karpova, T.S., M.M. Lepetit, and J.A. Cooper. 1993. Mutations that enhance the *cap2* null mutant phenotype in *Saccharomyces cerevisiae* affect the actin cytoskeleton, morphogenesis and pattern of growth. *Genetics* **135:** 693.

Kilmartin, J. 1981. Purification of yeast tubulin by self-assembly *in vitro*. *Biochemistry* **20:** 3629.

Kilmartin, J.V. and A.E.M. Adams. 1984. Structural rearrangements of tubulin and actin during the cell cycle of the yeast *Saccharomyces*. *J. Cell Biol.* **98:** 922.

Kilmartin, J., B. Wright, and C. Milstein. 1982. Rat monoclonal anti-tubulin antibodies derived by using a new nonsecreting rat cell line. *J. Cell Biol.* **93:** 576.

Kilmartin, J.V., S.L. Dyos, D. Kershaw, and J.T. Finch. 1993. A spacer protein in the *Saccharomyces cerevisiae* spindle pole body whose transcript is cell cycle-regulated. *J. Cell Biol.* **123:** 1175.

Kim, H.B., B.K. Haarer, and J.R. Pringle. 1991. Cellular morphogenesis in the *Saccharomyces cerevisiae* cell cycle: Localization of the *CDC3* gene product and the timing of events at the budding site. *J. Cell Biol.* **112:** 535.

Kim, J., P.O. Ljungdahl, and G.R. Fink. 1990. kem mutations affect nuclear fusion in *Saccharomyces cerevisiae. Genetics* **126:** 799.

Kingsbury, J. and D. Koshland. 1991. Centromere-dependent binding of yeast minichromosomes to microtubules *in vitro. Cell* **66:** 483.

Kirkpatrick, D. and F. Solomon. 1994. Overexpression of yeast homologs of the mammalian checkpoint gene RCC1 suppresses the class of α-tubulin mutations that arrest with excess microtubules. *Genetics* **137:** 381.

Kron, S.J., D.G. Drubin, D. Botstein, and J.A. Spudich. 1992. Yeast actin filaments display ATP-dependent sliding movement over surfaces coated with rabbit muscle myosin. *Proc. Natl. Acad. Sci.* **89:** 4466.

Kubler, E. and H. Riezman. 1993. Actin and fimbrin are required for the internalization step of endocytosis in yeast. *EMBO J.* **12:** 2855.

Kurihara, L.J., C.T. Beh, M. Latterich, R. Schekman, and M.D. Rose. 1994. Nuclear congression and membrane fusion: Two distinct events in the yeast karyogamy pathway. *J. Cell Biol.* **126:** 911.

Lechner, J. 1994. A zinc finger protein, essential for chromosome segregation, constitutes a putative DNA binding subunit of the *Saccharomyces cerevisiae* kinetochore complex, Cbf3. *EMBO J.* **13:** 5203.

Lechner, J. and J. Carbon. 1991. A 240kd multisubunit protein complex, CBF3, is a major component of the budding yeast centromere. *Cell* **64:** 717.

Lee, V.D. and B. Huang. 1993. Molecular cloning and centrosomal localization of human caltractin. *Proc. Natl. Acad. Sci.* **90:** 11039.

Leeuw, T., A. Fourest-Lieuvin, C. Wu, J. Chenevert, K. Clark, M. Whiteway, D.Y. Thomas, and E. Leberer. 1995. Pheromone response in yeast: Association of Bem1p with proteins of the MAP kinase cascade and actin. *Science* **270:** 1210.

Leung, D.W., E.Y. Chen, and D.V. Goeddel. 1989. A method for random mutagenesis of a defined DNA segment using a modified polymerase chain reaction. *Technique* **1:** 11.

Levine, B., A.J.G. Moir, V.B. Patchell, and S.V. Perry. 1990. The interaction of actin and dystrophin. *FEBS Lett.* **263:** 159.

Lewis, S.A., G. Tian, I.E. Vainberg, and N.J. Cowan. 1996. Chaperonin-mediated folding of actin and tubulin. *J. Cell Biol.* **132:** 1.

Li, Q. and K.A. Suprenant. 1994. Molecular characterization of the 77-kDa echinoderm microtubule-associated protein. Homology to the β-transducin family. *J. Biol. Chem.* **269:** 31777.

Li, R. and A.W. Murray. 1991. Feedback control of mitosis in budding yeast. *Cell* **66:** 519.

Li, R., Y. Zheng, and D.G. Drubin. 1995. Regulation of cortical actin cytoskeleton assembly during polarized cell growth in budding yeast. *J. Cell Biol.* **128:** 599.

Li, Y.-Y., E. Yeh, T. Hays, and K. Bloom. 1993. Disruption of mitotic spindle orientation in a yeast dynein mutant. *Proc. Natl. Acad. Sci.* **90:** 10096.

Lillie, S.H. and S.S. Brown. 1992. Suppression of a myosin defect by a kinesin-related gene. *Nature* **356:** 358.

———. 1994. Immunofluorescence localization of the unconventional myosin, Myo2p, and the putative kinesisn-related protein, Smy1p, to the same regions of polarized growth in *Saccharomyces cerevisiae. J. Cell Biol.* **125:** 825.

Liu, H. and A. Bretscher. 1989. Disruption of the single tropomyosin gene in yeast results in the disappearance of actin cables from the cytoskeleton. *Cell* **57:** 233.

———. 1992. Characterization of *TPM1* disrupted yeast cells indicates an involvement of tropomyosin in directed vesicular transport. *J. Cell Biol.* **118:** 285.

Machin, N.A., J.M. Lee, and G. Barnes. 1995. Microtubule stability in yeast: Characterization and dosage suppression of a benomyl-dependent tubulin mutant. *Mol. Biol. Cell* **6:** 1241.

Magdolen, V., D.G. Drubin, G. Mages, and W. Bandlow. 1993. High levels of profilin suppress the lethality caused by overproduction of actin in yeast cells. *FEBS Lett.* **316:** 41.

Marks, J., I.M. Hagan, and J.S. Hyams. 1986. Growth polarity and cytokinesis in fission yeast: The role of the cytoskeleton. *J. Cell Sci.* **5:** 229.

Marschall, L.G., R.L. Jeng, J. Mulholland, and T. Stearns. 1996. Analysis of Tub4p, a yeast γ-tubulin-like protein: Implications for microtubule organizing center function. *J. Cell Biol.* **134:** 443.

Matile, P. 1969. Prospects of yeast cytology. *Antonie Leeuwenhoek* (suppl.) **35:** 59.

McGough, A., M. Way, and D. DeRosier. 1994. Determination of the α-actinin-binding site on actin filaments by cryoelectron microscopy and image analysis. *J. Cell Biol.* **126:** 433.

McGrew, J.T., L. Goetsch, B. Byers, and P. Baum. 1992. Requirement for *ESP1* in the nuclear division of *Saccharomyces cerevisiae. Mol. Biol. Cell* **3:** 1443.

Meeks-Wagner, D., J.S. Wood, B. Garvik, and L.H. Hartwell. 1986. Isolation of two genes that affect mitotic chromosome transmission in *S. cerevisiae. Cell* **44:** 53.

Mejean, C., M.-C. Lebart, M. Boyer, C. Roustan, and Y. Benyamin. 1992. Localization and identification of actin structures involved in the filamin-actin interaction. *Eur. J. Biochem.* **209:** 555.

Meluh, P.B. and M.D. Rose. 1990. *KAR3*, a kinesin-related gene required for yeast nuclear fusion. *Cell* **60:** 1029.

Middleton, K. and J. Carbon. 1994. *KAR3*-encoded kinesin is a minus-end-directed motor that functions with centromere binding proteins (CBF3) on an in vitro yeast kinetochore. *Proc. Natl. Acad. Sci.* **91:** 7212.

Miklos, D., S. Caplan, D. Mertens, G. Hynes, Z. Pitluk, Y. Kashi, K. Harrison-Lavoie, S. Steneson, C. Brown, and B. Barrell. 1994. Primary structure and function of a second essential member of the heterooligomeric TCP1 chaperonin complex of yeast, TCP1β. *Proc. Natl. Acad. Sci.* **91:** 2743.

Miller, C.J. and E. Reisler. 1995. Role of charged amino acid pairs in subdomain-1 of actin in interactions with myosin. *Biochemistry* **34:** 2694.

Miller, C.J., P. Cheung, P. White, and E. Reisler. 1995. Actin's view of actomyosin interface. *Biophys. J.* (suppl.) **68:** 50.

Miller, C.J., T.C. Doyle, E. Bobkova, D. Botstein, and E. Reisler. 1996. Mutational analysis of the role of hydrophobic residues in the 338-348 helix on actin in actomyosin interactions. *Biochemistry* **35:** 3670.

Mirzayan, C., C.S. Copeland, and M. Snyder. 1992. The *NUF1* gene encodes an essential coiled-coil related protein that is a potential component of the yeast nucleoskeleton. *J. Cell Biol.* **116:** 1319.

Moir, D., S.E. Stewart, B.C. Osmond, and D. Botstein. 1982. Cold-sensitive cell-division-cycle mutants of yeast: Isolation, properties, and pseudoreversion studies. *Genetics* **100:** 547.

Moon, A.L., P.A. Janmey, K.A. Louie, and D.G. Drubin. 1993. Cofilin is an essential component of the yeast cortical cytoskeleton. *J. Cell Biol.* **120:** 421.

Muhua, L., T.S. Karpova, and J.A. Cooper. 1994. A yeast actin-related protein homologous to that in vertebrate dynactin complex is important for spindle orientation and nuclear migration. *Cell* **78:** 669.

Mulholland, J., D. Preuss, A. Moon, A. Wong, D. Drubin, and D. Botstein. 1994. Ultrastructure of the yeast actin cytoskeleton and its association with the plasma membrane. *J. Cell Biol.* **125:** 381.

Neff, N.F., J.H. Thomas, P. Grisafi, and D. Botstein. 1983. Isolation of the β-tubulin gene from yeast and demonstration of its essential function *in vivo*. *Cell* **33:** 211.

Ng, R. and J. Abelson. 1980. Isolation and sequence of the gene for actin in *Saccharomyces cerevisiae*. *Proc. Natl. Acad. Sci.* **77:** 3912.

Novick, P. and D. Botstein. 1985. Phenotypic analysis of temperature-sensitive yeast actin mutants. *Cell* **40:** 405.

Novick, P., B.C. Osmond, and D. Botstein. 1989. Suppressors of yeast actin mutations. *Genetics* **121:** 659.

Oakley, C.E. and B.R. Oakley. 1989. Identification of γ-tubulin, a new member of the tubulin superfamily encoded by *mipA* gene of *Aspergillus nidulans*. *Nature* **338:** 662.

Ohya, Y. and D. Botstein. 1994a. Diverse essential functions revealed by complementing yeast calmodulin mutants. *Science* **263:** 963.

———. 1994b. Structure-based systematic isolation of conditional-lethal mutations in the single yeast calmodulin gene. *Genetics* **138:** 1041.

Oliver, S.G., Q.J. van der Aart, M.L. Agostoni-Carbone, M. Aigle, L. Alberghina, D. Alexandraki, G. Antoine, R. Anwar, J.P. Ballesta, and P.E.A. Benit. 1992. The complete DNA sequence of yeast chromosome III. *Nature* **357:** 38.

Page, B.D. and M. Snyder. 1993. Chromosome segregation in yeast. *Ann. Rev. Microbiol.* **47:** 231.

Page, B.D., L.L. Satterwhite, M.D. Rose, and M. Snyder. 1994. Localization of the Kar3 kinesin heavy chain-related protein requires the Cik1 interacting protein. *J. Cell Biol.* **124:** 507.

Palmer, R.E., M., Koval, and D. Koshland. 1989. The dynamics of chromosome movement in the budding yeast *Saccharomyces cerevisiae*. *J. Cell Biol.* **109:** 3355.

Palmer, R.E., D.S. Sullivan, T. Huffaker, and D. Koshland. 1992. Role of astral microtubules and actin in spindle orientation and migration in the budding yeast, *Saccharomyces cerevisiae*. *J. Cell Biol.* **119:** 583.

Pasqualone, D. and T.C. Huffaker. 1994. *STU1:* A suppressor of a β-tubulin mutation, encodes a novel and essential component of the yeast mitotic spindle. *J. Cell Biol.* **127:** 1973.

Pellman, D., M. Bagget, H. Tu, and G.R. Fink, 1995. Two microtubule-associated proteins required for anaphase spindle movement in *Saccharomyces cerevisiae*. *J. Cell Biol.* **130:** 1373.

Peterson, J., Y. Zheng, L. Bender, A. Myers, R. Cerione, and A. Bender. 1994. Interactions between the bud emergence proteins Bem1p and Bem2p and Rho-type GTPases in yeast. *J. Cell Biol.* **127:** 1395.

Plamann, M., P.F. Minke, J.H. Tinsley, and K.S. Bruno. 1994. Cytoplasmic dynein and actin-related protein Arp1 are required for normal nuclear distribution in filamentous fungi. *J. Cell Biol.* **127:** 139.

Polaina, J. and J. Conde. 1982. Genes involved in the control of nuclear fusion during the sexual cycle of *Saccharomyces cerevisiae*. *Mol. Gen. Genet.* **1866:** 253.

Pringle, J.R., R.A. Preston, A.E.M. Adams, T. Stearns, D.G. Drubin, B.K. Haarer, and E.W. Jones. 1989. Fluorescence microscopy methods for yeast. *Methods Cell Biol.* **31:** 357.

Qadota, H., Y. Anraku, D. Botstein, and Y. Ohya. 1994. Conditional lethality of a yeast strain expressing human RHOA in place of RHO1. *Proc. Natl. Acad. Sci.* **91:** 9317.

Qadota, H., C.P. Python, S.B. Inoue, M. Arisawa, Y. Anraku, Y. Zheng, T. Watanabe, D.E. Levin, and Y. Ohya. 1996. Identification of yeast Rho1p GTPase as a regulatory subunit of 1,3-beta-glucan synthase. *Science* **272:** 279.

Raths, S., J. Rohrer, F. Crausaz, and H. Riezman. 1993. *end3* and *end4:* Two mutants defective in receptor-mediated and fluid-phase endocytosis in *Saccharomyces cerevisiae*. *J. Cell Biol.* **120:** 55.

Read, E.B., H.H. Okamura, and D.G. Drubin. 1992. Actin- and tubulin-dependent functions during *Saccharomyces cerevisiae* mating projection formation. *Mol. Biol. Cell* **3:** 429.

Reijo, R.A., E.M. Cooper, G.J. Beagle, and T.C. Huffaker. 1994. Systematic mutational analysis of the yeast β-tubulin gene. *Mol. Biol. Cell* **5:** 29.

Robinow, C.F. 1975. The preparation of yeasts for light microscopy. *Methods Cell Biol.* **11:** 1.

Robinow, C.F. and J. Marak. 1966. A fiber apparatus in the nucleus of the yeast cell. *J. Cell Biol.* **29:** 129.

Roof, D.M., P.B. Meluh, and M.D. Rose. 1992. Kinesin-related proteins required for assembly of the mitotic spindle. *J. Cell Biol.* **118:** 95.

Rose, M.D. and G.R. Fink. 1987. *KAR1:* A gene required for function of both intranuclear and extranuclear microtubules in yeast. *Cell* **48:** 1047.

Rose, M.D., L.M. Misra, and J.P. Vogel. 1989. *KAR2:* A karyogamy gene, is the yeast homolog of the mammalian Bip/GRP78 gene. *Cell* **57:** 1211.

Rose, M.D., B.R. Price, and G.R. Fink. 1986. *Saccharomyces cerevisiae* nuclear fusion requires prior activation by alpha-factor. *Mol. Cell. Biol.* **6:** 3490.

Rothstein, R.J. 1983. One-step gene disruption in yeast. *Methods Enzymol.* **101:** 202.

Rout, M.P. and J.V. Kilmartin. 1990. Components of the yeast spindle and spindle pole body. *J. Cell Biol.* **111:** 1913.

Salisbury, J.L., A.T. Baron, and M.A. Sanders. 1988. The centrin-based cytoskeleton of *Chlamydomonas reinhardtii:* Distribution in interphase and mitotic cells. *J. Cell Biol.* **107:** 635.

Salisbury, J.L., A. Baron, B. Surek, and M. Melkonian. 1984. Striated flagellar roots: Isolation and partial characterization of a calcium-modulated contractile organelle. *J. Cell Biol.* **99:** 962.

Saunders, W.S. and M.A. Hoyt. 1992. Kinesin-related proteins required for structural integrity of the mitotic spindle. *Cell* **70:** 451.

Saunders, W.S., D. Koshland, D. Eshel, I.R. Gibbons, and M.A. Hoyt. 1995. *Saccharomyces cerevisiae* kinesin- and dynein-related proteins required for anaphase chromosome segregation. *J. Cell Biol.* **128:** 617.

Schatz, P.J., F. Solomon, and D. Botstein. 1986a. Genetically essential and non-essential α-tubulin genes specify functionally interchangeable proteins. *Mol. Cell. Biol* **6:** 3722.

———. 1988. Isolation and characterization of conditional-lethal mutations in the *TUB1* α-tubulin gene of the yeast *Saccharomyces cerevisiae. Genetics* **120:** 681.

Schatz, P.J., L. Pillus, P. Grisafi, F. Solomon, and D. Botstein. 1986b. Two functional α-tubulin genes of the yeast *Saccharomyces cerevisiae* encode divergent proteins. *Mol. Cell. Biol.* **6:** 3711.

Schild, D., H.N. Ananthaswamy, and R.K. Mortimer. 1981. An endomitotic effect of a cell cycle mutation of Saccharomyces cerevisiae. *Genetics* **97:** 551.

Schopfer, L.M. and V. Massey. 1991. In *A study of enzymes* (ed. S.A. Kuby), p. 247. CRC Press, Boca Raton, Florida.

Schroer, T.A. 1994. Structure, function and regulation of cytoplasmic dynein. *Curr. Opin. Cell Biol.* **6:** 69.

Sethi, N., M.C. Monteagudo, D. Koshland, E. Hogan, and D.J. Burke. 1991. The *CDC20* gene product of *Saccharomyces cerevisiae:* A β-transducin homolog, is required for a subset of microtubule-dependent cellular processes. *Mol. Cell. Biol.* **11:** 5592.

Shortle, D., P. Novick, and D. Botstein. 1984. Construction and genetic characterization of temperature-sensitive mutant alleles of the yeast actin gene. *Proc. Natl. Acad. Sci.* **81:** 4889.

Simon, V.R., T.C. Swayne, and L.A. Pon. 1995. Actin-dependent mitochondrial motility in mitotic yeast and cell free systems: Identification of a motor activity on the mitochondrial surface. *J. Cell Biol.* **130:** 345.

Sivadon, P., F. Bauer, M. Aigle, and M. Crouzet. 1995. Actin cytoskeleton and budding pattern are altered in the yeast *rvs161* mutant: The Rvs161 protein shares common domains with the brain protein amphiphysin. *Mol. Gen. Genet.* **246:** 485.

Sloat, B.F., A. Adams, and J.R. Pringle. 1981. Roles of the *CDC24* gene product in cellular morphogenesis during the *Saccharomyces cerevisiae* cell cycle. *J. Cell Biol.* **89:** 395.

Smith, V., D. Botstein, and P.O. Brown. 1995. Genetic footprinting: A genomic strategy for determining a gene's function given its sequence. *Proc. Natl. Acad. Sci.* **92:** 6479.

Snyder, M. and R.W. Davis. 1988. *SPA1:* A gene important for chromosome segregation and other mitotic functions in *S. cerevisiae. Cell* **54:** 743.

Snyder, M., S. Gehrung, and B.D. Page. 1991. Studies concerning the temporal and genetic control of cell polarity in *Saccharomyces cerevisiae. J. Cell Biol.* **114:** 515.

Sobel, S.G. and M. Snyder. 1995. A highly divergent γ-tubulin gene is essential for cell growth and proper microtubule organization in *Saccharomyces cerevisiae. J. Cell Biol.* **131:** 1775.

Sorger, P.K., F. Severin, and A.A. Hyman. 1994. Factors required for the binding of reassembled yeast kinetochores to microtubules *in vitro. J. Cell Biol.* **127:** 995.

Spang, A., S. Geissler, K. Grein, and E. Schiebel. 1996. γ-Tubulin-like Tub4p of *Saccharomyces cerevisiae* is associated with the spindle pole body substructures that organize microtubules and is required for mitotic spindle formation. *J. Cell Biol.* **134:** 429.

Spang, A., I. Courtney, K. Grein, M. Matzner, and E. Schiebel. 1995. The Cdc31p-binding protein Kar1p is a component of the half bridge of the yeast spindle pole body. *J. Cell Biol.* **128:** 863.

Spudich, A. and J.A. Spudich. 1979. Actin in triton-treated cortical preparations of unfertilized and fertilized sea urchin eggs. *J. Cell Biol.* **82:** 212.

Stearns, T. 1988. "Genetic analysis of the yeast microtubule cytoskeleton." Ph.D. thesis. Massachusetts Institute of Technology, Cambridge.

———. 1990. The yeast microtubule cytoskeleton: Genetic approaches to structure and function. *Cell Motil. Cytoskeleton* **15:** 1.

———. 1995. The green revolution. *Curr. Biol.* **5:** 262.

Stearns, T. and D. Botstein. 1988. Unlinked noncomplementation: Isolation of new conditional-lethal mutations in each of the tubulin genes of *Saccharomyces cerevisiae. Genetics* **119:** 249.

Stearns, T. and M. Kirschner. 1994. In vitro reconstitution of centrosome assembly and function: The central role of γ-tubulin. *Cell* **76:** 623.

Stearns, T., L. Evans, and M. Kirschner. 1991. γ-Tubulin is a highly conserved component of the centrosome. *Cell* **65:** 825.

Stearns, T., M.A. Hoyt, and D. Botstein. 1990. Yeast mutants sensitive to antimicrotubule drugs define three genes that affect microtubule function. *Genetics* **124:** 251.

Stirling, D.A., K.A. Welch, and M.J.R. Stark. 1994. Interaction with calmodulin is required for the function of Spc110p, an essential component of the yeast spindle pole body. *EMBO J.* **13:** 4329.

Stott, K., K. Saito, D.J. Thiele, and V. Massey. 1993. Old yellow enzyme. The discovery of multiple isozymes and a family of related proteins. *J. Biol. Chem.* **268:** 6097.

Streiblova, E. and M. Girbardt. 1980. Microfilaments and cytoplasmic microtubules in cell division cycle mutants of *Schizosaccharomyces pombe. Can. J. Microbiol.* **26:** 250.

Strunnikov, A.V., J. Kingsbury, and D. Koshland. 1995. *CEP3* encodes a centromere protein of *Saccharomyces cerevisiae. J. Cell Biol.* **128:** 749.

Sullivan, D.S. and T.C. Huffaker. 1992. Astral microtubules are not required for anaphase B in *Saccharomyces cerevisiae. J. Cell Biol.* **119:** 379.

Sundberg, H.A., L. Goetsch, B. Byers, and T.N. Davis. 1996. Role of calmodulin and Spc110p interaction in the proper assembly of spindle pole body components. *J. Cell Biol.* **133:** 111.

Sweeney, F.P., M.J. Pocklington, and E. Orr. 1991. The yeast type II myosin heavy chain: Analysis of its predicted polypeptide sequence. *J. Muscle Res. Cell Motil.* **12:** 61.

Theriot, J.A. and T.J. Mitchison. 1993. The three faces of profilin. *Cell* **75:** 835.

Thomas, J.H. 1993. Thinking about genetic redundancy. *Trends Genet* **9:** 395.

Thomas, J.H., N.F. Neff, and D. Botstein. 1985. Isolation and characterization of mutations in the β-tubulin gene of *Saccharomyces cerevisiae. Genetics* **111:** 715.

Tian, G., Y. Huang, H. Rommelaere, J. Vandekerckhove, C. Ampe, and N.J. Cowan. 1996. Pathway leading to correctly folded β-tubulin. *Cell* **86:** 287.

Tilney, L.G., E.H. Egelman, D.J. DeRosier, and J.C. Saunder. 1983. Actin filaments, stereocilia, and hair cells of the bird cochlea. II. Packing of actin filaments in the stereocilia and in the cuticular plate and what happens to the organization when the stereocilia are bent. *J. Cell Biol.* **96:** 822.

Tkacz, J.S. and J.O. Lampen. 1972. Wall replication in *Saccharomyces* species: Use of fluorescein-conjugated concanavaline A to reveal the site of mannan insertion. *J. Gen. Microbiol.* **72:** 243.

Ursic, D. and M.R. Culbertson. 1991. The yeast homolog to mouse *Tcp-1* affects microtubule-mediated processes. *Mol. Cell. Biol.* **11:** 2629.

Ursic, D., J.C. Sedbrook, K.L. Himmel, and M.R. Culbertson. 1994. The essential yeast Tcp1 protein affects actin and microtubules. *Mol. Biol. Cell* **5:** 1065.

Vallen, E.A., M.A. Hiller, T.Y. Scherson, and M.D. Rose. 1992. Separate domains of *KAR1* mediate distinct functions in mitosis and nuclear fusion. *J. Cell Biol.* **117:** 1277.

Vallen, E.A., W. Ho, M. Winey, and M.D. Rose. 1994. Genetic interactions between *CDC31* and *KAR1*, two genes required for duplication of the microtubule organizing center in *Saccharomyces cerevisiae. Genetics* **137:** 407.

Vinh, D.B.-N. and D.G. Drubin. 1994. A yeast TCP-1-like protein is required for actin function in vivo. *Proc. Natl. Acad. Sci.* **91:** 9116.

Vinh, D.B., M.D. Welch, A.K. Corsi, K.F. Wertman, and D.G. Drubin. 1993. Genetic evidence for functional interactions between actin noncomplementing (Anc) gene products and actin cytoskeletal proteins in *Saccharomyces cerevisiae. Genetics* **135:** 275.

Vogel, J.P., L.M. Misra, and M.D. Rose. 1990. Loss of Bip/GRP78 function blocks translocation of secretory proteins in yeast. *J. Cell Biol.* **110:** 1885.

Vojtek, A., B. Haarer, J. Field, J. Gerst, T.D. Pollard, S. Brown, and M. Wigler. 1991. Evidence for a functional link between profilin and CAP in the yeast *Saccharomyces cerevisiae. Cell* **66:** 497.

Waddle, E.A., T.S. Karpova, R.H. Waterston, and J.A. Cooper. 1996. Movement of cortical actin patches in yeast. *J. Cell Biol.* **132:** 861.

Wang, T. and A. Bretscher. 1995. The *rho*-GAP encoded by BEM2 regulates cytoskeletal structure in budding yeast. *Mol. Biol. Cell.* **6:** 1011.

Weinstein, B. and F. Solomon. 1990. Phenotypic consequences of tubulin overproduction in *Saccharomyces cerevisiae:* Differences between alpha-tubulin and beta-tubulin. *Mol. Cell. Biol.* **10:** 5295.

Welch, M.D., D.A. Holtzman, and D.G. Drubin. 1994. The yeast actin cytoskeleton. *Curr. Opin. Cell Biol.* **6:** 110.

Welch, M.D., D.B.N. Vinh, H.H. Okamura, and D.G. Drubin. 1993. Screens for extragenic mutations that fail to complement *act1* alleles identify genes that are important for actin function in *Saccharomyces cerevisiae. Genetics* **135:** 265.

Wertman, K.F. and D.G. Drubin. 1992. Actin constitution: Guaranteeing the right to assemble. *Science* **258:** 759.

Wertman, K.F., D.G. Drubin, and D. Botstein. 1992. Systematic mutational analysis of the yeast *ACT1* gene. *Genetics* **132:** 337.

Winey, M. 1992. Spindle pole body of *Saccharomyces cerevisiae:* A model for genetic analysis of the centrosome cycle. In *The centrosome* (ed. V.I. Kalnins). Academic Press, San Diego.

Winey, M., M.A. Hoyt, C. Chan, L. Goetsch, D. Botstein, and B. Byers. 1993. *NDC1:* A nuclear periphery component required for yeast spindle pole body duplication. *J. Cell Biol.* **122:** 743.

Winey, M., C.L. Mamay, E.T. O'Toole, D.N. Mastronarde, T.H. Giddings, K.L. McDonald, and J.R. McIntosh. 1995. Three-dimensional ultrastructural analysis of the *Saccharomyces cerevisiae* mitotic spindle. *J. Cell Biol.* **129:** 1601.

Yanagida, M. 1987. Yeast tubulin genes. *Microbiol. Sci.* **4:** 115.

Yeh, E., R.V. Skibbens, J.W. Cheng, E.D. Salmon, and K. Bloom. 1995. Spindle dynamics and cell cycle regulation of dynein in the budding yeast *Saccharomyces cerevisiae. J. Cell Biol.* **130:** 687.

Zheng, Y., M.K. Jung, and B.R. Oakley. 1991. γ-Tubulin is present in *Drosophila melanogaster* and *Homo sapiens* and is associated with the centrosome. *Cell* **65:** 817.

Zheng, Y., M.L. Wong, B. Alberts, and T. Mitchison. 1995. Nucleation of microtubule assembly by a γ-tubulin-containing ring complex. *Nature* **378:** 578.

2

Protein Secretion, Membrane Biogenesis, and Endocytosis

Chris A. Kaiser, Ruth E. Gimeno, and David A. Shaywitz
Department of Biology
Massachusetts Institute of Technology
Cambridge, Massachusetts 02139

  87969-356-8/97 $5 + .00

I. INTRODUCTION

A growing yeast cell doubles its surface area every division cycle. The principal function of the yeast secretory pathway is to generate and deliver new membrane and protein to the growing surface of the bud. The pathway that newly synthesized proteins follow to the cell surface has the same outline in *Saccharomyces cerevisiae* as that originally defined in mammalian cells (Palade 1975). The principal steps in the pathway are outlined below and are diagrammed in Figure 1. The gene products responsible for the function of the secretory pathway are given at the end of this chapter in Table 1.

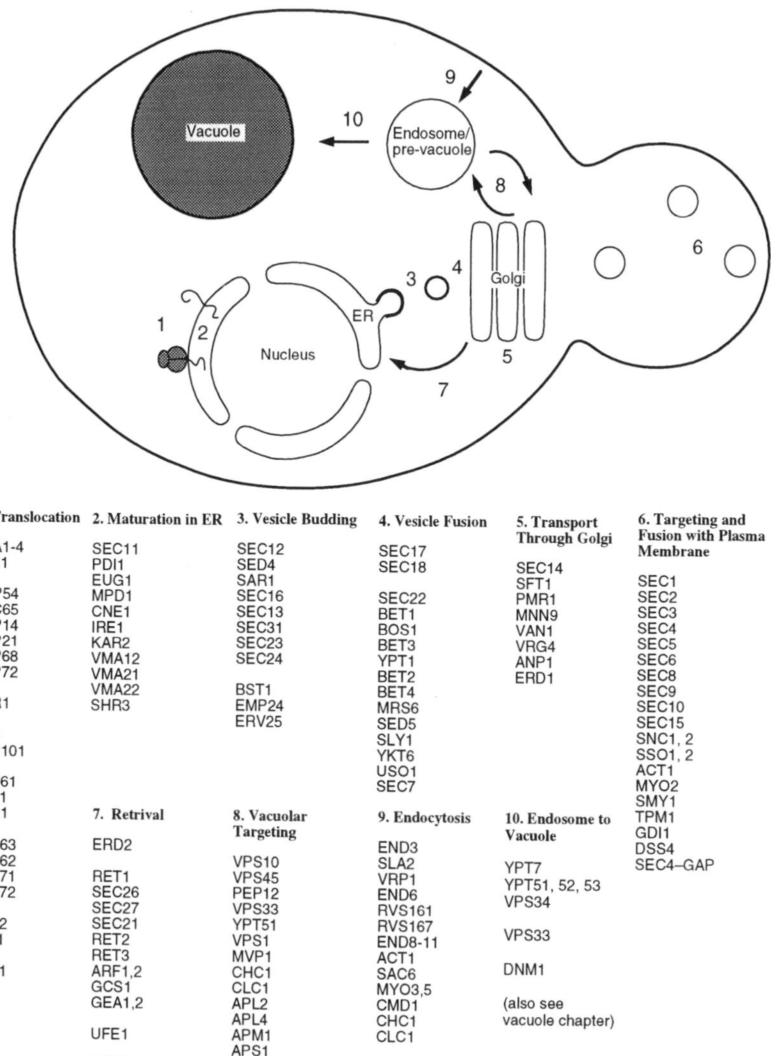

Figure 1 An overview of the different stages of the secretory pathway with the genes that are discussed in connection with each stage.

1. Secretory proteins are translocated from the cytosol into the lumen of the endoplasmic reticulum (ER). For some proteins, translocation occurs cotranslationally and for other proteins, translocation can occur posttranslationally, i.e., after completion of protein synthesis.
2. Secretory proteins assume a functional conformation within the lumen of the ER. This can involve covalent modifications of the pro-

tein such as cleavage of signal peptide, addition of N-linked and O-linked carbohydrate chains, disulfide bond formation, and addition of glycosylphosphatidylinositol (GPI) anchors. Catalyzed protein folding and, in some cases, assembly of multimeric protein complexes also take place in the ER lumen.

3. Secretory proteins are then packaged into 60-nm vesicles that bud from the ER.
4. These vesicles then deliver their contents to the Golgi by fusion with the Golgi membrane.
5. The Golgi apparatus is composed of three biochemically distinguishable compartments. Secretory proteins are carried through successive compartments of the Golgi probably by more rounds of vesicle formation and fusion.
6. Post-Golgi secretory vesicles 80–100 nm in diameter bud from the Golgi and then fuse with the plasma membrane of the bud. Targeting of these vesicles to the bud involves elements of the actin cytoskeleton.

The pathway also includes three branchpoints that enable protein sorting to different organelles to be achieved.

7. Resident ER proteins that have reached an early Golgi compartment are returned to the ER by retrograde vesicles that bud from the Golgi and fuse with the membrane of the ER.
8. In a late Golgi compartment, soluble and membrane vacuolar proteins are segregated from proteins destined for the plasma membrane.
9. Some integral membrane proteins that have reached the plasma membrane can be taken up by endocytosis. In the best-studied examples, the function of endocytosis appears to be the negative regulation of surface receptors and permeases.

A large number of different proteins are delivered to their final address by the secretory pathway. This number can be estimated from current knowledge of the approximately 6000 *S. cerevisiae* proteins. Functional or localization studies have shown that 277 of these genes are luminal or membrane proteins which are either at the cell surface or in one of the organelles along the secretory pathway (Yeast Protein Database [see Garrels 1995]). However, this number does not take into account all of the genes whose function or localization has yet to be explored. A better estimate can be obtained by considering the amino acid sequence of yeast proteins. By sequence alone, more than 1800 genes appear to encode proteins with one or more predicted transmembrane domains. Even allowing for the many integral membrane proteins in mitochondria and peroxisomes, it seems that from 10% to 20% of all yeast proteins are either residents or passengers of the secretory pathway.

To give an overview of the composition of the secretory pathway, the following are some of the types of proteins known to be in each organelle.

1. The ER contains membrane proteins responsible for the translocation of newly synthesized proteins across the ER membrane as well as luminal proteins involved in the folding and assembly of proteins within the ER. The ER membrane also contains proteins required for the formation of transport vesicles directed to the Golgi. In addition, the ER contains a large number of synthetases and transferases that carry out the synthesis of lipids and oligosaccharides in the ER. These enzymes are all integral membrane proteins.
2. The Golgi contains mannosyltransferases and proteases involved in the maturation of secretory proteins. Most of these proteins are type II integral membrane proteins (amino-terminal transmembrane anchors). The Golgi also contains an integral membrane Ca^{++} pump. This protein as well as other integral membrane Golgi proteins is required for normal maturation of proteins within the Golgi.
3. The lumen of the vacuole contains proteases and other hydrolytic enzymes. The vacuolar membrane contains a multisubunit H^+ ATPase that is responsible for acidification of the vacuole. The vacuole membrane also contains a phosphatase, a Ca^{++} pump, and transporters to concentrate amino acids in the vacuole lumen.
4. The plasma membrane carries a large number of integral membrane proteins that control the flux of small molecules and ions through the membrane. These include (i) permeases for cellular uptake of ammonia, amino acids, nucleotides, sugars, and other small molecules, (ii) H^+-transporting P-type ATPases and other ion pumps and transporters that set the electrical and ionic balance with the extracellular medium, and (iii) ATP-binding cassette (ABC)-type transporters that may be responsible for pumping compounds out of the cell. The plasma membrane also contains membrane-bound enzymes for the synthesis of cell wall chitin and glucan and proteases of unknown function. Finally, the plasma membrane contains proteins required for mating, which include the pheromone receptor proteins, agglutinins for cell adhesion, and an integral membrane protein required for plasma membrane fusion during mating.
5. In addition to the structural carbohydrates, chitin and glucan, the cell wall carries the soluble enzymes invertase, melibiase, glucanase, as-

paraginase, phosphatase, and proteases. These enzymes as well as mating pheromone and killer toxin are found trapped within the carbohydrate matrix of the cell wall and in the extracellular medium.

Most soluble and membrane proteins follow the same pathway to the cell surface. The existence of a single major pathway for protein secretion was demonstrated through the analysis of secretory pathway mutants. These *sec* mutants were first isolated as temperature-sensitive mutants that fail to express several different secreted proteins at the cell surface (Novick et al. 1980). When *sec* mutants are shifted to their restrictive temperature, the expansion of the cell surface stops abruptly, although protein synthesis continues for many hours. In more detailed examination of cell surface proteins, it was shown that no newly synthesized proteins could be detected at the surface after a secretory mutant block was established (Novick and Schekman 1983). Thus, for the plasma membrane and most surface proteins, there does not appear to be a significant alternative route for delivery of material to the cell surface other than the pathway defined by the *SEC* genes. Furthermore, no significant branches in the pathway are evident since mutations that block at each stage of the pathway—ER, Golgi, and secretory vesicles—all have the same effect of completely blocking surface growth.

Of all of the proteins that we now know to traverse the secretory pathway, three soluble proteins have been used for most of the in vivo and in vitro tests of secretory pathway function described in this chapter. These proteins are the secreted enzyme invertase, the vacuolar protease carboxypeptidase Y (CPY), and the secreted pheromone α-factor. This emphasis on three marker proteins is largely the result of historical precedent, and these soluble proteins now appear to represent a minor class of passengers in the secretory pathway, the vast majority of which are integral membrane proteins. Clearly, there are fundamental differences between the behavior of soluble and membrane proteins in the secretory pathway. A prominent example being the sorting of proteins to the vacuole and to the plasma membrane that occurs in a late Golgi compartment. When soluble proteins that are normally targeted to the vacuole are overproduced, the excess protein generally appears at the cell surface, indicating that transport to the plasma membrane is the default pathway for soluble proteins. In contrast, similar overproduction experiments indicate that transport to the vacuole is the default path for membrane proteins (Nothwehr and Stevens 1994). As the study of the function of the secretory pathway now has begun to emphasize issues of efficiency and fidelity, it will become increasingly important for the behavior of membrane proteins to be examined in parallel to the soluble marker proteins.

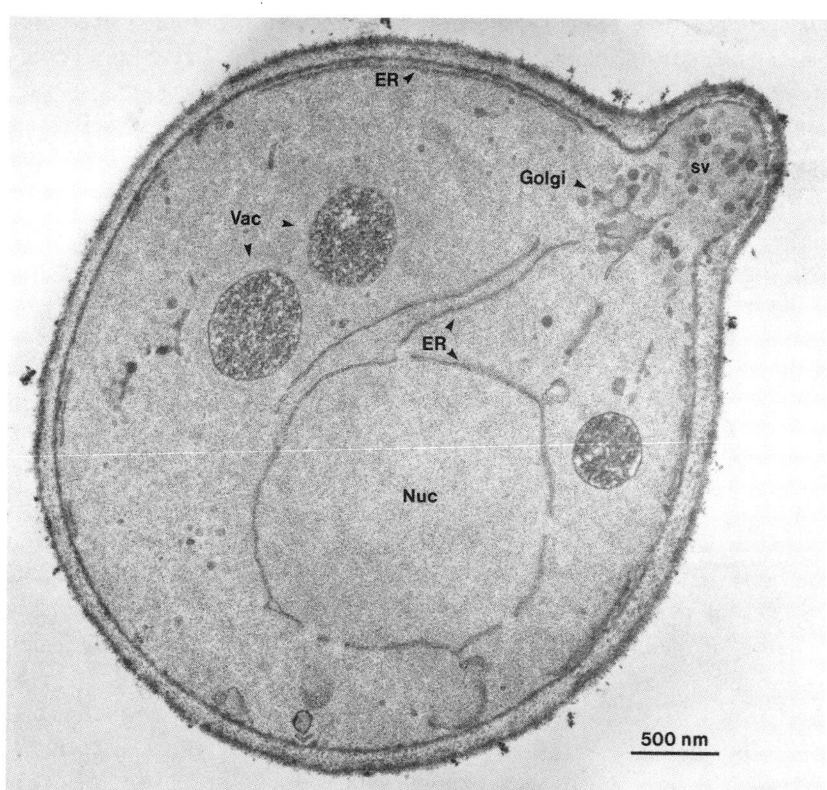

Figure 2 Electron micrograph of *S. cerevisiae* showing the organelles of the secretory pathway. The cell was fixed and stained with permanganate to highlight membrane structures. After embedding, thin sections were stained with uranyl acetate. (Nuc) Nucleus; (ER) endoplasmic reticulum (the nuclear envelope and a discontinuous, peripheral cisterna); (Golgi) fenestrated cisterna and surrounding vesicles; (sv) post-Golgi secretory vesicles; (Vac) vacuole.

Figure 2 shows an electron micrograph of an *S. cerevisiae* cell stained with $KMnO_4$ to highlight the major organelles of the secretory pathway. Localization of ER proteins such as Kar2p/BiP shows that the ER is composed of the nuclear envelope and a single discontinuous cisterna that lines the inner surface of the plasma membrane (Rose et al. 1989; Preuss et al. 1992). The secretory structures in *S. cerevisiae* occupy only a modest proportion of the cell interior as compared to the highly amplified membranes of the secretory organelles, typical of the specialized secretory cells used to study mammalian secretory pathway function. The low steady-state abundance of secretory organelles in *S. cerevisiae* is in part a consequence of the rapid transit of secretory proteins to the cell surface. The time from synthesis to appearance at the cell surface for typical secretory proteins is less than 10 minutes in *S. cerevisiae* as compared to more than 1 hour in a mammalian cell.

A fundamental difference between the biology of yeast cells and mammalian cells might also contribute to different morphologies of the secretory organelles. Under conditions of plentiful nutrients, yeast cells appear to be adapted to maximize growth rate. A rapidly dividing yeast cell doubles its surface area in 90 minutes, and under these conditions, it is likely that most vesicular transport between organelles is directed outward to the cell surface. Since retrograde vesicular traffic would tend to reduce the rate of surface growth, it is likely that few of the vesicles seen within a growing yeast cell are retrograde vesicles. Using the assumption that most vesicular traffic is in the forward direction, a simple accounting of the rate of plasma membrane growth compared to the size of transport vesicles yields the result that only two vesicles must bud from the ER every second to deliver the needed membrane to the cell surface. Given the brief lifetime of these vesicles, it is not surprising that few transport vesicles are evident in thin sections of wild-type yeast cells. In contrast, a mammalian cell that secretes a large quantity of a particular protein, such as a digestive enzyme or a peptide hormone, must do so without net surface growth so as to remain within the constraints of an epithelial cell layer. In such a cell, every forward transport step must be counterbalanced by an equivalent retrograde step. Even if vesicle shuttles perform with maximum efficiency, incorporating the secretory cargo only in forward steps and never in retrograde steps, twice the number of vesicles would be needed than for an equivalent growing cell.

II. PROTEIN TRANSLOCATION AND ASSEMBLY IN THE ENDOPLASMIC RETICULUM

A. Signal Sequences

The first insights into the mechanism of protein translocation into the ER came from the identification of hydrophobic signal sequences at the amino termini of precursors of secreted proteins (Blobel and Sabatini 1971; Milstein et al. 1972). Signal sequences of about 20 amino acid residues that direct secretory proteins to be translocated are present on precursors of almost all secreted proteins in both eukaryotes and prokaryotes. The case of invertase in *S. cerevisiae* provides a natural example of how the intracellular address of a protein can be specified by a signal sequence. The structural gene for invertase (*SUC2*) has two transcripts; the shorter transcript encodes a hydrophilic protein that resides in the cytosol and the longer transcript encodes the same protein but with a hydrophobic amino-terminal signal sequence of 20 amino acids. The product of the longer transcript is secreted to the cell surface (Carlson and Botstein 1980).

Comparison of the amino termini of a large number of secreted proteins reveals little conservation of primary sequence, but several general features of these sequences have been noted. Eukaryotic signal sequences usually have a short amino-terminal region that often carries a net positive charge, a hydrophobic region that consists of a segment of 7–15 residues that does not contain charged residues and is enriched in hydrophobic residues, and a signal peptidase cleavage region of three to seven polar residues (von Heijne 1985, 1990). Mutational studies to identify the essential components of signal sequences showed that usually a substantial deletion within the hydrophobic core of a signal sequence was required to eliminate function (Haguenauer-Tsapis and Hinnen 1984; Kaiser and Botstein 1986). In an extreme case, it was found that even after the entire hydrophobic core of the signal sequence of CPY was deleted, about 10% of the protein was still translocated into the ER (Blachly-Dyson and Stevens 1987). A direct experimental test of the limits of variation of signal sequences was performed by determining what kinds of sequences could functionally replace the signal sequence of invertase. About 20% of essentially random amino acid will direct secretion of invertase, showing that many different sequences have the capacity to target a protein to the translocation machinery (Kaiser et al. 1987). The random sequences that function as signal sequences are more hydrophobic than either random sequences that do not function as signal sequences or the amino termini of native cytosolic proteins. Presently, signal sequences are thought to be recognized either by the cytosolic signal recognition particle (SRP) complex or by the Sec63p complex in the ER membrane. Apparently the critical recognition events require little more than binding to a short hydrophobic peptide.

B. Cytosolic Factors

1. Hsp70s

The first major insight into the machinery required for translocation of proteins into the ER emerged from cell-free reconstitution of the translocation process. The key to a successful translocation assay, discovered simultaneously by three groups, is to use microsomal membranes, translation extract, and a secretory protein precursor all derived from *S. cerevisiae* (Hansen et al. 1986; Rothblatt and Meyers 1986; Waters and Blobel 1986). The assay involves expression of radioactively labeled prepro-α-factor by programming a translation extract with messenger RNA from the *MFA1* gene in the presence of cytosolic proteins and microsomal membranes. Translocation into the lumen of microsomal membranes could be detected by resistance of prepro-α-factor to exogenously added protease and acquisition of N-linked carbohydrate chains by pro-α-factor. Unlike the behavior of similar assays using components from mammalian cells, it was found that the processes of translation and translocation were not obligatorily coupled: Prepro-α-factor could first be translated and then be subsequently translocated into yeast microsomal membranes in the presence of cytosolic proteins after further protein synthesis was blocked by inhibitors. Not all precursor polypeptides can be translocated posttranslationally. For example, full-length preinvertase cannot be translocated after translation is completed. However, a portion of the molecule consisting of the amino-terminal 263 amino acid residues can be translocated posttranslationally (Hansen and Walter 1988).

The ability to translocate prepro-α-factor after translation allowed the dependence of translocation on cytosolic components to be tested. A factor that stimulated translocation was purified from crude cytosol and identified as 70-kD heat shock protein (Hsp70) (Chirico et al. 1988). Hsp70s are ATPases that have been implicated in a number of intracellular processes related to protein folding and disassembly or rearrangement of multimeric protein complexes (for review, see Georgopoulos et al. 1990). The function of Hsp70s in translocation is thought to be the maintenance of the precursor polypeptide in an unfolded state, necessary to engage properly with the translocation machinery. *S. cerevisiae* has two functionally distinct classes of cytosolic Hsp70s known as Ssa and Ssb proteins. The in vivo role of Hsp70s in translocation was examined by selectively depleting cells of the Ssa class of proteins. A triple deletion of *SSA1*, *SSA2*, and *SSA4* is inviable, but this strain can be kept alive by expression of *SSA3* from the *GAL1* promoter. When Ssa expression is repressed by growth on glucose, the cells begin to accumulate unprocessed precursors of secretory proteins as the endogenous level of Ssa protein falls (Deshaies et al. 1988).

Models for how Hsp70 proteins interact with their substrates are guided by studies of the *Escherichia coli* Hsp70 protein DnaK. DnaJ is one of the proteins that functions with DnaK in initiation of λ DNA replication and has been shown to stimulate the ATPase activity of DnaK

(Liberek et al. 1991). The *YDJ1* gene of *S. cerevisiae* encodes a homolog of DnaJ that appears to be important for translocation of certain precursor proteins into the ER (Caplan et al. 1992b). Ydj1p is a cytosolic protein that is modified by farnesylation, indicating that it is located on the cytosolic face of membranes (Caplan et al. 1992a). A direct interaction between Ydj1p and Ssa1p is indicated by the ability of Ydj1p to stimulate the ATPase activity of Ssa1p (Cyr et al. 1992; Ziegelhoffer et al. 1995) and by synthetic-lethal interactions between *ydj1* null alleles and temperature-sensitive *ssa1* alleles (Becker et al., 1996). Possible functions of Ydj1p are to catalyze the cycle of substrate peptide binding and release by Ssa proteins and to target Ssa proteins to the membranes where they perform their function.

2. *SRP and SRP Receptor*

The best-understood component of the eukaryotic translocation apparatus is SRP (for review, see Walter and Johnson 1994). SRP was discovered as a cytosolic factor that is needed to restore translocation competence to mammalian microsomal membranes that have been extracted with high salt to remove peripheral proteins (Walter and Blobel 1980). Mammalian SRP is a complex of a 7S RNA and six polypeptides. SRP recognizes signal sequences and selectively binds to ribosomes that are translating secretory proteins at the stage in chain elongation when the signal sequence emerges from the ribosome. The complex of nascent-chain, ribosome, and SRP can then engage the membrane by binding to the signal sequence receptor, which is composed of two subunits, SRα and SRβ (Gilmore et al. 1982). The nascent chain is then transferred to the membrane translocation apparatus and SRP is released for another cycle of ribosome binding. The SRP cycle is driven by GTP hydrolysis (Connolly and Gilmore 1986), and the hydrolysis appears to be catalyzed by at least three separate components: the 54-kD SRP subunit (SRP54), SRα, and SRβ (Bernstein et al. 1989; Römisch et al. 1989; Miller et al. 1995). The net action of SRP and the SRP receptor appears to be the efficient targeting of presecretory proteins to the ER membrane. These proteins are abundant constituents of mammalian secretory cells, and their presence accounts for the observation that most mammalian secretory proteins begin translocation across the ER membrane soon after the signal sequence emerges from the ribosome.

A yeast homolog of SRP has been discovered that is important but not essential for protein translocation in vivo. Comparison of mammalian

SRP54 with a distant homolog, the *ffh* gene of *E. coli*, identified highly conserved blocks of amino acids that allowed isolation of a yeast homolog (Hann et al. 1989). The product of *SRP54* is a component of a 16S cytosolic ribonucleoprotein complex that contains an RNA moiety encoded by *SCR1* (Felici et al. 1989; Hann and Walter 1991) and a homolog of the 19-kD SRP subunit encoded by *SEC65* (Hann et al. 1992; Stirling and Hewitt 1992). Purification of the yeast SRP complex by immunoaffinity chromatography has allowed the isolation of five additional subunits, four of which are homologous to mammalian SRP subunits and one of which appears to be unique to yeast (Brown et al. 1994). SRP function is not essential for yeast viability since strains with chromosomal deletions of either *SRP54* or *SEC65* grow slowly. Controlled inactivation of these genes has been achieved by repression of a form of *SRP54* regulated by the *GAL1* promoter or by use of a temperature-sensitive allele of *SEC65*. In both cases, defects in translocation into the ER were observed. Of the translocated proteins that have been examined, the effect of loss of SRP function is most severe for the integral membrane proteins dipeptidyl amino peptidase B (DPAP-B) and Pho8p; partial translocation defects were seen for the membrane protein Och1p and the luminal proteins invertase, prepro-α-factor, and Kar2p; and no defect was evident for the luminal proteins Gas1p, Pdi1p, and CPY (Hann and Walter 1991; Hann et al. 1992; Stirling and Hewitt 1992; Ng et al. 1996). These findings suggest that there are two avenues for protein translocation into the ER, one dependent on SRP and the other independent of SRP (Fig. 3). The SRP-independent pathway involves the Sec63p complex and is discussed below. Since SRP interacts with nascent chains early in translation, partitioning must be specified by sequences near the amino terminus of the nascent chain. Indeed, the severity of the translocation defect caused by SRP depletion for a given protein correlates with the apparent hydrophobicity of the signal sequences. Apparently, proteins with relatively hydrophobic signal sequences preferentially engage SRP, whereas sequences of lower hydrophobicity preferentially engage the Sec63p complex (Ng et al. 1996). The details of how signal sequence hydrophobicity controls this partitioning are not understood.

The SRP-independent translocation process can adaptively increase its capacity under conditions of SRP depletion, since the severity of the translocation defect diminishes after prolonged propagation of strains depleted of SRP (Ogg et al. 1992). This adaptation to reduced SRP levels could correspond to induced Kar2p expression, a luminal ATPase required for posttranslational translocation of presecretory proteins as de-

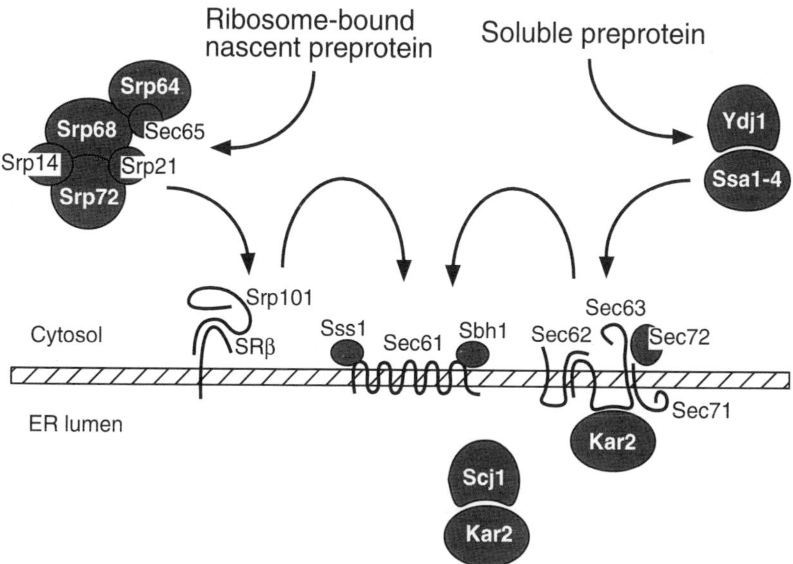

Figure 3 Soluble and membrane proteins involved in preprotein translocation across the ER membrane. The hypotheses that cotranslational translocation is mediated by SRP and that posttranscriptional translocation is mediated by cytosolic Hsp70 and the Sec63 complex have not been fully established.

scribed below, since depletion of Srp54p was found to increase the synthesis of other proteins likely to facilitate SRP-independent translocation, including Ssa1p, Ydj1p, and Kar2p (Arnold and Wittrup 1994).

The two subunits of the yeast SRP receptor have been identified by homology with their mammalian counterparts. *SRP101* encodes yeast SRα, a 69-kD peripheral membrane protein (Ogg et al. 1992) that is bound to the ER membrane by SRβ, a 30-kD integral membrane protein (Miller et al. 1995). Depletion of SRα has a similar effect on translocation as depletion of SRP subunits and strains with chromosomal deletion of *SRP101* are viable but grow slowly. A characteristic of mammalian in vitro translocation reactions is the capacity of SRP to arrest translation of nascent secretory proteins by binding to the signal sequence as it emerges from the ribosome. Subsequent interaction of SRP with the SRP receptor on the ER membrane transfers the nascent chain from SRP to the integral membrane translocation pore, and as translation continues, the nascent chain is translocated into the ER lumen (Walter and Blobel 1981). A predicted consequence of the ability of mammalian SRP to arrest translation and of the SRP receptor to release the nascent chain from this arrest is that depletion of the SRP receptor should give rise to unrelieved translational arrest of secretory proteins. However, depletion of SRα in yeast does not reduce translation of presecretory proteins, and it therefore can be concluded that yeast SRP is unlikely to cause a strong translation arrest (Ogg et al. 1992).

Important questions concerning the function of SRP in yeast still remain. Once an SRP-bound protein is brought to the membrane, the subsequent passage of the polypeptide chain across the membrane probably requires the Sec61p complex described below, but it is not known whether any of the other proteins that are required for posttranslational translocation are also needed. Furthermore, it is not understood how the nascent chain is transferred to the membrane translocation machinery. For this process to be efficient, a physical link between the translocation channel and SRP or SRP receptor is expected but has not yet been found. Clearly, an in vitro system that depends on exogenously added SRP for translocation into yeast microsomes will be a critical tool for addressing these questions.

C. Membrane and Luminal Factors

Membrane proteins required for protein translocation into the ER were first identified through the use of genetic screens. Fusion of a signal sequence to the cytosolic enzyme histidinol dehydrogenase directs this hybrid protein to the ER where it cannot function in histidine synthesis. Selection for mutants that misdirect some of the hybrid protein to the cytosol (where it is active) identified conditional mutations in three essential genes, *SEC61*, *SEC62*, and *SEC63* (Deshaies and Schekman 1987; Toyn et al. 1988; Rothblatt et al. 1989). A similar screen using a fusion of histidinol dehydrogenase to a membrane protein identified mutations in three additional genes, *SEC70*, *SEC71*, and *SEC72* (Green et al. 1992). Together, these gene products constitute most of the subunits of an ER membrane complex of seven polypeptides that in pure form is capable of translocating prepro-α-factor into phospholipid vesicles (Panzner et al. 1995). The full complex can be divided into two parts with different functions. The first, known as the Sec61p complex, is composed of three different polypeptides and probably forms the membrane channel through which the translocating polypeptide passes. The second, known as the Sec63p complex, is made up of four polypeptides and is thought to be critical for posttranslational translocation and to couple ATP hydrolysis by the luminal protein Kar2p to provide the energy to drive protein translocation across the membrane.

1. Sec61p Complex

SEC61 encodes a hydrophobic 52-kD ER protein that is predicted to have multiple membrane-spanning domains (Stirling et al. 1992). Evidence that Sec61p constitutes the core of a translocation channel comes from studies where a protein trapped in the membrane during translocation can be cross-linked to Sec61p (Müsch et al. 1992; Sanders et al. 1992). The Sec61p complex appears to be the central element of a conserved translocation machine, since it is similar in sequence to the *E. coli* translocation protein SecY (Stirling et al. 1992) and to a homologous mammalian protein that also can be cross-linked to translocating polypeptides (Görlich et al. 1992). Functional Sec61p is required for posttranslational translocation of prepro-α-factor into yeast microsomes (Panzner et al. 1995), whereas mammalian Sec61p is found associated with ribosome-bound nascent chains that have been brought to the membrane cotranslationally by SRP (Görlich and Rapoport 1993). Taken together, the experiments from yeast and mammalian cells situate Sec61p on both the SRP-dependent and SRP-independent pathways, a location consistent with its proposed role as the central translocation pore for both cotranslational and posttranslational translocation.

Mammalian Sec61p copurifies with two polypeptides, of approximately 9 kD each, that are referred to as the β and γ subunits of the complex (Görlich and Rapoport 1993; Hartmann et al. 1994). The yeast homolog of the γ-subunit is Sss1p which was isolated as a dosage-dependent suppressor of *sec61* mutations. Sss1p is similar in sequence to SecE protein, another component of the *E. coli* translocation apparatus (SecY, SecE, and SecG function as a complex in *E. coli*; a yeast homolog of SecG has not been identified). As expected for a component of the translocation channel, *SSS1* is an essential gene and depletion of Sss1p interferes with translocation of presecretory proteins: A severe defect in translocation of CPY was observed, whereas translocation of DPAP-B was hardly affected (Esnault et al. 1993). These relative effects on translocation of specific proteins are the inverse of those seen for depletion of SRP and suggest that Sss1p principally operates on an SRP-independent pathway. A yeast γ-subunit, encoded by *SBH1*, was identified by its homology with the mammalian protein and was found to copurify with yeast Sec61p and Sss1p (Panzner et al. 1995).

A second multimeric protein complex closely related to the Sec61p complex has been identified in yeast. This complex is composed of Ssh1p (a homolog of Sec61p), Sbh2p (a homolog of Sbh1p), and Sss1p (a component common to both complexes) (Finke et al. 1996). Like the complex that contains Sec61p, the Ssh1p complex is located in the ER membrane. Deletions of either *SSH1* or *SBH2* are viable, but exacerbate the growth and translocation defects of mutations in *SEC61* or *SBH1*, suggesting that the Ssh1p complex provides a translocation channel that acts in parallel with the Sec61p complex.

Immunoprecipitation experiments show that the Ssh1p complex associates with ER-bound ribosomes but not with the Sec63p complex. An attractive hypothesis is that the Sec61p complex, together with the Sec63p complex, is the principal channel for posttranslational translocation, and that the Ssh1p complex is the principle channel for cotranslational translocation. Development of an in vitro assay for translocation that depends on SRP would allow the proposed role of the Ssh1p complex in cotranslational translocation to be tested experimentally.

2. *Sec63p Complex*

SEC62 and *SEC63* are essential genes that encode integral membrane proteins required for protein translocation in vivo and in vitro (Deshaies and Schekman 1989; Rothblatt et al. 1989; Sadler et al. 1989). Antibody to Sec62p precipitates a complex of four proteins that includes Sec63p and is referred to here as the Sec63p complex (Deshaies et al. 1991). The two additional members of the complex are encoded by *SEC66/SEC71* and *SEC72* (Green et al. 1992; Feldheim et al. 1993; Kurihara and Silver 1993; Fang and Green 1994; Feldheim and Schekman 1994). Chromosomal deletions of *SEC71* and *SEC72* are viable but exhibit partial defects in translocation of a subset of presecretory proteins (Feldheim et al. 1993; Kurihara and Silver 1993; Fang and Green 1994; Feldheim and Schekman 1994). A form of the Sec63p complex that contains Sec63p, Sec71p, and Sec72p, but not Sec62p, can be purified from yeast microsomes and will confer the ability to translocate prepro-α-factor to proteoliposomes derived from *sec63* mutants (Brodsky and Schekman 1993). These findings suggest that the association of Sec62p with the complex is relatively weak and that Sec62p may not be required for the complex to function in vitro.

The Sec63p complex appears to reside close to the Sec61p complex in the ER membrane since Sec61p can be cross-linked to Sec62p (Deshaies et al. 1991), and both complexes copurify in an active form under conditions of mild detergent extraction (Panzner et al. 1995). The Sec63p complex is needed before the translocating chain associates with the Sec61p complex since mutations in either *SEC62* or *SEC63* block association of prepro-α-factor with Sec61p as detected in cross-linking studies

(Sanders et al. 1992). Moreover, the Sec63p complex has been implicated in recognition of signal sequences, clearly an early step in the translocation process. Mutations in *SEC62*, *SEC63*, or *SEC72* influence translocation of precursor proteins according to the composition of their signal sequence: Precursors with signal sequences of relatively low hydrophobicity are the most influenced when the function of the Sec63p complex is compromised (see Feldheim and Schekman 1994; Ng et al. 1996). An attractive possibility is that the Sec63p complex functions exclusively in an SRP-independent pathway to bring precursors to the translocation channel. This view is supported by the mutually exclusive associations of the Sec61p complex; some Sec61p complex extracted from yeast microsomes is bound to ribosomes (and therefore apparently engaged in cotranslational and probably SRP-dependent translocation), whereas a distinct pool of the Sec61p complex is found to be associated with the Sec63p complex (Panzner et al. 1995).

3. Kar2p/BiP

KAR2 encodes an essential yeast homolog of the Hsp70 family member, BiP, that is located in the lumen of the ER (Normington et al. 1989; Rose et al. 1989). Temperature-sensitive *kar2* mutations rapidly and completely block protein translocation in vivo, and membranes prepared from *kar2* mutants are defective for translocation in vitro (Vogel et al. 1990; Sanders et al. 1992). Direct binding of Kar2p to prepro-α-factor during the process of translocation has been demonstrated by cross-linking studies (Sanders et al. 1992). Translocation of prepro-α-factor into lipid vesicles containing purified Sec61p and Sec63p complexes is stimulated fivefold by inclusion of Kar2p in the lumen of these vesicles (Panzner et al. 1995). In vitro reconstitution of translocation in microsomes derived from *kar2* mutants shows that Kar2p is involved in both posttranslational and cotranslational translocation (Brodsky et al. 1995). Together, these findings suggest that a crucial aspect of the mechanism of translocation may be the binding of Kar2p to polypeptides as they emerge from the membrane into the ER lumen. A hypothesis related to the "Brownian ratchet" mechanism of translocation is that a translocating polypeptide would diffuse back and forth through a membrane channel until a sufficient length of the polypeptide had emerged into the lumen for Kar2p to bind to the polypeptide, thus restricting backward diffusion. Reiterations of this process would eventually draw the polypeptide into the ER lumen, using the energy of Kar2p binding to make the translocation process unidirectional (Simon et al. 1992).

A translocation mechanism driven by Kar2p binding would require a way for Kar2p to cycle between the free and polypeptide-bound states. Sec63p is the best candidate for a protein that could assist polypeptide binding and release by Kar2p. Sec63p was initially suspected to interact with Kar2p since a luminal segment of Sec63p has homology with DnaJ, an *E. coli* protein that acts in concert with DnaK, a member of the Hsp70 family to which Kar2p belongs (Sadler et al. 1989; Feldheim et al. 1992). Kar2p is found in association with the Sec63p complex isolated from detergent-solubilized microsomes by immunoprecipitation (Brodsky and Schekman 1993). Functional links between *KAR2* and *SEC63* are indicated by synthetic lethality between mutations in these two genes and by the ability of alleles of *KAR2* to suppress mutations in *SEC63* (Scidmore et al. 1993). Translocation of prepro-α-factor into microsomal vesicles depends on incorporation of Kar2p into the lumen of these vesicles, and the other Hsp70 proteins Ssa1p and DnaK will not substitute for Kar2p in this capacity (Brodsky et al. 1993). This specificity could be the consequence of the failure of Hsp70 proteins other than Kar2p to interact properly with the luminal domain of Sec63p. A second candidate for a protein that interacts with Kar2p is the DnaJ homolog, Scj1p. Null alleles of *SCJ1* are synthetically lethal with some, but not all, alleles of *KAR2*, and the portion of *SCJ1* that is homologous to DnaJ can function in place of the equivalent region of *SEC63*. However, a direct involvement of Scj1p in protein translocation has not been established (Schlenstedt et al. 1995).

In addition to its role in driving polypeptide translocation across the membrane, Kar2p has also been implicated in earlier steps in the translocation process. Prepro-α-factor arrested in the process of translocation can be cross-linked to Sec61p, but this association does not form when strains carrying certain alleles of *KAR2* are used in the preparation of microsomal membranes (Sanders et al. 1992). A similar defect in cross-linking of the precursor polypeptide is caused by the *sec63-1* mutation, which has been shown to interfere with the interaction between Sec63p and Kar2p (Lyman and Schekman 1995). These findings suggest that a polypeptide to be translocated posttranslationally interacts first with the Sec63p complex and is then transferred to the Sec61p complex and that Kar2p is required for either the interaction of the polypeptide with the Sec63p complex or the transfer of the polypeptide from the Sec63p complex to the Sec61p complex.

The proposed function of Kar2p as an ATPase that performs its function within the lumen of the ER implies that the ER membrane must contain transporters to carry ATP into the ER lumen from the cytosol. The ATP transporting activity was sought through the use of a biochemical assay for reconstituted ATP translocation into proteoliposomes. A single polypeptide with translocating activity purified from solubilized yeast microsomes was shown to be the product of the *SAC1* gene (Mayinger et al. 1995). Sac1p is an integral membrane protein located in ER and Golgi membranes consistent with this proposed function as an ER membrane transporter (Whitters et al. 1993). However, not all aspects of ATP transport into the ER are understood. Given that ATP in the ER lumen is probably necessary to sustain protein translocation into the ER, the ATP translocating activity should be essential. *sac1* deletion mutants are viable, except at low temperature, indicating that there must also be a Sac1p-independent mechanism for transport of ATP into the ER. In addition, *sac1* mutants display diverse phenotypes: *sac1* mutations were first identified as recessive suppressors of temperature-sensitive actin mutations (Novick et al. 1989), *sac1* mutations also suppress mutations in *SEC6*, *SEC9*, and *SEC14* (Cleves et al. 1989), and *sac1* mutants require more inositol for growth than do wild-type strains (Whitters et al. 1993). These results imply intriguing, but not easily explained, connections between ATP transport into the ER, the actin cytoskeleton, Golgi function, and lipid metabolism.

KAR2 was first identified in mutants that failed to fuse haploid nuclei during mating, a process known as karyogamy (Polaina and Conde 1982). The roles of Kar2p in karyogamy and in protein translocation are separable since the *kar2-1* allele shows no defect in translocation but has a strong karyogamy defect (Vogel et al. 1990). Initially, it was not obvious how Kar2p in the lumen of the ER would participate in the fusion of the outer surface of the nuclear envelope (which is also ER membrane) during nuclear fusion. It now appears that Kar2p may perform at least part of its function in karyogamy in conjunction with the Sec63p complex. Mutations in *SEC63*, *SEC71*, and *SEC72* (but not mutations in *SEC62*) cause karyogamy defects (Ng and Walter 1996). This finding implies that the Sec63p complex associated with Kar2p in the lumen could mediate bilayer fusion or that the Sec63p complexes could interact homotypically to dock one nuclear membrane against another in preparation for fusion. Mutations in *KAR2* appear to have a greater effect on karyogamy than mutations in *SEC63*, *SEC71*, or *SEC72* since *kar2-1* causes a unilateral defect (karyogamy fails if either parent carries the mutation), whereas mutations in components of the Sec63p complex cause bilateral defects (karyogamy fails only if both parents carry the mutation). The severity of the karyogamy defect in *kar2-1* mutants suggests that Kar2p may have additional functions in karyogamy that do not involve the Sec63p complex.

D. Protein Maturation in the ER

1. Signal Peptidase

As secretory proteins enter the lumen of the ER, their signal sequences are proteolytically removed. The enzyme responsible for signal sequence cleavage is a luminal protein complex known as signal peptidase. Signal peptidase isolated from *S. cerevisiae* is composed of four polypeptides of 25, 20, 18, and 13 kD (YaDeau et al. 1991; Fang et al. 1996; Mullins et al. 1996). The 18-kD subunit is encoded by *SEC11*, and temperature-sensitive *sec11* mutants fail to cleave signal sequences at the restrictive temperature (Bohni et al. 1988). The yeast complex is related to mammalian signal peptidase which is composed of five subunits, two of which are homologous to Sec11p (Shelness and Blobel 1990).

Comparison of a large number of known cleaved eukaryotic signal sequences indicates that signal peptidase cleavage usually occurs within a relatively polar region of the signal sequence, with small neutral or small polar residues at positions –1 and –3 with respect to the cleavage site (von Heijne 1983). An allele of *SUC2* that converts alanine to valine at position –1 of the cleavage site substantially reduces cleavage of the signal sequence (Schauer et al. 1985). Interestingly, this mutant invertase is transported out of the ER much more slowly than wild-type invertase. Similar defects in transport were observed for uncleaved signal sequences that were derived from random amino acid sequences, suggesting that amino-terminal hydrophobic peptides generally specify retention in the ER perhaps because of a structural similarity to unfolded proteins (Kaiser et al. 1987). *SEC11* was initially classified as a gene required for ER to Golgi transport; this transport defect can now be explained as an indirect consequence of retention in the ER of proteins with uncleaved signal sequences.

2. Protein Folding in the ER

Secretory proteins often contain disulfide bonds that form when the proteins enter the oxidizing environment of the ER lumen. A link between disulfide bond formation and protein export from the ER is evident from experiments in which the treatment of living yeast cells with the reducing agent DTT causes CPY, a protein whose native form contains five disulfide bonds, to accumulate in the ER (Winther and Breddam 1987; Jamsa et al. 1994). An oxidizing environment alone is not sufficient for the proper formation of disulfide bonds, and analysis of the enzyme protein disulfide isomerase (PDI) has shown that catalyzed rearrangement of disulfide bonds is essential for proper protein folding in the yeast ER. PDI isolated from yeast has the biochemical properties of mammalian PDI; the protein binds to peptides and catalyzes the refolding of pancreatic trypsin inhibitor (BPTI) under oxidizing conditions (LaMantia et al. 1991). The amino acid sequence of the yeast gene (*PDI1*) is 30% identical to that of mammalian PDI and contains two CXXC active site motifs as well as a carboxy-terminal HDEL signal for retention in the ER (Scherens et al. 1991). Deletion of *PDI1* is lethal, and when PDI is depleted from yeast cells by regulated repression of the gene, CPY accumulates within the ER (Tachibana and Stevens 1992). Initially, it was suspected that the essential function of PDI was actually not disulfide bond formation since mutation of both active sites to CXXS does not compromise cell viability but does disrupt the ability of the enzyme to catalyze folding of BPTI under oxidizing conditions (LaMantia and Lennarz 1993). This surprising result has been explained by subsequent work demonstrating that although the CXXS active site mutants will not catalyze dithiol oxidation, these mutants are fully functional for catalyzed reshuffling of disulfide bonds that have already formed (Laboissiere et al. 1995). This work also established the importance of the reshuffling activity for cell viability by showing that alteration of both active sites to SXXC is a lethal mutation and disrupts both dithiol oxidation and reshuffling activities.

Yeast carries two additional genes that appear to encode PDIs that are located in the ER. *EUG1* and *MPD1* share some sequence features with *PDI1*, and both genes suppress the lethality of a chromosomal deletion of *PDI1* when they are overexpressed, indicating that each is capable of carrying out the essential function of *PDI1* (Tachibana and Stevens 1992; Tachikawa et al. 1995). *EUG1* contains two active-site motifs of CXXS and *MDP1* contains a single active-site motif of CXXC. These variations in active-site configuration may reflect qualitative differences in function. Although overproduction of *EUG1* and *MPD1* can suppress the lethality of the *PDI1* deletion, in neither case is the normal rate of CPY transport from the ER fully restored. Further characterization of these different enzymes will be needed to evaluate possible differences in enzymatic mechanism or substrate specificity.

The role of Kar2p in ER protein maturation and exit from the ER has been difficult to evaluate because lethal *kar2* mutations block the translocation of proteins into the ER. One approach to circumvent this translocation block has been to take advantage of the reversible block in transport from the ER produced by treatment of yeast cells with the reducing agent DTT (Jamsa et al. 1994; Simons et al. 1995). After transport of

CPY from the ER is blocked by DTT, transport can be restored by dilution of the DTT. If this experiment is performed on a conditional *kar2* mutant that is shifted to the restrictive temperature as DTT is diluted, the transport of CPY is not restored. In cells where CPY transport is blocked because of defective Kar2p, the CPY protein is found in association with Kar2p in large protein aggregates (Simons et al. 1995). Together, these findings demonstrate that Kar2p function is essential to restore transport competence to reduced CPY and implies that Kar2p is normally an important chaperone for the maturation of ER proteins.

Calnexin is an abundant integral membrane protein of the mammalian ER that is involved in the retention of incorrectly folded proteins and unassembled protein complexes in the ER (Hammond and Helenius 1995). The yeast *CNE1* gene was identified by its homology to calnexin; Cne1p is 24% identical to mammalian calnexin and both proteins have amino-terminal luminal domains that bind calcium (Parlati et al. 1995). Cne1p has a transmembrane domain at the carboxyl terminus, and the protein is located in the ER membrane. Deletion mutants of *CNE1* grow normally and do not exhibit defects in secretion of acid phosphatase; however, Cne1p may be important for the retention of incorrectly folded proteins in the ER. Mutants of the α-factor receptor Ste2p and heterologously expressed mammalian α_1-antitrypsin are both retained in the ER in wild-type cells, but transport to the cell surface is partially restored in *cne1* deletions (Parlati et al. 1995). How *CNE1* participates in ER retention is not known.

3. *Regulation of ER Functions*

Proteins that function in the yeast secretory pathway usually are expressed constitutively throughout the cell cycle and under different growth conditions (Vahlensieck et al. 1995). A major exception to this generalization is the induction of ER resident proteins in response to agents that compromise the proper assembly of proteins in the ER. The logic underlying this regulation, often referred to as the unfolded protein response, appears to be analogous to induction of cytoplasmic chaperones in response to heat shock; i.e., catalysts of protein folding are induced to cope with an excess of incorrectly folded protein (Pelham 1986). Operationally, the unfolded protein response in the ER can be induced in yeast by treatment of cells with tunicamycin, 2-deoxyglucose, or 2-mercaptoethanol or by mutational inactivation of glycosylation (*sec53*) or signal sequence processing (*sec11*) (for review, see Shamu et al. 1994). The regulation of *KAR2* by the unfolded protein response has been extensively investigated, and a 22-nucleotide segment known as the unfolded response element (UPRE) has been identified in the *KAR2* promoter that is necessary and sufficient for regulation (Kohno et al. 1993). *FKB2* contains a similar promoter element that is necessary for its induction (Partaledis and Berlin 1993). *FKB2* encodes a luminal ER protein that has homology with peptidyl-prolyl isomerase, and although the function of this protein is not known, its inducibility by the unfolded protein response can be taken as evidence that it participates in protein folding in the ER. Similarly, the genes for protein disulfide isomerases, *PDI1* and *EUG1*, show the unfolded protein response and have promoter elements related to UPRE, although these sequences have not been shown directly to participate in regulation (Tachibana and Stevens 1992; Cox et al. 1993).

The *KAR2* promoter was fused to a *lacZ* reporter gene, and mutants defective in the unfolded protein response were identified by their failure to induce reporter expression. Mutations in *IRE1* block reporter induction in response to tunicamycin as well as induction of the wild-type *KAR2*, *PDI1*, and *EUG1* genes (Cox et al. 1993; Mori et al. 1993). *IRE1* encodes a transmembrane protein with an amino-terminal luminal domain and a cytosolic domain that has homology with serine/threonine kinases and appears to define a novel intracellular signal transduction pathway (Nikawa and Yamashita 1992). If Ire1p is located in the ER membrane, an attractive possibility would be that when the luminal domain senses the presence of unfolded protein in the ER lumen, the kinase is activated, producing a signal that ultimately leads to gene induction. Kinase activation is likely to involve oligomerization of the luminal domains since Ire1p monomers can be cross-linked to one another and truncations of the kinase domain can interfere with induction in a dominant fashion (Mori et al. 1993; Shamu and Walter 1996). The cytosolic domain includes both a kinase domain and a carboxy-terminal "tail" domain, both of which are required since mutation of conserved kinase residues or truncation of the tail disrupts the unfolded protein response (Mori et al. 1993; Shamu and Walter 1996). Interestingly, tail truncation mutants will complement kinase domain mutants, suggesting that when Ire1p oligomerizes, transphosphorylation of the tail by the kinase is part of the signaling mechanism (Shamu and Walter 1996).

HAC1 was identified as a gene that when overexpressed causes constitutive induction of the unfolded protein response, even in the absence of *IRE1*. Hac1p, a member of the basic-leucine zipper class of transcription factors, exhibits the properties expected for a transcription factor that acts on UPRE regulatory sites; deletion mutants of *HAC1* fail to induce the unfolded protein response and Hac1p binds specifically to segments of DNA containing a UPRE element (Cox and Walter 1996). Hac1p can only be detected in cells after induction of the unfolded protein response, although *HAC1* mRNA is produced constitutively. In an elegant set of experiments, Cox and Walter (1996) showed that the posttranscriptional regulation of Hac1p was based on a novel mRNA splicing mechanism. In uninduced cells, *HAC1* mRNA is not spliced and translation of this message produces a protein that is rapidly degraded by ubiquitin-dependent proteolysis. On induction of the unfolded protein response, *HAC1* mRNA is spliced, yielding a stable protein product with a carboxy-terminal peptide sequence different from that of the unstable form of the protein. The stable form of Hac1p, in turn, induces transcription at UPRE-containing promoters.

The regulated splicing of *HAC1* pre-mRNA was found not to depend on spliceosome function since splicing was not affected by pre-mRNA processing mutants. However, splicing of *HAC1* mRNA was disrupted by a particular allele of *RLG1*, a gene for a tRNA ligase required for pre-tRNA processing (Sidrauski et al. 1996). The unusual nature of *HAC1* pre-mRNA splicing is also evident in the intron sequence, which does not have consensus splice site or branchpoint sequences. *HAC1* splicing does depend on an intact *IRE1* gene, and homology between the Ire1p tail domain and a regulated RNA endonuclease suggests that Ire1p itself may participate in the processing of *HAC1* pre-mRNA. Involvement of Ire1p in RNA processing would imply that the protein spans the inner membrane of the nuclear envelope, transducing a signal from the ER lumen to the nuclear interior.

The amount of Kar2p in the ER is critical for setting the level of the unfolded protein response. Overproduction of Kar2p reduces the magnitude of the unfolded protein response, whereas reduction of the amount of Kar2p in the ER (by deletion of the carboxy-terminal HDEL sequence from Kar2p needed for efficient retention in the ER) induces the response in otherwise unstimulated cells (Hardwick et al. 1990; Kohno et al. 1993). When Kar2p retention in the ER fails, a compensatory increase in Kar2p expression is necessary for cell survival, since the combination of an *erd2* mutation (which prevents normal retrieval of Kar2p from the Golgi to ER) with an *ire1* mutation is lethal (Beh and Rose 1995).

4. Assembly and Processing Steps That Determine Protein Exit from the ER

In mammalian cells, membrane protein complexes composed of more than one subunit are usually assembled in the ER, and the formation of a completely assembled complex appears to be a prerequisite for transport out of the ER (Rose and Doms 1988). Examples of this type of regulation of exit from the ER are also found in yeast. The best-understood example of membrane protein complex that is assembled in the yeast ER is the H^+-translocating ATPase (V-ATPase) which is responsible for the acidification of the vacuole. The V-ATPase is composed of two parts, a V_0 integral membrane domain composed of at least four different polypeptides and a V_1 catalytic domain composed of at least six polypeptides (Graham et al. 1995). Although mutations in the V_1 subunits eliminate V-ATPase activity, the V_0 portion of the complex still assembles in these mutants and is transported to the vacuole membrane. In contrast, mutation of one of the V_0 subunits compromises the assembly of the remaining subunits and causes their rapid proteolytic degradation (Bauerle et al. 1993). Thus, it appears that all of the subunits of V_0 are needed for stable assembly of the complex.

Of particular interest for the study of protein assembly in the ER is the existence of genes whose products are not part of the V-ATPase but are required for the stable assembly of the V_0 complex. These include *VPH2/VMA12*, *VMA21*, and *VMA22*, all of which were identified by mutations that prevent the formation of functional V-ATPase and lead to degradation of subunits of the V_0 complex (Hirata et al. 1993; Hill and Stevens 1994, 1995). Vma12p and Vma21p are small integral membrane proteins located in the ER. Vma22p is a 22-kD peripheral membrane protein located on the cytosolic face of the ER membrane. Vma22p apparently binds to Vma12p since Vma22p association with the membrane is lost in *vma12Δ* cells (Hill and Stevens 1995). An attractive hypothesis is that Vma12p, Vma21p, Vma22p, and perhaps other as yet uncharacterized *VMA* gene products, form a complex in the ER membrane. The proposed function of this complex would be to assist in the insertion of V_0 subunits into the membrane and to bring about their proper assembly into a V_0 complex.

Another membrane complex that appears to require assembly in the ER is the permease required for high-affinity uptake of iron from the extracellular environment. The iron permease normally resides in the plasma membrane and is composed of at least two integral membrane proteins: Fet3p, a copper-containing oxidase with a single transmembrane domain, and Ftr1p, a polytopic integral membrane protein (Stearman et al. 1996). In *fet3* deletion strains, Ftr1p is retained in the ER instead of being transported to the cell surface. Similarly, in *ftr1* mutants, Fet3p fails to acquire copper and has aberrant glycosylation, both possible consequences of a failure to be transported through the secretory pathway (Stearman et al. 1996). Thus, it appears that the subunits of the complex are mutually dependent for transport from the ER. It is not known whether there are additional subunits of the permease complex or whether ER protein cofactors are needed for the proper assembly of the complex.

The plasma membrane permeases for uptake of amino acids appear to have special requirements for their transport from the ER. There are approximately 20 related amino acid permeases in yeast, each of which is a polytopic integral membrane protein that acts as a H^+/amino acid symporter (André 1995). A cofactor required for transport of permeases from the ER was identified in a screen for mutants resistant to toxic levels of histidine. Mutations in a single gene, *SHR3*, block the transport of a number of different permeases from the ER but do not affect transport of other soluble and membrane secretory proteins (Ljungdahl et al. 1992). Because of overlapping substrate specificities, it is difficult to define which of the permeases are affected by *shr3* mutations, but it is likely that at least the following permeases are affected: general amino acid permease (Gap1p), histidine permease (Hip1p), proline permease (Put4p), arginine permease (Can1p), lysine permease (Lyp1p), tryptophan permease (Scm2p), leucine permease (Tat1p), and dicarboxylic amino acid permease (Dip1p). Shr3p is a 24-kD polytopic integral ER membrane protein that is required either for the proper folding of amino acid permeases or for the selective loading of permeases into transport vesicles. The existence of an ER protein required for the proper transport of amino acid permeases suggests that there may be cofactors for the selective transport of other classes of membrane proteins.

Signal peptide cleavage by signal peptidase appears to be a general prerequisite for secretory protein exit from the ER (discussed above). For two luminal proteases, additional specialized intramolecular cleavages appear to be required for these proteins to exit the ER. The vacuolar protease proteinase B (PrB) is synthesized as a 76-kD precursor glycoprotein. The signal peptide is cleaved when the precursor enters the ER, an amino-terminal propeptide of 41 kD is removed within the ER lumen, and a carboxy-terminal peptide sequence is removed in two proteolytic cleavage steps within the vacuole lumen to finally generate the 31-kD mature enzyme (Moehle et al. 1989; Nebes and Jones 1991; Hirsch et al. 1992). Mutation of the PrB active site prevents cleavage of the amino-terminal propeptide, indicating that this is an intramolecular event (Nebes and Jones 1991; Hirsch et al. 1992). In these mutants, the uncleaved precursor PrB is not processed in the vacuole and the full-length precursor appears to remain in the ER, suggesting that the uncleaved amino-terminal propeptide can specify retention in the ER. However, the propeptide probably does not simply carry a determinant for ER retention, because under normal circumstances, the cleaved amino-terminal propeptide is itself transported to the vacuole where the propeptide is degraded (Hirsch et al. 1992). Possibly, autocleavage of the propeptide triggers a conformational change that renders the precursor protein competent for transport, or cleavage may actually occur within the retention sequence itself. A similar dependence on autoproteolytic cleavage for transport from the ER has been found for the Kex2p protease. Kex2p is a type I integral membrane protein, with an amino-terminal luminal domain synthesized with both a signal sequence and a proregion that are proteolytically processed within the ER. Mutation of the protease active site of Kex2p prevents processing of the propeptide, suggesting intramolecular processing, and the uncleaved precursor remains in the ER (Gluschankof and Fuller 1994). Again, these findings indicate that propeptide cleavage is needed to remove an ER retention signal.

A number of yeast plasma membrane proteins are modified at their carboxyl termini by addition of a glycosylphosphatidylinositol (GPI) lipid, which is added within the ER lumen. It is possible to rapidly block the formation of the GPI precursor by inositol starvation of an inositol auxotroph. This method was used to evaluate the effect of a failure in GPI modification on a major GPI-linked protein, Gas1p. On inositol starvation, unmodified Gas1p rapidly accumulates in the ER, but transport to the Golgi is restored once GPI synthesis resumes after inositol is provided to the cells (Doering and Schekman 1996). These results suggest that the GPI lipid may provide a signal needed for loading of Gas1p into budding transport vesicles. Alternatively, the hydrophobic carboxy-terminal peptide on Gas1p that is removed on GPI attachment could act to retain unmodified Gas1p in the ER.

5. *Protein Degradation in the ER*

Protein degradation in the ER can usefully be categorized as degradation of membrane proteins with sequences exposed to the cytosol and as degradation of soluble luminal proteins. The best-characterized substrate for ER protein degradation is HMG-CoA reductase, the first enzyme in the mevalonate pathway and an enzymatic activity encoded by two closely related membrane proteins, Hmg1p and Hmg2p (Basson et al. 1988). Hmg1p is stable, whereas Hmg2p is turned over with a half-life of about 1 hour (Hampton and Rine 1994). The turnover of Hmg2p appears to be related to homeostasis of the mevalonate pathway since pharmacological inhibition of HMG-CoA reductase reduces the rate of Hmg2p turnover. Hmg2p degradation does not occur in the vacuole since turnover occurs normally in mutants that are defective in vacuolar proteases (*pep4*Δ) and in transport from the ER to Golgi (*sec18*) (Hampton and Rine 1994). The protease that degrades Hmg2p has not yet been identified. Since Hmg2p is a polytopic integral membrane protein with substantial polypeptide segments on both the cytosolic and luminal sides of the membrane (Basson et al. 1988), proteolytic attack could conceivably come from either side of the membrane.

An important insight into the mechanism of ER membrane protein turnover has come from studies of genes in the ubiquitin pathway. The finding that the ubiquitin-conjugating enzyme Ubc6p is an integral membrane protein located in the ER suggested a possible role of Ubc6p in ER protein degradation. A role for Ubc6p in turnover of defective ER membrane proteins was established by the finding that loss-of-function mutations in *ubc6* can partially suppress missense mutations in *sec61* (Sommer and Jentsch 1993). In cells harboring a *sec61* mutation, both Sec61p and Sss1p are degraded. This degradation is prevented by deletion of *UBC6*, or *UBC7* (which encodes a cytosolic ubiquitin-conjugating enzyme), or by mutation of genes for proteasome subunits (Biederer et al. 1996). These findings indicate that incorrectly folded proteins in the ER membrane are recognized as substrates for ubiquitination and subsequent degradation by the proteasome. It is not yet clear whether just the cytosolic loops are subject to this type of degradation or whether the entire protein can be removed from the membrane and then be degraded by the ubiquitin/proteasome pathway.

Proteolysis within the ER lumen has been suggested as a mechanism to dispose of incorrectly folded proteins that are retained in the ER. Evidence for such degradation processes has been sought by examining the turnover of defective forms of luminal secretory proteins. Certain mutants of the soluble proteases CPY and proteinase A enter the ER, become glycosylated, and are then degraded. Degradation of these mutant proteins occurs even if transport from the ER is blocked by a *sec18* mutation, implying that proteolysis occurs prior to exit from the ER (Finger et al. 1993). The involvement of the proteasome in this degradation process was tested. Remarkably, mutants of either proteasome subunits (*pre1*, *cim3*, and *cim5*) or mutants of a ubiquitin-conjugating enzyme (*ubc7*) greatly slowed degradation (Hiller et al. 1996). The mutant CPY was proposed to be delivered to the proteasome by retrograde translocation of the misfolded glycoprotein from the ER lumen to the cytosol, since a form of the mutant protein that was both glycosylated and ubiquitinated and that appeared to reside on the cytosolic face of the ER could be detected (Hiller et al. 1996). In parallel experiments of similar design, a mutant form of human secretory protein α_1-antitrypsin that is unstable in human cells was shown also to be unstable when expressed in yeast cells (McCracken and Kruse 1993). Degradation of the mutant α_1-antitrypsin was slowed in *pre1 pre2* double mutants, indicating involvement of the proteasome (Werner et al. 1996). Together, these findings indicate that incorrectly folded luminal ER proteins are degraded by the proteasome in a manner similar to that of defective ER membrane proteins except that luminal proteins require the additional step of translocation from the lumen of the ER to the cytosol.

An in vitro assay that reproduces the cytosolic degradation of a luminal ER protein was developed from the observation that when prepro-α-factor is translocated into yeast microsomes under conditions where glycosylation is inhibited, the translocated protein is degraded by an ATP- and cytosol-dependent process (McCracken and Brodsky 1996). Degradation in this assay is slowed by compounds that selectively inhibit the proteasome or when cytosol is prepared from proteasome mutants. When cytosol from a proteasome mutant is used, unglycosylated pro-α-factor accumulates in the cytosol as an apparent intermediate in the degradation process (Werner et al. 1996). This in vitro retrograde translocation activity is possibly related to a previous finding that glycopeptides can be translocated out of the ER by an ATP- and cytosol-dependent activity (Römisch and Schekman 1992). Interestingly, pro-α-factor degradation in the cytosol was not affected by the ubiquitin-conjugating enzyme mutations *ubc4*, *ubc5*, *ubc6*, and *ubc7*, suggesting that in this case ubiquitination is not necessary for degradation by the proteasome. The Sec61p translocation channel has been implicated in a related retrograde translocation process in mammalian cells (Wiertz et al. 1996). It should now be possible to use the appropriate *sec* mutants to evaluate the role of each of the components of the translocation machinery in retrograde translocation of pro-α-factor.

The question of whether there is also a mechanism for degradation of incorrectly folded proteins within the lumen of the ER has not been resolved. Clearly, specific proteolytic cleavages do occur in the ER. These include cleavage of the signal peptide, cleavage of carboxy-terminal peptides coupled to GPI lipid attachment, and the autoproteolytic cleavages of proteinase B and Kex2p. But apart from these relatively limited proteolytic processes, more general degradative proteases have not been found in the ER. Experimental evidence against general proteolysis within the

ER comes from examination of the fate of thermally unstable forms of the bacteriophage λ repressor fused to invertase. In wild-type cells, these proteins are transported to the vacuole where the repressor moiety is readily degraded, but in mutants that block protein transport from the ER, the fusion proteins are stable (Hong et al. 1996). Thus, the potentially labile repressor sequences are not degraded when retained within the ER lumen. Since the discovery of degradation of ER proteins by the proteasome, it has been pointed out that it would make sense to exclude degradative proteases from the ER, a compartment primarily devoted to the folding and assembly of nascent polypeptide chains. By restricting proteolysis to the proteasome and the vacuole, proteins that are targeted for destruction can be segregated from proteins within the ER that are on the correct folding pathway but have not yet achieved a stable conformation.

III. PROTEIN TRANSPORT FROM ER TO GOLGI

In all eukaryotic cells, proteins are transported between organelles by cargo-carrying vesicles that bud from one membrane-bounded compartment and fuse with the next (Palade 1975). In the most fully understood examples of protein transport between organelles, budding of vesicle membrane is accompanied by the assembly onto the membrane of a coat that is derived from soluble cytosolic protein. This coat is thought to set the curvature of the vesicle membrane and to help to select the cargo molecules that will be accepted by the vesicle. Three different types of coated vesicles have been well characterized in yeast and mammalian cells. The first class of coated vesicles to be discovered have a cytosolic coat composed of an outer lattice of clathrin and four additional proteins that constitute the adapter protein (AP) complex (Pearse and Robinson 1990). Clathrin-coated vesicles in mammalian cells are responsible for transport steps directed to the lysosome either from the plasma membrane or from *trans*-Golgi compartment. A second type of vesicle, known as COPI, was discovered by cell-free reconstitution of transport between Golgi compartments. These vesicles are coated with a complex of seven proteins known as coatomer (Rothman 1994). A common mechanism is suggested by the sequence similarity of one of the coatomer proteins and one of the AP proteins (Duden et al. 1991). The involvement of COPI vesicles in transport through the Golgi has yet to be definitely established in vivo. Experiments in yeast have shown that COPI vesicles are involved in retrograde transport from the Golgi to ER

(Letourneur et al. 1994) and in budding from the ER (Bednarek et al. 1995). The third vesicle type carries protein from the ER to the Golgi and is coated with a set of proteins known as COPII (Barlowe et al. 1994).

In *S. cerevisiae*, vesicle transport between the ER and the Golgi has been divided into two distinct and experimentally separable steps: the formation of vesicles at the surface of the ER and the docking and fusion of vesicles with the surface of the Golgi. The dissection of ER to Golgi transport in yeast has been facilitated both by the discovery of conditional mutants that are defective in either vesicle formation or vesicle fusion and by in vitro reconstitution studies which have helped to define the essential components of the assembly and fusion machinery.

The first genetic screen for secretory (*sec*) mutants identified 23 complementation groups; analysis of the maturation of a marker protein revealed that a subset of the *sec* mutants was defective in ER to Golgi transport (Novick et al. 1981). Careful morphological analysis further divided the ER to Golgi mutants into a class required for the formation of vesicles and a class required for the consumption of vesicles; vesicle consumption mutants (including *sec17*, *sec18*, and *sec22*) accumulated 50-nm transport vesicles, whereas vesicle formation mutants (including *sec12*, *sec13*, *sec16*, and *sec23*) did not (Kaiser and Schekman 1990). Epistasis analysis confirmed that vesicle formation mutants function at an earlier stage than vesicle consumption mutants (Kaiser and Schekman 1990).

The second approach to dissection of ER to Golgi transport is based on the in vitro reconstitution of this transport step (Baker et al. 1988; Ruohola et al. 1988). In its simplest form, in vitro ER to Golgi transport assays measure addition of Golgi-specific carbohydrate modifications to radioactively labeled prepro-α-factor that has been posttranslationally translocated into the ER. Gently lysed cells or partially purified ER membranes can serve as a donor compartment, whereas the acceptor compartment is generally supplied by a membrane fraction enriched in Golgi membrane by differential centrifugation. In the presence of exogenously added cytosolic proteins and ATP, typically 25–50% of the input α-factor is modified in this system.

The basic in vitro assay has been modified to allow vesicle formation to be evaluated independently of vesicle fusion. The assay for vesicle budding begins with α-factor within rapidly sedimenting ER membranes and follows its conversion to a slowly sedimenting fraction that corresponds to free vesicles (Groesch et al. 1990; Rexach and Schekman 1991; Barlowe et al. 1994; Oka and Nakano 1994; Rexach et al. 1994). In this assay, mutants required for vesicle formation in vivo (such as *sec12* and *sec23*) were also found to be defective for vesicle budding in vitro. Similarly, the vesicle-accumulating mutant *sec18* was shown to be defective for vesicle fusion in vitro (Rexach and Schekman 1991; Oka and Nakano 1994; Rexach et al. 1994). Importantly, these studies of mutants establish a faithful correspondence between the in vitro transport assay and the events that occur in vivo.

Attempts to purify ER to Golgi transport vesicles from whole cells have failed probably because of the low abundance of these organelles. Pure ER to Golgi transport vesicles can be produced in quantity by in vitro budding reactions under conditions where vesicle fusion with an acceptor compartment is inhibited by either mutation or antibody (Groesch et al. 1990; Barlowe et al. 1994; Oka and Nakano 1994; Rexach et al. 1994). Purified ER-derived vesicles thus produced are functional transport intermediates since they are competent to fuse with Golgi membranes (Groesch et al. 1990; Barlowe et al. 1994; Oka and Nakano 1994; Rexach et al. 1994). ER-derived vesicles produced in vitro have uniform morphology (~60 nm diameter), and when formed under the appropriate conditions, they are encapsulated in a protein coat that can be visualized by electron microscopy after fixation and staining with tannic acid. This coat consists of a subset of the proteins required for vesicle formation in vivo and in vitro and has been termed COPII (Barlowe et al. 1994). Vesicle production in vitro appears to reproduce the normal selectivity in segregation of vesicle proteins from ER proteins that occurs before or during vesicle formation, since the vesicles that bud in vitro contain known vesicle docking factors, such as Sec22p and Bos1p, but do not contain resident ER proteins (Lian and Ferro-Novick 1993; Barlowe et al. 1994; Rexach et al. 1994). In addition, these vesicles have a characteristic set of 12 membrane proteins, termed ERV (ER-vesicle-associated) proteins, one of which has been shown to be Emp24p/Bst2p (Rexach et al. 1994; Schimmöller et al. 1995).

A. Formation of Vesicles at the ER: COPII-coated Vesicles

Vesicle formation from the ER in vitro requires both ER membranes and concentrated cytosolic protein. Five polypeptides—the Sec23p/Sec24p complex and the Sec13p/Sec31p complex, and a small GTP-binding protein, Sar1p—in pure form will satisfy this requirement for cytosolic protein in vesicle budding (Salama et al. 1993; Barlowe et al. 1994). These five proteins are present on completed vesicles and are abundant components of the COPII coat (Barlowe et al. 1994). The COPII coat is morphologically distinct from the clathrin coat and does not contain components of the coatomer (COPI) coats (Barlowe et al. 1994; Bednarek et al. 1995). In addition to these cytosolic factors, three additional proteins, Sec16p, Sec12p, and Sed4p, are also involved in COPII vesicle formation. Sec16p is tightly associated with the ER membrane and is also present on ER-derived vesicles produced in vitro (Espenshade et al. 1995). Sec16p binds to three cytosolic COPII coat components, Sec23p, Sec24p, and Sec31p, and may be a scaffold onto which these proteins assemble (Espenshade et al. 1995; Gimeno et al. 1996; D. Shaywitz and C. Kaiser, unpubl.). Two integral ER membrane proteins, Sec12p and Sed4p, regulate COPII vesicle formation, but they are not incorporated into completed vesicles (Barlowe et al. 1994; Gimeno et al. 1995).

It is important to note that the distinction between cytosolic and membrane-associated COPII coat components is defined by the behavior of the in vitro assay as carried out on ER membranes that have been extracted with urea to remove peripheral proteins and may not accurately reflect the distribution of these proteins in living cells. For example, Sec23p and Sar1p have been defined as cytosolic COPII coat components in the context of the in vitro assay, yet these proteins appear to be principally associated with membranes in crude cell extracts (Hicke and Schekman 1989; Nishikawa and Nakano 1991). It is also noteworthy that the in vitro budding assay has been most useful for study of essential factors that can be extracted from ER membranes and then be added back in soluble form. Additional peripheral membrane proteins such as Sec16p which are not easily extracted from ER membranes may have escaped detection by the budding assay in its current form.

Mammalian homologs of several COPII proteins have now been identified. Mammalian Sar1p, Sec13p, and Sec23p all localize to the transitional region of the ER, consistent with a role in vesicle formation (Orci et al. 1991; Kuge et al. 1994; Shaywitz et al. 1995). Furthermore, in mammalian cells, the transport of the marker protein VSV-G from the ER to the Golgi is inhibited both in vivo (by a dominant-negative mutant of hamster Sar1p) and in vitro (by anti-Sar1p antibodies), suggesting that the function of Sar1p, as well as the structure, has been evolutionarily conserved (Kuge et al. 1994). The demonstration that reciprocal human/yeast chimeras of Sec13p both complement the secretion defect of a temperature-sensitive *sec13* mutant provides further support for the conservation of COPII function through evolution (Shaywitz et al. 1995).

1. Sec23p and Sec24p: Cytosolic Vesicle Coat Proteins

SEC23 was initially identified as an ER to Golgi mutant in the original *sec* screen and was subsequently determined to participate in vesicle formation (Novick et al. 1981; Kaiser and Schekman 1990). The conservation of the temperature-sensitive *sec23* phenotype under in vitro conditions facilitated the biochemical isolation of Sec23p activity from wild-type cytosol (Baker et al. 1988; Hicke and Schekman 1989). Functional Sec23p was purified as a 300–400-kD complex that contains both the 85-kD Sec23p and a 105-kD protein designated Sec24p (Hicke et al. 1992; Barlowe et al. 1994); the precise stoichiometry of these two protein partners has not yet been established. Sec24p, like Sec23p, is essential and is required for ER to Golgi transport both in vitro (Hicke et al. 1992) and in vivo (R.E. Gimeno et al., unpubl.). Furthermore, the Sec23p/Sec24p complex represents one of the major components of the COPII vesicle coat and is required for its formation (Barlowe et al. 1994; Bednarek et al. 1995).

Not only do Sec23p and Sec24p interact with each other, both biochemically and genetically, but they also each exhibit genetic interactions with the genes encoding the other COPII components, *SEC13*, *SEC31*, *SEC16*, and *SAR1* (Kaiser and Schekman 1990; Oka and Nakano 1994). Sec16p has been shown to bind directly to both Sec23p and Sec24p; the two protein partners bind to different sites on Sec16p, and the binding of one subunit can occur in the absence of the other (Gimeno et al. 1996). Sec23p has also been shown to function as a GTPase-activating protein (GAP) for Sar1p, stimulating its GTPase activity by a factor of ten; Sec24p does not appear to affect the GAP activity of Sec23p (Yoshihisa et al. 1993). Although the GAP activity of Sec23p has also stimulated the generation of several different hypotheses regarding Sec23p function (Oka and Nakano 1994; Schekman and Orci 1996), the role of the Sec23p/Sec24p complex in the formation of COPII vesicles still remains to be established.

2. *Sec13p and Sec31p: Cytosolic Vesicle Coat Proteins*

SEC13 was identified in the original screen for secretion mutants and was shown to be required for the vesicle formation step of ER to Golgi transport (Novick et al. 1981; Kaiser and Schekman 1990). Immunodepletion experiments revealed a requirement for Sec13p in the in vitro transport assay (Pryer et al. 1993; Salama et al. 1993). Purification of Sec13p activity from wild-type cytosol revealed that Sec13p exists as a 700-kD complex with the 150-kD Sec31p (Pryer et al. 1993; Salama et al. 1993); the exact stoichiometry of Sec13p and Sec31p in this complex has not yet been established. Both *SEC13* and *SEC31* are essential genes required for vesicle transport in vivo and in vitro (Kaiser and Schekman 1990; Salama et al. 1993; Barlowe et al. 1994; Wuestehube et al. 1996).

The Sec13p/Sec31p complex, like the Sec23p/Sec24p complex, represents a major component of the COPII coat and is required for its formation (Barlowe et al. 1994; Bednarek et al. 1995).

Both *SEC13* and *SEC31* exhibit genetic interactions with the other COPII genes (*SEC23*, *SEC24*, *SAR1*, *SEC16*) (Kaiser and Schekman 1990; Gimeno et al. 1995; A. Frand and C. Kaiser, unpubl.). Structurally, Sec13p appears to consist almost entirely of six WD40 repeats, a motif first described in the β-subunit of trimeric G-proteins and subsequently shown to specify a β-propeller structure (Sondek et al. 1996). Sec13p appears to interact with a WD40-containing domain of Sec31p, raising the possibility of a homotypic interaction (D. Shaywitz and C. Kaiser, unpubl.). A different region of Sec31p has recently been shown to bind Sec16p, consistent with the proposed role of Sec16p as a COPII scaffold (D. Shaywitz and C. Kaiser, unpubl.).

The function of the Sec13p/Sec31p complex in the formation of COPII vesicles remains incompletely understood, a conclusion emphasized by recent data which suggest that Sec13p may not always be required for ER to Golgi transport. In particular, loss-of-function mutations in at least three genes (*BST1*, *EMP24*/*BST2*, and *ERV25*/*BST3*) have been isolated that bypass the requirement for Sec13p in ER to Golgi transport (Elrod-Erickson and Kaiser 1996). These mutations cause only subtle phenotypes on their own. Mutations in *BST1*, *BST2*, or *BST3* slow transport of a subset of secreted proteins from the ER to the Golgi apparatus and increase the rate at which ER-resident proteins Kar2p and Pdi1p reach the Golgi apparatus (Schimmöller et al. 1995; Stamnes et al. 1995; Elrod-Erickson and Kaiser 1996). This phenotype is consistent with a decrease in the fidelity of cargo sorting at the ER that could be explained if *BST1*, *BST2*, and *BST3* encode components of a checkpoint that monitors fidelity of cargo sorting and prevents the budding of immature or improperly coated vesicles (Elrod-Erickson and Kaiser 1996). Alternatively, since Emp24p/Bst2p and Erv25p/Bst3p are major integral membrane components of ER-derived transport vesicles, it has been suggested that Emp24p/Bst2p is a sorting receptor for a subset of cargo molecules (Schimmöller et al. 1995; Belden and Barlowe 1996).

Sec13p may also have functions apart from vesicle formation at the ER that are not shared by other COPII coat components. Sec13p and a Sec13p homolog, Seh1p, but not Sec31p are present in a subcomplex of the nuclear pore (Siniossoglou et al. 1996). However, *sec13* mutants are not defective in nuclear import, suggesting that the association of Sec13p with the pore may not be important for pore function. In contrast, Seh1p does appear to be involved in nuclear pore function, since *seh1* mutations genetically interact with other nuclear pore complex components (Siniossoglou et al. 1996).

3. Sec16p: An ER-associated Vesicle Coat Protein

SEC16 encodes an essential 240-kD hydrophilic protein required for the formation of ER to Golgi transport vesicles in vivo (Kaiser and Schekman 1990; Espenshade et al. 1995). Sec16p is tightly associated with the periphery of the ER and is also found on ER-derived transport vesicles that are produced in vitro (Espenshade et al. 1995). In contrast to the other COPII components, Sec16p cannot be extracted from membranes by urea, explaining why it has not needed to be added as a cytosolic factor to the in vitro transport assay (Espenshade et al. 1995). Finally, *SEC16* exhibits genetic interactions with the five other COPII genes (Nakano and Muramatsu 1989; Kaiser and Schekman 1990; Gimeno et al. 1995), and Sec16p has been shown to bind directly to the COPII subunits Sec23p, Sec24p, and Sec31p (Espenshade et al. 1995; Gimeno et al. 1996; D. Shaywitz and C. Kaiser, unpubl.). An attractive hypothesis is that Sec16p on the ER membrane organizes the recruitment and assembly of COPII coat components from the cytosol. Furthermore, the genetic interactions between *SEC16* and *SAR1* suggest that Sar1p could regulate the assembly of cytosolic coat components onto Sec16p (Nakano and Muramatsu 1989; Gimeno et al. 1995).

4. Sar1p: A GTPase Coupled to Vesicle Assembly

SAR1 was identified as a high-copy suppressor of mutations in *SEC12* and *SEC16* (Nakano and Muramatsu 1989). *SAR1* is an essential gene required for ER to Golgi transport, and it encodes a small GTP-binding protein most closely related to ARF, a GTP-binding protein that regulates assembly of the coatomer (COPI) coat (Oka et al. 1991; Barlowe et al. 1993). Sar1p is a component of the COPII vesicle coat and is required for its formation (Barlowe et al. 1994; Bednarek et al. 1995).

SAR1 exhibits genetic interactions with the five other COPII components. In addition, the activity of Sar1p has been shown to be modified by the COPII subunit Sec23p, which functions as a Sar1p-GAP, and by Sec12p, which functions to increase the rate of guanine nucleotide exchange by Sar1p. Sar1p has an intrinsic guanine-nucleotide off-rate (0.07 min^{-1}) and GTPase activity (0.0011 min^{-1}) similar to those of other small GTP-binding proteins (Barlowe et al. 1993). Sec12p and Sec23p increase these rates (respectively) by about one order of magnitude, and Sec12p and Sec23p together increase the rate of GTP hydrolysis by approximately 50-fold (Barlowe and Schekman 1993; Yoshihisa et al. 1993). Like Arf1p, Sar1p requires the presence of detergents or phospholipids for GTP binding (Barlowe et al. 1993); however, unlike other small GTP-binding proteins, including ARF, Sar1p does not appear to require lipid modifications for its function (Oka et al. 1991; Barlowe et al. 1993).

In vitro studies suggest that vesicle budding requires Sar1p in its GTP-bound form (Oka et al. 1991; Rexach and Schekman 1991; Barlowe et al. 1993, 1994; Oka and Nakano 1994). GTP hydrolysis is not required for vesicle formation since budding still occurs when nonhydrolyzable analogs of GTP are used, although under these conditions, greater quantities of Sec23p/Sec24p and Sec13p/Sec31p are needed (Barlowe et al. 1994). GTP hydrolysis by Sar1p is required for the overall transport process, as the vesicle fusion step is sensitive to nonhydrolyzable GTP analogs (Barlowe et al. 1994; Oka and Nakano 1994).

Sar1p is thought to function by regulating assembly and disassembly of the cytosolic COPII coat components. A simple model is that Sar1p in its GTP-bound state promotes assembly of the COPII coat, whereas GTP hydrolysis by Sar1p promotes disassembly of the COPII coat (Barlowe et al. 1994). This hypothesis is supported by in vitro data demonstrating that the COPII vesicle coat produced in the presence of GTP is unstable and readily dissociates from budded vesicles; however, when nonhydrolyzable analogs of GTP are substituted in the budding reaction, the resulting COPII coat is stabilized (Barlowe et al. 1994). These in vitro studies also indicate that COPII coat disassembly occurs in two distinct steps: First, Sar1p rapidly hydrolyzes GTP and dissociates from the vesicle, leaving a coat of Sec23p/Sec24p, Sec13p/Sec31p, and possibly Sec16p (Barlowe et al. 1994). Second, Sec23p/Sec24p and Sec13p/Sec31p dissociate from the vesicle (Barlowe et al. 1994); it has not been established whether the dissociation of Sec16p from the vesicle membrane ever occurs.

5. *Sec12p and Sed4p: ER Proteins That Regulate COPII Coat Formation*

Sec12p and Sed4p are integral membrane proteins that influence vesicle budding at the ER but are not themselves incorporated into vesicles. *SEC12*, an essential gene required for the formation of ER-derived vesicles, encodes a type II transmembrane protein that localizes to the ER but is absent from vesicles produced in vitro (Nakano et al. 1988; Nishikawa and Nakano 1993; Barlowe et al. 1994). The cytosolic domain of Sec12p is essential and stimulates guanine-nucleotide exchange by Sar1p, whereas the luminal domain appears to be less important since truncation of this domain does not greatly interfere with *SEC12* function (d'Enfert et al. 1991b; Barlowe and Schekman 1993). The cytosolic domain of Sec12p may also function to recruit Sar1p to the ER membrane since overexpression of Sec12p depletes Sar1p from the cytosol and increases the membrane-associated pool of Sar1p (d'Enfert et al. 1991a; Barlowe et al. 1993). Recruitment may not involve direct binding between Sec12p and Sar1p, since a stable association between these proteins has not been observed. However, if Sar1p in its GTP-bound state has a high affinity for the ER membrane, Sec12p may effect recruitment simply by stimulating nucleotide exchange. It appears that for Sar1p to be properly activated, the exchange activity of Sec12p must be in proximity to the ER membrane, since truncations of Sec12p that liberate a soluble exchange activity inhibit vesicle budding from the ER. This inhibition can be overcome by the addition of increased amounts of Sar1p (Barlowe et al. 1993).

SED4 encodes a type II transmembrane protein that is 45% identical to Sec12p in its cytosolic and transmembrane domains but has a divergent luminal domain (Hardwick et al. 1992); however, Sed4p does not appear to stimulate the nucleotide exchange activity of Sar1p. Sed4p is present on the ER, but it is not incorporated into vesicles (Gimeno et al. 1995). Although the deletion of *SED4* causes only minor growth and secretion defects in wild-type cells, a strain carrying an otherwise silent *sar1-5* mutation is rendered temperature-sensitive for growth and secretion when *SED4* is deleted (Gimeno et al. 1995). *SED4* exhibits genetic interactions with both *SAR1* and *SEC16*, and Sed4p has also been shown to bind to Sec16p; the functional significance of these interactions is the subject of ongoing investigation.

6. *Toward a Model for COPII Vesicle Formation*

Biochemical and molecular analyses of the proteins required for vesicle formation suggest an outline for the pathway for COPII coat assembly (Fig. 4). The first step in vesicle formation at the ER is thought to be guanine nucleotide exchange by Sar1p and recruitment of Sar1p to the ER membrane. This step is probably stimulated by Sec12p and may be facilitated by Sed4p. Sar1p in its GTP-bound membrane-associated form is thought to then recruit the cytosolic coat components Sec23p/Sec24p and Sec13p/Sec31p to Sec16p. This complex could then nucleate further polymerization of coat subunits to eventually form a vesicle bud.

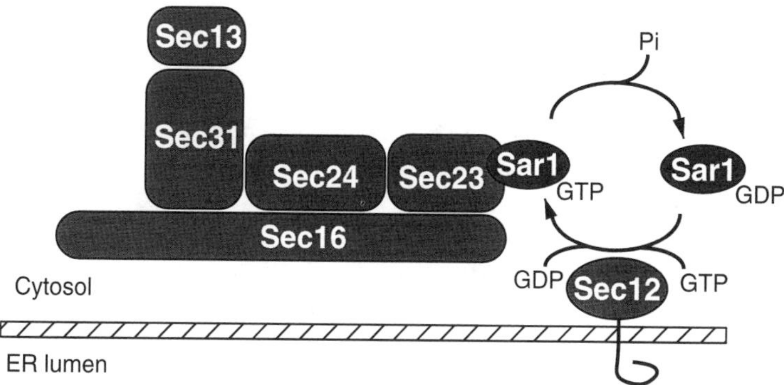

Figure 4 Components of the COPII vesicle coat complex. Completed vesicles do not contain Sec12p. The stoichiometry of individual proteins in the complex is not known.

The process of vesicle formation at the ER remains incompletely understood, and many fundamental questions remain. For example, how is the site of budding determined? What is the order of component assembly on the ER membrane? How is this assembly process regulated? Are there other factors required for vesicle assembly in vivo that, like Sec16p, have escaped detection by the in vitro assay? What is the relationship between the assembly of the coat structure and the deformation of the ER membrane from a flat surface into a vesicle bud? What is the mechanism of vesicle scission? Resolution of these questions will require both the refinement of the in vitro assay and the isolation of additional mutants.

7. *Cargo Sorting during COPII Vesicle Formation*

Considerable evidence has accumulated indicating that the entry of cargo molecules into ER-derived secretory vesicles reflects an important sorting decision made by the cell. For example, although discrete retention signals (which allow resident ER proteins to be retrieved from the Golgi) have been characterized, resident ER proteins lacking these signals still exit the ER at a rate much slower than that of actual secreted proteins (Munro and Pelham 1987; Hardwick et al. 1990). Furthermore, careful immunoelectron microscopy studies suggest that at least some secretory proteins are concentrated in ER-derived vesicles in mammalian cells (Balch et al. 1984; Mizuno and Singer 1993). Finally, COPII vesicles formed in vitro from either microsomes or nuclear envelope preparations are enriched for proteins known to enter the secretory pathway and seem to lack resident ER proteins (such as Kar2p, Sec12p, and Sec61p) (Barlowe et al. 1994; Bednarek et al. 1995). These data suggest that a distinction between resident and transported protein is made in the ER, at the level of packaging into transport vesicles.

If important sorting decisions occur during the formation of COPII vesicles, it is likely that the COPII proteins themselves participate in the selection process and communicate critical information about the selection of cargo to the budding vesicles, possibly through interactions with transmembrane proteins. Coat-mediated sorting is well documented during the formation of clathrin and COPI-mediated vesicles in mammalian cells (Schekman and Orci 1996). In these coats, sorting is mediated by binding of integral membrane cargo proteins or sorting receptors to subunits of the clathrin or the COPI coat (Cosson and Letourneur 1994; Ohno et al. 1995). Thus far, no binding between a COPII subunit and a cargo protein has been demonstrated, although it has been suggested that Sec23p may sample cargo proteins at the ER during the initial steps of vesicle formation (Schekman and Orci 1996).

The analysis of sorting during vesicle formation at the ER is complicated by the absence of a defined sorting signal or sorting receptor. No common signal required for exit from the ER has been defined, and mutants blocked in ER to Golgi transport did not reveal a candidate sorting receptor. One attractive hypothesis is that the sorting of integral membrane proteins is mediated by the interaction of these proteins with the COPII coat. In this scheme, the coat complex would collectively function as a "receptor surface," with many potential sites available for interaction with cargo membrane proteins. The sorting of soluble cargo molecules requires additional assumptions; one possibility is that soluble cargo binds to integral plasma membrane proteins and follows them through the secretory pathway. Alternatively, there may be a family of integral membrane proteins that function as "sorting receptors," binding, on the one hand, to COPII proteins and, on the other, to soluble cargo molecules. Emp24p/Bst2p has been suggested to be such a receptor. Emp24p is an integral ER membrane protein that was cloned because it is also a prominent component of ER-derived vesicles (Schimmöller et al. 1995). A chromosomal deletion of *EMP24* does not impair growth but slows ER to Golgi transport of a subset of cargo proteins (invertase and Gas1p), suggesting a role in segregating particular cargo molecules into the vesicles. However, direct binding of Emp24p to cargo proteins has not been observed.

Emp24p/Bst2p and a second ER membrane protein, Bst1p, were also identified because their deletion allows cells to grow in the absence of Sec13p (Elrod-Erickson and Kaiser 1996; see above). Strains deleted for Bst1p or Emp24p/Bst2p have a decreased fidelity of cargo protein sorting, evidenced by a kinetic defect in the export of a secreted protein (invertase) from the ER, and missorting of the normally ER-retained proteins Kar2p and Pdi1p and an invertase mutant defective in signal sequence cleavage (Elrod-Erickson and Kaiser 1996). Since *bst1* and *bst2* mutants affect both protein sorting and the requirement for a COPII coat component, Bst1p and Emp24p/Bst2p may constitute part of a checkpoint that monitors both protein sorting and COPII coat assembly, before allowing vesicle formation to proceed (Elrod-Erickson and Kaiser 1996).

8. *Sec20p and Tip1p: Additional Proteins Required for ER to Golgi Transport*

Sec20p and Tip1p are likely to act as a functional unit; however, the process in which they act is only poorly understood. *SEC20* was identified as a temperature-sensitive secretion mutant in the original *sec* mutant screen (Novick et al. 1980), and it has a mutant phenotype unique among ER to Golgi transport mutants: At the restrictive temperature, *sec20* mutants accumulate intermediate numbers of about 50-nm vesicles as well as ER membranes. Often, these vesicles are located in and extend the space between the peripheral ER and the plasma membrane (Kaiser and Schekman 1990). Sec20p is a type II integral ER membrane protein (Sweet and Pelham 1992). Overexpression of the cytosolic domain of Sec20p is toxic, and Tip1p was isolated because its overexpression suppresses this toxicity (Sweet and Pelham 1993). Tip1p is a peripheral membrane protein that localizes to the ER in a Sec20p-dependent manner. Furthermore, Tip1p binds to the cytosolic domain of Sec20p in coprecipitation experiments. Depletion of Tip1p causes morphological changes identical to those observed in *sec20* mutants (Sweet and Pelham 1993). Since *sec20* mutants do not show any strong genetic interactions with other ER to Golgi mutants, Sec20p and Tip1p may act together in a novel process, possibly affecting ER to Golgi transport indirectly.

B. Fusion of ER-derived Vesicles with the Golgi

Vesicles that have successfully budded from the ER must next dock and fuse with the appropriate target membrane, the *cis*-face of the Golgi. Ex-

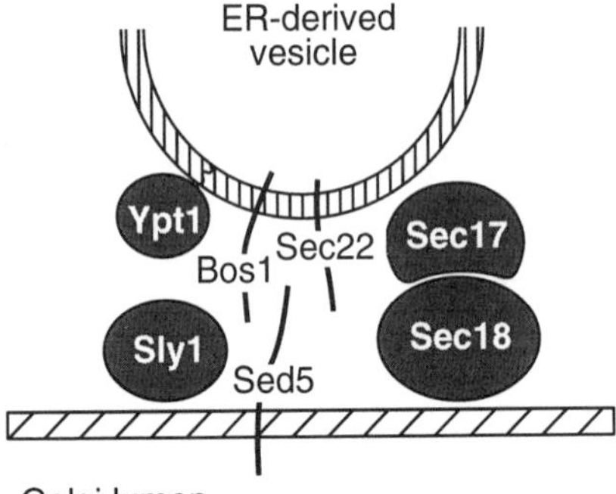

Figure 5 Components of the ER-derived vesicle fusion complex. Similar complexes are thought to function at each vesicle transport step in the pathway. The stoichiometry of individual proteins in the complex is not known.

tensive analysis of vesicle docking and fusion in yeast and mammalian cells has demonstrated both that a large number of gene products participate in this process and that many of the proteins involved have been remarkably conserved (diagrammed in Fig. 5) (for reviews, see Pryer et al. 1992; Bennett and Scheller 1993, 1994; Kaiser 1993; Ferro-Novick and Jahn 1994; Rothman 1994). Proteins in yeast apparently responsible for targeting ER-derived transport vesicles to their target membrane are, in many cases, closely related both to yeast proteins responsible for targeting Golgi-derived exocytic vesicles to the cell surface and to neuronal proteins involved in synaptic vesicle fusion.

In yeast, proteins involved in the docking and fusion of ER-derived transport vesicles include Sec22p, Bos1p, and (in some cases) Bet1p, proteins that are located on the vesicles and are homologous in sequence to the synaptic *v*esicle-*a*ssociated *m*embrane *p*rotein, VAMP (also known as synaptobrevin). The low-molecular-weight GTPase Ypt1p, which is homologous to the mammalian Rab proteins as well as to yeast Sec4p (involved in Golgi to plasma membrane transport; see below), is also associated with transport vesicles. Uso1p is homologous to the mammalian transcytosis-associated protein p115 and is a cytoplasmic factor required for the targeting and fusion of ER-derived vesicles. Other docking/fusion proteins are located on target membranes; one, Sed5p, is homologous to the neuronal presynaptic membrane protein, syntaxin. Also involved is Sly1p, a cytoplasmic protein that appears to associate tightly with Sed5p; Sly1p is homologous to the yeast Sec1p protein (involved in Golgi to plasma membrane transport, see below) as well as to the syntaxin-binding neuronal protein, n-Sec1. Mediating the binding of the vesicle-associated proteins and target-associated proteins are cytosolic Sec17p and Sec18p, known in mammalian cells as SNAP and NSF. Although the vesicle-associated and target-membrane associated docking/fusion proteins seem to vary throughout the cell, potentially contributing to the specificity required for accurate vesicular transport, the same Sec17p/Sec18p (or NSF/SNAP) complex appears to be utilized at every location and functions as a general fusion factor.

1. Sec17p and Sec18p: General Fusion Factors

Detailed morphological analysis of *sec* strains defective in ER to Golgi transport revealed a class of mutants that accumulated 60-nm vesicles at the nonpermissive temperature (*sec17*, *sec18*, *sec22*), presumably due to the failure of the vesicles to either locate or bind to their target membrane (Kaiser and Schekman 1990). The biochemical study of Sec17p and Sec18p, in particular, has been substantially facilitated by in vitro studies using components from mammalian cells.

In vitro reconstitution studies of intra-Golgi transport revealed the accumulation of uncoated vesicles on the target membranes when low concentrations of the cysteine-alkylating agent *N*-ethylmaleimide (NEM) were present (Balch et al. 1984). On the basis of this observation, an NEM-sensitive factor (NSF) was isolated and was found to be homologous to Sec18p. Remarkably, wild-type yeast cytosol could provide NSF activity, catalyzing the fusion of transport vesicles with target membranes in a mammalian cell-free system. The level of activity provided was higher in cytosol from a yeast strain expressing *SEC18* from a multicopy plasmid and not detectable in cytosol from a *sec18-1* mutant. These data together emphasize the high degree of functional conservation between the yeast and mammalian proteins.

Further analysis of NSF indicated that in addition to its demonstrated role in intra-Golgi transport, NSF is required for vesicular transport between the ER and the Golgi (Beckers et al. 1989) and also for vesicle fusion following receptor-mediated endocytosis (Diaz et al. 1989). Interestingly, although Sec18p was initially implicated in ER to Golgi transport (Novick et al. 1981), the assays used to make this determination identified only the first point in the secretory pathway when a protein is needed; subsequent kinetic analysis implicated Sec18p in at least four distinct stages of protein transport: between the ER and the Golgi, at two different intra-Golgi stages, and between the Golgi and the plasma membrane (Graham and Emr 1991). Thus, NSF/Sec18p appears to be required in both yeast and mammalian cells at every stage where protein movement is mediated by vesicular transport.

SEC18 is an essential gene that encodes a protein product with a predicted size of 84 kD (Eakle et al. 1988). Surprisingly, expression of *SEC18* both in vivo and in vitro results in the appearance of two protein products: the expected 84-kD protein and a slightly smaller 82-kD protein, which may result from the initiation of translation at a downstream AUG. Detailed biochemical analysis of mammalian NSF suggests that this factor, initially described as a tetramer (Tagaya et al. 1993), probably exists as a homotrimer (Whiteheart et al. 1994). Each individual NSF/Sec18p polypeptide subunit can be divided into three parts: an amino-terminal region (N) and two homologous ATP-binding domains (D1 and D2). Mutagenesis studies of NSF indicate that the ability of the D1 domain to hydrolyze ATP is required for NSF activity (and thus for fusion) and that the D2 domain is required for trimerization, but its ability to hydrolyze ATP is not absolutely required for NSF function (Whiteheart et al. 1994). These studies also demonstrated that NSF activity requires that all three subunits be functional; a mutation in a single subunit is as debilitating as the presence of the same mutation in every subunit.

NSF itself will only bind to membranes in the presence of cytosolic factors, designated SNAPs (*s*oluble *N*SF *a*ttachment *p*roteins) (Clary et al. 1990; Clary and Rothman 1990; Wilson et al. 1992). Three mammalian proteins with SNAP activity have been identified: α-SNAP and γ-SNAP, which have been found in every cell type examined, and β-SNAP, which appears to be brain-specific (Whiteheart et al. 1993). Each of the three SNAP proteins alone is sufficient to bind NSF to Golgi membranes, and each has been shown to interact directly with NSF (Clary et al. 1990).

Yeast cytosol can provide SNAP activity to a mammalian in vitro intra-Golgi transport assay. Cytosol from a *sec17* mutant is defective in SNAP activity, although this deficiency can be complemented by α-SNAP protein (Griff et al. 1992). *SEC17* is an essential gene, and it encodes a 34-kD protein product that is homologous to α-SNAP and that can provide SNAP activity to mammalian Golgi membranes (Griff et al. 1992). Sec17p, as well as α-, β-, and γ-SNAP, contains two *t*etratrico*p*eptide *r*epeats (TPRs), a motif also found in a number of genes encoding nuclear proteins, such as *nuc2*$^{+}$ and *CDC23* (Ordway et al. 1994). Both kinetic and morphological studies suggest that Sec17p is required in the vesicle consumption stage of protein transport (Kaiser and Schekman 1990; Griff et al. 1992). Coimmunoprecipitation experiments demonstrate that Sec17p and Sec18p are components of the same protein complex, and formation of this protein complex requires the presence of a saturable membrane receptor (Wilson et al. 1992). Since SNAPs will bind to mammalian Golgi membranes in the absence of NSF, and since α- and β-SNAP will compete for the same site on alkali-washed Golgi membranes, the presence of an integral membrane *SNA*P *re*ceptor (SNARE) was inferred.

2. *The SNARE Hypothesis*

NSF binds saturably and stoichiometrically to these SNAP/SNARE complexes, hydrolyzes ATP, and releases itself; the SNAP molecules apparently catalyze this reaction (Morgan et al. 1994). If ATP hydrolysis is blocked, however, such as through the use of a nonhydrolyzable ATP analog, the NSF/SNAP/SNARE complex can be extracted *en bloc* from Golgi membranes. Purification from bovine brain of the proteins comprising this complex revealed the presence of three major proteins besides NSF and SNAP: syntaxin, SNAP-25 (*syn*aptosome-*a*ssociated *p*rotein of 25 kD, unrelated to the SNAPs), and VAMP/synaptobrevin (Söllner et al. 1993). All three proteins were already known and had been localized: VAMPs/synaptobrevins are integral membrane proteins and are found on synaptic vesicles. Syntaxin is an integral membrane protein and is found at the presynaptic plasma membrane, whereas SNAP-25 is thought to be anchored to the presynaptic plasma membrane through the posttranslational fatty acylation of particular cysteine residues (Hess et al. 1992). Since NSF and SNAP are ubiquitous, and VAMP, syntaxin, and SNAP-25 are not, the so-called SNARE hypothesis was proposed (Söllner et al. 1993; for review, see Rothman 1994). In this scheme, NSF and SNAP universally mediate vesicular fusion; the specificity is imparted by the particular SNARE components, including those found on vesicles (v-SNAREs) and those found on the target membranes (t-SNAREs), both of which are predicted to vary at each stage of the transport process. This SNARE hypothesis offers a useful framework within which to consider the proteins that may provide specificity to ER to Golgi vesicular transport in *S. cerevisiae*, proteins such as the VAMP/synaptobrevin homologs Bet1p (also known as Sly12p), Bos1p, and Sec22p (also known as Sly2p) and the syntaxin homolog Sed5p.

3. *Bet1p, Bos1p, and Sec22p: Vesicle-associated Proteins*

BET1 is an essential gene. Conditional *bet1* mutants accumulate ER membranes and are blocked in the maturation of several marker secretory proteins, consistent with a defect in ER to Golgi transport (Newman and Ferro-Novick 1987). Bet1p is an 18-kD integral membrane protein residing on the cytoplasmic face of the ER (Newman et al. 1990, 1992a; Dascher et al. 1991); Bet1p is also found on ER-derived transport vesicles (Barlowe et al. 1994; Rexach et al. 1994), although not in every study (Lian and Ferro-Novick 1993).

During the effort to clone *BET1* by screening a genomic library for multicopy suppressors of *bet1*, *BOS1* (*b*et *o*ne *s*uppressor) was isolated (Newman et al. 1990). Bos1p and Bet1p are not equivalent; overexpression of Bos1p will suppress a *bet1-1* mutant but not the lethality associated with disruption of *BET1*. *BOS1* is also an essential gene, and it encodes a 27-kD integral membrane protein (Shim et al. 1991). Like Bet1p, Bos1p is found both on the cytoplasmic face of the ER and on ER-derived transport vesicles (Newman et al. 1992a; Lian and Ferro-Novick 1993; Rexach et al. 1994). In parallel to *bet*1 mutants, cells depleted of Bos1p fail to transport proteins to the Golgi, and accumulate ER membranes (Shim et al. 1991). Bos1p has also been shown to be required for the fusion competence of ER to Golgi transport vesicles (Lian and Ferro-Novick 1993).

Not only are both Bet1p and Bos1p members of the VAMP/synaptobrevin family, but they also interact genetically with a third VAMP/synaptobrevin family member, Sec22p. *BET1* and *BOS1* can each suppress both the growth and the secretory defects of a *sec22-3* mutant, and a *bet1 sec22-3* double mutant is inviable (Newman et al. 1990). *SEC22* was one of the 23 genes identified in the original *sec* mutant screen (Novick et al. 1980, 1981). *SEC22* is not an essential gene, but a *sec22* null mutant is both heat-sensitive and cold-sensitive, and under nonpermissive conditions, it is defective in ER to Golgi protein transport and accumulates ER membranes (Dascher et al. 1991; Ossig et al. 1991). Interestingly, *sec22* mutants still secrete some incompletely glycosylated invertase and form underglycoslylated mature CPY; these findings both suggest a bypass of Golgi compartments where outer-chain glycosylation takes place (Ossig et al. 1991). Underglycosylated mature CPY was also produced in Ypt1p-depleted strains suppressed by *SEC22* and *BET1* (Ossig et al. 1991). Subsequently, detailed morphological analysis revealed that *sec22* mutants accumulate ER-derived transport vesicles at the nonpermissive temperature, thus implicating Sec22p in the vesicle consumption stage of ER to Golgi transport (Kaiser and Schekman 1990). *SEC22* encodes an integral membrane protein of 25 kD that, like Bos1p and perhaps Bet1p, is found on ER to Golgi transport vesicles (Ossig et al. 1991; Barlowe et al. 1994; Rexach et al. 1994). *SEC22* is identical to *SLY2*

(Newman et al. 1992b), which was independently isolated as a high-copy suppressor of the loss of *YPT1* (Dascher et al. 1991). Although the suppression of *YPT1* loss by either *SEC22* or *BOS1* is quite weak, the overexpression of *SEC22* and *BOS1* together very efficiently suppresses the *YPT1* deficiency (Lian et al. 1994). Interestingly, the overexpression of *BET1*, *BOS1*, or *SEC22* partially suppresses the temperature-sensitive phenotype of *sec21-1* and *sec27-1*, mutant alleles of genes encoding the coatomer subunits γ-COP and β′-COP (Newman et al. 1990; Ossig et al. 1991; Duden et al. 1994). The molecular basis for this genetic interaction remains to be determined. Mammalian homologs of both *SEC22* and *BET1* have recently been characterized, and the predicted protein products are 32% and 21% identical, respectively, to their yeast counterparts (Hay et al. 1996).

4. Ypt1p: A Vesicle-associated Small GTPase

Originally identified as an open reading frame next to the actin gene (*ACT1*) (Gallwitz et al. 1983), *YPT1* is an essential gene, and it encodes a 23-kD protein that is one of the founding members of the Rab family of small-molecular-weight GTPases (Schmitt et al. 1986; Segev and Botstein 1987). Ypt1p shares the conserved domains characteristic of members of this family and has been shown to bind GTP (Schmitt et al. 1986). (The mechanism of action of small GTPases involved in secretion is considered in a separate section of this review.) Several different conditional *ypt1* mutants have been characterized including the cold-sensitive allele *ypt1-1* (Segev and Botstein 1987) and temperature-sensitive allele *ypt1^{ts}* (which represents a deliberately constructed N121I A161V double mutant) (Schmitt et al. 1988). At nonpermissive temperatures, both mutants accumulate incompletely glycoslylated invertase, consistent with a block in ER to Golgi transport (Schmitt et al. 1988; Segev et al. 1988). The transport of the vacuolar protein CPY was also monitored at the nonpermissive temperature in *ypt1-1* mutants, and it was found to accumulate in the p1 (ER) form (Bacon et al. 1989). Morphological studies of both mutants reveal an accumulation of ER-like membrane structures and small vesicles (Schmitt et al. 1988; Segev et al. 1988).

Since the addition of calcium to the culture medium partially rescues *ypt1^{ts}* mutants at the nonpermissive temperature, Ypt1p was hypothesized to be involved in the regulation of intracellular free calcium (Schmitt et al. 1988). However, the addition of calcium could not restore the activity of a *ypt1-2* mutant in an in vitro assay (Bacon et al. 1989);

additional in vitro studies indicated that the step involving calcium is distinguishable from the step involving Ypt1p (Baker et al. 1990). These data are consistent with the results of Rexach and Schekman (1991), which show that the stable attachment of transport vesicles to Golgi membranes requires the function of Ypt1p, whereas the fusion of these vesicles with the Golgi requires calcium. Significantly, Ypt1p appears not to be required for vesicle budding, and indeed, vesicles formed in the presence of defective Ypt1p (and thus unable to dock with Golgi membranes) regain their docking/fusion ability in the presence of exogenous Ypt1p, which the already formed vesicle presumably is able to acquire (Segev 1991).

Ypt1p and Sec4p in yeast, as well as Rab family proteins in mammalian cells, receive the posttranslational addition of a geranylgeranyl (GG) moiety, which in vitro studies suggest can be added to either or both of the two carboxy-terminal cysteines (Kinsella and Maltese 1991; Moores et al. 1991; for reviews, see Armstrong 1993; Omer and Gibbs 1994). This modification appears to mediate membrane attachment, although it is not clear whether this attachment facilitates insertion into the membrane lipid bilayer or mediates a specific interaction with an integral membrane protein, or perhaps both (Marshall 1993). Significantly, deletion or modification of the two carboxy-terminal cysteines renders both Ypt1p and Sec4p nonfunctional and unable to bind to membranes (Molenaar et al. 1988; Walworth et al. 1989). Geranylgeranylation of Ypt1p and Sec4p requires the activity of a prenyltransferase, an enzyme consisting of three subunits: component A, component B_α, and component B_β (for review, see Brown and Goldstein 1993). These proteins are encoded by *MRS6*, *BET4* (initially misidentified as the neighboring *MAD2* gene), and *BET2*, respectively (Rossi et al. 1991; Jiang et al. 1993; Seabra et al. 1993; Benito-Moreno et al. 1994; Jiang and Ferro-Novick 1994; Ragnini et al. 1994). Interestingly, if the geranygeranylated tail region of either Ypt1p or Sec4p is replaced with an appropriate transmembrane anchor, the chimeric protein is still functional (Ossig et al. 1995).

The involvement of Ypt1p-like proteins at several stages of transport in both yeast and mammalian cells suggested that perhaps they are responsible for the specificity of vesicle traffic; for example, perhaps one Ypt1p-like protein could direct ER-derived vesicles toward the early Golgi, and another might direct *trans*-Golgi-derived vesicles to fuse with the plasma membrane. However, when careful domain-exchange experiments were performed in yeast using Ypt1p and Sec4p (which is required for Golgi to plasma membrane transport), a chimera was found that could effectively complement both a *ypt1* mutant and a *sec4* mutant (Brennwald and Novick 1993; Dunn et al. 1993), Furthermore, a single *ypt1* mutant was shown to be defective in transport at two different stages, both between the ER and the *cis*-Golgi and between the *cis*-Golgi and the medial-Golgi (Jedd et al. 1995). These data, together with the observation that Ypt1p is not directly involved in the docking complex (see below), suggest that although Ypt1p may be necessary for accurate targeting, it is not the sole mediator of specificity in the secretory pathway.

Recent experiments have utilized site-directed mutagenesis to generate mutants of Ypt1p that appear to bind and sequester a GDP/GTP exchange factor; this still unidentified factor appears to be distinct from the Dss4p exchange protein and is required for ER to Golgi vesicular transport (Jones et al. 1995). Proteins that interact with both Ypt1p and Sec4p, including Dss4p, the mammalian exchange protein Mss4, and the GDP dissociation inhibitor protein Sec19p, are described below.

5. *Sly1p, Sed5p, and Sft1p: Target-membrane-associated Proteins*

In an effort to study proteins that might be involved in Ypt1p-related functions, a screen was performed to identify genes that suppress loss of *YPT1* (Dascher et al. 1991). Three genes that in high copy suppress the loss of *YPT1* are *SLY2* (identical to *SEC22*), *SLY12* (identical to *BET1*), and *SLY41*, which encodes an integral membrane protein somewhat homologous to chloroplast phosphate translocators. In contrast to *SLY2/SEC22* and *SLY12/BET1*, *SLY41* exhibits no known genetic interaction with any of the *SEC* genes, and its role (if any) in secretion remains obscure.

SLY1-20, a dominant mutation, was identified as a single-copy suppressor of the loss of *YPT1* (Dascher et al. 1991). *SLY1* itself is essential, and it encodes a 75-kD cytosolic protein homologous to Sec1p (a mediator of Golgi to plasma membrane transport) and Vps33p (a vacuolar sorting protein) (Dascher et al. 1991; Ossig et al. 1991; Aalto et al. 1992). Sly1p is required for ER to Golgi transport, as Sly1p-depleted cells accumulate core-glycosylated forms of invertase and CPY (Ossig et al. 1991). A neuronal Sly1p homolog, n-Sec1, binds tightly to neuronal syntaxin and inhibits its interaction with VAMP/synaptobrevin (Hata et al. 1993; Garcia et al. 1994; Pevsner et al. 1994). In yeast, Sly1p appears to associate tightly with the syntaxin homolog Sed5p (Søgaard et al. 1994), although the precise nature of this interaction remains to be established.

SED5 was initially isolated as a multicopy suppressor of an *ERD2* deletion (Hardwick and Pelham 1992). *SED5* is an essential gene, and it encodes a 39-kD integral membrane protein that may form homodimers and is homologous to neuronal syntaxin (Hardwick and Pelham 1992; Banfield et al. 1994). Sed5p-depleted cells exhibit a P1 block in CPY transport and accumulate ER membranes and vesicles, consistent with a block in ER to Golgi transport. By immunofluorescence, Sed5p is not found on the ER but on punctate structures throughout cytoplasm, presumed to be the Golgi. Besides interacting with Sly1p, Sed5p also has been shown to interact with the vesicular protein Sec22p under conditions when vesicle fusion is blocked, such as in a *sec18* mutant (Søgaard et al. 1994).

SFT1 was identified as a multicopy suppressor of a temperature-sensitive *sed5* mutant (Banfield et al. 1995). *SFT1* is an essential gene, and it encodes an 11-kD protein with a coiled-coil structure and a carboxy-terminal membrane anchor. By immunofluorescence, Sft1p appears to localize to the Golgi; the distribution of Sft1p partially, but not entirely, overlaps with that of Sed5p. Since Sft1p colocalizes well with the Golgi marker Mnt1p, and Sed5p colocalizes poorly with Mnt1p, Banfield et al. (1995) suggest that some Sft1p resides in a later compartment of the Golgi than the one occupied by Sed5p. This hypothesis would be consistent with the results of experiments following the maturation of marker proteins in temperature-sensitive *sft1* mutant strains under nonpermissive conditions, experiments which demonstrate that Sft1p is required for transport between early and late compartments of the Golgi. Curiously, Sft1p was also found in the "SNARE" complex (described below) formed during the docking of ER-derived vesicles in a *sec18* mutant strain and isolated by coimmunoprecipitation with Sed5p (Søgaard et al. 1994; Banfield et al. 1995). Banfield et al. (1995) propose that Sft1p functions as a t-SNARE for both ER-derived transport vesicles moving forward toward the *cis*-Golgi and for intra-Golgi vesicles moving backward toward the *cis*-Golgi. Further experiments will be required to characterize the role of Sft1p more precisely.

6. *Uso1p, Bet3p, and Sec7p: Additional Fusion Factors*

USO1 was identified in a conditional lethal mutant defective in ER to Golgi transport at nonpermissive temperatures (Nakajima et al. 1991). *USO1* is an essential gene that encodes a 200-kD protein found as a homodimer or homotrimer predominantly in the soluble fraction of wild-type cell lysates (Nakajima et al. 1991; Seog et al. 1994). One striking structural feature of Uso1p is a coiled-coil region required for function (Seog et al. 1994). Uso1p shares weak but significant sequence homology with mammalian transcytosis-associated protein (TAP), alternatively designated p115 (Barroso et al. 1995; Sapperstein et al. 1995). p115/TAP exists as a dimer, contains a coiled-coil domain, and appears to be required for vesicular docking in both intra-Golgi transport and transcytosis (Waters et al. 1992; Barroso et al. 1995; Sapperstein et al. 1995). p115/TAP is found in both the cytoplasm and associated with intracellular membranes and is present on transcytotic vesicles, on intra-Golgi transport vesicles, and on Golgi-derived secretory vesicles (Nakajima et al. 1991; Sztul et al. 1991; Waters et al. 1992; Gow et al. 1993; Barroso et al. 1995). The subcellular localization of Uso1p has yet to be established.

Both genetic and biochemical evidence points to a role for Uso1p in the docking or fusion stage of ER to Golgi vesicle transport. The temperature-sensitive growth defect of *uso1* cells is suppressed by *BOS1*, *BET1*, *SEC22* (all known ER to Golgi v-SNAREs), and *YKT6* (a putative ER to Golgi v-SNARE) (Sapperstein et al. 1996). Other suppressors include *YPT1* and the *sly1-20* mutation. Interestingly, all of the *uso1* suppressors, including *USO1* itself, also suppress the temperature-sensitive *ypt1-3* mutant; in addition, *ypt1-3* and *uso1* are synthetically lethal, further linking the function of the two gene products. Also like Ypt1p, Uso1p appears to be required for the formation of the docking complex, although it does not seem to participate directly in it (see below).

BET3 was initially identified in a screen for mutants demonstrating synthetic lethality with a *bet1* mutant (Rossi et al. 1995). *BET3* is an essential gene required for vesicular transport from the ER to the Golgi, and it encodes a 22-kD hydrophilic protein. Ultrastructural analysis reveals that *bet3* mutants at the restrictive temperature accumulate both small vesicles (~60 nm in size), characteristic of a block in ER to Golgi transport, and larger vesicles (~110 nm) similar in size to those seen in mutants that disrupt post-Golgi transport. Accumulation of vesicles of both sizes is suggestive of a defect in multiple stages of the secretory pathway; for example, both *bet2* and *sec19* mutants exhibit this phenotype and Bet2p and Sec19p operate at more than one step. Synthetic-lethal interactions observed between *bet3* and mutations affecting ER to Golgi transport (*bos1*, *sec21*, *sec22*, *ypt1*) and post-Golgi transport (*sec2*, *sec4*) support the hypothesis that *BET3* is required at more than one subcellular location. Although the precise role of Bet3p remains to be established, coimmunoprecipitation studies reveal that although Bet3p (like Uso1p and Ypt1p) does not appear to be part of the SNARE complex that forms in a *sec18* mutant, formation of this complex appears to require a functional Bet3p.

SEC7 is an essential gene that encodes a 230-kD hydrophilic protein that is found in both a soluble pool and associated with the cytoplasmic surface of Golgi membranes (Achstetter et al. 1988; Franzusoff and Schekman 1989; Franzusoff et al. 1991; Preuss et al. 1992). Although initial studies suggested that *sec7* mutants exhibit a defect in transport between Golgi compartments (Novick et al. 1981), additional backcrossing of the original mutant to a wild-type strain revealed a prominant defect in ER to Golgi transport, a phenotype apparently masked by the original strain background (Franzusoff and Schekman 1989). A role for Sec7p in ER to Golgi traffic is supported by in vitro data which demonstrate that in a reconstituted system (detailed below), the docking and fusion of ER-derived transport vesicles with the Golgi require Sec7p (Lupashin et al. 1996). In a different in vitro assay, antibodies to Sec7p were shown to block ER to Golgi transport, resulting in an accumulation of transport vesicles that contain Sec7p on their surface (Franzusoff et al. 1992). Since Sec7p is not required for the production of ER-derived vesicles (Barlowe et al. 1994), it remains unclear at what stage in the vesicle morphogenic pathway Sec7p associates with the vesicle, and additional experiments will be required to resolve the role of this important protein.

7. *Toward a Model for Fusion of ER-derived Transport Vesicles*

The vesicle delivery stage of ER to Golgi transport can be divided into three parts, based on data from in vitro studies. First, the COPII coat must be shed, an action apparently facilitated by the hydrolysis of GTP by Sar1p. Second, the vesicle must dock with the Golgi membrane, a process requiring SNARE components and Ypt1p. Finally, the docked vesicle fuses with the target membrane; calcium, ATP, and Sec18p are required for this step.

Several recent in vitro reconstitution studies, together with previously established genetic interactions, may help provide additional insight into the interrelationships among the various components of the vesicle docking and fusion machinery. For example, Rexach et al. (1994) took advantage of the ability of *ypt1* mutants to arrest vesicular transport at the docking stage to analyze the transport vesicles that accumulate in these mutants. In addition to Bet1p, Bos1p, and Sec22p, these vesicles also contained a number of unidentified integral membrane proteins, designated ERV (for their location on ER-derived vesicles). COPII subunits were also detected on these vesicles, although at much lower levels, and no apparent protein coats were visible by electron microscopy; however, it is not clear whether these vesicles had coats and lost them during the purification process or whether they never had coats at all. Since these uncoated vesicles were fusion-competent (if provided with fresh cytosol, membranes, and an ATP regeneration system) but lacked obvious coats, these data argue that the presence of a coat is not required for vesicles to be targeting-competent. In addition, although the transport vesicles are formed in the absence of Ypt1p, vesicle docking and/or fusion requires the presence of Ypt1p; presumably, the vesicles purified in this procedure acquire Ypt1p from the cytosol and/or the membranes used in the fusion assay.

COPII vesicles produced by the addition of purified COPII proteins to urea-washed membranes contain Sec22p and Bet1p, but not Ypt1p (Barlowe et al. 1994). Since it was previously demonstrated that vesicles formed by the addition of purified COPII proteins require Ypt1p to target and fuse with the Golgi (Salama et al. 1993), the presumption is that Ypt1p can be recruited to vesicles that have already been formed during the fusion assay.

A biochemical approach to identify yeast proteins that participate in the vesicle-docking complex has employed coimmunoprecipitation using anti-Sed5p antibodies (Søgaard et al. 1994). Sly1p, Sec17p, Bos1p, Sec22p, and Bet1p were recovered in the immune complexes, as well as three previously undescribed proteins designated p28, p26, and p14, in accordance with their apparent molecular weight. p26 was subsequently determined to be encoded by *YKT6* and is related to the synaptobrevin family. Interestingly, it is predicted not to have a proteinaceous membrane anchor at its carboxyl terminus; instead, it has a consensus site for farnesylation, which presumably provides a membrane anchor for the protein. The multiprotein complex is not observed in a wild-type strain (presumably because vesicles fuse shortly after docking) or in a *sec22* strain (presumably because docking does not occur); however, as discussed previously, Sly1p is always coimmunoprecipitated with Sed5p.

When the experiment was repeated using anti-Bos1p, the same multiprotein complex was observed in a *sec18* strain and not in either a *sec22* strain or a *sec17* strain (although in every instance, Bet1p was always found associated with Bos1p). The absence of complex formation in a *sec17* mutant suggests that Sec17p is required for stable vesicle docking in yeast. Perhaps most interestingly, although Ypt1p is not a member of this docking complex, docking fails to occur in *ypt1* mutants, suggesting that Ypt1p has a critical if indirect role in vesicle docking. A similar result was also observed using a conditional *uso1* mutant (Sapperstein et al. 1996); the multiprotein complex observed in a *sec18* strain was not seen in either a *uso1* strain or a *sec18 uso1* double-mutant strain, but it was seen in the double mutant in the presence of a Uso1p-encoding plasmid. These data argue that Uso1p is also required for the formation of the ER to Golgi SNARE complex. As also seen for Ypt1p, Uso1p was not observed to be present in the SNARE complex. A requirement for Uso1p in the fusion of ER-derived vesicles was also disclosed by in vitro reconstitution studies, as described below (Lupashin et al. 1996).

The protein composition of transport vesicles was also investigated by Lian and Ferro-Novick (1993), who immunoisolated fusion-competent vesicles using anti-Bos1p antibody and reported the presence of Bos1p, Sec22p, and Ypt1p on vesicles but, surprisingly, not Bet1p. In an additional study, Lian et al. (1994) observed that Bos1p and Sec22p physically associate on transport vesicles but not on ER membranes. Since this association does not occur in *ypt1^ts^* mutants, and since the overexpression of *BOS1* and *SEC22* together quite efficiently suppresses the loss of *YPT1*, Lian et al. (1994) propose that Ypt1p is responsible for the formation of a Bos1p/Sec22p complex, which itself is required for vesicle docking. These authors also suggest that Bet1p may mediate this association, since they observe an interaction between Bos1p and Bet1p on the ER.

One model emerging from these in vitro studies postulates that the core-docking interaction is between Sec22p, Bos1p, and perhaps Bet1p and Ykt6p on vesicles and Sed5p, Sly1p, and perhaps Sft1p on the Golgi. Sec17p may stabilize this complex and enable Sec18p to then bind and facilitate fusion. Ypt1p, as well as Uso1p and perhaps Sec7p and Bet3p, may have important, but still undefined, roles organizing the vesicular-docking proteins. Conversely, since the loss of Ypt1p can be suppressed by a mutation in Sly1p, Sly1p might inhibit the interaction between Sed5p and the vesicular-docking proteins, analogous to the inhibitory effect of n-Sec1, a neuronal Sly1p homolog, on syntaxin (a Sed5p homolog). However, such an argument, based as it is almost entirely on analogy, seems premature at this stage. Significantly, although Sly1p appears to be a member of the vesicle-docking complex in yeast, n-Sec1 was not identified in the corresponding complex in mammalian cells, suggesting that the two proteins need not have identical roles; alternatively, the discrepancy might reflect a difference in the way the two complexes were isolated. It is also curious that although SNAP-25 (unrelated to SNAP) is a component of the neuronal docking complex, a yeast SNAP-25 homolog implicated in ER to Golgi traffic has not yet been uncovered.

In an effort to define more precisely the mechanism of transport vesicle fusion, Lupashin et al. (1996) developed an in vitro assay utilizing highly purified Golgi membranes enriched for the *cis*-Golgi compartment. ER-derived transport vesicles could fuse with these membranes in a GTP- and ATP-dependent fashion; fusion also required wild-type cytosol. Incubation with cytosol prepared from temperature-sensitive mutant cells revealed the requirement in vesicle targeting and fusion for the following proteins: Sec7p, Ypt1p, Gdi1p/Sec19p, Sly1p, Uso1p, and Sec18p. Although a requirement for Sec17p was not detected by this assay, Sec17p was found to be already associated with the Golgi membranes. (Gdi1p/Sec19p, a GDP-dissociation inhibitor that can associate with both Ypt1p and Sec4p, is described below.)

It now appears that many, if not most, of the participants involved in the docking and fusion of ER-derived transport vesicles with the Golgi have been identified, and we are now challenged to resolve the mechanism by which these components interact and function.

8. *Small GTPases: Mechanistic Considerations*

One recurring theme in the discussion of vesicular transport has been the involvement of small GTPases at every stage of transport, in both budding (Sar1p, Arf1p) and targeting/fusion (Ypt1p, Sec4p). Although the presence of small GTPases in the secretory pathway has been well-established (for reviews, see Ferro-Novick and Novick 1993; Nuoffer and Balch 1994), the mechanism of action of these proteins remains the subject of intensive investigation. However, the conservation of structure among virtually all known small GTPases allows us to make informed guesses about the function of the particular small GTPases involved in secretion (Bourne et al. 1990, 1991).

Several GTPases, such as the bacterial translation elongation factor (EFTu), appear to contribute to intermolecular recognition by providing a time frame within which a particular molecular match must be established (Bourne 1988; Thompson 1988). This model might also represent a useful paradigm within which to consider the roles of Ypt1p and Sec4p. As discussed earlier, both proteins appear to be located on the surface of vesicles and to play an important part in the accurate targeting of these vesicles to the appropriate destination membrane. In both cases, however, it is possible to at least partially bypass the requirement for the small GTPase by overexpressing other molecules that also appear to be involved in targeting.

On the basis of the EFTu model, the initial association of, for example, the ER-derived transport vesicle with its target membrane would be expected to require the presence of Ypt1p-GTP, which might correctly position vesicle-docking proteins for optimal interaction with their targets. Once the initial recognition event occurs, there might be a race between the hydrolysis of GTP by Ypt1p and the dissociation of the vesicle. If Ypt1p does hydrolyze GTP first, then this might set up another competition, between the ability of Ypt1p-GDP to diffuse away from the complex (so that Sec17p and Sec18p can then stabilize the docking complex and catalyze fusion) and the propensity of the vesicle to dissociate from the Golgi before the docking complex is stabilized. This model, although at this point only speculative, is consistent with the known genetic and in vitro data, including the presence of both Ypt1p and Sec4p on transport vesicles (Goud et al. 1988; Lian and Ferro-Novick 1993) and the absence of Ypt1p in the docking complex isolated in a *sec18* mutant by Søgaard et al. (1994).

IV. RETRIEVAL OF ER PROTEINS FROM THE GOLGI

During the process of vesicle formation at the ER membrane, cargo proteins are segregated from resident ER proteins, but this segregation is not perfect and resident ER proteins do reach the Golgi at a slow rate. The cell's capacity to retain resident ER proteins is improved by mechanisms to return these proteins from early Golgi compartments to the ER. The best-understood retrieval mechanisms involve recognition of short peptide sequences on the proteins to be retrieved. One mechanism involves the recognition of the sequence HDEL at the carboxyl terminus of proteins located in the lumen of secretory organelles, a second mechanism involves recognition of the sequence KKXX at the carboxyl terminus of integral membrane proteins on the cytosolic side of the membrane, and a third mechanism involves recognition of a cytosolic double-arginine (RR) motif near the amino terminus of integral membrane proteins. In addition, there are less well characterized retrieval mechanisms for the integral membrane protein Sec12p (probably involving a determinant in the transmembrane domain) and for retrieval of the v-SNARE integral membrane proteins.

A. HDEL-mediated Recycling

The tetrapeptide HDEL was first recognized because of its appearance at the carboxyl terminus of soluble ER proteins, such as Kar2p (Pelham 1988), Eug1p (Tachibana and Stevens 1992), Kre5p (Meaden et al. 1990), Pdi1p (LaMantia et al. 1991), Trg1p (Gunther et al. 1991), and cyclophilin D (Frigerio and Pelham 1993). The two integral ER membrane proteins, Sec20p (Sweet and Pelham 1992) and Sed4p (Hardwick and Pelham 1992), also contain a carboxy-terminal HDEL located in the lumen since these proteins are of type II configuration. The HDEL sequence has been shown to be both necessary and sufficient for ER retention. Addition of HDEL to the carboxyl terminus of secreted proteins such as invertase or pro-α-factor confers ER localization on these proteins (Pelham et al. 1988; Dean and Pelham 1990). Conversely, deletion of the carboxy-terminal HDEL of Kar2p greatly increases the amount of Kar2p secreted into the medium, indicating escape from a retrieval mechanism (Hardwick et al. 1990). Deletion of HDEL from the integral membrane protein Sec20p also causes escape from retrieval back to the ER, but the protein in this case is transported to the vacuole where it is degraded (Sweet and Pelham 1992).

HDEL-mediated retention was shown to be saturated by the overproduction of an HDEL-bearing protein, implying the existence of a receptor for HDEL binding (Pelham et al. 1988; Hardwick and Pelham 1990). A screen for mutants that secrete elevated levels of HDEL-bearing proteins identified two candidate receptor genes: *ERD1* and *ERD2* (Pelham et al. 1988; Semenza et al. 1990). *ERD1* is required for proper Golgi function (Hardwick et al. 1990), and *ERD2* encodes an integral membrane protein located in the ER and Golgi with the expected properties for an HDEL receptor: Overexpression of Erd2p increases the capacity of the cells to retrieve HDEL-bearing proteins, whereas reduction of Erd2p decreases retrieval capacity (Semenza et al. 1990). Interestingly, *S. cerevisiae* does not retain proteins containing DDEL at the carboxyl terminus, but the related yeast, *K. lactis*, does. This difference was shown to be specified by Erd2p, since expression of the Erd2p homolog of *K. lactis* confers retention of DDEL-containing fusion proteins onto *S. cerevisiae* (Lewis et al. 1990).

HDEL is thought to function as a retrieval signal in the Golgi rather than as a retention mechanism in the ER. Initial evidence for a retrieval pathway came from the observation that HDEL-containing secretory proteins acquire carbohydrate modifications characteristic of early Golgi compartments but fractionate with the ER (Pelham 1988; Dean and Pelham 1990). The molecular characterization of Erd2p also suggests a function for Erd2p in the Golgi. Erd2p is a multiple membrane-spanning protein that is localized in the Golgi apparatus in both yeast (Semenza et al. 1990) and mammalian cells (Lewis and Pelham 1990). In mammalian cells, overexpression of KDEL-containing proteins results in the redistribution of Erd2 protein from the Golgi apparatus to the ER, suggesting recycling of the receptor-ligand complex to the ER (Lewis and Pelham 1992).

Although the intracellular retention of pro-α-factor with HDEL at the carboxyl terminus is quite efficient, only a small portion of the retained protein has Golgi-specific carbohydrate modifications (Dean and Pelham 1990), and overexpression of Erd2p entirely eliminates Golgi modifications on the retained protein (Semenza et al. 1990). These findings indicate that ER proteins may be retrieved from a compartment situated before the Golgi compartment where the first carbohydrate modification takes place, or that Erd2p can also slow the exit of proteins from the ER. Careful localization of mammalian Erd2 protein showed that it is present in several intracellular compartments beside the Golgi apparatus, including the ER, the intermediate compartment, and the *trans*-Golgi network, suggesting that Erd2p may act at multiple steps in the secretory pathway (Griffiths et al. 1994).

The mechanism by which Erd2p retrieves HDEL-containing proteins has not been established. An attractive possibility is that retrieval is mediated by vesicles whose coats assemble in response to interaction between Erd2p and HDEL-bearing proteins in a process analogous to receptor-mediated endocytosis. HDEL-mediated retrieval may be mediated by COPI-coated vesicles similar to the retrieval of KKXX-containing proteins, as discussed below. Temperature-sensitive mutants for two COPI coat components, Sec21p (γ-COP) and Ret2p (δ-COP), secrete significant amounts of Kar2p at the semipermissive temperature (A. Frand and C. Kaiser, unpubl.), and mutants in γ-COP fail to redistribute Erd2p to the ER at the nonpermissive temperature, a phenotype indicative of a failure to recycle Erd2p from the Golgi to the ER (Lewis and Pelham 1996). The suggestion that the cytosolic domains of Erd2p might carry a signal for coat assembly by analogy to receptor-mediated endocytosis prompted a mutant hunt for alleles of the human homolog of Erd2p that were defective for recycling to the ER (Townsley et al. 1993). An aspartate located in the extreme carboxy-terminal transmembrane domain of Erd2p was shown to be critical for retrieval and since charged residues in transmembrane domains often participate in protein-protein interactions within the membrane, it was suggested that retrieval of Erd2p involves interactions with itself or with other integral membrane proteins in the lipid bilayer (Townsley et al. 1993).

Deletion of *ERD2* is lethal and results in accumulation of Golgi-modified proteins, suggesting that Erd2p is important for transport of proteins through the Golgi apparatus (Semenza et al. 1990). The lethality of the *ERD2* deletion can be suppressed by overexpression of a variety of genes (termed *SED*) that have in common the property that their overexpression slows ER to Golgi transport (Hardwick et al. 1992). Possibly, deletion of *ERD2* causes an imbalance in Golgi traffic that can be counteracted by reducing the flux of proteins arriving from the ER. Alternatively, deletion of *ERD2* could be lethal because of a failure to maintain sufficient intracellular levels of essential HDEL-containing proteins (Townsley et al. 1994).

B. KKXX-mediated Recycling

1. KKXX-mediated Recycling: Role of COPI

A dilysine motif, KKXX or KXKXX, at the carboxyl terminus of type I ER membrane proteins is critical for targeting particular viral proteins to the ER in mammalian cells (Nilsson et al. 1989; Jackson et al. 1990). In yeast, at least four membrane-spanning proteins also contain a dilysine motif at their carboxyl termini: Wbp1p, a subunit of the *N*-oligosaccharyltransferase complex and a type I membrane protein (te Heesen et al. 1991; 1992); Vma21p, a multiple membrane-spanning protein required for assembly of the vacuolar ATPase complex (Hill and Stevens 1994); Emp47p, a type I membrane protein of unknown function (Schröder et al. 1995); and Erg25p, an enzyme in the ergosterol biosynthetic pathway (Bard et al. 1996). Both Wbp1p and Vma21p are located in the ER. Truncation of the KKXX sequence from Vma21p causes the protein to no longer be retained in the ER and to eventually be transported to the vacuole (Hill and Stevens 1994). In contrast, Wbp1p does not require the KKXX motif for retention in the ER, possibly because it is bound to additional, ER-retained proteins (Gaynor et al. 1994). Addition of KKXX to the carboxyl terminus of either multiple membrane-spanning proteins or type I integral membrane proteins confers ER localization (Gaynor et al. 1994; Letourneur et al. 1994; Townsley and Pelham 1994). As with HDEL-bearing proteins, retention of KKXX-bearing proteins appears to be the result of recycling from Golgi to ER since retained proteins are located in the ER but acquire Golgi-specific carbohydrate modification (Gaynor et al. 1994; Townsley and Pelham 1994). Fusion proteins bearing KKXX that are retained in the ER acquire Golgi modifications at the same rate as untagged control proteins, indicating that although addition of the KKXX peptide allows recycling, it does not interfere with exit from the ER (Gaynor et al. 1994).

The vesicle coat complex known as coatomer or COPI is a complex of seven proteins (α, β, β′, γ, δ, ε, and ζ-COP) originally identified as

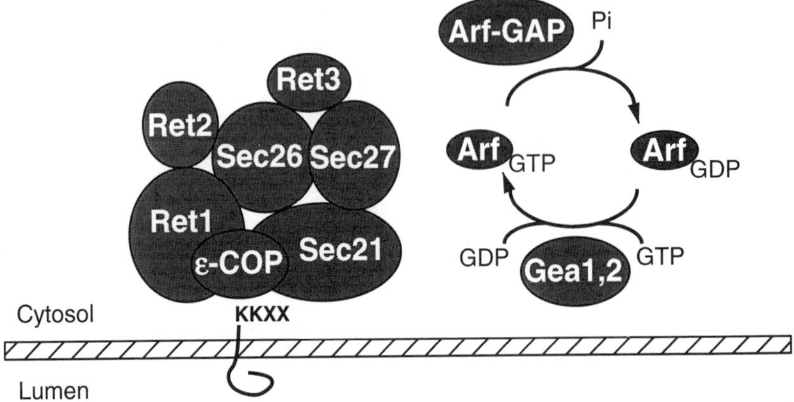

Figure 6 Components of the COPI protein vesicle coat complex. These proteins are required for the retrieval of proteins with the carboxy-terminal KKXX sequence from the Golgi to the ER. These proteins may also have a role in transport from the ER to Golgi and between compartments of the Golgi. The roles of Arf1p, Gea1p, Gea2p, and the putative ARF-GAP in COPI coat formation are inferred from mammalian studies but have not been demonstrated in yeast.

the major component of the coat of Golgi-derived vesicles in mammalian cells (Rothman 1994). A homologous complex was identified in yeast by purification of Sec21p (Hosobuchi et al. 1992). Sec21p itself is homologous to γ-COP (Stenbeck et al. 1992; Harter et al. 1996). Sequencing of the two largest subunits of the yeast complex identified proteins homologous to α-COP (Ret1p) and β-COP (Sec26p) (Duden et al. 1994; Gerich et al. 1995). Proteins homologous to δ-COP (Ret2p) and ζ-COP (Ret3p), as well as to α-COP (Ret1p), were identified in a screen for mutants defective in recycling of KKXX-containing proteins (Letourneur et al. 1994; Cosson et al. 1996). Finally, a homolog of β′-COP (Sec27p) was identified in a screen for mutants defective in ER to Golgi transport (Fig. 6) (Duden et al. 1994).

KKXX is present in the cytosol and might be expected to bind to retrieval factors. Indeed, a fusion of GST to the cytosolic domain of a KKXX-containing protein specifically binds to COPI (Cosson and Letourneur 1994). COPI function is critical for KKXX-mediated recycling since temperature-sensitive mutations in yeast coatomer subunit genes for α-COP (*RET1*), β′-COP (*SEC27*), γ-COP (*SEC21*), δ-COP (*RET2*), and ζ-COP (*RET3*) cause defects in the retention of KKXX fusion proteins at semipermissive temperatures (Letourneur et al. 1994; Cosson et al. 1996). Two of these mutations, *ret1* and *sec27*, also interfere with the ability of the COPI complex to bind to KKXX in vitro (Letourneur et al. 1994). Three subunits of coatomer, α-COP, β′-COP, and ε-COP, have been suggested to form a KKXX-binding site (Letourneur et al. 1994). These three proteins bind tightly to KKXX at salt concentrations that cause dissociation of the other subunits from the complex (Cosson and Letourneur 1994). γ-COP is likely to be in close proximity to KKXX since it can be cross-linked to KKXX-containing peptides (Harter et al. 1996).

COPI and the small GTP-binding protein, ARF, drive the assembly of coated vesicles from purified Golgi membranes in mammalian cells (Ostermann et al. 1993). An appealing model for KKXX-mediated recycling is that the binding of coatomer to the KKXX motif of Golgi proteins triggers formation of retrograde transport vesicles that deliver their contents back to the ER. Two yeast homologs of ARF, *ARF1* and *ARF2*, were identified by hybridization (Sewell and Kahn 1988; Stearns et al. 1990a). Arf1p and Arf2p are 96% identical and perform overlapping functions (Stearns et al. 1990a). Since both *ARF1* and *ARF2* show genetic interactions with *sec21* and *sec27* mutants (Stearns et al. 1990a; Duden et al. 1994), they may function together with yeast coatomer in KKXX-mediated recycling.

Formation and fusion of COPI-coated vesicles is regulated by nucleotide exchange and hydrolysis on ARF (Rothman 1994). Gea1p, a protein proposed to mediate nucleotide exchange on Arf1p and Arf2p, was identified in a screen for high-copy suppressors of dominant, cold-sensitive *ARF2* alleles designed to have increased affinity for a putative exchange factor (Peyroche et al. 1996). Gea1p is a component of a 500–600-kD complex that can stimulate the binding of GTP-γS to recombinant bovine myristoylated *ARF1* in vitro. Gea1p is 50% identical to another yeast gene product, Gea2p. Although neither *GEA1* nor *GEA2* is essential, deletion of both genes is lethal. When the double deletion strain is covered by a termperature-sensitive allele of *GEA1*, the strain exhibits an ER to Golgi secretion block at the nonpermissive temperature. *gea1* mutants are partially suppressed by multicopy *ARF1*, *ARF2*, or *SEC21*, consistent with a role for Gea1p in COPI vesicle formation. Gea1p, Gea2p, and ARNO, a human protein with nucleotide exchange activity for ARF, all contain a centrally located region of approximately 300 amino acids that is homologous to a domain of Sec7p. Since this domain is the only region common to all three proteins, it is likely to be responsible for the exchange activity. The intriguing possibility that Sec7p itself may function as a nucleotide exchange factor remains to be explored.

An ARF GTPase-activating protein (ARF-GAP) has been identified in mammalian cells (Cukierman et al. 1995). Yeast Gcs1p is homologous to mammalian ARF-GAP and stimulates the GTPase activity of yeast Arf1p (Ireland et al. 1994; Cukierman et al. 1995). Chromosomal deletions of *GCS1* are viable but fail to leave stationary phase at low temperatures (Ireland et al. 1994). The role of Gcs1p in the secretory pathway remains to be examined.

In addition to its role in retrieving ER proteins from the Golgi, the KKXX motif may also function in the localization of proteins to early Golgi compartments. Emp47p is a KKXX-containing protein that is mainly present in early Golgi compartments, but it can recycle to the ER (Schröder et al. 1995). The Golgi localization of Emp47p depends on a functional KKXX sequence but does not depend on α-COP (Schröder et al. 1995).

2. *Additional Roles for COPI-coated Vesicles*

It is possible that the COPI complex and ARF participate in cellular functions other than retrieval of KKXX-bearing proteins. In mammalian cells, COPI-coated vesicles have been implicated in forward transport through the Golgi apparatus (Rothman 1994), in transport from the ER to the Golgi apparatus (Peter et al. 1993; Aridor et al. 1995), and in endocytosis (Whitney et al. 1995; Aniento et al. 1996). In yeast, mutations of *ARF* genes or genes for COPI subunits cause defects in transport from the ER to the Golgi apparatus (Kaiser and Schekman 1990; Stearns et al. 1990a,b; Duden et al. 1994). Mutations affecting COPI usually give a weaker block in ER to Golgi transport than mutations affecting COPII, and it has been suggested that a failure of COPI mutants to recycle proteins properly could eventually lead to a failure of forward transport if essential transport factors must be recycled (Letourneur et al. 1994; Pelham 1994). Likely candidates for essential proteins that depend on COPI for recycling are the v-SNARE proteins Sec22p, Bet1p, and Bos1p, since mutations in their genes are exacerbated by the COPI mutations *sec21* and *sec27* (Newman et al. 1990; Ossig et al. 1991; Duden et al. 1994). Recent experiments showed that COPI-coated vesicles can bud from isolated yeast nuclei in vitro; the resulting vesicles contain v-SNAREs, such as Sec22p, Bet1p and Bos1p, but not the cargo protein α-factor (Bednarek et al. 1995). Thus, coatomer-coated vesicles may mediate a second pathway from the ER to the Golgi apparatus, possibly transporting distinct cargo proteins.

C. RR-mediated Retrieval

Paired arginine residues near the amino terminus of a mammalian type II transmembrane protein have been shown to confer ER localization on a plasma membrane protein, suggesting that this sequence acts much like KKXX to specify retrieval from a post-ER compartment (Schutze et al. 1994). The yeast ER protein, Bst1p, contains an amino-terminal RR motif, but the importance of this motif for retention of Bst1p in the ER has not been tested (Elrod-Erickson and Kaiser 1996).

D. Retrieval of Sec12p

Sec12p is mostly located in the ER, but it slowly acquires modifications characteristic of early Golgi compartments, suggesting that it is retained by a novel recycling mechanism since Sec12p does not contain a HDEL, KKXX, or RR motifs (Nakano et al. 1988). Screens for mutants that allow Sec12p to access late Golgi compartments identified three new genes, *RER1*, *RER2*, and *RER3* (Nishikawa and Nakano 1993; Boehm et al. 1994). *RER1* encodes a protein with four predicted membrane-spanning domains that is located in an early Golgi compartment (Boehm et al. 1994; Sato et al. 1995). Domain swapping experiments between Sec12p and another type II protein that is normally transported to the *trans*-Golgi compartment indicate that the transmembrane portion of Sec12p specifies *RER1*-dependent ER localization (Boehm et al. 1994). Even after deletion of *RER1*, significant quantities of Sec12p remain in the ER, suggesting additional retention or retrieval mechanisms for Sec12p. *RER1*-mediated recycling is likely distinct from HDEL- or KKXX-mediated recycling, since mutations in *rer1* do not affect the retrieval of a KKXX-containing fusion protein (Gaynor et al. 1994) and *rer1* mutant strains secrete only moderate amounts of the HDEL-bearing protein, Kar2p (Sato et al. 1995).

E. Fusion of Retrograde Transport Vesicles with the ER

It might be expected that vesicles that retrieve ER proteins from the Golgi apparatus use a fusion machinery similar to that employed by ER-derived vesicles to fuse with the Golgi apparatus. Ufe1p, an ER-localized protein related to the t-SNARE syntaxin, has the properties of a t-SNARE involved in fusion of retrograde vesicles with the ER. The temperature-sensitive *ufe1-1* allele prevents the proper recycling of Emp47p (a KKXX-bearing protein) and Erd2p to the ER. In addition, *ufe1-1* cells secrete Kar2p, a behavior consistent with a failure in recycling of the HDEL receptor (Lewis and Pelham 1996). The identification of Ufe1p as a t-SNARE for retrograde transport implies that v-SNAREs will also operate at this step, although specific v-SNAREs for retrieval from the Golgi have not been identified. Possibly, the v-SNAREs, Sec22p, Bet1p, and Bos1p operate in both forward and retrograde transport processes between the ER and Golgi (Lewis and Pelham 1996).

V. PROTEIN TRANSPORT THROUGH THE GOLGI

A. Inter-Golgi Transport

1. Compartmental Organization of the Golgi

In mammalian cells, the Golgi apparatus consists of a series of flattened stacks through which secretory proteins progressively pass. In *S. cerevisiae*, the morphology of the Golgi is considerably less well defined; careful immunoelectron microscopy suggests that the Golgi consists of single, isolated cisternae, typically not arranged in parallel stacks (Preuss et al. 1992). Nevertheless, secretory proteins appear to receive sequential modifications that are thought to occur in distinct Golgi compartments.

The existence of at least three individual compartments was established by the work of Graham and Emr (1991), who used conditional mutants and pulse-chase analysis to demonstrate that pro-α-factor and pro-CPY receive distinct carbohydrate modifications in compartments that can be separated either kinetically or by secretion mutant blocks. The *cis*-Golgi is defined by the acquisition of α-1,6 mannose residues (detected by antibody specific for this linkage); the medial Golgi is defined by the subsequent addition of α-1,3 mannose residues (also detected by antibody specific for this linkage); and the *trans*-Golgi is defined by the cleavage of pro-α-factor by Kex2p. Each transport step is presumed to occur by vesicular transport, since a mutation in *SEC18* was used to trap intermediates in each of the compartments. A *sec23* mutant was shown to block transport both between the ER and the *cis*-Golgi, and between the *cis* and medial Golgi; a similar result was recently obtained for a *ypt1* mutant (Jedd et al. 1995). Collectively, these data suggest that transport between the *cis* and medial Golgi is mechanistically similar to transport between the ER and the *cis*-Golgi; transport beyond the medial Golgi may require different components that remain to be specified.

Since the different compartments of the Golgi are defined by the modification of secretory proteins, it initially seemed reasonable to expect that these compartments might also be defined by the presence of the modifying enzyme. For example, the enzyme responsible for the α1,6 modification, Och1p, must be present in the *cis*-Golgi, whereas Kex2p, responsible for the cleavage of pro-α-factor, must be present in the *trans*-Golgi. However, the difficulty with using these enzymes as immunological markers for particular compartments is that the enzymes are not necessarily restricted to the first compartment in which they function. For example, Och1p is concentrated in the *cis*-Golgi, but it can access the medial and *trans*-Golgi compartments, from which it is thought to be subsequently retrieved (Harris and Waters 1996).

The enzyme responsible for one of the first steps of O-linked glycosylation, Kre2p/Mnt1p, has also been used as a Golgi marker, although the compartment in which it resides is not clear. Although some data suggest that it resides in the medial compartment (Lussier et al. 1995), other data argue that it is predominantly found colocalized with Och1p in the *cis*-Golgi (Chapman and Munro 1994). Immunoisolation of the *trans*-Golgi using anti-Kex2p antibodies has demonstrated that Kre2p/Mnt1p is largely absent from this compartment (Bryant and Boyd 1993).

Mnn1p, an enzyme responsible for outer-chain addition to both N-linked and O-linked sugar chains, and GDPase (an enzyme involved in salvage of luminal GDP released when mannose is transferred from GDP-mannose to the growing ologosaccharide of glycoproteins) represent additional Golgi compartment markers. However, although some cell fractionation studies suggest that Mnn1p and GDPase (which extensively colocalize) are present in at least two physically distinct compartments, one of which is the Kex2p-containing *trans*-Golgi (Graham et al. 1994), other experiments indicate that Mnn1p and GDPase do not extensively cofractionate with Kex2p (Whitters et al. 1994). The localization of Kex2p and its role in sorting at the *trans*-Golgi are discussed below.

As a whole, the data attempting to establish a correspondence between a specific Golgi compartment and a particular immunologic or enzymatic marker suggest that perhaps this approach represents a slight oversimplification. Rather than being defined by the presence or absence of one or another marker, Golgi compartments may be determined by the relative mixtures of Golgi proteins residing within individual cisternae (Whitters et al. 1994).

2. *Transport through the Golgi*

Although protein transport through the Golgi is thought to occur through vesicular intermediates, this hypothesis awaits experimental confirmation. One possible source of concern is the near absence of mutations that arrest protein maturation in the *cis* or medial Golgi. If transport between successive components of the Golgi requires the presence of unique targeting molecules, as suggested by the SNARE hypothesis, then Golgi-specific targeting proteins should have been identified by *sec* mutations. The only such candidate SNARE to date is Sft1p (described above), a Golgi protein that functions as a t-SNARE for ER-derived vesicles and possibly also for intra-Golgi vesicles that are returning to the *cis*-Golgi from a later compartment (Banfield et al. 1995).

Given the many mutations that block ER to Golgi transport and fusion of secretory vesicles with the plasma membrane, why have mutations not been identified that block protein traffic in the *cis* or medial compartments? One possibility is that such mutants could arise but have not yet been discovered; a priori, this possibility seems to be unlikely, given both the number of screens for *sec* mutants that have been performed and the number of genes involved in secretion that have already been identified. A second possibility is that a considerable amount of pathway redundancy may exist in the Golgi, so that even if a particular vesicle transport step is blocked, proteins would have ready access to other ways of reaching the next compartment; there is neither evidence for nor against this hypothesis.

A third possibility is that the sequential transport of secretory proteins through the Golgi may not be essential. Although ER-derived vesicles may be able to fuse only with the *cis*-Golgi, it is possible that vesicles derived from either the *cis* or medial Golgi may in fact be able to skip the next compartment if transport to that compartment is blocked. Such mutants might be able to successfully secrete proteins (as well as transport membrane to the cell surface), although these proteins would be expected to be only incompletely modified. In particular, some candidate mutants would be expected to exhibit the phenotype of incomplete Golgi-specific glycosylation.

Genes required for correct glycosylation in the Golgi that do not appear to be glycosylation enzymes themselves, fall into three classes: (1) Pmr1p; (2) Mnn9p, Anp1p, and Van1p; and (3) Erd1p and Van2p. Pmr1p is a Golgi-localized calcium ATPase; *pmr1* mutant strains secrete underglycosylated invertase (but which contains both α-1,6- and α-1,3-linked mannose residues) and exhibit genetic interactions with many different *sec* mutations (Antebi and Fink 1992). The effects of *pmr1* on the Golgi seem to be pleiotropic and can be partially reversed by the addition of calcium, suggesting that Pmr1p affects secretion indirectly by controlling the luminal Ca^{++} concentration of the Golgi.

Mnnp9, Anp1p, and Van1p are related type II integral membrane proteins (Chapman and Munro 1994; Yip et al. 1994; Kanik-Ennulat et al. 1995). Mutants for any one of these nonessential proteins secrete underglycosylated invertase and also exhibit a variety of different phenotypes, including, in all three cases, resistance to the toxic orthophosphate analog, vanadate. Once again, the major question here is whether the effects of these mutations on glycosylation in the Golgi reflect defects in the glycosylation enzymes themselves or perturbations in transport through the Golgi.

Van2p/Vrg4p and Erd1p both represent multiple membrane-spanning proteins that have not as yet been localized. *VAN2/VRG4* is an essential gene; conditional mutants secrete underglycosylated invertase, exhibit a kinetic defect in the transport of CPY, and secrete the resident ER protein Kar2p (Ballou et al. 1991; Poster and Dean 1996). *erd1* mutants also exhibit glycosylation defects and secrete Kar2p (Pelham et al. 1988; Hardwick et al. 1990). Again, the role of Van2p/Vrg4p and Erd1p in transport through the Golgi may be only indirect and remains to be resolved.

B. Sec14p: A Regulator of Golgi Phospholipid Composition

The organelles and vesicles that constitute the secretory pathway are composed, to a large extent, of various types of lipids, but the role of lipids in this pathway has remained obscure. Nevertheless, the asymmetric distribution of lipids across the organelles of the secretory pathway and, in particular, the increasing gradients of ergosterol and phosphatidylserine along the secretory pathway raise the possibility that the lipid content of particular membranes may play a critical part in protein sorting (for reviews, see Paltauf et al. 1992; Bretscher and Munro 1993; Pelham and Munro 1993). To date, however, the best evidence for a relationship between membrane lipid composition and secretion is based on the study of Sec14p.

SEC14 is essential for growth, and *sec14* mutants exhibit a secretion block at the level of the Golgi (Novick et al. 1981; Bankaitis et al. 1989). Sec14p is a hydrophilic 37-kD protein that in cell lysates behaves primarily as a soluble species (Bankaitis et al. 1989). However, a particulate component is also present, and immunolocalization and cell fractionation studies locate this portion of Sec14p to the Golgi (Cleves et al. 1991).

Sec14p is identical to Pit1p, a phosphatidylinositol (PI)/phosphatidylcholine (PC) transfer protein (Aitken et al. 1990; Bankaitis et al. 1990). Similar proteins have also been identified in mammalian cells and are characterized by their ability to transfer PI (and less efficiently, PC) from one lipid bilayer to another (for reviews, see Paltauf et al. 1992; Burgoyne 1994). Studies of *sec14* mutant strains have demonstrated that Sec14p is responsible for all of the detectable PI/PC transfer protein activity in *S. cerevisiae* (Aitken et al. 1990; Bankaitis et al. 1990).

In yeast, PC is synthesized through two distinct pathways. The first involves the methylation of phosphatidylethanolamine, and the second, termed the CDP-choline pathway, utilizes choline itself, either exogenously supplied or salvaged (for review, see Paltauf et al. 1992). The observation by Cleves et al. (1991) that a *sec14* null allele could be suppressed by the loss of function of proteins involved in the CDP-choline pathway for PC biosynthesis suggested that Sec14p functions to depress PC levels in the Golgi. Sec14p could conceivably accomplish this either by removing PC molecules from the Golgi and relocating them elsewhere or by negatively regulating synthesis or activity of an enzyme involved in PC biosynthesis. The ability of overexpressed Sec14p to decrease the rate of PC biosynthesis and the absence of any data demonstrating that PI/PC transfer proteins transport significant amounts of phospholipids in vivo both suggested that the second explanation was correct (Cleves et al. 1991; McGee et al. 1994).

Recently, Sec14p was shown to inhibit choline phosphate cytidylyltransferase (CCTase), the rate-limiting step of the CDP-choline pathway (Skinner et al. 1995). Interestingly, the inhibitor activity appears to be present only when Sec14p is bound to PC and not when bound to PI. Thus, Sec14p appears to be part of a feedback-inhibition pathway designed to maintain an optimal PI/PC ratio. Sec14p situated on the cytoplasmic surface of the Golgi could function as a sensor of phospholipid composition by periodically extracting either PI or PC phospholipids from the underlying membrane. If Sec14p extracts PC, then the Sec14p-PC complex will inhibit CCTase, thus decreasing the PC content of the phospholipid bilayer. Since Sec14p-PI has no effect on CCTase, the ability of Sec14p to inhibit PC production depends on the PI/PC ratio in the Golgi membrane.

Although appealing, this model still leaves unanswered the question of why a perturbation of PI/PC content should disrupt secretion. Also problematic is the observation that a rat PI/PC transfer protein, which has no structural homology with Sec14p, is able to complement a *sec14* temperature-sensitive mutant (although not a null allele), yet it does not appear to be able to inhibit CCTase at all (Skinner et al. 1993, 1995). Similarly, Sec14p is able to replace the mammalian PI/PC transfer protein in a neuroendocrine cell-based in vitro reconstitution assay and stimulate secretory vesicle formation (Ohashi et al. 1995). Another puzzling question is why secretion is halted rapidly at restrictive temperatures in a *sec14*^{ts} mutant, given the putative function of Sec14p. Undetermined as well is the function of several *sec14*^{ts} bypass suppressors that do not appear to be involved in PC synthesis, including two suppressors, *BSD1* and *BSD2*, defined by dominant mutations. Also isolated, as a recessive bypass suppressor of *sec14*^{ts}, was a mutation of *SAC1*, a mutation that in addition suppresses a temperature-sensitive mutation of the actin gene, *ACT1*.

Finally, if precise regulation of membrane phospholipid composition is essential for protein transport, then it is somewhat curious that other lipid transport or biosynthesis genes have not appeared in screens for secretion mutants.

VI. PROTEIN SORTING AT THE *TRANS*-GOLGI COMPARTMENT

Proteins that have reached the *trans*-Golgi have at least three possible fates: transport to the vacuole, transport to the plasma membrane via secretory vesicles, or residency in the *trans*-Golgi.

A. The Prevacuolar/Late Endosomal Sorting Compartment

In mammalian cells, soluble vacuolar proteins and their receptors are transported together to a prevacuolar/late endosomal sorting compartment. In this compartment, receptor and ligand dissociate, the receptor recycles back to the Golgi complex, and the cargo protein is transported to the vacuole (Kornfeld 1992). In yeast, soluble vacuolar proteins bind to a membrane protein receptor such as Vps10p in the *trans*-Golgi; the receptor and the protein ligand are then thought to be transported together to a prevacuolar compartment, where the protein ligand dissociates to be transported to the vacuole and the receptor is recycled back to the Golgi (Stack et al. 1995).

A prevacuolar compartment in yeast was first detected in cell fractionation studies where a compartment that contains the Golgi-modified forms of pro-CPY and pro-PrA (proteinase A) was physically separated from Golgi and vacuolar compartments (Vida et al. 1990, 1993). Entry of CPY into this compartment was shown to depend on a functional vacuolar sorting signal and Vps15p function, which situates this compartment on the vacuolar targeting pathway after sorting in the *trans*-Golgi. The prevacuolar compartment was found to cofractionate with α-factor that had been internalized by endocytosis, indicating that this compartment lies at the nexus of the endocytic and vacuolar pathways, as is the case in mammalian cells.

Evidence for a prevacuolar/late endosomal compartment also comes from the phenotypic characterization of mutants that are defective in vacuolar protein sorting (*vps* mutants). The class-E *vps* mutants accumulate soluble vacuolar proteins, such as CPY and PrA, and the vacuolar H^+ ATPase in a compartment adjacent to the vacuole as visualized by immunofluorescence microscopy (Raymond et al. 1992). This compartment also appears to contain endocytosed Ste3p and resident *trans*-Golgi proteins, such as Pep1p/Vps10p (Raymond et al. 1992; Piper et al. 1995). Thus, the so-called class-E compartment detected by microscopy appears to have a parallel composition to the prevacuolar compartment as isolated by subcellular fractionation. In a temperature-sensitive class-E *vps* mutant, the compartment is rapidly generated on shift to the restrictive temperature, suggesting that the compartment represents a physiological intermediate in the vacuolar protein-sorting pathways (Piper et al. 1995). The class-E compartment is not an intermediate step for all vacuolar proteins. For example, the vacuolar membrane protein alkaline phosphatase (ALP) does not accumulate in the class-E compartment, suggesting an alternative route to the vacuole (Raymond et al. 1992).

B. Sorting Signals for Vacuolar Proteins

Protein sorting to the vacuole is described in detail by Jones et al. (this volume) and has been reviewed recently (Stack et al. 1995); it will be discussed here only briefly. Soluble and integral membrane proteins are sorted to the vacuole by two apparently different mechanisms. Soluble vacuolar proteins generally require specific signal sequences for interaction with a luminal receptor protein for segregation to the vacuolar pathway, whereas membrane proteins appear to be transported to the vacuole without a need for particular vacuolar targeting signals (Roberts et al. 1992; Stack et al. 1995).

The vacuolar protein CPY requires a four-amino-acid peptide (QRPL) near the amino terminus of the pro-region for transport to the vacuole (Valls et al. 1987). The receptor for this signal is Pep1p/Vps10p, an integral membrane protein in the *trans*-Golgi with a large luminal domain. Importantly, Vps10p can be cross-linked to wild-type pro-CPY but not to a version of pro-CPY with a mutation in the sorting signal (Marcusson et al. 1994). PrA appears to be targeted to the vacuole by a similar mechanism. Initial experiments suggested that deletion of *VPS10* was not required for vacuolar targeting of PrA, but more recent studies have demonstrated direct binding of pro-PrA to Vps10p and have shown that Vps10p does participate in transport of PrA to the vacuole (Marcusson et al. 1994; Westphal et al. 1996). An even more general role for Vps10p in segregation of soluble proteins to the vacuole is suggested by experiments where invertase fused to a thermally unstable protein domain was shown to be transported to the vacuole by a Vps10p-dependent pathway (Hong et al. 1996). The finding that Vps10p recognizes unfolded polypeptide as a vacuolar targeting sequence suggests that Vps10p may also serve to direct incorrectly folded luminal proteins to the vacuole for their degradation.

C. Sorting Signals for *Trans*-Golgi Proteins

To function as a receptor for soluble vacuolar proteins, Vps10p must cycle through the *trans*-Golgi (Cereghino et al. 1995; Cooper and Stevens 1996). Other proteins with a topology similar to that of Vps10p that reside in the *trans*-Golgi are probably also localized by a recycling mechanism (Graham and Emr 1991). Vps10p and the α-factor processing enzymes Kex1p, Kex2p, and DPAP-A are all type I integral membrane proteins that have cytosolic domains of about 100 amino acid residues which are critical for localization of these proteins to the Golgi: Deletion of the cytosolic domain causes mislocalization of these proteins to the vacuole (Redding et al. 1991; Cooper and Bussey 1992; Roberts et al. 1992; Wilcox et al. 1992; Nothwehr et al. 1993; Cereghino et al. 1995). The signals required for sorting were investigated further, and it was noted that the cytosolic domains of Kex2p, DPAP-A, and Vps10p each contains a short sequence, rich in aromatic residues, that is similar to one of the motifs recognized to mediate clustering of mammalian membrane receptors in clathrin-coated pits (Trowbridge et al. 1993). Mutational studies demonstrated that this aromatic-rich motif is indeed required for efficient *trans*-Golgi localization of Vps10p, Kex2p, and DPAP-A (Wilcox et al. 1992; Nothwehr et al. 1993; Cereghino et al. 1995). Moreover, addition of this motif to a membrane protein that is normally transported to the vacuolar membrane caused this protein to be localized to the Golgi (Nothwehr et al. 1993).

The cytosolic domains of *trans*-Golgi proteins could cause retention by interacting with the sorting machinery which acts either to retain these proteins in the Golgi or to cause them to recycle to the Golgi from a prevacuolar compartment. Overexpression of Kex2p, DPAP-A, or Vps10p causes these proteins to be mislocalized to the vacuole, consistent with the idea that a saturable receptor is responsible for either the retention or recycling of these proteins. Overexpression of Kex2p also acts in *trans* to mislocalize a fusion protein containing the DPAP-A cytosolic domain to the vacuole, indicating that Kex2p and DPAP-A interact with a common sorting machinery (Roberts et al. 1992; Wilcox et al. 1992; Cereghino et al. 1995, Nothwehr et al. 1995).

D. Role of Clathrin and Related Proteins in Sorting at the *Trans*-Golgi Compartment

Clathrin-coated vesicles and buds are a prominent feature of the *trans*-Golgi in mammalian cells and have been implicated in sorting at this compartment (Kornfeld 1992). Clathrin-coated vesicles purified from yeast by differential centrifugation and gel filtration have a morphology similar to that of mammalian clathrin-coated vesicles and contain clathrin heavy- and light-chain molecules that can be released as typical clathrin triskelions (Mueller and Branton 1984; Payne and Schekman 1985). As with the mammalian protein, yeast clathrin triskelions can assemble into clathrin cages in vitro (Lemmon et al. 1988).

The genes encoding the clathrin heavy (*CHC1*) chain and light (*CLC1*) chain were isolated using antibodies raised against purified heavy chain (Payne and Schekman 1985) or by hybridization using a peptide sequence obtained from purified light chain (Silveira et al. 1990). Chc1p is 50% identical to mammalian clathrin heavy chain (Lemmon et al. 1991), whereas Clc1p has little sequence similarity to mammalian light chain but similar physical properties (Silveira et al. 1990). Deletion of *CHC1*, *CLC1*, or both genes causes slow growth or death depending on genetic background (Payne and Schekman 1985; Lemmon and Jones 1987; Payne et al. 1987; Silveira et al. 1990; Lemmon et al. 1991).

The function of clathrin in sorting of soluble vacuolar proteins was addressed by examination of both null and temperature-sensitive alleles of *CHC1* (Payne et al. 1988; Seeger and Payne 1992a). Null alleles of *chc1* exhibit normal vacuolar protein sorting, whereas a *chc1*^{ts} mutant causes both CPY and PrA to be secreted rather than transported to the vacuole. The difference in behavior between the temperature-sensitive and null alleles can be explained by a cellular adaptation to the loss of Chc1p and even for the *chc1*^{ts} mutant, transport to the vacuole is restored within 3 hours at the restrictive temperature (Seeger and Payne 1992a).

In addition to a role in transport of soluble vacuolar proteins, Chc1p is also required for the retention of *trans*-Golgi membrane proteins. This requirement was first revealed indirectly by the failure of *chc1* null mutants to process α-factor (Payne and Schekman 1989). It was subsequently shown that both *chc1*^{ts} and *chc1* deletion strains mislocalize DPAP-A and Kex2p to the plasma membrane, hence disrupting processing of α-factor in the Golgi (Seeger and Payne 1992b). Interestingly, *chc1* deletion strains are able to compensate for the defect in sorting of soluble vacuolar proteins, but a similar compensation is not observed for the defect in Golgi protein retention.

It has not yet been established whether clathrin is also required for transport of membrane proteins to the vacuole. A fraction of the vacuolar membrane ALP is present on the plasma membrane in both *chc1* deletion and *chc1*^{ts} mutants (Seeger and Payne 1992a,b). However, ALP appears on the plasma membrane only several hours after transfer of the *chc1*^{ts} mutant to the restrictive temperature, suggesting that the loss of clathrin function may bring about the mislocalization of ALP by an indirect mechanism (Seeger and Payne 1992b).

In mammalian cells, clathrin functions together with a heteromeric complex of four proteins called the adaptor protein (AP) complex. Two related AP complexes are found in mammalian cells: The AP1 complex, which contains γ, β1, μ1, and σ1 adaptin subunits, is located at the *trans*-Golgi compartment, and the AP2 complex, which contains α, β2, μ2, and σ2 adaptin subunits, is located at the plasma membrane. These adaptor complexes are thought to be crucial components of clathrin-coated vesicles and to both stimulate assembly of clathrin on the membrane and bind to the cytosolic tails of integral membrane cargo proteins (Robinson 1994). In yeast, proteins homologous to β1, μ1, and σ1 adaptins have been identified by searches of the protein database and by polymerase chain reaction (PCR) cloning strategies. Apl1p and Apl2p are related to the β subunits of AP1 and AP2 (Kirchhausen 1990; Rad et al. 1995), and Apm1p and Aps1p are most closely related to the μ1 and σ1 subunits (Phan et al. 1994; Stepp et al. 1995). Null alleles of *APL2*, *APM1*, or *APS1* do not produce either growth or vacuolar protein-sorting defects. However, these null alleles significantly exacerbate the growth and α-factor-processing defect of *chc1*^{ts} and *chc1* null mutations. Interestingly, *APM1* and *APS1* null alleles do not exacerbate the sorting defect of soluble vacuolar proteins exhibited by *chc1*^{ts}, and *APL2* null alleles even partially suppress this defect (Phan et al. 1994; Stepp et al. 1995). Both Apm1p and Aps1p cofractionate with clathrin-coated vesicles by gel filtration. This association depends on a functional clathrin heavy chain, indicating that Apm1p and Aps1p are indeed elements of an AP complex that coassembles with clathrin (Phan et al. 1994; Stepp et al. 1995).

Two GTP-binding proteins have been implicated in the formation of clathrin-coated vesicles in mammalian cells and could also function in conjunction with clathrin in yeast cells. ARF stimulates recruitment of mammalian AP complexes to the *trans*-Golgi compartment (Stamnes and Rothman 1993; Traub et al. 1993). The role of ARF proteins in sorting at the *trans*-Golgi compartment in yeast has not yet been evaluated. Dynamin is required for the pinching-off of clathrin-coated vesicles from the plasma membrane (Robinson 1994). A yeast protein homologous to dynamin, Vps1p, was identified in a screen for mutants defective in vacuolar protein sorting (Rothman et al. 1990). Null alleles of *VPS1* are completely defective in the normal transport of both soluble and membrane proteins from the *trans*-Golgi compartment to the vacuole, suggesting that Vps1p is a central component in the transport from the Golgi to the vacuole (Rothman et al. 1990; Nothwehr et al. 1995). These *vps1* mutants also cause Kex2p and other Golgi resident proteins to be transported to the vacuole (Wilsbach and Payne 1993; Nothwehr et al. 1995). Further analysis of the *vps1* phenotype gives a more complicated picture: In *vps1* null strains, Kex2p is first transported to the cell surface and then delivered to the vacuole by endocytosis (Nothwehr et al. 1995). Thus, Vps1p, like clathrin, appears to be required for transport of membrane proteins from the Golgi to the prevacuolar compartment. A failure to carry out this transport step appears to result in transport of Golgi membrane proteins to the plasma membrane instead. The difference in the final location of Kex2p in *chc1* mutants (plasma membrane) and *vps1* mutants (vacuole) can be explained by an independent requirement for clathrin in endocytosis that would prevent the uptake of Kex2p from the plasma membrane in $chc1^{ts}$ mutants. *MVP1*, a multicopy suppressor of a dominant-negative *vps1* mutation, encodes a 59-kD hydrophilic protein which colocalizes with Vps1p (Ekena and Stevens 1995). As with *vps1* mutants, null alleles of *MVP1* cause secretion of the soluble vacuolar proteins CPY and PrA, indicating that these genes function in the same process.

Most of the phenotypes associated with mutations in clathrin, Vps1p, or AP complex subunits are consistent with the view that these proteins function together in vesicular transport from the *trans*-Golgi to a prevacuolar compartment. Failure to carry out this transport step apparently leads to inappropriate packaging of Golgi resident proteins into secretory vesicles targeted to the plasma membrane. This mislocalization of Golgi membrane proteins in turn disrupts the proper sorting of soluble vacuolar proteins and also leads to their transport to the cell surface. The interaction between the cytosolic tails of receptor proteins and the AP complexes of clathrin vesicle coats in mammalian cells leads to the expectation that the cytosolic domains of Golgi proteins such as Kex2p and Vps10p should also interact with clathrin coats. If this were the case, the truncations of the cytosolic tails of Golgi proteins should have the same effect as defects in clathrin coat components. This prediction is not borne out since, for example, truncation of the cytosolic domain of Kex2p causes the protein to be transported directly to the vacuole, whereas mutation of clathrin causes Kex2p (with or without an intact cytosolic domain) to be transported to the plasma membrane (Redding et al. 1996). This discrepancy is not easily resolved, and more information is needed to explain the apparently complicated interaction between the yeast clathrin coat and the cytosolic domains of *trans*-Golgi membrane proteins.

E. Additional Proteins Involved in Transport from Golgi to Vacuole

A large number of mutants have been identified that are defective in sorting of soluble vacuolar proteins (*vps* mutants) and that are discussed in detail by Jones et al. (this volume). Morphological analysis of *vps* mutants suggests that transport from the *trans*-Golgi compartment to the vacuole involves at least two vesicular transport steps, and some of the *VPS* gene products have been assigned functions in either the formation or the fusion of these vesicles. Several of the *VPS* gene sequences are related to recognized components of the general vesicle fusion machinery. *PEP12/VPS6* encodes a protein homologous to syntaxins (Becherer et al. 1996), and Vps45p and Vps33p are homologous to the fusion component Sec1p (Banta et al. 1990; Wada et al. 1990; Cowles et al. 1994; Piper et al. 1994). Vps21p/Ypt51p and Ypt7p are homologous to members of the Rab family of GTPases (Horazdovsky et al. 1994). Consistent with the proposed role of these proteins in vesicle fusion, mutants in all of these genes lead to the accumulation of vesicles.

Perhaps surprisingly, no candidates for recognizable vesicle coat components have thus far been detected among *vps* mutants. Rather, many of these genes, such as *VPS3*, *VPS11*, *VPS16*, *VPS17*, *VPS27*, and *PEP7*, encode novel proteins with no known homologs. It is possible that transport from the *trans*-Golgi to the vacuole employs novel coats or a coat-independent mechanism of vesicle formation.

F. Vesicle Budding from the *Trans*-Golgi to the Plasma Membrane

Little is known of the mechanism by which proteins destined for the plasma membrane are sorted into secretory vesicles that bud from the *trans*-Golgi. Soluble proteins are thought to be packaged into secretory vesicles by default, since there are no known soluble resident proteins in the Golgi, and some soluble vacuolar proteins are secreted if they are overproduced or if vacuolar targeting sequences are removed. In contrast, membrane proteins do not appear to be packaged into secretory vesicles by default, since membrane proteins that have no known sorting signals are generally transported to the vacuole (Roberts et al. 1992). Conceivably, membrane proteins could require special signals to be loaded into secretory vesicles, but, such signals have not yet been identified. Temperature-sensitive mutations in the plasma membrane ATPase, Pma1p, are missorted to the vacuole at restrictive temperature (Chang and Fink 1995). The sorting defect of these mutants can be suppressed by overexpression of Ast1p, a novel, nonessential protein that associates with multiple intracellular membranes (Chang and Fink 1995). These mutants and Ast1p provide a starting point for the analysis of plasma membrane protein sorting at the *trans*-Golgi compartment.

By analogy to vesicle formation steps elsewhere in the secretory pathway, there is an expectation that formation of secretory vesicles will be coupled to assembly of a vesicle coat. No candidates for such a coat have yet been identified. Clathrin is unlikely to function as a secretory vesicle coat, since *chc1* null and *chc1*^{ts} mutants exhibit normal transport of soluble and integral membrane proteins to the plasma membrane (Payne and Schekman 1985; Payne et al. 1988; Seeger and Payne 1992a,b). The COPII coat component Sec23p is similarly not required for export of invertase after the early Golgi (Graham and Emr 1991).

An important question is whether secretory vesicles are all the same or whether different types of secretory vesicles exist. Secretory vesicles that accumulate in mutants defective in fusion with the plasma membrane (*sec1*, *sec4*, or *sec6*) have been examined by cell fractionation or by electron microscopy. Initial studies showed that the same secretory vesicle can contain different types of cargo molecules. Immunoelectron microscopy has shown that Pma1p and invertase localize to the same vesicles (Brada and Schekman 1988), and vesicles immunopurified with Pma1p antibody contain acid phosphatase (Holcomb et al. 1988). A recent study, however, provides evidence for at least two different classes of secretory vesicles. By separation of vesicles on shallow density gradients, two populations of secretory vesicles with different densities could be resolved (Harsay and Bretscher 1995). The denser vesicle population contains invertase, acid phosphatase, and the exoglucanase Exg1p, whereas the less dense population contains the majority of Pma1p, the synaptobrevin homolog Snc1p, and the endoglucanase Bgl2p. Both types of vesicles accumulate in three different fusion mutants (*sec1*, *sec4*, *sec6*), suggesting that different secretory vesicle populations use a common fusion machinery. If the two types of vesicles have different coats as suggested by their different densities, it may not be possible to find single mutations that prevent the formation of both types of vesicles. Thus, the existence of more than one type of secretory vesicle implies a functional redundancy in vesicle budding from the *trans*-Golgi that could explain why no *sec* mutant has been found that specifically acts at this step.

VII. PROTEIN TRANSPORT TO THE PLASMA MEMBRANE

A. Fusion of Golgi-derived Vesicles with the Plasma Membrane

The fusion of Golgi-derived vesicles with the plasma membrane is thought to parallel closely the fusion of ER-derived vesicles with the Golgi (for review, see Ferro-Novick and Jahn 1994). In both locations, fusion requires vesicle-specific and target-membrane-specific factors, a small-molecular weight GTPase, and the SNAP/NSF (Sec17p/Sec18p) complex. In contrast to transport between the ER and the Golgi, however, an assay has not yet been devised that reconstitutes Golgi to plasma membrane transport in vitro. As a consequence, the precise role of many of the gene products involved has been difficult to determine. Nevertheless, genetic interactions, morphological studies, and the examination of marker transport proteins such as invertase have implicated a number of proteins in Golgi to plasma membrane transport; cell-fractionation studies have helped determine whether these proteins are on vesicles, on the plasma membrane, or in the cytosol; and coimmunoprecipitation experiments have revealed interactions with other participating proteins. As a result of these studies, a general picture of vesicular transport between the Golgi and the plasma membrane has now begun to emerge.

Golgi-derived protein transport vesicles are approximately 100 nm in diameter, about twice as wide as ER-derived vesicles. Golgi-derived vesicles are targeted specifically to the tip of the developing bud, the site of exocytosis. In the current model, the fusion competence of a vesicle relies on the presence of the small-molecular-weight GTPase Sec4p, which is posttranslationally modified by geranlygeranlyation to facilitate its attachment to membranes. The localization of Sec4p is thought to be coupled to the phosphorylation status of the bound nucelotide (GDP or GTP) (Ferro-Novick and Novick 1993). Sec4p is found on the exterior of post-Golgi vesicles in the GTP-bound form; vesicular attachment to the plasma membrane is thought to be accompanied by the hydrolysis of Sec4p-GTP, possibly mediated by a Sec4p GAP on the cytoplasmic face of the plasma membrane. After fusion occurs, Sec4p-GDP is presumably extracted from the plasma membrane by Gdi1p (equivalent to Sec19p), a protein that also prevents the premature exchange of GDP for GTP. This exchange reaction eventually occurs, presumably on vesicles, and is probably mediated by the GDP dissociation stimulator, Dss4p.

The successful fusion of vesicles with the plasma membrane also requires the presence of two apparently distinct protein complexes: a SNARE complex, discussed in principle earlier, and a novel complex, containing at least eight different subunits. The plasma membrane SNARE complex consists of the vesicular proteins Snc1p/Snc2p (homologous to synaptobrevin and to Bos1p) and the membrane-associated proteins Sso1p/Sso2p (homologous to syntaxin and Sed5p), Sec9p (homologous to SNAP-25), and Sec1p (homologous to nSec1 and Sly1p). Sec17p and Sec18p (SNAP and NSF) are also needed for SNARE-mediated vesicular fusion. Although the general subcellular localization of SNARE components is consistent with a role in the determination of targeting specificity, the diffuse plasma membrane distribution of the SNARE molecules Sso1p/Sso2p and Sec9p suggests that other factors might also be involved.

One candidate suitably positioned to regulate vesicle targeting at the terminal stage of the secretory pathway is a large plasma-membrane-associated complex consisting of Sec6p, Sec8p, Sec15p, and at least five other proteins. Not only is this complex apparently required for vesicular traffic to the plasma membrane, but it also appears to be concentrated at the tips of developing buds, precisely the region where secretory vesicles are targeted. Furthermore, not only does this complex consist of at least three *SEC* gene products, but its formation requires the proper functioning of at least three other *SEC* gene products, Sec3p, Sec5p, and Sec10p, proteins that are not themselves members of the complex. The participation in this complex of at least six of the originally identified 23 essential genes required for secretion strongly suggests an important role for this complex in vesicular transport (TerBush and Novick 1995).

1. Sec4p

SEC4 is an essential gene, and it encodes a 24-kD Rab family protein 48% homologous to Ypt1p (Salminen and Novick 1987). Sec4p is required for vesicular transport between the Golgi and the plasma membrane; temperature-sensitive mutants of *SEC4* accumulate 100-nm post-Golgi vesicles under nonpermissive conditions (Novick et al. 1980, 1981). Sec4p is located both on the surface of post-Golgi secretory vesicles and on the cytosolic face of the plasma membrane. A small cytoplasmic pool, representing approximately 15% of the total Sec4p, is also present (Goud et al. 1988). As described earlier for Ypt1p, membrane attachment of Sec4p is mediated by the posttranslational modification of the carboxy-terminal region of the protein, and mutant proteins lacking this palmitylation site are unable to bind to membranes and behave as inactive proteins (Walworth et al. 1989; Jiang et al. 1993).

Like Ypt1p, Sec4p both binds GTP and has an intrinsic GTPase activity that is comparatively slow (0.0012/min; the equivalent figure for $p21^{ras}$ is 0.028/min (John et al. 1989; Walworth et al. 1989; Kabcenell et al. 1990). Sec4p has an inherent affinity for both GTP and GDP, although the dissociation rate for GTP (0.002/min) is about two orders of magnitude lower than that for GDP (0.21/min) (Kabcenell et al. 1990). Mutagenesis studies identified a mutation in Sec4p that partially interferes with its ability to hydrolyze GTP; the equivalent mutation in $p21^{ras}$ is oncogenic, but this mutation is lethal in Sec4p, producing a secretion defect characterized by the accumulation of post-Golgi vesicles (Walworth et al. 1992). This particular mutant *SEC4* gene behaves as a dominant loss-of-function allele and exhibits synthetic-lethal interactions with several other late-acting *sec* mutants (Walworth et al. 1992). The correlation of a *sec4* loss-of-function phenotype with decreased GTPase activity is in marked contrast to $p21^{ras}$ but is consistent with the involvement of Sec4p in a cycle, as originally proposed by Bourne (1988).

The involvement of Sec4p and Ypt1p in similar but distinct roles in the secretion pathway suggested that both of these proteins might contain a "destination" domain responsible for the specificity of vesicular targeting. However, as discussed previously, the creation of a chimeric Sec4p/Ypt1p molecule capable of fulfilling both Sec4p and Ypt1p functions, yet producing no missorting of vesicles, suggests that the primary source of specificity lies elsewhere (Brennwald and Novick 1993; Dunn et al. 1993).

2. *Gdi1p/Sec19p, Dss4p, Sec4p-GAP: Sec4p-modifying Proteins or Activities*

One class of protein that has been found to interact with members of the Rab family are the GDP-dissociation inhibitors (GDIs), proteins that both inhibit, in vitro, the dissociation of Rab-bound GDP and stimulate the release of prenylated GDP-bound Rab from membranes (see Ferro-Novick and Novick 1993; Novick and Garrett 1994; Nuoffer and Balch 1994). GDIs appear to be relatively nonspecific: Rab3a GDI, for example, will exhibit activity not only on many different mammalian Rab proteins, but also on yeast Sec4p (Garrett et al. 1993; Nuoffer and Balch 1994).

GDI1 was isolated from yeast on the basis of its homology with Rab3a GDI and *Drosophila* GDI (Garrett et al. 1994). The predicted protein sequence of 451 amino acids is 50% identical to both human and *Drosophila* GDIs. Gdi1p inhibits the dissociation of GDP from both Sec4p and Ypt1p in a dose-dependent fashion. It also stimulates the release of GDP-bound Sec4p (but not Ras2p) from yeast membranes. These data are consistent with properties of previously studied GDI proteins.

In an effort to determine where in the cell Gdi1p functions, a yeast strain was constructed in which Gdi1p could be depleted through the use of a regulated promoter (Garrett et al. 1994). Cells depleted of Gdi1p exhibit a partial block in secretion observed at several stages of protein transport. Several different forms of the marker protein invertase, representing various stages of glycosylation, were observed, suggesting a block between the ER and the Golgi and also at additional points later on in the pathway. Maturation of the vacuolar protein CPY was also examined, and both p1 and p2 intermediates were observed but no mature form, suggesting a slowing of the processing of CPY both before the Golgi and within it. Morphological studies of Gdi1p-depleted yeast revealed the exaggeration of ER structures, the accumulation of Golgi-related structures, and the build up of vesicles, mostly 100 nm in size, presumably Golgi-derived, but some 50 nm in size, consistent with an origin in the ER. As a whole, these data argue that Gdi1p is involved in multiple steps along the secretory pathway.

GDI1 was also shown to be the same as *SEC19*, one of the original secretion mutants (Novick et al. 1980; Garrett et al. 1994). Although GDI-depleted cells appear to be morphologically similar to *sec19* mutants at the restrictive temperature, *sec19* mutants are completely blocked in transport of CPY and invertase at the level of ER to Golgi transport (Novick et al. 1980; Stevens et al. 1982). These data are supported by in vitro reconstitution studies that reveal a requirement for Sec19p in the targeting or fusion of ER-derived vesicles (Lupashin et al. 1996). The detection of CPY and invertase transport defects later in the pathway in Gdi1p-depleted cells may reflect the residual Gdi1p present even after 10 hours of growth in glucose. In a sense, the Gdi1p-depleted yeast are comparable to yeast bearing a weak *sec19* allele. This example also serves as an important reminder that in studying mutations affecting progressive protein transport through the cell, it is only the first disruption in the pathway that is generally detected. Thus, if a protein affects multiple stages of the secretory pathway, its distal sites of actions might be easily overlooked.

Although Gdi1p/Sec19p exhibits limited homology with component A (Mrs6p) of the Sec4p and Ypt1p prenylation enzyme geranylgeranyltransferase II (GGTase II), it does not exhibit component-A function in a reconstitution assay of GGTase II function (Garrett et al. 1994; Jiang and Ferro-Novick 1994).

Studies of Ras-related proteins have also identified numerous factors that stimulate the exchange of GDP for GTP (for reviews, see Bourne et al. 1990, 1991; Downward 1992; Boguski and McCormick 1993). In a search for spontaneous suppressors of the temperature-sensitive *sec4-8* mutation, a GDP dissociation stimulator for Sec4p was identified as a dominant suppressor, *DSS4-1* (Moya et al. 1993). Although *DSS4* is not essential, a *dss4* null allele does exhibit synthetic-lethal interactions with several other late-acting *sec* mutants. Partially purified Dss4p is able to accelerate GDP dissociation by six to eight times; Dss4p also functions on Ypt1p, but only about half as efficiently. A mammalian Dss4p homolog, Mss4, has also been identified (Burton et al. 1993). Mss4 catalyzes GDP dissociation from Sec4p, Ypt1p, and mammalian Rab3a but has no effect on Ras2p. A second factor that exhibits exchange activity for Ypt1p has recently been identified; this factor appears to be distinct from Dss4p and is apparently essential for vesicular transport (Jones et al. 1995). The precise role of Dss4p, Mss4, and other exchange proteins in vesicular traffic remains to be established.

Studies of Ras and Ras-related proteins have also revealed the existence of GAPs (for reviews, see Bourne et al. 1990, 1991; Boguski and McCormick 1993). The GAP for $p21^{ras}$, for example, catalyzes the hydrolysis of $p21^{ras}$-GTP, accelerating the reaction by approximately five orders of magnitude (Gideon et al. 1992). At present, it is unclear whether GAPs represent simply negative regulators of Ras proteins or whether their GAP activity might be coupled to a role as a Ras protein effector. A membrane-associated, Sec4p-specific GAP activity has been located in yeast (Walworth et al. 1992). The protein or proteins responsible for this activity, which stimulates Sec4p GTP hydrolysis by about 3.5-fold, have not yet been isolated.

3. *Snc1p and Snc2p: Vesicle-associated Fusion Proteins*

Snc1p and Snc2p are VAMP/synaptobrevin homologs with predicted molecular weights of 13 kD localized to post-Golgi vesicles (Gerst et al. 1992; Protopopov et al. 1993). *SNC1* was initially identified as a suppressor of a null allele of the cyclase-associated protein (CAP), a protein required for responsiveness to Ras, for normal cell morphology and for cellular responses to nutrient extremes (Gerst et al. 1992). *SNC2* was cloned by degenerate PCR based on homology with Snc1p and mammalian synaptobrevins (Protopopov et al. 1993). Snc1p and Snc2p are 79% identical. Deletion of either *SNC1* or *SNC2* does not result in an obvious phenotype, but deletion of both genes renders the cell inviable on rich medium and capable of growing only on minimal medium at 30°C or lower. At 30°C, *SNC* null cells accumulate (rather than secrete) active invertase and accumulate 100-nm post-Golgi vesicles, phenotypes consistent with a defect in Golgi to plasma membrane transport. As discussed below, Snc proteins can form a complex with Sso proteins and Sec9p (Brennwald et al. 1994; Couve and Gerst 1994), analogous to the interaction observed in neurons between synaptobrevin, syntaxin, and SNAP-25.

4. *Sso1p, Sso2p, Sec9p, and Sec1p: Plasma-membrane-associated Proteins*

SSO1 and *SSO2* encode 39-kD syntaxin family proteins that share 72% amino acid identity (Aalto et al. 1993). Both genes were isolated as multicopy suppressors of *sec1*. Deletion of either *SSO1* or *SSO2* does not result in any apparent phenotype, but deletion of both is lethal. *SSO1* and *SSO2* exhibit a number of genetic interactions with other late-acting *SEC* genes, including *SEC9* and *SEC15*; indeed, Sso1p overexpression can efficiently suppress a *sec15* mutation and prevent the accumulation of 100-nm transport vesicles characteristic of these mutants. Conversely, depletion of Sso protein results in an accumulation of 100-nm vesicles. Sso proteins were shown by both immunofluorescence and cell-fractionation procedures to localize to the plasma membranes and to associate with Sec9p, as detailed below (Brennwald et al. 1994).

SEC9 is an essential gene, and it encodes a 74-kD protein (observed size: 106 kD) required for post-Golgi vesicular transport (Novick et al. 1981; Brennwald et al. 1994). The carboxy-terminal region of Sec9p is homologous to that of the neuronal SNAP-25 protein, a constituent of the SNARE complex; however, mammalian SNAP-25 cannot complement a *sec9-4* temperature-sensitive mutant. The importance of the carboxy-terminal domain of Sec9p is underscored by data indicating that an amino-terminal truncation of Sec9p can still complement *sec9-4* (Brennwald et al. 1994). However, it will not suppress a cold-sensitive *sec4* allele, in contrast to full-length Sec9p, suggesting that although the carboxy-terminal SNAP-25 domain of Sec9p is critical for Sec9p function, the amino-terminal domain does still have a facilitating role.

Sec9p is tightly associated with the periphery of the plasma membrane and is not found on transport vesicles (Brennwald et al. 1994). Immunofluorescence studies confirm its presence on the plasma membrane but do not reveal a particular association with the tip of growing buds, the site of membrane deposition. In the presence of a cross-linking agent, Sec9p coimmunoprecipitates with Sso1p/Sso2p and Snc1p/Snc2p but not in the presence of hydrolyzable ATP (Couve and Gerst 1994; Brennwald et al. 1994). These data are consistent with the hypothesis that Sso proteins, Snc proteins, and Sec9p form a SNARE complex whose formation is mediated by Sec17p and Sec18p.

SEC1 is an essential gene, and it encodes a 78-kD protein homologous to Sly1p, a constituent of the ER-derived vesicle-docking complex, and to n-Sec1, a component of the synaptic vesicle docking complex (Aalto et al. 1991, 1992; Egerton et al. 1993). Sec1p is required for protein transport between the Golgi and the plasma membrane (Novick and Schekman 1979; Novick et al. 1980). However, although *sec1* represents the first identified *sec* mutant, the role of Sec1p in vesicular transport remains incompletely understood, and further experiments will be required to establish the function of this obviously interesting protein.

5. *Sec2p: Additional Late-acting Factor*

SEC2 is an essential gene, and it encodes a protein with a predicted molecular weight of 84 kD (observed size is 105 kD) that functions at a post-Golgi stage of the secretory pathway (Nair et al. 1990). Alleles of *sec2* account for about half of the secretion-defective mutations isolated in the original *sec* screen, a probable consequence of the fact that carboxy-terminal truncation mutants of *SEC2* are temperature sensitive (Novick et al. 1980; Nair et al. 1990). *SEC2* exhibits strong genetic interactions with *SEC4* and *SEC15* (Nair et al. 1990). The amino-terminal region of Sec2p appears to contain a coiled-coil domain, although the relationship of this structural feature to the function of the protein remains unclear. Much of Sec2p is soluble, and under native conditions, it displays a molecular weight of 500–750 kD. The role of Sec2p in vesicular traffic between the Golgi and the plasma membrane remains the subject of continued investigation.

6. *The Sec6p/Sec8p/Sec15p Complex*

SEC6, *SEC8*, and *SEC15* are all essential genes, and they encode proteins with predicted molecular weights of 85 kD, 122 kD, and 105 kD, respectively (Salminen and Novick 1989; Bowser et al. 1992; Potenza et al. 1992); mammalian homologs of *SEC6* and *SEC8* have been isolated recently (Ting et al. 1995). All three genes were categorized as acting late in the secretory pathway on the basis of the accumulation of post-Golgi 100-nm vesicles in mutants at nonpermissive temperatures (Novick et al. 1980, 1981). Early studies of Sec8p and Sec15p suggested that these proteins were components of a 19.5S complex (Bowser and Novick 1991; Bowser et al. 1992). More recent work has demonstrated that Sec6p, Sec8p, and Sec15p are all components of a large protein complex, 1–2 million kD in size, which also contains five other prominent species whose genes have not yet been identified (TerBush and Novick 1995). This complex resides both in the cytosol and attached to the plasma membrane; both pools appear to have the same subunit composition, and there is no evidence for monomeric forms of Sec6p, Sec8p, or Sec15p.

Although both *SEC8* and *SEC15* exhibit strong genetic interactions with *SEC4*, Sec4p (in contrast to early indications; Bowser et al. 1992) does not appear to be a component of this complex (TerBush and Novick 1995). Moreover, mutations in *SEC4* do not appear to disrupt the formation of this complex. Mutations in *SEC1*, *SEC2*, *SEC8*, *SEC17*, or *SEC18* similarly do not affect the complex. Interestingly, although the protein products of *SEC3*, *SEC5*, and *SEC10* were not found to be components of this complex, mutations in any one of these genes resulted in the disappearance of several different subunits of this complex. Thus, these genes may have an important regulatory role in complex formation. Intriguingly, the Sec6p/Sec8p/Sec15p complex was shown by immunofluorescence studies to localize to the tips of small buds (TerBush and Novick 1995), consistent with a role in vesicle deposition at the plasma membrane surface and in contrast to the more diffuse plasma membrane staining seen for Sso1p/Sso2p and Sec9p (Brennwald et al. 1994). It is possible that the Sec6p/Sec8p/ Sec15p complex plays a critical part in the targeting of Golgi-derived vesicles to the correct location on the plasma membrane (TerBush and Novick 1995).

B. Role of the Actin Cytoskeleton in Vesicle Targeting

In actively growing yeast, both cell surface expansion and protein secretion occur almost exclusively at the developing bud; invertase, acid phosphatase, as well as **a**-agglutinin and other cell wall mannoproteins all initially appear at this location (Tkacz and Lampen 1972, 1973; Farkas et al. 1974; Field and Schekman 1980; Watzele et al. 1988). Immunoelectron microscopy studies indicate an accumulation of Golgi structures just below sites of bud emergence; both Golgi and ER structures appear to enter the daughter cell at an early stage and may account for much of the growth of the developing cell (Preuss et al. 1992).

The yeast actin cytoskeleton, consisting of both actin cables and actin patches, is polarized with respect to the new bud (Adams and Pringle 1984; Kilmartin and Adams 1984; Bretscher et al. 1994). Actin cables are generally oriented parallel to the bud axis and probably represent bundles of actin filaments. Actin patches assemble at the new bud site and then cluster at the tip of the developing bud; several patches are generally seen in the mother cell as well. The localization of both actin cables and actin patches is suggestive of a role in polarized cell growth.

The first concrete evidence linking actin and secretion was the demonstration that *act1* temperature-sensitive mutants accumulate intracellular invertase and amass post-Golgi vesicles (Novick and Botstein 1985). Since the secretion of external invertase is not altered in *act1* mutants, these data were interpreted as consistent with a slowing of vesicle transport to the plasma membrane.

The role of actin in protein transport was further explored by studies of *CAP1* and *CAP2*, which encode the two subunits of the yeast actin capping protein (Amatruda et al. 1992). Yeast either lacking or overexpressing both Cap1p and Cap2p are viable but lack detectable actin cables; an increased number of stained actin patches in the mother is also seen. Despite the lack of actin cables, however, polarized secretion appears to be normal in both null mutants and overexpression mutants. Moreover, *cap* mutants not only secrete invertase at a rate and level indistinguishable from those of wild-type cells, but they also fail to ac cumulate internal invertase, in contrast to the results obtained for *act1* mutants. Unfortunately, ultrastructural analysis for *cap* mutants was not performed, so it is not known whether vesicles accumulate in these cells. Nevertheless, these data suggest either that secretion does not require actin cables or that cables are still present in *cap* mutants but are not detectable by immunofluorescence.

The study of mutants that perturb the cytoskeleton has further defined its role in vesicular transport. For example, the actin cytoskeleton is dramatically disrupted by null alleles of the tropomyosin gene *TPM1* (Liu and Bretscher 1989). Yeast lacking *TPM1* are still viable but accumulate vesicles in the absence of a severe secretory defect (Liu and

Bretscher 1992). *tpm1* deletion mutants secrete invertase normally and also fail to accumulate invertase internally, in contrast to *act1* mutants. The vacuolar protein CPY also is processed normally in *tpm1*Δ mutants. However, **a**-agglutinin, a protein normally deposited at the tips of growing buds, is sometimes seen in scattered patches on *tpm1*Δ mother cells, suggesting a partial defect in localization. Accumulation of secretory vesicles was not observed in an epistasis test of *tpm1*Δ, *sec13*, and *tpm1*Δ, *sec18* double mutants, suggesting that Tpm1p acts later in the secretory pathway than the ER to Golgi step. However, *tpm1*Δ was not synthetically lethal with any of the *sec* mutants examined, including *sec1*, *sec4*, and *sec6*. The role of Tpm1p in vesicular transport is still unclear and remains the subject of continued investigation.

Class V myosins have been implicated in vesicular transport in a variety of organisms, including mouse, chicken, and yeast (Titus 1993). The yeast class V myosin, Myo2p, is essential for yeast cell viability and surface growth (Johnston et al. 1991). *myo2*^{ts} mutants are characteristically large and unbudded, apparently due to an inability to direct membrane growth to a new bud site. *myo2* mutants also exhibit a defective actin cytoskeleton; both actin cables and actin patches appear to be disrupted in these cells. Although *myo2* mutants secrete invertase normally, they also accumulate vesicles; these data were initially interpreted as representing a slowing in vesicle transport from the Golgi to the plasma membrane. However, careful measurements of the rate of protein secretion in *myo2* mutants showed no kinetic defect (Govindan et al. 1995); the cause of vesicle accumulation in *myo2* mutants thus continues to remain a mystery.

Genetic studies indicate that *myo2* is synthetically lethal with many mutations in late-acting *SEC* genes, suggesting a role for Myo2p in the transport of vesicles to the plasma membrane. The accumulation of secretory vesicles in *myo2* mutants is also reminiscent of many late-acting *sec* mutants. However, more detailed ultrastructural analysis of *myo2* mutants has revealed that the vesicles which accumulate at nonpermissive temperatures are concentrated in the mother cell; in contrast, late-acting *sec* mutants accumulate vesicles in arrested buds (Govindan et al. 1995). Epistasis analysis suggests that *myo2* operates at an earlier stage than *sec1* and *sec6*, the two *sec* mutants examined, since vesicles accumulate in the mother cell in *myo2 sec1* and *myo2 sec6* double mutants. At the same time, *myo2* mutants do not bypass the Sec6p-requiring step, as *myo2 sec6* mutants fail to secrete invertase. Significantly, the vesicles that accumulate in *myo2* mutant cells are not the results of endocytosis, since vesicles are still observed in a *myo2 end4* double mutant. *myo2* is also synthetically lethal with *tpm1*Δ, suggesting that both gene products may participate in vesicular transport.

Two high-copy suppressors of *myo2* have been isolated, *SMY1* and *SMY2* (Lillie and Brown 1992, 1994). Overexpression of *SMY2* also suppresses mutations in *SEC22*, *BET1*, *SEC16*, and *SPT15* (which encode the TATA-binding protein) (Lillie and Brown 1994). *SMY2* is nonessential, has no known homologies, and its function is not well understood. *SMY1* is also nonessential, but its protein product bears significant homology with the motor domains of the kinesin superfamily (Lillie and Brown 1992). This is surprising, because kinesins are microtubule-based motors, whereas myosins are actin-based. Furthermore, although microtubules are in a position to play a part in polarized secretion (as they emanate from the spindle pole body and extend into the bud), buds both form and grow normally in cells lacking microtubules (Huffaker et al. 1988; Jacobs et al. 1988).

*smy1*Δ is synthetically lethal with *myo2* mutations, and the products of both genes are localized to sites of active membrane growth, the tips of actively growing buds (Lillie and Brown 1992, 1994). Although both proteins are found in the same general area as actin patches, Smy1p and Myo2p do not appear to colocalize with actin (Lillie and Brown 1994). Both the structural features and the localization of Smy1p and Myo2p suggest that these proteins might act as a vesicular motor, responsible for the directed movement of secretory vesicles toward the growing bud. Curiously, when the distribution of Smy1p and Myo2p was examined in both *sec1* and *sec6* mutants at restrictive temperatures, the localization of both Smy1p and Myo2p was completely lost (Lillie and Brown 1994). These data were interpreted as evidence either that Smy1p and Myo2p are not associated with secretory vesicles or that the concentrated Myo2p/Smy1p staining at bud tips in wild-type cells does not represent groups of fusing secretory vesicles; further investigation will be required to resolve this important question.

A relationship between the secretory pathway and the actin cytoskeleton was also suggested by the isolation of *sac1* mutants, allele-specific loss-of-function suppressors of *act1*^{ts} mutants that can also function as bypass suppressors of a *sec14*Δ mutant (Cleves et al. 1989; Novick et al. 1989; Whitters et al. 1993). Other suppressors of *act1* identified do not suppress *sec14* mutants (Novick et al. 1989; Whitters et al. 1993). *sac1* mutants do not appear to be secretion-defective and do not accumulate vesicles at an appreciable rate (Novick et al. 1989). Genetic analysis reveals that *sac1* mutants weakly suppress *sec6* and *sec9* mutants, but they are synthetically lethal with *sec13* and synthetically harmful with *sec17*, sec*18*, *sec20*, *sec21*, and *sec23* (Cleves et al. 1989). Sac1p is a 71-kD integral membrane protein that appears to reside in both the ER and the Golgi (Whitters et al. 1993). Biochemical experiments indicate that Sac1p is involved in ATP transport into the ER (Mayinger et al. 1995), but the relationship of the function of Sac1p in either the late secretory pathway or the cytoskeleton has yet to be established.

Finally, as a caveat, it must be emphasized that the polarized movement of post-Golgi transport vesicles to the nascent bud site does not imply that these vesicles are solely responsible for membrane growth. For example, in mammalian cells, the transport of phosphatidylcholine and phosphatidylethanolamine to the plasma membrane was found to occur independently of metabolic energy, suggesting a mechanism distinct from vesicular flow (Sleight and Pagano 1983; Kaplan and Simoni 1985). Furthermore, in yeast, the transport of phosphatidylinositol and phosphatidylcholine to the plasma membrane was unaffected in secretion mutants representing successive steps of the pathway (Daum et al. 1986; Gnamusch et al. 1992; Paltauf et al. 1992). Thus, vesicular flow may constitute a necessary, but not sufficient, condition for membrane expansion, and the delivery of phospholipids and sterols to the growing plasma membrane may rely on alternative mechanisms of transport (see Paltauf et al. 1992).

VIII. ENDOCYTOSIS

A. Endocytosed Proteins

A number of different plasma membrane proteins are taken up by endocytosis and delivered to the vacuole. In some cases, these proteins are endocytosed constitutively in the absence of an external signal. Examples of constitutive endocytosis include the uptake of the mating pheromone receptors Ste2p and Ste3p in the absence of their ligands (Konopka et al. 1988; Reneke et al. 1988; Davis et al. 1993), the endocytosis of the **a**-factor transporter, Ste6p (Berkower et al. 1994; Kölling and Hollenberg 1994), and the internalization of a multidrug resistance protein, Pdr5p (Egner et al. 1995). Constitutive endocytosis of solutes also occurs, as demonstrated by the uptake of a fluorescent dye, lucifer yellow (Riezman 1985).

Receptor-mediated (or ligand-stimulated) endocytosis refers to the case where receptor and ligand are internalized in response to ligand binding. For example, when the mating pheromone α-factor is applied to *MAT***a** cells, the α-factor receptor Ste2p is taken up much more rapidly than in the absence of mating pheromone, and both the mating pheromone and its receptor are degraded in the vacuole (Chvatchko et al. 1986; Jenness and Spatrick 1986; Rohrer et al. 1993; Schandel and Jenness 1994). Internalization and degradation of the **a**-factor receptor Ste3p are similarly increased by the presence of **a**-factor (Davis et al. 1993). The carboxyl terminus of Ste3p is required for constitutive, but not ligand-stimulated, endocytosis, indicating that the two processes involve different parts of the receptor protein (Davis et al. 1993).

Endocytosis can be induced by environmental signals other than mating pheromones. Several plasma membrane permeases, including the uracil permease Fur4p, the general amino acid permease Gap1p, an inositol permease Itr1p, and the maltose transporter Mal1p, are turned over more rapidly under some growth conditions than under others (see Volland et al. 1994; Hein et al. 1995; Lai et al. 1995; Riballo et al. 1995). Degradation of some of these permeases occurs in the vacuole and is dependent on a functional endocytic pathway (Volland et al. 1994; Lai et al. 1995; Riballo et al. 1995).

B. Signals for Endocytosis

Proteins destined for endocytosis must have features that distinguish them from stable plasma membrane proteins. In mammalian cells, short peptide sequences in the cytosolic domain of proteins act as endocytosis signals. The known signals include a tyrosine-containing motif that directs the accumulation of receptors into clathrin-coated pits (Trowbridge et al. 1993), a dileucine (LL) motif (Letourneur and Klausner 1992), and a motif similar to the KKXX recycling signal (Itin et al. 1995). As yet, involvement of similar peptide signals in endocytosis in yeast has not been demonstrated.

A novel endocytic signal, ubiquitin, has been identified in *S. cerevisiae*. Ubiquitin is a 76-amino-acid protein that is covalently linked to lysine residues of target proteins by the action of ubiquitin-activating enzymes, ubiquitin-conjugating enzymes, and in some cases ubiquitin-protein ligases. Ubiquitination acts as a signal that regulates the fate of target proteins. The classical role of ubiquitin is to target proteins for degradation by the proteasome. Recent work on endocytosis in yeast suggests that ubiquitin can also tag proteins for endocytosis and subsequent degradation in the vacuole.

The role of ubiquitin in endocytosis has been shown most convincingly for α-factor-stimulated endocytosis of Ste2p (Hicke and Riezman 1996). A mutant defective in endocytic uptake (*end4*) accumulates ubiquitinated Ste2p. Ubiquitination is important for endocytosis of Ste2p, since mutants defective in ubiquitin-conjugating enzymes, most notably *ubc4*, block the uptake of α-factor (Hicke and Riezman 1996). Mutagenesis of the cytosolic domain of Ste2p identified five amino acids (DAKSS) that are necessary and sufficient for ligand-stimulated endocytosis (Rohrer et al. 1993). The central lysine in this motif is ubiquitinated in response to α-factor, and mutation of this lysine to arginine blocks both ubiquitination and α-factor internalization (Hicke and Riezman 1996). Mutations in adjacent serine residues both diminish α-factor internalization and block ubiquitination of Ste2p. Ubiquitin targets Ste2p to the vacuole rather than the proteasome, since mutations in vacuolar proteases, but not in proteasome subunits, block degradation of Ste2p (Hicke and Riezman 1996).

A growing body of evidence indicates that ubiquitin is a general signal for endocytosis of plasma membrane proteins. The **a**-factor receptor, Ste3p, has a sequence similar to the DAKSS motif of Ste2p (Rohrer et al. 1993) and may therefore be similarly ubiquitinated and endocytosed. Two constitutively endocytosed proteins, Ste6p and Pdr5p, accumulate in a ubiquitinated form in endocytosis mutants (Kölling and Hollenberg 1994; Egner and Kuchler 1996), and rapid degradation of Ste6p depends on the ubiquitin-conjugating enzymes Ubc4p and Ubc5p (Kölling and Hollenberg 1994). Constitutive endocytosis of Ste2p may also be due to ubiquitination, since some ubiquitinated Ste2p is present even in the absence of α-factor (Hicke and Riezman 1996). Furthermore, ubiquitination may function in the regulated uptake of permeases: The inositol-induced degradation of Itr1p is blocked by mutations in *ubc4* and *ubc5* (Robinson et al. 1996), and the rapid turnover of Gap1p and Fur4p is dependent on a ubiquitin-protein ligase, Rsp5p (Hein et al. 1995).

In contrast to endocytosis signals encoded in protein sequence, ubiquitin addition and removal can be controlled, giving additional capacity for regulation. The events that cause ubiquitination are still only poorly understood. In mammalian cells, ubiquitination of the signal transduction component IκBα occurs after phosphorylation (Hochstrasser 1996). Since Ste2p is hyperphosphorylated on serine residues upon ligand binding (Reneke et al. 1988), it has been proposed that phosphorylation of Ste2p triggers ubiquitination (Hochstrasser 1996). Other endocytosed proteins, such as Fur4p, are also phosphorylated (Volland et al. 1992, 1994). Some proteins that are ubiquitinated and degraded in a regulated manner, most notably cyclins, contain a nine-amino-acid sequence (termed "destruction box") which is required for ubiquitination and can confer ubiquitination on a test protein to which this sequence is fused, indicating that the destruction box sequence specifies ubiquitination (Glotzer et al. 1991). A sequence similar to the cyclin destruction box is present in the uracil permease, Fur4p, and a single-amino-acid substitution within this sequence significantly decreases regulated degradation of Fur4p (Galan et al. 1994).

C. Internalization at the Plasma Membrane

1. Role of Clathrin

Proteins to be endocytosed must be incorporated into endocytic vesicles. The paradigm for the formation of endocytic vesicles is the assembly of clathrin-coated vesicles on the plasma membrane of mammalian cells. As discussed above (see Section VI), clathrin-coated vesicles consist of clathrin heavy- and light-chain molecules that associate with the plasma membrane, along with a heteromeric adaptor complex consisting of four subunits. At the plasma membrane in mammalian cells, these four subunits are termed γ, β2, μ2, and σ2 adaptins (Robinson 1994). Proteins to be endocytosed are recruited to clathrin-coated pits by binding of their cytosolic domains to this adaptor complex (Trowbridge et al. 1993; see also Section VI). Coated pits mature into coated vesicles that pinch off the plasma membrane in a process dependent on the GTPase dynamin. Although the clathrin-mediated pathway is the best-understood endocytic pathway, clathrin-independent pathways also exist in mammalian cells (Lamaze and Schmid 1995).

The role of clathrin in endocytosis has been investigated through the use of mutants (Payne et al. 1988; Tan et al. 1993). Mutants in the clathrin heavy chain (*chc1* deletion or *chc1^{ts}*) show a twofold reduction in the internalization of α-factor and are defective in the uptake of the **a**-factor receptor, Ste3p (Payne et al. 1988; Tan et al. 1993), indicating that clathrin is important for endocytosis in yeast. However, since significant amounts of α-factor and Ste3p are internalized even in the absence of clathrin function, a clathrin-independent pathway must also exist. The uptake defects of *chc1^{ts}* mutants for both α-factor and Ste3p appeared rapidly after shift to the restrictive temperature, suggesting that the requirement for clathrin is direct (Tan et al. 1993).

Yeast genes homologous to mammalian plasma membrane AP complex subunits have been identified by homology with their mammalian counterparts (Phan et al. 1994; Rad et al. 1995). Deletions of either *APL1* or *APL2* (homologs of the mammalian β1 and β2 AP subunits) do not affect the internalization of α-factor, nor do they exacerbate the uptake defect of the *chc1^{ts}* mutant (Rad et al. 1995). Similarly, deletion of

APS2/YAP17 (a homolog of the σ2 AP subunit) has no detectable effect on the uptake of α-factor, even in a *chc1[ts]* mutant background (Phan et al. 1994). This lack of an effect of *apl* and *aps* mutations may be a consequence of functional redundancy among the numerous AP subunit homologs that are found in yeast. The phenotypes of single and multiple mutations in additional AP subunits may be revealing.

Three dynamin-like proteins, Vps1p, Dnm1p, and Mgm1p, have been identified in yeast. Vps1p is required for sorting at the *trans*-Golgi compartment (see Section VI), and Dnm1p has been implicated in endocytosis after the initial uptake step (see below). Deletion of either *VPS1* or *DNM1* has no obvious effect on the uptake of α-factor (Munn and Riezman 1994; Gammie et al. 1995; Nothwehr et al. 1995). Mgm1p was identified because of its role in mitochondrial genome maintenance, and its role in endocytosis has not yet been examined (Guan et al. 1993).

2. *Role of the Actin Cytoskeleton*

Multiple lines of evidence underscore the importance for endocytosis of actin and proteins related to the actin cytoskeleton. Temperature-sensitive alleles of actin (*act1-1* and *act1-2*) are completely defective in internalization of α-factor and of the fluid-phase endocytic marker, lucifer yellow (Kübler and Riezman 1993), and a new allele of actin (*end7-1*) was identified independently in a screen for mutants defective in α-factor uptake (Munn et al. 1995). Proteins with functions that can be related to the actin cytoskeleton are also important for endocytic uptake. Deletion of the actin bundling protein fimbrin (Sac6p) eliminates α-factor uptake (Kübler and Riezman 1993). Moreover, a block in α-factor internalization is caused by temperature-sensitive mutations in calmodulin (Kübler et al. 1994), a protein that partially colocalizes with actin and is required for maintenance of the actin cytoskeleton (Brockerhoff and Davis 1992). Furthermore, deletion of a type I myosin, Myo5p, causes a strong defect in α-factor internalization at elevated temperatures (Geli and Riezman 1996). Myo5p may functionally overlap with a second type I myosin, Myo3p: In a strain deleted for Myo3p, a temperature-sensitive mutation, *myo5-1*, causes a significant α-factor uptake defect even at 24°C (Geli and Riezman 1996). Not every cytoskeletal component is required for endocytic uptake: Strains bearing mutations affecting β-tubulin, the actin-binding protein Abp1p, or the type V myosin Myo2p show normal uptake of α-factor (Kübler and Riezman 1993; Kübler et al. 1994).

Four novel proteins, End3p, Sla2p/End4p, Vrp1p/End5p, and Rvs161p/End6p, were identified by screening a collection of temperature-sensitive mutants for defects in α-factor internalization (Raths et al. 1993; Munn et al. 1995). A fifth protein, Rvs167p, is homologous to End6p and was found to be also required for endocytic uptake (Munn et al. 1995). These five proteins also appear to be important for the proper function of the actin cytoskeleton since mutations in the corresponding genes affect actin organization. End3p contains sequence motifs that are commonly found in actin-binding proteins, such as an EF-hand Ca^{++}-binding site, a consensus sequence for binding phosphatidylinositol 4,5-biphosphate (PIP_2) and two repeated domains with sequence similarity to α-actinin (Bénédetti et al. 1994). One of the two repeated domains, but not the EF-hand motif and the PIP_2-binding site, is required for endocytic function of End3p (Bénédetti et al. 1994). The carboxy-terminal domain of Sla2p has homology with talin, a protein that links actin to plasma membrane proteins (Holtzman et al. 1993), and deletion of *SLA2* is synthetically lethal with mutations in the actin-binding protein Abp1p and the fimbrin homolog Sac1p (Holtzman et al. 1993). Mutations in *end6* fail to complement *ACT1* mutations (Munn et al. 1995), and the End6p homolog Rvs167p has been identified as an actin-binding protein (Amberg et al. 1995).

The function of the actin cytoskeleton in endocytosis has not yet been resolved at the molecular level. The internalization defect of temperature-sensitive mutants appears rapidly after shift to the nonpermissive temperature, suggesting that the requirement for the actin cytoskeleton is direct (Kübler and Riezman 1993; Bénédetti et al. 1994; Kübler et al. 1994; Munn and Riezman 1994; Munn et al. 1995). Possibly, the actin cytoskeleton provides the mechanical force necessary for the formation or pinching off of endocytic vesicles (Kübler and Riezman 1993). All of the mutants discussed above are completely defective in both α-factor internalization and fluid-phase endocytosis, indicating that the actin requirement may be a general feature of endocytic uptake in yeast. Actin appears to be similarly required for endocytosis in mammalian cells, where actin has been implicated in clathrin-mediated endocytosis of both plasma membrane proteins and fluid-phase markers (Sandvig and van Deurs 1990; Gottlieb et al. 1993). An intriguing link between the actin cytoskeleton and clathrin is suggested by the observation that calmodulin binds to the clathrin light chain (Silveira et al. 1990).

3. *Additional Functions Required for Endocytic Uptake*

Four additional genes, *END8*, *END9*, *END10*, and *END11*, required for uptake of α-factor and endocytosis of lucifer yellow were identified in a screen for mutants that render vacuolar H^+ ATPase defects lethal (Munn and Riezman 1994). The rationale for this screen was the observation that the cell depends on a functional endocytic pathway when acidification of the vacuole is compromised (Nelson and Nelson 1990; Munn and Riezman 1994). The identified alleles of *END8-11* are temperature-sensitive for growth, but the corresponding genes have not yet been cloned. It should be noted that additional classes of endocytic uptake mutants may exist since only those endocytosis mutants that are also temperature-sensitive for growth have been studied thus far.

D. Delivery of Internalized Proteins to the Vacuole

The intracellular events that occur after endocytic uptake are poorly understood. In mammalian cells, endocytosed proteins are delivered to early endosomes and then transported to late endosomes probably by vesicular transport (Gruenberg and Maxfield 1995). The early endosome is thought to be a recycling compartment from which proteins can be returned to the cell surface; the late endosome is likely identical to the prevacuolar compartment that also contains proteins from the *trans*-Golgi compartment destined for the vacuole (Gruenberg and Maxfield 1995)

Endocytosed proteins in yeast also traverse at least two separate compartments that may correspond to early and late endosomes, as demonstrated in cell-fractionation experiments using radioactively labeled α-factor (Singer-Krüger et al. 1993). These compartments are distinct from the plasma membrane and the vacuole and consist mostly of irregularly shaped, translucent vesicles with a diameter of 100–400 nm (Singer-Krüger et al. 1993).

Little is known about how vesicles form at the early endosome. If the early endosome is located before the intersection of the endocytic and vacuolar transport pathways, genes involved in vesicle formation at the early endosome should affect endocytosis but not vacuolar protein sorting. The dynamin homolog Dnm1p is likely to act at this step since *dnm1* mutants block endocytosis after internalization but do not affect vacuolar protein sorting (Gammie et al. 1995). Dnm1p, similar to the dynamin homolog Vps1p (see Section VI), lacks the proline-rich domain of mammalian dynamins and has only limited homology with dynamin in a pleckstrin homology domain (Gammie et al. 1995). Cells deleted for *DNM1* are viable and are not temperature-sensitive for growth (Gammie et al. 1995). Deletion of *DNM1* slows both the constitutive and the ligand-mediated delivery of Ste3p to the vacuole, whereas uptake of α-factor or endocytosis of lucifer yellow is not affected in *dnm1* deletion strains (Gammie et al. 1995). It could be that Dnm1p acts together with clathrin in the formation of coated vesicles at the early endosome. However, no compelling evidence exists that clathrin acts at this step since *chc1*^{ts} mutants do not affect the delivery of internalized Ste3p or internalized α-factor to the vacuole (Tan et al. 1993).

In mammalian cells, a subset of coatomer components (α-, β-, β′-, ε-, and ζ-COP, but not γ- and δ-COP) is associated with early endosomes, and at least β-COP is required for formation of transport vesicles from this compartment (Whitney et al. 1995; Aniento et al. 1996). The role of these coatomer subunits in endocytosis in yeast has not yet been examined.

The late endosomal compartment is probably the same as the prevacuolar compartment that also contains Golgi-derived vacuolar proteins (see Section VI). The class-E subset of *vps* mutants accumulates Golgi proteins and endocytosed Ste3p in a compartment adjacent to the vacuole that is thought to be an exaggerated form of the prevacuolar compartment (Raymond et al. 1992; Piper et al. 1995). Since the class-E compartment appears rapidly after shift to the nonpermissive temperature and since the accumulation of both Golgi proteins and Ste3p is reversible, the class-E compartment is likely a physiological intermediate in the endocytic and vacuolar targeting pathway (Piper et al. 1995). The prevacuolar compartment has been purified by subcellular fraction of wild-type cells, and it contains structures with a morphology similar to that of the early and late endosomes described above and cofractionates with internalized α-factor (Vida et al. 1993).

Additional evidence for a compartment common to the endocytic and vacuolar targeting pathways comes from the analysis of mutants defective in the delivery of endocytosed proteins to the vacuole. Three mutants, *end12*, *end13*, and *ren1*, that are defective in the delivery of α-factor to the vacuole are allelic to mutants defective in vacuolar protein sorting, *vps34*, *vps4*, and *vps2* (Davis et al. 1993; Munn and Riezman 1994), consistent with a common endocytic and vacuolar sorting pathway. *vps2* and *vps4* mutants accumulate the class-E compartment described above (Raymond et al. 1992) and may therefore be defective in vesicle formation from the prevacuolar/endosomal compartment. Although a large number of vacuolar protein sorting mutants have been identified, in most cases, it is unknown whether they also affect the endocytic pathway. It is likely that many more endocytosis mutants will emerge from the study of vacuolar targeting mutants.

Genes required for the delivery of endocytosed proteins to the vacuole have also been identified as homologs of Rab-like GTPases.

Four members of the Rab family of GTP-binding proteins, *YPT7*, *YPT51*, *YPT52*, and *YPT53,* were identified by PCR or genomic sequencing and were found to be required for transport of internalized α-factor to the vacuole (Wichmann et al. 1992; Singer-Krüger et al. 1994). All four Rab homologs are also required for the correct sorting of vacuolar proteases (Wichmann et al. 1992; Singer-Krüger et al. 1994). Ypt51p, Ypt52p, and Ypt53p are homologous to Rab5, a mammalian protein that localizes to early endosomes. The three proteins form a functionally overlapping group: Deletion of *YPT51* has the most severe endocytosis defect, and *ypt52* and/or *ypt53* mutations exacerbate the defects of *ypt51* mutants (Singer-Krüger et al. 1994). *YPT51* is identical to *VPS21*, a gene identified as a vacuolar protein-sorting mutant (Horazdovsky et al. 1994; Singer-Krüger et al. 1994); deletion of *vps21/ypt51* leads to accumulation of vesicles, consistent with a role for this protein in vesicle fusion (Horazdovsky et al. 1994; Singer-Krüger et al. 1994). Ypt7p is homologous to Rab7, a mammalian protein that localizes to late endosomes (Wichmann et al. 1992).

The stage in the endocytic pathway where Ypt51p and Ypt7p act was determined by cell fractionation. Cells deleted for *ypt51* accumulate internalized α-factor in a compartment with a density similar to that of early endosomes (Singer-Krüger et al. 1995), whereas cells deleted for *ypt7* accumulate internalized α-factor in a compartment with the properties of the late endosomes (Schimmöller and Riezman 1993). Given that Rab proteins are thought to generally function in vesicle fusion, these results are consistent with a requirement for Ypt51p, Ytp52p, and Ypt53p in fusion of vesicles derived from the early endosome with the prevacuolar/ late endosomal compartment and a requirement for Ypt7p in fusion of late endosome-derived vesicles with the vacuole.

IX. CONCLUSIONS

This review documents the enormous progress that has been made in the last decade in understanding the molecular mechanisms that underlie the secretory pathway. As in many areas of cell biology, the greatest strides have been made in areas where studies in yeast and in mammalian cells have reinforced each other. Genetic screens have yielded well over 100 yeast genes whose products act in the secretory or endocytic pathways. In most cases, it has been possible to define the precise site of action of these gene products through the use of detailed physiological assays. Robust cell-free assays have been developed for the processes of protein translocation through the membrane of the ER and for vesicle budding and fusion in ER to Golgi transport. These assays have defined the biochemical activities for a number of proteins as well as the specific requirements for nucleotide hydrolysis.

Key principles that have emerged from the examination of the secretory pathway of yeast and mammalian cells are (1) the final address of proteins that enter the secretory pathway is largely specified by short peptide sequences within the protein that determine interaction with particular components of the secretory machinery, (2) proteins are carried between organelles by transport vesicles, which bud from a donor organelle on assembly of a cytosolic protein coat, and which fuse with an acceptor organelle on assembly of a complex composed of proteins from both vesicle and acceptor membranes, and (3) the function of the macromolecular assemblies along the secretory pathway is usually coupled to nucleotide hydrolysis by the action of a recognizable ATPase or GTPase.

Important tools for future work—useful mutants, biochemical assays, and cell biological markers—are now available to allow every aspect of the function of the secretory pathway to be explored in greater detail. The next decade promises to yield new insights into the structure and regulation of the fundamental macromolecular machines of secretion: the translocation apparatus and transport vesicle coats. In addition, our understanding of the basic components of the secretory pathway now provides the opportunity to attack the more subtle questions of how selective sorting of cargo molecules occurs along the secretory pathway and how the function of the secretory pathway is integrated with cell growth and morphology.

ACKNOWLEDGMENTS

We would like to acknowledge Randy Schekman, whose pioneering work on many aspects of the yeast secretory pathway opened the way for much of the work described in this chapter. His high standards for rigor, imagination, and scientific openness have set the tone for the field and have been an inspiration for those of us who have worked with him. We also thank Greg Payne for help in identifying clathrin AP subunits, our colleagues in the Kaiser laboratory for valuable discussions, and Anne Paulin for preparation of the manuscript. Finally, we gratefully acknowledge the National Institutes of Health, the Lucille P. Markey Charitable Trust, and the Searle Scholars Program for supporting research in our laboratory.

Table 1 Yeast secretion genes

Protein	Homolog/Function	Properties	References
Cytosolic Factors for Translocation into the ER			
Srp54p	SRP subunit	60-kD hydrophilic protein; GTPase domain	Hann et al. (1989)
Sec65p	SRP subunit	19-kD hydrophilic protein required for assembly of Srp54p into SRP complex	Stirling et al. (1992)
Srp14p, Srp21p, Srp68p, Srp72p	SRP subunit	14-, 21-, 68-, 72-kD hydrophilic proteins; no known mammalian equivalent of Srp21p	Brown et al. (1994)
scR1	SRP RNA		Felici et al. (1989); Hann and Walter (1991)
Ssa1p-Ssa4p	Hsp70/DnaK homologs	four functionally redundant 70-kD cytosolic proteins	Werner-Washburne et al. (1987); Deshaies et al. (1988)
Ydj1p	DnaJ homolog	44-kD cytosolic protein, associates with Ssa1p-Ssa4p	Cyr et al. (1992)
Membrane Proteins for Translocation into the ER			
Srp101p	α-subunit of SRP receptor	69-kD peripheral membrane protein; GTPase domain	Ogg et al. (1992)
SRβ/YKL154w	β-subunit of SRP receptor	30-kD type II membrane protein; GTPase domain	Miller et al. (1995); Cherry et al. (*Saccharomyces* Genome Database, 1996)
Sec61p	translocation pore; homolog of *E. coli* SecY	53-kD polytopic integral membrane protein	Rothblatt et al. (1989); Stirling et al. (1992)
Sss1p	Sec61p complex subunit-γ	9-kD peripheral membrane protein	Esnault et al. (1993)
Sbh1p	Sec61p complex subunit-β	9-kD peripheral membrane protein	Panzner et al. (1995)
Sec63p	subunit of Sec63p complex	75-kD integral membrane protein; homology with *E. coli* DnaJ	Rothblatt et al. (1989); Deshaies et al. (1991)

Sec62p	subunit of Sec63p complex	32-kD integral membrane protein	Deshaies and Schekman (1989); Rothblatt et al. (1989)
Sec66p/Sec71p	subunit of Sec63p complex	31.5-kD type II integral membrane glycoprotein	Feldheim et al. (1993); Kurihara and Silver (1993)
Sec72p	subunit of Sec63p complex	22-kD peripheral membrane protein	Fang and Green (1994); Feldheim and Schekman (1994)
Scj1p	DnaJ homolog	45-kD luminal ER membrane protein, associates with Kar2p	Schlenstedt et al. (1995)
Kar2p	Hsp70/DnaK homolog (BiP)	78-kD luminal protein; ATPase domain; carboxy-terminal HDEL retrieval sequence	Rose et al. (1989)
Sac1p		71-kD integral membrane protein; located in ER and Golgi; ATP transporter	Novick et al. (1989); Mayinger et al. (1995)
Protein Maturation in the ER			
Sec11p	signal peptidase subunit	18-kD hydrophilic protein	Bohni et al. (1988)
Pdi1p	protein-disulfide isomerase	65-kD luminal glycoprotein; two active sites; carboxy-terminal HDEL retrieval sequence	LaMantia et al. (1991)
Eug1p	protein-disulfide isomerase	65-kD luminal glycoprotein; two active sites; carboxy-terminal HDEL retrieval sequence	Tachibana and Stevens (1992)
Mpd1p	protein-disulfide isomerase	36-kD luminal glycoprotein; one active site; carboxy-terminal HDEL retrieval sequence	Tachikawa et al. (1995)
Ire1p	regulator of luminal ER proteins	127-kD type I integral membrane protein; cytosolic Ser/Thr kinase domain	Cox et al. (1993); Mori et al. (1993)
Cne1p	calnexin	76-kD type I glycoprotein	Parlati et al. (1995)
Shr3p	assembly or transport of permeases	24-kD integral membrane protein	Ljungdahl et al. (1992)

(*Continued on following pages.*)

Table 1 (*Continued.*)

Protein	Homolog/Function	Properties	References
Vph2p/Vma12p	assembly of V-ATPase	25-kD integral membrane protein	Hirata et al. (1993)
Vma21p	assembly of V-ATPase	8.5-kD integral membrane protein; carboxy-terminal KKXX retrieval sequence	Hill and Stevens (1994)
Vma22p	assembly of V-ATPase	21-kD peripheral membrane protein	Hill and Stevens (1995)
Vesicle Formation at the ER			
Sec13p	COPII coat component	33-kD peripheral membrane protein present on ER to Golgi transport vesicles; WD40 repeats; forms a 700-kD complex with Sec31p	Pryer et al. (1993)
Sec31p	COPII coat component	150-kD hydrophilic protein present on ER to Golgi transport vesicles; WD40 repeats; forms a 700-kD complex with Sec13p	Salama et al. (1993); Wuestehube et al. (1996)
Sec23p	COPII coat component; GAP for Sar1p	84-kD peripheral membrane protein; located on ER to Golgi transport vesicles; forms an ~300-kD complex with Sec24p	Hicke and Schekman (1989); Hicke et al. (1992); Yoshihisa et al. (1993)
Sec24p	COPII coat component	105-kD hydrophilic protein; located on ER to Golgi transport vesicles; forms a ~300-kD complex with Sec23p	Hicke et al. (1992)
Sec16p	COPII coat component	240-kD peripheral ER membrane protein; located on ER to Golgi transport vesicles	Espenshade et al. (1995)
Sar1p	COPII coat component; Sar family GTPase	21-kD peripheral ER membrane protein; located on ER to Golgi transport vesicles	Nakano and Muramatsu (1989); Nishikawa and Nakano (1991)
Sec12p	exchange factor for Sar1p;	52-kD type II integral ER membrane protein	Nakano et al. (1988); d'Enfert et al. (1991b);

	homologous to Sed4p		Barlowe and Schekman (1993)
Sed4p	homologous to Sec12p	114-kD type II integral ER membrane protein; HDEL retrieval sequence; cytosolic domain binds to Sec16p	Hardwick et al. (1992); Gimeno et al. (1995)
Unclassified Proteins Required for ER to Golgi Transport			
Sec20p		44-kD type II integral ER membrane protein; HDEL retrieval sequence; binds to Tip1p	Sweet and Pelham (1992, 1993)
Tip1p		80-kD hydrophilic protein; binds to Sec20p; depends on Sec20p for ER binding	Sweet and Pelham (1993)
Sorting in the ER			
Emp24p/Bst2p		24-kD type I integral ER membrane protein; located in COPII vesicles; deletion suppresses *sec13*Δ	Schimmöller et al. (1995); Stamnes et al. (1995); Elrod-Erickson and Kaiser (1996)
Erv25p/Bst3p		25-kD type I integral ER membrane protein; located in COPII vesicles; binds to Emp24p; deletion suppresses *sec13*Δ	Belden and Barlowe (1996); Elrod-Erickson and Kaiser (1996)
Bst1p		118-kD polytopic integral membrane protein; located in the ER; XXRR retrieval sequence; deletion suppresses *sec13*Δ	Elrod-Erickson and Kaiser (1996)
Membrane Proteins Involved in ER Vesicle Docking or Fusion			
Bet1p/Sly12p	v-SNARE; synaptobrevin family	18-kD integral membrane protein; located in ER and possibly ER-derived transport vesicles	Newman et al. (1990); Dascher et al. (1991)
Bos1p	v-SNARE; synaptobrevin family	27-kD integral membrane protein; located in ER and ER-derived transport vesicles	Shim et al. (1991)

(*Continued on following pages.*)

Table 1 (*Continued.*)

Protein	Homolog/Function	Properties	References
Sec22p/Sly2p	v-SNARE; synaptobrevin family	25-kD integral membrane protein; located in ER and ER-derived transport vesicles	Ossig et al. (1991); Dascher et al. (1991); Newman et al. (1992a)
Ykt6p	v-SNARE; synaptobrevin family	23-kD hydrophilic protein; carboxy terminal farnesylation site	Søgaard et al. (1994); Cherry et al. (*Saccharomyces* Genome Database, 1996)
Sed5p	t-SNARE; syntaxin family	39-kD integral membrane protein; may form homodimers; located in Golgi	Hardwick and Pelham (1992)
Sft1p	t-SNARE	11-kD integral membrane protein; located in Golgi; has coiled-coil structure	Banfield et al. (1995)
Cytosolic Proteins Involved in ER Vesicle Docking or Fusion			
Sec17p	mammalian α-SNAP; general fusion factor	33-kD peripheral membrane protein	Griff et al. (1992)
Sec18p	mammalian NSF; general fusion factor	84-kD cytosolic/peripheral membrane protein; ATPase	Eakle et al. (1988); Wilson et al. (1989)
Sly1p	Sec1p	75-kD cytosolic protein; tightly associated with Sed5p	Ossig et al. (1991)
Uso1p	docking/fusion factor; homolog of p115/TAP	206-kD cytosolic (and possibly peripheral membrane) protein; homodimer or homotrimer; has coiled-coil domain	Nakajima et al. (1991)
Bet3p	docking/fusion factor	22-kD hydrophilic protein	Rossi et al. (1995)
Sec7p	docking/fusion factor	230-kD hydrophilic protein; contains highly charged, acidic domain near amino terminus; soluble and associated with cytosolic face of Golgi	Achstetter et al. (1988); Franzusoff et al. (1991)

Small GTP-binding Proteins and Related Factors Involved in Vesicle Docking or Fusion			
Ypt1p	Rab family GTPase	23-kD peripheral membrane protein; geranylgeranylated; located in ER-derived transport vesicles and the Golgi	Gallwitz et al. (1983); Segev (1991)
Sec4p	Rab family GTPase	24 kD; geranylgeranylated; located in post-Golgi vesicles and the plasma membrane	Salminen and Novick (1987)
Bet2p	β-subunit of component B of mammalian GGTase II; Dpr1p/Ram1p	37-kD hydrophilic protein	Rossi et al. (1991)
Bet4p	α-subunit of component B of mammalian GGTase II; Ram2p	35-kD hydrophilic protein; initially misidentified as Mad2p	Li et al. (1993, 1994); Jiang et al. (1993)
Mrs6p	component A of mammalian GGTase II; Rab escort protein (REP)	67-kD protein	Waldherr et al. (1993); Jiang and Ferro-Novick (1994)
Gdi1p/Sec19p	GDP-dissociation inhibitor; Rab3a GDI	51-kD protein; not localized, but by analogy to mammalian Rab-GDI, expected to be cytosolic	Garrett et al. (1994)
Dss4p	GDP-dissociation stimulator; Mss4p	17-kD hydrophillic protein associated with 100,000*g* pellet	Moya et al. (1993)
Sec4p-GAP	GTPase-activating protein	membrane-associated activity; protein not identified	Walworth et al. (1992)

(*Continued on following pages.*)

Table 1 (*Continued.*)

Protein	Homolog/Function	Properties	References
Retention or Retrieval of ER proteins			
(A) *COPI coat components*			
Ret1p/Sec33p	α-COP; coatomer coat component	136-kD hydrophilic protein; forms part of the 700-kD coatomer complex; four WD40 repeats near the amino terminus	Letourneur et al. (1994); Gerich et al. (1995); Wuestehube et al. (1996)
Sec26p	β-COP; coatomer coat component	109-kD hydrophilic protein; forms part of the 700-kD coatomer complex	Duden et al. (1994)
Sec27p	β′-COP; coatomer coat component	99-kD hydrophilic protein; forms part of the 700-kD coatomer complex; five WD40 repeats near the amino terminus	Duden et al. (1994)
Sec21p	γ-COP; coatomer coat component	105-kD hydrophilic protein; forms part of the 700-kD coatomer complex	Hosobuchi et al. (1992)
Ret2p	δ-COP; coatomer coat component	61-kD hydrophilic protein; homology with medium and small chains of clathrin adaptor complexes	Cosson et al. (1996)
YIL076W	ε-COP homolog	42-kD protein; function unknown	Cherry et al. (*Saccharomyces* Genome Database, 1996)
Ret3p	ζ-COP: coatomer coat component	21-kD hydrophilic protein; homology with medium and small chains of clathrin adaptor complexes	Cosson et al. (1996)
Arf1p,Arf2p	Arf family GTPase; coatomer coat component	21-kD, *N*-myristoylated proteins	Stearns et al. (1990b); Kahn et al. (1995)
Gea1p,Gea2p	putative nucleotide exchange factors for ARF; homologous to human ARF exchange protein, ARNO	159-kD, 166-kD proteins, respectively; 50% identical to each other; central region of each is homologous to a domain of Sec7p	Peyroche et al. (1996)

(B) *Other proteins implicated in retrieval from the Golgi apparatus*[a]			
Erd2p	Erd2p family; HDEL receptor	26-kD polytopic Golgi membrane protein	Semenza et al. (1990)
Ufe1p	t-SNARE; syntaxin family	41-kD integral ER membrane protein	Lewis and Pelham (1996)
Rer1p	retention of Sec12p	22-kD polytopic membrane protein	Nishikawa and Nakano (1993); Boehm et al. (1994)
Transport through the Golgi			
Sec14p	PI/PC transfer protein	35-kD peripheral membrane protein; located in Golgi; inhibitor of CCTase	Bankaitis et al. (1989)
Sft1p	t-SNARE located in Golgi	11-kD integral membrane protein; coiled-coil motif;	Banfield et al. (1995)
Pmr1p	Ca^{++} ATPase	105-kD polytopic integral membrane protein	Rudolph et al. (1989); Antebi and Fink (1992)
Mnn9p	Golgi function	46-kD type II membrane protein	Ballou et al. (1986); Yip et al. (1994)
Van1p	Golgi function	integral membrane glycoprotein; 61-kD apoprotein	Kanik-Ennulat and Neff (1990)
Van2p/Vrg4p	Golgi function	37-kD integral membrane protein	Kanik-Ennulat et al. (1995); Poster and Dean (1996)
Anp1p	Golgi function	58-kD type II integral membrane protein; located in the ER	Chapman and Munro (1994)
Erd1p	Golgi function	43-kD integral membrane protein	Hardwick et al. (1990)
Maturation and Processing in the Golgi			
Och1p	α-1,6 mannosyl transferase	66-kD type II integral membrane glycoprotein; *cis*-Golgi marker	Nakayama et al. (1991)
Mnn1p	α-1,3 mannosyl transferase	type II integral membrane glycoprotein; 85-kD apoprotein; medial Golgi marker	Yip et al. (1994)

(*Continued on following pages.*)

Table 1 (*Continued.*)

Protein	Homolog/Function	Properties	References
Kre2p/Mnt1p	α-1,2 mannosyl transferase	type II integral membrane glycoprotein; 85-kD apoprotein	Lussier et al. (1995)
Kex2p	furin-like processing protease	type I integral membrane glycoprotein; 77-kD apoprotein; cleaves at dibasic residues; *trans*-Golgi marker	Fuller et al. (1989)
Kex1p	prophermone processing protease	type I integral membrane glycoprotein; 80-kD apoprotein; cleaves basic carboxy-terminal residues; *trans*-Golgi marker	Dmochowska et al. (1987)
Ste13p/DPAP-A	prophermone processing protease	type II integral membrane glycoprotein; 107-kD apoprotein; cleaves basic carboxy-terminal residues; *trans*-Golgi marker	Suarez Rendueles and Wolf (1987); Nothwehr et al. (1993)
Transport from the *Trans*-Golgi Compartment to the Vacuole			
(A) *Components and regulators of clathrin-coated vesicles*			
Chc1p	clathrin heavy chain	187-kD hydrophilic protein; forms clathrin triskelions together with Clc1p	Payne et al. (1988); Lemmon et al. (1991); Tan et al. (1993)
Clc1p	clathrin light chain	38-kD hydrophilic protein; forms clathrin triskelions together with Chc1p	Silveira et al. (1990)
Apl1p/Yap80p	β-adaptin	80-kD hydrophilic protein	Kirchhausen (1990); Rad et al. (1995)
Apl2p	β-adaptin/AP-1 complex	82-kD hydrophilic protein; Golgi-specific phenotypes	Rad et al. (1995)
Apl3p/YBL037w	α-adaptin	115-kD hydrophilic protein	Cherry et al. (*Saccharomyces* Genome Database, 1996)
Apl4p/YPR029c	γ-adaptin/ AP-1 complex	94-kD hydrophilic protein	Cherry et al. (*Saccharomyces* Genome Database, 1996)

Apm1p/Yap54p	μ1-adaptin/ AP-1 complex	54-kD hydrophilic protein; component of an ~500-kD complex; Golgi-specific phenotypes	Nakayama et al. (1991); Stepp et al. (1995)
Apm2p	μ-adaptin	70-kD hydrophilic protein; component of an ~600-kD complex	Stepp et al. (1995)
Apm4p/YOL062c	μ-adaptin	55-kD hydrophilic protein	Cherry et al. (*Saccharomyces* Genome Database, 1996)
Aps1p/Yap19p	σ1-adaptin/AP-1 complex	18-kD hydrophilic protein; component of a 200–600-kD complex; Golgi-specific phenotypes	Nakai et al. (1993); Phan et al. (1994)
Aps2p/Yap17p	σ-adaptin	17-kD hydrophilic protein; component of a 200–600-kD complex	Kirchhausen et al. (1991); Phan et al. (1994)
Vps1p	dynamin-like GTPase	79-kD peripheral membrane protein	Rothman et al. (1990)
(B) *Additional proteins involved in transport from the* trans-*Golgi compartment to the vacuole*			
Pep1p/Vps10p		178-kD type I membrane protein; receptor for CPY	Marcusson et al. (1994)
Pep12p/Vps6p	t-SNARE; syntaxin-like	32-kD peripheral membrane protein	Becherer et al. (1996)
Vps45p	Sec1p-like	67-kD peripheral membrane protein	Cowles et al. (1994); Piper et al. (1994)
Vps33p/Slp1p	Sec1p-like	75-kD cytosolic protein; mutants have no detectable vacuole	Banta et al. (1990); Wada et al. (1990)
Mvp1p		60-kD hydrophilic protein; interacts genetically with Vps1p	Ekena and Stevens (1995)
Vesicular Transport from the Golgi to the Plasma Membrane			
Snc1p/Snc2p	v-SNARE; synaptobrevin family	13-kD integral membrane proteins; 79% identical to each other; located in post-Golgi vesicles	Protopopov et al. (1993)
Sso1p/Sso2p	t-SNARE; syntaxin family	39-kD integral membrane proteins; 72% identical to each other; located in plasma membrane	Aalto et al. (1993)

(*Continued on following pages.*)

Table 1 (*Continued.*)

Protein	Homolog/Function	Properties	References
Sec9p	mammalian SNAP-25	74-kD peripheral membrane protein; located in plasma membrane	Brennwald et al. (1994)
Sec1p	Sly1p, mammalian n-Sec1	83-kD hydrophilic protein associated with 12,000*g* and 100,000*g* pellets, but not easily extractable	Aalto et al. (1992); Egerton et al. (1993)
Sec2p		84-kD soluble protein; has coiled-coil domain; found in 500–750-kD complex	Nair et al. (1990)
Sec6p, Sec8p, Sec15p	components of Sec6p/Sec8p/Sec15p complex	85-, 122-, and 105-kD proteins; are all components of a large complex containing at least five other proteins; located in cytosol and the plasma membrane at the tips of growing buds	Salminen and Novick (1989); Bowser et al. (1992); Potenza et al. (1992); TerBush and Novick (1995)
Sec3p, Sec5p, Sec10p	assembly of Sec6p/Sec8p/Sec15p complex	mutants disrupt formation of the Sec6p/Sec8p/Sec15p complex	TerBush and Novick (1995)
Cytoskeleton-related Proteins Involved in Vesicular Transport			
Act1p	actin	42-kD; characteristically localized as both patches and cables	Shortle et al. (1982); Novick and Botstein (1985)
Tpm1p	tropomyosin	24-kD protein; colocalized with actin cables	Liu and Bretscher (1989)
Myo2p/Cdc66p	class V myosin	180-kD protein; localized to regions of active membrane growth	Johnston et al. (1991)
Smy1p	kinesin superfamily	74-kD protein; homology with motor domain of kinesin; localized to regions of active membrane growth	Lillie and Brown (1992)
Endocytic Uptake[b]			
End3p		40-kD hydrophilic protein; EF-hand Ca^{++}-binding motif; PIP_2-binding motif	Raths et al. (1993); Bénédetti et al. (1994)

Sla2p/End4p/ Mop2p	talin-like	109-kD peripheral plasma membrane protein	Holtzman et al. (1993); Raths et al. (1993); Na et al. (1995)
Vrp1p/End5p		83-kD proline-rich protein	Munn et al. (1995)
Rvs161p/End6p	Rvs167p; homology with amphyphysin	30-kD hydrophilic protein	Munn et al. (1995)
Rvs167p	Rvs161p; homology with amphyphysin	53-kD hydrophilic protein; SH3 domain; binds to actin	Amberg et al. (1995); Munn et al. 1995
Act1p/End7p	actin	42-kD cytoskeletal protein	Kübler and Riezman (1993); Munn et al. (1995)
Sac6p	fimbrin	72-kD hydrophilic protein; actin-bundling protein	Kübler and Riezman (1993)
Cmd1p	calmodulin	16-kD Ca^{++}-binding protein; Ca^{++}-binding sites not required for function in endocytosis	Kübler et al. (1994)
Myo5p	type I myosin; 87% identity to Myo3p	137-kD protein; SH3 domain near carboxyl terminus	Geli and Riezman (1996)
Myo3p	type I myosin; 87% identity to Myo5p	142-kD protein; SH3 domain near carboxyl terminus	Geli and Riezman (1996)
Delivery of Endocytosed Proteins to the Vacuole[c]			
Vps34p/End12p	PI3-kinase	95-kD peripheral membrane protein; acts in a complex with Vps15p	Herman and Emr (1990); Munn and Riezman (1994); Stack et al. (1995)
Vps21p/Ypt51p	Rab5p-like GTPase	23-kD peripheral membrane protein	Singer-Krüger et al. (1994); Horazdovsky et al. (1994)
Ypt52p	Rab5p-like GTPase	26-kD protein	Singer-Krüger et al. (1994)
Ypt53p	Rab5p-like GTPase	25-kD protein	Singer-Krüger et al. (1994)
Ypt7p	Rab7p-like GTPase	23-kD protein	Wichmann et al. (1992)
Dnm1p	dynamin-like GTPase	85-kD protein	Gammie et al. (1995)

[a]*RER2* and *RER3* are required for retention of Sec12p in the ER but have not been cloned (Nishikawa and Nakano 1993; Boehm et al. 1994).

[b]*END8, 9, 10, 11* are required for endocytic uptake but have not been cloned (Munn and Riezman 1994). Chc1p is required for endocytic uptake but is described as a component of clathrin-coated vesicles.

[c]*END13* (allelic to *VPS4*) and *REN1* (allelic to *VPS2*) are also required for delivery of endocytosed proteins to the vacuole, but they have not been cloned (Davis et al. 1993; Munn and Riezman 1994).

REFERENCES

Aalto, M.K., S. Jeränen, and H. Ronne. 1992. A family of proteins involved in intracellular transport. *Cell* **68:** 181.

Aalto, M.K., H. Ronne, and S. Keränen. 1993. Yeast syntaxins Sso1p and Sso2p belong to a family of related membrane proteins that function in vesicular transport. *EMBO J.* **12:** 4095.

Aalto, M.K., L. Ruohonen, K. Hosono, and S. Keränen. 1991. Cloning and sequencing of the yeast *Saccharomyces cerevisiae SEC1* gene localized on chromosome IV. *Yeast* **7:** 643.

Achstetter, T., A. Franzusoff, C. Field, and R. Schekman. 1988. SEC7 Encodes an unusual, high molecular weight protein required for membrane traffic from the yeast Golgi apparatus. *J. Biol. Chem.* **263:** 11711.

Adams, A.E.M. and J.R. Pringle. 1984. Relationship of actin and tubulin distribution to bud growth in wild-type and morphogenetic-mutant *Saccharomyces cerevisiae. J. Cell Biol.* **98:** 934.

Aitken, J.F., G.P. van Heusden, M. Temkin, and W. Dowhan. 1990. The gene encoding the phosphatidylinositol transfer protein is essential for cell growth. *J. Biol. Chem.* **265:** 7411.

Amatruda, J.F., D.J. Gattermeir, T.S. Karpova, and J.A. Cooper. 1992. Effects of null mutations and overexpression of capping protein on morphogenesis, actin distribution and polarized secretion in yeast. *J. Cell Biol.* **119:** 1151.

Amberg, D.C., E. Basart, and D. Botstein. 1995. Defining protein interactions with yeast actin *in vivo*. *Nat. Struct. Biol.* **2:** 28.

André, B. 1995. An overview of membrane transport proteins in *Saccharomyces cerevisiae. Yeast* **11:** 1575.

Aniento, F., F. Gu, R.G. Parton, and J. Gruenberg. 1996. An endosomal β-COP is involved in the pH-dependent formation of transport vesicles destined for late endosomes. *J. Cell Biol.* **133:** 29.

Antebi, A. and G.R. Fink. 1992. The yeast Ca^{2+}-ATPase homologue, *PMR1*, is required for normal Golgi function and localizes in a novel Golgi-like distribution. *Mol. Biol. Cell* **3:** 633.

Aridor, M., S.I. Bannykh, T. Rowe, and W.E. Balch. 1995. Sequential coupling between COPII and COPI vesicle coats in endoplasmic reticulum to Golgi transport. *J. Cell Biol.* **131:** 875.

Armstrong, J. 1993. Two fingers for membrane traffic. *Curr. Biol.* **3:** 33.

Arnold, C.E. and K.D. Wittrup. 1994. The stress response to loss of signal recognition particle function in *Saccharomyces cerevisiae. J. Biol. Chem.* **269:** 30412.

Bacon, R.A., A. Salminen, H. Ruohola, P. Novick, and S. Ferro-Novick. 1989. The GTP-binding protein Ypt1 is required for transport in vitro: The Golgi apparatus is defective in *ypt1* mutants. *J. Cell Biol.* **109:** 1015.

Baker, D., L. Hicke, M. Rexach, M. Schleyer, and R. Schekman. 1988. Reconstitution of *SEC* gene product-dependent intercompartmental protein transport. *Cell* **54:** 335.

Baker, D., L. Wuestehube, R. Schekman, D. Botstein, and N. Segev. 1990. GTP-binding Ypt1 protein and Ca^{2+} function independently in a cell-free protein transport reaction. *Proc. Natl. Acad. Sci.* **87:** 355.

Balch, W.E., B.S. Glick, and J.E. Rothman. 1984. Sequential intermediates in the pathway of intercompartmental transport in a cell-free system. *Cell* **39:** 525.

Ballou, L., R.A. Hitzeman, M.S. Lewis, and C.E. Ballou. 1991. Vanadate-resistant yeast mutants are defective in protein glycosylation. *Proc. Natl. Acad. Sci.* **88:** 3209.

Ballou, L., P. Gopal, B. Krummel, M. Tammi, and C.E. Ballou. 1986. A mutation that prevents glucosylation of the lipid-linked oligosaccharide precursor leads to underglycosylation of secreted yeast invertase. *Proc. Natl. Acad. Sci.* **83:** 3081.

Banfield, D., M.J. Lewis, and H.R.B. Pelham. 1995. A SNARE-like protein required for traffic through the Golgi complex. *Nature* **375:** 806.

Banfield, D.K., M.J. Lewis, C. Rabouille, G. Warren, and H.R. Pelham. 1994. Localization of Sed5, a putative vesicle targeting molecule, to the *cis*-Golgi network involves both its transmembrane and cytoplasmic domains. *J. Cell Biol.* **127:** 357.

Bankaitis, V.A., J.R. Aitken, A.E. Cleves, and W. Dowhan. 1990. An essential role for a phospholipid transfer protein in yeast Golgi function. *Nature* **347:** 561.

Bankaitis, V.A., D.E. Malehorn, S.D. Emr, and R. Greene. 1989. The *Saccharomyces cerevisiae SEC14* gene encodes a cytosolic factor that is required for transport of secretory proteins from the yeast Golgi complex. *J. Cell Biol.* **108:** 1271.

Banta, L.M., T.A. Vida, P.K. Herman, and S.D. Emr. 1990. Characterization of yeast Vps33p, a protein required for vacuolar protein sorting and vacuole biogenesis. *Mol. Cell. Biol.* **10:** 4638.

Bard, M., D.A. Bruner, C.A. Pierson, N.D. Lees, B. Biermann, L. Frye, C. Koegel, and R. Barbuch. 1996. Cloning and characterization of *ERG25*, the *Saccharomyces cerevisiae* gene encoding C-4 sterol methyl oxidase. *Proc. Natl. Acad. Sci.* **93:** 186.

Barlowe, C. and R. Schekman. 1993. *SEC12* encodes a guanine-nucleotide-exchange factor essential for transport vesicle budding from the ER. *Nature* **365:** 347.

Barlowe, C., C. d'Enfert, and R. Schekman. 1993. Purification and characterization of SAR1p, a small GTP-binding protein required for transport vesicle formation from the endoplasmic reticulum. *J. Biol. Chem.* **268:** 873.

Barlowe, C., L. Orci, T. Yeung, M. Hosobuchi, S. Hamamoto, N. Salama, M.F. Rexach, M. Ravazzola, M. Amherdt, and R. Schekman. 1994. COPII: A membrane coat formed by Sec proteins that drive vesicle budding from the endoplasmic reticulum. *Cell* **77:** 895.

Barroso, M., D.S. Nelson, and E. Sztul. 1995. Transcytosis-associated protein (TAP)/p115 is a general fusion factor required for binding of vesicles to acceptor membranes. *Proc. Natl. Acad. Sci.* **92:** 527.

Basson, M.E., M. Thorsness, M.J. Finer, R.M. Stroud, and J. Rine. 1988. Structural and functional conservation between yeast and human 3-hydroxy-3-methylglutaryl coenzyme A reductases, the rate-limiting enzyme of sterol biosynthesis. *Mol. Cell. Biol.* **8:** 3797.

Bauerle, C., M.N. Ho, M.A. Lindorfer, and T.H. Stevens. 1993. The *Saccharomyces cerevisiae VMA6* gene encodes the 36kDa subunit of the vacuolar membrane sector. *J. Biol. Chem.* **268:** 12749.

Becherer, K.A., S.E. Rieder, S.D. Emr, and E.W. Jones. 1996. Novel syntaxin homologue, Pep12p, required for the sorting of lumenal hydrolases to the lysosome-like vacuole in yeast. *Mol. Biol. Cell* **7:** 579.

Becker, J., W. Walter, W. Yan, and E.A. Craig. 1996. Functional interaction of cytosolic hsp70 and a DnaJ-related protein Ydj1p, in protein translocation in vivo. *Mol. Cell. Biol.* **16:** 4378.

Beckers, C.J.M., M.R. Block, B.S. Glick, J.E. Rothman, and W.E. Balch. 1989. Vesicular transport between the endoplasmic reticulum and the Golgi stack requires the NEM-sensitive fusion protein. *Nature* **339:** 397.

Bednarek, S.Y., M. Ravazzola, M. Hosobuchi, M. Amherdt, A. Perrelet, R. Schekman, and L. Orci. 1995. COPI- and COPII-coated vesicles bud directly from the endoplasmic reticulum in yeast. *Cell* **83:** 1183.

Beh, C.T. and M.D. Rose. 1995. Two redundant systems maintain levels of resident proteins within the yeast endoplasmic reticulum. *Proc. Natl. Acad. Sci.* **92:** 9820.

Belden, W.J. and C. Barlowe. 1996. Erv25p, a component of COPII-coated vesicles, forms a complex with Emp24p that is required for efficient endoplasmic reticulum to Golgi transport. *J. Biol. Chem.* **271:** 26939.

Bénédetti, H., S. Raths, F. Crausaz, and H. Riezman. 1994. The *END3* gene encodes a protein that is required for the internalization step of endocytosis and for actin cytoskeleton organization in yeast. *Mol. Biol. Cell* **5:** 1023.

Benito-Moreno, R.M., M. Miaczynska, B.E. Bauer, R.J. Schweyen, and A. Ragnaini. 1994. Mrs6p, the yeast homologue of the mammalian choroideraemia protein: Immunological evidence for its function as the Ypt1p Rab escort protein. *Curr. Genet.* **27:** 23.

Bennett, M.K. and R.H. Scheller. 1993. The molecular machinery for secretion is conserved from yeast to neurons. *Proc. Natl. Acad. Sci.* **90:** 2559.

———. 1994. A molecular description of synaptic vesicle membrane trafficking. *Annu. Rev. Biochem.* **63:** 63.

Berkower, C., D. Loayza, and S. Michaelis. 1994. Metabolic instability and constitutive endocytosis of *STE6*, the **a**-factor transporter of *Saccharomyces cerevisiae*. *Mol. Biol. Cell* **5:** 1185.

Bernstein, H.D., M.A. Portiz, K. Strub, P.J. Hoben, S. Brenner, and P. Walter. 1989. Model for signal sequence recognition from amino-acid sequence of 54k subunit of signal recognition particle. *Nature* **340:** 482.

Biederer, T., C. Volkwein, and T. Sommer. 1996. Degradation of subunits of the Sec61p complex, an integral component of the ER membrane, by the ubiquitin-proteasome pathway. *EMBO J.* **15:** 2069.

Blachly-Dyson, E. and T.H. Stevens. 1987. Yeast carboxypeptidase Y can be translocated and glycosylated without its amino-terminal signal sequence. *J. Cell Biol.* **104:** 1183.

Blobel, G. and D.D. Sabatini. 1971. Ribosome-membrane interaction in eukaryotic cells. In *Biomembranes* (ed. L.A. Manson), vol. 2, p. 193. Plenum Press, New York.

Boehm, J., H.D. Ulrich, R. Ossig, and H.D. Schmitt. 1994. Kex2-dependent invertase secretion as a tool to study the targeting of transmembrane proteins which are involved in ER–Golgi transport in yeast. *EMBO J.* **13:** 3696.

Boguski, M. and F. McCormick. 1993. Proteins regulating Ras and its relatives. *Nature* **366:** 643.

Bohni, P.C., R.J. Deshaies, and R. Schekman. 1988. *SEC11* is required for signal peptide processing and yeast cell growth. *J. Cell Biol.* **106:** 1035.

Bourne, H.R. 1988. Do GTPases direct membrane traffic in secretion? *Cell* **53:** 669.

Bourne, H.R., D.A. Sanders, and F. McCormick. 1990. The GTPase superfamily: A conserved switch for diverse cell functions. *Nature* **348:** 125.

———. 1991. The GTPase superfamily: Conserved structure and molecular mechanism. *Nature* **349:** 117.

Bowser, R. and P. Novick. 1991. Sec15 protein, an essential component of the exocytotic apparatus, is associated with the plasma membrane and with a soluble 19.5*S* particle. *J. Cell Biol.* **112:** 1117.

Bowser, R., H. Müller, B. Govindan, and P. Novick. 1992. Sec8p and Sec15p are components of a plasma membrane-associated 19.5*S* particle that may function downstream of Sec4p to control exocytosis. *J. Cell Biol.* **118:** 1041.

Brada, D. and R. Schekman. 1988. Coincident localization of secretory and plasma membrane proteins in organelles of the yeast secretory pathway. *J. Bacteriol.* **170:** 2775.

Brennwald, P. and P. Novick. 1993. Interactions of three domains distinguishing the Ras-related GTP-binding proteins Ypt1 and Sec4. *Nature* **362:** 560.

Brennwald, P., B. Kearns, K. Champion, S. Keränen, V. Bankaitis, and P. Novick. 1994. Sec9 is a SNAP-25-like component of a yeast SNARE complex that may be the effector of Sec4 function in exocytosis. *Cell* **79:** 245.

Bretscher, A., B. Drees, E. Harsay, D. Schott, and T. Wang. 1994. What are the basic functions of microfilaments? Insights from studies in budding yeast. *J. Cell Biol.* **126:** 821.

Bretscher, M.S. and S. Munro. 1993. Cholesterol and the Golgi apparatus. *Science* **261:** 1280.

Brockerhoff, S.E. and T.N. Davis. 1992. Calmodulin concentrates at regions of cell growth in *Saccharomyces cerevisiae*. *J. Cell Biol.* **118:** 619.

Brodsky, J.L. and R. Schekman. 1993. A Sec63p-Bip complex from yeast is required for protein translocation in a reconstituted system. *J. Cell Biol.* **123:** 1355.

Brodsky, J., L. Goeckeler, and R. Schekman. 1995. BiP and Sec63p are required for both co- and posttranslational translocation into the yeast endoplasmic reticulum. *Proc. Natl. Acad. Sci.* **92:** 9643.

Brodsky, J.L., S. Hamamoto, D. Feldheim, and R. Schekman. 1993. Reconstitution of protein translocation from solubilized yeast membranes reveals topologically distinct roles for BiP and cytosolic Hsc70. *J. Cell Biol.* **120:** 95.

Brown, J.D., B.C. Hann, K.F. Medzihradzky, M. Niwa, A.L. Burlingame, and P. Walter. 1994. Subunits of the *Saccharomyces cerevisiae* signal recognition particle required for its functional expression. *EMBO J.* **13:** 4390.

Brown, M.S. and J.L. Goldstein. 1993. Mad Bet for Rab. *Nature* **366:** 14.

Bryant, N.J. and A. Boyd. 1993. Immunoisolation of Kex2p-containing organelles from yeast demonstrates colocalisation of three processing proteinases to a single Golgi compartment. *J. Cell Sci.* **106:** 815.

Burgess, T.L. and R.B. Kelly. 1987. Constitutive and regulated secretion of proteins. *Annu. Rev. Cell Biol.* **3:** 243.

Burgoyne, R.D. 1994. Phosphoinositides in vesicular traffic. *Trends Biochem. Sci.* **19:** 55.

Burton, J., D. Roberts, M. Montaldi, P. Novick, and P. De Camilli. 1993. A mammalian guanine-nucleotide-releasing protein enhances function of yeast secretory protein Sec4. *Nature* **361:** 464.

Caplan, A.J., D.M. Cyr, and M.G. Douglas. 1992a. Ydj1p facilitates polypeptide translocation across different intracellular membranes by a conserved mechanism. *Cell* **71:** 1143.

Caplan, A., J. Tsai, P. Casey, and M. Douglas. 1992b. Farnesylation of Ydj1p is required for function at elevated growth temperatures in *S. cerevisiae*. *J. Biol. Chem.* **267:** 18890.

Carlson, M. and D. Botstein. 1982. Two differentially regulated mRNAs with different 5′ ends encode secreted with intracellular forms of yeast invertase. *Cell* **28:** 145.

Cereghino, J.L., E.G. Marcusson, and S.D. Emr. 1995. The cytoplasmic tail domain of the vacuolar protein sorting receptor Vps10p and a subset of VPS gene products regulate receptor stability, function, and localization. *Mol. Biol. Cell* **6:** 1089.

Chang, A. and G.R. Fink. 1995. Targeting of the yeast plasma membrane [H^+]ATPase: A novel gene *AST1* prevents mislocalization of mutant ATPase to the vacuole. *J. Cell Biol.* **128:** 39.

Chapman, R.E. and S. Munro. 1994. The functioning of the yeast Golgi apparatus requires an ER protein encoded by *ANP1*, a member of a new family of genes affecting the secretory pathway. *EMBO J.* **13:** 4896.

Cherry, J.M., C. Adler, C. Ball, S. Dwight, S. Chervitz, G. Juvik, S. Weng, and D. Botstein. 1996. *Saccharomyces* Genome Database (http://genome-www.stanford.edu/ Saccharomyces/SacchDB4.6.1).

Chirico, W.J., M.G. Waters, and G. Blobel. 1988. 70K heat shock related proteins stimulate protein translocation into microsomes. *Nature* **332:** 805.

Chvatchko, Y., I. Howald, and H. Riezman. 1986. Two yeast mutants defective in endocytosis are defective in pheromone response. *Cell* **46:** 355.

Clary, D.O. and J.E. Rothman. 1990. Purification of three related peripheral membrane proteins needed for vesicular transport. *J. Biol. Chem.* **265:** 10109.

Clary, D.O., I.C. Griff, and J.E. Rothman. 1990. SNAPs, a family of NSF attachment proteins involved in intracellular membrane fusion in animals and yeast. *Cell* **61:** 709.

Cleves, A.E., P.J. Novick, and V.A. Bankaitis. 1989. Mutations in the *SAC1* gene suppress defects in yeast Golgi and yeast actin function. *J. Cell Biol.* **109:** 2939.

Cleves, A.E., T.P. McGee, E.A. Whitters, K.M. Champion, J.R. Aitken, W. Dowhan, M. Goebl, and V.A. Bankaitis. 1991. Mutations in the CDP-choline pathway for phospholipid biosynthesis bypass the requirement for an essential phospholipid transfer protein. *Cell* **64:** 789.

Connolly, T. and R. Gilmore. 1986. Formation of a functional ribosome-membrane junction during translocation requires the participation of a GTP-binding protein. *J. Cell Biol.* **103:** 2253.

Cooper, A. and H. Bussey. 1992. Yeast Kex1p is a Golgi-associated membrane protein: Deletions in a cytoplasmic targeting domain result in mislocalization to the vacuolar membrane. *J. Cell Biol.* **119:** 1459.

Cooper, A.A. and T.H. Stevens. 1996. Vps10p cycles between the late-Golgi and prevacuolar compartments in its function as the sorting receptor for multiple yeast vacuolar hydrolases. *J. Cell Biol.* **133:** 529.

Cosson, P. and F. Létournèur. 1994. Coatomer interaction with di-lysine endoplasmic reticulum retention motifs. *Science* **263:** 1629.

Cosson, P., C. Démollière, S. Hennecke, R. Duden, and F. Letourneur. 1996. δ- and ζ-COPI, two coatomer subunits homologous to clathrin-associated proteins, are involved in ER retrieval. *EMBO J.* **15:** 1792.

Couve, A. and J.E. Gerst. 1994. Yeast Snc proteins complex with Sec9: Functional interactions between putative SNARE proteins. *J. Biol. Chem.* **269:** 23391.

Cowles, C.R., S.D. Emr, and B. Horazdovsky. 1994. Mutations in the *VPS45* gene, a *SEC1* homologue, result in vacuolar protein sorting defects and accumulation of membrane vesicles. *J. Cell Sci.* **107:** 3449.

Cox, J.S. and P. Walter. 1996. A novel mechanism for regulating activity of a transcription factor that controls the unfolded protein response. *Cell* **87:** 391.

Cox, J.S., C.E. Shamu, and P. Walter. 1993. Transcriptional induction of genes encoding endoplasmic reticulum resident proteins requires a transmembrane protein kinase. *Cell* **73:** 1197.

Cukierman, E., I. Huber, M. Rotman, and D. Cassel. 1995. The *ARF1* GTPase-activating protein: Zinc finger motif and Golgi complex localization. *Science* **270:** 1999.

Cyr, D., X. Lu, and M. Douglas. 1992. Regulation of Hsp70 function by a eukaryotic DnaJ homolog. *J. Biol. Chem.* **267:** 20927.

Dascher, C., R. Ossig, D. Gallwitz, and H.D. Schmitt. 1991. Identification and structure of four yeast genes (*SLY*) that are able to suppress the functional loss of *YPT1*, a member of the RAS superfamily. *Mol. Cell Biol.* **11:** 872.

Daum, G., E. Heidorn, and F. Paltauf. 1986. Intracellular transfer of phospholipids in the yeast, *Saccharomyces cerevisiae*. *Biochim. Biophys. Acta* **878:** 93.

Davis, N.G., J.L. Horecka, and G.F. Sprague, Jr. 1993. *cis*- and *trans*-acting functions required for endocytosis of the yeast pheromone receptors. *J. Cell Biol.* **122:** 53.

Dean, N. and H.R.B. Pelham. 1990. Recycling of proteins from the Golgi compartment to the ER in yeast. *J. Cell Biol.* **111:** 369.

d'Enfert, C., L.J. Wuestehube, T. Lila, and R. Schekman. 1991a. Sec12p-dependent membrane binding of the small GTP-binding protein Sar1p promotes formation of transport vesicles from the ER. *J. Cell Biol.* **114:** 663.

d'Enfert, C., C. Barlowe, S.-I. Nishikawa, A. Nakano, and R. Schekman. 1991b. Structural and functional dissection of a membrane glycoprotein required for vesicle budding from the endoplasmic reticulum. *Mol. Cell. Biol.* **11:** 5727.

Deshaies, R.J. and R. Schekman. 1987. A yeast mutant defective at an early stage in import of secretory protein precursors into the endoplasmic reticulum. *J. Cell Biol.* **105:** 633.

———. 1989. Sec62 encodes a putative membrane protein required for protein translocation into the yeast endoplasmic reticulum. *J. Cell Biol.* **109:** 2653.

Deshaies, R.J., S.L. Sanders, D.A. Feldheim, and R. Schekman. 1991. Assembly of yeast Sec proteins involved in translocation into the endoplasmic reticulum into a membrane-bound multisubunit complex. *Nature* **349:** 806.

Deshaies, R.J., B.D. Koch, M. Werner-Washburne, E.A. Craig, and R. Schekman. 1988. A subfamily of stress proteins facilitates translocation of secretory and mitochondrial precursor polypeptides. *Nature* **332:** 800.

Diaz, R., L.S. Mayorga, P.J. Weidman, J.E. Rothman, and P.D. Stahl. 1989. Vesicle fusion following receptor-mediated endocytosis requires a protein active in Golgi transport. *Nature* **339:** 398.

Dmochowska, A., D. Dignard, D. Henning, D.Y. Thomas, and H. Bussey. 1987. Yeast *KEX1* gene encodes a putative protease with a carboxypeptidase B-like function involved in killer toxin and α-factor precursor processing. *Cell* **50:** 573.

Doering, T.L. and R. Schekman. 1996. GPI anchor attachment is required for Gas1p transport from the endoplasmic reticulum in COP II vesicles. *EMBO J.* **15:** 182.

Downward, J. 1992. Ras regulation: Putting back the GTP. *Curr. Biol.* **2:** 329.

Duden, R., G. Griffiths, R. Frank, P. Argos, and T.E. Kreis. 1991. β-COP, a 110 kd protein associated with non-clathrin-coated vesicles and the Golgi complex, shows homology to β-adaptin. *Cell* **64:** 649.

Duden, R., M. Hosobuchi, S. Hamamoto, M. Winey, B. Byers, and R. Schekman. 1994. Yeast β- and β′-coat proteins (COP). *J. Biol. Chem.* **269:** 24486.

Dunn, B., T. Stearns, and D. Botstein. 1993. Specificity domains distinguish the Ras-related GTPases Ypt1 and Sec4. *Nature* **362:** 563.

Eakle, K.A., M. Bernstein, and S.D. Emr. 1988. Characterization of a component of the yeast secretion machinery: Identification of the *SEC18* gene product. *Mol. Cell Biol.* **8:**

4098.

Egerton, M., J. Zueco, and A. Boyd. 1993. Molecular characterization of the *SEC1* gene of *Saccharomyces cerevisiae:* Subcellular distribution of a protein required for yeast protein secretion. *Yeast* **9:** 703.

Egner, R. and K. Kuchler. 1996. The yeast multidrug transporter Pdr5 of the plasma membrane is ubiquinated prior to endocytosis and degradation in the vacuole. *FEBS Lett.* **378:** 177.

Egner, R., Y. Mahe, R. Pandjaitan, and K. Kuchler. 1995. Endocytosis and vacuolar degradation of the plasma membrane-localized Pdr5 ATP-binding cassette multidrug transporter in *Saccharomyces cerevisiae. Mol. Cell. Biol.* **15:** 5879.

Ekena, K. and T.H. Stevens. 1995. The *Saccharomyces cerevisiae MVP1* gene interacts with *VPS1* and is required for vacuolar protien sorting. *Mol. Cell. Biol.* **15:** 1671.

Elrod-Erickson, M.J. and C.A. Kaiser. 1996. Genes that control the fidelity of endoplasmic reticulum to Golgi transport identified as suppressors of vesicle budding mutations. *Mol. Biol. Cell* **7:** 1043.

Esnault, Y., M.O. Blondel, R.J. Deshaies, R. Schekman, and F. Kepes. 1993. The yeast *SSS1* gene is essential for secretory protein translocation and encodes a highly conserved protein of the endoplasmic reticulum. *EMBO J.* **12:** 4083.

Espenshade, P., R.E. Gimeno, E. Holzmacher, P. Teung, and C.A. Kaiser. 1995. Yeast *SEC16* gene encodes a multidomain vesicle coat protein that interacts with Sec23p. *J. Cell Biol.* **131:** 311.

Fang, H. and N. Green. 1994. Nonlethal *sec71-1* and *sec72-1* mutations eliminate proteins associated with the Sec63p-BiP complex from *S. cerevisiae. Mol. Biol. Cell* **5:** 933.

Fang, H., S. Panzner, C. Mullins, E. Hartmann, and N. Green. 1996. The homologue of mammalian SPC12 is important for efficient signal peptidase activity in *Saccharomyces cerevisiae. J. Biol. Chem.* **271:** 16460.

Farkas, V., J. Kovarik, A. Kosinova, and S. Bauer. 1974. Autoradiographic study of mannan incorporation into the growing cell walls of *Saccharomyces cerevisiae. J. Bacteriol.* **117:** 265.

Feldheim, D. and R. Schekman. 1994. Sec72p contributes to the selective recognition of signal peptides by the secretory polypeptide translocation complex. *J. Cell Biol.* **126:** 935.

Feldheim, D., J. Rothblatt, and R. Schekman. 1992. Topology and functional domains of Sec63p, an endoplasmic reticulum membrane protein required for secretory protein translocation. *Mol. Cell. Biol.* **12:** 3288.

Feldheim, D., K. Yoshimura, A. Admon, and R. Schekman. 1993. Structural and functional characterization of Sec66p, a new subunit of the polypeptide translocation apparatus in the yeast endoplasmic reticulum. *Mol. Biol. Cell* **4:** 931.

Felici, F., G. Cesareni, and J.M.X. Hughes. 1989. The most abundant small cytoplasmic RNA of *Saccharomyces cerevisiae* has an important function required for normal cell growth. *Mol. Cell Biol.* **9:** 3260.

Ferro-Novick, S. and R. Jahn. 1994. Vesicle fusion from yeast to man. *Nature* **370:** 191.

Ferro-Novick, S. and P. Novick. 1993. The role of GTP-binding proteins in transport along the exocytic pathway. *Annu. Rev. Cell Biol.* **9:** 575.

Field, C. and R. Schekman. 1980. Localized secretion of acid phosphatase reflects the pattern of cell surface growth in *Saccharomyces cerevisiae. J. Cell Biol.* **86:** 123.

Finger, A., M. Knop, and D.H. Wolf. 1993. Analysis of two mutated vacuolar proteins reveals a degradation pathway in the endoplasmic reticulum or a related compartment of yeast. *Eur. J. Biochem.* **218:** 565.

Finke, K., K. Plath, S. Panzer, S. Prehn, T.A. Rapoport, E. Hartmann, and T. Sommer. 1996. A second trimeric complex containing homologs of the Sec61p complex functions in protein transport across the ER membrane of *S. cerevisiae*. *EMBO J.* **15:** 1482.

Franzusoff, A. and R. Schekman. 1989. Functional compartments of the yeast Golgi apparatus are defined by the *sec7* mutation. *EMBO J.* **8:** 2695.

Franzusoff, A., E. Lauzé, and K.E. Howell. 1992. Immuno-isolation of Sec7p-coated transport vesicles from the yeast secretory pathway. *Nature* **355:** 173.

Franzusoff, A., K. Redding, J. Crosby, R.S. Fuller, and R. Schekman. 1991. Localization of components involved in protein transport and processing through the yeast Golgi apparatus. *J. Cell Biol.* **112:** 27.

Frigerio, G. and H.R.B. Pelham. 1993. A *Saccharomyces cerevisiae* cyclophilin resident in the endoplasmic reticulum. *J. Mol. Biol.* **233:** 183.

Fuller, R.S., A.J. Brake, and J. Thorner. 1989. Intracellular targeting and structural conservation of a prohormone-processinf endoprotease. *Science* **246:** 482.

Galan, J.M., C. Volland, D. Urban-Grimal, and R. Haguemauer-Tsapis. 1994. The yeast plasma membrane uracil permease is stabilized against stress induced degradation by a point mutation in a cyclin-like "destruction box". *Biochem. Biophys. Res. Commun.* **201:** 769.

Gallwitz, D., C. Donath, and C. Sander. 1983. A yeast gene encoding a protein homologous to the human *c-has/bas* proto-oncogene product. *Nature* **306:** 704.

Gammie, A.E., L.J. Kurihara, R.B. Vallee, and M.D. Rose. 1995. *DNM1*, a dynamin-related gene, participates in endosomal trafficking in yeast. *J. Cell Biol.* **130:** 553.

Garcia, E.P., E. Gatti, M. Butler, J. Burton, and P. DeCamilli. 1994. A rat brain Sec1 homolog related to Rop and *UNC18* interacts with syntaxin. *Proc. Natl. Acad. Sci.* **91:** 2003.

Garrels, J.I. 1995. YPD—A database for the proteins of *Saccharomyces cerevisiae* (http://quest7.proteome.com/YPDhome.html). *Nucleic Acids Res.* **24:** 46.

Garrett, M.D., J.E. Zahner, C.M. Cheney, and P.J. Novick. 1994. *GDI1* encodes a GDP dissociation inhibitor that plays an essential role in the yeast secretory pathway. *EMBO J.* **13:** 1718.

Garrett, M.D., A.K. Kabcenell, J.E. Zahner, K. Kaibuchi, T. Sasaki, Y. Takai, C.M. Cehney, and P.J. Novick. 1993. Interaction of Sec4 with GDI proteins from bovine brain, *Drosophila melanogaster* and *Saccharomyces cerevisiae:* Conservation of GDI membrane dissociation activity. *FEBS Lett.* **331:** 233.

Gaynor, E.C., S. te Heesen, T.R. Graham, M. Aebi, and S.D. Emr. 1994. Signal-mediated retrieval of membrane protein from the Golgi to the ER in yeast. *J. Cell Biol.* **127:** 653.

Geli, M.I. and H. Riezman. 1996. Role of type I myosins in receptor-mediated endocytosis in yeast. *Science* **272:** 533.

Georgopoulos, C., D. Ang, K. Liberek, and M. Zylicz. 1990. Properties of the *Escherichia coli* heat shock proteins and their role in bacteriophage λ growth. In *Stress proteins in biology and medicine* (ed. R.I. Morimoto et al.), p. 191. Cold Spring Harbor Laboratory Press, Cold Spring Harbor, New York.

Gerich, B., L. Orci, H. Tschochner, F. Lottspeich, M. Ravazzola, M. Amherdt, F. Wieland, and C. Harter. 1995. Non-clathrin-coat protein α is a conserved subunit of coatomer and in *Saccharomyces cerevisiae* is essential for growth. *Proc. Natl. Acad. Sci.* **92:** 3229.

Gerst, J.E., L. Rodgers, M. Riggs, and M. Wigler. 1992. *SNC1*, a yeast homolog of the synaptic vesicle-associated membrane protein/synaptobrevin gene family: Genetic interaction with the *RAS* and *CAP* genes. *Proc. Natl. Acad. Sci.* **89:** 4338.

Gideon, P., J. John, M. Frech, A. Lautwein, R. Clark, J.E. Scheffler, and A. Wittinghofer. 1992. Mutational and kinetic analysis of the GTPase-activating protein (GAP)-p21 interaction: The C-terminal domain of GAP is not sufficient for full activity. *Mol. Cell. Biol.* **12:** 2050.

Gilmore, R., P. Walter, and G. Blobel. 1982. Protein translocation across the endoplasmic reticulum. II. Isolation and characterization of the signal recognition particle receptor. *J. Cell Biol.* **95:** 470.

Gimeno, R.E., P. Espenshade, and C.A. Kaiser. 1995. *SED4* encodes a yeast endoplasmic reticulum protein that binds Sec16p and participates in vesicle formation. *J. Cell Biol.* **131:** 325.

———. 1996. COPII coat subunit interactions: Sec24p and Sec23p bind to adjacent regions of Sec16p. *Mol. Biol. Cell* **7:** 1815.

Glotzer, M., A.W. Murray, and M.W. Kirschner. 1991. Cyclin is degraded by the ubiquitin pathway. *Nature* **349:** 132.

Gluschankof, P. and R.S. Fuller. 1994. A C-terminal domain conserved in precursor processing proteases is required for intramolecular N-terminal maturation of pro-Kex2 protease. *EMBO J.* **13:** 2280.

Gnamusch, E., C. Kalaus, C. Hrastnik, F. Paltauf, and G. Daum. 1992. Transport of phospholipids between subcellular membranes of wild-type yeast cells and of the phosphatidylinositol transfer protein-deficient strain *Saccharomyces cerevisiae sec14*. *Biochim. Biophys. Acta* **1111:** 120.

Görlich, D. and T.A. Rapoport. 1993. Protein translocation into proteoliposomes reconstituted from purified components of the endoplasmic reticulum membrane. *Cell* **75:** 615.

Görlich, D., S. Prehn, E. Hartmann, K.U. Kalies, and T.A. Rapoport. 1992. A mammalian homolog of Sec16p and SecYp is associated with ribosomes and nascent polypeptides during translocation. *Cell* **71:** 489.

Gottlieb, T.A., I.E. Ivanov, M. Adesnik, and D.D. Sabatini. 1993. Actin microfilaments play a critical role in endocytosis at the apical but not the basolateral surface of polarized epithelial cells. *J. Cell Biol.* **120:** 695.

Goud, B., A. Salminen, N.C. Walworth, and P.J. Novick. 1988. A GTP-binding proteins required for secretion rapidly associates with secretory vesicles and the plasma membrane in yeast. *Cell* **53:** 753.

Govindan, B., R. Bowser, and P. Novick. 1995. The role of Myo2, a yeast class V myosin, in vesicular transport. *J. Cell Biol.* **128:** 1055.

Gow, A., D. Nelson, S. Rahim, and E. Sztul. 1993. A coat protein required for transcytotic traffic exists as a multimeric complex. *Eur. J. Cell Biol.* **61:** 184.

Graham, L.A., K.J. Hill, and T.H. Stevens. 1995. *VMA8* encodes a 32-kDa V_1 subunit of the *Saccharomyces cerevisiae* vacuolar H^+-ATPase required for function and assembly of the enzyme complex. *J. Biol. Chem.* **270:** 15037.

Graham, T.R. and S.D. Emr. 1991. Compartmental organization of Golgi-specific protein modification and vacuolar protein sorting events defined in a yeast *sec18* (NSF) mutant. *J. Cell Biol.* **114:** 207.

Graham, T.R., M. Seeger, G.S. Payne, V.L. MacKay, and S.D. Emr. 1994. Clathrin-dependent localization of α1,3 mannosyltransferase to the Golgi complex of *Saccharomyces cerevisiae*. *J. Cell Biol.* **127:** 667.

Green, N., H. Fang, and P. Walter. 1992. Mutants in three novel complementation groups inhibit membrane protein insertion into and soluble protein translocation across the endoplasmic reticulum membrane of *Saccharomyces cerevisiae*. *J. Cell Biol.* **116:** 597.

Griff, I.C., R. Schekman, J.E. Rothman, and C.A. Kaiser. 1992. The yeast *SEC17* gene product is functionally equivalent to mammalian α-SNAP protein. *J. Biol. Chem.* **267:** 12106.

Griffiths, G., M. Ericsson, J. Krijnse-Locker, T. Nilsson, B. Goud, H.-D. Söling, B.L. Tang, S.H. Wong, and W. Hong. 1994. Localization of the lys, asp, glu, leu tetrapeptide receptor to the Golgi complex and the intermediate compartment in mammalian cells. *J. Cell Biol.* **127:** 1557.

Groesch, M.E., H. Ruohola, R. Bacon, G. Rossi, and S. Ferro-Novick. 1990. Isolation of a functional vesicular intermediate that mediates ER to Golgi transport in yeast. *J. Cell Biol.* **111:** 45.

Gruenberg, J. and F.R. Maxfield. 1995. Membrane transport in the endocytic pathway. *Curr. Opin. Cell Biol.* **7:** 552.

Guan, K., L. Farh, T.K. Marshall, and R.J. Deschenes. 1993. Normal mitochondrial structure and genome maintenance in yeast requires the dynamin-like product of the *MGM1* gene. *Curr. Genet.* **24:** 141.

Gunther, R., C. Brauer, B. Janetzky, H.H. Forster, I.M. Ehbrecht, L. Lehle, and H. Kuntzel. 1991. The *Saccharomyces cerevisiae TRG1* gene is essential for growth and encodes a lumenal endoplasmic reticulum glycoprotein involved in the maturation of vacuolar carboxypeptidase. *J. Biol. Chem.* **266:** 24557.

Haguenauer-Tsapis, R. and A. Hinnen. 1984. A deletion that includes the signal peptidase cleavage site impairs processing, glycosylation, and secretion of cell surface yeast acid phosphatase. *Mol. Cell. Biol.* **4:** 2668.

Hammond, C. and A. Helenius. 1995. Quality control in the secretory pathway. *Curr. Opin. Cell Biol. 7:* 523.

Hampton, R.Y. and J. Rine. 1994. Regulated degradation of HMG-CoA reductase, and integral membrane protein of the endoplasmic reticulum, in yeast. *J. Cell Biol.* **125:** 299.

Hann, B.C. and P. Walter. 1991. The signal recognition particle in *S. cerevisiae*. *Cell* **67:** 131.

Hann, B.C., M.A. Poritz, and P. Walter. 1989. *Saccharomyces cerevisiae* and *Schizosaccharomyces pombe* contain a homologue to the 54-kD subunit of the signal recognition particle that in *S. cerevisiae* is essential for growth. *J. Cell Biol.* **109:** 3223.

Hann, B.C., C.J. Stirling, and P. Walter. 1992. *SEC65* gene product is a subunit of the yeast signal recognition particle required for its integrity. *Nature* **356:** 532.

Hansen, W. and P. Walter. 1988. Prepro-carboxypeptidase Y and a truncated form of preinvertase, but not full-length invertase, can be posttranslationally translocated across microsomal membranes from *Saccharomyces cerevisiae*. *J. Cell Biol.* **106:** 1075.

Hansen, W., P.D. Garcia, and P. Walter. 1986. *In vitro* protein translocation across the yeast endoplasmic reticulum: ATP-dependent posttranslational translocation of the prepro-α-factor. *Cell* **45:** 397.

Hardwick, K.G. and H.R.B. Pelham. 1990. ERS1 a seven transmembrane domain protein from *Saccharomyces cerevisiae*. *Nucleic Acids Res.* **18:** 2177.

———. 1992. *SED5* encodes a 39-kD integral membrane protein required for vesicular transport between the ER and the Golgi complex. *J. Cell Biol.* **119:** 513.

Hardwick, K.G., J.C. Boothroyd, A.D. Rudner, and H.R.B. Pelham. 1992. Genes that allow yeast cells to grow in the absence of the HDEL receptor. *EMBO J.* **11:** 4187.

Hardwick, K.G., M.J. Lewis, J. Semenza, N. Dean, and H.R.B. Pelham. 1990. *ERD1*, a yeast gene required for the retention of luminal endoplasmic reticulum proteins, affects glycoprotein processing in the Golgi apparatus. *EMBO J.* **9:** 623.

Harris, S.L. and M.G. Waters. 1996. Localization of a yeast early Golgi mannosyltransferase, Och1p, involves retrograde transport. *J. Cell Biol.* **132:** 985.

Harsay, E. and A. Bretscher. 1995. Parallel secretory pathways to the cell surface in yeast. *J. Cell Biol.* **131:** 297.

Harter, C., J. Pavel, F. Coccia, E. Draken, S. Wegehingel, H. Tschochner, and F. Wieland. 1996. Nonclathrin coat protein γ, a subunit of coatomer, binds to the cytoplasmic dilysine motif of membrane proteins of the early secretory pathway. *Proc. Natl. Acad. Sci.* **93:** 1902.

Hartmann, E., T. Sommer, S. Prehn, D. Görlich, S. Jentsch, and T. Rapoport. 1994. Evolutionary conservation of components of the protein translocation complex. *Nature* **367:** 654.

Hata, Y., C.A. Slaughter, and T.C. Südhof. 1993. Synaptic vesicle fusion complex contains *unc-18* homologue bound to syntaxin. *Nature* **366:** 347.

Hay, J.C., H. Hirling, and R.H. Scheller. 1996. Mammalian vesicle trafficking proteins of the endoplasmic reticulum and Golgi apparatus. *J. Biol. Chem.* **271:** 5671.

Hein, C., J.-Y. Springael, C. Volland, R. Haguenauer-Tsapis, and B. André. 1995. *NPI1*, an essential yeast gene involved in induced degradation of Gap1 and Fur4 permeases, encodes the Rsp5 ubiquitin-protein ligase. *Mol. Microbiol.* **18:** 77.

Herman, P.K. and S.D. Emr. 1990. Characterization of *VPS34*, a gene required for vacuolar protein sorting and vacuole segregation in *Saccharomyces cerevisiae. Mol. Cell. Biol.* **10:** 6742.

Hess, D.T., T.M. Slater, M.C. Wilson, and J.H.P. Skene. 1992. The 25 kDa synaptosomal-associated protein SNAP-25 is the major methionine-rich polypeptide in rapid axonal transport and a major substrate for palmitoylation in adult CNS. *J. Neurosci.* **12:** 4634.

Hicke, L. and H. Riezman. 1996. Ubiquitination of a yeast plasma membrane receptor signals its ligand-stimulated endocytosis. *Cell* **84:** 277.

Hicke, L. and R. Schekman. 1989. Yeast Sec23p acts in the cytoplasm to promote protein transport from the endoplasmic reticulum to the Golgi complex *in vivo* and *in vitro*. *EMBO J.* **8:** 1677.

Hicke, L., T. Yoshihisa, and R. Schekman. 1992. Sec23p and a novel 105-kDa protein function as a multimeric complex to promote vesicle budding and protein transport from the endoplasmic reticulum. *Mol. Biol. Cell* **3:** 667.

Hill, K.J. and T.H. Stevens. 1994. Vma21p is a yeast membrane protein retained in the endoplasmic reticulum by a di-lysine motif and is required for the assembly of the vacuolar H^+-ATPase complex. *Mol. Biol. Cell* **5:** 1039.

———. 1995. Vma22p is a novel endoplasmic reticulum-associated protein required for assembly of the yeast vacuolar H^+-ATPase complex. *J. Biol. Chem.* **270:** 22329.

Hiller, M.M., A. Finger, M. Schweiger, and D.H. Wolf. 1996. ER degradation of a misfolded luminal protein by the cytosolic ubiquitin-proteasome pathway. *Science* **273:** 1725.

Hirata, R., N. Umemoto, M.N. Ho, Y. Ohya, T.H. Stevens, and Y. Anraku. 1993. VMA12 is essential for the assembly of the vacuolar H^+-ATPase subunits onto the vacuolar membrane in *Saccharomyces cerevisiae. J. Biol. Chem.* **268:** 961.

Hirsch, H.H., H.H. Schiffer, H. Muller, and D.H. Wolf. 1992. Biogenesis of the yeast vacuole (lysosyme). Mutation in the active site of the vacuolar serine proteinase yscB abolishes proteolytic maturation of its 73-kDa precursor to the 41.5-kDa pro-enzyme and a newly detected 41-kDa peptide (erratum appears in *Eur. J. Biochem.* [1992] **205:** 1217). *Eur. J. Biochem.* **203:** 641.

Hochstrasser, M. 1996. Protein degradation or regulation: Ub the judge. *Cell* **84:** 813.

Holcomb, C.L., W.J. Hansen, T. Etcheverry, and R. Schekman. 1988. Secretory vesicles externalize the major plasma membrane ATPase in yeast. *J. Cell Biol.* **106:** 641.

Holtzman, D.A., S. Yang, and D.G. Drubin. 1993. Synthetic-lethal interactions identify two novel genes, *SLA1* and *SLA2*, that control membrane cytoskeleton assembly in *Saccharomyces cerevisiae. J. Cell Biol.* **122:** 635.

Hong, E., A.R. Davidson, and C.A. Kaiser. 1996. A pathway for targeting soluble misfolded proteins to the yeast vacuole. *J. Cell Biol.* **135:** 623.

Horazdovsky, B.F., G.R. Busch, and S.D. Emr. 1994. *VPS21* encodes a rab5-like GTP binding protein that is required for the sorting of yeast vacuolar proteins. *EMBO J.* **13:** 1297.

Hosobuchi, M., T. Kreis, and R. Schekman. 1992. *SEC21* is a gene required for ER to Golgi protein transport that encodes a subunit of a yeast coatomer. *Nature* **360:** 603.

Huffaker, T.C., J.H. Thomas, and D. Botstein. 1988. Diverse effects of β-tubulin mutations on microtubule formation and function. *J. Cell Biol.* **106:** 1997.

Ireland, L.S., G.C. Johnston, M.A. Drebot, N. Dhillon, A.J. DeMaggio, M.F. Hoekstra, and R.A. Singer. 1994. A member of a novel family of yeast "Zn-finger" proteins mediates the transition from stationary phase to cell proliferation. *EMBO J.* **13:** 3812.

Itin, C., F. Kappeler, A.D. Linstedt, and H.P. Hauri. 1995. A novel endocytosis signal related to the KKXX ER-retrieval signal. *EMBO J.* **14:** 2250.

Jackson, M.R., T. Nilsson, and P.A. Peterson. 1990. Identification of a consensus motif for retention of transmembrane proteins in the endoplasmic reticulum. *EMBO J.* **9:** 3153.

Jacobs, C.W., A.E.M. Adams, P.J. Szaniszlo, and J.R. Pringle. 1988. Functions of microtubules in the *Saccharomyces cerevisiae* cell cycle. *J. Cell Biol.* **107:** 1409.

Jamsa, E., M. Simonen, and M. Makarow. 1994. Selective retention of secretory proteins in the yeast endoplasmic reticulum by treatment of cells with reducing agent. *Yeast* **10:** 355.

Jedd, G., C. Richardson, R. Litt, and N. Segev. 1995. The Ypt1 GTPase is essential for the first two steps of the yeast secretory pathway. *J. Cell Biol.* **131:** 583.

Jenness, D.D. and P. Spatrick. 1986. Down regulation of the α-factor pheromone receptor in *S. cerevisiae. Cell* **46:** 345.

Jiang, Y. and S. Ferro-Novick. 1994. Identification of yeast component A: Reconstitution of the geranylgeranyltransferase that modifies Ypt1p and Sec4p. *Proc. Natl. Acad. Sci.* **91:** 4377.

Jiang, Y., G. Rossi, and S. Ferro-Novick. 1993. Bet2p and Mad2p are components of a prenyltransferase that adds geranylgeranyl onto Ypt1p and Sec4p. *Nature* **366:** 84.

John, J., I. Schlichting, E. Schiltz, P. Rösch, and A. Wittinghofer. 1989. C-terminal truncation of $p21^H$ preserves crucial kinetic and structural properties. *J. Biol. Chem.* **264:** 13086.

Johnston, G.C., J.A. Prendergast, and R.A. Singer. 1991. The *Saccharomyces cerevisiae MYO2* gene encodes an essential myosin for vectorial transport of vesicles. *J. Cell Biol.* **113:** 539.

Jones, S., R.J. Litt, C.J. Richardson, and N. Segev. 1995. Requirement of nucleotide exchange factor for Ypt1 GTPase mediated protein transport. *J. Cell Biol.* **130:** 1051.

Kabcenell, A.K., B. Goud, J.K. Northrup, and P.J. Novick. 1990. Binding and hydrolysis of guanine nucleotides by Sec4p, a yeast protein involved in the regulation of vesicular traffic. *J. Biol. Chem.* **265:** 9366.

Kahn, R.A., J. Clark, C. Rulka, T. Stearns, C.-J. Zhang, P.A. Randazzo, T. Terui, and M. Cavenaugh. 1995. Mutational analysis of *Saccharomyces cerevisiae ARF1. J. Biol. Chem.* **270:** 143.

Kaiser, C. 1993. Protein transport from the endoplasmic reticulum to the Golgi apparatus. In *Mechanisms of intracellular trafficking and processing of proproteins* (ed. Y.P. Loh), p. 79. CRC Press, Boca Raton, Florida.

Kaiser, C.A. and D. Botstein. 1986. Secretion-defective mutations in the signal sequence for *Saccharomyces cerevisiae* invertase. *Mol. Cell. Biol.* **6:** 2382.

Kaiser, C.A. and R. Schekman. 1990. Distinct sets of *SEC* genes govern transport vesicle formation and fusion early in the secretory pathway. *Cell* **61:** 723.

Kaiser, C.A., D. Preuss, P. Grisafi, and D. Botstein. 1987. Many random sequences functionally replace the secretion signal sequence of yeast invertase. *Science* **235:** 312.

Kanik-Ennulat, C. and N. Neff. 1990. Vanadate-resistant mutants of *Saccharomyces cerevisiae* show alterations in protein phosphorylation and growth control. *Mol. Cell. Biol.* **10:** 898.

Kanik-Ennulat, C., E. Montalvo, and N. Neff. 1995. Sodium orthovanadate-resistant mutants of *Saccharomyces cerevisiae* show defects in Golgi-mediated protein glycosylation, sporulation and detergent resistance. *Genetics* **140:** 933.

Kaplan, M.R. and R.D. Simoni. 1985. Intracellular transport of phosphatidylcholine to the plasma membrane. *J. Cell Biol.* **101:** 441.

Kilmartin, J.V. and A.E.M. Adams. 1984. Structural rearrangements of tubulin and actin during the cell cycle of the yeast *Saccharomyces. J. Cell Biol.* **98:** 922.

Kinsella, B.T. and W.A. Maltese. 1991. Rab GTP-binding proteins implicated in vesicular transport are isoprenylated *in vitro* at cysteines within a novel carboxyl-terminal motif. *J. Biol. Chem.* **266:** 8540.

Kirchhausen, T. 1990. Indentification of a putative yeast homolog of the mammalian β chains of the clathrin-associated protein complexes. *Mol. Cell. Biol.* **10:** 6089.

Kirchhausen, T., A.C. Davis, S. Frucht, B.O. Greco, G.S. Payne, and B. Tubb. 1991. AP17 and AP19, the mammalian small chains of the clathrin-associated protein complexes show homology to Yap17p, their putative homolog in yeast. *J. Biol. Chem.* **266:** 11153.

Kohno, K., K. Normington, J. Sambrook, M.-J. Gething, and K. Mori. 1993. The promoter region of the yeast *KAR2* (BiP) gene contains a regulatory region domain that responds to the presence of unfolded proteins in the endoplasmic reticulum. *Mol. Cell Biol.* **13:** 877.

Kölling, R. and C.P. Hollenberg. 1994. The ABC-transporter Ste6 accumulates in the plasma membrane in a ubiquinated form in endocytosis mutants. *EMBO J.* **13:** 3261.

Konopka, J.B., D.D. Jenness, and L.H. Hartwell. 1988. The C-terminus of the *S. cerevisiae* α-pheromone receptor mediates and adaptive response to pheromone. *Cell* **54:** 609.

Kornfeld, S. 1992. Structure and function of the mannose 6-phosphate/insulin-like growth factor II receptors. *Annu. Rev. Biochem.* **61:** 307.

Kübler, E. and H. Riezman. 1993. Actin and fimbrin are required for the internalization step of endocytosis in yeast. *EMBO J.* **12:** 2855.

Kübler, E., F. Schimmöller, and H. Riezman. 1994. Calcium-independent calmodulin requirement for endocytosis in yeast. *EMBO J.* **13:** 5539.

Kuge, O., C. Dascher, L. Orci, T. Rowe, M. Amherdt, H. Plutner, M. Ravazzola, G. Tanigawa, J.E. Rothman, and W.E. Balch. 1994. Sar1 promotes vesicle budding from the endoplasmic reticulum but not Golgi compartments. *J. Cell Biol.* **125:** 51.

Kurihara, T. and P. Silver. 1993. Suppression of a Sec63 mutation identifies a novel component of the yeast endoplasmic reticulum translocation apparatus. *Mol. Biol. Cell* **4:** 919.

Laboissiere, M.C., S.L. Sturley, and R.T. Raines. 1995. The essential function of protein-disulfide isomerase is to unscramble non-native disulfide bonds. *J. Biol. Chem.* **270:** 28006.

Lai, K., C.P. Bolognese, S. Swift, and P. McGraw. 1995. Regulation of inositol transport in *Saccharomyces cerevisiae* involves inositol-induced changes in permease stability and endocytic degradation in the vacuole. *J. Biol. Chem.* **270:** 2525.

LaMantia, M. and W.J. Lennarz. 1993. The essential function of yeast protein disulfide isomerase does not reside in its isomerase activity. *Cell* **74:** 899.

LaMantia, M., T. Miura, H. Tachikawa, A. Kaplan, W.J. Lennarz, and T. Mizunaga. 1991. Glycosylation site binding protein and protein disulfide isomerase are identical and essential for cell viability in yeast. *Proc. Natl. Acad. Sci.* **88:** 4453.

Lamaze, C. and S.L. Schmid. 1995. The emergence of clathrin-independent pinocytic pathways. *Curr. Opin. Cell Biol.* **7:** 573.

Lemmon, S.K. and E.W. Jones. 1987. Clathrin requirement for normal growth of yeast. *Science* **238:** 504.

Lemmon, S.K., V.P. Lemmon, and E.W. Jones. 1988. Characterization of yeast clathrin and anticlathrin heavy chain monoclonal antibodies. *J. Cell. Biochem.* **36:** 329.

Lemmon, S.K., A. Pellicena-Palle, K. Conley, and C.L. Freund. 1991. Sequence of the clathrin heavy chain from *Saccharomyces cerevisiae* and requirement of the COOH terminus for clathrin function. *J. Cell Biol.* **112:** 65.

Letourneur, F. and R.D. Klausner. 1992. A novel di-leucine motif and a tyrosine-based motif independently mediate lysosomal targeting and endocytosis of CD3 chains. *Cell* **69:** 1143.

Letourneur, F., E.C. Gaynor, S. Hennecke, C. Démollière, R. Duden, S.D. Emr, H. Riezman, and P. Cosson. 1994. Coatomer is essential for retrieval of dilysine-tagged proteins to the endoplasmic reticulum. *Cell* **79:** 1199.

Lewis, M.J. and H.R.B. Pelham. 1990. A human homologue of the yeast HDEL receptor. *Nature* **348:** 162.

———. 1992. Ligand-induced redistribution of human KDEL receptor from the Golgi complex to the endoplasmic reticulum. *Cell* **68:** 353.

———. 1996. SNARE-mediated retrograde traffic from the Golgi complex to the endoplasmic reticulum. *Cell* **85:** 205.

Lewis, M.J., D.J. Sweet, and H.R.B. Pelham. 1990. The *ERD2* gene determines the specificity of the luminal ER protein retention system. *Cell* **61:** 1359.

Li, R., C. Havel, J.A. Watson, and A.W. Murray. 1993. The mitotic feedback control gene *MAD2* encodes the α-subunit of a prenyltransferase (see correction). *Nature* **366:** 82.

———. 1994. The mitotic feedback control gene *MAD2* encodes the α-subunit of a prenyltransferase (correction). *Nature* **371:** 438.

Lian, J.P. and S. Ferro-Novick. 1993. Bos1p, an integral membrane protein of the endoplasmic reticulum to Golgi transport vesicles, is required for their fusion competence. *Cell* **73:** 735.

Lian, J.P., S. Stone, Y. Jiang, P. Lyons, and S. Ferro-Novick. 1994. Ypt1p implicated in v-SNARE activation. *Nature* **372:** 698.

Liberek, K., J. Marszalek, D. Ang, C. Georgopoulos, and M. Zylicz. 1991. *Escherichia coli* DnaJ and GrpE heat shock proteins jointly stimulate ATPase activity of DnaK. *Proc. Natl. Acad. Sci.* **88:** 2874.

Lillie, S.H. and S.S. Brown. 1992. Suppression of a myosin defect by a kinesin-related gene. *Nature* **356:** 358.

———. 1994. Immunofluorescence localization of the unconventional myosin, Myo2p, and the putative kinesin-related protein, Smy1p, to the same regions of polarized growth in *Saccharomyces cerevisiae*. *J. Cell Biol.* **125:** 825.

Liu, H. and A. Bretscher. 1989. Disruption of the single tropomyosin gene in yeast results in the disappearance of actin cables from the cytoskeleton. *Cell* **57:** 233.

———. 1992. Characterization of *TPM1* disrupted yeast cells indicates an involvement of tropomyosin in directed vesicular transport. *J. Cell Biol.* **118:** 285.

Ljungdahl, P.O., C.J. Gimeno, C.A. Styles, and G.R. Fink. 1992. *SHR3:* A novel component of the secretory pathway specifically required for localization of amino acid permeases in yeast. *Cell* **71:** 463.

Lupashin, V.V., S. Hamamoto, and R.W. Schekman. 1996. Biochemical requirements for the targeting and fusion of ER-derived transport vesicles with purified yeast Golgi membranes. *J. Cell Biol.* **132:** 277.

Lussier, M., A.M. Sdicu, T. Ketela, and H. Bussey. 1995. Localization and targeting of the *Saccharomyces cerevisiae* Kre2p/Mnt1p α-1,2-mannosyltransferase to a medial-Golgi compartment. *J. Cell Biol.* **131:** 913.

Lyman, S.K. and R. Schekman. 1995. Interaction between BiP and Sec63p is required for the completion of protein translocation into the ER of *Saccharomyces cerevisiae*. *J. Cell Biol.* **131:** 1163.

Marcusson, E.G., B.F. Horazdovsky, J.L. Cereghino, E. Gharakhanian, and S.D. Emr. 1994. The sorting receptor for yeast vacuolar carboxypeptidase Y is encoded by the *VPS10* gene. *Cell* **77:** 579.

Marshall, C.J. 1993. Protein prenylation: A mediator of protein-protein interactions. *Science* **259:** 1865.

Mayinger, P., V.A. Bankaitis, and D.I. Meyer. 1995. Sac1p mediates the adenosine triphosphate transport into yeast endoplasmic reticulum that is required for protein translocation. *J. Cell Biol.* **131:** 1377.

McCracken, A.A. and J.L. Brodsky. 1996. Assembly of ER-associated protien degradation in vitro: Dependence on cytosol, calnexin, and ATP. *J. Cell Biol.* **132:** 291.

McCracken, A.A. and K.B. Kruse. 1993. Selective protein degradation in the yeast exocytic pathway. *Mol. Biol. Cell* **4:** 729.

McGee, T.P., H.B. Skinner, E.A. Whitters, S.A. Henry, and V.A. Bankaitis. 1994. A phosphatidylinositol transfer protein controls the phosphatidylcholine content of yeast Golgi membranes. *J. Cell Biol.* **124:** 273.

Meaden, P., K. Hill, J. Wagner, D. Slipetz, S.S. Sommer, and H. Bussey. 1990. The yeast *KRE5* gene encodes a probable endoplasmic reticulum protein required for (1→6)-β-D-glucan synthesis and normal cell growth. *Mol. Cell. Biol.* **10:** 3013.

Miller, J.D., S. Tajima, L. Lauffer, and P. Walter. 1995. The β subunit of the signal recognition particle receptor is a transmembrane GTPase that anchors the α subunit, a peripheral membrane GTPase, to the endoplasmic reticulum membrane. *J. Cell Biol.* **128:** 273.

Milstein, C., G.G. Brownlee, T.M. Harrison, and M.D. Mathews. 1972. A possible precursor of immunoglobulin light chains. *Nat. New Biol.* **239:** 117.

Mizuno, M. and S.J. Singer. 1993. A soluble secretory protein is first concentrated in the endoplasmic reticulum before transfer to the Golgi apparatus. *Proc. Natl. Acad. Sci.* **90:** 5732.

Moehle, C.M., C.K. Dixon, and E.W. Jones. 1989. Processing pathway for protease B of *Saccharomyces cerevisiae*. *J. Cell Biol.* **108:** 309.

Molenaar, C.M.T., R. Prange, and D. Gallwitz. 1988. A carboxyl-terminal cysteine residue is required for palmitic acid binding and biological activity of the *ras*-related yeast *YPT1* protein. *EMBO J.* **7:** 971.

Moores, S.L., M.D. Schaber, S.D. Mosser, E. Rands, M.B. O'Hara, V.M. Garsky, M.S. Marshall, D.L. Pompliano, and J.B. Gibbs. 1991. Sequence dependence of protein isoprenylation. *J. Biol. Chem.* **266:** 14603.

Morgan, A., R. Dimaline, and R.D. Burgoyne. 1994. The ATPase activity of N-ethylmaleimide-sensitive fusion protein (NSF) is regulated by soluble NSF attachment proteins. *J. Biol. Chem.* **269:** 29347.

Mori, K., W. Ma, M. Gething, and J. Sambrook. 1993. A transmembrane protein with a cdc^{2+}/CDC28-related kinase activity is required for signaling from the ER to the nucleus. *Cell* **74:** 743.

Moya, M., D. Roberts, and P. Novick. 1993. *DSS4-1* is a dominant suppressor of *sec4-8* that encodes a nucleotide exchange protein that aids Sec4p function. *Nature* **361:** 460.

Mueller, S.C. and D. Branton. 1984. Identification of coated vesicles in *Saccharomyces cerevisiae*. *J. Cell Biol.* **98:** 341.

Munn, A.L. and H. Riezman. 1994. Endocytosis is required for the growth of vaculoar H^+-ATPase-defective yeast: Identification of six new *END* genes. *J. Cell Biol.* **127:** 373.

Munn, A.L., B.J. Stevenson, M.I. Geli, and H. Riezman. 1995. *end5*, *end6*, and *end7:* Mutations that cause actin delocalization and block the internalization step of endocytosis in *Saccharomyces cerevisiae*. *Mol. Biol. Cell* **6:** 1721.

Munro, S. and H.R.B. Pelham. 1987. A C-terminal signal prevents secretion of luminal ER proteins. *Cell* **48:** 899.

Müsch, A., M. Weidman, and T. Rapaport. 1992. Yeast Sec proteins interact with polypeptides traversing the endoplasmic reticulum membrane. *Cell* **69:** 343.

Na, S., M. Hincapie, J.H. McCusker, and J.E. Haber. 1995. *MOP2* (*SLA2*) affects the abundance of the plasma membrane H^+-ATPase of *Saccharomyces cerevisiae*. *J. Biol. Chem.* **270:** 6815.

Nair, J., H. Müller, M. Peterson, and P. Novick. 1990. Sec2 protein contains a coiled-coil domain essential for vesicular transport and a dispensable carboxy terminal domain. *J. Cell Biol.* **110:** 1897.

Nakai, M., T. Takada, and T. Endo. 1993. Cloning of the *YAP19* gene encoding a putative yeast homolog of AP19, the mammalian small chain of the clathrin-assembly proteins. *Biochim. Biophys. Acta* **1174:** 282.

Nakajima, H., A. Hirata, Y. Ogawa, T. Yonehara, K. Yoda, and M. Yamasaki. 1991. A cytoskeleton-related gene, *USO1*, is required for intracellular protein transport in *Saccharomyces cerevisiae*. *J. Cell Biol.* **113:** 245.

Nakano, A. and M. Muramatsu. 1989. A novel GTP-binding protein, Sar1p, is involved in transport from the endoplasmic reticulum to the Golgi apparatus. *J. Cell Biol.* **109:** 2677.

Nakano, A., D. Brada, and R. Schekman. 1988. A membrane glycoprotein, Sec12p, required for protein transport from the endoplasmic reticulum to the Golgi apparatus in yeast. *J. Cell Biol.* **107:** 851.

Nakayama, Y., M. Goebl, B.O. Greco, S. Lemmon, E.P. Chow, and T. Kirchhausen. 1991. The medium chains of the mammalian clathrin-associated proteins have a homolog in yeast. *Eur. J. Biochem.* **202:** 569.

Nebes, V.L. and E.W. Jones. 1991. Activation of the proteinase B precursor of the yeast *Saccharomyces cerevisiae* by autocatalysis and by an internal sequence. *J. Biol. Chem.* **266:** 22851.

Nelson, H. and N. Nelson. 1990. Disruption of genes encoding subunits of yeast vacuolar H(+)-APTase causes conditional lethality. *Proc. Natl. Acad. Sci.* **87:** 3503.

Newman, A.P. and S. Ferro-Novick. 1987. Characterization of new mutants in the early part of the yeast secretory pathway isolated by a [^{3}H]mannose suicide selection. *J. Cell Biol.* **105:** 1587.

Newman, A.P., M.E. Groesch, and S. Ferro-Novick. 1992a. Bos1p, a membrane protein required for ER to Golgi transport in yeast, co-purifies with the carrier vesicles and with Bet1p and the ER membrane. *EMBO J.* **11:** 3609.

Newman, A.P., J. Shim, and S. Ferro-Novick. 1990. *BET1*, *BOS1*, and *SEC22* are members of a group of interacting yeast genes required for transport from the endoplasmic reticulum to the Golgi complex. *Mol. Cell. Biol.* **10:** 3405.

Newman, A.P., J. Graf, P. Mancini, G. Rossi, J.P. Lian, and S. Ferro-Novick. 1992b. *SEC22* and *SLY2* are identical (letter). *Mol. Cell. Biol.* **12:** 3663.

Ng, D.T.W. and P. Walter. 1996. ER membrane protein complex required for nuclear fusion. *J. Cell Biol.* **132:** 499.

Ng, D.T.W., J.D. Brown, and P. Walter. 1996. Signal sequences specify the targeting route to the endoplasmic reticulum membrane. *J. Cell Biol.* **134:** 269.

Nikawa, J.I. and S. Yamashita. 1992. *IRE1* encodes a putative protein kinase containing a membrane-spanning domain and is required for inositol protophy in *Saccharomyces cerevisiae*. *Mol. Microbiol.* **6:** 1441.

Nilsson, T., M. Jackson, and P.A. Peterson. 1989. Short cytoplasmic sequences serve as retention signals for transmembrane proteins in the endoplasmic reticulum. *Cell* **58:** 707.

Nishikawa, S. and A. Nakano. 1991. The GTP-binding Sar1 protein is localized to the early compartment of the yeast secretory pathway. *Biochim. Biophys. Acta* **1093:** 135.

———. 1993. Identification of a gene required for membrane protein retention in the early secretory pathway. *Proc. Natl. Acad. Sci.* **90:** 8179.

Normington, K., K. Kohno, Y. Kozutsumi, M.J. Gething, and J. Sambrook. 1989. *S. cerevisiae* encodes an essential protein homologous in sequence and function to mammalian BiP. *Cell* **57:** 1223.

Nothwehr, S.F. and T.H. Stevens. 1994. Sorting of membrane proteins in the yeast secretory pathway. *J. Biol. Chem.* **269:** 10185.

Nothwehr, S.F., E. Conibear, and T.H. Stevens. 1995. Golgi and vacuolar membrane proteins reach the vacuole in *vps1* mutant yeast cells via the plasma membrane. *J. Cell Biol.* **129:** 35.

Nothwehr, S.F., C.J. Roberts, and T.H. Stevens. 1993. Membrane protein retention in the yeast Golgi apparatus: Dipeptidyl aminopeptidase A is retained by a cytoplasmic signal containing aromatic residues. *J. Cell Biol.* **121:** 1197.

Novick, P. and D. Botstein. 1985. Phenotypic analysis of temperature-sensitive yeast actin mutants. *Cell* **40:** 405.

Novick, P. and M.D. Garrett. 1994. No exchange without receipt. *Nature* **369:** 18.

Novick, P. and R. Schekman. 1979. Secretion and cell-surface growth are blocked in a temperature-sensitive mutant of *Saccharomyces cerevisiae. Proc. Natl. Acad. Sci.* **76:** 1858.

———. 1983. Export of major cell surface proteins is blocked in yeast secretory mutants. *J. Cell Biol.* **96:** 541.

Novick, P., S. Ferro-Novick, and R. Schekman. 1981. Order of events in the yeast secretory pathway. *Cell* **25:** 461.

Novick, P., C. Field, and R. Schekman. 1980. Identification of 23 complementation groups required for post-translational events in the yeast secretory pathway. *Cell* **21:** 205.

Novick, P., B.C. Osmond, and D. Botstein. 1989. Suppressors of yeast actin mutations. *Genetics* **121:** 659.

Nuoffer, C. and W.E. Balch. 1994. GTPases: Multifunctional molecular switches regulating vesicular traffic. *Annu. Rev. Biochem.* **63:** 949.

Ogg, S., M. Poritz, and P. Walter. 1992. The signal recognition particle receptor is important for growth and protein secretion in *Saccharomyces cerevisiae. Mol. Biol. Cell* **3:** 895.

Ohashi, M., K.J. de Vries, R. Frank, G. Snoek, V. Bankaitis, K. Wirtz, and W.B. Huttner. 1995. A role for phosphatidylinositol transfer protein in secretory vesicle formation. *Nature* **377:** 544.

Ohno, H., J. Stewart, M.C. Fournier, H. Bosshart, I. Rhee, S. Miyatake, T. Saito, A. Gallusser, T. Kirchhausen, and J.S. Bonifacino. 1995. Interaction of tyrosine-based sorting signals with clathrin-associated proteins. *Science* **269:** 1872.

Oka, T. and A. Nakano. 1994. Inhibition of GTP hydrolysis by Sar1p causes accumulation of vesicles that are a functional intermediate of the ER-to-Golgi transport in yeast. *J. Cell Biol.* **124:** 425.

Oka, T., S. Nishikawa, and A. Nakano. 1991. Reconstitution of GTP-binding Sar1 protein function in ER Golgi transport. *J. Cell Biol.* **114:** 671.

Omer, C.A. and J.B. Gibbs. 1994. Protein prenylation in eukaryotic microorganisms: Genetics, biology, and biochemistry. *Mol. Microbiol.* **11:** 219.

Orci, L., M. Ravazzola, P. Meda, C. Holcomb, H.-P. Moore, L. Hicke, and R. Schekman. 1991. Mammalian Sec23p homologue is restricted to the endoplasmic reticulum transitional cytoplasm. *Proc. Natl. Acad. Sci.* **88:** 8611.

Ordway, R.W., L. Pallanck, and B. Ganetzky. 1994. A TPR domain in the SNAP secretory proteins. *Trends Biochem. Sci.* **19:** 530.

Ossig, R., W. Laufer, H.D. Schmitt, and D. Gallwitz. 1995. Functionality and specific membrane localization of transport GTPases carrying C-terminal membrane anchors of synaptobrevin-like proteins. *EMBO J.* **14:** 3645.

Ossig, R., C. Dascher, H.-H. Trepte, H.D. Schmitt, and D. Gallwitz. 1991. The yeast *SLY* gene products, suppressors of defects in the essential GTP-binding Ypt1 protein, may act in endoplasmic reticulum-to-Golgi transport. *Mol. Cell Biol.* **11:** 2980.

Ostermann, J., L. Orci, K. Tani, M. Amerdt, M. Ravazzola, Z. Elazar, and J.E. Rothman. 1993. Stepwise assembly of functionally active transport vesicles. *Cell* **75:** 1015.

Palade, G. 1975. Intracellular aspects of the process of protein secretion. *Science* **189:** 347.

Paltauf, F., S.D. Kohlwein, and S.A. Henry. 1992. Regulation and compartmentalization of lipid synthesis in yeast. In *Molecular and cellular biology of the yeast* Saccharomyces: *Gene expression* (ed. J.R. Broach et al.), p. 415. Cold Spring Harbor Laboratory Press, Cold Spring Harbor, New York.

Panzner, S., L. Dreier, E. Hartmann, S. Kostka, and T.A. Rapoport. 1995. Posttranslational protein transport in yeast reconstituted with a purified complex of Sec proteins and Kar2p. *Cell* **81:** 561.

Parlati, F., M. Dominguez, J.J.M. Bergeron, and D.Y. Thomas. 1995. *Saccharomyces cerevisiae CNE1* encodes and endoplasmic reticulum (ER) membrane protein with sequence similarity to calnexin and calreticulin and functions as a constituent of the ER quality control apparatus. *J. Biol. Chem.* **270:** 244.

Partaledis, J.A. and V. Berlin. 1993. The *FKB2* gene of *Saccharomyces cerevisiae*, encoding the immunosuppressant-binding protein FKBP-13, is regulated in response to accumulation of unfolded proteins in the endoplasmic reticulum. *Proc. Natl. Acad. Sci.* **90:** 5450.

Payne, G.S. and R. Schekman. 1985. A test of clathrin function in protein secretion and cell growth. *Science* **230:** 1009.

———. 1989. Clathrin: A role in the intracellular retention of a Golgi membrane protein. *Science* **245:** 1358.

Payne, G.S., D. Baker, E. can Tuinen, and R. Schekman. 1988. Protein transport to the vacuole and receptor-mediated endocytosis by clathrin heavy chain-deficient yeast. *J. Cell Biol.* **106:** 1453.

Payne, G.S., T.B. Hasson, M.S. Hasson, and R. Schekman. 1987. Genetic and biochemical characterization of clathrin-deficient *Saccharomyces cerevisiae. Mol. Cell. Biol.* **7:** 3888.

Pearse, B.M.F. and M.S. Robinson. 1990. Clathrin, adaptors, and sorting. *Annu. Rev. Cell Biol.* **6:** 151.

Pelham, H.R.B. 1986. Speculations on the functions of the major heat shock and glucose regulated proteins. *Cell* **46:** 959.

———. 1988. Evidence that luminal ER proteins are sorted from secreted proteins in a post-ER compartment. *EMBO J.* **7:** 913.

———. 1994. About turn for the COPS? *Cell* **79:** 1125.

Pelham, H.R.B. and S. Munro. 1993. Sorting of membrane proteins in the secretory pathway. *Cell* **75:** 603.

Pelham, H.R.B., K.G. Hardwick, and M.J. Lewis. 1988. Sorting of soluble ER proteins in yeast. *EMBO J.* **7:** 1757.

Peter, F., H. Plutner, H. Zhu, T.E. Kreis, and W.E. Balch. 1993. β-COP is essential for transport of protein from the endoplasmic reticulum to the Golgi *in vitro*. *J. Cell Biol.* **122:** 1155.

Pevsner, J., S.-C. Hsu, J.E.A. Braun, N. Calakos, A.E. Ting, M.K. Bennett, and R.H. Scheller. 1994. Specificity and regulation of a synaptic vesicle docking complex. *Neuron* **13:** 353.

Peyroche, A., S. Paris, and C.L. Jackson. 1996. Nucleotide exchange on ARF mediated by yeast Gea1 protein. *Nature* **384:** 479.

Phan, H.L., J.A. Finlay, D.S. Chu, P.K. Tan, T. Kirchhausen, and G.S. Payne. 1994. The *Saccharomyces cerevisiae APS1* gene encodes a homolog of the small subunit of the mammalian clathrin AP-1 complex: Evidence for functional interaction with clathrin at the Golgi complex. *EMBO J.* **13:** 1706.

Piper, R.C., E.A. Whitters, and T.H. Stevens. 1994. Yeast Vps45p is a Sec1p-like protein required for the consumption of vacuole-targeted, post-Golgi transport vehicles. *Eur. J. Cell Biol.* **65:** 305.

Piper, R.C., A.A. Coope, H. Yang, and T.H. Stevens. 1995. *VPS27* controls vacuolar and endocytic traffic through a prevacuolar compartment in *Saccharomyces cerevisiae. J. Cell Biol.* **131:** 603.

Polaina, J. and J. Conde. 1982. Genes involved in the control of nuclear fusion during the sexual cycle of *Saccharomyces cerevisiae. Mol. Gen. Genet.* **186:** 253.

Poster, J.B. and N. Dean. 1996. The yeast *VRG4* gene is required for normal Golgi functions and defines a new family of related genes. *J. Biol. Chem.* **271:** 3837.

Potenza, M., R. Bowser, H. Müller, and P. Novick. 1992. *SEC6* encodes an 85 kDa soluble protein required for exocytosis in yeast. *Yeast* **8:** 549.

Preuss, D., J. Mulholland, A. Franzusoff, N. Segev, and D. Botstein. 1992. Characterization of the *Saccharomyces* Golgi complex through the cell cycle by immunoelectron microscopy. *Mol. Biol. Cell* **3:** 789.

Protopopov, V., B. Govindan, P. Novick, and J.E. Gerst. 1993. Homologs of the synaptobrevin/VAMP family of synaptic vesicle proteins function on the late secretory pathay in *S. cerevisiae. Cell* **74:** 855.

Pryer, N.K., L.J. Wuestehube, and R. Schekman. 1992. Vesicle-mediated protein sorting. *Annu. Rev. Biochem.* **61:** 471.

Pryer, N.K., N.R. Salama, R. Schekman, and C.A. Kaiser. 1993. Cytostolic Sec13p complex is required for vesicle formation from the endoplasmic reticulum *in vitro. J. Cell Biol.* **120:** 865.

Rad, M.R., H.L. Phan, L. Kirchrath, P.K. Tan, T. Kirchhausen, C.P. Hollenberg, and G.S. Payne. 1995. *Saccharomyces cerevisiae* Apl2p, a homologue of the mammalian clathrin AP β subunit, plays a role in clathrin-dependent Golgi functions. *J. Cell Sci.* **108:** 1605.

Ragnani, A., R. Teply, M. Waldherr, A. Voskova, and R.J. Schweyen. 1994. The yeast protein Mrs6p, a homologue of the rabGDI and human choroideraemia proteins, affects cytoplasmic and mitochondrial functions. *Curr. Genet.* **26:** 308.

Raths, S., J. Rohrer, F. Crausaz, and H. Riezman. 1993. *end3* and *end4:* Two mutants defective in receptor-mediated and fluid-phase endocytosis in *Saccharomyces cerevisiae. J. Cell Biol.* **120:** 55.

Raymond, C.K., I. Howald-Stevenson, C.A. Vater, and T.H. Stevens. 1992. Morphological classification of the yeast vacuolar protein sorting mutants: Evidence for a prevacuolar compartment in class E *vps* mutants. *Mol. Biol. Cell* **3:** 1389.

Redding, K., C. Holcomb, and R.S. Fuller. 1991. Immunolocalization of Kex2 protease identifies a putative late Golgi compartment in the yeast *Saccharomyces cerevisiae. J. Cell Biol.* **113:** 527.

Redding, K., M. Seeger, G.S. Payne, and R.S. Fuller. 1996. The effects of clathrin inactivation on localization of Kex2 protease are independent of the TGN localization signal in the cytosolic tail of Kex2p. *Mol. Biol. Cell* **7:** 1667.

Reneke, J.E., K.J. Blumer, W.E. Courchesne, and J. Thorner. 1988. The carboxy-terminal segment of the yeast α-factor receptor is a regulatory domain. *Cell* **55:** 221.

Rexach, M.F. and R.W. Schekman. 1991. Distinct biochemical requirements for the budding, targeting, and fusion of ER-derived transport vesicles. *J. Cell Biol.* **114:** 219.

Rexach, M.F., M. Latterich, and R.W. Schekman. 1994. Characteristics of endoplasmic reticulum-derived transport vesicles. *J. Cell Biol.* **126:** 1133.

Riballo, E., M. Herweijer, D. Wolf, and R. Lagunas. 1995. Catabolite inactivation of the yeast maltose transporter occurs in the vacuole after internalization by endocytosis. *J. Bacteriol.* **177:** 5622.

Riezman, H. 1985. Endocytosis in yeast: Several of the yeast secretory mutants are defective in endocytosis. *Cell* **40:** 1001.

Roberts, C.J., S.F. Nothwehr, and T.H. Stevens. 1992. Membrane protein sorting in the yeast secretory pathway: Evidence that the vacuole may be the default compartment. *J. Cell Biol.* **119:** 69.

Robinson, K.S., K. Lai, T.A. Cannon, and P. McGraw. 1996. Inositol transport in *Saccharomyces cerevisiae* is regulated by transcriptional and degradative endocytic mechanisms during the growth cycle that are distinct from inositol-induced regulation. *Mol. Biol. Cell* **7:** 81.

Robinson, M.S. 1994. The role of clathrin, adaptors and dynamin in endocytosis. *Curr. Opin. Cell Biol.* **6:** 538.

Rohrer, J., H. Bénédetti, B. Zanolari, and H. Riezman. 1993. Identification of a novel sequence mediating regulated endocytosis of the G protein-coupled α-pheromone receptor in yeast. *Mol. Biol. Cell* **4:** 511.

Römisch, K. and R. Schekman. 1992. Distinct processes mediate glycoprotein and glycopeptide export from the endoplasmic reticulum in *Saccharomyces cerevisiae. Proc. Natl. Acad. Sci.* **89:** 7227.

Römisch, K., J. Webb, J. Herz, S. Prehn, R. Frank, M. Vingron, and B. Dobberstein. 1989. Homology of the 54K protein of signal recognition particle, docking protein, and two *E. coli* proteins with putative GTP-binding domains. *Nature* **340:** 478.

Rose, J. and R.W. Doms. 1988. Regulation of protein export from the endoplasmic reticulum. *Annu. Rev. Cell Biol.* **4:** 257.

Rose, M.D., L.M. Misra, and J.P. Vogel. 1989. *KAR2*, a karyogamy gene, is the yeast homolog of the mammalian BiP/*GRP78* gene. *Cell* **57:** 1211.

Rossi, G., Y. Jiang, A.P. Newman, and S. Ferro-Novick. 1991. Dependence of Ypt1 and Sec4 membrane attachment on Bet2. *Nature* **351:** 158.

Rossi, G., K. Kolstad, S. Stone, F. Palluault, and S. Ferro-Novick. 1995. *BET3* encodes a novel hydrophilic protein that acts in conjunction with yeast SNAREs. *Mol. Biol. Cell* **6:** 1769.

Rothblatt, J.A. and D.I. Myers. 1986. Secretion in yeast: Reconstitution of the translocation and glycosylation of α-factor and invertase in a homologous cell-free system. *Cell* **44:** 619.

Rothblatt, J.A., R.J. Deshaies, S.L. Sanders, G. Daum, and R. Schekman. 1989. Multiple genes are required for proper insertion of secretory proteins into the endoplasmic reticulum in yeast. *J. Cell Biol.* **109:** 2641.

Rothman, J.E. 1994. Mechanisms of intracellular protein transport. *Nature* **372:** 55.

Rothman, J.H., C.K. Raymond, G. Teresa, P.J. O'Hara, and T.H. Stevens. 1990. A putative GTP binding protein homologues to interferon-inducible Mx proteins performs an essential function in yeast protein sorting. *Cell* **61:** 1063.

Rudolph, H.K., A. Antebi, G.R. Fink, C.M. Buckley, T.E. Dorman, J. Le Vitre, L.S. Davidow, J.I. Mao, and D.T. Moir. 1989. The yeast secretory pathway is perturbed by mutations in *PMR1*, a member of a Ca^{2+} ATPase family. *Cell* **58:** 133.

Ruohola, H., A.K. Kabcenell, and S. Ferro-Novick. 1988. Reconstitution of protein transport from the endoplasmic reticulum to the Golgi complex in yeast: The acceptor Golgi compartment is defective in the *sec23* mutant. *J. Cell Biol.* **107:** 1465.

Sadler, I., A. Chiang, T. Kurihara, J. Rothblatt, J. Way, and P. Silver. 1989. A yeast gene important for protein assembly into the endoplasmic reticulum and the nucleus has homology to DnaJ, an *Escherichia coli* heat shock protein. *J. Cell Biol.* **109:** 2665.

Salama, N.R., T. Yeung, and R.W. Schekman. 1993. The Sec13p complex and reconstitution of vesicle budding from the ER with purified cytosolic proteins. *EMBO J.* **12:** 4073.

Salminen, A. and P.J. Novick. 1987. A *ras*-like protein is required for a post-Golgi event in yeast secretion. *Cell* **49:** 527.

———. 1989. The Sec15 protein responds to the function of the GTP binding protein, Sec4, to control vesicular traffic in yeast. *J. Cell Biol.* **109:** 1023.

Sanders, S.L., K.M. Whitfield, J.P. Vogel, M.D. Rose, and R.W. Schekman. 1992. Sec61p and BiP directly facilitate polypeptide translocation into the ER. *Cell* **69:** 353.

Sandvig, K. and B. van Deurs. 1990. Selective modulation of the endocytic uptake of ricin and fluid phase markers without alteration in transferrin endocytosis. *J. Biol. Chem.* **265:** 6382.

Sapperstein, S.K., V.V. Lupashin, H.D. Schmitt, and M.G. Waters. 1996. Assembly of the ER to Golgi SNARE complex requires Uso1p. *J. Cell Biol.* **132:** 755.

Sapperstein, S.K., D.M. Walter, A.R. Grosvenor, J.E. Heuser, and M.G. Waters. 1995. p115 is a general vesicular transport factor related to the yeast endoplasmic reticulum to Golgi transport factor Uso1p. *Proc. Natl. Acad. Sci.* **92:** 522.

Sato, K., S. Nishikawa, and A. Nakano. 1995. Membrane protein retrieval from the Golgi apparatus to the endoplasmic reticulum (ER): Characterization of the *RER1* gene product as a component involved in ER localization of Sec12p. *Mol. Biol. Cell* **6:** 1459.

Schandel, K.A. and D.D. Jenness. 1994. Direct evidence for ligand-induced internalization of the yeast α-factor pheromone receptor. *Mol. Cell. Biol.* **14:** 7245.

Schauer, I., S. Emr, C. Gross and R. Schekman. 1985. Invertase signal and mature sequence substitutions that delay intercompartmental transport of active enzyme. *J. Cell Biol.* **100:** 1664.

Schekman, R. and L. Orci. 1996. Coat proteins and vesicle budding. *Science* **271:** 1526.

Scherens, B., E. Dubois, and F. Messenguy. 1991. Determination of the sequence of the yeast YCL313 gene localized on chromosome III. Homology with the protein disulfide isomerase (PDI gene product) of other organisms. *Yeast* **7:** 185.

Schimmöller, F. and H. Riezman. 1993. Involvement of Ypt7p, a small GTPase, in traffic from late endosome to the vacuole in yeast. *J. Cell Sc.* **106:** 823.

Schimmöller, F., B. Singer-Krüger, S. Schröder, U. Krüger, C. Barlowe, and H. Riezman. 1995. The absence of Emp24p, a component of ER-derived COPII-coated vesicles, causes a defect in transport of selected proteins to the Golgi. *EMBO J.* **14:** 1329.

Schlenstedt, G., S. Harris, B. Risse, R. Lill, and P.A. Silver. 1995. A yeast DnaJ homologue, Scj1p, can function in the endoplasmic reticulum with BiP/Kar2p via a conserved domain that specifies interactions with Hsp70s. *J. Cell Biol.* **129:** 979.

Schmitt, H.D., M. Puzicha, and D. Gallwitz. 1988. Study of a temperature-sensitive mutant of the *ras*-related *YPT1* gene product in yeast suggests a role in the regulation of intracellular calcium. *Cell* **53:** 635.

Schmitt, H.D., P. Wagner, E. Pfaff, and D. Gallwitz. 1986. The *ras*-related *YPT1* gene product in yeast: A GTP-binding protein that might be involved in microtubule organization. *Cell* **47:** 410.

Schröder, S., F. Schimmöller, B. Singer-Krüger, and H. Riezman. 1995. The Golgi-localization of yeast Emp47p depends on its di-lysine motif but is not affected by the *ret1-1* mutation in α-COP. *J. Cell Biol.* **131:** 895.

Schutze, M.-P., P.A. Peterson, and M.R. Jackson. 1994. An N-terminal double-arginine motif maintains type II membrane proteins in the endoplasmic reticulum. *EMBO J.* **13:** 1696.

Scidmore, M.A., H.H. Okamura, and M.D. Rose. 1993. Genetic interactions between *KAR2* and *SEC63*, encoding eukaryotic homologues of DnaK and DnaJ in the endoplasmic reticulum. *Mol. Biol. Cell* **4:** 1145.

Seabra, M.C., M.S. Brown, and J.L. Goldstein. 1993. Retinal degeneration in choroideremia: Deficiency of Rab geranylgeranyltransferase. *Science* **259:** 377.

Seeger, M. and G.S. Payne. 1992a. A role for clathrin in the sorting of vacuolar proteins in the Golgi complex of yeast. *EMBO J.* **11:** 2811.

———. 1992b. Selective and immediate effects of clathrin heavy chain mutations on Golgi membrane protein retention in *Saccharomyces cerevisiae*. *J. Cell Biol.* **118:** 531.

Segev, N. 1991. Mediation of the attachment or fusion step in vesicular transport by the GTP-binding Ypt1 protein. *Science* **252:** 1553.

Segev, N. and D. Botstein. 1987. The *ras*-like yeast *YPT1* gene is itself essential for growth, sporulation, and starvation response. *Mol. Cell Biol.* **7:** 2367.

Segev, N., J. Mulholland, and D. Botstein. 1988. The yeast GTP-binding *YPT1* protein and a mammalian counterpart are associated with the secretion machinery. *Cell* **52:** 915.

Semenza, J.C., K.G. Hardwick, N. Dean, and H.R.B. Pelham. 1990. *ERD2*, a yeast gene required for the receptor-mediated retrieval of luminal ER proteins from the secretory pathway. *Cell* **61:** 1349.

Seog, D.-H., M. Kito, K. Yoda, and M. Yamasaki. 1994. Uso1 protein contains a coiled-coil rod region essential for protein transport from the ER to the Golgi apparatus in *Saccharomyces cerevisiae*. *J. Biochem.* **116:** 1341.

Sewell, J.L. and R.A. Kahn. 1988. Sequences of the bovine and yeast ADP-ribosylation factor and comparison to other GTP-binding proteins. *Proc. Natl. Acad. Sci.* **85:** 4620.

Shamu, C. and P. Walter. 1996. Oligomerization and phosphoralation of the Ire1p kinase during intracellular signaling from the endoplasmic reticulum to the nucleus. *EMBO J.* **15:** 3028.

Shamu, C.W., J.S. Cox, and P. Walter. 1994. The unfolded-protein-response pathway in yeast. *Trends Cell Biol.* **4:** 56.

Shaywitz, D.A., L. Orci, M. Ravazzola, A. Swaroop, and C.A. Kaiser. 1995. Human SEC13Rp functions in yeast and is located on transport vesicles budding from the endoplasmic reticulum. *J. Cell Biol.* **128:** 769.

Shelness, G.S. and G. Blobel. 1990. Two subunits of canine signal peptidase complex are homologous to yeast SEC11 complex. *J. Biol. Chem.* **265:** 9512.

Shim, J., A.P. Newman, and S. Ferro-Novick. 1991. The *BOS1* gene encodes an essential 27-kD putative membrane protein that is required for vesicular transport from the ER to the Golgi complex in yeast. *J. Cell Biol.* **113:** 55.

Shortle, D., J.E. Haber, and D. Botstein. 1982. Lethal disruption of the yeast actin gene by integrative DNA transformation. *Science* **217:** 371.

Sidrauski, C., J.S. Cox, and P. Walter. 1996. tRNA ligase is required for regulated mRNA splicing in the unfolded protein response. *Cell* **87:** 405.

Silveira, L.A., D.H. Wong, F.R. Masiarz, and R. Schekman. 1990. Yeast clathrin has a distinctive light chain that is important for cell growth. *J. Cell Biol.* **111:** 1437.

Simon, S.M., C.S. Peskin, and G.F. Oster. 1992. What drives the translocation of proteins? *Proc. Natl. Acad. Sci.* **89:** 3370.

Simons, J.F., S. Ferro-Novick, M.D. Rose, and A. Helenius. 1995. BiP/Kar2p serves as a molecular chaperone during carboxypeptidase Y folding in yeast. *J. Cell Biol.* **130:** 41.

Singer-Krüger, B., H. Stenmark, and M. Zerial. 1995. Yeast Ypt51p and mammalian Rab5: Counterparts with similar function in the early endocytic pathway. *J. Cell Sci.* **108:** 3509.

Singer-Krüger, B., R. Frank, F. Crausaz, and H. Riezman. 1993. Partial purification and characterization of early and late endosomes from yeast. *J. Biol. Chem.* **268:** 14376.

Singer-Krüger, B., H. Stenmark, A. Düsterhöft, P. Philippsen, J.-S. Yoo, D. Gallwitz, and M. Zerial. 1994. Role of three Rab5-like GTPases, Ypt51p, Ypt52p, and Ypt53p, in the endocytic and vacuolar protein sorting pathways of yeast. *J. Cell Biol.* **125:** 283.

Siniossoglou, S., C. Wimmer, M. Rieger, V. Doye, H. Tekotte, C. Weise, S. Emig, A. Segref, and E.C. Hurt. 1996. A novel complex of nucleoporins, which includes Sec13p and a Sec13p homolog, is essential for normal nuclear pores. *Cell* **84:** 265.

Skinner, H.B., J.G.J. Alb, E.A. Whitters, G.M.J. Helmkamp, and V.A. Bankaitis. 1993. Phospholipid transfer activity is relevant to but not sufficient for the essential function of the yeast *SEC14* gene product. *EMBO J.* **12:** 4775.

Skinner, H.B., T.P. McGee, C.R. McMaster, M.R. Fry, R.B. Bell, and V.A. Bankaitis. 1995. The *Saccharomyces cerevisiae* phosphatidylinositol-transfer protein effects a ligand-dependent inhibition of choline-phosphate cytidylyltransferase activity. *Proc. Natl. Acad. Sci.* **92:** 112.

Sleight, R.G. and R.E. Pagano. 1983. Rapid appearance of newly synthesized phosphatidylethanolamine at the plasma membrane. *J. Biol. Chem.* **258:** 9050.

Søgaard, M., K. Tani, R.R. Ye, S. Geromanos, P. Tempst, T. Kirchhausen, J. Rothman, and T. Söllner. 1994. A Rab protein is required for the assembly of SNARE complexes in the docking of transport vesicles. *Cell* **78:** 937.

Söllner, T., S.W. Whiteheart, M. Brunner, H. Erdjument-Bromage, S. Geromanos, P. Tempst, and J.E. Rothman. 1993. SNAP receptors implicated in vesicle targeting and fusion. *Nature* **362:** 318.

Sommer, T. and S. Jentsch. 1993. A protein translocation defect linked to ubiquitin conjugation at the endoplasmic reticulum. *Nature* **365:** 176.

Sondek, J., A. Bohm, D.G. Lambright, H.E. Hamm, and P.B. Sigler. 1996. Crystal structure of a GA protein βγ dimer at 2.1 Å resolution. *Nature* **379:** 369.

Stack, J.H., B. Horazdovsky, and S.D. Emr. 1995. Receptor-mediated protein sorting to the vacuole in yeast: Roles for a protein kinase, a lipid kinase and GTP-binding proteins. *Annu. Rev. Cell Dev. Biol.* **11:** 1.

Stamnes, M.A. and J.E. Rothman. 1993. The binding of AP-1 clathrin adaptor particles to Golgi membranes requires ADP-ribosylation factor, a small GTP-binding protein. *Cell* **73:** 999.

Stamnes, M.A., M.W. Craighead, M.H. Hoe, N. Lampen, S. Geromanos, P. Tempst, and J.E. Rothman. 1995. A integral membrane component of coatomer-coated transport vesicles defines a family of proteins involved in budding. *Proc. Natl. Acad. Sci.* **92:** 8011.

Stearman, R., D.S. Yuan, Y. Yamaguchi-Iwai, R.D. Klausner, and A. Dancis. 1996. A permease-oxidase complex involved in high-affinity iron uptake in yeast. *Science* **271:** 1552.

Stearns, T., R.A. Kahn, D. Botstein, and M.A. Hoyt. 1990a. ADP ribosylation factor is an essential protein in *Saccharomyces cerevisiae* and is encoded by two genes. *Mol. Cell. Biol.* **10:** 6690.

Stearns, T., M.C. Willingham, D. Botstein, and R.A. Kahn. 1990b. ADP-ribosylation factor is functionally and physically associated with the Golgi complex. *Proc. Natl. Acad. Sci.* **87:** 1238.

Stenbeck, G., R. Schreiner, D. Herrmann, S. Auerbach, F. Lottspeich, J.E. Rothman, and F.T. Wieland. 1992. γ-COP, a coat subunit of non-clathrin-coated vesicles with homology to Sec21p. *FEBS Lett.* **314:** 195.

Stepp, J.D., A. Pellicena-Palle, S. Hamilton, T. Kirchhausen, and S.K. Lemmon. 1995. A late Golgi sorting function for *Saccharomyces cerevisiae* Apm1p, but not for Apm2p, a second yeast clathrin AP medium chain-related protein. *Mol. Biol. Cell* **6:** 41.

Stevens, T., B. Esmon, and R. Schekman. 1982. Early stages in the yeast secretory pathway are required for transport of carboxypeptidase Y to the vacuole. *Cell* **30:** 439.

Stirling, C.J. and E.W. Hewitt. 1992. The *S. cerevisiae SEC65* gene encodes a components of the yeast signal recognition particle with homology to human SRP19. *Nature* **356:** 534.

Stirling, C.J., J. Rothblatt, M. Hosobuchi, R. Deshaies, and R. Schekman. 1992. Protein translocation mutants defective in the insertion of integral membrane proteins into the endoplasmic reticulum. *Mol. Biol. Cell* **3:** 129.

Suarez Rendueles, P. and D.H. Wolf. 1987. Identification of the structural gene for dipeptidyl aminopeptidase yscV (DAP2) of *Saccharomyces cerevisiae. J. Bacteriol.* **169:** 4041.

Sweet, D.J. and H.R.B. Pelham. 1992. The *Saccharomyces cerevisiae SEC20* gene encodes a membrane glycoprotein which is sorted by the HDEL retrieval system. *EMBO J.* **11:** 423.

———. 1993. The *TIP1* gene of *Saccharomyces cerevisiae* encodes an 80 kDa cytoplasmic protein that interacts with the cytoplasmic domain of Sec20p. *EMBO J.* **12:** 2831.

Sztul, E., A. Kaplin, L. Saucan, and G. Palade. 1991. Protein traffic between distinct plasma membrane domains: Isolation and characterization of vesicular carriers involved in transcytosis. *Cell* **64:** 81.

Tachibana, C. and T.H. Stevens. 1992. The yeast *EUG1* gene encodes an endoplasmic reticulum protein that is functionally related to protein disulfide isomerase. *Mol. Cell. Biol.* **12:** 4601.

Tachikawa, H., Y. Takeuchi, W. Funahashi, T. Miura, X.-D. Gao, D. Fujimoto, T. Mizunaga, and K. Onodera. 1995. Isolation and characterization of a yeast gene, *MPD1*, the overexpression of which suppresses inviability cause by protein disulfide isomerase depletion. *FEBS Lett.* **369:** 212.

Tagaya, M., D.W. Wilson, M. Brunner, N. Arango, and J.E. Rothman. 1993. Domain structure of an N-ethylmaleimide-sensitive fusion protein involved in vesicular transport. *J. Biol. Chem.* **268:** 2662.

Tan, P.K., N.G. Davis, G.F. Sprague, and G.S. Payne. 1993. Clathrin facilitates the internalization of seven transmembrane segment receptors for mating pheromones in yeast. *J. Cell Biol.* **123:** 1707.

te Heesen, S., B. Janetsky, L. Lehle, and M. Aebi. 1992. The yeast *WBP1* is essential for oligosaccharyl transferase activity in vivo and in vitro. *EMBO J.* **11:** 2071.

te Heesen, S., R. Rauhut, R. Aebersold, J. Abelson, M. Aebi, and M.W. Clark. 1991. An essential 45kDa yeast transmembrane protein reacts with anti-nuclear pore antibodies:

Purification of the protein, immunolocalization and cloning of the gene. *Eur. J. Cell Biol.* **56:** 8.

TerBush, D.R. and P. Novick. 1995. Sec6, Sec8, and Sec15 are components of a multi-subunit complex which localizes to small bud tips in *Saccharomyces cerevisiae*. *J. Cell Biol.* **130:** 299.

Thompson, R.C. 1988. EFTu provides an internal kinetic standard for translational accuracy. *Trends Biochem. Sci.* **13:** 91.

Ting, A.E., C.D. Hazuka, S.-C. Hsu, M.D. Kirk, A.J. Bean, and R.H. Scheller. 1995. rSec6 and rSec8, mammalian homologs of yeast proteins essential for secretion. *Proc. Natl. Acad. Sci.* **92:** 9613.

Titus, M.A. 1993. From fat yeast and nervous mice to brain myosin-V. *Cell* **75:** 9.

Tkacz, J.S. and J.O. Lampen. 1972. Wall replication in *Saccharomyces* species: Use of fluorescein-conjugated concanavalin A to reveal the site of mannan insertion. *J. Gen. Microbiol.* **72:** 243.

———. 1973. Surface distribution of invertase on growing *Saccharomyces* cells. *J. Bacteriol.* **113:** 1073.

Townsley, F.M. and H.R.B. Pelham. 1994. The KKXX signal mediates retrieval of membrane proteins from the Golgi to the ER in yeast. *Eur. J. Cell Biol.* **64:** 211.

Townsley, F.M., G. Frigerio, and H.R.B. Pelham. 1994. Retrieval of HDEL proteins is required for growth of yeast cells. *J. Cell Biol.* **127:** 21.

Townsley, F.M., D.W. Wilson, and H.R.B. Pelham. 1993. Mutational analysis of the human KDEL receptor: Distinct structural requirements for Golgi retention, ligand binding and retrograde transport. *EMBO J.* **12:** 2821.

Toyn, J., A.R. Hibbs, P. Sanz, J. Crowe, and D.I. Meyer. 1988. *In vivo* and *in vitro* analysis of ptl1, a yeast *ts* mutant with a membrane-associated defect in protein translocation. *EMBO J.* **7:** 4347.

Traub, L.M., J.A. Ostrom, and S. Kornfeld. 1993. Biochemical dissection of AP-1 recruitment onto Golgi membranes. *J. Cell Biol.* **123:** 561.

Trowbridge, I.S., J.F. Collawn, and C.R. Hopkins. 1993. Signal-dependent membrane protein trafficking in the endocytic pathway. *Annu. Rev. Cell Biol.* **9:** 129.

Vahlensieck, Y., H. Riezman, and B. Meyhack. 1995. Transcriptional studies on yeast *SEC* genes provide no evidence for regulation at the transcriptional level. *Yeast* **11:** 901.

Valls, L.A., C.P. Hunter, J.H. Rothman, and T.H. Stevens. 1987. Protein sorting in yeast: The localization determinant of yeast vacuolar carboxypeptidase Y resides in the propeptide. *Cell* **48:** 887.

Vida, T.A., T.R. Graham, and S.D. Emr. 1990. *In vitro* reconstitution of intercompartmental protein transport to the yeast vacuole. *J. Cell Biol.* **111:** 2871.

Vida, T.A., G. Huyer, and S.D. Emr. 1993. Yeast vacuolar proenzymes are sorted in the late Golgi complex and transported to the vacuole via a prevacuolar endosome-like compartment. *J. Cell Biol.* **121:** 1245.

Vogel, J.P., L.M. Misra, and M.D. Rose. 1990. Loss of Bip/Grp78 function blocks translocation of secretory proteins in yeast. *J. Cell Biol.* **110:** 1885.

Volland, C., C. Garnier, and R. Haguenauer-Tsapis. 1992. *In vivo* phosphorylation of the yeast uracil permease. *J. Biol. Chem.* **267:** 23767.

Volland, C., D. Urban-Grimal, G. Geraud, and R. Haguenauer-Tsapis. 1994. Endocytosis and degradation of the yeast uracil permease under adverse conditions. *J. Biol. Chem.* **269:** 9833.

von Heijne, G. 1983. Patterns of amino acids near signal sequence cleavage sites. *Eur. J. Biochem.* **133:** 17.

———. 1985. Signal sequences. The limits of variation. *J. Mol. Biol.* **184:** 99.

———. 1990. The signal peptide. *J. Membr. Biol.* **115:** 195.

Wada, Y., K. Kitamoto, T. Kanbe, K. Tanaka, and Y. Anraku. 1990. The *SLP1* gene of *Saccharomyces cerevisiae* is essential for vacuolar morphogenesis and function. *Mol. Cell. Biol.* **10:** 2214.

Waldherr, M., A. Ragnani, R.J. Schweyer, and M.S. Boguski. 1993. MRS6—Yeast homologue of the choroideraemia gene (letter). *Nat. Genet.* **3:** 193.

Walter, P. and G. Blobel. 1980. Purification of a membrane-associated protein complex required for protein translocation across the endoplasmic reticulum. *Proc. Natl. Acad. Sci.* **77:** 7112.

———. 1981. Translocation of proteins across the endoplasmic reticulum. III. Signal recognition protein (SRP) causes signal sequence-dependent and site-specific arrest of chain elongation that is released by microsomal membranes. *J. Cell Biol.* **91:** 557.

Walter, P. and A.E. Johnson. 1994. Signal sequence recognition and protein targeting to the endoplasmic reticulum membrane. *Annu. Rev. Cell Biol.* **10:** 87.

Walworth, N.C., B. Goud, A.K. Kabcenell, and P.J. Novick. 1989. Mutational analysis of *SEC4* suggests a cyclical mechanism for the regulation of vesicular traffic. *EMBO J.* **8:** 1685.

Walworth, N.C., P. Brennwald, A.K. Kabcenell, M. Garrett, and P. Novick. 1992. Hydrolysis of GTP by Sec4 protein plays an important role in vesicular transport and is stimulated by a GTPase-activating protein in *Saccharomyces cerevisiae. Mol. Cell. Biol.* **12:** 2017.

Waters, M. and G. Blobel. 1986. Secretory protein translocation in a yeast cell-free system can occur post-translationally and requires ATP hydrolysis. *J. Cell Biol.* **102:** 1543.

Waters, M.G., D.O. Clary, and J.E. Rothman. 1992. A novel 115-kD peripheral membrane protein is required for intercisternal transport in the Golgi stack. *J. Cell Biol.* **118:** 1015.

Watzele, M., F. Klis, and W. Tanner. 1988. Purification and characterization of the inducible **a** agglutinin of *Saccharomyces cerevisiae. EMBO J.* **7:** 1483.

Werner, E.D., J.L. Brodsky, and A.A. McCracken. 1996. Proteasome-dependent endoplasmic reticulum-associated protein degradation: An unconventional route to a familiar fate. *Proc. Natl. Acad. Sci.* **93:** 1397.

Werner-Washburne, M., D.E. Stone, and E.A. Craig. 1987. Complex interactions among members of an essential subfamly of hsp70 genes in *Saccharomyces cerevisiae. Mol. Cell. Biol.* **7:** 2568.

Westphal, V., E.G. Marcusson, J.R. Winther, S.D. Emr, and H.B. van den Hazel. 1996. Multiple pathways for vacuolar sorting of yeast proteinase A. *J. Biol. Chem.* **271:** 11865.

Whiteheart, S.W., K. Rossnagel, S.A. Buhrow, M. Brunner, R. Jaenicke, and J.E. Rothman. 1994. N-ethylmaleimide-sensitive fusion protein: A trimeric ATPase whose hydrolysis of ATP is required for membrane fusion. *J. Cell Biol.* **126:** 945.

Whiteheart, S.W., I.C. Griff, M. Brunner, D.O. Clary, T. Mayer, S.A. Buhrow, and J.E. Rothman. 1993. SNAP family of NSF attachment proteins includes a brain-specific isoform. *Nature* **362:** 353.

Whitney, J.A., M. Gomez, D. Sheff, T.E. Kreis, and I. Mellman. 1995. Cytoplasmic coat proteins involved in endosome function. *Cell* **83:** 703.

Whitters, E.A., T.P. McGee, and V.A. Bankaitis. 1994. Purification and characterization of a late Golgi compartment from *Saccharomyces cerevisiae*. *J. Biol. Chem.* **269:** 28106.

Whitters, E.A., A.E. Cleves, T.P. McGee, H.B. Skinner, and V.A. Bankaitis. 1993. Sac1p is an integral membrane protein that influences the cellular requirement for phospholipid transfer protein function and inositol in yeast. *J. Cell Biol.* **122:** 79.

Wichmann, H., L. Hengst, and D. Gallwitz. 1992. Endocytosis in yeast: Evidence for the involvement of a small GTP-binding protein (Ypt7p). *Cell* **71:** 1131.

Wiertz, E.J., D. Tortorella, M. Bogyo, J. Yu, W. Mothes, T.R. Jones, T.A. Rapoport, and H.L. Ploegh. 1996. Sec61-mediated transfer of a membrane protein from the endoplasmic reticulum ot the proteasome for destruction. *Nature* **384:** 432.

Wilcox, C.A., K. Redding, R. Wright, and R.S. Fuller. 1992. Mutation of a tyrosine localization signal in the cytosolic tail of yeast Kex2 protease disrupts Golgi retention and results in default transport to the vacuole. *Mol. Biol. Cell* **3:** 1353.

Wilsbach, K. and G.S. Payne. 1993. Vps1p, a member of the dynamin GTPase family, is necessary for Golgi membrane protein retention in *Saccharomyces cerevisiae*. *EMBO J.* **12:** 3049.

Wilson, D.W., S.W. Whiteheart, M. Wiedman, M. Brunner, and J.E. Rothman. 1992. A multisubunit particle implicated in membrane fusion. *J. Cell Biol.* **117:** 531.

Wilson, D.W., C.A. Wilcox, G.C. Flynn, E. Chen, W.-J. Kuang, W.J. Henzel, M.R. Block, A. Ullrich, and J.E. Rothman. 1989. A fusion protein required for vesicle-mediated transport in both mammalian cells and yeast. *Nature* **339:** 355.

Winther, J.R. and K. Breddam. 1987. The free sulfhydryl group (cys 341) of carboxypeptidase Y: Functional effects of mutational substitutions. *Carlsberg Res. Commun.* **52:** 263.

Wuestehube, L.J., R. Duden, A. Eun, S. Hamamoto, P. Korn, R. Ram, and R. Schekman. 1996. New mutants of *Saccharomyces cerevisiae* affected in the transport of proteins from the endoplasmic reticulum to the Golgi complex. *Genetics* **142:** 393.

YaDeau, J., C. Klein, and G. Blobel. 1991. Yeast signal peptidase contains a glycoprotein and the Sec11 gene product. *Proc. Natl. Acad. Sci.* **88:** 517.

Yip, C.L., S.K. Welch, F. Klebl, G. Teresa, P. Seidel, F.J. Grant, P.J. O'Hara, and V.L. MacKay. 1994. Cloning and analysis of the *Saccharomyces cerevisiae MNN9* and *MNN1* genes required for complex glycosylation of secreted proteins. *Proc. Natl. Acad. Sci.* **91:** 2723.

Yoshihisa, T., C. Barlowe, and R. Schekman. 1993. Requirement for a GTPase-activating protein in vesicle budding from the endoplasmic reticulum. *Science* **259:** 1466.

Zieglhoffer, T., P. Lopez-Buesa, and E.A. Craig. 1995. The dissociation of ATP from the hsp70 of *Saccharomyces cerevisiae* is stimulated by both Ydj1p and peptide substrates. *J. Biol. Chem.* **270:** 10412.

3

Biogenesis of Yeast Wall and Surface Components

Peter Orlean
Department of Biochemistry
University of Illinois at Urbana-Champaign
Urbana, Illinois 61801

I. INTRODUCTION

The cell wall preserves the osmotic integrity of *Saccharomyces cerevisiae* and defines the morphology of the yeast cell during budding growth and during the developmental processes of mating, sporulation, and pseudohypha formation. The approximately 200-nm-wide wall endows the cell with osmotic stability and is also the framework to which biologically active proteins such as cell adhesion molecules or hydrolytic enzymes are attached or within which they act. The wall makes up 15–30% of the dry weight of the vegetative cell, and its major components, by weight, are mannoproteins (40%) and β-1,3 and β-1,6 glucans (50–60%). Chitin makes up 1–2% of the wall and is found mostly in the division septum. Some surface glycoproteins also receive a glycosyl phosphatidylinositol (GPI) membrane anchor. The plasma membrane is enriched in inositol phosphoceramides, which are cotransported to the cell surface with GPI-anchored proteins.

Yeast cell wall biogenesis is fascinating in its own right as an example of cellular morphogenesis, since the wall must expand during cell growth and a division septum must be synthesized at a specific time and place in the cell division cycle. Furthermore, yeast has proven to be a rewarding model system in which to explore biosynthetic pathways that are obviously conserved among eukaryotes, such as asparagine-linked glycosylation. However, it is emerging that seemingly specialized aspects of yeast wall biogenesis also have their parallels in other organisms. Thus, chitin-synthase-related proteins are present both in rhizobia and in vertebrates, the pathway leading to β-1,6 glucan assembly recalls the synthesis of extracellular matrix polysaccharides in plants, and analogies can be drawn between the formation of focal adhesions at the surface of mammalian cells and the formation of a bud in yeast. To regulate wall biogenesis, yeast uses familiar mechanisms; for example, it transmits signals through MAP kinase cascades, and it regulates β-1,3 glucan synthase with a GTPase. Indeed, connections are starting to be made between cell wall synthesis and fields that up to now have been explored separately, such as signal transduction, control of bud placement and emergence, and the roles of intermediate filaments (for discussion, see Cid et al. 1995; Lew et al., this volume). From a practical standpoint, *S. cerevisiae* also provides the model for wall synthesis in pathogenic fungi. Interference with wall biogenesis can lead to lysis and death of *S. cerevisiae* cells, and studies in this yeast should identify steps in cell wall assembly that are good targets for antifungal agents.

Following the pioneering work of Ballou (for review, see Ballou 1982, 1990), who isolated mutants that failed to express certain carbohydrate epitopes on their surface, a variety of strategies have been used to identify cell wall synthesis mutants and genes involved in wall assembly. Such studies are yielding an increasingly detailed picture of how a yeast cell synthesizes its wall components and have also revealed that the synthesis of much wall material is carried out along the secretory pathway. Furthermore, pairs or families of genes can be involved in the synthesis of certain wall polymers. In some cases, this has led to great redundancy among the enzymes that make a certain component, whereas in others, obviously related proteins have assumed specialized functions in the synthesis of a particular wall polymer.

Wall biogenesis is presented here in the context of the successive sites at which components are believed to be formed. We proceed from precursor formation in the cytoplasm to the secretory-pathway-based synthesis of glycoproteins, GPI anchors, and β-1,6 glucan, then to the synthesis of chitin and β-1,3 glucan at the plasma membrane, and finally, to the outside of the cell, where components are incorporated into the wall. Wall synthesis in vegetative cells is described first and then the modifications and additions made to these processes during the developmental processes of mating and sporulation. Examples are then given of how wall biogenesis can be regulated by a protein kinase C cell integrity pathway, which is in turn responsive to environmental signals and to growth signals given at the start of a round of budding. A list of the genes and mutants discussed is presented in the Appendix (see Section XI).

II. WALL COMPOSITION AND ISOLATION OF WALL MUTANTS

A. Overview of Wall and Surface Components

The major wall components of *S. cerevisiae* are mannoproteins, β-glucans, and chitin, and the plasma membrane is enriched in inositol phosphoceramides. For comprehensive reviews of cell wall structure and composition, see Arnold (1981), Fleet (1991), and Ruiz-Herrera (1992).

S. cerevisiae glycoproteins can bear both *N*-glycosidically linked carbohydrate chains and short oligosaccharides consisting of two to five α-1,2-linked mannose residues *O*-glycosidically linked to serine or threonine. The N-linked chains can be extended by the addition of an "outer chain" containing up to 200 mannose residues in a structure that consists of a backbone of α-1,6-linked mannose chains with α-1,2-mannose- and α-1,3 mannose-containing side chains (for review, see Ballou 1990; Herscovics and Orlean 1993). Such hyperglycosylated molecules in the cell wall are referred to as mannan or mannoprotein. Some of these glycoproteins can also receive a GPI membrane anchor at their carboxyl terminus, a structure that can participate in the cross-linking of glycoproteins to cell wall β-1,6 glucan. Cell surface mannoproteins can be anchored in the plasma membrane or be soluble periplasmic enzymes, as is invertase, but many are cross-linked to protein via disulfide bonds or linked to carbohydrate through glycosidic bonds. Mannoproteins can be extracted into hot citrate buffer or released by treatment with sulfhydryl reagents or upon digestion of the wall with β-glucanases (see Section VI.A). Many of the wall and surface glycoproteins that have been studied have no known catalytic activity, although the phenotypes that arise when such proteins are absent, or when their expression is induced, indicate that they have roles in morphogenesis and cell adhesion, examples of the latter being the mating agglutinins and proteins involved in flocculation (Lipke and Kurjan 1992; Stratford 1992a). Both *N*- and *O*-glycosylation in *S. cerevisiae* are initiated in the endoplasmic reticulum (ER), and glycoproteins are also found in secretory compartments, either as cargo or as resident membrane proteins. Methods for the analysis of yeast glycoproteins are discussed by Ballou (1990), Orlean et al. (1991), and Trimble and Atkinson (1992).

β-glucans make up 30–60% of the dry weight of the wall of stationary-phase cells, and three populations can be defined empirically on the basis of their solubility in acid and alkali (Fleet 1991). Approximately 35% of the dry weight of the wall is an acid- and alkali-insoluble branched β-1,3 glucan with an average degree of polymerization of 1500 residues, containing 3% of β-1,6 glucosidic interchain linkages (Manners et al. 1973a). A second fraction, contributing about 20% to the dry weight of the wall, is alkali-soluble, but it is otherwise of essentially the same composition and size as the alkali-insoluble fraction. These two fractions differ only in their degree of cross-linking to chitin, which renders the glucan alkali-insoluble (Hartland et al. 1994; see Section VI.B). Approximately 5% of the dry weight of the cell wall is an alkali-insoluble polymer with an average degree of polymerization of 140, 14% of whose β-1,6-linked glucan core consists of β-1,3-linked branch residues, which are in turn extended with β-1,6-linked side chains (Manners et al. 1973b). This polymer can become cross-linked to certain wall glycoproteins (see Section VI.A). β-1,3 glucan is the major structural component of the cell wall. Proteolytic digestion of the cell wall leaves behind a densely interwoven mesh of microfibrils (Kopecká et al. 1974), and when protoplasts are allowed to regenerate their cell wall, fibrillar nets are also formed that are crystalline aggregates of strands themselves composed of three intertwined helical glucan chains (Kopecká and Kreger 1986).

Chitin, a homopolymer of β-1,4-linked *N*-acetylglucosamine (GlcNAc) residues, represents 1–2% of the dry weight of the wall of vegetative cells. Chitin makes up the ring at the base of the emerging bud, forms the primary septum, and is also laid down during thickening of the secondary septa. A small amount of chitin is also deposited in the lateral wall.

Wall composition varies in response to the different developmental programs followed during budding growth, mating, sporulation and, presumably, during pseudohyphal growth. Cell wall β-glucan and mannoprotein compositions also vary with the growth phase of vegetative cells (McMurrough and Rose 1967; Valentin et al. 1987; De Nobel et al. 1991). Nevertheless, assessment of mannose/glucose ratios in cell walls is a sensitive way to detect mutants containing low levels of either mannan or glucan in their walls. Normally, the mannose/glucose ratio is about 1, but this number can vary between 0.16 and 3.9 in mutants (Klis 1994; Ram et al. 1994). Indeed, the phenotypes of cell wall synthesis mutants reveal that *S. cerevisiae* can tolerate large reductions in the levels of certain wall components. Strategies for identifying cell wall mutants are described below.

B. Isolation of Mutants and Genes in Cell Wall Synthesis

Mutants altered in wall structure or defective in the biosynthesis of wall components have been isolated in a variety of ways, and many of the genes involved have been cloned by complementation of a mutant phenotype. Other genes have been isolated following protein purification or by screening for in vitro overexpression of a given enzyme activity. Subsequent study of null or conditional alleles of such genes has shed light on the function of their products in vivo. Screens for gene-dosage-dependent resistance to inhibitors have also led to the isolation of genes involved in the synthesis of wall components.

Many mutants have been identified by using agents that bind to specific cell wall components. Some of these agents also have deleterious effects on the cells. Use of antisera raised to specific determinants in surface mannans led to the isolation of nonconditional *mnn* mutants that failed to be agglutinated. Among these, mutants defective in the attachment of mannobiosylphosphate additionally fail to bind the cationic dye Alcian Blue (for review, see Ballou 1990). A screen for lack of cell surface expression of α-mating agglutinin led to the isolation of candidate GPI anchoring mutants (Benghezal et al. 1995). Fluorescent dyes such as

Calcofluor White and Congo Red are used to stain β-linked polysaccharides, and in *S. cerevisiae*, these dyes appear to stain chitin specifically (Elorza et al. 1983; Vannini et al. 1983). The binding of these dyes to chitin seems to have a specific effect on chitin synthesis in vivo. Treatment of *S. cerevisiae* cells with Calcofluor White or Congo Red causes the cells to aggregate, form aberrantly thick septa, and ultimately cease growth, although bud emergence is not affected (Roncero and Durán 1985; Vannini et al. 1993). Spheroplasts allowed to regenerate cell wall polymers in the presence of Calcofluor White make considerably more chitin, whereas glucan synthesis is unaffected (Roncero and Durán 1985). The effect of Calcofluor White was attributed to binding of the dye to nascent chitin chains as they are being extruded into the cell wall. Dye binding presumably interferes with the hydrogen bonding of polysaccharide chains to form crystalline microfibrils: This self-assembly process may normally limit the rate of GlcNAc polymerization by chitin synthase (Elorza et al. 1983; Roncero and Durán 1985). A screen for mutants resistant to the effects of Calcofluor White led to the isolation of strains deficient in chitin synthesis in vivo (Roncero et al. 1988). Given that Calcofluor White perturbs wall structure and halts growth of wild-type cells, mutations affecting other aspects of wall assembly might exacerbate the effects of this dye, and indeed a screen for Calcofluor White hypersensitivity has led to the isolation of mutants in the synthesis of a wide array of cell wall components (Ram et al. 1994).

A number of killer toxins bind specifically to certain cell wall polysaccharides, and screens for cells that lack these receptors, or which are somehow modified in the structure of their walls, have proven to be powerful means with which to identify genes in cell wall assembly. For example, a screen for resistance to *S. cerevisiae* K1 killer toxin, a pore-forming protein whose receptor is β-1,6 glucan (Hutchins and Bussey 1983; Bussey 1991), has led to the isolation of many genes involved in the formation of this glucan (see Section IV.E), and resistance to *Hansenula mrakii* killer toxins has allowed the identification of genes implicated in β-1,3 glucan synthesis (see Section V.B). *Kluyveromyces lactis* K1 killer toxin binds to GlcNAc-containing carbohydrates, and its α-subunit has exochitinase activity: Resistance to this toxin is conferred by a mutation that drastically lowers the chitin content of the wall (Butler et al. 1992; Takita and Castilho-Valavicius 1993).

Inhibitors have been exploited in various ways to isolate genes involved in cell surface assembly. For example, selections for mutants resistant to orthovanadate yielded mutants with defects in mannan synthesis (Ballou et al. 1991; Chi et al. 1996). A related approach is to screen colonies of cells transformed with multicopy genomic DNA libraries for those that are resistant to inhibitors because of increased dosage with the gene encoding the inhibitor's target protein (Rine et al. 1983). Proteins involved in *N*-glycosylation and β-1,3 glucan synthesis have been identified in this way in screens for resistance to the antibiotics tunicamycin and echinocandin. Mutants isolated as hypersensitive to the immunosuppressives FK506 and cyclosporin A have also made possible the isolation of genes involved in β-1,3 glucan synthesis.

Screens using colony autoradiography have been designed to detect defects in the incorporation of a specific radioactively labeled precursor into a given cell surface component. In an elegant study, Huffaker and Robbins (1982, 1983) used [^{3}H]mannose suicide selection to enrich for mutants defective in *a*sparagine-*l*inked *g*lycosylation (*alg*), which in consequence did not incorporate damaging amounts of [^{3}H]mannose into their long mannan outer chains. Chitin-deficient mutants have been isolated by screening colonies of mutagenized cells for those defective in [^{3}H]GlcN incorporation into chitin (Bulawa 1992) and GPI anchoring mutants by screening for defects in [^{3}H]inositol incorporation into protein (Leidich et al. 1994).

Purification of proteins by monitoring their enzymatic activity has allowed the corresponding genes to be cloned and the phenotypes of null or conditional mutants to be explored. In related approaches, banks of temperature-sensitive mutants have been screened by in vitro assay for mutants with low activity of enzymes in an essential process, for example, in the dolichol pathway of *N*-glycosylation (Roos et al. 1994). Procedures in which lysed colonies are assayed on filter replicas for glycosyltransferase or hydrolase activity have led to the isolation of mutants lacking an in vitro activity or, if colonies had previously been transformed with multicopy libraries, of genes conferring enzyme overproduction. Isolation of in vitro mutants can often make subsequent detection of enzyme overproduction easier by lowering the background signal. Genes encoding chitin synthases I and II (Bulawa et al. 1986; Silverman et al. 1988), dolichyl phosphate mannose synthase (Orlean et al. 1988), a β-1,3 glucanase (Nebrada et al. 1986; Kuranda and Robbins 1987), and chitinase (Kuranda and Robbins 1987) have been isolated using such approaches.

Screens for secretion defects have led to the identification of two mutants with global blocks in glycosylation and glycolipid anchoring (Ferro-Novick et al. 1984). Osmotic fragility should be an obvious consequence of defective cell wall assembly, and indeed mutants have been isolated that lyse unless given osmotic support (Venkov et al. 1974;

Cabib and Durán 1975; Blagoeva et al. 1991; Payton et al. 1991; Song et al. 1992; Shimizu et al. 1994). Some mutants isolated by other means have proven to be osmotically fragile, sometimes only in combination with other cell wall synthesis mutations. Furthermore, a number of mutations that confer osmotic fragility are in regulatory genes, for example, *PKC1* (Levin and Bartlett-Heubusch 1992; Paravicini et al. 1992), suggesting that multiple processes in cell wall assembly must be impaired in order to compromise cell integrity.

Once these "primary screens" for mutants and genes in cell wall synthesis have borne fruit, further genes involved in the process can often be identified using screens for multicopy suppressors or synthetically lethal mutations. Furthermore, an increasing number of genes, especially those encoding potential glycosyltransferases, are being found to have homologs elsewhere in the yeast genome, allowing the function of such genes to be explored as well. Once a number of mutants have been isolated that are defective in the synthesis of a specific wall component, such strains may be found as a group to be unusually sensitive or resistant to some agent, offering a simple screening strategy for further mutants.

III. SUPPLY OF PRECURSORS AND CYTOPLASMIC GLYCOSYLATION

A. Sugar Donors

1. Sugar Nucleotides and Dolichyl-phosphate-linked Sugars

The major monosaccharides found in yeast glycoproteins, glycolipids, and polysaccharides are mannose (Man), glucose (Glc), and *N*-acetylglucosamine (GlcNAc). The nucleotide sugars GDPMan, UDPGlcNAc, and UDPGlc, themselves formed in the cytoplasm, serve as donors in cytoplasmic reactions leading to protein glycosylation. Chitin and β-1,3 glucan synthases use UDPGlcNAc and UDPGlc, respectively, as donors in vitro and presumably in vivo. These donors may be cytoplasmic, but the possibility can be entertained that the UDP sugars used by polysaccharide synthases may be concentrated in specialized compartments (see Section V.A.3.f). Sugar nucleotides also participate in luminal reactions: GDPMan is used in glycosylation reactions in post-ER secretory compartments, and UDPGlc may be a donor in the synthesis of β-1,6 glucan in secretory compartments, as well as in transient reglucosylation of protein in the ER (see Sections IV.A.2.b and IV.E.2.b). A system that transports GDPMan into the lumen of Golgi compartments has been characterized (see Section IV.A.2.e). Not all of the genes encoding enzymes leading to nucleotide sugar synthesis have been isolated, but mutations in known genes involved in the synthesis of sugar donors perturb wall synthesis, morphogenesis, and secretion.

The gene encoding UDPGlc pyrophosphorylase, *UGP1*, was identified following amino acid sequence comparisons, and it is essential for viability (Daran et al. 1995). UDPGlc is the sugar donor in β-glucan, glycogen, and dolichyl phosphate glucose (Dol-P-Glc) synthesis. Depletion of UDPGlc pyrophosphorylase activity leads to a perturbation in cell wall assembly, reflected in a decrease in β-glucan content and in increased resistance of cells to lysis by β-1,3-glucanase (Daran et al. 1995).

Dol-P-Glc, synthesized by transfer of Glc from UDPGlc to Dol-P, donates the three glucose residues in the completion of the dolichyl pyrophosphatase (Dol-PP)-linked precursor oligosaccharide in *N*-glycosylation. Two mutants, *alg5* and *dpg1*, are defective in glucosylation in vivo and lack in vitro Dol-P-Glc synthetic activity (Huffaker and Robbins 1982; Ballou et al. 1986). Neither of these nonallelic mutations confers an obvious growth defect, indicating that the Dol-P-Glc-requiring steps in *N*-glycosylation are dispensable in *S. cerevisiae* and also that Dol-P-Glc has no significant role in the synthesis of wall β-glucan. The *ALG5* gene, cloned by complementation of the temperature sensitivity of an *alg5 wbp1* double mutant is a candidate for the structural gene (te Heesen et al. 1994), for it endows *Escherichia coli* transformants with Dol-P-Glc synthetic activity, and the protein it encodes corresponds in size to a 35-kD protein that can be photoaffinity-labeled with a UDPGlc analog (Palamarczyk et al. 1990). Neither the nature nor the role of the protein defective in the *dpg1* mutant is known.

Two phosphoglucose isomerases effect the conversion of Glc-6-P to Fru-6-P and hence are required for Man and GlcNAc supply. One of these proteins is encoded by the *CDC30* gene and is a high-K_m phosphoglucose isomerase. Some alleles of *cdc30* mutants have a temperature-sensitive phosphoglucose isomerase activity, and the cells arrest as unbudded G_1 cells, a phenotype attributed to a lack of Man and GlcNAc needed for cytokinesis and cell separation (i.e., cell wall formation). The *PGI1* gene encodes the low-K_m phosphoglucose isomerase (Dickinson 1991).

Mannose is formed upon conversion of Fru-6-P to Man-6-P by phosphomannose isomerase. A mutant defective in this enzyme, *pmi40*, was isolated as a temperature-sensitive osmotically fragile mutant, whose phenotype could be rescued by the inclusion of mannose in the culture medium (Payton et al. 1991). *pmi40* has *N*-glycosylation and secretion defects, consistent with an inability to make the precursor GDPMan. Disruption of the cloned *PMI40* gene yields mannose auxotrophs (Smith et al. 1992). *PMI40* expression is influenced by carbon source and is regulated at the transcriptional level, and maybe posttranslationally as well.

Phosphomannomutase is encoded by the *SEC53* gene, which is defective in the class-B secretion mutant *sec53*, which seemed to be blocked at an early stage in protein translocation into the ER (Ferro-Novick et al. 1984; Kepes and Schekman 1988). Alleles of *sec53* were also isolated as *alg4* in a [^{3}H]mannose suicide screen for glycosylation mutants (Huffaker and Robbins 1983). *sec53* mutants are blocked in *N*-glycosyation, *O*-mannosylation, GPI anchoring, and in the formation of mannosylinositol phosphoceramides (Huffaker and Robbins 1983; Kepes and Schekman 1988; Conzelmann et al. 1990; Orlean et al. 1991), a pleiotropic effect due to a block in the synthesis of GDPMan. The secretion defect in this mutant is presumably due to global block in all protein glycosylation, resulting in the accumulation of malfolded, secretion-incompetent proteins in the ER.

Dolichyl phosphate mannose (Dol-P-Man), which is synthesized by transfer of Man from GDPMan to Dol-P, donates Man residues in three of the glycosylation pathways in the ER. The gene for Dol-P-Man synthase, *DPM1*, was isolated using a colony screen to detect transformants that overproduced enzyme activity (Orlean et al. 1988), and it encodes a 30-kD protein, a size in agreement with that obtained for the purified yeast enzyme (Haselbeck 1989). Cell fractionation and indirect immunofluorescence studies show that Dpm1p is localized in the ER (Preuss et al. 1991). Disruption of *DPM1* is lethal, and temperature-sensitive *dpm1* mutants were used to show that Dpm1p participates in *N*-glycosylation, *O*-mannosylation, and GPI anchoring (Orlean 1990). In contrast to *sec53*, *dpm1* mutants are not blocked in the secretion of invertase (Orlean 1992).

DPM1 was also isolated in a screen for genes that at high copy allow cells to grow in the absence of Erd2p, the receptor for proteins bearing the carboxy-terminal amino acid sequence HDEL, the retention signal for luminal ER proteins (Hardwick et al. 1992). Although not compromised for growth, cells overexpressing *DPM1* show extensive fragmentation of their ER into small vesicles. This disruption of the ER may perturb membrane trafficking in such a way as to bypass the need for Erd2p, but it is not clear whether it is the catalytic activity of Dpm1p or some other property of the protein that causes suppression.

GlcNAc synthesis is initiated by glutamine: Fru-6-P amidotransferase, which makes GlcN-6-P. The gene for this enzyme, *GFA1*, was isolated by complementation of the glucosamine auxotroph *gcn1* isolated by Ballou and co-workers (Whelan and Ballou 1975; Ballou et al. 1977;

Watzele and Tanner 1989). When *gcn1* cells are deprived of GlcN, they proceed through several division cycles and form strings of two to eight undivided cells before succumbing to "glucosamineless death" (Ballou et al. 1977). Amidotransferase deficiency will, in addition to blocking *N*-glycosylation and GPI anchoring, block chitin synthesis and affect wall morphogenesis and may also influence transcription by interfering with *O*-GlcNAc attachment (see Section III.C). In *Aspergillus* and *Blastocladiella* species, the amidotransferase may be regulated by phosphorylation, the enzyme in its phosphorylated state being sensitive to feedback inhibition by UDPGlcNAc (Etchebehere and Da Costa Maia 1989; Borgia 1992). In *S. cerevisiae*, expression of *GFA1* is under pheromonal control (see Section VIII.A). *AGM1*, encoding GlcNAc-phosphate mutase, was identified on account of its ability to complement the growth defects of mutants deficient in phosphoglucomutase (Hofmann et al. 1994). Haploid cells harboring a disrupted *AGM1* gene stop growing after five generations and, like *gcn1* mutants, form strings of undivided cells.

2. *Dolichyl Phosphate*

Yeast dolichol is a mixture of molecules containing 14–18 isoprene units (Jung and Tanner 1973). De novo synthesis of dolichol involves the condensation of farnesyl pyrophosphate with 11–15 isopentenyl pyrophosphate units to form polyprenyl pyrophosphate, a reaction catalyzed by *cis*-prenyl transferase (Adair and Cafmeyer 1987). Dol-P is formed by CTP-dependent dolichol kinase (Szkopinska et al. 1988), a very hydrophobic protein encoded by the *SEC59* gene, which is defective in the class-B *sec* mutant *sec59* (Ferro-Novick et al. 1984; Bernstein et al. 1989; Heller et al. 1992). *sec59* cells have a phenotype similar to that of *sec53* and are blocked in *N*-glycosylation, *O*-mannosylation, and GPI anchoring (Conzelmann et al. 1990; Orlean 1990), due to the depletion of Dol-P which participates in all three pathways as a carrier lipid or sugar donor. It is not clear how Dol-P becomes depleted in *sec59* membranes as rapidly as it does, since Dol-P is regenerated after this polyisoprenoid has donated the sugar residue or oligosaccharide it carries. Indeed, how dolichol kinase participates in cellular metabolism presents a conundrum, for Dol-PP is the likely endproduct of the dolichol biosynthetic pathway, and free dolichol is not believed to be an intermediate. For a discussion of the models for the metabolic role of dolichol kinase, see Heller et al. (1992). The enzyme could be involved in a pathway in which Dol-P is dephosphorylated on one side of the ER membrane, after which free dolichol is translocated back across the membrane and then rephosphorylated at the cytoplasmic surface by dolichol kinase. Alternatively, dolichol kinase may serve to mobilize preexisting dolichol pools, and it also remains possible that phosphorylation of dolichol is an obligatory step in de novo synthesis of Dol-P.

B. Ceramide

The sphingolipids inositol phosphoceramide and mannosylinositol- and diinositol-phosphoceramide, which are synthesized in compartments of the secretory pathway (Puoti et al. 1991), become localized in the plasma membrane (Patton and Lester 1991). Furthermore, the lipid portion of many GPI anchors contains a type of ceramide (Conzelmann et al. 1992). Yeast ceramide contains the long-chain-base phytosphingosine, which is amide-linked to a C_{26} fatty acid. The first step in sphingolipid synthesis, condensation of serine with palmitoyl CoA, is catalyzed by a transferase likely to consist of related subunits encoded by *LCB1* and *LCB2/SCS1* (Buede et al. 1991; Nagiec et al. 1994; Zhao et al. 1994). Mutations in either gene alone result in sphingosine auxotrophy (Pinto et al. 1992); however, extragenic suppressors of *lcb1*-deficient strains can be isolated that grow without making detectable sphingolipids (Dickson et al. 1990). These strains compensate for the *lcb1* defect by making novel inositol glycerophospholipids consisting of diacylglycerol esterified with one C_{26} and one medium-length fatty acid chain (Lester et al. 1993). The protein encoded by the *SLC1* suppressor gene is a variant acyltransferase with an altered substrate specificity that allows it to use a C_{26} fatty acid instead of $C_{16/18}$ fatty acids to acylate the *sn*-2 position of inositol-containing glycerolipids (Nagiec et al. 1993). The new diacylglycerol is presumably used as a substrate for subsequent mannose and phosphoinositol transfer, and the resulting lipids serve as functional analogs of sphingolipids. The fact that this bypass mechanism involves incorporation of a C_{26} fatty acid indicates that the long-chain fatty acid is a critical substituent of sphingolipids. Some of yeast's fatty-acid-elongating activity is localized in the outer mitochondrial membrane (Bessoule et al., 1988), suggesting that there may be a transport mechanism to make the C_{26} fatty acid available for incorporation into ceramide, a process that occurs in the ER.

SLC1-suppressed *lcb1* cells cannot grow at 37°C, at low pH, or in high salt (Patton et al. 1992), indicating that the phytosphingosine portion of yeast ceramide has a key role as well. These authors' findings suggested a requirement for sphingolipids in maintenance of the cells' permeability barrier to protons or in proton efflux. The finding that

LCB2/SCS1 is a second-site suppressor of the Ca^{++}-sensitive phenotype of a strain deficient in Csg2p, a membrane protein required for growth at high Ca^{++} concentrations, suggested a relationship between sphingolipid metabolism and Ca^{++} homeostasis. Alternatively, or in addition, sphingolipid metabolism may be regulated by Ca^{++} (Zhao et al. 1994).

C. Cytoplasmic Glycosylation

Most of the glycosylation pathways in eukaryotes are based in the endomembrane system and lead to the covalent modification of secretory proteins. A major exception is the attachment of O-linked GlcNAc residues to both cytoplasmic and nuclear proteins (Hart et al. 1989). *O*-GlcNAc has been proposed to serve a regulatory role as an alternative to phosphorylation. O-linked GlcNAc has been reported to occur in *S. cerevisiae* (Hart et al. 1989), and such residues have been found on certain proteins in a fraction enriched in DNA-binding proteins isolated from extracts of *Aspergillus oryzae* (Machida and Jigami 1994).

A cytoplasmic protein that shows homology with rabbit phosphoglucomutase becomes modified by covalent attachment of Glc-1-P, and a UDPGlc:glycoprotein glucose-1-phosphotransferase has been detected in *S. cerevisiae* extracts). Covalently linked mannose is also present on phosphoglucomutase and may be the acceptor for Glc-1-P attachment (Marchase et al. 1993). This leads to the prediction that a cytoplasmically oriented protein:mannosyltransferase exists and that Glc-1-P transfer may be blocked in mutants such as *sec53* and *dpm1*.

IV. ASSEMBLY OF SURFACE COMPONENTS ALONG THE SECRETORY PATHWAY

The secretory pathway is heavily involved in the biosynthesis of cell surface components and in their transport to the cell surface. Five biosynthetic pathways can be distinguished. *N*-glycosylation, *O*-mannosylation, and GPI anchoring result in the covalent attachment of carbohydrate or a glycolipid to protein in the ER, and the resulting structures are modified or extended in later secretory compartments. Some proteins can receive all three types of modification. β-1,6 glucan synthesis is also initiated in the ER and may involve the direct glucosylation of protein; further steps in this pathway are also are associated with later secretory compartments. The fifth pathway, also started in the ER and completed in later compartments, is the formation of inositol phosphoceramides. Glycosylation in yeast has been reviewed by Ballou (1982, 1990), Tanner and Lehle (1987), Kukuruzinska et al. (1987), and Herscovics and Orlean (1993).

A. *N*-glycosylation

1. Biosynthesis in the ER

a. Synthesis of the Dol-PP-linked Precursor. In *S. cerevisiae*, as in other eukaryotes, *N*-glycosylation starts with the assembly of the precursor oligosaccharide $GlcNAc_2Man_9Glc_3$ on the polyisoprenoid carrier lipid Dol-PP (Fig. 1). This occurs stepwise in a series of reactions catalyzed by glycosyltransferases that are believed to be localized in the membrane of the ER: Certainly, oligosaccharide transfer to a protein is accomplished before the protein exits the ER. The pathway for lipid-linked oligosaccharide assembly is highly conserved between yeast and mammalian cells. The first seven sugars are transferred from the sugar nucleotides UDPGlcNAc and GDPMan, yielding Dol-PP-$GlcNAc_2Man_5$, and at this point, the heptasaccharide is believed to be translocated into the ER lumen, where four mannoses are transferred from Dol-P-Man and three glucoses are transferred from Dol-P-Glc.

Mutants in the assembly of the Dol-PP-linked precursor oligosaccharide have been isolated, and the steps at which oligosaccharide assembly is blocked in these strains have been identified in radiolabeling experiments with [^{3}H]GlcN or [^{3}H]Man. Thus, *alg1* accumulates mainly Dol-PP-linked $GlcNAc_2$; *alg2*, Dol-PP-$GlcNAc_2Man_2$; *alg3*, mainly Dol-PP-$GlcNAc_2Man_5$; *alg9*, Dol-PP-$GlcNAc_2Man_6$; *alg5* and *alg6*, Dol-PP-$GlcNAc_2Man_9$; and *alg8*, Dol-PP-$GlcNAc_2Man_9$Glc (Huffaker and Robbins 1982, 1983; Runge et al. 1984; Runge and Robbins 1986; Verostek et al. 1991; Aebi et al. 1996; Burda et al. 1996). Epistasis relationships between such mutants have also provided genetic evidence for the ordering of steps in the assembly of the Dol-PP-linked oligosaccharide shown in Figure 1. For example, Δ*alg3* is epistatic to Δ*alg9*, indicating that Alg3p-dependent addition of the α-1,3-linked mannose to Dol-PP-$GlcNAc_2Man_5$ must precede addition of further mannose residues (Burda et al. 1996). Furthermore, the truncated oligosaccharides that accumulate in mannosylation-defective *alg3* and *alg9* cells do not receive any glucose residues, even though the mannose residue to which they normally become attached is present, indicating that glucose addition requires the complete $GlcNAc_2Man_9$ structure (Verostek et al. 1993a,b; Burda et al. 1996). Truncated, glucose-free oligosaccharides can, however, be transferred to protein.

The phenotypes of *alg* mutants revealed that the early steps in the pathway, such as those blocked in the *alg4* (*sec53*), *alg1*, and *alg2* mutants, are essential, whereas *alg3*, *alg9*, and the later-acting *alg5*, *dpg1*, *alg6*, and *alg8* mutants, in which addition of glucose residues to the lipid-linked oligosaccharide is blocked, have no obvious growth

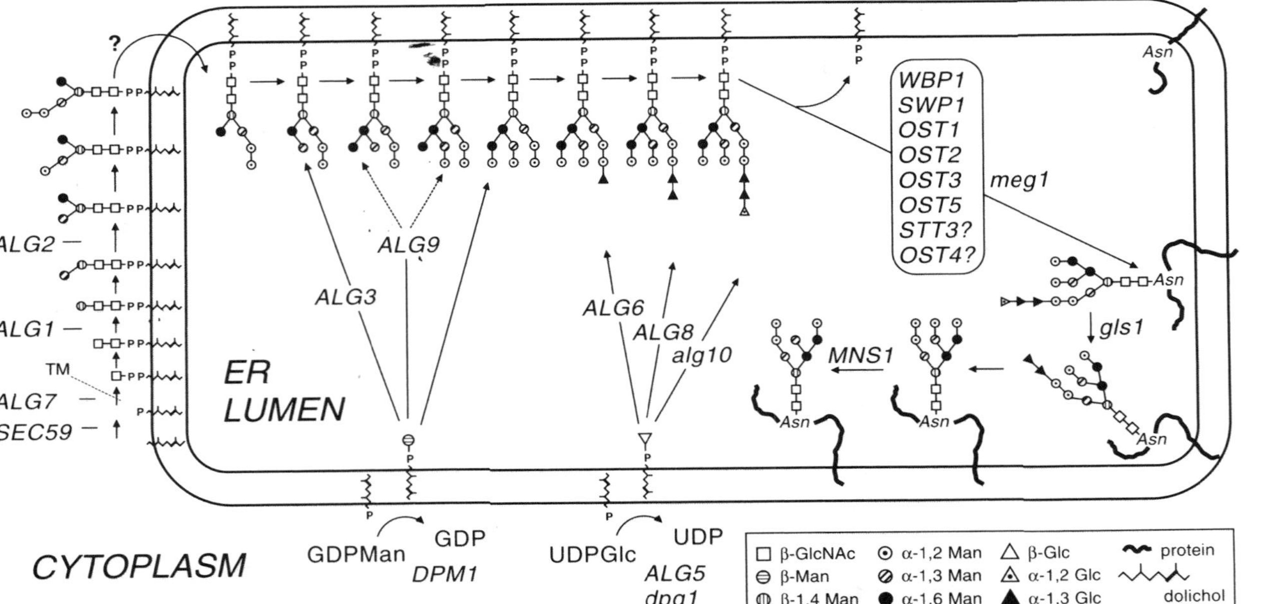

Figure 1 Assembly of the dolichol-linked precursor oligosaccharide in *N*-glycosylation and its transfer to protein in the endoplasmic reticulum (ER), showing the genetically defined steps. Cloned genes are labeled in uppercase letters, and steps defined by mutations are in lowercase letters. Sugar addition at the cytoplasmic face of the ER membrane is directly from a sugar nucleotide. The postulated transmembrane translocation of the Dol-PP-linked heptasaccharide is indicated by a question mark. Dol-P-Man and Dol-P-Glc are believed to be formed at the cytoplasmic face of the membrane, but used in luminal reactions.

defect, but transfer nonglucosylated oligosaccharides to protein. Mutations blocking transfer of the oligosaccharide to protein are lethal. The apparent "viability cutoff" before the *alg3* step may reflect a requirement for a minimum oligosaccharide structure to permit its efficient transfer to protein and to allow initiation of mannan outer-chain synthesis by Och1p (see Section IV.A.2.c). Whether mutations affecting the formation of Dol-PP-GlcNAc$_2$Man$_3$ and Dol-PP-GlcNAc$_2$Man$_4$ will be lethal remains to be seen.

The *alg3* step coincides with the stage in lipid-linked oligosaccharide assembly at which GDPMan is replaced by Dol-P-Man as donor and at which Dol-PP-GlcNAc$_2$Man$_5$ is believed to be translocated across the ER membrane in order to be a substate for Dol-P-Man-dependent elongation reactions. It has long been speculated that specific "flippases" exist for Dol-P-Man and Dol-PP-GlcNAc$_2$Man$_5$. In elegantly conceived experiments, Haselbeck and Tanner (1982, 1983) obtained evidence that purified Dol-P-Man synthase itself might catalyze the translocation of its product across membranes, but other proteins are now presumed to be responsible (Schutzbach and Zimmerman 1992). A candidate membrane protein-mediated Dol-P-Man translocating system has been detected in mouse liver microsomes (Rush and Waechter 1995). Mutations in a Dol-P-Man translocator should be lethal, providing there is an essential Dol-P-Man-requiring reaction that is exclusively luminal. Candidates for such reactions are those involved in GPI anchor assembly and in *O*-mannosylation of serine or threonine (see Sections IV.B and IV.D). The *alg3* mutant, in which the four Dol-P-Man-derived mannoses are not added, is not defective in Dol-PP-GlcNAc$_2$Man$_5$ translocation, because GlcNAc$_2$Man$_5$ is found on protein (Huffaker and Robbins 1983; Verostek et al. 1993a,b; Aebi et al. 1996), nor are *alg3* cells defective in Dol-P-Man translocation, for they carry out luminal Dol-P-Man-dependent *O*-mannosylation. It is likely that *alg3* mutants are deficient in a Dol-P-Man-dependent mannosyltransferase (Verostek et al. 1991). If a Dol-PP-oligosaccharide "flippase" indeed exists, it must be able to translocate shorter Dol-PP-linked oligosaccharides as well, for in *alg2* mutants, some Man$_{1-2}$GlcNAc$_2$ is found on protein (Jackson et al. 1989). If "flipping" is indeed protein-mediated, then the "flippase" gene should be an essential one.

A number of the *ALG* genes have been cloned, and with the exception of *ALG4/SEC53* (see Section III.A.1), all encode transmembrane proteins. The essential *ALG7* gene, which encodes the first enzyme of assembly pathway, was isolated by screening transformants for those resistant to the inhibitor tunicamycin on account of increased dosage with the inhibitor's target enzyme, the GlcNAc-1-P transferase (Rine et al. 1983; Barnes et al. 1984). *ALG7* shows a complex transcription pattern that is influenced by sequences in the gene's 3′-untranslated region (Kukuruzinska and Robbins, 1987). Strains with deletions in the 3′-untranslated region have lowered GlcNAc-1-P transferase activity and are hypersensitive to tunicamycin (Kukuruzinska and Lennon 1995). Alg7p activity is influenced by levels of phosphatidylinositol (PI) in the membrane: Membranes from inositol-starved inositol auxotrophs show decreased GlcNAc-1-P transferase activity, which is relieved by the addition of PI (but not by inositol phosphoceramides), suggesting that PI functions as a cofactor in the transferase reaction (Hanson and Lester 1982). Indeed, an early effect of inositol starvation is inhibition of mannan synthesis (Hanson and Lester 1980).

ALG1 and *ALG2* were cloned by complementation of their temperature sensitivity. Alg1p (Albright and Robbins 1990) was shown indeed to be the enzyme that transfers mannose from GDPMan to Dol-PP-$GlcNAc_2$, for this β-mannosyltransferase was expressed in *E. coli* transformants harboring the *ALG1* gene (Couto et al. 1984). Alg2p is required for mannosylation of Dol-PP-$GlcNAc_2Man_2$ to give Dol-PP-$GlcNAc_2$ Man_3 (Jackson et al. 1993).

Nonconditional *alg* mutations severely affect cell growth when combined in strains harboring a temperature-sensitive allele of *wbp1*, which causes defective oligosaccharide transfer to protein, and such double mutants fail to grow at 30°C. This finding has been exploited to clone the late-stage *ALG6* and *ALG8* genes by complementation (Stagljar et al. 1994; Reiss et al. 1996). Alg6p and Alg8p are potential glucosyltransferases that respectively transfer the first and second α-1,3-linked glucoses to Dol-PP-$GlcNAc_2Man_9$. These two proteins are clearly related, but nonetheless each seems to require a very specific acceptor structure. Screens for synthetic lethality with *wbp1* subsequently yielded mutants in two new complementation groups, *alg9* and *alg10*, which accumulate Dol-PP-$GlcNAc_2Man_6$ and Dol-PP-$GlcNAc_2Man_9Glc_2$, respectively (Zufferey et al. 1995). *ALG9* has been cloned, but it is not yet known whether its product is involved in the addition of an α-1,2- or α-1,6-linked mannose to Dol-PP-$GlcNAc_2Man_6$ (Burda et al. 1996). *ALG3* was isolated by complementation of the temperature sensitivity that results when *alg3* is combined with *stt3*, a mutation that also affects oligosaccharide transfer to protein (Aebi et al. 1996).

b. Oligosaccharide Transfer to Protein. Oligosaccharyltransferase (OTase), a complex of proteins, catalyzes the transfer of $GlcNAc_2Man_9$

Glc_3 from Dol-PP to the amide group of the asparagine residue in the tripeptide "sequon" Asn-X-Ser/Thr, X being any amino acid except proline (Lehle and Bause 1984). A survey of the N-linked sites characterized up to that point indicated that there was a bias in favor of the use of Asn-X-Thr sites in yeast glycoproteins (Moelhle et al. 1987), but it is not yet clear whether this reflects a substrate preference on the part of OTase (Basco et al. 1993). Not all *N*-glycosylation sequons in a protein receive an oligosaccharide chain, and the structures that are subsequently elaborated at these sites can differ. Thus, the invertase monomer has 14 sequons, of which 8 are almost always glycosylated and 5 are glycosylated no more than half the time, such that on average in wild-type cells, 9–10 sites on invertase receive an oligosaccharide (Reddy et al. 1988). Whether a given site is used, i.e., is a good substrate for OTase, presumably reflects the nature of the amino acids near that site and the local conformation the peptide adopts (Reddy et al. 1988). Consistent with this, the presence of a proline immediately carboxy-terminal to the hydroxy amino acid in the sequon prevents glycosylation of that site (Roitsch and Lehle 1989).

The size of the Dol-PP-linked oligosaccharide can influence the number of chains transferred to protein: Full-length $GlcNAc_2Man_9Glc_3$ is transferred more efficiently to acceptor peptides in vitro than are shorter, nonglucosylated oligosaccharides (Trimble et al. 1980; Sharma et al. 1981). The shorter oligosaccharides made in mutants blocked in lipid-linked oligosaccharide assembly can also be transferred to protein in vivo (Huffaker and Robbins 1983). Thus, $GlcNAc_2Man_{1\text{-}2}$ is found on protein in *alg2*; $GlcNAc_2Man_5$ in *alg3* and *dpm1*; $GlcNAc_2Man_6$ in Δ*alg9*; $GlcNAc_2Man_9$ in *alg5*, *alg6*, and *dpg1*; and $GlcNAc_2Man_9Glc$ in *alg8* (Ballou et al. 1986; Runge and Robbins 1986; Jackson et al. 1989; Verostek et al. 1991; Muñoz et al. 1994; Burda et al. 1996). In general in such mutants, fewer sites become glycosylated, but the efficiency with which truncated oligosaccharides are transferred varies with the protein. Relatively little reduction in efficiency of transfer of $GlcNAc_2Man_5$ to protein is seen with invertase (Verostek et al. 1993a,b), whereas with Exg1p, mutants unable to make glucosylated $GlcNAc_2Man_9$ showed more pronounced underglycosylation. Moreover, the presence of one Glc residue on $GlcNAc_2Man_9$ led to a bias in transfer of that chain toward one specific sequon in Exg1p (Muñoz et al. 1994). The glucose residues on the Dol-PP-linked saccharides, in general, enhance glycosylation of N-linked sites, and they may do so by adopting a specific conformation that contributes to recognition of the oligosaccharide by the OTase complex (Alvarado et al. 1991). The importance of the Glc residues, however, is clear from the growth defect seen when mutations abolishing Glc attachment (*alg5*, *alg6*, and *alg8*) are combined with a mutation that impairs oligosaccharide transfer to protein (Stagljar et al. 1994).

At least eight proteins are required for oligosaccharyl transfer to protein, and the genes that encode them have been cloned. Up to six of these proteins (Wbp1p, Swp1p, Ost1p, Ost2p, Ost3p, and Ost5p) are present as a complex in purified OTase (Kelleher and Gilmore 1994; Knauer and Lehle 1994; Pathak et al. 1995; Silberstein and Gilmore 1996; Reiss et al. 1997), and the products of two further genes (Ost4p and Ost3p) are also required for oligosaccharyl transfer in vivo and may yet prove to be part of the OTase complex. Five of the eight cloned genes required for oligosaccharyl transfer are essential and three are not, but mutations in all of these genes result in transfer of fewer oligosaccharide chains to protein in vivo, as well as defects in OTase activity in vitro. Genetic evidence in the form of a number of instances of multicopy suppression and synthetic interaction indicates that many of the OTase proteins interact with one another. The following are the properties and possible functions of these proteins.

Wbp1p was originally identified as an ER membrane protein, whose gene was then cloned and found to be essential. Temperature-sensitive *wbp1* alleles are defective in oligosaccharide transfer to protein in vivo and in OTase activity in vitro (te Heesen et al. 1992). Wbp1p has a carboxy-terminal KKXX sequence that mediates retrieval of this type I membrane protein to the ER from an early Golgi compartment containing Och1p (Gaynor et al. 1994; Section IV.A.2.c). A second essential gene, *SWP1*, was isolated as a multicopy suppressor of the *wbp1-2* mutant (te Heesen et al. 1993). The allele specificity of this suppression suggests that Wbp1p and Swp1p interact with each other in vivo, consistent with the fact these proteins copurify in the OTase complex (Kelleher and Gilmore 1994; Knauer and Lehle 1994). Swp1p shows homology with mammalian ribophorin II (Kelleher and Gilmore 1994). Sequence information obtained from three further subunits of the purified OTase complex was used to design oligonucleotide probes, which were used to isolate the corresponding genes, *OST1* (*NLT1*), *OST2*, and *OST3*. Ost1p is homologous to mammalian ribophorin I (Pathak et al. 1995; Silberstein et al. 1995b). *OST2*, which also gives multicopy suppression of *wbp1-2*, encodes a protein showing similarity to the vertebrate DAD1 ("defender against apoptotic death") protein (Silberstein et al. 1995a). *OST3* is a nonessential gene. However, *ost3* disruptants show "biased underglycosylation" in that soluble glycoproteins (like carboxypeptidase Y) that use the signal recognition particle (SRP)-independent pathway are almost completely *N*-glycosylated, whereas membrane proteins (including Wbp1p), which use the SRP-dependent translocation process, are underglycosylated (Karaoglu et al. 1995). This raises the intriguing possibility that Ost3p may help position the OTase complex for better *N*-glycosylation of proteins that use the SRP-dependent translocation pathway. *OST5*, the gene for the sixth 9-kD subunit of the purified OTase complex was identified on account of the synthetic phenotype of the *ost5* mutation with Δ*alg5* (Reiss et al. 1997). Ost5p is not an essential protein, and its depletion results in relatively mild hypoglycosylation and modest effects on in vitro OTase activity. *OST5* shows specific genetic interactions with two other OTase genes: Overexpression of *OST5* suppresses *ost1* mutations, suggestive of an interaction between Ost5p and Ost1p, and Δ*ost5* is lethal in combination with the *stt3* mutation (see below).

Two further proteins, not detectable in the purified OTase complex, seem to be involved in oligosaccharide transfer. Mutations in the *OST4* gene, isolated in a screen of a bank of temperature-sensitive mutants for those defective in vitro in the *N*-glycosylation of an ^{125}I-labeled acceptor peptide, as well as in a screen for mutants resistant to orthovanadate, lead to defects in oligosaccharyl transfer in vivo and in vitro (Roos et al. 1994; Chi et al. 1996). Ost4p is a hydrophobic 36-amino-acid protein that may lie entirely within the hydrophobic core of the membrane. *ost4* null mutants are slow growing at 25°C but inviable at 37°C and show a marked reduction in oligosaccharide transfer in vivo. The function of the very small Ost4p and Ost5p OTase subunits is unclear, but such proteins might contribute to the stability or assembly of the membrane protein complex (Reiss et al. 1997). *ost4* mutants show resistance to orthovanadate, a phenotype normally associated with mutants affected in the function of the secretory pathway at the level of the Golgi complex (see Section IV.2.c), but it is not clear whether a reduced ability of cells to transfer oligosaccharides to asparagine also leads to orthovanadate resistance. *ost4* mutants are also subject to multicopy suppression by a gene designated *MEG1* (Roos et al. 1994; Chi et al. 1996), but the nature and function of this gene's product are not yet known. Mutations in Stt3p were identified in a screen for mutants synthetically lethal with *wbp1* (Zufferey et al. 1995), and they seem to affect the substrate specificity of OTase, for very little oligosaccharide transfer occurs when incomplete precursor oligosaccharide is available, whereas transfer of the full-length precursor is scarcely affected. Indeed, the combination of the nonconditional *stt3-3* mutation with deletions in *ALG3*, *ALG5*, and *ALG8* results in inviability. Stt3p was not among the proteins in the purified OTase complex but may yet prove to be a component. *STT3* is an essential gene.

Possible roles for its product are in substrate recognition or in the correct assembly of the OTase complex (Zufferey et al. 1995). *stt3* mutants were originally identified in a screen for sensitivity to staurosporine, an inhibitor of protein kinase C. Pkc1p has regulatory roles in cell wall biogenesis, and *pkc1* mutations result in osmotic fragility (Levin and Bartlett-Heubusch 1992; Paravicini et al. 1992; Roemer et al. 1994) (Section IX). The fact that a defect in *N*-glycosylation exacerbates the effect of a Pkc1p deficiency is consistent with the notion that *N*-glycosylation has a key role in wall biogenesis.

The availability of many, if not all, of the OTase genes should allow the interactions between the OTase subunits, hence the structural organization of the complex to be defined genetically and biochemically. The challenge will be to establish which components of OTase are responsible for the complex's various functions, namely, recognition of the acceptor peptide, recognition of the Dol-PP-linked precursor oligosaccharide, and catalysis of oligosaccharyl transfer.

2. *Processing and Maturation of N-linked Oligosaccharides*

a. Processing Glycosidases. After transfer of the 14-sugar oligosaccharide to protein, the structure is trimmed by processing glycosidases in the ER or in an intermediate compartment that forms behind the secretion block in the *sec18* mutant (Esmon et al. 1984). The terminal α-1,2 glucose is removed by glucosidase I (Kilker et al. 1981; Bause et al. 1986), and the two α-1,3 glucoses are removed by glucosidase II (Saunier et al. 1982). The *gls1* mutant, which retains all three glucose residues, grows and extends its mannan outer chains normally (Esmon et al. 1984). A single, specific α-1,2 mannose residue is then cleaved off, leaving a single isomer of the structure $GlcNAc_2Man_8$ (Byrd et al. 1982). This mannosidase has been purified (Jelenik-Kelley and Herscovics 1988; Ziegler and Trimble 1991), and the gene encoding it, *MNS1*, has been cloned (Camirand et al. 1991). Null mutants in *MNS1* are not defective in growth or in mannan chain formation (Puccia *et al.* 1993).

b. Transient Reglucosylation of N-linked Oligosaccharides. In various eukaryotes, a UDPGlc:glycoprotein glucosyltransferase reglucosylates the glucose-free N-linked oligosaccharides generated by glucosidases I and II. This transferase is a component of a proposed "quality control" mechanism that allows partially folded or misfolded proteins to be retained in the ER until they adopt their native conformation (Hebert et al. 1995). The mechanism involves the deglucosylation and reglucosylation of $GlcNAc_2Man_9$ by the opposing activities of glucosidase II and UDPGlc:glycoprotein glucosyltransferase. The latter enzyme, however, acts only on denatured proteins, and proteins bearing the resulting monoglucosylated N-linked chains are in turn bound and retained in the ER by the membrane-associated chaperone calnexin. Once the glycoprotein attains its native conformation, it ceases to be a substrate for UDPGlc:glycoprotein glucosyltransferase and can exit the ER. UDPGlc: glycoprotein glucosyltransferase activity has not been detected in *S. cerevisiae*, but the enzyme has been purified from *Schizosaccharomyces pombe* (Fernández et al. 1994), and its gene, *gpt1*$^+$, has been cloned (Fernández et al. 1996). Disruption of *gpt1*$^+$ is not lethal, and it does not lead to any reduction in growth rates, although the disruptants are somewhat smaller than wild-type cells. Nevertheless, UDPGlc:glycoprotein glucosyltransferase is a stress-response protein, for synthesis of *gpt1*$^+$ mRNA is induced by heat shock and upon inhibition of *N*-glycosylation, conditions that impair protein folding in the ER (Fernández et al. 1996). These findings support the notion that the transferase is involved in monitoring the completion of protein folding in the ER, although the mechanism is dispensable for cell viability. Intriguingly, Gpt1$^+$p shows sequence similarity to Kre5p, an ER protein required for the initiation of β-1,6 glucan synthesis in *S. cerevisiae* (see Section IV.E.2.b).

c. Outer-chain Elaboration and Candidate Mannosyltransferases. The structure of the mannan outer chain was investigated with the aid of *mnn* mutants isolated by Ballou and co-workers (for review, see Ballou 1982, 1990). These workers' efforts, and those of Trimble and colleagues, have now led to a consensus for the structure of mature yeast N-linked carbohydrates (Fig. 2) (Ballou 1990; Trimble and Atkinson 1992). The $GlcNAc_2Man_8$ structure can be matured in different ways, depending on the glycoprotein and on the individual N-linked sites within it. Three of the four saccharides in the vacuolar glycoprotein carboxypeptidase Y retain a "mature core" of 9–13 mannose residues, and 1 is extended to $Man_{8\text{-}12}$ (Trimble et al. 1983). Some sites in the secreted protein invertase and in cell wall proteins also retain a $Man_{9\text{-}13}$ core, whereas other chains are extended with branched mannan outer chains containing up to 200 residues. Whether an N-linked chain is extended or not depends on local peptide conformation, which dictates whether the saccharide is sterically accessible to the mannosyltransferases that catalyze chain elaboration and mannosyl phosphate addition. Those chains that are not extended also tend to be those that are inaccessible to, hence resistant to, the deglycosylating enzyme endo-β-*N*-acetylglucosaminidase H (Trimble et al. 1983; Ziegler et al. 1988; Basco et al. 1993).

Elaboration of the mannan outer chain is affected by mutations in a large number of genes, but of the genes cloned, few have been shown to encode mannosyltransferases. Studies on the extent of glycosylation that occurs in mutants blocked at different stages along the secretory pathway, and on the localization of individual transferases, indicate that outer-chain extension occurs in successive Golgi compartments (Fig. 2) (Franzusoff and Schekman 1989; Franzusoff 1992; Graham and Emr 1991; Gaynor et al. 1994; Lussier et al. 1995a). Indeed, the glycosylation state of reporter proteins has long been exploited to monitor their progress along the secretory pathway.

Formation of the outer chains is initiated by the Och1p α-1,6 mannosyltransferase (Nakanishi-Shindo et al. 1993; Romero et al. 1994). The gene encoding this membrane enzyme was cloned by complementation of the temperature-sensitive *och1* mutant, which had been isolated following [^{3}H]mannose suicide enrichment (Nagasu et al. 1992; Nakayama et al. 1992). Och1p is predicted to be a type II membrane protein that is normally localized in, and indeed defines, an early Golgi compartment (Chapman and Munro 1994; Gaynor et al. 1994). Its localization mecha-

Figure 2 Elaboration of N-linked carbohydrate chains in successive compartments of the Golgi complex. The marker proteins Och1p, Kre2p, Mnn1p, and Kex2p define distinct Golgi compartments in which the indicated carbohydrate additions and proteolytic cleavages are carried out by resident enzymes. Depending on the glycoprotein and the individual N-linked sites within it, the $GlcNAc_2Man_8$ structure formed in the ER can undergo maturation in one of two ways. A $GlcNAc_2Man_{9\text{-}13}$ "mature core" is formed at certain sites in secretory proteins and on all vacuolar glycoproteins, whereas a branched mannan outer chain of up to 200 mannoses is formed at certain sites in secretory proteins. In the major maturation pathway, the α-1,2-linked Man is removed from the $GlcNAc_2Man_8$ core by Mns1p, but this residue is retained in the minor pathway. WT denotes the structure formed in wild-type cells. (*) α-1,2-linked Man that is added to the outer-chain-initiating α-1,6-linked Man (possibly by M_2MT-1; Lewis and Ballou 1991), and that may serve as a stop signal that prevents elongation of the outer chain. (#) Additional sites at which Man-1-P is transferred from GDPMan to the 6 position. The effects of *mnn* mutations on the carbohydrate structure are indicated, uppercase letters being used to denote cloned genes. The figure is based on structural work by Ballou (1982, 1990) and by Trimble and Atkinson (1992), and on models presented by Franzusoff and Schekman (1989), Graham and Emr (1991), Franzusoff (1992), Gaynor et al. (1994), and Lussier et al. (1995a).

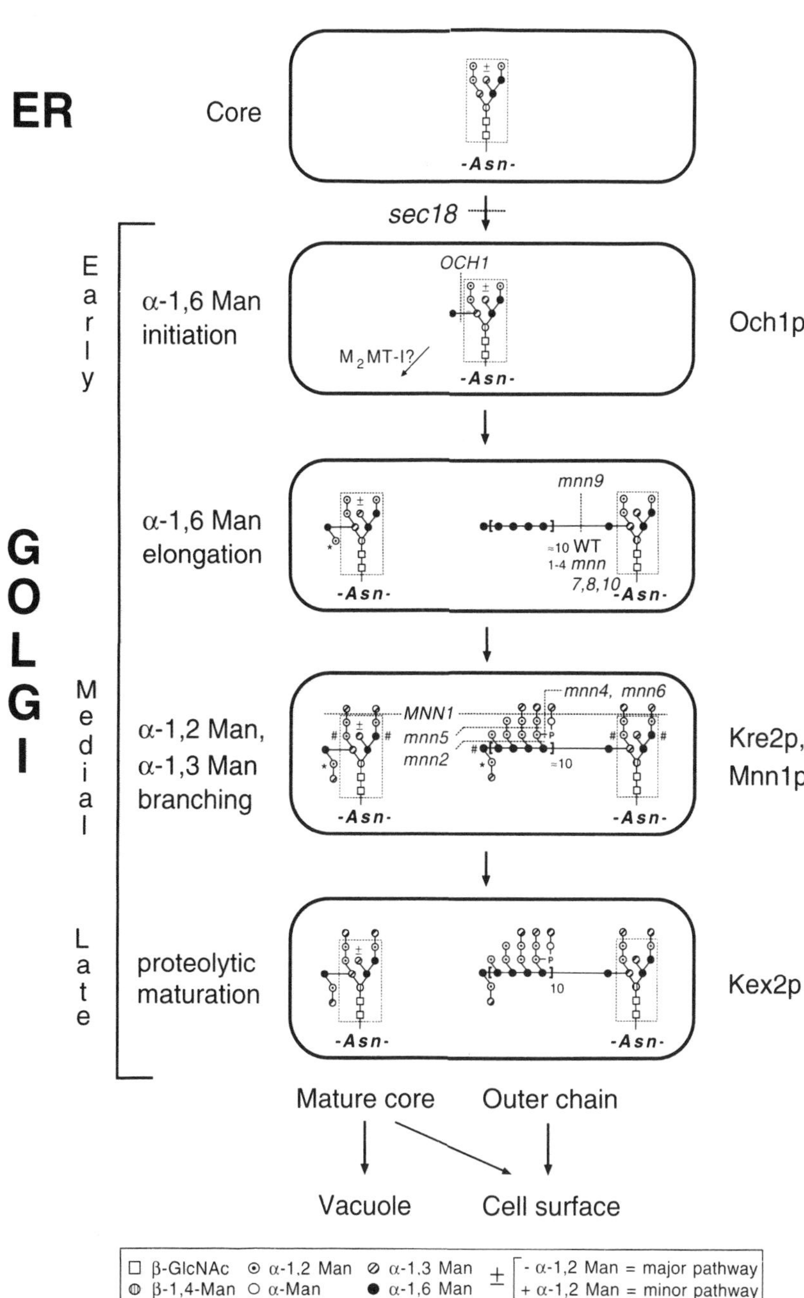

Figure 2 (See facing page for legend.)

nism, however, may involve movement to the *trans*-Golgi and subsequent retrieval to the earlier compartment (Harris and Waters 1996). *och1* disruptants are viable, lack any α-1,6-polymannose outer chains, but are temperature-sensitive, indicating that outer-chain addition per se is required for growth at elevated temperature. At nonpermissive temperature, *och1* mutants halt growth with small buds (Nagasu et al. 1992). Och1p may not be able to initiate outer-chain formation on the truncated N-linked saccharides made in the *alg1* and *alg2* mutants, explaining why the latter mutations are lethal. The structures made in *alg3* cells and in mutants blocked at later stages in oligosaccharide assembly, however, can be extended into more or less normal outer chains.

Addition of an α-1,2-linked mannose to the α-1,6-linked outer-chain initiating mannose residue may serve as a stop signal that prevents elongation of the outer chain at certain sites and allows formation instead of "mature core" stuctures (Ballou 1990). A candidate for this enzyme, M_2MT-1, which transfers to α-1,6 mannose-containing acceptors, has been purified (Lewis and Ballou 1991). A mannosyltransferase that elongates the α-1,6 backbone has yet to be identified, although Mnn10p may be a candidate (see Section IV.A.2.d).

A number of nonconditional *mnn* mutants fail to make specific linkages during the formation and extension of branches off the outer-chain backbone, but not all of these have been shown to be defective in mannosyltransferase activity. However, since a given type of glycosidic linkage can occur in many places in mannan, multiple transferases may exist, and an absence of one of them may simply not be noticeable in a given assay. The *mnn2* mutant does not form branches off the α-1,6 outer-chain backbone, being blocked in the addition of the α-1,2-linked mannose residue on which the side chain is extended. Neither the *mnn2* mutant nor cells overexpressing *MNN2* show any change in detectable α-1,2 mannosyltransferase activity (Devlin and Ballou 1990). Mnn2p may yet prove to catalyze mannosyl transfer or instead it may be a regulatory protein. *mnn5* mutants are defective in the addition of the second α-1,2 mannose residue to the side chain. Subsequent addition of the terminal α-1,3 mannose to the α-1,2 side branch is blocked in the *mnn1* mutant, extracts of which also lack in vitro α-1,3 mannosyltransferase activity. The *mnn1* mutation also blocks α-1,3 mannose addition to *O*-glycosidically linked mannose trisaccharides (see Section IV.D.3). Mnn1p is a type II membrane protein (Yip et al. 1994) with a short cytoplasmic tail and is likely to encode the α-1,3 mannosyltransferase because antiserum raised to the *MNN1* gene product precipitates α-1,3 mannosyltransferase activity and because *mnn1* null mutants lack this enzyme activity (Graham et al. 1994). Mnn1p is normally localized in the functional equivalents of the medial and *trans*-Golgi (Graham et al. 1994) but mislocalized to the plasma membrane in clathrin heavy-chain-deficient strains, although how clathrin influences the localization of this mannosyltransferase is not yet clear.

The mannotriosyl side branches in the outer chain, as well as mannoses in the mature core, can further be modified by the transfer of mannosylphosphate units from GDPMan. The resulting phosphodiester-linked branches can themselves be extended by Mnn1p with an α-1,3-linked mannose. Mannosylphosphate addition is reduced in the *mnn4* and *mnn6* mutants, which fail to bind the cationic dye Alcian Blue, but whether these mutations affect a transferase, or are in regulatory proteins, is not yet known (Ballou 1982, 1990). The *mnn3* mutant exhibits a general shortening in the mannan outer chain, and its *O*-linked manno-oligosaccharides are also affected. Again, it is not known whether this mutation is in a regulatory or a mannosyltransferase gene.

A family of nine potential mannosyltransferases has been identified whose members affect the extent of mannosylation of N-linked, and in one case, also O-linked, chains (Lussier et al. 1997). The gene encoding the first member, Kre2p/Mnt1p, was isolated by complementation of the K1 killer-toxin-resistant mutant *kre2* and also following purification of an α-1,2 mannosyltransferase activity (Haüsler and Robbins 1992; Haüsler et al. 1992; Hill et al. 1992). The protein is localized in the Golgi (Chapman and Munro 1994; Lussier et al. 1995a). *kre2/mnt1* null mutants are blocked in the elongation of O-linked mannobiose chains, and the N-linked outer chains in these strains are also decreased in size. Eight further genes encoding the Kre2p/Mnt1p-related proteins Ktr1p, Ktr2p, Ktr3p, Ktr4p, Ktr5p, Ktr6p, Ktr7p, and Yur1p, have been identified on the basis of sequence homology, and all encode type II membrane proteins with a short cytoplasmic amino terminus and a conserved luminal domain (Hill et al. 1992; Lussier et al. 1993, 1997; Mallet et al. 1994). Of these, Ktr1p, Ktr2p, and Yur1p have in vitro mannosyltransferase activity and are localized in the Golgi complex (Lussier et al. 1996). None of these proteins seem to be involved in *O*-mannosylation in vivo, but *ktr1*, *ktr2*, and *yur1* disruptants show partial resistance to K1 killer toxin, indicating alterations in the cell surface, in turn suggesting that these proteins may be redundant mannosyltransferases that all participate in *N*-glycosylation in the Golgi (Lussier et al. 1996).

d. Glycosylation Defects in Mutants Affected in the Function of the Secretory Pathway. Residence time in a given compartment may influence the extent of outer-chain glycosylation of a protein. The hyperglycosylated forms of proteins seen in certain *sec* mutants may arise because these proteins become trapped in the accumulated secretory vesicles along with substrate transporters and mannosyltransferases that aberrantly glycosylate available acceptor glycans. However, outer-chain extension is strongly affected by mutations in genes that are likely to be involved in the organization and functioning of the secretory pathway, as opposed to protein export per se. Mutants such as *mnn7/8*, *mnn9*, and *mnn10*, identified on account of their reduced ability to be recognized by antiserum against the unbranched α-1,6 backbone of the outer chain, were the first *mnn* mutants to be isolated that had obvious growth defects, among these, osmotic fragility (Ballou 1982). The *mnn9* mutant, which showed the most severe underglycosylation, made N-linked chains in which the outer-chain-initiating α-1,6 mannose had been attached, but not further elongated. Addition to that core structure of a few α-1,2- and α-1,3-linked mannoses gave a $Man_{13}GlcNAc_2$ structure (Tsai et al. 1984; Ballou 1990). One possible function for Mnn9p is that it is the α-1,6 mannosyltransferase that elongates the backbone of the outer chain; however, recent findings suggest that the glycosylation phenotype may be an indirect one, due to perturbation of certain functions of the secretory pathway.

Mutants severely affected in outer-chain extension have additional phenotypes: Alleles of *mnn7/8*, *mnn9*, and *mnn10* were among the *vrg* mutants isolated in a screen for sodium orthovanadate resistance (Ballou et al. 1991), and these strains also showed enhanced hygromycin B sensitivity. These studies, and those of Kanik-Ennulat et al. (1995) with further vanadate-resistant *van* mutants, showed that the degree of resistance is correlated with the severity of the glycosylation defect in the mutants. Although orthovanadate-resistant cells underglycosylate invertase, they are not perturbed in the secretion of the protein (Ballou et al. 1991; Kanik-Ennulat et al. 1995). *MNN9* (Yip et al. 1994) is a member of a gene family. One other member, *VAN1*, was isolated by complementation of a vanadate-resistant mutant (Kanik-Ennulat et al. 1995), and the third, *ANP1*, complemented *gem3*, a mutant isolated in a screen for mutants that failed to retain a reporter protein in an early Golgi compartment (Chapman and Munro 1994). All three genes encode type II membrane proteins. Anp1p, which shows 41% and 20% similarity, respectively, to Van1p and Mnn9p, is localized in the ER when expressed from a 2μ vector. The subcellular localization of the other two proteins is not yet known. Null alleles of the three genes are viable, but *mnn9* disruptants grow particularly slowly and are morphologically abnormal and osmotically fragile (Yip et al. 1994). *anp1* and *van1* mutants also show underglycosylation of invertase, but not to the degree *mnn9* does, and of the two, *anp1* was the least severely affected. *anp1 van1* double mutants show synthetic lethality (Chapman and Munro 1994). This finding suggests that Anp1p and Van1p are not mannosyltransferases required for outer-chain elongation, for null mutants in outer-chain-initiating Och1p grow at 25°C. Although formally Mnn9p, as well as Anp1p and Van1p, could be mannosyltransferases, another possibility is that proteins in the Mnn9p family are involved in the function and organization of the secretory pathway (Ballou et al. 1991; Chapman and Munro 1994; Yip et al. 1994; Kanik-Ennulat et al. 1995). The vanadate resistance conferred by defects in Mnn9p, Van1p, and Anp1p is consistent with this notion and suggests some possible functions for these proteins (Kanik-Ennulat et al. 1995). Vanadate toxicity may be due to the formation of a vanadate-derived molecule in the late ER or early Golgi, which in turn can inhibit ATP-dependent processes. If Mnn9p, Van1p, and Anp1p were transporters, mutations in them could prevent access of vanadate to its luminal site of toxification or cause the derivative to be diluted below the level at which it is normally toxic to a key luminal target. Such mutations would also affect glycosylation by preventing the supply of substrates and cofactors for luminal mannosyltransferases. Were Mnn9p, Van1p, and Anp1p to influence vesicle targeting, mutations in these proteins may result in the opening of a pathway to the plasma membrane that skips the later-stage Golgi compartments, thus eliminating toxic vanadate from the cells but, in so doing, bypassing the mannosyltransferases that reside in those compartments (Ballou et al. 1991).

The hygromycin B sensitivity of the *mnn* outer-chain mutants seems to be a consequence of underglycosylation, for mutants blocked in the assembly and transfer of the precursor oligosaccharide to protein also show this sensitivity (Dean 1995). In this case, alterations in *N*-glycosylation of one or more proteins lead either to an enhanced ability to take up hygromycin B or to a reduced ability to detoxify it.

A close coupling of *N*-glycosylation and secretion is emphasized by the fact that mutations in three further proteins that affect the function of the secretory pathway in ways distinct from Mnn9p, Van1p, and Anp1p also lead to underglycosylation. Mutations in the essential *VRG4* (*VAN2*) gene cause defects in protein sorting in the Golgi apparatus and in retrieval of HDEL-bearing proteins from the Golgi to the ER, and *vrg4* mutants show altered membrane morphology as assessed by electron microscopy (Poster and Dean 1996). *vrg4* mutants are also orthovanadate-resistant and severely impaired in outer mannan chain elongation (Ballou et al. 1991; Kanik-Ennulat et al. 1995; Poster and Dean 1996). These diverse defects suggest that Vrg4p, which is localized to the Golgi (Poster and Dean 1996), is important for normal Golgi functions, and unlikely to be a mannosyltransferase. Interestingly, though, Vrg4p resembles a *Leishmania donovanii* protein that has been proposed to be involved in GDPMan transport in the parasite's Golgi (Section IV.A.2.e). Mutation of nonessential *ERD1* leads to defects in the retention of luminal ER proteins in that compartment, and *erd1* mutants scarcely elongate the N-linked chains of invertase (Hardwick et al. 1990). Since *erd1* cells seem to be affected in various Golgi functions and mannosylation events, Erd1p is unlikely to be a mannosyltransferase. Mutants defective in nonessential *PMR1*, a Ca^{++} ATPase localized in a Golgi-like compartment, are, like *mnn9*, impaired in outer-chain elongation (Antebi and Fink 1992), indicating that alterations in Ca^{++} balance influence Golgi glycosylation events.

A relationship between *N*-glycosylation, secretory pathway function, and polarized growth is also suggested by the phenotype of the *mnn10* mutant (also known as *bed1*). In addition to having a severe defect in N-linked outer-chain extension and being orthovanadate-resistant (Ballou 1990; Ballou et al. 1991; Dean and Poster 1996), the *mnn10/bed1* mutant is synthetically lethal with overproduction of mitotic cyclins (Montdésert and Reed 1996). *mnn10/bed1* disruptants are delayed in bud emergence, enlarged, and round, and they show delocalized chitin deposition. The mutation seems to disrupt the coordination between nuclear division and budding by delaying budding but allowing mitosis to proceed and was proposed to do so by affecting the targeting of secretory vesicles during polarized growth (Montdésert and Reed 1996). Mnn10p resembles an α-1,2 galactosyltransferase from *S. pombe* Golgi membranes, and two proposals, based on the assumption that the protein is indeed a glycosyltransferase, have been made for the protein's function. The first is that Mnn10p may be an outer-chain-elongating α-1,6 mannosyltransferase, the similarity between the bakers' yeast and fission yeast proteins reflecting in this case a possible requirement for an acceptor oligosaccharide of similar structure, rather than a shared requirement for UDPGal as substrate (Dean and Poster 1996). An alternative proposal is that Mnn10p is a galactosyltransferase that may modify a protein or lipid that affects the functioning of the secretory pathway and indirectly, in turn, mannan outer-chain extension (Montdésert and Reed 1996). However, disruption of *GAL10*, which encodes UDP-galactose-4-epimerase, the enzyme that interconverts UDPGlc and UDPGal, does not give rise to the *mnn10/*

bed1 phenotype in cells grown on glucose (Montdésert and Reed 1996), suggesting that if Mnn10p is indeed a galactosyltransferase, another UDPGal epimerase exists. Modifications of proteins or lipids with galactose must be very rare in *S. cerevisiae:* The only place where galactose has so far been reported to occur is in the GPI anchor of a cAMP-binding protein (Müller et al. 1992). It is not yet clear where Mnn10p functions in *S. cerevisiae*, for the protein has been reported to be present predominantly both in the ER (Montdésert and Reed 1996) and in the Golgi (Dean and Poster 1996). It would be interesting to establish whether mutations in early and late steps in *N*-glycosylation, or in secretory pathway function, also show synthetic lethality with cyclin overproduction.

e. GDPMan Transport. GDPMan, which is synthesized in the cytoplasm, must be transported into the lumen of Golgi compartments in order to serve as mannosyl donor in the extension of mannan chains. Import of GDPMan into the lumen of the Golgi in *S. cerevisiae* is proposed to involve an antiport mechanism analogous to that used in mammalian cells (Hirschberg and Snider 1987). The model is that GDPMan enters the lumen via a specific membrane-bound carrier protein and donates its mannose residue to an endogenous acceptor, releasing GDP, which is in turn hydrolyzed to GMP, whose exit from the Golgi lumen is coupled to the import of further GDPMan. This model has received strong support from studies by Hirschberg and colleagues, who identified a key predicted component of this antiport system, the GDPase, which is indeed localized in a Golgi fraction (Abeijon et al. 1989; Yanagisawa et al. 1990). Disruption of the cloned *GDA1* gene yielded viable haploids, but Golgi vesicles isolated from Δ*gda1* strains showed a fivefold reduction in the rate at which GDPMan was transported into them (Berninsone et al. 1994). *gda1* disruptants showed a marked reduction in Golgi mannosylation events, not only those leading to maturation of N-linked chains, but also those leading to the extension of O-linked saccharides and the formation of mannosylinositol phosphoceramides (Abeijon et al. 1993). N-linked chains on carboxypeptidase Y were not elongated beyond the core glycosylated ER stage, and invertase was underglycosylated and had a large proportion of core, N-linked saccharides, although this effect was not as severe as that seen in *pmr1* and *mnn9* mutants, where the outer chain is virtually absent from invertase. These general mannosylation defects were consistent with a reduced availability of GDPMan for a host of different Golgi mannosyltransfer reactions. Blockage of mannosylation in *gda1* disruptants is not complete, and the antiporter GMP may also be generated in the mutant's Golgi vesicles when GDPMan donates mannosyl phosphate to mannan, or by nonspecific phosphatases (Berninsone et al. 1994). The effect of a GDPMan deficiency on the formation of mannose-containing structures is presumably a complex one, influenced by the K_m values different mannosyltransferases have for GDPMan and by the possibility that different Golgi compartments differ in the composition and size of their complement of mannosyltransferases.

A candidate for the GDPMan transporter has emerged in Vrg4p (Van2p) (Section IV.A.2.d). Vrg4p resembles the *L. donovanii* LPG2 protein, which is required for synthesis of this parasite's mannose-containing lipophosphoglycan (Poster and Dean 1996). *LPG2* mutants are defective in vitro in GDPMan transport, suggesting that this protein is closely involved in GDPMan transport into the lumen of the parasite's Golgi (Ma et al. 1997) and, in turn, that Vrg4p may have the same role in *S. cerevisiae.* Should Vrg4p indeed be a GDPMan transporter, this would explain the outer-chain elongation defect in *vrg4* mutants but would also suggest a requirement for GDPMan-dependent Golgi mannosylation events for Golgi functions such as protein sorting and retrieval of proteins to the ER. The critical Golgi mannosylation reactions may not be those affecting N-linked carbohydrates, for the *mnn9* mutant, which fails to extend its outer mannan chains (Section IV.A.2.d), does not show the defect in retention of ER proteins that *vrg4* cells do (Poster and Dean 1996).

3. *Functions of N-linked Chains*

a. Folding, Multimerization, and Efficient Secretion of Protein. Studies on individual *N*-glycosylated proteins are yielding a general consensus for roles for N-linked saccharides. A primary one is to prevent the malfolding and aggregation of newly translocated proteins in the ER lumen and hence promote secretion competence of proteins, some of which may be required for cell viability. This is apparent from a comparison of the phenotypes of the *sec53* and *sec59* mutants with that of the *dpm1* mutant: The former two show a global block in secretion (see Section III.A), whereas *dpm1* cells do not. *dpm1* cells differ from *sec53* and *sec59* cells only in their ability to attach N-linked oligosaccharides—mainly truncated ones—to protein, suggesting that cells must attach N-linked chains of some minimum size to protein, perhaps in some minimum number, in order to promote secretion competence of sufficient proteins (Orlean 1992). Furthermore, double mutants harboring both the *wbp1* mutation and a mutation in the luminal ER protein Kar2p (BiP) have a very slow growth rate at a temperature permissive for growth of single *wbp1* or *kar2* mutants (te Heesen and Aebi 1994). Defects in Kar2p, a chaperone that normally stabilizes the folding-competent state of a protein, are therefore exacerbated in cells impaired in *N*-glycosylation. N-linked chains also promote multimerization and, in one case, influence catalytic activity of a protein. The results of studies with a number of different proteins are recounted below. The temperature sensitivity of the *och1* disruptant indicates that the ability to extend a mannan outer chain on an N-linked oligosaccharide is a function required for growth of *S. cerevisiae* at high temperature, although it is not known what key function these structures fulfill.

The stabilities of glycosylated and unglycosylated forms of invertase to heat- or denaturant-induced unfolding do not differ significantly, but the unglycosylated protein tends to aggregate. The role of N-linked glycosylation was proposed in this case to enhance the solubility of the protein, and so allow newly translocated invertase to fold into its native state in the ER lumen, rather than form nonspecific aggregates (Schülke and Schmid 1988). Glycosylation of invertase also promotes multimerization (Chu et al. 1983). Secreted invertase normally forms an octamer of core glycosylated subunits that is itself a complex of four invertase dimers. Multimerization occurs in the lumen of the ER, the N-linked saccharides are extended in later secretory compartments, and octameric invertase is secreted into the periplasmic space. When invertase is formed in the presence of tunicamycin, secretion of the unglycosylated invertase that is formed is delayed, and rather than octamers, invertase dimers are formed, some of which are secreted into the medium. This indicates that the N-linked chains stabilize multimers or promote their formation. Octamer formation may be required for rapid secretion, and glycosylation-dependent multimerization presumably also helps retain invertase in the periplasmic space (Esmon et al. 1987). Mel1p (α-galactosidase) and Pho3p (acid phosphatase) also form multimers, whose large sizes are likely to aid in retaining such proteins in the periplasmic space (Esmon et al. 1987). A comparison of the stability of the octamers formed by the *mnn1 mnn9* mutant (which has 8–11 carbohydrate chains, each with 10 mannose residues) with those formed by the *mnn1 mnn9 dpg1* mutant (which has 4–7 carbohydrate chains, each with 10 mannoses) showed that the more extensively glycosylated invertase molecules from *mnn1 mnn9* formed the more stable octamers (Tammi et al. 1987). Oligosaccharides at some sites in the protein may be more important than those at other sites in promoting oligomerization (Reddy et al. 1990).

In the case of carboxypeptidase Y (encoded by *PRC1*), blockage of *N*-glycosylation with tunicamycin, or elimination of the four *N*-glycosylation sites by mutation, does not prevent sorting of the protein to its correct destination, the vacuole (Schwaiger et al. 1982; Winther et al. 1991); however, the rate of transport of the protein to the vacuole was reduced, the more so at higher temperatures (Winther et al. 1991). In Prc1p, an N-linked chain at Asn-87 is of specific importance for efficient transport, for a protein glycosylated at this site alone is transported to the vacuole at the normal rate. Interestingly, the carbohydrate chain at Asn-87 is smaller than the chains at the other three sites and is not phosphorylated (Trimble et al. 1983). Elimination of the three N-linked glycosylation sites of pro-α-factor (MFα1p) delayed transport of the protein from the ER to the Golgi, again indicating that the N-linked chains promote solubility and inhibit aggregation of this protein in the ER lumen (Caplan et al. 1991). No specific roles could be assigned to the individual chains in pro-α-factor. Pho5p, alkaline phosphatase, has 12 N-linked sites that are all core-glycosylated. In experiments in which different numbers of sites in different positions along the protein were eliminated by site-specific mutagenesis, secreted alkaline phosphatase activity was only detected when the protein bore at least six N-linked chains (Riederer and Hinnen 1991). The position of the chains along the protein also influenced the amount of enzyme activity recovered. Since the underglycosylated variants of Pho5p were not secreted, not enzymically active, and not degraded within the cell, it was concluded that the hypoglycosylated forms were accumulating in a malfolded, inactive state. This malfolding was temperature-dependent, for it was relieved upon incubation of the cells at lower temperature. In the case of the exo-β-1,3-glucanase Exg1p, the nature of an N-linked chain has been shown to influence enzyme activity. Exg1p can be glycosylated at both its N-linked sites, but outer-chain elongation occurs at Asn-325 only (Basco et al. 1994). The glycoform bearing the elongated chain, which represents 10% of secreted β-1,3-glucanase activity, showed half the affinity for the high-molecular-weight substrate laminarin as did glycoforms bearing core N-linked chains at either or both sequons, but affinity for a low-molecular-weight substrate was unaltered. An explanation is that the N-linked chain hinders binding of the enzyme to high-molecular-weight substrates.

b. Cell Cycle Progression. Inhibition of *N*-glycosylation gives rise to additional phenotypes that suggest a role for *N*-glycosylation in cell cycle progression. When tunicamycin, which inhibits Alg7p, the first enzyme in Dol-PP-linked oligosaccharide assembly, is added to logarithmically growing cells, the cells complete their division cycle, separate, and halt growth as unbudded cells (Arnold and Tanner 1982). Although these effects resemble a G_1 cell cycle arrest, the cells have a postreplicative G_2 content of DNA but are impaired in subsequent nuclear division (Vai et al. 1987). Tunicamycin therefore affects both bud emergence and nuclear division, but the reason for the latter effect is not clear. *alg1* cells also show an apparent G_1 arrest upon shift to their nonpermissive temperature (Klebl et al. 1984), and OTase mutants might also be expected to show this arrest phenotype, but this has not been recorded. It is possible, then, that cell cycle progression requires one or more key proteins to be *N*-glycosylated. Consistent with this notion, the growth defect of *alg1* point mutants, but not their glycosylation defect, can be suppressed by overexpression of G_1 cyclins, suggesting that a cell cycle checkpoint sensitive to levels of *N*-glycosylation exists but that it can be overcome in cells with lowered *N*-glycosylating capacity (Benton et al. 1996). *S. cerevisiae* cells also enlarge greatly their ability to carry out *N*-glycosylation during the G_1 phase. Thus, glucose-starved "G_0" cells, in which *ALG7*, *ALG1*, *ALG2*, *ALG5*, *ALG6*, *ALG8*, *WBP1*, and *SWP1* mRNAs had become depleted, showed a rapid induction of all these mRNAs when cells exited G_0 and growth resumed in the presence of glucose (Kukuruzinska and Lennon 1994; Lennon et al. 1995). This induction took place when protein synthesis was blocked, suggesting that expression of genes of the dolichol pathway is an early and presumably important response to a growth stimulus. Elevated levels of *N*-glycosylation are presumably required to promote efficient protein secretion and to meet the demand for glycoproteins during wall expansion and bud growth. As discussed in Section IV.A.2.d, coordination of glycosylation and secretory pathway function, on the one hand, with cell cell division, on the other, is also suggested by the finding that mutants defective in the putative Mnn10p glycosyltransferase proceed with nuclear division but are delayed in budding (Montdésert and Reed 1996).

B. GPI Anchors

1. GPI Structure and Identification of GPI Anchoring Mutants

Various glycoproteins in *S. cerevisiae* receive a GPI anchor at their carboxyl terminus. These structures are attached to protein in the ER, whereupon the proteins transit the secretory pathway and end up anchored in the external leaflet of the plasma membrane. GPI-anchored proteins in *S. cerevisiae* were first detected in experiments in which yeast proteins were radioactively labeled in vivo with [^{3}H]inositol, which becomes covalently attached to protein as part of the GPI anchor (Conzelmann et al. 1988). The structure of the core glycan of the GPI anchor isolated from protein is the same as that found in all other eukaryotes studied and consists of protein-CO-NH-CH_2-CH_2-PO_4-6-Man-α-1,2-Man-α-1,6-Man-α-1,4-GlcN-α-1,6-myo-inositol-1-PO_4-lipid (Fig. 3) (Fankhauser et al. 1993). A fourth α-1,2-linked mannose branches from the core α-1,2 mannose and to this fourth residue may be added either an α-1,2-linked or an α-1,3-linked mannose (Fankhauser et al. 1993). The phospholipid moiety can be a base-labile C_{26}/C_{26} diacylglycerol or a base-stable ceramide (Conzelmann et al. 1992).

Known yeast GPI-anchored proteins include Gas1p/Ggp1p (Nuoffer et al. 1991; Vai et al. 1991), the Agα1 mating agglutinin (Wojciechowicz et al. 1993), and a cAMP-binding protein (Müller et al. 1992). Six [^{3}H]inositol-labeled proteins can be separated by two-dimensional gel electrophoresis (Vai et al. 1990). Increasing numbers of proteins, such as the Aga1p agglutinin, Kre1p (involved in β-1,6 glucan synthesis [Boone et al. 1990]), cell wall proteins (Van der Vaart et al. 1995), and possibly Flo1p (involved in flocculation [Teunissen et al. 1993; Watari et al. 1994]), are being inferred to be GPI-anchored from the deduced amino acid sequence of the carboxyl terminus of the protein. These proteins are often serine- and threonine-rich, hence possibly heavily *O*-mannosylated as well.

The pleiotropic glycosylation mutants *sec53*, *sec59*, and *dpm1* are blocked in the incorporation of [^{3}H]inositol into protein, hence in GPI anchoring, at nonpermissive temperature, but none of the *alg* mutations affect GPI anchoring (Conzelmann et al. 1990; Orlean 1990). The temperature-sensitive mutants *gpi1*, *gpi2*, and *gpi3* (*spt14*), which have a specific defect in GPI anchoring, were isolated using a screen for colonies of mutagenized cells blocked in the incorporation of [^{3}H]inositol into protein (Leidich et al. 1994, 1995). A screen for Calcofluor resistance yielded *cwh6,* allelic to *gpi3 (spt14)* (Vossen et al. 1995). The finding that the α-mating agglutinin (Agα1p) is GPI-anchored was exploited in a screen for mutants defective in cell surface expression of this glycoprotein and led to isolation of candidate GPI anchoring mutants (Benghezal et al. 1995). The *gaa1* mutant, originally identified as endocytosis-defective *end2*, turned out to be defective in attachment of the GPI precursor to protein (Hamburger et al. 1995).

2. *Biosynthesis and Attachment to Protein*

a. Assembly of the GPI Precursor Glycolipid. Yeast GPI anchors are preassembled in a pathway whose steps are believed to take place in the

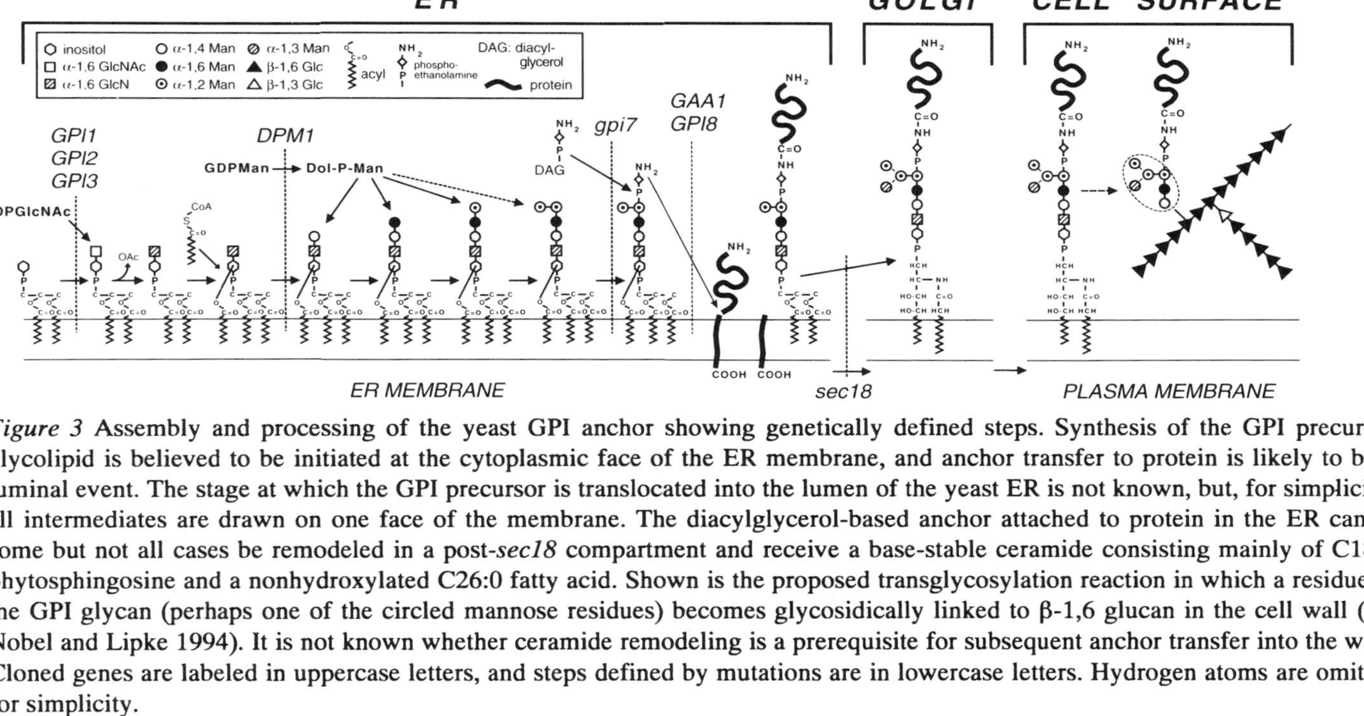

Figure 3 Assembly and processing of the yeast GPI anchor showing genetically defined steps. Synthesis of the GPI precursor glycolipid is believed to be initiated at the cytoplasmic face of the ER membrane, and anchor transfer to protein is likely to be a luminal event. The stage at which the GPI precursor is translocated into the lumen of the yeast ER is not known, but, for simplicity, all intermediates are drawn on one face of the membrane. The diacylglycerol-based anchor attached to protein in the ER can in some but not all cases be remodeled in a post-*sec18* compartment and receive a base-stable ceramide consisting mainly of C18:0 phytosphingosine and a nonhydroxylated C26:0 fatty acid. Shown is the proposed transglycosylation reaction in which a residue in the GPI glycan (perhaps one of the circled mannose residues) becomes glycosidically linked to β-1,6 glucan in the cell wall (De Nobel and Lipke 1994). It is not known whether ceramide remodeling is a prerequisite for subsequent anchor transfer into the wall. Cloned genes are labeled in uppercase letters, and steps defined by mutations are in lowercase letters. Hydrogen atoms are omitted for simplicity.

membrane of the ER, and indeed, the completed glycolipid is transferred to protein in the ER, for proteins that accumulate in the *sec18* mutant already bear a GPI anchor (Conzelmann et al. 1988).

Early steps in yeast GPI synthesis have been detected in a cell-free system, the first two being transfer of GlcNAc from UDPGlcNAc to an acceptor PI yielding GlcNAc-PI, and deacetylation of GlcNAc-PI to GlcN-PI (Costello and Orlean 1992). At least three gene products are required for GlcNAc-PI synthesis, for the *gpi1*, *gpi2*, and *gpi3* mutants all lack in vitro GlcNAc-PI synthetic activity (Leidich et al. 1994, 1995; Schönbächler et al. 1995). The genetic interactions observed raise the possibility that Gpi1p, Gpi2p, and Gpi3p interact with one another in a complex: Thus, no pairwise combination of these three mutations yields viable haploids, and overexpression of *GPI2* partially suppresses the growth defect of *gpi1* cells at 37°C (Leidich et al. 1994, 1995). *GPI2* and *GPI3* (*SPT14,* see below) are essential genes (Leidich et al. 1995), whereas *gpi1* disruptants are viable, lack in vitro GlcNAc-PI-synthetic activity, but are temperature-sensitive for growth (Leidich and Orlean 1996). Gpi1p, Gpi2p, and Gpi3p are all membrane proteins. Of the three, Gpi3p/Spt14p may be catalytic, for this protein shows amino acid sequence similarity to certain bacterial glycosyltransferases that recognize UDP sugars (Vossen et al. 1995), although it does not resemble Alg7p or chitin synthases. Gpi3p/Spt14p is 44% identical to the human Pig-A protein, mutations in which abolish GPI anchoring and GlcNAc-PI synthesis and lead to the clinical condition paroxysmal nocturnal hemoglobinuria (Takeda et al. 1993; Schönbächler et al. 1995; Vossen et al. 1995).

The first mannose residue originates from Dol-P-Man, since GPI anchoring is blocked in the *dpm1* mutant (Orlean 1990). Evidence for an additional step preceding mannosylation came from experiments in which the *dpm1* mutant was pulse-labeled with [^{3}H]inositol or [^{3}H]GlcN. At nonpermissive temperature, *dpm1* (and *sec53* and *sec59* cells) accumulates GlcN-PI with a fatty acyl group esterified to an inositol hydroxyl, which renders the structure resistant to hydrolysis by PI-specific phospholipase C. In vitro studies show that this inositol acyl group originates from fatty acyl CoA (Orlean 1990; Costello and Orlean 1992). The acyl group is preserved on subsequent GPI intermediates, for the full-length GPI precursor is also inositol-acylated (Sipos et al. 1994). However, a deacylation step must subsequently occur, for the mature, base-labile GPI anchors on protein are susceptible to PI-specific phospholipase C hydrolysis (Conzelmann et al. 1990). On the basis of results obtained with a trypanosomal cell-free system, the second and third mannoses in the core glycan are likely to originate from Dol-P-Man (Menon et al. 1990). The fourth, α-1,2-linked branching mannose, seems to be added to the yeast GPI precursor in the ER as well (Sipos et al. 1995). Two isolates in which Agα1p expression at the cell surface is abolished accumulate GPI precursors whose structures suggest that these strains may be defective in the addition of the first and third mannoses, respectively, to the GPI core (Benghezal et al. 1995). The *S. cerevisiae* genome contains two open reading frames encoding proteins resembling the mammalian PIG-B protein (Takahashi et al. 1996), which is required for addition of the α-1,2-linked mannose to the growing GPI. Intriguingly, all three PIG-B-type proteins show local similarity to Alg9p (C.H. Taron and P. Orlean, unpubl.), a candidate α-1,2 mannosyltransferase of the N-linked pathway, indicating a relationship between the Dol-P-Man-utilizing mannosyltransferases of the *N*-glycosylation and GPI-anchoring pathways. Whether these similarities reflect the fact that these proteins are all α-1,2 mannosyltransferases is an open question. Three mutants that show a defect in the addition to the first mannose to the GPI precursor have been described, but these strains are also affected in elongation of N-linked chains, hence unlikely to be deficient in the α-1,4 mannosyltransferase. Intriguingly, although these mutants, *gpi4*, *gpi6*, and *gpi9*, seem to be defective in Dol-P-Man supply, they complement the known mutants affecting this process, *sec53*, *sec59*, *dpm1*, and *pmi40* (Benghezal et al. 1995).

The phosphoethanolamine (EtN-P) moiety that is added to the third mannosyl residue of the GPI precursor is likely to originate from phosphatidylethanolamine (PE). *S. cerevisiae* mutants disrupted in the ethanolamine phosphotransferase (*EPT1*) and choline phosphotransferase (*CPT1*) genes, and thus unable to make CDP-ethanolamine, nevertheless make GPI anchors but do not incorporate exogenously added [^{3}H]ethanolamine into them, indicating that CDP-ethanolamine does not serve as EtN-P donor. Since *ept1/cpt1* disruptants can still make PE by decarboxylation of phosphatidylserine, PE is the probable EtN-P donor (Menon and Stevens 1992). The *gpi7* mutant is defective in EtN-P addition in vivo (Benghezal et al. 1995), and a candidate for an *S. cerevisiae* protein required for this step has been identified because of its resemblance to PIG-F protein, which is involved in EtN-P addition during completion of human GPIs (Inoue et al. 1993). The *S. cerevisiae* PIG-F homolog is essential for viability and functionally equivalent to its human counterpart, for PIG-F rescues the lethal null mutation in its *S. cerevisiae* counterpart (C.H. Taron and P. Orlean, unpubl.). Despite the identification of a number of genes involved specifically in GPI assembly and attachment, biochemical activities remain to be demonstrated for these genes' products.

b. Transfer of the GPI Anchor to Protein. Proteins that receive a GPI anchor have both an amino-terminal secretion signal and a carboxy-terminal signal for GPI attachment. Anchor transfer to protein is believed to take place in a transamidation reaction, involving cleavage of a carboxy-terminal hydrophobic sequence and concomitant formation of an amide linkage between the EtN-P of the GPI and the new carboxy-terminal amino acid (Udenfriend and Kodukula 1995). Studies with model mammalian proteins and yeast Gas1p/Ggp1p have defined the sequence requirements for GPI addition. The attachment signal consists of a stretch of hydrophobic amino acids at the protein's carboxyl terminus, followed by a short stretch of more hydrophilic residues, and an attachment site consisting of three consecutive residues with small side chains, designated ω, ω+1, and ω+2. Peptide cleavage occurs between the ω and ω+1 residues, and the GPI anchor is attached to the ω residue. The ω+1 and ω+2 amino acids are immediately carboxy-terminal to the cleavage site, and of the two, the ω+2 site has the more stringent requirement for a small side chain (Nuoffer et al. 1991, 1993).

The *gaa1* and *gpi8* mutants both accumulate the complete GPI anchor precursor at nonpermissive temperature and exhibit synthetic lethality (Benghezal et al. 1995; Hamburger et al. 1995). These findings suggest that anchor transfer to protein may be carried out by a protein complex, as is oligosaccharyl transfer to asparagine (see Section IV.A.1.b). *GPI8*, an essential gene, encodes a protein resembling a plant vacuolar asparaginyl endopeptidase that has transamidase activity in vitro (Benghezal et al. 1996), a finding consistent with the notion that GPI attachment proceeds by transamidation. *GAA1*, likewise essential, encodes a protein localized to the ER (Hamburger et al. 1995). Overexpression of Gaa1p leads to improved GPI attachment to forms of Gas1p/Ggp1p with mutated, suboptimal anchor attachment sites (Hamburger et al. 1995; see below).

c. Anchor Remodeling and Modification. The GPI anchor precursors that can be identified by in vivo and in vitro radioactive labeling experiments are all base-labile (Costello and Orlean 1992; Sipos et al. 1994), indicating that the yeast GPI anchor precursor is assembled on a PI with ester-linked fatty acyl chains. The diacylglyerol moiety of the GPI anchor attached to Gas1p/Ggp1 contains two C_{26} fatty acids (Fankhauser et al.

1993). Whether the GPI precursor is assembled on a specialized PI with extra long-chain fatty acids (rather than on dipalmitoylglyceryl-based PI, the most abundant species in yeast) or whether fatty acid remodeling occurs at a later stage is not known.

In pulse-chase experiments with the *sec18* mutant, which is blocked in transport from the ER to the Golgi, it was shown that many protein-bound GPI anchors can also undergo a remodeling reaction in which the diacylglycerol is replaced with a ceramide (Conzelmann et al. 1992). The GPI-anchored proteins that accumulate in *sec18* cells at nonpermissive temperature have base-labile anchors. After return of the *sec18* cells to their permissive temperature, the proportion of proteins with base-stable, ceramide-containing anchors increases, although not all proteins receive the ceramide. The remodeling reaction, which must occur in a compartment beyond the *sec18* block, may involve transfer of the entire phosphoinositol glycan protein "headgroup" to ceramide, by analogy with the formation of inositol phosphoceramide from PI (Becker and Lester 1980). Anchor remodeling is prevented in the presence of an inhibitor of ceramide synthesis (Horvath et al. 1994). The acceptor ceramide seems to be a novel one not found in any of the major yeast inositol phosphoceramides (Fankhauser et al. 1993). Addition of the fifth mannose to the branch off the trimannosyl core of protein-bound anchors takes place in the Golgi and, intriguingly, is not affected in the *mnn9* mutant, which drastically affects elongation of N-linked chains (Sipos et al. 1995; see Section IV.A.2.d).

3. *Functions of GPIs and Phenotypes of GPI Anchoring Mutants*

GPI synthesis and attachment to protein are essential functions in *S. cerevisiae*, for the GPI anchoring mutants so far identified all show temperature-sensitive growth either on standard media or on Calcofluor-White-containing medium (Leidich et al. 1994, 1995; Benghezal et al. 1995; Hamburger et al. 1995; Vossen et al. 1995). So far, no individual protein that normally becomes GPI-anchored has been shown to be essential for viability; however, GPIs have key roles in the intracellular transport of certain proteins and in cell wall assembly, and inducible cleavage of these anchors may be a way in which cell surface expression of certain proteins can be regulated.

Attachment of a GPI is critical for efficient intracellular transport of the class of proteins that normally receive an anchor. Studies using Gas1p/Ggp1p as a model protein reveal that abolition of GPI anchor attachment to Gas1p/Ggp1p, caused by shifting the *gaa1* or *gpi3/spt14* mutants to nonpermissive temperature or by mutating the protein's carboxy-terminal signal cleavage site, results in a defect in the transport of the protein to the Golgi, whereas deletion of the entire carboxy-terminal hydrophobic domain leads to secretion of the truncated protein into the medium (Nuoffer et al. 1991; Hamburger et al. 1995). GPI anchoring and transport of Gas1p/Ggp1p from the ER is rapidly abolished upon starvation of inositol auxotrophs for inositol, perhaps because GPI precursor synthesis is unusually sensitive to interruptions in phosphatidylinositol supply (Doering and Schekman 1996). Transit and modification of non-GPI-anchored proteins along the secretory pathway is not affected in the GPI anchoring mutants tested (Leidich et al. 1994; Benghezal et al. 1995; Hamburger et al. 1995; Schönbächler et al. 1995), and inositol starvation, which rapidly and specifically inhibits transport of GPI-anchored proteins, has a much delayed effect on the transport of nonanchored proteins such as invertase and Prc1p (Doering and Schekman 1996).

The requirement of GPI anchors for transport of the proteins that normally receive them correlates with the finding that the transport of such proteins is dependent on ceramide biosynthesis. Thus, treatment of cells with myriocin, an inhibitor of ceramide synthesis, severely retards transport of GPI-anchored proteins from the ER to the Golgi, but it does not affect transport of other secretory proteins. An attractive explanation for this finding is that GPI-anchored proteins and ceramides are normally cotransported, the role of the ceramides being to promote the clustering of GPI-anchored proteins, perhaps in a specialized membrane microdomain, and so make them competent for transport to the Golgi (Horvath et al. 1994). A myriocin-resistant mutant was isolated in which normal protein transport, but not ceramide synthesis, was restored. Whether the bypass mechanism is due to synthesis of a functional replacement for ceramide (see Section III.B) is not known.

GPI anchoring per se and arrival of GPI-anchored proteins at the plasma membrane are also necessary for normal cell wall assembly. When grown at their semipermissive temperatures, *gpi1* and *gpi2* strains are enlarged, misshapen, and show a separation defect (Leidich and Orlean 1996). Moreover, the GPI anchoring mutants so far tested all show hypersensitivity to Calcofluor White (Benghezal et al. 1995; Vossen et al. 1995). These wall-synthetic defects may in part be explained by the fact that the GPI on some glycoproteins serves as an intermediate in the transfer of the protein to an acceptor glycan in the cell wall (see Section VI.A.2.b). The finding that inositol starvation of inositol auxotrophs leads to rapid inhibition of GPI anchoring (Doering and Schekman 1996) and mannan synthesis (Hanson and Lester 1980) is consistent with a block in transport and subsequent cross-linking of mannoproteins into the wall. A GPI anchoring defect also affects the function of Gas1p/Ggp1p, which has its own role in cell wall synthesis. *gas1/ggp1* disruptants have morphological defects similar to those manifested in the *gpi1* and *gpi2* mutants, and cells are resistant to lysis by the β-1,3-glucanase zymolyase (Popolo et al. 1993b). Moreover, a mutant allele of *GAS1/GGP1*, *cwh52*, is hypersensitive to Calcofluor White and has a severe defect in in vivo β-1,3 glucan synthesis (Ram et al. 1995; Section V.B.2.b). The presence of the GPI anchor is necessary for this protein's function, for a mutant Gas1p/Ggp1p lacking its carboxy-terminal hydrophobic domain was secreted into the medium and failed to complement the growth defect of the *gas1/ggp1* disruptant (Nuoffer et al. 1991).

GPIs can also be a target for cleavage by phospholipase and perhaps other processing enzymes, which are induced, for example, by nutritional upshift. Thus, the GPI-anchored forms of a small subset of proteins are cleaved by an endogenous GPI-specific phospholipase that is activated when spheroplasts are transferred from lactate- to glucose-containing medium (Müller and Bandlow 1993; Bandlow et al. 1995; Müller et al. 1996). The in vivo function of the proteins so processed, which include a cAMP-binding protein, although not Gas1p/Ggp1p, is not known, nor is the basis for the selectivity of cleavage of these GPI-anchored proteins clear. It is possible that GPI cleavage from a protein is accompanied by a conformational change, which in turn may regulate the function of the protein. Indeed, GPI cleavage from the cAMP-binding protein results in an increase in this protein's affinity for its ligand (Müller and Bandlow 1994). Parallels have been drawn between glucose-induced processing of GPI-anchored proteins in yeast and the insulin-induced lipolytic release of certain GPI-anchored proteins from the surface of adipocytes (Bandlow et al. 1995; Müller et al. 1996). One possible consequence of the latter is that processing generates a GPI-derived glycan that is somehow internalized and functions as an intracellular messenger to mediate some of the effects of insulin. A second processing step, which follows lipolytic cleavage and removes the protein-attached anchor phosphoinositol glycan and perhaps some carboxy-terminal amino acids, has been demonstrated for the yeast cAMP-binding protein (Müller et al. 1996). Whether the GPI glycopeptide that is generated has any physiological role in *S. cerevisiae* remains to be established.

GPI anchoring mutants defective in GlcNAc-PI synthesis also have a transcriptional phenotype. This emerged from the finding that the gene defective in the *gpi3* and *cwh6* mutants had previously been identified as

SPT14, whose protein product resembles that of the human Pig-A gene (Leidich et al. 1995; Schönbächler et al. 1995; Vossen et al. 1995). *spt14* mutants were identified as extragenic suppressors of the histidine auxotrophy of the *his4-917* transposon insertion mutation (Fassler et al. 1991). The *gpi1* and *gpi2* mutants also suppress the *his4-917* mutation (Leidich et al. 1995), but none of the three proteins resemble known transcriptional activators. One possible link between the GlcNAc-PI synthesis defect and the transcriptional phenotype of these mutants may be through altered partitioning of UDPGlcNAc between the GlcNAc-PI synthetic machinery and a protein:*O*-GlcNAc transferase that modifies one or more proteins that in turn influence *HIS4* expression.

Competition for UDPGlcNAc presumably also occurs between the GPI-synthetic and *N*-glycosylation pathways, and indeed, this is seen in vitro. Membranes from the *gpi1*, *gpi2*, and *gpi3* mutants make no GlcNAc-PI in vitro but show greatly enhanced synthesis of Dol-PP-$GlcNAc_{1,2}$ compared with wild-type membranes, which make mostly GlcNAc-PI and GlcN-PI (Leidich et al. 1994, 1995). An explanation is that in wild-type membranes in vitro, GlcNAc-PI formation competes favorably with Dol-PP-$GlcNAc_{1,2}$ synthesis for UDPGlcNAc. Whether this competition is significant in vivo is not known.

C. Inositol Phosphoceramides

Three major classes of inositol phosphoceramides (IPCs), which show diversity in their long-chain base and in the chain length and degree of hydroxylation of their fatty acid, are found in *S. cerevisiae* (Smith and Lester 1974). IPCs are made by transfer of inositol phosphate to ceramides (Becker and Lester 1980). One of the IPC species can be mannosylated, GDPMan being the donor in vitro (Becker and Lester 1980), and the resulting mannosylinositol phosphoceramide (MIPC) can receive a further phosphoinositol on the mannose residue, yielding $M(IP)_2C$. Experiments with various secretion mutants, in which yeast IPCs were pulse-labeled with [^{3}H]inositol, showed that formation of MIPC, $M(IP)_2C$, and one class of IPC is prevented in *sec* mutants defective in budding of secretory vesicles from the ER and in the fusion of such vesicles with the Golgi membrane. The likely explanation for these findings is that transport of the IPC substrate to its site of mannosylation is dependent on vesicular flow between the ER and Golgi (Puoti et al. 1991). Formation of MIPC and $M(IP)_2C$ shows a severe decrease in GDPase-deficient, Δ*gda1* cells, indicating that mannosylation indeed takes place in the lumen of a Golgi compartment (Abeijon et al. 1993).

Candidate genes for potential IPC:GDPMan mannosytransferase(s) have not yet been identified, and the functions of MIPC and $M(IP)_2C$ are unknown. IPCs are ultimately delivered to the cell surface, where they are enriched in the plasma membrane (Patton and Lester 1991). The key general role for ceramides in cotransport of GPI-anchored proteins has been recounted above (see Section IV.B.3).

D. *O*-Mannosylation

1. Structure and Occurrence

Many secretory proteins, including those that receive N-linked sugars and a GPI anchor, are modified by the attachment of O-glycosidically oligosaccharides containing between one and five mannose residues to serine and threonine residues (Sentandreu and Northcote 1968; Tanner and Lehle 1987). Although a common modification of fungal proteins, *O*-mannosylation has so far only been detected in mammals in brain proteoglycan (Krusius et al. 1986). Yeast proteins that are heavily, sometimes exclusively, *O*-mannosylated are in most cases either structural components of the wall or involved in some way in wall formation. These proteins include a structural cell wall protein (Frevert and Ballou 1985), the mating agglutinins (see Section VIII.A.1), the Fus1p cell fusion protein (McCaffrey et al. 1987; Truehart et al. 1987; Truehart and Fink 1989), Kex2 protease (Fuller et al. 1989), chitinase (Kuranda and Robbins 1991), the Gas1p/Ggp1p (Nuoffer et al. 1991; Vai et al. 1991), the secreted heat shock protein Hsp150p (Russo et al. 1992), Kre1p and Kre9p involved in β-1,6 glucan synthesis (Boone et al. 1990; Brown and Bussey 1993), and Flo1p (Teunissen et al. 1993; Watari et al. 1994). Further proteins likely to be *O*-mannosylated include small cell wall proteins (van der Vaart et al. 1995) and the related Msb2p and Hkr1p which are, respectively, a multicopy suppressor of *cdc24* (Bender and Pringle 1992) and a possible regulator of β-glucan synthesis (Kasahara et al. 1994b). Most of these proteins contain stretches of amino acids rich in serine and threonine and may bear many O-linked chains. The exact contribution of these chains to the mass of a protein may be hard to assess using SDS-PAGE, since the proteins may bind relatively little SDS and hence migrate aberrantly slowly. There seems to be no attachment consensus for *O*-mannosylation, although studies using synthetic peptides as acceptors indicate that a nearby proline residue can favor *O*-mannosylation in vitro (Lehle and Bause 1984; Strahl-Bolsinger and Tanner 1991). The existence of multiple protein:*O*-mannosyltransferases suggests an explanation for the lack of an apparent attachment consensus: These transferases may differ in their acceptor specificities and collectively recognize a range of different peptide sequences.

2. *Biosynthesis in the Endoplasmic Reticulum*

Early biochemical studies by Tanner and co-workers revealed that the mannose residue linked to protein is transferred from Dol-P-Man but that the subsequent mannose units are transferred from GDPMan (Tanner and Lehle 1987). The former finding is borne out by the fact that *O*-mannosylation is completely abolished when the temperature-sensitive Dol-P-Man synthase mutant is shifted to its nonpermissive temperature (Orlean 1990). *O*-mannosylation is initiated in the ER, for O-linked mannose and mannobiose become attached to protein in *sec18* cells after shift to nonpermissive temperature (Haselbeck and Tanner 1983). *S. cerevisiae* contains a family of seven Dol-P-Man:protein *O*-mannosyltransferases (Gentzsch et al. 1995a; Gentzsch and Tanner 1996). The gene for the first of these, *PMT1*, was cloned following purification of an enzyme by monitoring mannosyl transfer to a specific acceptor peptide (Sharma et al. 1991; Strahl-Bolsinger and Tanner 1991; Strahl-Bolsinger et al. 1993). *pmt1* disruptants were viable, but they lacked the in vitro *O*-mannosyltransferase activity used to monitor enzyme purification; however, as Δ*pmt1* strains still contained about half the total amount of O-linked mannose found in wild-type cells (Strahl-Bolsinger et al. 1993), further transferases had to be present. A second transferase was duly detected in Δ*pmt1* strains by using a different acceptor peptide (Gentzsch et al. 1995b). This activity is encoded by the *PMT2* gene, whose product is 31% identical to Pmt1p (Lussier et al. 1995b). Single *pmt1* and *pmt2* null mutants show reduced levels of *O*-mannosylation, reduced growth rates, and partial resistance to K1 killer toxin, and these effects are cumulative, for Δ*pmt1* Δ*pmt2* double disruptants have a severe growth defect, are more killer-resistant, and show lower levels of *O*-mannosylation than either *pmt* mutant alone. These observations indicate that *PMT1* and *PMT2* interact genetically, that O-linked manno-oligosaccharides have a role in normal growth, and that the O-linked saccharides contribute to the structure that serves as receptor for killer toxin. Overexpression of either *PMT1* or *PMT2* in a *pmt1 pmt2* double disruptant in each case enhances the growth rate of vegetative cells and restores killer sensitivity, indicating that Pmt1p and Pmt2p are functionally homologous (Lussier et al. 1995b). However, it seems that Pmt1p and Pmt2p can also form a heterodimer, for Pmt2p copurifies with Pmt1p on an anti-Pmt1p immunoaffinity column, and simultaneous overexpression of

Pmt1p and Pmt2p leads to a threefold increase in in vitro protein:*O*-mannosyltransferase activity, whereas overexpression of the individual Pmt proteins does not (Gentzsch et al. 1995a). Whether these two proteins normally function as a heterodimer in vivo is not yet clear.

Δ*pmt2* Δ*pmt2* strains are still capable of *O*-mannosylation and contain residual protein *O*-mannosyltransferase activity under the conditions used to detect Pmt2p activity, and indeed, two more related sequences, *PMT3* and *PMT4*, have been characterized and three further *PMT* genes also exist (Immervoll et al. 1995; Gentzsch and Tanner 1996). Single *pmt3* and *pmt4* disruptants grow well, and Δ*pmt4* or Δ*pmt3* Δ*pmt4* cells show a slight decrease in *O*-mannosylation, whereas Δ*pmt3* cells *O*-mannosylate normally. Δ*pmt1* Δ*pmt2* Δ*pmt4* and Δ*pmt2* Δ*pmt3* Δ*pmt4* strains are inviable, leading to the conclusion that *S. cerevisiae* must need to carry out a minimum level of *O*-mannosylation in general or that one or more yeast proteins are critically dependent on *O*-mannosylation to fulfill some essential function (Gentzsch and Tanner 1996). The roles of *PMT5*, *PMT6*, and *PMT7* remain to be explored. The defects in viable strains deficient in multiple *PMT* genes indicate that *O*-mannosylation of protein is required for maintenance of cell wall integrity, which in turn permits normal budding and cell separation. Δ*pmt2* Δ*pmt4*, Δ*pmt2* Δ*pmt4*, and Δ*pmt1* Δ*pmt2* Δ*pmt3* strains require osmotic support, and even then grow slowly and form clumps of cells that are often multinucleate. Analysis of cell wall hydrolysates of the Δ*pmt2* Δ*pmt3* disruptant, which shows a milder degree of osmotic fragility, reveals a 20% decrease in the glucan fraction, suggesting a role for O-linked mannoproteins in retaining glucan in the cell wall, which may in turn help maintain wall integrity (Gentzsch and Tanner 1996). However, the glycoprotein-β-glucan linkages that have so far been demonstrated do not involve O-linked mannose directly (see Section VI.A.2).

One explanation for the existence of multiple protein:*O*-mannosyltransferases in yeast is that they recognize different acceptor peptides and that collectively the Pmtp family allows *S. cerevisiae* to *O*-mannosylate a wider range of proteins (Lussier et al. 1995b; Finck et al. 1996; Gentzsch and Tanner 1996). In vivo evidence that this is so comes from comparing the effects of the *pmt1 pmt2* and Δ*pmt4* deletions on two *O*-mannosylated proteins. Thus, *O*-mannosylation of chitinase is significantly lowered in a Δ*pmt1* Δ*pmt2* strain, but little affected in a Δ*pmt4* mutant, whereas the opposite is true for Gas1p/Ggp1p, *O*-mannosylation of which is affected only in Δ*pmt4* (Gentzsch and Tanner 1996). Furthermore, the existence of a mutant that is defective in vivo and in vitro in mannosylation of a specific threonine residue in heterologously expressed human insulin-like growth factor (Finck et al. 1996) is also consistent with the notion that different peptides are recognized by different *O*-mannosyltransferases, although it is not known whether this particular mutant is defective in a *PMT* gene. Different acceptor preferences are also seen in vitro. Pmt1p distributes mannose evenly between serine and threonine, whereas Pmt2p, assayed in extracts of Δ*pmt1* cells, shows a strong preference for mannosylating serine residues (Gentzsch et al. 1995b). It is possible that the Pmt proteins are able to form a range of different heterodimers, which in turn might broaden the range of acceptors that could be efficiently *O*-mannosylated. How the Pmt proteins are organized in the ER membrane and positioned to recognize a nascent secretory protein are open questions.

3. Elongation of O-linked Chains

The second, α-1,2-linked mannose residue may be added to the initiating O-linked mannose in a compartment intermediate between the ER and Golgi, possibly that defined by the *sec18* block, for O-linked mannobiose is found on protein in *sec18* cells after shift to nonpermissive temperature (Haselbeck and Tanner 1983). In vitro, the mannosyl donor for the addition of mannoses 2 through 5 is GDPMan (Tanner and Lehle 1987), although this sugar nucleotide is not believed to be transported into the ER lumen. Since longer O-linked mannose chains are found on protein in other secretion mutants, such as *sec20* (Zueco et al. 1986), *sec6*, and *sec7* (Kuranda and Robbins 1991), it is possible that hyperextension of O-linked chains is carried out by mannosyltransferases en route to later compartments, and which, along with sugar nucleotide transporters, are trapped in the secretory vesicles where they act on any mannosyl acceptors present. However, Δ*gda1* strains, which are impaired in GDPMan transport into the Golgi, nevertheless contain large amounts of O-linked mannobiose (Abeijon et al. 1993), raising the possibility that addition of the second mannose is indeed a specialized and localized process that may require additional proteins. GDPMan transport is not totally abolished in Δ*gda1* cells, and regardless of which explanation holds, the extension of O-linked chains may also be influenced by the amount of free GDPMan in a given compartment, by the K_m values of different mannosyltransferases for this substrate, and by the K_i values of the enzyme for GDP (Abeijon et al. 1993). As would expected, mutants such as Δ*pmr1*, which is perturbed in the normal functioning of the secretory pathway and in N-linked outer-chain extension, also show diminished extension of O-linked chains (Abeijon et al. 1993).

Attachment of the third mannose residue requires the Kre2p/Mnt1p α-1,2 mannosyltransferase (Haüsler et al. 1992), an enzyme localized to the medial Golgi compartment (Chapman and Munro 1994; Lussier et al. 1995a). Other members of the Kre2p/Mnt1p family do not seem to be involved in *O*-mannosylation (Lussier et al. 1996). The O-linked trisaccharide can be elongated with one or two α-1,3-linked mannoses, the first of which is added by Mnn1p (Graham et al. 1994; Yip et al. 1994).

4. *Functions of O-linked Chains*

The O-linked mannose chains of yeast proteins are concentrated in the serine- and threonine-rich stretches that characterize these proteins. The effects of these many short O-linked chains may be to confer a stiff and extended conformation on those parts of the protein or to protect or stabilize the protein (Jentoft 1990; Stratford 1994). As with N-linked chains, O-linked chains collectively are important to the cell, but whether the modification makes any contribution at all to function presumably depends on the individual protein, although no protein has yet been shown to need all, or any subset, of its O-linked chains. Thus, in the case of Gas1p/Ggp1p, neither deletion nor duplication of the serine-rich and threonine-rich region affects the function of that protein in vivo (Gatti et al. 1994), nor does deletion of the serine-rich and threonine-rich region of Kex2p affect the catalytic activity of this protein (Fuller et al. 1989). Furthermore, O-linked manno-oligosaccharides do not seem to have a role in cell-cell adhesion mediated by the **a**-mating agglutinin (Cappellaro et al. 1994; see Section VII.A).

E. β-1,6 Glucan

1. *β-1,6 Glucan-Protein Complexes*

The major class of β-1,6-linked glucan is the alkali-insoluble fraction, although β-1,6 linkages are found in other wall fractions as well (see Section II.A). β-1,6 glucan can also be covalently associated with mannan, for treatment of alkali-soluble β-glucan with β-1,3-glucanase releases material consisting of both mannan and β-1,6-linked glucan (Fleet and Manners 1977). In vitro assays for the biosynthesis of β-1,6 glucan have yet to be developed, but β-1,6 glucan synthesis can be detected in in vivo radioactive labeling experiments. Thus, a *glc1 pgm2 GAL81 GAL82* strain, grown on glucose, can take up [^{14}C]galactose and convert it to UDP[^{14}C]Glc (Tkacz 1984), and the *pgm2* and *glc1* mutations, respectively, prevent catabolism of radioactively labeled hexose and diversion of UDPGlc into glycogen. Treatment of such [^{14}C]galactose-labeled cells with endo-β-1,3-glucanase released radioactively labeled macromolecules that could be separated into different size classes by gel-filtration chromatography. Three fractions were converted to lower-molecular-weight products by β-1,6-glucanase, indicating that their radioactivity was present in β-1,6 glucan chains, and one of these β-1,6-glucan-containing fractions was also reduced in size upon treatment with endo-β-*N*-acetylglucosaminidase H. Radioactive labeling of this material was also inhibited by tunicamycin. These results indicate that at least one class of β-1,6-glucan-containing molecule is covalently associated with *N*-glycosylated protein. Moreover, in some cases, β-1,6 glucan can be directly attached to protein (Montijn et al. 1994), but some mannoproteins become cross-linked to β-1,6 glucan in the cell wall through their carbohydrate (see Section VI.A.2). However, it is not yet clear whether the β-1,6-glucan-containing fractions identified in various ways have their origin in the same biosynthetic pathway or whether some are the products of separate biosynthetic routes (Klis 1994). Genetic analyses focusing on alkali-insoluble β-1,6 glucan have revealed that the biosynthesis of this component proceeds along the secretory pathway in a process that may parallel the synthesis of extracellular matrix polysaccharides of higher plants (Roemer et al. 1994).

2. *Mutants and Genes in β-1,6 Glucan Synthesis*

a. K1 Killer Toxin Resistance as a Selection for Mutants in β-1,6 Glucan Synthesis. A key finding, which led to the development of an elegant screen for mutants in β-1,6 glucan synthesis, was that β-1,6 glucan is a critical component of the cell surface receptor that is bound by K1 killer toxin (Hutchins and Bussey 1983). Many of the mutants selected on the basis of resistance to this toxin, "*kre*" mutants, have defects in alkali-insoluble β-1,6 glucan synthesis and show altered proportions of β-1,3-linked and β-1,6-linked residues in this cell wall fraction, changes that can be detected by ^{13}C-labeled nuclear magnetic resonance (NMR) spectroscopy (Boone et al. 1990). Some mutations also affect *N*- and *O*-glycosylation of protein. A mutant defective in the synthesis of alkali-insoluble, predominantly β-1,6-linked glucan was also isolated on account of its hypersensitivity to a lytic β-1,3-glucanases (Shiota et al. 1985), and K1 killer-toxin-resistant mutants were among those displaying Calcofluor White sensitivity (Ram et al. 1994).

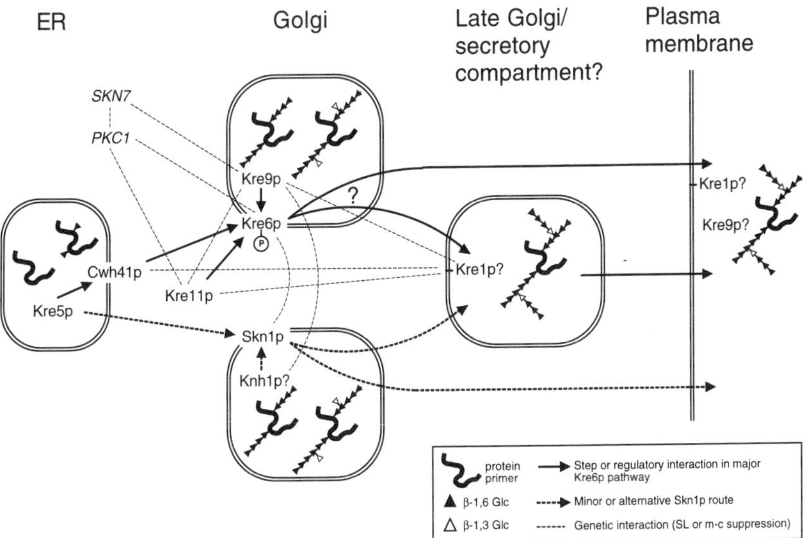

Figure 4 Suggested pathway for secretory-pathway-based synthesis of β-1,6 glucan. Bold arrows indicate steps or regulatory interactions in the major "Kre6p pathway." The order of events is based on epistasis relationships between mutants in the genes involved and on the subcellular localization of some of the proteins. Kre1p and Kre9p may function in secretory compartments or outside the plasma membrane, or both. In many cases, synthetic lethality or a synthetic phenotype arises when two mutations affecting steps in this pathway are combined. Dashed arrows indicate the possible minor or alternative route involving Skn1p and Knh1p, homologs respectively of Kre6p and Kre9p. Dashed lines indicate additional genetic interactions, synthetic lethality (SL) or multiply suppression (m-c). Symbols suggest the relative size and the composition of the growing, branched β-1,6 glucan.

b. kre *Mutants and Interactions between* KRE *Genes*. The phenotypes conferred by *kre* mutations and by mutations in genes that show genetic interactions with *KRE* genes are allowing a pathway to be charted for the synthesis of an alkali-insoluble β-1,6 glucan consisting of a core β-1,6 glucan, with β-1,3-linked side branches in turn extended by further β-1,6-linked glucose residues (Fig. 4). Some mutants have a lowered content of β-1,6 glucan, whereas in others, a polymer is made that is smaller or that lacks the β-1,6-linked side branches. Intriguingly, many of the gene products involved are proteins that reside in secretory compartments or transit the pathway, but some are cytoplasmic and seem to regulate the activity of secretory pathway-based proteins. These proteins are discussed below in the order in which they appear to contribute to β-1,6 glucan synthesis.

The *kre5* mutation is epistatic to all *kre* mutations so far isolated (Boone et al. 1990). *kre5* disruptants are viable but contain virtually no alkali-insoluble β-1,6 glucan (Meaden et al. 1990), although they seem to make normal levels of β-1,3 glucan. Δ*kre5* cells grow very slowly, and although often enlarged, multiply budded, and defective in cell separation, they are not osmotically fragile. These results indicated that β-1,6 glucan is very important for normal vegetative growth, although second-site suppressors could be isolated that allowed somewhat faster growth of *kre5* disruptants without restoring detectable synthesis of β-1,6 glucan. Kre5p is a hydrophilic *N*-glycosylated protein that is likely to function in the ER for it has the ER retention signal "HDEL" at its carboxyl terminus: Deletion of HDEL yielded a protein that failed to complement the *kre5* null mutant, consistent with the notion that Kre5p normally functions in the ER.

A second ER protein, the *N*-glycosylated integral membrane protein Cwh41p, also has a role in β-1,6 glucan synthesis. *cwh41* disruptants make 50% of normal amounts of β-1,6 glucan and show increased resistance to K1 killer toxin as well as Calcofluor White hypersensitivity, but, like Δ*kre5* cells, they are unaffected in β-1,3 glucan synthesis (Jiang et al. 1996). *KRE5*, however, is epistatic to *CWH41*, for Δ*cwh41* Δ*kre5* cells have the same phenotype as Δ*kre5* cells. On the other hand, *CWH41* shows strong genetic interactions with *KRE6* and *KRE1*, whose products are involved in steps in β-1,6 glucan synthesis that occur, respectively, in the Golgi and at the cell surface (see below). Thus, Δ*cwh41* Δ*kre6* mutants are not viable, whereas the effects of combining Δ*cwh41* with Δ*kre1* are additive and lead to a 75% reduction in β-1,6 glucan and a very slow growth phenotype. These interactions indicate that Cwh41p indeed has a role in β-1,6 glucan synthesis, but it is distinct from those of Kre6p and Kre1p, for overexpression of *KRE6* or *KRE1* does not suppress the *cwh41* mutation (Jiang et al. 1996).

kre6 null mutants show about a 50% reduction in alkali-insoluble β-1,6 glucan (Roemer and Bussey 1991; Roemer et al. 1993) and are slow growing and morphologically abnormal. Electron micrographs of thin sections of these cells' walls also reveal ultrastructural alterations (Roemer et al. 1994), although the cells are osmotically stable. *kre6* disruptants are also resistant to *H. mrakii* HM-1 killer toxin (Kasahara et al. 1994a). The finding that β-glucan synthesis was reduced only to 50% of wild-type levels in *kre6* null mutants prompted the search for a functional homolog of Kre6p required for the synthesis of the residual β-glucan and resulted in the identification of *SKN1* as a dose-dependent suppressor of Δ*kre6*. Skn1p is 66% identical to Kre6p. In contrast to the *kre6* disruptants, Δ*skn1* cells grow at wild-type rates and make normal levels of β-1,6 glucan that is structurally indistinguishable from that made in wild-type cells (Roemer et al. 1993). Together, the functional homologs Kre6p and Skn1p are responsible for the bulk of alkali-insoluble β-1,6 glucan synthesis. Thus, double *kre6 skn1* disruptants are inviable in some strain backgrounds and alive but extremely slow growing in others, and their levels of β-1,6 glucan are considerably lower than they are in Δ*kre6* cells: The double mutants made no more than 10% of wild-type amounts of β-1,6 glucan. This residual β-1,6 glucan polymer is apparently altered in structure and smaller in size than the polymer in wild-type cells. Whether this residual β-1,6 glucan is a Kre5p-dependent "primer," or the product of other unidentified enzymes, is not yet clear. As with Δ*kre5*, the slow-growth phenotype of Δ*kre6* Δ*skn1* double disruptants can partially be suppressed without restoration of β-1,6 glucan synthesis (Roemer et al. 1993). Since deletion of Kre6p has a substantial effect on β-1,6 glucan synthesis that cannot be compensated by Skn1p, whereas deletion of *SKN1* has little effect in a *KRE6* background and its role in β-1,6 glucan synthesis is only uncovered when *KRE6* is deleted, Kre6p may normally make most of the alkali-insoluble β-1,6 glucan. The finding that single disruptions of *KRE6* or *SKN1* do not affect the actual structure of the β-1,6 glucan suggests that the products of these genes act in parallel pathways in the synthesis of this β-glucan (Roemer et al. 1993). *kre6* disruptants, and to a lesser extent, *skn1* disruptants, have in vitro β-1,3 glucan synthase activities that are lowered to as little as one third of wild-type levels, although no further decrease is seen in membranes from double disruptants (Roemer et al. 1993; Kasahara et al. 1994a). In vivo levels of β-1,3 glucan, however, are not affected in these strains, so the contribution that Kre6p and Skn1p make to β-1,3 glucan synthesis is unclear.

Both Kre6p and Skn1p receive N-linked sugars. Kre6p has been localized to the Golgi apparatus, whereas the subcellular location of Skn1p remains to be ascertained. Consistent with the involvement of these two proteins in glucan metabolism is the presence in their carboxy-terminal domains, which are predicted to be luminally oriented, of homology with proteins predicted to bind β-glucans, including a short stretch of amino acids proposed to be the active site of a β-1,3-glucanase. Furthermore, both proteins can be phosphorylated on their cytoplasmic domains (Roemer et al. 1994), suggesting that they may be regulated by this means. *KRE6* also shows genetic interactions with the protein kinase C (Pkc1p) pathway, which has roles in regulating cell wall synthesis (see Section IX). Δ*kre6* is synthetically lethal with Δ*pkc*, and *KRE6* partially suppresses the lysis defect of Δ*pkc1* strains; however, the mechanism for this suppression is not clear, because it does not involve restoration of β-1,6 glucan synthesis. It seems unlikely that phosphorylation of Kre6p is regulated by Pkc1p, for Kre6p can be phosphorylated in Δ*pkc1* strains (Roemer et al. 1994).

kre11 null mutants also show a 50% reduction in β-1,6 glucan but little apparent growth defect, and although the β-1,6 glucan made in Δ*kre11* cells is somewhat smaller than that made in wild-type cells, it contains a similar proportion of β-1,3 and β-1,6 linkages (Brown et al. 1993b). Δ*kre11* Δ*kre6* double mutants are inviable, whereas Δ*kre11* Δ*skn1* strains resemble *kre11* disruptants (Roemer et al. 1993). Kre11p is a soluble, cytoplasmic protein that does, however, contain the short sequence RXGG, present in the putative UDPGlc-binding domain of glycogen synthase, suggesting that Kre11p may bind this sugar nucleotide, the presumed donor in glucan synthesis (Brown et al. 1993b). Kre11p may serve to regulate the activity of a membrane-associated complex that makes β-1,6 glucan.

Kre1p functions downstream from this putative complex. *kre1* mutants make 40% of wild-type levels of a β-1,6-linked glucan that has 1,3 branchpoints, but these mutants seem to be unable to extend β-1,6 glucan chains on these branches. Levels of β-1,6 glucan in *kre1 kre6* double mutants are the same as those in *kre6* single mutants (Boone et al. 1990). Δ*kre11* Δ*kre1* double disruptants have a severe growth defect: Their β-1,6 glucan structures, which could only be analyzed in strains containing spontaneous second-site suppressors of the slow-growth phenotype, revealed a more severe deficiency in β-1,6 glucan synthesis than that seen in Δ*kre11* or Δ*kre1* alone (a 75–80% reduction vs. 40–50% in the single mutants). The small amount of β-1,6 glucan in Δ*kre11* Δ*kre1* double disruptants was of the size made in the Δ*kre1* mutant and contained the same ratio of β-1,6 to β-1,3 linkages as that strain; hence, the cells were unable to extend the β-1,3 branches with β-1,6 glucan. Results of combining the Δ*kre1* and Δ*kre11* mutations indicate that Kre11p must precede Kre1p in the synthesis of this β-glucan (Brown et al. 1993b). Kre1p has both an amino-terminal secretion signal and a potential carboxy-terminal site for GPI anchor attachment and is also *O*-mannosylated (Roemer and Bussey 1995) and may thus normally function at the cell surface, possibly by elongating the β-1,6 glucan chains on β-1,3 side branches. However, were Kre1p to function outside the plasma membrane, it would be unable to use UDPGlc as glycosyl donor. Instead, it may catalyze a transglycosylation reaction.

Two further secretory proteins, Kre9p (Brown and Bussey 1993) and

Knh1p (Dijkgraaf et al. 1996), which show 46% identity at the amino acid level, have roles in β-1,6 glucan synthesis. Both are serine- and threonine-rich *O*-mannosylated proteins that can be secreted into the medium if overproduced. *kre9* disruptants make no more than 20% of normal levels of β-1,6 glucan, are slow growing and β-1,3-glucanase-sensitive, and show morphological abnormalities (Brown and Bussey 1993): The morphological aberrations may be due solely to a defect in β-1,6 glucan synthesis, but the possibility cannot be ruled out that *KRE9* is involved in further processes. The residual β-1,6 glucan in Δ*kre9* cells is less than half the average size of the wild-type material and is also altered in structure. Kre9p affects β-1,6 glucan synthesis downstream from Kre5p, as Δ*kre9* Δ*kre5* mutants grow no more slowly than do Δ*kre5* strains. Combination of the *kre9* disruption with Δ*kre6* or Δ*kre11* results in inviability in the strain background tested, although whether the three gene products interact directly is not known. Δ*kre9* Δ*skn1* mutants, however, grow no worse than Δ*kre9* cells do. Δ*kre9* Δ*kre1* mutants also proved to be inviable, indicating that Kre1p can normally act on the polymer made in Δ*kre9*: The inviability of this double disruptant was attributed to the cumulative lowering of β-1,6 glucan to a level below that required to maintain viability. In contrast to the consequences of the *kre9* disruption, deletion of *knh1* has no discernible effect on β-1,6 glucan synthesis, K1 killer toxin resistance, or the cells' growth rate, nor does combination of Δ*knh1* with Δ*kre1*, Δ*kre6*, or Δ*kre11* yield an exacerbated phenotype. However, combination of Δ*knh1* with Δ*kre9* results in synthetic lethality, indicating that both growth and β-1,6 glucan synthesis in each of the single null mutants are dependent on the presence of the other gene's product. Furthermore, overexpression of *KNH1* suppresses the β-1,6 glucan synthetic and growth defects of Δ*kre9* cells. Taken together, these findings indicate that Knh1p is a functional homolog of Kre9p (Dijkgraaf et al. 1996); however, neither the role of these two proteins in β-1,6 glucan synthesis nor their precise cellular location is known.

The growth defect of *kre9* disruptants, but not of Δ*knh1* cells, is relieved by the multicopy suppressor *SKN7* (Brown et al. 1993a; Dijkgraaf et al. 1996). *SKN7* does not restore the β-1,6 glucan deficiency of Δ*kre9* cells, but it does have some impact on the cell wall, giving partial restoration of the morphological defects and allowing retention of cell wall proteins that are more rapidly shed from Δ*kre9* cells into the medium. *SKN7* also fails to rescue the lethal Δ*kre9* Δ*knh1* double mutant (Dijkgraaf et al. 1996), a finding consistent with the notion that *SKN7* suppresses the growth defect, but not the β-1,6 glucan defect, in Δ*kre9* cells. *SKN7*, however, does not suppress other mutants in β-1,6 glucan synthesis and is therefore not a general bypass suppressor of a β-1,6 glucan deficiency. *skn7*-disrupted cells have no discernible phenotype, and combination of Δ*skn7* with Δ*kre9*, Δ*kre6*, Δ*skn1*, Δ*kre11*, or Δ*kre1* did not lead to any further effect on growth and β-1,6 glucan levels than those seen with the singly disrupted strains.

Intriguingly, Skn7p has regions resembling the receiver domain in the response-regulating partner of a bacterial two-component signal transduction pathway, as well as regions resembling the DNA-binding domain of heat shock transcription factors and a transcriptional activation domain (Brown et al. 1993a; 1994). Skn7p activates transcription of a reporter gene and is predominantly located in the nucleus. Moreover, the effects of amino acid substitutions at the conserved aspartate residue that can be phosphorylated in this class of protein influenced the biological activity of the protein (Brown et al. 1994). Skn7p therefore has the potential to regulate the expression of genes involved in aspects of cell wall synthesis, although neither the upstream sensor kinase to which this regulator presumably responds nor the signal that is sensed has been identified.

A pathway leading to the synthesis of the bulk of alkali-insoluble β-1,6 glucan can now be suggested (Fig. 4): It includes all of the above proteins, except the Kre6p and Kre9p homologs, single mutations in which give no discernible β-1,6 glucan synthesis defect. Synthesis of a core β-1,6 glucan is initiated in the ER and involves Kre5p, but whether initiation is by direct glucosylation of a protein primer and perhaps formation of the protein-glucose linkage which has been described (Montijn et al. 1994; see Section VI.A.2.a) is not known. Cwh41p acts either downstream from Kre5p or perhaps as an auxiliary component of a Kre5p-containing complex. The core glucan formed in the ER is completed in the Golgi with extension of the β-1,6 backbone and addition of β-1,3 branches. Transmembrane Kre6p may be a regulator that couples cytoplasmic signals with β-glucan synthesis in the Golgi lumen, or it could make the glucan there itself and be regulated on both sides of the Golgi membrane by luminal Kre9p and by cytoplasmic Kre11p. Kre6p may be phosphorylated on its cytoplasmic domain. The Kre6p homolog Skn1p can function in parallel and partially substitute for Kre6p, but its role is normally a minor one compared with that of Kre6p. Kre1p is involved in the extension of β-1,6 chains at a later stage, but whether this occurs in a late-stage secretory compartment, at, or outside the plasma membrane is unknown. Kre9p, which can be secreted from the cells, may act at the cell surface.

This scheme leads to a number of predictions, for example, that the secretory vesicles that accumulate in late-stage *sec* mutants should contain β-1,6 glucan. Furthermore, a mechanism for UDPGlc transport into the Golgi would be predicted to exist, and mutations such as *mnn9* and *pmr1*, which affect mannan chain elongation, might also affect the size or branching of β-1,6 glucan. The roles of Skn1p and Knh1p, functional homologs of Kre6p and Kre9p, respectively, are not yet clear. One possibility is that there are further, although normally very minor, pathways for β-1,6 glucan synthesis. Another possibility, which emerges from the finding that levels of transcription of the *KNH* gene are higher in cells grown in galactose than in cells grown in glucose-containing medium, is that Knh1p and Skn1p have specialized functions in β-1,6 glucan synthesis that depend on the cells' carbon source (Dijkgraaf et al. 1996).

The severe growth defects of Δ*kre5* strains and of Δ*kre1* Δ*kre11* and Δ*kre6* Δ*skn1* double mutants can be partially relieved by second-site suppressor mutations without restoration of β-1,6 glucan synthesis or killer toxin sensitivity. *S. cerevisiae* can therefore compensate for the β-1,6 glucan deficiency, although the mechanism is unknown.

V. PLASMA MEMBRANE: CHITIN AND β-1,3 GLUCAN SYNTHESIS

The polysaccharides chitin and β-1,3 glucan are synthesized at the plasma membrane and are structural components of the wall, chitin being a key component of the division septum (for review, see Cabib et al. 1982; Cabib 1987; Bulawa 1993) and β-1,3 glucan making up a densely interwoven fibrillar mesh that envelops the cell (Kopecka et al. 1974; Fleet 1991). As seen with glycosylation and β-1,6 glucan synthesis, families of related, sometimes partially redundant, proteins are involved in both chitin and β-1,3 glucan synthesis, along with other proteins that may participate in polysaccharide synthetic complexes or serve as regulators.

A. Chitin Synthesis

1. Chitin Synthesis as a Model for Morphogenesis

Quantitatively a minor component of the wall of vegetative cells, much of the cell's chitin is made at a specific place and at a specific time in the division cycle. At the end of the G_1 phase, chitin starts to be deposited at the cell surface, outside the plasma membrane, in a ring that defines the neck between the mother cell and the bud that emerges there. Following growth of the bud and separation of a nucleus into the daughter cell, the plasma membrane invaginates, and outside it, a thin disk of chitin is extended inward from the original chitin ring. This structure, the primary septum, eventually closes the gap between mother and daughter cell. The primary septum is then thickened on both sides by the deposition of the secondary septa consisting of glucan and mannoprotein. Cell separation ensues, permitted by the dissolution of some of the chitin in the primary septum by the Cts1p endochitinase. The chitin ring and most of the primary septum remain as the bud scar on the mother cell. A small amount of chitin may be present in the birth scar on the daughter cell, and some chitin is also laid down in the lateral wall of the daughter cell following septation.

Delocalized, mislocalized, and excessive deposition of chitin can occur in cell division cycle mutants such as *cdc24* and *cdc3*, emphasizing the fact that chitin synthesis is normally highly regulated (Sloat et al. 1981; Roberts et al. 1983; Slater et al. 1985). Synthesis of chitin is important for the maintenance of cell integrity, for when growing cells are treated with the chitin synthase inhibitor polyoxin D, the primary septum is not formed, and lysis can occur at the junction between mother and daughter cells (Bowers et al. 1974). Indeed, formation of a minimum amount of chitin is essential for viability (Shaw et al. 1991).

Chitin synthases use UDPGlcNAc as sugar donor and catalyze the reaction $2n$ UDPGlcNAc→[GlcNAc-β-1,4-GlcNAc]$_n$ + $2n$ UDP. In vitro activity of chitin synthases from *S. cerevisiae* and many fungi is stimulated or made detectable by pretreatment of enzyme preparations with a protease such as trypsin. Chitin synthase activity is present in the plasma membrane (Kang et al. 1985), where synthesis of the polysaccharide presumably occurs, as well as in "chitosomes," intracellular vesicles that were proposed to deliver the enzyme to the plasma membrane (Bartnicki-Garcia et al. 1978; Schekman and Brawley 1979; Florez-Martinez and Schwencke 1988; Leal-Morales et al. 1988, 1994). Recent studies have revealed that chitin synthases indeed transit the secretory pathway but that chitosomes are likely to be an endocytically derived compartment (Chuang and Schekman 1996; Ziman et al. 1996). Until 1986, all chitin synthesis in *S. cerevisiae* was assumed to be due to one enzyme, and although two further enzymes are now known to be present in yeast, the early studies with this major in vitro chitin synthetic activity yielded insights into the membrane topography and possible mechanism of polysaccharide synthesis. Thus, chitin synthesis seems to be a transmembrane process across the plasma membrane, since UDPGlcNAc is probably available only in the cytoplasm, whereas chitin is extracellular. Experiments on chitin synthase in spheroplast membranes support this notion.

Chitin synthase activity of spheroplast membranes could be influenced by nonpenetrating agents only after the spheroplasts had been lysed, suggesting that the enzyme's catalytic domain is exposed toward the cytoplasm. Upon incubation of isolated plasma membranes with UDPGlcNAc, the chitin that was formed was found on the extracytoplasmic side of the membranes: Chitin synthases can therefore catalyze the vectorial synthesis of polysaccharide by extruding a growing chitin chain across the plasma membrane (Cabib et al. 1983). Solubilized, partially purified chitin synthase (Chs1p) makes chitin chains of an average length of about 100 GlcNAc units and does so by a processive mechanism, remaining bound to its product until it has completed the chain (Kang et al. 1984). Newly extruded chitin chains are believed subsequently to hydrogen-bond together to form crystalline chitin in the wall.

Genes involved in chitin synthesis have been isolated through the use of colony screens that detect in vitro chitin synthetic activity or by the use of screens for mutants that make low levels of chitin in vivo. These two types of approaches have complemented one another, for each has yielded up genes that the other approach would have been unlikely to detect. Three genes have now been identified whose products are likely to catalyze chitin synthesis, and a number of other presumably noncatalytic proteins affect chitin synthesis in vivo and in vitro.

2. *Isolation of Mutants and Genes in Chitin Synthesis*

a. In Vitro Activity Screens. Genetic analysis of chitin synthesis started with the isolation of the *CHS1* gene. In a landmark study, Bulawa et al. (1986) used an autoradiographic screen of autolyzed colonies to detect mutants defective in the synthesis of chitin from UDP[^{14}C]GlcNAc. The *chs1* mutants so isolated had no obvious growth defect, but they lacked in vitro proteolytically activatable chitin synthetic activity. "Reversal" of the screen, in which the autoradiographic assay was carried out on colonies of the *chs1* mutant transformed with a multicopy library, led to detection of transformants that overexpressed chitin synthase activity. Good evidence was obtained that the gene conferring overproduction, *CHS1*, was the structural gene for a chitin synthase. Chitin synthetic activity of a Chs1p–β-galactosidase fusion protein could be immunoprecipitated with anti-β-galactosidase serum. Furthermore, expression of the *S. cerevisiae CHS1* gene in *S. pombe*, a fungus believed to contain neither chitin nor chitin synthase activity, endowed the fission yeast transformants with trypsin-activatable chitin synthase activity (Bulawa et al. 1986). However, *S. pombe* is now recognized to make a chitin-linked polymer and to contain both an in vitro chitin synthetic activity (Sietsma and Wessels 1990) and a chitin synthase-related gene (Bowen et al. 1992; I. MacAllister and P. Orlean, unpubl.), and thus the heterologously expressed protein could, in principle, have been an activator of an endogenous enzyme. Recent findings, though, strongly indicate that the Chs proteins indeed catalyze chitin synthesis.

Disruption of *CHS1*, surprisingly, yielded viable haploids which although they lacked apparent trypsin-stimulated chitin synthase activity, nevertheless contained levels of chitin indistinguishable from their wild-type siblings. Vegetatively growing Δ*chs1* cells had a subtle phenotype (see later), but neither mating nor sporulation was affected in the disruptants. Clearly, another chitin synthase had to be at work in Δ*chs1* cells, and incubations of Δ*chs1* membranes with UDP[^{14}C]GlcNAc indeed led to the identification of chitin synthetic activity. One activity ("chitin synthase 2"; Sburlati and Cabib 1986) was stimulated upon treatment with proteases, and gave highest activity with Co^{++}, whereas the other ("chitin synthase II"; Orlean 1987) was inhibited by trypsin and gave highest activity with Mg^{++}. It was subsequently established that these two activities are due to two different enzymes: chitin synthase II and chitin synthase III, respectively (Bulawa and Osmond 1990).

CHS2, encoding proteolytically activated chitin synthase II was cloned using the colony assay screen on Δ*chs1* cells transformed with a multicopy library (Silverman et al. 1988). Chs2p, which shows 42% identity to Chs1p in its carboxy-terminal two thirds, has important roles in growth and development. Disruption of *CHS2*, although in some cases lethal (Silverman et al. 1988), yields viable, but morphologically abnormal, haploids that nevertheless contain nearly wild-type levels of chitin (Bulawa and Osmond 1990; Baymiller and McCullough 1993). Membranes of Δ*chs2* cells had a trypsin-sensitive chitin synthase activity, chitin synthase III (Bulawa and Osmond 1990). Screens for mutants defective in chitin synthesis in vivo led to identification of genes governing the bulk of chitin synthesis.

b. Screens for In Vivo Chitin Deficiency. A screen for resistance to Calcofluor White led to the isolation of strains deficient in chitin synthesis in vivo, some of which contained no more than 10% of normal levels of chitin ("*cal*" mutants; Roncero et al. 1988). Using an autoradiographic approach, in which colonies of mutagenized cells were screened for those defective in [^{3}H]glucosamine incorporation into chitin, Bulawa (1992) isolated chitin-synthesis-deficient "*csd*" mutants. A number of the genes identified by these two procedures are allelic, and the two approaches combined yielded four complementation groups of nonconditional mutants that make no more than 20% of normal amounts of chitin. Of these, *cal1/csd2*, *cal2/csd4*, and *cal3* are defective in vitro in chitin synthase III activity, whereas *csd3* mutants have normal in vitro chitin synthase III activity. *cal1/csd2* mutants are defective in a gene encoding a protein with sequence similarity to Chs1p and Chs2p, and this protein is therefore the presumed catalytic subunit of chitin synthase III. The nomenclature used here for genes involved in chitin synthesis is essentially that recently adopted by a number of workers in the field. The *CAL1/CSD2* and *CAL3* loci are renamed *CHS3* and *CHS5*, respectively, but the *CAL2/CSD4* and *CSD3* loci will be referred to as *CHS4/CSD4* and *CSD3*, respectively.

In some strain backgrounds, *chs3*, *csd3*, and *chs4/csd4* confer temperature sensitivity for growth on low-osmolarity media. This effect is alleviated by inclusion of osmotic support in the medium or by introduction of the *SSD-v1* allele of the polymorphic *SSD1/SRL* gene (Kim et al. 1994), which encodes a protein phosphatase that suppresses a number of mutations affecting growth control (Bulawa 1992, 1993). Suppression of the temperature sensitivity of these strains is in neither case accompanied by a restoration of chitin synthesis. *csd1* is the only chitin synthesis-deficient mutant that is temperature-sensitive for growth: One possibility is that these strains are defective in UDPGlcNAc supply (Bulawa 1993).

Alleles of *CHS3* were also isolated by complementation of mutants resistant to *K. lactis* K1 killer toxin (Takita and Castilho-Valavicius 1993; Butler et al. 1992). The α-subunit of this toxin has chitinase activity and may kill cells by binding to and hydrolyzing chitin, and resistance to the toxin could therefore be due to the absence of its receptor from the mutants. A screen for mutants defective in the formation of the dityrosine layer of the ascospore wall (see Section VIII.B) also yielded a mutant allele of *CHS3*, *dit101* (Pammer et al. 1992).

3. Chitin Synthases: Regulation and Function

a. General Properties of Chitin Synthases. Chs1p, Chs2p, and Chs3p are membrane proteins that are clearly homologous. *S. cerevisiae* Chs1p and Chs2p are closely related to each other, but both show a lower level of similarity (28% identity over 180 amino acids) to Chs3p. The similarities between the Chs proteins are most pronounced in their carboxy-terminal halves, and indeed there is striking variation in the proteins' amino-terminal halves: For example, the pI values of the amino-terminal thirds of Chs1p and Chs2p are 4.5 and 10.4, respectively (Silverman 1989). The functions of these dissimilar domains is unknown, and one third of Chs1p's amino-terminal region and one quarter of Chs2p's amino-terminal region can be deleted without abolishing enzyme activity or in vivo function (Bulawa et al. 1986; Silverman et al. 1988; Ford et al. 1996; Uchida et al. 1996). Although chitin synthases can be activated in vitro by treatment with proteases, there is as yet no evidence that proteolytically processed forms of these proteins are ever generated in vivo, and no specific activating proteases have been identified. Chs2p and Chs3p enter the secretory pathway via the ER (Chuang and Schekman 1996; Ziman et al. 1996), and Chs1p presumably does so as well.

In addition to the demonstration of chitin synthesis by Chs1p after electrophoresis in a nondenaturing polyacrylamide gel (Kang et al. 1984), evidence has been accumulating that the other Chs proteins catalyze polymerization of GlcNAc. Site-specific mutagenesis of Chs2p has identified amino residues important for in vitro activity of Chs2p in the protein's conserved carboxy-terminal domain, and the same residues are conserved in other proteins implicated in the synthesis of β-1,4 glycosidic linkages (Nagahashi et al. 1995). Among these, strikingly, are the *Rhizobium* NodC protein (Bulawa 1992; Geremia et al. 1994), which synthesizes a chito-oligosaccharide involved in signaling between bacterium and legume during nodulation, and the developmentally regulated *Xenopus laevis* DG42 protein, which, after translation in a cell-free system, also synthesizes a chito-oligosaccharide from UDPGlcNAc (Semino and Robbins 1995). Intriguingly, the chitin made in vivo, and by chitin synthases I and III in vitro, is uniform in size. The average chain length of in vivo chitin is about 100–120 GlcNAc residues, as is the in vitro product of purified Chs1p (Kang et al. 1984), and the chitin made in vitro by Chs3p in a microsomal fraction was, on average, 170 GlcNAc residues in length (Orlean 1987).

The three chitin synthases all have relatively high K_m values for UDPGlcNAc, in the range of 0.5–0.8 mM (Kang et al. 1984; Sburlati and Cabib 1986; Orlean 1987), and in some cases, the enzymes show differential sensitivities in vitro to nikkomycins and polyoxins (Gaughran et al. 1994), competitive inhibitors that are structural analogs of UDPGlcNAc (Debono and Gordee 1994; Georgopapadakou and Tkacz 1995). The in vitro effects of these inhibitors are consistent with their effects on septum formation in vivo (Bowers et al. 1974), but as the uptake of these compounds is hindered by competing peptides in the growth medium, the clinical effectiveness of these drugs is limited.

b. Chitin Synthase I. In in vitro assays of membrane preparations from wild-type cells, chitin synthase I shows the highest specific activities of the three chitin synthases (Sburlati and Cabib 1986; Orlean 1987; Valdivieso et al. 1991; Bulawa 1992). However, little if any chitin synthase I activity can ever be detected in membranes from vegetative cells unless the membranes are preincubated with protease (Orlean 1987). Levels of total Chs1p activity, measured after proteolytic treatment, are the same in cultures harvested in the logarithmic and stationary phases of growth and show no change during the vegetative growth cycle in synchronized cells (Orlean 1987; Choi et al. 1994b), and levels of Chs1p do not fluctuate significantly during the cell cycle (Ziman et al. 1996). Levels of Chs1p therefore do not seem to be regulated by synthesis and degradation, and the catalytic potential of Chs1p in vegetative cells remains mostly, if not completely, latent, consistent with the fact that *chs1* disruptants are not obviously defective in chitin synthesis in vivo. Levels of *CHS1* mRNA, however, show modest fluctuation during the cell cycle (Pammer et al. 1992).

Some cells in Δ*chs1* cultures accumulate small refractile buds, whose appearance is correlated with acidification of the growth medium (Cabib et al. 1989). These apparently lysed buds have a small hole at the center of the birth scar. This phenotype is relieved by provision of osmotic support or by maintaining the pH of the medium above 5; the number of lysed buds in a culture is also lowered in the presence of the chitinase inhibitor allosamidin or in Δ*chs1* cells whose chitinase gene (*CTS1*) has been deleted. These findings led to the notion that Chs1p repairs wall damage caused at low pH by chitinase, a periplasmic enzyme with a low pH optimum. Whether cell wall damage itself or acidic culture conditions activate chitin synthase I is not known. Recessive mutations in the *SCS1* gene suppress the lysis phenotype of Δ*chs1* strains, indicating that lysis is not exclusively due to chitinase (Cabib et al. 1992).

c. Chitin Synthase II. Chitin synthase II, most easily detected in a Δ*chs1* background, is also activated in vitro by treatment with proteases, but total activities are about 5–15% of that of chitin synthase I. Although proteolysis is a potential mechanism by which Chs2p could be activated at a specific time and place in the cell, proteolytic activation may be strictly an in vitro phenomenon. Indeed, trypsin activatability in vitro does not correlate with in vivo functionality of certain Chs2p truncations. Thus, certain forms of Chs2p amino-terminally truncated beyond the protein's dispensable amino-terminal quarter lose in vivo function but retain high in vitro activity (Ford et al. 1996; Uchida et al. 1996). Although these deletions could somehow prevent interactions of Chs2p with an activator whose effects are mimicked by trypsin, this is not a general explanation, for some Chs2p truncations with in vitro activity show but modest stimulation by trypsin (Ford et al. 1996). Under certain conditions, Chs2p can catalyze chitin synthesis in vitro without prior trypsin treatment (Uchida et al. 1996), raising the possibility that Chs2p functions in vivo in an unprocessed form.

Chitin synthase II activity is detectable only in membranes from growing cells (Bulawa 1993). Levels of both Chs2p and of chitin synthase II activity measured after trypsin treatment fluctuate during the vegetative growth cycle, showing a peak at the time the primary septum is formed, and a decrease after cytokinesis, consistent with rapid turnover of the protein (Choi et al. 1994b; Chuang and Schekman 1996). *CHS2* mRNA levels also peak just before primary septum formation, then drop (Pammer et al. 1992). Levels of Chs2p therefore seem to be tightly controlled by a cell-cycle-regulated synthesis and degradation mechanism. This mechanism operates in concert with one that very precisely determines the temporal and spatial localization of the protein (see below).

From measurements of chitin levels in Δ*chs1* Δ*chs3* cells, it seems that Chs2p is responsible for the synthesis of 5–10% of the chitin in vegetative cells. Disruption of *CHS2*, which is lethal under some conditions and in some strain backgrounds, yields cells that are enlarged and often elongated, and which also have a separation defect (Bulawa and Osmond 1990; Shaw et al. 1991). In some media, Δ*chs2* cells make twice as much chitin as their wild-type siblings. The phenotypes of Δ*chs2* strains are the same regardless of whether or not *CHS1* is present. Thin-section electron micrographs reveal that Δ*chs2* cells have thickened, amorphous septa that lack a primary septum. The latter structure, however, is detectable as a thin layer of wheat-germ agglutinin-gold-labeled material in the otherwise chitin-deficient septa of Δ*chs1* Δ*chs3* cells (Shaw et al. 1991). These observations, which suggest that a major function of Chs2p is to synthesize the primary septum, correlate well with the localization of Chs2p.

Chs2p, strikingly, is localized to the neck between mother cell and bud at the end of mitosis only (Chuang and Schekman 1996), and the protein's placement in the plasma membrane there seems to be what regulates chitin deposition. Chitin synthase II activity is found both in the plasma membrane and in chitosomes (Sburlati and Cabib 1986; Leal-Morales et al. 1994), and Chs2p enters the ER and is delivered to the plasma membrane via the secretory pathway (Chuang and Schekman 1996). Chs2p's specific localization at the mother-bud junction is likely to be dictated by polarized secretion at this stage in the cell cycle (Chuang and Schekman 1996). Once there, the protein may become sequestered and organized into a ring structure by the septins, components of the 10-nm neck filaments that form a ring under the plasma membrane. Rapid degradation of Chs2p at the end of mitosis requires delivery of the protein to the vacuole, for degradation is blocked in both *end4* mutants, which are defective in the internalization step in endocytosis, and in *pep4* mutants, which are defective in maturation of a variety of vacuolar hydrolases (Chuang and Schekman 1996).

Interestingly, *CHS2* disruptants are sometimes multinucleate, a defect that would not be expected to arise upon loss of the primary septum, since nuclear segregation normally precedes septation (Bulawa 1993). Indeed, the "strings" of cells made by glucosamine auxotrophs, which are unable to make any GlcNAc-containing macromolecules, are mononucleate (Section III.A.1). It is therefore possible that Chs2p has further roles in cell division. Overexpression of *CHS2* causes flocculation, but it is not known whether this is accompanied by changes in chitin levels or wall organization (Sudoh et al. 1991).

d. Chitin Synthase III and Proteins Influencing Its Activity. Chitin synthase III is the major, if not only, activity that is measured in yeast membranes without prior treatment with trypsin, and indeed, such activity is trypsin-sensitive (Orlean 1987; Bulawa and Osmond 1990; Valdivieso et al. 1991), as well as sensitive to certain detergents (Orlean 1987; Choi et al. 1994a). This activity presumably represents endogenously active enzyme. After detergent extraction of Δ*chs1* Δ*chs2* membranes and elimination of active chitin synthase III, it is possible to detect a pool of trypsin-stimulated chitin synthase activity. Trypsin treatment also restores some chitin synthase activity to detergent-extracted Δ*chs4/csd4* membranes, but not to Δ*chs5* membranes, suggesting that proteolysis mimics the effects of Chs4p/Csd4p, a candidate activator of chitin synthase III, and indicating that chitin synthase III activity is dependent on Chs3p and Chs5p (Choi et al. 1994a). Whether the endogenously active form of Chs3p is the full-length protein, or a form generated by proteolytic processing in vivo, remains to be established.

Highest specific activities of chitin synthase III are measured in membranes from logarithmically growing cultures, and activity declines sharply as cultures enter the stationary phase (Orlean 1987). Expression of *CHS3* fluctuates modestly during the cell cycle, levels of *CHS3* mRNA being highest after primary septum formation, somewhat after

CHS2 mRNA levels peak (Pammer et al. 1992). However, Chs3p is synthesized throughout the cell cycle and steady-state levels of Chs3p do not show cell cycle regulation, indicating that, in contrast to Chs2p, levels of Chs3p are not regulated by synthesis and degradation, but rather, they are regulated posttranslationally (Choi et al. 1994b; Chuang and Schekman 1996; Ziman et al. 1996).

The phenotypes of *CHS3* disruptants show that Chs3p, which makes 90–95% of the chitin in the cell wall (Shaw et al. 1991; Bulawa 1992), makes the chitin ring that is formed upon bud emergence and which remains in the bud scar, as well as the chitin that is deposited in the lateral wall. Evidence for the latter role came from the finding that the increased and delocalized synthesis of chitin seen upon shift of the *cdc3* and *cdc24* mutants to nonpermissive temperature was abolished when these strains also harbored a *chs3* mutation (Shaw et al. 1991). The accumulation of chitin in these *cdc* mutants was suggested to be due to multiple cycles of chitin deposition in the wall of cells blocked in cell division. Chs3p-dependent chitin synthesis can also occur at other times and places, namely, during mating and during synthesis of the chitosan layer of the ascospore wall (Section VIII.A.2.a). The chitin made by Chs3p can also be cross-linked to alkali-insoluble β-1,3 glucan (Section VI.B).

Chs3p becomes localized to sites of polarized growth in a cell-cycle-dependent fashion. During the budding cycle, Chs3p appears first in a ring at the cell surface where a bud is about to emerge, a localization consistent with the protein's role in formation of the chitin ring. The protein remains in a ring in small-budded cells in the neck joining mother and bud and then disappears from the neck of medium-budded cells, only to reappear in the neck region of large-budded cells undergoing cytokinesis (Chuang and Schekman 1996; Santos and Snyder 1997). This highly specific temporal and spatial localization of Chs3p suggests that localized chitin synthesis by Chs3p is governed by specific placement of the protein, rather than by localized activation of a uniformly distributed, latent enzyme (Chuang and Schekman 1996).

In addition to its specific localization to the neck region, Chs3p is also detectable in patches in the cytoplasm at different stages of the cell cycle. Indeed, up to 50% of the protein is present in a unique intracellular vesicular fraction (Chuang and Schekman 1996; Ziman et al. 1996), probably corresponding to chitosomes (Bartnicki-Garcia et al. 1978; Schekman and Brawley 1979; Florez-Martinez and Schwencke 1988; Leal-Morales et al. 1988, 1994). Chs3p, like Chs2p, enters the secretory pathway in the ER, where it receives N-linked carbohydrate chains, before being conveyed to the plasma membrane in parallel with Chs2p (Chuang and Schekman 1996; Santos and Snyder 1997). Chs3p may then be organized or incorporated into a ring structure by the 10-nm neck filaments. Although Chs3p transits the secretory pathway, the chitosomal fraction that contains Chs3p does not correspond to the ER, Golgi, or vacuole; rather, chitosomes are derived from the plasma membrane by endocytosis. Thus, formation of chitosomes is blocked in *end4* mutants and therefore dependent on the internalization step in endocytosis. The fate of Chs3p after endocytosis contrasts markedly with that of Chs2p. Chs2p is rapidly taken to the vacuole and degraded, and little is detected in chitosomes by immunostaining, whereas Chs3p persists in chitosomes after endocytosis (Chuang and Schekman 1996). Although the distibution of Chs3p between plasma membrane and chitosomes does not vary during the cell cycle, the possibility was raised that Chs3p could be mobilized from its chitosomal pool to a future site of budding. This recycling mechanism might permit cell-cycle-regulated targeting of Chs3p to its site of action before the secretory pathway becomes polarized (Chuang and Schekman 1996; Ziman et al. 1996).

Overexpression of *CHS3* leads to little if any increase in chitin synthase III activity, indicating that other components are limiting for the reaction (Valdivieso et al. 1991; Bulawa 1992; Choi et al. 1994a). Chs4p/Csd4p and Chs5p, which resemble neither chitin synthases nor known proteases, are required for chitin synthase III activity in vivo and in vitro, and Csd3p is required for chitin synthesis in vivo. Chs5p, Myo2p, and possibly Chs4p/Csd4p are necessary for correct localization of Chs3p to the cell periphery.

chs4/csd4 mutants have the same phenotype as *chs3* mutants, and like *CHS3*, *CHS4/CSD4* is not an essential gene (Bulawa 1992, 1993). Overexpression of *CHS4/CSD4* results in a three- to tenfold increase in in vitro chitin synthase III activity, although simultaneous overexpression of *CHS3* does not further increase chitin synthase III activity. Chs4p/Csd4p may function as a limiting subunit of a complex containing Chs3p (Bulawa 1993), but it also interacts, directly or indirectly, with one of the septin components of the 10-nm neck filaments (Longtine et al. 1996). Chs4p/Csd4p may therefore have an intriguing dual role in vegetative cells in which it not only helps localize Chs3p, but also triggers catalytic activity. A gene encoding a Chs4p/Csd4p homolog, *SHC1*, is expressed during sporulation, but not during exponential growth, and Shc1p may be required for chitosan synthesis during sporulation (Bulawa 1993) (Section VIII.B.2).

chs5 mutants make about 20% of normal levels of chitin (Roncero et al. 1988) and lack in vitro chitin synthase III activity (Valdivieso et al. 1991). Interestingly, the *CHS5* gene, which is nonessential, encodes a protein showing resemblance to the human large neurofilament (NF-H) protein (Santos and Durán 1992; unpublished data cited by Cid et al. 1995). Although Chs5p is limiting for chitin synthase III activity in vitro, overexpression of *CHS5* does not lead to an increase in enzyme activity. Chs5p is required for localization of Chs3p to its specific sites on the cell surface, for in Δ*chs5* cells, Chs3p is not detected at the cell periphery, but rather, accumulates in cytoplasmic patches. In *CHS5* cells, both Chs5p and Chs3p colocalize to the same cytoplasmic compartment, suggesting that Chs5p is involved in targeting Chs3p to its correct location at the cell surface, where chitin synthesis is activated, perhaps by Chs4p/Csd4p (Santos and Snyder 1997). It is not yet known how Chs5p is involved in targeting of Chs3p, nor whether Chs5p targets only Chs3p being delivered afresh via the secretory pathway, nor whether it also targets Chs3p that is being recycled from an endocytic compartment.

Targeting of the secretory vesicles that contain Chs3p is also dependent on the actin cytoskeleton and proteins affecting its polarity, including the myosin Myo2p. *myo2* mutants are perturbed in actin polarization and show delocalized deposition of chitin (Johnston et al. 1991). The latter is likely to be due to a lack of polarization of Chs3p, for upon shift of a temperature-sensitive *myo2* mutant to nonpermissive temperature, Chs3p is no longer localized on the cell periphery but accumulates in cytoplasmic patches (Santos and Snyder 1997). Localization of Chs5p, however, is not affected in *myo2* mutants.

Csd3p, a hydrophilic protein, is required for chitin synthesis in vivo but not for chitin synthase III activity in vitro (Bulawa 1992, 1993). Δ*csd3* cells are viable, ruling out a role for Csd3p in UDPGlcNAc synthesis. Possible roles for Csd3p may be to localize Chs3p or to recycle the protein from chitosomes to the site on the plasma membrane where it becomes activated (Ziman et al. 1996). Mixed membranes of *csd3* cells may show in vitro chitin synthase activity because Chs3p, physically separated from its activators in vivo, may be reconstituted with its activators in vitro.

e. Mutants Lacking Two or More Chitin Synthases. The presence of either Chs2p or Chs3p is required for cell viability. Thus, Δ*chs1*/Δ*chs2* and Δ*chs1*/Δ*chs3* double disruptants are viable, but triple mutants are not (Bulawa and Osmond 1990; Shaw et al. 1991; Bulawa 1992). The consequences of losing all three chitin synthases were examined using a Δ*chs1*

Δ*chs2* Δ*chs3* strain harboring *CHS2* under the control of the *GAL1* promoter. Upon shift from galactose to glucose, chitin synthase II activity dropped rapidly within one doubling time, and cells developed defects in separation and nuclear segregation, became greatly enlarged in some cases, and then died (Shaw et al. 1991). Since these phenotypes were not rescued by introduction of *CHS1*, Chs1p cannot substitute for either of the other chitin synthases. Assuming that chitin synthesis is the only function of Chs2p and Chs3p, then a minimum amount of the polysaccharide must be made to ensure viability (Shaw et al. 1991).

The above studies reveal that each chitin synthase makes chitin at different times during the cell cycle and at different places in the cell. Chs1p, required for normal budding in acidic media, may make chitin to repair damage caused at low pH. Chs2p forms the primary septum but may have further roles in cell division, and Chs3p makes the chitin ring formed on bud emergence, as well as the chitin in the lateral wall.

f. Chitin-synthase-related Proteins in Other Eukaryotes. Other fungi also have a repertoire of chitin synthases (Bowen et al. 1992), some of which may have roles analogous to those of the three *S. cerevisiae* enzymes and others of which may have specialized functions in hyphal extension and branching. *Candida albicans* has at least three chitin synthases, resembling both Chs1p/Chs2p and Chs3p types (Au-Young and Robbins 1990; Chen-Wu et al. 1992; Sudoh et al. 1993), whereas at least six chitin synthase genes are present in *Aspergillus* species (Motoyama et al. 1994; Mellado et al. 1995). Genes encoding chitin-synthase-related proteins have now been identified in *Xenopus*, zebrafish, and mice and all are expressed only for a short time during embryonic development (Semino et al. 1996). What might these proteins be making in vertebrates? One possibility, based on the finding that the *Xenopus* DG42 makes chito-oligosaccharides in vitro (Semino and Robbins 1995), is that these proteins make chito-oligosaccharides that function as signals at defined stages in embryonic development, or which serve as primers for hyaluronic acid synthesis (Semino et al. 1996; Varki 1996). Another possibility is that DG42 synthesizes hyaluronic acid (Meyer and Kreil 1996), a linear polymer of alternating GlcNAc and D-glucuronic acid residues that is a key component of the extracellular matrix in animals. Either way, chitin-synthase-related enzymes are likely to have important roles in vertebrate development, and the existence of these proteins needs to be taken into account when chitin synthase inhibitors are considered for use as antifungal agents (Semino and Robbins 1995).

B. β-1,3 Glucan Synthesis

1. Enzymology

Although β-1,3 glucan is found in vivo in cell wall fractions that contain mixtures of β-1,3 and β-1,6 glucosidic linkages, so far, only linear β-1,3 glucan chains have been synthesized in in vitro systems. The single apparent β-1,3 glucan synthase activity in *S. cerevisiae* membranes uses UDPGlc as donor, and its product is a β-1,3 glucan with a minimum chain length of 60–80 glucose residues (Shematek et al. 1980). About two thirds of the β-1,3 glucan synthase activity is found in plasma membrane fractions, the remainder in a lower-density fraction distinct from chitosomes (Shematek et al. 1980; Leal-Morales et al. 1988). Elegant experiments by Cabib and co-workers revealed that β-1,3 glucan synthase activity is regulated by a GTP-binding component. In vitro glucosyl transfer is stimulated by GTP and its nonhydrolyzable analogs (Shematek and Cabib 1980; Notario et al. 1982), and the activity can be dissociated into a membrane-bound and a GTP-binding component (Kang and Cabib 1986; Mol et al. 1994). GTP stimulation is a general regulatory mechanism, for the effect is also seen with β-1,3 glucan synthases from other fungi, including the pathogen *C. albicans* (Orlean 1982; Staniszlo et al. 1985). β-1,3 glucan synthase is inhibited in vitro by acylated cyclic hexapeptides of the echinocandin group and by papulacandins, acylated derivatives of β-1,4 galactosylglucose (Debono and Gordee 1994; Georgopapadakou and Tkacz 1995): Both classes of compounds have antifungal activity in vivo. β-1,3 glucan synthase activity is sensitive to treatments that perturb the membrane environment, and the lipophilic nature of these molecules may contribute to their inhibitory activity (Ko et al. 1994).

2. Mutants and Genes Affecting Synthesis of β-1,3 Glucan

Mutations that affect formation of alkali-insoluble β-1,3 glucan and lower in vitro β-1,3 glucan synthase activity have been isolated in screens for osmotic fragility (Song et al. 1992). Genes involved in β-1,3 glucan synthesis have been identified through use of screens for resistance to various agents, as well as, intriguingly, for hypersensitivity to the immunosuppressives cyclosporin A and FK506. Some mutations, although affecting primarily β-1,3 glucan formation, also affect the synthesis of β-1,6 glucan and chitin. Components of a possible β-1,3 glucan-synthesizing complex, as well as possible regulatory proteins, have been identified.

a. Components of β*-1,3 Glucan Synthase and Its GTP-binding Activator.* Screens for mutations conferring hypersensitivity to FK506, cyclosporin A, and Calcofluor White, or causing resistance to echinocandin and papulacandin, all led to identification of a gene first called *FKS1* (Douglas et al. 1994a,b; Eng et al. 1994; Castro et al. 1995; Ram et al. 1995). This gene was also cloned following partial purification of β-1,3 glucan synthase (Inoue et al. 1995). *fks1* mutants show a 75% reduction in β-1,3 glucan, and Δ*fks1* cells grow slowly and have reduced in vitro β-1,3 glucan synthase activity. Mutants in *fks1* also show hypersensitivity to the chitin synthase inhibitor nikkomycin Z (El-Sherbeini and Clemas 1995a), suggesting that such cells have become more dependent on chitin for maintenance of their wall. In the presence of lethal concentrations of Calcofluor White, *fks1* mutants give a cell cycle arrest after forming a small bud (Ram et al. 1995). Consistent with the latter phenotype, Fks1p, a protein with 16 potential transmembrane helices, is localized to the tip of growing buds (Qadota et al. 1996), suggesting that Fks1p is required for bud expansion (Ram et al. 1995). Moreover, levels of *FKS1* transcript are highest in late G_1/early S phase (Mazur et al. 1995; Ram et al. 1995). Whether Fks1p, which lacks a cleavable signal sequence, reaches the plasma membrane via the secretory pathway is unknown.

FKS2, encoding a protein 88% identical to Fks1p, was identified by virtue of its cross-hybridization with *FKS1* (Mazur et al. 1995) and because Fks2p copurifies with Fks1p (Inoue et al. 1995). *fks2* disruptants have no defect in vegetative growth, but they have a severe one in sporulation. Double *fks1 fks2* disruptants, however, are inviable. Despite the close similarity of Fks2p to Fks1p, *FKS2* expression is regulated differently, and each protein may be a component of one of two different β-1,3 glucan synthetic complexes that have partially overlapping functions (Inoue et al. 1995; Mazur et al. 1995). *FKS2* expression is induced by Ca^{++} and by mating pheromone and is dependent on calcineurin, a Ca^{++}/calmodulin-dependent phosphoprotein phosphatase that consists of the Cna1p, Cna2p, and Cnb1p subunits and is inhibited by FK506. In accord with this, calcineurin mutants are synthetically lethal with *fks1*, and *FKS2* transcription is inhibited by FK506 (Mazur et al. 1995; Ram et al. 1995). Hypersensitivity of *fks1* mutants to FK506 seems to arise because transcription of *FKS2*, which can partially substitute for *FKS1*, is Ca^{++}- and calcineurin-dependent. Thus, when calcineurin is mutated or inhibited, the burden of synthesizing β-1,3 glucan to survive falls on Fks1p, and mutations in this protein become lethal (Mazur et al. 1995). Collectively, these results suggest that Fks2p-dependent β-1,3 glucan synthesis is in part controlled by Ca^{++}/calcineurin-regulated phosphorylation and dephosphorylation, but exactly how Fks2p activation is effected is not yet known.

The regulatory GTP-binding subunit that can be purified from membrane preparations containing β-1,3 glucan synthase activity has been identified as the essential Rho1p, a member of a family of GTPases (Drgonová et al. 1996; Mazur and Baginsky 1996; Qadota et al. 1996). Membranes from *rho1* mutants have very low levels of the GTP-binding component of β-1,3 glucan synthase and a thermolabile β-1,3 glucan synthase activity that can be fully restored, in a GTP-dependent fashion, by the addition of purified or recombinant Rho1p (Drgonová et al. 1996; Qadota et al. 1996). Furthermore, membranes from cells overexpressing a constitutively active, GTP-locked allele of *RHO1* have GTP-independent β-1,3 glucan synthase activity (Qadota et al. 1996). Specific inactivation of Rho1p by ADP ribosylation in turn inactivates β-1,3 glucan synthase activity in membranes from both Δ*fks1* and Δ*fks2* strains, suggesting that Rho1p may normally regulate the activity of β-1,3 glucan synthase complexes containing either Fks1p or Fks2p (Mazur and Baginsky 1996). Rho1p is isoprenylated at its carboxyl terminus (Yamochi et al. 1994), and the lipid, which presumbly mediates attachment of the protein to the membrane, is important for the function of Rho1p in vivo. Thus, mutations in Cdc43p, a subunit of a geranylgeranyltransferase that may modify Rho1p, result in lowered in vitro β-1,3 glucan synthase activity, and mutations in the *cwg2*$^+$ gene, which encodes the *S. pombe* counterpart of Cdc43p, cause osmotic fragility and lowered β-1,3 glucan synthase activity in fission yeast (Diaz et al. 1993).

Rho1p coimmunoprecipitates with Fks1p (Mazur and Baginsky 1996; Qadota et al. 1996), and these two proteins could indeed potentially interact in vivo, for indirect immunofluorescence studies reveal that Fks1p and Rho1p are colocalized at the site of bud emergence and at the tip of the growing bud (Qadota et al. 1996). Consistent with a requirement for Rho1p for maintenance of cell integrity, conditional *rho1* mutants are hypersensitive to echinocandin and Calcofluor White (Qadota et al. 1996), as well as osmotically fragile, ceasing growth with small buds that lyse at their tips (Qadota et al. 1994; Yamochi et al. 1994). These findings lead to the following model for the activation of β-1,3 glucan synthase for cell wall synthesis at bud emergence. Exchange of GDP for GTP on isoprenylated Rho1p allows the protein to bind to the transmembrane components of β-1,3 glucan synthase, including Fks1p and Fks2p. Binding of Rho1p-GTP may expose a catalytic site on the glucan synthase complex to the substrate UDPGlc, upon which β-1,3 glucan chains are synthesized, and perhaps simultaneously extruded through the membrane, by analogy with the formation of chitin (Drgonová et al. 1996).

Intriguingly, Rho1p has two further functions as a regulator of cell wall synthesis and morphogenesis (see Section IX). Rho1p activates protein kinase C (Nonaka et al. 1995; Kamada et al. 1996), which in turn regulates a pathway that leads to cell wall synthesis in response to low osmolarity. Pkc1p, however, is not directly involved in the activation of β-1,3 glucan synthase (Drgonová et al. 1996; Qadota et al. 1996). Furthermore, Rho1p may also be involved in the organization of the actin cytoskeleton at the bud tip (Yamochi et al. 1994).

b. Further Genes Involved in β-1,3 Glucan Synthesis. Four other genes have been implicated in β-1,3 glucan synthesis. *KNR4* was cloned by complementation of the *knr4* mutant, selected for its resistance to *H. mrakii* killer toxin (Hong et al. 1994b). The gene's product was also identified as nuclear Smi1p which binds matrix association regions in DNA (Fishel et al. 1993). *knr4* disruptants are viable, but they have much lower levels of both β-1,3 and β-1,6 glucan, and β-1,3 glucan synthase activity is also reduced (Hong et al. 1994a,b). Intriguingly, Δ*knr4* cells have elevated levels of chitin. In some strain backgrounds, *knr4* null mutants show a temperature-sensitive cell cycle arrest in S phase as tiny-budded cells (Fishel et al. 1993). Knr4p/Smi1p could have a regulatory role in expression of certain cell wall biosynthetic genes.

Mutations in *GNS1* lead to resistance to the semisynthetic papulacandin analog L-733,560 (El-Sherbeini and Clemas 1995b). *gns1* mutants grow slowly, and haploid *gns1* cells show a bilateral mating defect. Homozygous *gns1* diploids have a severe sporulation defect, and these diploids are also osmotically fragile. Membranes from *gns1* cells have in vitro β-1,3 glucan synthetic activity that is 10% of wild-type levels. *GNS1* interacts genetically with *FKS1*, for Δ*gns1* Δ*fks1* strains grow more slowly than either single null mutant and show even lower in vitro β-1,3 glucan synthase activity. Gns1p is a membrane protein, and since it confers L-733,560 sensitivity on the cells, one possibility is that it is a drug-binding subunit of the β-1,3 glucan synthase complex.

HKR1 was identified as an essential gene that when overexpressed confers resistance to *H. mrakii* HM-1 killer toxin and causes a 2.5-fold increase in alkali-insoluble β-1,3 glucan, although levels of in vitro β-1,3 glucan synthase activity are unchanged (Kasahara et al. 1994a,b). Hkr1p is a serine- and threonine-rich protein with a transmembrane domain toward its carboxyl terminus that is followed by a potential cytoplasmic

Ca^{++}- binding domain. Insertion of *LEU2* at a site immediately 5′ to the sequence encoding the protein's transmembrane and carboxy-terminal cytoplasmic domains creates a viable *hkr1* mutant, which, however, contains lower levels of cell wall β-1,3 glucan and has lowered in vitro β-1,3 glucan synthase activity (Yabe et al. 1996). These results implicate Hkr1p in β-1,3 glucan synthesis. Assuming that the *hkr1* mutant makes a truncated Hkr1p lacking its Ca^{++}-binding domain, then this domain is not required for vegetative growth, but it is needed for normal levels of β-1,3 glucan synthesis. Hkr1p shows 33% identity to the product of the nonessential *MSB2* gene, a multicopy suppressor of the bud emergence defect of *cdc24* (Bender and Pringle 1992). Overexpression of *HKR1*, though, does not suppress *cdc24* (Kasahara et al. 1994b). However, the *hkr1* mutants were in some cases enlarged and deformed, and some haploids showed a departure from the expected axial budding pattern (Yabe et al. 1996). Hkr1p, detectable only when overexpressed, is localized mainly at the cell surface and presumably functions there.

The GPI-anchored Gas1p/Ggp1p cell surface glycoprotein is also involved in the formation of β-1,3 glucan. A mutant allele of *GAS1/GGP1*, *cwh52*, shows Calcofluor White hypersensitivity and a 75% decrease in levels of cell wall β-1,3 glucan and β-1,6 glucan (Ram et al. 1995). Levels of in vitro β-1,3 glucan synthase activity, however, are normal. Δ*gas1/ggp1* cells show resistance to β-1,3-glucanase, although thin-section electron micrographs of their walls do not reveal major alterations (Popolo et al. 1993b). In the presence of lethal concentrations of Calcofluor White, *gas1/ggp1/cwh52* cells halt growth after forming a small bud (Ram et al. 1995), indicating that Gas1p/Ggp1p is normally required for bud growth at that stage in the cell cycle. Indeed, expression of *GAS1/GGP1* is highest in G_1/early S phase cells, and, consistent with the notion that transcription of the gene is cell-cycle-controlled, *GAS1/GGP1* has several cell cycle regulatory elements in its 5′ upstream region (Popolo et al. 1993a; Ram et al. 1995). A *C. albicans* counterpart of *GAS1/GGP1* has been identified whose expression is inducible at alkaline pH, conditions that elicit hypha formation (Saporito-Irwin et al. 1995). *CaPHR1* is required for normal morphogenesis of *C. albicans*, since disruption of this gene yields cells incapable of forming either buds or normal hyphae at high pH. CaPhr1p becomes GPI-anchored in *S. cerevisiae* and complements the defects of an *S. cerevisiae* Δ*gas1/ggp1* strain (Vai et al. 1996).

Whether Knr4p, Gns1p, Hkr1p, and Gas1p/Ggp1p all directly affect Fks1p- and Fks2p-dependent β-1,3 glucan synthesis, or whether they act in additional pathways, is not clear. Following synthesis at the plasma membranes, β-1,3 glucan chains become incorporated into the growing wall, a process that may involve reactions such as that catalyzed by the Bgl2p "endoglucanase," which joins two β-1,3-linked chains via a β-1,6 linkage (see Section VII.A).

VI. INCORPORATION OF COMPONENTS INTO THE WALL

Components elaborated in compartments of the secretory pathway and delivered to the plasma membrane are incorporated into the wall outside the cell in various ways. Some glycoproteins remain soluble (or periplasmic) or loosely attached through disulfide bonds, whereas others become cross-linked to β-1,6 glucan. β-1,3 glucan becomes cross-linked to chitin.

A. Wall Glycoproteins

1. Extraction and Identification

Enzymes that can be assayed in intact cells without disruption of the plasma membrane and that can be rendered soluble by dissolution, even simple ''cracking'' of the wall, are defined as periplasmic (Arnold 1981). Many proteins—some of which may be legitimate wall components and others of which may be contaminating proteins—can be extracted from isolated cell walls into urea or solubilized by heating walls with SDS; approximately 30% of this material may be present in larger complexes that can be dissociated by treatment with thiol reagents (Valentin et al. 1984). Subsequent treatment of SDS-extracted cell walls with β-glucanases yields a different spectrum of proteins, containing N- and O-linked carbohydrates, that are heterogeneous in size and often of much higher average molecular weight, on account of their long N-linked mannan outer chains (Shibata et al. 1983; Pastor et al. 1984; Valentin et al. 1984; Fleet 1991; van Rinsum et al. 1991). Indeed, isolation and characterization of the very high-molecular-weight material is made difficult by this hyperglycosylation (much of this material will not enter SDS-polyacrylamide gels). Use of the *mnn9* mutant, which is defective in the synthesis of polymannose outer chains (Section IV.A.2.d), results in the formation of smaller, underglycosylated versions of these molecules (Frevert and Ballou 1985; Montijn et al. 1994).

GPI-anchored proteins, radioactively labeled in vivo with [^{3}H]inositol, can be extracted from slurries of broken cells into Triton X-114 (Conzelmann et al. 1988; Fankhauser and Conzelmann 1991). Treatment of the extracted proteins cleaves their GPI anchor and, in a few cases, notably that of Gas1p/Ggp1p; the protein becomes sufficiently hydrophilic to shift to the aqueous phase on subsequent partitioning between Triton X-114 and buffer. However, the majority of yeast GPI-anchored proteins remain too hydrophobic to be separated in a Triton X-114 partitioning, even though their GPI anchors are cleavable by PI-PLC (Conzelmann et al. 1990).

2. *Cross-Linkage*

Early studies of yeast cell wall composition by chemical fractionation suggested the existence of glucomannan-protein complexes (Kessler and Nickerson 1959; Korn and Northcote 1960). The facts that glycoproteins can be released from the wall with β-1,3 glucanase and that the extractable mannoproteins also contain glucose suggest that some cell wall proteins may be covalently cross-linked to β-glucan. Evidence has been obtained for two types of linkages.

a. Protein-linked, Glucose-containing Side Chains. Four β-1,3-glucanase-releasable proteins that can be separated by SDS-PAGE are recognized by antiserum raised against β-1,6 glucan (van Rinsum et al. 1991; Montijn et al. 1994). In addition to N-linked chains, these proteins have a new class of covalently attached glucose-containing side chain. These chains contain roughly equal amounts of β-1,6-linked glucose and α-1,6-linked mannose residues and are attached directly to protein (Montijn et al. 1994). Neither the nature of the linkage between the glucose-containing chains and protein nor the cellular site of attachment of these chains is yet known, although there is a precedent for a glucosylasparagine linkage in mammalian cells (Schreiner et al. 1994). It has been speculated that these glucose-containing side chains become attached to protein in the ER or Golgi and that their attachment, perhaps by Kre5p, Kre6p, or Skn1p, may be an early step in β-1,6 glucan assembly (Montijn et al. 1994). The glucose-containing chains may also have an indirect cross-linking role in the cell wall. Since the proteins that bear these chains were released upon treatment with β-1,3-glucanase, the possibility was raised that the β-1,6 glucose units in the side chains had become extended with β-1,3 glucan chains. The latter, in turn, could be interwoven with β-1,3 glucan fibrils, thus anchoring the glucomannoprotein in the wall (Montijn et al. 1994). Consistent with this, when synthesis of alkali-insoluble glucan is inhibited by aculeacin A (an echinocandin-class inhibitor of β-1,3 glucan synthase), wall proteins that are normally only released by Zymolyase are instead secreted into the growth medium (Valentin et al. 1986). Furthermore, the susceptibility of cells to spheroplasting by β-1,3-glucanase is enhanced upon treatment with the drug bleomycin and is accompanied by the release of cell wall mannoproteins into the medium (Beaudouin et al. 1993). Bleomycin seems to affect anchoring of proteins in the wall, and since this drug is known to disrupt glycosidic linkages in DNA, it may be acting on wall glycans involved in cross-linking.

b. Transfer of Mannoproteins from GPI Anchors to β-1,6 Glucan. Studies initiated with the α-mating agglutinin Agα1p have revealed that GPI-anchored mannoproteins can become covalently cross-linked to β-1,6 glucan through the glycan portion of the GPI, after cleavage of the anchor lipid. Mature Agα1p is extractable from whole cells by treatment with glucanases, suggesting that the protein is cross-linked to wall glucan (Hauser and Tanner 1989; Schreuder et al. 1993). However, both Agα1p and Aga1p are secretory glycoproteins whose hydrophobic carboxy-terminal domains contain the signals for GPI attachment. The finding that deletion of these carboxy-terminal putative anchor attachment sites caused both Agα1p and Aga1p to be secreted into the growth medium (Roy et al. 1991; Wojciechowicz et al. 1993; van Berkel et al. 1994) led to the suggestion that GPI-anchored forms of these proteins are normally intermediates whose transport to the plasma membrane precedes their cross-linkage to wall glucan (De Nobel and Lipke 1994; Lu et al. 1994). This notion was substantiated by kinetic analyses of Agα1p processing in pulse-chase experiments. The appearance of high-molecular-weight, GPI-anchored Agα1p at the cell surface was followed by the formation of a soluble, periplasmic form lacking inositol and the diacylglycerol components of a GPI anchor but retaining phosphoethanolamine and one or more mannoses of the anchor glycan. This form was in turn converted to the glucanase-extractable cell wall form (Lu et al. 1995).

The glucanase-releasable cell wall form of Agα1p, but not the plasma membrane or periplasmic forms, is recognized by antiserum against β-1,6 glucan (Lu et al. 1995), suggesting that Agα1p becomes cross-linked to β-1,6 glucan. Support for this notion came from experiments on the localization and β-1,6 glucan composition of Agα1p made in various *kre* mutants, which are defective in β-1,6 glucan synthesis (see Section IV.E.2.b). Less Agα1p was associated with the wall in these strains, and more secreted into the medium, an effect most pronounced in *kre5* cells, which showed the most severe defect in β-1,6 glucan synthesis (Lu et al. 1995). Moreover, the β-1,6 glucan component of Agα1p made in the *kre1* and *kre9* strains seemed to contain the shorter chains characteristic of the β-1,6 glucan made in those mutants. Recognition of Agα1p by antiserum to β-1,6 glucan was abolished when the carboxy-terminal GPI anchor attachment site was deleted, indicating that the GPI anchor is required for cross-linking of Agα1p to β-1,6 glucan. Cross-linking could occur in a transglycosylation reaction between β-1,6 glucan and the reducing end of the residual GPI glycan remaining after cleavage of the inositol phospholipid portion (De Nobel and Lipke 1994), although the generation of a soluble, periplasmic form of the protein suggests that the transfer is indirect. Linkage between β-1,6 glucan and Agα1p as well as cell wall mannoproteins of both *S. cerevisiae* and *C. albicans* is indeed through a phosphodiester linkage (Kapteyn et al. 1995, 1996), consistent with the fact that the residual GPI glycan contains a phosphodiester link between ethanolamine and mannose. In addition, the β-1,6 glucan was itself shown to be attached to β-1,3 glucan (Kapteyn et al. 1996), and it was proposed that this heteropolymer with phosphodiester-linked mannoprotein is the alkali-soluble, protein-bound β-1,6- and β-1,3-glucan-containing macromolecule described by Fleet and Manners (1977).

A number of other glucanase-extractable cell wall proteins have now been identified that may be cross-linked in the same way (Klis 1994; van der Vaart et al. 1995), and, as with Agα1p, the extent to which these proteins are incorporated into the cell wall decreases the more severely the cells are defective in β-1,6 glucan synthesis (Jiang et al. 1996). Should this mode of cross-linking, which so far has no counterpart in mammalian cells, be critical for the correct integration of certain glycoproteins into the cell wall, then the transglycosylation reaction presents an attractive target for antifungal therapy. The morphological abnormalities and Calcofluor White sensitivity of GPI anchoring mutants may in part be a consequence of their inability to cross-link key proteins into the wall.

The existence of a further mechanism for cross-linking of formerly GPI-anchored proteins to cell wall glucan is suggested by studies with a cAMP-binding protein. Following lipolytic cleavage of this protein's GPI anchor, a second cleavage event occurs that seems to involve proteolysis at or near the protein's carboxyl terminus, since no residual GPI-derived inositol or ethanolamine can be detected on the protein (Müller et al. 1996; see Section IV.B.3). The doubly processed form of the protein is detected as a periplasmic species and as a detergent-insoluble higher-molecular-weight form that can be released from the cell wall by treatment with a β-1,3-glucanase-containing preparation, suggesting that the protein had become covalently linked to a wall polymer. Since the processed protein is devoid of all GPI components, the linkage formed cannot be the phosphodiester described above: Given that the mature cAMP-binding protein terminates in asparagine, the possibility was raised that the protein could be attached to β-glucan via a glucosylasparagine linkage (see Section VI.A.2.a) (Müller et al. 1996). The basis for the selectivity of this double-processing mechanism for only certain GPI-anchored proteins is not known, although it seems to be restricted to those proteins whose lipolytic cleavage and subsequent incorporation into the cell wall is induced by nutritional shift (Müller et al. 1996).

c. Protein-Carbohydrate Interactions. Some cell wall polysaccharides are bound tightly to certain proteins in a manner independent of the protein's sequence, whereas some proteins have specific sequences that bind specific polysaccharides. Bgl2p, an endo-β-1,3-glucanase and abundant cell wall glycoprotein, binds tightly to β-glucan in both native and heat-denatured states, presumably by hydrogen bonding (Klebl and Tanner 1989; Mrsa et al. 1992, 1993). β-glucan is able to bind tightly to other proteins in vitro, suggesting that no specific protein sequence is a binding determinant (Mrsa et al. 1992). Such binding may be important in the anchoring of bona fide wall proteins, but the same interactions might interfere with the secretion of heterologously expressed proteins. However, not all secreted proteins are affected: The periplasmic enzymes invertase and acid phosphatase do not bind β-glucan, even in their deglycosylated state (Mrsa et al. 1992).

Chitinase, Cts1p, has a domain at its carboxyl terminus that binds chitin with high affinity (Kuranda and Robbins 1991), and which is analogous to the noncatalytic polysaccharide-binding domains in certain cellulases. This domain is not required for catalysis by Cts1p; rather, it may be required for localization of a portion of Cts1p to the chitin that joins mother cell and bud (see Section VII.B).

3. Functions of Wall Glycoproteins

a. Porosity. Some wall proteins, such as glycosidases (see Section VII) are active in wall metabolism, and some have roles in sexual agglutination (see Section VIII.A.1) and flocculation, but en masse, wall glycoproteins also influence the porosity of the wall to high-molecular-weight molecules. Intact cells of many strains resist spheroplasting by β-1,3-glucanases unless they have been pretreated with protease (often present in preparations of lytic enzymes such as Zymolyase) and a thiol reagent (Kitamura 1982; Valentin et al. 1984; Zlotnik et al. 1984). This suggests that the outer layer of the wall is covered in mannoprotein, which must be removed to make the underlying β-1,3 glucan accessible. The β-1,3 glucan of isolated walls can be solubilized by protease-free Zymolyase, indicating that the inner surface of the cell wall is not covered by mannoprotein (Zlotnik et al. 1984).

Yeast glycoproteins, or heterologously expressed glycoproteins, 200–400 kD in size, can be secreted through the wall into the medium or retained within the wall to varying extents (De Nobel and Barnett 1991). The porosity of the wall is determined by wall mannoproteins and can be estimated by comparing the relative sensitivities of cells to lysis by high-molecular-weight polycationic DEAE-dextran and poly-L-lysine. Porosity is highest in logarithmically growing cells but much lower in stationary-phase cells (De Nobel et al. 1991), and it also varies during the cell cycle, being highest during the early stages of bud growth. Changes in porosity correlate with different rates of incorporation of glucanase-soluble wall proteins (De Nobel et al. 1991). β-glucanase-releasable mannoproteins from late exponential phase cells are more extensively *N*-glycosylated than proteins from cells in early logarithmic phase (Valentin et al. 1987), and *mnn* mutations reducing the size and degree of branching of mannan outer chains have more porous walls, indicating that the carbohydrate is a major determinant of porosity (De Nobel et al. 1990). Cross-linking either to β-glucan fractions or to other proteins via disulfide bridges also has a role: The porosity-limiting mannoproteins are mainly the glucanase-soluble ones, and the proteins in the less porous walls of late logarithmic phase cells have a much higher content of disulfide cross-links (Valentin et al. 1987). Perturbations in β-glucan synthesis may also enhance the passage of macromolecules through the wall; thus, mutants resistant to the glucan synthase inhibitor aculeacin A show higher transformation efficiencies with circular DNA (Gallego et al. 1993). Protease treatment of intact yeast cells also results in higher transformation efficiencies (Brzobohaty and Kovác 1986).

b. Flocculation. Important in the brewing process, flocculation involves the formation of large aggregates of cells but is unrelated to sexual agglutination (for review, see Calleja 1987; Stratford 1992a). Flocculation seems to be due to the interaction between sugar-binding lectin-like proteins on flocculent strains and the outer mannan chain side branches of glycoproteins on other cells (Miki et al. 1982; Stratford 1992b). Candidate lectin molecules have been identified (Shankar and Umesh-Kumar 1994; Straver et al. 1994). Mannose-specific flocculation is expressed in cells containing the dominant *FLO1* gene, but the phenomenon is subject to suppression in certain strains. One suppressor mutation is in *SFL1*, which encodes domains homologous to Myc protein and to a yeast heat shock transcription factor and which may be involved in the expression of genes involved in flocculation and, indeed, in normal cell surface assembly (Fujita et al. 1989). *FLO1* encodes a very large serine- and threonine-rich cell wall protein that may become GPI-anchored (Teunissen et al. 1993; Watari et al. 1994). It is not clear whether Flo1p is the lectin itself, an extended "scaffold" subunit for a lectin, or the lectin's receptor or whether it somehow regulates expression of the lectin (Klis 1994; Stratford 1994).

B. Chitin–β-1,3 Glucan Linkages

Extraction of yeast walls with hot alkali leaves behind an insoluble fraction containing β-1,3 and β-1,6 glucan and chitin. The glucan component of this fraction can be made alkali-soluble by prior treatment with chitinase, indicating that chitin had rendered the β-glucan insoluble (Mol and Wessels 1987). Pulse-chase experiments with [^{14}C]Glc using synchronized cells show that preformed, soluble β-glucan is a precursor of alkali-insoluble glucan (Hartland et al. 1994). Up to 40% of the radioactive label that first appeared in alkali-soluble β-glucan was converted to alkali-insoluble material during the chase period. Whether cells were pulse-labeled early or later in their budding cycle, the conversion continued at a reasonably steady rate for some time, suggesting that the process is not restricted to any stage in the cell cycle. The extent to which alkali-soluble glucan can be rendered insoluble correlates with the amount of chitin made in the cells. Thus, chitin-deficient *chs3* cells contain much lower levels of alkali-insoluble glucan, and correspondingly elevated levels of alkali-soluble glucan, compared with wild-type cells (Roncero et al. 1988). Pulse-chase experiments with *chs3* mutants show that much less alkali-insoluble material was made in wild-type cells, whereas in *cdc24* cells, which show increased and delocalized synthesis of chitin after shift to nonpermissive temperature, almost 80% of alkali-soluble β-glucan became insoluble (Hartland et al. (1994). It therefore seems that the two β-glucan fractions differ from each other only in the degree to which they are cross-linked to chitin.

The chitin–β-1,3 glucan linkage was characterized in oligosaccharides obtained following digestion of cell walls with both β-1,3-glucanase and chitinase and consists of a β-1,4 linkage between the terminal reducing residue of a chitin chain and the nonreducing end of a β-1,3 glucan chain (Kollár et al. 1995). The linkage oligosaccharide cannot be isolated from Δ*chs3* cells, indicating that the chitin involved in the cross-link is that synthesized by chitin synthase III. It was estimated that 40–50% of chitin chains are cross-linked to glucan. However, there is only one chitin-glucan bond for every 8000 hexose units in the wall, so these cross-links, although rare, profoundly affect the solubility of highly polymerized glucans. Given that most chitin is incorporated into the bud wall late in its cell cycle, whereas alkali-soluble β-glucan seems to be made continuously, it was proposed that in buds at least, the chitin is attached to preexisting glucan (Kollár et al. 1995). The cross-linking reaction, which presumably takes place outside the plasma membrane, may involve a transglycosylation reaction from a newly formed chitin chain. Alternatively, if chitin chains grow from their reducing end and retain an activated GlcNAc residue there, the complete chitin chain could be transferred to glucan (Kollár et al. 1995).

The cross-linking reaction is a potential target for an antifungal agent, although an inhibitory or fungicidal effect may require simultaneous inhibition or perturbation of synthesis of other wall components. Certainly, combination of the Δ*chs3* mutation with the *cdc24* mutation, which results in excessive and delocalized chitin deposition, gives cells that are sensitive to hypotonic shock, indicating that cross-linking becomes important for maintenance of wall integrity when wall synthesis is abnormal and imbalanced (Hartland et al. 1994).

VII. CELL WALL ENZYMES: GLYCOSIDASES

S. cerevisiae makes a variety of hydrolases with specificity toward some of the β-glycosidic linkages found in wall components. All glycosidases studied so far are secretory glycoproteins that end up in the periplasm or become cell-wall-associated.

A. β-1,3 Glucanases

Both exo-β-1,3 glucanase activities, which hydolyze glycosidic linkages at the nonreducing ends of β-1,3 glucans, and endo-β-1,3-glucanase activities, which cleave within the chains, have been detected in periplasmic and wall fractions of *S. cerevisiae* cells (Hien and Fleet 1983; Fleet 1991). Some of the chromatographically separable glucanase activities may be due to differently glycosylated forms of the same gene product, as is the case with Exg1p (for review, see Larriba et al. 1995). The in vitro substrate specificities of these enzymes suggest that they could function during wall growth, for example, by "nicking" the microfibrillar glucan net to permit the insertion of new polysaccharide chains. So far, only two β-1,3-glucanase genes have been isolated because they encode previously characterized enzyme activities, but two further β-1,3-glucanase genes have been identified in other ways. Further enzymes have subsequently been identified in null mutants lacking the other activities. We focus here on the four β-1,3-glucanase genes that have been cloned and disrupted, three of which encode members of a protein family.

Exg1p is the major secreted exo-β-1,3-glucanase, three variants of which can be separated, bearing either one or two N-linked chains. One of the chains can be elongated to yield the third glycoform (Hernández et al. 1992; Larriba et al. 1995). *EXG1* (*BGL1*) was cloned by screening colonies of cells transformed with genomic DNA libraries for expression of activity that hydrolyzed the fluorogenic substrate 4-methylumbelliferyl-β-D-glucoside (Nebrada et al. 1986; Kuranda and Robbins 1987). Exg1p-deficient strains show no obvious defects in growth and development (Nebrada et al. 1986), but they are more sensitive to K1 killer toxin and show a small increase in β-1,6 glucan levels, whereas overexpression of *EXG1* leads to toxin resistance and a reduction in β-1,6 glucan (Jiang et al. 1995). Exg1p therefore has a role in cell wall glucan metabolism and perhaps preferentially degrades the β-1,3 side branches on β-1,6 glucan. Expression of *EXG1* is modulated through the Pbs2p-Hog1p MAP kinase signaling cascade involved in osmoregulation (see Section IX).

EXG2, identified along with *EXG1* because it also conferred β-1,3-glucanase activity on an *exg1*-deficient strain (Nebrada et al. 1986), encodes a protein similar in sequence to Exg1p (J. Correa, unpubl., cited by Larriba et al. 1995). In contrast to Exg1p, Exg2p has a carboxy-terminal GPI anchor attachment site, deletion of which leads to secretion of the protein into the medium. Null mutations in *EXG2* cause no discernible growth defect. Analysis of *exg1* null mutants has led to the identification of two further β-glucanase activities (Cenamor et al. 1987).

SSG1/SPR1, encoding a third member of the Exg1p family, was cloned using peptide sequence information obtained from an exo-β-1,3-glucanase synthesized during sporulation (San Segundo et al. 1993) and was also identified as a gene expressed during a late stage of sporulation (Muthukumar et al. 1993). The amino acid sequence of Ssg1p/Spr1p is 64% identical to Exg1p. This protein makes a small but dispensable contribution to ascus formation: The onset of ascus formation is delayed in homozygous *ssg1/spr1* disruptants, but the ascospores eventually formed appear to be normal. Exg1p and Exg2p do not substitute for Ssg1p/Spr1p, because disruption in an *ssg1/spr1* background of either or both *EXG* genes has no further effect on the sporulation phenotype.

Bgl2p is an abundant Zymolyase-extractable cell wall protein that shows sequence similarity to plant β-1,3-glucanases, has endo-β-1,3-glucanase activity, and binds to β-glucans and chitin (Klebl and Tanner 1989; Mrsa et al. 1992, 1993). *bgl2* disruptants show no discernible effect on vegetative growth, mating, or sporulation, but overproduction of Bgl2p retards growth, an effect that may be lethal if expression exceeds three times wild-type levels (Klebl and Tanner 1989). Bgl2p levels were increased two- to threefold in cells harboring alleles of *pkc1*. Such cells are slow growing, osmotically fragile, and contain 30% of wild-type levels of alkali-insoluble glucan, which may account for the strain's osmotic fragility (Shimizu et al. 1994). Whether any effects of *BGL2* overexpression are indeed due to the protein's catalytic activity is not yet known, since disruption of *BGL2* in a Δ*pkc1* background suppressed the slow-growth phenotype but not the osmotic fragility of the mutant. It was proposed that Pkc1p may be a negative regulator of *BGL2* expression.

In addition to catalyzing β-1,3 glucan hydrolysis, Bgl2p can carry out a glycosyl transfer reaction and join two β-1,3-linked glucan chains, reducing end to nonreducing end via a β-1,6 glucosidic linkage. The mechanism involves cleavage of laminaribiose from the reducing end of a β-1,3-linked chain to create an enzyme-bound "activated donor" chain. The latter is then transferred to C_6 of the glucosyl residue at the non-reducing end of an acceptor β-1,3 glucan chain (Yu et al. 1993; Goldman et al. 1995). Studies with purified Bgl2p and soluble β-1,3 glucan show that at low concentration of acceptor glucan, mainly the hydrolysis step is detected, because water is the predominant acceptor. The donor chain may arise in vivo by de novo synthesis of β-1,3 glucan or it could be released upon cleavage of preexisting β-1,3 glucan by endoglucanase. The existence of this novel cross-linking activity suggests a way in which β-1,3 glucan could be incorporated into existing wall polysaccharide and allow wall expansion. In pulse-chase experiments in which cell wall β-glucan was specifically labeled, radioactively labeled laminaribiose was released from prelabeled β-glucan and could therefore indeed have been generated in vivo in a Bgl2p-like glucosyltransfer reaction (Coen et al. 1994).

B. Chitinase

Endochitinase is a periplasmic enzyme, but much is secreted into the medium when cells are grown in nutrient-rich YPD medium (Correa et al. 1982; Elango et al. 1982; Kuranda and Robbins 1991). *CTS1* was cloned by overexpression of activity toward the fluorogenic substrate 4-methylumbelliferyl-tri-*N*-acetylchitotriose, and it encodes a secretory protein with a number of domains (Kuranda and Robbins 1987, 1991). The catalytic domain shows significant homology with a cucumber chitinase produced in response to invasion by pathogens. Next is a central, serine- and threonine-rich region that is heavily *O*-mannosylated (Orlean 1990), and the protein's carboxyl terminus contains a noncatalytic high-affinity chitin-binding domain. *cts1* disruptants form large aggregates of cells joined together at their chitin-containing septa, showing that Cts1p has a key role in cell separation (Kuranda and Robbins 1991), and indeed, treatment of wild-type cells with the potent chitinase inhibitor dimethylallosamidin leads to the same cell separation defect seen in Δ*cts1* strains (Sakuda et al. 1990). Interestingly, the shape of the aggregates formed by haploid Δ*cts1* cells was dependent on whether the disruptee normally displayed an axial or a random budding pattern. Since the cell separation defect of Δ*cts1* strains was only partially complemented by Cts1p lacking its chitin-binding domain, this domain is normally important for the localization of the enzyme in the septal region of the wall.

Studies with synchronized cells show that *CTS1* expression is periodic and that transcription of the gene is regulated by Ace2p, a zinc-finger-containing transcription factor (Dohrmann et al. 1992). Mutation of *ACE2* virtually abolishes *CTS1* transcription, and *ace2* mutants show a cell separation defect. Ace2p is homologous to the cell-cycle-regulated transcriptional activator Swi5p, which regulates expression of *HO* with a periodicity the same as that achieved in Ace2p-governed *CTS1* transcription. Swi5p-regulated transcription of *HO* is asymmetrical in that *HO* is expressed in the mother cell resulting from a mitotic division and not in the daughter. Should *CTS1* likewise be expressed in only one of the mitotic progeny, digestion of the division septum from one side only may result in the bud scar remaining on the mother cell (Dohrmann et al. 1992).

VIII. MATING AND SPORULATION

A. Mating

Conjugation between haploids is preceded by formation of the mating projection or shmoo, a process accompanied by change in cell wall composition and ultrastructure. The signaling events involved in this alteration in cell polarity are reviewed by Sprague and Thorner (1992). Many mutants affected in the synthesis of surface components mate, although with varying degrees of efficiency, and some of the observed changes in morphology, in wall composition, and in expression of wall synthesis genes in pheromone-treated wild-type cells either are not absolutely required for mating or have no obvious role in the process. We focus here on the glycoproteins and extra chitin that are deposited at the surface of the projection.

1. Glycoprotein Agglutinins

The peptide mating pheromones **a**- and α-factor induce the expression of glycoprotein agglutinins at the surface of haploid cells of the opposite mating type (for review, see Yanagishima 1984; Lipke and Kurjan 1992). The α-agglutinin, Agα1p, is a secretory protein with 12 potential *N*-glycosylation sites, possible O-linked mannose, and a GPI anchor (Terrance et al. 1987; Hauser and Tanner 1989; Lipke et al. 1989). The latter structure is required for eventual cross-linking of the protein in the cell wall (Lu et al. 1994; see Section VI.A.2.b), where it is localized in the outer fibrillar layer of the wall (Cappellaro et al. 1994). *N*-glycosylation, however, is not required for function, since deglycosylated Agα1p retains full biological activity (Terrance et al. 1987; Hauser and Tanner 1989). The region of Agα1p defined by deletion analysis as important for agglutination shows weak homology with immunoglobulin fold sequences (Wojciechowicz et al. 1993).

The **a**-mating agglutinin consists of two components, the Aga1p anchorage subunit and the small Aga2p-binding subunit. Aga1p is an exclusively *O*-mannosylated secretory protein with a potential GPI attachment site (Roy et al. 1991). Unlike the other two agglutinin genes, expression of *AGA1* is pheromone-inducible in both **a** and α cells: Aga1p may therefore be a wall glycoprotein present in both **a** and α cells: Its heavy *O*-mannosylation may allow the protein to serve as an extended scaffold to which inducible Aga2p becomes attached in **a** cells. Aga2p is an 18-kD, *O*-mannosylated protein that can be released from the wall by treatment with mercaptoethanol and which forms a 1:1 complex with Aga1p in vitro (Orlean et al. 1986; Watzele et al. 1988; Cappellaro et al. 1991). Aga2p is joined to Aga1p by two disulfide bonds, and mutation of the two cysteines in Aga2p that are involved causes the protein to be released to the medium and abolishes agglutination (Cappellaro et al. 1994). Aga2p retains its biological activity after deglycosylation (Watzele et al. 1988; Cappellaro et al. 1991). The biologically active portion of Aga2p has been localized to ten amino acids at the protein's carboxyl terminus. When normally glycosylated, this ten-amino-acid peptide can inhibit the interaction between Aga2p and Agα1p at concentrations four to five times lower than the unglycosylated peptide, indicating a modest role for the O-linked sugars (Cappellaro et al. 1994). Altogether though, it seems that mating agglutination mediated by Agα1p and Aga2p is due mainly to protein-protein interactions.

Fus1p, also induced by mating pheromones, is required for the cytoplasmic fusion that occurs after agglutination of mating partners (McCaffrey et al. 1987; Truehart et al. 1987). Fus1p is a membrane-spanning *O*-mannosylated protein that localizes to the tip of the shmoo (Truehart and Fink 1989). Although cell fusion requires dissolution of the walls of the two haploids, Fus1p is not believed to be a hydrolase, and other lytic enzymes involved in cell fusion have not been identified. One intriguing role for Fus1p is to "sense" the presence of the partner, transmit this information to the cytoplasm, and trigger the mechanism that allows fusion (Truehart and Fink 1989).

2. *Polysaccharide Synthesis*

*a. Chitin. MAT***a** haploids treated with α-factor show a three- to fourfold increase in chitin content, and the extra chitin is deposited diffusely in the shmoo (Schekman and Brawley 1979). Chs3p is responsible for most if not all of this extra chitin synthesis, since Δ*chs3* and Δ*chs4/csd4* haploids fail to make extra chitin in response to α-factor, and *csd3* and *chs5* point mutants show some, although much reduced, synthesis of chitin in response to the pheromone (Orlean 1987; Roncero et al. 1988; Valdivieso et al. 1991; Bulawa 1993). Thus, neither Chs1p nor Chs2p can substitute for Chs3p in this process, nor indeed is Chs1p required, for Δ*chs1* cells do show pheromone-induced chitin synthesis (Bulawa et al. 1986; Orlean 1987). In *MAT***a** haploids treated with α-factor, Chs3p is localized predominantly at the periphery of the mating projection, whereas in Δ*chs5* strains, Chs3p is found mainly in cytoplasmic patches, a redistribution similar to that seen in vegetatively growing Δ*chs5* cells (Santos and Snyder 1997). Chs5p is therefore required for targeting of Chs3p to the shmoo, but, since *chs5* mutants have a severe mating defect, Chs5p must have additional roles in cell fusion. A plausible role would be the targeting of transport compartments that contain proteins involved in cell fusion, such as Fus1p. *chs3* and *chs4/csd4* mutants are able to mate (Roncero et al. 1988; Bulawa 1992), but *chs5* mutants show a severe decrease in mating efficiency (Santos and Snyder 1997). Interestingly, the increase in chitin synthesis in response to pheromone is not required for projection formation, for shmoos are made by the chitin-deficient Δ*chs3* and Δ*csd4* mutants; however, in the *SSD-v1*-suppressible strain background, shmoo morphology is abnormal (Bulawa 1993).

The extra chitin synthesis in pheromone-treated haploids is not obviously correlated with levels of transcription of *CHS* genes or levels of chitin synthase activity. By in vitro assay, pheromone-treated cells do not show an increase in specific chitin synthase III activity, nor do levels of *CHS3* mRNA change (Orlean 1987; Choi et al. 1994b). Levels of *CHS1* mRNA, on the other hand, rise 4–15-fold upon treatment of both *MAT***a** and *MAT*α haploids with pheromone from cells of opposite mating type, and indeed, *CHS1* has four pheromone response elements in an upstream DNA sequence that confer hormone inducibility on *CHS1* (Appeltauer and Achstetter 1989). Chitin synthase I activities double in *MAT***a** haploids treated with α-factor (Schekman and Brawley 1979; Orlean 1987), but virtually all this activity is detectable only after trypsin treatment. Treatment of *MAT***a** cells with α-factor results in a decrease in *CHS2* mRNA, followed by a decrease in chitin synthase II activity (Pammer et al. 1992; Choi et al. 1994b). How chitin synthase III is stimulated in response to pheromone and whether Chs1p and Chs2p make undetectable contributions to the synthesis of extra chitin, or have other functions as well, are unknown.

The additional chitin that is made in response to mating pheromone requires that more precursor GlcNAc be available, and indeed endogenous GlcNAc synthesis is stimulated in α-factor-treated *MAT***a** haploids (Orlean et al. 1985). Furthermore, as envisaged by Ballou (1982), the expression of genes responsible for GlcNAc synthesis can be regulated. Thus, the promoter regions of the *GFA1* amidotransferase and *AGM1* GlcNAc phosphate mutase genes contain, respectively, six and three copies of the pheromone response element in their promoter regions, and levels of *GFA1* mRNA rise two- to threefold upon treatment of *MAT***a** haploids with α-factor (Watzele and Tanner 1989; Hoffmann et al. 1994). Synthesis of extra chitin may proceed at the expense of glycogen (Watzele and Tanner 1989).

b. β-glucan. *MAT***a** haploids treated with α-factor develop an increased susceptibility to lysis by β-glucanases and show a decrease in mannose to glucose ratios from 1 to about 0.66, indicating changes in the deposition and organization of β-glucan in the cell wall (Lipke et al. 1976). None of the genes implicated in β-1,3 glucan synthesis have been shown to be required for pheromone-induced morphogenesis. However, expression of *FKS2* is highly induced in response to mating pheromone, although in contrast to *CHS1*, this occurs after a delay of 45–60 minutes. *FKS2* induction is also dependent on calcineurin (Mazur et al. 1995). Fks2p may have a role in wall remodeling during mating, but this role can be bypassed or compensated for because Δ*fks2* strains form normal shmoos and do not lose viability upon pheromone treatment. Of the mutants affected in β-1,6 glucan synthesis, only Δ*kre9* strains show a marked decrease in mating efficiency. Growth of Δ*kre9 MAT***a** cells is arrested by α-factor, but shmoos are not formed, indicating a role for Kre9p in this polarized growth process (Brown and Bussey 1993).

B. Sporulation

1. Structure of the Ascospore Wall and Mutations Affecting Its Formation

The ascospore wall, which endows the spore with resistance to various adverse conditions, is formed after the second meiotic division. Wall components are deposited between a double membrane that surrounds the haploid nucleus to create a four-layered structure (Byers 1981). The inner two layers consist of β-glucan and mannan (Kathoda et al. 1984; Briza et al. 1988), components similar to those found in the vegetative cell wall, but the next two are peculiar to the ascospore. The third layer consists mainly of chitosan, a polymer of β-1,4-linked glucosamine, and the outer layer is an insoluble macromolecule containing LL-dityrosine and DL-dityrosine and protein, but which is otherwise of unknown composition (Briza et al. 1986, 1988, 1990b).

A variety of mutations affecting the synthesis of components of the vegetative cell wall cause sporulation defects in homozygous diploids. Inhibition of glucosamine synthesis by starvation of *gcn1/gfa1* diploids prevents formation of the outer two layers of the ascospore wall, indicating the importance of glucosamine-containing polymers in ascospore wall assembly (Whelan and Ballou 1975; Ballou et al. 1977). The same phenotype is caused by tunicamycin, an inhibitor of *N*-glycosylation (Weinstock and Ballou 1987). Further mutants with sporulation defects include vanadate-resistant (*mnn*) mutants (Ballou 1982), GPI anchoring mutants (*gpi1*, *gpi2*, and *gpi3*; Leidich and Orlean 1996), β-1,3 glucan synthesis mutants (Δ*fks2* [Mazur et al. 1995] and *gns1* [El-Sherbeini and Clemas 1995b]), and the *lcb1* mutant defective in ceramide synthesis (Buede et al. 1991). Abolition of *N*-glycosylation and GPI anchoring may alter the structure or compromise the function and localization of glycoproteins normally incorporated into the ascospore wall. One such protein is sporulation-specific Sps100p, a protein that is potentially heavily *N*- and *O*-glycosylated and that at an early stage in ascospore wall formation, contributes to the resistance of spores to ether (Law and Segall 1988).

Mutations causing defects in ascospore maturation were identified by screening a strain that can undergo haploid meiosis and form two spores for mutants that formed visible asci, but which failed to show the natural fluorescence imparted by the outer dityrosine-containing layer (Briza et al. 1990a). These *dit* mutants have shed light on the biosynthesis of the chitosan and dityrosine components of the ascospore wall (Pammer et al. 1992; Briza et al. 1994).

2. *Chitosan Synthesis*

Spores made by *dit101* mutants lack their dityrosine and chitosan layers and fail to make chitosan but are viable, although they are sensitive to lytic enzymes. Formation of the dityrosine layer is therefore dependent on the presence of the underlying chitosan layer. *DIT101* proved to be identical to *CHS3*, and chitin synthase III is therefore required for synthesis of chitosan in the ascospore wall (Pammer et al. 1992). *CHS3* expression is elevated during sporulation (Choi et al. 1994b), but Chs4p/Csd4p, which is required for chitin synthesis in vegetative cells, does not seem to be involved in chitosan synthesis, for Δ*chs4/csd4* strains make a chitosan layer (Bulawa 1993). However, a Chs4p/Csd4p homolog exists and is encoded by *SHC1*, whose expression rises sharply during sporulation. It has been proposed that Shc1p modulates the activity and perhaps localization of Chs3p during ascospore wall synthesis (Bulawa 1993).

The requirement for Chs3p for chitosan synthesis indicates that chitin is the biosynthetic precursor of chitosan and that chitin is deacetylated subsequent to synthesis. Chitin deacetylase activity has been detected in *Mucor rouxii* (Davis and Bartnicki-Garcia 1984) and in sporulating *S. cerevisiae* diploids (Mishra et al. 1997). *S. cerevisiae* contains two genes, *CDA1* and *CDA2*, that encode soluble proteins similar to the *Mucor* deacetylase (Kafetzopoulos et al. 1993), and whose expression is sporulation-specific (Christodoulidou et al. 1996; Mishra et al. 1997). Both *CDA1* and *CDA2* conferred in vitro chitin deacetylase activity when expressed in vegetative cells under the control of an inducible promoter, and approximately 20% of the chitin in these cells was converted to chitosan in vivo. Moreover, in membranes from vegetative cells synthesizing chitin in vitro, the presence of chitin deacetylase expressed from a plasmid led to deacetylation of the in vitro chitin (Mishra et al. 1997). Since microfibrillar, hydrogen-bonded chitin resists enzymatic deacetylation, the deacetylases presumably act on freshly formed chitin and may indeed become physically associated with chitin synthase III.

3. *Formation of the Dityrosine Layer*

The *dit1* and *dit2* mutants both make ascospores that lack dityrosine and show low resistance to various treatments, although the spores are formed at the normal rate and are fully viable under normal conditions (Briza et al. 1990a). Expression of *DIT1* and *DIT2* is restricted to sporulating diploids (Briza et al. 1990a). The function of the *DIT1* and *DIT2* gene products was investigated by expressing the genes in vegetative cells and led to the proposal of a pathway for dityrosine synthesis. The first step is the conversion of L-tyrosine into soluble, L-tyrosine-containing precursors by sporulation-specific Dit1p. These are in turn cross-linked to form LL-dityrosine-containing molecules, a step requiring both the sporulation-specific cytochrome P450 Dit2p and the nonspecific NADPH-cytochrome P450 reductase Cpr1p. This precursor is further metabolized and then incorporated into the ascospore wall, after which some of the LL-dityrosine is epimerized to DL-dityrosine (Briza et al. 1994).

IX. REGULATION: A PROTEIN KINASE C CELL INTEGRITY PATHWAY AND MULTIPLE ROLES OF THE Rho1p GTPase

Wall biogenesis is regulated by cell cycle controls and developmental programs, but the process is also responsive to environmental conditions. During vegetative growth, for example, bud site selection, bud emergence, and polarized growth of the bud, which are under the control of specific genes (Lew et al., this volume), dictate subsequent localized deposition of wall components. Superimposed on the morphogenetic events during budding growth is the need for the cell to preserve its osmotic integrity. For both purposes to be achieved, synthesis of individual wall components and their assembly into the wall must be highly controlled. Indeed, severe defects in the synthesis of specific wall components can result in osmotic fragility, but from the existence of mutants lacking much of a given wall component and from the occurrence of suppression of some mutants' severe growth phenotype without restoration of their wall defect, it seems that *S. cerevisiae* can also tolerate or compensate for considerable changes in wall composition. A number of instances of suppression involve interactions between cell wall biosynthesis genes and genes encoding protein kinases and phosphatases in the protein kinase C (PKC) signaling pathway. Moreover, mutations in components of the PKC pathway result in osmotic fragility, indicating that this pathway has a key regulatory role in maintaining cell integrity, perhaps by coordinating the activities of several wall synthetic activities.

The two-component Skn7p pathway (see Section IV.E.2.b) and the Pbs2p-Hog1p osmosensing pathway have parallel regulatory roles. Protein phosphorylation pathways and their relationships to morphogenesis have been discussed by Cid et al. (1995).

The PKC pathway illustrates the complexity of regulatory interactions involved in maintaining wall integrity and allowing bud morphogenesis. This pathway responds to a mechanical environmental signal, as well as to activation of the cyclin-dependent kinase Cdc28p at the Start point in the G_1 phase of the cell cycle, which marks commitment to a new round of cell division. Furthermore, Pkc1p itself shares the activating GTPase Rho1p with the β-1,3 glucan synthase complex, as well as with a mechanism that organizes the actin cytoskeleton at the bud tip, exemplifying how the syntheses of cell wall components, as well as the generation of cell polarity, can be coordinated with one another.

Disruption of the yeast PKC homolog Pkc1p results in a temperature-sensitive lysis phenotype. When osmotic support is removed, cells rapidly lyse at the bud tip, indicating that the preexisting wall had become weakened (Levin and Bartlett-Heubusch 1992; Paravicini et al. 1992). Mutations downstream from *pkc1* in the MAP kinase branch of the PKC pathway—in *MKK1/MKK2* (which encode a pair of MAP kinase kinase homologs) and in *MPK1* (a MAP kinase that may phosphorylate transcription factors)—also give a temperature-sensitive lysis defect (Torres et al. 1991; Irie et al. 1993; Lee et al. 1993). Δ*pkc1* cells grown in the presence of osmotic support show a 30–70% reduction in total β-glucan, as well as a slight reduction in their mannan content (Roemer et al. 1994; Shimizu et al. 1994). These defects in wall polysaccharide synthesis, which may in turn affect the extent of cross-linking of wall components, may cause osmotic fragility.

Mutations affecting the synthesis of different surface components can show synthetic lethality with PKC pathway mutants, and in some cases, the wild-type alleles of such genes can also suppress the lysis defect of Δ*pkc1* cells. β-1,6 glucan synthesis (see Section IV.E.2.b) provides an illustration. Thus, the Δ*kre6* and Δ*kre11* mutations, which affect β-1,6 glucan synthesis and its regulation, are lethal in combination with Δ*pkc1*, as is the Δ*kre2/mnt1* mutation, which affects *O*-mannosylation. Δ*kre6* is also synthetically lethal with Δ*mpk1* and Δ*mkk1*/Δ*mkk2* (Roemer et al. 1994). The basis for this lethality is not known, but it may reflect a weakening of the cell wall. Of the above interacting genes, only *KRE6* partially suppresses the osmotic fragility of a *pkc1* disruptant, but the suppressees show no detectable increase in their β-1,6 glucan content nor any obvious change in cell wall composition, and thus *KRE6* is not acting as a suppressor by promoting β-1,6 glucan synthesis (Roemer et al. 1994). *KRE6* does not suppress Δ*mpk1* or Δ*mkk1*/Δ*mkk2*, suggesting that *KRE6* does not exert its effect via the MAP kinase branch of the PKC pathway but rather through a second regulatory branch. However, it does not seem that Pkc1p directly regulates Kre6p because overexpression of *PKC1* does not suppress the growth defects of Δ*kre6*, and Kre6p can be phosphorylated in Δ*pkc1* strains (Roemer et al. 1994).

SKN7, whose product functions in a two-component signaling pathway involved in the regulation of β-1,6 glucan synthesis (see Section IV.E.2.b), also interacts with *PKC1*. Δ*skn7* Δ*pkc1* strains are inviable even in the presence of osmotic support, and *SKN7* suppresses the lysis defect of Δ*pkc1* strains but not of Δ*mkk1* Δ*mkk2* double mutants. The Skn7p pathway therefore does not act through the MAP kinase branch of the Pkc1p pathway, rather it may operate in parallel to coordinate extracellular signals that regulate assembly of the wall (Brown et al. 1994).

Cell wall metabolism is also influenced through a second MAP kinase cascade, the Pbs2p-Hog1p pathway, which signals stress induced by high osmolarity and causes the accumulation of cytoplasmic glycerol (Brewster et al. 1993). Overexpression of the Pbs2p MAP kinase kinase leads to enhanced expression of the *EXG1* exoglucanase gene and to killer toxin resistance, suggesting lowered synthesis of β-1,6 glucan. In contrast, Δ*exg1* and Δ*pbs2* strains are hypersensitive to killer toxin, presumably on account of synthesizing more receptor β-1,6 glucan. Pbs2p-regulated expression of *EXG1*, and perhaps of other genes involved in cell wall glucan metabolism, is antagonized by Ptc1p phosphatase. A mutant allele of *PTC1*, *cwh47*, is Calcofluor-White-sensitive and killer-toxin-resistant, and disruption of *PTC1* results in activation of *EXG1* expression (Jiang et al. 1995). Whether further aspects of wall synthesis are regulated through this high-osmolarity-sensing pathway is not yet known. Together, however, the Pbs2p-Hog1p and Pkc1p pathways would allow the cell to regulate cell wall synthesis in response to both high and low osmolarity (Gustin et al. 1995).

What are the stresses that activate signaling through these pathways? It seems that stretching of the plasma membrane has a role in activation of Pkc1p (Kamada et al. 1995). Cells normally resist low osmolarity on account of having an intact wall, and if necessary, they respond to low osmolarity by strengthening the wall. If cell wall synthesis is impaired, or if the wall-augmentation signal cannot be sent (as is the case in Δ*pkc1* mutants), the weakened wall will no longer prevent lysis. This weakness might normally be sensed as stress on the plasma membrane. Indeed, the PKC signaling pathway is activated by an amphipathic drug that generates membrane stretch by inserting preferentially into the cytoplasmic leaflet of the plasma membrane (Kamada et al. 1995). Furthermore, heat shock, which would increase membrane fluidity, also results in activation of PKC.

Pkc1p-mediated signaling of wall synthesis through the MAP kinase pathway also seems to be triggered at the point in the cell cycle when cells initiate budding (Marini et al. 1996). This possibility emerged from the demonstration that a mutant *pkc1* allele exacerbates the temperature-sensitive phenotype conferred by mutations in the cyclin-dependent kinase Cdc28p, which functions at Start in the cell cycle. Furthermore, overexpression of the G_1 cyclin Cln2p, an activator of Cdc28p, suppresses the mutant *pkc1* allele, albeit incompletely, consistent with Cdc28p-dependent activation of the partially functional, mutant Pkc1p. Cdc28p activation at Start is also closely correlated with a burst of synthesis of diacylglycerol, a known activator of PKCs. These findings led to the proposal for a pathway in which activation of Cdc28p stimulates a phospholipase C that apparently acts on phosphatidylcholine to release diacylglycerol, which in turn activates Pkc1p (Marini et al. 1996). The net result of this pathway is to ensure that the cell is primed to expand its bud and maintain the integrity of its wall during growth.

Activation of Pkc1p is also achieved by the Rho1p GTPase, a protein that also regulates two further pathways involved in the maintenance of cell integrity and polarity. Mutants with conditional defects in *RHO1* are blocked in signaling through the MAP kinase branch of the PKC pathway and show a temperature-sensitive lysis defect (Qadota et al. 1994; Yamochi et al. 1994) that can be suppressed by overexpression of *PKC1* or by a constitutively active form of the kinase (Nonaka et al. 1995; Kamada et al. 1996). Pkc1p and Rho1p indeed interact with each other: Two-hybrid interactions between Rho1p-GTP and Pkc1p show that this occurs in vivo (Nonaka et al. 1995), and Rho1p and Pkc1p can be coimmunoprecipitated in a GTP-dependent fashion (Kamada et al. 1996). Moreover, the GTP-bound form of Rho1p confers on Pkc1p the ability to respond to the activating cofactor phosphatidylserine and phosphorylate a substrate protein (Kamada et al. 1996). Interestingly, the region of Pkc1p that Rho1p interacts with is the predicted diacyglycerol-binding domain of Pkc1p (Nonaka et al. 1995; Kamada et al. 1996). This finding raises the question of whether, how, and when Pkc1p activity could be stimulated by both GTPase and diacylglycerol.

Rho1p-GTP also activates β-1,3 glucan synthase (see Section V.B.2.a), but this effect must be mediated through a pathway that does not directly involve Pkc1p, since the lethality of the *pkc1* disruption is suppressed in the presence of osmotic stabilizer, whereas *rho1* disruptants are not rescued (Yamochi et al. 1994). Moreover, neither a constitutively active form of Pkc1p nor overexpression of *PKC1* restores β-1,3 glucan synthase activity to *rho1* membranes, and Pkc1p fails to suppress the lysis defect of all but one allele of *rho1*. *pkc1* mutants show at best a partial defect in β-1,3 glucan synthase activity, which, when seen, is associated with the catalytic fraction of the enzyme preparations (Drgonová et al. 1996; Qadota et al. 1996). These findings lead to a model in which Rho1p has a dual role in coordinating wall synthesis for the maintenance of cell integrity. Thus, a signal generated by membrane stress in cells with weakened walls stimulates guanyl nucleotide exchange on Rho1p, whereupon GTP-bound Rho1p directly stimulates β-1,3 glucan synthesis and also activates Pkc1p, which in turn leads to expression of genes required for the formation of supplemental wall components (Drgonová et al. 1996; Kamada et al. 1996).

The third function of Rho1p is as a regulator of polarized growth by organizing the cytoskeleton at the site of budding. The protein's localization closely matches that of actin patches on the inner surface of the plasma membrane at the tips of growing buds and in the neck between mother cell and bud before cytokinesis (Yamochi et al. 1994). RhoA, the mammalian counterpart of Rho1p, has a role in organizing the actin stress fibers that extend from focal adhesions, small patches on the plasma membrane (for review, see Takai et al. 1995). Human RhoA can partially substitute for yeast Rho1p (Qadota et al. 1994), inviting a parallel to be drawn between the roles of RhoA and Rho1p in polarized growth processes of focal contact formation in mammalian cells and bud formation in yeast (for discussion, see Bussey 1996). In yeast, the actin cytoskeleton would serve to direct secretory vesicles and their cargo of cell wall precursors and synthetic enzymes to the growing bud (see Section V.A.2.d).

The regulatory networks radiating from and connecting Pkc1p and Rho1p are beginning to reveal how cell wall biogenesis can be controlled. We now await further links to be forged between these networks, the mechanisms governing bud emergence (see Lew et al., this volume), those triggering the formation of individual polysaccharides such as chitin.

X. WHAT'S NEXT?

Since the cell surface was last reviewed in these volumes (Ballou 1982), molecular genetic approaches have led to the discovery of many genes involved in cell wall biogenesis. Pathways involved in wall synthesis have been identified or charted in greater detail, further wall components have been detected, and cross-links between wall macromolecules, long predicted to occur, have been demonstrated. It is also clear that the synthesis of wall components is integrated with the secretory pathway and with intracellular signaling pathways and that many of the processes involved in yeast wall biogenesis follow variations on themes familiar in eukaryotic cell biology.

In the near future, we can expect the identification of many more genes involved in cell wall biogenesis. Genes that encode homologs of known proteins will be identified in the complete sequence of the *S. cerevisiae* genome and the function of their products will be explored systematically. New genes, and many more genetic interactions between wall biogenesis genes, will be uncovered in screens for multicopy suppression and synthetic lethality, perhaps superimposed on broad screens for cell wall synthesis defects such as Calcofluor White sensitivity. Genes for predicted or conjectural proteins such as cross-linking enzymes and ''flippases'' should eventually be flushed from their hiding places in the genome.

A major challenge will be to establish the functions of the gene products implicated in wall biogenesis. Indeed, catalytic activity has so far been demonstrated in surprisingly, if not embarrassingly, few cases. Likewise, we have few biochemical details of the protein-protein interactions and regulatory roles for proteins that have been revealed by genetic analyses, nor do we know in many cases where in the cell these proteins act and how they get there. We would like to know, for example, exactly how chitin synthesis is triggered and whether and how it is coordinated with the Rho1p-dependent morphogenetic pathways.

A further challenge will be to exploit what has been learned in *S. cerevisiae* to identify processes in cell wall assembly that, when inhibited, compromise the viability or virulence of pathogenic fungi. Work with *S. cerevisiae* has highlighted the fact that there is redundancy among wall synthetic proteins and that yeast can tolerate considerable reductions in the amounts of certain wall components. Nevertheless, candidate lethal targets can be explored systematically, and among them may be polysaccharide synthases and cross-linking enzymes. Will it be possible to discover or design specific inhibitors of such proteins, and will it be necessary to inhibit more than one synthase? Virulence may, among other factors, be conferred by particular carbohydrate structures that promote adhesion to host cells: Could the mannosyltransferases that make the key structures be inhibited or, given the redundancy among outer-chain mannosyltransferases, would it be more practical to use specific oligosaccharides as competitors?

The complexity underlying the biogenesis of cell surface carbohydrates in yeast is likely to be amplified in mammalian cells. Indeed, the degree of gene duplication and redundancy seen among yeast Golgi proteins implicated in elongation and branching of N-linked outer mannan chains presages enormous genetic complexity in the generation of complex carbohydrate structures in mammalian cells. Numerous glycosyltransferase genes may lurk in the mammalian genome, and should some prove redundant, it will not be easy to uncover the function of their products. Moreover, as has been shown for yeast Chs1p, readily assayable enzyme activity does not necessarily mean that a glycosyltransferase is a dominant one in vivo.

Finally, as in yeast, Golgi glycosylation events in mammalian cells are likely to be strongly influenced by proteins that are involved in aspects of Golgi and secretory pathway function but that are not themselves glycosyltransferases.

Fascinating in their own right, as well as for practical reasons, studies of yeast cell wall biosynthesis have already contributed much information about conserved biosynthetic pathways and will continue to do so. It will be intriguing to see how processes first characterized in yeast, such as chitin and β-1,6 glucan synthesis, are used or have been adapted by multicellular organisms.

ACKNOWLEDGMENTS

I acknowledge Clinton Ballou for his many contributions to genetic and structural analysis of the yeast cell wall. I am very grateful to my mentors, David Kerridge, Widmar Tanner, and Phil Robbins, who gave me the privilege of working with them, and who taught me so much about this field. I thank the many colleagues in this field who generously shared preprints and unpublished results with me, and I very much appreciate Phil Robbins's thoughtful comments on this chapter. I thank my own students for their enthusiastic contributions to this field. Research in my laboratory was supported by National Institutes of Health grant GM-46220 and by America Cancer Society grant JFRA-459.

XI. Appendix: Genes involved in cell wall biogenesis

Gene or mutant[a]	Other names	Product/Biochemical defect	Remarks/Mutant phenotype[b]	References
Precursor Supply				
UGP1		UDPGlc pyrophosphorylase	Ugp1p depletion leads to lowered β-glucan in vivo	Daran et al. (1995)
ALG5		Dol-P-Glc synthase	nonconditional, SL with *wbp1*	Huffaker and Robbins (1983); te Heesen (1994)
dpg1		Dol-P-Glc synthesis defect	not allelic to *alg5*	Ballou et al. (1986)
CDC30		phosphoglucose isomerase	high K_m enzyme; ts; G_1 arrest	Dickinson (1991)
PGI1		phosphoglucose isomerase	low K_m enzyme	Dickinson (1991)
PMI40		phosphomannose isomerase	ts lytic; mannose remedial	Payton et al. (1991); Smith et al. (1992)
SEC53	*alg4*	phosphomannomutase	ts *sec*; blocked in *N*-, *O*-, GPI	Conzelmann et al. (1988); Kepes and Schekman (1988); Orlean et al. (1991)
DPM1		Dol-P-Man synthase	ts; affects *N*-, blocks *O*-, *GPI*	Orlean et al. (1988); Haselbeck (1989); Orlean (1990)
GFA1	*gcn1*	Fru-6-P:glutamine amido-transferase	GlcN auxotrophs; pheromone-induced	Whelan and Ballou (1975); Watzele and Tanner (1989)
AGM1		GlcNAc-P mutase	pheromone-induced	Hoffmann et al. (1994)
SEC59		dolichol kinase	ts *sec*; blocked in *N*-, *O*-, GPI	Bernstein et al. (1989); Heller et al. (1992)
LCB1		serine palmitoyl transferase	sphingosine auxotroph	Buede et al. (1991)
LCB2	*SCS1*	serine palmitoyl transferase	Lcb1p homolog	Nagiec et al. (1994); Zhao et al. (1994)
SLC1		fatty acyl transferase	mutant allele suppresses *lcb*	Nagiec et al. (1993)

Secretory Pathway: Mutants and Genes Affecting N-glycosylation; GPI Anchoring, O-mannosylation, and Pathway Function				
ALG7		Dol-PP: GlcNAc-1-P transferase	tunicamycin-sensitive growth-related expression	Barnes et al. (1984); Kukuruzinska and Lennon (1995)
ALG1		mannosyltransferase	ts; accumulates Dol-PP-$GlcNAc_2$	Huffaker and Robbins (1982); Couto et al. (1984); Albright and Robbins (1990)
ALG2		mannosyltransferase?	ts; accumulates Dol-PP-$GlcNAc_2Man_2$	Jackson et al. (1993)
ALG3		mannosyltransferase?	nonconditional; accumulates Dol-PP-$GlcNAc_2Man_5$	Huffaker and Robbins (1983); Verostek et al. (1991); Aebi et al. (1996)
ALG6		glucosyltransferase?	nonconditional; accumulates Dol-PP-$GlcNAc_2Man_9$	Runge et al. (1984); Reiss et al. (1996)
ALG8		glucosyltransferase?	nonconditional; accumulates Dol-PP-$GlcNAc_2Man_9Glc$	Runge and Robbins (1986); Stagljar et al. (1994)
ALG9		mannosyltransferase?	*alg9* SL with *wbp1*: accumulates Dol-PP-$GlcNAc_2Man_6$	Zufferey et al. (1995); Burda et al. (1996)
alg10		mannosyltransfer	SL with *wbp1*: accumulates Dol-PP-$GlcNAc_2Man_9Glc_2$	Zufferey et al. (1995)
WBP1		OTase subunit	SL with nonconditional *alg*	te Heesen et al. (1992)
SWP1		OTase subunit	homology with ribophorin II	te Heesen et al. (1993)
OST1	*NLT1*	OTase subunit	homology with ribophorin I	Pathak et al. (1995); Silberstein et al. (1995b)
OST2		OTase subunit	resembles defender against apoptotic death" protein	Silberstein et al. (1995a)
OST3		OTase subunit	nonessential; Δ*ost3* shows "biased underglycosylation"	Karaoglu et al. (1995)

(*Continued on following pages.*)

XI. Appendix: (Continued.)

Gene or mutant[a]	Other names	Product/Biochemical defect	Remarks/Mutant phenotype[b]	References
OST4		36-amino-acid protein OTase subunit?	Δ*ost4* grows slowly at 25°C, inviable at 37°C	Chi et al. (1996)
OST5		OTase subunit	SL Δ*alg5*; high-copy suppression of *ost1*	Reiss et al. (1997)
STT3		nonstoichiometric OTase subunit?	SL *wbp1*, Δ*alg3*, Δ*alg5*, Δ*ost5* Δ*alg8*; staurosporine-sensitive	Zufferey et al. (1995)
meg1			in vitro defect in oligo-saccharyl transfer	Roos et al. (1994)
gls1		glucosidase I?	defective in Glc trimming of Asn-$GlcNAc_2Man_9Glc_3$	Esmon et al. (1984)
MNS1		α-1,2 mannosidase		Camirand et al. (1991)
OCH1		outer-chain-initiating α-1,6 mannosyltransferase	Δ*och1* ts; Och1p defines early Golgi compartment	Nakayama et al. (1992); Nakanishi-Shindo et al. 1993; Chapman and Munro (1994); Gaynor et al. (1994)
MNN1		α-1,3 mannosyltransferase	medial and *trans*-Golgi; mannosylates N- and O-chains	Graham et al. (1994); Yip et al. (1994)
MNN2		no α-1,2 branches off α-1,6 outer-chain backbone	transferase or regulator	Devlin and Ballou (1990)
mnn3		general shortening of N- and O-linked chains		Ballou (1990)
mnn4		mannosylphosphate addition		Ballou (1990)
mnn5		second α-1,2 Man not added to α-1,2 side chain		Ballou (1990)
mnn6		mannosylphosphate addition		Ballou (1990)

mnn7,8	defective outer mannan chain extension	osmotically fragile; vanadate-resistant	Ballou (1990)
KRE2/MNT1	α-1,2 mannosyltransferase	elongation of O- and N-linked chains; medial Golgi	Hill et al. (1992); Haüsler et al. (1992)
KTR1	mannosyltransferase	Kre2p/Mnt1p homolog; *N*- only; Golgi	Hill et al. (1992); Lussier et al. (1996)
KTR2	mannosyltransferase	Kre2p/Mnt1p homolog; *N*- only; Golgi	Lussier et al. (1993, 1996)
KTR3	mannosyltransferase?	Kre2p/Mnt1p homolog; *N*- only	Mallet et al. (1994)
KTR4	mannosyltransferase?	Kre2p/Mnt1p homolog; *N*- only	
KTR5,6,7	mannosyltransferase?	Kre2p/Mnt1p homolog; *N*-only	Lussier et al. (1997)
YUR1	mannosyltransferase	Kre2p/Mnt1p homolog; *N*- only; Golgi	Hill et al. (1992); Lussier et al. (1996)
GDA1	GDPase	role in GDPMan transport into Golgi vesicles; general decrease in mannosylation	Abeijon et al. (1989, 1993)
MNN9	severe defect in outer mannan chain extension	null alleles viable but slow growing; vanadate-resistant; mannosyltransferase; transport or vesicle targeting?	Ballou et al. 1991; Yip et al. 1994
MNN10 *BED1*	defective in polarized secretion; mannosyltransferase?	homolog of *S. pombe* galacto-syltransferase	Dean and Poster (1996); Mondésert and Reed (1996)
VAN1	underglycosylates	Mnn9p homolog; vanadate-resistant	Kanik-Ennulat et al. (1995)
ANP1	underglycosylates	Mnn9p homolog; vanadate-resistant; Δ*anp1* SL Δ*van1*; ER-localized?	Chapman and Munro (1994)
VRG4 VAN2	defective outer mannan chain elongation	Golgi-localized; essential; defective in Golgi to ER retrieval	Ballou et al. (1991); Kanik-Ennulat et al. (1995); Poster and Dean (1996)

(*Continued on following pages.*)

XI. Appendix: (Continued.)

Gene or mutant[a]	Other names	Product/Biochemical defect	Remarks/Mutant phenotype[b]	References
ERD1		underglycosylates	ER-retention-defective	Hardwick et al. (1990)
PMR1		Ca^{++} ATPase	Golgi defect: impairs elongation of N- and O-linked chains	Antebi and Fink (1992)
GPI1		GlcNAc-PI synthesis	Δ*gpi1* ts; SL *gpi2* and *gpi3*	Leidich et al. (1994); Leidich and Orlean (1996)
GPI2		GlcNAc-PI synthesis	essential; SL *gpi1* and *gpi3*	Leidich et al. (1995)
GPI3	*SPT14*, *CWH6*	GlcNAc-PI synthesis	essential; possible glycosyl-transferase: CFW-sensitive	Leidich et al. (1995); Schönbächler et al. (1995); Vossen et al. (1995)
gpi4,5,6,9		mannosylation of GPI precursor	ts or ts on CFW; *gpi4,6,9* affect *N*-	Benghezal et al. (1995)
gpi7		phosphoethanolamine addition to GPI precursor	ts on CFW; yeast PIG-F homolog essential	Benghezal et al. (1995); Inoue et al. (1993); C. Taron and P. Orlean (unpubl.)
GPI8		GPI anchor attachment; aminopeptidase	accumulates complete GPI precursor; ts on CFW	Benghezal et al. (1995, 1996)
GAA1	*END2*	GPI anchor attachment	accumulates complete GPI SL with *gpi8*	Hamburger et al. (1995)
PMT1		Dol-P-Man: protein *O*-mannosyl-transferase	*O*- down to half in Δ*pmt1*	Strahl-Bolsinger et al. (1993)
PMT2		Dol-P-Man: protein *O*-mannosyl-transferase	Pmt1p homolog; Δ*pmt1*/Δ*pmt2* severe *O*- and growth defects	Lussier et al. (1995b)
PMT3,4,5,6,7		Dol-P-Man: protein *O*-mannosyl-transferases?	Pmt1p homologs; Δ*pmt1*/Δ*pmt2*/Δ*pmt3*/Δ*pmt4* osmotically remediable	Gentzsch et al. (1995a); Immervoll et al. (1995); Gentzsch and Tanner (1996)

Secretory Pathway: Mutants and Genes Affecting β-1,6 Glucan Synthesis			
KRE5	virtually no alkali-insoluble β-1,6 glucan	epistatic to all *kre* mutants; severe growth defect; ER protein; resembles *S. pombe* UDPGlc: glycoprotein glucosyltransferase	Meaden et al. (1990); Fernández et al. (1996)
CWH41	alkali-insoluble β-1,6 glucan reduced 50%	ER protein, SL Δ*kre6*, 75% reduction in β-1,6 glucan in Δ*cwh41* Δ*kre1*	Jiang et al. (1996)
KRE6	alkali-insoluble β-1,6 glucan reduced 50%	slow growing; SL Δ*pkc1*; *KRE6* suppresses Δ*pkc1*; Golgi membrane; phosphorylated	Roemer and Bussey (1991); Roemer et al. (1993, 1994)
SKN1	Kre6p homolog; normal β-1,6 glucan, but 10% of normal in Δ*skn1*/Δ*kre6*	suppresses Δ*kre6*; Δ*skn1*/Δ*kre6* dead or severe growth defect; minor role in β-1,6 glucan synthesis?	Roemer et al. (1993)
KRE11	alkali-insoluble β-1,6 glucan reduced by 50%	little growth defect; SL Δ*kre6*; cytoplasmic; UDPGlc binding?	Brown et al. (1993b); Roemer et al. (1993)
KRE9	alkali-insoluble β-1,6 glucan reduced by 80%	slow growth; SL Δ*kre6*, Δ*kre11*; Ser/Thr-rich secretory protein	Brown and Bussey (1993)
KNH1	Kre9p homolog; normal β-1,6 glucan	Δ*knh1* SL with Δ*kre9*	Jiang et al. (1996)
SKN7	transcriptional activator	suppressor of Δ*kre9*, Δ*pkc1*; two-component receiver domain	Brown et al. (1993a, 1994)

(Continued on following pages.)

XI. Appendix: (Continued.)

Gene or mutant[a]	Other names	Product/Biochemical defect	Remarks/Mutant phenotype[b]	References
KRE1		alkali-insoluble β-1,6 glucan reduced by 60%	cell surface, GPI-anchored? Δ*kre1*/Δ*kre11* severe growth defect and 20% β-1,6 glucan	Boone et al. (1990); Brown et al. (1993b)
Plasma Membrane: Genes Involved in Chitin and β-1,3 Glucan Synthesis				
CHS1		chitin synthase I	Δ*chs1* makes normal amounts of chitin; repair at low pH? pheromone-responsive expression, but no chitin made	Kang et al. (1984); Bulawa et al. (1986); Orlean (1987); Appeltauer and Achstetter (1989); Cabib et al. (1989)
CHS2		chitin synthase II	Chs1p homolog; makes 5–10% of chitin in cell; responsible for chitin in primary septum?	Silverman et al. (1988); Silverman (1989); Bulawa and Osmond (1990)
CHS3	*CAL1*, *CSD2*, *DIT101*, *KTI1*	chitin synthase III	Chs1p-related; makes 90–95% of chitin: chitin ring, lateral wall chitin, chitin induced by α-factor, chitosan; CFW-resistant; Δ*chs1*/Δ*chs2*/Δ*chs3* inviable	Orlean (1987); Bulawa and Osmond (1990); Shaw et al. (1991); Valdivieso et al. (1991); Bulawa (1992); Pammer et al. (1992); Takita and Castilho-Valavicius (1993)
CHS5		chitin reduced by 80%	required for targeting of Chs3p and for chitin synthase III activity in vitro; CFW-resistant; resembles neurofilament NF-H protein	Roncero et al. (1988); Valdivieso et al. (1991); Santos and Durán (1992); Santos and Snyder (1997)
CSD3		chitin reduced by 90%	required for localization of Chs3p? CFW-resistant	Bulawa (1992, 1993)

CHS4/ CSD4	*CAL2*	chitin reduced by 90%	limiting subunit of complex containing Chs3p; CFW-resistant	Bulawa (1992, 1993)
FKS1	*ETG1, CWH53, GSC1, PBR1*	β-1,3 glucan synthase subunit	hypersensitive to FK506, CFW, resistant to echinocandin; activated by Rho1p	Douglas et al. (1994b); Castro et al. (1995); Inoue et al. (1995); Mazur et al. (1995); Ram et al. (1995)
FKS2	*GSC2*	β-1,3 glucan synthase subunit	Fks1p homolog; calcineurin-dependent expression; SL Δ*fks1*	Inoue et al. (1995); Mazur et al. (1995)
RHO1		GTPase	activating subunit of Fks1p β-1,3 glucan synthase and Pkc1p; ts *rho1* osmotically fragile and CFW-hyper-sensitive	Qadota et al. (1994, 1996); Yamochi et al. (1994;) Drgonová et al. (1996); Kamada et al. (1996)
CDC43		geranylgeranyltransferase subunit	may modify Rho1p; low in vitro β-1,3 glucan synthase in *cdc43*	Diaz et al. (1993)
KNR4	*SMI1*	low in vivo and in vitro β-1,3 and β-1,6 glucan synthesis	*knr4* restant to HM-k9 toxin Δ*knr4* has elevated chitin; nuclear, regulatory protein	Fishel et al. (1993); Hong et al. (1994b)
GNS1		in vitro β-1,3 glucan synthase defect	papulacandin-resistant; drug-binding subunit of β-1,3 glucan synthase complex? Δ*fks1*/Δ*gns1* has severe growth defect	El-Sherbeini and Clemas (1995b)

(*Continued on following pages.*)

XI. Appendix: (Continued.)

Gene or mutant[a]	Other names	Product/Biochemical defect	Remarks/Mutant phenotype[b]	References
HKR1		low in vivo and in vitro β-1,3 glucan synthesis	essential gene; overexpression confers HM-1 killer resistance and raises in vivo β-1,3 glucan; Hkr1p is Ser/Thr-rich membrane protein with cytoplasmic EF-hand	Kasahara et al. (1994b); Yabe et al. (1996)
GAS1/CWH52 GGP1		in vivo β-1,3- and β-1,6 glucan reduced by 75%	GPI-anchored, *O*- and *N*-glycosylated surface protein; CFW-hypersensitive	Nuoffer et al. (1991); Vai et al. (1991); Ram et al. (1995)
Cell Wall Hydrolases				
EXG1/BGL1		exo-β-1,3 glucanase	Δ*exg1* K1-killer toxin-sensitive; *EXG1* overexpression gives K1 resistance; Exg1p secreted, *N*-glycosylated	Nebrada et al. (1986); Kuranda and Robbins (1987); Hernández et al. (1992)
EXG2		exo-β-1,3 glucanase	Exg1p homolog; GPI anchor attachment site	Larriba et al. (1995)
SSG1	*SPR1*	exo-β-1,3 glucanase	sporulation-specific Exg1p homolog	San Segundo et al. (1993); Muthukumar et al. (1993)
BGL2		endo-β-1,3 glucanase	Δ*bgl2* no obvious phenotype; high overexpression lethal; expression regulated via Pkc1p; Exg1p catalyzes glycosyl transfer	Klebl and Tanner (1989); Shimizu et al. (1994); Goldman et al. (1995)

CTS1	endochitinase	cell-separation defect in Δ*cts1*; Ser/Thr-rich secreted protein; chitin-binding domain; Ace2p regulated-expression	Kuranda and Robbins (1987, 1991); Dohrman et al. (1992)
Mating and Sporulation			
AGα1	α-agglutinin	Agα1p glycosylated, GPI-anchored, then cross-linked to wall	Hauser and Tanner (1989); Lipke et al. (1989); Lu et al. (1994)
AGA1	**a**-agglutinin anchorage subunit	*O*-mannosylated; GPI attachment sequence	Roy et al. (1991)
AGA2	**a**-agglutinin binding subunit	Agα2p small, *O*-mannosylated; disulfide-linked to Aga1p	Orlean et al. (1986); Watzele et al. (1988); Cappellaro et al. (1991, 1994)
FUS1	cell fusion	*O*-mannosylated; transmembrane	McCaffrey et al. (1987); Truehart et al. (1987); Truehart and Fink (1989)
SHC1	Csd4p homolog	sporulation-specific expression	Bulawa (1993)
CDA1	chitin deacetylase	sporulation-specific expression; forms chitosan	Christodoulidou et al. (1996); Mishra et al. (1997)
CDA2	chitin deacetylase	sporulation-specific expression; forms chitosan	Christodoulidou et al. (1996); Mishra et al. (1997)
DIT1	dityrosine layer formation	sporulation-specific expression; converts L-tyrosine to precursor	Briza et al. (1990a, 1994)
DIT2	cytochrome P450 dityrosine layer formation	sporulation-specific expression; forms LL-dityrosine	Briza et al. (1990a, 1994)

(*Continued on following pages.*)

XI. Appendix: (Continued.)

Gene or mutant[a]	Other names	Product/Biochemical defect	Remarks/Mutant phenotype[b]	References
CPR1	NADPH	cytochrome P450 reductase dityrosine layer formation	nonsporulation-specific acts with Dit2p	Briza et al. (1994)
Regulation and Low- and High-osmolarity-sensing Pathways				
SSD-v1	*SRK1*	protein phosphatase	suppresses various mutations affecting growth control, also ts of *chs3*, *csd3*, *csd4*	Bulawa (1993); Kim et al. (1994)
PKC1		protein kinase C	Δ*pkc1* ts lytic, low β-glucan, SL Δ*kre6*, Δ*kre11*, Δ*kre2*/*mnt1*; Pkc1p activated by Rho1p-GTP	Levin and Bartlett Heubusch (1992); Paravicini et al. (1992); Roemer et al. (1994); Shimizu et al. (1994); Nonaka et al. 1995; Kamada et al. (1996)
MKK1 + *MKK2*		MAP kinase kinase homologs	Δ*mkk1*/Δ*mkk2* ts lytic and SL Δ*kre6*; act downstream of Pkc1p	Irie et al. (1993); Roemer et al. (1994)
MPK1	*SLT2*	MAP kinase	ts lysis; SL Δ*kre6*; acts downstream from Mkk1,2p	Torres et al. (1991); Lee et al. (1993); Roemer et al. (1994)
PBS2		MAP kinase kinase	acts in Pbs2p-Hog1p high-osmolarity-sensing pathway regulates expression of *EXG1*	Brewster et al. (1993); Jiang et al. (1995)
PTC1	*CWH47*	Ser/Thr protein phosphatase	negative regulator of Pbs2p	Jiang et al. (1995)

[a]Genes and mutations are listed in the approximate order in which they are discussed in the text and in which they may function in a biosynthetic pathway. Gene designation is given, unless locus is defined only by a mutation.

[b]Abbreviations used are ts, temperature-sensitive growth; SL, synthetically lethal with; *N*-, *N*-glycosylation; *O*-, O-mannosylation; GPI, GPI anchoring; CFW, Calcofluor White.

REFERENCES

Abeijon, C., P. Orlean, P.W. Robbins, and C.B. Hirschberg. 1989. Topography of glycosylation in yeast: Characterization of GDP-mannose transport and lumenal guanosine diphosphatase activities in Golgi-like vesicles. *Proc. Natl. Acad. Sci.* **86:** 6935.

Abeijon, C., K. Yanagisawa, E.C. Mandon, A. Häusler, K. Moremen, C.B. Hirschberg, and P.W. Robbins. 1993. Guanosine diphosphatase is required for protein and sphingolipid glycosylation in the Golgi lumen of *Saccharomyces cerevisiae. J. Cell Biol.* **122:** 307.

Adair, W.L., Jr. and N. Cafmeyer. 1987. Characterization of the *Saccharomyces cerevisiae cis*-prenyltransferase required for dolichyl phosphate biosynthesis. *Arch. Biochem. Biophys.* **259:** 589.

Aebi, M., J. Gassenhuber, H. Domdey, and S. te Heesen. 1996. Cloning and characterization of the *ALG3* gene of *Saccharomyces cerevisiae. Glycobiology* **6**: 439.

Albright, C.F. and P.W. Robbins. 1990. The sequence and transcript heterogeneity of the yeast gene *ALG1*, an essential mannosyltransferase involved in *N*-glycosylation. *J. Biol. Chem.* **265:** 7042.

Alvarado, E., T. Nukada, T. Ogawa, and C.E. Ballou. 1991. Conformation of the glucotriose unit in the lipid-linked oligosaccharide precursor for protein glycosylation. *Biochemistry* **30:** 881.

Antebi, A. and G.R. Fink. 1992. The yeast Ca^{2+}-ATPase homologue, PMR1, is required for normal Golgi function and localizes in a novel Golgi-like distribution. *Mol. Biol. Cell* **3:** 633.

Appeltauer, U. and T. Achstetter. 1989. Hormone-induced expression of the *CHS1* gene from *Saccharomyces cerevisiae. Eur. J. Biochem.* **181:** 243.

Arnold, E. and W. Tanner. 1982. An obligatory role of protein glycosylation in the life cycle of yeast cells. *FEBS Lett.* **148:** 49.

Arnold, W.N. 1981. Physical aspects of the yeast cell envelope. In *Yeast cell envelopes, biochemistry, biophysics, and ultrastructure* (ed. W.N. Arnold), vol. 1, p. 25. CRC Press, Boca Raton, Florida.

Au-Young, J. and P.W. Robbins. 1990. Isolation of a chitin synthase gene (*CHS1*) from *Candida albicans* by expression in *Saccharomyces cerevisiae. Mol. Microbiol.* **4:** 197.

Ballou, C.E. 1982. Yeast cell wall and cell surface. In *The molecular biology of the yeast* Saccharomyces: *Metabolism and gene expression* (ed. J.N. Strathern et al.), p. 335. Cold Spring Harbor Laboratory, Cold Spring Harbor Laboratory, New York .

———. 1990. Isolation, characterization, and properties of *Saccharomyces cerevisiae mnn* mutants with nonconditional protein glycosylation defects. *Methods Enzymol.* **185:** 440.

Ballou, C.E., S.K. Maitra, J.W. Walker, and W.L. Whelan. 1977. Developmental defects associated with glucosamine auxotrophy in *Saccharomyces cerevisiae. Proc. Natl. Acad. Sci.* **74:** 4351.

Ballou, L., R.A. Hitzeman, M.S. Lewis, and C.E. Ballou. 1991. Vanadate-resistant yeast mutants are defective in protein glycosylation. *Proc. Natl. Acad. Sci.* **88:** 3209.

Ballou, L., P. Gopal, B. Krummel, M. Tammi, and C.E. Ballou. 1986. A mutation that prevents glucosylation of the lipid-linked oligosaccharide precursor leads to underglycosylation of secreted yeast invertase. *Proc. Natl. Acad. Sci.* **83:** 3081.

Bandlow, W., S. Wied, and G. Müller. 1995. Glucose induces amphiphilic to hydrophilic conversion of a subset of glycosyl-phosphatidylinositol-anchored ectoproteins in yeast.

Arch. Biochem. Biophys. **324:** 300.

Barnes, G., W.J. Hansen, C.L. Holcomb, and J. Rine. 1984 Asparagine-linked glycosylation in *Saccharomyces cerevisiae*: Genetic analysis of an early step. *Mol. Cell. Biol.* **4:** 2381.

Bartnicki-Garcia, S., C.E. Bracker, E Reyes, and J. Ruiz-Herrera. 1978. Isolation of chitosomes from taxonomically diverse fungi and synthesis of chitin microfibrils *in vitro*. *Exp. Mycol.* **2:** 173.

Basco, R.D., M.D. Muñoz, L.M. Hernández, C. Vazquez de Aldana, and G. Larriba. 1993. Reduced efficiency in the glycosylation of the first sequon of *Saccharomyces cerevisiae* exoglucanase leads to the synthesis and secretion of a new glycoform of the molecule. *Yeast* **9:** 221.

Basco, R.D., L.M. Hernández, M.D. Muñoz, I. Olivero, E. Andaluz, F. Del Rey, and G. Larriba. 1994. Selective elongation of the oligosaccharide attached to the second potential glycosylation site of yeast exoglucanase: Effects on the activity and properties of the enzyme. *Biochem. J.* **304:** 917.

Bause, E., R. Erkens, J. Schweden, and L. Jaenicke. 1986. Purification and characterization of trimming glucosidase I from *Saccharomyces cerevisiae. FEBS Lett.* **206:** 208.

Baymiller, J. and J.E. McCollough. 1993. Identification of a *Saccharomyces cerevisiae* mutation that allows cells to grow without chitin synthase 1 or 2. *Curr. Genet.* **23:** 102.

Beaudouin, R., S.T. Lim, J.-A. Steide, M. Powell, J. Mc.Koy, A.J. Pramanik, E. Johnson, C.W. Moore, and P.N. Lipke. 1993. Bleomycin affects cell wall anchorage of mannoproteins in *Saccharomyces cerevisiae. Antimicrob. Agents Chemother.* **37:** 1264.

Becker, G.W. and R.L. Lester. 1980. Biosynthesis of phosphoinositol-containing sphingolipids from phosphatidylinositol by a membrane preparation from *Saccharomyces cerevisiae. J. Bacteriol.* **142:** 747.

Bender, A. and J.R. Pringle. 1992. A Ser/Thr-rich multicopy suppressor of a *cdc24* bud emergence defect. *Yeast* **8:** 315.

Benghezal, M., P.N. Lipke, and A. Conzelmann. 1995. Identification of six complementation classes involved in the biosynthesis of glycosylphosphatidylinositol anchors in *Saccharomyces cerevisiae. J. Cell Biol.* **130:** 1333.

Benghezal, M., A. Benachour, S. Rusconi, M. Aebi, and A. Conzelmann. 1996. Yeast Gpi8p is essential for GPI anchor attachment onto proteins. *EMBO J.* **15**: 6575.

Benton, B.K., S.D. Plump, J. Roos, W.J. Lennarz, and F.R. Cross. 1996. Overexpression of *S. cerevisiae* G_1 cyclins restores the viability of *alg1 N*-glycosylation mutants. *Curr. Genet.* **29:** 106.

Berninsone, P., J.J. Miret, and C.B. Hirschberg. 1994. The Golgi guanosine diphosphatase is required for transport of GDP-mannose into the lumen of *Saccharomyces cerevisiae* Golgi vesicles. *J. Biol. Chem.* **269:** 207.

Bernstein, M., F. Kepes, and R. Schekman. 1989. *SEC59* encodes a membrane protein required for core glycosylation in *Saccharomyces cerevisiae. Mol. Cell. Biol.* **9:** 1191.

Bessoule, J.-J., R. Lessire, M. Rigoulet, B. Guerin, and C. Cassagne. 1988. Localization of the synthesis of the very-long-chain fatty acid in mitochondria from *Saccharomyces cerevisiae. Eur. J. Biochem.* **177:** 207.

Blagoeva, J., G. Stoev, and P. Venkov. 1991. Glucan structure in a fragile mutant of *Saccharomyces cerevisiae. Yeast* **7:** 455.

Boone, C., S.S. Sommer, A. Hensel, and H. Bussey. 1990. Yeast *KRE* genes provide evidence for a pathway of cell wall β-glucan assembly. *J. Cell Biol.* **110:** 1833.

Borgia, P.T. 1992. Roles of the *orlA*, *tsE*, and *bimG* genes of *Aspergillus nidulans* in chitin synthesis. *J. Bacteriol.* **174:** 384.

Bowen, A.R., J.L. Chen-Wu, M. Momany, R. Young, P.J. Szaniszlo, and P.W. Robbins. 1992. Classification of fungal chitin synthases. *Proc. Natl. Acad. Sci.* **89:** 519.

Bowers, B., G. Levin, and E. Cabib. 1974. Effect of polyoxin D on chitin synthesis and septum formation in *Saccharomyces cerevisiae*. *J. Bacteriol.* **119:** 564.

Brewster, J.L., T. de Valoir, N.D. Dwyer, E. Winter, and M.C. Gustin. 1993. An osmosensing signal transduction pathway in yeast. *Science* **259:** 1760.

Briza, P., M. Eckerstorfer, and M. Breitenbach. 1994. The sporulation-specific enzymes encoded by the *DIT1* and *DIT2* genes catalyze a two-step reaction leading to a soluble LL-dityrosine-containing precursor of the yeast spore wall. *Proc. Natl. Acad. Sci.* **91:** 4524.

Briza, P., M. Breitenbach, A. Ellinger, and J. Segall. 1990a. Isolation of two developmentally regulated genes involved in spore wall maturation in *Saccharomyces cerevisiae*. *Genes Dev.* **4:** 1775.

Briza, P., A. Ellinger, G. Winkler, and M. Breitenbach. 1988. Chemical composition of the yeast ascospore wall. The second outer layer consists of chitosan. *J. Biol. Chem.* **263:** 11569.

———. 1990b. Characterization of a D,L,-dityrosine-containing macromolecule from yeast ascospore walls. *J. Biol. Chem.* **265:** 15118.

Briza, P., G. Winkler, H. Kalchhauser, and M. Breitenbach. 1986. Dityrosine is a prominent component of the yeast ascospore wall. A proof of its structure. *J. Biol. Chem.* **261:** 4288.

Brown, J.L. and H. Bussey. 1993. The yeast *KRE9* gene encodes an O glycoprotein involved in cell surface β-glucan assembly. *Mol. Cell. Biol.* **13:** 6346.

Brown, J.L., H. Bussey, and R.C. Stewart. 1994. Yeast Skn7p functions in a eukaryotic two-component regulatory pathway. *EMBO J.* **13:** 5186.

Brown, J.L., S. North, and H. Bussey. 1993a. *SKN7*, a yeast multicopy suppressor of a mutation affecting cell wall β-glucan assembly, encodes a product with domains homologous to prokaryotic two-component regulators and to heat shock transcription factors. *J. Bacteriol.* **175:** 6908.

Brown, J.L., Z. Kossaczka, B. Jiang, and H. Bussey. 1993b. A mutational analysis of killer toxin resistance in *Saccharomyces cerevisiae* identifies new genes involved in cell wall (1→6)-β-glucan synthesis. *Genetics* **133:** 837.

Brzobohaty, B. and L. Kovác. 1986. Factors enhancing genetic transformation of intact yeast cells modify cell wall porosity. *J. Gen. Microbiol.* **132:** 3089.

Buede, R., C. Rinker-Schaffer, W.J. Pinto, R.L. Lester, and R.C. Dickson. 1991. Cloning and characterization of *LCB1*, a *Saccharomyces* gene required for biosynthesis of the long-chain base component of sphingolipids. *J. Bacteriol.* **173:** 4325.

Bulawa, C.E. 1992. *CSD2*, *CSD3*, and *CSD4*, genes required for chitin synthesis in *Saccharomyces cerevisiae:* The *CSD2* gene product is related to chitin synthases and to developmentally regulated proteins in *Rhizobium* species and *Xenopus laevis*. *Mol. Cell. Biol.* **12:** 1764.

———. 1993. Genetics and molecular biology of chitin synthesis in fungi. *Annu. Rev. Microbiol.* **47:** 505.

Bulawa, C.E. and B.C. Osmond. 1990. Chitin synthase I and chitin synthase II are not required for chitin synthesis *in vivo* in *Saccharomyces cerevisiae*. *Proc. Natl. Acad. Sci.* **87:** 7424.

Bulawa, C.E., M. Slater, E. Cabib, J. Au-Young, A. Sburlati, W.L. Adair, Jr., and P.W.

Robbins. 1986. The *S. cerevisiae* structural gene for chitin synthase is not required for chitin synthesis *in vivo*. *Cell* **46:** 213.

Burda, P., S. te Heesen, A. Brachat, A. Wach, A. Düsterhöft, and M. Aebi. 1996. Stepwise assembly of the lipid-linked oligosaccharide in the endoplasmic reticulum of *Saccharomyces cerevisiae*: Identification of the *ALG9* gene encoding a putative mannosyl transferase. *Proc. Natl. Acad. Sci.* **93:** 7160.

Bussey, H. 1991. K1 killer toxin, a pore-forming protein from yeast. *Mol. Microbiol.* **5:** 2339.

———. 1996. Cell shape determination: A pivotal role for Rho. *Science* **272:** 224.

Butler, A.R., V.J. Martin, J.H. White, and M.J.R. Stark. 1992. Genetic and biochemical analysis of the *Kluyveromyces lactis* toxin mode of action. *Yeast* **8:** S319.

Byers, B. 1981. Cytology of the yeast life cycle. In *The molecular biology of the yeast* Saccharomyces: *Life cycle and inheritance* (ed. J.N. Strathern et al.), p. 59. Cold Spring Harbor Laboratory, Cold Spring Harbor, New York.

Byrd, J.C., A.L. Tarentino, F. Maley, P.H. Atkinson, and R.B. Trimble. 1982. Glycoprotein synthesis in yeast. Identification of $Man_8GlcNAc_2$ as an essential intermediate in oligosaccharide processing. *J. Biol. Chem.* **257:** 14657.

Cabib, E. 1987. The synthesis and degradation of chitin. *Adv. Enzymol. Relat. Areas Mol. Biol.* **59:** 59.

Cabib, E. and A. Durán. 1975. Simple and sensitive procedure for screening yeast mutants that lyse at nonpermissive temperatures. *J. Bacteriol.* **124:** 1604.

Cabib, E., B. Bowers, and R.L. Roberts. 1983. Vectorial synthesis of a polysaccharide by isolated plasma membranes. *Proc. Natl. Acad. Sci.* **80:** 3318.

Cabib, E., R. Roberts, and B. Bowers. 1982. Synthesis of the yeast cell wall and its regulation. *Annu. Rev. Biochem.* **51:** 763.

Cabib, E., S.J. Silverman, and J.A. Shaw. 1992. Chitinase and chitin synthase 1: Counterbalancing activities in cell separation of *Saccharomyces cerevisiae*. *J. Gen. Microbiol.* **138:** 97.

Cabib, E., A. Sburlati, B. Bowers, and S.J. Silverman. 1989. Chitin synthase 1, an auxiliary enzyme for chitin synthesis in *Saccharomyces cerevisiae*. *J. Cell Biol.* **108:** 1665.

Calleja, G.B. 1987. Cell aggregation. In *The yeasts*, 2nd edition (ed. A.H. Rose and J.S. Harrison), vol. 2, p. 165. Academic Press, New York.

Camirand, A., A. Heysen, B. Grondin, and A. Herscovics. 1991. Glycoprotein biosynthesis in *Saccharomyces cerevisiae*. Isolation and characterization of the gene encoding a specific processing a-mannosidase. *J. Biol. Chem.* **266:** 15120.

Caplan, S., R. Green, J. Rocco, and J. Kurjan. 1991. Glycosylation and structure of the yeast *MFα1* α-factor precursor is important for efficient transport through the secretory pathway. *J. Bacteriol.* **173:** 627.

Cappellaro, C., C. Baldermann, R. Rachel, and W. Tanner. 1994. Mating type-specific cell-cell recognition of *Saccharomyces cerevisiae:* Cell wall attachment and active sites of a- and α-agglutinin. *EMBO J.* **13:** 4737.

Cappellaro, C., K. Hauser, V. Mrsa, M. Watzele, G. Watzele, C. Gruber, and W. Tanner. 1991. *Saccharomyces cerevisiae* **a**- and α-agglutinin: Characterization of their molecular interaction. *EMBO J.* **10:** 4081.

Castro, C., J.C. Ribas, M.H. Valdivieso, R. Varona, F. del Rey, and A. Duran. 1995. Papulacandin B resistance in budding and fission yeasts: Isolation and characterization of a gene involved in (1,3)β-D-glucan synthesis in *Saccharomyces cerevisiae*. *J. Bacteriol.* **177:** 5732.

Cenamor, R., M. Molina, J. Galdona, M. Sánchez, and C. Nombela. 1987. Production and secretion of *Saccharomyces cerevisiae* β-glucanases: Differences between protoplast and periplasmic enzymes. *J. Gen. Microbiol.* **133:** 619.

Chapman, R.E. and S. Munro. 1994. The functioning of the yeast Golgi apparatus requires an ER protein encoded by *ANP1*, a member of a new family of genes affecting the secretory pathway. *EMBO J.* **13:** 4896.

Chen-Wu, J.L., J. Zwicker, A.R. Bowen, and P.W. Robbins. 1992. Expression of chitin synthase genes during yeast and hyphal growth phases of *Candida albicans. Mol. Microbiol.* **6:** 497.

Chi, J.H., J. Roos, and N. Dean. 1996. The *OST4* gene of *Saccharomyces cerevisiae* encodes an unusually small protein required for normal levels of oligosaccharyltransferase activity. *J. Biol. Chem.* **271:** 3132.

Choi, W.-J., A. Sburlati, and E. Cabib. 1994a. Chitin synthase 3 from yeast has zymogenic properties that depend on both the *CAL1* and the *CAL3* genes. *Proc. Natl. Acad. Sci.* **91:** 4727.

Choi, W.-J., B. Santos, A. Durán, and E. Cabib. 1994b. Are yeast chitin synthases regulated at the transcriptional or the posttranslational level? *Mol. Cell. Biol.* **14:** 7685.

Christodoulidou, A., V. Bouriotis, and G. Thireos. 1996. Two sporulation-specific chitin deacetylase-encoding genes are required for the ascopore wall rigidity of *Saccharomyces cerevisiae. J. Biol. Chem.* **271:** 31420.

Chu, F.K., W. Watorek, and F. Maley. 1983. Factors affecting the oligomeric structure of yeast external invertase. *Arch. Biochem. Biophys.* **223:** 543.

Chuang, J.S. and R.W. Schekman. 1996. Differential trafficking and timed localization of two chitin synthase proteins, Chs2p and Chs3p. *J. Cell Biol.* **135**: 597.

Cid, V.J., A. Durán, F. del Rey, M.P. Snyder, C. Nombela, and M. Sanchez. 1995. Molecular basis of cell integrity and morphogenesis in *Saccharomyces cerevisiae. Microbiol. Rev.* **59:** 345.

Coen, M.L., C.G. Lerner, J.O. Capobianco, and R.C. Goldman. 1994. Synthesis of yeast cell wall glucan and evidence for glucan metabolism in a *Saccharomyces cerevisiae* whole cell system. *Microbiology* **140:** 2229.

Conzelmann, A., C. Fankhauser, and C. Desponds. 1990. Myoinositol gets incorporated into numerous membrane glycoproteins of *Saccharomyces cerevisiae*; incorporation is dependent on phosphomannomutase (*SEC53*). *EMBO J.* **9:** 653.

Conzelmann, A., A. Puoti, R.L. Lester, and C. Desponds. 1992. Two different types of lipid moieties are present in glycophosphoinositol-anchored membrane proteins of *Saccharomyces cerevisiae. EMBO. J.* **11:** 457.

Conzelmann, A., H. Riezman, C. Desponds, and C. Bron. 1988. A major 125-kD membrane glycoprotein of *Saccharomyces cerevisiae* is attached to the lipid bilayer through an inositol-containing phospholipid. *EMBO J.* **7:** 2233.

Correa, J.U., N. Elango, I. Polacheck, and E. Cabib. 1982. Endochitinase, a mannan-associated enzyme from *Saccharomyces cerevisiae. J. Biol. Chem.* **257:** 1392.

Costello, L.C. and P. Orlean. 1992. Inositol acylation of a potential glycosyl phosphoinositol anchor precursor from yeast requires acyl coenzyme A. *J. Biol. Chem.* **267:** 8599.

Couto, J.R., T.C. Huffaker, and P.W. Robbins. 1984. Cloning and expression in *Escherichia coli* of a yeast mannosyltransferase from the asparagine-linked glycosylation pathway. *J. Biol. Chem.* **259:** 378.

Daran, J.M., N. Dallies, D. Thines-Sempoux, V. Paquet, and J. François. 1995. Genetic and biochemical characterization of the *UGP1* gene encoding the UDP-glucose pyrophosphorylase from *Saccharomyces cerevisiae*. *Eur. J. Biochem.* **233:** 520.

Davis, L.L. and S. Bartnicki-Garcia. 1984. Chitosan synthesis by the tandem action of chitin synthetase and chitin deacetylase from *Mucor rouxii*. *Biochemistry* **23:** 1065.

Dean, N. 1995. Yeast glycosylation mutants are sensitive to aminoglycosides. *Proc. Natl. Acad. Sci.* **92:** 1287.

Dean, N. and J.B. Poster. 1996. Molecular and phenotypic analysis of the *S. cerevisiae MNN10* gene identifies a family of related glycosyltransferases. *Glycobiology* **6:** 73.

Debono, M. and R.S. Gordee. 1994. Antibiotics that inhibit fungal cell wall development. *Annu. Rev. Microbiol.* **48:** 471.

De Nobel, H. and P.N. Lipke. 1994. Is there a role for GPIs in yeast cell-wall assembly? *Trends Cell Biol.* **4:** 42.

De Nobel, J.G. and J.A. Barnett. 1991. Passage of molecules through yeast cell walls: A brief essay-review. *Yeast* **7:** 313.

De Nobel, J.G., F.M. Klis, J. Priem, T. Munnik, and H. Van Den Ende. 1990. The glucanase-soluble mannoproteins limit cell wall porosity in *Saccharomyces cerevisiae*. *Yeast* **6:** 491.

De Nobel, J.G., F.M. Klis, A. Ram, H. Van Unen, J. Priem, T. Munnik, and H. Van Den Ende. 1991. Cyclic variations in the permeability of the cell wall of *Saccharomyces cerevisiae*. *Yeast* **7:** 589.

Devlin, C. and C.E. Ballou. 1990. Identification and characterization of a gene and protein required for glycosylation in the yeast Golgi. *Mol. Microbiol.* **4:** 1993.

Diaz, M., Y. Sanchez, T. Bennett, C.R. Sun, C. Godoy, F. Tamanoi, A. Durán, and P. Perez. 1993. The *Schizosaccharomyces pombe cwg2*$^{+}$ gene codes for the β-subunit of a geranylgeranyltransferase type I required for β-glucan synthesis. *EMBO J.* **12:** 5245.

Dickinson, J.R. 1991. Biochemical and genetic studies on the function of, and relationship between, the *PGI1*- and *CDC30*-encoded phosphoglucose isomerases in *Saccharomyces cerevisiae*. *J. Gen. Microbiol.* **137:** 765.

Dickson, R.C., G.B. Wells, A. Schmidt, and R.L. Lester. 1990. Isolation of mutant *Saccharomyces cerevisiae* strains that survive without sphingolipids. *Mol. Cell. Biol.* **10:** 2176.

Dijkgraaf, G.J.P., J.L. Brown, and H. Bussey. 1996. The *KNH1* gene of *Saccharomyces cerevisiae* is a functional homolog of *KRE9*. *Yeast* **12:** 683.

Doering, T.L. and R. Schekman. 1996. GPI anchor attachment is required for Gas1p transport from the endoplasmic reticulum in COP II vesicles. *EMBO J.* **15:** 182.

Dohrmann, P.R., G. Butler, K. Tamai, S. Dorland, J.R. Greene, D.J. Thiele, and D.J. Stillman. 1992. Parallel pathways of gene regulation: Homologous regulators *SWI5* and *ACE2* differentially control transcription of *HO* and chitinase. *Genes Dev.* **6:** 93.

Douglas, C.M., J.A. Marrinan, W. Li, and M.B. Kurtz. 1994a. A *Saccharomyces cerevisiae* mutant with echinocandin-resistant 1,3-β-D-glucan synthase. *J. Bacteriol.* **176:** 5686.

Douglas, C.M, F. Foor, J.A. Marrinan, N. Morin, J.B. Nielsen, A.M. Dahl, P. Mazur, W. Baginsky, W. Li, M. El-Sherbeini, J.A. Clemas, S.M. Mandala, B.R. Frommer, and M.B. Kurtz. 1994b. The *Saccharomyces cerevisiae FKS1* (*ETG1*) gene encodes an integral membrane protein which is a subunit of 1,3-β-D-glucan synthase. *Proc. Natl. Acad. Sci.* **91:** 12907.

Drgonová, J., T. Drgon, K. Tanaka, R. Kollár, G.-C. Chen, R.A. Ford, C.S.M. Chan, Y. Takai, and E. Cabib. 1996. Rho1p, a yeast protein at the interface between cell polarization and morphogenesis. *Science* **272:** 277.

Elango, N., J.U. Correa, and E. Cabib. 1982. Secretory character of yeast chitinase. *J. Biol. Chem.* **257:** 1398.

El-Sherbeini, M. and J.A. Clemas. 1995a. Nikkomycin Z supersensitivity of an echinocandin-resistant mutant of *Saccharomyces cerevisiae. Antimicrob. Agents Chemother.* **39:** 200.

———. 1995b. Cloning and characterization of *GNS1*: A *Saccharomyces cerevisiae* gene involved in synthesis of 1,3-β-glucan in vitro. *J. Bacteriol.* **177:** 3227.

Elorza, M.V., H. Rico, and R. Sentandreu. 1983. Calcofluor white alters the assembly of chitin fibrils in *Saccharomyces cerevisiae. J. Gen. Microbiol.* **129:** 1577.

Eng, W.-K., L. Faucette, M.M. McLaughlin, R. Cafferkey, Y. Koltin, R.A. Morris, P.R. Young, R.K. Johnson, and G.P. Livi. 1994. The yeast *FKS1* gene encodes a novel membrane protein, mutations in which confer FK506 and cyclosporin A hypersensitivity and calcineurin-dependent growth. *Gene* **151**: 61.

Esmon, B., P.C. Esmon, and R. Schekman. 1984. Early steps in processing of yeast glycoproteins. *J. Biol. Chem.* **259:** 10322.

Esmon, P.C., B.E. Esmon, I.E. Schauer, A. Taylor, and R. Schekman. 1987. Structure, assembly, and secretion of octameric invertase. *J. Biol. Chem.* **262:** 4387.

Etchebehere, L.C. and J.C. Da Costa Maia. 1989. Phosphorylation-dependent regulation of amidotransferase during the development of *Blastocladiella emersonii. Arch. Biochem. Biophys.* **272:** 301.

Fankhauser, C. and A. Conzelmann. 1991. Purification, biosynthesis, and cellular localization of a major 125-kDa glycophosphatidylinositol-anchored membrane glycoprotein of *Saccharomyces cerevisiae. Eur. J. Biochem.* **195:** 439.

Fankhauser, C., S.W. Homans, J.E. Thomas-Oates, M.J. McConville, C. Desponds, A. Conzelmann, and M.A.J. Ferguson. 1993. Structures of glycosylphosphatidylinositol membrane anchors from *Saccharomyces cerevisiae. J. Biol. Chem.* **268:** 26365.

Fassler, J.S., W. Gray, J. P. Lee, G. Yu, and G. Gingerich. 1991. The *Saccharomyces cerevisiae SPT14* gene is essential for normal expression of the yeast transposon, Ty, as well as for expression of the *HIS4* gene and several genes in the mating pathway. *Mol. Gen. Genet.* **230:** 310.

Fernández, F.S., S.E. Trombetta, U. Hellman, and A.J. Parodi. 1994. Purification to homogeneity of UDP-glucose: Glycoprotein glucosyltransferase from *Schizosaccharomyces pombe* and apparent absence of the enzyme from *Saccharomyces cerevisiae. J. Biol. Chem.* **269:** 30701.

Fernández, F.S., M. Jannatipour, U. Hellman, L.A. Rokeach, and A.J. Parodi. 1996. A new stress protein: Synthesis of *Schizosaccharomyces pombe* UDP-Glc:glycoprotein glucosyltransferase mRNA is induced by stress conditions but the enzyme is not essential for cell viability. *EMBO J.* **15:** 705.

Ferro-Novick, S., P. Novick, C. Field, and R. Schekman. 1984. Yeast secretory mutants that block the formation of active cell surface enzymes. *J. Cell Biol.* **98:** 35.

Finck, M., N. Bergmann, B. Jansson, and J.F. Ernst. 1996. Defective threonine-linked glycosylation of human insulin-like growth factor in mutants of the yeast *Saccharomyces cerevisiae. Glycobiology* **6**: 313.

Fishel, B.R., A.O. Sperry, and W.T. Garrard. 1993. Yeast calmodulin and a conserved nuclear protein participate in the *in vivo* binding of a matrix association region. *Proc. Natl. Acad. Sci.* **90:** 5623.

Fleet, G.H. 1991. Cell walls. In *The yeasts*, 2nd edition (ed. A.H. Rose and J.S. Harrison), vol. 4, p. 199. Academic Press, New York.

Fleet, G.H. and D.J. Manners. 1977. The enzymic degradation of an alkali-soluble glucan from the cell walls of *Saccharomyces cerevisiae*. *J. Gen. Microbiol.* **98:** 315.

Flores-Martinez, A. and J. Schwencke. 1988. Chitin synthetase activity is bound to chitosomes and to the plasma membrane in protoplasts of *Saccharomyces cerevisiae*. *Biochim. Biophys. Acta* **946:** 328.

Ford, R.A., J.A. Shaw, and E. Cabib. 1996. Yeast chitin synthases 1 and 2 consist of a non-homologous and dispensible N-terminal region and of a homologous moiety essential for function. *Mol. Gen. Genet.* **252**: 420.

Franzusoff, A. 1992. Beauty and the yeast: compartmental organization of the secretory pathway. *Semin. Cell Biol.* **3:** 309.

Franzusoff, A. and R. Schekman. 1989. Functional compartments of the yeast Golgi apparatus are defined by the *sec7* mutation. *EMBO J.* **8:** 2695.

Frevert, J. and C.E. Ballou. 1985. *Saccharomyces cerevisiae* structural cell wall mannoprotein. *Biochemistry* **24:** 753.

Fujita, A., Y. Kikuchi, S. Kuhara, Y. Misumi, S. Matsumoto, and H. Kobayashi. 1989. Domains of the SFL1 protein of yeasts are homologous to Myc oncoproteins or yeast heat-shock transcription factor. *Gene* **85:** 321.

Fuller, R.S., A. Brake, and J. Thorner. 1989. Yeast prohormone processing enzyme (*KEX2* gene product) is a Ca^{2+}-dependent serine protease. *Proc. Natl. Acad. Sci.* **86:** 1434.

Gallego, C., C. Casas, and E. Herrero. 1993. Increased transformation levels in intact cells of *Saccharomyces cerevisiae* Aculeacin A-resistant mutants. *Yeast* **9:** 523.

Gatti, E., L. Popolo, M. Vai, N. Rota, and L. Alberghina. 1994. *O*-linked oligosaccharides in yeast glycosyl phosphatidylinositol-anchored protein gp115 are clustered in a serine-rich region not essential for its function. *J. Biol. Chem.* **269:** 19695.

Gaughran, J.P., M.H. Lai, D.R. Kirsch, and S.J. Silverman. 1994. Nikkomycin Z is a specific inhibitor of *Saccharomyces cerevisiae* chitin synthase isozyme Chs3 in vitro and in vivo. *J. Bacteriol.* **176:** 5857.

Gaynor, E.C., S. te Heesen, T.R. Graham, M. Aebi, and S.D. Emr. 1994. Signal-mediated retrieval of a membrane protein from the Golgi to the ER in yeast. *J. Cell Biol.* **127:** 653.

Gentzsch, M. and W. Tanner. 1996. The *PMT* gene family: Protein *O*-glycosylation in *Saccharomyces cerevisiae* is vital. *EMBO J.* **15**: 5752.

Gentzsch, M., T. Immervoll, and W. Tanner. 1995a. Protein *O*-glycosylation in *Saccharomyces cerevisiae:* The protein *O*-mannosyltransferases Pmt1p and Pmt2p function as heterodimer. *FEBS Lett.* **377:** 128.

Gentzsch, M., S. Strahl-Bolsinger, and W. Tanner. 1995b. A new Dol-P-Man:protein *O*-D-mannosyltransferase activity from *Saccharomyces cerevisiae*. *Glycobiology* **5:** 77.

Georgopapadakou, N.H. and J.S. Tkacz. 1995. The fungal cell wall as a drug target. *Trends Microbiol.* **3:** 98.

Geremia, R.A., P. Mergaert, D. Geelen, M. Van Montagu, and M. Holsters. 1994. The NodC protein of *Azorhizobium caulinodans* is an *N*-acetyglucosaminyltransferase. *Proc. Natl. Acad. Sci.* **91:** 2669.

Goldman, R.C., P.A. Sullivan, D. Zakula, and J.O. Capobianco. 1995. Kinetics of β-1,3 glucan interaction at the donor and acceptor sites of the fungal glucosyltransferase encoded by the *BGL2* gene. *Eur. J. Biochem.* **227:** 372.

Graham, T.R. and S.D. Emr. 1991. Compartmental organization of Golgi-specific protein modification and vacuolar protein sorting events defined in a yeast *sec18* (*NSF*) mutant. *J. Cell Biol.* **114:** 207.

Graham, T.R., M. Seeger, G.S. Payne, V.L. MacKay, and S.D. Emr. 1994. Clathrin-dependent localization of α 1,3 mannosyltransferase to the Golgi complex of *Saccharomyces cerevisiae. J. Cell Biol.* **127:** 667.

Gustin, M.C., K. Davenport, and M. Sohaskey. 1995. Osmosensing signal transduction pathways. *Yeast* **11:** S17.

Hamburger, D., M. Egerton, and H. Riezman. 1995. Yeast Gaa1p is required for attachment of a completed GPI anchor onto proteins. *J. Cell Biol.* **129:** 629.

Hanson, B.A. and R.L. Lester. 1980. Effects of inositol starvation on phospholipid and glycan syntheses in *Saccharomyces cerevisiae. J. Bacteriol.* **142:** 79.

———. 1982. Effect of inositol starvation on the in vitro syntheses of mannan and *N*-acetylglucosaminylpyrophosphoryldolichol in *Saccharomyces cerevisiae. J. Bacteriol.* **151:** 334.

Hardwick, K.G., J.C. Boothroyd, A.D. Rudner, and H.R.B. Pelham. 1992. Genes that allow yeast cells to grow in the absence of the HDEL receptor. *EMBO J.* **11:** 4187.

Hardwick, K.G., M.J. Lewis, J. Semenza, N. Dean and H.R.B. Pelham. 1990. *ERD1*, a yeast gene required for the retention of luminal endoplasmic reticulum proteins, affects glycoprotein processing in the Golgi apparatus. *EMBO J.* **9:** 623.

Harris, S.L. and M.G. Waters. 1996. Localization of a yeast early Golgi mannosyltransferase, Och1p, involves retrograde transport. *J. Cell Biol.* **132:** 985.

Hart, G.W., R.S. Haltiwanger, G.D. Holt, and W.G. Kelly. 1989. Glycosylation in the nucleus and cytoplasm. *Annu. Rev. Biochem.* **58:** 841.

Hartland, R.P., C.A. Vermeulen, F.M. Klis, J.H. Sietsma, and J.G.H. Wessels. 1994. The linkage of (1,3)-β-glucan to chitin during cell wall assembly in *Saccharomyces cerevisiae. Yeast* **10:** 1591.

Haselbeck, A. 1989. Purification of GDP mannose: dolichyl-phosphate *O*-β-D-mannosyltransferase from *Saccharomyces cerevisiae. Eur. J. Biochem.* **181:** 663.

Haselbeck, A. and W. Tanner. 1982. Dolichyl phosphate-mediated mannosyl transfer through liposomal membranes. *Proc. Natl. Acad. Sci.* **79:** 1520.

———. 1983. *O*-Glycosylation in *Saccharomyces cerevisiae* is initiated at the endoplasmic reticulum. *FEBS Lett.* **158:** 335.

———. 1984. Further evidence for dolichyl phosphate-mediated glycosyl translocation through membranes. *FEMS Microbiol. Lett.* **21:** 305.

Hauser, K. and W. Tanner. 1989. Purification of the inducible α-agglutinin of *S. cerevisiae* and molecular cloning of the gene. *FEBS Lett.* **255:** 290.

Häusler, A. and P.W. Robbins 1992. Glycosylation in *Saccharomyces cerevisiae:* Cloning and characterization of an α-1,2-mannosyltransferase structural gene. *Glycobiology* **2:** 77.

Häusler, A., L. Ballou, C.E. Ballou, and P.W. Robbins. 1992. Yeast glycoprotein biosynthesis: *MNT1* encodes an α-1,2-mannosyltransferase involved in *O*-glycosylation. *Proc. Natl. Acad. Sci.* **89:** 6846.

Hebert, D.N., B. Foellmer, and A. Helenius. 1995. Glucose trimming and reglucosylation determine glycoprotein association with calnexin in the endoplasmic reticulum. *Cell* **81:** 425.

Heller, L., P. Orlean, and W.L. Adair, Jr. 1992. *Saccharomyces cerevisiae sec59* cells are deficient in dolichol kinase activity. *Proc. Natl. Acad. Sci.* **89:** 7013.

Hernández, L.M., I. Olivero, E. Alvarado, and G. Larriba. 1992. Oligosaccharide structures of the major exoglucanase secreted by *Saccharomyces cerevisiae*. *Biochemistry* **31:** 9823.

Herscovics, A. and P. Orlean. 1993. Glycoprotein biosynthesis in yeast. *FASEB J.* **7:** 540.

Hien, N.H. and G. H. Fleet. 1983. Separation and characterization of six (1→3)-β-glucanases from *Saccharomyces cerevisiae*. *J. Bacteriol.* **156:** 1204.

Hill, K., C. Boone, M. Goebl, R. Puccia, A.-M. Sdicu, and H. Bussey. 1992. Yeast *KRE2* defines a new gene family encoding probable secretory proteins, and is required for the correct *N*-glycosylation of proteins. *Genetics* **130:** 273.

Hirschberg, C.B. and Snider, M.D. 1987. Topography of glycosylation in the rough endoplasmic reticulum and Golgi apparatus. *Annu. Rev. Biochem.* **56:** 63.

Hofmann, M., E. Boles, and F.K. Zimmermann. 1994. Characterization of the essential yeast gene encoding *N*-acetylglucosamine-phosphate mutase. *Eur. J. Biochem.* **221:** 741.

Hong, Z., P. Mann, K.J. Shaw, and B. DiDomenico. 1994a. Analysis of β-glucans and chitin in a *Saccharomyces cerevisiae* cell wall mutant using high-performance liquid chromatography. *Yeast* **10:** 1083.

Hong, Z., P. Mann, N.H. Brown, L.E. Tran, K.J. Shaw, R.S. Hare, and B. DiDomenico. 1994b. Cloning and characterization of *KNR4*, a yeast gene involved in (1,3)-β-glucan synthesis. *Mol. Cell. Biol.* **14:** 1017.

Horvath, A., C. Sütterlin, U. Manning-Krieg, N.R. Movva, and H. Riezman. 1994. Ceramide synthesis enhances transport of GPI-anchored proteins to the Golgi apparatus in yeast. *EMBO J.* **13:** 3687.

Huffaker, T.C. and P.W. Robbins. 1982. Temperature-sensitive yeast mutants deficient in asparagine-linked glycosylation. *J. Biol. Chem.* **257:** 3203.

———. 1983. Yeast mutants deficient in protein glycosylation. *Proc. Natl. Acad. Sci.* **80:** 7466.

Hutchins, K. and H. Bussey. 1983. Cell wall receptor for yeast killer toxin: Involvement of (1→6)-β-D-Glucan. *J. Bacteriol.* **154:** 161.

Immervoll, T., M. Gentzsch, and W. Tanner. 1995. *PMT3* and *PMT4*, two new members of the protein-O-mannosyltransferase gene family of *Saccharomyces cerevisiae*. *Yeast* **11:** 1345.

Inoue, N., T. Kinoshita, T. Orii, and J. Takeda. 1993. Cloning of a human gene, PIG-F, a component of glycosyl-phosphatidylinositol anchor biosynthesis, by a novel expression cloning strategy. *J. Biol. Chem.* **268:** 6882.

Inoue, S.B., N. Takewaki, T. Takasuka, T. Mio, M. Adachi, Y. Fujii, C. Miyamoto, M. Arisawa, Y. Furuichi, and T. Watanabe. 1995. Characterization and gene cloning of 1,3-β-D-glucan synthase from *Saccharomyces cerevisiae*. *Eur. J. Biochem.* **231:** 845.

Irie, K., M. Takase, K.S. Lee, D.E. Levin, H. Araki, K. Matsumoto, and Y. Oshima. 1993. *MKK1* and *MKK2*, which encode *Saccharomyces cerevisiae* mitogen-activated protein kinase-kinase homologs, function in the pathway mediated by protein kinase C. *Mol. Cell. Biol.* **13:** 3076.

Jackson, B.J., M.A. Kukuruzinska, and P.W. Robbins. 1993. Biosynthesis of asparagine-linked oligosaccharides in *Saccharomyces cerevisiae:* The *alg2* mutation. *Glycobiology* **3:** 357.

Jackson, B.J., C.D. Warren, B. Bugge, and P.W. Robbins. 1989. Synthesis of lipid-linked oligosaccharides in *Saccharomyces cerevisiae*: $Man_2GlcNAc_2$ and $Man_1GlcNAc_2$ are transferred from dolichol to protein *in vivo*. *Arch. Biochem. Biophys.* **272:** 203.

Jelinek-Kelly, S. and A. Herscovics. 1988. Glycoprotein biosynthesis in *Saccharomyces cerevisiae*: Purification of the α-mannosidase which removes one specific mannose residue from $Man_9GlcNAc$. *J. Biol. Chem.* **263:** 14757.

Jentoft, N. 1990. Why are proteins *O*-glycosylated? *Trends Biochem. Sci.* **15:** 291.

Jiang, B., A.F.J. Ram, J. Sheraton, F.M. Klis, and H. Bussey. 1995. Regulation of cell wall β-glucan assembly: *PTC1* negatively affects *PBS2* action in a pathway that includes modulation of *EXG1* transcription. *Mol. Gen. Genet.* **248:** 260.

Jiang, B., J. Sheraton, A.F.J. Ram, G.J.P. Dijkgraaf, F.M. Klis, and H. Bussey. 1996. *CWH41* encodes a novel endoplasmic reticulum membrane *N*-glycoprotein involved in β1,6-glucan assembly. *J. Bacteriol.* **178:** 1162.

Johnston, G.C., J.A. Prendergast, and R.A. Singer. 1991. The *Saccharomyces cerevisiae MYO2* gene encodes an essential myosin required for vectorial transport of vesicles. *J. Cell Biol.* **113**: 539.

Jung, P. and W. Tanner. 1973. Identification of the lipid intermediate in yeast mannan biosynthesis. *Eur. J. Biochem.* **37:** 1.

Kafetzopoulos, D., G. Thireos, J.N. Vournakis, and V. Bouriotis. 1993. The primary structure of a fungal chitin deacetylase reveals the function for two bacterial gene products. *Proc. Natl. Acad. Sci.* **90:** 8005.

Kamada, Y., U.S. Jung, J. Piotrowski, and D.E. Levin. 1995. The protein kinase C-activated MAP kinase pathway of *Saccharomyces cerevisiae* mediates a novel aspect of the heat shock response. *Genes Dev.* **9:** 1559.

Kamada, Y., H. Qadota, C.P. Python, Y. Anraku, Y. Ohya, and D.E. Levin. 1996. Activation of yeast protein kinase C by Rho1 GTPase. *J. Biol. Chem.* **271:** 9193.

Kang, M.S. and E. Cabib. 1986. Regulation of fungal cell wall growth: A guanine nucleotide-binding, proteinaceous component required for activity of (1→3)-β-D-glucan synthase. *Proc. Natl. Acad. Sci.* **83:** 5808.

Kang, M.S., J. Au-Young, and E. Cabib. 1985. Modification of yeast plasma membrane density by Concanavalin A attachment. Application to study of chitin synthetase distribution. *J. Biol. Chem.* **260:** 12680.

Kang, M.S., N. Elango, E. Mattia, J. Au-Young, P.W. Robbins, and E. Cabib. 1984. Isolation of chitin synthetase from *Saccharomyces cerevisiae*. Purification of an enzyme by entrapment in the reaction product. *J. Biol. Chem.* **259:** 14966.

Kanik-Ennulat, C., E. Montalvo, and N. Neff. 1995. Sodium orthovanadate-resistant mutants of *Saccharomyces cerevisiae* show effects in Golgi-mediated protein glycosylation, sporulation and detergent resistance. *Genetics* **140:** 933.

Kapteyn, J.C., R.C. Montijn, G.J.P. Dijkgraaf, H. van den Ende, and F.M. Klis. 1995. Covalent association of β-1,3-glucan with β-1,6-glucosylated mannoproteins in cell walls of *Candida albicans*. *J. Bacteriol.* **177:** 3788.

Kapteyn, J.C., R.C. Montijn, E. Vink, J. de la Cruz, A. Llobell, J.E. Douwes, H. Shimoi, P.N. Lipke, and F.M. Klis. 1996. Retention of *Saccharomyces cerevisiae* cell wall proteins through a phosphodiester-linked β-1,3-/β-1,6-glucan heteropolymer. *Glycobiology* **6:** 337.

Karaoglu, D., D.J. Kelleher, and R. Gilmore. 1995. Functional characterization of Ost3p. Loss of the 34-kD subunit of the *Saccharomyces cerevisiae* oligosaccharyltransferase results in biased underglycosylation of acceptor subtrates. *J. Cell. Biol.* **130:** 567.

Kasahara, S., S.B. Inoue, T. Mio, T. Yamada, T. Nakajima, E. Ichishima, Y. Furuichi, and H. Yamada. 1994a. Involvement of cell wall β-glucan in the action of HM-1 killer toxin. *FEBS Lett.* **348:** 27.

Kasahara, S., H. Yamada, T. Mio, Y. Shiratori, C. Miyamoto, T. Yabe, T. Nakajima, E.

Ichishima, and Y. Furuichi. 1994b. Cloning of the *Saccharomyces cerevisiae* gene whose overexpression overcomes the effects of HM-1 killer toxin, which inhibits β-glucan synthesis. *J. Bacteriol.* **176:** 1488.

Katohda, S., K. Konno, Y. Sasaki, K. Suzuki, and S. Sakamoto. 1984. Isolation and composition of the spore wall of *Saccharomyces cerevisiae. Agric. Biol. Chem.* **48:** 895.

Kelleher, D.J. and R. Gilmore. 1994. The *Saccharomyces cerevisiae* oligosaccharyltransferase is a protein complex composed of Wbp1p, Swp1p, and four additional polypeptides. *J. Biol. Chem.* **269:** 12908.

Kepes, F. and R. Schekman. 1988. The yeast *SEC53* gene encodes phosphomannomutase. *J. Biol. Chem.* **263:** 9155.

Kessler, G. and W.J. Nickerson. 1959. Glucomannan-protein complexes from cell walls of yeasts. *J. Biol. Chem.* **234:** 2281.

Kilker, R.D, Jr., B. Saunier, J.S. Tkacz, and A. Herscovics. 1981. Partial purification from *Saccharomyces cerevisiae* of a soluble glucosidase which removes the terminal glucose from the oligosaccharide $Glc_3Man_9GlcNAc_2$. *J. Biol. Chem.* **256:** 5299.

Kim, Y.-J., L. Francisco, G.-C. Chen, E. Marcotte, and C.S.M. Chan. 1994. Control of cellular morphogenesis by the Ipl2/Bem2 GTPase-activating protein: Possible role of protein phosphorylation. *J. Cell Biol.* **127:** 1381.

Kitamura, K. 1982. A protease that participates in yeast cell wall lysis during zymolyase digestion. *Agric. Biol. Chem.* **46:** 2093.

Klebl, F. and W. Tanner. 1989. Molecular cloning of a cell wall exo-β-1,3-glucanase from *Saccharomyces cerevisiae. J. Bacteriol.* **171:** 6259.

Klebl, F., T. Huffaker, and W. Tanner. 1984. A temperature-sensitive *N*-glycosylation mutant of *Saccharomyces cerevisiae* that behaves like a cell cycle mutant. *Exp. Cell Res.* **150:** 309.

Klis, F.M. 1994. Cell wall assembly in yeast. *Yeast* **10:** 851.

Knauer, R. and L. Lehle. 1994. The *N*-oligosaccharyltransferase complex from yeast. *FEBS Lett.* **344:** 83.

Ko, Y.-T., D.J. Frost, C.-T. Ho, R.D. Ludescher, and B.P. Wasserman. 1994. Inhibition of yeast (1,3)-β-glucan synthase by phospholipase A_2 and its reaction products. *Biochim. Biophys. Acta* **1193:** 31.

Kollár, R., E. Petráková, G. Ashwell, P.W. Robbins, and E. Cabib. 1995. Architecture of the yeast cell wall. The linkage between chitin and β(1→3)-glucan. *J. Biol. Chem.* **270:** 1170.

Kopecká, M. and D.R. Kreger. 1986. Assembly of microfibrils in vivo and in vitro from (1→3)-β-D-glucan synthesized by protoplasts of *Saccharomyces cerevisiae. Arch. Microbiol.* **143:** 387.

Kopecká, M., H.J. Phaff, and G.H. Fleet. 1974. Demonstration of a fibrillar component in the cell wall of the yeast *Saccharomyces cerevisiae* and its chemical nature. *J. Cell Biol.* **62,** 66.

Korn, E.D. and D.H. Northcote. 1960. Physical and chemical properties of polysaccharides and glycoproteins of the yeast-cell wall. *Biochem. J.* **75:** 12.

Krusius, T., J. Finne, R.K. Margolis, and R.U. Margolis. 1986. Identification of an *O*-glycosidic mannose-linked sialylated tetrasaccharide and keratan sulfate oligosaccharides in the chondroitin sulfate proteoglycan of brain. *J. Biol. Chem.* **261:** 8237.

Kukuruzinska, M.A, and K. Lennon. 1994. Growth-related coordinate regulation of the early *N*-glycosylation genes in yeast. *Glycobiology* **4:** 437.

———. 1995. Diminished activity of the first *N*-glycosylation enzyme, dolichol-P-dependent *N*-acetylglucosamine-1-P transferase (GPT), gives rise to mutant phenotypes in yeast. *Biochim. Biophys. Acta* **1247:** 51.

Kukuruzinska, M.A. and P.W. Robbins. 1987. Protein glycosylation in yeast: Transcript heterogeneity of the *ALG7* gene. *Proc. Natl. Acad. Sci.* **84:** 2145.

Kukuruzinska, M.A., M.L.E. Bergh, and B.J. Jackson. 1987. Protein glycosylation in yeast. *Annu. Rev. Biochem.* **56:** 915.

Kuranda, M.J. and P.W. Robbins. 1987. Cloning and heterologous expression of glycosidase genes from *Saccharomyces cerevisiae. Proc. Natl. Acad. Sci.* **84:** 2585.

———. 1991. Chitinase is required for cell separation during growth of *Saccharomyces cerevisiae. J. Biol. Chem.* **266:** 19758.

Larriba, G., E. Andaluz, R. Cueva, and R.D. Basco. 1995. Molecular biology of yeast exoglucanases. *FEMS Microbiol. Lett.* **125:** 121.

Law, D.T.S. and Segall, J. 1988. The *SPS100* gene of *Saccharomyces cerevisiae* is activated late in the sporulation process and contributes to spore wall maturation. *Mol. Cell. Biol.* **8:** 912.

Leal-Morales, C.A., C.E. Bracker, and S. Bartnicki-Garcia. 1988. Localization of chitin synthetase in cell-free homogenates of *Saccharomyces cerevisiae*: Chitosomes and plasma membrane. *Proc. Natl. Acad. Sci.* **85:** 8516.

———. 1994. Subcellular localization, abundance and stability of chitin synthetases 1 and 2 from *Saccharomyces cerevisiae. Microbiology* **140:** 2207.

Lee, K.S., K. Irie, Y. Gotoh, Y. Watanabe, H. Araki, E. Nishida, K. Matsumoto, and D.E. Levin. 1993. A yeast mitogen-activated protein kinase homolog (Mpk1p) mediates signaling by protein kinase C. *Mol. Cell. Biol.* **13:** 3067.

Lehle, L. and E. Bause. 1984. Primary structural requirements for *N*- and *O*-glycosylation of yeast mannoproteins. *Biochim. Biophys. Acta* **799:** 246.

Leidich, S.D. and P. Orlean. 1996. Gpi1, a *Saccharomyces cerevisiae* protein that participates in the first step in GPI anchor synthesis. *J. Biol. Chem.* **271:** 27829.

Leidich, S.D., D.A. Drapp, and P. Orlean. 1994. A conditionally lethal yeast mutant blocked at the first step in glycosyl phosphatidylinositol anchor synthesis. *J. Biol. Chem.* **269:** 10193.

Leidich, S.D., Z. Kostova, R.R. Latek, L.C. Costello, D.A Drapp, W. Gray, J.S. Fassler, and P. Orlean. 1995. Temperature-sensitive yeast GPI anchoring mutants *gpi2* and *gpi3* are defective in the synthesis of *N*-acetylglucosaminyl phosphatidylinositol: Cloning of the *GPI2* gene. *J. Biol. Chem.* **270:** 13029.

Lennon, K., R. Pretel, J. Kesselheim, S. te Heesen, and M.A. Kukuruzinska. 1995. Proliferation-dependent differential regulation of the dolichol pathway genes in *Saccharomyces cerevisiae. Glycobiology* **5**: 633.

Lester, R.L., G.B. Wells, G. Oxford, and R.C. Dickson. 1993. Mutant strains of *Saccharomyces cerevisiae* lacking sphingolipids synthesize novel inositol glycerophospholipids that mimic sphingolipid structures. *J. Biol. Chem.* **268:** 845.

Levin, D.E. and E. Bartlett-Heubusch. 1992. Mutants in the *S. cerevisiae PKC1* gene display a cell cycle-specific osmotic stability defect. *J. Cell Biol.* **116:** 1221.

Lewis, M.S. and C.E. Ballou. 1991. Separation and characterization of two α1,2-mannosyltransferase activities from *Saccharomyces cerevisiae. J. Biol. Chem.* **266:** 8255.

Lipke, P.N. and J. Kurjan. 1992. Sexual agglutination in budding yeasts: Structure, function, and regulation of adhesion glycoproteins. *Microbiol. Rev.* **56:** 180.

Lipke, P.N., A. Taylor, and C.E. Ballou. 1976. Morphogenic effects of α-factor on *Sac-*

charomyces cerevisiae **a** cells. *J. Bacteriol.* **127**: 610.

Lipke, P.N., D. Wojciechowicz, and J. Kurjan. 1989. *AGα1* is the structural gene for the *Saccharomyces cerevisiae* α-agglutinin, a cell surface glycoprotein involved in cell-cell interactions during mating. *Mol. Cell. Biol.* **9:** 3155.

Longtine, M.S., D.J. DeMarini, M.L. Valencik. O.S. Al-Awar, H. Fares, C. De Virgilio, and J.R. Pringle. 1996. The septins: Roles in cytokinesis and other processes. *Curr. Opin. Cell Biol.* **8**: 106.

Lu, C.-F., J. Kurjan, and P.N. Lipke. 1994. A pathway for cell wall anchorage of *Saccharomyces cerevisiae* α-agglutinin. *Mol. Cell. Biol.* **14:** 4825.

Lu, C.-F., R.C. Montijn, J.L. Brown, F. Klis, J. Kurjan, H. Bussey, and P.N. Lipke. 1995. Glycosyl phosphatidylinositol-dependent cross-linking of α-agglutinin and β1,6-glucan in the *Saccharomyces cerevisiae* cell wall. *J. Cell Biol.* **128:** 333-340.

Lussier, M., A. Camirand, A.-M. Sdicu, and H. Bussey. 1993. *KTR2:* A new member of the *KRE2* mannosyltransferase gene family. *Yeast* **9:** 1057.

Lussier, M., A.-M. Sdicu, A. Camirand, and H. Bussey. 1996. Functional characterization of the *YUR1, KTR1*, and *KTR2* genes as members of the yeast *KRE2/MNT1* mannosyltransferase gene family. *J. Biol. Chem.* **271:** 11001.

Lussier, M., A.-M. Sdicu, T. Ketela, and H. Bussey. 1995a. Localization and targeting of the *Saccharomyces cerevisiae* Kre2p/Mnt1p α1,2-mannosyltransferase to a *medial*-Golgi compartment. *J. Cell Biol.* **131:** 913.

Lussier, M., M. Gentzsch, A.-M. Sdicu, H. Bussey, and W. Tanner. 1995b. Protein *O*-glycosylation in yeast. The *PMT2* gene specifies a second protein *O*-mannosyltransferase that functions in addition to the *PMT1*-encoded activity. *J. Biol. Chem.* **270:** 2770.

Lussier, M., A.-M. Sdicu, E. Winnett, D.H. Vo, J. Sheraton, A. Düsterhöft, R.K. Storms, and H. Bussey. 1997. Completion of the *Saccharomyces cerevisiae* genome sequence allows identification of *KTR5*, *KTR6*, and *KTR7* and definition of the nine-membered *KRE2/MNT1* mannosyltransferase gene family in this organism. *Yeast* **13**: 267.

Ma, D., D.G. Russell, S.M. Beverley, and S.J. Turco. 1997. Golgi GDP-mannose uptake requires *Leishmania LPG2*. A member of a eukaryotic family of putative nucleotide-sugar transporters. *J. Biol. Chem.* **272:** 3799.

Machida, M. and Y. Jigami. 1994. Glycosylated DNA-binding proteins from filamentous fungus, *Aspergillus oryzae*: Modification with *N*-acetylglucosamine monosaccharide through an *O*-glycosidic linkage. *Biosci. Biotech. Biochem.* **58:** 344.

Mallet, L., F. Bussereau, and M. Jacquet. 1994. Nucleotide sequence analysis of an 11.7 kb fragment of yeast chromosome II including *BEM1*, a new gene of the WD-40 repeat family and a new member of the *KRE2/MNT1* family. *Yeast* **10:** 819.

Manners, D.J., A.J. Masson, and J.C. Patterson. 1973a. The structure of a β-(1→3)-D-glucan from yeast cell walls. *Biochem. J.* **135:** 19.

Manners, D.J., A.J. Masson, J.C. Patterson, H. Björndal, and B. Lindberg. 1973b. The structure of a β-(1→6)-D-glucan from yeast cell walls. *Biochem. J.* **135:** 31.

Marchase, R.B., P. Bounelis, L.M. Brumley, N. Dey, B. Browne, D. Auger, T.A. Fritz, P. Kulesza, and D.M. Bedwell. 1993. Phosphoglucomutase in *Saccharomyces cerevisiae* is a cytoplasmic glycoprotein and the acceptor for a Glc-phosphotransferase. *J. Biol. Chem.* **268:** 8341.

Marini, N.J., E. Meldrum, B. Buehrer, A.V. Hubberstey, D.E. Stone, A. Traynor-Kaplan, and S.I. Reed. 1996. A pathway in the yeast cell division cycle linking protein kinase C (Pkc1) to activation of Cdc28 at START. *EMBO J.* **15:** 3040.

Mazur, P. and W. Baginsky. 1996. *In vitro* activity of 1,3-β-D-glucan synthase requires the GTP-binding protein Rho1. *J. Biol. Chem.* **271:** 14604.

Mazur, P., N. Morin, W. Baginsky, M. El-Sherbeini, J.A. Clemas, J.B. Nielsen, and F. Foor. 1995. Differential expression and function of two homologous subunits of yeast 1,3-β-D-glucan synthase. *Mol. Cell. Biol.* **15:** 5671.

McCaffrey, G., F.J. Clay, K. Kelsay, and G.F. Sprague, Jr. 1987. Identification and regulation of a gene required for cell fusion during mating of the yeast *Saccharomyces cerevisiae. Mol. Cell. Biol.* **7:** 2680.

McMurrough, I. and A.H. Rose. 1967. Effect of growth rate and substrate limitation on the composition and structure of the cell wall of *Saccharomyces cerevisiae. Biochem. J.* **105:** 189.

Meaden, P., K. Hill, J. Wagner, D. Slipetz, S.S. Sommer, and H. Bussey. 1990. The yeast *KRE5* gene encodes a probable endoplasmic reticulum protein required for (1→6)-β-D-glucan synthesis and normal cell growth. *Mol. Cell. Biol.* **10:** 3013.

Mellado, E., A. Aufauvre-Brown, C.A. Specht, P.W. Robbins, and D.W. Holden. 1995. A multigene family related to chitin synthase genes of yeast in the opportunistic pathogen *Aspergillus fumigatus. Mol. Gen. Genet.* **246:** 353.

Menon, A.K. and V.L. Stevens. 1992. Phosphatidylethanolamine is the donor of the ethanolamine residue linking a glycosylphosphatidylinositol anchor to protein. *J. Biol. Chem.* **267:** 15277.

Menon, A.K., S. Mayor, and R.T. Schwarz. 1990. Biosynthesis of glycosylphosphatidylinositol lipids in *Trypanosoma brucee:* Involvement of mannosyl-phosphoryldolichol as the mannose donor. *EMBO J.* **9:** 4249.

Meyer, M.F. and G. Kreil. 1996. Cells expressing the DG42 gene from early *Xenopus* embryos synthesize hyaluronan. *Proc. Natl. Acad. Sci.* **93:** 4543.

Miki, B.L.A., N.H. Poon, A.P. James, and V.L. Seligy. 1982. Possible mechanism for flocculation interactions governed by gene *FLO1* in *Saccharomyces cerevisiae. J. Bacteriol.* **150:** 878.

Mishra, C., C.E. Semino, K.J. McCreath, H. de la Vega, C.A. Specht, and P.W. Robbins. 1997. Cloning and expression of two chitin deacetylase genes in *Saccharomyces cerevisiae. Yeast* (in press).

Moehle, C.M., R. Tizard, S.K. Lemmon, J. Smart, and E.W. Jones. 1987. Protease B of the lysosomelike vacuole of the yeast *Saccharomyces cerevisiae* is homologous to the subtilisin family of serine proteases. *Mol. Cell. Biol.* **7:** 4390.

Mol, P.C. and J.G.H. Wessels. 1987. Linkages between glucosaminoglycan and glucan determine alkali-insolubility of the glucan in walls of *Saccharomyces cerevisiae. FEMS Microbiol. Lett.* **41:** 95.

Mol, P.C., H.-M. Park, J.T. Mullins, and E. Cabib. 1994. A GTP-binding protein regulates the activity of (1-3)-β-glucan synthase, an enzyme directly involved in yeast cell wall morphogenesis. *J. Biol. Chem.* **269:** 31267.

Mondésert, G. and S.I. Reed. 1996. *BED1*, a gene encoding a galactosyltransferase homologue, is required for polarized growth and efficient bud emergence in *Saccharomyces cerevisiae. J. Cell Biol.* **132:** 137.

Montijn, R.C., J. van Rinsum, F.A. van Schagen, and F.M. Klis. 1994. Glucomannoproteins in the cell wall of *Saccharomyces cerevisiae* contain a novel type of carbohydrate side chain. *J. Biol. Chem.* **269:** 19338.

Motoyama, T., N. Kojima, H. Horiuchi, A. Ohta, and M. Takagi. 1994. Isolation of a chitin synthase gene (*chsC*) of *Aspergillus nidulans. Biosci. Biotech. Biochem.* **58:**

2254.

Mrsa, V., F. Klebl, and W. Tanner. 1993. Purification and characterization of the *Saccharomyces cerevisiae BGL2* gene product, a cell wall endo-β-1,3-glucanase. *J. Bacteriol.* **175:** 2102.

Mrsa, V., T. Ugarkovic, and S. Barbaric. 1992. Binding of *Saccharomyces cerevisiae* extracellular proteins to glucane. *Arch. Biochem. Biophys.* **296:** 569.

Müller, G. and W. Bandlow. 1993. Glucose induces lipolytic cleavage of a glycolipidic plasma membrane anchor in yeast. *J. Cell Biol.* **122:** 325.

———. 1994. Lipolytic membrane release of two phosphatidylinositol-anchored cAMP receptor proteins in yeast alters their ligand-binding parameters. *Arch. Biochem. Biophys.* **308:** 504.

Müller, G., E. Gross, S. Wied, and W. Bandlow. 1996. Glucose-induced sequential processing of a glycosyl-phosphatidylinositol-anchored ectoprotein in *Saccharomyces cerevisiae. Mol. Cell. Biol.* **16:** 442.

Müller, G., K. Schubert, F. Fiedler, and W. Bandlow. 1992. The cAMP-binding ectoprotein from *Saccharomyces cerevisiae* is membrane-anchored by glycosyl-phosphatidylinositol. *J. Biol. Chem.* **267:** 25337.

Muñoz, M.D., L.M. Hernández, R. Basco, E. Andaluz, and G. Larriba. 1994. Glycosylation of yeast exoglucanase sequons in *alg* mutants deficient in the glucosylation steps of the lipid-linked oligosaccharide. Presence of glucotriose unit in Dol-PP-$GlcNAc_2Man_9Glc_3$ influences both glycosylation efficiency and selection of *N*-linked sites. *Biochim. Biophys. Acta* **1201:** 361.

Muthukumar, G., S.-H. Suhng, P.T. Magee, R.D. Jewell, and D.A. Primerano. 1993. The *Saccharomyces cerevisiae SPR1* gene encodes a sporulation-specfic exo-1,3-β-glucanase which contributes to ascospore thermoresistance. *J. Bacteriol.* **175:** 386.

Nagahashi, S., M. Sudoh, N. Ono, R. Sawada, E. Yamaguchi, Y. Uchida, T. Mio, M. Takagi, M. Arisawa, and H. Yamada-Okabe. 1995. Characterization of chitin synthase 2 of *Saccharomyces cerevisiae*. Implication of two highly conserved domains as possible catalytic sites. *J. Biol. Chem.* **270:** 13961.

Nagasu, T., Y-I. Shimma, Y. Nakanishi, J. Kuromitsu, K. Iwama, K.-I. Nakayama, K. Suzuki, and Y. Jigami. 1992. Isolation of new temperature-sensitive mutants of *Saccharomyces cerevisiae* deficient in mannose outer chain elongation. *Yeast* **8:** 535.

Nagiec, M.M., G.B. Wells, R.L. Lester, and R.C. Dickson. 1993. A suppressor gene that enables *Saccharomyces cerevisiae* to grow without making sphingolipids encodes a protein that resembles an *Escherichia coli* fatty acyltransferase. *J. Biol. Chem.* **268:** 22156.

Nagiec, M.M., J.A. Baltisberger, G.B. Wells, R.L. Lester, and R.C. Dickson. 1994. The *LCB2* gene of *Saccharomyces* and the related *LCB1* gene encode subunits of serine palmitoyltransferase, the initial enzyme in sphingolipid synthesis. *Proc. Natl. Acad. Sci.* **91:** 7899.

Nakanishi-Shindo, Y., K. Nakayama, A. Tanaka, Y. Toda, and Y. Jigami. 1993. Structure of the *N*-linked oligosaccharides that show the complete loss of α-1,6-polymannose outer chain from *och1*, *och1 mnn1*, and *och1 mnn1 alg3* mutants of *Saccharomyces cerevisiae*. *J. Biol. Chem.* **268:** 26338.

Nakayama, K., T. Nagasu, Y. Shimma, J. Kuromitsu, and Y. Jigami. 1992. *OCH1* encodes a novel membrane bound mannosyltransferase: Outer chain elongation of asparagine-linked oligosaccharides. *EMBO J.* **11:** 2511.

Nebreda, A.R., T.G. Villa, J.R. Villanueva, and F. del Rey. 1986. Cloning of genes related to exo-β-glucanase production in *Saccharomyces cerevisiae:* Characterization of an exo-β-glucanase structural gene. *Gene* **47:** 245.

Nonaka, H., K. Tanaka, H. Hirano, T. Fujiwara, H. Kohno, M. Umikawa, A. Mino, and Y. Takai. 1995. A downstream target of *RHO1* small GTP-binding protein is *PKC1*, a homolog of proten kinase C, which leads to activation of the MAP kinase cascade in *Saccharomyces cerevisiae. EMBO J.* **14:** 5931.

Notario, V., H. Kawai, and E. Cabib. 1982. Interaction between yeast β-(1→3)glucan synthetase and activating phosphorylated compounds. A kinetic study. *J. Biol. Chem.* **257:** 1902.

Nuoffer, C., A. Horvath, and H. Riezman. 1993. Analysis of the sequence requirements for glycosylphosphatidylinositol anchoring of *Saccharomyces cerevisiae* Gas1 protein. *J. Biol. Chem.* **268:** 10558.

Nuoffer, C., P. Jenö, A. Conzelmann, and H. Riezman. 1991 Determinants for glycophospholipid anchoring of the *Saccharomyces cerevisiae* GAS1 protein to the plasma membrane. *Mol. Cell. Biol.* **11:** 27.

Orlean, P.A.B. 1982. (1,3)-β-D-glucan synthase from budding and filamentous cultures of the dimorphic fungus *Candida albicans. Eur. J. Biochem.* **127:** 397.

———. 1987. Two chitin synthases in *Saccharomyces cerevisiae. J. Biol. Chem.* **262:** 5732.

———. 1990. Dolichol phosphate mannose synthase is required in vivo for glycosyl phosphatidylinositol membrane anchoring, O mannosylation, and N glycosylation of protein in *Saccharomyces cerevisiae. Mol. Cell. Biol.* **10:** 5796.

———. 1992. Enzymes that recognize dolichols participate in three glycosylation pathways and are required for protein secretion. *Biochem. Cell Biol.* **70:** 438.

Orlean, P., C. Albright, and P.W. Robbins. 1988. Cloning and sequencing of the yeast gene for dolichol phosphate mannose synthase, an essential protein. *J. Biol. Chem.* **263:** 17499.

Orlean, P., E. Arnold, and W. Tanner. 1985. Apparent inhibition of glycoprotein synthesis by *S. cerevisiae* mating pheromones. *FEBS Lett.* **184:** 313.

Orlean, P., M.J. Kuranda, and C.F. Albright. 1991. Analysis of glycoproteins from *Saccharomyces cerevisiae. Methods Enzymol.* **194:** 682.

Orlean, P., H. Ammer, M. Watzele, and W. Tanner. 1986. Synthesis of an *O*-glycosylated cell surface protein induced in yeast by α factor. *Proc. Natl. Acad. Sci.* **83:** 6263.

Palamarczyk, G., R. Drake, B. Haley, and W.J. Lennarz. 1990. Evidence that the synthesis of glucosylphosphodolichol in yeast involves a 35-kDa membrane protein. *Proc. Natl. Acad. Sci.* **87:** 2666.

Pammer, M., P. Briza, A. Ellinger, T. Schuster, R. Stucka, H. Feldmann, and M. Breitenbach. 1992. *DIT101* (*CSD2, CAL1*), a cell cycle-regulated yeast gene required for synthesis of chitin in cell walls and chitosan in spore walls. *Yeast* **8:** 1089.

Paravicini, G., M. Cooper, L. Friedli, D.J. Smith, J.-L. Carpentier, L.S. Klig, and M.A. Payton. 1992. The osmotic integrity of the yeast cell requires a functional *PKC1* gene product. *Mol. Cell. Biol.* **12:** 4896.

Pastor, F.I.J., E. Valentin, E. Herrero, and R. Sentandreu. 1984. Structure of the *Saccharomyces cerevisiae* cell wall: Mannoproteins released by Zymolyase and their contribution to wall architecture. *Biochim. Biophys. Acta* **802:** 292.

Pathak, R., C.S. Parker, and B. Imperiali. 1995. The essential yeast *NLT1* gene encodes the 64 kDa glycoprotein subunit of the oligosaccharyl transferase. *FEBS Lett.* **362:** 229.

Patton, J.L. and R.L. Lester. 1991. The phosphoinositol sphingolipids of *Saccharomyces cerevisiae* are highly localized in the plasma membrane. *J. Bacteriol.* **173:** 3101.

Patton, J.L., B. Srinivasan, R.C. Dickson, and R.L. Lester. 1992. Phenotypes of sphingolipid-dependent strains of *Saccharomyces cerevisiae*. *J. Bacteriol.* **174:** 7180.

Payton, M.A., M. Rheinnecker, L.S. Klig, M. DeTiani, and E. Bowden. 1991. A novel *Saccharomyces cerevisiae* secretory mutant possesses a thermolabile phosphomannose isomerase. *J. Bacteriol.* **173:** 2006.

Pinto, W.J., G.W. Wells, and R.L. Lester. 1992. Characterization of enzymatic synthesis of sphingolipid long-chain bases in *Saccharomyces cerevisiae:* Mutant strains exhibiting long-chain-base auxotrophy are deficient in serine palmitoyltransferase activity. *J. Bacteriol.* **174:** 2575.

Popolo, L., P. Cavadini, M. Vai, and L. Alberghina. 1993a. Transcript accumulation of the *GGP1* gene, encoding a yeast GPI-anchored glycoprotein, is inhibited during arrest in the G_1 phase and during sporulation. *Curr. Genet.* **24:** 382.

Popolo, L., M. Vai, E. Gatti, S. Porello, P. Bonfante, R. Balestrini, and L. Alberghina. 1993b. Physiological analysis of mutants indicates involvement of the *Saccharomyces cerevisiae* GPI-anchored protein gp115 in morphogenesis and cell separation. *J. Bacteriol.* **175:** 1879.

Poster, J.B. and N. Dean. 1996. The yeast *VRG4* gene is required for normal Golgi functions and defines a new family of related genes. *J. Biol. Chem.* **271:** 3837.

Preuss, D., J. Mulholland, C.A. Kaiser, P. Orlean, C. Albright, M.D. Rose, P.W. Robbins, and D. Botstein. 1991. Structure of the yeast endoplasmic reticulum: localization of ER proteins using immunofluorescence and immunoelectron microscopy. *Yeast* **7:** 891.

Puccia, R., B. Grondin, and A. Herscovics. 1993. Disruption of the processing α-mannosidase gene does not prevent outer chain synthesis in *Saccharomyces cerevisiae*. *Biochem. J.* **290:** 21.

Puoti, A., C. Desponds, and A. Conzelmann. 1991. Biosynthesis of mannosylinositolphosphoceramide in *Saccharomyces cerevisiae* is dependent on genes controlling the flow of secretory vesicles from the endoplasmic reticulum to the Golgi. *J. Cell Biol.* **113:** 515.

Qadota, H., Y. Anraku, D. Botstein, and Y. Ohya. 1994. Conditional lethality of a yeast strain expressing human *RHOA* in place of *RHO1*. *Proc. Natl. Acad. Sci.* **91:** 9317.

Qadota, H., C.P. Python, S.B. Inoue, M. Arisawa, Y. Anraku, Y. Zheng, T. Watanabe, D.E. Levin, and Y. Ohya. 1996. Identification of yeast Rho1p GTPase as a regulatory subunit of 1,3-β-glucan synthase. *Science* **272:** 279.

Ram, A.F.J., S.S.C. Brekelmans, L.J.W.M. Oehlen, and F.M. Klis. 1995. Identification of two cell cycle regulated genes affecting the β1,3-glucan content of cell walls in *Saccharomyces cerevisiae*. *FEBS Lett.* **358:** 165.

Ram, A.F.J., A. Wolters, R. ten Hoopen, and F.M. Klis. 1994. A new approach for isolating cell wall mutants in *Saccharomyces cerevisiae* by screening for hypersensitivity to Calcofluor White. *Yeast* **10:** 1019.

Reddy, A.V., R. MacColl, and F. Maley. 1990. Effect of oligosaccharides and chloride on the oligomeric structures of external, internal, and deglycosylated invertase. *Biochemistry* **29:** 2482.

Reddy, V.A., R.S. Johnson, K. Biemann, R.S. Williams, F.D. Ziegler, R.B. Trimble, and F. Maley. 1988. Characterization of the glycosylation sites in yeast external invertase. I. *N*-linked oligosaccharide content of the individual sequons. *J. Biol. Chem.* **263:** 6978.

Reiss, G., S. te Heesen, R. Gilmore, R. Zufferey, and M. Aebi. 1997. A specific screen for oligosaccharyltransferase mutations identifies the 9 kDa *OST5* protein required for optimal activity *in vivo* and *in vitro*. *EMBO J.* (in press).

Reiss, G., S. te Heesen, J. Zimmerman, P.W. Robbins, and M. Aebi. 1996. Isolation of the *ALG6* locus of *Saccharomyces cerevisiae* required for glucosylation in the *N*-linked glycosylation pathway. *Glycobiology* **6**: 493.

Riederer, M.A. and A. Hinnen. 1991. Removal of *N*-glycosylation sites of the yeast acid phosphatase severely affects protein folding. *J. Bacteriol.* **173:** 3539.

Rine, J., W. Hansen, E. Hardeman, and R.W. Davis. 1983. Targeted selection of recombinant clones through gene dosage effects. *Proc. Natl. Acad. Sci.* **80:** 6750.

Roberts, R.L., B. Bowers, M.L. Slater, and E. Cabib. 1983. Chitin synthesis and localization in cell division cycle mutants of *Saccharomyces cerevisiae*. *Mol. Cell. Biol.* **3:** 922.

Roemer, T. and H. Bussey. 1991. Yeast β-glucan synthesis: *KRE6* encodes a predicted type II membrane protein required for glucan synthesis *in vivo* and for glucan synthase activity *in vitro*. *Proc. Natl. Acad. Sci.* **88:** 11295.

———. 1995. Yeast Kre1p is a cell surface *O*-glycoprotein. *Mol. Gen. Genet.* **249**: 209.

Roemer, T., S. Delaney, and H. Bussey. 1993. *SKN1* and *KRE6* define a pair of functional homologs encoding putative membrane proteins involved in β-glucan synthesis. *Mol. Cell. Biol.* **13:** 4039.

Roemer, T., G. Paravicini, M.A. Payton, and H. Bussey. 1994. Characterization of the yeast (1→6)-β-glucan biosynthetic components, Kre6p and Skn1p, and genetic interactions between the *PKC1* pathway and extracellular matrix assembly. *J. Cell Biol.* **127:** 567.

Roitsch, T. and L. Lehle. 1989. Structural requirements for protein *N*-glycosylation. *Eur. J. Biochem.* **181:** 525.

Romero, P.A., B. Sleno, and A. Herscovics. 1994. Glycoprotein biosynthesis in *Saccharomyces cerevisiae*. Partial purification of the α-1,6-mannosyltransferase that initiates outer chain synthesis. *Glycobiology* **4:** 135.

Roncero, C. and A. Durán. 1985. Effect of Calcofluor White and Congo Red on fungal cell wall morphogenesis: In vivo activation of chitin polymerization. *J. Bacteriol.* **163:** 1180.

Roncero, C., M.H. Valdivieso, J.C. Ribas, and A. Durán. 1988. Isolation and characterization of *Saccharomyces cerevisiae* mutants resistant to Calcofluor White. *J. Bacteriol.* **170:** 1945.

Roos, J., R. Sternglanz, and W.J. Lennarz. 1994. A screen for yeast mutants with defects in the dolichol-mediated pathway for *N*-glycosylation. *Proc. Natl. Acad. Sci.* **91:** 1485.

Roy, A., C.F. Lu, D.L. Marykwas, P.N. Lipke, and J. Kurjan. 1991. The *AGA1* product is involved in cell surface attachment of the *Saccharomyces cerevisiae* cell adhesion glycoprotein **a**-agglutinin. *Mol. Cell. Biol.* **11:** 4196.

Ruiz-Herrera, J. 1992. *Fungal cell wall: Structure, synthesis, and assembly.* CRC Press, Boca Raton, Florida.

Runge, K.W. and P.W. Robbins. 1986. A new yeast mutation in the glucosylation steps of the asparagine-linked glycosylation pathway. Formation of a novel asparagine-linked oligosaccharide containing two glucose residues. *J. Biol. Chem.* **261:** 15582.

Runge, K.W., T.C. Huffaker, and P.W. Robbins. 1984. Two yeast mutations in glucosylation steps of the asparagine glycosylation pathway. *J. Biol. Chem.* **259:** 412.

Rush, J.S. and C.J. Waechter. 1995. Transmembrane movement of a water-soluble analogue of mannosylphosphoryldolichol is mediated by an endoplasmic reticulum protein. *J. Cell Biol.* **130:** 529.

Russo, P., N. Kalkinnen, H. Sareneva, J. Paakkola, and M. Makarow. 1992. A heat shock gene from *Saccharomyces cerevisiae* encoding a secretory glycoprotein. *Proc. Natl. Acad. Sci.* **89:** 3671.

Sakuda, S., Y. Nishimoto, M. Ohi, M. Watanabe, S. Takayama, A. Isogai, and Y. Yamada. 1990. Effects of demethylallosamidin, a potent yeast chitinase inhibitor, on the cell division of yeast. *Agric. Biol. Chem.* **54:** 1333.

San Segundo, P., J. Correa, C.R. Vazquez de Aldana, and F. del Rey. 1993. *SSG1*, a gene encoding a sporulation-specific 1,3-β-glucanase in *Saccharomyces cerevisiae*. *J. Bacteriol.* **175:** 3823.

Santos, B. and A. Durán. 1992. Cloning of *CAL3*, a *Saccharomyces cerevisiae* gene involved in Calcofluor sensitivity and chitin synthesis. *Yeast* **8:** S522.

Santos, B. and M. Snyder. 1997. Targeting of chitin synthase 3 to polarized growth sites in yeast requires Chs5p and Myo2p. *J. Cell Biol.* **136**: 95.

Saporito-Irwin, S.M., C.E. Birse, P.S. Sypherd, and W.A. Fonzi. 1995. *PHR1*, a pH-regulated gene of *Candida albicans*, is required for morphogenesis. *Mol. Cell. Biol.* **15:** 601.

Saunier, B., R.D. Kilker, Jr., J.S. Tkacz, A Quaroni,, and A. Herscovics. 1982. Inhibition of *N*-linked complex oligosaccharide formation by 1-deoxynojirimycin, an inhibitor of processing glucosidases. *J. Biol. Chem.* **257:** 14155.

Sburlati, A. and E. Cabib. 1986. Chitin synthetase 2, a presumptive participant in septum formation in *Saccharomyces cerevisiae*. *J. Biol. Chem.* **261:** 15147.

Schekman, R. and V. Brawley. 1979. Localized deposition of chitin on the yeast cell surface in response to mating pheromone. *Proc. Natl. Acad. Sci.* **76:** 645.

Schönbächler, M., A. Horvath, J. Fassler, and H. Riezman. 1995. The yeast *spt14* gene is homologous to the human *PIG-A* gene and is required for GPI anchor synthesis. *EMBO J.* **14:** 1637.

Schreiner, R., E. Schnabel, and F. Wieland. 1994. Novel *N*-glycosylation in eukaryotes: Laminin contains the linkage unit β-glucosylasparagine. *J. Cell Biol.* **124:** 1071.

Schreuder, M.P., S. Brekelmans, H. van den Ende, and F.M. Klis. 1993. Targeting of a heterologous protein to the cell wall of *Saccharomyces cerevisiae*. *Yeast* **9:** 399.

Schülke, N. and F.X. Schmid. 1988. Effect of glycosylation on the mechanism of renaturation of invertase from yeast. *J. Biol. Chem.* **263:** 8832.

Schutzbach, J. S. and J.W. Zimmerman. 1992. Yeast dolichyl-phosphomannose synthase: Reconstitution of enzyme activity with phospholipids. *Biochem. Cell Biol.* **70:** 460.

Schwaiger, H., A. Hasilik, K. von Figura, A. Wiemken, and W. Tanner. 1982. Carbohydrate-free carboxypeptidase Y is transferred into the lysosome-like yeast vacuole. *Biochem. Biophys. Res. Commun.* **104:** 950.

Semino, C.E. and P.W. Robbins. 1995. Synthesis of "Nod"-like chitin oligosaccharides by the *Xenopus* developmental protein DG42. *Proc. Natl. Acad. Sci.* **92:** 3498.

Semino, C.E., C.A. Specht, A. Raimondi, and P.W. Robbins. 1996. Homologs of the *Xenopus* developmental gene *DG42* are present in zebrafish and mouse and are involved in the synthesis of Nod-like chitin oligosaccharides during early embryogenesis. *Proc. Natl. Acad. Sci.* **93:** 4548.

Sentandreu, R. and Northcote, D.H. 1968. The structure of a glycopeptide isolated from the yeast cell wall. *Biochem. J.* **109:** 419.

Shankar, C.S. and S. Umesh-Kumar. 1994. A surface lectin associated with flocculation in brewing strains of *Saccharomyces cerevisiae*. *Microbiology* **140:** 1097.

Sharma, C.B., C. D'Souza, and A.D. Elbein. 1991. Partial purification of a mannosyltransferase involved in the *O*-mannosylation of glycoproteins from *Saccharomyces cerevisiae. Glycobiology* **1:** 367.

Sharma, C.B., L. Lehle, and W. Tanner. 1981. *N*-glycosylation of yeast proteins. Characterization of the solubilized oligosaccharyl transferase. *Eur. J. Biochem.* **116:** 101.

Shaw, J.A., P.C. Mol, B. Bowers, S.J. Silverman, M.H. Valdivieso, A. Durán, and E. Cabib. 1991. The function of chitin synthases 2 and 3 in the *Saccharomyces cerevisiae* cell cycle. *J. Cell. Biol.* **114:** 111.

Shematek, E.M., and E. Cabib. 1980. Biosynthesis of the yeast cell wall. II. Regulation of β-(1→3)glucan synthetase by ATP and GTP. *J. Biol. Chem.* **255:** 895.

Shematek, E.M., J.A. Braatz, and E. Cabib. 1980. Biosynthesis of the yeast cell wall. I. Preparation and properties of β-(1→3)glucan synthetase. *J. Biol. Chem.* **255:** 888.

Shibata, N., K. Mizugami, K. Takano, and S. Suzuki. 1983. Isolation of mannan-protein complexes from viable cells of *Saccharomyces cerevisiae* X2180-1A wild type and *Saccharomyces cerevisiae* X2180-1A-5 mutant strains by the action of Zymolyase-60,000. *J. Bacteriol.* **156:** 552.

Shimizu, J., K. Yoda, and M. Yamasaki. 1994. The hypo-osmolarity-sensitive phenotype of the *Saccharomyces cerevisiae hpo2* mutant is due to a mutation in *PKC1*, which regulates expression of β-glucanase. *Mol. Gen. Genet.* **242:** 641.

Shiota, M., T. Nakajima, A. Satoh, M. Shida, and K. Matsuda. 1985. Comparison of β-glucan structures in a cell wall mutant of *Saccharomyces cerevisiae* and the wild type. *J. Biochem.* **98:** 1301.

Sietsma, J.H. and J.G.H. Wessels. 1990. The occurrence of glucosaminoglycan in the wall of *Schizosaccharomyces pombe. J. Gen. Microbiol.* **136:** 2261.

Silberstein, S. and R. Gilmore. 1996. Biochemistry, molecular biology, and genetics of the oligosaccharyltransferase. *FASEB J.* **10:** 849.

Silberstein, S., P.G. Collins, D.J. Kelleher, and R. Gilmore. 1995a. The essential *OST2* gene encodes the 16-kD subunit of the yeast oligosaccharyltransferase, a highly conserved protein expressed in diverse eukaryotic organisms. *J. Cell. Biol.* **131:** 371.

Silberstein, S., P.G. Collins, D.J. Kelleher, P.J. Rapiejko, and R. Gilmore. 1995b. The α subunit of the *Saccharomyces cerevisiae* oligosaccharyltransferase complex is essential for vegetative growth of yeast and is homologous to mammalian ribophorin I. *J. Cell Biol.* **128:** 525.

Silverman, S.J. 1989. Similar and different domains of chitin synthases 1 and 2 of *S. cerevisiae*: two isozymes with distinct functions. *Yeast* **5:** 459.

Silverman, S.J., A. Sburlati, M.L. Slater, and E. Cabib. 1988. Chitin synthase 2 is essential for septum formation and cell division in *Saccharomyces cerevisiae. Proc. Natl. Acad. Sci.* **85:** 4735.

Sipos, G., A. Puoti, and A. Conzelmann. 1994. Glycosylphosphatidylinositol membrane anchors in *Saccharomyces cerevisiae:* Absence of ceramides from complete precursor glycolipids. *EMBO J.* **13:** 2789.

———. 1995. Biosynthesis of the side chain of yeast glycosylphosphatidylinositol anchors is operated by novel mannosyltransferases located in the endoplasmic reticulum and the Golgi apparatus. *J. Biol. Chem* **270:** 19709.

Slater, M.L., B. Bowers, and E. Cabib. 1985. Formation of septum-like structures at locations remote from the budding sites in cytokinesis-defective mutants of *Saccharomyces cerevisiae. J. Bacteriol.* **162:** 763.

Sloat, B.F., A. Adams, and J.R. Pringle. 1981. Roles of the *CDC24* gene product in cellular morphogenesis during the *Saccharomyces cerevisiae* cell cycle. *J. Cell Biol.* **89:** 395.

Smith, D.J., A. Proudfoot, L. Friedli, L.S. Klig, G. Paravicini, and M.A. Payton. 1992. *PMI40*, an intron-containing gene required for early steps in yeast mannosylation. *Mol. Cell. Biol.* **12:** 2924.

Smith, S.W. and R.L. Lester. 1974. Inositol phosphorylceramide, a novel substance and the chief member of a major group of yeast sphingolipids containing a single inositol phosphate. *J. Biol. Chem.* **249:** 3395.

Sprague, G.F., Jr. and J.W. Thorner. 1992. Pheromone response and signal transduction during the mating process of *Saccharomyces cerevisiae.* In *The molecular and cellular biology of the yeast* Saccharomyces: *Gene expression* (ed. E.W. Jones et al.), p. 657. Cold Spring Harbor Laboratory Press, Cold Spring Harbor, New York.

Song, M.R., D.W. Lee, S.W. Park, K.S. Bae, and H.M. Park. 1992. Isolation and characterization of *Saccharomyces cerevisiae* mutants deficient in (1→3)-β-D-glucan synthase. *Korean J. Appl. Microbiol. Biotechnol.* **20:** 642.

Stagljar, I., S. te Heesen, and M. Aebi. 1994. New phenotype of mutations deficient in glucosylation of the lipid-linked oligosaccharide: Cloning of the *ALG8* locus. *Proc. Natl. Acad. Sci.* **91:** 5977.

Staniszlo, P.J., M.S. Kang, and E. Cabib. 1985. Stimulation of β(1→3)glucan synthetase of various fungi by nucleoside triphosphates: Generalized regulatory mechanism for cell wall biosynthesis. *J. Bacteriol.* **161:** 1188.

Strahl-Bolsinger, S. and W. Tanner. 1991. Protein *O*-glycosylation in *Saccharomyces cerevisiae.* Purification and characterization of the dolichyl-phosphate-D-mannose-protein *O*-D-mannosyltransferase. *Eur. J. Biochem.* **196:** 185.

Strahl-Bolsinger, S., T. Immervoll, R. Deutzmann, and W. Tanner. 1993. *PMT1*, the gene for a key enzyme of protein *O*-glycosylation in *Saccharomyces cerevisiae. Proc. Natl. Acad. Sci.* **90:** 8164.

Stratford, M. 1992a. Yeast flocculation: Reconciliation of physiological and genetic viewpoints. *Yeast* **8:** 25.

———. 1992b. Yeast flocculation: Receptor definition by *mnn* mutants and concanavalin A. *Yeast* **8:** 635.

———. 1994. Another brick in the wall? Recent developments concerning the yeast cell envelope. *Yeast* **10:** 1741.

Straver, M.H., V.M. Traas, G. Smit, and J.W. Kijne. 1994. Isolation and partial purification of mannose-specific agglutinin from brewer's yeast involved in flocculation. *Yeast* **10:** 1183.

Sudoh, M., H. Shimada, M. Arisawa, K. Yano, and M. Takagi. 1991. Yeast flocculation mediated by overexpression of the chitin synthase 2 gene in *Saccharomyces cerevisiae. Agric. Biol. Chem.* **55:** 2901.

Sudoh, M., S. Nagahashi, M. Doi, A. Ohta, M. Takagi, and M. Arisawa. 1993. Cloning of the chitin synthase 3 gene from *Candida albicans* and its expression during yeast-hyphal transition. *Mol. Gen. Genet.* **241:** 351.

Szkopinska, A., L. Nowak, E. Swiezewska, and G. Palamarczyk. 1988. CTP-dependent lipid kinases of yeast. *Arch. Biochem. Biophys.* **266:** 124.

Takahashi, M., N. Inoue, K. Ohishi, Y. Maeda, N. Nakamura, Y. Endo, T. Fujita, J. Takeda, and T. Kinoshita. 1996. PIG-B, a membrane protein of the endoplasmic reticulum with a large lumenal domain, is involved in transferring the third mannose of the GPI anchor. *EMBO J.* **15**: 4254.

Takai, Y., T. Sasaki, K. Tanaka, and H. Nakanishi. 1995. Rho as a regulator of the cytoskeleton. *Trends Biochem. Sci.* **20:** 227.

Takeda, J., T. Miyata, K. Kawagoe, Y. Iida, Y. Endo, T. Fujita, M. Takahashi, T. Kitani, and T. Kinoshita. 1993. Deficiency of the GPI anchor caused by a somatic mutation of the *PIG-A* gene in paroxysmal nocturnal hemoglobinuria. *Cell* **73:** 703.

Takita, M.A. and B. Castilho-Valavicius. 1993. Absence of cell wall chitin in *Saccharomyces cerevisiae* leads to resistance to *Kluyveromyces lactis* killer toxin. *Yeast* **9:** 589.

Tammi, M., L. Ballou, A. Taylor, and C.E. Ballou. 1987. Effect of glycosylation on yeast invertase oligomer stability. *J. Biol. Chem.* **262:** 4395.

Tanner, W. and L. Lehle. 1987. Protein glycosylation in yeast. *Biochim. Biophys. Acta* **906:** 81.

te Heesen, S. and M. Aebi. 1994. The genetic interaction of *kar2* and *wbp1* mutations. Distinct functions of binding protein BiP and *N*-linked glycosylation in the processing pathway of secreted proteins in *Saccharomyces cerevisiae. Eur. J. Biochem.* **222:** 631.

te Heesen, S., B. Janetzky, L. Lehle, and M. Aebi. 1992. The yeast *WBP1* is essential for oligosaccharyltransferase activity *in vivo* and *in vitro. EMBO J.* **11:** 2071.

te Heesen, S., R. Knauer, L. Lehle, and M. Aebi. 1993. Yeast Wbp1p and Swp1p form a protein complex essential for oligosaccharyltransferase activity. *EMBO J.* **12:** 279.

te Heesen, S., L. Lehle, A. Weissmann, and M. Aebi. 1994. Isolation of the *ALG5* locus encoding the UDP-glucose:dolichyl-phosphate glucosyltransferase from *Saccharomyces cerevisiae. Eur. J. Biochem.* **224:** 71.

Terrance, K., P. Heller, Y.-S. Wu, and P.N. Lipke. 1987. Identification of glycoprotein components of α-agglutinin, a cell adhesion protein from *Saccharomyces cerevisiae. J. Bacteriol.* **169:** 475.

Teunissen, A.W.R.H., E. Holub, J. van der Hucht, J. A. van den Berg, and H.Y. Steensma. 1993. Sequence of the open reading frame of the *FLO1* gene from *Saccharomyces cerevisiae. Yeast* **9:** 423.

Tkacz, J.S. 1984. *In vivo* synthesis of β-1,6-glucan in *Saccharomyces cerevisiae.* In *Microbial cell wall synthesis and autolysis* (ed. C. Nombela), p. 287. Elsevier, Amsterdam.

Torres, L., H. Martin, M.I. Garcia-Saez, J. Arroyo, M. Molina, M. Sanchez, and C. Nombela. 1991. A protein kinase gene complements the lytic phenotype of *Saccharomyces cerevisiae lyt2* mutants. *Mol. Microbiol.* **5:** 2845.

Trimble, R.B. and P.H. Atkinson. 1992. Structural heterogeneity in the $Man_{8-13}GlcNAc$ oligosaccharides from log-phase *Saccharomyces* yeast: A one- and two-dimensional ^1H NMR spectroscopic study. *Glycobiology.* **2:** 57.

Trimble, R.B., F. Maley, and F.K. Chu. 1983. Glycoprotein biosynthesis in yeast. Protein conformation affects processing of high mannose oligosaccharides on carboxypeptidase Y and invertase. *J. Biol. Chem.* **258:** 2562.

Trimble, R.B., F. Maley, and A.L. Tarentino. 1980. Characterization of large oligosaccharide-lipids synthesized *in vitro* by microsomes from *Saccharomyces cerevisiae. J. Biol. Chem.* **255:** 10232.

Trueheart, J. and G.R. Fink. 1989. The yeast cell fusion protein FUS1 is *O*-glycosylated and spans the plasma membrane. *Proc. Natl. Acad. Sci.* **86:** 9916.

Trueheart, J., J.D. Boeke, and G.R. Fink. 1987. Two genes required for cell fusion during yeast conjugation: Evidence for a pheromone-induced surface protein. *Mol. Cell. Biol.* **7:** 2316.

Tsai, P.-K., J. Frevert, and C.E. Ballou. 1984. Carbohydrate structure of *Saccharomyces cerevisiae mnn9* mannoprotein. *J. Biol. Chem.* **259:** 3805.

Uchida, Y., O. Shimmi, M. Sudoh, M. Arisawa, and H. Yamada-Okabe. 1996. Characterization of chitin synthase 2 of *Saccharomyces cerevisiae*. II. Both full-size and processed enzymes are active for chitin synthesis. *J. Biochem.* **119**: 659.

Udenfriend, S. and K. Kodukula. 1995. How glycosyl-phosphatidylinositol-anchored membrane proteins are made. *Annu. Rev. Biochem.* **64:** 563.

Vai, M., L. Popolo, and L. Alberghina. 1987. Effect of tunicamycin on cell cycle progression in budding yeast. *Exp. Cell Res.* **171:** 448.

Vai, M., E. Gatti, E. Lacanà, L. Popolo, and L. Alberghina. 1991. Isolation and deduced amino acid sequence of the gene encoding gp115, a yeast glycophospholipid-anchored protein containing a serine-rich region. *J. Biol. Chem.* **266:** 12242.

Vai, M., I. Orlandi, P. Cavadini, L. Alberghina, and L. Popolo. 1996. *Candida albicans* homologue of *GGP1/GAS1* gene is functional in *Saccharomyces cerevisiae* and contains the determinants for glycosylphosphatidylinositol attachment. *Yeast* **12:** 361.

Vai, M., L. Popolo, R. Grandori, E. Lacanà, and L. Alberghina. 1990. The cell cycle modulated glycoprotein GP115 is one of the major yeast proteins containing glycosylphosphatidylinositol. *Biochim. Biophys. Acta* **1038:** 277.

Valdivieso, M.H., P.C. Mol, J.A. Shaw, E. Cabib, and A. Durán. 1991. *CAL1*, a gene required for activity of chitin synthase 3 in *Saccharomyces cerevisiae*. *J. Cell Biol.* **114:** 101.

Valentin, E., E. Herrero, and R. Sentandreu. 1986. Incorporation of mannoproteins into the walls of aculeacin A-treated yeast cells. *Arch. Microbiol.* **146:** 214.

Valentin, E., E. Herrero, F.I.J. Pastor, and R. Sentandreu. 1984. Solubilization and analysis of mannoprotein molecules from the cell wall of *Saccharomyces cerevisiae*. *J. Gen. Microbiol.* **130:** 1419.

Valentin, E., E. Herrero, H. Rico, F. Miragall, and R. Sentandreu. 1987. Cell wall mannoproteins during the population growth phases in *Saccharomyces cerevisiae*. *Arch. Microbiol.* **148:** 88.

van Berkel, M.A.A., L.H.P. Caro, R.C. Montijn, and F.M. Klis. 1994. Glucosylation of chimeric proteins in the cell wall of *Saccharomyces cerevisiae*. *FEBS Lett.* **349:** 135.

van der Vaart, J.M., L.H.P. Caro, J.W. Chapman, F.M. Klis, and C.T. Verrips. 1995. Identification of three mannoproteins in the cell wall of *Saccharomyces cerevisiae*. *J. Bacteriol.* **177:** 3104.

Vannini, G.L., F. Poli, A. Donini, and S. Pancaldi. 1983. Effects of Congo red on wall synthesis and morphogenesis in *Saccharomyces cerevisiae*. *Plant Sci. Lett.* **31:** 9.

van Rinsum, J., F.M. Klis, and H. van den Ende. 1991. Cell wall glucomannoproteins of *Saccharomyces cerevisiae mnn9*. *Yeast* **7:** 717.

Varki, A. 1996. Does DG42 synthesize hyaluronan or chitin? A controversy about oligosaccharides in vertebrate development. *Proc. Natl. Acad. Sci.* **93:** 4523.

Venkov, P.V., A.A. Hadjiolov, E. Battaner, and D. Schlessinger. 1974. *Saccharomyces cerevisiae:* Sorbitol-dependent fragile mutants. *Biochem. Biophys. Res. Commun.* **56:** 599.

Verostek, M.F., P.H. Atkinson, and R.B. Trimble. 1991. Structure of *Saccharomyces cerevisiae alg3, sec18* mutant oligosaccharides. *J. Biol. Chem.* **266:** 5547.

———. 1993a. Glycoprotein biosynthesis in the *alg3 Saccharomyces cerevisiae* mutant. I. Role of glucose in the initial glycosylation of invertase in the endoplasmic reticulum. *J. Biol. Chem.* **268:** 12095.

———. 1993b. Glycoprotein biosynthesis in the *alg3 Saccharomyces cerevisiae* mutant. II. Structure of novel $Man_{6-10}GlcNAc_2$ processing intermediates on secreted invertase.

J. Biol. Chem. **268:** 12104.

Vossen, J.H., A.F.J. Ram, and F.M. Klis. 1995. Identification of *SPT14 / CWH6* as the yeast homologue of hPIG-A, a gene involved in the biosynthesis of GPI anchors. *Biochim. Biophys. Acta* **1243:** 549.

Watari, J., Y. Takata, M. Ogawa, H. Sahara, S. Koshino, M.-L. Onnela, U. Airaksinen, R. Jaatinen, M. Penttila, and S. Keränen. 1994. Molecular cloning and analysis of the yeast flocculation gene *FLO1*. *Yeast* **10:** 211.

Watzele, G. and W. Tanner. 1989. Cloning of the glutamine: Fructose-6-phosphate amidotransferase gene from yeast. Pheromonal regulation of its transcription. *J. Biol. Chem.* **264:** 8753.

Watzele, M., F. Klis, and W. Tanner. 1988. Purification and characterization of the inducible **a** agglutinin of *Saccharomyces cerevisiae*. *EMBO. J.* **7:** 1483.

Weinstock, K.G. and C.E. Ballou. 1987. Tunicamycin inhibition of epispore formation in *Saccharomyces cerevisiae*. *J. Bacteriol.* **169:** 4384.

Whelan, W.L. and C.E. Ballou. 1975. Sporulation in D-glucosamine auxotrophs of *Saccharomyces cerevisiae:* Meiosis with defective ascospore wall formation. *J. Bacteriol.* **125:** 1545.

Winther, J.R., T.H. Stevens, and M.C. Kielland-Brandt. 1991. Yeast carboxypeptidase Y requires glycosylation for efficient intracellular transport, but not for vacuolar sorting, *in vivo* stability, or activity. *Eur. J. Biochem.* **197:** 681.

Wojciechowicz, D., C.-F. Lu, J. Kurjan, and P.N. Lipke. 1993. Cell surface anchorage and ligand-binding domains of the *Saccharomyces cerevisiae* cell adhesion protein α-agglutinin, a member of the immunoglobulin superfamily. *Mol. Cell. Biol.* **13:** 2554.

Yabe, T., T. Yamada-Okabe, S. Kasahara, Y. Furuichi, T. Nakajima, E. Ichishima, M. Arisawa, and H. Yamada-Okabe. 1996. *HKR1* encodes a cell surface protein that regulates both cell wall β-glucan synthesis and budding pattern in the yeast *Saccharomyces cerevisiae*. *J. Bacteriol.* **178:** 477.

Yamochi, W., K. Tanaka, H. Nonaka, A, Maeda, T. Musha, and Y. Takai. 1994. Growth site localization of Rho1 small GTP-binding protein and its involvement in bud formation in *Saccharomyces cerevisiae*. *J. Cell Biol.* **125:** 1077.

Yanagashima, N. 1984. Mating systems and sexual interactions in yeast. In *Encyclopedia of plant physiology*, New Series (ed. H.F. Linskens and J.H. Heslop-Harrison), vol. 17, p. 402. Springer Verlag, Berlin.

Yanagisawa, K., D. Resnick, C. Abeijon, P.W. Robbins, and C.B. Hirschberg. 1990. A guanosine diphosphatase enriched in Golgi vesicles of *Saccharomyces cerevisiae*. Purification and characterization. *J. Biol. Chem.* **265:** 19351.

Yip, C.L., S.K. Welch, F. Klebl, T. Gilbert, P. Seidel, F.J. Grant, P.J. O'Hara, and V.L. MacKay. 1994. Cloning and analysis of the *Saccharomyces cerevisiae MNN9* and *MNN1* genes required for complex glycosylation of secreted protteins. *Proc. Natl. Acad. Sci.* **91:** 2723.

Yu, L., R. Goldman, P. Sullivan, G.F. Walker, and S.W. Fesik. 1993. Heteronuclear NMR studies of ^{13}C-labeled yeast cell wall b-glucan oligosaccharides. *J. Biomol. NMR* **3:** 429.

Zhao, C., T. Beeler, and T. Dunn. 1994. Suppressors of the Ca^{2+}-sensitive yeast mutant (*csg2*) identify genes involved in sphingolipid biosynthesis. Cloning and characterization of *SCS1*, a gene required for serine palmitoyltransferase activity. *J. Biol. Chem.* **269:** 21480.

Ziegler, F.D. and R.B. Trimble. 1991. Glycoprotein biosynthesis in yeast: Purification and characterization of the endoplasmic reticulum processing α-mannosidase. *Glycobiology* **1:** 605.

Ziegler, F.D., F. Maley, and R.B. Trimble. 1988. Characterization of the glycosylation sites in yeast external invertase. II. Location of the endo-β-*N*-acetylglucosaminidase H-resistant sequons. *J. Biol. Chem.* **263:** 6986.

Ziman, M., J.S. Chuang, and R.W. Schekman. 1996. Chs1p and Chs3p, two proteins involved in chitin synthesis, populate a compartment of the *Saccharomyces cerevisiae* endocytic pathway. *Mol. Biol. Cell* **7**: 1909.

Zlotnik, H., M.P. Fernandez, B. Bowers, and E. Cabib. 1984. *Saccharomyces cerevisiae* mannoproteins form an external cell wall layer that determines wall porosity. *J. Bacteriol.* **159:** 1018.

Zueco, J., S. Mormeneo, and R. Sentandreu. 1986. Temporal aspects of the *O*-glycosylation of *Saccharomyces cerevisiae* mannoproteins. *Biochim. Biophys. Acta* **884:** 93.

Zufferey, R., R. Knauer, P. Burda, I. Stagljar, S. te Heesen, L. Lehle, and M. Aebi. 1995. *STT3*, a highly conserved protein required for yeast oligosaccharyl transferase activity *in vivo*. *EMBO J.* **14:** 4949.

4

Biogenesis and Function of the Yeast Vacuole

Elizabeth W. Jones, Gene C. Webb, and Mark A. Hiller
Carnegie Mellon University
Pittsburgh, Pennsylvania 15213

Yeast III © 1997 Cold Spring Harbor Laboratory Press 0-87969-356-8/97 $5 + .00

I. INTRODUCTION

A defining characteristic of all eukaryotic cells is the presence of distinct endomembrane organelles. These compartments allow a cell to carry out competing processes in separate environments suited to the particular requirements of each process (e.g., biosynthesis vs. degradation). The presence of endomembrane organelles requires not only the formation and maintenance of the lipid components that delineate the different compartments, but also the synthesis and targeting of proteins that serve either structural or enzymatic functions in the organelles. The vacuole of *Saccharomyces cerevisiae* is a large organelle, comprising as much as 25% of the cellular volume (Wiemken and Durr 1974). There are some indications that vacuolar morphology is correlated with the cell's position in the cell cycle (Hartwell 1970; Wiemken et al. 1970; Jones et al. 1993); logarithmically growing cells often have a multilobed vacuole, with interconnections between lobes indicative of a reticulum, whereas stationary-phase cells and cells in G_1 typically have a single large vacuole that is relatively less dense (Hartwell 1970; Wiemken et al. 1970; Preston et al. 1989; Pringle et al. 1989; Jones et al. 1993). The vacuole is a very dynamic organelle. When cells are starved of a carbon source, the vacuolar lobes rapidly fuse to give a single large vacuole that is less dense than prior to fusion; it will be phase bright, whereas the reticular vacuole is not. Refeeding results in a rapid return to the denser reticular form (Pringle et al. 1989).

The yeast vacuole shares features with lysosomes. It contains an ensemble of hydrolases, including endo- and exoproteases, ribonuclease(s), endo- and exopolyphosphatases, α-mannosidase, trehalase, and alkaline phosphatase (Table 1) (Wiemken et al. 1979). It is an acidic compartment, maintained at pH 6.2 (Preston et al. 1989), with a proton gradient generated and maintained by a V-type ATPase (Anraku 1987). The vacuole also serves as a reservoir for a number of small molecules, including phosphate and polyphosphate, amino acids, *S*-adenosylmethionine, allantoin and allantoate, several divalent cations (Ca^{++}, Mg^{++}, Mn^{++}), and other ions and small molecules (Wiemken et al. 1979; Cooper 1982; Serrano 1991).

II. DELIVERY TO THE VACUOLE

The vacuole receives input from several different pathways. Many newly synthesized proteins reach the vacuole via the secretory pathway. There are some indications that some vacuolar membrane proteins may travel a route separate from that transited by soluble proteins, except for steps at the very beginning and very end of the pathway for soluble proteins. A select few enzymes reach the vacuole via an extrasecretory cytoplasm-to-vacuole pathway. In addition, if the secretory pathway is blocked, some membrane proteins may travel to the vacuole via the plasma membrane and the endocytic pathway. No evidence excludes travel via the plasma membrane as a route normally followed by some proteins, as is the case for acid phosphatase in animal cells (Braun et al. 1989; Gottschalk et al. 1989). The vacuolar ATPase is a special case. The membrane sector is synthesized via the secretory pathway, whereas the catalytic sector is synthesized in the cytoplasm and associates reversibly with the membrane sector in a regulated fashion.

Proteins are also delivered to the vacuole for degradation. Some membrane proteins proceed there directly from their normal locations in the Golgi cisternae, presumably by a normal delivery pathway, others from the plasma membrane via (often selective) endocytosis, and still others via autophagy. Soluble proteins may be selectively imported along a newly described vesicular pathway or via autophagy. In addition, the α-factor pheromone is delivered via receptor-mediated endocytosis.

Ions and small molecules can obviously reach the vacuole via fluid-phase endocytosis. Amino acids, purines, ions, and some other small molecules may also be transported through the plasma membrane and, in a second step, into the vacuole.

A. Secretory Pathway

Most vacuolar hydrolytic enzymes are synthesized as inactive precursors, whether they reach the vacuole via the secretory pathway or the extra-

Table 1 Vacuolar hydrolases

Hydrolase	Abbrev.	Gene (ORF aa)	Activity	Molecular mass (kD)	Glycosylation	Route to vacuole	Precursor processed by	Function
Soluble hydrolases								
Proteinase A	PrA	*PEP4* (405)	aspartyl endoproteinase	42	N-linked (2)	secretory	self(PrA), PrB	hydrolase precursor processing especially as initiator of cascade; protein degradation
Proteinase B	PrB	*PRB1* (635)	serine endoproteinase	31	O-linked; precursor-1 N-linked	secretory	PrA,PrB	hydrolase precursor processing; protein degradation
Carboxypeptidase Y	CpY	*PRC1* (532)	serine carboxypeptidase	61	N-linked (4)	secretory	PrA, PrB	peptide metabolism
Carboxypeptidase S	CpS	*CPS1* (576)	metallo (Zn^{++}) carboxypeptidase	74, 77	N-linked (2–3)	secretory	PrB, (PrA?)	peptide metabolism
Aminopeptidase I	ApI	*LAP4* (514)	metallo (Zn^{++}) aminopeptidase	50[a]	none	extra-secretory	PrB	
Aminopeptidase Y	ApY	*APE3* (537)	metallo (Zn^{++}) aminopeptidase activated by Co^{++}	70, 75	N-linked (5–8)	secretory	PrB	

α-Mannosidase	αMS	*AMS1* (1083)	α-mannosidase	73, 107 s.u. 560	none	extra-secretory	PrA[b]	degradation of glycoconjugates
Trehalase				220	N-linked	secretory	PrA[b]	
RNase(s)							PrA[b]	
Endopolyphosphatase							PrA[b]	
Exopolyphosphatase	ScPPX2						PrA[b]	
Exopolyphosphatase	ScPPX3							
Vacuolar glycoprotein	Vgp72			72, 90–150	N-linked	secretory	PrA[b]	
Integral membrane hydrolases								
Alkaline phosphatase	AlP	*PHO8* (566)	nonspecific alkaline phosphatase	72	N-linked (2)	secretory	PrA, PrB	removes phosphate from phosphopeptides
Dipeptidylaminopeptidase	DPAPB	*DAP2* (818)	type IV dipeptidase	120	N-linked (6–7)	secretory	none	

[a]See text.
[b]*PEP4*-dependent.

secretory cytoplasm-to-vacuole pathway. The vacuolar hydrolases that follow the secretory route undergo posttranslational modifications in the form of glycosylation and peptide cleavages while transiting through these compartments; those imported via the extrasecretory pathway may undergo proteolytic cleavages but they are not glycosylated (Figs. 1 and 2 and Table 1) (for review, see Jones et al. 1989; Jones 1991b; Jones and Murdock 1994).

1. Membrane Enzymes

The route for delivery of enzymes of the vacuolar membrane has been investigated for two enzymes: DPAPB (dipeptidylaminopeptidase B) and AlP (alkaline phosphatase). Both enzymes are classified as type II membrane proteins, possessing short amino-terminal cytoplasmic domains, subterminal *trans*-membrane domains, and substantial carboxy-terminal luminal domains. Both are glycoproteins. DPAPB and AlP reach the vacuole via the endoplasmic reticulum (ER) and the Golgi, for both proteins receive glycosyl modifications catalyzed by ER and Golgi enzymes and are blocked in transit to the vacuole by mutations that block passage from the ER to the Golgi membranes (*sec18*) or within the Golgi (*sec7*). Neither protein is affected by mutations, like *sec1*, that block late steps in the secretory pathway (Klionsky and Emr 1989; Roberts et al. 1989). If the direct route to the vacuole is blocked by the *vps1* mutation, AlP can still reach the vacuole, albeit with slower kinetics. The protein passes via the late stages in the secretory pathway to the plasma membrane, from which it is taken up by endocytosis and delivered to the vacuole (Nothwehr et al. 1995). Whether DPAPB follows a similar route in the *vps1* mutant has not been examined.

AlP passes to the vacuole as a zymogen that must be activated by proteolytic cleavage(s) at the carboxyl terminus (Jones et al. 1982; Kaneko et al. 1985, 1987; Klionsky and Emr 1989). The cleavage does not occur in the *pep4* mutant (Jones et al. 1982; Klionsky and Emr 1989). The evidence suggests that either PrA or PrB can catalyze activating cleavages, for AlP activity is produced when PrA is absent but PrB is present (during *PRB1*-dependent phenotypic lag in growth of *pep4* mutant spore clones) (Zubenko et al. 1982), as well as when PrA is present but PrB is absent (in *prb1* mutants) (A. Heinzer et al., unpubl.), but not when both are absent (in the *pep4* mutant after emerging from phenotypic lag). DPAPB, in contrast, does not require or undergo proteolytic cleavages (Roberts et al. 1989).

2. *Soluble Hydrolases*

a. PrA. PrA is a glycoprotein that is translocated into the ER and traverses the ER and Golgi en route to the vacuole (Klionsky et al. 1988). Three proteolytic cleavages can occur during posttranslational maturation of the precursor. After scission of the signal peptide (Klionsky et al. 1988), probably between amino acids 22 and 23 (Wolff et al. 1996), the product polypeptide undergoes an autocatalytic cleavage, probably between amino acids 67 and 68, and probably in the vacuole, generating an enzymatically active species (Rupp et al. 1991; van den Hazel et al. 1992; Woolford et al. 1993; Wolff et al. 1996). The mature amino terminus at residue 76 is generated by PrB (Jones et al. 1989; Hirsch et al. 1992; van den Hazel et al. 1992; Woolford et al. 1993). The ultimate precursor has been detected in the *prb1*Δ mutant but not kinetically.

b. PrB. PrB is a glycoprotein. All of the precursors to protease B carry one N-linked glycosyl side chain; the mature enzyme carries only O-linked glycosylation (Mechler et al. 1982; Moehle et al. 1987, 1989). Analyses of *sec7, sec11, sec14,* and *sec18* mutants, as well as the glycosylation patterns, indicate that Prb1p passes through the ER and Golgi en route to the vacuole (Mechler et al. 1988; Moehle et al. 1989; V. Nebes and E. Jones, unpubl.). The rapidity of the kinetics suggests transit directly from the Golgi to the endosome and vacuole, although passage via the plasma membrane has not been rigorously excluded (Moehle et al. 1989). Four proteolytic cleavages occur during posttranslational maturation of Prb1p. Removal of the signal peptide by signal peptidase (V. Nebes and E. Jones, unpubl.) and autocatalytic scission of a large amino-terminal propeptide between residues 280 and 281 occur in the ER (Mechler et al. 1988; Moehle et al. 1989; Nebes and Jones 1991; Hirsch et al. 1992). The third cleavage, which converts a 40-kD intermediate to a 37-kD species by removal of a carboxy-terminal peptide, is catalyzed by PrA, presumably in the vacuole (Moehle et al. 1989). The final cleavage, which yields mature enzyme, may result from PrB cleavage or autocatalysis (Moehle et al. 1989; Nebes and Jones 1991). The fourth cleavage is presumed to occur in the vacuole and removes the N-linked glycosyl chain together with its peptide. This final PrB-catalyzed cleavage is not dependent on prior cleavage by PrA as evidenced by *PRB1*-dependent phenotypic lag (Zubenko et al. 1982).

c. CpY. CpY is a glycoprotein bearing four N-linked glycosyl side chains. Analysis of the *sec1*, *sec7*, and *sec18* mutants indicates that Prc1p

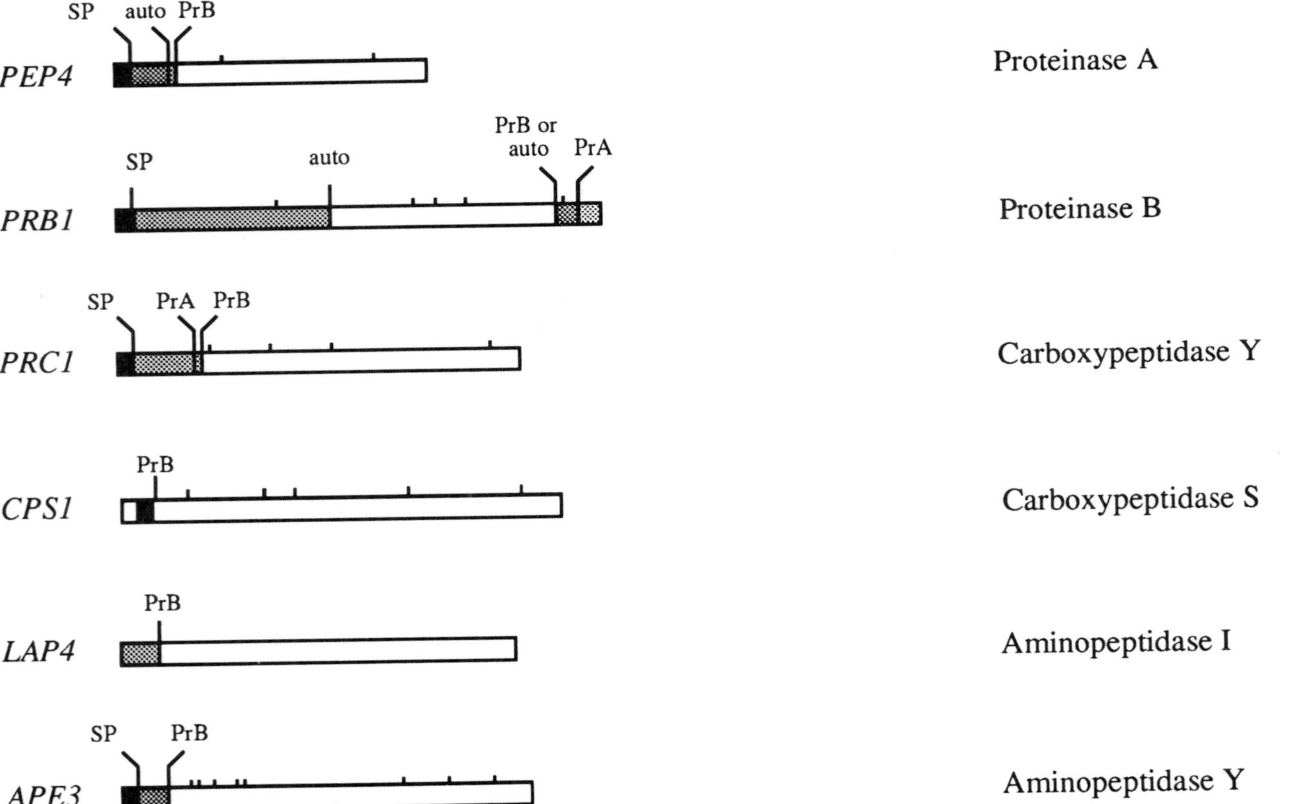
SP
auto
PrB
PEP4
Proteinase A
PrB or auto
SP
auto
PrA
PRB1
Proteinase B
SP
PrA
PrB
PRC1
Carboxypeptidase Y
PrB
CPS1
Carboxypeptidase S
PrB
LAP4
Aminopeptidase I
SP
PrB
APE3
Aminopeptidase Y

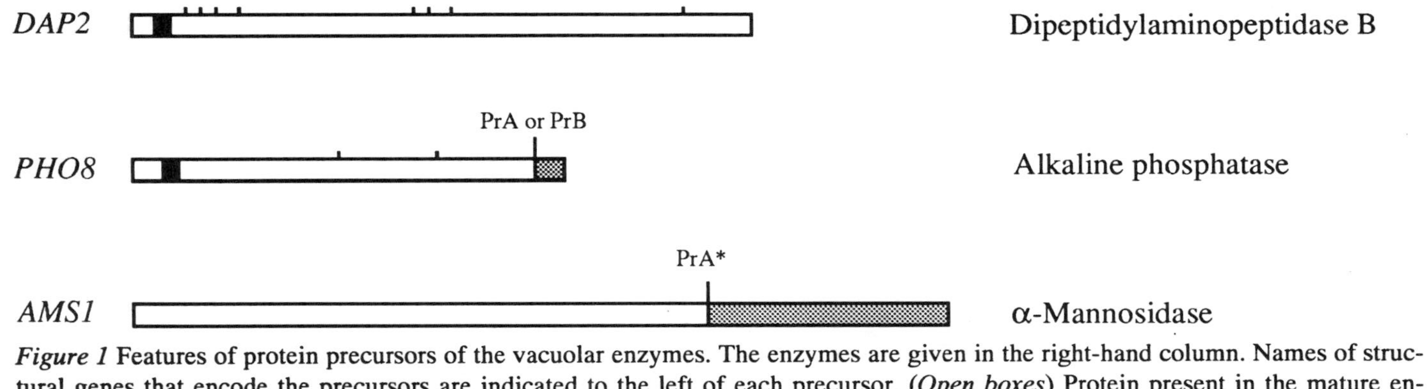

Figure 1 Features of protein precursors of the vacuolar enzymes. The enzymes are given in the right-hand column. Names of structural genes that encode the precursors are indicated to the left of each precursor. (*Open boxes*) Protein present in the mature enzyme; (*closed boxes*) hydrophobic peptides that function either as signal sequences or transmembrane domains; (*lightly shaded boxes*) peptides removed during processing. Sites of proteolytic cleavage and the enzymes that catalyze them are indicated above each precursor. Cleavage enzymes: (PrA) proteinase A; (PrB) proteinase B; (SP) signal peptidase; (auto) autocatalytic. PrA* indicates that cleavage does not occur in a *pep4* mutant; cleavage might be PrA- and/or PrB-dependent. Short lines above cartoons of precursors correspond to potential sites of *N*-glycosylation for precursors that transit the secretory pathway; such sites are not so identified for Lap4p and Ams1p, which follow an extrasecretory pathway.

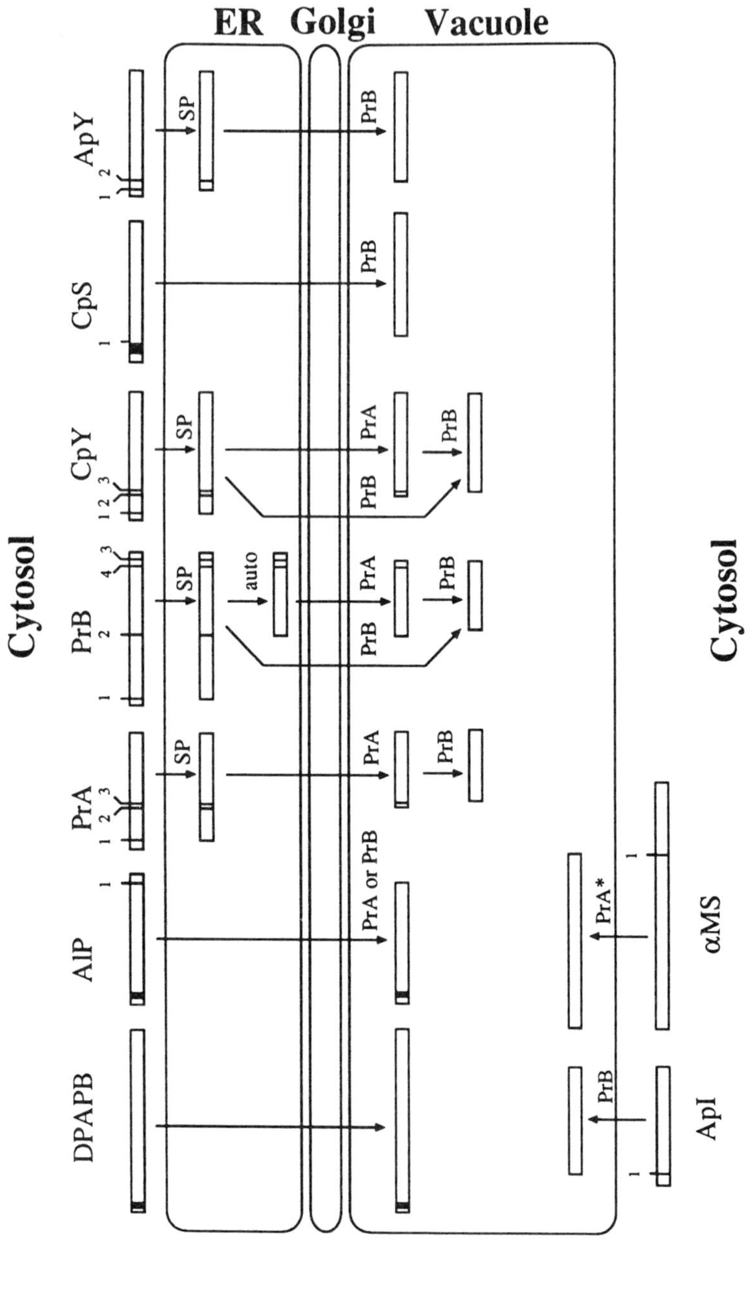

Figure 2 (See facing page for legend.)

passes through the ER and Golgi but not the late steps in the secretory pathway en route to the vacuole, in agreement with the glycosylation patterns (Stevens et al. 1982). Three proteolytic cleavages occur during maturation of Prc1p, all on the amino-terminal side. After signal peptidase cleavage in the ER (Blachly-Dyson and Stevens 1987), PrA catalyzes removal of an amino-terminal peptide, probably by cleavage between residues 105 and 106, in the vacuole (Hasilik and Tanner 1978a; Hemmings et al. 1981; Stevens et al. 1982; Mechler et al. 1987; Sorensen et al. 1994); PrB generates the mature amino terminus at residue 112 of Prc1p (Mechler et al. 1987; Hirsch et al. 1992; Sorensen et al. 1994). The final two cleavages need not be ordered; either suffices to produce active enzyme (Zubenko et al. 1979, 1982; Mechler et al. 1987; Sorensen et al. 1994).

d. ApY. ApY is a glycoprotein with eight potential acceptor sites for N-linked glycosylation; five to seven sites are typically used (Nishizawa et al. 1994; Yasuhara et al. 1994). Preferred substrates of the enzyme carry basic residues. Because the activity is markedly stimulated by Co^{++} and because disruption of the encoding gene *APE3* results in the absence of an aminopeptidase activity responsive to Co^{++} (Nishizawa et al. 1994; Yasuhara et al. 1994), it seems likely that ApY corresponds to ApCo (Achstetter et al. 1982), despite the discrepancy in the molecular masses reported (see Table 1). ApY is inferred to reach the vacuole via the ER, Golgi, endosome route because it is *N*-glycosylated, although direct tests using *sec* mutants have not been done. Ape3p appears to undergo two proteolytic cleavages, the first by signal peptidase and the second catalyzed by PrB. The latter apparently generates the mature amino terminus at residue 57 and activates the zymogen (Nishizawa et al. 1994; Yasuhara et al. 1994).

Figure 2 Events and intracellular locations of processing events in maturation of vacuolar hydrolase precursors. Hydrolases: (DPAPB) dipeptidylaminopeptidase B; (AlP) alkaline phosphatase; (PrA) proteinase A; (PrB) proteinase B; (CpY) carboxypeptidase Y; (CpS) carboxypeptidase S; (ApY) aminopeptidase Y; (ApI) aminopeptidase I; (αMs) α-mannosidase. Numbers over precursors identify sites of cleavage and the order in which cleavages occur. (*Down arrows*) Events and locations of cleavages along the secretory pathway; (*up arrows*) processing along the extrasecretory pathway. (*Closed boxes*) Transmembrane domains. Cleavage enzymes: (SP) signal peptidase; (PrA) proteinase A; (PB) proteinase B; (auto) autocatalytic; (PrA*) PrA-dependent, which does not distinguish between PrA, PrB, or both.

e. CpS. CpS is a glycoprotein. It has five potential acceptor sites for N-linked glycosylation, of which two to three on average are used (Bordallo et al. 1991; Spormann et al. 1991, 1992). Analyses of *sec18* and *sec7* mutants, as well as the glycosylation patterns, indicate that Cps1p passes through the ER and Golgi en route to the vacuole. It has not been excluded that Cps1p travels to the vacuole by way of the plasma membrane and endocytotic vesicles, since *sec* mutants blocked for late steps were not analyzed (Spormann et al. 1992).

Interestingly enough, Cps1p is a type II glycoprotein and travels to the vacuole as a membrane protein. Within the vacuole, a proteolytic cleavage adjacent to the membrane renders the enzyme soluble. The cleavage is primarily catalyzed by PrB (Spormann et al. 1992). The unprocessed CpS precursor is active, in contrast to all of the other precursors described above that undergo processing (Spormann et al. 1992).

CpS is processed to the mature, soluble form (in the vacuole) with a half-time of about 20 minutes, in contrast to shorter half-times of about 6 minutes for the other hydrolases (normalized to 30°C) (Hasilik and Tanner 1978a; Klionsky et al. 1988; Klionsky and Emr 1989; Moehle et al. 1989). The fact that CpS shows slower kinetics and is synthesized via a type II membrane glycoprotein precursor raises the interesting possibility that CpS, like lysosomal acid phosphatase in fibroblasts (also synthesized via a type II membrane glycoprotein precursor), might reach the vacuole via the plasma membrane (Braun et al. 1989; Gottschalk et al. 1989). Analysis of synthesis in a *sec1* or *sec4* mutant at the restrictive temperature would clarify the issue.

f. Trehalase. Vacuolar (acidic) trehalase is delivered to the vacuole via the secretory pathway, for it is a glycoprotein and its synthesis is impaired in *sec18* and *sec7* mutants (Harris and Cotter 1988; Mittenbuhler and Holzer 1988, 1991). It must be synthesized as a zymogen that requires proteolytic activation, for it is not synthesized in a *pep4* mutant (Harris and Cotter 1987). Nothing more has been reported on its synthesis.

g. Vgp72. Nishizawa et al. (1990) have described synthesis of a peripheral vacuolar membrane glycoprotein of unknown function. It is delivered rapidly and directly to the vacuole via the secretory pathway, as evidenced by its N-linked glycosyl side chains and analysis of *sec1*, *sec7*, and *sec18* mutants. Unlike other vacuolar glycoproteins, forms apparently bearing outer chain glycosylation are delivered to the vacuole as well as a form bearing the modestly processed core side chains. The Vpg72 precursor undergoes Sec11p- and Pep4p-dependent proteolytic cleavages (Nishizawa et al. 1990).

B. Extrasecretory Pathway

1. αMS

α-mannosidase is a peripheral membrane protein located within the vacuole (Yoshihisa et al. 1988; Yoshihisa and Anraku 1990). It contains seven potential acceptor sites for N-linked glycosylation but is not a glycoprotein (Yoshihisa et al. 1988; Yoshihisa and Anraku 1990). It continues to be synthesized in *sec* mutants held at temperatures that block the secretory pathway (Yoshihisa and Anraku 1990), and no signal sequence is predicted from the sequence of *AMS1* (Yoshihisa and Anraku 1989). These observations led the authors to infer that αMS enters the vacuole by an extrasecretory pathway.

Analysis of delivery of *AMS1-SUC2* and *PRC1-AMS1* fusion proteins to the vacuole confirms two inferences: (1) The presence of the Ams1p sequence at the amino terminus of a protein results in absence of glycosylation and exclusion from the secretory pathway and (2) the putative glycosylation acceptor sites on Ams1p can accept glycosyl side chains when the protein passes through the secretory pathway (Yoshihisa et al. 1988; Yoshihisa and Anraku 1990). These observations reinforce the conclusion that there is an extrasecretory route to the vacuole.

αMS is synthesized as a 107-kD polypeptide that is proteolytically processed to a 73-kD polypeptide in a *PEP4*-dependent manner with a half-time of 10 hours. Since both species appear to be enzymatically active, the cleavage is not required for activation. The evidence suggests that the cleavage occurs in the carboxy-terminal half of the molecule, since the 73-kD species has a blocked amino terminus, presumably via a cytoplasmic reaction, even though the 73-kD species resides in the vacuole (Yoshihisa et al. 1988; Yoshihisa and Anraku 1990).

The results in toto thus make it very clear that Ams1p reaches the vacuole, where it is processed, by a route other than the secretory pathway. A cautionary flag is possibly raised by the extraordinary long half-time for delivery to the vacuole and the fact that synthesis of αMS is glucose-repressed (Opheim 1978) or requires transfer to 37°C (see Yoshihisa and Anraku 1990).

The extraordinarily long 10-hour half-time for delivery of αMS to the vacuole, and the fact that synthesis of αMS is glucose-repressed (Opheim 1978), lead us to speculate along the following lines. αMS will be synthesized rapidly as cells approach stationary phase, for its synthesis will be released from glucose repression. Stationary-phase cells turn over substantially more protein per hour than do logarithmically growing cells (Halvorson 1958a,b). If cells were to enhance the rate of autophagy as cells approach stationary phase, they could incorporate hexameric (6 x 107 kD) αMS into vacuoles via autophagy, allowing conversion of some of the 107-kD polypeptides to 73-kD + 31-kD species within the (also active) hexamer (Yoshihisa et al. 1988). This hypothesis would account for the finding of 73-, 31-, and 107-kD units within the hexamer, which the assumption of a 107→73 kD processing pathway does not. In addition, this import mechanism would ensure that αMS is present at low levels in the cytoplasm during logarithmic growth, where it could degrade glycoconjugates derived from misfolded glycoproteins expelled from the ER (Hiller et al. 1996), but is incorporated into vacuoles when the cells most need the ability to degrade and recycle glycoconjugates, namely, in stationary phase.

2. *ApI*

ApI is a soluble, vacuolar hydrolase. The encoding gene, *LAP4*, contains four potential acceptor sites for N-linked glycosylation; the amino-terminal amino acid of purified ApI corresponds to residue 46 of the open reading frame (ORF) (Chang and Smith 1989).

ApI is not a glycoprotein (Klionsky et al. 1992a), earlier reports notwithstanding (Metz and Rohm 1976). It continues to enter the vacuole and to be proteolytically processed in *sec* mutants held at temperatures that block the secretory pathway (Klionsky et al. 1992a). Thus, ApI appears to enter the vacuole by an extrasecretory pathway.

Lap4p is proteolytically processed in the vacuole with a half-time of 45 minutes. The cleavage is catalyzed by PrB and is required for enzyme activity. There are some indications that maturation may occur in two steps with cleavage first by PrA, then PrB, although the kinetic data are unclear at this point. Thus, the apparent requirement for PrA for activation may be a manifestation of the requirement for PrA in processing of the penultimate PrB precursor (Trumbly and Bradley 1983; Klionsky et al. 1992a; Segui-Real et al. 1995).

Overproduction of αMS or ApI from high-copy plasmids results in accumulation of the unprocessed polypeptides in the cytoplasm (Yoshihisa and Anraku 1990; Klionsky et al. 1992a) and for ApI, at least, increases the half-time for proteolytic processing. That ApI and αMS may share the same pathway or components is indicated by the observation that overproduction of αMS lengthens the half-time for ApI processing (Klionsky et al. 1992a). Recent experiments indicate that some components may participate both in this extrasecretory pathway and in vacuolar delivery via the secretory pathway (Harding et al. 1995; see also Sections VI.C.2 and VII), on the one hand, and be required both for extrasecretory import into the vacuole and for autophagy, on the other hand (Harding et al. 1996).

C. Other Hydrolases

RNase(s) is present in vacuoles (Wiemken et al. 1979). Synthesis of the activity(s) is *PEP4*-dependent (Jones et al. 1982). Nothing more has been reported. Yeast vacuoles contain endo- and exopolyphosphatases (Wiemken et al. 1979). Recently, Wurst et al. (1995) inferred the existence of two vacuolar exopolyphosphatases, ScPPX2 and ScPPX3, with synthesis of ScPPX2 being *PEP4*-dependent. Within this work, the authors report the existence of an endopolyphosphatase whose synthesis is also *PEP4*-dependent.

D. Pro Sequences and Chaperones

Many of the precursors to vacuolar hydrolases contain pro sequences that are removed through proteolysis. What functions do these pro regions serve?

The propeptides of precursors to AlP, PrA, PrB, CpY, trehalase, ApY, and ApI clearly act to inhibit their own enzymatic activities (Hemmings et al. 1981; Jones et al. 1982; Harris and Cotter 1987; Kaneko et al. 1987; Klionsky et al. 1992a; Yasuhara et al. 1994; Wolff et al. 1996). We assume that this is true for the RNase(s) and exo- and endopolyphosphatases whose activities depend on *PEP4* function. It is equally clear that the propeptides of precursors to CpS and αMS do not inhibit their own enzymatic activities: αMS is active as both 107-kD and 73-kD polypeptides and CpS is active as a precursor in the *pep4* and *prb1* mutants (Yoshihisa et al. 1988; Spormann et al. 1992). For CpS, the transmembrane peptide clearly functions to attach the protein to the membrane and is assumed to direct the protein to the vacuole, although this has not been demonstrated; the vacuole is the default compartment for membrane proteins (Roberts et al. 1992; Wilcox et al. 1992; Nothwehr and Stevens 1994). A role has not been assigned or even suggested for the severed peptide for αMS.

The amino-terminal propeptides of the PrA and PrB precursors have been shown to be necessary in vivo for production of the respective active enzymes (Nebes and Jones 1991; van den Hazel et al. 1993). The propeptides normally function in *cis* but can function in *trans*, albeit less efficiently. They are inferred to catalyze conformational changes (folding) needed for production of active, stable enzymes; i.e., they function as intramolecular chaperones (Nebes and Jones 1991; van den Hazel et al. 1993). In vitro data have been presented that imply a similar role for the propeptide of the CpY precursor (Winther and Sorensen 1991).

A propeptide of about 4-kD is removed from the carboxyl terminus of the AlP precursor in a *PEP4*-dependent fashion in the vacuole (Jones et al. 1982; Klionsky and Emr 1989). The propeptide must contain approximately 35 amino acids. Because a truncated *PH08* clone lacking 22 carboxy-terminal residues can complement the *pep4* mutant for AlP activity, although the full-length *PH08* clone cannot (Kaneko et al. 1985, 1987), it is unlikely that the propeptide of Pho8p functions as an intramolecular chaperone, since active enzyme must be being produced from the truncated clone, presumably without proteolytic processing. No information is available on possible chaperone functions for propeptides of the other hydrolase precursors. Recently, Kar2p/BiP has been shown to be required as a molecular chaperone in folding of the CpY precursor (Simons et al. 1995).

III. FUNCTIONS OF THE HYDROLASES

Several of the hydrolases are synthesized as inactive precursors that must be activated by proteolytic cleavages. A key enzyme in the activation is PrA, for it initiates the activation cascade by autocatalytic activation of its own zymogen in the vacuole. PrA activity is normally essential for production of PrB activity from its precursor. The two enzymes between them catalyze all of the proteolytic cleavages required for maturation of vacuolar hydrolases (Fig. 2 and Table 1) (for review, see Jones 1991b; Jones and Murdock 1994). Thus, directly or indirectly PrA is required for activation and/or maturation of most of the vacuolar hydrolases. Not surprisingly, if PrA activity is completely eliminated by mutation (*pep4*Δ), cells are pleiotropically deficient for many hydrolase activities, including PrA, PrB, CpY, vacuolar AlP, RNase(s), ApI, vacuolar trehalase, one exopolyphosphatase, and one endopolyphosphatase (Jones et al. 1982; Trumbly and Bradley 1983; Harris and Cotter 1987; Wurst et al. 1995). PrA participates in protein degradation in growing or starved vegetative cells and in sporulating cells (Zubenko and Jones 1981; Teichert et al.

1989). Whether the failure of *pep4Δ* homozygotes to sporulate arises from the deficiency of protein degradation that they manifest or from some more complex consequence of the pleiotropy has not been disentangled.

PrB also participates in protein degradation in growing or starved vegetative cells and in sporulating cells (Wolf and Ehmann 1979; Zubenko and Jones 1981; Teichert et al. 1989) and is the key enzyme for degradation during autophagy (Takeshige et al. 1992). Indeed, it was the mutational loss of PrB activity that allowed detection of autophagy in yeast cells (Takeshige et al. 1992). Loss of PrB activity reduces sporulation; whether sporulation is eliminated by loss of PrB may depend on genetic background (Wolf and Ehmann 1979; Zubenko and Jones 1981; Teichert et al. 1989).

If both PrA and PrB activities are eliminated by mutation, either in double mutants or through the pleiotropic effects of the *pep4Δ* mutation, the cells die fairly rapidly if nitrogen starvation is imposed (Teichert et al. 1989; A. Heinzer and E. Jones, unpubl.).

Both carboxypeptidases are involved in metabolism of externally supplied peptides (Jones 1977; Wolf and Weiser 1977; Wolf and Ehmann 1979; Teichert et al. 1989); CpY (and presumably CpS) degrades internally generated peptides as well (Zubenko and Jones 1981).

Kinetic studies have implicated αMS in degradation of glycoconjugates (Kuranda and Robbins 1987). It seems possible that it may carry out this role in the cytoplasm in logarithmically growing cells and in the vacuole at later growth stages and in starving cells. AlP dephosphorylates phosphopeptides in vitro; the *pho8* mutant lacks this activity (Donella-Deana et al. 1993). The vacuolar location of AlP would appear to restrict AlP to a degradative and recycling function rather than any regulatory role as a protein phosphatase. Although probable roles can be deduced for the other hydrolases, evidence is lacking.

IV. VACUOLE FUNCTION

A. Transport and Storage of Ions and Other Molecules

A major difference between the yeast vacuole and the mammalian lysosome is the vacuole's ability to sequester high concentrations of small molecules and ions, making it more like a plant vacuole. The vacuole stores ions, amino acids, purines, polyamines, allantoin, and allantoate and *S*-adenosylmethionine (see Svihla et al. 1963, 1964; Nakamura and Schlenk 1974; Wiemken and Durr 1974; Schwencke and De Robichon-Szulmajster 1976; Zacharski and Cooper 1978; Wiemken et al. 1979; Messenguy et al. 1980; Ohsumi and Anraku 1981; Kitamoto et al. 1988a; Kakinuma et al. 1992). The ions sequestered include Ca^{++} (Ohsumi and Anraku 1983; Dunn et al. 1994), Mg^{++} (Okorokov et al. 1978), Mn^{++} (Okorokov et al. 1977; Nieuwenhuis et al. 1981; Theuvenet et al. 1986), Sr^{++} (Nieuwenhuis et al. 1981; Ohsumi and Anraku 1983; Theuvenet et al. 1986; Avery and Tobin 1992), Zn^{++} (Bilinski and Miller 1983; White and Gadd 1986), Fe^{++} (Raguzzi et al. 1988), Li^{+} (Perkins and Gadd 1993), phosphate (Urech et al. 1978), K^{+} (Eilam et al. 1985; Ohsumi et al. 1988), Cl^{-} (Wada et al. 1992a), and Cd^{++} (as a Cd-glutathione conjugate) (Li et al. 1996). Ions and small molecules may accumulate to high concentrations; the Ca^{++} concentration in the vacuole can reach the millimolar range (Halachmi and Eilam 1989), the arginine concentration 400–500 mM (Kitamoto et al. 1988a).

1. Vacuolar Calcium Accumulation

Ca^{++} transport is the best studied of all vacuolar transport processes, except for proton transport. Ca^{++} transport into the vacuole is dependent on a proton gradient. The apparent ATP requirement for Ca^{++} transport in isolated vacuoles is thought to reflect a need for ATP hydrolysis by the proton pumping vacuolar ATPase (V-ATPase) to generate the proton gradient (Ohsumi and Anraku 1983; Dunn et al. 1994). In agreement with this observation, disruption of genes for subunits of the V-ATPase eliminates the ability of the vacuole to sequester Ca^{++} and causes an increase in cytoplasmic calcium concentration (Ohya et al. 1991; Halachmi and Eilam 1993). When Ca^{++} is added to acidified vacuoles, the pH gradient is dissipated, indicating that proton antiport is the mechanism for transport of Ca^{++} (Ohsumi and Anraku 1983; Belde et al. 1993). It is estimated that two protons exit the vacuole for every Ca^{++} ion that enters the vacuole; K^{+} ions exit simultaneously (Nieuwenhuis et al. 1981; Ohsumi and Anraku 1983). The rate of transport is thought to be high enough to allow the vacuole to balance the rate of Ca^{++} accumulation into the cell and thereby keep cytoplasmic concentrations low (Dunn et al. 1994).

Table 2 summarizes proteins involved in vacuolar Ca^{++} homeostasis. Recent experiments have identified two genes that affect Ca^{++} homeostasis and have led to the hypothesis that Ca^{++} transport across the vacuolar membrane may be accomplished by a Ca^{++}-ATPase together with a Ca^{++}/H^{+} exchanger (Cunningham and Fink 1994, 1996; Pozos et al. 1996). Both gene products, Pmc1p and Vcx1p/Hum1p, localize to the vacuole membrane by immunofluorescence (Cunningham and Fink

Table 2 Vacuole-related proteins important for Ca^{++} homeostasis

SGD name	Other names	Proteins homolog	Size (aa)	Properties
Pmc1p		Ca^{++}ATPase (plasma membrane family)	1173	vacuole integral membrane protein
Pmr1p	Bsd1p	Ca^{++}ATPase	950	Golgi integral membrane protein; transports Ca^{++} into the Golgi complex
	Ssc1p	(Sarc/ER family)		
Cnb1p	Ycn2p	calcineurin	174	regulatory subunit of calcineurin, type-2B protein phosphatase
Cna1p	Cmp1p	calcineurin	553	catalytic subunit of calcinuerin, type-2B protein serine/threonine phosphatase; cytoplasmic
Cna2p	Cmp2p	calcineurin	604	catalytic subunit of calcineurin, type-2B protein serine/threonine phosphatase; redundant with Cna1p; cytoplasmic
Vcx1p	Hum1p	$Na^+/Ca^{++}/K^+$ antiporter; Na^+/Ca^{++}	411	Ca^{++}/H^+ exchanger

1994, 1996; Pozos et al. 1996). Pmc1p has homology with a plasma membrane Ca^{++}-ATPase. A *pmc1* null mutation causes a Ca^{++}-sensitive phenotype and results in less Ca^{++} per cell than is present in wild type (Cunningham and Fink 1994). These are the phenotypes expected from a defect in vacuolar storage of Ca^{++}, and they resemble phenotypes caused by mutations that alter V-ATPase subunits. A second gene, *VCX1/HUM1*, has homology with a mammalian retinal rod Na^+/Ca^{++}, K^+ exchanger and a mammalian cardiac Na^+/Ca^{++} exchanger (Cunningham and Fink 1996; Pozos et al. 1996). Vacuoles purified from *vcx1/hum1* mutants exhibit no vacuolar H^+/Ca^{++} exchange function, and Ca^{++} will not dissipate the proton gradient in acidified vacuoles (Cunningham and Fink 1996; Pozos et al. 1996). However, whole-cell Ca^{++} accumulation is only slightly reduced in *vcx1/hum1* strain. Unlike mutants lacking V-ATPase activity or the *pmc1* mutant, a *vcx1/hum1* strain is able to grow on a medium containing elevated levels of Ca^{++} (Cunningham and Fink 1996; Pozos et al. 1996). It is not completely clear how Pmc1p and

Vcx1p/Hum1p function together to sequester Ca^{++} in the vacuole and indications are that calcineurin may regulate each transporter (Cunningham and Fink 1994, 1996; Pozos et al. 1996).

2. *Vacuolar Polyphosphate*

Phosphate is stored in the vacuole both as inorganic phosphate and as polymers of polyphosphate. Vacuolar polyphosphate is thought to serve as a counterion for the accumulation of positively charged molecules such as Ca^{++} and arginine (Urech et al. 1978; Durr et al. 1979). It has been demonstrated in vitro that polyphosphate can bind Ca^{++}, Mg^{++}, and *S*-adenosylmethionine (Miller 1984; Dunn et al. 1994). In permeablized cells, the extent to which vacuoles accumulate Ca^{++} is proportional to polyphosphate content (Dunn et al. 1994). Mutants with defects in vacuolar acidification have little or no phosphate in the vacuole; for this reason, proton exchange is thought to be the mechanism for transport of phosphate into the vacuole, although this has not been tested using purified vacuoles (Beauvoit et al. 1991). A second proposed function of vacuolar polyphosphate is pH regulation, as the breakdown of polyphosphate can lower pH by producing protons (Pick et al. 1990). Increasing the pH of the growth medium causes an increase in cytoplasmic pH, followed by production of short phosphate chains from vacuolar polyphosphate, and restoration of neutral cytoplasmic pH (Castro et al. 1995). When phosphate in the medium is limiting, vacuolar phosphate can act as a reserve and be mobilized (Gillies et al. 1981; Campbell-Burk and Shulman 1987; Greenfield et al. 1987; Westenberg et al. 1989). Cells shifted from a low phosphate to a high phosphate growth medium (and hence, at the time of the shift, cells efficient at scavenging phosphate from the medium) rapidly increase vacuolar phosphate levels. This is presumably due to vacuolar accumulation of excess inorganic phosphate not needed by the cell. Stationary-phase cells accumulate inorganic phosphate and have polyphosphate with more residues per chain than do cells in logarithmic growth (Greenfield et al. 1987).

3. *Amino Acid Storage*

Ten different amino acids can accumulate to high concentrations in the vacuole, and indeed, pools for these amino acids are in the vacuole (see Wiemken et al. 1979; Messenguy et al. 1980; Kitamoto et al. 1988a). The basic amino acids, in particular including ornithine and citrulline, can accumulate to very high levels. Arginine, for example, is present in vacuoles at 60–70 mM when cells are grown in minimal medium or rich yeast extract-peptone dextrose (YPD) medium and at 400–500 mM when arginine is the nitrogen source (Kitamoto et al. 1988a). The concentrations of various amino acids found in the vacuole during growth in YPD and other nitrogen sources have been reported (Messenguy et al. 1980; Kitamoto et al. 1988a). Transport into vacuoles of arginine, lysine, histidine, leucine, tyrosine, isoleucine, glutamine, tryptophan, phenylalanine, and asparagine is dependent on the pH gradient generated by the V-ATPase (Ohsumi and Anraku 1981; Sato et al. 1984b). Seven different transport systems have been biochemically characterized (Sato et al. 1984a,b), including those that primarily transport arginine; arginine and lysine; histidine; phenylalanine and tryptophan; tyrosine; glutamine and asparagine; and leucine and isoleucine.

4. *Transport of Small Peptides*

A role for the vacuole in glutathione *S*-conjugate-mediated cytoplasmic detoxification of Cd^{++} and other toxic molecules has been demonstrated. A vacuolar transporter, encoded by *YCF1*, is able to transport glutathione *S*-conjugates across the vacuolar membrane. Ycf1p is a member of the ATP-binding cassette (ABC) transporter family. *ycf1*Δ strains are sensitive for growth in medium with increased levels of Cd^{++} and a glutathione *S*-conjugate precursor 1-chloro-2,4-dinitrobenzene (CDNB) (Szczypka et al. 1994; Li et al. 1996). A recent study, primarily of *Schizosaccharomyces pombe* mutants but including the *S. cerevisiae* *ycf1*Δ mutant, has implicated Ycf1p and glutathione in detoxification and transport of purine precursors into the vacuole and ultimately in production of the red pigment of *ade2* and *ade1* mutants (Chaudhuri et al. 1997). In a study of glutathione *S*-conjugate transport into vacuolar membrane vesicles by Li et al. (1996), two MgATP-dependent activities were described. The first is dependent on Ycf1p and vanadate-sensitive ATP hydrolysis, and the second is dependent on the $\psi\Delta$ generated by the V-ATPase but independent of the ΔH. The activity is present in *ycf1*Δ strains and is mediated by an unknown gene product.

5. *Functions of Sequestration*

The vacuole serves diverse functions through sequestration. At least some of the stored molecules and ions can be mobilized by the cell. The amount of phosphate and the chain length of polyphosphates in the vacuole are dynamic. This store can be mobilized as needed for pH regulation and as a source of phosphate; it has the potential to serve as an energy store (Wurst et al. 1995). Stored iron can be mobilized for production of cytochromes under iron-limiting conditions (Raguzzi et al. 1988). There is one report of inositol-triphosphate-stimulated release of Ca^{++} from the vacuole (Belde et al. 1993), although there is no demonstrated role in *S. cerevisiae* for release of calcium from the vacuole. Amino acids can be mobilized from the vacuole and used as nitrogen sources when cells are subjected to nitrogen starvation (Whitney and Magasanik 1973; Kitamoto et al. 1988a). When this occurs, there is a counterflow of K^+ so that vacuolar turgor pressure is maintained (Kitamoto et al. 1988a). Amino acids also exit the vacuole in response to α-factor-mediated G_1 arrest (Sumrada and Cooper 1978). A surplus of one amino acid can alter the composition of the vacuolar pool (Kitamoto et al. 1988a). Thus, the vacuolar amino acid pools are very dynamic; we have little insight into this phenomenon. What the signals may be that control polyphosphate breakdown and phosphate release or mobilization of arginine and iron are unknown, as are the mechanisms of transmittal of the signal, the mechanisms of transport of the ions and molecules out of the vacuole, and the counterflow of potassium ions.

It is equally clear that vacuolar storage can serve a detoxification function. The vacuole detoxifies metal ions such as Ca^{++} or Cd^{++} (as glutathione *S*-conjugates) by keeping cytosolic concentrations low. Similarly, cells may detoxify purine precursors (also likely as glutathione *S*-conjugates) by removing them from the cytosol.

B. Vacuolar Acidification

The vacuolar-type proton pumping ATPase (V-ATPase) is a multiprotein complex responsible for generating and maintaining the acidic pH of the vacuole. The initial characterization of V-ATPase activity revealed an MgATP hydrolysis activity enriched in a purified vacuole fraction (Kakinuma et al. 1981). The activity is distinct from the ATPase activities of the mitochondria and the plasma membrane. Neither vanadate, an inhibitor of the plasma membrane ATPase, nor oligomycin, an inhibitor of the mitochondrial ATPase, affects the ATPase activity associated with the vacuole. *N,N'*-dicyclohexylcarbodiimide (DCCD) is an inhibitor of both the mitochondrial ATPase and vacuolar ATPase, whereas Bafilomycin A_1 and concanamycin A are specific inhibitors of the V-ATPase (Kakinuma et al. 1981; Kinashi et al. 1984; Bowman et al. 1988). ATP is able to stimulate transport of protons into sealed vacuoles in a DCCD-sensitive manner, indicating that the enzyme is a proton pumping ATPase responsible for vacuolar acidification (Kakinuma et al. 1981). Purification of the active complex from vacuoles using glycerol gradients further separates the V-ATPase activity from the mitochondria and plasma membrane ATPase activities (Uchida et al. 1985). This purified complex is also a starting point for the identification of proteins that make up the V-ATPase (Uchida et al. 1985, 1988b; Kane et al. 1989; Hirata et al. 1990). Table 3 summarizes the proteins that comprise the V-ATPase and/or are required for assembly of the V-ATPase.

The V-ATPase is composed of two sectors: the V_0 sector resides in the vacuolar membrane; the V_1 sector is the catalytic sector. The V_1 sector associates reversibly with the vacuolar membrane, in a regulated manner, and is composed of seven polypeptides of 69, 60, 54, 42, 32, 27, and 14 kD. Tfp1p (69 kD) and Vma2p (60 kD) provide catalytic function; Tfp1p contains a nucleotide-binding site (Uchida et al. 1988a; Hirata et al. 1990) and Vma2p contains a predicted regulatory binding site (Nelson et al. 1989). A proposed structure based on studies of the bovine V-ATPase has three copies of Tfp1p and Vma2p each in the V_1 sector (Arai et al. 1988). The other polypeptides in the V_1 sector are Vma13p (54 kD; Ho et al. 1993b), Vma5p (42 kD; Beltran et al. 1992; Ho et al. 1993a), Vma8p (32 kD; Graham et al. 1995; Nelson et al. 1995), Vma4p (27 kD; Foury 1990; Ho et al. 1993a), and Vma7p (14 kD; L.A. Graham et al. 1994; Nelson et al. 1994).

The V_0 sector is composed of six polypeptides of 100, 36, 23, 17, 17, and 13 kD, of which Vph1p (100 kD; Manolson et al. 1992b), Ppa1p (23 kD; Apperson et al. 1990), Vma3p (17 kD; Nelson and Nelson 1989; Umemoto et al. 1990), and Tfp3p (17 kD; Shih et al. 1990; Umemoto et al. 1991) are integral membrane subunits. Although Vma6p (36 kD; Bauerle et al. 1993) and Vma10p (13 kD; Supekova et al. 1995) are judged to be peripheral membrane proteins, they are thought to be part of the V_0 sector, along with the integral membrane proteins, because they remain associated with the vacuolar membrane under conditions that remove the V_1 sector and because of their roles in V-ATPase assembly (Bauerle et al. 1993). Vma3p is a hydrophobic polypeptide defined as the proteolipid subunit, as it can be extracted from the vacuole membrane into a chloroform/methanol mixture. It will bind the V-ATPase inhibitor DCCD. It is thought that six copies form the pore in the membrane through which the protons are transported (Umemoto et al. 1990). Tfp3p and Ppa1p appear to be proteolipid subunits on the basis of sequence homology with Vma3p and proteolipid subunits from other species. Although the amino acid similarity between Vma3p and Tfp3p, in particular, is high, expression of either from a high-copy plasmid will not allow functional complementation of a null mutation in the other. Disruption of either single gene alone results in a complete loss of ATPase function,

Table 3 Proteins involved in vacuolar acidification

SGD name	Other names	Amino acids (kD)	Localization, role	References
V_1 sector (cytoplasmic):				
Tfp1p	Vma1p Cls8p	617 (69)	V_1 sector, peripheral, catalytic (A subunit)	Hirata et al. (1990); Shih et al. (1988); Ohya et al. (1991)
Vma2p	Vat2p VatPbp	517 (60)	V_1 sector, peripheral, nucleotide bending (B subunit)	Yamashiro et al. (1990); Nelson et al. (1989)
Vma4p	Vat5p	233 (27)	V_1 sector, peripheral (E subunit)	Foury (1990); Ho et al. (1993a)
Vma5p	VatCp Vat3p	392 (42)	V_1 sector, peripheral (C subunit)	Ho et al. (1993a); Beltran et al. (1992)
Vma7p		118 (14)	V_1 sector, peripheral (F subunit)	L.A. Graham et al. (1994); Nelson et al. (1995)
Vma8p		256 (32)	V_1 sector, peripheral (D subunit)	Graham et al. (1995); Nelson et al. (1995)
Vma13p	Cls11p	478 (54)	V_1 sector, peripheral	Ho et al. (1993b); Ohya et al. (1991)
V_0 sector (membrane):				
Vma6p		345 (36)	V_0 sector, peripheral, required for assembly (D subunit of membrane sector)	Bauerle et al. (1993)
Vma10p		114 (13)	V_0	Supekova et al. (1995)
Tfp3p	Vma11p Cls9p Vma16p	164 (17)	V_0 integral? needed for Cup5p assembly	Umemoto et al. (1991); Ohya et al. (1991); Shih et al. (1990)
Ppa1p		213 (23)	V_0 integral? proteolipid subunit	Apperson et al. (1990)
Cup5p	Vma3p Cls7p Gef2p VatPcp	159 (17)	V_0 integral, proteolipid subunit (subunit C of membrane sector)	Umemoto et al. (1990); Ohya et al. (1991); Nelson and Nelson (1989)
Vph1p		840 (100)	V_0 integral, vacuole-specific	Manolson et al. (1992b)
Stv1p		890 (110)	V_0 integral, not in vacuole membrane	Manolson et al. (1994)
Assembly factors for V-ATPase:				
Vph2p	Vma12p Cls10p	215 (25)	ER membrane	Hirata et al. (1993); Ohya et al. (1991); Bachhawat et al. (1993)
Vma21p		77 (8)	ER membrane	Hill and Stevens (1994)
Vma22p	Vph6p	181 (23)	ER, peripheral	Hill and Stevens (1995); Hemenway et al. (1995)

and the double mutant has the same phenotype as each single mutant. Thus, despite high sequence homology between the two, each may have a distinct function in the V-ATPase. Tfp3p has not yet been shown to purify with the V-ATPase (Umemoto et al. 1991). Ppa1p also has not been shown to purify with the V-ATPase, and its assignment to the V_1 sector is based solely on homology with other proteolipid subunits.

VPH1 and *STV1* encode integral membrane subunits with multiple predicted membrane-spanning domains. The two have extensive sequence homology and appear to be functional homologs with different subcellular localizations. On the basis of immunofluorescence microscopy, Vph1p resides in the vacuole, whereas the localization of Stv1p is reminiscent of a Golgi protein, although the overlap is not complete. Manolson et al. (1992b, 1994) have suggested that the V-ATPase is present on at least two different organellar membranes in the cell, that *STV1* and *VPH1* encode organelle-specific 100-kD subunits, and that all other V-ATPase subunits are held in common.

A purified and enzymatically active complex was used for the initial characterization of the polypeptides that make up the V-ATPase. The polypeptides that are bona fide subunits of the V-ATPase both localize to the vacuole or vacuolar vesicles and cofractionate with Bafilomycin A_1-sensitive ATPase activity. The pH of the wild-type vacuole is 6.2 as determined using the fluorescence of 6-carboxyfluorescein (6CF); 6CF fluorescence is pH-dependent (Preston et al. 1989). This pH is similar to the vacuolar pH inferred from ^{31}P nuclear magnetic resonance (NMR) studies of vacuolar phosphate (Nicolay et al. 1983; Greenfield et al. 1987). Strains containing null mutations in genes encoding various subunits have no Bafilomycin A_1-sensitive ATPase activity, vacuoles purified from the strains are unable to transport protons, and the vacuoles have neutral pH (see, e.g., Hirata et al. 1990; Umemoto et al. 1990; Yamashiro et al. 1990).

Defects in vacuole acidification caused by mutations that disrupt V-ATPase function result in strains with a wide range of phenotypes. The mutant strains grow more slowly than corresponding wild-type strains on rich medium (YEPD). They will not grow at all on medium buffered to pH 7.5, and it is speculated that this phenotype reflects the importance of the vacuole for regulation of cytoplasmic pH (Yamashiro et al. 1990). Additionally, mutant strains will not grow on medium containing a nonfermentable carbon source, medium without inositol, or medium with elevated levels of many ions, including Ca^{++}, Mn^{++}, Zn^{++}, Sr^{++}, Fe^{++}, Na^{+}, and Cu^{++} (Ohya et al. 1991; for further references, see Section IV). Vacuoles from these strains do not accumulate the red pigment that results from an *ade2* mutation, and as a consequence, they do not attain the red colony phenotype normally associated with *ade2* containing strains (Foury 1990). Some of these defects are attributed to the importance of the proton gradient for vacuolar accumulation of ions and other small molecules (see above). It is not at all clear why other phenotypes, such as the growth defect on a nonfermentable carbon source, result from a defect in vacuolar acidification. Taken together, these results may indicate that vacuolar acidification has diverse roles in cell physiology. Several synthetic lethal interactions between mutations that affect vacuolar acidification and mutations that affect other cellular processes have been characterized. Mutations that result in endocytosis defects are lethal in combination with mutations in genes that alter vacuolar acidification (Munn and Riezman 1994). Blocking function of the protein phosphatase calcineurin in strains with vacuolar acidification defects is lethal (Garrett-Engele et al. 1995; Hemenway et al. 1995; Tanida et al. 1995).

In almost every case examined, the disruption of genes encoding subunits of the V-ATPase causes the phenotypes listed above. The exceptions are *VPH1* and *STV1*. *vph1* null mutants weakly exhibit some of the phenotypes, whereas *stv1* null mutants manifest none of these mutant phenotypes. The *vph1 stv1* double-mutant strain behaves like strains with null mutations in other V-ATPase-encoding genes. For example, *vph1*Δ strains grow slowly on YPD buffered to pH 7.5, whereas *tfp1*Δ strains do not grow at all on the same medium. The *vph1*Δ *stv1*Δ double mutant is also unable to grow on YPD buffered to pH 7.5 (Manolson et al. 1994). The facts that the *tfp1*Δ, *vph1*Δ, and *stv1*Δ single mutants all have different phenotypes and that the *vph1*Δ *stv1*Δ double mutant has mutant phenotypes like those of the *tfp1*Δ mutant are support for the hypothesis that *STV1* and *VPH1* encode organelle-specific V-ATPase subunits, whereas *TFP1* encodes a subunit common to all V-ATPases. The fact that the phenotype of the *vph1*Δ *stv1*Δ double mutant is like that of the *tfp1*Δ mutant further suggests that there are no other V-ATPases in the yeast cell, in agreement with the finding that there are no other homologs of *VPH1* and *STV1* in the yeast genome. Expression of *STV1* from a high-copy plasmid causes vacuolar localization of Stv1p and assembly of Stv1p into the Bafilomycin A_1-sensitive vacuolar V-ATPase and can partially rescue the phenotypes associated with mutations in *VPH1* (Manolson et al. 1994). This finding not only confirms that Vph1p and Stv1p are functional homologs, but also suggests that Stv1p normally resides in a compartment upstream of Vph1p in the vacuolar branch of the secretory pathway or the secretory pathway itself.

Blocking acidification of mammalian lysosomes causes secretion of hydrolases (Mellman et al. 1986). This has been tested in yeast using mutations in genes encoding subunits of the V-ATPase. Strains completely deficient in vacuolar acidification have near wild-type hydrolase activity as judged from the intracellular accumulation of mature CpY and PrA; thus, V-ATPase function is not essential for hydrolase function (Umemoto et al. 1990; Yamashiro et al. 1990). However, 20–25% of CpY and PrA are secreted in the *vma2Δ* mutant, as measured in pulse-chase experiments, whereas the kinetics of processing of the internal pool of precursors is not affected. Wild-type cells secrete 4–8% CpY (Yamashiro et al. 1990). *vma3Δ* mutant strains secrete some CpY and PrA and convert CpY to a mature form with slightly slower kinetics than wild type (Umemoto et al. 1990). Similarly, treating cells with Bafilomycin A_1 results in the secretion of some CpY and PrA (Banta et al. 1988). Other results indicate that the degree of the defect in targeting and processing is in part dependent on the specific mutation used to block vacuolar acidification. The kinetics of CpY processing is slowed in *vma4Δ* mutants but is more dramatically affected in *vma3Δ* mutants (Morano and Klionsky 1994). The hydrolase CpS behaves like CpY, but the membrane protein AlP is not affected by either mutation. Evidence exists that CpY accumulates in the Golgi in a *vma3Δ* strain (Yaver et al. 1993). None of these experiments specifically addresses the role of vacuolar acidification in sorting within the Golgi or in precursor processing, because all of these experiments employed mutations that would abrogate all V-ATPase activity, regardless of location, as would the Bafilomycin A_1. To delineate the role of vacuolar acidification per se in these processes, the *vph1Δ* mutant should be used.

Recent experiments by Kane (1995) indicate that assembly of the V_1 complex onto the vacuolar membrane is regulated. Removal of glucose from the growth medium causes the V_1 complex to come off the vacuolar membrane as judged from immunofluorescence and biochemical fractionation. The effect is reversible, as adding glucose back to the culture induces the V_1 complex to move back onto the vacuolar membrane. In both cases, the effect is rapid, occurring in less than 5 minutes. Protein synthesis is not required, as the results are the same in the presence or absence of cycloheximide. The same effect is observed when the carbon source is shifted from glucose to raffinose, galactose, or glycerol/ethanol. After glucose is added back to the culture, the V_1 subunit is again localized to the vacuolar membrane. One possibility is that V-ATPase activity, and thereby vacuolar acidification, is regulated by movement of the V_1 subunit onto the vacuole membrane. The mechanisms of disassembly and reassembly promise to be exciting future research.

C. Protein Turnover

A yeast cell often needs to alter its state in response to regulated developmental pathways or to environmental cues that originate in other cells or from changing or changed metabolic conditions. Quick responses can be achieved by altering the interface between the cell and the environment by selectively changing the complement of plasma membrane proteins through selective endocytosis and degradation and/or by altering regulatory circuitry via selected, targeted molecules by stabilizing markedly unstable proteins or by destabilizing hitherto stable proteins. Examples of both kinds of regulatory modulations are known in yeast (for reviews and further details, see Jones and Murdock 1994; Hochstrasser 1995).

Leaving aside the molecules that might be viewed by cells as "recyclables"—"by-product" peptides, abnormal proteins, and unassembled subunits of protein complexes or supramolecular assemblies—the unstable proteins fall into interesting categories: They are key regulatory components, functioning to activate or inhibit key enzymes of the cell cycle, such as Cdc28p kinase, or they are sensors at the plasma membrane, detecting environmental changes via receptors and transporters (in or out) and allowing for an appropriate response through a signal transduction pathway or directly through altering transcription patterns. The sensors are degraded in the vacuole after endocytosis; the intracellular components are degraded via the proteasome. These proteolyses allow for very rapid responses and adaptations to environmental change.

1. Vegetative Cells

Yeast cells growing exponentially in a minimal medium degrade protein at rates of 0.5–2% per hour (Halvorson 1958b; Betz and Weiser 1976; Lopez and Gancedo 1979; Chiang et al. 1996). For vegetative cells, about 35% of the turnover of short-lived proteins and 45% of the turnover of long-lived proteins can be attributed to vacuolar proteolysis (Teichert et al. 1989).

Short-lived proteins degraded in compartments other than the vacuole (and the mitochondrion, which will not be discussed) include cell-cycle-related proteins such as Sci1p, an inhibitor of the Cdc28p kinase, and the G_1 and M cyclins; mating-related proteins, including the pheromone response protein, Gpa1p (the α-subunit of the heterotrimeric G protein) and the *MATα2* repressor and *MATα1* transcriptional activator proteins; the transcriptional activator protein for general control of amino acid biosynthesis, Gcn4p; unassembled proteins, such as ribosomal proteins and the α-subunit of fatty acid synthase (FAS), when they are surplus to requirements; and misfolded proteins exported from the ER for degradation (Hochstrasser and Varshavsky 1990; Chen et al. 1993; Egner et al. 1993; Kornitzer et al. 1994; Madura and Varshavsky 1994; Schwob et al. 1994; Deshaies et al. 1995; Seufert et al. 1995; Yaglom et al. 1995; Hiller et al. 1996). In other words, the short-lived proteins degraded, presumably by the proteasome, are proteins regulating transcription (α1, α2, a1, Gcn4p), the response to external triggers (Gpa1p), and the cell cycle (Sci1p, cyclins), as well as orphan or defective proteins.

Short-lived "proteins" degraded in the vacuole include membrane proteins delivered directly from membranes of the secretory pathway and by endocytosis from the plasma membrane; peptides derived from processing events in the secretory pathway or, like α-factor, taken up by endocytosis; and orphan or defective proteins. The membrane proteins that are degraded in the vacuole are presented in Table 4, as well as whether delivery for degradation is constitutive, triggered, and/or related to growth stage.

The type II glycoproteins Och1p, which adds the initial α-1,6-linked mannosyl residue to glycosyl side chains in the *cis*-Golgi, and Ste13p (DPAPA), which removes dipeptides from the α-factor precursor in the *trans*-Golgi, and the type I glycoproteins Kex2p and Kex1p, an endopeptidase and carboxypeptidase, respectively, involved in α-factor processing in the *trans*-Golgi, are all degraded in the vacuole (Cooper and Bussey 1992; Roberts et al. 1992; Wilcox et al. 1992; Harris and Waters 1996). Mutant forms of the plasma membrane ATPase and Wbp1p, a type I ER membrane subunit of the oligosaccharyltransferase, are both delivered directly to the vacuole and degraded there (Gaynor et al. 1994; Chang and Fink 1995).

A number of proteins are selectively removed from the plasma membrane in a regulated manner by endocytosis and are degraded in the vacuole. The best studied example is Ste2p, the G-protein-coupled receptor for the pheromone α-factor. Ste2p is internalized in a clathrin-dependent manner in the absence of ligand; the rate is stimulated severalfold by pheromone binding (Jenness and Spatrick 1986; Tan et al. 1993). The receptor becomes phosphorylated, which positively regulates ubiquitination of the cytoplasmic tail of the receptor; ubiquitination is required for endocytosis (Hicke and Riezman 1996). The receptor is trans-

Table 4 Membrane protein degradation

Protein	Gene product	Stimulus	*PEP4*-dependent	Endocytosis required	Ubiquitination[a]	
					observed	required for endocytosis
Receptors						
α-factor receptor	Ste2p	constitutive; ligand triggered	yes	yes	yes	yes
a-factor receptor	Ste3p	constitutive; ligand triggered	yes	yes		
PM permeases						
uracil	Fur4p	constitutive;	yes	yes	yes	yes
		stationary phase;	yes	yes	yes	yes
		starvation	yes	yes	yes	yes
inositol	Itr1p	inositol addition;	yes	yes?		yes
		stationary phase	yes	yes		?
maltose	Mal61p	glucose addition;	yes	yes		yes?
		N starvation	yes	yes		
galactose	Gal2p	glucose addition	yes			

general aa	Gap1p	Gln or NH_3 addition		yes?		yes
glutamine	Gnp1p	NH_3 addition				yes
allantoate	Dal5p	NH_3 addition				
allantoin	Dal4p	Asn addition				
PM pumps						
a-factor	Ste6p	constitutive	yes	yes	yes	yes
multidrug	Pdr5p	constitutive	yes	yes	yes	yes
Golgi proteins						
kexin	Kex2p	overexpressed; high temperature	yes	no		
mannosyl-transferase	Och1p	constitutive	yes	no		
DPAPA	Ste13p	overexpression	yes	no		
Kex1p	Kex1p	overexpression	yes	no		

Ste2p (Davis et al. 1993; Rohrer et al. 1993; Tan et al. 1993; Schandel and Jenness 1994; Hicke and Riezman 1996); Ste3p (Davis et al. 1993; Tan et al. 1993); Fur4p (Volland et al. 1994; Hein et al. 1995; Galan et al. 1996); Itr1p (Lai et al. 1995; Robinson et al. 1996); Mal61p (Riballo et al. 1995; Medintz et al. 1996; C. Michels, pers. comm.); Gal2p (Chiang et al. 1996); Gap1p (Grenson 1983; Hein et al. 1995); Gnp1p (Grenson and Dubois 1982); Dal5p (Grenson 1983; Grenson and Dubois 1982); Dal4p (Cooper 1982; Cooper and Sumrada 1983); Ste6p (Berkower et al. 1994; Kolling and Hollenberg 1994); Pdr5p (Egner et al. 1995; Egner and Kuchler 1996); Kex2p (Wilcox et al. 1992); Och1p (Harris and Waters 1996); DPAPA (Roberts et al. 1992); Kex1p (Cooper and Bussey 1992).

[a]Observed directly or inferred from effects of *ubc* or *npi1* mutants.

ported through the endocytic pathway to the vacuole, where it is degraded by vacuolar proteases (in a *PEP4*-dependent manner) not by the proteasome (Singer and Riezman 1990; Raths et al. 1993; Schimmoller and Riezman 1993; Schandel and Jenness 1994; Hicke and Riezman 1996).

The other plasma membrane proteins listed in Table 4 share one or more characteristics with Ste2p; a protein may demonstrably be internalized, show *PEP4*-dependent degradation, require endocytosis for degradation, be stable in *ubc* and/or *npi1* mutants (functions required for or associated with ubiquitination), and become ubiquitinated, all signs suggesting that the proteins are taken up into the cell by endocytosis and delivered to the vacuole and degraded there. The details for a given protein can be found in the references in Table 4.

The α-factor tridecapeptide is taken up by receptor-mediated endocytosis and hydrolyzed in the vacuole, since cleavage does not occur in a *pep4* strain. The degradation appears to be a normal component of the adaptation and recovery mechanism, for *pep4* mutants are slowed in recovery from G_1 arrest (Singer and Riezman 1990). Numerous peptides are taken up and utilized by cells. For Cbz-Gly-Leu and Cbz-Phe-Leu, at least, cleavage must occur in the vacuole, since mobilization of the Leu from these peptides requires CpY and/or CpS (Wolf and Ehmann 1981; Jones 1991a and unpubl.).

2. *Stationary Phase*

Resting cells show higher rates of protein turnover than growing cells (Halvorson 1958a,b), and rates of protein degradation increase severalfold as cells enter stationary phase (Bakalkin et al. 1976; Teichert et al. 1989). Proteolytic activities increase as cells reach stationary phase; the levels of vacuolar proteases, particularly PrB, rise substantially (Betz and Weiser 1976; Hansen et al. 1977; Moehle and Jones 1990). Transcription of *UBI4*, encoding polyubiquitin, also increases markedly at this time (Finley et al. 1987). Thus, cells in stationary phase seem to be well endowed for both proteasomal and vacuolar proteolysis. Ubiquitinated proteins accumulate in the vacuoles of a mutant deficient for PrA and PrB but not in wild-type vacuoles (Simeon et al. 1992), implicating the vacuole in turnover of ubiquitinated proteins during stationary phase. The accumulated ubiquitinated proteins remain pelletable after treatment with 0.1 M Na_2CO_3 at pH 11.5 (Simeon et al. 1992). Whether this means that the ubiquitinated proteins are all integral membrane proteins, perhaps not unexpected considering recent observations about ubiquitination of endocytosed membrane proteins (see above), or might include aggregated proteins is not known.

3. *Transitions*

a. Changes in Carbon Source. The complement of enzymes present and the levels of enzymes of carbon metabolism vary with the carbon and energy source used for growth. As cells exhaust the glucose in a medium, they derepress synthesis of the respiratory components and enzymes they will need for the new metabolic state. These include enzymes of the tricarboxylic acid cycle, the glyoxylate shunt, synthetic pathways for reserve carbohydrates, gluconeogenic enzymes, and mitochondrial components (Polakis and Bartley 1965; Polakis et al. 1965; Duntze et al. 1969; Perlman and Mahler 1974; Wales et al. 1980; François et al. 1991).

If cells are transferred from acetate, ethanol, pyruvate, or oleic acid to a glucose-containing medium, there is a large increase (16x) in the rate of degradation of long-lived proteins (Chiang et al. 1996). The levels of some enzymes, including enzymes of gluconeogenesis and reserve carbohydrate biosynthesis, mitochondrial components, and peroxisomal enzymes decrease more rapidly than expected from dilution (Witt et al. 1966; Ferguson et al. 1967; Chapman and Bartley 1968; Gancedo 1971; Gancedo and Schwerzmann 1976; François et al. 1991). Fructose bisphosphatase (FBPase), PEP carboxykinase (PEPCK), the cytoplasmic form of malate dehydrogenase (cMDH), isocitrate lyase, and thiolase are rapidly inactivated. For FBPase, cMDH, and thiolase, the enzyme protein is known to be degraded (Hagele et al. 1978; Neeff et al. 1978; Funayama et al. 1980; Chiang and Schekman 1991; Chiang et al. 1996). The phenomenon was named *catabolite inactivation* by Holzer (1976). Glucose also triggers catabolite inactivation of the fermentative capacity for maltose and galactose. For both of these inducible systems, the effect is on transport of the sugar; the permease is degraded in each case (Riballo et al. 1995; Chiang et al. 1996; Medintz et al. 1996). The vacuolar ApI also undergoes catabolite inactivation and degradation (Frey and Rohm 1979).

Upon catabolite inactivation, degradation of the maltose permease, Mal61p, is catalyzed by vacuolar proteases in the vacuole, not by the proteasome; Mal61p reaches the vacuole via endocytosis, for in mutants lacking the vacuolar hydrolases or defective in endocytosis, Mal61p is not degraded (Riballo et al. 1995; Medintz et al. 1996). Gal2p, the galactose permease, appears to share a similar fate upon catabolite inactivation, although the evidence is less complete (Chiang et al. 1996).

Chiang and Schekman (1991) and Chiang et al. (1996) reported that the degradation of FBPase that takes place when glucose is added to cells growing on poorer carbon sources occurs in the vacuole. A short period after glucose exposure, the FBPase is transferred from the cytoplasm into the vacuole in a process that appears to require synthesis of a protein that must pass through the Golgi complex, at least (and presumably to the vacuole) to facilitate transfer. Degradation of FBPase is impaired in a *pep4* mutant, implicating vacuolar hydrolases in the degradation (Funaguma et al. 1985; Chiang and Schekman 1991; Chiang et al. 1996). Two lines of evidence suggest that FBPase is taken into a vesicular intermediate and then transferred to the vacuole. Huang and Chiang (1997) have purified a novel vesicular component that contains FBPase. FBPase becomes incorporated into this vesicular fraction only after cells are shifted to a glucose-containing medium from media containing poorer carbon sources. In addition, Hoffman and Chiang (1996) have obtained a large number of mutants defective in degradation of FBPase. These *vid* mutants are defective in importing FBPase into vacuoles. In some of the mutants, FBPase is present in vesicles that never fuse with the vacuoles (Hoffman and Chiang 1996).

Evidence has also been adduced, however, to suggest that degradation of FBPase is effected by the cytoplasmic proteasome after ubiquitination of FBPase (Schafer et al. 1987; Teichert et al. 1989; Schork et al. 1994, 1995). It seems likely that both sets of investigators are correct. A fraction of the FBPase is nearly perfectly stable in the *pep4* mutant (~50% in Chiang and Schekman 1991), but the rest is quite unstable. All of the data could be accounted for if the unstable 50% were degraded by the proteasome in the cytosol and the remaining 50% were degraded in the vacuole by vacuolar hydrolases. The distribution of degradation between the two systems might vary with the experimental regimen and reflect kinetics of the processes involved. Analysis of relevant double and/or triple mutants should be instructive.

b. Changes in Nitrogen Source. The complement of enzymes and the levels of enzymes of nitrogen metabolism will vary with the nitrogen source available to the cell. Good nitrogen sources such as glutamine, asparagine, and ammonia repress synthesis of degradative enzymes for a variety of nitrogenous compounds, including arginine, urea, allantoin, allantoate, GABA, glutamate, and proline and for anabolic enzymes such as glutamine synthetase; the enzymes are expressed at higher levels on poor nitrogen sources such as proline (for review, see Magasanik 1992). By analogy, this phenomenon is known as nitrogen catabolite repression.

It affects not only the enzymes required for catabolism of these compounds, but also many of the permeases that transport them into cells, including the general amino acid permease, Gap1p, and permeases for allantoin, Dal4p; allantoate, Dal5p; urea, Dur3p; proline, Put4p; arginine Can1p; and GABA, Uga4p (Daugherty et al. 1993; Coffman et al. 1994).

If cells are shifted from a poor nitrogen source to a good nitrogen source, transport via Gap1p or the inducible allantoin permease, Dal4p, decreases very rapidly (Cooper and Sumrada 1983; Grenson 1983; Stanbrough and Magasanik 1995). Again, by analogy, this is referred to as nitrogen catabolite inactivation. The inactive Gap1p has been shown to be degraded, albeit at rates slower than inactivation (Hein et al. 1995; Stanbrough and Magasanik 1995). The responsible proteolytic system has not been identified. Gap1p is neither inactivated nor degraded in the *npi1* (*mut2*) mutant, which is defective in ubiquitin ligase (Grenson 1983; Hein et al. 1995). By analogy with the fate of other membrane proteins whose activities are being modulated, we feel confident in predicting that Gap1p is being taken as a ubiquitinated protein into the vacuole by endocytosis and degraded there. Similarly, we expect loss of transport capacity for glutamine, proline, allantoate, allantoin, and GABA upon addition of a good nitrogen source (Grenson and Dubois 1982; Grenson 1983) to result in ubiquitination, endocytosis, and degradation of the relevant permeases.

4. *Starvations*

a. Carbon or Nitrogen Starvation. If cells are transferred from a glucose ammonia medium to a medium lacking a usable carbon or nitrogen source, the rate of protein degradation increases and may reach 3–4% per hour (Lopez and Gancedo 1979; Wolf and Ehmann 1979; Takeshige et al. 1992). Vacuolar hydrolases have a major role in degradation under these conditions, with PrB accounting for at least 60% of the degradation (Wolf and Ehmann 1979; Takeshige et al. 1992). Carbon or nitrogen starvation triggers autophagy (discussed below).

The only specific protein known to be inactivated and degraded upon carbon starvation is NADP-dependent glutamate dehydrogenase (Mazon 1978; Mazon and Hemmings 1979). However, this degradation appears not to be effected by vacuolar hydrolases, since *pep4*, *prb1*, and *prc1* mutations do not affect the degradation (Wolf and Ehmann 1979; Hemmings et al. 1980; Mechler and Wolf 1981).

Nitrogen starvation of cells deficient for PrA and PrB results in accumulation of substantial amounts of large ubiquitinated proteins in vacuoles, whereas wild-type cells accumulate lesser amounts of somewhat smaller ubiquitinated proteins (Simeon et al. 1992). Whether these all correspond to endocytosed membrane proteins has not been examined.

If cells growing on an acetate-based medium are transferred to a nitrogen-free acetate-based medium, protein is degraded at a rate of 2–3% per hour (Betz and Weiser 1976; Zubenko and Jones 1981; Teichert et al. 1989). The protein degradation, while taking place in sporulation medium, is a response to the starvation regimen per se, rather than to sporulation, for ploidy and *MAT* genotype are without influence on the phenomenon (Betz and Weiser 1976; Zubenko and Jones 1981; but see Hopper et al. 1974). This sporulation medium can be seen to trigger autophagy in strains deficient in PrB activity; autophagy is the response to starvation not "sporulation," for haploids and petite strains, neither of which can sporulate, manifest the response (Takeshige et al. 1992). It seems likely that the autophagy is nonselective, since up to 70% of all vegetative proteins present in the cells are degraded in this regimen (Hopper et al. 1974; Klar and Halvorson 1975; Betz and Weiser 1976; Zubenko and Jones 1981).

b. Analog-containing Proteins. Incorporation of amino acid analogs into proteins results in increased protein degradation. Initial rates of degradation of canavanine-containing proteins are comparable for wild-type and protease-deficient cells ($Pra^{-}Prb^{-}$). However, within 1 hour, the rate falls off in the mutant cells to about 65% of that in wild-type cells (Teichert et al. 1989). Thus, about 35% of the abnormal proteins are being degraded by vacuolar hydrolases. Since Wolf and his colleagues have demonstrated that degradation of analog-containing proteins is impaired in mutants with defective proteasome subunits, we presume that 65% of analog-containing proteins are degraded by the proteasome (Heinemeyer et al. 1991, 1993; Hilt et al. 1993).

V. AUTOPHAGY

In the process of autophagy, organelles and cytosol are taken up into the lysosome for degradation. Autophagy in mammalian cells has been separated into microautophagy and macroautophagy. During macroautophagy, cellular components become surrounded by an invagination of the rough ER to become an autophagosome, which then fuses with the lysosomal membrane to give an autolysosome. Degradation of the autosomal membrane and its contents follows. During microautophagy, cellular components are incorporated into lysosomes by an invagination of the lysosomal membrane. Macroautophagy is generally thought to be nonselective, although it can be selective under certain conditions.

When wild-type *S. cerevisiae* cells are starved of carbon or nitrogen or sulfur, they show increased rates of protein degradation. If mutant cells lacking PrB are subjected to the same starvation conditions, autophagic bodies are found in the vacuoles (Takeshige et al. 1992). The process appears to be nonselective since ribosomes, mitochondria, membranes of the ER, lipid and glycogen granules, membrane vesicles, and cytosolic enzymes have all been found in vacuoles under these conditions. Such autophagic bodies are not seen in wild-type cells under the same conditions, presumably because they are degraded as rapidly as they are formed. This inference is strengthened by the finding that wild-type cells starved of nitrogen will form autophagic bodies if PMSF (phenylmethylsulfonyl fluoride), which covalently inactivates serine proteases such as PrB, is present. Washout of the PMSF by transfer to rich medium results in the disappearance of the autophagic bodies within a few hours, presumably via degradation by newly synthesized PrB (Takeshige et al. 1992).

Methylotrophic yeasts like *Pichia pastoris* and *Hansenula polymorpha* synthesize large peroxisomes when grown on methanol. When methanol is replaced by glucose as the carbon source in both yeasts, the peroxisomes and the enzymes within them are selectively degraded by vacuolar hydrolases and, in the case of *P. pastoris*, the related cytosolic enzyme formate dehydrogenase is degraded selectively in the same organelle (Veenhuis et al. 1983; Tuttle et al. 1993; Tuttle and Dunn 1995). Mitochondrial enzymes are not affected (Tuttle et al. 1993). Peroxisomal alcohol oxidase is not degraded unless it is within peroxisomes (van der Klei et al. 1991), suggesting that a peroxisomal membrane protein is the recognition factor in this selective autophagy. The process resembles microautophagy for the peroxisomes are engulfed by invaginations of the vacuolar membrane and are not enclosed in an additional surrounding membrane before engulfment; they do not appear to be autophagosomes (Tuttle and Dunn 1995). Because cycloheximide prevents engulfment, it is inferred that a protein, possibly a receptor, must be synthesized to allow engulfment.

If the methanol is replaced by ethanol for *P. pastoris*, peroxisomes are selectively degraded by vacuolar hydrolases, but cytosolic formate dehydrogenase is not. The peroxisomes appear to be in autophagosomes, although the source of the surrounding membrane is unknown; mitochondria are not engulfed (Tuttle and Dunn 1995). This appears to be selective macroautophagy. Cycloheximide does not affect the process.

Apparently any glucose regimen that triggers catabolite inactivation of FBPase (and thiolase if peroxisomes are present) also triggers autophagy in *pep4* cells (Chiang et al. 1996). The autophagy seems to be selective, since all of the FBPase and thiolase protein are degraded, but only a small fraction of the cytosolic enzymes are taken into vacuoles and degraded (Chiang and Schekman 1991; Takeshige et al. 1992; Chiang et al. 1996). However, autophagy toward FBPase will appear more selective than it really is if any FBPase in the cytoplasm is also being degraded by the proteasome.

Mutants defective in degradation of FBPase have been obtained and so far have defined 20 genes (Hoffman and Chiang 1996). Many of the mutants appear to be defective in importing FBPase into vacuoles; a few appear to transfer FBPase into vesicles that never fuse with vacuoles (Hoffman and Chiang 1996). Mutants unable to import the vacuolar hydrolase ApI into vacuoles via the extrasecretory route have also been obtained (Harding et al. 1995) as have mutants unable to undergo autophagy (Tsukada and Ohsumi 1993; Thumm et al. 1994). Recent studies have revealed that functional overlap exists between the genes required for autophagy (*AUT*) and the genes required for cytoplasm-to-vacuole targeting (*CVT*). Representative mutants for 7 of 9 *AUT* genes are also Cvt$^-$. Similarly, representative mutants for 10 of 12 *CVT* genes are also Aut$^-$. Three of the genes, *AUT5/CVT17*, *AUT10/CVT10*, and *AUT9/CVT7*, are common to both pathways (Harding et al. 1996). Further work should help clarify whether there are multiple routes for importing proteins directly into vacuoles, to what extent the routes overlap or are redundant, and whether the routes used for incorporating normal vacuolar constituents such as ApI and αMS are also used for selective autophagy.

VI. HYDROLASE TRAFFICKING TO THE YEAST VACUOLE

A. The Early Secretory Pathway

Several lines of evidence indicate that most of the soluble and membrane-associated vacuolar hydrolases proceed to the vacuole via the secretory pathway. First, in the case of CpY, the most extensively studied of all the vacuolar hydrolases, it has been shown that preproCpY contains a canonical signal sequence that directs the nascent polypeptide into the lumen of the ER, where it is then cleaved from the protein (Blachly-

Dyson and Stevens 1987). This signal sequence has been shown to be sufficient to direct a heterologous polypeptide into the lumen of the ER, as it can functionally replace the signal sequence of the normally secreted enzyme invertase (Johnson et al. 1987). Second, it has also been demonstrated that proCpY, like secreted proteins, interacts with the ER-localized chaperone protein BiP (Kar2p), indicating its presence in the ER (Vogel et al. 1990; te Heesen and Aebi 1994; Chaudhuri et al. 1995; Simons et al. 1995). This interaction may facilitate the folding of the nascent polypeptide and/or be part of a "quality control" mechanism prior to release of the protein from the ER. Third, CpY has been demonstrated to be a glycoprotein, containing four N-linked carbohydrate side chains (Hasilik and Tanner 1978b; Stevens et al. 1982; Johnson et al. 1987; Klionsky et al. 1988; Franzusoff and Schekman 1989). Since N-linked protein glycosylation takes place only in the lumens of the ER and Golgi membranes, this again demonstrates that proCpY is routed to the vacuole via the secretory path. In addition, as specific mannosyltransferase enzymes are known to be spatially restricted to specific compartments of the secretory pathway and to catalyze the addition of carbohydrate moieties via specific and unique linkages, the types of carbohydrate linkages seen on a secretory pathway mannoprotein give an unambiguous signature of which secretory compartments have been encountered. In this way, it can be demonstrated that proCpY encounters not only the ER, where the core glycosylation event takes place, but also the Golgi compartments that contain the core unit trimming functions and α-1,6-mannosyltransferase (*cis*-Golgi compartment) and α-1-3-mannosyltransferase (medial Golgi compartment) (Klionsky et al. 1988; T.R. Graham et al. 1994). Fourth, kinetic experiments on CpY synthesis and processing demonstrate that the hydrolase passes through compartments identifiable as the ER and Golgi complex en route to the vacuole (Vida et al. 1993). Finally, the investigation of CpY production in various mutants that are blocked at translocation into the ER (Deshaies and Schekman 1987), transport between the ER and Golgi (Stevens et al. 1982), and between different compartments of the Golgi apparatus (Stevens et al. 1982; Franzusoff and Schekman 1989; Banfield et al. 1995) demonstrates that proCpY is transported along the secretory pathway prior to arrival at the vacuole. Importantly, these same mutations have the same effects on vacuolar hydrolase precursors as they do on the progression of normally secreted proteins. This is consistent with the CpY precursor being transported along the secretory pathway with normally secreted proteins. Equally important is the further observation that mutations in genes that function in transport from the Golgi apparatus to the plasma membrane have no effect on CpY synthesis (Stevens et al. 1982), indicating that in the late Golgi complex, distal to both the α-1-3-mannosyltransferase and the Kex2p-containing compartment (see below), vacuolar proteins and secreted proteins follow separate routes.

Although the data are most complete for CpY, similar results have been found with other vacuolar hydrolases. Specifically, precursor forms of PrA have been demonstrated to enter the ER early in its synthesis (Klionsky et al. 1988), to be glycoproteins with carbohydrate linkages indicating transport through the *trans*-Golgi apparatus (Meussdoerffer et al. 1980; Mechler et al. 1982), and whose progress is blocked by mutations in early-acting *SEC* genes (Klionsky et al. 1988; Mechler et al. 1988), but not those functioning in transport between the Golgi complex and the plasma membrane (Rothman et al. 1986).

The earliest precursor form of PrB contains a functional signal sequence that is cleaved in the ER (V. Nebes and E. Jones, unpubl.). Precursor forms of PrB contain one N-linked glycosyl side chain; mature PrB contains none, but since mature PrB is a glycoprotein (Ulane and Cabib 1976; Kominami et al. 1981; Mechler et al. 1982; Moehle et al. 1989), the mannosyl residues are inferred to be O-linked (for discussion, see Moehle et al. 1987). Because both N- and O-linked carbohydrates are added in the secretory pathway, PrB must transit the secretory pathway (Mechler et al. 1982, 1988; Haselbeck and Tanner 1983; Moehle et al. 1989). Like CpY and PrA, *sec* mutations that block translocation into the ER and transport between the ER and Golgi compartment, or between Golgi compartments, block the synthesis of PrB (Mechler et al. 1988; Moehle et al. 1989). Data are not available demonstrating what effect, if any, late-acting *sec* mutations have on PrB synthesis.

A similar picture emerges for the type II integral membrane vacuolar hydrolases, alkaline phosphatase (AlP) and DPAPB. For AlP, the subamino-terminal transmembrane domain (amino acids 34–59) can act like an uncleaved signal sequence when put in place of the invertase signal sequence, indicating that it may function to direct the membrane protein to the ER (Klionsky and Emr 1990). Furthermore, it has been demonstrated that precursor AlP is inserted in the ER membrane and is transported along the secretory path as an integral membrane protein (Klionsky and Emr 1989). AlP contains N-linked glycosyl side chains with structures consistent with passage through both the ER and Golgi apparatus up through the *trans*-Golgi compartment (Onishi et al. 1979; Clark et al. 1982; Klionsky and Emr 1989). In addition, *sec* mutations that block transport between the ER and Golgi block the synthesis of AlP (Klionsky and Emr 1989). However, a mutation in *SEC7* that blocks transport between specific Golgi compartments only slows the synthesis of the protein (Klionsky and Emr 1989); this is unlike what is seen for the precursor to CpY (Stevens et al. 1982). This difference in behavior may indicate a difference in AlP progression from that seen for the soluble hydrolases. Mutations in late-acting *SEC* genes have no effect on AlP synthesis (Klionsky and Emr 1989).

Less is known about DPAPB in the early secretory pathway, yet what is known is consistent with that discussed above. DPAPB is an integral membrane glycoprotein, indicating passage through the secretory pathway (Roberts et al. 1989). However, data on the type of carbohydrate linkages are not available (Roberts et al. 1989). Mutations in *SEC* genes required for ER to Golgi apparatus transport block the synthesis of DPAPB (Roberts et al. 1989). As is seen with AlP, however, the *sec7* mutation leads only to a partial retardation of synthesis. Again, the significance of this difference is not presently known. The functions of late-acting *SEC* genes are not required for synthesis of DPAPB (Roberts et al. 1989).

As noted above, at steps in the secretory pathway distal to the α-1-3-mannosyltransferase and Kex2p compartments, secreted and vacuolar proteins demonstrate marked differences in genetic requirements for further progress along the secretory pathway. Whereas secreted proteins rely on late *SEC* gene functions for transport from the Golgi apparatus to the plasma membrane and subsequent exocytosis, vacuolar proteins are transported to the vacuole independent of these late *SEC* functions and require cellular functions not needed by secreted proteins, indicating that secreted and vacuolar proteins move through separate compartments once beyond the late Golgi apparatus and that the cell contains functions required specifically for trafficking between the Golgi apparatus and the vacuole. These vacuole-trafficking-specific functions have been the research focus of several laboratories, including our own.

B. Trafficking between the Golgi Apparatus and the Endosome

Mutants defective in transport between the Golgi apparatus and the vacuole have been isolated in several different laboratories using different paradigms. The first screen that surfaced mutants defective in vacuolar transport was actually designed to isolate protease-deficient mutants. Mutants isolated as deficient in CpY defined 16 complementation groups designated *prc1* and *pep1-pep15*, signifying peptidase-deficient (Jones 1977). *pep16* and *pep21* were later similarly defined. Many of these mutations were pleiotropic, as initially reported, and later proved to be defective in Golgi to vacuole transport.

Two groups sought mutants that secreted vacuolar hydrolases: One group used a *prc1::SUC2* fusion, which contains ER and vacuolar targeting information from the CpY precursor fused to invertase with selection for Suc$^+$ (invertase-secreting) colonies (Bankaitis et al. 1986; Robinson et al. 1988), and the second group selected for mutant derivatives of a *pep4* mutant (which is unable to process the CpY precursor; Hemmings et al. 1981) able to use *N*-Cbz-L-phenylalanyl-L-leucine as a leucine source (Rothman and Stevens 1986). The former selection identified complementation groups *vpt1-vpt33* and the latter identified *vpl1-vpl19.* These two groups later fused their mutant nomenclature to give complementation groups *vps1-vps46* (Robinson et al. 1988; Rothman et al. 1989a; Raymond et al. 1992a). *vpt, vpl,* and *vps* refer to *v*acuole *p*rotein *t*argeting or *l*ocalization or *s*orting, respectively.

Mutants were also isolated that displayed abnormal *va*cuolar *m*orphology microscopically; a preliminary screen identified mutants that accumulated less red pigment (in an *ade1* background) or less chloroquine (Wada et al. 1992b). These mutants were labeled *vam1-vam9.*

Screens for mutants defective in either fluid-phase or receptor-mediated endocytosis or for synthetic lethality with mutations in a gene for a V-ATPase subunit also surfaced mutants defective in the Golgi to vacuole pathway (Chvatchko et al. 1986; Raths et al. 1993; Munn and Riezman 1994; Munn et al. 1995). These mutants were called *end1-end13.*

The final systematic screen was carried out by Preston et al. (1992), who used the pH-dependent fluorescence of 6-carboxyfluorescein to isolate mutants unable to form an acidic vacuole. Complementation groups *vph1-vph9* were defined.

Extensive overlap exists among the genes defined in these six screens, as well as with the broader screen for calcium-sensitive mutants (Ohya et al. 1986). The following sections discuss those genes for which a significant body of information exists and any prevailing themes that are beginning to emerge from this branch of the secretory pathway. In the tables summarizing genes and functions, the name used in the Saccharomyces Genome Database is given primacy; all other names known to us are listed as well.

Figure 3 summarizes what is known or conjectured about the sites of operation of gene products in steps between the TGN and the vacuole. The proteins known to participate in transport between the TGN and the endosome are summarized in Table 5, and those that function between the endosome and the vacuole are summarized in Table 6.

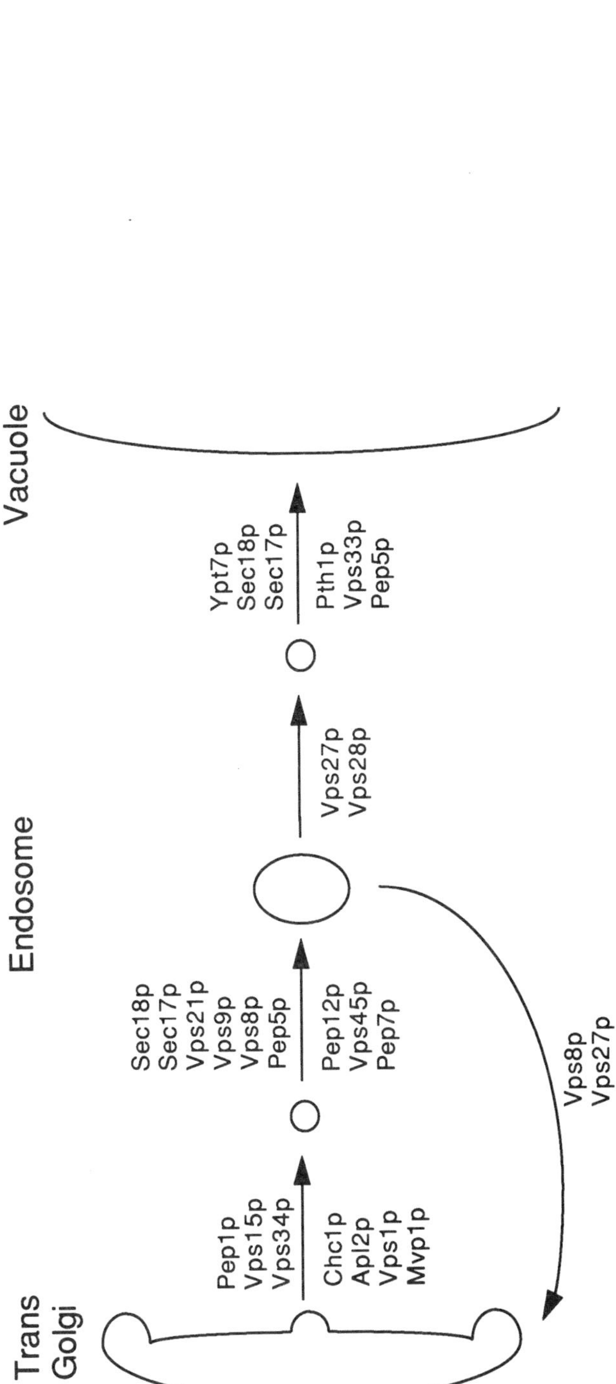

Figure 3 Compartments and gene products involved in delivery of hydrolase precursors from the late Golgi to the vacuole. Precursors to soluble hydrolases are incorporated into vesicles that fuse with the endosome. Transport between the endosome and the vacuole is assumed to occur via vesicles. Recycling from the endosome to the late Golgi is inferred from the properties of mutants thought to be defective in the process. Gene products are placed adjacent to the step(s) in which they participate. See the text for the identity and functions of the gene products.

Table 5 Proteins involved in transport of precursors for soluble vacuolar hydrolases from late Golgi (TGN?) to (late?) endosome

SGD name	Other names	Homolog	Size (aa)	Properties
Vesicle budding:				
Pep1p	Vps10p Vpt1p		1577	CpY sorting receptor in late Golgi; type I integral membrane protein 166-aa cytoplasmic tail, 1300-aa luminal domain
Vps15p	Vpt15p Vpl19p Vac4p Grd8p		1455	myristoylated Ser/Thr kinase
Vps34p	Vpt29p Vp17p Pep15p		875	PtdIns 3-kinase
Chc1p		clathrin heavy chain	1653	presumed vesicle coat protein
Apl2p		β-adaptin, Ap-1 complex	726	large subunit of clathrin-associated protein complex
Vps1p	Bpt26p Vpl1p Spo15p Lam1p Grd1p	dynamin	704	member of dynamin family of GTPases
Mvp1p			511	associates with Vps1p; itself required for vacuolar sorting
Vesicle docking and/or fusion:				
Sec18p		NSF homolog; general fusion factor	758	cytosolic and peripheral membrane protein; ATPase
(Sec17p)		α-SNAP general fusion factor	292	peripheral membrane protein
Pep12p	Vps6 Vpt13 Vpl6	t-SNARE; syntaxin family	288	integral membrane protein; carboxy-terminal TMD; located in endosome

Vps45p	Vpl28p Stt10p	Sec1 homolog	577	cytosolic and peripheral membrane protein; interacts genetically with Pep12p and Pep7p
Pep7p	Vps19p Vpt19p Vp121p Vac1p	cysteine region homologous to part of Vps27p, Fab1p, 2 *S.c.* ORFs	514	cytosolic and peripheral membrane protein; interacts genetically with Vps45p and Pep7p; cysteine-rich region aa are essential for function; highly similar regions are present in Vps27p, Fab1p, as well as other ORFs in *S.c., H.s., C.e., M.m., P.d.*
Vps21p	Ypt51p Vpt21p Vps12p Vpt12p	rab5 homolog	210	small GTP-binding protein; geranylgeranylated; geranylgeranylation required for membrane association; also involved in endocytosis postvesicle internalization
Ypt52p		rab5 homolog	234	small GTP-binding protein; also involved in endocytosis
Ypt53p		rab5 homolog	220	small GTP-binding protein
Vps9p	Vpt9p Vpl31p	domain related to inhibitors of Ras	451	
Vps8p		see below	1176	interacts with Vps21p; also see below
Pep5p	Vps11p Vpt11p Vpl9p End1p Vam1p	*M.m., H.s.* ESTs	1029	peripheral membrane protein; a *pep5* allele interacts genetically with a missense allele of *VPS8*
Endosome to Golgi recycling:				
Vps8p	Vpt8p Vpl18p Fun15p	*H.s.* EST	1176	hydrophilic protein; in its absence, Vps10p goes through to the vacuole
Vps27p	Vpt27p Vpl27p Grd11p		623	see Table 6
Vps28p	Vpt28p	*H.s.* ESTs	242	see Table 6

Table 6 Proteins involved in transport of precursors for soluble vacuolar hydrolases from late endosome to the vacuole

SGD name	Other names	Homolog	Size (aa)	Properties
Exit from the endosome:				
Vps27p	Vpt27p Vpl23p Grd11p Ssv7p		623	hydrophilic protein; has cysteine-rich putative zinc finger essential for function, similar to ORFs in *S.c., H.s., M.m., C.e.*
Vps28p	Vpt28p Vpll3p Lpe4p	*H.s.* ESTs	242	soluble, hydrophilic protein; no known relatives
Docking/fusion at the vacuole:				
Pth1p	Vam3p	syntaxin	283	predicted carboxy-terminal TMD, operates downstream from Pep12p
Ypt7p	Ast4p Vam4p	rab7	208	required for homotypic fusion event in vacuole inheritance, for endosome-endosome fusion, and for fusion of endosomes to vacuoles when expressed from high-copy plasmid
Vps33p?	Pep14p Slp1p Vpt33p Vpl25p Vam5p Cls14p Met27p	Sec1p	691	soluble when expressed from high-copy plasmid
Vam7p?		SNAP-23; SNAP-25	316	hydrophilic protein; heptad repeat motif

1. Sorting

Sorting of both soluble and membrane vacuolar hydrolase precursors appears to take place in a late Golgi compartment. The best evidence concerns CpY. CpY passes through compartments containing the α-1,3-mannosyltransferase and Kex2p protease, indicating that the sorting event takes place at or distal to the compartment(s) containing these functions (Klionsky et al. 1988; Graham and Emr 1991). Because the Kex2p-containing compartment is thought to be distal to the α-1,3-mannosyltransferase compartment, these results indicate that CpY travels through the compartments containing α-1,3-mannosyltransferase and Kex2p and that sorting to the vacuole could take place within the Kex2p compartment. By comparison of functional genetic landmarks along the secretory pathway, the Kex2p-containing compartment is often characterized as the yeast equivalent to the TGN (Graham and Emr 1991; Wilcox and Fuller 1991). Because it is thought that α-factor, which requires Kex2p function for maturation, is packaged into secretory vesicles subsequent to cleavage by Kex2p (see Graham and Emr 1991), and CpY does not enter the late secretory pathway, it is likely that the sorting event takes place in this Kex2p compartment. Although both DPAPA (Ste13p) and Kex1p act subsequent to Kex2p activity in α-factor maturation, they are both thought to be in the same compartment as Kex2p (Bryant and Boyd 1993; Nothwehr et al. 1993). Therefore, there appears to be no present need to hypothesize more distal Golgi compartments on the basis of these Golgi apparatus activities.

Additional findings that tie the sorting events to these late Golgi compartments are found in the work on *vps1* mutants. These mutants not only missort both soluble and membrane-associated vacuolar hydrolases to the plasma membrane, but also fail to retain both Kex2p and a DPAPA reporter construct in the late Golgi complex, suggesting that the integrity of this late Golgi compartment is compromised in *vps1* mutants and is therefore crucial for vacuolar hydrolase sorting (Nothwehr et al. 1995; Vater et al. 1992).

It was reported that overexpression of CpY led to the secretion of the hydrolase precursor in a Golgi-modified form, indicating that the sorting mechanism for this precursor was saturable and might involve a receptor that recognized the precursor form of CpY (Stevens et al. 1986). Since late-acting *SEC* genes were required for the secretion of the CpY precursor, the CpY precursor is exiting the cell via the normal secretory pathway (Stevens et al. 1986). Early reports concluded that Pep1p/Vps10p, a type I transmembrane glycoprotein found in the TGN, was the proCpY-specific receptor (Marcusson et al. 1994). In mutants deleted for the

PEP1/VPS10 gene, P2 CpY was selectively secreted to the plasma membrane, whereas the PrA and PrB precursors were trafficked and matured nearly as well as in a wild-type strain (Marcusson et al. 1994). More recent work on *VPS10*, however, indicates that it may function in the sorting or trafficking process of multiple vacuolar hydrolases (Cooper and Stevens 1996; Westphal et al. 1996). Because Vps10p appears to interact physically with proteins destined for the vacuole, it may still prove to be a receptor but lack the ligand specificity originally attributed to it (Marcusson et al. 1994; Cooper and Stevens 1996; Westphal et al. 1996). Consistent with its postulated role as a sorting receptor, it was determined that mutations and deletions affecting the cytoplasmic tail of Vps10p result in relocation of Vps10p from the Golgi membranes to the vacuole, where Vps10p is degraded in a *PEP4*-dependent manner, and secretion of P2 CpY (Cereghino et al. 1995). It was further noted that overexpression of Vps10p led to its accumulation in the vacuole in addition to its presence in the Golgi complex, suggesting that the protein could traverse part of or the entire route to the vacuole. Interestingly, *VTH1* and *VTH2* are apparently redundant homologs of *VPS10*, suggesting that a family of receptors with unknown but perhaps overlapping cargo "specificity" may function in sorting proteins toward the vacuole from the TGN (Cooper and Stevens 1996; Westphal et al. 1996).

Unlike mammalian lysosomal protein sorting, which for many proteins involves a mannose-6-phosphate receptor (for review, see Hille-Rehfeld 1995), sorting of both soluble and membrane vacuolar hydrolases does not require glycosylation (Clark et al. 1982; Mechler et al. 1982; Schwaiger et al. 1982; Stevens et al. 1982, 1986; Johnson et al. 1987; Klionsky et al. 1988; Klionsky and Emr 1989, 1990; Moehle et al. 1989; Nebes and Jones 1992) and therefore the same mechanism is not being utilized in yeast; rather, the vacuolar sorting signals reside in the polypeptide. Deletion analysis demonstrated that the sorting determinant in P2 CpY resides in the propeptide between amino acids 20 and 31; point mutations that resulted in secretion identified the sequence QRPL, amino acids 24–27, in the propeptide of the zymogen (Valls et al. 1987, 1990).

The precursor to PrA traverses the compartment containing the α-1,3-mannosyltransferase; passage through the Kex2p compartment has not been examined (Klionsky et al. 1988). It has been assumed that because precursor PrA is a soluble vacuolar enzyme that is proteolytically processed like the CpY precursor and takes about the same amount of time as CpY to travel to the vacuole (Klionsky and Emr 1990), that they therefore travel the same path to the vacuole. In favor of this model are the observations that not only do secretory pathway mutants block precursor forms of PrA as they do precursor forms of CpY (Graham and Emr 1991), but also many of the same mutations (*pep*, *vps*, *vam*) shown to affect proCpY sorting also affect proPrA sorting and processing (Jones 1977; Rothman and Stevens 1986; Robinson et al. 1988). As for the CpY precursor, overexpression of *PEP4*, the structural gene for the PrA precursor, leads to secretion of precursor forms of PrA via the late secretory pathway (Rothman et al. 1986). Unlike CpY, however, the form of PrA secreted under conditions of overexpression is not the ER, Golgi apparatus, or vacuolar form, suggesting some novel processing (Rothman et al. 1986). It is not clear if this novel processing takes place internal or external to the cell.

The vacuolar sorting signal in the PrA precursor has been more difficult to identify than that in the CpY precursor. Experiments with fusion constructs between the pro region of proPrA (amino acids 1–76) and invertase showed that this pro region allowed retention of invertase within the cell to almost the same extent as full-length fusions, suggesting that most of the sorting information lies in the pro region. However, further analysis was confounded because large deletions led to unstable proteins (Klionsky et al. 1988). Very small deletions in the pro region that removed five amino acids at a time at amino acid positions 36–40, 57–61, or 68–72 resulted in stable proteins, but they had no effect on movement of proPrA between the Golgi complex and vacuole (Klionsky et al. 1988). Therefore, these 15 amino acids do not appear to contain the sorting signal.

Recent data on the function of *VPS10*, *VTH1*, and *VTH2* suggest that, in part, these proteins may act as receptors for proPrA (Cooper and Stevens 1996; Westphal et al. 1996). Less is known of both the *cis* and *trans* sorting requirements of the PrB precursor. As for CpY, mutations in *SEC18*, *SEC7*, or *SEC14* lead to a block in transit of the PrB precursor, indicating that the two hydrolases travel the same route through the ER and Golgi complex (Moehle et al. 1989). This is further supported by the observation that, as for CpY and PrA, the PrB precursor is affected in similar ways by *pep* and *vps* mutations, although, as for PrA, the precursor forms of PrB are not secreted but accumulate within the cell in an uncharacterized location (Robinson et al. 1988; Woolford et al. 1990; Preston et al. 1991; Becherer et al. 1996). Overexpression of *PRB1*, the structural gene for the PrB precursor, does not result in secretion of forms of PrB. To date, no sorting receptor for PrB has been reported. Because fusion proteins between invertase and either the prepro region of precursor PrB or the mature and post regions are both found in the vacuole, it may be the case that precursor PrB contains multiple vacuolar targeting sequences (D. Murdock et al., unpubl.).

It should be noted that although the above listed observations are consistent with both precursor forms of both PrA and PrB traversing the same compartments as the CpY precursor en route to the vacuole, nearly all of the *PEP*, *VPS*, and *VAM* mutants examined to date for sorting of the three hydrolases show reproducible differences in the amounts of the three hydrolase precursors secreted at the plasma membrane. In all cases, P2 CpY is secreted in the greatest amounts, with precursors to PrA and PrB being secreted to lesser extents. The obvious exception is *PEP4*, which is the structural gene for PrA and which is required for processing but not sorting of several precursors to vacuolar hydrolases (Jones et al. 1989). If it is accepted that all three hydrolases are sorted at the same site in the late Golgi complex, and subsequently traverse the same path to the vacuole, then it suggests that the mechanism of sorting of the PrA and PrB precursors is significantly different from that utilized by the CpY precursor or that a putative receptor(s) interacts with the PrA and PrB presursor ligands with significantly different dynamics. Further work on both PrA and PrB will have to be undertaken to resolve this apparent inconsistency.

The sorting event for AlP takes place distal to the α-1,3-mannosyltransferase compartment of the Golgi (Klionsky and Emr 1989). Also like the soluble hydrolases, delivery of AlP to the vacuole does not normally depend on the action of late-acting *SEC* genes (*SEC1* and *SEC4*) (Klionsky and Emr 1989; Nothwehr et al. 1995).

In an attempt to locate a *cis*-acting sorting signal, fusions to invertase were utilized as in the experiments on CpY and PrA. These experiments demonstrated that the first 52 amino acids of AlP, which contain most of the potential transmembrane domain (amino acids 34–59), were sufficient to confer vacuolar localization to the fusion protein (Klionsky and Emr 1990). If these amino-terminal 52 amino acids were replaced by the 20-amino-acid signal sequence from Prc1p, that fraction of the protein which received the Golgi-associated α-1-3-mannosyl residues was secreted (Klionsky and Emr 1990). This observation was taken to demonstrate that the first 52 amino acids were also necessary for vacuole localization. It was inferred that specific vacuolar targeting signals are present in the first 52 amino acids. However, as will be seen below for the case of DPAPB, it is possible that all that is required to target membrane proteins to the vacuole is a transmembrane domain that contains no specific targeting information.

Studies on the *cis*-acting sorting requirements of DPAPB were very elegantly executed using domain swapping between DPAPB and the close structural and functional homolog, DPAPA, found in the late Golgi complex (Roberts et al. 1992). Retention of DPAPA in the Golgi complex is dependent on sequences found in the cytoplasmic domain; in the absence of these sequences, the protein was delivered to the vacuole (Roberts et al. 1992). Work on sorting of DPAPB determined that the cytoplasmic domain was not required for proper sorting to the vacuole and, furthermore, that when the transmembrane domain of DPAPB was replaced with that of DPAPA (cytoplasmic domain of DPAPB, *trans*-membrane domain of DPAPA, luminal domain of DPAPB), the chimeric protein was correctly sorted to the vacuole. This result indicates that the transmembrane domain did not by itself direct targeting to the vacuole. Finally, it was shown that fusion of the luminal domain of the protein to the signal sequence of a secreted protein resulted in secretion of the luminal domain at the plasma membrane, indicating that the luminal domain did not by itself encode sorting information. This led Roberts et al. (1992) to conclude that the default destination for all secretory pathway membrane proteins is the vacuole and that no specific vacuolar targeting information for membrane proteins exists (for review, see Nothwehr and Stevens 1994). The information presented above for AlP neither confirms nor refutes this model. Also consistent with the default model is the observation that DPAPA (normally found in the late Golgi compartment) transited to the vacuole when its Golgi-specific retrieval signals in the cytoplasmic domain were removed or changed. To invoke a specific vacuolar targeting mechanism, one must infer that DPAPA contains cryptic vacuolar sorting signals and that signals for vacuolar targeting in DPAPB and perhaps AlP are redundant and present in both the cytoplasmic and transmembrane domains.

The effect of pH on vacuolar sorting has been investigated using different means to perturb the vacuole, secretory pathway, and/or cellular pH, each with varying and often unkown degrees of specificity. Early on, it was observed that weak bases and proton ionophores caused the mislocalization of precursors of CpY and PrA (see Banta et al. 1988; Rothman et al. 1989b) However, since these reagents presumably cause global cellular collapse of pH gradients, causal relationships are very difficult to interpret but suggest that pH could play a part in sorting. The observation that Bafilomycin A_1, an inhibitor specific for the V-type H^+-

ATPases (Bowman et al. 1988), perturbed the sorting of the CpY, PrA, and PrB precursors, but not the AlP precursor, suggested more specifically that the acidic pH of particular cellular compartment(s) was important for sorting of soluble hydrolases (Banta et al. 1988; Klionsky and Emr 1989).

Caution must be used in interpreting the site of action of Bafilomycin A_1 and therefore the site of this theorized pH perturbation, because V-type or vacuolar-type ATPases, despite the name, are not limited to vacuoles; they are found on other cellular organelles (Moriyama and Nelson 1989). It is well established that the yeast vacuole contains a V-type ATPase whose function is inhibited by Bafilomycin A_1 (Banta et al. 1988; Manolson et al. 1992a,b). In addition, however, the identification of a structural and functional homolog to the vacuole's 100-kD ATPase subunit that localizes to a nonvacuole, non-ER site in the secretory pathway indicates that this type of proton pump is found elsewhere in the secretory path and would presumably be affected by Bafilomycin A_1 (Manolson et al. 1994). Therefore, although Bafilomycin A_1 leads to a collapse of the vacuole's pH gradient, it likely also leads to perturbation of any pH gradients produced by other V-type ATPases at other sites along the secretory path. The data present in the literature on Bafilomycin A_1 do not allow the differentiation of the various compartments.

Genetic manipulations, viz disruption and/or deletion of genes known to encode subunits of the V-type proton pump, have also been used to study the effects of pH perturbation on hydrolase sorting. In most cases studied, the disruption or deletion of genes encoding either soluble or membrane subunits leads to perturbation of sorting and processing of CpY, PrA, PrB, and, in contrast to the Bafilomycin A_1 results, AlP precursors (Nelson and Nelson 1990; Yamashiro et al. 1990; Klionsky et al. 1992b; Morano and Klionsky 1994; Yaver et al. 1993). Precursor forms for the soluble hydrolases were secreted; both precursor and mature forms of the hydrolases were present intracellularly, indicating that sorting was partially perturbed, but properly sorted precursors were processed normally. This suggests that the altered pH in these mutants affects sorting rather than processing of the hydrolase precursors. This conclusion is further supported by the observation that the internal pool of hydrolase precursors is kinetically stable, indicating that the defect is not a slowing of processing (Klionsky et al. 1992b). Interestingly, fractionation experiments in V-type ATPase mutant strains indicate that CpY- and AlP-invertase fusion proteins cofractionate with Kex2p on sucrose gradients, as if the fusion proteins were blocked in departure from the Golgi complex (Yaver et al. 1993). As these studies were carried out using strains disrupted or deleted for genes encoding nucleotide-binding subunits of the proton pump, the defects seen could be secondary effects, due to long-term disabling of pump activity. Experiments utilizing ATPase subunit-encoding genes expressed from the inducible/repressible *GAL* promoter indicated that the effect on CpY is a primary effect due to the loss of pump activity (Morano and Klionsky 1994). In these same experiments, however, AlP was only perturbed after multiday growth under conditions where the catalytic subunit was not expressed (Morano and Klionsky 1994). Therefore, the defects observed in AlP synthesis in ATPase subunit mutants are likely to be secondary.

If it is true that the effects seen on soluble hydrolases in the ATPase subunit mutants are due to effects on sorting, it is difficult to see how altering the pH of the target organelle (the vacuole) would affect sorting, particularly in light of the evidence that an endosomal compartment is present between the Golgi complex and vacuole (see below). This positions the presumed site of action of the proton-pumping ATPase two compartments away from the site of the sorting event. A more plausible explanation is that a second nonvacuolar V-type ATPase is present in the late Golgi compartment or TGN. As noted above, the identification of a structural and functional homolog of the 100-kD V-ATPase subunit in a non-ER, nonvacuolar location in the secretory pathway is consistent with this hypothesis (Manolson et al. 1994). Moreover, the fact that high-copy expression of Stv1p (the nonvacuolar 100-kD V-ATPase subunit) complements the deficiency for Vph1p (the vacuolar 100-kD V-ATPase subunit) and results in incorporation of Stv1p into the vacuolar V-ATPase implies that Stv1p is a subunit of a second V-ATPase in a compartment upstream of the vacuole.

2. *Vesicle Budding*

a. VPS15 *and* VPS34. In an invaluable attempt to bring order to the many *VPS/PEP* complementation groups, Stevens and colleagues (Raymond et al. 1992a) classified mutants in all of the known *VPS* complementation groups by both vacuole morphology and protein localization. The *vps15* and *vps34* mutants were placed in class D with other mutants that gave a similar vacuole morphology (single large vacuolar compartment), but it was noted that *vps15* and *vps34* appeared to define a subset of the class-D mutants because they accumulated abnormal Golgi membranes in the form of Berkeley bodies (Banta et al. 1988), secreted the greatest fraction of P2 CpY of all the *vps* mutants (>95%) (Robinson et al. 1988), manifested punctate cytoplasmic staining for AlP in addition to the vacuole membrane staining seen in other class-D mutants, and gave a higher fraction of vacuoleless (avacuolar? avacuolate?) daughter cells than other class-D mutants (Raymond et al. 1992a). *VPS15* and *VPS34* have been reviewed elsewhere (Conibear and Stevens 1995; Stack et al. 1995b).

VPS15 was cloned and sequenced and was found to contain significant similarity to the serine/threonine family of protein kinases in the amino-terminal third of the predicted 166-kD protein (Herman et al. 1991b). Changes in highly conserved residues in the protein kinase domain led to defects in vacuolar sorting and autophosphorylation, indicating that kinase activity is important for both functions (Herman et al. 1991b). Vps15p was shown to be on the cytoplasmic face of the Golgi membranes and is modified at its amino terminus with myristic acid, whose transferase is found only in the cytoplasm (Herman et al. 1991a,b). Loss of Vps15p function leads to the secretion of more than 95% of CpY in a Golgi-modified form as well as to a block in processing of the PrA and PrB precursors (Robinson et al. 1988; Herman et al. 1991a). Analysis of a *vps15*^{ts} mutant led to the observation that although both P2 CpY and P2 PrA were immediately affected by the loss of Vps15p function, the integral membrane hydrolase AlP was unaffected, indicating a direct function of Vps15p apparently only in soluble hydrolase sorting or targeting (Herman et al. 1991a). Furthermore, even though a block in P2 CpY transport was immediately apparent, at very early time points following a shift to the restrictive temperature, the CpY precursor accumulated internally in a Golgi-modified form, only later being secreted, suggesting that the cell's ability to sort precursor CpY from secreted proteins was still functional. The fact that the internally accumulated P2 CpY was delivered to the vacuole upon a reshift to the permissive temperature argues that it was held in a bona fide secretory compartment (Herman et al. 1991a). Kinetic studies on the movement of P2 CpY from the Golgi to the yeast endosome en route to the vacuole (see below) indicate that *vps15* mutations block the movement of P2 CpY from the Golgi complex (Vida et al. 1993). In total, the above phenotypic and localization data suggest that Vps15p functions downstream from the receptor-binding event in the late Golgi complex but prior to departure from the Golgi complex.

VPS34 was cloned and sequenced and was found to encode a hydrophilic 95-kD peripheral membrane protein thought to be present in a large multiprotein complex (Herman and Emr 1990). Protein sequence similarity was noted between Vps34p and the catalytic subunit of bovine brain phosphoinositide 3-kinase (PI 3-kinase). Both in vitro and in vivo observations demonstrated that Vps34p is a phosphatidylinositol-specific PI 3-kinase (PtdIns 3-kinase) (see Schu et al. 1993; Stack and Emr 1994).

Vps34p also proved to contain a protein serine/threonine kinase activity responsible for an autophosphorylation event (Stack and Emr 1994).

The loss of Vps34p function leads to the secretion of more than 95% of CpY in a Golgi-modified form, as well as a block in the processing of precursor forms of both PrA and PrB (Herman and Emr 1990). Analysis of a *vps34*^{ts} mutant showed a rapid decrease in cellular PtdIns 3-kinase activity upon a shift to the restrictive temperature, suggesting that the sorting defect observed in *vps34* mutants is directly related to the loss of PtdIns 3-kinase activity (Stack et al. 1995b); soluble hydrolases are directly affected, and the membrane hydrolase AlP is unaffected (Stack et al. 1995b).

Numerous genetic and biochemical interactions between *VPS15* and *VPS34* or their products have surfaced. Overexpression of the *VPS34* lipid kinase partially suppresses the pleiotropic defects exhibited by some *vps15* mutants, strains deleted for *VPS15* are defective in Vps34p PI 3-kinase activity, and a temperature-conditional allele of *VPS15* leads to temperature-conditional lipid kinase activity (Stack et al. 1993, 1995b). In addition, Vps34p was found solely in the soluble fractions in strains deleted for *VPS15*, indicating that Vps15p is necessary for binding Vps34p to cellular membranes (Stack et al. 1993). Vps15p and Vps34p were shown to be members of a hetero-oligomeric protein complex in a 1:1 stoichiometric ratio in immunoprecipitation and cross-linking experiments (Stack et al. 1993). Other as yet unidentified proteins were also present in this complex, suggesting that Vps15p and Vps34p do not function alone in carrying out their respective functions. Because the membrane-bound form of Vps34p contains greater lipid kinase activity than does free Vps34p, the authors suggested that Vps15p is responsible for activating the lipid kinase activity of Vps34p. Studies of *vps15* mutants indicate that Vps15p kinase activity is not required for association of Vps15p and Vps34p but is required for activation of Vps34p (Stack et al. 1993). Together, these data indicate that Vps34p operates downstream from Vps15p and that this communication is part of a chain of events that lead to the departure of soluble hydrolases from the Golgi complex. Presumably, these activities are both downstream from a hydrolase receptor-binding event.

b. Coat Proteins and Related Functions. Because vesicle-mediated transport has been shown to be the mechanism by which proteins transit along the secretory path (for review, see Schekman and Orci 1996; Kaiser et al., this volume), it was expected that a similar mechanism would be employed for transport out of the Golgi complex. This expectation has been confirmed in kinetic studies of P2 CpY delivery to the vacuole (Vida et al. 1993); other soluble and membrane hydrolases are presumed to proceed similarly. Coat proteins are thought to participate in such vesiculation events. Coat proteins are hypothesized to be responsible for generating the physical force that forms the highly curved membranes of vesicles, and they may also participate in concentrating and sorting cargo (for review, see Barlowe et al. 1994; Kreis and Pepperkok 1994; Harter 1995; Kreis et al. 1995; Mallabiabarrena and Malhotra 1995; Schekman and Orci 1996). Yeast contain at least three different vesicle coat complexes, including COPI, COPII, and clathrin (for review, see Wilsbach and Payne 1993; Salama and Schekman 1995; Stack et al. 1995a). Although there is no evidence that COPI or COPII function in movement of vacuolar hydrolases from the Golgi complex, there is evidence for the involvement of clathrin (for a brief review, see Stack and Emr 1993; Conibear and Stevens 1995; Stack et al. 1995a). Studies of a strain bearing a temperature-conditional allele of the clathrin heavy-chain gene (*CHC1*) showed that following a 5-minute incubation at the restrictive temperature, approximately half of the CpY produced was secreted from the cell in a Golgi-modified form, whereas the remainder was held inside the cell also in a Golgi-modified form (Seeger and Payne 1992a). Transit of the PrA precursor to the vacuole was also affected but not as severely. This was in contrast to the AlP precursor, which showed no perturbation in transit to the vacuole. This constellation of phenotypes (severe CpY effects, less severe PrA[PrB] effects, and no effect on AlP) was similar to that seen in *vps15/vps34* mutants as well as other vacuole-specific mutants (see below) and suggests that clathrin functions in the same process as these other gene products. Studies on CpY demonstrated that P2 CpY is secreted in a *SEC1*-dependent manner in the clathrin mutant at the restrictive temperature (Seeger and Payne 1992a).

The hypothesis that clathrin is the coat protein responsible for soluble hydrolase departure from the Golgi complex is consistent with the above data. However, the loss of clathrin function has only a temporary effect. Three hours after the loss of clathrin function in *chc1*^{ts} strains or in strains deleted for *CHC1*, transit of P2 CpY to the vacuole was unperturbed (Payne et al. 1988; Seeger and Payne 1992a). This observation appears to mean one of two possibilities. The cells may "adapt" to the loss of clathrin and invoke a mechanism that is as efficient as the presumed clathrin-dependent mechanism in delivering the soluble hydrolases. This could be perhaps a de novo mechanism, utilization of an alternative pathway, or coatless transport/vesiculation (see Rothman and Warren 1994; Taylor et al. 1994; Misteli and Warren 1995a,b; Warren et al. 1995). Alternatively, the effects of the loss of clathrin function on soluble hydrolase transit were only a secondary, and in this way a temporary defect. The observation that the imposition of the defect in vacuolar trafficking in the conditional *chc1* mutant is severe and immediate and quickly reversed upon shifting back to the permissive temperature suggests that the cells are adapting by some means to the loss of clathrin (Seeger and Payne 1992a). Otherwise, one might expect that either the imposition or reversal might be slow.

Of the four clathrin-associated proteins known in yeast, only one, Apl2p, appears to impinge upon vacuolar hydrolase trafficking (Phan et al. 1994; Rad et al. 1995). *APL2* was identified in yeast through protein sequence similarity with a mammalian clathrin-associated protein (Rad et al. 1995). In a strain deleted for the *APL2* gene and carrying the temperature-conditional allele of *CHC1*, the P2 CpY sorting defect was partially suppressed, suggesting that the cells may already utilize a clathrin-independent pathway in the absence of Apl2p.

In addition to its role in soluble vacuolar hydrolase trafficking, clathrin is also known to be necessary for retention of late Golgi proteins (Kex1p, Kex2p, and DPAPA) (Payne and Schekman 1989; Seeger and Payne 1992b; Whitters et al. 1994; for review, see Wilsbach and Payne 1993). It has been hypothesized that the defect in Golgi retention is due to the loss of the same function required for delivery of soluble vacuolar proteins; i.e., the loss of functional trafficking out of the Golgi is responsible for a perturbation in the presumed recycling of late Golgi proteins between the TGN and endosome and thus leads to mislocalization of the late Golgi proteins to the plasma membrane (for review, see Wilsbach and Payne 1993; Kaiser et al., this volume).

VPS1 encodes an 80-kD cytoplasmic protein with significant similarity to the dynamin family of proteins (Rothman et al. 1990). The *vps1* mutant missorts soluble hydrolase precursors and secretes P2 CpY (Robinson et al. 1988; Rothman et al. 1990; Latterich and Watson 1991; Raymond et al. 1992b; Vater et al. 1992; Nothwehr et al. 1995). The *vps1* mutant alone of all the mutants thought to be defective in the pathway between the late Golgi and the endosome missorts the precursor of the membrane protein AlP to the plasma membrane, from which it is delivered to the vacuole for subsequent processing by endocytosis (Nothwehr et al. 1995). A subset of the dynamin family of proteins has been shown to be involved in vesicle-mediated transport in both mammals and flies (Damke et al. 1994; Hinshaw and Schmid 1995; Takei et al. 1995), suggesting that a similar function is feasible for Vps1p. The mammalian and fly proteins are thought to function in the scission of nascent vesicles formed through the action of clathrin from the source organelle. Immunofluorescence studies indicate that Vps1p may localize to the Golgi complex and that this distribution is perturbed by secretory path mutations that alter Golgi structure (Rothman et al. 1990; Vater et al. 1992). Furthermore, electron microscopic studies on *vps1* mutants showed the presence of Berkeley bodies, indicating a possible role of Vps1p in departure from the Golgi (Banta et al. 1988; Rothman et al. 1990). Because *vps1* mutants present a vacuole morphology somewhat different from that of *vps15* or *vps34* mutants, which are also defective in departure from the Golgi (the *vps1* mutant has a single large vacuole surrounded by small vacuoles, and the *vps15* and *vps34* mutants have a single large vacuole), this suggests that *vps1* may operate at additional steps in the pathway(s) leading to the vacuole (Raymond et al. 1992a).

Early reports on in-vitro-synthesized Vps1p (identified as Spo15p) suggested that Vps1p was a microtubule-binding protein (Yeh et al. 1991), as is seen for dynamin (Obar et al. 1990; Shpetner and Vallee 1992). However, subsequent work has demonstrated that microtubules are unlikely to be involved in the in vivo functioning of Vps1p (Vater et al. 1992), suggesting that Vps1p binding to microtubules is either an in vitro artifact or an unutilized capability provided by evolutionarily conserved sequence similarity. In this regard, it should be noted that the region of highest similarity is in the amino-terminal GTP-binding domain (Rothman et al. 1990), whereas the carboxy-terminal domain, which is thought to be utilized in protein-protein interactions, is highly divergent (Rothman et al. 1990; Vater et al. 1992). The probable functional interactions of Vps1p and clathrin (see below) make this a particularly appealing model. Indeed, this type of functional coupling of dynamin and clathrin has been observed in mammalian cells (Takei et al. 1996).

Many of the phenotypes manifested by clathrin heavy-chain mutants (vacuolar protein missorting and Golgi protein retention defects) are also seen in *vps1* mutants. For this reason, it has been suggested that these two proteins may operate together in departure of proteins from the Golgi complex (for review, see Wilsbach and Payne 1993; Conibear and Stevens 1995). However, there are differences between the mutants. *vps1* mutants abrogate direct delivery of the AlP precursor to the vacuole, whereas *chc1* mutants do not. *chc1* mutants are defective in endocytosis of serpentine receptors (Tan et al. 1993), whereas *vps1* mutants are not. Most of the data are consistent with the view that clathrin and Vps1p function together at the same step in production of vesicles from the late Golgi for delivery to the prevacuolar compartment. Synthetic phenotypes of *chc1 vps1* double mutants are consistent with this hypothesis (see Nothwehr et al. 1995).

MVP1 was identified as a high-copy suppressor of a dominant-negative allele of *VPS1* (Ekena and Stevens 1995). *MVP1* encodes a hydrophilic 60-kD protein with no similarity to any protein in the sequence databases. Mvp1p is thought to function in trafficking hydrolases to the vacuole, since the *mvp1* null mutant is defective in P2 CpY sorting. However, the precise function of Mvp1p and its relationship to Vps1p function are unknown.

3. *The Yeast Endosome*

The transit of mammalian lysosomal proteins through an endosomal compartment en route to the lysosome is an established delivery route (for review, see Kornfeld 1987; Kornfeld and Mellman 1989). Substantial evidence has accumulated to indicate that the same is true in *S. cerevisiae.* Studies utilizing buoyant density centrifugation to separate membrane compartments during kinetic studies of P2 CpY transport and processing indicate the presence of a compartment of intermediate density through which Golgi-modified CpY must pass prior to maturation in the vacuole (Vida et al. 1993). This compartment is kinetically located after the Golgi complex but before the vacuole. Quantitation of the amounts of CpY in each compartment and estimations from in vitro processing assays indicate that between 35% and 45% of P2 CpY is found in this compartment, with the remainder thought to localize in the Golgi (Vida et al. 1990, 1993). Furthermore, the ability of P2 CpY to progress to this intermediate compartment depends on *VPS15* function, one of the genes required for exit from the Golgi complex as discussed above (Vida et al. 1993). These data indicate that the CpY precursor transits one or more compartments between the TGN, where sorting occurs, and the vacuole, where processing by PrA is normally thought to occur.

The most striking evidence for a prevacuolar compartment comes from studies of class-E vacuolar trafficking mutants. Class-E mutants include the *vps2, 4, 20, 22, 24, 25, 27, 28, 31, 32, 36,* and *37* mutants; all accumulate a prominent organelle distinct from the vacuole (Raymond et al. 1992a; Piper et al. 1995; Rieder et al. 1996). This compartment contains significant amounts of CpY, PrA, and PrB (Raymond et al. 1992a), suggesting that it is an exaggerated form of a compartment that in wild-type cells is transited by soluble vacuolar hydrolases. Studies carried out on conditional *vps27* mutants confirm that this compartment is likely to be a true intermediate compartment, as accumulation, sorting, and processing defects associated with loss of Vps27p function are reversible (Piper et al. 1995). This exaggerated compartment has been demonstrated to contain a functional V-type proton pumping ATPase and contains much of the 60-kD V-ATPase subunit. The class-E compartment is acidic, and the adjacent vacuole is neutral (Raymond et al. 1992a; Cereghino et al. 1995; Piper et al. 1995; Rieder et al. 1996).

Studies of CpY precursor processing in the *vps27* and *vps28* class-E mutants indicated that the compartment in which CpY accumulates is competent to carry out the PrA-dependent maturation cleavages (Piper et al. 1995; Rieder et al. 1996). For this reason, this compartment contains not only active PrA, but also active CpY (and presumably PrB). In wild-type cells, the PrA- and PrB-dependent cleavage steps of hydrolase processing are thought to take place in or very close to delivery of the hydrolase precursor to the vacuole. This therefore means that in the class-E mutants, a prevacuolar compartment that normally contains little active hydrolase becomes charged with active hydrolases. This conclusion is supported by the finding that immediately following the inactivation of Vps27p function in a *vps27* conditional mutant, there is a short delay before processing of P2 CpY begins. This delay is not seen at later time points or in strains containing null alleles of *VPS27* (Piper et al. 1995); it is interpreted to be the time required for this compartment to accumulate active processing hydrolase(s).

The above discussions of vacuolar hydrolase trafficking through the yeast late endosome has centered on the soluble hydrolases, particularly CpY. However, as seen at multiple earlier steps, the precursor to the integral membrane hydrolase AlP often demonstrates very different requirements for vacuolar trafficking. This again appears to be the case at the late endosome. As noted by both Raymond et al. (1992a) and Davis et al. (1993), class-E mutants demonstrate relatively normal AlP immunofluorescent staining of the dominant vacuolar compartment, suggesting that AlP may transit to the vacuole via a route not including the late endosome. In *ypt7* mutants at steady state, PrA is found in its mature, presumably active, form, whereas AlP is found almost entirely in its precursor form (see Wichmann et al. 1992), suggesting that the two hydrolases are in separate compartments. Immunoprecipitation experiments also demonstrated a severe defect in processing of AlP in *ypt7* mutants.

These data, together with the studies discussed above, suggest that precursors to the soluble hydrolases CpY, PrA, and PrB are transported together to the vacuole via the late endosome, whereas the AlP precursor is delivered via a separate route. In class-E mutants, the soluble hydrolases accumulate in an aberrantly large late endosome where they initially undergo maturation with delayed kinetics but ultimately reach near wild-type levels at steady state, whereas the AlP precursor is still transported to the vacuole but is not efficiently matured because of lack of vacuolar PrA.

4. Vesicle Docking/Fusion at the Late Endosome

As noted above, transport from the TGN to the endosome is thought to occur by vesicle-mediated transport, because fractionation profiles of precursor CpY demonstrate transport via buoyant compartments (Vida et al. 1993). The involvement of Sec18p in this step further supports this conclusion (Graham and Emr 1991). Sec18p has been demonstrated to be a functional homolog of mammalian NSF, a protein that functions in vesicle fusion events (Wilson et al. 1989). Because Sec18p has been demonstrated to operate at all steps of vesicle transport from the ER to the plasma membrane (Graham and Emr 1991), and specifically to operate at the vesicle docking stage at the individual steps (Wilson et al. 1989), this was taken as an early indication that transport to the endosome would utilize the same mechanism of vesicle fusion seen at other steps in the secretory pathway. This postulated mechanism has been designated the "SNARE" (SNAP receptor; SNAP is a soluble NSF attachment protein) hypothesis (for review, see Ferro-Novick and Jahn 1994; Rothman 1994; Rothman and Warren 1994; Südhof 1995; Kaiser et al., this volume). Although strict biochemical demonstration of the utilization of a SNARE mechanism at the Golgi to endosome step has not been reported, genetic searches have yielded a collection of presumed functional homologs to various SNARE proteins that operate in eukaryotic cells from yeast to humans (for review, see Ferro-Novick and Jahn 1994; see Table 7). An expected participant in NSF-catalyzed reactions is α-SNAP. Sec17p has been demonstrated to be functionally equivalent to SNAP in the secretory pathway (Griff et al. 1992). It is reasonable to expect that Sec17p will be required in the vacuolar branch both because of the nature of the fusion event and because it is the sole protein in the yeast genome related to SNAP; the test has not been carried out.

One of the key components of the SNARE mechanism is the interaction of integral membrane proteins, designated v-SNAREs and t-SNAREs, that is thought to impart the vesicle targeting specificity between the cargo carrying transport vesicle (v-SNARE) and the intended target organelle (t-SNARE) (for review, see Ferro-Novick and Jahn 1994; Rothman 1994; Rothman and Warren 1994; Südhof 1995; Kaiser

Table 7 Sites of action of proposed yeast SNARE mechanism components

Transport step	v-SNAREs	t-SNAREs	SNAP-25	Sec1p	NSF	αSNAP	rab
ER to Golgi complex	Bos1p (Sec22p)	Sed5p	Bet1p?	Sly1p	Sec18p	Sec17p	Ypt1p
Intra-Golgi	?	Sft1p	?	?	?	?	?
TGN to plasma membrane	Snc1p, Snc2p	Sso1p, Sso2p	Sec9p	Sec1p	Sec18p	Sec17p	Sec4p
TGN to endosome	?	Pep12p	?	Vps45p	Sec18p	Sec17p	Vps21p
Endosome to vacuole	?	Pth1p	Vam7p	Vps33p	Sec18p	Sec17p	Ypt7p

Bos1p (Shim et al. 1991); Sec22p (Lian and Ferro-Novick 1993); Sed5p (Hardwick and Pelham 1992; Banfield et al. 1994); Bet1p (Newman et al. 1990); Sly1p (Ossig et al. 1991); Sec18p (Eakle et al. 1988; Graham and Emr 1991); Sec17p (Griff et al. 1992); Ypt1p (Segev et al. 1988); Sft1p (Banfield et al. 1995); Snc1p, Snc2p (Protopopov et al. 1993); Sso1p, Sso2p (Aalto et al. 1993; Jantti et al. 1994); Sec9p (Brennwald et al. 1994); Sec1p (Aalto et al. 1991, 1992); Sec4p (Goud et al. 1988; Kabcenell et al. 1990), Pep12p (Becherer et al. 1996); Vps45p (Cowles et al. 1994; Piper et al. 1994); Vps21p (Horazdovsky et al. 1994; Singer-Kruger et al. 1994); Pth1p/Vam3p (A. Srivastava and E. Jones, unpubl.); Vam7p (Wada and Anraku 1992; A. Srivastava and E. Jones, unpubl.); Slp1p/Vps33p/Pep14p (Jones 1977; Banta et al. 1990; Wada et al. 1990); and Ypt7p (Wichmann et al. 1992).

et al., this volume). Pep12p has the properties expected for a v-SNARE of the endosome. Pep12p shows up to 25% identity and 49% similarity to members of the syntaxin protein family. It is a cytoplasmically oriented integral membrane protein with a carboxy-terminal transmembrane domain. It fractionates predominantly with the yeast endosome (Becherer et al. 1996).

Ultrastructural analysis via electron microscopy revealed the accumulation of 40–50-nm vesicles in *pep12* mutants, suggesting a defect in a vesicle consumption step in the vacuolar pathway (Becherer et al. 1996), consistent with Pep12p functioning in a vesicle docking/fusion step. Loss of function for Pep12p prevents maturation and vacuolar delivery of precursors to all soluble vacuolar hydrolases. Upon transfer to the restrictive temperature, a *pep12^{ts}* mutant manifested a block in maturation of precursor forms of CpY, PrA, and PrB, but not of AlP, again suggesting that the AlP precursor transits to the vacuole via a separate route (S. Rieder and S. Emr, pers. comm.). Because *pep12* mutants share many phenotypic characteristics with *vps15* and *vps34* mutants (Raymond et al. 1992a), including vacuolar morphology and levels of precursor CpY secretion, and as both Vps15p and Vps34p have been demonstrated to function between the late Golgi and the endosome, it seems likely that Pep12p also functions at this step. The fact that *vps15* and *vps34* mutants accumulate expanded Golgi complex structures with high penetrance and *pep12* mutants do so rarely, if at all (Banta et al. 1988; Becherer et al. 1996), and that only the *pep12* mutant accumulates vesicles, suggests that Pep12p may function downstream from the Vps15p/ Vps34p complex, consistent with their proposed roles in docking/fusion and budding, respectively, of vesicles containing precursors to soluble hydrolases.

Sec1p or its homologs participate in all known syntaxin-related docking/fusion events involving vesicular transport. Although the function of members of this family of proteins is unknown, both genetic and physical interactions between members of this family and syntaxins in several systems suggest that Sec1p family members participate in SNARE complex function (for a brief review, see Ferro-Novick and Jahn 1994; Südhof 1995; Kaiser et al., this volume). Vps45p and the *vps45* mutant have the properties expected of the Sec1p homolog for this fusion event.

The loss of Vps45p function leads to vacuolar morphology and inheritance patterns identical to those seen for *pep12* mutants, leading to the conclusion that Vps45p may operate at the same step of transport as Pep12p (Raymond et al. 1992a; Cowles et al. 1994; Piper et al. 1994). Because *vps45* mutants accumulate 40–60-nm vesicles and not Golgi structures (Cowles et al. 1994; Piper et al. 1994), Vps45p is thought to operate at a vesicle consumption stage of transport, as is hypothesized for Pep12p. Genetic interaction between *PEP12* and *VPS45* has been seen through synthetic phenotypes in *pep12^{ts} vps45^{ts}* double mutants (C. Burd and S. Emr, pers. comm.) and suppression of *vps45^{ts}* phenotypes by high-copy *PEP12* (E. Whitters, per. comm.). Localization data on Vps45p are also consistent with the protein's functioning between the TGN and late endosome compartments (Cowles et al. 1994; Piper et al. 1994).

It is likely that Pep7p also functions at the step of vesicle docking/fusion at the endosome. *pep7* mutants have a single large vacuole, accumulate 40–60-nm vesicles in the cytoplasm, secrete P2 CpY, and fail to sort and process precursors to soluble but not membrane vacuolar hydrolases to the vacuole (Webb et al. 1997b). All of these characteristics of the *pep7* mutant are strikingly similar to the phenotypes of *vps45* and *pep12* mutants.

Pep7p is found free in the cytoplasm and peripherally associated with membranous structures, as if it cycles on and off a membrane. The *pep7-20^{ts}* allele is partially suppressed by *PEP12* and *VPS45* in high copy but not by high-copy *PTH1/VAM3* or *VPS33/PEP14*, genes that encode other vacuolar pathway syntaxin and Sec1p homologs (Webb et al. 1997a). These results suggest that Pep7p, Pep12p, and Vps45p function together in a docking/fusion complex at the endosomal membrane.

One of the protein participants involved in docking/fusion events is a small GTP-binding protein of the rab family. The precise function of these GTPases at these various events is not known. However, it is thought that they facilitate the interaction of the structural SNARE components but do not themselves enter into this complex (for review, see Ferro-Novick and Jahn 1994; Rothman 1994; Südhof 1995). Three structural homologs of rab5, namely, Vps21p/Ypt51p, Ypt52p, and Ypt53p, and a rab7 homolog, namely, Ypt7p, have been implicated in the vacuolar pathway. Analysis of the *vps21* (Horazdovsky et al. 1994) and the *ypt51*, *ypt52*, and *ypt53* (Singer-Kruger et al. 1994) set of mutants implicates these rab5 homologs in the Golgi to endosome step. The dominant protein of the three seems to be Vps21p, since the *vps21* null mutant has a single large vacuole, accumulates vesicles, has a severe defect in sorting precursors to soluble vacuolar hydrolases, and secretes P2 CpY. Some of Vps21p is geranylgeranylated; the geranylgeranylated protein is associated with the membrane. Cytoplasmic Vps21p does not carry the lipid modification (Horazdovsky et al. 1994) The *ypt52* and *ypt53* null mutants are essentially normal; the *ypt51 ypt52* double mutant is more severely affected than either single mutant, and the triple mutant is even more severely affected. Thus, Vps21p/Ypt51p, Ypy52p, and Ypt53p seem to compose a functionally redundant set of three GTPases. Loss of function of the three GTPases leads to a set of phenotypes similar to those manifested by *vps15* and *vps34* mutants, and strikingly similar to those seen in both *pep12* and *vps45* mutants (Banta et al. 1988; Raymond et al. 1992a; Horazdovsky et al. 1994; Singer-Kruger et al. 1994), suggesting that Vps21p might participate in a vesicle docking step in transport between the TGN and the late endosome.

Mutations in *YPT7* cause markedly different vacuolar morphology and hydrolase maturation phenotypes as compared to *vps21* mutations (Wichmann et al. 1992; Schimmoller and Riezman 1993). The role of *YPT7* is discussed below.

Vps9p is thought to be involved in regulating the fusion step at the endosome (Burd et al. 1996). The *vps9* mutant, like other mutants blocked at this step, has a large vacuole, accumulates vesicles, secretes P2 CpY, and has a severe sorting defect for soluble hydrolase precursors. Processing of the AlP precursor appears to be normal but maturation of the CpS precursor is slowed in the *vps9^ts^* mutant after transfer to the restrictive temperature. The CpS precursor is a type II integral membrane protein that is cleaved near the membrane within the vacuolar lumen to give the soluble hydrolase. Since both AlP and CpS travel to the vacuole as integral membrane proteins that are processed by PrA and/or PrB, it is somewhat surprising that the two proteins behave differently in the *vps9* mutant and implies that they may travel to the vacuole by different routes (Burd et al. 1996). The Vps9p sequence contains three motifs that are all present in two human proteins that inhibit Ras; these authors propose that these are G-protein-binding sequences. Burd et al. (1996) propose that Vps9p binds to Vps21p to regulate its activity.

Vps8p is a hybrophilic protein that is found associated with membranes and in the soluble fraction. In a wild-type cell, about 30% pellets with the P13 fraction, 55% with the P100 fraction, the rest being soluble (Chen and Stevens 1996; Horazdovsky et al. 1996). Deletion/disruption of *VPS8* results in a class-D mutant phenotype: The *vps8*Δ mutant secretes most CpY antigen as P2 CpY, is temperature-sensitive, has a single large vacuole, and is inefficiently inherited (Horazdovsky et al. 1996). (Chen and Stevens [1996] report that the *vps8*Δ mutant has a normal, class-A vacuole; we, like Horazdovsky et al. [1996], classify it as class D [C. Woolford and E. Jones, unpubl.].) Because of membership of the *vps8*Δ mutant in class D, interactions were sought between Vps8p and the products of other class-D genes, *VPS45, PEP12*, and *VPS21*. In the *vps21*Δ mutant, the intracellular distribution of Vps8p was altered, for 80% of Vps8p was in the soluble fraction; the distribution was like wild type in the *vps45*Δ and *pep12*Δ mutants. These data suggest that Vps21p directly or indirectly is responsible for the association of Vps8p with membranes (Horazdovsky et al. 1996). The authors suggest that Vps8p participates in the Golgi to endosome step in the pathway and further suggest that Vps8p may modulate Vps21p function, since overexpression of Vps21p can partially suppress the *vps8*Δ mutation.

Pep5p is a 116-kD peripheral membrane protein that is enriched in purified vacuoles (the fraction also contains endosomes and transport vesicles). The *pep5* mutant secretes P2 CpY and is deficient for PrA, PrB, CpY, and AlP. It is a class-C mutant (no discernible vacuole) (Banta et al. 1988; Robinson et al. 1988; Dulic and Riezman 1989; Woolford et al. 1990). A *pep5::TRP1* insertion allele is partially suppressed by a missense *vps8* allele, *vps8-200*, resulting in restoration of substantial amounts of the soluble hydrolases and the presence of a vacuole, albeit fragmented (because *pep5* is epistatic to *pep8*, the *pep5*Δ *vps8*Δ double mutant lacks a vacuole; the presence of an abnormal vacuole in the suppressed strain indicates suppression of the *pep5* defect in vacuole formation. The fact that *vps8* and *pep5* alleles interact genetically, we take as prima facie evidence that Pep5p and Vps8p interact and have thus placed Pep5p provisionally at the docking/fusion step at the endosome along with Vps8p (C. Woolford and E. Jones, unpubl.).

In summary, the analysis of SNARE component homologs seen to operate in transport to the vacuole indicates that one set (Pep12p, Vps45p, and Vps21p) probably operates together at a single step of vesicle docking/fusion in the transport of precursor forms of CpY, PrA, and PrB at the endosome. Pep7p, Vps8p, and Vps9p are novel participants in the complex. If the parallel to other complexes involved in vesicle trafficking in the yeast secretory pathway continues to hold, we can expect to identify genes that encode homologs to the v-SNARE(s) and the SNAP-25 t-SNARE, as well as other possibly novel components.

Two additional correlations can be made about the gene products that are thought to function uniquely between the late Golgi and the late endosome: For all save the CpY receptor and clathrin (namely, Vps15p, Vps34p, Pep12p, Vps45p, Pep7p, and Vps21p), loss of function results in a single large vacuole that is inherited inefficiently (class D) (Weisman et al. 1990; Shaw and Wickner 1991; Raymond et al. 1992a; Weisman and Wickner 1992), and, for all those null mutants tested (*pep12, pep7, vps21/ypt51*, and by implication *vps45*), the vacuole is not acidic (Preston et al. 1989, 1992; Rothman et al. 1989b; Raymond et al. 1992a;

Singer-Kruger et al. 1994; Horazdovsky et al. 1996) The implication of the latter observation is that the vacuolar ATPase is not being properly assembled. We know this to be true for the *pep12* and *vps3* mutants, for the preponderance of the 60-kD and 69-kD V-ATPase subunits is not assembled onto the vacuolar membrane but remains in the cytoplasm (Rothman et al. 1989b; M. Manolson et al., unpubl.). It will be interesting to see whether the vacuolar morphology of null mutants will continue to serve as a valid guide to the likely position within the pathway in which a gene product functions.

5. *Recycling from the Endosome to the TGN*

Vps8p is a hydrophilic protein that is found in association with membranes and particularly the membrane fraction that contains Golgi membranes, transport vesicles, and endosomes. The *vps8* null mutant secretes 70% of CpY as P2 CpY and, peculiarly enough, appears to secrete PrA rather than P2 PrA (Chen and Stevens 1996). Maturation of the AlP precursor is slightly slowed, probably because the levels of processing hydrolases in the vacuole are low (a conditional mutant was not used). Vacuole morphology is normal. The most revealing phenotype of the *vps8* null mutant is the fate of Vps10p; in the absence of Vps8p, Vps10p transits all the way to the vacuole, where it is cleaved. This is interpreted to indicate that Vps8p functions in recycling the Vps10p receptor, and possibly other components, back to the late Golgi to allow further rounds of sorting. The *vps8* mutant is also defective in retention of Golgi proteins.

The *vps27* and *vps28* mutants accumulate the Vps10p receptor in the exaggerated endosome, the class-E compartment, where the receptor is proteolyzed (Cereghino et al. 1995; Piper et al. 1995); the mutants secrete P2 CpY. Since the *pep4*Δ mutation, which results in elimination of nearly all vacuolar hydrolases (see Jones 1991b), does not reduce the amount of P2 CpY secreted, these authors suggest that P2 CpY secretion results from failure to recycle the receptor and not to low levels of Vps10p due to proteolysis (Cereghino et al. 1995).

C. Transport between the Late Endosome and the Vacuole

The evidence that trafficking between the TGN and endosome is a vesicle-mediated transfer involving SNARE complexes is largely circumstantial, based on the accumulation of 40–60-nm vesicles in some mutants, sequence similarity between some of the gene products required for the Golgi to endosome steps and gene products required for secretion and exocytosis in mammalian and yeast cells, and kinetic analyses of the processing and processing defects, particularly in conditional mutants. For step(s) between the late endosome and the vacuole, the evidence is weaker. Nevertheless, a start has been made. Proteins that participate in transport between the endosome and the vacuole are summarized in Table 6.

1. *Exit from the Endosome*

Two mutants, *vps27* and *vps28*, appeared to be blocked in exit from the endosome (Piper et al. 1995; Rieder et al. 1996). Both mutants have been classified as class-E mutants, for they accumulate a novel prevacuolar compartment distinct from the vacuole but adjacent to it. This class-E compartment is thought to be an enlarged late endosome (see above). Studies carried out on conditional *vps27* mutants indicate that the class-E compartment accumulates and gradually becomes charged with hydrolases at the restrictive temperature. The accumulation, sorting, and processing defects associated with loss of Vps27p function are reversible (Piper et al. 1995). Thus, the *vps27* mutant seems to be blocked in exit from the endosome.

The prevacuolar compartment present in class-E mutants has also been shown to accumulate proteins and dyes brought into the cell by endocytosis. As mentioned earlier, the pheromone receptors Ste2p and Ste3p are taken up by endocytosis and delivered to the vacuole where they are degraded by vacuolar proteases (Davis et al. 1993). In the *vps27* and *vps28* mutants, Ste2p and Ste3p accumulated in the class-E compartment adjacent to the nucleus (Piper et al. 1995), and as demonstrated for vacuolar hydrolases, this accumulation was reversible in a conditional *vps27* mutant. The lipophilic styryl dye FM 4-64 assembles into the plasma membrane and is delivered to the vacuole by endocytosis (Vida and Emr 1995). In the *vps27* and *vps28* mutants, FM 4-64 accumulates in the class-E compartment rather than the vacuole (Vida and Emr 1995; Rieder et al. 1996). These observations suggest that the endocytic pathway to the vacuole and the TGN to vacuole pathway converge at the late endosome or that the late endosome is downstream from that event.

It should be noted that although *vps28* mutants are classified as class E, as analyzed by electron microscopy, the compartment accumulated is significantly different in the *vps28* mutant from that seen in other class-E mutants. The *vps28* compartment appears as stacks of curved cisternal membranes (Rieder et al. 1996). It is not clear if this is a permutation of the class-E compartment seen in other sorting mutants, or perhaps represents a compartment through which both membrane and soluble hydrolases pass en route to the vacuole that is downstream from the late endosome, but upstream of the vacuole.

It has been demonstrated several times that late Golgi proteins (Kex2p and Vps10p) or reporter constructs (A-AlP) accumulate in the endosome in class-E mutants (Raymond et al. 1992a; Cereghino et al. 1995; Piper et al. 1995; Rieder et al. 1996). This is not entirely surprising, because it has been hypothesized that late Golgi complex proteins might cycle between the TGN and the first compartment in the vacuolar pathway (for review, see Wilsbach and Payne 1993). Because of this accumulation phenotype, it has been suggested that class-E mutations not only block forward movement toward the vacuole, but also block movement out of the endosome back to the TGN (recycling pathway). Consistent with the hypothesis of the accumulation of the TGN proteins in the late endosome where active PrA and CpY accumulate (see above) is the observation that both Vps10p and Kex2p are very unstable in these mutants and undergo partial (Vps10p) or complete (Kex2p) proteolysis (Cereghino et al. 1995; Piper et al. 1995). Overexpression of *VPS10* in the *vps27* and *vps28* mutants substantially reduced the amount of P2 CpY secreted, which was interpreted to mean that the two mutants were defective in recycling the CpY receptor (Piper et al. 1995). An alternative explanation is that the receptor is being inactivated by proteolysis in the class-E compartment so that it cannot be recycled and that by supplying extra receptor, the loss by proteolysis becomes irrelevant. Since it is thought that under wild-type conditions, the late endosome does not contain active hydrolases, the hypothesized cycling of late Golgi proteins through this compartment would not generally lead to their encountering these destructive enzymes. The proteolytic cleavage of Vps10p in class-E mutants has been demonstrated to be PrA-dependent and to not take place in the vacuole (Cereghino et al. 1995). Stabilization of Vps10p in class-E mutants by deletion of *PEP4* (PrA structural gene) did not lead to an improvement in the mutants' ability to sort P2 CpY, supporting the hypothesis that the *vps27* and *vps28* mutants are not able to recycle Golgi apparatus proteins back to the TGN (Cereghino et al. 1995).

2. *Vesicle Docking/Fusion at the Vacuole*

The data of Graham and Emr (1991) suggest that Sec18p may not be required for delivery of hydrolases from the endosome to the vacuole, because when the *sec18* mutant was incubated at the permissive temperature to label uniformly all CpY species, a substantial fraction of P2 CpY was able to transit to and be processed in the vacuole after the cells were transferred to the restrictive temperature, as if Sec18p were not needed for the late steps. However, Haas and Wickner (1996) found that Sec18p is required for fusion of vacuoles in their in vitro reaction. Thus, whether or not Sec18p is required for fusion of endosome-derived vesicles remains an open question. The involvement of Sec17p in docking/fusion of vesicles at the vacuole has not been examined. As indicated above, there is no homolog to Sec17p by sequence in the yeast genome, and Haas and Wickner (1996) have demonstrated its essentiality for vacuole fusion in vitro, suggesting that it will be required in fusion of endosome-derived vesicles at the vacuole.

Pth1p/Vam3p (*P*EP *t*welve *h*omolog) has at least some of the properties expected for a t-SNARE of the vacuole. Pth1p shows up to 34% identity and 57% similarity to the syntaxin protein family (A. Srivastava and E. Jones, unpubl.). Deletion mutations in this gene cause kinetic defects in processing of precursors to CpY, PrB, PrA, and AlP; at steady state, CpY, PrB, and PrA are present as their mature forms but the AlP precursor remains unprocessed. Thus, the consequences of loss of Pth1p are less severe than for loss of Pep12p for the soluble hydrolases but not the membrane hydrolases. In contrast to the *pep12* mutant, the *pth1* mutant secretes little P2 CpY, suggesting that Pth1p may operate deeper into the vacuolar pathway than does Pep12p (A. Srivastava and E. Jones, unpubl.). Consistent with this hypothesis is the observation that the *pep12 pth1* double mutant secretes nearly all P2 CpY, i.e., the *pep12Δ* mutation is epistatic to *pth1Δ* for the P2 CpY secretion phenotype. The *pth1* mutant has a fragmented vacuole, and the *pep12* mutant has a single large vacuole. The *pep12 pth1* double mutant has a fragmented vacuole (A. Srivastava and E. Jones, unpubl.). These observations make sense if the epistasis of *pep12* reflects events in a biosynthetic and kinetic pathway, with the prior block being epistatic, whereas the epistasis of *pth1* for vacuole morphology reflects a defect in fusion of vacuoles at the last step in the assembly pathway for the organelle in the single and double mutants. Interestingly, the *pth1* mutant does not process the ApI precursor. This finding implicates *PTH1* in the cytoplasm-to-vacuole targeting pathway.

Ypt7p is a member of the rab7 family of GTPases. The *ypt7* mutant, like the *pth1* mutant, has a fragmented vacuole and low levels of P2 CpY secretion (Schimmoller and Riezman 1993; Wichmann et al. 1992). Vacuolar hydrolase maturation in *ypt7* mutants shows a significant lag that has been interpreted as evidence of slow maturation kinetics in the late endosomal or subsequent compartment en route to the vacuole

(Wichmann et al. 1992). Fractionation profiles of endocytosed proteins in *ypt7* mutants indicate that Ypt7p operates in transport between the late endosome and vacuole (Schimmoller and Riezman 1993). These results would appear to place Ypt7p as the rab GTPase at the vacuolar docking/fusion complex with Pth1p. This finding is consistent with the observation that Ypt7p is required for in vitro fusion of vacuoles (Haas et al. 1995) and that in both the *ypt7* and *pth1* mutants, AlP is found almost entirely in its precursor form (Wichmann et al. 1992; A. Srivastava and E. Jones, unpubl.). It is quite noteworthy that *ypt7* and *pth1* have such profound effects on maturation of the AlP precursor but not on those of soluble hydrolases. Certainly, it reinforces the conclusion that soluble hydrolase precursors and the precursor to AlP travel to the vacuole by different routes. One wonders whether there is any significance to the fact that the only mutations that affect both membrane and soluble hydrolase precursor trafficking, namely, *vps1*, *ypt7*, and *pth1*, operate at the beginning and end of the pathway.

A second vacuolar pathway homolog of Sec1p exists, namely, Vps33p. Analysis of the *vps33* mutant reveals a set of phenotypes very different from those seen in *vps45* mutants. This suggests that the two presumed homologs operate at different steps of transport to the vacuole. Cells lacking Vps33p function show extreme defects in precursor hydrolase sorting and processing, as well as the absence of a recognizable vacuolar compartment (Banta et al. 1988; Wada et al. 1990; Raymond et al. 1992a; Jacquemin-Faure et al. 1994). Analysis of a conditional allele of *VPS33* appears to indicate that the primary function of Vps33p is indeed transport of precursor hydrolases in the vacuolar pathway and that the loss of a discernible vacuolar structure is a secondary effect (Banta et al. 1990). It is possible that one or more blocks in transport of vesicle components at key positions leading to the vacuole could secondarily lead to the loss of the vacuole compartment. The loss of an identifiable vacuolar compartment may be responsible for a broad range of phenotypes observed in *vps33* mutants (Ohya et al. 1986; Kitamoto et al. 1988b; Wada et al. 1992b; Jacquemin-Faure et al. 1994). Because of the above-noted overlap of the endosomal pathway with the vacuolar biosynthetic pathway, the effects of *vps33* mutations on endocytic traffic could be either due to a primary defect in a common step in transport or secondarily due to the loss of active hydrolase or the entire vacuolar compartment (Dulic and Riezman 1990). No information directly ties Vps33p to any particular docking/fusion step in the vacuolar pathway. However, Vps33p remains the best candidate for the Sec1p homolog participating in the docking/fusion step at the vacuole.

The *vam7* mutant has a fragmented vacuole phenotype and has lower CpY levels than wild type in immunoblots (Wada and Anraku 1992; Wada et al. 1992b). Lowered intracellular CpY is commonly correlated with secretion of P2 CpY; no kinetic or secretion data have been reported for this mutant, however. Vam7p shows limited sequence similarity to human SNAP-25 and SNAP-23 proteins, making Vam7p a candidate for a SNAP-25 t-SNARE. The weakness of the mutant phenotype makes it a better candidate for a SNAP-25 homolog at a step deep in the pathway, i.e., at the vacuole, rather than at the endosome. Thus, we may already have identified a syntaxin t-SNARE homolog (Pth1p), a Sec1p homolog (Vps33p), a SNAP-25 t-SNARE homolog (Vam7p), and rab protein (Ypt7p) for a docking/fusion step at the vacuole.

As mentioned above, *pep5::TRP1* alleles are partially suppressed by *vps8-200*. Maturation of the soluble hydrolases PrA, PrB, and CpY is restored; maturation of the membrane hydrolase AlP is not (C. Woolford and E. Jones, unpubl.). This finding suggests that the soluble hydrolase precursors and their product hydrolases are not present in the same compartments as the AlP precursor. The simplest interpretation of these findings is that Pep5p is required for docking/fusion steps at both the endosome and the vacuole and that, in the suppressed strain, Pep5p function is restored for fusion at the endosome but not for fusion at the vacuole. In this interpretation, soluble hydrolases are being matured in an endosomelike compartment and have not progressed beyond the endosome.

The ApI precursor is not matured in either the *pep5*Δ mutant or the *pep5::TRP1 vps8-200* suppressed strain (C. Woolford and E. Jones, unpubl.). This finding implicates Pep5p in the cytoplasm-to-vacuole targeting pathway and, together with the similar finding for the *pth1*Δ mutant, suggests that the TGN-to-vacuole pathway and the cytoplasm-to-vacuole pathway may share a common final docking/fusion step at the vacuole. This step may also be required for vacuole inheritance (see Section VII).

The *pth1*, *vps33*, *vam7*, and *ypt7* mutants do not have completely comparable phenotypes. It is not yet clear whether Pth1p, Vps33p, Vam7p, and Ypt7p operate at the same step of transport. It may be an oversimplification to think that mutants defective for the same step will have comparable phenotypes, because it assumes that each protein operates at only one step. This assumption is presumably valid for proteins that are hypothesized to act as either v-SNAREs or t-SNAREs, since they are thought to impart location specificity. It is clear though that other components of SNARE complexes need not be specific for a particular step (for review, see Ferro-Novick and Jahn 1994; Rothman 1994).

Therefore, the lack of common phenotypes may not exclude the possibility of two genes functioning at one or more common steps; rather, it may indicate that one or more of the genes function at one or more additional steps. The converse situation, where mutations in two or more genes lead to very similar phenotypes, would appear to be fairly strong evidence for the genes functioning at the same step or steps, as is interpreted here for *PEP12*, *VPS45*, *PEP7*, and *VPS21/YPT52/YPT53*.

3. *Intersection of the Endocytic and Golgi to Vacuole Pathways*

Membrane proteins, peptides, and some lipophilic dyes are taken into the cell by endocytosis (for a more extensive discussion of endocytosis, see Kaiser et al., this volume). Data also demonstrate that endocytosed proteins (α-factor and its receptor Ste2p, and **a**-factor receptor Ste3p) are delivered from the plasma membrane to the vacuole via at least two separate intermediate compartments that, by definition, would be considered "endosomes" (Singer and Riezman 1990; Singer-Kruger et al. 1993). An early compartment (early endosome) is seen as peripherally located multiple small membrane compartments, whereas a kinetically ordered later compartment (late endosome) is seen as a larger perivacuolar membrane compartment. Incubation of cells at 15°C leads to uptake and loading of the early endosome, with little transport to the late endosome (Singer and Riezman 1990; Schimmoller and Riezman 1993). This block in transport is reversible, suggesting that the accumulated compartment is a bona fide endosomal intermediate (Schimmoller and Riezman 1993; Singer-Kruger et al. 1993).

As paths from both the TGN and plasma membrane to the vacuole utilize intermediate compartments, the obvious question becomes: Are any of these compartments held in common between the two pathways? The most striking evidence for a common prevacuolar compartment comes from studies of class-E vacuolar trafficking mutants. Class-E mutants include the *vps2, 4, 20, 22, 24, 25, 27, 28, 31, 32, 36*, and *37* mutants; all accumulate a prominent organelle distinct from the vacuole (Raymond et al. 1992a; Piper et al. 1995; Rieder et al. 1996). This class-E compartment contains significant amounts of CpY (Raymond et al. 1992a), PrA, and PrB (see Raymond et al. 1992a), suggesting that it is an exaggerated form of a compartment which in wild-type cells is transited by soluble vacuolar hydrolases. Studies carried out on conditional *vps27* mutants confirm that this compartment is likely to be a true intermediate compartment, as accumulation, sorting, and processing defects associated with loss of Vps27p function are reversible (Piper et al. 1995). Studies of

CpY precursor processing in the *vps27* and *vps28* class-E mutants indicated that the compartment in which CpY accumulates is competent to carry out the PrA- and PrB-dependent maturation cleavages (Piper et al. 1995; Rieder et al. 1996). For this reason, this compartment contains not only active PrA and PrB, but also active CpY. In wild-type cells, the PrA-dependent cleavage steps of hydrolase processing are thought to take place in the vacuole. This therefore means that in the class-E mutants, a prevacuolar compartment that normally contains little active hydrolase becomes charged with active hydrolases. This conclusion is supported by the finding that immediately following the inactivation of Vps27p function in a *vps27* conditional mutant, there is a short delay before processing of P2 CpY begins. This delay is not seen at later time points or in strains containing null alleles of *VPS27* (Piper et al. 1995). This delay is interpreted to be the time required for this compartment to accumulate active processing hydrolase(s).

The prevacuolar compartment present in class-E mutants has also been shown to accumulate proteins and dyes brought into the cell by endocytosis. The class-E gene, *VPS2*, was separately identified in a screen for mutants that stabilize the **a**-factor receptor, Ste3p (Davis et al. 1993), which was shown to be endocytosed and degraded in the vacuole. The phenotypes manifested by *vps2* mutants suggest that blocks internal to the cell affect the cell's ability to clear the Ste3p receptor from the plasma membrane. This can mean either that Vps2p acts in two places (internal and at the cell surface) or that internal blocks lead secondarily to a "back up" of protein onto the cell surface. *vps2* mutants were demonstrated to accumulate Ste3p and Ste2p in a compartment next to but distinct from the vacuole (Davis et al. 1993). *vps27* mutants showed the same accumulation of endocytosed Ste3p and Ste2p in the class-E compartment (Piper et al. 1995), and as demonstrated for vacuolar hydrolases, this accumulation was reversible in a conditional *vps27* mutant. A third class-E mutant, *vps28*, has also been demonstrated to accumulate Ste3p as well as endocytosed dyes in a compartment next to the vacuole (Rieder et al. 1996). The class-E gene, *VPS4*, was also identified in a mutant hunt of genes affecting endocytosis (Munn and Riezman 1994). Additionally, *pep5*, *vps34*, and *vps21* vacuolar pathway mutants have also been identified in genetic hunts for endocytic mutants (Dulic and Riezman 1989; Munn and Riezman 1994; Singer-Kruger et al. 1994). Unlike the class-E mutants, the significance of these potentially common components are not well understood, except to suggest further that paths from the TGN and plasma membrane to the vacuole are closely entwined.

The morphological similarities between the compartments in class-E mutants that accumulate vacuolar and endocytic proteins argue that they are indeed the same compartment and, furthermore, that this compartment is the late endosome. The demonstration that this is indeed the case comes from studies of proteolysis of endocytosed proteins in these mutants. Since the class-E compartment contains active CpY and PrA (see above), it would be predicted that if endocytosed proteins are degraded in a PrA- and/or CpY-dependent manner, delivery to this compartment would not preclude proteolysis and would, to the contrary, demonstrate PrA-dependent proteolysis. This is indeed what was found for α-factor in *ypt7* mutants (Schimmoller and Riezman 1993) and for Ste2p and Ste3p in *vps2* mutants (Davis et al. 1993). Similarly, all studies that demonstrated, by immunofluoresence, accumulation of pheromone receptors in the late endosome used strains lacking vacuolar hydrolases (*pep4* mutants), for accumulation cannot be seen in *PEP4* strains (Davis et al. 1993; Piper et al. 1995; Rieder et al. 1996). The observation that this proteolysis does not take place on proteins endocytosed at 15°C indicates that the reaction does not take place in the early endosome (Singer and Riezman 1990).

Early work on CpY transport to the vacuole suggested that the endosomal compartment that accumulates at 15°C overlaps with the intermediate compartment through which P2 CpY passes (Vida et al. 1993). However, in light of subsequent work discussed above, it appears more likely that the early and late endosomes cofractionate on the sucrose density gradients presented in these studies.

The overall picture then is that the endocytic and vacuolar transport pathways converge in the vicinity of the late endosome. However, there are still a number of observations at odds with this conclusion. Among them are the fact that the *ypt7* mutant is severely defective in maturation of the AlP precursor but only kinetically deficient in maturation of precursors for soluble hydrolases (Wichmann et al. 1992) and that the class-E mutants, which seem to be blocked in exit from the late endosome, are not blocked for transport of the AlP precursor to the vacuole (Raymond et al. 1992a).

The "SNARE" hypothesis was developed to account for the specificity of vesicle fusion events along the secretory pathway. Protein participants include v- and t-SNAREs, NSF, α-SNAP, SNAP-25, a rab protein, and a Sec1p homolog. That the same or similar mechanisms are employed both in exocytosis of synaptic vesicles during synaptic transmission and at all steps in the secretory pathway from the ER to the plasma membrane and that the proteins carrying out like functions at the several steps are highly similar in sequence facilitate recognition of steps and pathways employing this mechanism. Table 7 shows clearly how helpful this paradigm has been in analysis of the secretory pathway and its vacuolar branch. Although we have not yet identified all of the proteins predicted by this paradigm, the outline is already quite clear. Transport between the TGN and the endosome and between the endosome and the vacuole almost certainly proceeds via vesicular carriers.

The entries in Table 7, however, even when the table is completed, will be only a partial list of participating proteins. As described previously in this section, genetic analysis has already identified a number of novel protein components for the vacuolar pathway that have as yet no counterparts in either the secretory or synaptic transmission pathways. Table 8 summarizes information on a set of additional gene products and the properties of the corresponding mutants. Because the genes have been identified and their sequences are known, we already know that these genes also encode novel components without homologs in the secretory or synaptic pathways.

In Table 9 are listed genes that have been identified by mutation but not yet tied to an ORF in the yeast genome. Possibly some of these correspond to missing SNARE components. Some, however, are bound to be novel.

D. A Special Case: Assembly of the V-ATPase

The V-ATPase of the vacuole membrane consists of a catalytic sector, V_1, and a membrane sector, V_o. There are seven subunits associated with the V_1 sector and six subunits associated with the V_o sector (see Table 3 and references therein). The catalytic sector associates reversibly with the membrane sector in response to the nutritional state of the cell (Kane 1995).

The catalytic sector is synthesized and assembled in the cytoplasm and can then assemble onto the vacuolar membrane. With a single exception, deletion of a gene encoding a subunit of the catalytic sector results in synthesis and stable accumulation of the remaining V_1 subunits in the cytoplasm; they do not assemble onto the vacuolar membrane (Kane et al. 1992; Ho et al. 1993a; Doherty and Kane 1993). Deletion of *VMA13* (54-kD V_1 subunit) abrogates vacuolar ATPase activity but does not affect assembly of the V_1 sector or its association with the vacuolar membrane (Ho et al. 1993b). Subcomplexes containing various subunits of the V_1 sector can be detected in the cytoplasm of wild-type yeast as well as

Table 8 Other proteins involved in vacuolar transport

SGD name	Other names	Homolog	Size (aa)	Properties of protein and mutant
Pep8p	Vps26p Vpt4p Grd6p	H<β>58 *M.m.*	379	peripheral membrane protein; immunogold EM localizes it to the vacuole membrane but also possibly to endosomal membrane (Bachhawat et al. 1994) *mutant*: Class A (normal) vacuole; P2 CpY not processed, little secreted; PrA and PrB affected less, AlP not at all.
Pep11p	Vps29p Vpt6p	limited homol. M6P and interferon receptor	282	hydrophilic protein; two exons *mutant*: Class A vacuole; P2 CpY secreted, some matured internally; PrA and PrB affected somewhat, AlP not at all; *pep8* interacts genetically with *pep1* (Banta et al. 1988; Raymond et al. 1992a; D. Friedrichsen and E. Jones, unpubl.)
Vps35p	Vpt7p Pep2p Grd9p	*C.e.*, *M.m.*, *Sch.p.* ORFs	937	hydrophilic protein associated with a membrane; released by salt and detergent, may be in a complex (Paravicini et al. 1992) *mutant*: Class A vacuole; P2 CpY secreted, small amount of P2 PrA secreted, most processed to PrA internally; PrB and AlP processing somewhat slowed (Paravicini et al. 1992)
Vps17p	Pep21p Vpt3p		551	hydrophilic peripheral membrane protein on cytoplasmic face; phosphorylated *mutant*: Class B (fragmented) vacuole; P2 CpY secreted; PrA and PrB maturation slowed two-fold but proteins are not secreted; AlP processed (Köhrer and Emr 1993)
Pep3p	Vps18p Vpt18p Vpl21p Vam8p Cls18p	*D.m.* deep orange, *C.e.* ORF	918	peripheral membrane protein; present in purified vacuole fractions (which also contain endosomes and transport vesicles) (Preston et al. 1991; Robinson et al. 1991) *mutant*: Class C (no discernible) vacuole; P2 CpY secreted; cells lack mature PrA, PrB, CpY, AlP (Robinson et al. 1988, 1991; Preston et al. 1991)
Vps16p	Vph4p Vam9p Vpt16p Cls17p	*M.m.* EST	797	hydrophilic protein found in a particulate fraction, fractionates away from cellular membranes, probably a large protein complex; if overproduced some fraction is soluble (Horazdovsky and Emr 1993) *mutant*: Class C vacuole; secretes P2 CpY; P2 PrA is retained intracellularly (Horazdovsky and Emr 1993)
Vps3p	Pep6p Vpt17p Vpl3p		1011	hydrophilic protein; partly soluble, partly pelletable *mutant*: Class D (single large nonacidic) vacuole; P2 CpY secreted; in *vps3*^{ts} mutant takes 60 min incubation at 37°C to get substantial P2 CpY secretion (slow inactivation or lag in seeing consequences of inactivation?); defective in vacuole inheritance (Raymond et al. 1990)

Table 9 Mutants: Genes involved in vacuole biogenesis not yet cloned

SGD name	Other names	SGD name	Other names
Class A (wild type) vacuoles		Class E (enlarged prevacuolar compartment next to vacuole)	
Vps13p	Vpt2p	Vps2p	Vpt14p
Vps30p	Vpt30p		Vpl2p
Vps38p	Vpl17p		Ren1p
Vps44p	Vpl27p		Grd7p
Vps46p	Vpl30p	Vps4p	Vpt10p
			Vpl4p
Class B (fragmented) vacuole			Grd13p
Pep10p	Vps5p	Vps20p	Vpt20p
	Vpt5p		Vpl10p
Vps39p	Vpl18p/22p	Vps22p	Vpt22p
	Vam6p		Vpl14p
Vps41p	Vpl20p	Vps23p	Vpt23p
	Vam2p		Vpl15p
	Fet2p	Vps24p	Vpt24p
Vps43p	Vpl24p		Vpl26p
Vph6p		Vps25p	Vpt25p
Vph7p			Vpl12p
Vph9p		Vps30p	Vpt30p
		Vps31p	Vpt31p
		Vps32p	Vpt32p
Class D (single, large vacuole)		Vps36p	Vpl11p
Vph5p			Vac3p
Vph8p			Grd12p

in various mutants (Doherty and Kane 1993; Tomashek et al. 1996). Deletion of genes for V_0 subunits has no effect on synthesis and assembly of the V_1 subunits and their subcomplexes, although the complexes do not assemble on the vacuolar membrane (Kane et al. 1992; Manolson et al. 1992b; Ho et al. 1993a; Doherty and Kane 1993; Tomashek et al. 1996).

Deletion of a gene encoding a subunit of the membrane sector results in failure of the other V_0 subunits to accumulate in the vacuolar membrane (Kane et al. 1992; Bauerle et al. 1993). It seems likely that the V_0 subunits are being degraded under these conditions by analogy with what happens in the *vma21* and *vma22* mutants (Hill and Stevens 1994, 1995).

Kinetic experiments suggested that the V_0 sector undergoes assembly in the ER (Doherty and Kane 1993). Recent work has demonstrated that three proteins, Vph2p/Vma12p, Vma21p, and Vma22p/Vph6p, all of which are located in the ER, are required for assembly of the V_0 sector of the V-ATPase; they are not subunits of the V-ATPase (Hirata et al. 1993; Bachhawat et al. 1993; Hill and Stevens 1994, 1995; Hemenway et al. 1995). Vph2p/Vma12p and Vma21p are each predicted to contain two *trans*-membrane domains; both are integral membrane proteins (Hirata et al. 1993; Bachhawat et al. 1993; Hill and Stevens 1994). Vma22p is hydrophilic and a peripheral ER protein (Hill and Stevens 1995). Its association with the ER is dependent on the presence of Vma12p (Hill and Stevens 1995).

Several observations suggest that the function(s) of the three assembly factors may be highly specific, possibly limited to assembly of the V_0 sector of the V-ATPase. None of the genes is essential, and deletion of either *VMA21* or *VMA22* (*VMA12* was not tested) has no effect on synthesis or localization of the plasma membrane ATPase or vacuolar AlP. In addition, although deletion of *VPH2/VMA12* affects the ability of cells to grow on nonfermentable carbon sources or at pH 7.5 and renders them sensitive to a number of cations, the effects seen are no more severe than those that result from loss of subunits of the catalytic sector of the V-ATPase (Ohya et al. 1991; Hirata et al. 1993; Bachhawat et al. 1993; Hill and Stevens 1994, 1995; Hemenway et al. 1995).

Unlike what is seen for another vacuolar integral membrane protein, AlP, mutations that disrupt the TGN to vacuole pathway for soluble vacuolar hydrolases also result in a nonacidic vacuole, presumably owing to failure to deliver the assembled V_0 sector to the vacuole (Preston et al. 1989, 1992; Rothman et al. 1989b; Raymond et al. 1992a; Singer-Kruger et al. 1994; Horazdovsky et al. 1996). Indeed, in the class-E mutants, the V-ATPase appears to accumulate in the class-E compartment, which is thought to be an enlarged endosome (Raymond et al. 1992a).

To sum up, the V_0 and V_1 sectors are assembled independently. The data are consistent with self-assembly of the V_1 sector; at least three ER-localized proteins are required for assembly of the V_0 sector. The V_0 sector seems to follow the pathway taken by the soluble hydrolase precursors. There is no evidence bearing on whether delivery of the V_0 sector to the vacuole requires targeting information or is by default.

E. Extrasecretory Pathway

Both αMS and ApI reach the vacuole by an extrasecretory route (Yoshihisa and Anraku 1990; Klionsky et al. 1992a). The kinetics for delivery of the two proteins are very different: The half-time for processing of αMS is approximately 10 hours and for processing of ApI, 30–45 minutes (Yoshihisa and Anraku 1990; Klionsky et al. 1992a). Klionsky and colleagues have made a very good start at genetically analyzing the import pathway for ApI by isolating mutants that accumulate the ApI precursor. Eight *cvt* (*c*ytoplasm to *v*acuole *t*argeting) complementation groups were identified in the initial screen, two of which proved to be *VPS39* and *VPS41* (Harding et al. 1995). Harding et al. (1996) and Scott et al. (1996) investigated the overlap between *cvt* mutants and mutants unable to undergo autophagy (Tsukada and Ohsumi 1993; Thumm et al. 1994) and found that most of the *cvt* mutants were also defective in autophagy and vice versa. Seven genes proved to have been identified in both the *cvt* screen and an autophagy screen, and two genes were identified in all three screens (*cvt2/apg7; cvt5/apg8; cvt7/apg9/aut9; cvt10/apg1/aut3*; *cvt11/apg15*; *cvt12/apg14;* and *cvt17/aut5*) (Harding et al. 1996; Scott et al. 1996). Despite the fact that shared gene products are used in the ApI uptake pathway and the autophagy pathway, the two pathways operate differently. The *cvt* pathway is constitutive and is kinetically much faster than the autophagy pathway. The autophagy pathway is induced (derepressed?) by starvation for any one of a number of nutrients, for example, nitrogen or carbon (Takeshige et al. 1992; Scott et al. 1996). The *cvt* pathway has a limited capacity; when *LAP4* (encoding ApI) is overexpressed, a substantial portion of proApI remains in the cytoplasm. Upon nitrogen starvation, essentially all of proApI is taken into vacuoles and matured. Moreover, if cells that are overexpressing *LAP4* and have accumulated proApI in the cytoplasm are then starved for nitrogen, the cytoplasmic proApI is taken into the vacuole (by autophagy) and matured. Thus, the autophagic pathway has a substantially greater capacity than the *cvt* pathway. Both pathways are greatly slowed if the vacuolar H^+-ATPase is nonfunctional.

Recent evidence suggests that proApI is imported into vacuoles via a vesicular intermediate: proApI rapidly oligomerizes in the cytosol to form a dodecamer that binds to a nonvacuolar membrane; proApI can be recovered in a vesicle fraction from the cytoplasm (D. Klionsky, pers. comm.). In addition, in one *cvt* mutant, proApI is present in intravacuolar vesicles (Harding et al. 1995; D. Klionsky, pers. comm.).

The pro region of proApI is predicted to contain two α-helices separated by a β-turn. The targeting determinant seems to reside in the more amino-terminal of the two α-helices, for mutations that change amino acids in this helix result in failure to target. The position and spacing of charges in this amphipathic helix are important for its targeting function (Oda et al. 1996).

As further work brings us information on the gene products identified by the *cvt/aut/apg* mutations, it will be very interesting to see which functions are shared by the *cvt* and autophagy pathways as well as which functions are not and how the two pathways are regulated to achieve cellular purposes.

VII. VACUOLE INHERITANCE

Time-lapse phase-contrast photography carried out by Rudolph Müller of the Zentral Institut für Mikrobiologie und Experimentelle Therapie, Jena, in 1958 revealed that the vacuole is a dynamic organelle that blebs off smaller vacuoles that can re-fuse with the parent organelle or be distributed to the bud (the film was shown at the 1984 International Yeast Meeting in Sussex). Wiemken et al. (1970), using electron microscopy, described vacuolar dynamics in cells as a function of the cell cycle; the large vacuole of unbudded cells shrinks and fragments into small vacuoles as bud initiation occurs and cell density increases. Small vacuoles move into the bud, followed by fusion and expansion of the vacuole as cell density decreases.

Vacuolar segregation begins with elongation of the vacuole toward and/or into the emerging bud as a tubule or line of vesicles (Weisman and Wickner 1988; Raymond et al. 1990; Gomes de Mesquita et al. 1991). This alignment is referred to as the vacuolar segregation structure. It is present in buds with a length 0.25 that of the bud length seen at cytokinesis and precedes entry of the nucleus into the bud neck and mitosis (Raymond et al. 1990; Gomes de Mesquita et al. 1991; Cardenas and Heitman 1995). The *TOR2-1* mutation results in a reversal of the order; the vacuole enters after the nucleus and nuclear division, even though *TOR2-1* does not affect inheritance of the vacuole. Jones et al. (1993), using video microscopy, observed vacuolar segregation to the bud after nuclear segregation, a finding at odds with all other reports cited above. They did, however, observe fusion of vacuoles in the bud after nuclear segregation, as was previously reported (Wiemken et al. 1970).

A number of the mutants isolated for their defects in processing and sorting of hydrolase precursors have been reported to be defective in segregation of the vacuole into the bud (Herman and Emr 1990; Raymond et al. 1990, 1992a; Weisman et al. 1990). These include *vps3, vps9, vps15, vps21, vps34, vps45, pep7*, and *pep12*. All of these mutants possess single large vacuoles (class D) and are defective in steps between the late Golgi and the vacuole (see Table 5).

Mutants defective in vacuolar inheritance have been identified in a number of screens (Weisman et al. 1990; Shaw and Wickner 1991; Weisman and Wickner 1992; Nicolson et al. 1995; Gomes de Mesquita et al. 1996; Wang et al. 1996). All of the screens have identified mutants with defects in vacuole inheritance that are a secondary consequence of aberrant vacuole morphology and vacuolar hydrolase delivery. These mutants fell into previously defined complementation groups: *pep7* (*vac1*), *vps34*, *vps15* (*vac4*), all class-D mutants, and *vps36* (*vac3*, class E). *vac5* proved to be an allele of *PHO80*, which encodes a cyclin that activates the cyclin-dependent kinase, Pho85p (Nicolson et al. 1995, 1996). It seems likely that the effect of *vac5* on vacuole inheritance will prove to be indirect.

A number of mutants have emerged from the screens that seem to be defective solely in vacuole inheritance. These include *vac2* (Shaw and Wickner 1991), *vac6* and *vac7* (Gomes de Mesquita et al. 1996), and *vac8-vac12* (Wang et al. 1996). Wang et al. (1996) have examined vacuolar morphology in the inheritance-defective mutants. Morphology can be normal (*vac8*, *vac9*, *vac10*), a single round vacuole like class-D sorting-defective mutants (*vac2*, *vac5*, *vac6*, *vac11*, *vac12*), or a very large round vacuole like that seen in the *fab1* mutant (*vac7*, *fab1*) (Yamamoto et al. 1995). Sorting and processing of hydrolase precursors via the secretory pathway and the endocytic and autophagic pathways are normal in the *vac* mutants. However, the *vac8* mutant is defective in targeting of ApI to the vacuole via the *cvt* pathway (Klionsky et al. 1992a). *vac8* seems not to be allelic to previously identified *CVT* genes.

Weisman and her collaborators have also examined vacuolar segregation in mutants with defects in genes encoding proteins of the cytoskeleton (Hill et al. 1996). They reported that vacuolar inheritance is defective in several *act1* mutants. The defect can be specific to vacuoles (mitochondria also rely on the actin cytoskeleton for inheritance, but cells can be defective for inheritance of one but not the other in both directions). Mutations in a gene for a class V myosin, *MYO2*, can also impair vacuole inheritance. Similarly, a mutation that affects a residue in profilin important for actin binding results in a Vac$^-$ phenotype. In keeping with these observations, actin cables appear to colocalize with vacuolar membranes and Myo2p appears to colocalize with the tip of the segregation structure by immunofluorescence (Hill et al. 1996). Thus, there is very good evidence that vacuole inheritance involves the actin cytoskeleton. It will be very interesting to see whether the seven new genes, *VAC2* and *VAC6-VAC12*, will encode proteins previously known to be associated with the actin cytoskeleton or novel proteins specific for vacuolar inheritance.

An in vitro system has been developed that carries out at least some of the reactions involved in vacuolar inheritance. The morphological correlates that have been seen in vitro are the formation of vacuole-derived tubules and vesicles and the fusion of vacuoles with attendant maturation and thereby activation of the AlP precursor (precursor supplied by one parent vacuole and active processing protease by the other parent vacuole). The fusion reaction allows quantitation (Conradt et al. 1992, 1994; Haas et al. 1994). Four different steps have been delineated in the overall reaction. The overall reaction requires ATP, vacuoles, and cytosol and is inhibited by Bafilomycin A_1, which inhibits the V-ATPases, and the proton ionophore CCCP, indicating a requirement for a pH gradient and electrochemical potential across the vacuolar membrane. GTPγS and mastaparan both inhibit fusion, indicating that heterotrimeric G proteins participate. Inhibition by neomycin implicates phosphatidylinositol polyphosphates in the process (Haas et al. 1994), and inhibition by microcystin-LR, which specifically inhibits protein phosphatases of types 1 and 2A, indicates that a dephosphorylation is required (Conradt et al. 1994).

Using the *ypt7* mutant, which has a fragmented vacuole (Wichmann et al. 1992), Haas et al. (1995) demonstrated that Ypt7p must be present on both partner vacuoles for fusion to occur. The presence of excess Ypt7p in the cytosol is not able to overcome the lack of Ypt7p on one of the participating vacuoles. Antibodies to Ypt7p can inhibit fusion. Addition of excess Ypt7p but not the related small G proteins Vps21p or Ypt1p, which participate in other vesicle fusion reactions, overcame the inhibition. Vacuoles from the *vac2* mutant are unable to fuse even in the presence of wild-type cytoplasm (Haas et al. 1994).

Recently, Haas and Wickner (1996) showed that Sec18p (NSF) and Sec17p (α-SNAP) are required for fusion of vacuoles in vitro. Sec17p and Sec18p are needed early in the fusion reaction, for fusion is resistant to addition of either antibody after a short incubation. Sec17p is released from the vacuoles before fusion occurs (Mayer et al. 1996). ATP, Ypt7p, the pH gradient, heterotrimeric G protein(s), and protein phosphatase are required for a longer period, apparently acting later in the sequence (Mayer et al. 1996).

It was reported recently that a heterodimer, which was purified from the cytosol and which contains thioredoxin and the cytoplasmic inhibitor of PrB, IB2, is required for vacuolar fusion in vitro (Xu and Wickner 1996; Xu et al. 1997), and indeed, strains deleted for *TRX1* and *TRX2* or for *PBI2* proved to have a defect in the pattern and extent of inheritance of vacuoles.

Clearly, the in vitro reactions can only partially recapitulate the phenomenon of vacuolar inheritance, for the actin cytoskeleton is left out of the equation in vitro. Nevertheless, a good beginning has been made.

One study has examined the inheritance of the vacuole and its components during meiosis (Roeder and Shaw 1996). These authors reported that some vesicular components may be almost entirely excluded from ascospores, whereas other components may be at least partially included. The latter observation confirmed the initial reports of phenotypic lag attributed to inclusion of PrB in ascospores (Zubenko et al. 1982).

VIII. SUMMARY AND FUTURE DIRECTIONS

We have made considerable progress in delineating the skeletons of the pathways that deliver proteins to the vacuole. Transport is via vesicles and involves proteins homologous to proteins of the secretory and synaptic transmission pathways. It is equally clear that another route must exist and must be sought, for the AlP precursor reaches the vacuole when all known pathways are blocked; this route seems to depart from the TGN using Vps1p and rejoins the endosomal pathway somewhere in the vicinity of the vacuole. At least two apparently independent cytoplasm-to-vacuole delivery routes exist (ApI and FDPase); both seem to employ vesicular carriers. That followed by ApI appears to join the endosomal pathway at a step near the vacuole. Whether the FDPase pathway will also converge is unknown. It is worth noting that delineation of the pathway skeletons is no mean achievement and that by comparison to knowledge of the lysosomal targeting pathway in animal (or plant) cells, where little is known beyond the M6P receptor, clathrin, and adaptor complexes, looms large.

We should soon complete the assigning of proteins already identified genetically to pathway steps, although additional components will undoubtedly continue to be identified by genetic means. The most pressing and demanding challenges will be biochemical, however, for we need to put biochemical flesh (molecular mechanism) onto the pathway skeletons. We will also need to delineate more clearly how the various pathways (endosomal, endocytic, cytoplasm to vacuole, vacuole inheritance) converge or overlap.

Interesting questions await with respect to the V-ATPase(s). Where in the pathway is the second (Stv1p-containing) V-ATPase? How are the two ATPases targeted to their respective organelles? Is one by default and the other by targeting and/or retrieval? How do the assembly factors function in assembling the V_0 sector of the vacuolar V-ATPase? Do they also function to assemble the V_0 sector of the second V-ATPase? How is the V_1 sector assembled? And how is assembly/disassembly of the V_1 sector on and off the V_0 sector regulated?

There are still numbers of questions about vacuolar function that remain unaddressed. How are fluxes of small molecules in and out of vacuoles regulated? What are the signals and how are they read? How is the signal(s) that triggers autophagy generated and read? The vacuole is in some fashion involved in osmoregulation and in Fe^{++} uptake/metabolism. What are the phenomena and how are they effected and regulated? Why does loss of vacuole function interfere with aerobic carbon metabolism? Does it make the cell sensitive to various cations (beyond Ca^{++})? The next decade promises to be as interesting as the last one has been!

ACKNOWLEDGMENTS

We thank numerous colleagues for supplying unpublished observations and the members of our laboratory for helpful discussions. Research in this laboratory was supported by grants DK-18090 and GM-2973 from the National Institutes of Health and by an Academic Development Program predoctoral fellowship from Merck Research Laboratories (M.A.H.).

REFERENCES

Aalto, M.K., S. Keranen, and H. Ronne. 1992. A family of proteins involved in intracellular transport. *Cell* **68:** 181.

Aalto, M.K., H. Ronne, and S. Keranen. 1993. Yeast syntaxins Sso1p and Sso2p belong to a family of related membrane proteins that function in vesicular transport. *EMBO J.* **12:** 4095.

Aalto, M.K., L. Ruohonen, K. Hosono, and S. Keranen. 1991. Cloning and sequencing of the yeast *Saccharomyces cerevisiae SEC1* gene localized on chromosome IV. *Yeast* **7:** 643.

Achstetter, T., C. Ehmann, and D.H. Wolf. 1982. Aminopeptidase Co, a new yeast peptidase. *Biochem. Biophys. Res. Commun.* **109:** 341.

Anraku, Y. 1987. Unveiling the mechanism of ATP-dependent energization of yeast vacuolar membranes: Discovery of a third type of H^+-translocating ATPase. In *Bioenergetics: Structure and function of energy transducing systems* (ed. T. Ozawa, and S. Papa), p. 249. Japan Scientific Societies Press, Tokyo.

Apperson, M., R.E. Jensen, K. Suda, C. Witte, and M.P. Yaffe. 1990. A yeast protein, homologous to the proteolipid of the chromaffin granule proton-ATPase, is important for cell growth. *Biochem. Biophys. Res. Commun.* **168:** 574.

Arai, H., G. Terres, S. Pink, and M. Forgac. 1988. Topography and subunit stoichiomety of the coated vesicle proton pump. *J. Biol. Chem.* **263:** 8796.

Avery, S.V. and J.M. Tobin. 1992. Mechanisms of strontium uptake by laboratory and brewing strains of *Saccharomyces cerevisiae*. *Appl. Environ. Microbiol.* **58:** 3883.

Bachhawat, A.K., J. Suhan, and E.W. Jones. 1994. The yeast homolog of H<β>58, a mouse gene essential for embryogenesis, performs a role in the delivery of proteins to the vacuole. *Genes Dev.* **8:** 1379.

Bachhawat, A.K., M.F. Manolson, D.G. Murdock, J.D. Garman, and E.W. Jones. 1993. The *VPH2* gene encodes a 25 kDa protein required for activity of the yeast vacuolar H^+-ATPase. *Yeast* **9:** 175.

Bakalkin, G., S. Kalnov, A. Zubatov, and V. Luzikov. 1976. Degradation of total cell protein at different stages of *Saccharomyces cerevisiae* yeast growth. *FEBS Lett.* **63:** 218.

Banfield, D., M. Lewis, and H. Pelham. 1995. A SNARE-like protein required for traffic through the Golgi complex. *Nature* **375:** 806.

Banfield, D.K., M.J. Lewis, C. Rabouille, G. Warren, and H.R.B. Pelham. 1994. Localization of Sed5, a putative vesicle targeting molecule, to the *cis*-Golgi network involves both its transmembrane and cytoplasmic domains. *J. Cell Biol.* **127:** 357.

Bankaitis, V.A., L.M. Johnson, and S.D. Emr. 1986. Isolation of yeast mutants defective in protein targeting to the vacuole. *Proc. Natl. Acad. Sci.* **83:** 9075.

Banta, L.M., J.S. Robinson, D.J. Klionsky, and S.D. Emr. 1988. Organelle assembly in yeast: Characterization of yeast mutants defective in vacuolar biogenesis and protein sorting. *J. Cell Biol.* **107:** 1369.

Banta, L.M., T.A. Vida, P.K. Herman, and S.D. Emr. 1990. Characterization of yeast Vps33p, a protein required for vacuolar protein sorting and vacuole biogenesis. *Mol. Cell. Biol.* **10:** 4638.

Barlowe, C., L. Orci, T. Yeung, M. Hosobuchi, S. Hamamoto, N. Salama, M.F. Rexach, M. Ravazzola, M. Amherdt, and R. Schekman. 1994. COPII: A membrane coat formed by Sec proteins that drive vesicle budding from the endoplasmic reticulum. *Cell* **77:** 895.

Bauerle, C., M.N. Ho, M.A. Lindorfer, and T.H. Stevens. 1993. The *Saccharomyces cerevisiae VMA6* gene encodes the 36-kDa subunit of the vacuolar H^+-ATPase membrane sector. *J. Biol. Chem.* **268:** 12749.

Beauvoit, B., M. Rigoulet, G. Raffard, P. Canioni, and B. Guerin. 1991. Differential sensitivity of the cellular compartments of *Saccharomyces cerevisiae* to protonophoric uncoupler under fermentative and respiratory energy supply. *Biochemistry* **30:** 11212.

Becherer, K.A., S.E. Rieder, S.D. Emr, and E.W. Jones. 1996. Novel syntaxin homologue, Pep12p, required for the sorting of lumenal hydrolases to the lysosome-like vacuole in yeast. *Mol. Biol. Cell* **7:** 579.

Belde, P.J., J.H. Vossen, G.W. Borst-Pauwels, and A.P. Theuvenet. 1993. Inositol 1,4,5-trisphosphate releases Ca^{2+} from vacuolar membrane vesicles of *Saccharomyces cerevisiae*. *FEBS Lett.* **323:** 113.

Beltran, C., J. Kopecky, Y.C. Pan, H. Nelson, and N. Nelson. 1992. Cloning and mutational analysis of the gene encoding subunit C of yeast vacuolar H^+-ATPase. *J. Biol. Chem.* **267:** 774.

Berkower, C., D. Loayza, and S. Michaelis. 1994. Metabolic instability and constitutive endocytosis of *STE6*, the **a**-factor transporter of *Saccharomyces cerevisiae*. *Mol. Biol. Cell* **5:** 1185.

Betz, H. and U. Weiser. 1976. Protein degradation and proteinases during yeast sporulation. *Eur. J. Biochem.* **62:** 65.

Bilinski, C.A. and J.J. Miller. 1983. Translocation of zinc from vacuole to nucleus during yeast meiosis. *Can. J. Genet. Cytol.* **25:** 415.

Blachly-Dyson, E. and T.H. Stevens. 1987. Yeast carboxypeptidase Y can be translocated

and glycosylated without its amino-terminal signal sequence. *J. Cell Biol.* **104:** 1183.
Bordallo, J., C. Bordallo, S. Gascon, and P. Suarez-Rendueles. 1991. Molecular cloning and sequencing of genomic DNA encoding yeast vacuolar carboxypeptidase yscS. *FEBS Lett.* **283:** 27.
Bowman, E.J., A. Siebers, and K. Altendorf. 1988. Bafilomycins: A class of inhibitors of membrane ATPases from microorganisms, animal cells, and plant cells. *Proc. Natl. Acad. Sci.* **85:** 7972.
Braun, M., A. Waheed, and K. von Figura. 1989. Lysosomal acid phosphatase is transported to lysosomes via the cell surface. *EMBO J.* **8:** 3633.
Brennwald, P., B. Kearns, K. Champion, S. Keranen, V. Bankaitis, and P. Novick. 1994. Sec9 is a SNAP-25-like component of a yeast SNARE complex that may be the effector of Sec4 function in exocytosis. *Cell* **79:** 245.
Bryant, N. and A. Boyd. 1993. Immunoisolation of Kex2p-containing organelles from yeast demonstrates colocalisation of three processing proteinases to a single Golgi compartment. *J. Cell Sci.* **106:** 815.
Burd, C.G., P.A. Mustol, P.V. Schu, and S.D. Emr. 1996. A yeast protein related to a mammalian Ras-binding protein, Vps9p, is required for localization of vacuolar proteins. *Mol. Cell. Biol.* **16:** 2369.
Campbell-Burk, S.L. and R.G. Shulman. 1987. High-resolution NMR studies of *Saccharomyces cerevisiae*. *Annu. Rev. Microbiol* **41:** 595.
Cardenas, M.E. and J. Heitman. 1995. FKBP12-rapamycin target TOR2 is a vacuolar protein with an associated phosphatidylinositol 4-kinase activity. *EMBO J.* **14:** 5892.
Castro, C.D., A.J. Meehan, A.P. Koretsky, and M.M. Domach. 1995. In situ 31P nuclear magnetic resonance for observation of polyphosphate and catabolite responses of chemostat-cultivated *Saccharomyces cerevisiae* after alkalinization. *Appl. Environ. Microbiol.* **61:** 4448.
Cereghino, J.L., E.G. Marcusson, and S.D. Emr. 1995. The cytoplasmic tail domain of the vacuolar protein sorting receptor Vps10p and a subset of *VPS* gene products regulate receptor stability, function, and localization. *Mol. Biol. Cell* **6:** 1089.
Chang, A. and G.R. Fink. 1995. Targeting of the yeast plasma membrane H^+-ATPase: A novel gene *AST1* prevents mislocalization of mutant ATPase to the vacuole. *J. Cell Biol.* **128:** 39.
Chang, Y.H. and J.A. Smith. 1989. Molecular cloning and sequencing of genomic DNA encoding aminopeptidase I from *Saccharomyces cerevisiae*. *J. Biol. Chem.* **264:** 6979.
Chapman, C. and W. Bartley. 1968. The kinetics of enzyme changes in yeast under conditions that cause the loss of mitochondria. *Biochem. J.* **107:** 455.
Chaudhuri, B., N.S. Delany, and C. Stephen. 1995. The unfolded-protein-response element discriminates misfolding induced by different mutant pro-sequences of yeast carboxypeptidase y. *Biochem. Biophys. Res. Commun.* **209:** 31.
Chaudhuri, B., S. Inhavale, and A.K. Bachhawat. 1997. *apd1*$^+$, a gene required for red pigment formation in *ade6* mutants of *Schizosaccharomyces pombe*, encodes an enzyme required for glutathione biosynthesis: A role for glutathione and a glutathione-conjugate pump. *Genetics* **145:** 75.
Chen, P., P. Johnson, T. Sommer, S. Jentsch, and M. Hochstrasser. 1993. Multiple ubiquitin-conjugating enzymes participate in the in vivo degradation of the yeast *MAT*α*2* repressor. *Cell* **74:** 357.
Chen, Y.-J. and T. Stevens. 1996. The *VPS8* gene is required for localization and trafficking of the CPY sorting receptor in *Saccharomyces cerevisiae*. *Eur. J. Cell Biol.* **70:**

289.

Chiang, H.L. and R. Schekman. 1991. Regulated import and degradation of a cytosolic protein in the yeast vacuole. *Nature* **350:** 313.

Chiang, H.L., R. Schekman, and S. Hamamoto. 1996. Selective uptake of cytosolic, peroxisomal, and plasma membrane proteins into the yeast lysosome for degradation. *J. Biol. Chem.* **271:** 9934.

Chvatchko, Y., I. Howald, and H. Riezman. 1986. Two yeast mutants defective in endocytosis are defective in pheromone response. *Cell* **46:** 355.

Clark, D.W., J.S. Tkacz, and J.O. Lampen. 1982. Asparagine-linked carbohydrate does not determine the cellular location of yeast vacuolar nonspecific alkaline phosphatase. *J. Bacteriol.* **152:** 865.

Coffman, J.A., H.M. El Berry, and T.G. Cooper. 1994. The *URE2* protein regulates nitrogen catabolic gene expression through the GATAA-containing UAS_{NTR} element in *Saccharomyces cerevisiae. J. Bacteriol.* **176:** 7476.

Conibear, E. and T.H. Stevens. 1995. Vacuolar biogenesis in yeast: Sorting out the sorting proteins. *Cell* **83:** 513.

Conradt, B., A. Haas, and W. Wickner. 1994. Determination of four biochemically distinct, sequential stages during vacuole inheritance in vitro. *J. Cell Biol.* **126:** 99.

Conradt, B., J. Shaw, T. Vida, S. Emr, and W. Wickner. 1992. In vitro reactions of vacuole inheritance in *Saccharomyces cerevisiae. J Cell Biol.* **119:** 1469.

Cooper, A. and H. Bussey. 1992. Yeast Kex1p is a Golgi-associated membrane protein: Deletions in a cytoplasmic targeting domain result in mislocalization to the vacuolar membrane. *J. Cell Biol.* **119:** 1459.

Cooper, A.A. and T.H. Stevens. 1996. Vps10p cycles between the late-Golgi and prevacuolar compartments in its function as the sorting receptor for multiple yeast vacuolar hydrolases. *J. Cell Biol.* **133:** 529.

Cooper, T.G. 1982. Transport in *Saccharomyces cerevisiae.* In *The molecular biology of the yeast* Saccharomyces: *Metabolism and gene expression* (ed. J.N. Strathern et al.), p. 399. Cold Spring Harbor Laboratory, Cold Spring Harbor, New York.

Cooper, T.G. and R.A. Sumrada. 1983. What is the function of nitrogen catabolite repression in *Saccharomyces cerevisiae*? *J. Bacteriol.* **155:** 623.

Cowles, C.R., S.D. Emr, and B.F. Horazdovsky. 1994. Mutations in the *VPS45* gene, a *SEC1* homologue, result in vacuolar protein sorting defects and accumulation of membrane vesicles. *J. Cell Sci.* **107:** 3449.

Cunningham, K.W. and G.R. Fink. 1994. Calcineurin-dependent growth control in *Saccharomyces cerevisiae* mutants lacking *PMC1*, a homolog of plasma membrane Ca^{2+} ATPases. *J. Cell Biol.* **124:** 351.

———. 1996. Calcineurin inhibits VCX1-dependent H^+/Ca^{2+} exchange and induces Ca^{2+} ATPases in *Saccharomyces cerevisiae. Mol. Cell. Biol.* **16:** 2226.

Damke, H., T. Baba, D.E. Warnock, and S.L. Schmid. 1994. Induction of mutant dynamin specifically blocks endocytic coated vesicle formation. *J. Cell Biol.* **127:** 915.

Daugherty, J.R., R. Rai, H.M. El Berry, and T.G. Cooper. 1993. Regulatory circuit for responses of nitrogen catabolic gene expression to the *GLN3* and *DAL80* proteins and nitrogen catabolite repression in *Saccharomyces cerevisiae. J. Bacteriol.* **175:** 64.

Davis, N.G., J.L. Horecka, and G.F. Sprague, Jr. 1993. *cis*- and *trans*-acting functions required for endocytosis of the yeast pheromone receptors. *J. Cell Biol.* **122:** 53.

Deshaies, R.J. and R. Schekman. 1987. A yeast mutant defective at an early stage in import of secretory protein precursors into the endoplasmic reticulum. *J. Cell Biol.* **105:**

633.

Deshaies, R.J., V. Chau, and M. Kirschner. 1995. Ubiquitination of the G_1 cyclin Cln2p by a Cdc34p-dependent pathway. *EMBO J.* **14:** 303.

Doherty, R.D. and P.M. Kane. 1993. Partial assembly of the yeast vacuolar H^+-ATPase in mutants lacking one subunit of the enzyme. *J. Biol. Chem.* **268:** 16845.

Donella-Deana, A., S. Ostojic, L.A. Pinna, and S. Barbaric. 1993. Specific dephosphorylation of phosphopeptides by the yeast alkaline phosphatase encoded by *PHO8* gene. *Biochim. Biophys. Acta* **1177:** 221.

Dulic, V. and H. Riezman. 1989. Characterization of the *END1* gene required for vacuole biogenesis and gluconeogenic growth of budding yeast. *EMBO J.* **8:** 1349.

———. 1990. *Saccharomyces cerevisiae* mutants lacking a functional vacuole are defective for aspects of the pheromone response. *J. Cell Sci.* **97:** 517.

Dunn, T., K. Gable, and T. Beeler. 1994. Regulation of cellular Ca^{2+} by yeast vacuoles. *J. Biol. Chem.* **269:** 7273.

Duntze, W., W. Neumann, J. Gancedo, W. Atzpodien, and H. Holzer. 1969. Studies on the regulation and localization of the glyoxylate enzymes in *Saccharomyces cerevisiae. Eur. J. Biochem.* **10:** 83.

Durr, M., K. Urech, T. Boller, A. Wiemken, J. Schwencke, and M. Nagy. 1979. Sequestration of arginine by polyphosphate in vacuoles of yeast (*Saccharomyces cerevisiae*). *Arch. Microbiol.* **121:** 169.

Eakle, K.A., M. Bernstein, and S.D. Emr. 1988. Characterization of a component of the yeast secretion machinery: Identification of the *SEC18* gene product. *Mol. Cell. Biol.* **8:** 4098.

Egner, R. and K. Kuchler. 1996. The yeast multidrug transporter Pdr5 of the plasma membrane is ubiquitinated prior to endocytosis and degradation in the vacuole. *FEBS Lett.* **378:** 177.

Egner, R., Y. Mahe, R. Pandjaitan, and K. Kuchler. 1995. Endocytosis and vacuolar degradation of the plasma membrane-localized Pdr5 ATP-binding cassette multidrug transporter in *Saccharomyces cerevisiae. Mol. Biol. Cell* **15:** 5879.

Egner, R., M. Thumm, M. Straub, A. Simeon, H.J. Schuller, and D.H. Wolf. 1993. Tracing intracellular proteolytic pathways. Proteolysis of fatty acid synthase and other cytoplasmic proteins in the yeast *Saccharomyces cerevisiae. J. Biol. Chem.* **268:** 27269.

Eilam, Y., H. Lavi, and N. Grossowicz. 1985. Active extrusion of potassium in the yeast *Saccharomyces cerevisiae* induced by low concentrations of trifluoperazine. *J. Gen. Microbiol.* **131:** 2555.

Ekena, K. and T.H. Stevens. 1995. The *Saccharomyces cerevisiae MVP1* gene interacts with *VPS1* and is required for vacuolar protein sorting. *Mol. Cell. Biol.* **15:** 1671.

Ferguson, J.J., Jr., M. Boll, and H. Holzer. 1967. Yeast malate dehydrogenase enzyme inactivation in catabolite repression. *Eur. J. Biochem.* **1:** 21.

Ferro-Novick, S. and R. Jahn. 1994. Vesicle fusion from yeast to man. *Nature* **370:** 191.

Finley, D., E. Ozaynak, and A. Varshavsky. 1987. The yeast polyubiquitin gene is essential for resistance to high temperature, starvation, and other stresses. *Cell* **48:** 1035.

Foury, F. 1990. The 31-kDa polypeptide is an essential subunit of the vacuolar ATPase in *Saccharomyces cerevisiae. J. Biol. Chem.* **265:** 18554.

François, J., M.-J. Neves, and H.-G. Hers. 1991. The control of trehalose biosynthesis in *Saccharomyces cerevisiae:* Evidence for a catabolite inactivation and repression of trehalose-6-phosphate synthase and trehalose-6-phosphate phosphatase. *Yeast* **7:** 575.

Franzusoff, A. and R. Schekman. 1989. Functional compartments of the yeast Golgi apparatus are defined by the *sec7* mutation. *EMBO J.* **8:** 2695.

Frey, J. and K.-H. Rohm. 1979. Glucose induced inactivation of aminopeptidase I in *Saccharomyces cerevisiae. FEBS Lett.* **100:** 261.

Funaguma, T., Y. Toyoda, and J. Sy. 1985. Catabolite inactivation of fructose 1,6-bisphosphatase and cytoplasmic malate dehydrogenase in yeast. *Biochem. Biophys. Res. Commun.* **130:** 467.

Funayama, S., J.M Gancedo, and C. Gancedo. 1980. Turnover of yeast fructose-bisphosphatase in different metabolic conditions. *Eur. J. Biochem.* **109:** 61.

Galan, J.M., V. Moreau, B. Andre, C. Volland, and R. Haguenauer-Tsapis. 1996. Ubiquitination mediated by the Npi1p/Rsp5p ubiquitin-protein ligase is required for endocytosis of the yeast uracil permease. *J. Biol. Chem.* **271:** 10946.

Gancedo, C. 1971. Inactivation of fructose 1,6-diphosphatase by glucose in yeast. *J. Bacteriol.* **107:** 401.

Gancedo, C. and J. Schwerzmann. 1976. Inactivation by glucose of phosphoenolpyruvate carboxykinase from *Saccharomyces cerevisiae. Arch. Microbiol.* **109:** 221.

Garrett-Engele, P., B. Moilanen, and M.S. Cyert. 1995. Calcineurin, the Ca^{2+}/calmodulin-dependent protein phosphatase, is essential in yeast mutants with cell integrity defects and in mutants that lack a functional vacuolar H^+-ATPase. *Mol. Cell. Biol.* **15:** 4103.

Gaynor, E.C., S. te Heesen, T.R. Graham, M. Aebi, and S.D. Emr. 1994. Signal-mediated retrieval of a membrane protein from the Golgi to the ER in yeast. *J. Cell Biol.* **127:** 653.

Gillies, R.J., K. Ugurbil, J.A. Den Hollander, and R.G. Shulman. 1981. 31P NMR studies of intracellular pH and phosphate metabolism during cell division cycle of *Saccharomyces cerevisiae. Proc. Natl. Acad. Sci.* **78:** 2125.

Gomes de Mesquita, D.S., R. ten Hoopen, and C.L. Woldringh. 1991. Vacuolar segregation to the bud of *Saccharomyces cerevisiae:* An analysis of morphology and timing in the cell cycle. *J. Gen. Microbiol.* **137:** 2447.

Gomes de Mesquita, D.S., H.B. van den Hazel, J. Bowman, and C.L. Woldringh. 1996. Characterization of new vacuolar segregation mutants, isolated by screening for loss of proteinase B self-activation. *Eur. J. Cell Biol.* **71:** 237.

Gottschalk, S., A. Waheed, B. Schmidt, P. Laidler, and K. von Figura. 1989. Sequential processing of lysosomal acid phosphatase by a cytoplasmic thiol proteinase and a lysosomal aspartyl proteinase. *EMBO J.* **8:** 3215.

Goud, B., A. Salminen, N. Walworth, and P. Novick. 1988. A GTP-binding protein required for secretion rapidly associates with secretory vesicles and the plasma membrane in yeast. *Cell* **53:** 753.

Graham, L.A., K.J. Hill, and T.H. Stevens. 1994. *VMA7* encodes a novel 14-kDa subunit of the *Saccharomyces cerevisiae* vacuolar H^+-ATPase complex. *J. Biol. Chem.* **269:** 25974.

———. 1995. *VMA8* encodes a 32-kDa V_1 subunit of the *Saccharomyces cerevisiae* vacuolar H^+-ATPase required for function and assembly of the enzyme complex. *J. Biol. Chem.* **270:** 15037.

Graham, T.R. and S.D. Emr. 1991. Compartmental organization of Golgi-specific protein modification and vacuolar protein sorting events defined in a yeast *sec18* (NSF) mutant. *J. Cell Biol.* **114:** 207.

Graham, T.R., M. Seeger, G.S. Payne, V.L. MacKay, and S.D. Emr. 1994. Clathrin-dependent localization of α-1,3 mannosyltransferase to the Golgi complex of *Sac-*

charomyces cerevisiae. *J. Cell Biol.* **127:** 667.

Greenfield, N.J., M. Hussain, and J. Lenard. 1987. Effects of growth state and amines on cytoplasmic and vacuolar pH, phosphate and polyphosphate levels in *Saccharomyces cerevisiae:* A 31P-nuclear magnetic resonance study. *Biochim. Biophys. Acta* **926:** 205.

Grenson, M. 1983. Inactivation-reactivation process and repression of permease formation regulate several ammonia sensitive permeases in the yeast *Saccharomyces cerevisiae*. *Eur. J. Biochem.* **133:** 135.

Grenson, M. and E. Dubois. 1982. Pleiotropic deficiency in nitrogen-uptake systems and derepression of nitrogen-catabolic enzymes in *npr1* mutants of *Saccharomyces cerevisiae*. *Eur. J. Biochem.* **121:** 643.

Griff, I.E., R. Schekman, J.E. Rothman, and C.A. Kaiser. 1992. The yeast SEC17 gene product is functionally equivalent to mammalian α-SNAP protein. *J. Biol. Chem.* **267:** 12106.

Haas, A. and W. Wickner. 1996. Homotypic vacuole fusion requires Sec17p (yeast α-SNAP) and Sec18p (yeast NSF). *EMBO J.* **15:** 3296.

Haas, A., B. Conradt, and W. Wickner. 1994. G-protein ligands inhibit in vitro reactions of vacuole inheritance. *J. Cell Biol.* **126:** 87.

Haas, A., D. Scheglmann, T. Lazar, D. Gallwitz, and W. Wickner. 1995. The GTPase Ypt7p of *Saccharomyces cerevisiae* is required on both partner vacuoles for the homotypic fusion step of vacuole inheritance. *EMBO J.* **14:** 5258.

Hagele, E., J. Neefe, and D. Mecke. 1978. The malate dehydrogenase isoenzymes of *Saccharomyces cerevisiae*. *Eur. J. Biochem.* **83:** 67.

Halachmi, D. and Y. Eilam. 1989. Cytosolic and vacuolar Ca^{2+} concentrations in yeast cells measured with the Ca^{2+}-sensitive fluorescence dye indo-1. *FEBS Lett.* **256:** 55.

———. 1993. Calcium homeostasis in yeast cells exposed to high concentrations of calcium. Roles of vacuolar H^+-ATPase and cellular ATP. *FEBS Lett.* **316:** 73.

Halvorson, H. 1958a. Intracellular protein and nucleic acid turnover in resting yeast cells. *Biochim. Biophys. Acta* **27:** 255.

———. 1958b. Studies on protein and nucleic acid turnover in growing cultures of yeast. *Biochim. Biophys. Acta* **27:** 267.

Hansen, R., R. Switzer, H. Hinze, and H. Holzer. 1977. Effects of glucose and nitrogen source on the levels of proteinases, peptidases, and proteinase inhibitors in yeast. *Biochim. Biophys. Acta* **196:** 103.

Harding, T.M., A. Hefner-Gravink, M. Thumm, and D.J. Klionsky. 1996. Genetic and phenotypic overlap between autophagy and the cytoplasm to vacuole protein targeting pathway. *J. Biol. Chem.* **271:** 17621.

Harding, T.M., K.A. Morano, S.V. Scott, and D.J. Klionsky. 1995. Isolation and characterization of yeast mutants in the cytoplasm to vacuole protein targeting pathway. *J. Cell Biol.* **131:** 591.

Hardwick, K.G. and H.R.B. Pelham. 1992. *SED5* encodes a 39-kD integral membrane protein required for vesicular transport between the ER and the Golgi complex. *J. Cell Biol.* **119:** 513.

Harris, S.D. and D.A. Cotter. 1987. Vacuolar (lysosomal) trehalase of *Saccharomyces cerevisiae*. *Curr. Microbiol.* **15:** 247.

———. 1988. Transport of yeast vacuolar trehalase to the vacuole. *Can. J. Microbiol.* **34:** 835.

Harris, S.L. and G. Waters. 1996. Localization of a yeast early Golgi mannosyltransferase, Och1p, involves retrograde transport. *J. Cell Biol.* **132:** 985.

Harter, C. 1995. COP-coated vesicles in intracellular protein transport. *FEBS Lett.* **369:** 89.

Hartwell, L. 1970. Periodic density fluctuation during the yeast cell cycle and the selection of synchronous cultures. *J. Bacteriol.* **104:** 1280.

Haselbeck, A. and W. Tanner. 1983. O-glycosylation in *Saccharomyces cerevisiae* is initiated at the endoplasmic reticulum. *FEBS Lett.* **158:** 335.

Hasilik, A. and W. Tanner. 1978a. Biosynthesis of the vacuolar yeast glycoprotein carboxypeptidase Y. Conversion of precursor into the enzyme. *Eur. J. Biochem.* **85:** 599.

———. 1978b. Carbohydrate moiety of carboxypeptidase Y and perturbation of its biosynthesis. *Eur. J. Biochem.* **91:** 567.

Hein, C., J.-Y. Springael, C. Volland, R. Haguenauer-Tsapis, and B. Andre. 1995. *NPI1*, an essential yeast gene involved in induced degradation of Gap1 and Fur4 permeases, encodes the Rsp5 ubiquitin-protein ligase. *Mol. Microbiol.* **18:** 77.

Heinemeyer, W., A. Gruhler, V. Mohrle, Y. Mahe, and D.H. Wolf. 1993. *PRE2*, highly homologous to the human major histocompatibility complex-linked *RING10* gene, codes for a yeast proteasome subunit necessary for chrymotryptic activity and degradation of ubiquitinated proteins. *J. Biol. Chem.* **268:** 5115.

Heinemeyer, W., A. Simeon, H.H. Hirsch, H.H. Schiffer, U. Teichert, and D.H. Wolf. 1991. Lysosomal and non-lysosomal proteolysis in the eukaryotic cell: Studies on yeast. *Biochem. Soc. Trans.* **19:** 724.

Hemenway, C.S., K. Dolinski, M.E. Cardenas, M.A. Hiller, E.W. Jones, and J. Heitman. 1995. *vph6* mutants of *Saccharomyces cerevisiae* require calcineurin for growth and are defective in vacuolar H^+-ATPase assembly. *Genetics* **141:** 833.

Hemmings, B., G.S. Zubenko, and E.W. Jones. 1980. Proteolytic inactivation of the NADP-dependent glutamate dehydrogenase in proteinase-deficient mutants of *Saccharomyces cerevisiae*. *Arch. Biochem. Biophys.* **202:** 657.

Hemmings, B.A., G.S. Zubenko, A. Hasilik, and E.W. Jones. 1981. Mutant defective in processing of an enzyme located in the lysosome-like vacuole of *Saccharomyces cerevisiae*. *Proc. Natl. Acad. Sci.* **78:** 435.

Herman, P.K. and S.D. Emr. 1990. Characterization of *VPS34*, a gene required for vacuolar protein sorting and vacuole segregation in *Saccharomyces cerevisiae*. *Mol. Cell. Biol.* **10:** 6742.

Herman, P.K., J.H. Stack, and S.D. Emr. 1991a. A genetic and structural analysis of the yeast Vps15 protein kinase: Evidence for a direct role of Vps15p in vacuolar protein delivery. *EMBO J.* **10:** 4049.

Herman, P.K., J.H. Stack, J.A. DeModena, and S.D. Emr. 1991b. A novel protein kinase homolog essential for protein sorting to the yeast lysosome-like vacuole. *Cell* **64:** 425.

Hicke, L. and H. Riezman. 1996. Ubiquitination of a yeast plasma membrane receptor signals its ligand-stimulated endocytosis. *Cell* **84:** 277.

Hill, K.J. and T.H. Stevens. 1994. Vma21p is a yeast membrane protein retained in the endoplasmic reticulum by a di-lysine motif and is required for the assembly of the vacuolar H^+-ATPase complex. *Mol. Biol. Cell* **5:** 1039.

———. 1995. Vma22p is a novel endoplasmic reticulum-associated protein required for assembly of the yeast vacuolar H^+-ATPase complex. *J. Biol. Chem.* **270:** 22329.

Hill, K.L., N.L. Catlett, and L.S. Weisman. 1996. Actin and myosin function in directed vacuole movement during cell division in *Saccharomyces cerevisiae*. *J. Cell Biol.* **135:** 1535.

Hille-Rehfeld, A. 1995. Mannose 6-phosphate receptors in sorting and transport of lysosomal enzymes. *Biochim. Biophys. Acta* **1241:** 177.

Hiller, M.M., A. Finger, M. Schweiger, and D.H. Wolf. 1996. ER degradation of a misfolded luminal protein by the cytosolic ubiquitin-proteasome pathway. *Science* **273:** 1725.

Hilt, W., C. Enenkel, A. Gruhler, T. Singer, and D.H. Wolf. 1993. The *PRE4* gene codes for a subunit of the yeast proteasome necessary for peptidylglutamyl-peptide-hydrolyzing activity. Mutations link the proteasome to stress- and ubiquitin-dependent proteolysis. *J. Biol. Chem.* **268:** 3479.

Hinshaw, J.E. and S.L. Schmid. 1995. Dynamin self-assembles into rings suggesting a mechanism for coated vesicle budding [see comments]. *Nature* **374:** 190.

Hirata, R., Y. Ohsumk, A. Nakano, H. Kawasaki, K. Suzuki, and Y. Anraku. 1990. Molecular structure of a gene, *VMA1*, encoding the catalytic subunit of H^+-translocating adenosine triphosphatase from vacuolar membranes of *Saccharomyces cerevisiae*. *J. Biol. Chem.* **265:** 6726.

Hirata, R., N. Umemoto, M.N. Ho, Y. Ohya, T.H. Stevens, and Y. Anraku. 1993. *VMA12* is essential for assembly of the vacuolar H^+-ATPase subunits onto the vacuolar membrane in *Saccharomyces cerevisiae*. *J. Biol. Chem.* **268:** 961.

Hirsch, H.H., H.H. Schiffer, and D.H. Wolf. 1992. Biogenesis of the yeast vacuole (lysosome). Proteinase yscB contributes molecularly and kinetically to vacuolar hydrolase-precursor maturation. *Eur. J. Biochem.* **207:** 867.

Ho, M.N., K.J. Hill, M.A. Lindorfer, and T.H. Stevens. 1993a. Isolation of vacuolar membrane H^+-ATPase-deficient yeast mutants; the *VMA5* and *VMA4* genes are essential for assembly and activity of the vacuolar H^+-ATPase. *J. Biol. Chem.* **268:** 221.

Ho, M.N., R. Hirata, N. Umemoto, Y. Ohya, A. Takatsuki, T.H. Stevens, and Y. Anraku. 1993b. *VMA13* encodes a 54-kDa vacuolar H^+-ATPase subunit required for activity but not assembly of the enzyme complex in *Saccharomyces cerevisiae*. *J. Biol. Chem.* **268:** 18286.

Hochstrasser, M. 1995. Ubiquitin, proteasomes, and the regulation of intracellular protein degradation. *Curr. Opin. Cell Biol.* **7:** 215.

Hochstrasser, M. and A. Varshavsky. 1990. In vivo degradation of a transcriptional regulator: The yeast α2 repressor. *Cell* **61:** 697.

Hoffman, M. and H.-L. Chiang. 1996. Isolation of degradation-deficient mutants defective in the targeting of fructose-1,6-bisphosphatase into the vacuole for degradation in *Saccharomyces cerevisiae*. *Genetics* **143:** 1555.

Holzer, H. 1976. Catabolite inactivation in yeast. *Trends Biochemi. Sci.* **1:** 178.

Hopper, A., P. Magee, S. Welch, M. Friedman, and B. Hall. 1974. Macromolecule synthesis and breakdown in relation to sporulation and meiosis in yeast. *J. Bacteriol.* **119:** 619.

Horazdovsky, B.F. and S.D. Emr. 1993. The *VPS16* gene product associates with a sedimentable protein complex and is essential for vacuolar protein sorting in yeast. *J. Biol. Chem.* **268:** 4953.

Horazdovsky, B.F., G.R. Busch, and S.D. Emr. 1994. *VPS21* encodes a rab5-like GTP binding protein that is required for the sorting of yeast vacuolar proteins. *EMBO J.* **13:** 1297.

Horazdovsky, B.F., C. Cowles, P. Mustol, M. Holmes, and S.D. Emr. 1996. A novel RING finger protein, Vps8p, functionally interacts with the small GTPase, Vps21p, to facilitate soluble vacuolar protein localization. *J. Biol. Chem.* **271:** 33607.

Huang, P.-H. and H.-L. Chiang. 1997. Identification of novel vesicles in the cytosol to lysosome degradation pathway. *J. Cell Biol.* (in press).

Jacquemin-Faure, I., D. Thomas, J. Laporte, C. Cibert, and Y. Surdin-Kerjan. 1994. The vacuolar compartment is required for sulfur amino acid homeostasis in *Saccharomyces cerevisiae*. *Mol. Gen. Genet.* **244:** 519.

Jantti, J., S. Keranen, J. Toikkanen, E. Kuismanen, C. Ehnholm, H. Soderlund, and V.M. Olkkonen. 1994. Membrane insertion and intracellular transport of yeast syntaxin Sso2p in mammalian cells. *J. Cell Sci.* **107:** 3623.

Jenness, D. and P. Spatrick. 1986. Down regulation of the α-factor pheromone receptor in *Saccharomyces cerevisiae*. *Cell* **46:** 345.

Johnson, L.M., V.A. Bankaitis, and S.D. Emr. 1987. Distinct sequence determinants direct intracellular sorting and modification of a yeast vacuolar protease. *Cell* **48:** 875.

Jones, E.W. 1977. Proteinase mutants of *Saccharomyces cerevisiae*. *Genetics* **85:** 23.

———. 1991a. Tackling the protease problem in *Saccharomyces cerevisiae*. *Methods Enzymol.* **194:** 428.

———. 1991b. Three proteolytic systems in the yeast *Saccharomyces cerevisiae*. *J. Biol. Chem.* **266:** 7963.

Jones, E.W. and D.G. Murdock. 1994. Proteolysis in the yeast vacuole. In *Cellular proteolytic systems* (ed. A.J. Ciechanover and A.L. Schwartz), p. 115. Wiley-Liss, New York.

Jones, E.W., G.S. Zubenko, and R.R. Parker. 1982. *PEP4* gene function is required for expression of several vacuolar hydrolases in *Saccharomyces cerevisiae*. *Genetics* **102:** 665.

Jones, E.W., C.A. Woolford, C.M. Moehle, J.A. Noble, and M.I. Innis. 1989. Genes, zymogens, and activation cascades of yeast vacuolar proteases. In *Cellular proteases and control mechanisms: Proceedings of a Glaxo-UCLA Colloquium on Cellular Proteases and Control Mechansims* (ed. T.E. Hugli), p. 141. A.R. Liss, New York.

Jones, H.D., M. Schliwa, and D.G. Drubin. 1993. Video microscopy of organelle inheritance and motility in budding yeast. *Cell Motil. Cytoskeleton* **25:** 129.

Kabcenell, A.K., B. Goud, J.K. Northup, and P.J. Novick. 1990. Binding and hydrolysis of guanine nucleotides by Sec4p, a yeast protein involved in the regulation of vesicular traffic. *J. Biol. Chem.* **265:** 9366.

Kakinuma, Y., N. Masuda, and K. Igarashi. 1992. Proton potential-dependent polyamine transport by vacuolar membrane vesicles of *Saccharomyces cerevisiae*. *Biochim. Biophys. Acta* **1107:** 126.

Kakinuma, Y., Y. Ohsumi, and Y. Anraku. 1981. Properties of H^+-translocating adenosine triphosphatase in vacuolar membranes of *Saccharomyces cerevisiae*. *J. Biol. Chem.* **256:** 10859.

Kane, P.M. 1995. Disassembly and reassembly of the yeast vacuolar H^+-ATPase *in vivo*. *J. Biol. Chem.* **270:** 17025.

Kane, P.M., C.T. Yamashiro, and T.H. Stevens. 1989. Biochemical characterization of the yeast vacuolar H^+-ATPase. *J. Biol. Chem.* **264:** 19236.

Kane, P.M., M.C. Kuehn, I. Howald-Stevenson, and T.H. Stevens. 1992. Assembly and targeting of peripheral and integral membrane subunits of the yeast vacuolar H^+-ATPase. *J. Biol. Chem.* **267:** 447.

Kaneko, Y., Y. Tamai, A. Toh-e, and Y. Oshima. 1985. Transcriptional and post-translational control of *PHO8* expression by *PHO* regulatory genes in *Saccharomyces cerevisiae*. *Mol. Cell. Biol.* **5:** 248.

Kaneko, Y., N. Hayashi, A. Toh-e, I. Banno, and Y. Oshima. 1987. Structural characteristics of the *PHO8* gene encoding repressible alkaline phosphatase in *Saccharomyces cerevisiae*. *Gene* **58:** 137.

Kinashi, H., K. Someno, and K. Sakaguchi. 1984. Isolation and characterization of concanamycins A, B, and C. *J. Antibiot.* **37:** 1333.

Kitamoto, K., K. Yoshizawa, Y. Ohsumi, and Y. Anraku. 1988a. Dynamic aspects of vacuolar and cytosolic amino acid pools of *Saccharomyces cerevisiae*. *J. Bacteriol.* **170:** 2683.

———. 1988b. Mutants of *Saccharomyces cerevisiae* with defective vacuolar function. *J. Bacteriol.* **170:** 2687.

Klar, A. and H. Halvorson. 1975. Proteinase activities of *Saccharomyces cerevisiae*. *J. Bacteriol.* **124:** 863.

Klionsky, D.J. and S.D. Emr. 1989. Membrane protein sorting: Biosynthesis, transport and processing of yeast vacuolar alkaline phosphatase. *EMBO J.* **8:** 2241.

———. 1990. A new class of lysosomal/vacuolar protein sorting signals. *J. Biol. Chem.* **265:** 5349.

Klionsky, D.J., L.M. Banta, and S.D. Emr. 1988. Intracellular sorting and processing of a yeast vacuolar hydrolase: Proteinase A propeptide contains vacuolar targeting information. *Mol. Cell. Biol.* **8:** 2105.

Klionsky, D.J., R. Cueva, and D.S. Yaver. 1992a. Aminopeptidase I of *Saccharomyces cerevisiae* is localized to the vacuole independent of the secretory pathway. *J. Cell Biol.* **119:** 287.

Klionsky, D.J., H. Nelson, N. Nelson, and D.S. Yaver. 1992b. Mutations in the yeast vacuolar ATPase result in the mislocalization of vacuolar proteins. *J. Exp. Biol.* **172:** 83.

Köhrer, K. and S.D. Emr. 1993. The yeast *VPS17* gene encodes a membrane-associated protein required for the sorting of soluble vacuolar hydrolases. *J. Biol. Chem.* **268:** 559.

Kolling, R. and C.P. Hollenberg. 1994. The ABC-transporter Ste6 accumulates in the plasma membrane in a ubiquitinated form in endocytosis mutants. *EMBO J.* **13:** 3261.

Kominami, E., H. Hoffschulte, and H. Holzer. 1981. Purification and properties of proteinase B from yeast. *Biochim. Biophys. Acta* **661:** 124.

Kornfeld, S. 1987. Trafficking of lysosomal enzymes. *FASEB J.* **1:** 462.

Kornfeld, S. and I. Mellman. 1989. The biogenesis of lysosomes. *Annu. Rev. Cell Biol.* **5:** 483.

Kornitzer, D., B. Raboy, R.G. Kulka, and G.R. Fink. 1994. Regulated degradation of the transcription factor Gcn4. *EMBO J.* **13:** 6021.

Kreis, T.E. and R. Pepperkok. 1994. Coat proteins in intracellular membrane transport. *Curr. Opin. Cell Biol.* **6:** 533.

Kreis, T.E., M. Lowe, and R. Pepperkok. 1995. COPs regulating membrane traffic. *Annu. Rev. Cell Dev. Biol.* **11:** 677.

Kuranda, M. and P. Robbins. 1987. Cloning and heterologous expression of glycosidase genes from *Saccharomyces cerevisiae*. *Proc. Natl. Acad. Sci.* **84:** 2585.

Lai, K., C.P. Bolognese, S. Swift, and P. McGraw. 1995. Regulation of inositol transport in *Saccharomyces cerevisiae* involves inositol-induced changes in permease stability and endocytic degradation in the vacuole. *J. Biol. Chem.* **270:** 2525.

Latterich, M. and M.D. Watson. 1991. Isolation and characterization of osmosensitive vacuolar mutants of *Saccharomyces cerevisiae*. *Mol. Microbiol.* **5:** 2417.

Li, Z.S., M. Szczypka, Y.P. Lu, D.J. Thiele, and P.A. Rea. 1996. The yeast cadmium factor protein (YCF1) is a vacuolar glutathione S-conjugate pump. *J. Biol. Chem.* **271:** 6509.

Lian, J.P. and S. Ferro-Novick. 1993. Bos1p, an integral membrane protein of the endoplasmic reticulum to Golgi transport vesicles, is required for their fusion competence. *Cell* **73:** 735.

Lopez, S. and J. Gancedo. 1979. Effect of metabolic condition on protein turnover in yeast. *Biochem. J.* **178:** 769.

Madura, K. and A. Varshavsky. 1994. Degradation of Gα by the N-end rule pathway. *Science* **265:** 1454.

Magasanik, B. 1992. Regulation of nitrogen utilization. In *The molecular and cellular biology of the yeast* Saccharomyces: *Gene expression* (ed. E.W. Jones et al.), p. 283. Cold Spring Harbor Laboratory Press, Cold Spring Harbor, New York.

Mallabiabarrena, A. and V. Malhotra. 1995. Vesicle biogenesis: The coat connection (comment). *Cell* **83:** 667.

Manolson, M.F., D. Proteau, and E.W. Jones. 1992a. Evidence for a conserved 95–120 kDa subunit associated with and essential for activity of V-ATPases. *J. Exp. Biol.* **172:** 105.

Manolson, M.F., B. Wu, D. Proteau, B.E. Taillon, B.T. Roberts, M.A. Hoyt, and E.W. Jones. 1994. *STV1* gene encodes functional homologue of 95-kDa yeast vacuolar H^+-ATPase subunit Vph1p. *J. Biol. Chem.* **269:** 14064.

Manolson, M.F., D. Proteau, R.A. Preston, A. Stenbit, B.T. Roberts, M.A. Hoyt, D. Preuss, J. Mulholland, D. Botstein, and E.W. Jones. 1992b. The *VPH1* gene encodes a 95-kDa integral membrane polypeptide required for in vivo assembly and activity of the yeast vacuolar H^+-ATPase. *J. Biol. Chem.* **267:** 14294.

Marcusson, E.G., B.F. Horazdovsky, J.L. Cereghino, E. Gharakhanian, and S.D. Emr. 1994. The sorting receptor for yeast vacuolar carboxypeptidase Y is encoded by the *VPS10* gene. *Cell* **77:** 579.

Mayer, A., W. Wickner, and A. Haas. 1996. Sec18p (NSF) driven release of Sec17p (α-SNAP) can precede docking and fusion of yeast vacuoles. *Cell* **85:** 83.

Mazon, M. 1978. Effect of glucose starvation on the nicotinamide adenine dinucleotide phosphate-dependent glutamate dehydrogenase of yeast. *J. Bacteriol.* **133:** 780.

Mazon, M. and B. Hemmings. 1979. Regulation of *Saccharomyces cerevisiae* nicotinamide adenine dinucleotide phosphate-dependent glutamate dehydrogenase by proteolysis during carbon starvation. *J. Bacteriol.* **139:** 686.

Mechler, B. and D.H. Wolf. 1981. Analysis of proteinase A function in yeast. *Eur. J. Biochem.* **121:** 47.

Mechler, B., H. Muller, and D.H. Wolf. 1987. Maturation of vacuolar (lysosomal) enzymes in yeast: Proteinase yscA and proteinase yscB are catalysts of the processing and activation event of carboxypeptidase yscY. *EMBO J.* **6:** 2157.

Mechler, B., H.H. Hirsch, H. Muller, and D.H. Wolf. 1988. Biogenesis of the yeast lysosome (vacuole): Biosynthesis and maturation of proteinase yscB. *EMBO J.* **7:** 1705.

Mechler, B., M. Muller, H. Muller, F. Meussdoerffer, and D.H. Wolf. 1982. In vivo biosynthesis of the vacuolar proteinases A and B in the yeast *Saccharomyces cerevisiae. J. Biol. Chem.* **257:** 11203.

Medintz, I., H. Jiang, E.-K. Han, W. Cui, and C.A. Michels. 1996. Characterization of the glucose-induced inactivation of maltose permease in *Saccharomyces cerevisiae. J. Bacteriol.* **178:** 2245.

Mellman, I., R. Fuchs, and A. Helenius. 1986. Acidification of the endocytic and exocytic pathways. *Annu. Rev. Biochem.* **55:** 663.

Messenguy, F., D. Colin, and J.P. ten Have. 1980. Regulation of compartmentation of amino acid pools in *Saccharomyces cerevisiae* and its effects on metabolic control. *Eur. J. Biochem.* **108:** 439.

Metz, G. and K.-H. Rohm. 1976. Yeast aminopeptidase. I. Chemical composition and catalytic properties. *Biochim. Biophys. Acta* **429:** 933.

Meussdoerffer, F., P. Tortora, and H. Holzer. 1980. Purification and properties of proteinase A from yeast. *J. Biol. Chem.* **255:** 12087.

Miller, J.J. 1984. In vitro experiments concerning the state of polyphosphate in the yeast vacuole. *Can. J. Microbiol* **30:** 236.

Misteli, T. and G. Warren. 1995a. Mitotic disassembly of the Golgi apparatus in vivo. *J. Cell Sci.* **108:** 2715.

———. 1995b. A role for tubular networks and a COP I-independent pathway in the mitotic fragmentation of Golgi stacks in a cell-free system. *J. Cell Biol.* **130:** 1027.

Mittenbuhler, K. and H. Holzer. 1988. Purification and characterization of acid trehalase from the yeast *suc2* mutant (erratum in *J. Biol. Chem.* [1988] **263:** 16512). *J. Biol. Chem.* **263:** 8537.

———. 1991. Characterization of different forms of yeast acid trehalase in the secretory pathway. *Arch. Microbiol.* **155:** 217.

Moehle, C.M. and E.W. Jones. 1990. Consequences of growth media, gene copy number, and regulatory mutations on the expression of the *PRB1* gene of *Saccharomyces cerevisiae. Genetics* **124:** 39.

Moehle, C.M., C.K. Dixon, and E.W. Jones. 1989. Processing pathway for protease B of *Saccharomyces cerevisiae. J. Cell Biol.* **108:** 309.

Moehle, C.M., R. Tizard, S.K. Lemmon, J. Smart, and E.W. Jones. 1987. Protease B of the lysosomelike vacuole of the yeast *Saccharomyces cerevisiae* is homologous to the subtilisin family of serine proteases. *Mol. Cell. Biol.* **7:** 4390.

Morano, K.A. and D.J. Klionsky. 1994. Differential effects of compartment deacidification on the targeting of membrane and soluble proteins to the vacuole in yeast. *J. Cell Sci.* **107:** 2813.

Moriyama, Y. and N. Nelson. 1989. H^+-translocating ATPase in Golgi apparatus. Characterization as vacuolar H^+-ATPase and its subunit structures. *J. Biol. Chem.* **264:** 18445.

Munn, A.L. and H. Riezman. 1994. Endocytosis is required for the growth of vacuolar H^+-ATPase-defective yeast: Identification of six new *END* genes. *J. Cell Biol.* **127:** 373.

Munn, A.L., B.J. Stevenson, M.I. Geli, and H. Riezman. 1995. *end5, end6,* and *end7:* Mutations that cause actin delocalization and block the internalization step of endocytosis in *Saccharomyces cerevisiae. Mol. Biol. Cell* **6:** 1721.

Nakamura, K.D. and F. Schlenk. 1974. Active transport of exogenous S-adenosylmethionine and related compounds into cells and vacuoles of *Saccharomyces cerevisiae. J. Bacteriol.* **120:** 482.

Nebes, V.L. and E.W. Jones. 1991. Activation of the proteinase B precursor of the yeast *Saccharomyces cerevisiae* by autocatalysis and by an internal sequence. *J. Biol. Chem.* **266:** 22851.

———. 1992. N-linked glycosylation of proteinase B precursors of the yeast *Saccharomyces cerevisiae* is not required for proper targeting or processing of the enzyme. *Yeast* **8:** 353.

Neeff, J., E. Hagele, J. Nauhaus, U. Heer, and D. Mecke. 1978. Evidence for catabolite degradation in the glucose-dependent inactivation of cytoplasmic malate dehydrogenase. *Eur. J. Biochem.* **87:** 489.

Nelson, H. and N. Nelson. 1989. The progenitor of ATP synthases was closely related to the current vacuolar H^+-ATPase. *FEBS Lett.* **247:** 147.

———. 1990. Disruption of genes encoding subunits of yeast vacuolar H^+-ATPase causes conditional lethality. *Proc. Natl. Acad. Sci.* **87:** 3503.

Nelson, H., S. Mandiyan, and N. Nelson. 1989. A conserved gene encoding the 57-kDa subunit of the yeast vacuolar H^+-ATPase (erratum in *J. Biol. Chem.* [1989] **264:** 5313). *J. Biol. Chem.* **264:** 1775.

———. 1994. The *Saccharomyces cerevisiae VMA7* gene encodes a 14-kDa subunit of the vacuolar H^+-ATPase catalytic sector. *J. Biol. Chem.* **269:** 24150.

———. 1995. A bovine cDNA and a yeast gene (*VMA8*) encoding the subunit D of the vacuolar H^+-ATPase. *Proc. Natl. Acad. Sci.* **92:** 497.

Newman, A.P., J. Shim, and S. Ferro-Novick. 1990. *BET1*, *BOS1*, and *SEC22* are members of a group of interacting yeast genes required for transport from the ER to the Golgi complex. *Mol. Cell. Biol.* **10:** 3405.

Nicolay, K., W.A. Scheffers, P.M. Bruinenberg, and R. Kaptein. 1983. In vivo 31P NMR studies on the role of the vacuole in phosphate metabolism in yeasts. *Arch. Microbiol.* **134:** 270.

Nicolson, T.A., B. Conradt, and W. Wickner. 1996. A truncated form of the Pho80 cyclin of *Saccharomyces cerevisiae* induces expression of a small cytosolic factor which inhibits vacuole inheritance. *J. Bacteriol.* **178:** 4047.

Nicolson, T.A., L.S. Weisman, G.S. Payne, and W.T. Wickner. 1995. A truncated form of the Pho80 cyclin redirects the Pho85 kinase to disrupt vacuole inheritance in *S. cerevisiae*. *J. Cell Biol.* **130:** 835.

Nieuwenhuis, B.J., C.A. Weijers, and G.W. Borst-Pauwels. 1981. Uptake and accumulation of Mn^{2+} and Sr^{2+} in *Saccharomyces cerevisiae*. *Biochim. Biophys. Acta* **649:** 83.

Nishizawa, M., T. Yasuhara, T. Nakai, Y. Fujiki, and A. Ohashi. 1994. Molecular cloning of the aminopeptidase Y gene of *Saccharomyces cerevisiae*. Sequence analysis and gene disruption of a new aminopeptidase. *J. Biol. Chem.* **269:** 13651.

Nishizawa, S., N. Umemoto, Y. Ohsumi, A. Nakano, and Y. Anraku. 1990. Biogenesis of vacuolar membrane glycoproteins of yeast *Saccharomyces cerevisiae*. *J. Biol. Chem.* **265:** 7440.

Nothwehr, S.F. and T.H. Stevens. 1994. Sorting of membrane proteins in the yeast secretory pathway. *J. Biol. Chem.* **269:** 10185.

Nothwehr, S.F., E. Conibear, and T.H. Stevens. 1995. Golgi and vacuolar membrane proteins reach the vacuole in *vps1* mutant yeast cells via the plasma membrane. *J. Cell Biol.* **129:** 35.

Nothwehr, S.F., C.J. Roberts, and T.H. Stevens. 1993. Membrane protein retention in the yeast Golgi apparatus: Dipeptidyl aminopeptidase A is retained by a cytoplasmic signal containing aromatic residues. *J. Cell Biol.* **121:** 1197.

Obar, R.A., C.A. Collins, J.A. Hammarback, H.S. Shpetner, and R.B. Vallee. 1990. Molecular cloning of the microtubule-associated mechanochemical enzyme dynamin reveals homology with a new family of GTP-binding proteins (comments). *Nature* **347:** 256.

Oda, M., S. Scott, A. Hefner-Gravink, A. Caffarelli, and D. Klionsky. 1996. Identification of a cytoplasm to vacuole targeting determinant in aminopeptidase I. *J. Cell Biol.* **132:** 999.

Ohsumi, Y. and Y. Anraku. 1981. Active transport of basic amino acids driven by a proton motive force in vacuolar membrane vesicles of *Saccharomyces cerevisiae. J. Biol. Chem.* **256:** 2079.

———. 1983. Calcium transport driven by a proton motive force in vacuolar membrane vesicles of *Saccharomyces cerevisiae. J. Biol. Chem.* **258:** 5614.

Ohsumi, Y., K. Kitamoto, and Y. Anraku. 1988. Changes induced in the permeability barrier of the yeast plasma membrane by cupric ion. *J. Bacteriol.* **170:** 2676.

Ohya, Y., Y. Ohsumi, and Y. Anraku. 1986. Isolation and characterization of Ca^{2+}-sensitive mutants of *Saccharomyces cerevisiae. J. Gen. Microbiol.* **132:** 979.

Ohya, Y., N. Umemoto, I. Tanida, A. Ohta, H. Iida, and Y. Anraku. 1991. Calcium-sensitive *cls* mutants of *Saccharomyces cerevisiae* showing a Pet^- phenotype are ascribable to defects of vacuolar membrane H^+-ATPase activity. *J. Biol. Chem.* **266:** 13971.

Okorokov, L.A., S.B. Letrikevich, L.P. Lichko, and E.V. Mel'nikova. 1978. Vacuolar pool of magnesium in cells of the yeast *Saccharomyces cerevisiae. Biol. Bull. Acad. Sci. USSR* **5:** 638.

Okorokov, L.A., L.P. Lichko, V.M. Kadomtseva, V.P. Kholodenko, V.T. Titovsky, and I.S. Kulaev. 1977. Energy-dependent transport of manganese into yeast cells and distribution of accumulated ions. *Eur. J. Biochem.* **75:** 373.

Onishi, H.R., J.S. Tkacz, and J.O. Lampen. 1979. Glycoprotein nature of yeast alkaline phosphatase. Formation of active enzyme in the presence of tunicamycin. *J. Biol. Chem.* **254:** 11943.

Opheim, D. 1978. α-Mannosidase of *Saccharomyces cerevisiae*. Characterization and modulation of activity. *Biochim. Biophys. Acta* **524:** 121.

Ossig, R., C. Dascher, H.-H. Trepte, H.D. Schmitt, and D. Gallwitz. 1991. The yeast *SLY* gene products, suppressors of defects in the essential GTP-binding Ypt1 protein, may act in endoplasmic reticulum-to-Golgi transport. *Mol. Cell. Biol.* **11:** 2980.

Paravicini, G., B.F. Horazdovsky, and S.D. Emr. 1992. Alternative pathways for the sorting of soluble vacuolar proteins in yeast: A *vps35* null mutant missorts and secretes only a subset of vacuolar hydrolases. *Mol. Biol. Cell* **3:** 415.

Payne, G.S. and R. Schekman. 1989. Clathrin: A role in the intracellular retention of a Golgi membrane protein. *Science* **245:** 1358.

Payne, G.S., D. Baker, E. van Tuinen, and R. Schekman. 1988. Protein transport to the vacuole and receptor-mediated endocytosis by clathrin heavy chain-deficient yeast. *J. Cell Biol.* **106:** 1453.

Perkins, J. and G.M. Gadd. 1993. Accumulation and intracellular compartmentation of lithium ions in *Saccharomyces cerevisiae* (erratum in *FEMS Microbiol. Lett.* [1993] **110:** 349). *FEMS Microbiol. Lett.* **107:** 255.

Perlman, P. and H. Mahler. 1974. Derepression of mitochondria and their enzymes in yeast: Regulatory aspects. *Arch. Biochem. Biophys.* **162:** 248.

Phan, H.L., J.A. Finlay, D.S. Chu, P.K. Tan, T. Kirchhausen, and G.S. Payne. 1994. The *Saccharomyces cerevisiae APS1* gene encodes a homolog of the small subunit of the mammalian clathrin AP-1 complex: Evidence for functional interaction with clathrin at the Golgi complex. *EMBO J.* **13:** 1706.

Pick, U., M. Bental, E. Chitlaru, and M. Weiss. 1990. Polyphosphate-hydrolysis—A protective mechanism against alkaline stress? *FEBS Lett.* **274:** 15.

Piper, R.C., E.A. Whitters, and T.H. Stevens. 1994. Yeast Vps45p is a Sec1p-like protein required for the consumption of vacuole-targeted, post-Golgi transport vesicles. *Eur. J.*

Cell Biol. **65:** 305.

Piper, R.C., A.A. Cooper, H. Yang, and T.H. Stevens. 1995. *VPS27* controls vacuolar and endocytic traffic through a prevacuolar compartment in *Saccharomyces cerevisiae. J. Cell Biol.* **131:** 603.

Polakis, E.S. and W. Bartley. 1965. Changes in the enzyme activities of *Saccharomyces cerevisiae* during aerobic growth on different carbon sources. *Biochem. J.* **97:** 284.

Polakis, E.S., W. Bartley, and G. Meek. 1965. Changes in the activities of respiratory enzymes during the aerobic growth of yeast on different carbon sources. *Biochem. J.* **97:** 298.

Pozos, T.C., I. Sekler, and M.S. Cyert. 1996. The product of *HUM1*, a novel yeast gene, is required for vacuolar Ca^{2+}/H^+ exchange and is related to mammalian Na^+/Ca^{2+} exchangers. *Mol. Cell. Biol.* **16:** 3730.

Preston, R.A., R.F. Murphy, and E.W. Jones. 1989. Assay of vacuolar pH in yeast and identification of acidification-defective mutants. *Proc. Natl. Acad. Sci.* **86:** 7027.

Preston, R.A., P.S. Reinagel, and E.W. Jones. 1992. Genes required for vacuolar acidity in *Saccharomyces cerevisiae. Genetics* **131:** 551.

Preston, R.A., M.F. Manolson, K. Becherer, E. Weidenhammer, D. Kirkpatrick, R. Wright, and E.W. Jones. 1991. Isolation and characterization of *PEP3*, a gene required for vacuolar biogenesis in *Saccharomyces cerevisiae. Mol. Cell. Biol.* **11:** 5801.

Pringle, J.R., R.A. Preston, A.E.M. Adams, T. Stearns, D.G. Drubin, B.K. Haarer, and E.W. Jones. 1989. Fluorescence microscopy methods for yeast. *Methods Cell Biol.* **31:** 357.

Protopopov, V., B. Govindan, P. Novick, and J.E. Gerst. 1993. Homologs of the synaptobrevin/VAMP family of synaptic vesicle proteins function on the late secretory pathway in *S. cerevisiae. Cell* **74:** 355.

Rad, M.R., H.L. Phan, L. Kirchrath, P.K. Tan, T. Kirchhausen, C.P. Hollenberg, and G.S. Payne. 1995. *Saccharomyces cerevisiae* Apl2p, a homologue of the mammalian clathrin AP β subunit, plays a role in clathrin-dependent Golgi functions. *J. Cell Sci.* **108:** 1605.

Raguzzi, F., E. Lesuisse, and R.R. Crichton. 1988. Iron storage in *Saccharomyces cerevisiae. FEBS Lett.* **231:** 253.

Raths, S., J. Rohrer, F. Crausaz, and H. Riezman. 1993. *end3* and *end4:* Two mutants defective in receptor-mediated and fluid-phase endocytosis in *Saccharomyces cerevisiae. J. Cell Biol.* **120:** 55.

Raymond, C.K., I. Howald-Stevenson, C.A. Vater, and T.H. Stevens. 1992a. Morphological classification of the yeast vacuolar protein sorting mutants: Evidence for a prevacuolar compartment in class E *vps* mutants. *Mol. Biol. Cell* **3:** 1389.

Raymond, C.K., P.J. O'Hara, G. Eichinger, J.H. Rothman, and T.H. Stevens. 1990. Molecular analysis of the yeast *VPS3* gene and the role of its product in vacuolar protein sorting and vacuolar segregation during the cell cycle. *J. Cell Biol.* **111:** 877.

Raymond, C.K., C.J. Roberts, K.E. Moore, I. Howald, and T.H. Stevens. 1992b. Biogenesis of the vacuole in *Saccharomyces cerevisiae. Int. Rev. Cytol.* **139:** 59.

Riballo, E., M. Herweijer, D. Wolf, and R. Lagunas. 1995. Catabolite inactivation of the yeast maltose transporter occurs in the vacuole after internalization by endocytosis. *J. Bacteriol.* **177:** 5622.

Rieder, S.E., L.M. Banta, K. Kohrer, J.M. McCaffery, and S.D. Emr. 1996. Multilamellar endosome-like compartment accumulates in the yeast *vps28* vacuolar protein sorting mutant. *Mol. Biol. Cell* **7:** 985.

Roberts, C.J., S.F. Nothwehr, and T.H. Stevens. 1992. Membrane protein sorting in the yeast secretory pathway: Evidence that the vacuole may be the default compartment. *J. Cell Biol.* **119:** 69.

Roberts, C.J., G. Pohlig, J.H. Rothman, and T.H. Stevens. 1989. Structure, biosynthesis, and localization of dipeptidyl aminopeptidase B, an integral membrane glycoprotein of the yeast vacuole. *J. Cell Biol.* **108:** 1363.

Robinson, J.S., T.R. Graham, and S.D. Emr. 1991. A putative zinc finger protein, *Saccharomyces cerevisiae* Vps18p, affects late Golgi functions required for vacuolar protein sorting and efficient α-factor prohormone maturation. *Mol. Cell. Biol.* **11:** 5813.

Robinson, J.S., D.J. Klionsky, L.M. Banta, and S.D. Emr. 1988. Protein sorting in *Saccharomyces cerevisiae:* Isolation of mutants defective in the delivery and processing of multiple vacuolar hydrolases. *Mol. Cell. Biol.* **8:** 4936.

Robinson, K.S., K. Lai, T.A. Cannon, and P. McGraw. 1996. Inositol transport in *Saccharomyces cerevisiae* is regulated by transcriptional and degradative endocytic mechanisms during the growth cycle that are distinct from inositol-induced regulation. *Mol. Biol. Cell* **7:** 81.

Roeder, A. and J. Shaw. 1996. Vacuole partitioning during meiotic division in yeast. *Genetics* **144:** 445.

Rohrer, J., H. Benedetti, B. Zanolari, and H. Riezman. 1993. Identification of a novel sequence mediating regulated endocytosis of the G protein-coupled α-pheromone receptor in yeast. *Mol. Biol. Cell* **4:** 511.

Rothman, J.E. 1994. Mechanisms of intracellular protein transport. *Nature* **372:** 55.

Rothman, J.E. and G. Warren. 1994. Implications of the SNARE hypothesis for intracellular membrane topology and dynamics. *Curr. Biol.* **4:** 220.

Rothman, J.H. and T.H. Stevens. 1986. Protein sorting in yeast: Mutants defective in vacuole biogenesis mislocalize vacuolar proteins into the late secretory pathway. *Cell* **47:** 1041.

Rothman, J.H., I. Howald, and T.H. Stevens. 1989a. Characterization of genes required for protein sorting and vacuolar function in the yeast *Saccharomyces cerevisiae. EMBO J.* **8:** 2057.

Rothman, J.H., C.P. Hunter, L.A. Valls, and T.H. Stevens. 1986. Overproduction-induced mislocalization of a yeast vacuolar protein allows isolation of its structural gene. *Proc. Natl. Acad. Sci.* **83:** 3248.

Rothman, J.H., C.K. Raymond, T. Gilbert, P.J. O'Hara, and T.H. Stevens. 1990. A putative GTP binding protein homologous to interferon-inducible Mx proteins performs an essential function in yeast protein sorting. *Cell* **61:** 1063.

Rothman, J.H., C.T. Yamashiro, C.K. Raymond, P.M. Kane, and T.H. Stevens. 1989b. Acidification of the lysosome-like vacuole and the vacuolar H^+-ATPase are deficient in two yeast mutants that fail to sort vacuolar proteins. *J. Cell Biol.* **109:** 93.

Rupp, S., H.H. Hirsch, and D.H. Wolf. 1991. Biogenesis of the yeast vacuole (lysosome). Active site mutation in the vacuolar aspartate proteinase yscA blocks maturation of vacuolar proteinases. *FEBS Lett.* **293:** 62.

Salama, N.R. and R.W. Schekman. 1995. The role of coat proteins in the biosynthesis of secretory proteins. *Curr. Opin. Cell Biol.* **7:** 536.

Sato, T., Y. Ohsumi, and Y. Anraku. 1984a. An arginine/histidine exchange transport system in vacuolar-membrane vesicles of *Saccharomyces cerevisiae. J. Biol. Chem.* **259:** 11509.

———. 1984b. Substrate specificities of active transport systems for amino acids in vacuolar-membrane vesicles of *Saccharomyces cerevisiae*. Evidence of seven independent proton/amino acid antiport systems. *J. Biol. Chem.* **259:** 11505.

Schafer, W., H. Kalisz, and H. Holzer. 1987. Evidence for non-vacuolar proteolytic catabolite inactivation of yeast fructose-1,6-bisphosphatase. *Biochim. Biophys. Acta* **925:** 150.

Schandel, K.A. and D.D. Jenness. 1994. Direct evidence for ligand-induced internalization of the yeast α-factor pheromone receptor. *Mol. Cell. Biol.* **14:** 7245.

Schekman, R. and L. Orci. 1996. Coat proteins and vesicle budding. *Science* **271:** 1526.

Schimmoller, F. and H. Riezman. 1993. Involvement of Ypt7p, a small GTPase, in traffic from late endosome to the vacuole in yeast. *J. Cell Sci.* **106:** 823.

Schork, S., M. Thumm, and D.H. Wolf. 1995. Catabolite inactivation of fructose-1,6-bisphosphatase of *Saccharomyces cerevisiae*. Degradation occurs via the ubiquitin pathway. *J. Biol. Chem.* **270:** 26446.

Schork, S.M., G. Bee, M. Thumm, and D.H. Wolf. 1994. Catabolite inactivation of fructose-1,6-bisphosphatase in yeast is mediated by the proteasome. *FEBS Lett.* **349:** 270.

Schu, P.V., K. Takegawa, M.J. Fry, J.H. Stack, M.D. Waterfield, and S.D. Emr. 1993. Phosphatidylinositol 3-kinase encoded by yeast *VPS34* gene essential for protein sorting. *Science* **260:** 88.

Schwaiger, H., A. Hasilik, K. von Figura, A. Wiemken, and W. Tanner. 1982. Carbohydrate-free carboxypeptidase Y is transferred into the lysosome-like yeast vacuole. *Biochem. Biophys. Res. Commun.* **104:** 950.

Schwencke, J. and H. De Robichon-Szulmajster. 1976. The transport of *S*-adenosyl-L-methionine in isolated yeast vacuoles and spheroplasts. *Eur. J. Biochem.* **65:** 49.

Schwob, E., T. Bohm, M.D. Mendenhall, and K. Nasmyth. 1994. The B-type cyclin kinase inhibitor $p40^{SIC1}$ controls the G_1 to S transition in *S. cerevisiae*. *Cell* **79:** 233.

Scott, S., A. Hefner-Gravink, K. Morano, T. Noda, Y. Ohsumi, and D. Klionsky. 1996. Cytoplasm-to-vacuole targeting and autophagy employ the same machinery to deliver proteins to the yeast vacuole. *Proc. Natl. Acad. Sci.* **93:** 12304.

Seeger, M. and G.S. Payne. 1992a. A role for clathrin in the sorting of vacuolar proteins in the Golgi complex of yeast. *EMBO J.* **11:** 2811.

———. 1992b. Selective and immediate effects of clathrin heavy chain mutations on Golgi membrane protein retention in *Saccharomyces cerevisiae*. *J. Cell Biol.* **118:** 531.

Segev, N., J. Mulholland, and D. Botstein. 1988. The yeast GTP-binding YPT1 protein and a mammalian counterpart are associated with the secretion machinery. *Cell* **52:** 915.

Segui-Real, B., M. Martinez, and I. Sandoval. 1995. Yeast aminopeptidase I is post-translationally sorted from the cytosol to the vacuole by a mechanism mediated by its tripartite N-terminal extension. *EMBO J.* **14:** 5476.

Serrano, R. 1991. Transport across yeast vacuolar and plasma membrane. In *The molecular and cellular biology of the yeast* Saccharomyces: *Genome dynamics, protein synthesis, and energetics* (ed. J.R. Broach et al.), p. 523. Cold Spring Harbor Laboratory Press, Cold Spring Harbor, New York.

Seufert, W., B. Futcher, and S. Jentsch. 1995. Role of a ubiquitin-conjugating enzyme in degradation of S- and M-phase cyclins. *Nature* **373:** 78.

Shaw, J.M. and W.T. Wickner. 1991. *vac2:* A yeast mutant which distinguishes vacuole segregation from Golgi-to-vacuole protein targeting. *EMBO J.* **10:** 1741.

Shih, C.K., J. Kwong, E. Montalvo, and N. Neff. 1990. Expression of a proteolipid gene from a high-copy-number plasmid confers trifluoperazine resistance to *Saccharomyces cerevisiae*. *Mol. Cell. Biol.* **10:** 3397.

Shih, C.K., R. Wagner, S. Feinstein, C. Kanik-Ennulat, and N. Neff. 1988. A dominant trifluoroperazine resistance gene from *Saccharomyces cerevisiae* has homology with F0F1 ATP synthase and confers calcium-sensitive growth. *Mol. Cell. Biol.* **8:** 3094.

Shim, J., A.P. Newman, and S. Ferro-Novick. 1991. The *BOS1* gene encodes an essential 27-kD putative membrane protein that is required for vesicular transport from the ER to the Golgi complex in yeast. *J. Cell Biol.* **113:** 55.

Shpetner, H.S. and R.B. Vallee. 1992. Dynamin is a GTPase stimulated to high levels of activity by microtubules. *Nature* **355:** 733.

Simeon, A., I.J. van der Klei, M. Veenhuis, and D.H. Wolf. 1992. Ubiquitin, a central component of selective cytoplasmic proteolysis, is linked to proteins residing at the locus of non-selective proteolysis, the vacuole. *FEBS Lett.* **301:** 231.

Simons, J.F., S. Ferro-Novick, M.D. Rose, and A. Helenius. 1995. BiP/Kar2p serves as a molecular chaperone during carboxypeptidase Y folding in yeast. *J. Cell Biol.* **130:** 41.

Singer, B. and H. Riezman. 1990. Detection of an intermediate compartment involved in transport of α-factor from the plasma membrane to the vacuole in yeast. *J. Cell Biol.* **110:** 1911.

Singer-Kruger, B., R. Frank, F. Crausaz, and H. Riezman. 1993. Partial purification and characterization of early and late endosomes from yeast. Identification of four novel proteins. *J. Biol. Chem.* **268:** 14376.

Singer-Kruger, B., H. Stenmark, A. Dusterhoft, P. Philippsen, J.S. Yoo, D. Gallwitz, and M. Zerial. 1994. Role of three rab5-like GTPases, Ypt51p, Ypt52p, and Ypt53p, in the endocytic and vacuolar protein sorting pathways of yeast. *J. Cell Biol.* **125:** 283.

Sorensen, S.O., H.B. van den Hazel, M.C. Kielland-Brandt, and J.R. Winther. 1994. pH-dependent processing of yeast procarboxypeptidase Y by proteinase A in vivo and in vitro. *Eur. J. Biochem.* **220:** 19.

Spormann, D.O., J. Heim, and D.H. Wolf. 1991. Carboxypeptidase yscS: Gene structure and function of the vacuolar enzyme. *Eur. J. Biochem.* **197:** 399.

———. 1992. Biogenesis of the yeast vacuole (lysosome). The precursor forms of the soluble hydrolase carboxypeptidase yscS are associated with the vacuolar membrane. *J. Biol. Chem.* **267:** 8021.

Stack, J.H. and S.D. Emr. 1993. Genetic and biochemical studies of protein sorting to the yeast vacuole. *Curr. Opin. Cell Biol.* **5:** 641.

———. 1994. Vps34p required for yeast vacuolar protein sorting is a multiple specificity kinase that exhibits both protein kinase and phosphatidylinositol-specific PI 3-kinase activities. *J. Biol. Chem.* **269:** 31552.

Stack, J.H., B. Horazdovsky, and S.D. Emr. 1995a. Receptor-mediated protein sorting to the vacuole in yeast: Roles for a protein kinase, a lipid kinase and GTP-binding proteins. *Ann. Rev. Cell Dev. Biol.* **11:** 1.

Stack, J., D. DeWald, K. Takegawa, and S. Emr. 1995b. Vesicle-mediated protein transport: Regulatory interactions between the Vps15 protein kinase and the Vps34 PtdIns 3-kinase essential for protein sorting to the vacuole in yeast. *J. Cell Biol.* **129:** 321.

Stack, J.H., P.K. Herman, P.V. Schu, and S.D. Emr. 1993. A membrane-associated complex containing the Vps15 protein kinase and the Vps34 PI 3-kinase is essential for protein sorting to the yeast lysosome-like vacuole. *EMBO J.* **12:** 2195.

Stanbrough, M. and B. Magasanik. 1995. Transcriptional and posttranslational regulation of the general amino acid permease of *Saccharomyces cerevisiae*. *J. Bacteriol.* **177:** 94.

Stevens, T., B. Esmon, and R. Schekman. 1982. Early stages in the yeast secretory pathway are required for transport of carboxypeptidase Y to the vacuole. *Cell* **30:** 439.

Stevens, T.H., J.H. Rothman, G.S. Payne, and R. Schekman. 1986. Gene dosage-dependent secretion of yeast vacuolar carboxypeptidase Y. *J. Cell Biol.* **102:** 1551.

Südhof, T.C. 1995. The synaptic vesicle cycle: A cascade of protein-protein interactions. *Nature* **375:** 645.

Sumrada, R. and T.G. Cooper. 1978. Control of vacuole permeability and protein degradation by the cell cycle arrest signal in *Saccharomyces cerevisiae. J. Bacteriol.* **136:** 234.

Supekova, L., F. Supek, and N. Nelson. 1995. The *Saccharomyces cerevisiae VMA10* is an intron-containing gene encoding a novel 13-kDa subunit of vacuolar H^+-ATPase. *J. Biol. Chem.* **270:** 13726.

Svihla, G., J.L. Dainko, and F. Schlenk. 1963. Ultraviolet microscopy of purine compounds in the yeast vacuole. *J. Bacteriol.* **85:** 399.

———. 1964. Ultraviolet microscopy of the vacuole of *Saccharomyces cerevisiae* during sporulation. *J. Bacteriol.* **88:** 449.

Szczypka, M.S., J.A. Wemmie, W.S. Moye-Rowley, and D.J. Thiele. 1994. A yeast metal resistance protein similar to human cystic fibrosis transmembrane conductance regulator (CFTR) and multidrug resistance-associated protein. *J. Biol. Chem.* **269:** 22853.

Takei, K., O. Mundigl, L. Daniell, and P. De Camilli. 1996. The synaptic vesicle cycle: A single vesicle budding step involving clathrin and dynamin. *J. Cell Biol.* **133:** 1237.

Takei, K., P.S. McPherson, S.L. Schmid, and P. De Camilli. 1995. Tubular membrane invaginations coated by dynamin rings are induced by GTP-γS in nerve terminals (comments). *Nature* **374:** 186.

Takeshige, K., M. Baba, S. Tsuboi, T. Noda, and Y. Ohsumi. 1992. Autophagy in yeast demonstrated with proteinase-deficient mutants and conditions for its induction. *J. Cell Biol.* **119:** 301.

Tan, P.K., N.G. Davis, G.F. Sprague, and G.S. Payne. 1993. Clathrin facilitates the internalization of seven transmembrane segment receptors for mating pheromones in yeast. *J. Cell Biol.* **123:** 1707.

Tanida, I., A. Hasegawa, H. Iida, Y. Ohya, and Y. Anraku. 1995. Cooperation of calcineurin and vacuolar H^+-ATPase in intracellular Ca^{2+} homeostasis of yeast cells. *J. Biol. Chem.* **270:** 10113.

Taylor, T.C., M. Kanstein, P. Weidman, and P. Melancon. 1994. Cytosolic ARFs are required for vesicle formation but not for cell-free intra-Golgi transport: Evidence for coated vesicle-independent transport. *Mol. Biol. Cell* **5:** 237.

te Heesen, S. and M. Aebi. 1994. The genetic interaction of *kar2* and *wbp1* mutations. Distinct functions of binding protein BiP and N-linked glycosylation in the processing pathway of secreted proteins in *Saccharomyces cerevisiae. Eur. J. Biochem.* **222:** 631.

Teichert, U., B. Mechler, H. Muller, and D.H. Wolf. 1989. Lysosomal (vacuolar) proteinases of yeast are essential catalysts for protein degradation, differentiation, and cell survival. *J. Biol. Chem.* **264:** 16037.

Theuvenet, A.P., B.J. Nieuwenhuis, J. van de Mortel, and G.W. Borst-Pauwels. 1986. Effect of ethidium bromide and DEAE-dextran on divalent cation accumulation in yeast. Evidence for an ion-selective extrusion pump for divalent cations. *Biochim. Biophys. Acta* **855:** 383.

Thumm, M., R. Egner, B. Koch, M. Schlumpberger, M. Straub, M. Veenhuis, and D.H.

Wolf. 1994. Isolation of autophagocytosis mutants of *Saccharomyces cerevisiae. FEBS Lett.* **349:** 275.

Tomashek, J.J., J.L. Sonnenburg, J.M. Artimovich, and D.J. Klionsky. 1996. Resolution of subunit interactions and cytoplasmic subcomplexes of the yeast vacuolar proton-translocating ATPase. *J. Biol. Chem.* **271:** 10397.

Trumbly, R. and G. Bradley. 1983. Isolation and characterization of aminopeptidase mutants of *Saccharomyces cerevisiae. J. Bacteriol.* **156:** 36.

Tsukada, M. and Y. Ohsumi. 1993. Isolation and characterization of autophagy-defective mutants of *Saccharomyces cerevisiae. FEBS Lett.* **333:** 169.

Tuttle, D.L. and W.A. Dunn, Jr. 1995. Divergent modes of autophagy in the methylotrophic yeast *Pichia pastoris. J. Cell Sci.* **108:** 25.

Tuttle, D.L., A.S. Lewin, and W.A. Dunn, Jr. 1993. Selective autophagy of peroxisomes in methylotrophic yeasts. *Eur. J. Cell Biol.* **60:** 283.

Uchida, E., Y. Ohsumi, and Y. Anraku. 1985. Purification and properties of H^+-translocating, Mg^{2+}-adenosine triphosphatase from vacuolar membranes of *Saccharomyces cerevisiae. J. Biol. Chem.* **260:** 1090.

———. 1988a. Characterization and function of catalytic subunit α of H^+-translocating adenosine triphosphatase from vacuolar membranes of *Saccharomyces cerevisiae.* A study with 7-chloro-4-nitrobenzo-2-oxa-1,3-diazole. *J. Biol. Chem.* **263:** 45.

———. 1988b. Purification of yeast vacuolar membrane H^+-ATPase and enzymological discrimination of three ATP-driven proton pumps in *Saccharomyces cerevisiae. Methods Enzymol.* **157:** 544.

Ulane, R.E. and E. Cabib. 1976. The activation system of chitin synthetase from *Saccharomyces cerevisiae. J. Biol. Chem.* **251:** 3367.

Umemoto, N., Y. Ohya, and Y. Anraku. 1991. *VMA11*, a novel gene that encodes a putative proteolipid, is indispensable for expression of yeast vacuolar membrane H^+-ATPase activity. *J. Biol. Chem.* **266:** 24526.

Umemoto, N., T. Yoshihisa, R. Hirata, and Y. Anraku. 1990. Roles of the *VMA3* gene product, subunit c of the vacuolar membrane H^+-ATPase on vacuolar acidification and protein transport. A study with *VMA3*-disrupted mutants of *Saccharomyces cerevisiae. J. Biol. Chem.* **265:** 18447.

Urech, K., M. Durr, T. Boller, A. Wiemken, and J. Schwencke. 1978. Localization of polyphosphate in vacuoles of *Saccharomyces cerevisiae. Arch. Microbiol.* **116:** 275.

Valls, L.A., J.R. Winther, and T.H. Stevens. 1990. Yeast carboxypeptidase Y vacuolar targeting signal is defined by four propeptide amino acids. *J. Cell Biol.* **111:** 361.

Valls, L.A., C.P. Hunter, J.H. Rothman, and T.H. Stevens. 1987. Protein sorting in yeast: The localization determinant of yeast vacuolar carboxypeptidase Y resides in the propeptide. *Cell* **48:** 887.

van den Hazel, H.B., M.C. Kielland-Brandt, and J.R. Winther. 1992. Autoactivation of proteinase A initiates activation of yeast vacuolar zymogens. *Eur. J. Biochem.* **207:** 277.

———. 1993. The propeptide is required for in vivo formation of stable active yeast proteinase A and can function even when not covalently linked to the mature region. *J. Biol. Chem.* **268:** 18002.

van der Klei, I., W. Harder, and M. Veenhuis. 1991. Selective inactivation of alcohol oxidase in two peroxisome-deficient mutants of the yeast *Hansenula polymorpha. Yeast* **7:** 813.

Vater, C.A., C.K. Raymond, K. Ekena, I. Howald-Stevenson, and T.H. Stevens. 1992.

The *VPS1* protein, a homolog of dynamin required for vacuolar protein sorting in *Saccharomyces cerevisiae*, is a GTPase with two functionally separable domains. *J. Cell Biol.* **119:** 773.

Veenhuis, M., A. Douma, W. Harder, and M. Osumi. 1983. Degradation and turnover of peroxisomes in the yeast *Hansenula polymorpha* induced by selective inactivation of peroxisomal enzymes. *Arch. Microbiol.* **134:** 193.

Vida, T.A. and S.D. Emr. 1995. A new vital stain for visualizing vacuolar membrane dynamics and endocytosis in yeast. *J. Cell Biol.* **128:** 779.

Vida, T.A., T.R. Graham, and S.D. Emr. 1990. In vitro reconstitution of intercompartmental protein transport to the yeast vacuole. *J. Cell Biol.* **111:** 2871.

Vida, T.A., G. Huyer, and S.D. Emr. 1993. Yeast vacuolar proenzymes are sorted in the late Golgi complex and transported to the vacuole via a prevacuolar endosome-like compartment. *J. Cell Biol.* **121:** 1245.

Vogel, J.P., L.M. Misra, and M.D. Rose. 1990. Loss of BiP/GRP78 function blocks translocation of secretory proteins in yeast. *J. Cell Biol.* **110:** 1885.

Volland, C., D. Urban-Grimal, G. Geraud, and R. Haguenauer-Tsapis. 1994. Endocytosis and degradation of the yeast uracil permease under adverse conditions. *J. Biol. Chem.* **269:** 9833.

Wada, Y. and Y. Anraku. 1992. Genes for directing vacuolar morphogenesis in *Saccharomyces cerevisiae*. II. *VAM7*, a gene for regulating morphogenic assembly of the vacuoles. *J. Biol. Chem.* **267:** 18671.

Wada, Y., Y. Ohsumi, and Y. Anraku. 1992a. Chloride transport of yeast vacuolar membrane vesicles: A study of in vitro vacuolar acidification. *Biochim. Biophys. Acta* **1101:** 296.

———. 1992b. Genes for directing vacuolar morphogenesis in *Saccharomyces cerevisiae.* I. Isolation and characterization of two classes of *vam* mutants. *J. Biol. Chem.* **267:** 18665.

Wada, Y., K. Kitamoto, T. Kanbe, K. Tanaka, and Y. Anraku. 1990. The *SLP1* gene of *Saccharomyces cerevisiae* is essential for vacuolar morphogenesis and function. *Mol. Cell. Biol.* **10:** 2214.

Wales, D., T. Cartledge, and D. Lloyd. 1980. Effects of glucose repression and anaerobiosis on the activities and intracellular distribution of tricarboxylic acid cycle and associated enzymes in *Saccharomyces carlsbergensis*. *J. Gen. Microbiol.* **116:** 93.

Wang, Y.-X., H. Zhao, T.M. Harding, D.S. Gomes de Mesquita, C.L. Woldringh, D.J. Klionsky, A.L. Munn, and L.S. Weisman. 1996. Multiple classes of yeast mutants are defective in vacuole partitioning yet target vacuole proteins correctly. *Mol. Biol. Cell* **7:** 1375.

Warren, G., T. Levine, and T. Musteli. 1995. Mitotic dissasembly of the mammalian Golgi apparatus. *Trends Cell Biol.* **5:** 413.

Webb, G.C., M. Hoedt, L.J. Poole, and E.W. Jones. 1997a. Genetic interactions between a *pep7* mutation and the *PEP12* and *VPS45* genes: Evidence for a novel SNARE component in transport between the *S. cerevisiae* Golgi complex and endosome. *Genetics* (in press).

Webb, G.C., J-.Q. Zhang, S.J. Garlow, A. Wesp, H. Riezman, and E.W. Jones. 1997b. Pep7p, a novel protein involved in vesicle-mediated transport between the yeast Golgi and endosome. *Mol. Biol. Cell* **8**: (in press).

Weisman, L.S. and W. Wickner. 1988. Intervacuole exchange in the yeast zygote: A new pathway in organelle communication. *Science* **241:** 589.

———. 1992. Molecular characterization of *VAC1*, a gene required for vacuole inheritance and vacuole protein sorting. *J. Biol. Chem.* **267:** 618.

Weisman, L.S., S.D. Emr, and W.T. Wickner. 1990. Mutants of *Saccharomyces cerevisiae* that block intervacuole vesicular traffic and vacuole division and segregation. *Proc. Natl. Acad. Sci.* **87:** 1076.

Westenberg, B., T. Boller, and A. Wiemken. 1989. Lack of arginine- and polyphosphate-storage pools in a vacuole-deficient mutant (*end1*) of *Saccharomyces cerevisiae. FEBS Lett.* **254:** 133.

Westphal, V., E.G. Marcusson, J.R. Winther, and S.D. Emr. 1996. Multiple pathways for vacuolar sorting of yeast proteinase A. *J. Biol. Chem.* **271:** 11865.

White, C. and G.M. Gadd. 1986. Uptake and cellular distribution of copper, cobalt and cadmium in strains of *Saccharomyces cerevisiae* cultured on elevated concentrations of these metals. *FEMS Microbiol. Ecol.* **38:** 277.

Whitney, P. and B. Magasanik. 1973. The induction of arginase in *Saccharomyces cerevisiae. J. Biol. Chem.* **248:** 6197.

Whitters, E.A., T.P. McGee, and V. Bankaitis. 1994. Purification and characterization of a late Golgi compartment from *Saccharomyces cerevisiae. J. Biol. Chem.* **269:** 28106.

Wichmann, H., L. Hengst, and D. Gallwitz. 1992. Endocytosis in yeast: Evidence for the involvement of a small GTP-binding protein (Ypt7p). *Cell* **71:** 1131.

Wiemken, A. and M. Durr. 1974. Characterization of amino acid pools in the vacuolar compartment of *Saccharomyces cerevisiae. Arch. Microbiol.* **101:** 45.

Wiemken, A., P. Matile, and H. Moor. 1970. Vacuolar dynamics in synchronously budding yeast. *Arch. Mikrobiol.* **70:** 89.

Wiemken, A., M. Schellenberg, and U. Urech. 1979. Vacuoles: The sole compartments of digestive enzymes in yeast (*Saccharomyces cerevisiae*)? *Arch. Microbiol.* **123:** 23.

Wilcox, C.A. and R.S. Fuller. 1991. Posttranslational processing of the pro-hormone cleaving Kex protease in the *Saccharomyces cerevisiae* secretory pathway. *J. Cell Biol.* **115:** 297.

Wilcox, C.A., K. Redding, R. Wright, and R.S. Fuller. 1992. Mutation of a tyrosine localization signal in the cytosolic tail of yeast Kex2 protease disrupts Golgi retention and results in default transport to the vacuole. *Mol. Biol. Cell* **3:** 1353.

Wilsbach, K. and G.S. Payne. 1993. Dynamic retention of TGN membrane proteins in *Saccharomyces cerevisiae. Trends Cell Biol.* **3:** 426.

Wilson, D.W., C.A. Wilcox, G.C. Flynn, E. Chen, W.J. Kuang, W.J. Henzel, M.R. Block, A. Ullrich, and J.E. Rothman. 1989. A fusion protein required for vesicle-mediated transport in both mammalian cells and yeast. *Nature* **339:** 355.

Winther, J.R. and P. Sorensen. 1991. Propeptide of carboxypeptidase Y provides a chaperone-like function as well as inhibition of the enzymatic activity. *Proc. Natl. Acad. Sci.* **88:** 9330.

Witt, I., R. Kronau, and H. Holzer. 1966. Isoenzym der malatdehydrogenase und ihre regulation in *Saccharomyces cerevisiae. Biochim. Biophys. Acta* **128:** 63.

Wolf, D.H. and C. Ehmann. 1979. Studies on a proteinase B mutant of yeast. *Eur. J. Biochem.* **98:** 375.

———. 1981. Carboxypeptidase S- and carboxypeptidase Y-deficient mutants of *Saccharomyces cerevisiae. J. Bacteriol.* **147:** 418.

Wolf, D.H. and U. Weiser. 1977. Studies on a carboxypeptidase Y mutant of yeast and evidence for a second carboxypeptidase activity. *Eur. J. Biochem.* **73:** 553.

Wolff, A.M., N. Din, and J.G.L. Petersen. 1996. Vacuolar and extracellular maturation of

Saccharomyces cerevisiae proteinase A. *Yeast* **12:** 823.

Woolford, C.A., C.K. Dixon, M.F. Manolson, R. Wright, and E.W. Jones. 1990. Isolation and characterization of *PEP5*, a gene essential for vacuolar biogenesis in *Saccharomyces cerevisiae*. *Genetics* **125:** 739.

Woolford, C.A., J.A. Noble, J.D. Garman, M.F. Tam, M.A. Innis, and E.W. Jones. 1993. Phenotypic analysis of proteinase A mutants. Implications for autoactivation and the maturation pathway of the vacuolar hydrolases of *Saccharomyces cerevisiae*. *J. Biol. Chem.* **268:** 8990.

Wurst, H., T. Shiba, and A. Kornberg. 1995. The gene for a major exopolyphosphatase of *Saccharomyces cerevisiae*. *J. Bacteriol.* **177:** 898.

Xu, Z. and W. Wickner. 1996. Thioredoxin is required for vacuole inheritance in *Saccharomyces cerevisiae*. *J. Cell Biol.* **132:** 787.

Xu, Z., A. Mayer, E. Muller, and W. Wickner. 1997. A heterodimer of thioredoxin and IB2 cooperates with Sec18p (NSF) to promote vacuole inheritance. *J. Cell Biol.* **136:** 299.

Yaglom, J., M.H.K. Linskens, S. Sadis, D.M. Rubin, B. Futcher, and D. Finley. 1995. p34^{Cdc28}-mediated control of Cln3 cyclin degradation. *Mol. Cell. Biol.* **15:** 731.

Yamamoto, A., D.B. DeWald, I.V. Boronenkov, R.A. Anderson, S.D. Emr, and D. Koshland. 1995. Novel PI(4)P 5-kinase homologue, Fab1p, essential for normal vacuole function and morphology in yeast. *Mol. Biol. Cell* **6:** 525.

Yamashiro, C.T., P.M. Kane, D.F. Wolczyk, R.A. Preston, and T.H. Stevens. 1990. Role of vacuolar acidification in protein sorting and zymogen activation: A genetic analysis of the yeast vacuolar proton-translocating ATPase. *Mol. Cell. Biol.* **10:** 3737.

Yasuhara, T., T. Nakai, and A. Ohashi. 1994. Aminopeptidase Y, a new aminopeptidase from *Saccharomyces cerevisiae*. Purification, properties, localization, and processing by protease B. *J. Biol. Chem.* **269:** 13644.

Yaver, D.S., H. Nelson, N. Nelson, and D.J. Klionsky. 1993. Vacuolar ATPase mutants accumulate precursor proteins in a pre-vacuolar compartment. *J. Biol. Chem.* **268:** 10564.

Yeh, E., R. Driscoll, M. Coltrera, A. Olins, and K. Bloom. 1991. A dynamin-like protein encoded by the yeast sporulation gene *SPO15*. *Nature* **349:** 713.

Yoshihisa, T. and Y. Anraku. 1989. Nucleotide sequence of *AMS1*, the structure gene of vacuolar α-mannosidase of *Saccharomyces cerevisiae*. *Biochem. Biophys. Res. Commun.* **163:** 908.

———. 1990. A novel pathway of import of α-mannosidase, a marker enzyme of vacuolar membrane, in *Saccharomyces cerevisiae*. *J. Biol. Chem.* **265:** 22418.

Yoshihisa, T., Y. Ohsumi, and Y. Anraku. 1988. Solubilization and purification of α-mannosidase, a marker enzyme of vacuolar membranes in *Saccharomyces cerevisiae*. *J. Biol. Chem.* **263:** 5158.

Zacharski, C.A. and T.G. Cooper. 1978. Metabolite compartmentation in *Saccharomyces cerevisiae*. *J. Bacteriol.* **135:** 490.

Zubenko, G.S. and E.W. Jones. 1981. Protein degradation, meiosis, and sporulation in proteinase-deficient mutants of *Saccharomyces cerevisiae*. *Genetics* **97:** 45.

Zubenko, G.S., A.P. Mitchell, and E.W. Jones. 1979. Septum formation, cell division, and sporulation in mutants of yeast deficient in protease B. *Proc. Natl. Acad. Sci.* **76:** 2395.

Zubenko, G.S., F.J. Park, and E.W. Jones. 1982. Genetic properties of mutations at the *PEP4* locus in *Saccharomyces cerevisiae*. *Genetics* **102:** 679.

5

The Nucleus and Nucleocytoplasmic Transport in *Saccharomyces cerevisiae*

Susan R. Wente
Department of Cell Biology and Physiology
Washington University School of Medicine
St. Louis, Missouri, 63110

Susan M. Gasser
Swiss Institute for Experimental Cancer Research
ISREC, 155, chemin des Boveresses
Epalinges, Switzerland CH-1066

Avrom J. Caplan
Department of Cell Biology and Anatomy
Mount Sinai School of Medicine
New York, New York 10029

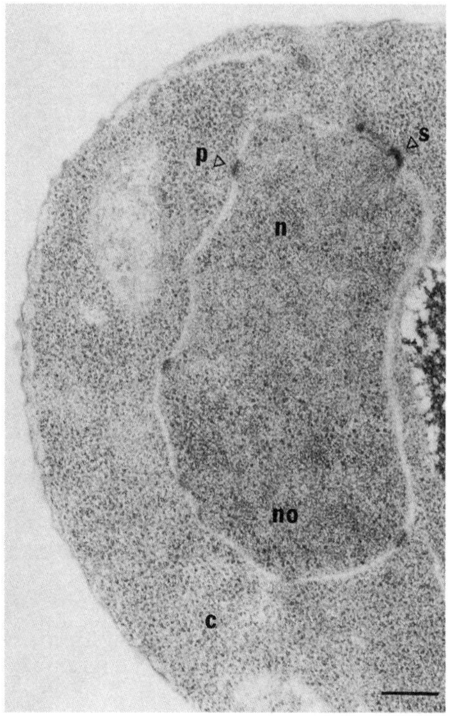

Figure 1 Thin-section electron micrograph of an *S. cerevisiae* nucleus. The nuclear envelope forms the unstained border between the cytoplasmic and nuclear compartments. (c) Cytoplasm; (n) nucleus/chromatin; (no) nucleolus; (s) spindle pole body; (p) nuclear pore complex. Bar, 0.25 μm.

I. INTRODUCTION

Compartmentation of cell functions is one of the biological features that most clearly differentiates eukaryotic cells from prokaryotic cells. In contrast to the single cytoplasmic compartment of a typical prokaryote, a eukaryotic cell contains numerous membranous organelles that are remarkably distinct both functionally and structurally. The yeast *Saccharomyces cerevisiae* has served as an important model system for studying organellar structure and function (see Pon and Schatz 1991; Serrano 1991; and other chapters in this volume). A striking merger of molecular genetic, biochemical, and morphological analyses has similarly advanced our understanding of the yeast nucleus, such that studies of yeast are making important contributions to the understanding of this organelle as well.

The structure of the nucleus is classically defined by its morphology.

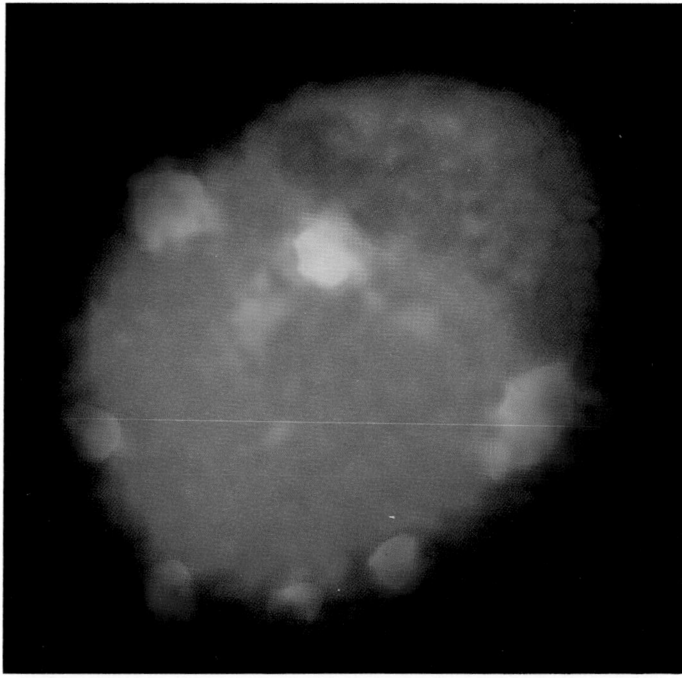

Figure 2 Organization of subcompartments of the yeast nucleus as visualized by indirect immunofluorescence. Wild-type diploid cells were prepared for immunofluorescence by a brief incubation with β-glucanase followed by fixation with 3.7% formaldehyde in growth medium. Immunodetection of Rap1p (Shore and Nasmyth 1987) and Nop1p (Henríquez et al. 1990) using affinity-purified rabbit anti-Rap1p antibodies and a mouse monoclonal antibody specific for Nop1p was performed as described by Gotta et al. (1996). (*Green*) Anti-Rap1p (FITC); (*blue*) anti-Nop1p (CY5). Ethidium bromide (*red*) was used as a general DNA stain. Shown is a merger of the three images captured on a Zeiss Axiovert 100 LSM confocal microscope using a 100X Plan-Apochromat objective. The blue merged with red appears violet. Bar, 2 μm.

Figures 1 and 2 show views of the yeast nucleus as obtained by electron and fluorescence microscopy, respectively, and illustrate several aspects of interphase nuclear organization. There are at least two functional compartments within the nucleus, one containing the majority of the chromatin (approximately two-thirds by volume; n in Fig. 1; stained red with ethidium bromide in Fig. 2) and one containing the nucleolus (approximately one-third by volume; no in Fig. 1; stained violet with anti-Nop1p antibodies in Fig. 2), where rDNA is transcribed and ribosomes are assembled (Woolford and Warner 1991; Mélèse and Xue 1995).

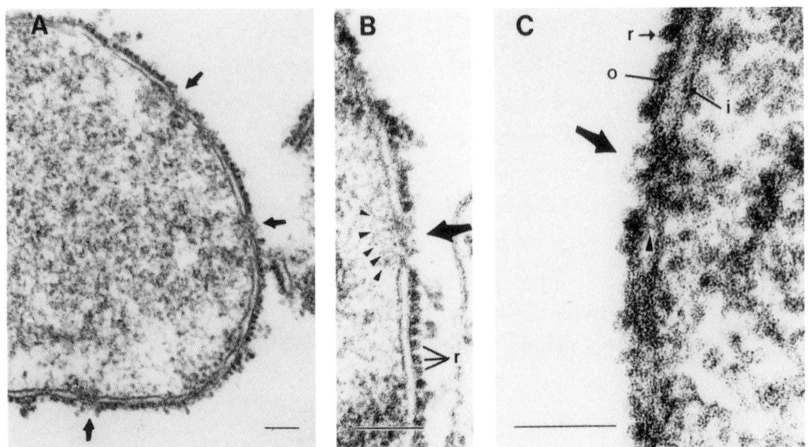

Figure 3 Thin-section electron micrographs of isolated yeast nuclei. Arrows denote NPCs; arrowheads in *B* highlight the nucleoplasmic side of an NPC; arrowhead in *C* points toward the pore membrane region. (r) Ribosomes; (o) outer nuclear membrane; (i) inner nuclear membrane. Bars, 0.1 μm. (Courtesy of Dr. J. Allen.)

Shown by green fluorescence in Figure 2 is the immunodetection of *r*epressor *a*ctivator *p*rotein 1 (Rap1p) (Shore and Nasmyth 1987), an abundant nuclear protein that binds to the repetitive sequences found at all yeast telomeres (Walmsley et al. 1984; see below). The focal staining pattern of Rap1p reflects the positioning of telomeric sequences in wild-type cells (Gotta et al. 1996), as discussed further below. The nucleolus is usually quite distinct in electron micrographs of yeast nuclei (see also Figs. 8A,B and 9B) as an electron-dense, crescent-shaped structure that occupies one side of the nucleus and associates closely with the nuclear envelope. The cytoplasm and nucleoplasm are separated by the two lipid bilayer membranes of the nuclear envelope (clear boundary in Fig. 1; visualized as the two membranes in Fig. 3). The inner nuclear membrane apposes the chromatin and the nucleolus. The outer nuclear membrane is continuous with the endoplasmic reticulum (ER), so that the intermembrane space is continuous with the lumen of the ER. Indeed, in yeast cells, the outer nuclear membrane represents the bulk of the rough ER (see Kaiser et al., this volume) and is studded with ribosomes (Fig. 3). Two different, structurally complex organelles penetrate the double lipid bilayer—the nuclear pore complexes (NPCs: p in Fig. 1; arrows in Fig. 3) and the spindle pole bodies (SPBs: s in Fig. 1).

The goal of many recent studies has been to define these nuclear structures in terms of their functional composition. This review provides a general description of yeast nuclear organization and of the strategies employed to identify factors required for nuclear structure and function. We place special emphasis on how the nuclear envelope mediates selective transport between the cytoplasm and nucleoplasm. Other valuable reviews on the yeast nucleus (Osborne and Silver 1993) and general nuclear structure (Gerace and Burke 1988; Miller et al. 1991; Forbes 1992) are also available. SPBs are discussed in more detail by Botstein et al. (this volume).

II. INTRANUCLEAR ARCHITECTURE

A. Genome and Chromatin Organization

The yeast haploid genome contains 1.4×10^7 base pairs divided among 16 chromosomes that range in size from 230 to more than 1000 kbp. Although the yeast genome contains more duplicated genes and gene families than originally expected (~10% of coding sequences share more than 90% identity with another yeast gene; K.H. Wolfe, pers. comm.), it has little simple- or middle-repetitive DNA and consists largely (≥70%) of coding sequence (see, *inter alia*, Dujon et al. 1994; Johnston et al. 1994). One of the few examples of repetitive DNA is the 300 plus or minus 75 bp of an irregular $TG_{1\text{-}3}$ repeat found at the ends (telomeres) of all yeast chromosomes, which can also be found in shorter stretches separating sequential subtelomeric elements (for review, see Louis 1995). Subtelomeric elements called Y′(which can be either 5.4 or 6.7 kb in length) and X (which carries a conserved core element of 500 bp) are usually found immediately adjacent to the more simple telomeric repeat. Although the sequences of Y′and core X elements are highly conserved, their presence and copy number vary among chromosomes and strains (for review, see Louis 1995). The other major repeated sequences in yeast are the family of large retrotransposon-like elements called Ty (for review, see Boeke and Sandmeyer 1991; Voytas and Boeke 1993) and the 9-kb ribosomal DNA repeat. The rDNA repeat is present in 50–70 tandem copies, covering nearly 500 kb on chromosome XII, and it encodes the 35S precursor and the 5S structural RNAs of ribosomes (Petes 1979; Woolford and Warner 1991).

Like all eukaryotic genomes, yeast DNA is organized at the most basic level into the nucleosome, the core of which is defined by 146 bp wrapped in two turns around a complex composed of two copies of each of the four core histones, H2A, H2B, H3, and H4 (Thomas and Furber 1976). Histones H3 and H4 are among the most highly conserved proteins known, and even the amino-terminal tails, which are not necessary for forming the core nucleosome, reveal highly conserved patterns of sequence and residue-specific modifications (Davie et al. 1981; for review, see Grunstein 1990). Nucleosomal core particles are separated by a stretch of linker DNA, whose length can differ from species to species, in different tissues, and even within different domains of a single chromosome (Bradbury et al. 1981). In isolated nuclei, this linker DNA is more sensitive to attack by micrococcal nuclease, which allows definition of the nucleosomal repeat. In *S. cerevisiae*, the repeat is about 165 bp (Lohr et al. 1977; Nelson and Fangman 1979), shorter than the repeat unit found in mammalian cells, which ranges between 180 and 200 bp (Bradbury et al. 1981). Other lower eukaryotes, such as *Aspergillus* and *Neurospora*, also typically show a shorter nucleosomal repeat, although closer analysis in yeast has shown that linker size can vary significantly in different regions of the genome. For example, in the *URA3* gene alone, some nucleosomes are packed immediately next to one another (300 bp for two nucleosomes), whereas others are precisely positioned at 180-bp intervals (Bernardi et al. 1992). The spacing has been shown to be influenced by three parameters: (1) the nature of the DNA sequence itself, (2) chromatin folding, and (3) boundary elements, which can be promoter regions, terminators, or origins of replication (for review, see Thoma 1992). The physiological significance of the variations in nucleosomal repeat sizes is not understood despite extensive study. Early experiments to determine whether nucleosomes were specifically positioned with respect to sequence suggested that the majority were not. Nevertheless, it is clear that the translational and rotational setting of certain individual nucleosomes with respect to strategically located regulatory sequences does have an important role in transcriptional control (see below).

It has been proposed that the shorter repeat unit of the nucleosome in yeast may reflect the absence of histone H1, a slightly larger, linker histone, which binds the nucleosome in a 1:1 ratio at the point where the DNA enters and leaves the core particle. Although early attempts to identify a yeast equivalent to H1 identified a mitochondrial protein (Certa et al. 1984), the yeast genome project has revealed a nuclear protein that shares significant homology with histone H1, although it has two, rather than one, globular domains (Landsman 1996; Ushinsky et al. 1997). The gene product is nonessential for mitotic growth, although disruption slightly increases expression of a plasmid-borne *CYC1-lacZ* fusion (Ushinsky et al. 1997). Heterologous expression of an animal H1 in yeast results in growth inhibition and transcriptional repression at substoichiometric levels of expression (Linder and Thoma 1994), yet its presence does not alter the nucleosome repeat unit, suggesting that animal histone H1 is not sufficient to determine nucleosomal positioning. Other typical chromatin proteins such as the HMG (high-mobility group) family are also found in yeast, and at least four genes encoding HMG1-like proteins have been cloned (Haggren and Kolodrubetz 1988; Kolodrubetz and Burgum 1990; Kruger and Herskowitz 1991). These highly charged proteins are essential for viability and are thought to modulate gene expression by altering either local nucleosome positioning (Tremethick and Drew 1993) or stability (Kruger and Herskowitz 1991). The effect of these HMG1-like proteins on the compaction of the nucleosomal fiber has not yet been established, although mutation of one such gene, *SIN1*, influences mitotic chromosome stability (Kruger and Herskowitz 1991). No homologs of HMGI/Y, proteins that bind preferentially to the AT-rich, late-replicating, Giemsa-dark bands of facultative heterochromatin in human and mouse mitotic chromosomes (Saitoh and Laemmli 1994), have been reported for *S. cerevisiae*.

One role proposed for histone H1 is the stabilization of a 30-nm solenoidal structure formed from the nucleosomal filament (Finch and Klug 1976; Thoma et al. 1979). Like mammalian chromatin, yeast nucleosomes have been observed to fold spontaneously into a solenoidal structure in a salt-dependent manner (Rattner et al. 1982; Lowary and Widom 1989), although the solenoidal fiber makes up only a small proportion of the DNA observed in electron micrographic spreads (Rattner et al. 1982). No protein has yet been identified in yeast that stabilizes the solenoid into a fiber-like structure, yet the level of chromatin compaction in yeast suggests an equivalent degree of organization, at least for regions that are not actively transcribed. A simple calculation based on the volume of the nucleus and the length of DNA found within suggests that the necessary degree of compaction for a human genome in a typical nucleus of 10 μm in diameter is only two- to threefold higher than that for a haploid yeast genome in a nucleus of slightly under 2 μm in diameter. Supporting this, in situ hybridization on flattened yeast nuclei using probes from chromosome XVI suggested a compaction ratio of approximately 80 (Guacci et al. 1994), which is two- to threefold less than the ratios estimated using similar methods for regions of the interphase mammalian genome (Lawrence et al. 1988; Trask et al. 1989; van den Engh et al. 1992). The significance of these values is limited, however, because of the number of probes used, the resolution of fluorescence microscopy, and the complications inherent in correctly measuring distances in nonoriented three-dimensional structures (Dietzel et al. 1995). Finally, one cannot exclude that chromatin organization is altered by the in situ hybridization treatment itself. It is clear that heterochromatic regions of the mammalian genome remain in a fully condensed state throughout interphase, at a level of compaction unlikely to be found in the yeast genome.

The higher levels of chromosomal organization include all levels of folding beyond the 30-nm fiber and are commonly thought to require the participation of nonhistone chromosomal proteins in both interphase and metaphase cells. The most widely accepted model for higher-order organization is the radial-loop structure, in which the 30-nm solenoid folds into loops ranging in size from 5 to 100 kb, with the base of each loop anchored by interactions with nonhistone proteins (Laemmli et al. 1978). Evidence for loop organization derives from observations of lampbrush chromosomes (for review, see Callan 1986), transcriptional domains in polytene chromosomes, and the halos of DNA loops that extend from proteinaceous residual structures in both histone-depleted chromosomes (Paulson and Laemmli 1977) and interphase nuclei treated with either high- or low-salt extraction techniques (Cook and Brazell 1975; Benyajati and Worcel 1976).

B. Nuclear Skeletal Structures

The proteinaceous structures that remain after extraction of membranes and histones are variously called scaffolds, nuclear matrices, cages, nucleo- or karyoskeletons, or the nuclear matrix–pore complex–lamina fraction (for review, see Gasser et al. 1989). The composition of these various substructures depends largely on the protocols used for nuclear isolation and for histone depletion, and only in the case of the nuclear lamina is it clear that the proteins isolated as a nuclear substructure actually form an analogous structure in intact cells (for review, see Nigg 1992). Nevertheless, it is noteworthy that nuclear scaffolds or matrices can be isolated from both fission and budding yeast nuclei after a stabilization step involving both $CuSO_4$ and a temperature shift to 37°C (Cardenas et al. 1990). However, no salt-resistant, perinuclear lamin meshwork was observed by electron microscopy of yeast nuclei after extraction (Cardenas et al. 1990) or envelope purification (Strambio-de-Castillia et al. 1995), consistent with the apparent lack of true lamin homologs in yeast, as judged from antibody screening and thorough scanning of the sequence data bank (G.-C. Alghizi and S.M. Gasser, unpubl.).

The development of a low-ionic-strength method using lithium 3′,5′ diiodosalicylate (LIS) for the extraction of histones allowed the identification of a class of AT-rich, scaffold-associated DNA sequences (SARs or MARs) (for review, see Laemmli et al. 1992). This mapping technique was readily applicable to isolated yeast nuclei, and in contrast to higher eukaryotic systems, in which the function of scaffold-attachment regions is largely unclear, those identified in yeast coincided with well-characterized functional elements. In particular, autonomous replication sequences (ARSs), centromeres, and the transcriptional silencers that flank the silent mating-type loci were all found to be scaffold attached (Amati and Gasser 1988, 1990; Hofmann et al. 1989). In addition, competitive binding studies using *ARS1* and *CEN3* sequences showed that *ARS1* can compete for *CEN* binding to the nuclear scaffold, but not the reverse, defining at least two classes of SARs in yeast (Amati and Gasser 1988).

The finding that yeast ARS elements, which serve as origins of replication, are associated with a residual nuclear structure is consistent with evidence gathered over many years indicating that DNA synthesis itself may be linked to a nuclear substructure, (see, e.g., Dingman 1974; Dijkwel et al. 1979; Pardoll et al. 1980; Vogelstein et al. 1980). Replication forks detected by nascent, pulse-labeled DNA, by the presence of the polymerase α-primase complex, or by the presence of DNA replication intermediates are associated with the high-salt nuclear-matrix fraction (Berezney and Coffey 1975; Wanka et al. 1977; Dijkwel et al. 1979), as well as with matrices prepared under other conditions (Jackson and Cook 1986, 1988; Vaughn et al. 1990). The matrix association of nascent DNA was also observed for high-salt-extracted matrices from yeast nuclei (Potashkin et al. 1984). Despite the difficulty of identifying replication origins in higher eukaryotic cells, there is strong evidence that origins of replication are located 5′ of the chick α-globin gene and 3′ of the Chinese hamster DHFR gene, and both are found close to scaffold attachment sites (for review, see Dijkwel and Hamlin 1995), similar to what has been observed in yeast. Deletion analyses in yeast have shown that the sequence requirements for scaffold binding are similar, but not always identical, to those for the initiation of DNA replication (Amati et al. 1990). In particular, both activities require oligo(dA) sequences and an overall strand asymmetry in base composition, but ARS activity requires a less extensive A/T-rich region and 11-bp ARS consensus, which is not essential for scaffold association.

It is still unclear what proteins mediate interaction with the scaffold association in yeast, although it is noteworthy that Orc2p, the DNA-binding subunit of the *o*rigin *r*ecognition *c*omplex (ORC), and Cdc6p, which appears to interact with the ORC, are both recovered in the nuclear scaffold fraction under conditions of extraction that remove 95% of the nuclear protein (P. Pasero and S.M. Gasser, unpubl.). In mammalian nuclei, a large coiled-coil protein called NuMA (*n*uclear matrix *m*itotic *a*pparatus protein) is required for nuclear decondensation after metaphase and appears to form an integral part of the nuclear matrix fraction (Compton and Cleveland 1994). A closely related protein called Nuf1p was discovered in yeast but proved to be identical to the p110 component of the SPB (Mirzayan et al. 1992; Kilmartin et al. 1993). DNA topoisomerase II is a highly conserved protein that binds preferentially to a narrow minor groove of oligo(dT)- or oligo(dA)-rich DNA, most notably in the context of SAR fragments (Adachi et al. 1989). Topoisomerase II has been recovered in scaffolds of a variety of species (Berrios et al. 1985; Earnshaw et al. 1985; Cardenas et al. 1990), and both yeast genetics and reconstitution studies using *Xenopus* extracts suggest that this enzyme is essential for chromosome condensation. The coupling of a cold-sensitive (*cs*) *top2* mutation with a cold-sensitive β-tubulin mutation in fission yeast resulted in a mitotic arrest with elongated, noncondensed chromosomes, suggesting a role for topoisomerase II in proper chromosome condensation (Uemura et al. 1987). Consistent with this interpretation, depletion of topoisomerase II from *Xenopus* egg extracts in vitro resulted in extracts that failed to convert nuclei into condensed mitotic chromosomes (Adachi et al. 1991; Hirano and Mitchison 1993). The condensation activity could be restored by the addition of purified yeast topoisomerase II. Although it cannot be assumed that the same protein(s) both promotes and maintains the highly condensed structure of mammalian mitotic chromosomes, tight association of topoisomerase II with chromosomes is consistent with a role in one function or the other (Gasser et al. 1986).

C. Mitotic Chromosome Condensation

The less-soluble nonhistone chromosomal proteins were originally proposed to condense and maintain the highly compacted (10^4-fold) structure of mammalian mitotic chromosomes (Laemmli et al. 1978). In the case of *S. cerevisiae*, evidence as to whether the chromosomes actually condense during mitosis has been both difficult to obtain and contradictory (Wintersberger et al. 1975; Peterson and Ris 1976; Gordon 1977), and only one report has convincingly identified individual yeast chromosomes in Giesma-stained whole-cell spreads (Galeotti and Williams 1978). Recently, studies using fluorescence in situ hybridization (FISH) have determined a chromosome condensation ratio for metaphase nuclei that is approximately twice that observed in interphase nuclei (Guacci et al. 1994), suggesting that a modest but reproducible degree of mitotic chromosome condensation does occur in *S. cerevisiae*.

In addition, important new light has been shed on the mechanism of chromosome condensation in yeast and other organisms by the discovery of a family of proteins, showing strong structural homology from yeast to humans, that are involved in this process (for review, see Hirano 1995). This group of proteins is called SMC after the *S. cerevisiae* gene *SMC1* (*s*tability of *m*itotic *c*hromosomes), the first member with a defined function (Strunnikov et al. 1993). A second *S. cerevisiae* gene, *SMC2*, has since been characterized (Strunnikov et al. 1995), and sequence analysis has revealed at least three other family members in the yeast genome. Similarly, *Schizosaccharomyces pombe* has three known *SMC* genes, two of which, called *cut3* and *cut14* (for *c*hromosomes *u*ntimely *t*orn; Saka et al. 1994), are involved in chromosome segration, and one of which, *rad18*, is involved in DNA repair (Lehmann et al. 1995). Homologs have also been cloned from chicken, where the protein corresponds to SCII (scaffold protein II), the second most abundant scaffold protein recovered from metaphase chromosomes (Saitoh et al. 1994); from *C. elegans* (Chuang et al. 1994), and from *Xenopus*, where two SMC-like proteins, XCAP-C and XCAP-E, were recovered as abundant nonhistone proteins bound tightly to chromosomes condensed in vitro in *Xenopus* egg extracts (Hirano and Mitchison 1994).

All members of the SMC family have striking structural similarity. They are large proteins (1000–1500 amino acids) that contain a globular head domain with a nucleotide-binding motif, followed by two extended stretches of α-helix, followed by a globular tail domain of 100–200 amino acids that includes a conserved motif also implicated in nucleotide binding. The three-domain structure with conserved internal sequences is reminiscent of mechanochemical motor proteins like kinesin.

In *S. cerevisiae* strains harboring various mutant alleles of *SMC1*, and in an *smc2* temperature-sensitive mutant, the stability of both natural chromosomes and minichromosomes is reduced (Strunnikov et al. 1993, 1995). The *S. pombe cut3* and *cut14* mutants arrest with noncondensed chromosomes at the restrictive temperature and show aberrant mitotic segregation (Saka et al. 1994). Notably, in the *smc2* mutant at nonpermissive temperature, the distances measured by FISH between pairs of hybridized probes were reproducibly twofold larger than those in mitotically arrested *SMC*$^+$ cells (Strunnikov et al. 1995). Thus, although it is not yet clear whether the primary role of the SMCs in yeast is in sister chromatid segregation or in condensation, the two may be intimately related, and even the small degree of condensation measured by in situ hybridization in yeast may be important for proper mitotic division.

D. Chromatin Organization, Subnuclear Domains, and Transcriptional Regulation

1. Nucleosome Positioning in Relation to Transcription

Unlike higher eukaryotic genomes, yeast chromatin generally shows little differential sensitivity to DNase I, and the vast majority of both transcribed and nontranscribed sequences show very similar accessibility to nucleases (Lohr and Hereford 1979). This has led to the proposal that the majority of the yeast genome is in a transcriptionally poised state, whether or not the genes are actively transcribed (Sledziewski and Young 1982; for review, see Pérez-Ortin et al. 1989). Nevertheless, distinct differences exist in the patterns of DNase I and micrococcal-nuclease-hypersensitive sites when active and repressed states are compared. Nuclease-hypersensitive sites often correlate with sites of transcription initiation and with binding sites for the sequence-specific *trans*-activators (see, e.g., Nasmyth 1982; Sledziewski and Young 1982; Bergman and Kramer 1983; Lohr and Hopper 1985), and increased sensitivity at these sites usually correlates with induction of transcription.

Contradictory results have been published as to the fate of nucleosomes in highly transcribed yeast sequences. In some cases, the regular nucleosomal pattern is clearly altered, although the nucleosomes or "half-nucleosomes" remain closely associated with the transcription unit (Lohr 1983; Lee and Garrard 1991). The degree of alteration in the nucleosomal pattern is almost certainly correlated with the level of transcription. For example, the induction of an integrated *GAL* promoter resulted in the loss of nucleosome positioning (but not the loss of nucleosomes) on an adjacent transcription unit (Cavalli and Thoma 1993), whereas the same promoter on a high-copy-number plasmid induced only an enhanced susceptibility to micrococcal nuclease, with no loss of nucleosome positions (Fedor and Kornberg 1989). The higher copy number implies a lower level of induction per copy, and this is thought to explain the contrasting results. Importantly, the boundaries of a particular nucleosomal organization, such as the phased nucleosomes at the *HSP82* locus, appear to correlate precisely with the transcription unit itself, starting at the upstream activating sequences and extending to the termination of transcription (Szent-Györgyi et al. 1987). In this way, each yeast gene, whether its nucleosomes are positioned or not, acts as a chromatin microenvironment in which the boundaries of the transcription unit provide the boundaries of the nucleosomal alterations. This is in sharp contrast to the chromatin domains found in mammalian cells, which can include up to 100 kb of flanking DNA displaying fluctuations in nuclease sensitivity indicative of the gene's potential for transcription.

In yeast, once a highly transcribed gene has been switched off by the removal of its activator, a delay of 10–50 minutes can ensue prior to restoration of the typical inactive chromatin structure, which in the case of the *GAL10* promoter involves a nucleosome positioned over the TATA box (Cavalli and Thoma 1993). This delay may reflect a requirement for the passage of a replication fork for remodeling of the nucleosomes from an active to a repressed pattern. A similar requirement for DNA replication (again presumably to allow nucleosomal assembly following DNA synthesis) was observed for the restoration of a repressed chromatin state at the silent mating-type loci after a shift from restrictive to permissive temperature for a mutation in *SIR3* (Miller and Nasmyth 1984). It is important to note, however, that passage through S phase is not necessary to derepress silent genes after inactivation of the silencer factor.

Further evidence that positioned nucleosomes can repress transcription in yeast has come from chromatin analysis on the yeast *TRP1/ARS1* minichromosome carrying the operator for the α2 repressor. It was first shown that the minichromosome carrying the α2 operator was packaged into stably positioned nucleosomes in cells expressing the α2 repressor (Roth et al. 1990). In a subsequent study, binding of the α2 repressor in the promoters of the *STE6* and *BAR1* genes was shown to position nucleosomes both translationally and rotationally over sequences adjoining the operator (Shimizu et al. 1991). The stable positioning of this nucleosome is thought to help repress **a**-cell-specific promoters in α cells.

2. *Repressed Chromatin and Subnuclear Domains*

Despite the observation that the majority of the yeast genome appears to be in an open or transcriptionally poised state, two classes of sites show a position-dependent transcriptional repression that correlates with a less accessible chromatin structure. The first of these are the transcriptionally repressed mating-type loci on chromosome III (*HML* and *HMR*), which in vivo show a reduced accessibility to the yeast HO endonuclease and to an ectopically expressed *Escherichia coli dam* methylase. In lysed cells as well, exogenous restriction enzymes cut less readily in the repressed domain than in flanking regions (for review, see Laurenson and Rine 1992). Similarly, DNA polymerase II genes positioned adjacent to the poly($TG_{1\text{-}3}$) tracts at yeast telomeres were found to succumb to a heritable, but reversible, transcriptional inactivation that also correlates with reduced accessibility to an ectopically expressed *dam* methylase

(Gottschling et al. 1990; Aparicio et al. 1991; Gottschling 1992). This phenomenon is part of a telomeric position effect, or TPE, which, like position-effect variegation in *Drosophila*, appears to spread into adjacent euchromatin from a transcriptionally inactive repetitive element, in this case the TG_{1-3} repeat at the telomere (Renauld et al. 1993).

Nucleation of the repressed chromatin state at telomeres requires the *cis*-acting telomeric repeat, to which bind multiple copies of *r*epressor *a*ctivator *p*rotein 1 (Rap1p) (Longtine et al. 1989; Gottschling et al. 1990; Gilson et al. 1993b; Liu et al. 1994; Stavenhagen and Zakian 1994; Buck and Shore 1995). At *HML* and *HMR*, short *cis*-acting silencer elements serve the same purpose and bind *trans*-acting factors needed for the establishment of silencing, namely, Rap1p (Shore and Nasmyth 1987), Abf1p (*A*RS *b*inding *f*actor 1; Buchman et al. 1988), and the *o*rigin *r*ecognition *c*omplex (ORC; Micklem et al. 1993), a six-polypeptide complex that recognizes the ARS consensus (Bell et al. 1993). In addition to these *cis*-acting sequences and their ligands, both the telomere position-effect and mating-type silencing require the "silent information regulator" genes, *SIR2*, *SIR3*, and *SIR4*; the *NAT1/ARD1* amino-terminal acetylase; and the intact amino termini on histones H3 and H4 (for review, see Laurenson and Rine 1992). *SIR1* is necessary for repression at *HML* and *HMR* but not for repression of genes inserted near the TG_{1-3} tracts at telomeres.

Several lines of evidence suggest that Sir3p and Sir4p are structural components of repressed chromatin. First, these proteins were shown to bind selectively to the amino termini of histones H3 and H4 in vitro, and mutations in the histone amino termini that disrupt this interaction and derepress silencing in vivo also lead to a redistribution of Sir3p throughout the nucleus (Kayne et al. 1988; Thompson et al. 1994b; Hecht et al. 1995). Second, overexpression of Sir4p, or domains of Sir4p, derepresses both mating-type and telomeric silencing (Marshall et al. 1987; Cockell et al. 1995), whereas overexpression of Sir3p extends repression inward from a marked telomere (Renauld et al. 1993). Finally, Sir3p and Sir4p interact in two-hybrid assays with themselves, with each other, and with Rap1p (Chien et al. 1991; Moretti et al. 1994), and Sir4p and Rap1p coprecipitate in a DNase-insensitive complex from yeast nuclear extracts (Cockell et al. 1995).

3. Telomeres and Functional Nuclear Compartments

As demonstrated in Figure 2 for Rap1p, the silencing factors Rap1p, Sir3p, and Sir4p are not randomly distributed throughout the wild-type yeast cell nucleus, but instead they localize in approximately ten foci, many of which are found near the nuclear periphery in G_1-phase nuclei (Palladino et al. 1993). The three proteins are interdependent for localization to these foci, and a number of mutations that disrupt silencing also result in the dispersed staining of all three proteins (Palladino et al. 1993; Cockell et al. 1995; Hecht et al. 1995). A double immunofluorescence/ FISH protocol has shown that 75% of the foci detectable by anti-Rap1p, anti-Sir3p, and anti-Sir4p antibodies in wild-type cells coincide precisely with a probe that detects the Y′ subtelomeric repeat, which is found immediately adjacent to the short $TG_{1\text{-}3}$ stretch (Gotta et al. 1996). This indicates that both Sir proteins and Rap1p are present in high concentrations at their major sites of action (i.e., at telomeres), which are themselves clustered. However, because Rap1p staining (presumably indicating telomeres) remains clustered in a *rap1-21* mutant that is fully derepressed, the clustering of telomeres is not sufficient for the establishment or maintenance of repression (Cockell et al. 1995).

Most studies to date on telomere positioning have utilized fluorescence microscopy, the resolution of which (i.e., ~200 nm) is insufficient to judge whether the foci of telomeres, Rap1p, or Sir proteins are directly apposed to the nuclear envelope. In silencing-competent yeast cells, approximately 70% of telomeres (or Rap1p foci) are found within a peripheral zone measuring one fifth of the nuclear radius, which comprises 50% of the spherical nuclear volume (Klein et al. 1992; Gotta et al. 1996), although this fraction may vary through the cell cycle. Rap1p foci do not appear to be directly associated with nuclear pore complexes, because double immunostaining with anti-Rap1p and anti-pore-protein antibodies primarily labels nonoverlapping foci at the nuclear periphery, with only 14% direct coincidence. In addition, Rap1p immunofluorescence in a strain carrying a mutation in *nup133*, which provokes the clustering of nuclear pores on one side of the nucleus (Doye et al. 1994), has a wild-type focal pattern, suggesting that Rap1p/telomere complexes are positioned independently of the pore complexes (cited in Gotta et al. 1996). Although Sir4p has a coiled-coil domain in its essential carboxy-terminal region, which has been likened to the α-helical domain of a nuclear lamin (Diffley and Stillman 1989), Sir4p staining does not form the ring-like pattern typical for staining of nuclear lamins, suggesting that it does not form a stable link to a peripheral nuclear substructure. Whether there exist proteins stably associated with the nuclear envelope that are responsible for the clustering of yeast telomeres remains to be seen.

The juxtaposition of multiple telomeres may represent a subnuclear organization of the genome that facilitates the maintenance and replication of telomeres as well as transcriptional repression of telomere-proximal sequences. A nonrandom distribution of telomeres and centromeres in the nucleus is not unique to *S. cerevisiae* (for review, see Gilson et al. 1993a). For instance, telomeres appear to be adjacent to the nuclear envelope in both polytene nuclei and embryonic cells of *Drosophila* (Mathog et al. 1984), and cell-cycle-dependent positioning of telomeres has been observed in *S. pombe* and mammalian tissue culture cells (Funabiki et al. 1993; Vourc'h et al. 1993). In most species, telomeres cluster at the nuclear envelope in the "bouquet" stage of meiosis, and in *S. pombe*, dramatic movement of the zygotic nucleus is led by telomeres clustered at the SPB (Chikashige et al. 1994).

Other evidence for higher-order organization within yeast nuclei was obtained using immunological probes for the nucleolus and the SPB. In this way, Yang et al. (1989) have shown that the nucleolus generally lies opposite the SPB in vegetatively growing *S. cerevisiae* cells. Moreover, during mitosis, the nucleolar DNA remains closely associated with the bulk of the genomic DNA, as if an attachment or structure is maintained, unlike the situation in *S. pombe*, where the nucleolus trails behind the segregating chromosomes (Granot and Snyder 1991).

Recent studies have shown that the proximity of a silencer-flanked construct to a telomere influences the efficiency with which it is repressed (Thompson et al. 1994a; Maillet et al. 1996), consistent with the idea that the higher-order organization of chromosomes within the nucleus can influence transcriptional states. It was also observed that by enlarging the pool of Sir3p or Sir4p that is not bound to telomeric sites, either by overexpressing the Sir proteins or by mutations of *RAP1*, it is possible to repress efficiently at nontelomeric sites in the genome (Lustig et al. 1996; Maillet et al. 1996; Marcand et al. 1996). Thus, it appears likely that the subnuclear organization of telomeres and the resulting unequal distribution of chromatin-binding factors are capable of modulating transcription patterns in yeast. This lends weight to the notion that nuclear organization has a distinct function within the cell (Cremer et al. 1993).

III. THE NUCLEAR ENVELOPE

A. Nuclear Envelope Division, Fusion, and Biogenesis

During the "closed" mitosis of yeast, the nuclear envelope and resident NPCs apparently remain intact. This is in sharp contrast to the "open" mitosis of vertebrate cells, which is accompanied by a vesiculation of the nuclear envelope and reversible disassembly of the NPCs (Wiese and

Wilson 1993). Several techniques have been used to analyze nuclear membrane dynamics during budding and division, including electron microscopy (Byers 1981), immunofluorescence using antibodies that recognize nuclear envelope proteins in fixed cells (Fig. 4) (Davis and Fink 1990; Copeland and Snyder 1993), video-enhanced DIC microscopy on living cells (Yeh et al. 1995), and fluorescence microsopy on living cells after staining with a lipophilic dye (Koning et al. 1993). As the newly formed bud enlarges, the nucleus migrates to a position proximal to the bud site (Fig. 4a). This is followed by extension of the nucleus into the bud, which involves a very elongated protrusion of nuclear membrane through the mother-bud neck (Fig. 4d). Penetration of the nuclear envelope occurs before any DAPI staining (chromatin) is seen crossing the neck. The envelope continues to elongate rapidly during cell division, coincident with extension of the spindle, and it remains intact until the time of cytokinesis. The forces producing the ultimate scission of the nuclear envelope remain unknown.

The molecular components that mediate nuclear migration and spindle orientation are partially understood, and, not surprisingly, both processes involve microtubule function. Both spindle and astral (cytoplasmic) microtubules are attached to the SPBs (for review, see Winey and Byers 1992; Page and Snyder 1993; Kilmartin 1994; Botstein et al., this volume). Using drugs such as nocodazole that inhibit microtubule polymerization (Jacobs et al. 1988) and mutants defective in tubulin (Huffaker et al. 1988; Sullivan and Huffaker 1992), nuclear migration to the mother-bud neck was shown to require astral microtubules, because in their absence both migration and spindle orientation failed to occur. However, once the nucleus has migrated to the bud neck, it can be stably maintained in that position independent of microtubule function (Jacobs et al. 1988).

During or after nuclear migration, the spindle becomes oriented such that one pole penetrates the mother-bud neck prior to anaphase, so that subsequent chromosome (and nuclear envelope) migration occurs along the proper axis. In addition to the astral microtubules themselves, this process requires dynein (Eshel et al. 1993; Li et al. 1993), the novel protein Jnm1p (McMillan and Tatchell 1994), and actin (Palmer et al. 1992; Sullivan and Huffaker 1992). The role of dynein (Dhc1p), a motor protein, in nuclear migration and spindle mutation was explored by Yeh et al. (1995). Nuclear migration to a position proximal to the neck was normal in *dhc1* mutants, although spindle orientation and penetration of the nucleus into the neck were impaired. Localization of dynein fusion proteins both to SPBs and to astral microtubules further supports the con-

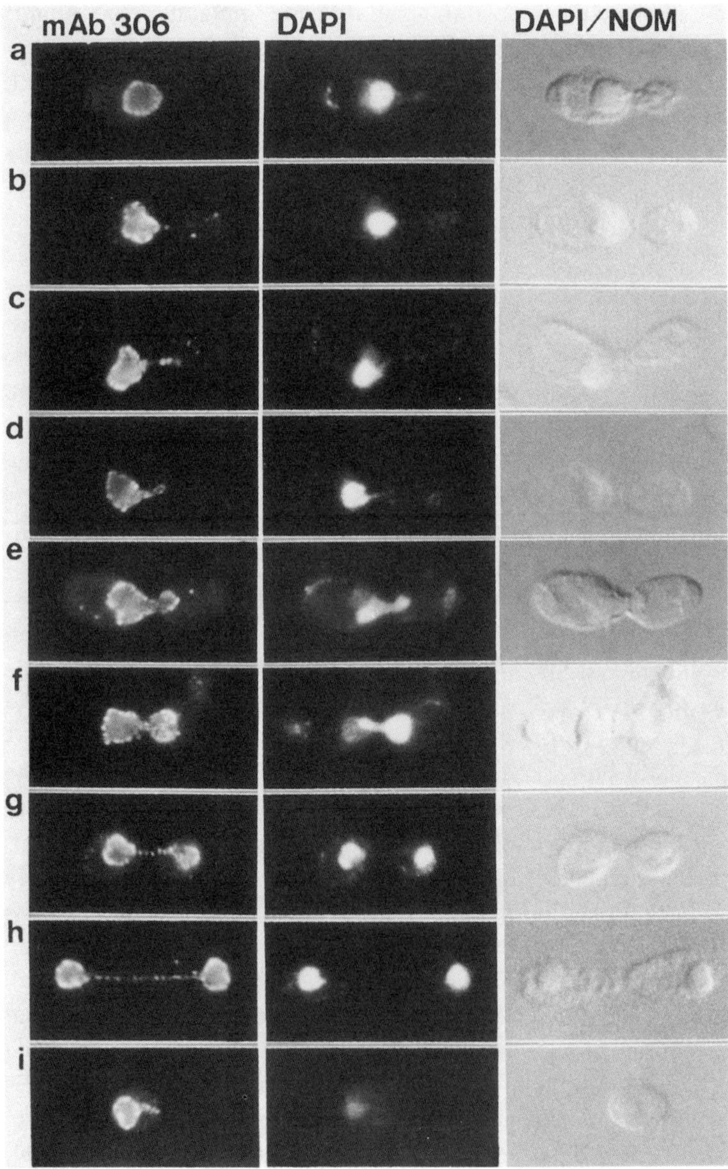

Figure 4 Nuclear envelope dynamics during budding and mitosis in *S. cerevisiae*. Indirect immunofluorescence was performed using a monoclonal antibody (mAB 306) against the Nup1p nucleoporin (*left panels*). DAPI staining marks the DNA (*middle panels*). Whole cells/nuclei are shown in the right column by differential interference contrast optics plus DAPI fluorescence. (Reprinted, with permission, from Davis and Fink 1990 [copyright Cell Press].)

nection between motor function on microtubules and nuclear movement, although the precise mechanism of dynein action remains unclear.

The role of actin is also not well understood, although mutation in an actin-like protein related to the dynactin complex polypeptide Arp1 (*ACT5/ACT3*; Clark and Meyer 1994; Muhua et al. 1994) causes impairment in nuclear migration and spindle orientation. The dynactin complex is known to interact with dynein itself in higher eukaryotic systems (Gill et al. 1991; Schroer and Sheetz 1991), and *act5 dhc1* double mutants have phenotypes similar to those of the single mutants. Deletion of the actin capping protein gene (*CAP1*) also led to impaired nuclear migration and, in combination with an *act5* mutation, caused more severe nuclear migration and spindle orientation defects than did either mutation alone (Muhua et al. 1994). These data suggest cooperation or crosstalk between the actin- and microtubule-based cytoskeletal networks during nuclear movement and spindle orientation. Since actin is polarized in the bud, it seems likely that microtubule-based nuclear movement is given directionality by anchoring to actin-based structures residing in the bud itself.

During mating, nuclei from the two haploid cells migrate together such that their nuclear envelopes are juxtaposed with their SPBs adjacent to one another. Subsequent fusion of the nuclear envelopes (karyogamy) forms the diploid nucleus. Elegant genetic strategies have identified two classes of genes required for karyogamy (*KAR* genes: Conde and Fink 1976; Polaina and Conde 1982; Kurihara et al. 1994; for review, see Rose 1996). Wild-type SPBs and cytoplasmic microtubules are required for mating (Rose 1991). It is therefore not surprising that one class of genes identified in these screens encodes SPB components (e.g., Kar1p) and microtubule-based motors (e.g., the kinesin homolog Kar3p) (see Meluh and Rose 1990; Vallen et al. 1992) that are required for nuclear movement. A second class of *KAR* gene products may have direct roles in nuclear fusion. Microscopic analysis of *kar2, kar5, kar7*, and *kar8* mutants reveals that the haploid nuclear membranes are closely abutting but not fused (Rose et al. 1989; Kurihara et al. 1994). *KAR2* encodes a lumenal protein of the ER and is homologous to BiP, a molecular chaperone of the Hsp70 class. *KAR7* (also called *SEC71*) encodes an integral membrane protein of the ER (M. Rose, pers. comm.). Given the continuity of the outer nuclear membrane with the ER (see Section I), the localization of these proteins is consistent with their apparent involvement in nuclear membrane fusion.

Nuclear envelope biogenesis probably occurs via membrane flow from the ER, but this process is not well understood. Normally regulated growth can be perturbed, however, by overexpression of genes encoding several different ER membrane proteins. For example, a tenfold transcriptional induction of the *HMG1* gene (one of two genes encoding HMG-CoA reductase, which catalyzes an early step in steroid biosynthesis), results in the proliferation of "karmellae," or stacked membrane pairs surrounding the nucleus (Wright et al. 1988). Up to ten extra membrane pairs may be formed, although the karmellae do not completely surround the nucleus. Small regions of the original nuclear envelope remain and contain the NPCs and SPBs. The karmellae are not inherited by daughter cells upon cell division. Overexpression in yeast of rat cytochrome b_5 (Vergères et al. 1993), the chicken lamin B receptor (another integral nuclear membrane protein; Smith and Blobel 1994), or *Candida maltosa* cytochrome P450 (Schunck et al. 1991) also results in the formation of karmellae. However, this phenotype is not associated with all ER integral membrane proteins, as overexpression of *SEC63* (whose product is required for polypeptide translocation into the ER) does not stimulate formation of karmellae (R. Wright, pers. comm.), and overexpression of *HMG2* (which encodes the other isozyme of HMG-CoA reductase) produces patterns of membrane proliferation distinct from those seen with *HMG1* (Koning et al. 1996). How overexpression of *HMG1* stimulates the formation of karmellae is unclear, although the region of Hmg1p responsible for the effect is thought to reside in the lumen of the ER (Parrish et al. 1995), suggesting that the machinery for membrane proliferation also resides therein.

B. Communication Across the Nuclear Envelope

The nuclear envelope separates the synthesis of RNA from that of proteins both spatially and temporally. Cell viability depends on the continual exchange of macromolecules and metabolites across this double bilayer. A series of elegant studies in vertebrate cells utilizing microinjection of tracer molecules and cell-free transport systems has led to the conclusion that the nuclear pore complex is the only site for both inward and outward nuclear transport (Paine et al. 1975; Feldherr et al. 1984; Dworetzky and Feldherr 1988). Nuclear pores are formed where the inner and outer nuclear membranes fuse (see Fig. 3). The membrane spanning the pore itself is termed the pore membrane. A supramolecular assembly (the NPC) resides within the pore and is represented in thin-section electron micrographs as a dense mass approximately 100 nm in diameter. Small molecules and metabolites diffuse through the aqueous channels of the NPCs (which have an effective diameter of ~9 nm and should allow the passage of globular proteins up to ~65 kD). However, macromolecules ranging in size from 20 kD to greater than 100 MDa traverse the envelope in an energy-dependent, receptor-mediated manner (for review, see Gerace and Burke 1988; Dingwall 1991; Miller et al. 1991; Akey 1992; Forbes 1992; Davis 1995).

1. Nuclear Pore Complex Structure

All eukaryotic NPCs are thought to be architecturally similar, and NPCs from all organisms studied have an eightfold symmetry and filamentous networks radiating outward from their cytoplasmic and nucleoplasmic faces (see the yeast NPCs in Fig. 3B,C). The octagonal symmetry of the yeast NPC is especially evident in the enlarged insets in Figure 5 (Rout and Blobel 1993). A combination of high-resolution cryoelectron microscopy and transmission scanning electron microscopy techniques with nuclear envelope preparations from *Xenopus* oocytes and mammalian cells has revealed the major structural components of the vertebrate NPC (Unwin and Milligan 1982; Akey 1989; Ris 1991; Goldberg and Allen 1992, 1993; Hinshaw et al. 1992; Akey and Radermacker 1993; Ris and Malecki 1993), and three-dimensional reconstructions have resolved its basic framework to a level of approximately 10 nm (Hinshaw et al. 1992; Akey and Radermacker 1993). A schematic drawing of the proposed NPC structure is shown in Figure 6 (Rout and Wente 1994). The filamentous structures (CF and NB) extend from both the cytoplasmic and nuclear faces, with the former attached to particles (P) and the latter forming a basket-like structure. The overall architecture is based on eight equally spaced spokes (S) that are sandwiched between two rings (CR and NR). The spokes are also attached to an inner spoke ring (IS) that encompasses a central plug (CP), and the spokes appear to traverse the pore membrane for connection to an outer ring (OS) in the nuclear envelope (NE) lumen. In these higher eukaryotic cells, well-defined nuclear lamina (L) and nuclear envelope lattice (NEL) structures also seem to have connections to the NPC. Such fine details of structure are not clear for the yeast NPC at the level of resolution afforded by standard electron microscopy technology.

On the basis of its size in electron micrographs (Reichelt et al. 1990), it is estimated that the molecular mass of the vertebrate NPC is approximately 1.25×10^8 daltons. Measurements by Rout and Blobel (1993) of the electron-dense mass of isolated NPCs and of NPCs in isolated nuclei suggest that the yeast NPC is smaller than that of vertebrate cells (~30 nm in thickness vs. the ~70–80 nm of *Xenopus* NPCs). These investigators have proposed that the smaller dimensions may reflect the absence

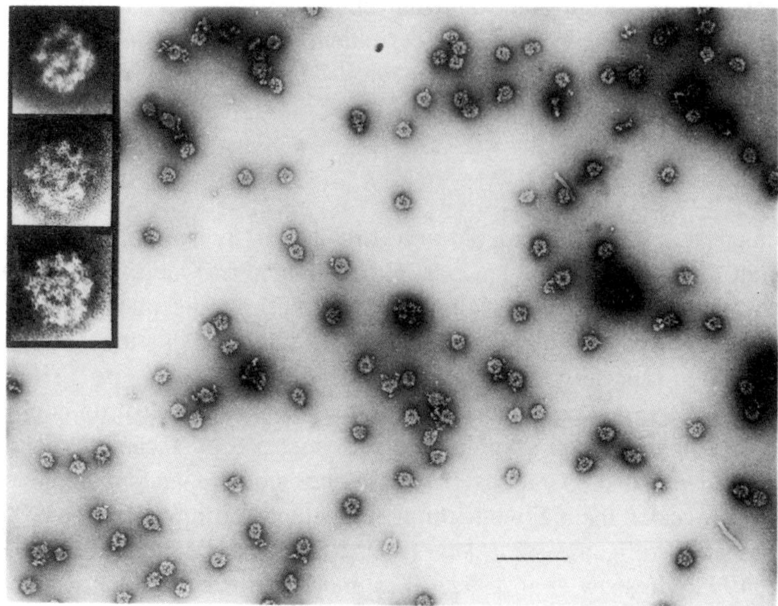

Figure 5 Electron micrograph of negatively stained particles from a highly enriched, detergent-solubilized yeast NPC fraction. Bar, 0.5 μm. (*Inset*) Three examples at high magnification (diameter of each particle is 90 nm). (Reprinted, with permission, from Rout and Blobel 1993.)

of cytoplasmic and/or nuclear NPC ring structures in yeast (Rout and Blobel 1993). Some of this difference may be due either to the loss of NPC components during the biochemical fractionation of the NPCs and nuclei or to less extensive uranyl acetate staining in the nuclear thin sections. The apparent size difference may also reflect actual differences in protein composition between yeast and other eukaryotic NPCs. Pinpointing the differences and similarities between yeast and vertebrate NPC structure and composition will be key to understanding nucleocytoplasmic trafficking in all cells.

2. *Nuclear Pore Complex Composition*

The large size of the NPC, in combination with its myriad of transport functions, implies that a large number of different polypeptides comprise its structure; some recent estimates suggest that there are at least 50 or more different NPC proteins (nucleoporins) (Reichelt et al. 1990; Rout and Blobel 1993). Several recent reviews have described in depth the current status of attempts to define NPC polypeptide composition (Panté

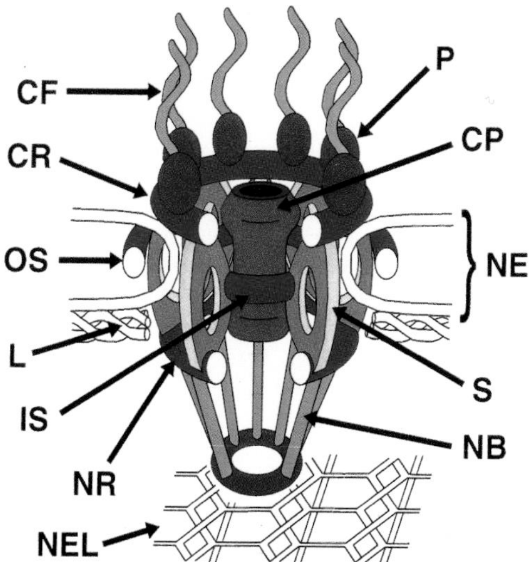

Figure 6 Schematic model of vertebrate NPC structure. The cytoplasmic face is uppermost, and sections of facing portions have been removed for clarity. (CF) cytoplasmic filament; (P) particles; (CR) cytoplasmic ring; (CP) central plug; (OS) outer spoke ring; (NE) nuclear envelope; (L) lamina; (IS) inner spoke ring; (S) spoke; (NR) nuclear ring; (NB) nuclear basket; (NEL) nuclear envelope lattice. Structures are drawn approximately to scale with the pore through the membrane of ~90 nm and the diameter of the OS ~120 nm. (Reprinted, with permission, from Rout and Wente 1994.)

and Aebi 1994; Rout and Wente 1994; Davis 1995). The molecular characterization of the first two yeast nucleoporins, Nup1p and Nsp1p, was reported in 1990 (Davis and Fink 1990; Nehrbass et al. 1990). In the short six years since then, the catalog of known yeast nucleoporins has grown to include 20 different polypeptides (Table 1). Immunological, biochemical, and genetic strategies have all been successfully employed to clone the genes encoding these proteins.

The first yeast nucleoporins were identified immunochemically. A number of monoclonal antibodies generated independently by several laboratories bind in a polyspecific manner to multiple vertebrate nucleoporins (Davis and Blobel 1986; Park et al. 1987; Snow et al. 1987). The structural basis of this recognition is conserved epitopes in domains of FG, FXFG, or GLFG peptide repeats shared by the respective nucleoporins (see below; for review, see Rout and Wente 1994; Davis 1995). Several of these monoclonal antibodies also possess extensive interspecies cross-reactivity and recognize homologous regions in

Table 1 Yeast nuclear pore complex proteins

NPC protein[a]	Synonyms	Motifs[b]	Essential?	Effect of mutant[c] perturbations on			References
				import	export	NPC or NE structure	
Gle2p (40)			no, ts	no	yes	yes	1
Nic96p			yes	yes	no	yes	2–4
Nsp1p (86)		FXFG	yes	yes	no	yes	5–7
Nup1p (113)		FXFG	yes or ts	yes	yes	yes	8–10
Nup2p (78)		FXFG	no			n.d.	11
Nup49p	Nsp49p	GLFG	yes	yes	yes	n.d.	12–15
Nup57p		GLFG	yes	n.d.	n.d.	n.d.	2,12
Nup82p			yes	no	yes	no	16,17
Nup84p			no, ts	no	yes	yes	18
Nup85p	Rat9p		no, ts	no	yes	yes	18,19
Nup100p		GLFG	no			no	12,20
Nup116p	Nsp116p	GLFG	no, ts	n.d.	yes	yes	12,21
Nup120p	Rat2p		no, ts	no	yes	yes	22,23
Nup133p	Rat3p		no, ts	no	yes	yes	15,24,25
Nup145p	Rat10p	GLFG	yes	yes	yes	yes	20,26

Nup157p			no			no	27
Nup159p	Rat7p	FG	yes	no	yes	yes	28,29
Nup170p	Nle3p		no			yes	27,30
Nup188p			no	no	no	yes	4,23,31
Pom152p		transmemb.	no			no	32
Seh1p (39)	Sec13 homolog		no	n.d.	no	n.d.	18

References: (1) Murphy et al. 1996; (2) Grandi et al. 1995a; (3) Grandi et al. 1993; (4) Zabel et al. 1996; (5) Hurt 1988; (6) Nehrbass et al. 1990; (7) Mutvei et al. 1992; (8) Davis and Fink 1990; (9) Belanger et al. 1994; (10) Bogerd et al. 1994; (11) Loeb et al. 1993; (12) Wente et al. 1992; (13) Wimmer et al. 1992; (14) Schlenstedt et al. 1993; (15) Doye et al. 1994; (16) Grandi et al. 1995b; (17) Hurwitz and Blobel 1995; (18) Siniossoglou et al. 1996; (19) Goldstein et al. 1996; (20) Wente and Blobel 1994; (21) Wente and Blobel 1993; (22) Heath et al. 1995; (23) Aitchison et al. 1995a; (24) Pemberton et al. 1995; (25) Li et al. 1995; (26) Fabre et al. 1994; (27) Aitchison et al. 1995b; (28) Gorsch et al. 1995; (29) Kraemer et al. 1995; (30) Kenna et al. 1996; (31) Nehrbass et al. 1996; (32) Wozniak et al. 1994.

[a]Nomenclature for NPC proteins: (Gle) GLFG lethal; (Nic) Nucleoporin-interacting components; (Nsp) nucleoskeletal-like protein; (Nup) nuclear pore protein; (Pom) pore membrane protein; (Seh) Sec13p homolog, with numerical assignment reflecting either predicted molecular mass (kD) or genetic identification (with the predicted molecular mass following in parentheses). Synonyms are other names also found in the literature (Rat, for mRNA trafficking; Nle, for *nup1* lethal).

[b]The polypeptides contain: (GLFG) multiple repeats of glycine-leucine-phenylalanine-glycine, separated by uncharged spacers; (FXFG) multiple repeats of phenylalanine-X-phenylalanine-glycine, separated by highly charger spacers; (FG) multiple repeats of phenylalanine-glycine; (Transmemb.) span(s) of hydrophobic sequence characteristic of transmembrane domains.

[c]"Mutant" refers to a variety of non-wild-type configurations including viable/conditional null alleles, conditional temperature- or cold-sensitive alleles; *GAL*-regulated overexpression in wild-type cells, or *GAL*-regulated depletion in the background of a lethal mull mutation. (n.d.) Not determined. Blank boxes indicate the gene is not essential and a conditional background appropriate for assaying transport has not identified.

nucleoporins of yeast (among other organisms) (Aris and Blobel 1989; Davis and Fink 1990; Carmo-Fonseca et al. 1991; Wente et al. 1992). This cross-reactivity has been exploited by immunoscreening expression libraries to isolate genes encoding six *S. cerevisiae* proteins (Davis and Fink 1990; Wente et al. 1992; Loeb et al. 1993; Wente and Blobel 1994), and a seventh gene was identified using a polyclonal antibody raised against yeast nuclear fractions (Hurt 1988).

In biochemical approaches, numerous attempts have been made to isolate NPCs and their constituent proteins from yeast nuclei (Hurt et al. 1988; Aris and Blobel 1988, 1989; Allen and Douglas 1989; Cardenas et al. 1990). Recently, two distinct strategies have proven effective. First, protocols that yield remarkably enriched fractions either of nuclear envelopes (Strambio-de-Castillia et al. 1995) or of detergent-solubilized NPC particles (Rout and Blobel 1993) have been devised. The strategy for NPCs is based on detergent lysis of purified *Saccharomyces uvarum* nuclei followed by sequential sedimentation steps. A representative field of such NPCs is shown in Figure 5. Approximately 80 distinct polypeptides can be resolved electrophoretically and chromatographically from such preparations (Rout and Blobel 1993; Wozniak et al. 1994). Second, subcomplexes of nucleoporins have been purified by an affinity chromatographic approach. Hurt and co-workers found that by fusing the coding sequence of Nsp1p to five tandem repeats of the IgG-binding region of protein A, they could copurify at least four other nucleoporins (Nic96p, Nup82p, Nup49p, and Nup57p) with the protein A–Nsp1p (Grandi et al. 1993, 1995a,b). After sequencing peptides from several of these NPC-associated proteins, the correponding genes have been identified either by database searches of the yeast genome or by conducting the polymerase chain reaction (PCR) with degenerate oligonucleotides encoding the amino acid sequences (Grandi et al. 1993, 1995a,b; Wozniak et al. 1994; Aitchison et al. 1995a,b; Hurwitz and Blobel 1995; Kraemer et al. 1995; Pemberton et al. 1995).

Finally, several genetic approaches have also led to the isolation of multiple yeast nucleoporin genes (for review, see Doye and Hurt 1995). One strategy has employed screens for new mutations that are lethal in combination with a mutation in a previously known nucleoporin gene. Such "synthetic lethal" screens are modeled after that described by Bender and Pringle (1991) and may identify genes that are redundant in function or whose protein products physically interact. Such a screen with an *nsp1*^{ts} mutant has successfully identified mutant alleles of multiple different genes encoding bona fide nucleoporins (*NUP49, NUP116, NUP145*, and *NUP84*) (Wimmer et al. 1992; Fabre et al. 1994; Siniossoglou et al. 1996). Similarly, a screen with *nup49*^{ts} identified *NUP133* (Doye et al. 1994); a screen with *nup1*^{ts} has identified up to seven different complementation groups, two of which are *srp1* (a nuclear import factor; see below) and *nup170* (Belanger et al. 1994; Kenna et al. 1996); and a screen with a *pom152* null yielded *nic96*, *nup170*, and *nup188* mutants (Aitchison et al. 1995b).

Another fruitful genetic strategy has been to screen banks of temperature- or cold-sensitive mutants directly for those that are transport-defective. In situ hybridization and immunofluorescence microscopy have determined the location of polyadenylated RNA in the arrested cells, and strains that accumulate poly(A) RNA in the nucleus at the non-permissive temperature were isolated (Amberg et al. 1992; Kadowaki et al. 1992, 1994b; Noble and Guthrie 1996). Thus far, from the screen conducted by Cole and co-workers, five different *rat* (for m*R*NA *t*rafficking) complementation groups have been shown to involve mutations in genes encoding NPC proteins (Table 1) (Gorsch et al. 1995; Heath et al. 1995; Li et al. 1995; Goldstein et al. 1996; C. Cole, pers. comm.). It is striking that in many cases, the same nucleoporins have been identified independently by both genetic and biochemical/immunological strategies (Table 1: Nup49p, Nup116p, Nup120p, Nup133p, Nup145p, and Nup159p).

Yeast nucleoporins can be classified in part by the primary structure domains and peptide motifs that they contain (Rout and Wente 1994). First, the central regions of Nup1p (Davis and Fink 1990), Nup2p (Loeb et al. 1993), and Nsp1p (Nehrbass et al. 1990) contain multiple repeats of "FXFG" peptides separated by highly charged spacer sequences (29, 15, and 22 repeats, respectively). A second group including Nup49p (Wente et al. 1992; Wimmer et al. 1992), Nup57p (Grandi et al. 1995a), Nup100p (Wente et al. 1992), Nup116p (Wente et al. 1992; Wimmer et al. 1992), and Nup145p (Fabre et al. 1994; Wente and Blobel 1994) lacks the multiple FXFG repeats; instead, their amino-terminal regions have multiple repeats of "GLFG" separated by uncharged spacers (13, 9, 29, 33, and 12 repeats, respectively). The FXFG region from Nup1p and the GLFG region from Nup116p are shown as examples in Figure 7. Recently, a yeast nucleoporin (Nup159p) with an extensive region of FG repeats separated by both uncharged and charged spacers has been characterized (Gorsch et al. 1995). There are also multiple nucleoporins that lack any of these repetitive motifs (Table 1).

The prevalence of repetitive motifs in nucleoporins is intriguing. These regions may serve as domains that assemble into the filamentous structures of the NPC (Kraemer et al. 1994, 1995; Panté et al. 1994; Wu et al. 1995). Such a location would be required if the repetitive-motif region of these nucleoporins serves a role in the docking step of transport (see Section IV.C.2). Many of these repetitive regions appear to have redundant roles, as deletion of the FXFG region from Nsp1p or Nup2p (Nehrbass et al. 1990; Loeb et al. 1993) or deletion of the GLFG region from Nup49p, Nup57p, Nup100p, or Nup145p (Wente et al. 1992; Wente and Blobel 1994; Grandi et al. 1995a; Iovine et al. 1995) has no effect on cell viability. However, the GLFG region of Nup116p may have a unique function, as its deletion does confer a loss-of-function phenotype (Iovine et al. 1995) (see Section IV.C.2).

One known protein with a transmembrane segment, Pom152p, is associated with the yeast NPC and thus localized in the pore membrane

Nup116p

		TTSATTG
205	GLFG	QKPAT
214	GMFG	TGTGSG
224	GGFG	SGATNST
235	GLFG	SSTNLSGNSAFGANKPATSG
259	GLFG	NTTNNPTNGTNNT
276	GLFG	QQNSNTNG
288	GLFG	QQQNSFGANNVSNG
306	GAFG	QVNRGAFPQQQTQQGSG
327	GIFG	QSNANANG
339	GAFG	QQQGTG
349	ALFG	AKPASG
359	GLFG	QSAGSKAFGMNTNPTGTTG
382	GLFG	QTNQQQSGG
395	GLFG	QQQNSNAG
407	GLFG	QNNQSQNQS
420	GLFG	QQNSSNAFGQPQQQG
439	GLFG	SKPAG
448	GLFG	QQQGASTFASGNAQNNSIFGQNNQQQQSTG
482	GLFG	QQNNQSQSQPG
497	GLFG	QTNQNNNQPFGQNGLQQPQQNN
523	SLFG	AKPTGFGNT
536	GLFS	NSTTNQSNGISGNNLQQQSG
560	GLFQ	NKQQPASG
572	GLFG	SKPSNTVGG
585	GLFG	NNQVANQNNPASTSG
604	GLFG	SKPATG
614	SLFG	GTNSTAPNASSG
630	GIFG	SNNASNTAATTNST
648	GLFG	NKPVGAGASTSAG
665	GLFG	NNNNSSLNNSNGST
683	GLFG	SNNTSQSTNAG
698	GLFQ	NNTSTNTSGG
712	GLFS	QPSQPMAQSQNALQQQQQQRLQIQNNNPYGTNE

Figure 7 (*See facing page for legend and part 2 of art.*)

(Wozniak et al. 1994). In addition, three of the GLFG nucleoporins (Nup100p, Nup116p, and Nup145p) have peptide octamers in their non-GLFG regions that may confer their affinity for homopolymeric poly(G) RNA (Fabre et al. 1994). The role of this affinity in NPC structure and function remains to be determined.

As noted above, the NPCs of all eukaryotic organisms are believed to be functionally similar. Comparison of the amino acid sequences of the known yeast nucleoporins to those reported from vertebrates has revealed some striking similarities. However, to date, there is only one report of functional complementation by heterologous expression of a

Nup1p

		EPKKDKESIVLP
343	TVGFDFIKD	NETPSKKTSPKATSSAGAVF
362	KSSVEMGKT	DKSTKTAEAP
381	TLSFNFSQK	ANKTKAVDNTVPS
403	TTLFNFGGK	STDVTSA
419	SQPFKFGKT	SEKSENHTESDAPPKST
445	APIFSFGKQ	EENGDEGDDENEPKRKRRLPVSEDTNT
481	KPLFDFGKT	GDQKETKKGESEKDASG
507	KPSFVFGAS	DKQAEG
522	TPLFTFGKK	ADVTSNIDS
540	SAQFTFGKA	ATAKETHTKPSETPATIVK
568	KPTFTFGQS	TSENKISEGSA
588	KPTFSFSKS	EEERKSSPSNEAA
611	KPSFSFPGK	PVDVQAPTDDKTL
633	KPTFSFTEP	AQKDSSVVSEPK
654	KPSFTFASS	KPSQP
668	KPLFSFGKS	DAAKEPPGS
686	NTSFSFTKP	PANETDKRPT
705	PPSFTFGGS	TTNNTTTTST
724	KPSFSFGAP	ESMKSTASTAAANTEKL
750	SNGFSFTKF	NHNKEKSNS
768	PTSFFDGSA	SSTPIPVLGKPTDATGNTTS
797	KSAFSFGTA	NTNGTNASAN
816	STSFSFNAP	ATGNGTTTTSNTSGTNI
842	AGTFNVGKP	DQSIASGNTNGA
863	GSAFGFSSS	GTAATGAASN
882	QSSFNFGNN	GAGGLNPFTSATS
904	STNANAGLF	NKPPSTNAQNVNV
926	PSAFNFTGN	NSTPGG
941	GSVFNMNGN	

Figure 7 The "FXFG" repeat region of Nup1p and the "GLFG" repeat region of Nup116p. The amino acid sequence of each region has been aligned such that the consensus repeat motifs are shown in the middle column with the spacer sequences in the right column. Amino acid positions in the context of the full-length proteins are designated in the left column.

vertebrate nucleoporin gene in a yeast mutant: Rat Nup155p complements a *nup170 pom152* double-mutant strain (Aitchison et al. 1995b). Moreover, the precise sublocalization of any yeast nucleoprotein to a distinct structure of an NPC (i.e., filaments, cytoplasmic face, nucleoplasmic face) also awaits advances in yeast immunoelectron microscopic resolution.

3. *Nucleoporin Transport Function*

Although many of the nucleoporins are essential NPC components, defining their precise roles in nuclear import, nuclear export, or both has proven difficult. Part of the difficulty is due to the intrinsic overlap in the structural and enzymatic requirements of the import and export processes (see Section IV.C). A single nucleoporin may also be involved both in transport and in assembly of the NPC and/or nuclear envelope pore. To evaluate the nucleoporin roles, mutant strains have been assayed for perturbations of nuclear protein import, poly(A) RNA export, and NPC/nuclear envelope morphology.

Both in vitro and in vivo assays can be used to monitor transport capacity. Protein import can be assayed in vitro by monitoring the nuclear accumulation of a fluorescently labeled substrate in semi-intact spheroplasts (Schlenstedt et al. 1993). More commonly, an in vivo assay is employed. One such assay monitors by immunofluorescence microscopy whether endogenous nuclear proteins have been mislocalized (Nehrbass et al. 1990; Loeb et al. 1995). A second in vivo import assay relies on the induced overexpression of a reporter protein and measurement of its nuclear accumulation by immunofluorescence microscopy after shifting a conditional nucleoporin mutant to restrictive conditions (see, e.g., Gu et al. 1992; Nehrbass et al. 1993; Bogerd et al. 1994; Doye et al. 1994; Fabre et al. 1994). Data obtained using in vivo assays with reporter proteins must be interpreted cautiously. First, as export of the induced mRNA is also required, a lack of nuclear protein accumulation may be due to a primary defect in mRNA export. Second, the half-lives of the reporter proteins have not been assessed. If they have slow turnover times in the nucleus, any nuclear accumulation from before the shift to nonpermissive conditions may persist and be aberrantly scored as import. An alternative in vivo assay monitors the import of a green fluorescent protein reporter (Shulga et al. 1996). This protocol may prove quite powerful because it also measures perturbations of import rates. To measure nuclear export capacity, the location of poly(A) RNA can be monitored by in situ hybridization as described in Section IV.C.4 (with nuclear accumulation reflecting a block in export) (Amberg et al. 1992; Forrester et al. 1992; Kadowaki et al. 1992). Finally, the transport function of a particular nucleoporin may also be suggested by determining its nearest-neighbor protein-protein interactions (see Section IV.C.2).

With both in vitro and in vivo transport assays, a rapid and penetrant phenotype would provide key evidence that the mutation has a direct effect on transport (and thus that the affected protein has a direct role in transport). The best example to date is the analysis of a *nup159* temperature-sensitive strain: Export is blocked very rapidly (within 15 minutes), whereas import remains apparently unaffected (Gorsch et al. 1995). A transport defect might also be the indirect consequence of an unusual structural perturbation of the NPC and/or nuclear envelope, such as those seen in cells lacking Nup116p (Figs. 8 and 9A) (Wente and Blobel 1993). The results of such analyses to date of nucleoporin function are summarized in Table 1.

Two other phenotypes have also been reported to correlate with a lack of nuclear transport. Some yeast mutants that are defective for nuclear transport coincidentally have altered nucleolar morphology (for review, see Schneiter et al. 1995). As noted above, the nucleolus in wild-type yeast cells is a single crescent-shaped organelle. Fragmentation and unfolding of nucleolar structure into multiple intranuclear foci have been observed with mutant alleles of *MTR2*, *RNA1*, *PRP20*, and *SRP1* (all discussed in more detail below) (Aebi et al. 1990; Kadowaki et al. 1994a,b; Yano et al. 1994; Loeb et al. 1995). In addition, transport mutants can also accumulate unspliced pre-tRNA, pre-rRNA, and/or pre-mRNA (see, e.g., Traglia et al. 1989; Aebi et al. 1990; Kadowaki et al. 1993, 1994b; Sharma et al. 1996). Both of these phenotypes are probably pleiotropic defects resulting from the probable dependence of nucleolar structural integrity and the splicing process on effective transport of splicing and ribosomal components.

4. Nuclear Pore Complex Biogenesis

In addition to mediating the translocation of macromolecules to and from the nucleus, individual nucleoporins may be required for pore and NPC assembly. A requisite step in the biogenesis of an NPC is the formation of the pore through the nuclear envelope, presumably via a regulated fusion of the outer and inner nuclear membranes. On the basis of measurements of total pore number per cell, yeast NPC formation appears to peak twice during the cell cycle, in early G_1 and in mitosis (Jordan et al. 1977). Some basal level of NPC turnover probably also occurs indepen-

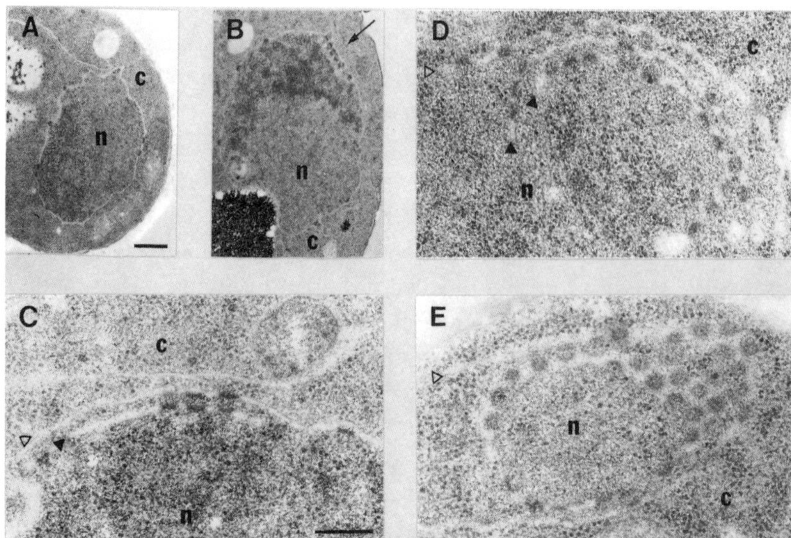

Figure 8 Thin-section electron micrographs of intranuclear structures found in cells lacking the Nup116p nucleoporin. These annulate lamellae-like structures are characterized by double membranes studded with NPC structures. They are present at the permissive (this figure, all panels) and nonpermissive (arrow, Fig. 9A) growth temperatures. (*A*) Cross section through the nucleus of a wild-type yeast cell. (*B*) Cross section through the nucleus of a cell lacking Nup116p. Arrow points to a section of intranuclear "annulate lamella" at low magnification. (*C,D*) Cross sections through two (*C*) or three (*D*) stacked double lipid bilayers, each studded with NPCs, at higher magnification. (*Open arrowheads*) Outermost envelopes; (*solid arrowheads*) intranuclear bilayers. (*E*) Tangential section along an invagination of intranuclear annulate lamella. (*A,B*) 0.5 μm; (*C–E*) 0.25 μm. (c) Cytoplasm; (n) nucleoplasm.

dent of the cell cycle stage. Membrane fusion may also be required for the disassembly of a nuclear pore and the restoration of an uninterrupted double-membrane structure. The mechanism by which these membrane fusion events occur has not been elucidated. Given the common requirements for nuclear membrane fusion events in karyogamy and ER homotypic fusion (Kurihara et al. 1994; Latterich and Schekman 1994; Latterich et al. 1995; for review, see Wilson 1995), it is interesting to speculate that the membrane fusion step in NPC biogenesis may require the same proteins.

Integral membrane proteins localized specifically to the membrane surrounding the NPC (such as Pom152p in yeast) are thought to serve as anchors for peripheral nucleoporins (Gerace et al. 1982; Gerace and Foisner 1994). However, it has not been determined whether Pom152p

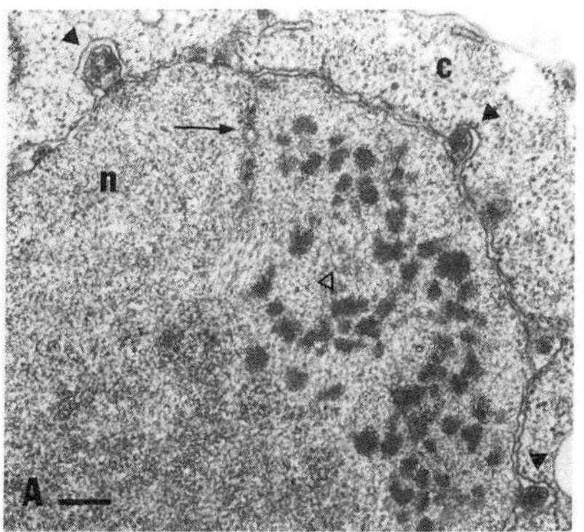

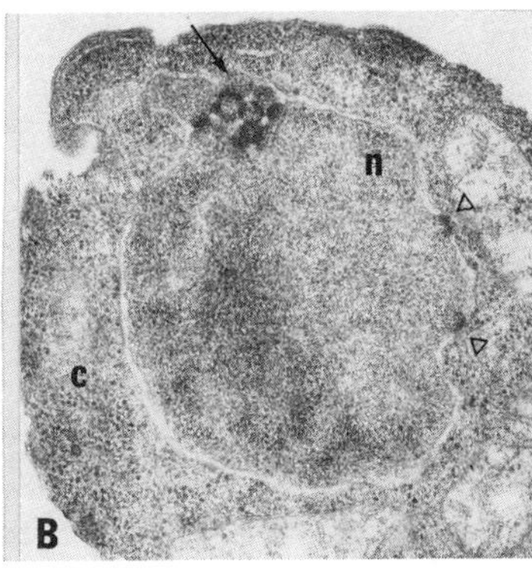

Figure 9 Thin-section electron micrographs of aberrant NPC structures. (*A*) Herniations that form in a temperature-dependent manner in cells lacking the Nup116p nucleoporin (*closed arrowheads*); inner nuclear membrane invagination forming intranuclear "annulate lamellae" (*arrow*); and intranuclear electron-dense particles possibly representing nucleolar fragmentation (*open arrowhead*). (Reprinted, with permission, from Wente and Blobel 1993.) (*B*) Clusters of NPCs formed constitutively in cells lacking the amino-terminal region of the Nup145p nucleoporin (*arrow*). The nucleus also contains apparently intact single NPCs (*arrowheads*). Bar, 0.25 μm. (n) Nucleoplasm; (c) cytoplasm. (Reprinted, with permission, from Wente and Blobel 1994.)

itself serves as such an anchor or what its nearest neighbors are in terms of peripheral NPC proteins. Also intriguing is the possible role of other nuclear envelope integral membrane proteins. For example, *ndc1* mutant cells were originally characterized for their aberrant SPB assembly; however, Ndc1p has been localized over the entire nuclear envelope surface by immunofluorescence microscopy (Winey et al. 1993). SPB duplication in *ndc1*-arrested cells is blocked at the point of insertion into the nuclear membranes. It remains to be seen whether Ndc1p is required for a similar step in formation of the nuclear pore.

To maintain nuclear integrity, formation of the nuclear pore and assembly of the peripheral nucleoporins must be coincident. In wild-type yeast cells, no cytoplasmic pool of the peripheral nucleoporins from either the GLFG family (Wente et al. 1992) or the FXFG family (Aris and Blobel 1989) has been detected. Thus, it is presumed that these nucleoporins are rapidly assembled into NPCs. However, no studies have directly addressed the biosynthesis of any individual NPC polypeptide and its incorporation into the yeast NPC.

Recently, insights into the dynamics of NPC structure and assembly have been afforded by the morphological analysis of nucleoporin mutant strains. A striking array of different NPC-specific perturbations have been observed. Nuclear migration defects observed in *nup1* mutant strains (Bogerd et al. 1994) are also accompanied by extensive perturbations of the nuclear envelope. Thin-section electron micrographs of the *nup1* cells show extensive lobulation of the nucleus and massive projections of the nuclear envelope (Bogerd et al. 1994). These events presumably require an enormous proliferation of nuclear/ER membrane. In contrast, depletion of Nsp1p (a related FXFG nucleoporin) results in a coincident decrease in NPC number but without any apparent perturbations of the remaining NPCs or the nuclear envelope (Mutvei et al. 1992). Revealing the structural basis of these mutant perturbations may provide insight into the pathways for NPC and nuclear envelope biogenesis.

Two different NPC/nuclear envelope structural perturbations are observed in temperature-sensitive mutant cells that lack the GLFG nucleoporin Nup116p (Wente and Blobel 1993). At permissive growth temperature (23°C), the only ultrastructural differences between wild-type cells and mutant cells are unusual invaginations of the inner nuclear membrane (Fig. 8). These invaginations are usually observed in cross section as stacks of intranuclear double membranes that lie parallel to the outermost nuclear envelope. These intranuclear membranes are studded with densities that resemble NPCs. Similar structures have also been observed when *NUP170* is overexpressed using the *GAL10* promoter in a *pom152* null mutant (Aitchison et al. 1995b). Such invaginations have not been observed in wild-type yeast cells, yet they are morphologically very similiar in structure to the intranuclear annulate lamellae that have been found in other eukaryotic cells (Kessel 1983). It has been proposed that annulate lamellae may serve as an NPC storage form in cells poised to divide rapidly (Stafstrom and Staehelin 1984). Therefore, it is interesting to speculate that the *nup116* mutation or *NUP170* overexpression in the absence of Pom152p may affect the assembly pathway for NPCs such that an accumulation of NPC precursors results in the formation of structures resembling intranuclear annulate lamellae.

At nonpermissive growth temperature (37°C), the intranuclear annulate lamellar structures still exist (Fig. 9A), but there is also a dramatically different perturbation of the Nup116p-depleted NPCs in the nuclear envelope. Although the mutant NPCs still appear to be in place and to be membrane-anchored, a double-membrane seal continuous with the inner and outer nuclear membranes forms over the cytoplasmic faces of the NPCs (Fig. 9A). Electron-dense material accumulates between the cytoplasmic faces of these NPCs and the membrane seal, resulting in "herniations" of the nuclear envelope around individual NPCs. This apparent "sealing off" of NPCs prevents nucleocytoplasmic trafficking, blocking the export of polyadenylated RNA from the nucleus (Wente and Blobel 1993). Although such herniated structures are not seen in wild-type yeast cells, a very similar structure has been reported in a single micrograph of a chimpanzee zygote nuclear envelope (Maul 1977). The absence of Nup116p may affect the stability of intact NPCs, the pore membrane, and/or the surrounding nuclear envelope.

Freeze-fracture and thin-section electron microscopic analyses have revealed that wild-type NPCs are distributed over the nuclear surface at a density of 15 NPC/μm^2 (e.g., see Figs. 1 and 3) (Severs et al. 1976; Jordan et al. 1977). However, NPCs are excluded from areas where the vacuolar and nuclear membranes abut (Severs et al. 1976). Mutations in several nucleoporin genes can alter this distribution. Clusters of NPC-like structures in "aggregated" patches of nuclear envelope are observed in cells deleted for the amino-terminal half of Nup145p (Fig. 9B) (Wente and Blobel 1994), in *nup120* mutant cells (Aitchison et al. 1995a; Heath et al. 1995), and in *nup84* and *nup85* mutant cells (Goldstein et al. 1996; Siniossoglou et al. 1996). In cells lacking Nup133p (Doye et al. 1994; Li et al. 1995; Pemberton et al. 1995), or in cells with an undefined mutant allele of *NUP159* (Gorsch et al. 1995), the NPCs are concentrated in a single patch of nuclear envelope (appearing as a linear cluster in cross section). Interestingly, in *gle2* temperature-sensitive mutants, the NPCs cluster in a temperature-dependent manner at 37°C (Murphy et al. 1996). Vertebrate NPCs are also found in localized patches in certain specialized cell types such as in spermatocytes (Fawcett 1981). The formation of clusters may be due to changes in the attachment of NPCs (and/or of the nuclear envelope) to the intranuclear architecture. Two basic alternatives for the pathway of NPC cluster formation are possible. If NPCs are assembled at just one site in the membrane, the clusters may arise from the inability of these newly assembled NPCs to migrate away. In contrast, if NPCs are inserted randomly over the surface, clusters would form via aggregation of NPCs diffusing in the nuclear envelope. In no cell type is it known where NPCs assemble into the intact nuclear envelopes of interphase cells.

IV. NUCLEOCYTOPLASMIC TRANSPORT

A. Signals for Transport

For macromolecules too large to diffuse freely through the nuclear pore (see Section III.B), transport across the nuclear envelope in either direction is an extremely selective process (Forbes 1992; Davis 1995; Gerace 1995; Görlich and Mattaj 1996). Following the paradigm of the signal hypothesis for transport across the ER membrane (Blobel and Dobberstein 1975), amino acid sequences that are both necessary and sufficient for the nuclear localization of proteins have been identified (Dingwall et al. 1982; Hall et al. 1984; Kalderon et al. 1984; Lanford and Butel 1984; Silver et al. 1984; for review, see Dingwall and Laskey 1991). Vertebrate nuclear localization signals (NLSs) have been grouped into two classes, those similar to the "single cluster" of SV40 large T antigen (PKKKRKV) and those like the signal in nucleoplasmin and termed "bipartite," which contain two essential and interdependent basic clusters (2–4 residues) separated by 10–18 spacer residues. Such NLSs are similar in yeast and vertebrate proteins (see Osborne and Silver 1993; see below).

NLSs have been defined in yeast proteins by deletion and site-directed mutagenesis analyses and by construction of gene fusions that express short putative NLS peptide sequences fused to reporter proteins (usually β-galactosidase). Single NLS sequence spans have been found in the amino-terminal regions of the transcriptional repressor α2p (Hall et al. 1984), the transcriptional activator Gal4p (Silver et al. 1984), histone H2B (Moreland et al. 1987), and ribosomal protein L29 (Moreland et al. 1985). However, further studies revealed that α2p actually has two independent NLS sequences, the second one being in the homeodomain (Hall et al. 1990). Ribosomal proteins L29 (Underwood and Fried 1990) and L25 (Schaap et al. 1991) were both also shown to have two independent NLS sequences.

The two NLS sequences in α2p may be functionally distinct (Hall et al. 1990): In 30% of cells expressing an α2p with only a single NLS, the protein was found to accumulate at the nuclear envelope. Likewise, a mutation that impairs nuclear transport of the *URA2* gene product also leads to its accumulation at the nuclear envelope (Nagy et al. 1989). It is not clear why some nuclear proteins have two independent NLS sequences, but some possible explanations are discussed by Garcia-Bustos et al. (1991). The spacing between the two NLS sequences of ribosomal protein L3 suggests that they might constitute a single bipartite NLS (Fried 1992), the caveat being that in a truly bipartite NLS, each individual part should be insufficient to localize a protein by itself. The ability of each of the L3 NLS sequences to mediate import may be a function of their fusion to a heterologous protein (Fried 1992). Indeed, Nelson and Silver (1989) pointed out that the efficiency of nuclear localization by an NLS must be considered in the context of the reporter protein's tertiary structure.

Although there is no true consensus among the yeast NLSs themselves or in comparison to the vertebrate NLSs, all of the yeast NLSs analyzed do involve basic residues and are not cleaved after transport; they are located at various points in proteins (for review, see Osborne and Silver 1993). Furthermore, import signals from divergent phyla appear to function universally, as the SV40 large T antigen NLS can target proteins to the yeast nucleus (Nelson and Silver 1989). It is also possible for a non-NLS protein to be cotransported into the nucleus by its interaction with an NLS-containing protein (Moreland et al. 1985).

The export of macromolecules from the nucleus is also a signal-dependent event; however, most of the transported particles are remarkably complex. It is generally believed that RNA molecules are exported as RNA-protein (RNP) complexes (Izaurralde and Mattaj 1995). Thus, the signals for nuclear export may reside within particular amino acid sequences (see below), and the mechanism of RNA export in essence becomes that of protein export. These RNPs contain multiple polypeptides, and it has been difficult to determine which of the bound proteins is/are responsible for selective export and whether signals residing in the RNAs themselves may also play a part. In vertebrate cells, the export of U snRNAs (small nuclear RNAs) requires a saturable factor that binds to their 5′ monomethyl cap structure (Jarmolowski et al. 1994; Izaurralde et al. 1995). Whether the cap structure is required for yeast nuclear export has not yet been reported.

Recently, amino acid sequences that are necessary and sufficient for mediating the export of proteins from the nucleus have been characterized in mammalian cells (Gerace 1995). The nuclear export signals (NESs) for protein kinase inhibitor (Wen et al. 1995) and the HIV-1 Rev protein (Fischer et al. 1995) bear no similarity to NLSs. Both are short stretches (10 and 9 residues, respectively) that are enriched in leucine and other hydrophobic amino acids. When either of two leucines (underlined) or the isoleucine in the protein kinase inhibitor NES (LALKLA GLDI) is changed by mutagenesis, nuclear export is diminished (Wen et al. 1995). The NES in the Rev protein is functional in yeast (Stutz et al. 1995), and the analysis of similar NESs in yeast proteins has now begun (see Section IV.C.4).

In summary, both nuclear import and export are signal-mediated processes; as discussed further below (Section IV.C), recognition of the NLS or NES by an appropriate receptor is the first of a sequence of steps resulting in transport.

B. Regulated Nuclear Import

The bulk of nucleocytoplasmic trafficking, such as the import of polymerases to their DNA substrates and the export of mRNAs in RNP complexes to their sites of cytoplasmic translation, appears to be constitutive. However, nuclear transport can also be precisely regulated such that access to the genetic information is restricted. A number of different mechanisms have been characterized for controlling the entry of a given protein into the nucleus, including proteolytic processing, ligand binding, and posttranslational modifications (for review, see Dingwall 1991; Osborne and Silver 1993). Typically, these mechanisms result in the release of a particular protein from a "cytoplasmic anchor" or the unmasking of a constitutive NLS. Such regulation is common in complex organisms, where spatial and temporal developmental stages, as well as responses to hormones, are governed by transcription factors that can activate large numbers of genes.

In yeast, mating-type switching is triggered in mother cells by the action of the HO endonuclease, whose expression is dependent on various factors, among them Swi5p. Swi5p is a zinc finger transcription factor (85 kD) whose action is in part controlled by regulating its entry into the nucleus in a cell-cycle-dependent manner (Nasmyth et al. 1990). *SWI5* is transcribed during the S, G_2, and M phases of the cell cycle and Swi5p accumulates in the cytoplasm. In early G_1, however, Swi5p rapidly translocates to the nucleus, where it activates *HO* gene transcription with the help of Swi4p and Swi6p (Nasmyth et al. 1990).

The means by which Swi5p nuclear entry is regulated has been dissected by Moll et al. (1991). Cytoplasmic Swi5p is phosphorylated at three serine residues near its NLS sequence by the Cdc28p kinase. This net increase in negative charge is proposed to abrogate the NLS function and result in the observed cytoplasmic retention. Subsequent nuclear translocation of Swi5p would therefore require dephosphorylation at these sites, thus "unmasking" the NLS. Cell-cycle-regulated nuclear localization has also been documented for Ace2p (a transcriptional regulator) (Dohrmann et al. 1992) and for Cdc46p and Cdc47p (factors that regulate DNA replication) (Hennessy et al. 1990; Dalton and Whitbread 1995). In fission yeast, nuclear retention of the Cdc2p kinase (the equivalent of Cdc28p in *S. cerevisiae*) is itself regulated by association with the Cdc13p cyclin protein (Booher et al. 1989).

A recent study by Loeb et al. (1995) has highlighted the temporal requirement for import of cell cycle regulators that allow proper cell cycle progression. A conditional mutation in the gene encoding the NLS-receptor protein Srp1p (see Section IV.C.1) blocks nuclear import and arrests cell division at the G_2/M transition. The *srp1* import defect may directly block the timely degradation of mitotic cyclins by inhibiting the nuclear localization of an essential cell cycle regulator (Loeb et al. 1995).

It is interesting that mutations in at least three nucleoporin genes also result in arrest at the large-budded stage: For the essential *NUP1* (Davis and Fink 1990) and *NUP49* (Wente et al. 1992) genes, this has been observed with the growth of dissected null-mutant spores, and for *NUP116* (Wente and Blobel 1993), it has been observed during growth arrest at 37°C of a temperature-sensitive null strain. Such cell cycle arrest may prove to be a common phenotype for import-compromised strains.

C. Mechanisms of Nuclear Trafficking

1. Cytosolic Factors for Import

The overall process of nuclear import can be divided into at least two steps: NLS-dependent binding at the NPC, followed by nucleotide-dependent transport through the pore (Newmeyer and Forbes 1988; Richardson et al. 1988). Each of these steps is mediated by multiple proteins. A cytosolic receptor is required for recognition of NLSs, distinct cytosolic components facilitate docking at the nuclear envelope, and additional soluble factors carry out the subsequent postdocking energy-dependent translocation step through the NPC (for review, see Moore and Blobel 1994a; Powers and Forbes 1994; Adam 1995; Melchior and Gerace 1995; and see below). Although a number of schemes have been used in attempts to identify yeast import factors directly (see Section IV.C.1.a), most progress to date has come from identifying the yeast homologs of import factors isolated originally in vertebrate biochemical systems (as described below in Section IV.C.1.b).

a. Strategies for Identifying Yeast Import Factors. A variety of biochemical and genetic strategies have been used in attempts to identify a yeast NLS receptor. For example, proteins that specifically bind wild-type but not mutant NLS peptides have been purified. One of these, Nsr1p, is a 67-kD protein with RNA-binding motifs (Lee and Mélèse 1989; Lee et al. 1991). Nsr1p localizes to the nucleolus and probably has a direct role in rRNA processing and ribosome biogenesis (Lee et al. 1992). Since assembly of ribosomes involves shuttling of ribosomal proteins in and out of the nucleus, Nsr1p may act to integrate these processes (Xue and Mélèse 1994). However, a direct role for Nsr1p in NLS-mediated nuclear import has not been demonstrated.

A different NLS-binding protein of 70 kD (Nbp70) is localized to the nucleus (Silver et al. 1989; Stochaj et al. 1991, 1993). Because antiserum against Nbp70 blocks the in vitro binding of NLS-bearing proteins to yeast and animal cell nuclei, Nbp70 may indeed be the same as the bona fide NLS receptor (Srp1p) described below (Stochaj and Silver 1992). The apparent molecular mass of Srp1p is similar (60 kD) (Yano et al. 1992; Görlich et al. 1994), and both are phosphoproteins (Stochaj and Silver 1992; Azuma et al. 1995). However, Srp1p is not localized predominantly in the nucleus but rather at the nuclear rim and in the cytoplasm (Yano et al. 1992; Aitchison et al. 1995a). Whether Nbp70 is related to Srp1p will be clarified when the gene encoding Nbp70 is characterized.

Several genetic strategies to isolate nuclear import mutants have been based on mistargeting of essential mitochondrial proteins to the nucleus. Specifically, when any of the mitochondrial proteins, cytochrome c_1, iso-1-cytochrome *c*, or the F_1b subunit of the ATPase, is directed to the nucleus by fusing it to an NLS sequence, virtually all of the NLS-linked protein is transported to the nucleus (Sadler et al. 1989; Bossie et al. 1992; Gu et al. 1992). Under these conditions, the mitochondrial function of the cells is diminished such that they do not grow on glycerol, a nonfermentable carbon source. Defective nuclear import presumably allows the NLS-linked mitochondrial proteins to be imported instead into the mitochondria and thus restores growth on glycerol.

These selection schemes have identified a variety of genes (*NPL* and *NIP*). Some appear to have only indirect links to nuclear transport, but they may have essential roles in nuclear structure and/or assembly. *NIP1* encodes an essential cytoplasmic protein that is associated with 40S ribosomes and may be required for translation (Gu et al. 1992; D. Goldfarb, pers. comm.). The first gene to be cloned from the *NPL* screen (*NPL1*) is allelic to *SEC63* (Sadler et al. 1989), which encodes a component of the translocation apparatus in the ER membrane (Rothblatt et al. 1989; Sanders et al. 1992). It is unclear whether Sec63p also functions directly in nucleocytoplasmic transport, or if the import deficiency of *npl1-1* is due to general perturbations of nuclear assembly. Recently, Ng and Walter (1996) have reported that several *sec63* mutant alleles produce severe karyogamy defects, suggesting that the import defects of the *npl1-1* mutant may be related to perturbations of NPC or nuclear envelope biogenesis, as described above in Section III.B.4. Another gene cloned from this screen, *NPL3*, encodes a 45-kD protein with two RNA-binding motifs and a carboxy-terminal region with repeats of the tetrapeptide RGGY (Bossie et al. 1992). Because Npl3p shuttles between the nuclear and cytoplasmic compartments (Flach et al. 1994), it may function in the export of poly(A) RNA from the nucleus (see Section IV.C.4).

Recently, a new screen for nuclear import mutants has been undertaken by Silver and colleagues. Pools of temperature-sensitive mutants arrested at the nonpermissive temperature were examined for their ability to accumulate Npl3p in the nucleus. One of the mutants isolated can be complemented by the gene that encodes a protein homologous to the vertebrate import factor p10/B2/NTF2 (see below; Corbett and Silver 1996) and thus is indeed defective in a nuclear import factor. This type of screen may prove to be very fruitful for identifying biochemically intractable import factors, as a similar screen for export mutants has been very successful (see Section IV.C.4).

b. Identification of Homologs of Vertebrate Import Factors. The definitive identification of yeast proteins as import factors has to date come largely come from comparison to their vertebrate homologs (Table 2). Several of the yeast and vertebrate factors have multiple names (separated by slashes in Table 2) resulting from their independent isolation by different investigators and the current lack of consensus to nomenclature. Although most of the yeast proteins had already been linked to nuclear, NPC, and/or transport functions, several were originally identified in seemingly unrelated investigations. Studies of vertebrate nuclear import have made great strides in the past several years due largely to the development of a reliable in vitro nuclear protein import assay by Adam and Gerace (Adam et al. 1990). In this assay, nuclear import in permeabilized tissue culture cells is dependent on cytosol and energy, which allows the purification of soluble import factors. Although a similar system with yeast semi-intact cells is available, it has not yet been used to purify transport factors (Schlenstedt et al. 1993).

Analysis of the putative *Xenopus* NLS receptor revealed that it is 44% identical in amino acid sequence to the previously identified yeast *SRP1* gene product (Görlich et al. 1994; Adam 1995; Moroianu et al. 1995a; Weis et al. 1995). Yeast Srp1p has since been shown to bind NLS sequences, and *srp1* mutants are defective in nucleocytoplasmic trafficking (Azuma et al. 1995; Loeb et al. 1995), confirming the role of this protein in nuclear import. The localization of Srp1p in the cytoplasm and at the nuclear rim (presumably associated with NPCs) by immunofluorescence microscopy is also consistent with its putative NLS receptor function (Yano et al. 1992). Interestingly, mutant alleles of *SRP1* were first isolated as suppressors of RNA polymerase I (*rpa190*) mutations (hence, the nomenclature, with its unfortunate potential for confusion with the signal recognition particle, or SRP, of ER protein translocation) (McCusker et al. 1991; Yano et al. 1992). Srp1p has recently been renamed Kap60p (for *ka*ryo*p*herin of 60 kD) by Blobel and co-workers (Enenkel et al. 1995), but it will be referred to as Srp1p in this chapter. *SRP1* has also been isolated as a multicopy suppressor of a *cse1* chromosome segrega-

Table 2 Nuclear transport factors

Yeast gene[a]	Essential?	Subcellular localization[b]	Transport activity[c] import	Transport activity[c] export	Role in transport[d]	Vertebrate homolog[a]	References
SRP1/KAP60	yes	nuclear rim + cytoplasm	yes	no	NLS receptor	NLS receptor/importin/ Srp1/karyopherin α	1–8
KAP95/ IMP90H/RSL1	yes	nuclear rim + cytoplasm	yes	no	import docking	p97/Imp90/karyo- pherin β	8–14
GSP1/CNR1 RNA1	yes	nuclear + faint cytoplasm	yes	yes	import post-docking GTPase	Ran/TC4	15–19
GSP2/CNR2	no	n.d.	n.d.	n.d.	same as Gsp1p?	Ran/TC4	15–19
RNA1	yes	cytoplasm	yes	yes	GAP for Gsp1p	RanGAP1	20–23
YRB1/CST20	yes	nuclear rim + cytoplasm	yes	yes	Gsp1p-binding stimula- tor of GTP hydrolysis	RanBP1	22,24,25
NTF2	yes	nuclear rim	yes	no	import post-docking	p10/B2/NTF2	26–29
RIP1	no	NPC + faint nuclear	n.d.	yes	NES receptor	Rab/hRIP	30–32
GLE1/BRR3/ RSS1	yes	NPC	no	yes	RNA export mediator with an NES	n.d.	33–35
PRP20/MTR1/ SRM1	yes	nuclear	no	yes	GEF for Gsp1p	RCC1	36–42

References: (1) Yano et al. 1992, 1994; (2) Görlich et al. 1994, 1995b; (3) Adam 1995; (4) Moroianu et al. 1995a; (5) Weis et al. 1995; (6) Azuma et al. 1995; (7) Loeb et al. 1995; (8) Enenkel et al. 1995; (9) Adam and Adam 1994; (10) Chi et al. 1995; (11) Görlich et al. 1995a; (12) Radu et al. 1995a; (13) Iovine et al. 1995; (14) Koepp et al. 1996; (15) Moore and Blobel 1993, 1994a; (16) Melchior et al. 1993; (17) Kadowaki et al. 1993; (18) Belhumeur et al. 1993; (19) Schlenstedt et al. 1995b; (20) Hartwell 1967; (21) Hopper et al. 1990; (22) Bischoff et al. 1995a,b; (23) Corbett et al. 1995; (24) Schlenstedt et al. 1995a; (25) Ouspenski et al. 1995; (26) Moore and Blobel 1994b; (27) Paschal and Gerace 1995; (28) P. Silver, pers. comm.; (29) Nehrbass and Blobel 1996; (30) Fritz et al. 1995; (31) Bogerd et al. 1995; (32) Stutz et al. 1995; (33) Murphy and Wente 1996; (34) Del Priore et al. 1996; (35) T. Awadhy and C. Guthrie, pers. comm.; (36) Bischoff and Ponstingl 1991a,b; (37) Aebi et al. 1990; (38) Vijayraghavan et al. 1989; (39) Amberg et al. 1993; (40) Kadowaki et al. 1993; (41) Dasso 1993; (42) Cheng et al. 1995.

[a]Most of the yeast and vertebrate factors have multiple names, which are separated in the table by slashes.

[b]Subcellular localization as determined by immunofluorescence microscopy.

[c]Import and export activities as determined using conditional mutant strains or inferred from the function of the vertebrate homolog.

[d](GTPase) GTP hydrolyzing enzmye; (GEF) guanine-nucleotide-exchange protein; (GAP) GTPase-activating protein; (NLS) nuclear localization sequence.

tion mutation (Xiao et al. 1993). Both this and the genetic interaction with *rpa190* could be pleiotropic effects resulting from effects on nuclear transport (Yano et al. 1994).

After binding of an NLS-bearing protein to the NLS receptor, a second component is required for docking at the nuclear envelope (Adam and Adam 1994; Chi et al. 1995; Görlich et al. 1995b; Radu et al. 1995b). *KAP95* (for karyopherin of 95 kD)/*RSL1* (for R*NA1* *s*ynthetic *l*ethal) encodes a putative yeast homolog of this docking factor, with 34% sequence identity to the human protein (Görlich et al. 1995b; Enenkel et al. 1995). *KAP95/RSL1* is essential (Iovine et al. 1995), and, as predicted from the biochemical role of its vertebrate homolog, a temperature-sensitive *kap95/rsl1* mutant has diminished nuclear import capacity in vivo, although export is not detectably affected (Koepp et al. 1996). *kap95/rsl1* mutations are also synthetically lethal with an *rna1-1* mutation (Koepp et al. 1996), thus genetically linking two essential components of nuclear transport (see discussion of Rna1p below). Immunofluorescence microscopy with anti-Kap95p antibodies results in a staining pattern similar to that for Srp1p (Koepp et al. 1996; M. Rout, pers. comm.; M.K. Iovine and S.R. Wente, unpubl.). Kap95p and Srp1p can be purified from yeast cytosol as a heterodimeric complex, and when expressed and purified from bacterial cells, they form a complex that is active for targeting NLS proteins to the nuclear envelopes of permeabilized mammalian cells (Enenkel et al. 1995).

Subsequent to docking of an NLS-containing protein at the NPC via interactions with Srp1p and Kap95p, translocation through the NPC requires a small Ras-like GTP-binding protein, called Ran or TC4 in vertebrate cells (Melchior et al. 1993; Moore and Blobel 1993; for review, see Moore and Blobel 1994a). The yeast homologs (Gsp1p/Cnr1p and Gsp2p/Cnr2p) are a striking 88% identical to the mammalian proteins (Belhumeur et al. 1993; Kadowaki et al. 1993). *GSP1* and *GSP2* (for *G*TP-binding *s*uppressor of P*RP20*) were isolated as multicopy suppressors of a temperature-sensitive mutation in *PRP20*, which also encodes a nuclear transport factor (see below) (Belhumeur et al. 1993). These genes were independently identified by their high DNA sequence similarity to Ran and named *CNR1* and *CNR2* (for *c*onserved *n*uclear *R*as homologs) (Kadowaki et al. 1993); in this review, the *GSP* nomenclature will be used. *GSP1* and *GSP2* encode polypeptides of 219 and 220 amino acids, respectively, that are more than 95% identical, but only *GSP1* is essential for cell growth (Belhumeur et al. 1993; Kadowaki et al. 1993). In agreement with the in vitro data from vertebrate systems, overexpression of a mutant Gsp1p that is "locked" in the GTP-bound state blocks both import and export (Schlenstedt et al. 1995b). Therefore, GTP hydrolysis is necessary for the translocation step of nuclear transport. In addition, a requirement for ATP has been reported with in vitro import assays (Newmeyer and Forbes 1988; Richardson et al. 1988; Moore and Blobel 1992), but nuclear transport factors or NPC components that hydrolyze ATP have not yet been identified.

Several additional cytoplasmic factors that may regulate Ran/TC4 activity have been isolated from vertebrate cells, including a GTPase-activating protein (RanGAP1) (Bischoff et al. 1995a), a stimulator of the GTPase activity in the presence of the GAP (RanBP1) (Bischoff et al. 1995b), and a small homodimer (p10/B2/NTF2) that may influence Ran's affinity for the NPC and/or the NLS receptor and the docking factor (Moore and Blobel 1994b; Paschal and Gerace 1995; Nehrbass and Blobel 1996). Surprisingly, the apparent yeast homolog of human RanGAP1, with 26% sequence identity, is encoded by *RNA1* (Bischoff et al. 1995a), a temperature-sensitive allele of which was identified nearly 30 years ago on the basis of its general inhibition of RNA processing (Hartwell 1967; Hutchison et al. 1969). Subsequent studies by Hopper et al. (1990) have pinpointed Rna1p location to the cytoplasm and have, more recently, definitively shown that it is a GAP required for nuclear import (Corbett et al. 1995). The *gsp1* and *rna1-1* mutants also have RNA export defects (Shiokawa and Pogo 1974; Amberg et al. 1992; Schlenstedt et al. 1995b), and thus these proteins may have direct roles in nuclear export as well (see Section IV.C.4).

The vertebrate RanBP1 directly interacts with Ran (Coutavas et al. 1993), and in a similar manner, the putative yeast homolog Yrb1p (for *y*east *R*an *b*inding protein) (Schlenstedt et al. 1995a)/Cst20p (for *c*hromosome *st*ability) (Ouspenski et al. 1995) copurifies with the GTP-bound form of Gsp1p (Schlenstedt et al. 1995a). *YRB1/CST20* is an essential gene that encodes a cytoplasmic/nuclear rim-localized protein with 50% identity to the mouse RanBP1. Although temperature-sensitive mutations in *YRB1* inhibit both nuclear protein import and RNA export, the import block is primary (Schlenstedt et al. 1995a). Rna1p and Yrb1p, along with Prp20p (a nucleoplasmic GDP-GTP exchange factor; see Section IV.C.4), probably regulate nuclear transport by modulating the nucleotide-bound state of Gsp1p.

The yeast genome-sequencing project identified a predicted yeast protein with 46% identity to the human p10/B2/NTF2 import factor (Moore and Blobel 1994b; Paschal and Gerace 1995). A mutant allele of this gene has also been identified in a screen for nuclear import mutants (see above; Corbett and Silver 1996). Unlike the *gsp1*, *rna1-1*, and *yrb1* mutant cells, this mutant is not defective for the export of poly(A) RNA (Corbett and Silver 1996). Ntf2p localizes predominantly at the nuclear envelope in yeast cells (Corbett and Silver 1996; Nehrbass and Blobel 1996). In vitro experiments with bacterially expressed import factors have suggested that Ntf2p binds to GDP-Gsp1p and may function to coordinate the Gsp1p-dependent association and disassociation of the import factors with the NPC (Nehrbass and Blobel 1996; Paschal et al. 1996). A model for how nuclear import may proceed is discussed below.

Overall, there is an extraordinary similarity in the primary structures of many import factors between very distantly related organisms. However, it has been reported that yeast cytosol does not have the capacity to mediate nuclear import in permeabilized vertebrate cells (Adam et al. 1990). Therefore, some intrinsic difference(s) between species must exist. Nevertheless, the conservation of so many elements of this system means that genetic studies of their function in yeast will be powerfully synergistic with the biochemical studies in vertebrate systems.

2. *Interactions of Transport Factors with Nucleoporins*

The docking and translocation steps of nuclear import require direct interactions of transport factors with NPC proteins. Extensive analyses of vertebrate nucleoporins have highlighted possible interactions with the cytosolic import factors (Finlay et al. 1987, 1991; Yoneda et al. 1987; Dabauvalle et al. 1988a,b, 1990; Featherstone et al. 1988; Finlay and

Table 3 Transport factor/nucleoprotein interactions

Transport factor	NPC protein	Interaction criteria	References
Srp1p	Nup1p	two-hybrid; genetic;[a] coimmunoprecipitation	1
	Nup2p	coimmunoprecipitation	1
Kap95p	Nup116p	two-hybrid; blot overlay;[b] coimmunoprecipitation	2,3
	Nup1p	in vitro[b]	4
Rat karyopherin β[c]	Nup159p	blot overlay[b]	5
Gsp1p	Nup2p	two-hybrid	6
Rna1p	Nup1p	genetic[a]	7
Gle1p	Nup100p	two-hybrid	8

References: (1) Belanger et al. 1994; (2) Iovine et al. 1995; (3) M. Iovine and S. Wente, unpubl.; (4) Rexach and Blobel 1995; (5) Kraemer et al. 1995; (6) Dingwall et al. 1995; (7) Bogerd et al. 1994; (8) Murphy and Wente (1996).

[a]Mutant alleles are synthetically lethal.

[b]With bacterially expressed proteins.

[c]Rat karyopherin β is the homolog of yeast Kap95p.

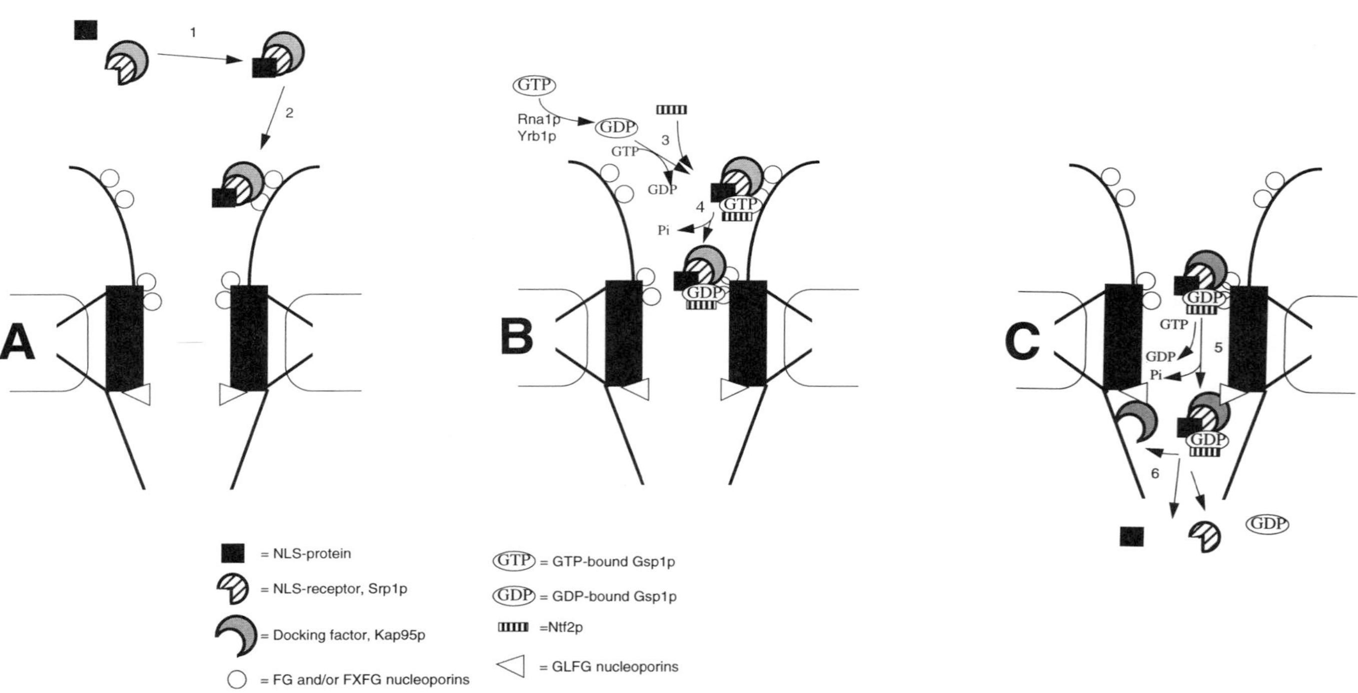
1
2
A
GTP
Rna1p
Yrb1p
GDP
GTP
3
GDP
4
GTP
Pi
GDP
B
GDP
GTP
5
GDP
Pi
GDP
6
GDP
C
= NLS-protein
= NLS-receptor, Srp1p
= Docking factor, Kap95p
= FG and/or FXFG nucleoporins
GTP = GTP-bound Gsp1p
GDP = GDP-bound Gsp1p
=Ntf2p
= GLFG nucleoporins

Figure 10 Model for nuclear protein import. (*A*) Docking step of nuclear import. NLS-containing proteins in the cytoplasm bind to the NLS receptor-docking factor (Srp1p-Kap95p) heterodimer (*1*). Kap95p mediates binding of the trimeric complex on the cytoplasmic filamentous structures of the NPC (*2*). The nucleoporins required for this interaction are believed to contain FG and/or FXFG repeats. However, substructural localization to the cytoplasmic face of the NPC in yeast has so far only been reported for Nup159p (Kraemer et al. 1995). Although the localization of FG/FXFG versus GLFG nucleoporins is shown as asymmetrical in the diagram, each class may in fact be symmetrically positioned and on both faces. (*B*) Translocation step of nuclear import. Inward movement of the trimeric complex along a nucleoporin scaffold of repeat-containing proteins requires GTP hydrolysis by Gsp1p (*4*) (see Koepp and Silver 1996). The exchange of GDP to GTP on Gsp1p may be mediated by Ntf2p (*3*). The presence of Rna1p and Yrb1p in the cytoplasm suggests that the steady-state form of the unbound Gsp1p is the GDP form. (*C*) Recycling of import factors. After an undetermined number of translocation/GTP hydrolysis steps (*5*), the import complex reaches the nucleoplasmic face of the NPC. The mechanism of complex disassembly is unknown, but recent studies suggest that only Kap95p stays associated with the NPC via interactions with GLFG nucleoporins (*6*). Export of Kap95p, Srp1p, and Gsp1p is required for further rounds of import and may be based both on diffusive export and on the active NES-based pathway diagrammed in Fig. 11.

Forbes 1990; Sterne-Marr et al. 1992), and several relevant complexes have been reported, including p10/B2/NTF2 with rat p62 (Paschal and Gerace 1995), karyopherin α/karyopherin β with rat Nup98p (Radu et al. 1995a), and Ran/TC4 with RanBP2 (Yokoyama et al. 1995). Genetic and biochemical methods have detected similar interactions in yeast (Table 3). Interestingly, all of the NPC proteins in Table 3 contain FXFG, GLFG, or FG repeats (see Table 1), and Kap95p binding to Nup116p specifically requires the Nup116p GLFG region (Iovine et al. 1995). However, the Gsp1p interaction with Nup2p is mediated by carboxy-terminal Gsp1p-binding sites that lie outside the nucleoporin FXFG region (Dingwall et al. 1995). Such Gsp1p-binding sites are also found in several other proteins, including Yrb1p and Yrb2p (Hartmann and Görlich 1995; Schlenstedt et al. 1995a; Dingwall et al. 1995).

The Srp1p-Nup1p, Srp1p-Nup2p, and Kap95p-Nup116p interactions can occur in an NLS-independent manner (Belanger et al. 1994; Iovine et al. 1995), and each may be required for a distinct phase of the import process. During import, the vertebrate NLS receptor and docking factor move through the NPC and must then be recycled back to the cytoplasm (Görlich et al. 1995a; Moroianu et al. 1995b). It is presumed that import in yeast cells proceeds similarly, as diagrammed in Figure 10 (Section IV.C.3). The interactions with Nup1p and Nup2p may be required primarily for movement of Srp1p from the cytoplasm to the nuclear surface. After NLS-ligand binding to Srp1p, Gsp1p-dependent translocation through the NPC could be mediated by the Nup2p Gsp1p-binding sites and the Rna1p-Srp1p-Nup1p interactions. In contrast, several studies support the model that the Kap95p-Nup116p interaction may be required for the recycling of Kap95p from the nucleoplasm to the cytoplasm after a round of import. First, a temperature-sensitive mutation in *KAP95* that abolishes the interaction with Nup116p causes the mutant Kap95p to accumulate at the NPC/nuclear envelope (M. Iovine and S. Wente, unpubl.). In addition, the putative vertebrate homolog of Nup116p (rat Nup98p/*Xenopus* p97) is localized on the nucleoplasmic face of the NPC (Radu et al. 1995b), and its depletion from *Xenopus* NPCs/nuclei limits nuclear import capacity in terms of nuclear growth and DNA replication (Powers et al. 1995). Further studies are necessary to clarify when during transport these particular interactions of transport factors with nucleosporins are required.

3. *Model for Nuclear Protein Import*

The model for nuclear protein import presented in Figure 10 diagrams the probable roles of the factors described in Sections IV.C.1.b and

IV.C.2 (for additional details, see recent reviews by Moore and Blobel 1994a; Gerace 1995; Melchior and Gerace; 1995; Görlich and Mattaj 1996; Koepp and Silver 1996). This model also integrates the results from recent in vitro studies with recombinant yeast import factors and nucleoporins (Rexach and Blobel 1995; Nehrbass and Blobel 1996). After translation, an NLS-containing protein binds in the cytoplasm to the heterodimeric Srp1p-Kap95p complex. This trimer docks on the cytoplasmic NPC filaments by interacting with FG- or FXFG-repeat-containing nucleoporins (Rexach and Blobel 1995). Translocation through the aqueous channel of the NPC is proposed to proceed by a succession of energy-dependent disassociation-reassociation steps with nucleoporins. The cumulative effect of these rounds of binding reactions is to propel the NLS protein–Srp1p–Kap95p trimer forward along a nucleoporin scaffold to the nucleoplasm.

The energy requirement for this movement is due at least in part to GTP hydrolysis by Gsp1p. Additional sources of energy (ATP hydrolysis) may also be required but have not been well-defined. The localization of Rna1p and Yrbp1 in the cytoplasm suggests that cytoplasmic Gsp1p is mainly in the GDP-bound state. Ntf2p may aid in regulating the nucleotide-bound state of Gsp1p (Nehrbass and Blobel 1996; Paschal et al. 1996). The GTP is then hydrolyzed by Gsp1p, causing the import complex to move to the next binding site and completion of a step in the translocation process. The precise role of GTP hydrolysis by Gsp1p requires further analysis (Koepp and Silver 1996). How many rounds of association and GTP hydrolysis are required to reach the nucleoplasm is not known.

At the nucleoplasmic face of the NPC, the import complex appears to disassemble. Srp1p, Gsp1p, Ntf2p, and the NLS-containing protein are released into the nucleoplasm, but Kap95p apparently stays associated with the NPC (Görlich et al. 1995a; Moroianu et al. 1995b). Continuation of import requires that Gsp1p, Srp1p, Ntf2p, and Kap95p be efficiently returned to the cytoplasm. The recycling of Kap95p may require an interaction with the GLFG region of Nup116p (M. Iovine and S. Wente, unpubl.). Some Srp1p recycling may be consistent with U snRNA export (Görlich et al. 1996) and Gsp1p may be small enough (30 kD) to diffuse efficiently, or it may be recycled via a Gsp1p requirement in RNA export. The presence of a guanine nucleotide exchange protein (GEF) (Prp20p/Srm1p) in the nucleus is thought to generate the GTP-bound form of Gsp1p, presumably for mediation of RNA export (see Section IV.C.4). Although the overall model for protein import is already quite complex, many aspects remain unresolved. The in vitro assays in vertebrate systems have detected clear requirements for rate-limiting factors, but regulatory factors also presumably exist; for example, Srp1p may be an in vivo substrate for a unique protein kinase (Azuma et al. 1995). Furthermore, most of the transport factor–nucleoporin interactions have only been documented in vitro (Table 3). Yeast genetic technology should prove very valuable for testing the current import model, as mutants defective in the relevant factors should be able to block transport at intermediate steps.

4. *Analysis of Export Requirements*

Export of proteins and RNP particles from the nucleus also takes place via the NPCs and is a facilitated and signal-dependent process (for review, see Gerace 1995; Izaurralde and Mattaj 1995). As described above, protein export from the nucleus can be mediated by NESs. The best characterized vertebrate mediator of RNA export is the HIV-1 Rev protein. An intron in unspliced HIV-1 RNAs contains a Rev response element which serves as a specific binding site for Rev (Daly et al. 1989; Malim et al. 1989; Zapp and Green 1989; Böhnlein et al. 1991; Malim and Cullen 1991; Zapp et al. 1991). The RNA-binding domain of Rev is distinct from its NES region, and thus the NES directs export of the protein with the bound RNA transported coincidentally (Meyer and Malim 1994). Because function of the Rev protein in yeast is dependent on its NES sequence (Stutz et al. 1995), NES-based export machinery appears to be functionally conserved across species. The model in Figure 11 diagrams a proposed mechanism for NES-driven export based on the studies described below.

A candidate for an NES receptor has been isolated in several independent two-hybrid screens. Using the NES region of HIV-1 Rev (Fischer et al. 1995) as bait, a yeast protein (Rip1p for *R*ev-*i*nteracting *p*rotein) (Stutz et al. 1995) and a related mammalian protein (Rab/hRIP1 for *R*ev/Rex-*a*ctivation-domain *b*inding or *R*ev-*i*nteracting *p*rotein) (Bogerd et al. 1995; Fritz et al. 1995) were isolated. Both the yeast and mammalian proteins contain multiple FG repeats typical of nucleoporins, and localization at the immunofluorescence level of the yeast Rip1p is primarily at or near the NPCs (Stutz et al. 1995). *RIP1* is not an essential gene and its deletion does not appear to affect mRNA export, but its deletion or overexpression does diminish Rev's activity in yeast (Stutz et al. 1995). This implies a perturbation of the export of Rev protein itself and its bound RNA.

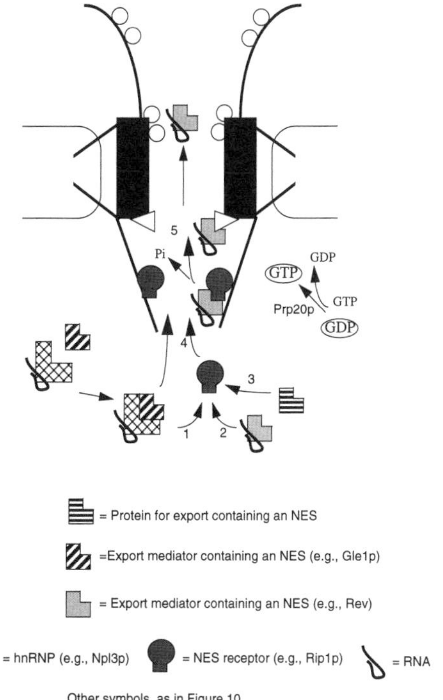

Figure 11 Model for nuclear export. Molecules bound for export can either contain an NES or be bound to a protein that contains an NES. Other non-NES-based export pathways probably also exist (Fischer et al. 1995). The diagram illustrates three different NES-containing substrates for export: (*1*) Gle1p serving as an adaptor to hnRNP proteins (meant to be provocative and not based on experimental data); (*2*) Rev serving as an adaptor to Rev-response-element-containing RNAs; and (*3*) a NES-containing protein itself. The NES regions are necessary and sufficient for interaction with Rip1p, thus defining Rip1p as a possible NES receptor (*4*). The two arrows at point 4 indicate that Rip1p may be localized both in the nucleoplasm and at the NPC. (Immunofluorescence studies reported to date have not resolved this question.) Moreover, as described in the text, nucleoporins Nup100p and Nup159p can also interact with NES regions. If export proceeds in a fashion similiar to import, the NES complex may move through the NPC via sequential association-disassociation binding steps with repeat-containing nucleoporins. Rip1p with its multiple FG repeats could facilitate the first step in this process. The translocation steps are also believed to require GTP hydrolysis by Gsp1p (*5*), with the GTP-bound form of Gsp1p being generated by Prp20p (a GEF) in the nucleoplasm. How Gsp1p interacts with the export complex is unknown. If Gsp1p accompanies the export complex through the NPC, it would coincidentally be recycled for mediating protein import (Fig. 10).

In addition, a cellular protein with a NES whose mutation blocks RNA export was recently identified. Gle1p is a novel and essential 62-kD yeast protein that localizes predominantly at the NPCs (Murphy and Wente 1996). *GLE1* (for *G*LFG *le*thal) was identified in a genetic screen for novel mutations that were lethal in combination with a null allele of *NUP100*. Mutation of the NES in Gle1p prevents mRNA export from the nucleus in a temperature-dependent manner (Murphy and Wente 1996). Another allele of the same gene was identified independently in a screen of cold-sensitive mutants for RNA export mutants (*brr3*) (Noble and Guthrie 1996; T. Awabdy and C. Guthrie, pers. comm.; see below). *GLE1* has also been isolated as a high-copy suppressor of a temperature-sensitive *rat7/nup159* mutant (*RSS1*, for *r*at *s*even *s*uppressor) (Del Priore et al. 1996). The two-hybrid assay showed that Gle1p interacts with both Rip1p and Nup100p (Table 3) (Murphy and Wente 1996). Rev also binds to Rip1p, to the GLFG region of Nup100p, and to the FG-containing fragment of Nup159p (Stutz et al. 1995). These properties suggest that Gle1p is an RNA-export factor and that Rev may mediate viral RNA export by mimicking the function of Gle1p. The mechanism of mRNA export may involve sequence-specific RNA binding by Gle1p or, more likely, direct targeting of RNP molecules by Gle1p-binding hnRNP proteins followed by translocation through the NPC via Rip1p and Nup100p interactions (Fig. 11). It is not known whether Rip1p accompanies the exported substrates to the cytoplasm.

The fact that *RIP1* and *NUP100* are not essential suggests that other (and perhaps redundant) factors are also involved in NES recognition and nuclear export. Both Rna1p and Gsp1p have both import and export "activity" as assessed by studies of conditional mutant strains (see above and Table 2). However, the export defects of these mutants may merely be indirect consequences of their failure to import required export factors, and further studies are required to pinpoint their export functions.

Because quantitative in vitro assays for nuclear export have not yet been developed, genetic approaches are proving to be particularly useful in identifying other potential export factors (as described above for Gle1p). Collections of temperature-sensitive and cold-sensitive mutant strains have been screened directly for blocks in poly(A) RNA export (Amberg et al. 1992; Kadowaki et al. 1992, 1994b; Noble and Guthrie 1996). In yeast, pre-mRNAs are typically methylated at the 5′end and contain short poly(A) tails. In situ hybridization with oligonucleotides of polydeoxy T detects the location of such poly(A) tails in whole cells. At the fluorescence microscopy level, the signal in wild-type cells is diffuse and cytoplasmic, because the poly(A) RNA molecules are exported rapidly from the nucleus. In RNA export mutants, the poly(A) RNA accumulates in the nucleus. These screens have identified *rat* (for m*R*NA *t*rafficking) (Amberg et al. 1992), *mtr* (for *m*RNA *tr*ansport) (Kadowaki et al. 1992), and *brr* (for *b*ad *r*esponse to *r*efrigeration) (Noble and Guthrie 1996) mutants. As discussed above, numerous *RAT* genes encode NPC proteins (Table 1: *RAT2/NUP120*, *RAT3/NUP133*, *RAT7/NUP159*, *RAT9/NUP85*, *RAT10/NUP145*) (Gorsch et al. 1995; Heath et al. 1995; Li et al. 1995; Goldstein et al. 1996; C. Cole, pers. comm.). Aside from the interaction of Nup159p with Rev in the two-hybrid assay (see above), the roles of these nucleoporins in RNA export have not been defined.

The *mtr1* mutations proved to be allelic to *prp20/srm1* mutations (Kadowaki et al. 1993), and, indeed, multiple *prp20* alleles show rapid mRNA export defects (Amberg et al. 1993). *PRP20* was first identified in a screen for RNA-splicing mutants (Vijayraghavan et al. 1989) and is homologous to the human *RCC1* gene (Aebi et al. 1990). The 30% identity in sequence between the yeast and mammalian proteins is sufficient to allow functional complementation between species. In particular, expression of the human *RCC1* gene can rescue a *prp20* mutation (Fleischmann et al. 1991), and, conversely, yeast *PRP20* can complement a mutation of *RCC1* (tsBN2) in a hamster cell line (Ohtsubo et al. 1991). Interestingly, this *RCC1* mutation was identified in a screen for hamster cell mutants defective in progression through the cell cycle (for review, see Dasso 1993), and some have speculated that RCC1 and Ran/TC4 may have additional cellular functions besides mediating nuclear transport (Dasso 1993; Sazer 1996). RCC1 appears to function in nuclear transport via its interactions with Ran/TC4 as a guanine nucleotide exchange protein (GEF) (Bischoff and Ponstingl 1991a,b; Cheng et al. 1995). This implies that a Gsp1p-mediated GTP hydrolysis step may also be required for RNA export.

The most abundant RNA-binding proteins in vertebrate cells are the hnRNP proteins. A subclass of vertebrate hnRNP proteins are known to shuttle rapidly to and from the nucleus, suggesting that these proteins are involved directly or indirectly in RNA export (Piñol-Roma and Dreyfuss 1992, 1993). The yeast Npl3p, identified in a selection scheme for import mutants (see Section IV.C.1.a; Bossie et al. 1992), is a member of the hnRNP family, and it shuttles rapidly between the nucleus and cytoplasm (Flach et al. 1994). This same protein was identified independently as a nucleolar-associated protein (Nop3p) required for pre-rRNA processing (Russell and Tollervey 1992). Mutations in *NOP3/NPL3* not only confer import defects (see Section IV.C.1.a), but also cause accumulation of mRNA in the nucleus at the nonpermissive temperature (Flach et al. 1994; Wilson et al. 1994; Russell and Tollervey 1995; Singleton et al. 1995). This suggests that Nop3p/Npl3p may facilitate export while accompanying the mRNA to the cytoplasm. In support of this model, a temperature-sensitive mutation in the RNA recognition motifs of Npl3p/Nop3p blocks both RNA export and export of the protein itself (Lee et al. 1996). Npl3p/Nop3p may act as an adaptor to the export machinery (Fig. 11).

V. SUMMARY AND PROSPECTS

In recent years, numerous yeast nuclear envelope and NPC components have been defined. The overlap between biochemical and genetic studies in various species has been particularly exciting. Aside from the continued cataloging of factors required for nuclear physiology, the next challenge will be to resolve how these proteins assemble into functional structures. How is the intranuclear architecture established and maintained? What components of the nuclear envelope have an active role in nuclear migration, nuclear division, and NPC or SPB assembly?

In addition, many aspects of the mechanisms of nucleocytoplasmic trafficking remain to be dissected at a finer level. Separating the import and export processes will continue to be a daunting task, as both traverse the same portal (the NPC) and may use common soluble factors. The fact that many components have dual roles in nuclear import and export is clear from Tables 1 and 2. To complicate matters further, studies with vertebrate and yeast cells have also suggested that different, independent import and export pathways may operate depending on the type of macromolecule being transported, for example, in the export of U snRNAs versus 5S rRNA (Jarmolowski et al. 1994) and the import of RNA-binding proteins (Aitchison et al. 1996; Pollard et al. 1996). With many assays and strategies now available, yeast cell biologists are poised to contribute importantly to a deeper understanding of nuclear physiology during the next several years.

ACKNOWLEDGMENTS

The authors thank Drs. Jerry Lyn Allen, Laura Davis, Monica Gotta, and Michael Rout for contributing Figures 3, 4, 2, and 5, respectively. They also greatly appreciate the sharing of unpublished results by Drs. Mark Rose, Robyn Wright, Pamela Silver, Michael Rout, Charles Cole, David Goldfarb, Mark Winey, and K.H. Wolfe. A.J.C. thanks Dr. Michael

Douglas for his support during the writing of this manuscript and for help in organizing the initial drafts. S.R.W. is indebted to Dr. Günter Blobel, in whose laboratory (with the technical assistance of Eleana Sphica and the Rockefeller University Electron Microscopy Facility) the experiments for Figures 1 and 8 were conducted. S.R.W. is supported by a grant from the National Institutes of Health (R01-GM-51219-02). S.M.G. is supported by the Swiss National Science Fund.

REFERENCES

Adachi, Y., E. Kaes, and U.K. Laemmli. 1989. Preferential, cooperative binding of topoisomerase II to scaffold associated regions. *EMBO J.* **8:** 3997.

Adachi, Y., M. Luke, and U.K. Laemmli. 1991. Chromosome assembly in vitro. Topoisomerase II is required for condensation. *Cell* **64:** 137.

Adam, E.J.H. and S.A. Adam. 1994. Identification of cytosolic factors required for nuclear location sequence-mediated binding to the nuclear envelope. *J. Cell Biol.* **125:** 547.

Adam, S.A. 1995. The importance of importin. *Trends Cell Biol.* **5:** 189.

Adam, S.A., R. Sterne-Marr, and L. Gerace. 1990. Nuclear protein import in permeabilized mammalian cells requires soluble cytoplasmic factors. *J. Cell Biol.* **111:** 807.

Aebi, M., M.W. Clark, U. Vijayraghavan, and J. Abelson. 1990. A yeast mutant, *PRP20*, altered in mRNA metabolism and maintenance of the nuclear structure, is defective in a gene homologous to the human gene *RCC1* which is involved in the control of chromosome condensation. *Mol. Gen. Genet.* **224:** 72.

Aitchison, J.D., G. Blobel, and M.P. Rout. 1995a. Nup120p: A yeast nucleoporin required for NPC distribution and mRNA transport. *J. Cell Biol.* **131:** 1659.

———. 1996. Kap104p: A karyopherin involved in nuclear transport of messenger RNA binding proteins. *Science* **274:** 624.

Aitchison, J.D., M.P. Rout, M. Marelli, G. Blobel, and R.W. Wozniak. 1995b. Two novel related yeast nucleoporins Nup170p and Nup157p: Complementation with the vertebrate homologue Nup155p and functional interactions with the yeast nuclear pore-membrane protein Pom152p. *J. Cell Biol.* **131:** 1133.

Akey, C.W. 1989. Interactions and structure of the nuclear pore complex revealed by cryo-electron microscopy. *J. Cell Biol.* **109:** 955.

———. 1992. The nuclear pore complex: A macromolecular transporter. In *Nuclear trafficking* (ed. C. Feldherr), p. 31. Academic Press, New York.

Akey, C.W. and M. Radermacher. 1993. Architecture of the *Xenopus* nuclear pore complex revealed by three-dimensional cryo-electron microscopy. *J. Cell Biol.* **122:** 1.

Allen, J.L. and M.G. Douglas. 1989. Organization of the nuclear pore complex in *Saccharomyces cerevisiae*. *J. Ultrastruct. Mol. Struct. Res.* **102:** 95.

Amati, B.B. and S.M. Gasser. 1988. Chromosomal ARS and CEN elements bind specifically to the yeast nuclear scaffold. *Cell* **54:** 967.

———. 1990. *Drosophila* scaffold-attached regions bind nuclear scaffolds and can function as ARS elements in both budding and fission yeasts. *Mol. Cell. Biol.* **10:** 5442.

Amati, B., L. Pick, T. Laroche, and S.M. Gasser. 1990. Nuclear scaffold attachment stimulates, but is not essential for ARS activity in *Saccharomyces cerevisiae:* Analysis of the *Drosophila ftz* SAR. *EMBO J.* **9:** 4007.

Amberg, D.C., A.L. Goldstein, and C.N. Cole. 1992. Isolation and characterization of *RAT1:* An essential gene of *Saccharomyces cerevisiae* required for the efficient nucleocytoplasmic trafficking of mRNA. *Genes Dev.* **6:** 1173.

Amberg, D.C., M. Fleischmann, I. Stagljar, C.N. Cole, and M. Aebi. 1993. Nuclear PRP20 protein is required for mRNA export. *EMBO J.* **12:** 233.

Aparicio, O.M., B.L. Billington, and D.E. Gottschling. 1991. Modifiers of position effect are shared between telomeric and silent mating-type loci in *S. cerevisiae. Cell* **66:** 1279.

Aris, J.P. and G. Blobel. 1988. Identification and characterization of a yeast nucleolar protein that is similar to a rat liver nucleolar protein. *J. Cell Biol.* **107:** 17.

———. 1989. Yeast nuclear envelope proteins cross react with an antibody against mammalian pore complex proteins. *J. Cell Biol.* **108:** 2059.

Azuma, Y., M.M. Tabb, L. Vu, and M. Nomura. 1995. Isolation of a yeast protein kinase that is activated by the protein encoded by *SRP1* (Srp1p) and phosphorylates Srp1p complexed with nuclear localization signal peptides. *Proc. Natl. Acad. Sci.* **92:** 5159.

Belanger, K.D., M.A. Kenna, S. Wei, and L.I. Davis. 1994. Genetic and physical interactions between Srp1p and nuclear pore complex proteins Nup1p and Nup2p. *J. Cell Biol.* **126:** 619.

Belhumeur, P., A. Lee, R. Tam, T. DiPaolo, N. Fortin, and M.W. Clark. 1993. *GSP1* and *GSP2*, genetic suppressors of the *prp20-1* mutant in *Saccharomyces cerevisiae:* GTP-binding proteins involved in the maintenance of nuclear organization. *Mol. Cell. Biol.* **13:** 2152.

Bell, S.P., R. Kobayashi, and B. Stillman. 1993. Yeast origin recognition complex functions in transcription silencing and DNA replication. *Science* **262:** 1844.

Bender, A. and J.R. Pringle. 1991. Use of a screen for synthetic lethal and multicopy suppressee mutants to identify two new genes involved in morphogenesis in *Saccharomyces cerevisiae. Mol. Cell. Biol.* **11:** 1295.

Benyajati, C. and A. Worcel. 1976. Isolation, characterization, and structure of the folded interphase genome of *Drosophila melanogaster. Cell* **9:** 393.

Berezney, R. and D.S. Coffey. 1975. Nuclear protein matrix: Association with newly synthesized DNA. *Science* **189:** 291.

Bergman, L.W. and R.A. Kramer. 1983. Modulation of chromatin structure associated with derepression of the acid phosphatase gene of *Saccharomyces cerevisiae. J. Biol. Chem.* **258:** 7223.

Bernardi, F., M. Zatchej, and F. Thoma. 1992. Species specific protein-DNA interactions may determine the chromatin units of genes in *S. cerevisiae* and in *S. pombe. EMBO J.* **11:** 1177.

Berrios, M., N. Osheroff, and P.A. Fisher. 1985. In situ localization of DNA topoisomerase II, a major polypeptide component of the *Drosophila* nuclear matrix fraction. *Proc. Natl. Acad. Sci.* **82:** 4142.

Bischoff, F.R. and H. Ponstingl. 1991a. Catalysis of guanine nucleotide exchange on Ran by the mitotic regulator RCC1. *Nature* **354:** 80.

———. 1991b. Mitotic regulator protein RCC1 is complexed with a nuclear *ras*-related polypeptide. *Proc. Natl. Acad. Sci.* **88:** 10830.

Bischoff, F.R., H. Krebber, T. Kempf, I. Hermes, and H. Ponstingl. 1995a. Human RanGTPase-activating protein RanGAP1 is a homologue of yeast Rna1p involved in mRNA processing and transport. *Proc. Natl. Acad. Sci.* **92:** 1749.

Bischoff, F.R., H. Krebber, E. Smirnova, W. Dong, and H. Ponstingl. 1995b. Co-activation of RanGTPase and inhibition of GTP dissociation by Ran-GTP binding protein RanBP1. *EMBO J.* **14:** 705.

Blobel, G. and B. Dobberstein. 1975. Transfer of proteins across membranes. II. Reconstitution of functional rough microsomes from heterologous components. *J. Cell Biol.* **67:** 852.

Boeke, J.D. and S.B. Sandmeyer. 1991. Yeast transposable elements. In *The molecular and cellular biology of the yeast* Saccharomyces: *Genome dynamics, protein synthesis, and energetics* (ed. J.R. Broach et al.), vol. 1, p. 193. Cold Spring Harbor Laboratory Press, Cold Spring Harbor, New York.

Bogerd, H.P., R.A. Fridell, S. Madore, and B.R. Cullen. 1995. Identification of a novel cellular cofactor for the Rev/Rex class of retroviral regulatory proteins. *Cell* **82:** 485.

Bogerd, A.M., J.A. Hoffman, D.C. Amberg, G.R. Fink, and L.I. Davis. 1994. *nup1* mutants exhibit pleiotropic defects in nuclear pore complex function. *J. Cell Biol.* **127:** 319.

Böhnlein, E., J. Berger, and J. Hauber. 1991. Functional mapping of the human immunodeficiency virus type 1 Rev RNA binding domain: New insights into the domain structure of Rev and Rex. *J. Virol.* **65:** 7051.

Booher, R.N., C.E. Alfa, J.S. Hyams, and D.H. Beach. 1989. The fission yeast cdc2/cdc13/suc1 protein kinase: Regulation of catalytic activity and nuclear localization. *Cell* **58:** 485.

Bossie, M.A., C. DeHoratius, G. Barcelo, and P.A. Silver. 1992. A mutant nuclear protein with similarity to RNA binding proteins interferes with nuclear import in yeast. *Mol. Biol. Cell* **3:** 875.

Bradbury, E.M., N. Maclean, and H.R. Matthews. 1981. *DNA, chromatin and chromosomes.* Wiley, New York.

Buchman, A.R., W.J. Kimmerly, J. Rine, and R.D. Kornberg. 1988. Two DNA-binding factors recognize specific sequences at silencers, upstream activating sequences, autonomously replicating sequences, and telomeres in *Saccharomyces cerevisiae. Mol. Cell. Biol.* **8:** 210.

Buck, S.W. and D. Shore. 1995. Action of a RAP1 carboxy-terminal silencing domain reveals an underlying competition between *HMR* and telomeres in yeast. *Genes Dev.* **9:** 370.

Byers, B. 1981. Cytology of the yeast life cycle. In *The molecular biology of the yeast* Saccharomyces: *Life cycle and inheritance* (ed. J.N. Strathern et al.), p. 59. Cold Spring Harbor Laboratory, Cold Spring Harbor, New York.

Callan, H.G. 1986. *Lampbrush chromosomes.* Springer-Verlag, Berlin.

Cardenas, M.E., T. Laroche, and S.M. Gasser. 1990. The composition and morphology of yeast nuclear scaffolds. *J. Cell Sci.* **96:** 439.

Carmo-Fonseca, M., H. Kern, and E.C. Hurt. 1991. Human nucleoporin p62 and the essential yeast nuclear pore protein NSP1 show sequence homology and a similar domain organization. *Eur. J. Cell Biol.* **55:** 17.

Cavalli, G. and F. Thoma. 1993. Chromatin transitions during activation and repression of galactose-regulated genes in yeast. *EMBO J.* **12:** 4603.

Certa, U., M. Colavito-Shepanski, and M. Grunstein. 1984. Yeast may not contain histone H1: The only known "histone H1-like" protein in *Saccharomyces cerevisiae* is a mitochondrial protein. *Nucleic Acids Res.* **12:** 7975.

Cheng, Y., J.E. Dahlberg, and E. Lund. 1995. Diverse effects of the guanine nucleotide exchange factor RCC1 on RNA transport. *Science* **267:** 1807.

Chi, N.C., E.J.H. Adam, and S.A. Adam. 1995. Sequence and characterization of cytoplasmic nuclear protein import factor p97. *J. Cell Biol.* **130:** 265.

Chien, C.-T., P.L. Bartel, R. Sternglanz, and S. Fields. 1991. The two-hybrid system: A method to identify and clone genes for proteins that interact with a protein of interest. *Proc. Natl. Acad. Sci.* **88:** 9578.

Chikashige, Y., D. Ding, H. Funabiki, T. Haraguchi, S. Mashiko, M. Yanagida, and Y. Hiraoka. 1994. Telomere-led premeiotic chromosome movement in fission yeast. *Science* **264:** 270.

Chuang, P.-T., D.G. Albertson, and B.J. Meyer. 1994. DPY-27: A chromosome condensation protein homolog that regulates *C. elegans* dosage compensation through association with the X chromosome. *Cell* **79:** 459.

Clark, S.W. and D.I. Meyer. 1994. *ACT3:* A putative centractin homologue in *S. cerevisiae* is required for proper orientation of the mitotic spindle. *J. Cell Biol.* **127:** 129.

Cockell, M., F. Palladino, T. Laroche, G. Kyrion, C. Liu, A.J. Lustig, and S.M. Gasser. 1995. The carboxy termini of Sir4 and Rap1 affect Sir3 localization: Evidence for a multicomponent complex required for yeast telomeric silencing. *J. Cell Biol.* **129:** 909.

Compton, D.A. and D.W. Cleveland. 1994. NuMA, a nuclear protein involved in mitosis and nuclear reformation. *Curr. Opin. Cell Biol.* **6:** 343.

Conde, J. and G.R. Fink. 1976. A mutant of *Saccharomyces cerevisiae* defective for nuclear fusion. *Proc. Natl. Acad. Sci.* **73:** 3651.

Cook, P.R. and I.A. Brazell. 1975. Supercoils in human DNA. *J. Cell Sci.* **19:** 261.

Copeland, C.S. and M. Snyder. 1993. Nuclear pore complex antigens delineate nuclear envelope dynamics in vegetative and conjugating *Saccharomyces cerevisiae. Yeast* **9:** 235.

Corbett, A.H. and P.A. Silver. 1996. The *NTF2* gene encodes an essential, highly conserved protein that functions in nuclear transport *in vivo. J. Biol. Chem.* **271:** 18477.

Corbett, A.H., D.M. Koepp, G. Schlenstedt, M.S. Lee, A.K. Hopper, and P.A. Silver. 1995. Rna1p, a Ran/TC4 GTPase activating protein, is required for nuclear import. *J. Cell Biol.* **130:** 1017.

Coutavas, E., M. Ren, J.D. Oppenheim, P. D'Eustachio, and M.G. Rush. 1993. Characterization of proteins that interact with the cell-cycle regulatory protein Ran/TC4. *Nature* **366:** 585.

Cremer, T., A. Kurz, R. Zirbel, S. Dietzel, B. Rinke, E. Schröck, M.R. Speicher, U. Mathieu, A. Jauch, P. Emmerich, H. Scherthan, T. Ried, C. Cremer, and P. Lichter. 1993. Role of chromosome territories in the functional compartmentalization of the cell nucleus. *Cold Spring Harbor Symp. Quant. Biol.* **58:** 777.

Dabauvalle, M.-C., R. Benavente, and N. Chaly. 1988a. Monoclonal antibodies to a M_r 68000 pore complex glycoprotein interfere with nuclear protein uptake in *Xenopus* oocytes. *Chromosoma* **97:** 193.

Dabauvalle, M.-C., K. Loos, and U. Scheer. 1990. Identification of a soluble precursor complex essential for nuclear pore assembly in vitro. *Chromosoma* **100:** 56.

Dabauvalle, M.-C., B. Schulz, U. Scheer, and R. Peters. 1988b. Inhibition of nuclear accumulation of kayophilic proteins in living cells by microinjection of the lectin wheat germ agglutinin. *Exp. Cell Res.* **174:** 291.

Dalton, S. and L. Whitbread. 1995. Cell cycle-regulated nuclear import and export of Cdc47, a protein essential for initiation of DNA replication in budding yeast. *Proc. Natl. Acad. Sci.* **92:** 2514.

Daly, T.J., K.S. Cook, G.S. Gray, T.E. Maione, and J.R. Rusche. 1989. Specific binding of HIV-1 recombinant Rev protein to the Rev-responsive element *in vitro*. *Nature* **342:** 816.

Dasso, M. 1993. RCC1 in the cell cycle: the regulator of chromosome condensation takes on new roles. *Trends Biochem. Sci.* **13:** 96.

Davie, J.R., C.A. Saunders, J.M. Walsh, and S.C. Weber. 1981. Histone modifications in the yeast *S. cerevisiae. Nucleic Acids Res.* **9:** 3205.

Davis, L.I. 1995. The nuclear pore complex. *Annu. Rev. Biochem.* **64:** 865.

Davis, L.I. and G. Blobel. 1986. Identification and characterization of a nuclear pore complex protein. *Cell* **45:** 699.

Davis, L.I. and G.R. Fink. 1990. The *NUP1* gene encodes an essential component of the yeast nuclear pore complex. *Cell* **61:** 965.

Del Priore, V., C.A. Snay, A. Bahr, and C.N. Cole. 1996. The product of the *Saccharomyces cerevisiae RSS1* gene, identified as a high-copy suppressor of the *rat7-1* temperature-sensitive allele of the *RAT7/NUP159* nucleoporin, is required for efficient mRNA export. *Mol. Biol. Cell* **7:** 1601.

Dietzel, S., E. Weilandt, R. Eils, C. Münkel, C. Cremer, and T. Cremer. 1995. Three-dimensional distribution of centromeric or paracentromeric heterochromatin of chromosomes 1, 7, 15 and 17 in human lymphocyte nuclei studied with light microscopic axial tomography. *Bioimaging* **3:** 121.

Diffley, J.F.X. and B. Stillman. 1989. Transcriptional silencing and lamins. *Nature* **342:** 24.

Dijkwel, P.A. and J.L. Hamlin. 1995. Origins of replication and the nuclear matrix: The DHFR domain as a paradigm. *Int. Rev. Cytol.* **162A:** 455.

Dijkwel, P.A., L.H.F. Mullenders, and F. Wanka. 1979. Analysis of the attachment of replicating DNA to a nuclear matrix in mammalian interphase nuclei. *Nucleic Acids Res.* **6:** 219.

Dingman, C.W. 1974. Bidirectional chromosome replication: Some topological considerations. *J. Theor. Biol.* **43:** 187.

Dingwall, C. 1991. Transport across the nuclear envelope: Enigmas and explanations. *BioEssays* **13:** 213.

Dingwall, C. and R.A. Laskey. 1991. Nuclear targeting sequences—A consensus? *Trends Biochem. Sci.* **16:** 478.

Dingwall, C., S. Kandels-Lewis, and B. Séraphin. 1995. A family of Ran binding proteins that includes nucleoporins. *Proc. Natl. Acad. Sci.* **92:** 7525.

Dingwall, C., S.V. Sharnick, and R.A. Laskey. 1982. A polypeptide domain that specifies migration of nucleoplasmin into the nucleus. *Cell* **30:** 449.

Dohrmann, P.R., G. Butler, K. Tamai, S. Dorland, J.R. Greene, D.J. Thiele, and D.J. Stillman. 1992. Parallel pathways of gene regulation: Homologous regulators *SWI5* and *ACE2* differentially control transcription of *HO* and chitinase. *Genes Dev.* **6:** 93.

Doye, V. and E.C. Hurt. 1995. Genetic approaches to nuclear pore structure and function. *Trends Genet.* **11:** 235.

Doye, V., R. Wepf, and E.C. Hurt. 1994. A novel nuclear pore protein Nup133p with distinct roles in poly(A)$^+$ RNA transport and nuclear pore distribution. *EMBO. J.* **13:** 6062.

Dujon, B., D. Alexandraki, B. André, W. Ansorge, V. Baladron, J.P.G. Ballestra, A. Banrevi, P.A. Bolle, M. Bolotin-Fukuhara, P. Bossier, and 98 others. 1994. Complete DNA sequence of yeast chromosome XI. *Nature* **369:** 371.

Dworetzky, S.I. and C.M. Feldherr. 1988. Translocation of RNA-coated gold particles through the nuclear pores of oocytes. *J. Cell Biol.* **106:** 575.

Earnshaw, W.C., B. Halligan, C.A. Cooke, M.M.S. Heck, and L.F. Liu. 1985. Topoisomerase II is a structural component of mitotic chromosome scaffolds. *J. Cell Biol.* **100:** 1706.

Enenkel, C., G. Blobel, and M. Rexach. 1995. Identification of a yeast karyopherin heterodimer that targets import substrate to mammalian nuclear pore complexes. *J. Biol. Chem.* **270:** 16499.

Eshel, D., L.A. Urrestarazu, S. Vissers, J.-C. Jauniaux, J.C. van Vliet-Reedijk, R.J. Planta, and I.R. Gibbons. 1993. Cytoplasmic dynein is required for normal nuclear segregation in yeast. *Proc. Natl. Acad. Sci.* **90:** 11172.

Fabre, E., W.C. Boelens, C. Wimmer, I.W. Mattaj, and E.C. Hurt. 1994. Nup145p is required for nuclear export of mRNA and binds homopolymeric RNA *in vitro* via a novel conserved motif. *Cell* **78:** 275.

Fawcett, D.W. 1981. *The cell*, 2nd edition. W.B. Saunders, Philadelphia.

Featherstone, C., M.K. Darby, and L. Gerace. 1988. A monoclonal antibody against the nuclear pore complex inhibits nucleocytoplasmic transport of protein and RNA *in vivo*. *J. Cell Biol.* **107:** 1289.

Fedor, M.J. and R.D. Kornberg. 1989. Upstream activation sequence-dependent alteration of chromatin structure and transcription activation of the yeast *GAL1-GAL10* genes. *Mol. Cell. Biol.* **9:** 1721.

Feldherr, C.M., E. Kallenbach, and N. Schultz. 1984. Movement of a karyophilic protein through the nuclear pores of oocytes. *J. Cell Biol.* **99:** 2216.

Finch, J.T. and A. Klug. 1976. Solenoidal model for superstructure in chromatin. *Proc. Natl. Acad. Sci.* **73:** 1897.

Finlay, D.R. and D.J. Forbes. 1990. Reconstitution of biochemically altered nuclear pores: Transport can be eliminated and restored. *Cell* **60:** 17.

Finlay, D.R., D.D. Newmeyer, T.M. Price, and D.J. Forbes. 1987. Inhibition of in vitro nuclear transport by a lectin that binds to nuclear pores. *J. Cell Biol.* **104:** 189.

Finlay, D.R., E. Meier, P. Bradley, J. Horecka, and D.J. Forbes. 1991. A complex of nuclear pore proteins required for pore function. *J. Cell Biol.* **114:** 169.

Fischer, U., J. Huber, W.C. Boelens, I.W. Mattaj, and R. Lührmann. 1995. The HIV-1 Rev activation domain is a nuclear export signal that accesses an export pathway used by specific cellular RNAs. *Cell* **82:** 475.

Flach, J., M. Bossie, J. Vogel, A. Corbett, T. Jinks, D.A. Willins, and P.A. Silver. 1994. A yeast RNA-binding protein shuttles between the nucleus and the cytoplasm. *Mol. Cell. Biol.* **14:** 8399.

Fleischmann, M., M.W. Clark, W. Forrester, M. Wickens, T. Nishimoto, and M. Aebi. 1991. Analysis of yeast *prp20* mutations and functional complementation by the human homologue *RCC1*, a protein involved in the control of chromosome condensation. *Mol. Gen. Genet.* **227:** 417.

Forbes, D.J. 1992. Structure and function of the nuclear pore complex. *Annu. Rev. Cell Biol.* **8:** 495.

Forrester, W., F. Stutz, M. Rosbash, and M. Wickens. 1992. Defects in mRNA 3′-end formation, transcription initiation, and mRNA transport associated with the yeast mutation *prp20:* Possible coupling of mRNA processing and chromatin structure. *Genes Dev.* **6:** 1914.

Fried, H.M. 1992. Transport of ribosomal proteins and rRNA, tRNA, and snRNA. In

Nuclear trafficking (ed. C.M. Feldherr), p. 291. Academic Press, New York.

Fritz, C.C., M.L. Zapp, and M.R. Green. 1995. A human nucleoporin-like protein that specifically interacts with HIV Rev. *Nature* **376:** 530.

Funabiki, H., I. Hagan, S. Uzawa, and M. Yanagida. 1993. Cell cycle-dependent specific positioning and clustering of centromeres and telomeres in fission yeast. *J. Cell Biol.* **121:** 961.

Galeotti, C.L. and K.L. Williams. 1978. Giemsa staining of mitotic chromosomes in *Kluyveromyces lactis* and *Saccharomyces cerevisiae. J. Gen. Microbiol.* **104:** 337.

Garcia-Bustos, J.F., P. Wagner, and M.N. Hall. 1991. Nuclear import substrates compete for a limited number of binding sites. Evidence for different classes of yeast nuclear import receptors. *J. Biol. Chem.* **266:** 22303.

Gasser, S.M., B.B. Amati, M.E. Cardenas, and J.F.-X. Hofmann. 1989. Studies on scaffold attachment sites and their relation to genome function. *Int. Rev. Cytol.* **119:** 57.

Gasser, S.M., T. Laroche, J. Falquet, E. Boy de la Tour, and U.K. Laemmli. 1986. Metaphase chromosome structure: Involvement of topoisomerase II. *J. Mol. Biol.* **188:** 613.

Gerace, L. 1995. Nuclear export signals and the fast track to the cytoplasm. *Cell* **82:** 341.

Gerace, L. and B. Burke. 1988. Functional organization of the nuclear envelope. *Annu. Rev. Cell Biol.* **4:** 335.

Gerace, L. and R. Foisner. 1994. Integral membrane proteins and dynamic organization of the nuclear envelope. *Trends Cell Biol.* **4:** 127.

Gerace, L., Y. Ottaviano, and C. Kondor-Koch. 1982. Identification of a major polypeptide of the nuclear pore complex. *J. Cell Biol.* **95:** 826.

Gill, S.R., T.A. Schroer, I. Szilak, E.R. Steuer, M.P. Sheetz, and D.W. Cleveland. 1991. Dynactin, a conserved, ubiquitously expressed component of an activator of vesicle motility mediated by cytoplasmic dynein. *J. Cell Biol.* **115:** 1639.

Gilson, E., T. Laroche, and S.M. Gasser. 1993a. Telomeres and the functional architecture of the nucleus. *Trends Cell Biol.* **3:** 128.

Gilson, E., M. Roberge, R. Giraldo, D. Rhodes, and S.M. Gasser. 1993b. Distortion of the DNA double helix by RAP1 at silencers and multiple telomeric binding sites. *J. Mol. Biol.* **231:** 293.

Goldberg, M.W. and T.D. Allen. 1992. High resolution scanning electron microscopy of the nuclear envelope: Demonstration of a new, regular, fibrous lattice attached to the baskets of the nucleoplasmic face of the nuclear pores. *J. Cell Biol.* **119:** 1429.

———. 1993. The nuclear pore complex: Three-dimensional surface structure revealed by field emission, in-lens scanning electron microscopy, with underlying structure uncovered by proteolysis. *J. Cell Sci.* **106:** 261.

Goldstein, A.L., C.A. Snay, C.V. Heath, and C.N. Cole. 1996. Pleiotropic nuclear defects associated with a conditional allele of the novel nucleoporin Rat9p/Nup85p. *Mol. Biol. Cell* **7:** 917.

Gordon, C.N. 1977. Chromatin behaviour during the mitotic cell cycle of *Saccharomyces cerevisiae. J. Cell Sci.* **24:** 81.

Görlich, D. and I.W. Mattaj. 1996. Nucleocytoplasmic transport. *Science* **271:** 1513.

Görlich, D., S. Prehn, R.A. Laskey, and E. Hartmann. 1994. Isolation of a protein that is essential for the first step of nuclear protein import. *Cell* **79:** 767.

Görlich, D., F. Vogel, A.D. Mills, E. Hartmann, and R.A. Laskey. 1995a. Distinct functions for the two importin subunits in nuclear protein import. *Nature* **377:** 246.

Görlich, D., S. Kostka, R. Kraft, C. Dingwall, R. Laskey, E. Hartmann, and S. Prehn.

1995b. Two different subunits of importin cooperate to regulate nuclear localization sequences and bind them to the nuclear envelope. *Curr. Biol.* **5:** 383.

Görlich, D., R. Kraft, S. Kostka, F. Vogel, E. Hartmann, R.A. Laskey, I.W. Mattaj, and E. Izuarralde. 1996. Importin provides a link between nuclear protein import and U snRNA export. *Cell* **87:** 21.

Gorsch, L.C., T.C. Dockendorff, and C.N. Cole. 1995. A conditional allele of the novel repeat-containing yeast nucleoporin *RAT7/NUP159* causes both rapid cessation of mRNA export and reversible clustering of nuclear pore complexes. *J. Cell Biol.* **129:** 939.

Gotta, M., T. Laroche, A. Formenton, L. Maillet, H. Scherthan, and S.M. Gasser. 1996. The clustering of telomeres and colocalization with Rap1, Sir3 and Sir4 proteins in wild-type *Saccharomyces cerevisiae*. *J. Cell Biol.* **134:** 1349.

Gottschling, D.E. 1992. Telomere-proximal DNA in *Saccharomyces cerevisiae* is refractory to methyltransferase activity *in vivo*. *Proc. Natl. Acad. Sci.* **89:** 4062.

Gottschling, D.E., O.M. Aparicio, B.L. Billington, and V.A. Zakian. 1990. Position effect at *S. cerevisiae* telomeres: Reversible repression of PolII transcription. *Cell* **63:** 751.

Grandi, P., V. Doye, and E.C. Hurt. 1993. Purification of NSP1 reveals complex formation with "GLFG" nucleoporins and a novel nuclear pore protein NIC96. *EMBO J.* **12:** 3061.

Grandi, P., N. Schlaich, H. Tekotte, and E.C. Hurt. 1995a. Functional interaction of Nic96p with a core nucleoporin complex consisting of Nsp1p, Nup49p, and a novel protein Nup57p. *EMBO J.* **14:** 76.

Grandi, P., S. Emig, C. Weise, F. Hucho, T. Pohl, and E.C. Hurt. 1995b. A novel nuclear pore protein Nup82p which specifically binds to a fraction of Nsp1p. *J. Cell Biol.* **130:** 1263.

Granot, D. and M. Snyder. 1991. Segregation of the nucleolus during mitosis in budding and fission yeast. *Cell Motil. Cytoskeleton* **20:** 47.

Grunstein, M. 1990. Histone function in transcription. *Annu. Rev. Cell Biol.* **6:** 643.

Gu, Z., R.P. Moerschell, F. Sherman, and D.S. Goldfarb. 1992. *NIP1*, a gene required for nuclear transport in yeast. *Proc. Natl. Acad. Sci.* **89:** 10355.

Guacci, V., E. Hogan, and D. Koshland. 1994. Chromosome condensation and sister chromatid pairing in budding yeast. *J. Cell Biol.* **125:** 517.

Haggren, W. and D. Kolodrubetz. 1988. The *Saccharomyces cerevisiae ACP2* gene encodes an essential HMG1-like protein. *Mol. Cell. Biol.* **8:** 1282.

Hall, M.N., L. Hereford, and I. Herskowitz. 1984. Targeting of *E. coli* β-galactosidase to the nucleus in yeast. *Cell* **36:** 1057.

Hall, M.N., C. Craik, and Y. Hiraoka. 1990. Homeodomain of yeast repressor α2 contains a nuclear localization signal. *Proc. Natl. Acad. Sci.* **87:** 6954.

Hartmann, E. and D. Görlich. 1995. A Ran-binding motif in nuclear pore proteins. *Trends Cell Biol.* **5:** 192.

Hartwell, L.H. 1967. Macromolecule synthesis in temperature-sensitive mutants of yeast. *J. Bacteriol.* **93:** 1662.

Heath, C.V., C.S. Copeland, D.C. Amberg, V. Del Priore, M. Snyder, and C.N. Cole. 1995. Nuclear pore complex clustering and nuclear accumulation of poly(A)$^+$ RNA associated with mutation of the *Saccharomyces cerevisiae RAT2/NUP120* gene. *J. Cell Biol.* **131:** 1677.

Hecht, A., T. Laroche, S. Strahl-Bolsinger, S.M. Gasser, and M. Grunstein. 1995. Histone H3 and H4 N-termini interact with SIR3 and SIR4 proteins: A molecular model for the formation of heterochromatin in yeast. *Cell* **80:** 583.

Hennessy, K.M., C.D. Clark, and D. Botstein. 1990. Subcellular localization of yeast *CDC46* varies with the cell cycle. *Genes Dev.* **4:** 2252.

Henríquez, R., G. Blobel, and J. Aris. 1990. Isolation and sequencing of *NOP1*. A yeast gene encoding a nucleolar protein homologous to a human autoimmune antigen. *J. Biol. Chem.* **265:** 2209.

Hinshaw, J.E., B.O. Carragher, and R.A. Milligan. 1992. Architecture and design of the nuclear pore complex. *Cell* **69:** 1133.

Hirano, T. 1995. Biochemical and genetic dissection of mitotic chromosome condensation. *Trends Biochem. Sci.* **20:** 357.

Hirano, T. and T.J. Mitchison. 1993. Topoisomerase II does not play a scaffolding role in the organization of mitotic chromosomes assembled in *Xenopus* extracts. *J. Cell Biol.* **120:** 601.

———. 1994. A heterodimeric coiled-coil protein required for mitotic chromosome condensation *in vitro*. *Cell* **79:** 449.

Hofmann, J.F.-X., T. Laroche, A.H. Brand, and S.M. Gasser. 1989. RAP-1 factor is necessary for DNA loop formation in vitro at the silent mating type locus *HML*. *Cell* **57:** 725.

Hopper, A.K., H.M. Traglia, and R.W. Dunst. 1990. The yeast *RNA1* gene product necessary for RNA processing is located in the cytosol and apparently excluded from the nucleus. *J. Cell Biol.* **111:** 309.

Huffaker, T.C., J.H. Thomas, and D. Botstein. 1988. Diverse effects of β-tubulin mutations on microtubule formation and function. *J. Cell Biol.* **106:** 1997.

Hurt, E.C. 1988. A novel nucleoskeletal-like protein located at the nuclear periphery is required for the life cycle of *Saccharomyces cerevisiae*. *EMBO. J.* **7:** 4323.

Hurt, E.C., A. McDowall, and T. Schimmang. 1988. Nucleolar and nuclear envelope proteins of the yeast *Saccharomyces cerevisiae*. *Eur. J. Cell Biol.* **46:** 554.

Hurwitz, M.E. and G. Blobel. 1995. NUP82 is an essential yeast nucleoporin required for poly(A)$^+$ RNA export. *J. Cell Biol.* **130:** 1275.

Hutchison, H.T., L.H. Hartwell, and C.S. McLaughlin. 1969. Temperature-sensitive yeast mutant defective in ribonucleic acid production. *J. Bacteriol.* **99:** 807.

Iovine, M.K., J.L. Watkins, and S.R. Wente. 1995. The GLFG repetitive region of the nucleoporin Nup116p interacts with Kap95p, an essential yeast nuclear import factor. *J. Cell Biol.* **131:** 1699.

Izaurralde, E. and I.W. Mattaj. 1995. RNA export. *Cell* **81:** 153.

Izaurralde, E., J. Lewis, C. Gamberi, A. Jarmolowski, C. McGuigan, and I.W. Mattaj. 1995. A cap-binding protein complex mediating U snRNA export. *Nature* **376:** 709.

Jackson, D.A. and P.R. Cook. 1986. Replication occurs at a nucleoskeleton. *EMBO J.* **5:** 1403.

———. 1988. Visualization of a filamentous nucleoskeleton with a 23 nm axial repeat. *EMBO J.* **7:** 3667.

Jacobs, C.W., A.E.M. Adams, P.J. Szaniszlo, and J.R. Pringle. 1988. Functions of microtubules in the *Saccharomyces cerevisiae* cell cycle. *J. Cell Biol.* **107:** 1409.

Jarmolowski, A., W.C. Boelens, E. Izaurralde, and I.W. Mattaj. 1994. Nuclear export of different classes of RNA is mediated by specific factors. *J. Cell Biol.* **124:** 627.

Johnston, M., S. Andrews, R. Brinkman, J. Cooper, H. Ding, J. Dover, Z. Du, A. Favello, L. Fulton, S. Gattung, and 25 others. 1994. Complete nucleotide sequence of *Saccharomyces cerevisiae* chromosome VIII. *Science* **265:** 2077.

Jordan, E.G., N.J. Severs, and D.H. Williamson. 1977. Nuclear pore formation and the cell cycle in *Saccharomyces cerevisiae*. *Exp. Cell Res.* **104:** 446.

Kadowaki, T., Y. Zhao, and A.M. Tartakoff. 1992. A conditional yeast mutant defective in mRNA transport from nucleus to cytoplasm. *Proc. Natl. Acad. Sci.* **89:** 2312.

Kadowaki, T., M. Hitomi, S. Chen, and A.M. Tartakoff. 1994a. Nuclear mRNA accumulation causes nucleolar fragmentation in yeast *mtr2* mutant. *Mol. Biol. Cell* **5:** 1252.

Kadowaki, T., D. Goldfarb, L.M. Spitz, A.M. Tartakoff, and M. Ohno. 1993. Regulation of RNA processing and transport by a nuclear guanine nucleotide release protein and members of the Ras superfamily. *EMBO J.* **12:** 2929.

Kadowaki, T., S. Chen, M. Hitomi, E. Jacobs, C. Kumagai, S. Liang, R. Schneiter, D. Singleton, J. Wisniewska, and A.M. Tartakoff. 1994b. Isolation and characterization of *Saccharomyces cerevisiae* mRNA transport-defective (*mtr*) mutants. *J. Cell Biol.* **126:** 649.

Kalderon, D., B.L. Roberts, W.D. Richardson, and A.E. Smith. 1984. A short amino acid sequence able to specify nuclear location. *Cell* **39:** 499.

Kayne, P.S., U.-J. Kim, M. Han, J.R. Mullen, F. Yoshizaki, and M. Grunstein. 1988. Extremely conserved histone H4 N-terminus is dispensable for growth but essential for repressing the silent mating loci in yeast. *Cell* **55:** 27.

Kenna, M.A., J.G. Petranka, J.L. Reilly, and L.I. Davis. 1996. Yeast Nle3p/Nup170p is required for normal stoichiometry of FG nucleoporins within the nuclear pore complex. *Mol. Cell. Biol.* **16:** 2025.

Kessel, R.G. 1983. The structure and function of annulate lamellae: Porous cytoplasmic and intranuclear membranes. *Int. Rev. Cytol.* **82:** 181.

Kilmartin, J.V. 1994. Genetic and biochemical approaches to spindle function and chromosome segregation in eukaryotic microorganisms. *Curr. Opin. Cell Biol.* **6:** 50.

Kilmartin, J.V., S.L. Dyos, D. Kershaw, and J.T. Finch. 1993. A spacer protein in the *Saccharomyces cerevisiae* spindle pole body whose transcript is cell cycle-regulated. *J. Cell Biol.* **123:** 1175.

Klein, F., T. Laroche, M.E. Cardenas, J.F.-X. Hofmann, D. Schweizer, and S.M. Gasser. 1992. Localization of RAP1 and topoisomerase II in nuclei and meiotic chromosomes of yeast. *J. Cell Biol.* **117:** 935.

Koepp, D.M. and P.A. Silver. 1996. A GTPase controlling nuclear trafficking: Running the right way or walking RANdomly. *Cell* **87:** 1.

Koepp, D.M., D.H. Wong, A.H. Corbett, and P.A. Silver. 1996. Dynamic localization of the nuclear import receptor and its interactions with transport factors. *J. Cell Biol.* **133:** 1163.

Kolodrubetz, D. and A. Bergum. 1990. Duplicated *NHP6* genes of *Saccharomyces cerevisiae* encode proteins homologous to bovine high mobility protein. *J. Biol. Chem.* **265:** 3234.

Koning, A.J., C.J. Roberts, and R.L. Wright. 1996. Different subcellular localization of *Saccharomyces cerevisiae* HMG-CoA reductase isozymes at elevated levels corresponds to distinct endoplasmic reticulum membrane proliferations. *Mol. Biol. Cell* **7:** 769.

Koning, A.J., P.Y. Lum, J.M. Williams, and R. Wright. 1993. $DiOC_6$ staining reveals organelle structure and dynamics in living yeast cells. *Cell Motil. Cytoskeleton* **25:** 111.

Kraemer, D.M., C. Strambio-de-Castillia, G. Blobel, and M.P. Rout. 1995. The essential yeast nucleoporin NUP159 is located on the cytoplasmic side of the nuclear pore complex and serves in karyopherin-mediated binding of transport substrate. *J. Biol. Chem.* **270:** 19017.

Kraemer, D., R.W. Wozniak, G. Blobel, and A. Radu. 1994. The human CAN protein, a putative oncogene product associated with myeloid leukemogenesis, is a nuclear pore complex protein that faces the cytoplasm. *Proc. Natl. Acad. Sci.* **91:** 1519.

Kruger, W. and I. Herskowitz. 1991. A negative regulator of *HO* transcription, SIN1 (SPT2), is a nonspecific DNA-binding protein related to HMG1. *Mol. Cell. Biol.* **11:** 4135.

Kurihara, L.J., C.T. Beh, M. Latterich, R. Schekman, and M.D. Rose. 1994. Nuclear congression and membrane fusion: Two distinct events in the yeast karyogamy pathway. *J. Cell Biol.* **126:** 911.

Laemmli, U.K., E. Käs, L. Poljak, and Y. Adachi. 1992. Scaffold-associated regions: *cis*-acting determinants of chromatin structural loops and functional domains. *Curr. Opin. Genet. Dev.* **2:** 275.

Laemmli, U.K., S.M. Cheng, K.W. Adolph, J.R. Paulson, J.A. Brown, and W.R. Baumbach. 1978. Metaphase chromosome structure: the role of nonhistone proteins. *Cold Spring Harbor Symp. Quant. Biol.* **42:** 351.

Landsman, D. 1996. Histone H1 in *Saccharomyces cerevisiae* a double mystery solved? *Trends Biochem. Sci.* **21:** 287.

Lanford, R.E. and J.S. Butel. 1984. Construction and characterization of an SV40 mutant defective in nuclear transport of T antigen. *Cell* **37:** 801.

Latterich, M. and R. Schekman. 1994. The karyogamy gene *KAR2* and novel proteins are required for ER membrane fusion. *Cell* **78:** 87.

Latterich, M., K.-U. Fröhlich, and R. Schekman. 1995. Membrane fusion and the cell cycle: Cdc48p participates in the fusion of ER membranes. *Cell* **82:** 885.

Laurenson, P. and J. Rine. 1992. Silencers, silencing, and heritable transcriptional states. *Microbiol. Rev.* **56:** 543.

Lawrence, J.B., C.A. Villnave, and R.H. Singer. 1988. Sensitive, high-resolution chromatin and chromosome mapping in situ: Presence and orientation of two closely integrated copies of EBV in a lymphoma line. *Cell* **52:** 51.

Lee, M.-S. and W.T. Garrard. 1991. Transcription-induced nucleosome "splitting:" An underlying structure for DNase I sensitive chromatin. *EMBO J.* **10:** 607.

Lee, M.S., M. Henry, and P.A. Silver. 1996. A protein that shuttles between the nucleus and the cytoplasm is an important mediator of RNA export. *Genes Dev.* **10:** 1233.

Lee, W.-C. and T. Mélèse. 1989. Identification and characterization of a nuclear localization sequence-binding protein in yeast. *Proc. Natl. Acad. Sci.* **86:** 8808.

Lee, W.-C., Z. Xue, and T. Mélèse. 1991. The *NSR1* gene encodes a protein that specifically binds nuclear localization sequences and has two RNA recognition motifs. *J. Cell Biol.* **113:** 1.

Lee, W.-C., D. Zabetakis, and T. Mélèse. 1992. NSR1 is required for pre-rRNA processing and for the proper maintenance of steady-state levels of ribosomal subunits. *Mol. Cell. Biol.* **12:** 3865.

Lehmann, A.R., M. Walicka, D.J.F. Griffiths, J.M. Murray, F.Z. Watts, S. McCready, and A.M. Carr. 1995. The *rad18* gene of *Schizosaccharomyces pombe* defines a new subgroup of the SMC superfamily involved in DNA repair. *Mol. Cell. Biol.* **15:** 7067.

Li, O., C.V. Heath, D.C. Amberg, T.C. Dockendorff, C.S. Copeland, M. Snyder, and C.N. Cole. 1995. Mutation or deletion of the *Saccharomyces cerevisiae RAT3/NUP133* gene causes temperature-dependent nuclear accumulation of poly(A)$^+$ RNA and constitutive clustering of nuclear pore complexes. *Mol. Biol. Cell* **6:** 401.

Li, Y.-Y., E. Yeh, T. Hays, and K. Bloom. 1993. Disruption of mitotic spindle orientation in a yeast dynein mutant. *Proc. Natl. Acad. Sci.* **90:** 10096.

Linder, C. and F. Thoma. 1994. Histone H1 expressed in *Saccharomyces cerevisiae* binds to chromatin and affects survival, growth, transcription and plasmid stability but does not change nucleosomal spacing. *Mol. Cell. Biol.* **14:** 2822.

Liu, C., X. Mao, and A.J. Lustig. 1994. Mutational analysis defines a C-terminal tail domain of RAP1 essential for telomeric silencing in *Saccharomyces cerevisiae. Genetics* **138:** 1025.

Loeb, J.D.J., L.I. Davis, and G.R. Fink. 1993. *NUP2*, a novel yeast nucleoporin, has functional overlap with other proteins of the nuclear pore complex. *Mol. Biol. Cell* **4:** 209.

Loeb, J.D.J., G. Schlenstedt, D. Pellman, D. Kornitzer, P.A. Silver, and G.R. Fink. 1995. The yeast nuclear import receptor is required for mitosis. *Proc. Natl. Acad. Sci.* **92:** 7647.

Lohr, D. 1983. The chromatin structure of an actively expressed, single copy yeast gene. *Nucleic Acids Res.* **11:** 6755.

Lohr, D. and L. Hereford. 1979. Yeast chromatin is uniformly digested by DNase I. *Proc. Natl. Acad. Sci.* **76:** 4285.

Lohr, D. and J.E. Hopper. 1985. The relationship of regulatory proteins and DNase I hypersensitive sites in the yeast *GAL1-10* genes. *Nucleic Acids Res.* **13:** 8409.

Lohr, D., R.R. Kovacic, and K.E. Van Holde. 1977. Quantitative analysis of the digestion of yeast chromatin by staphylococcal nuclease. *Biochemistry* **16:** 463.

Longtine, M.S., N.M. Wilson, M.E. Petracek, and J. Berman. 1989. A yeast telomere binding activity binds to two related telomere sequence motifs and is indistinguishable from RAP1. *Curr. Genet.* **6:** 225.

Louis, E.J. 1995. The chromosome ends of *Saccharomyces cerevisiae. Yeast* **11:** 1553.

Lowary, P.T. and J. Widom. 1989. Higher-order structure of *Saccharomyces cerevisiae* chromatin. *Proc. Natl. Acad. Sci.* **86:** 8266.

Lustig, A.J., C. Liu. C. Zhang, and J.P. Hanish. 1996. Tethered Sir3p nucleates silencing at telomeres and internal loci in *Saccharomyces cerevisiae. Mol. Cell. Biol.* **16:** 2483.

Maillet, L., C. Boscheron, M. Gotta, S. Marcand, E. Gilson, and S.M. Gasser. 1996. Evidence for silencing compartments within the yeast nucleus: A role for telomere proximity and Sir protein concentration in silencer-mediated repression. *Genes Dev.* **10:** 1796.

Malim, M.H. and B.R. Cullen. 1991. HIV-1 structural gene expression requires the binding of multiple Rev monomers to the viral RRE: Implications for HIV-1 latency. *Cell* **65:** 241.

Malim, M.H., S. Böhnlein, J. Hauber, and B.R. Cullen. 1989. Functional dissection of the HIV-1 Rev *trans*-activator: Derivation of a *trans*-dominant repressor of Rev function. *Cell* **58:** 205.

Marcand, S., S.W. Buck, P. Moretti, E. Gilson, and D. Shore. 1996. Silencing of genes at nontelomeric sites in yeast is controlled by sequestration of silencing factors at telomeres by Rap1 protein. *Genes Dev.* **10:** 1297.

Marshall, M., D. Mahoney, A. Rose, J.B. Hicks, and J.R. Broach. 1987. Functional domains of *SIR4*, a gene required for position effect regulation in *Saccharomyces cerevisiae. Mol. Cell. Biol.* **7:** 4441.

Mathog, D., M. Hochstrasser, Y. Gruenbaum, H. Saumweber, and J. Sedat. 1984. Characteristic folding pattern of polytene chromosomes in *Drosophila* salivary gland nuclei.

Nature **308:** 414.

Maul, G.G. 1977. The nuclear and the cytoplasmic pore complex: Structure, dynamics, distribution, and evolution. *Int. Rev. Cytol.* (suppl.) **6:** 75.

McCusker, J.H., M. Yamagishi, J.M. Kolb, and M. Nomura. 1991. Suppressor analysis of temperature-sensitive RNA polymerase I mutations in *Saccharomyces cerevisiae:* Suppression of mutations in a zinc-binding motif by transposed mutant genes. *Mol. Cell. Biol.* **11:** 746.

McMillan, J.N. and K. Tatchell. 1994. The *JNM1* gene in the yeast *Saccharomyces cerevisiae* is required for nuclear migration and spindle orientation during the mitotic cell cycle. *J. Cell Biol.* **125:** 143.

Melchior, F. and L. Gerace. 1995. Mechanisms of nuclear protein import. *Curr. Opin. Cell Biol.* **7:** 310.

Melchior, F., B. Paschal, J. Evans, and L. Gerace. 1993. Inhibition of nuclear protein import by nonhydrolyzable analogues of GTP and identification of the small GTPase Ran/TC4 as an essential transport factor. *J. Cell Biol.* **123:** 1649.

Mélèse, T. and Z. Xue. 1995. The nucleolus: An organelle formed by the act of building a ribosome. *Curr. Opin. Cell Biol.* **7:** 319.

Meluh, P.B. and M.D. Rose. 1990. *KAR3*, a kinesin-related gene required for yeast nuclear fusion. *Cell* **60:** 1029.

Meyer, B.E. and M.H. Malim. 1994. The HIV-1 Rev *trans*-activator shuttles between the nucleus and the cytoplasm. *Genes Dev.* **8:** 1538.

Micklem, G., A. Rowley, J. Harwood, K. Nasmyth, and J.F.X. Diffley. 1993. Yeast origin recognition complex is involved in DNA replication and transcriptional silencing. *Nature* **366:** 87.

Miller, A.M. and K.A. Nasmyth. 1984. Role of DNA replication in the repression of silent mating type loci in yeast. *Nature* **312:** 247.

Miller, M., M.K. Park, and J.A. Hanover. 1991. Nuclear pore complex: Structure, function, and regulation. *Physiol. Rev.* **71:** 909.

Mirzayan, C., C.S. Copeland, and M. Snyder. 1992. The *NUF1* gene encodes an essential coiled-coil related protein that is a potential component of the yeast nucleoskeleton. *J. Cell Biol.* **116:** 1319.

Moll, T., G. Tebb, U. Surana, H. Robitsch, and K. Nasmyth. 1991. The role of phosphorylation and the CDC28 protein kinase in cell cycle-regulated nuclear import of the *S. cerevisiae* transcription factor SWI5. *Cell* **66:** 743.

Moore, M.S. and G. Blobel. 1992. The two steps of nuclear import, targeting to the nuclear envelope and translocation through the nuclear pore, require different cytosolic factors. *Cell* **69:** 939.

———. 1993. The GTP-binding protein Ran/TC4 is required for protein import into the nucleus. *Nature* **365:** 661.

———. 1994a. A G protein involved in nucleocytoplasmic transport: The role of Ran. *Trends Biochem. Sci.* **19:** 211.

———. 1994b. Purification of a Ran-interacting protein that is required for protein import into the nucleus. *Proc. Natl. Acad. Sci.* **91:** 10212.

Moreland, R.B., H.G. Nam, L.M. Hereford, and H.M. Fried. 1985. Identification of a nuclear localization signal of a yeast ribosomal protein. *Proc. Natl. Acad. Sci.* **82:** 6561.

Moreland, R.B., G.L. Langevin, R.H. Singer, R.L. Garcea, and L.M. Hereford. 1987. Amino acid sequences that determine the nuclear localization of yeast histone 2B. *Mol. Cell. Biol.* **7:** 4048.

Moretti, P., K. Freeman, L. Coodly, and D. Shore. 1994. Evidence that a complex of SIR proteins interacts with the silencer and telomere-binding protein RAP1. *Genes Dev.* **8:** 2257.

Moroianu, J., G. Blobel, and A. Radu. 1995a. Previously identified protein of uncertain function is karyopherin α and together with karyopherin β docks import substrate at nuclear pore complexes. *Proc. Natl. Acad. Sci.* **92:** 2008.

Moroianu, J., M. Hijikata, G. Blobel, and A. Radu. 1995b. Mammalian karyopherin α1β and α2β heterodimers: α1 or α2 subunit binds nuclear localization signal and β subunit interacts with peptide repeat-containing nucleoporins. *Proc. Natl. Acad. Sci.* **92:** 6532.

Muhua, L., T.S. Karpova, and J.A. Cooper. 1994. A yeast actin-related protein homologous to that in vertebrate dynactin complex is important for spindle orientation and nuclear migration. *Cell* **78:** 669.

Murphy, R. and S.R. Wente. 1996. An RNA export mediator with an essential nuclear export signal. *Nature* **383:** 357.

Murphy, R., J.L. Watkins, and S.R. Wente. 1996. *GLE2*, a *Saccharomyces cerevisiae* homologue of the *Schizosaccharomyces pombe* export factor *RAE1*, is required for nuclear pore complex structure and function. *Mol. Biol. Cell* **7:** (in press).

Mutvei, A., S. Dihlmann, W. Herth, and E.C. Hurt. 1992. NSP1 depletion in yeast affects nuclear pore formation and nuclear accumulation. *Eur. J. Cell Biol.* **59:** 280.

Nagy, M., M. Le Gouar, S. Potier, J.-L. Souciet, and G. Hervé. 1989. The primary structure of the aspartate transcarbamylase region of the *URA2* gene product in *Saccharomyces cerevisiae:* Features involved in activity and nuclear localization. *J. Biol. Chem.* **264:** 8366.

Nasmyth, K.A. 1982. The regulation of yeast mating-type chromatin structure by *SIR:* An action at a distance affecting both transcription and transposition. *Cell* **30:** 567.

Nasmyth, K., G. Adolf, D. Lydall, and A. Seddon. 1990. The identification of a second cell cycle control on the *HO* promoter in yeast: Cell cycle regulation of SWI5 nuclear entry. *Cell* **62:** 631.

Nehrbass, U. and G. Blobel. 1996. Role of the nuclear transport factor p10 in nuclear import. *Science* **272:** 120.

Nehrbass, U., E. Fabre, S. Dihlmann, W. Herth, and E.C. Hurt. 1993. Analysis of nucleo-cytoplasmic transport in a thermosensitive mutant of nuclear pore protein NSP1. *Eur. J. Cell Biol.* **62:** 1.

Nehrbass, U., M.P. Rout, S. Maguire, G. Blobel, and R.W. Wozniak. 1996. The yeast nucleoporin Nup188p interacts genetically and physically with the core structures of the nuclear pore complex. *J. Cell Biol.* **133:** 1153.

Nehrbass, U., H. Kern, A. Mutvei, H. Horstmann, B. Marshallsay, and E.C. Hurt. 1990. NSP1: A yeast nuclear envelope protein localized at the nuclear pores exerts its essential function by its carboxy-terminal domain. *Cell* **61:** 979.

Nelson, M. and P.A. Silver. 1989. Context affects nuclear protein localization in *Saccharomyces cerevisiae. Mol. Cell. Biol.* **9:** 384.

Nelson, R.G. and W.L. Fangman. 1979. Nucleosome organization of the yeast 2-μm DNA plasmid: A eukaryotic minichromosome. *Proc. Natl. Acad. Sci.* **76:** 6515.

Newmeyer, D.D. and D.J. Forbes. 1988. Nuclear import can be separated into distinct steps in vitro: Nuclear pore binding and translocation. *Cell* **52:** 641.

Ng, D.T.W. and P. Walter. 1996. ER membrane protein complex required for nuclear fusion. *J. Cell Biol.* **132:** 499.

Nigg, E. 1992. Assembly-disassembly of the nuclear lamina. *Curr. Opin. Cell Biol.* **4:**

105.
Noble, S.M. and C. Guthrie. 1996. Identification of novel genes required for yeast pre-mRNA splicing by means of cold-sensitive mutations. *Genetics* **143:** 67.
Ohtsubo, M., T. Yoshida, H. Seino, H. Nishitani, K.L. Clark, G.F. Sprague, Jr., M. Frasch, and T. Nishimoto. 1991. Mutation of the hamster cell cycle gene *RCC1* is complemented by the homologous genes of *Drosophila* and *S. cerevisiae*. *EMBO J.* **10:** 1265.
Osborne, M.A. and P.A. Silver. 1993. Nucleocytoplasmic transport in the yeast *Saccharomyces cerevisiae*. *Annu. Rev. Biochem.* **62:** 219.
Ouspenski, I.I., U.W. Mueller, A. Matynia, S. Sazer, S.J. Elledge, and B.R. Brinkley. 1995. Ran-binding protein-1 is an essential component of the Ran/RCC1 molecular switch system in budding yeast. *J. Biol. Chem.* **270:** 1975.
Page, B.D. and M. Snyder. 1993. Chromosome segregation in yeast. *Annu. Rev. Microbiol.* **47:** 231.
Paine, P.L., L.C. Moore, and S.B. Horowitz. 1975. Nuclear envelope permeability. *Nature* **254:** 109.
Palladino, F., T. Laroche, E. Gilson, A. Axelrod, L. Pillus, and S.M. Gasser. 1993. SIR3 and SIR4 proteins are required for the positioning and integrity of yeast telomeres. *Cell* **75:** 543.
Palmer, R.E., D.S. Sullivan, T. Huffaker, and D. Koshland. 1992. Role of astral microtubules and actin in spindle orientation and migration in the budding yeast *Saccharomyces cerevisiae*. *J. Cell Biol.* **119:** 583.
Panté, N. and U. Aebi. 1994. Towards understanding the three-dimensional structure of the nuclear pore complex at the molecular level. *Curr. Opin. Struct. Biol.* **4:** 187.
Panté, N., R. Bastos, I. McMorrow, B. Burke, and U. Aebi. 1994. Interactions and three-dimensional localization of a group of nuclear pore complex proteins. *J. Cell Biol.* **126:** 603.
Pardoll, D.M., B. Vogelstein, and D.S. Coffey. 1980. A fixed site of DNA replication in eucaryotic cells. *Cell* **19:** 527.
Park, M.K., M. D'Onofrio, M.C. Willingham, and J.A. Hanover. 1987. A monoclonal antibody against a family of nuclear pore proteins (nucleoporins): O-linked *N*-acetylglucosamine is part of the immunodeterminant. *Proc. Natl. Acad. Sci.* **84:** 6462.
Parrish, M.L., C. Sengstag, J.D. Rine, and R.L. Wright. 1995. Identification of the sequences in HMG-CoA reductase required for karmellae assembly. *Mol. Biol. Cell* **6:** 1535.
Paschal, B.M. and L. Gerace. 1995. Identification of NTF2, a cytosolic factor for nuclear import that interacts with nuclear pore complex protein p62. *J. Cell Biol.* **129:** 925.
Paschal, B.M., C. Delphin, and L. Gerace. 1996. Nucleotide-specific interaction of Ran/TC4 with nuclear transport factors NTF2 and p97. *Proc. Natl. Acad. Sci.* **93:** 7679.
Paulson, J.R. and U.K. Laemmli. 1977. The structure of histone-depleted metaphase chromosomes. *Cell* **12:** 817.
Pemberton, L.F., M.P. Rout, and G. Blobel. 1995. Disruption of the nucleoporin gene *NUP133* results in clustering of nuclear pore complexes. *Proc. Natl. Acad. Sci.* **92:** 1187.
Perez-Ortin, J.E., E. Matallana, and L. Franco. 1989. Chromatin structure in yeast. *Yeast* **5:** 219.
Peterson, J.B. and H. Ris. 1976. Electron-microscopic study of the spindle and chromosome movement in the yeast *Saccharomyces cerevisiae*. *J. Cell Sci.* **22:** 219.

Petes, T.D. 1979. Meiotic mapping of yeast ribosomal deoxynucleic acid on chromosome XII. *J. Bacteriol.* **138:** 185.

Piñol-Roma, S. and G. Dreyfuss. 1992. Shuttling of pre-mRNA-binding proteins between nucleus and cytoplasm. *Nature* **355:** 730.

———. 1993. hnRNP proteins: Localization and transport between the nucleus and the cytoplasm. *Trends Cell Biol.* **3:** 151.

Polaina, J. and J. Conde. 1982. Genes involved in the control of nuclear fusion during the sexual cycle of *Saccharomyces cerevisiae. Mol. Gen. Genet.* **186:** 253.

Pollard, V.W., W.M. Michael, S. Nakielny, M.C. Siomi, F. Wang, and G. Dreyfuss. 1996. A novel receptor-mediated nuclear protein import pathway. *Cell* **86:** 985.

Pon, L. and G. Schatz. 1991. Biogenesis of yeast mitochondria. In *The molecular and cellular biology of the yeast* Saccharomyces: *Genome dynamics, protein synthesis, and energetics* (ed. J.R. Broach et al.) vol. 1, p. 333. Cold Spring Harbor Laboratory Press, Cold Spring Harbor, New York.

Potashkin, J.A., R.F. Zeigel, and J.A. Huberman. 1984. Isolation and initial characterization of residual nuclear structures from yeast. *Exp. Cell Res.* **153:** 374.

Powers, M.A. and D.J. Forbes. 1994. Cytosolic factors in nuclear transport: What's importin? *Cell* **79:** 931.

Powers, M.A., C. Macaulay, F.R. Masiarz, and D.J. Forbes. 1995. Reconstituted nuclei depleted of a vertebrate GLFG nuclear pore protein, p97, import but are defective for nuclear growth and replication. *J. Cell Biol.* **128:** 721.

Radu, A., G. Blobel, and M.S. Moore. 1995a. Identification of a protein complex that is required for nuclear protein import and mediates docking of import substrate to distinct nucleoporins. *Proc. Natl. Acad. Sci.* **92:** 1769.

Radu, A., M.S. Moore, and G. Blobel. 1995b. The peptide repeat domain of nucleoporin Nup98 functions as a docking site in transport across the nuclear pore complex. *Cell* **81:** 215.

Rattner, J.B., C. Saunders, J.R. Davie, and B.A. Hamkalo. 1982. Ultrastructural organization of yeast chromatin. *J. Cell Biol.* **92:** 217.

Reichelt, R., A. Holzenburg, E.L. Buhle, Jr., M. Jarnik, A. Engel, and U. Aebi. 1990. Correlation between structure and mass distribution of the nuclear pore complex and of distinct pore complex components. *J. Cell Biol.* **110:** 883.

Renauld, H., O.M. Aparicio, P.D. Zierath, B.L. Billington, S.K. Chhablani, and D.E. Gottschling. 1993. Silent domains are assembled continuously from the telomere and are defined by promoter distance and strength, and by SIR3 dosage. *Genes Dev.* **7:** 1133.

Rexach, M. and G. Blobel. 1995. Protein import into nuclei: Association and dissociation reactions involving transport substrate, transport factors, and nucleoporins. *Cell* **83:** 683.

Richardson, W.D., A.D. Mills, S.M. Dilworth, R.A. Laskey, and C. Dingwall. 1988. Nuclear protein migration involves two steps: Rapid binding at the nuclear envelope followed by slower translocation through nuclear pores. *Cell* **52:** 655.

Ris, H. 1991. The three-dimensional structure of the nuclear pore complex as seen by high voltage electron microscopy and high resolution low voltage scanning electron microscopy. *Electron Microscopy Society of America Bull.* **21:** 54.

Ris, H. and M. Malecki. 1993. High-resolution field emission scanning electron microscope imaging of internal cell structures after Epon extraction from sections: A new approach to correlative ultrastructural and immunocytochemical studies. *J. Struct. Biol.*

111: 148.

Rose, M.D. 1991. Nuclear fusion in yeast. *Annu. Rev. Microbiol.* **45:** 539.

———. 1996. Nuclear fusion in the yeast *Saccharomyces cerevisiae Annu. Rev. Cell Dev. Biol.* **12:** 663.

Rose, M.D., L.M. Misra, and J.P. Vogel. 1989. *KAR2*, a karyogamy gene, is the yeast homolog of the mammalian BiP/GRP78 gene. *Cell* **57:** 1211.

Roth, S.Y., A. Dean, and R.T. Simpson. 1990. Yeast α2 repressor positions nucleosomes in *TRP1/ARS1* chromatin. *Mol. Cell. Biol.* **10:** 2247.

Rothblatt, J.A., R.J. Deshaies, S.L. Sanders, G. Daum, and R. Schekman. 1989. Multiple genes are required for proper insertion of secretory proteins into the endoplasmic reticulum in yeast. *J. Cell Biol.* **109:** 2641.

Rout, M.P. and G. Blobel. 1993. Isolation of the yeast nuclear pore complex. *J. Cell Biol.* **123:** 771.

Rout, M.P. and S.R. Wente. 1994. Pores for thought: Nuclear pore complex proteins. *Trends Cell Biol.* **4:** 357.

Russell, I.D. and D. Tollervey. 1992. NOP3 is an essential yeast protein which is required for pre-rRNA processing. *J. Cell Biol.* **119:** 737.

———. 1995. Yeast Nop3p has structural and functional similarities to mammalian pre-mRNA binding proteins. *Eur. J. Cell Biol.* **66:** 293.

Sadler, I., A. Chiang, T. Kurihara, J. Rothblatt, J. Way, and P. Silver. 1989. A yeast gene important for protein assembly into the endoplasmic reticulum and the nucleus has homology to DnaJ, an *Escherichia coli* heat shock protein. *J. Cell Biol.* **109:** 2665.

Saitoh, N., I.G. Goldberg, E.R. Wood, and W.C. Earnshaw. 1994. ScII: An abundant chromosome scaffold protein is a member of a family of putative ATPases with an unusual predicted tertiary structure. *J. Cell Biol.* **127:** 303.

Saitoh, Y. and U.K. Laemmli. 1994. Metaphase chromosome structure: Bands arise from a differential folding path of the highly AT-rich scaffold. *Cell* **76:** 609.

Saka, Y., T. Sutani, Y. Yamashita, S. Saitoh, M. Takeuchi, Y. Nakaseko, and M. Yanagida. 1994. Fission yeast cut3 and cut14, members of a ubiquitous protein family, are required for chromosome condensation and segregation in mitosis. *EMBO J.* **13:** 4938.

Sanders, S.L., K.M. Whitfield, J.P. Vogel, M.D. Rose, and R.W. Schekman. 1992. Sec61p and BiP directly facilitate polypeptide translocation into the ER. *Cell* **69:** 353.

Sazer, S. 1996. The search for the primary function of the Ran GTPase continues. *Trends Cell Biol.* **6:** 81.

Schaap, P.J., J. van't Reit, C.L. Woldringh, and H.A. Raué. 1991. Identification and functional analysis of the nuclear localization signals of ribosomal protein L25 from *Saccharomyces cerevisiae. J. Mol. Biol.* **221:** 225.

Schlenstedt, G., E. Hurt, V. Doye, and P.A. Silver. 1993. Reconstitution of nuclear protein transport with semi-intact yeast cells. *J. Cell Biol.* **123:** 785.

Schlenstedt, G., D.H. Wong, D.M. Koepp, and P.A. Silver. 1995a. Mutants in a yeast Ran binding protein are defective in nuclear transport. *EMBO. J.* **14:** 5367.

Schlenstedt, G., C. Saavedra, J.D.J. Loeb, C.N. Cole, and P.A. Silver. 1995b. The GTP-bound form of the yeast Ran/TC4 homologue blocks nuclear protein import and appearance of poly(A)$^+$ RNA in the cytoplasm. *Proc. Natl. Acad. Sci.* **92:** 225.

Schneiter, R., T. Kadowaki, and A.M. Tartakoff. 1995. mRNA transport in yeast: Time to reinvestigate the functions of the nucleolus. *Mol. Biol. Cell* **6:** 357.

Schroer, T.A. and M.P. Sheetz. 1991. Two activators of micro-tubule based vesicle transport. *J. Cell Biol.* **115:** 1309.

Schunck, W.-H., F. Vogel, B. Gross, E. Kärgel, S. Mauersberger, K. Köpke, C. Gengnagel, and H.G. Müller. 1991. Comparison of two cytochromes P-450 from *Candida maltosa:* Primary structures, substrate specificities and effects of their expression in *Saccharomyces cerevisiae* on the proliferation of the endoplasmic reticulum. *Eur. J. Cell Biol.* **55:** 336.

Serrano, R. 1991. Transport across yeast vacuolar and plasma membranes. In *The molecular and cellular biology of the yeast* Saccharomyces; *Genome dynamics, protein synthesis, and energetics* (ed. J.R. Broach et al.), vol. 1, p. 523. Cold Spring Harbor Laboratory Press, Cold Spring Harbor, New York.

Severs, N.J., E.G. Jordan, and D.H. Williamson. 1976. Nuclear pore absence from areas of close association between nucleus and vacuole in synchronous yeast cultures. *J. Ultrastruct. Res.* **54:** 374.

Sharma, K., E. Fabre, H. Tekotte, E.C. Hurt, and D. Tollervey. 1996. Yeast nucleoporin mutants are defective in pre-tRNA splicing. *Mol. Cell. Biol.* **16:** 294.

Shimizu, M., S.Y. Roth, C. Szent-Györgyi, and R.T. Simpson. 1991. Nucleosomes are positioned with base pair precision adjacent to the α2 operator in *Saccharomyces cerevisiae. EMBO J.* **10:** 3033.

Shiokawa, K. and A.O. Pogo. 1974. The role of cytoplasmic membranes in controlling the transport of nuclear messenger RNA and initiation of protein synthesis. *Proc. Natl. Acad. Sci.* **71:** 2658.

Shore, D. and K. Nasmyth. 1987. Purification and cloning of a DNA binding protein from yeast that binds to both silencer and activator elements. *Cell* **51:** 721.

Shulga, N., P. Roberts, Z. Gu, L. Spitz, M.M. Tabb, M. Nomura, and D.S. Goldfarb. 1996. In vivo nuclear kinetics in *Saccharomyces cerevisiae:* A role for heat shock protein 70 during targeting and translocation. *J. Cell Biol.* **135:** 329.

Silver, P., L.P. Keegan, and M. Ptashne. 1984. Amino terminus of the yeast *GAL4* gene product is sufficient for nuclear localization. *Proc. Natl. Acad. Sci.* **81:** 5951.

Silver, P., I. Sadler, and M.A. Osborne. 1989. Yeast proteins that recognize nuclear localization sequences. *J. Cell Biol.* **109:** 983.

Singleton, D.R., S. Chen, M. Hitomi, C. Kumagai, and A.M. Tartakoff. 1995. A yeast protein that bidirectionally affects nucleocytoplasmic transport. *J. Cell Sci.* **108:** 265.

Siniossoglou, S., C. Wimmer, M. Rieger, V. Doye, H. Tekotte, C. Weise, S. Emig, A. Segref, and E.C. Hurt. 1996. A novel complex of nucleoporins, which includes Sec13p and a Sec13p homolog, is essential for normal nuclear pores. *Cell* **84:** 265.

Sledziewski, A. and E.T. Young. 1982. Chromatin conformational changes accompany transcriptional activation of a glucose-repressed gene in *Saccharomyces cerevisiae. Proc. Natl. Acad. Sci.* **79:** 253.

Smith, S. and G. Blobel. 1994. Colocalization of vertebrate lamin B and lamin B receptor (LBR) in nuclear envelopes and in LBR-induced membrane stacks of the yeast *Saccharomyces cerevisiae. Proc. Natl. Acad. Sci.* **91:** 10124.

Snow, C.M., A. Senior, and L. Gerace. 1987. Monoclonal antibodies identify a group of nuclear pore complex glycoproteins. *J. Cell Biol.* **104:** 1143.

Stafstrom, J.P. and L.A. Staehelin. 1984. Are annulate lamellae in the *Drosophila* embryo the result of overproduction of nuclear pore components? *J. Cell Biol.* **98:** 699.

Stavenhagen, J.B. and V.A. Zakian. 1994. Internal tracts of telomeric DNA act as silencers in *Saccharomyces cerevisiae. Genes Dev.* **8:** 1411.

Sterne-Marr, R., J.M. Blevitt, and L. Gerace. 1992. O-linked glycoproteins of the nuclear pore complex interact with a cytosolic factor required for nuclear protein import. *J. Cell Biol.* **116:** 271.

Stochaj, U., and P.A. Silver. 1992. A conserved phosphoprotein that specifically binds nuclear localization sequences is involved in nuclear import. *J. Cell Biol.* **117:** 473.

Stochaj, U., M. Osborne, T. Kurihara, and P. Silver. 1991. A yeast protein that binds nuclear localization signals: Purification, localization, and antibody inhibition of binding activity. *J. Cell Biol.* **113:** 1243.

Stochaj, U., M.A. Bossie, K. van Zee, A.M. Whalen, and P.A. Silver. 1993. Analysis of conserved binding proteins for nuclear localization sequences. *J. Cell Sci.* **104:** 89.

Strambio-de-Castillia, C., G. Blobel, and M.P. Rout. 1995. Isolation and characterization of nuclear envelopes from the yeast *Saccharomyces. J. Cell Biol.* **131:** 19.

Strunnikov, A.V., E. Hogan, and D. Koshland. 1995. *SMC2*, a *Saccharomyces cerevisiae* gene essential for chromosome segregation and condensation, defines a subgroup within the SMC family. *Genes Dev.* **9:** 587.

Strunnikov, A.V., V.L. Larionov, and D. Koshland. 1993. *SMC1:* An essential yeast gene encoding a putative head-rod-tail protein is required for nuclear division and defines a new ubiquitous protein family. *J. Cell Biol.* **123:** 1635.

Stutz, F., M. Neville, and M. Rosbash. 1995. Identification of a novel nuclear pore-associated protein as a functional target of the HIV-1 Rev protein in yeast. *Cell* **82:** 495.

Sullivan, D.S. and T.C. Huffaker. 1992. Astral microtubules are not required for anaphase B in *Saccharomyces cerevisiae. J. Cell Biol.* **119:** 379.

Szent-Györgyi, C., D.B. Finkelstein, and W.T. Garrard. 1987. Sharp boundaries demarcate the chromatin structure of a yeast heat-shock gene. *J. Mol. Biol.* **193:** 71.

Thoma, F. 1992. Nucleosome positioning. *Biochim. Biophys. Acta* **1130:** 1.

Thoma, F., T. Koller, and A. Klug. 1979. Involvement of histone H1 in the organization of the nucleosome and of the salt dependent superstructure of chromatin. *J. Cell Biol.* **83:** 403.

Thomas, J.O. and V. Furber. 1976. Yeast chromatin structure. *FEBS Lett.* **66:** 274.

Thompson, J.S., L.M. Johnson, and M. Grunstein. 1994. Specific repression of the yeast silent mating type locus *HMR* by an adjacent telomere. *Mol. Cell. Biol.* **14:** 446.

Thompson, J.S., X. Ling, and M. Grunstein. 1996. Histone H3 amino terminus is required for telomeric and silent mating locus repression in yeast. *Nature* **369:** 245.

Traglia, H.M., N.S. Atkinson, and A.K. Hopper. 1989. Structural and functional analyses of *Saccharomyces cerevisiae* wild-type and mutant *RNA1* genes. *Mol. Cell Biol.* **9:** 2989.

Trask, B., D. Pinkel, and G. van den Engh. 1989. The proximity of DNA sequences in interphase cell nuclei is correlated to genomic distance and permits ordering of cosmids spanning 250 kilobase pairs. *Genomics* **5:** 710.

Tremethick, D.J. and H.R. Drew. 1993. High mobility group proteins 14 and 17 can space nucleosomes *in vitro*. *J. Biol. Chem.* **268:** 11389.

Uemura, T., H. Ohkura, Y. Adachi, K. Morino, K. Shiozaki, and M. Yanagida. 1987. DNA topoisomerase II is required for condensation and separation of mitotic chromosomes in *S. pombe*. *Cell* **50:** 917.

Underwood, M.R. and H.M. Fried. 1990. Characterization of nuclear localizing sequences derived from yeast ribosomal protein L29. *EMBO J.* **9:** 91.

Unwin, P.N.T. and R.A. Milligan. 1982. A large particle associated with the perimeter of the nuclear pore complex. *J. Cell Biol.* **93:** 63.

Ushinsky, S.C., H. Bussey, A. Ahmed, Y. Wang, J. Friesen, B.A. Williams, and R.K. Storms. 1997. Histone H1 in *Saccharomyces cerevisiae. Yeast* (in press).

Vallen, E.A., M.A. Hiller, T.Y. Scherson, and M.D. Rose. 1992. Separate domains of *KAR1* mediate distinct functions in mitosis and nuclear fusion. *J. Cell Biol.* **117:** 1277.

van den Engh, G., R. Sachs, and B.J. Trask. 1992. Estimating genomic distance from DNA sequence location in cell nuclei by a random walk model. *Science* **257:** 1410.

Vaughn, J.P., P.A. Dijkwel, L.H.F. Mullenders, and J.L. Hamlin. 1990. Replication forks are associated with the nuclear matrix. *Nucleic Acids Res.* **18:** 1965.

Vergères, G., T.S.B. Yen, J. Aggeler, J. Lausier, and L. Waskell. 1993. A model system for studying membrane biogenesis: Overexpression of cytochrome b_5 in yeast results in marked proliferation of the intracellular membrane. *J. Cell Sci.* **106:** 249.

Vijayraghavan, U., M. Company, and J. Abelson. 1989. Isolation and characterization of pre-mRNA splicing mutants of *Saccharomyces cerevisiae. Genes Dev.* **3:** 1206.

Vogelstein, B., D.M. Pardoll, and D.S. Coffey. 1980. Supercoiled loops and eucaryotic DNA replication. *Cell* **22:** 79.

Vourc'h, C., D. Taruscio, A.L. Boyle, and D.C. Ward. 1993. Cell cycle-dependent distribution of telomeres, centromeres, and chromosome-specific subsatellite domains in the interphase nucleus of mouse lymphocytes. *Exp. Cell Res.* **205:** 142.

Voytas, D.F. and J.D. Boeke. 1993. Yeast retrotransposons and tRNAs. *Trends Genet.* **12:** 421.

Walmsley, R.W., C.S.M. Chan, B.-K. Tye, and T.D. Petes. 1984. Unusual DNA sequences associated with the ends of yeast chromosomes. *Nature* **310:** 157.

Wanka, F., L.H.F. Mullenders, A.G.M. Bekers, L.J. Pennings, J.M.A. Aelen, and J. Eygensteyn. 1977. Association of nuclear DNA with a rapidly sedimenting structure. *Biochem. Biophys. Res. Commun.* **74:** 739.

Weis, K., I.W. Mattaj, and A.I. Lamond. 1995. Identification of hSRP1α as a functional receptor for nuclear localization sequences. *Science* **268:** 1049.

Wen, W., J.L. Meinkoth, R.Y. Tsien, and S.S. Taylor. 1995. Identification of a signal for rapid export of proteins from the nucleus. *Cell* **82:** 463.

Wente, S.R. and G. Blobel. 1993. A temperature-sensitive *NUP116* null mutant forms a nuclear envelope seal over the yeast nuclear pore complex thereby blocking nucleocytoplasmic traffic. *J. Cell Biol.* **123:** 275.

Wente, S.R. and G. Blobel. 1994. *NUP145* encodes a novel yeast glycine-leucine-phenylalanine-glycine (GLFG) nucleoporin required for nuclear envelope structure. *J. Cell Biol.* **125:** 955.

Wente, S.R., M.P. Rout, and G. Blobel. 1992. A new family of yeast nuclear pore complex proteins. *J. Cell Biol.* **119:** 705.

Wiese, C. and K.L. Wilson. 1993. Nuclear membrane dynamics. *Curr. Opin. Cell Biol.* **5:** 387.

Wilson, K.L. 1995. NSF-independent fusion mechanisms. *Cell* **81:** 475.

Wilson, S.M., K.V. Datar, M.R. Paddy, J.R. Swedlow, and M.S. Swanson. 1994. Characterization of nuclear polyadenylated RNA-binding proteins in *Saccharomyces cerevisiae. J. Cell Biol.* **127:** 1173.

Wimmer, C., V. Doye, P. Grandi, U. Nehrbass, and E.C. Hurt. 1992. A new subclass of nucleoporins that functionally interact with nuclear pore protein NSP1. *EMBO J.* **11:** 5051.

Winey, M. and B. Byers. 1992. Spindle pole body of *Saccharomyces cerevisiae:* A model for genetic analysis of the centrosome cycle. In *The centrosome* (ed. V. Kalins), p. 197.

Academic Press, Orlando, Florida.

Winey, M., M.A. Hoyt, C. Chan, L. Goetsch, D. Botstein, and B. Byers. 1993. *NDC1:* A nuclear periphery component required for yeast spindle pole body duplication. *J. Cell Biol.* **122:** 743.

Wintersberger, U., M. Binder, and P. Fischer. 1975. Cytogenetic demonstration of mitotic chromosomes in the yeast *Saccharomyces cerevisiae. Mol. Gen. Genet.* **142:** 13.

Woolford, J.L., Jr. and J.R. Warner. 1991. The ribosome and its synthesis. In *The molecular and cellular biology of the yeast* Saccharomyces: *Genome dynamics, protein synthesis, and energetics* (ed. J.R. Broach et al.), p. 587. Cold Spring Harbor Laboratory Press, Cold Spring Harbor, New York.

Wozniak, R.W., G. Blobel, and M.P. Rout. 1994. POM152 is an integral membrane protein of the pore membrane domain of the yeast nuclear envelope. *J. Cell Biol.* **125:** 31.

Wright, R., M., Basson, L. D'Ari, and J. Rine. 1988. Increased amounts of HMG-CoA reductase induce "karmellae": A proliferation of stacked membrane pairs surrounding the yeast nucleus. *J. Cell Biol.* **107:** 101.

Wu, J., M.J. Matunis, D. Kraemer, G. Blobel, and E. Coutavas. 1995. Nup358, a cytoplasmically exposed nucleoporin with peptide repeats, Ran-GTP binding sites, zinc fingers, a cyclophilin A homologous domain, and a leucine-rich region. *J. Biol. Chem.* **270:** 14209.

Xiao, Z., J.T. McGrew, A.J. Schroeder, and M. Fitzgerald-Hayes. 1993. *CSE1* and *CSE2*, two new genes required for accurate mitotic chromosome segregation in *Saccharomyces cerevisiae. Mol. Cell. Biol.* **13:** 4691.

Xue, Z. and T. Mélèse. 1994. Nucleolar proteins that bind NLSs: A role in nuclear import or ribosome biogenesis? *Trends Cell Biol.* **4:** 414.

Yang, C.H., E.J. Lambie, J. Hardin, J. Craft, and M. Snyder. 1989. Higher order structure is present in the yeast nucleus: Autoantibody probes demonstrate that the nucleolus lies opposite the spindle pole body. *Chromosoma* **98:** 123.

Yano, R., M.L. Oakes, M.M. Tabb, and M. Nomura. 1994. Yeast Srp1p has homology to armadillo/plakoglobin/β-catenin and participates in apparently multiple nuclear functions including the maintenance of the nucleolar structure. *Proc. Natl. Acad. Sci.* **91:** 6880.

Yano, R., M. Oakes, M. Yamaghishi, J.A. Dodd, and M. Nomura. 1992. Cloning and characterization of *SRP1*, a suppressor of temperature-sensitive RNA polymerase I mutations, in *Saccharomyces cerevisiae. Mol. Cell. Biol.* **12:** 5640.

Yeh, E., R.V. Skibbens, J.W. Cheng, E.D. Salmon, and K. Bloom. 1995. Spindle dynamics and cell cycle regulation of dynein in the budding yeast, *Saccharomyces cerevisiae. J. Cell Biol.* **130:** 687.

Yokoyama, N., N. Hayashi, T. Seki, N. Panté, T. Ohba, K. Nishii, K. Kuma, T. Hayashida, T. Miyata, U. Aebi, M. Fukui, and T. Nishimoto. 1995. A giant nucleopore protein that binds Ran/TC4. *Nature* **376:** 184.

Yoneda, Y., N. Imamoto-Sonobe, M. Yamaizumi, and T. Uchida. 1987. Reversible inhibition of protein import into the nucleus by wheat germ agglutinin injected into cultured cells. *Exp. Cell Res.* **173:** 586.

Zabel, U., V. Doye, H. Tekotte, R. Wepf, P. Grandi, and E.C. Hurt. 1996. Nic96p is required for nuclear pore formation and functionally interacts with a novel nucleoporin, Nup188p. *J. Cell Biol.* **133:** 1141.

Zapp, M.L. and M.R. Green. 1989. Sequence-specific RNA binding by the HIV-1 Rev protein. *Nature* **342:** 714.

Zapp, M.L., T.J. Hope, T.G. Parslow, and M.R. Green. 1991. Oligomerization and RNA binding domains of the type 1 human immunodeficiency virus Rev protein: A dual function for an arginine-rich binding motif. *Proc. Natl. Acad. Sci.* **88:** 7734.

6
Peroxisomes

Paul B. Lazarow
Department of Cell Biology and Anatomy
Mount Sinai School of Medicine
New York, New York 10029

Wolf-H. Kunau
Institut für Physiologische Chemie
Medizinische Fakultät, Ruhr-Universität
Bochum, 44780 Bochum, Germany

I. INTRODUCTION

Peroxisomes are the major group in the microbody family of organelles, which also includes glyoxysomes, glycosomes, and the catalase-free microbodies. The "peroxisome" was defined by de Duve (de Duve and Baudhuin 1966) on the basis of biochemical function: It contains at least one enzyme that produces H_2O_2 (an oxidase) together with catalase, which decomposes H_2O_2 either catalatically ($2H_2O_2 \rightarrow 2H_2O + O_2$) or peroxidatically ($RH_2 + H_2O_2 \rightarrow R + 2H_2O$). This combination of enzymes allows the peroxisome to carry out respiration, an essential cell function in an aerobic world. Peroxisomal respiration does not conserve energy as ATP and may have arisen in evolution before the more elaborate mitochondrial respiratory system involving oxidative phosphorylation (de Duve 1969). A second major peroxisomal function is the β-oxidation of fatty acids, which is conserved (Kunau et al. 1996) among peroxisomes of evolutionarily diverse organisms, including animals (van den Bosch et al. 1992; Reddy and Mannaerts 1994), plants (Beevers 1979; Tolbert 1981), and fungi (Tanaka et al. 1982; Kunau et al. 1987). Interestingly, mitochondria also catalyze β-oxidation in animals but not in plants or fungi (Kunau et al. 1987). A third important cell function to which the peroxisome contributes in plant and fungal cells is the glyoxylate cycle, which is required for the net conversion of two-carbon compounds (and of fat, which is catabolized to acetyl-CoA) to carbohydrate. The "glyoxysome" of castor bean was given this name because it contains all of the enzymes of the glyoxylate cycle (Beevers 1982). The glyoxysome also catalyzes H_2O_2-based respiration and fatty acid β-oxidation and is thus also a peroxisome.

In contrast to the "peroxisome," the "microbody" is defined morphologically: It is an organelle with one membrane, a homogeneous electron-dense matrix, and sometimes a crystalloid inclusion (de Duve and Baudhuin 1966). All peroxisomes are microbodies, and "peroxisome" and "microbody" are sometimes used interchangeably. An unusual microbody in trypanosomes contains all of the enzymes catalyzing glycolysis, which are in the cytosol in most other organisms (Opperdoes and Michels 1993). This "glycosome" generally lacks catalase but does contain some other enzymes in common with peroxisomes. Moreover, glycosomal and peroxisomal proteins share common topogenic information (Sommer and Wang 1994). *Neurospora crassa* contains two types of microbodies: One contains β-oxidation enzymes but no catalase, whereas the other catalyzes peroxisomal respiration (Wanner and Theimer 1982; Kionka and Kunau 1985).

Peroxisomes and/or related microbodies are nearly ubiquitous in eukaryotic cells, and they demonstrate remarkable biochemical specializations in various cell types. In addition to the examples already noted, other peroxisome specializations are related to their essential roles in the utilization of diverse carbon and nitrogen sources by eukaryotic microorganisms. For example, *Hansenula polymorpha* peroxisomes contain methanol oxidase and amine oxidase, which allow this yeast to grow on methanol or on methylamine (Veenhuis and Harder 1991; Veenhuis 1992). In addition, mammalian peroxisomes carry out essential functions in lipid biosynthesis (van den Bosch et al. 1992; Reddy and Mannaerts 1994). Reflecting these diverse and specialized roles, the abundance of peroxisomes and peroxisomal enzymes is closely regulated throughout nature. In mammals, peroxisomes are controlled by diet, hormones, and environmental temperature. In eukaryotic microorganisms, they are induced when needed to assimilate a carbon or a nitrogen source. The size of the individual peroxisomes and the number of peroxisomes per cell are separately regulated.

Because the peroxisome and related microbodies were among the last organelles to be discovered, and because most of the mapping of the intracellular sites of metabolism had been completed before peroxisomes were investigated, our knowledge of peroxisome biochemistry is probably still fragmentary and incomplete. Because "mitochondrial fractions" generally contain peroxisomes, some peroxisomal enzymes (e.g., catalase) were once thought to belong to mitochondria. Other, as yet unrecognized, misassignments of peroxisomal enzymes to other organelles may also have occurred. It is entirely possible, even likely, that additional peroxisomal functions remain to be discovered.

Interest in peroxisomes has increased in recent years for several reasons, including the discovery that mammalian peroxisomes are involved in lipid metabolism (Lazarow and de Duve 1976); the fact that compounds causing peroxisome proliferation in rodents are carcinogenic (Reddy et al. 1980); the finding of huge inducible peroxisomes in diverse yeast species other than *Saccharomyces cerevisiae* (Tanaka et al. 1982; Gould et al. 1992; Veenhuis 1992); and the recognition that many inherited, lethal human disorders are caused by defects in peroxisome function and/or assembly (Schutgens et al. 1986; Lazarow and Moser 1995). Interest in *S. cerevisiae* peroxisomes, the focus of the present chapter, was stimulated by the discovery that these peroxisomes could be induced by growth on oleic acid (Kunau et al. 1987; Veenhuis et al. 1987). Additional information on peroxisome biology and pathology can be found in several recent reviews (van den Bosch et al. 1992; Subramani 1993; Lazarow and Moser 1995) and in the proceedings of recent international symposia (Latruffe 1993a,b; van Roermund et al. 1995; Reddy et al. 1996).

II. DISCOVERY, OCCURRENCE, AND MORPHOLOGY OF *S. CEREVISIAE* PEROXISOMES

Peroxisomes were found in *S. cerevisiae* in 1968 by Charlotte Avers and her group, who employed a combination of analytical cell fractionation and electron microscopic methods. Spherical organelles with a homogeneous matrix, surrounded by a single membrane (Avers and Federman 1968), showed a positive cytochemical staining reaction for catalase (Hoffmann et al. 1970). These catalase-containing particles sedimented more slowly than cytochrome-*c*-oxidase-containing particles (mitochondria) (Avers and Federman 1968) and cosedimented with glycolate oxidase and malate synthase (two other peroxisomal enzymes) (Szabo and Avers 1969). The presence of the H_2O_2-producing oxidase together with the H_2O_2-destroying catalase satisfied de Duve's definition of the peroxisome (de Duve and Baudhuin 1966). The presence of malate synthase and (apparently) isocitrate lyase, key enzymes of the glyoxylate cycle, demonstrated a kinship with *Tetrahymena* peroxisomes (Müller et al. 1968) and plant glyoxysomes (Beevers 1982). In addition, Perlman and Mahler (1970) reported the presence of citrate synthase, another enzyme of the glyoxylate cycle, in the peroxisomes. The existence of catalase-containing peroxisomes in *S. cerevisiae* was reconfirmed by Parish (1975) but then was questioned by Susani et al. (1976), who thought that the catalase was in vacuoles. Remarkably, that was essentially the end of research on peroxisomes in *S. cerevisiae* for 11 years.

With the benefit of hindsight, the early observations on *S. cerevisiae* peroxisomes seem even more remarkable because they succeeded despite then unknown complications. In particular, peroxisomes are fragile and easily damaged during cell fractionation: There is more than one isozyme of catalase and of malate synthase, and at least one isozyme of each is in a cell compartment other than peroxisomes; expression of peroxisomal enzymes is subject to glucose repression, so that the abundance of peroxisomes in cells grown on glucose depends on where in the growth curve the harvesting is carried out; and even when cells are grown on ethanol or other nonfermentable carbon sources, the abundance of peroxisomes is rather low.

The discovery that peroxisomes contain an inducible enzyme system for the β-oxidation of fatty acids both in rat liver (Lazarow and de Duve 1976) and in the alkane-degrading yeast *Candida tropicalis* (Kawamoto et al. 1978) suggested additional enzymes that might be present in *S. cerevisiae* peroxisomes. In addition, an active peroxisomal degradation of fatty acids was predicted by Monson and Hayes (1982) on the basis of metabolic labeling studies. Subsequently, the β-oxidation enzymes were indeed found in *S. cerevisiae* peroxisomes, and it was also observed that these enzymes could be induced by growth of cells in the presence of fatty acids (Kunau et al. 1987; Veenhuis et al. 1987; Skoneczny et al. 1988). This enzyme induction was accompanied by a proliferation of peroxisomes that was observed morphologically (Veenhuis et al. 1987). These and many subsequent biochemical and morphological analyses have confirmed unambiguously the original observations by Avers and her co-workers.

S. cerevisiae cells grown on glucose each contains at least one and generally a few small peroxisomes (Thieringer et al. 1991; Skoneczny and Rytka 1996), which are often found at the periphery of the cell, as illustrated in Figure 1A. These peroxisomes are bounded by the canonical single membrane and have a diameter of approximately 0.1–0.2 μm. As described further below, peroxisomes are somewhat larger in cells grown on glycerol or on ethanol (Fig. 1B) and are both larger and more abundant in cells grown in the presence of oleic acid (Fig. 1C). The peroxisomes are sometimes found in clusters. Paracrystalline inclusions have not been observed in *S. cerevisiae* peroxisomes.

Although peroxisomes generally present circular profiles in electron microscopic thin sections, tubular peroxisomes are also observed. Peroxisomes can be interconnected via these tubular elements into a peroxisomal reticulum (Lazarow et al. 1980). A striking example occurs in the mouse preputial gland, in which all peroxisomes are connected to form a single compartment (Gorgas 1984). Tubular elements have also been observed in *S. cerevisiae*, suggesting the presence of some peroxisomal reticulum in this organism as well (Erdmann and Blobel 1995).

III. PEROXISOME PROLIFERATION AND REGULATION

A. Organelle Proliferation and Enzyme Induction

A characteristic feature of peroxisome biology is that the abundance of the organelle and its enzymes within any one cell type is highly variable and closely regulated. This applies to mammals, plants, and eukaryotic microorganisms. For example, in rat liver, peroxisomal activities may be

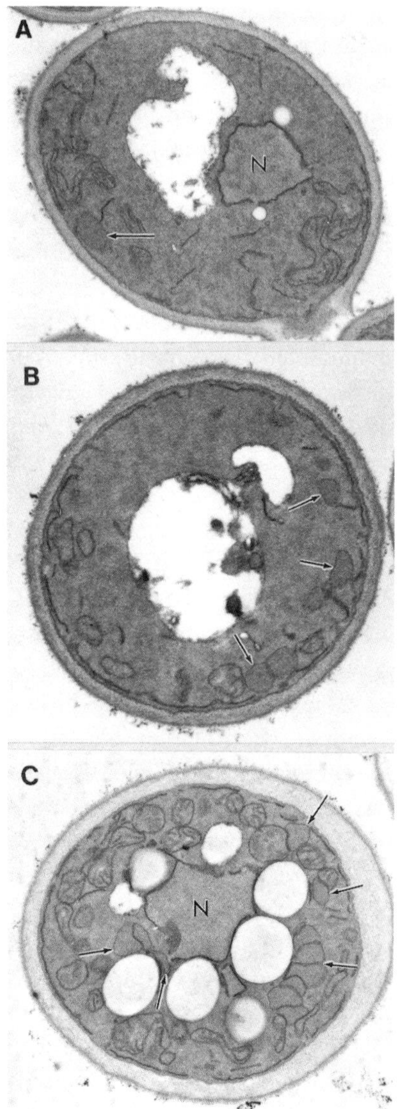

Figure 1 Electron microscopic observation of peroxisomes in *S. cerevisiae* cells grown on glucose (*A*), ethanol (*B*), oleic acid (*C*). (N) Nucleus. Arrows indicate peroxisomes. (Micrographs courtesy of M. Veenhuis.)

induced by high-fat diets or a cold environment, and the organelle proliferates as much as five- to tenfold in response to hypolipidemic drugs or xenobiotic compounds. Such proliferation is accompanied by variable inductions of different enzyme activities. The β-oxidation enzymes increase by 10–100-fold, other enzymes such as catalase merely double, and still others decrease in activity (Lazarow 1977; Nemali et al. 1988). The proliferation of peroxisomes and selective induction of peroxisomal enzymes are especially pronounced in certain yeast species that can live on distinct carbon and nitrogen sources, including such unusual compounds as methanol, alkanes, alkylamines, fatty acids, D-amino acids, and uric acid (Tanaka et al. 1982; Veenhuis and Harder 1991). In every case, the proliferated peroxisomes contain at least some of the enzymes required by the yeast to catabolize that carbon or nitrogen source.

As mentioned above, oleic acid induces both the proliferation of peroxisomes and the induction of enzymes for catabolism at fatty acids in *S. cerevisiae* (Kunau et al. 1987; Veenhuis et al. 1987; Skoneczny et al. 1988; Thieringer et al. 1991). Both the size and number of peroxisomes increase substantially upon such induction (Fig. 1); oleate-grown cells contain many peroxisomes with diameters up to 0.5 μm. These peroxisomes are often found in clusters, sometimes near the cell surface but also throughout the cytoplasm. Such proliferation is prevented by growth under anaerobic conditions. Other carbon sources, including ethanol and acetate, and a variety of nitrogen sources do not appear to cause *S. cerevisiae* peroxisomes to proliferate, although there is some increase in peroxisomal size and enzyme content (Veenhuis et al. 1987; Thieringer et al. 1991). *S. cerevisiae*, in contrast to many other yeasts, seems unable to grow on methanol, amines, or uric acid as sole carbon source, and, to date, xenobiotic compounds appear to be without effect on peroxisomes in this organism.

The regulation of expression of peroxisomal proteins in *S. cerevisiae* is complex. The entire organelle and most of its enzymes (including catalase and the β-oxidation enzymes) are subject to glucose repression and derepression (see below). Some peroxisomal enzymes (e.g., malate synthase and malate dehydrogenase) can be induced by nonfermentable carbon sources (e.g., acetate and ethanol). Many enzymes, including those catalyzing the complete β-oxidation cycle, are induced by fatty acids with chain lengths from 8 to 22 carbons (Skoneczny et al. 1988). For example, the activities of acyl-CoA oxidase and thiolase are extremely low (sometimes barely detectable) in the presence of ≥2% glucose. They increase about 20-fold when cells are grown on a nonfermentable carbon source such as ethanol or glycerol and then increase by another factor of 20 when cells are grown on oleic acid. Measurements of catalase activity are complicated by the existence of two isozymes in wild-type cells: catalase A (encoded by the *CTA1* gene) (Cohen et al. 1988) and catalase T (encoded by the *CTT1* gene) (Hartig and Ruis 1986). Catalase A is in peroxisomes and catalase T is in the cytosol

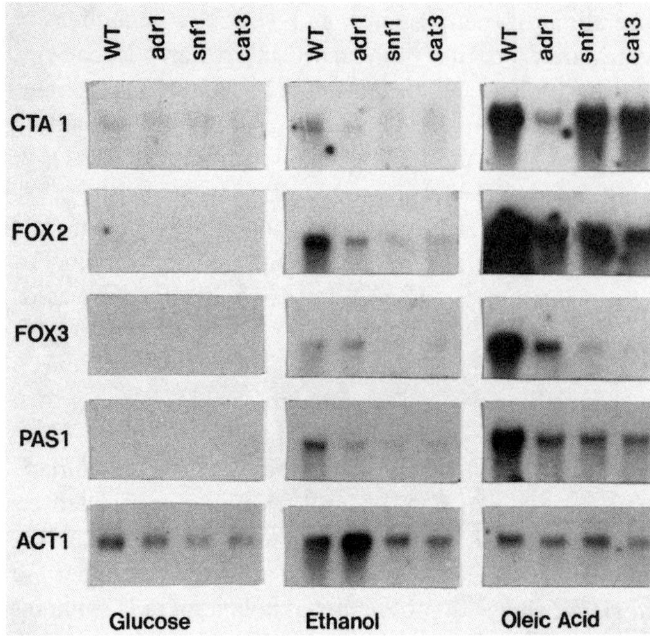

Figure 2 Abundance of mRNAs of genes encoding peroxisomal catalase (*CTA1*), fatty acid β-oxidation enzymes (*FOX*), a protein required for peroxisome assembly (*PAS1*), and (as a control) actin (*ACT1*). Wild-type cells and cells defective in one of the regulatory genes (*adr1, snf1*, or *snf4*) were grown overnight with the indicated carbon source. (Reprinted, with permission, from Simon et al. 1992.)

(Skoneczny et al. 1988; van der Klei et al. 1990). In cells containing only catalase A, there is no detectable catalase activity in cells grown on 10% glucose, 53 units/mg protein in cells grown on ethanol, and 253 units/mg protein in cells grown on oleate (Skoneczny et al. 1988). In contrast, catalase T is not induced by oleate.

B. Transcriptional Regulation

1. Cis-*acting Regulatory Elements*

Considerable evidence has accumulated demonstrating that the repression, derepression, and induction of genes encoding peroxisomal enzymes are strongly controlled at the transcriptional level. Many such genes have now been cloned, which has permitted both Northern blot analyses of mRNA abundance (Fig. 2) and the identification and characterization of the upstream *cis*-acting regulatory elements (Fig. 3).

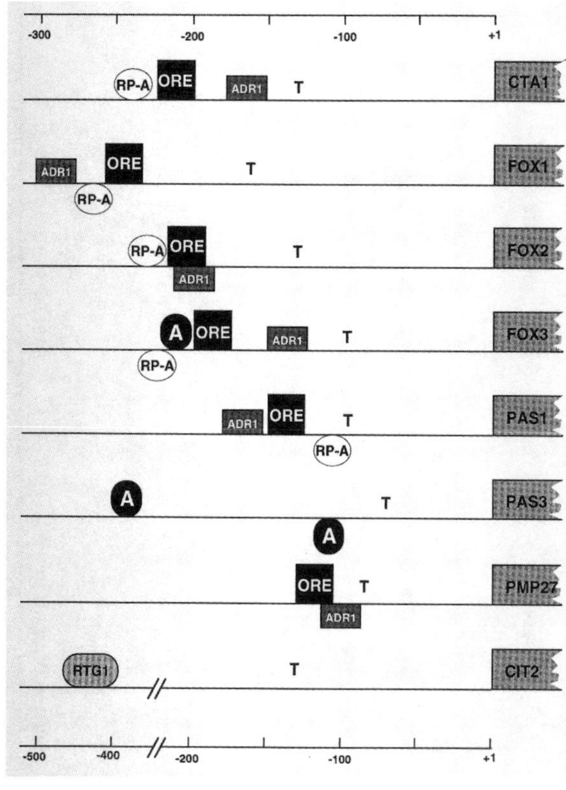

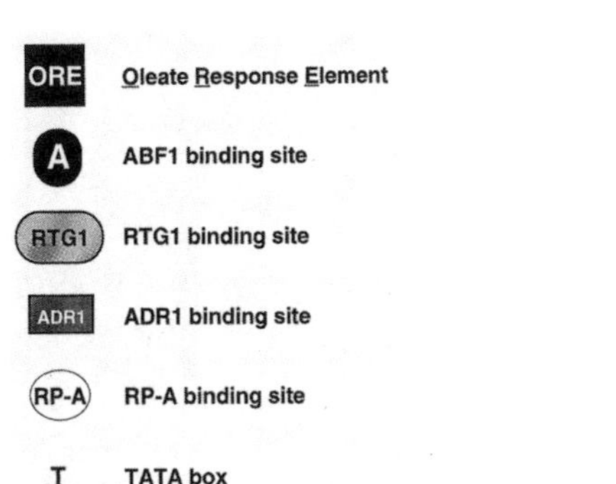

Figure 3 Cis-acting regulatory elements. (Reprinted, with permission, from Einerhand et al. 1992.)

Cis elements responsible for the glucose repression have been reported for *FOX1/POX1* and *FOX3/POT1*, which encode enzymes catalyzing β-oxidation reactions (Table 1A). In the case of *FOX1*, two glucose-responsive elements have been identified: One consists of an in-

Table 1 Peroxisomal matrix and membrane proteins

Gene	Enzyme name	Function	Mass (kD)	PTS	Note
(A) *Matrix proteins*[a,b]					
CTA1	catalase A	H_2O_2 degradation	58	redundant	c
FOX1/POX1	acyl-CoA oxidase	β-oxidation	84	?	c
FOX2	multifunctional protein	β-oxidation	99	SKL	c
FOX3/POT1	3-oxoacyl-CoA thiolase	β-oxidation	45	PTS2	c
CAT2/YCAT1	carnitine acetyltransferase	acetyl group transport	73	AKL	c
CIT2	citrate synthase 2	glyoxylate cycle	52	SKL	c
MDH2	malate dehydrogenase 2	glyoxylate cycle	41	?	c
MDH3	malate dehydrogenase 3	oxidation of NADH	37	SKL	c
MLS1	malate synthase	glyoxylate cycle	63	SKL	c
DAL7	malate synthase	allantoin catabolism	63	SKL	d
YBR1444/YBR204C	lipase motif	?	43	SKL	d
SPS19	?	sporulation	31	SKL	d
PAS6	?	peroxisome biogenesis	68	redundant	c
FAA2	acyl-CoA synthetase	β-oxidation	83	?	c
PMP60/YBR222C	acyl-CoA synthetase (?)	β-oxidation	60	SKL	c
AAT2	aspartate aminotransferase	?	45	AKL	c
PAS7/PEB1	PTS2 receptor	peroxisome biogenesis	42	first 55 amino acids	c

(B) *Membrane proteins*[b]

Gene	Enzyme name	Function	Mass (kD)	Disposition
PMP27	?	peroxisome proliferation and size control	27	see text
PAS3	?	peroxisome biogenesis	51	integral
PAS11	?	peroxisome biogenesis	46	integral
PAS20	?	peroxisome biogenesis	43	integral
PAS21	?	peroxisome biogenesis	43	integral
PXA1/PAL1/SSH2	human adrenoleukodystrophy protein ortholog	ABC-transporter	100	integral
YKL741	similar to PXA1 and human PMP70	ABC-transporter	100	integral
PAS2/UBC10[e]	ubiquitin-conjugating enzyme	peroxisome biogenesis	21	peripheral

[a]Proteins transported into the peroxisome (matrix proteins and proteins located at the inner face of the peroxisomal membrane).
[b]See text for references.
[c]Localization demonstrated.
[d]Localization presumed.
[e]Discussed in Section VI.E.2.c.

verted repeat and the other a direct repeat (Wang et al. 1992, 1994). In the case of *FOX3*, the glucose-responsive element has been shown to consist of overlapping Abf1p and RP-A sites (Einerhand et al. 1991, 1992). The RP-A site (= *CAR1* URS = FUF1) also has been identified by sequence analysis in *CTA1, FOX1/POX1, FOX2,* and *PAS1* (Fig. 3).

Two synergistically acting control elements are involved in derepression and induction of genes encoding peroxisomal enzymes. An Adr1p-binding element is required for the derepression of catalase A (Simon et al. 1991), and such sites have also been identified by sequence analysis in the upstream regions of other genes for peroxisomal proteins (Fig. 3). In addition, an oleate-response element or "β-oxidation box" (Einerhand et al. 1991) or "peroxisome box" (Filipits et al. 1993) mediates the specific induction of these genes by fatty acids. Oleate-response elements have been identified by DNA footprinting, band-shift experiments, mutational analysis, and tests of promoter function (using reporter genes) in the upstream regions of *CTA1* (Filipits et al. 1993), *FOX1* (Wang et al. 1994), and *FOX3* (Table 2) (Einerhand et al. 1991, 1993). The consensus sequence of this element is CGG NNN TNA (where N = any nucleotide) (Filipits et al. 1993). This consensus sequence is generally found as imperfect inverted or direct repeats and sometimes, in addition, as single elements. In the case of *FOX3*, one half of the repeat alone (i.e., a single consensus sequence) gives a weak response (Einerhand et al. 1993). The orientation of the oleate-response elements can be reversed in *CTA1* without altering their function (Filipits et al. 1993). Computer searches have identified both presumed Adr1p-binding sites and apparent oleate-response elements in the promoter regions of additional peroxisomal genes (Table 2; Fig. 3) (Einerhand et al. 1992; Simon et al. 1992).

The oleate-response element appears to be conserved in other yeasts. The *C. tropicalis FOX2* homolog (*HDE*) is regulated by oleate when expressed in *S. cerevisiae*. A 52-nucleotide sequence within its upstream region is required for the oleate response, and this sequence contains a nonanucleotide, CGGTTATTA, that is similar to the oleate-response element of *S. cerevisiae* (Sloots et al. 1991). Similar sequences are found upstream of other oleate-induced genes of *C. tropicalis* (Table 2).

2. Trans-*acting Transcription Factors*

Intensive efforts are currently under way in a number of laboratories to identify the *trans*-acting factors that interact with the *cis* elements described above.

Three proteins (Abf1p, RP-A, and Car80p) are essential for mainten-

Table 2 Oleate-response element

	Position	Sequence	Position	Note
1. *Saccharomyces cerevisiae*				
Catalase A (*CTA1*)	- 209	C G G C T T T A A	- 201	a
	- 186	C G G A G T T A A	- 194	a
Acyl-CoA oxidase (*FOX1/POX1*)	- 251	A A C G G C T A T T A	- 241	a
	- 277	T T C G G T C A T T A	- 267	a
Multifunctional protein (*FOX2*)	- 204	C G G G G A T C A	- 194	b
	- 231	C G G G G G T T A	- 223	b
	- 250	C G G T G T T A A	- 258	b
	- 272	C G G G G A A T T	- 264	b
3-Oxoacyl-CoA thiolase (*FOX3/POT1*)	- 188	C G G G G A T A A	- 180	a
	- 166	C G G T G T T A A	- 174	a
Malate synthase (*MLS1*)	- 332	A C C G G C T A G T A	- 322	b
Sporulation protein (*SPS19*)		C G G C G T T T A		b
		C G G C G C G T A		b
Consensus		C G G N N N T N A		
2. *Candida tropicalis*				
Multifunctional protein (*HDE*)	- 355	C G G T T A T T A	- 347	a
Acyl-CoA oxidase (*POX4*)	- 359	C G G T T A T T C	- 351	b
Catalase (*CATL*)	- 695	T G G T T A T T A	- 687	b

[a]Found by functional tests.
[b]Discovered by computer search.

ance of glucose repression. Abf1p is an abundant yeast phosphoprotein that binds to the consensus DNA sequence 5′-RTCRYNNNNNACG-3′ (Abf1p-binding site) (R = purine, Y = pyrimidine, N = any nucleotide). It is an essential multifunctional protein; depending on the context of the binding site, Abf1p acts as a repressor or an activator of transcription, or it is involved in DNA replication (Einerhand et al. 1995). RP-A is *r*eplication *p*rotein *A* (= BUF = binding URS1 factor [Luche et al. 1992]). RP-A is a hetero-oligomer encoded by three essential genes (Brill and Stillman 1991). The action of RP-A on *FOX3* requires Car80p/Ume6p (Einerhand et al. 1995). RP-A was found to bind to the RP-A-binding sites in the other peroxisomal genes tested (*CTA1, FOX1, FOX2,* and *PAS1*). Interestingly, a number of other genes whose products are known to be involved in general glucose repression (*TUP1, CYC8/ SSN6, RGR1, GRR1, MIG1, HXK1, HXK2, CID1, CAT4, REG1*) are not required for glucose repression of *FOX1*, according to genetic tests (Stanway et al. 1995).

Three proteins, Adr1p, Snf1p, and Snf4p, are known to function as global regulators of derepression of glucose-repressible genes and have also been shown to act as positive regulators of genes encoding peroxisomal proteins, including *CTA1, FOX2*, and *FOX3* (Fig. 2) (Simon et al. 1991, 1992; Navarro and Igual 1994). Indeed, mutants with defects in these regulatory proteins have been isolated in a selection procedure for peroxisome biogenesis mutants (van der Leij et al. 1992). Adr1p is a DNA-binding transcriptional activator (Simon et al. 1991), and Snf1p is a cytosolic serine/threonine protein kinase (Celenza and Carlson 1986). Snf4p is physically associated with Snf1p and is required for maximal kinase activity. Signal transduction from cytosolic Snf1p to nuclear Adr1p appears to involve Snf2p, Snf5p, and Snf6p, which function in a large multisubunit complex (Laurent et al. 1991), together with Swi1p and Swi3p (Peterson and Herskowitz 1992), none of which binds DNA (Fig. 4). These regulatory proteins are required not only for derepression of peroxisomal genes in ethanol-grown cells, but also for optimal induction of these genes by oleic acid (see Fig. 2). Expression of reporter constructs driven only by the oleate-response elements are reduced in a *snf1* background (Filipits et al. 1993). Peroxisomal enzymes are not induced in *adr1, snf1*, or *snf4* backgrounds, and peroxisomes were not detected by immunofluorescence in *snf1* or *snf4* mutants (Simon et al. 1992). Moreover, yeast do not grow on oleate if *SNF1, SNF2, SNF5*, or *SNF6* is disrupted (Einerhand et al. 1992). The carboxy-terminal half of Adr1p is critical for the response to oleate (Simon et al. 1995). It is important to note that there are significant quantitative differences in the sensitivities

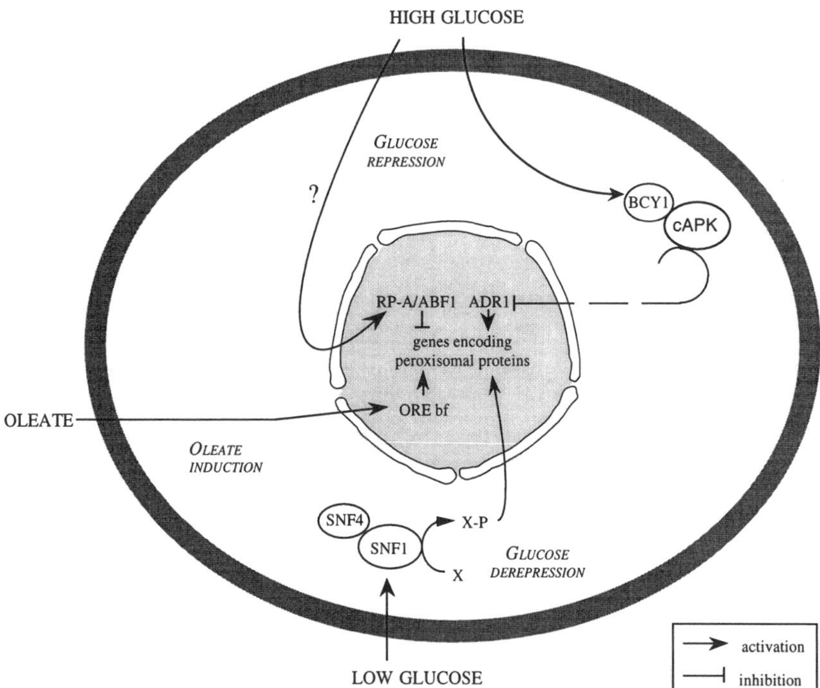

Figure 4 Signal transduction pathways influencing peroxisomal protein expression and assembly in *S. cerevisiae*. (Reprinted, with permission, from Einerhand et al. 1992.)

of *CTA1, FOX2*, and *FOX3* to regulation by Adr1p, Snf1p, and Snf4p. For example, *CTA1* depends more on the action of Adr1p, whereas *FOX3* is very sensitive to disruption of the *SNF* genes (Simon et al. 1992; Navarro and Igual 1994). Interestingly, *FOX1* is not much affected by either Adr1p or Snf1p (Stanway et al. 1995), but a protein with an apparent mass of approximately 110 kD that binds with high affinity to the oleate-response element of *FOX1* has been purified by oligonucleotide-affinity chromatography and its gene identified (Y. Luo and G. Small, pers. comm.). It contains a fungal Zn(2)-Cys(6) binuclear cluster domain near its amino terminus and appears to be a phosphoprotein. A protein of similar size, which forms dimers, binds to the oleate-response element of *FOX3* (H. Tabak, pers. comm.). What may be the same gene was also identified as *PIP2* in a genetic screen for factors required for oleate induction of *CTA1* (H. Ruis, pers. comm.). In *pip2* cells, neither peroxisomes nor peroxisomal fatty-acid-oxidizing enzymes can be induced by oleate.

Two additional regulatory genes, *RTG1* and *RTG2*, were identified by their roles in "retrograde" communication to the nucleus of defective mitochondrial function (Liao and Butow 1993). Impairment of mitochondrial tricarboxylic acid cycle function (e.g., by disruption of *CIT1*, encoding mitochondrial citrate synthase, or *MDH1*, encoding mitochondrial malate dehydrogenase) causes the up-regulation of *CIT2* (encoding peroxisomal citrate synthase) as well as the up-regulation of other genes encoding glyoxylate cycle enzymes. Rtg1p, a basic helix-loop-helix transcription factor, and Rtg2p act via an upstream activation site in the *CIT2* gene (Liao and Butow 1993). Further studies have shown that these two regulatory proteins are also required for the full induction of several other peroxisomal proteins (catalase, acyl-CoA oxidase, and a membrane protein designated Pmp27p) by oleate (Chelstowska and Butow 1995).

Rodent peroxisomes and their β-oxidation enzymes are induced by a high-fat diet and by a variety of xenobiotic chemicals ("peroxisomal proliferators"). This induction is mediated by a peroxisome proliferator-activated receptor (PPAR) (Bieri 1993). No PPAR homolog has been identified in *S. cerevisiae.* Moreover, there is no apparent similarity between the *S. cerevisiae* oleate-response element and the mammalian PPAR-response element. Thus, it would appear that there may be conservation of regulatory elements mediating the response to fatty acids among yeasts but not between yeasts and mammals.

3. *Summary*

The control of expression of peroxisomal enzymes and the control of proliferation of the entire organelle are obviously complex. Many of the genes involved are regulated by the general glucose repression and derepression system. In addition, a specific induction of some genes by oleic acid is mediated by a conserved upstream activating sequence and one or more *trans*-acting factors. More information on these factor(s) can be expected in the near future. Much further research is required on the extent of coordinated regulation of sets of genes encoding peroxisomal enzymes and on the relationship between control of enzymes and control of organelle proliferation.

IV. ISOLATION AND PROPERTIES OF PEROXISOMES

A. Isolation, Density, and Fragility

The purification of peroxisomes from *S. cerevisiae* (for review, see Zinser and Daum 1995) follows the same principles as the isolation of peroxisomes from other organisms (Lazarow et al. 1991). Peroxisomes are first induced by growth on oleic acid. After removal of the cell wall by enzymatic digestion, the spheroplasts are broken by homogenization with a Dounce or Potter-Elvehjem homogenizer or sometimes lysed by osmotic shock. A significant difference is that the homogenization medium for *S. cerevisiae* should be slightly acidic (pH ~6.0), whereas a pH of 7.0–7.4 is used with rat liver. An appropriate osmotic strength is provided by including sorbitol (typically 0.6 M) in the medium. A crude organelle fraction, consisting mostly of mitochondria and peroxisomes, is then prepared by differential centrifugation. Gentle handling is required during these steps to avoid mechanical damage to the peroxisomes and thus to minimize leakage of their enzymes into the suspending medium.

The peroxisomes are then separated from other organelles by equilibrium density centrifugation of the crude organelle fraction in a gradient of Nycodenz (Fig. 5) (Lewin et al. 1990) or sucrose (McCammon et al. 1990a). Either discontinuous gradients, as in the papers just cited, or continuous gradients (Höhfeld et al. 1991; Thieringer et al. 1991) can be employed. Carefully chosen step gradients are very useful for preparative purposes, whereas continuous gradients can give better resolution and are more useful for analytical purposes. Peroxisomes have a greater equilibrium density than do mitochondria (see below) but are smaller. As a result, peroxisomes tend to follow the mitochondria down the gradient and then pass them when the mitochondria stop at their equilibrium density. Thus, there is a risk that peroxisomes may get trapped in a layer of mitochondria that is compacted on a density shelf. It should also be noted that a bell-shaped curve of peroxisome abundance versus density can be broken artifactually into apparently separate peaks by discontinuous density steps (de Duve 1965). Centrifugation in vertical rotors is preferable (to minimize hydrostatic pressure, which can damage organelles), but swinging bucket rotors are also satisfactory.

The distributions of organelles in the gradient are determined by measurement of their marker enzymes, as described by de Duve (1965). The purity of the organelles can then be estimated from calculations of the relative specific activities of the marker enzymes in the various fractions (for examples of such purity calculations, see Thieringer et al. 1991; Zinser et al. 1991). Some cross-contamination of peroxisomes and mitochondria is frequently observed. This can be tolerated if taken into account in the interpretation, or it can be minimized by careful use of linear gradients (Höhfeld et al. 1991; Thieringer et al. 1991).

The equilibrium density of peroxisomes depends on the growth medium. In particular, peroxisomes are less dense (and therefore less well

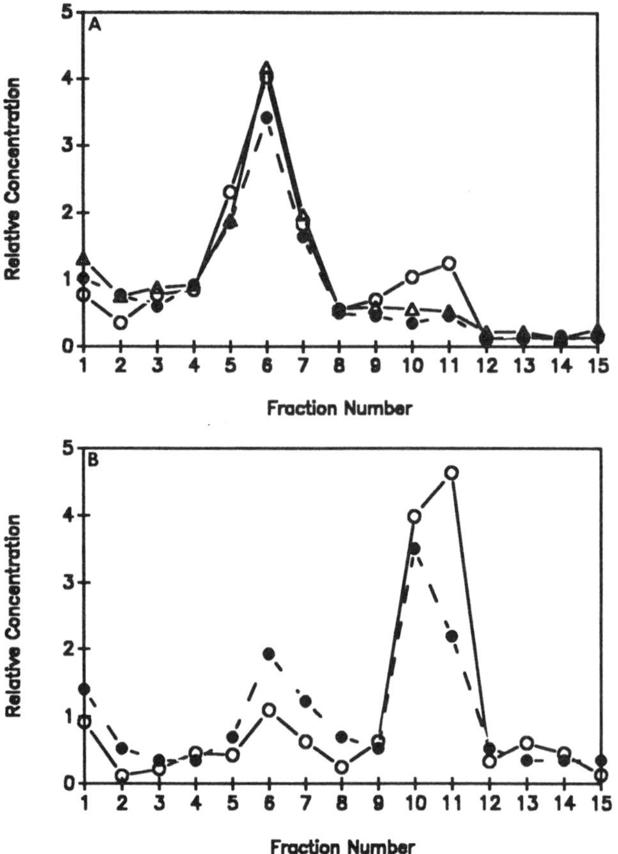

Figure 5 Separation of peroxisomes and mitochondria by equilibrium density centrifugation of a crude organelle fraction in a discontinuous Nycodenz step gradient. (*A*) Relative concentrations of protein (*open triangle*), the mitochondrial marker enzyme cytochrome *c* oxidase (*closed circle*), and malate dehydrogenase (*open circle*) (for discussion of the multiple MDH isozymes, see Section V.B). (*B*) Relative concentrations of the peroxisomal marker enzymes catalase (*open circle*) and acyl-CoA oxidase (*closed circle*). (Reprinted, with permission, from Lewin et al. 1990.)

separated from mitochondria upon centrifugation) when grown on glycerol than when gown on a combination of glycerol and oleic acid (McCammon et al. 1990a). Similarly, a time-dependent increase in peroxisome density from 1.15 g/ml to 1.21 g/ml was observed after shifting *S. cerevisiae* cells from glucose-containing medium to oleic-acid-containing medium (Erdmann and Blobel 1995). In addition, the absolute density difference between peroxisomes and mitochondria is greater in

Nycodenz (1.17–1.21 g/ml vs. 1.10–1.12 g/ml) than in sucrose (1.20–1.21 g/ml vs. 1.17 g/ml). Thus, a greater separation of peroxisomes from mitochondria can be obtained in Nycodenz than in sucrose. A purification of peroxisomes of 23–31-fold relative to the homogenate can be accomplished (Thieringer et al. 1991).

Despite care and practice in purifying peroxisomes by the above methods, some leakage of peroxisomal enzymes into the surrounding medium inevitably occurs. For example, a variety of studies have found 15–49% of the thiolase, 15–35% of the acyl-CoA oxidase, and 10–70% of the catalase in the postorganelle supernatant during the initial differential centrifugation (Erdmann et al. 1989; McCammon et al. 1990a; Höhfeld et al. 1991; Thieringer et al. 1991; van der Leij et al. 1993; Erdmann 1994; Marzioch et al. 1994). In addition, peroxisomal enzymes are sometimes found near the top of density gradients; this is usually due to leakage of enzymes from peroxisomes during resuspension of the crude organelle pellet. Such leakage must be taken into consideration when interpreting experimental results, or it can be avoided by using the digitonin permeabilization procedure described below.

B. Subfractionation

There are two well-documented methods for the subfractionation of peroxisomes, and additional methods have also been reported. In the case of rat liver peroxisomes, matrix proteins (the soluble proteins inside the organelle) can be released by stirring the purified organelles for 24 hours in dilute, slightly alkaline pyrophosphate buffer (10 mM, pH 8) at 4°C (Leighton et al. 1969). After such treatment, the peroxisomes appear by electron microscopy to lack content (except that they retain their paracrystalline urate oxidase inclusion); nevertheless, the membranes appear in thin sections to be closed, intact circles. Enzyme assays and SDS-PAGE indicate that essentially all content proteins have been released into the medium. In addition, unsealed sheets of rat liver peroxisome membranes containing integral membrane proteins can be prepared by treating the purified organelles for 30 minutes with 0.1 M Na_2CO_3, pH 11 (Fujiki et al. 1982b). This alkaline treatment removes peripheral membrane proteins as well as content proteins. This method is also applicable to other organelle membranes (Fujiki et al. 1982a) and has been widely applied, including to *S. cerevisiae* peroxisomes (McCammon et al. 1990a; Erdmann and Blobel 1995).

There are reports in the literature of the rupture of *S. cerevisiae* peroxisomes and release of content proteins (to obtain a membrane fraction) by brief treatment with pH 8 buffer. In our experience, a 1-hour treatment releases some content proteins, but such release is far from quantitative; for instance, considerable thiolase can be retained. So far as we are aware, the longer time of incubation, with stirring, that is highly effective with the rat liver organelles has not been tried with *S. cerevisiae* peroxisomes.

There are also reports of the release of matrix proteins from peroxisomes by hypotonic shock. These reports should be interpreted with caution, for two reasons. First, peroxisomes appear to be relatively insensitive to such shock, presumably due to the fact that, as isolated, they are permeable to solutes as large as sucrose (see Section IV.D). This insensitivity explains the successful isolation of peroxisomes from spheroplasts lysed by hypotonic shock (McCammon et al. 1990a). Second, different matrix proteins leak from peroxisomes to quite different extents (Alexson et al. 1985). Clearly, peroxisomes are fragile, and some leakage of content proteins routinely occurs when they are pelleted by centrifugation, resuspended, diluted in another medium, or repelleted. Perhaps such leakage is greater in hypotonic media, and increasing the pH to 8 may well also increase protein leakage (recall that pH 6 appears to be optimal for purification of *S. cerevisiae* peroxisomes).

Salt washes also cause the release of proteins from peroxisomes (Erdmann and Blobel 1995; Zinser and Daum 1995). Among the β-oxidation enzymes located in the matrix, thiolase is released at low salt, whereas acyl-CoA oxidase and the multifunctional protein require 0.25 M KCl. Some of the unidentified proteins released at 0.25 M KCl may be peripheral membrane proteins, whereas others are probably more difficult to extract matrix proteins. Careful controls with proteins whose intraperoxisomal location is already well established are essential to future studies of differential protein release from *S. cerevisiae* and other peroxisomes. Ultimately, immunoelectron microscopy on intact peroxisomes appears to be the most reliable method for distinguishing a peripheral membrane localization from a matrix localization. Electron microscopy is also recommended for evaluation of attempts to prepare membrane fractions.

1. Matrix Proteins

Purified *S. cerevisiae* peroxisomes contain at least 21 proteins visible by SDS-PAGE (Kunau et al. 1993). Among these are the β-oxidation enzymes, as well as catalase A, malate dehydrogenase, and citrate synthase. Most of these 21 proteins are extracted by low or high salt and are probably present in the matrix, although a few might be peripheral membrane proteins. The soluble enzymes that have been found in *S. cerevisiae* peroxisomes are listed in Table 1A. The table includes several enzymes that are predicted to be in peroxisomes because they contain the peroxisomal targeting sequence SKL (described in detail later in this chapter) at their carboxyl termini. *S. cerevisiae* peroxisomes also contain two proteins of unknown function that cross-react with an antiserum against human sterol carrier protein 2 (Tahotna et al. 1993). They have molecular masses of 35 kD and 58 kD.

2. *Membrane Components*

a. Proteins. Membranes have been isolated from peroxisomes by means of carbonate extraction (see above). Such membranes contain the phospholipids and integral membrane proteins, some of which have been identified (Table 1B). In total, about 30 polypeptides were discerned by sequential high-performance liquid chromatography (HPLC) and SDS-PAGE analyses of highly purified peroxisomal membranes (Erdmann and Blobel 1995), including abundant integral membrane proteins with apparent molecular masses of 32, 31, and 24 kD (McCammon et al. 1990a).

The gene (*PMP27*) that encodes the last of these proteins has been cloned and sequenced; its product has a predicted mass of 27 kD (Erdmann and Blobel 1995; Marshall et al. 1995). Pmp27p has several hydrophobic patches and is firmly associated with the membrane, but it might not span the bilayer. Deletion of *PMP27* has no apparent effect on peroxisome structure or function in cells growing on glucose, acetate, or ethanol. However, the deletion causes the formation of giant peroxisomes in oleate-induced cells, and both segregation of peroxisomes to buds and growth are impaired. Overexpression of Pmp27p causes the proliferation of normal-size peroxisomes, many of which are elongated in shape early in induction. These data indicate that Pmp27p is not required for the normal division and segregation of peroxisomes; however, it may be involved in proliferation of the organelle and control of peroxisome size (Erdmann and Blobel 1995; Marshall et al. 1995).

Limited information is also available for several other peroxisomal integral membrane proteins. Pas3p is essential for peroxisome biogenesis (Höhfeld et al. 1991). It contains one membrane-spanning domain and a large cytosolic domain; its function is presently unknown. Its abundance is so low that it is not visible among total peroxisomal proteins by Coomassie blue staining, although it is readily detected by immunoblotting. Pas11p (W.-H. Kunau, unpubl.), Pas20p (which contains an SH3 domain; R. Erdmann; L.Y. Elgersma; both pers. comm.), and Pas21p (H. Tabak, pers. comm.) are also integral proteins of the peroxisome membrane, present in low abundance, that are required for peroxisome biogenesis. These proteins are candidates to participate in the translocation of newly synthesized proteins into the organelle.

Interestingly, two members of the ABC transporter family are found in the *S. cerevisiae* peroxisome membrane. Pxa1p/Pal1p/Ssh2p is most similar to the human ABC transporter that is defective in adrenoleukodystrophy (Shani et al. 1995; Swartzman et al. 1996). Ykl741p is most similar (H. Tabak, pers. comm.) to the human PMP70, an ABC transporter that has been found to be defective in some Zellweger syndrome patients (Gartner et al. 1992).

b. Lipids. The phospholipid composition of the peroxisome membrane has been reported to be 48% phosphatidylcholine, 23% phosphatidylethanolamine, 16% phosphatidylinositol, and 4.5% phosphatidylserine (Zinser et al. 1991). This resembles the composition of the nuclear and vacuolar membranes but differs from the compositions of the plasma membrane and mitochondrial membranes (Table 3). The peroxisome membrane fraction also contained 7% cardiolipin, almost half as much as the mitochondrial inner membrane. The ratio of ergosterol to phospholipid in the peroxisomal membrane was 0.39, compared to 3.31 in the plasma membrane and 0.20 in the mitochondrial membranes (Zinser et al. 1991).

C. Cell Fractionation by Selective Permeabilization of Membranes with Digitonin

The plasma membrane and intracellular membranes of animal cells can be selectively permeabilized by digitonin on the basis of their cholesterol content (de Duve 1965; Wanders et al. 1984). Digitonin forms a stoichiometric complex with cholesterol, producing holes in the membrane, and the concentration of digitonin needed to permeabilize a membrane is inversely proportional to the abundance of sterol in that membrane. Thus, the plasma membrane, which is richest in cholesterol (Colbeau et al. 1971), is permeabilized by low concentrations of digitonin; this releases cytosolic proteins from the cells. The lysosome membrane contains less cholesterol and requires a higher concentration of digitonin to allow release of lysosomal enzymes. The mitochondrion and peroxisome, which contain little if any cholesterol in their membranes, require the highest concentrations of digitonin for permeabilization (de Duve 1965;

Table 3 Phospholipid compositions of subcellular fractions of *S. cerevisiae*

Subcellular fraction	% of total phospholipid[a]						
	PtdCho	PtdEtn	PtdIns	PtdSer	CL	PA	others
Plasma membrane	16.8	20.3	17.7	33.6	0.2	3.9	6.9
Secretory vesicles	35.0	22.3	19.1	12.9	0.7	1.2	8.8
Vacuoles	46.5	19.4	18.3	4.4	1.6	2.1	7.7
Nucleus	44.6	26.9	15.1	5.9	<1.0	2.2	4.3
Peroxisomes	48.2	22.9	15.8	4.5	7.0	1.6	n.d.
Light microsomes	51.3	33.4	7.5	6.6	0.4	0.2	0.5
Mitochondria	40.2	26.5	14.6	3.0	13.3	2.4	n.d.
Inner mitochondrial membrane	38.4	24.0	16.2	3.8	16.1	1.5	n.d.
Outer mitochondrial membrane	45.6	32.6	10.2	1.2	5.9	4.4	n.d.

[a](PtdCho) Phosphatidylcholine; (PtdEtn) phosphatidylethanolamine; (PtdIns) phosphatidylinositol; (PtdSer) phosphatidylserine; (CL) cardiolipin; (PA) phosphatidic acid; (others) other phospholipids; n.d. indicates not detectable. (Reprinted, with permission, from Zinser et al. 1991.)

Wanders et al. 1984); in these cases, the digitonin may be functioning as a nonspecific detergent. Thus, the intracellular locations of enzymes can be determined by analyzing the extent of their release from cells as a function of the digitonin concentration applied. An important advantage of this method is that it avoids mechanical breakage of the organelles and the attendant artifactual release of organelle proteins.

S. cerevisiae does not contain cholesterol, but it does contain a related sterol, ergosterol (Longley et al. 1968; Rattray et al. 1975). As mentioned above, the ergosterol/phospholipid ratio in peroxisome membranes is much lower than in plasma membranes but higher than in mitochondrial membranes (Zinser et al. 1991). This permits the selective permeabilization of these membranes by treatment of spheroplasts with progressively increasing concentrations of digitonin (Zhang et al. 1993b), as illustrated in Figure 6. The three peroxisomal enzymes thiolase, catalase, and acyl-CoA oxidase are consistently released together, and their pattern of release can be reproducibly distinguished from the release of enzymes from other compartments. The data suggest that these three enzymes are all found within peroxisomes in wild-type *S. cerevisiae* to the extent of at least 93% (Zhang et al. 1993b). This contrasts strikingly with the substantial and variable leakage of these same proteins during classical homogenization and cell fractionation, as discussed above. Thus, the digitonin method is more reliable than classical cell fractionation for quantitative estimation of how much of an enzyme is in yeast peroxi-

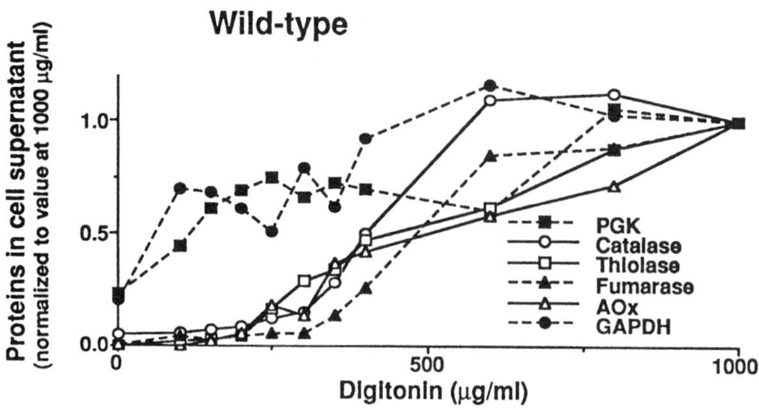

Figure 6 Release of enzymes from *S. cerevisiae* spheroplasts by permeabilization with increasing concentrations of digitonin. (*Closed squares and circles*) Cytosolic enzymes; (*open symbols*) peroxisomal enzymes; (*closed triangles*) mitochondrial enzyme. (Reprinted, with permission, from Zhang et al. 1993b.)

somes. The digitonin method has proved to be very effective in studying the subcellular distribution of gene products required for peroxisome biogenesis (Zhang and Lazarow 1995) and for studying protein topogenesis (Zhang and Lazarow 1996).

D. Membrane Permeability

From the available literature, the permeability of the peroxisome membrane to small solutes appears to depend both on the organism studied and on the in vivo or in vitro methods employed. In the case of rat liver, isolated peroxisomes are permeable to the substrates of its enzymes (e.g., glycolate or urate) and to the solutes used to form the density gradients used in cell-fractionation experiments (de Duve 1965; de Duve and Baudhuin 1966); the entry of sucrose into peroxisomes during centrifugation explains the high equilibrium density of peroxisomes. This permeability of the rat liver peroxisome membrane has been reported to be due to the presence of a nonselective pore-forming protein (perhaps with an M_r = 22,000), which allows the free passage of NAD, ATP, coenzyme A, and smaller metabolites (Van Veldhoven et al. 1987). This is analogous to the function of porin in mitochondrial outer membranes (Benz 1990).

In contrast, the peroxisomes of methylotrophic yeasts (*H. polymorpha* and *Candida boidinii*) appear to have tight membranes in vivo. These peroxisomes appear to be somewhat more acidic than the cytosol (pH

5.8–6.0 vs. pH 7.1) according to nuclear magnetic resonance (NMR) spectroscopy of living *H. polymorpha* cells (Nicolay et al. 1987) and to immunocytochemical evidence for weak base accumulation (Waterham et al. 1990); they contain a membrane protein that cross-reacts with an antibody against the mitochondrial proton-translocating ATPase (Douma et al. 1990). These data suggest that the peroxisomal membrane is impermeable to protons in vivo, which argues against the presence of open pores. Instead, there might be carriers or specific channels for the transport of metabolites. Indeed, the *C. boidinii* peroxisomal membrane contains a 47-kD protein that resembles mitochondrial solute-carrier proteins, including the ATP/ADP translocator (Jank et al. 1993). Nevertheless, *H. polymorpha* peroxisome membranes, like their rat liver counterparts, contain a pore-forming protein (M_r = 31,000) (Sulter et al. 1993). Moreover, the high equilibrium density of these peroxisomes when sedimented in sucrose gradients suggests that they, like the rat liver organelles, are permeable to sucrose in vitro. Taken together, the data suggest the possibility that the porins of both rat and yeast peroxisomes may be regulated, being mostly closed in vivo and stuck open in vitro. It also remains possible that there are true species differences and that the rat liver peroxisome pores are open in vivo.

The permeability of the peroxisome membrane in *S. cerevisiae* has been investigated by a combination of genetics and biochemical assays (van Roermund et al. 1995). NAD is reduced to NADH within peroxisomes during the β-oxidation of fatty acids. The NADH might diffuse out of peroxisomes for reoxidation in the cytosol if it can permeate through the peroxisome membrane; if it cannot exit, it would require reoxidation within the organelle, perhaps by malate dehydrogenase (MDH). van Roermund et al. (1995) tested the effect of deleting the *MDH3* gene, which encodes a peroxisomal isozyme of MDH (Steffan and McAlister-Henn 1992). The deletion abolished peroxisomal fatty acid oxidation in vivo, and 3-hydroxyacyl-CoA, the metabolic intermediate that immediately precedes the NAD-requiring step, accumulated. In contrast, cell extracts of this strain containing broken peroxisomes and added NAD showed normal β-oxidation in vitro, indicating that all of the required enzyme activities except MDH are present within peroxisomes in the *mdh3* cells. The data thus suggest that NADH cannot pass through the peroxisome membrane and so must be reoxidized within the organelle (Fig. 7).

In a similar fashion, the mechanism of egress of acetyl-CoA, the end product of peroxisomal β-oxidation, was also tested (van Roermund et al. 1995). If the acetyl-CoA could not pass through the peroxisomal mem-

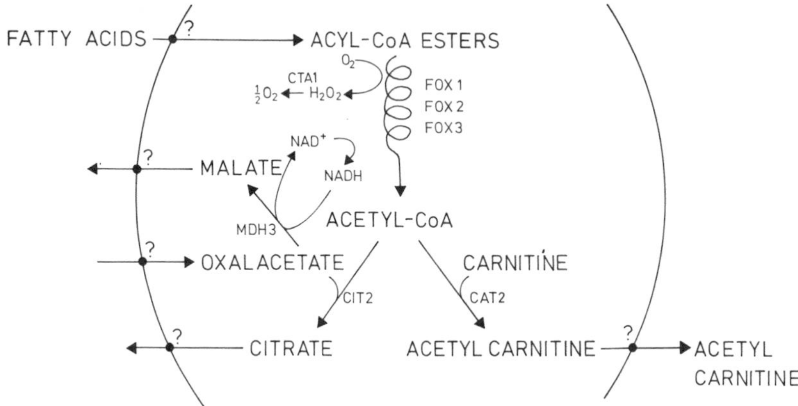

Figure 7 Metabolism of fatty acids in *S. cerevisiae* peroxisomes. Question marks indicate the unknown mechanisms by which metabolites cross the peroxisome membrane (see text, Section IV.D).

brane, acetyl moieties might be transferred to the much smaller molecule, carnitine (Atomi et al. 1993), or condensed with oxaloacetate to form citrate. Disruption of *CAT2* (*YCAT*), which encodes carnitine acetyltransferase (Elgersma et al. 1995), had no discernible effect on peroxisomal β-oxidation in vivo. Disruption of *CIT2*, which encodes the peroxisomal isozyme of citrate synthase (Lewin et al. 1990), likewise had no discernible effect. However, simultaneous disruption of both genes completely prevented peroxisomal ß-oxidation in vivo. In contrast, cell-free extracts of the double-disruption strain, with broken membranes, retained the ability to ß-oxidize fatty acids. These data suggest that the peroxisome has two alternative methods for the further metabolism and export of acetyl-CoA, with regeneration of free coenzyme A in the peroxisome (Fig. 7). When both mechanisms are inoperative, acetyl-CoA accumulates, bringing fatty acid oxidation to a halt.

In summary, the available data allow for two alternative interpretations. The *S. cerevisiae* peroxisomal membrane could have pores (perhaps regulated) that are large enough for the passage of citrate, acetylcarnitine, malate, and oxaloacetate (M_r = 114–203) but not big enough for NADH or acetyl-CoA (M_r = 646–1059). Alternatively (or in addition), there could be carriers or specific channels for the smaller metabolites. Irrespective of which interpretation is correct, there must be a mechanism by which peroxisomes acquire the larger cofactors, NAD and coenzyme A.

V. FUNCTIONS OF PEROXISOMES IN *S. CEREVISIAE*

A. β-oxidative Degradation of Fatty Acids

During the 1970s and 1980s, the peroxisomal fatty acid β-oxidation system was investigated intensively in rat liver (Hashimoto 1982; Osmundsen 1982), castor bean (Beevers 1982), and *C. tropicalis* (Tanaka et al. 1982). In contrast to animals in which fatty acid degradation occurs both in peroxisomes and in mitochondria, plants and fungi degrade fatty acids exclusively in microbodies (Kunau et al. 1996). The observation in the late 1980s that *S. cerevisiae* grows (although not efficiently) on oleate with the concomitant induction of a complete set of peroxisomal β-oxidation enzymes left little doubt that fatty acids are degraded in the peroxisomes in this yeast as well. Conversion of fatty acids to water-soluble products and CO_2 was recently demonstrated directly (Shani et al. 1995; van Roermund et al. 1995). Mutants defective in fatty acid oxidation (*fox* mutants) were identified and used to clone the genes encoding the proteins involved (see Table 1A), including an acyl-CoA oxidase (*FOX1*/*POX1*), a multifunctional protein (*FOX2*), and a 3-oxoacyl-CoA thiolase (*FOX3*/*POT1*) (Figs. 7 and 8).

As in all peroxisomal β-oxidation systems, the first reaction of the β-oxidation cycle in *S. cerevisiae* is catalyzed by an acyl-CoA oxidase rather than an acyl-CoA dehydrogenase (Veenhuis et al. 1987; Skoneczny et al. 1988). The corresponding gene, *FOX1*/*POX1*, encodes a predicted polypeptide of 748 amino acids (84 kD) (Dmochowska et al. 1990). The oligomerization state of Fox1p has not yet been determined, but the acyl-CoA oxidases of both *C. tropicalis* and *Yarrowia lipolytica* are octameric flavoproteins with FAD as prosthetic group (Kunau et al. 1996).

The second and third reactions of the *S. cerevisiae* β-oxidation cycle are catalyzed by a multifunctional protein, as in the microbodies of other species (Fig. 8). Thus, this enzyme possesses both 2-enoyl-CoA hydratase and 3-hydroxyacyl-CoA dehydrogenase activities. Surprisingly, however, the stereochemical course of the two reactions in *S. cerevisiae* is opposite to that found in their mammalian, plant, and bacterial counterparts (Kunau et al. 1996). 2-Trans-enoyl-CoA esters are converted via D- instead of L-3-hydroxyacyl-CoA esters into 3-oxoacyl-CoA intermediates (Fig. 8). This means that the enoyl-CoA hydratase is a hydratase 2 and that the 3-hydroxyacyl-CoA dehydrogenase is specific for D-isomers. The stereochemical properties of the *S. cerevisiae* multifunctional protein are shared by those of *C. tropicalis*, *Y. lipolytica*, and the filamentous fungus *N. crassa* (Kunau et al. 1996). Originally, three other activities, enoyl-CoA hydratase 1, L-3-hydroxyacyl-CoA dehydrog-

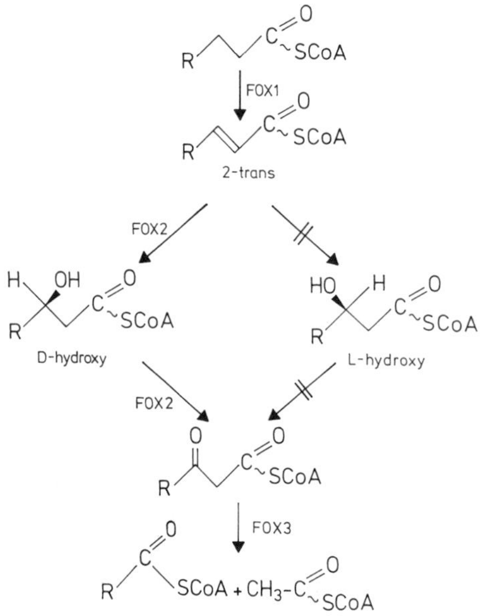

Figure 8 β-oxidation reactions in peroxisomes.

enase, and 3-hydroxyacyl-CoA epimerase, had erroneously been assigned to all four of these fungal multifunctional proteins.

FOX2, which encodes the multifunctional protein, was cloned by functional complementation of the *fox2* mutant (Hiltunen et al. 1992). Its open reading frame predicts a polypeptide of 900 amino acids (98.7 kD). The first and second thirds of the predicted polypeptide show extensive sequence similarity to each other, suggesting that a partial gene duplication occurred during evolution of the gene. Similar sequence stretches have been found in the corresponding genes of *C. tropicalis* and *N. crassa* (Kunau et al. 1996). These three fungal proteins exhibit no significant sequence similarity to the multifunctional proteins either of mammalian or plant peroxisomes or of several bacterial species (Kunau et al. 1996). The multifunctional protein of *S. cerevisiae* and a truncated version of it lacking the 271 carboxy-terminal amino acids were purified to apparent homogeneity from cells overexpressing the proteins (Hiltunen et al. 1992). The truncated protein possessed only the D-3-hydroxyacyl-CoA dehydrogenase activity.

The fourth and last reaction of the β-oxidation cycle is catalyzed in *S. cerevisiae* by a typical peroxisomal β-keto-acyl-Co-A thiolase. Unlike cytosolic and mitochondrial thiolases, peroxisomal thiolases are homodimeric proteins (Kunau et al. 1996). The thiolase gene of *S.*

cerevisiae has been cloned (Erdmann 1989; Igual et al. 1991), and the protein has been purified to apparent homogeneity from wild-type and overproducing yeast cells (Erdmann and Kunau 1994). The isolated protein began with the fifth or sixth amino acid predicted by the open reading frame of its gene. It is not yet known whether this observation is due to limited proteolysis during purification or reflects processing of the protein during or after import into the peroxisome. The purified protein was crystallized and its structure was determined at a resolution of 2.8 Å (Mathieu et al. 1994); its 1.8-Å structure is almost completed (R. Wierenga, pers. comm.).

Proteins likely to be acyl-CoA synthetases (necessary for the initial esterification to coenyzme A of imported fatty acids) have also been identified in *S. cerevisiae* peroxisomes (Table 1A; E. Hettema and H. Tabak; R. Erdmann and G. Blobel; both pers. comm.). The degradation of unsaturated fatty acids has not been studied in *S. cerevisiae* to the same extent as in *C. tropicalis*. However, the occurrence of the two auxiliary enzymatic activities, 2,4-dienoyl-CoA reductase and 2,3-enoyl-CoA isomerase, required to metabolize intermediates with *cis*-double bonds at either even-numbered or odd-numbered carbon atoms has been established. These activities could be separated both from the multifunctional protein and from each other and therefore seem to represent monofunctional enzymes (W.-H. Kunau, unpubl.).

B. Glyoxylate Cycle?

An important biochemical function is the formation of glucose from smaller metabolites, when needed. Whereas 3- or 4-carbon compounds such as lactate or succinate can be converted into sugars by cytosolic enzymes that mostly just reverse the reactions of glycolysis, gluconeogenesis from 2-carbon compounds such as ethanol or acetate requires an additional metabolic cycle, the glyoxylate cycle (Fig. 9A). Five enzymes catalyze the formation of one molecule of succinate from two molecules of acetyl-CoA. Isocitrate lyase (ICL), which forms glyoxylate, and malate synthase (MS), which consumes glyoxylate, are the key enzymes. They are found in peroxisomes in diverse species, sometimes with and sometimes without the other enzymes of the glyoxylate cycle.

For example, some seeds efficiently convert stored fat to sugar upon germination. The fat is metabolized to succinate within peroxisomes (glyoxysomes) by the sequential action of β-oxidation and the glyoxylate cycle; the further conversion of succinate to sugar occurs elsewhere in the cell. These seed peroxisomes contain all five enzymes of the glyoxy-

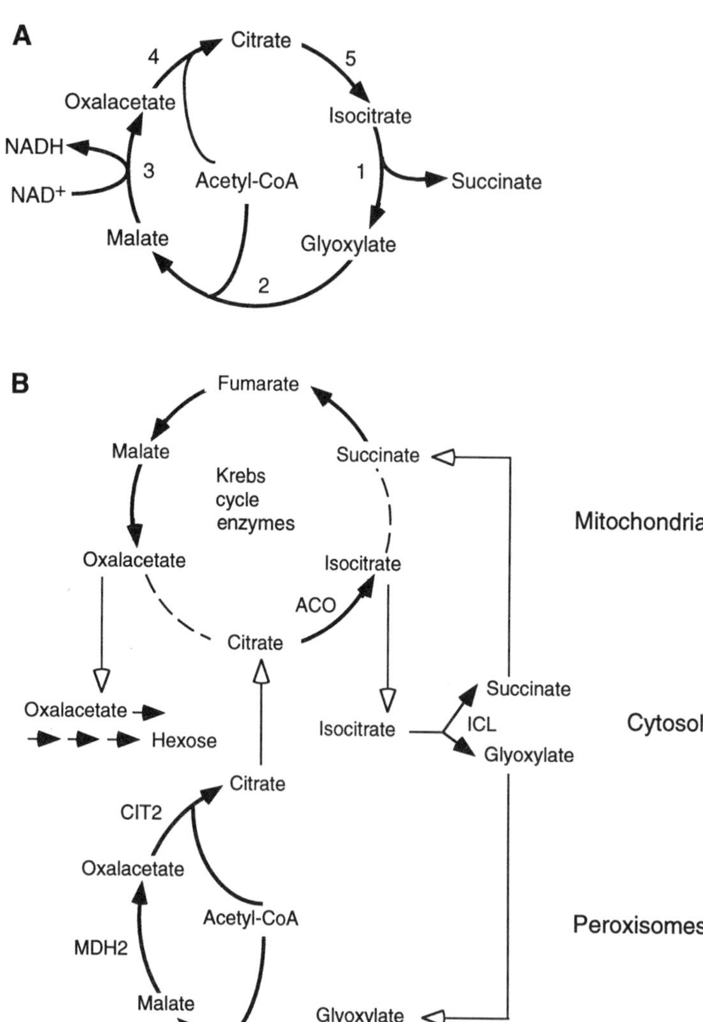

Figure 9 Glyoxylate cycle in *S. cerevisiae*. (*A*) Five steps, catalyzed by (*1*) isocitrate lyase, (*2*) malate synthase, (*3*) malate dehydrogenase, (*4*) citrate synthase, and (*5*) aconitase. (*B*) Apparent compartmentation of glyoxylate cycle isozymes and the additional reactions converting acetyl-CoA to hexose in *S. cerevisiae*. (*Thick lines*) Reactions; (*thin lines*) transport of metabolites between intracellular compartments; (*dashed lines*) reactions not part of this pathway.

late cycle (Beevers 1979). Three of these activities (malate dehydrogenase [MDH], citrate synthase [CS], and aconitase) are also present in mitochondria, where they participate in the Krebs cycle. In contrast, in the yeast *C. tropicalis* (Fahimi et al. 1982) and the protozoan *Tetrahymena pyriformis* (Müller et al. 1968), the peroxisomes contain only the two key enzymes of the glyoxylate cycle, ICL and MS. MDH, CS, and aconitase are present in mitochondria, and the glyoxylate cycle operates by the metabolic cooperation of these two organelles.

In *S. cerevisiae*, the two key glyoxylate cycle enzymes are present and are induced by growth on acetate, ethanol, or oleate. MS is present in peroxisomes, together with isozymes of MDH and CS, but, curiously, ICL apparently is not. *S. cerevisae* contains at least two malate synthase genes, *MLS1* and *DAL7* (Hartig et al. 1992), both of which encode proteins ending in the peroxisomal targeting sequence SKL. *MLS1* is induced by growth on acetate, ethanol, or oleic acid (it contains an oleate-response element), and peroxisomes from oleate-grown *S. cerevisiae* contain malate synthase activity (McCammon et al. 1990a). ICL was reported to be partly associated with peroxisomes (Szabo and Avers 1969), but subsequent investigations by cell fractionation have found this enzyme exclusively in the cytosol in *S. cerevisiae* (Parish 1975; Douglas et al. 1979; McCammon et al. 1990a; Thieringer et al. 1991). Although this might have been due to leakage of ICL from fragile peroxisomes damaged during cell homogenization, subsequent studies using the more gentle digitonin permeabilization procedure (see above; Zhang et al. 1993b) have confirmed a cytosolic localization for ICL (J.W. Zhang and P.B. Lazarow, unpubl.).

S. cerevisiae peroxisomes appear to contain two isozymes of MDH. One of these is the product of *MDH2*. Although it lacks a recognizable peroxisomal targeting sequence (see Section VI.B), some Mdh2p was found in peroxisomes of oleate-grown cells with an isozyme-specific antibody, and peroxisome-associated MDH activity was reduced in an *mdh2* mutant strain (McCammon et al. 1990a). Peroxisomes also contain the MDH isozyme encoded by *MDH3* (Steffan and McAlister-Henn 1992). Mdh3p ends with SKL and was demonstrated to be in peroxisomes by cell fractionation of wild-type and *mdh3* deletion strains (van Roermund et al. 1995). Mdh3p is required for growth on oleate, and, as discussed above, this isozyme is thought to reoxidize NADH formed within peroxisomes during β-oxidation. Mdh2p and Mdh3p are induced 30-fold and 10-fold, respectively, by growth on acetate (Steffan and McAlister-Henn 1992), and *MDH2* is essential for growth on acetate, whereas *MDH3* is not. Mdh1p is in mitochondria and is not induced by acetate. Taken together, these data suggest that Mdh2p is probably the isozyme most important for glyoxylate cycle function and that it operates, at least in part, in peroxisomes.

S. cerevisiae peroxisomes also contain (Lewin et al. 1990) an isozyme of citrate synthase, Cit2p (M_r = 50,000) (Kispal and Srere 1991), which ends in SKL. Another isozyme, Cit1p, is in mitochondria. Cit2p is induced as much as 30-fold when mitochondrial function is compromised, as in [rho^0] yeast (Liao et al. 1991). Deletion of *CIT2* causes massive accumulation of fat droplets when cells are grown on oleate, indicating that Cit2p is essential for fatty acid metabolism, presumably by its contribution to the glyoxylate cycle (Lewin et al. 1990).

Aconitase (the product of *ACO1*) is present in mitochondria, where it is required for the Krebs cycle (Gangloff et al. 1990). Aconitase activity also has been found in the high-speed supernatant (cytosol), although this might have been due to leakage of enzyme from mitochondria during cell fractionation (Duntze et al. 1969). Extreme leakage could also explain the claim that all malate synthase is cytosolic in the same cells (Duntze et al. 1969). Deletion of *ACO1* abolishes all aconitase activity (Gangloff et al. 1990), suggesting that Aco1p is the only aconitase isozyme. There is no evidence for a peroxisomal localization of aconitase in *S. cerevisiae*.

Taken together, the data indicate that three of the five glyoxylate cycle enzymes are present in peroxisomes. What are the functional implications? The data suggest that the glyoxylate cycle may operate in *S. cerevisiae* by the metabolic cooperation of peroxisomes, mitochondria, and cytosol, as shown in Figure 9B. This would be similar to the metabolic cooperation of cell compartments in *T. pyriformis* and *C. tropicalis*, except that the distribution of enzymes in *S. cerevisiae* is different. On the other hand, *S. cerevisiae* is capable of growing on acetate even when peroxisomes do not assemble (Erdmann et al. 1989). Moreover, MDH is believed to reduce oxaloacetate to malate in the process of regenerating NAD for β-oxidation (van Roermund et al. 1995), whereas it must oxidize malate during the glyoxylate cycle. These observations suggest the need for further experiments to clarify the compartmentation of glyoxylate cycle function during growth on two-carbon compounds such as acetate and ethanol and during growth on oleate.

C. Respiration

S. cerevisiae displays a classic peroxisomal respiration in which hydrogen peroxide is formed by oxidases and decomposed by catalase. The peroxisomes contain both acyl-CoA oxidase, the first enzyme in the β-oxidation spiral (see above), and glycolate oxidase (Szabo and Avers 1969); the H_2O_2 formed within peroxisomes is decomposed inside the organelle by catalase A, encoded by *CTA1* (Cohen et al. 1988). Additional cellular protection against H_2O_2 is provided in the cytosol by catalase T, encoded by *CTT1* (Hartig and Ruis 1986). In some strains, there appears to be further protection against H_2O_2 by mitochondrial cytochrome *c* peroxidase (van der Leij et al. 1992), but this was ineffective in other strains (Zhang et al. 1993a).

Neither peroxisomal nor mitochondrial enzymes are present to any appreciable extent in anaerobically grown yeast. Peroxisomal enzyme induction in response to gradually increasing oxygen in the medium is sigmoidal, with half-maximal increases in the range of 1.0–1.2 μM O_2 (Rogers and Stewart 1973). This contrasts with the mitochondrial enzymes, which are half-maximally induced at 0.2 μM O_2 or less.

D. Other Functions

1. Contribution to Allantoin Catabolism

Glyoxylate is a by-product when allantoin is catabolized as a nitrogen source. Glyoxylate accumulation in the cell is toxic, and it is removed (via condensation with acetyl-CoA) by an isozyme of malate synthase encoded by *DAL7* (Hartig et al. 1992). This gene is part of a coordinately regulated cluster encoding the enzymes required to convert allantoin to urea. Dal7p ends in the peroxisomal targeting sequence SKL and is likely to be located in peroxisomes, although this has not yet been tested experimentally. The two genes encoding malate synthase have totally different regulatory controls: *DAL7* is induced by growth on glucose + allantoin, which causes strong carbon catabolite repression of *MLS1;* conversely, *DAL7* is repressed by growth on rich medium containing acetate, which induces the expression of *MLS1*. Peroxisomes also contribute to allantoin degradation in some animal species, in more elaborate ways (Scott et al. 1969).

2. Transamination

Transamination is a common peroxisomal function in diverse species. Aspartate aminotransferase, encoded by *AAT2*, ends in AKL (a functional variant of SKL) and is located in peroxisomes in *S. cerevisiae* (R. Wanders, pers. comm.).

3. Possible Roles in Spore Formation and Karyogamy

A gene that is expressed only during sporulation, *SPS19*, encodes a protein ending in the peroxisomal targeting sequence SKL (Coe et al.

Table 4 Peroxisome assembly mutants of *S. cerevisiae*

Mutants	Alleles	Phenotype				References	Comments
			enzyme	enzyme distribution			
		peroxisomes[a]	amount	catalase[b]	thiolase[c]		
pas1	4	absent[d]	normal	c	c	1,2,6,9	=*peb2*, AAA family[e]
pas2	1	absent	normal	c	c	1,2,8	=*ubc10*, ubiquitin conjugating enzyme
pas3	4	absent[f]	normal	c	c	1,2,3,7	Integral membrane protein
pas4	1	1–2	normal	p+c	p+c	2,3,9	Ring finger protein, required for peroxisome proliferation
pas5	1	absent		n.d.	n.d.	2,3,9	Ring finger protein
pas6	2	absent[f]	normal	c	c	2,3	Contains both PTS2 and SKL sequences
pas7	2	present	normal	p	c	2,3,12	=*peb1*, PTS2 receptor
pas8	3	absent	normal	c	c	2,3,10	AAA family[e]; possible transmembrane protein
pas9	1	absent	normal	c	c	2,16	peripheral membrane protein
pas10	5	reticulum	normal	c	p	2,3,4,11,13	PTS1 receptor
pas11	1	absent	normal	c	c	2,15	integral membrane protein
pas12	1	absent	normal	c	c	2,9,16	Prenylated protein
pas13	1	absent	normal	c	c	2,16	=*pas3*
pas14	4	1–2	low	c	p	3,17	=*snf1*
pas15		1–2	low	c	–	3,17	=*pas14* =*snf1*

pas16		1–2	low	c	–	3	
pas17		few	low	c	–	3,17	=*pas 14* =*snf1*
pas18		1–2	low	c	–	3	
pas19						4,17	=*adr1*
pas20		absent	normal	c	c	4,15	contains SH3 domain, integral membrane protein
pas21		absent	normal	c	c	4,15	integral membrane protein
pas22		present[h]	normal	p+c	p+c	4,15	DNA J family
peb1	5	clusters	normal	p	c	5,14	=*pas7*, PTS2 receptor
peb2	1	absent[d]	normal	c	c	5	=*pas1*
peb3	2	present	low	p	p	5	
peb4	1	absent[d]	normal	c	c	5	
peb5	1	clusters	normal	c	p	5	

References: (1) Erdmann et al. 1989; (2) Kunau and Hartig 1992; (3) van der Leij et al. 1992; (4) Elgersma et al. 1993; (5) Zhang et al. 1993a,b; (6) Erdmann et al. 1991; (7) Höhfeld et al. 1991; (8) Wiebel and Kunau 1992; (9) Kunau et al. 1993; (10) Voorn-Brouwer et al. 1993; (11) van der Leij et al. 1993; (12) Marzioch et al. 1994; (13) Brocard et al. 1994; (14) Zhang and Lazarow 1995, 1996; (15) Rehling et al. 1996a; (16) W.-H. Kunau, unpubl.; (17) H. Tabak, unpubl.

[a]Absent means no morphologically recognizable peroxisomes.

[b]Catalase activity in the *pas* mutants includes the peroxisomal (p) isozyme, catalase A, which predominates after oleate induction, and the cytosolic (c) isozyme, catalase T. In mutants *pas14–pas18*, the residual catalase activity may be predominantly catalase T. In the *peb* mutants, only catalase A is present.

[c](–) Denotes no thiolase detected by immunoblotting.

[d]Peroxisome membrane ghosts detected by immunogold labeling.

[e]See text.

[f]Membrane structures were seen in the original mutant but not in the subsequent gene knockout strain.

[g](n.d.) Not determined.

[h]Normal and double membranes.

[?]References 1–5, mutant isolation.

[?]References 6–17, gene and/or protein characterization.

1994). Sps19p is a 31-kD protein that belongs to a family of short-chain, nonmetalloenzyme dehydrogenases. Peroxisomal Fox2p, the multifunctional protein involved in β-oxidation, is also a member of this family. Deletion of *SPS19* (together with another sporulation-specific gene, *SPS18*) does not prevent the formation of viable spores, but the efficiency of sporulation is reduced and the spores are much less resistant to ether. These results suggest that a peroxisomal enzyme may contribute in some fashion to the synthesis of a spore-wall component (Coe et al. 1994).

CAR1, a fungal homolog of *PAF-1* (a human gene encoding a peroxisomal membrane protein essential for peroxisome biogenesis; Shimozawa et al. 1992), is essential not only for peroxisome assembly, but also for karyogamy (nuclear fusion) in *Podospora anserina* (Berteaux-Lecellier et al. 1995). In filamentous ascomycetes, karyogamy occurs as part of the process leading to meiosis and sexual sporulation. There is some similarity of the human Paf-1 and the fungal Car1 proteins to *S. cerevisiae* Pas4p and Pas5p (Table 4), including the presence of a RING-finger motif in each. It will be of considerable interest to determine whether a yeast homolog of Paf-1 is involved in karyogamy.

VI. PEROXISOME BIOGENESIS

A. Overview

The major features of peroxisome biogenesis, illustrated in Figure 10, appear to be the same in animals, plants, and several species of yeasts (Lazarow and Fujiki 1985; Subramani 1993; Lazarow and Moser 1995). The genes encoding peroxisomal proteins are all located in the nucleus; there is no DNA in peroxisomes (Kamiryo et al. 1989). The RNA transcripts, after processing and export, are translated on free polyribosomes; this includes mRNAs encoding soluble matrix proteins (found inside the organelle) as well as mRNAs encoding integral membrane proteins. The newly synthesized proteins are targeted to, and imported into, preexisting peroxisomes. Thus, the organelles increase in size. Phospholipids, which are required for the elaboration of the peroxisomal membrane as the organelle grows, are synthesized in the endoplasmic reticulum and in mitochondria (Zinser et al. 1991; Paltauf et al. 1992) and conveyed to peroxisomes by an unknown mechanism. New peroxisomes form by the division of preexisting peroxisomes. Peroxisomes are thought to undergo fusion as well as fission (Poole et al. 1970; de Duve 1973; Lazarow et al. 1982). In certain cell types, some or even all peroxisomes may be interconnected into a peroxisomal reticulum (Lazarow et al. 1980, 1982).

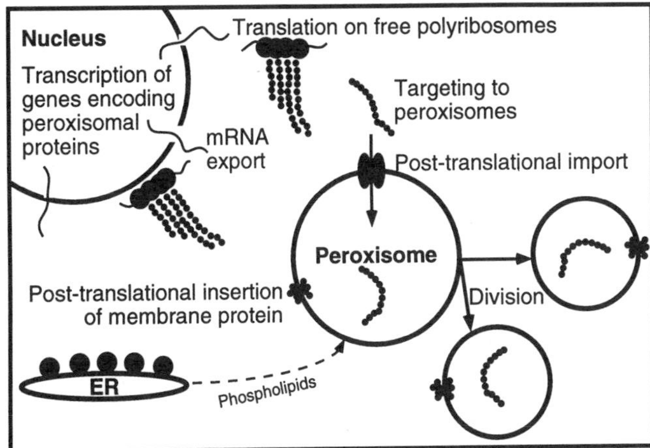

Figure 10 Overview of peroxisome biogenesis. (Reprinted, with permission, from Lazarow 1995.)

Peroxisomes are destroyed by autophagy (sequestration of the entire organelle within a membrane-bound compartment that fuses with lysosomes [or the vacuole], leading to digestion of proteins and lipids) (de Duve and Baudhuin 1966; Dunn 1994). In the case of normal rat liver, autophagic turnover is random, and peroxisomes have an average half-life of 1.5 days (Poole 1971). In *H. polymorpha* and *Pichia pastoris*, peroxisomes may be selectively degraded upon change of carbon source (Veenhuis et al. 1983; Tuttle et al. 1993). Autophagy occurs in *S. cerevisiae* as well (Takeshige et al. 1992), and mutants defective in this process have been described previously (Tsukada and Ohsumi 1993; Thumm et al. 1994). Mutants unable to autophagocytose peroxisomes have been isolated in *H. polymorpha* (Titorenko et al. 1995) and *P. pastoris* (Tuttle and Dunn 1995).

Although both matrix and integral membrane proteins are imported (or inserted) into preexisting peroxisomes posttranslationally, the mechanisms of their targeting and import or insertion are distinct. As discussed further below, three types of topogenic sequences have been found for soluble matrix proteins, and thus far, the integral membrane proteins seem not to make use of any of them. Moreover, translocation of matrix proteins through the peroxisomal membrane has requirements not shared by the integral membrane proteins. This is seen clearly in fibroblasts of Zellweger syndrome patients, in which membrane proteins appear to assemble normally into peroxisomal membranes, but the matrix proteins are not imported. This results in the presence in the fibroblasts of nearly empty membrane ghosts of peroxisomes, which can be recognized by immunochemical methods but not by simple electron microscopy (Santos et al. 1988). There are at least ten complementation groups among the patients with inherited disorders of peroxisome biogenesis (prototype = Zellweger syndrome) (Shimozawa et al. 1993; Slawecki et al. 1995). In at least eight of the ten complementation groups, peroxisome ghosts are present (for summary, see Purdue and Lazarow 1994). These general features of peroxisome biogenesis, which were established for animals, plants, and other yeasts, are presumed to apply to *S. cerevisiae*.

B. Targeting of Proteins to Peroxisomes

At least three types of peroxisomal targeting sequences (PTSs) direct matrix proteins to peroxisomes; two of these have been well characterized (de Hoop and Ab 1992; Subramani 1993; Purdue and Lazarow 1994). PTS1 is a carboxy-terminal tripeptide, typically SKL, but sometimes a conservative variant thereof. PTS2 is an amino-terminal 11–16-amino-acid oligopeptide containing conserved pairs of amino acids (RL and H/Q-L) with a characteristic spacing of five amino acids between them. The third type of PTS consists of internal, sometimes redundant, sequences, which have not yet been characterized in detail.

All three PTSs appear to function in *S. cerevisiae*. PTS1 (SKL or AKL) (Gould et al. 1990) has been found at the carboxyl terminus of a number of *S. cerevisiae* proteins (see Table 1A). Many of these proteins have been shown to be in peroxisomes; in other cases, the presumed peroxisomal localization remains to be tested (Table 1A). In contrast, *S. cerevisiae* thiolase (Fox3p/Pot1p) is directed to peroxisomes by the amino-terminal oligopeptide, MSQ<u>RL</u>QSIKD<u>HL</u>VLSA, which contains a functional PTS2 (Erdmann 1994; Glover et al. 1994b).

A PTS1-related sequence, SSNSKF, functions as a PTS1 at the carboxyl terminus of catalase A. In addition, this protein contains additional internal topogenic information that is sufficient to direct it to the peroxisomes when the carboxy-terminal PTS1 is removed (Kragler et al. 1993). Acyl-CoA oxidase and Peb1p are other interesting cases. In rat liver, acyl-CoA oxidase is directed to peroxisomes by a carboxy-terminal SKL. In *C. tropicalis*, there are three acyl-CoA oxidase genes, none of whose products ends in SKL. The one protein among these that has been investigated, Pox4p, is targeted to peroxisomes by either of two internal, redundant sequences (Small et al. 1988; Kamiryo et al. 1989). The *S. cerevisiae* acyl-CoA oxidase (Fox1p/Pox1p) likewise lacks a carboxy-terminal SKL and is thought to be targeted to peroxisomes in a manner similar to that of *C. tropicalis* Pox4p. Peb1p, a protein required for the import of thiolase into peroxisomes, is itself targeted to peroxisomes by its amino-terminal 55 amino acids, which do not resemble PTS2 (Zhang and Lazarow 1996).

The topogenesis of peroxisomal membrane proteins is less well understood. None of the known integral membrane proteins in the *S. cerevisiae* organelle contains a recognizable PTS of the types known to function for matrix proteins. The first 40 amino acids of Pas3p are sufficient to convert a passenger protein into a peroxisomal integral membrane protein (W.-H. Kunau, unpubl.). These 40 amino acids contain a hydrophobic membrane-spanning domain surrounded by positively charged amino acids. The peroxisomal membrane protein PMP47 of *C. boidinii* is properly sorted to peroxisomal membranes in *S. cerevisiae*, indicating that the topogenic information is conserved (McCammon et al. 1990b). A hydrophilic loop of 20 amino acids is necessary and sufficient for this targeting (Dyer et al. 1996).

C. Import of Oligomeric Proteins as well as of Unfolded (or Loosely Folded) Aposubunits

Careful kinetic experiments have shown that catalase A is normally imported into *S. cerevisiae* peroxisomes in the form of apo-monomers. Inside the organelle, each monomer acquires a heme group and assembles into a tetramer (Zimniak et al. 1976). This is the same manner by which rat liver catalase enters peroxisomes and assembles (Lazarow and de Duve 1973). Similarly, methanol oxidase, a homo-octameric flavoprotein of *H. polymorpha*, is imported as a monomer with peroxisomes of *S. cerevisiae* (Distel et al. 1987). Until recently, it has been assumed, by analogy with the situation for mitochondria, that all peroxisomal proteins must be unfolded for import.

Recent data from two studies, however, suggest that oligomerized proteins can also enter *S. cerevisae* peroxisomes. One study used the bacterial enzyme, chloramphenicol acetyltransferase (ChAT), whose mature form is a homotrimer. This enzyme is normally found in the cytosol when expressed in eukaryotic cells, but it can be targeted to peroxisomes by the addition of an SKL to the carboxyl terminus (ChAT-SKL). McNew and Goodman (1994) demonstrated that ChAT, tagged with a hemagglutinin (HA) epitope but lacking any PTS, was also imported into *S. cerevisiae* peroxisomes if, and only if, it was coexpressed with ChAT-SKL. This import is presumably explained by the formation in the cytosol of mixed trimers of HA-ChAT and ChAT-SKL, which are then coimported into peroxisomes. The authors suggested that trimeric, fully folded ChAT might enter peroxisomes through pores (either static or dynamic) or by membrane internalization (McNew and Goodman 1994). Similarly, thiolase was shown to be capable of entering peroxisomes as a dimer (Glover et al. 1994a). Truncated thiolase, lacking its PTS2 but epitope-tagged, could enter peroxisomes when coexpressed with full-length thiolase. Immunoprecipitation experiments suggested that the thiolase heterodimer remained together during translocation.

There has not yet been a formal test of whether such oligomers are truly folded during the import process. However, peroxisomes are presumably capable of importing folded oligomers, at least in cultured animal cells, because they are able to import appropriately tagged gold particles (Walton et al. 1995).

D. *S. cerevisiae* Mutants Defective in Peroxisome Formation

A variety of screening and selection procedures have been used to identify *S. cerevisiae* mutants defective in peroxisome assembly (*pas*) or peroxisome biogenesis (*peb*) (see Table 4). Erdmann et al. (1989) screened for mutants that had lost the ability to grow on oleic acid but retained the ability to grow on acetate. These mutants (32 out of 130,000 colonies screened by replica plating) were then screened by electron microscopy, identifying four that contained no recognizable peroxisomal structures. Cell fractionation studies showed that peroxisomal enzymes (catalase and thiolase) were present in the cytosol (in active form), rather than in peroxisomes. The four mutants fell into three complementation groups (*pas1–pas3*). As expected, other mutants identified in this screen had defects in individual fatty acid oxidation enzymes (*fox* mutants). Additional mutants with reduced amounts of the β-oxidation enzymes (presumably regulatory mutants) were not further characterized. In subsequent experiments, mutants that grew on acetate but not on oleate were analyzed next by cell fractionation for the sedimentability of peroxisomal enzymes, and nine additional *pas* complementation groups were identified (Table 4, *pas4–pas12*) (Kunau and Hartig 1992). (The original *pas13* turned out to be allelic to *pas3* [Kunau et al. 1993].)

van der Leij et al. (1992) devised a selection procedure based on the fact that peroxisomal β-oxidation produces H_2O_2 as a side product, which is toxic to the cells if not decomposed by catalase. These authors inhibited catalase (both isozymes) with a specific inhibitor, 3-amino-1,2,4-triazole (3-AT) and observed that this prevented the cells from growing on lauric acid. As expected, 3-AT did not prevent the cells from growing on maltose, presumably because little H_2O_2 was produced. Mutants unable to utilize fatty acids (including peroxisome assembly mutants) should not produce much H_2O_2 even in the presence of lauric acid, and so would be expected to grow despite the inhibition of catalase as long as they were provided with another carbon source (e.g., maltose). Of EMS-generated mutants that survived this positive selection procedure, 1535 were screened for their ability to grow on glycerol but not on oleate, and 36 passed this screen. These 36 were analyzed by cell fractionation, and in 16, catalase and/or thiolase was mislocalized to the cytosol or reduced in abundance. These 16 mutants belonged to 12 complementation groups, many of which were identical to those identified by Kunau and co-workers. *pas3–pas8* and *pas10* proved to be defective in peroxisome biogenesis per se, whereas *pas14–pas19* proved to be regulatory mutants (Table 4). (Subsequently, *pas15* and *pas17* turned out to be allelic to *pas14* [note added in proof to van der Leij et al. 1992].)

Zhang et al. (1993a) independently devised a positive selection procedure to obtain additional peroxisome biogenesis (*peb*) mutants in *S. cerevisiae*. This scheme was also based on the principle of exploiting the toxicity of H_2O_2 produced during peroxisomal fatty acid oxidation, but it employed a yeast strain in which both catalase genes had been mutated. Moreover, the requirement for utilization of glycerol was included in the selection strategy. Cells were mutagenized and grown on medium containing oleic acid plus glycerol as a second carbon source. Colonies that grew well were screened by immunofluorescence (with simplified methods) for missing (or altered) peroxisomes. Approximately 1700 of the 10^6 mutagenized cells grew into large colonies, and, of these, 11 had striking and reproducible changes in thiolase immunofluorescence. These mutants fell into five *peb* complementation groups, of which four are bona fide peroxisome biogenesis mutants and one (*peb3*) is probably a regulatory mutant. The *peb* mutants were cross-complemented with *pas1–pas6* and four were different (Table 4); one of these (*peb1*) was subsequently shown to be allelic to *pas7*. Two of the four biogenesis mutants showed striking clustering of peroxisomes.

Another powerful selection procedure was based on the expression of the bleomycin resistance protein, fused to luciferase (Elgersma et al. 1993). The fusion protein was imported into peroxisomes; sequestered in this compartment, it had only a weak effect in defending the cells against the toxic phleomycin (bleomycin analog) ligand. In peroxisome-assembly mutants, the fusion protein was mislocalized to the cytosol, where the bleomycin domain effectively protected the cell against toxicity. Of 800 phleomycin-resistant colonies, 84 were unable to grow on oleate, but they were able to grow on glycerol; 30 of these were tested for peroxisomal protein import and all were defective. These mutants reidentified many of the existing *pas* complementation groups as well as three new groups (*pas20–pas22*). Taking all of these data together, there are now 15–17 groups of *pas* or *peb* mutants that are defective in peroxisome biogenesis per se and 4 or 5 complementation groups of regulatory mutants that affect peroxisomes.

E. Pathways of Matrix Protein Import

1. *Mutant Phenotypes Suggest a Three-branched Import Pathway*

Phenotypic analysis of the peroxisome biogenesis mutants of *S. cerevisiae* has been informative. Certain mutants (*pas7/peb1, pas10, peb5*) contain morphologically recognizable peroxisomes (or unusual membrane structures in the case of *pas10*; van der Leij et al. 1992), normal expression of all peroxisomal proteins tested, and selective defects in the import of matrix proteins (Table 4). For example, *pas7/peb1* cells fail to import thiolase into peroxisomes (leaving it in the cytosol) but do import acyl-CoA oxidase, the multifunctional Fox2p (hydratase-dehydrogenase), catalase A, and transgenically expressed luciferase (Zhang et al. 1993a,b; Marzioch et al. 1994). Thiolase differs from these other proteins in that it is targeted to peroxisomes by an amino-terminal PTS2, suggesting that Pas7p/Peb1p functions specifically in the import of PTS2-targeted proteins (Fig. 11A).

In contrast, *pas10* cells fail to import Fox2p, luciferase, and catalase A into peroxisomes but do import thiolase (van der Leij et al. 1993), and *peb5* cells fail to import catalase A but do import thiolase as well as acyl-CoA oxidase (Zhang et al. 1993a,b). The nonimported proteins all contain a carboxy-terminal PTS1, suggesting that Pas10p and Peb5p function specifically in the import of PTS1-targeted proteins (Fig. 11A).

It is noteworthy that the only protein known to be imported normally into peroxisomes in both *pas7/peb1* and *peb5* mutants is acyl-CoA oxidase, which is targeted by neither a PTS1 nor a PTS2, but instead has redundant internal PTSs (at least in *C. tropicalis*) (Small et al. 1988; Kamiryo et al. 1989; see above). Taken together, the data suggest that there are three branches in the pathway for the import of matrix proteins into *S. cerevisiae* peroxisomes (Fig. 11A).

In addition, mutants in many other complementation groups lack morphologically recognizable peroxisomes and have all tested matrix proteins in the cytosol (*pas1/peb2, pas2, 3, 5, 6, 8, 9, 11, 12, 20, 21*, and

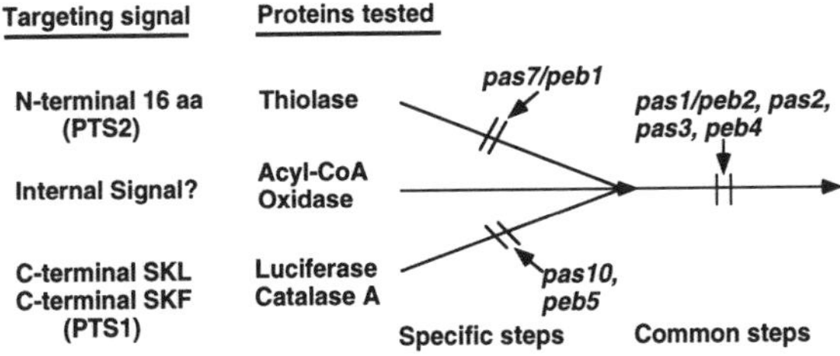

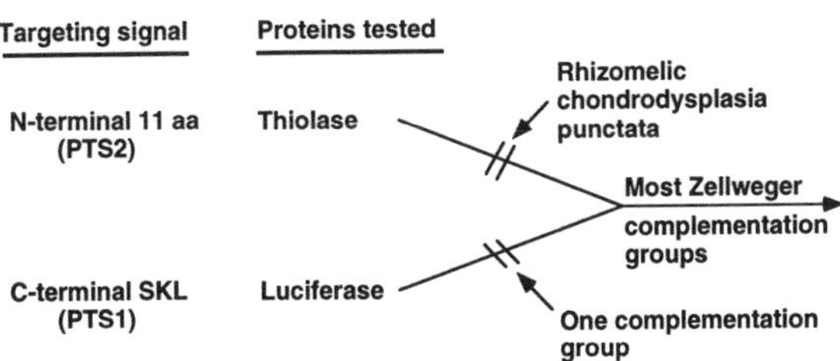

Figure 11 Branched pathways of peroxisomal protein import, deduced from analysis of mutants. (Modified, with permission, from Lazarow 1995.)

peb4). (The original *pas3* and *pas6* mutant isolates contained some type of peroxisomal membrane structures, but strains harboring disruptions of these genes do not.) These mutants might be defective in assembling peroxisomal membrane proteins, leading to the gradual loss of peroxisomes as the cells grow and divide, or they might be defective in importing most or all matrix proteins through the membranes (as in Zellweger fibroblasts), rendering the peroxisomes unrecognizable. As described below, some of these mutants do indeed contain peroxisomal membrane ghosts; these are likely to be defective in shared components that are required for the import of all matrix proteins (Fig. 11A).

A similar variety of phenotypes have been observed among human patients with inherited disorders involving peroxisome biogenesis. Fibroblasts from patients in most complementation groups (including Zellweger syndrome, neonatal adrenoleukodystrophy, and infantile Refsum's disease) contain peroxisomal membranes but fail to import any tested protein (Fig. 11B). One patient is selectively defective in the import of PTS1-targeted proteins. In contrast, patients with rhizomelic chondrodysplasia punctata are defective in thiolase import (and also lack two other enzymes with unknown topogenesis) (Fig. 11B).

2. *Hypothesis of Multiple Receptors and a Shared Translocation Machinery*

One simple interpretation of the branched pathway in Figure 11 is that multiple receptors exist, one for each type of topogenic information, together with a shared translocation machinery (Purdue and Lazarow 1994). Recent data strongly support this interpretation.

a. Receptors. PAS10 encodes a PTS1 receptor. Pas10p is a hydrophilic, 69-kD protein containing eight tetratricopeptide repeats and no predicted membrane-spanning domains (van der Leij et al. 1993). It interacts with a PTS1-containing peptide (–KNIESKL) in the yeast two-hybrid assay (Brocard et al. 1994), and it is homologous to *P. pastoris* Pas8p, which has been shown to be a receptor for PTS1 (McCollum et al. 1993; Terlecky et al. 1995). Homologs have also been identified in *H. polymorpha* (van der Klei et al. 1995), *Y. lipolytica* (Szilard et al. 1995), and human cells (Dodt et al. 1995; Fransen et al. 1995; Wiemer et al. 1995), all of which also interact with SKL. Interestingly, these proteins have been reported in a variety of intracellular locations, including the cytosol, the outer surface of the peroxisome, firmly embedded in the peroxisome membrane, and inside the peroxisome. Taken together, these data indicate that *S. cerevisiae* Pas10p is a PTS1 receptor, but its site(s) and manner of action require further study.

PAS7/PEB1 encodes a PTS2 receptor. Pas7p/Peb1p is a 42-kD hydrophilic protein with no predicted transmembrane segment (Marzioch et al. 1994; Zhang and Lazarow 1995). It contains six WD repeats that comprise the large majority of the protein, with the exception of the first 55 amino acids. WD repeats are thought to fold into β sheets, separated by turns, which may self-associate; the motif has been identified in 27 proteins with diverse functions (Neer et al. 1994). Pas7p/Peb1p binds specifically to the PTS2 of thiolase (its amino-terminal 16 amino acids), according to affinity-binding experiments, immunoprecipitation, and the two-hybrid method (Rehling et al. 1996b; Zhang and Lazarow 1996). A mutation that is known to reduce thiolase topogenesis in vivo (leucine to arginine at residue 5) also reduces thiolase's interaction with Pas7p/Peb1p in vitro. This demonstrates a functional relationship between PTS2 structure, function, and binding to Pas7p/Peb1p (Zhang and Lazarow 1996). Another mutation in thiolase's PTS2 (arginine to glycine at residue 4) makes thiolase import into peroxisomes temperature-sensitive in vivo, and the temperature-sensitive phenotype is rescued by overexpression of *PAS7/PEB1* (W.-H. Kunau, unpubl.). As with Pas10p and its orthologs in other organisms (see above), there are conflicting data concerning the precise site of function Pas7p/Peb1p (Rachubinski and Subramani 1995). It has been found exclusively inside peroxisomes (Zhang and Lazarow 1995, 1996), as well as predominantly outside peroxisomes in the cytosol (Marzioch et al. 1994); its location and function are under active investigation.

b. Peroxisomal Membrane Proteins. Of the *PAS* and *PEB* gene products whose structures are known, only a few have been demonstrated to be integral membrane proteins (see Table 4). Pas3p (50.6 kD) has a predicted amino-terminal membrane-spanning domain and is not extracted from peroxisomes with sodium carbonate (Höhfeld et al. 1991), indicating that it is embedded in the membrane (Fujiki et al. 1982a). Its susceptibility to protease suggests that most of the protein is cytoplasmically disposed. Because Pas3p has no sequence similarity to other proteins in the databases, we can only speculate about its function: By virtue of its location, it could be a receptor, a docking protein, or a component of the translocation machinery. The same possibilities apply to Pas11p, which has two predicted membrane-spanning domains and is an integral membrane protein by the test of carbonate nonextractability (W.-H. Kunau, unpubl.). Of particular interest is Pas8p, which has two predicted ATP-binding sites as well as two predicted membrane-spanning domains (Voorn-Brouwer et al. 1993). If Pas8p is in fact a membrane protein, it would be a good candidate for a component of the translocation apparatus, which has been shown to require ATP, at least in animal cells (Imanaka et al. 1987; Wendland and Subramani 1993a).

Another peroxisomal integral membrane protein, Pmp27p, which was identified by reverse genetics, rather than by complementation of a mutant, is not required for peroxisome biogenesis. However, disruption of this gene causes the formation of giant peroxisomes during growth on oleate, suggesting that it may be involved in peroxisome proliferation and the control of peroxisome size (Erdmann and Blobel 1995; Marshall et al. 1995).

c. Other Possible Components of the Translocation Machinery Used by All Matrix Proteins. Some of the other *PAS* gene products have interesting or provocative biochemical functions, but their precise roles in the import of matrix proteins remain mysterious. For example, *PAS1* encodes a hydrophilic 117-kD protein with two predicted ATP-binding sites (Erdmann et al. 1991). It (together with Pas8p) is a member of the AAA protein family (distinct from the ABC protein family) (Kunau et al. 1993). The AAA proteins include Sec18p of *S. cerevisiae* and NEM-sensitive factor (NSF) of CHO cells, which have essential roles in membrane fusion events in the secretory pathway, as well as other proteins that contain two ATP-binding sites but have totally different functions. The abundance of Pas1p is very low and its intracellular location is not known.

Another intriguing protein is Pas2p/Ubc10p (21.1 kD), which is a member of the ubiquitin-conjugating enzyme family (Wiebel and Kunau 1992). It lacks predicted membrane-spanning domains, but it cofractionates with the peroxisomal membrane and appears to be a peripheral membrane protein. Degradation of a factor that inhibits peroxisome biogenesis has been suggested as a possible function. Other *PAS* gene products with distinguishing biochemical features include Pas4p and Pas5p (Liu et al. 1996), which contain zinc fingers of the C3HC4 type ("ring fingers"), and Pas12p, which is prenylated (Kunau et al. 1993). The only SKL-containing protein thus far identified among the *PAS* gene products is Pas6p, which in addition contains a PTS2 and is homologous to *H. polymorpha* Per3p (P. Rehling and W.-H. Kunau, unpubl.).

F. Peroxisome Ghosts and the Origin of Peroxisomes

The central premise of peroxisome biogenesis, namely, that new peroxisomes form by division from preexisting peroxisomes, has been challenged by the discovery of *S. cerevisiae* mutants such as *pas1/peb2* and *pas2*, which appear, by morphological criteria, to be entirely lacking in peroxisomes (Erdmann et al. 1989; Zhang et al. 1993a) but which nevertheless recover a normal population of peroxisomes when transformed with a wild-type *PAS1/PEB2* or *PAS2* gene, respectively (Erdmann et al. 1991; Wiebel and Kunau 1992). From where do these peroxisomes arise? One possibility is that the mutants contain empty membrane ghosts of peroxisomes, similar to those detected in fibroblasts from Zellweger syndrome patients (Santos et al. 1988, 1992; Suzuki et al. 1992; Wendland and Subramani 1993b; Motley et al. 1994). Such ghosts could regain import competence upon reintroduction of the wild-type gene, refill with matrix proteins, and carry on with growth and division.

To look for such ghosts in *S. cerevisiae*, a good membrane marker was required. Hemagglutinin epitope-tagged Pas3p was found to sort efficiently to the peroxisome membrane and could be detected by immunoblotting, immunofluorescence, and immunogold labeling (Fig. 12) (Purdue and Lazarow 1995). Enhanced clustering of peroxisomes occurred at the expression level employed. This peroxisomal membrane

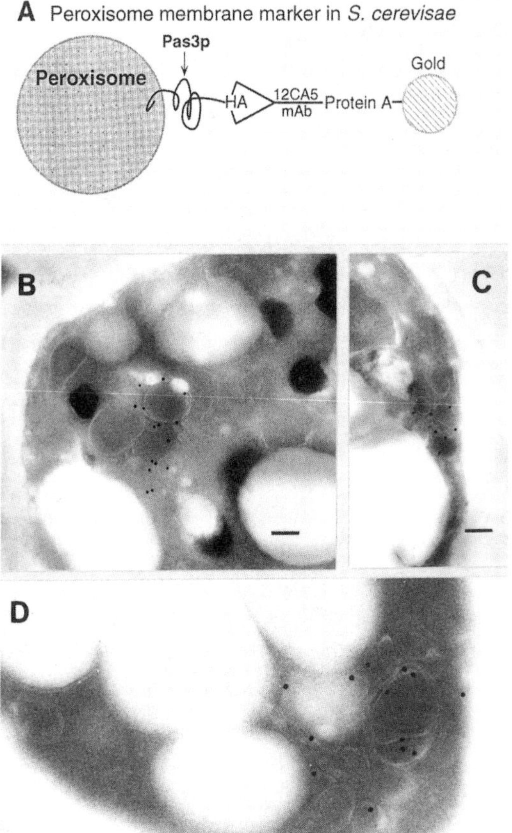

Figure 12 Identification of peroxisomal membrane ghosts in *S. cerevisiae* mutants by immunoelectron microscopy. (*A*) The peroxisomal transmembrane protein Pas3p was tagged with the hemagglutinin (HA) epitope and detected with monoclonal antibody 12CA5 and protein-A coupled to gold beads. (*B,C*) Wild-type *S. cerevisiae*. (*D*) Mutant *peb2*, in which peroxisomes had appeared to be missing altogether. (Modified, with permission, from Purdue and Lazarow 1995.)

marker was expressed in two mutants, *pas1/peb2* and *peb4*, in which peroxisomes appeared to be completely absent (Erdmann et al. 1989; Zhang et al. 1993a) and peroxisomal membrane ghosts were identified (Fig. 12) (Purdue and Lazarow 1995). The ghosts presumably provide the structural basis for reforming peroxisomes when the missing wild-type gene is introduced into the mutant.

VII. CONCLUDING REMARKS AND OUTLOOK

Enormous progress has been made in cloning and characterizing both the genes that encode peroxisomal proteins and genes whose products are required for peroxisome biogenesis. At the present rate of this work, and with the recent completion of the *S. cerevisiae* genome sequencing project, we can expect to have a largely complete inventory of relevant genes in the not too distant future. In particular, the discovery of additional gene products that contain recognizable peroxisomal targeting sequences has implied that these proteins are likely to be located within peroxisomes. This has suggested new functions for the organelle that must be tested experimentally. More such discoveries and follow-up investigations can be expected. Lagging a little behind, but probably not for long, is identification of the set of genes involved in regulating peroxisome proliferation and the abundance of individual peroxisomal enzymes.

Not unexpectedly, because it is more difficult, work on the function of the *PAS* and *PEB* gene products is proceeding more slowly. Two of these gene products are receptors for peroxisomal targeting sequences: Pas10p is a PTS1 (SKL) receptor, and Pas7p/Peb1p is a PTS2 receptor. Major challenges for the field are the elucidation of the physiological functions of the other genes, as well as elucidation of the mechanisms for targeting and insertion of peroxisomal membrane proteins.

As might be expected, the ease of testing structure/function relationships by deleting (or modifying) genes in *S. cerevisiae* is proving to be enormously valuable for studies of enzyme function, gene regulation, and peroxisome formation. For example, deletion of *MDH3* suggested that NADH must be reoxidized within peroxisomes and that malate and oxaloacetate probably shuttle in and out carrying the reducing equivalents. Deletion of the *RTG* genes unexpectedly showed a pronounced effect on peroxisome proliferation, thereby identifying crucial regulatory proteins. Modification of the presumed active site of Pas2p confirmed that ubiquitin conjugation is essential for peroxisome assembly (for reasons that remain mysterious). It can be confidently anticipated that further applications of these methods will give many important new insights.

This chapter has focused on *S. cerevisiae*, but it should be noted that other yeasts are proving to be equally valuable for studies of peroxisome biogenesis. These include *P. pastoris* (Gould et al. 1992), *H. polymorpha* (Veenhuis 1992), *Y. lipolytica* (Tanaka et al. 1982), and *C. boidinii*. The advantages of these other yeasts include the ability to induce peroxisome proliferation with other carbon sources, such as methanol, and to cause so much proliferation that a large fraction of the cell volume is occupied by the peroxisomes. Recent articles review peroxisome biogenesis in these organisms (Purdue and Lazarow 1994; Rachubinski and Subramani 1995; McNew and Goodman 1996).

It is exciting, although no longer surprising, that the yeast mutants which are defective in peroxisome biogenesis have been found to be good cellular models for human inherited diseases, including Zellweger syndrome. The studies on yeast are likely to provide a basic understanding of the normal functioning of the peroxisome and of the ways that this can fail. These studies will provide insights into the nature of the defects in the human disorders. Identification of the relevant yeast genes is likely to aid in the identification and characterization of the homologous human genes, and perhaps eventually lead to useful therapeutic interventions.

ACKNOWLEDGMENTS

We thank Ms. Kara Lukin for help in gathering references and Mr. Julio Robles for word processing. The work of the authors was supported by National Institutes of Health grant DK19394 to P.B.L. and DFG grants KU329-17-1 and KU329-17-2 to W.H.K.

REFERENCES

Alexson, S.E.H., Y. Fujiki, H. Shio, and P.B. Lazarow. 1985. Partial disassembly of peroxisomes. *J. Cell Biol.* **101:** 294.

Atomi, H., M. Ueda, J. Suzuki, Y. Kamada, and A. Tanaka. 1993. Presence of carnitine acetyltransferase in peroxisomes and in mitochondria of oleic acid-grown *Saccharomyces cerevisiae. FEMS Microbiol. Lett.* **112:** 31.

Avers, C.J. and M. Federman. 1968. The occurrence in yeast of cytoplasmic granules which resemble microbodies. *J. Cell Biol.* **37:** 555.

Beevers, H. 1979. Microbodies in higher plants. *Annu. Rev. Plant Physiol.* **30:** 159.

———. 1982. Glyoxysomes in higher plants. *Ann. N.Y. Acad. Sci.* **386:** 243.

Benz, R. 1990. Biophysical properties of porin pores from mitochondrial outer membrane of eukaryotic cells. *Experientia* **46:** 131.

Berteaux-Lecellier, V., M. Picard, C. Thompson-Coffe, D. Zickler, A. Panvier-Adoutte, and J.-M. Simonet. 1995. A nonmammalian homolog of the *PAF1* gene (Zellweger syndrome) discovered as a gene involved in caryogamy in the fungus *Podospora anserina*. *Cell* **81:** 1043.

Bieri, F. 1993. Peroxisome proliferators and cellular signaling pathways: A review. *Biol. Cell* **77:** 43.

Brill, S.J. and B. Stillman. 1991. Replication factor-A from *Saccharomyces cerevisiae* is encoded by three essential genes coordinately expressed in S phase. *Genes Dev.* **5:** 1589.

Brocard, C., F. Kragler, M.M. Simon, T. Schuster, and A. Hartig. 1994. The tetratricopeptide repeat-domain of the PAS10 protein of *Saccharomyces cerevisiae* is essential for binding the peroxisomal targeting signal-SKL. *Biochem. Biophys. Res. Commun.* **204:** 1016.

Celenza, J.L. and M. Carlson. 1986. A yeast gene that is essential for release from glucose repression encodes a protein kinase. *Science* **233:** 1175.

Chelstowska, A. and R.A. Butow. 1995. *RTG* genes in yeast that function in communication between mitochondria and the nucleus are also required for expression of genes encoding peroxisomal proteins. *J. Biol. Chem.* **270:** 18141.

Coe, J.G.S., L.E. Murray, and I.W. Dawes. 1994. Identification of a sporulation-specific promoter regulating divergent transcription of two novel sporulation genes in *Saccharomyces cerevisiae. Mol. Gen. Genet.* **244:** 661.

Cohen, G., W. Rapatz, and H. Ruis. 1988. Sequence of the *Saccharomyces cerevisiae CTA1* gene and amino acid sequence of catalase A derived from it. *Eur. J. Biochem.* **176:** 159.

Colbeau, A., J. Nachbaur, and P.M. Vignais. 1971. Enzymic characterization and lipid composition of rat liver subcellular membranes. *Biochim. Biophys. Acta* **249:** 462.

de Duve, C. 1965. The separation and characterization of subcellular particles. *Harvey Lect.* **59:** 49.

———. 1969. Evolution of the peroxisome. *Ann. N.Y. Acad. Sci.* **168:** 369.

———. 1973. Biochemical studies on the occurrence, biogenesis and life history of mammalian peroxisomes. *J. Histochem. Cytochem.* **21:** 941.

de Duve, C. and P. Baudhuin. 1966. Peroxisomes (microbodies and related particles). *Physiol. Rev.* **46:** 323.

de Hoop, M.J. and G. Ab. 1992. Import of proteins into peroxisomes and other microbodies. *Biochem. J.* **286:** 657.

Distel, B., M. Veenhuis, and H.F. Tabak. 1987. Import of alcohol oxidase into peroxisomes of *Saccharomyces cerevisiae. EMBO J.* **6:** 3111.

Dmochowska, A., D. Dignard, R. Maleszka, and D.Y. Thomas. 1990. Structure and transcriptional control of the *Saccharomyces cerevisiae POX1* gene encoding acyl-coenzyme A oxidase. *Gene* **88:** 247.

Dodt, G., N. Braverman, C. Wong, A. Moser, H.W. Moser, P. Watkins, D. Valle, and S.J. Gould. 1995. Mutations in the PTS1 receptor gene, *PXR1*, define complementation group 2 of the peroxisome biogenesis disorders. *Nat. Genet.* **9:** 115.

Douglas, M., D. Finkelstein, and R.A. Butow. 1979. Analysis of products of mitochondrial protein synthesis in yeast: Genetic and biochemical aspects. *Methods Enzymol.* **56:** 58.

Douma, A.C., M. Veenhuis, H.R. Waterham, and W. Harder. 1990. Immunocytochemical demonstration of the peroxisomal ATPase of yeasts. *Yeast* **6:** 45.

Dunn, W.A., Jr. 1994. Autophagy and related mechanisms of lysosome-mediated protein degradation. *Trends Cell Biol.* **4:** 139.

Duntze, W., D. Neumann, J.M. Gancedo, W. Atzpodien, and H. Holzer. 1969. Studies on the regulation and localization of the glyoxylate cycle enzymes in *Saccharomyces cerevisiae. Eur. J. Biochem.* **10:** 83.

Dyer, J.M., J.A. McNew, and J.M. Goodman. 1996. The sorting sequence of the peroxisomal integral membrane protein PMP47 is contained within a short hydrophilic loop. *J. Cell Biol.* **133:** 269.

Einerhand, A.W.C., W.T. Kos, B. Distel, and H.F. Tabak. 1993. Characterization of a transcriptional control element involved in proliferation of peroxisomes in yeast in response to oleate. *Eur. J. Biochem.* **214:** 323.

Einerhand, A.W.C., I. Van Der Leij, W.T. Kos, B. Distel, and H.F. Tabak. 1992. Transcriptional regulation of genes encoding proteins involved in biogenesis of peroxisomes in *Saccharomyces cerevisiae*. *Cell Biochem. Funct.* **10:** 185.

Einerhand, A.W.C., T.M. Voorn-Brouwer, R. Erdmann, W.-H. Kunau, and H.F. Tabak. 1991. Regulation of transcription of the gene coding for peroxisomal 3-oxoacyl-CoA thiolase of *Saccharomyces cerevisiae*. *Eur. J. Biochem.* **200:** 113.

Einerhand, A.W.C., W. Kos, W.C. Smart, A.J. Kal, H.F. Tabak, and T.G. Cooper. 1995. The upstream region of the *FOX3* gene encoding peroxisomal 3-oxoacyl-coenzyme A thiolase in *Saccharomyces cerevisiae* contains ABF1- and replication protein A-binding sites that participate in its regulation by glucose repression. *Mol. Cell. Biol.* **15:** 3405.

Elgersma, Y., M. van den Berg, H.F. Tabak, and B. Distel. 1993. An efficient positive selection procedure for the isolation of peroxisomal import and peroxisome assembly mutants of *Saccharomyces cerevisiae*. *Genetics* **135:** 731.

Elgersma, Y., C.W.T. van Roermund, R.J.A. Wanders, and H.F. Tabak. 1995. Peroxisomal and mitochondrial carnitine acetyltransferases of *Saccharomyces cerevisiae* are encoded by a single gene. *EMBO J.* **14:** 3472.

Erdmann, R. 1989. "Isolierung und Charakterisierung peroxisomaler Mutanten von *Saccharomyces cerevisiae* und deren Verwendung zur Untersuchung der biogese von Peroxisomen." Ph.D. thesis, Ruhr-Universitat, Bochum, Germany.

———. 1994. The peroxisomal targeting signal of 3-oxoacyl-CoA thiolase from *Saccharomyces cerevisiae*. *Yeast* **10:** 935.

Erdmann, R. and G. Blobel. 1995. Giant peroxisomes in oleic acid-induced *Saccharomyces cerevisiae* lacking the peroxisomal membrane protein Pmp27p. *J. Cell Biol.* **128:** 509.

Erdmann, R. and W.-H. Kunau. 1994. Purification and immunolocalization of the peroxisomal 3-oxoacyl-CoA thiolase from *Saccharomyces cerevisiae*. *Yeast* **10:** 1173.

Erdmann, R., M. Veenhuis, D. Mertens, and W.-H. Kunau. 1989. Isolation of peroxisome-deficient mutants of *Saccharomyces cerevisiae*. *Proc. Natl. Acad. Sci.* **86:** 5419.

Erdmann, R., F.F. Wiebel, A. Flessau, J. Rytka, A. Beyer, K.-U. Fröhlich, and W.-H. Kunau. 1991. *PAS1*, a yeast gene required for peroxisome biogenesis, encodes a member of a novel family of putative ATPases. *Cell* **64:** 499.

Fahimi, H.D., A. Reinicke, M. Sujatta, S. Yokota, M. Özel, F. Hartig, and K. Stegmeier. 1982. The short- and long-term effects of bezafibrate in the rat. *Ann. N.Y. Acad. Sci.* **386:** 111.

Filipits, M., M.M. Simon, W. Rapatz, B. Hamilton, and H. Ruis. 1993. A *Saccharomyces cerevisiae* upstream activating sequence mediates induction of peroxisome proliferation by fatty acids. *Gene* **132:** 49.

Fransen, M., C. Brees, E. Baumgart, J.C.T. Vanhooren, M. Baes, G.P. Mannaerts, and P.P. Van Veldhoven. 1995. Identification and characterization of the putative human peroxisomal C-terminal targeting signal import receptor. *J. Biol. Chem.* **270:** 7731.

Fujiki, Y., A.L. Hubbard, S. Fowler, and P.B. Lazarow. 1982a. Isolation of intracellular membranes by means of sodium carbonate treatment: Application to endoplasmic reticulum. *J. Cell Biol.* **93:** 97.

Fujiki, Y., S. Fowler, H. Shio, A.L. Hubbard, and P.B. Lazarow. 1982b. Polypeptide and phospholipid composition of the membrane of rat liver peroxisomes: Comparison with endoplasmic reticulum and mitochondrial membranes. *J. Cell Biol.* **93:** 103.

Gangloff, S.P., D. Marguet, and G.J.-M. Lauquin. 1990. Molecular cloning of the yeast mitochondrial aconitase gene (*ACO1*) and evidence of a synergistic regulation of expression by glucose plus glutamate. *Mol. Cell. Biol.* **10:** 3551.

Gartner, J., H. Moser, and D. Valle. 1992. Mutations in the 70K peroxisomal membrane protein gene in Zellweger syndrome. *Nat. Genet.* **1:** 16.

Glover, J.R., D.W. Andrews, and R.A. Rachubinski. 1994a. *Saccharomyces cerevisiae* peroxisomal thiolase is imported as a dimer. *Proc. Natl. Acad. Sci.* **91:** 10541.

Glover, J.R., D.W. Andrews, S. Subramani, and R.A. Rachubinski. 1994b. Mutagenesis of the amino targeting signal of *Saccharomyces cerevisiae* 3-ketoacyl-CoA thiolase reveals conserved amino acids required for import into peroxisomes *in vivo*. *J. Biol. Chem.* **269:** 7558.

Gorgas, K. 1984. Peroxisomes in sebaceous glands. V. Complex peroxisomes in the mouse preputial gland: Serial sectioning, and three-dimensional reconstruction studies. *Anat. Embryol.* **169:** 261.

Gould, S.J., D. McCollum, A.P. Spong, J.A. Heyman, and S. Subramani. 1992. Development of the yeast *Pichia pastoris* as a model organism for a genetic and molecular anal ysis of peroxisome assembly. *Yeast* **8:** 613.

Gould, S.J., G.-A. Keller, M. Schneider, S.H. Howell, L.J. Garrard, J.M. Goodman, B. Distel, H. Tabak, and S. Subramani. 1990. Peroxisomal protein import is conserved between yeast, plants, insects and mammals. *EMBO J.* **9:** 85.

Hartig, A. and H. Ruis. 1986. Nucleotide sequence of the *Saccharomyces cerevisiae CTT1* gene and deduced amino acid sequence of yeast catalase T. *Eur. J. Biochem.* **160:** 487.

Hartig, A., M.M. Simon, T. Schuster, J.R. Daugherty, H.S. Yoo, and T.G. Cooper. 1992. Differentially regulated malate synthase genes participate in carbon and nitrogen metabolism of *S. cerevisiae. Nucleic Acids Res.* **20:** 5677.

Hashimoto, T. 1982. Individual peroxisomal β-oxidation enzymes. *Ann. N.Y. Acad. Sci.* **386:** 5.

Hiltunen, J.K., B. Wenzel, A. Beyer, R. Erdmann, A. Fosså, and W.-H. Kunau. 1992. Peroxisomal multifunctional β-oxidation protein of *Saccharomyces cerevisiae*. Molecular analysis of the *FOX2* gene and gene product. *J. Biol. Chem.* **267:** 6646.

Hoffmann, H.-P., A. Szabo, and C.J. Avers. 1970. Cytochemical localization of catalase activity in yeast peroxisomes. *J. Bacteriol.* **104:** 581.

Höhfeld, J., M. Veenhuis, and W.-H. Kunau. 1991. *PAS3*, a *Saccharomyces cerevisiae* gene encoding a peroxisomal integral membrane protein essential for peroxisome biogenesis. *J. Cell Biol.* **114:** 1167.

Igual, J.C., E. Matallana, C. Gonzalez-Bosch, L. Franco, and J.E. Pérez-Ortin. 1991. A new glucose-repressible gene identified from the analysis of chromatin structure in deletion mutants of yeast *SUC2* locus. *Yeast* **7:** 379.

Imanaka, T., G.M. Small, and P.B. Lazarow. 1987. Translocation of acyl-CoA oxidase into peroxisomes requires ATP hydrolysis but not a membrane potential. *J. Cell Biol.* **105:** 2915.

Jank, B., B. Habermann, R.J. Schweyen, and T.A. Link. 1993. PMP47, a peroxisomal homologue of mitochondrial solute carrier proteins. *Trends Biochem. Sci.* **18:** 427.

Kamiryo, T., Y. Sakasegawa, and H. Tan. 1989. Expression and transport of *Candida tropicalis* peroxisomal acyl-coenzyme A oxidase in the yeast *Candida maltosa*. *Agric. Biol. Chem.* **53:** 179.

Kawamoto, S., C. Nozaki, A. Tanaka, and S. Fukui. 1978. Fatty acid β-oxidation system in microbodies of *n*-alkane-grown *Candida tropicalis*. *Eur. J. Biochem.* **83:** 609.

Kionka, C. and W.-H. Kunau. 1985. Inducible β-oxidation pathway in *Neurospora crassa*. *J. Bacteriol.* **161:** 153.

Kispal, G. and P.A. Srere. 1991. Studies on yeast peroxisomal citrate synthase. *Arch. Biochem. Biophys.* **286:** 132.

Kragler, F., A. Langeder, J. Raupachova, M. Binder, and A. Hartig. 1993. Two independent peroxisomal targeting signals in catalase A of *Saccharomyces cerevisiae*. *J. Cell Biol.* **120:** 665.

Kunau, W.-H. and A. Hartig. 1992. Peroxisome biogenesis in *Saccharomyces cerevisiae*. *Antonie Leeuwenhoek* **62:** 63.

Kunau, W.-H., V. Dommes, and H. Schulz. 1996. β-Oxidation of fatty acids in mitochondria, peroxisomes, and bacteria: A century of continued progress. *Prog. Lipid Res.* (in press).

Kunau, W.-H., A. Beyer, T. Franken, K. Götte, M. Marzioch, J. Saidowsky, A. Skaletz-Rorowski, and F.F. Wiebel. 1993. Two complementary approaches to study peroxisome biogenesis in *Saccharomyces cerevisiae*: Forward and reversed genetics. *Biochimie* **75:** 209.

Kunau, W.-H., C. Kionka, A. Ledebur, M. Mateblowski, M. Moreno de la Garza, U. Schultz-Borchard, R. Thieringer, and M. Venehuis. 1987. *β-Oxidation systems in eukaryotic microorganisms* (ed. H.D. Fahimi and H. Sies), p. 128. Springer-Verlag, Heidelberg.

Latruffe, N., Ed. 1993a. Cellular aspects related to peroxisomes. *Biol. Cell* **77:** 1.

———. 1993b. Peroxisomes: Biochemistry, molecular biology and genetic diseases. *Biochimie* **75:** 141.

Laurent, B.C., M.A. Treitel, and M. Carlson. 1991. Functional interdependence of the yeast SNF2, SNF5, and SNF6 proteins in transcriptional activation. *Proc. Natl. Acad. Sci.* **88:** 2687.

Lazarow, P.B. 1977. Three hypolipidemic drugs increase hepatic palmitoyl-coenzyme A oxidation in the rat. *Science* **197:** 580.

———. 1995. Peroxisome structure, function, and biogenesis—Human patients and yeast mutants show strikingly similar defects in peroxisome biogenesis. *J. Neuropathol. Exp. Neurol.* **54:** 720.

Lazarow, P.B. and C. de Duve. 1973. The synthesis and turnover of rat liver peroxisomes. V. Intracellular pathway of catalase synthesis. *J. Cell Biol.* **59:** 507.

———. 1976. A fatty acyl-CoA oxidizing system in rat liver peroxisomes: Enhancement by clofibrate, a hypolipidemic drug. *Proc. Natl. Acad. Sci.* **73:** 2043.

Lazarow, P.B. and Y. Fujiki. 1985. Biogenesis of peroxisomes. *Annu. Rev. Cell Biol.* **1:** 489.

Lazarow, P.B. and H.W. Moser. 1995. Disorders of peroxisome biogenesis. In *The metabolic basis of inherited disease* (ed. C.R. Scriver et al.), p. 2287. McGraw-Hill, New York.

Lazarow, P.B., H. Shio, and M. Robbi. 1980. Biogenesis of peroxisomes and the peroxisome reticulum hypothesis. In *31st Mosbach Colloquium: Biological chemistry of organelle formation* (ed. T. Bucher et al.), p. 187. Springer-Verlag, New York.

Lazarow, P.B., M. Robbi, Y. Fujiki, and L. Wong. 1982. Biogenesis of peroxisomal proteins *in vivo* and *in vitro*. *Ann. N.Y. Acad. Sci.* **386:** 285.

Lazarow, P.B., R. Thieringer, G. Cohen, T. Imanaka, and G.M. Small. 1991. Protein import into peroxisomes *in vitro*. *Methods Cell Biol.* **34**: 304.

Leighton, F., B. Poole, P.B. Lazarow, and C. de Duve. 1969. The synthesis and turnover of rat liver peroxisomes. I. Fractionation of peroxisome proteins. *J. Cell Biol.* **41:** 521.

Lewin, A.S., V. Hines, and G.M. Small. 1990. Citrate synthase encoded by the *CIT2* gene of *Saccharomyces cerevisiae* is peroxisomal. *Mol. Cell. Biol.* **10:** 1399.

Liao, X. and R. Butow. 1993. *RTG1* and *RTG2:* Two yeast genes required for a novel path of communication from mitochondria to the nucleus. *Cell* **72:** 61.

Liao, X., W.C. Small, P.A. Srere, and R.A. Butow. 1991. Intramitochondrial functions regulate nonmitochondrial citrate synthase (*CIT2*) expression in *Saccharomyces cerevisiae. Mol. Cell Biol.* **11:** 38.

Liu, Y., K. Gu, and C. Dieckmann. 1996. Independent regulation of full-length and 5′-truncated *PAS5* mRNAs in *Saccharomyces cerevisiae. Yeast* **12:** 135.

Longley, R.P., A.H. Rose, and B.A. Knights. 1968. Composition of the protoplast membrane from *Saccharomyces cerevisiae. Biochem. J.* **108:** 401.

Luche, R.M., W.C. Smart, and T.G. Cooper. 1992. Purification of the heteromeric protein binding to the URS1 transcriptional repression site in *Saccharomyces cerevisiae. Proc. Natl. Acad. Sci.* **89:** 7412.

Marshall, P.A., Y.I. Krimkevich, R.H. Lark, J.M. Dyer, M. Veenhuis, and J.M. Goodman. 1995. Pmp27 promotes peroxisomal proliferation. *J. Cell Biol.* **129:** 345.

Marzioch, M., R. Erdmann, M. Veenhuis, and W.-H. Kunau. 1994. *PAS7* encodes a novel yeast member of the WD-40 protein family essential for import of 3-oxoacyl-CoA thiolase, a PTS2-containing protein, into peroxisomes. *EMBO J.* **13:** 4908.

Mathieu, M., J.P. Zeelen, R.A. Pauptit, R. Erdmann, W.-H. Kunau, and R.K. Wierenga. 1994. The 2.8 Å crystal structure of peroxisomal 3-ketoacyl-CoA thiolase of *Saccharomyces cerevisiae*: A five-layered αβαβα structure constructed from two core domains of identical topology. *Structure* **2:** 797.

McCammon, M.T., M. Veenhuis, S.B. Trapp, and J.M. Goodman. 1990a. Association of glyoxylate and beta-oxidation enzymes with peroxisomes of *Saccharomyces cerevisiae. J. Bacteriol.* **172:** 5816.

McCammon, M.T., C.A. Dowds, K. Orth, C.R. Moomaw, C.A. Slaughter, and J.M. Goodman. 1990b. Sorting of peroxisomal membrane protein PMP47 from *Candida boidinii* into peroxisomal membranes of *Saccharomyces cerevisiae. J. Biol. Chem.* **265:** 20098.

McCollum, D., E. Monosov, and S. Subramani. 1993. The *pas8* mutant of *Pichia pastoris* exhibits the peroxisomal protein import deficiencies of Zellweger syndrome cells—The PAS8 protein binds to the COOH-terminal tripeptide peroxisomal targeting signal, and is a member of the TPR protein family. *J. Cell Biol.* **121:** 761.

McNew, J.A. and J.M. Goodman. 1994. An oligomeric protein is imported into peroxisomes in vivo. *J. Cell Biol.* **127:** 1245.

———. 1996. The targeting and assembly of peroxisomal proteins: Some old rules do not apply. *Trends Biochem. Sci.* **21:** 54.

Monson, K.D. and J.M. Hayes. 1982. Biosynthetic control of the natural abundance of carbon 13 at specific positions within fatty acids in *Saccharomyces cerevisiae*. Isotopic fractionation in lipid synthesis as evidence for peroxisomal regulation. *J. Biol. Chem.* **257:** 5568.

Motley, A., E. Hettema, B. Distel, and H. Tabak. 1994. Differential protein import deficiencies in human peroxisome assembly disorders. *J. Cell Biol.* **125:** 755.

Müller, M., J.F. Hogg, and C. de Duve. 1968. Distribution of tricarboxylic acid cycle enzymes and glyoxylate cycle enzymes between mitochondria and peroxisomes in

Tetrahymena pyriformis. J. Biol. Chem. **243:** 5385.

Navarro, B. and J.C. Igual. 1994. *ADR1* and *SNF1* mediate different mechanisms in transcriptional regulation of yeast *POT1* gene. *Biochem. Biophys. Res. Commun.* **202:** 960.

Neer, E.J., C.J. Schmidt, R. Nambudripad, and T.F. Smith. 1994. The ancient regulatory-protein family of WD-repeat proteins. *Nature* **371:** 297.

Nemali, M.R., N. Usuda, M.K. Reddy, K. Oyasu, T. Hashimoto, T. Osumi, M.S. Rao, and J.K. Reddy. 1988. Comparison of constitutive and inducible levels of expression of peroxisomal β-oxidation and catalase genes in liver and extrahepatic tissues of rat. *Cancer Res.* **48:** 5316.

Nicolay, K., M. Veenhuis, A.C. Douma, and W. Harder. 1987. A ^{31}P NMR study of the internal pH of yeast peroxisomes. *Arch. Microbiol.* **147:** 37.

Opperdoes, F.R. and P.A.M. Michels. 1993. The glycosomes of the *Kinetoplastida*. *Biochimie* **75:** 231.

Osmundsen, H. 1982. Peroxisomal β-oxidation of long fatty acids: Effects of high fat diets. *Ann. N.Y. Acad. Sci.* **386:** 13.

Paltauf, F., S.D. Kohlwein, and S.A. Henry. 1992. Regulation and compartmentalization of lipid synthesis in yeast. In *The molecular and cellular biology of the yeast* Saccharomyces*: Gene expression* (ed. E.W. Jones et al.), vol. 2, p. 415. Cold Spring Harbor Laboratory Press, Cold Spring Harbor, New York.

Parish, R.W. 1975. The isolation and characterization of peroxisomes (microbodies) from baker's yeast, *Saccharomyces cerevisiae. Arch. Microbiol.* **105:** 187.

Perlman, P.S. and H.R. Mahler. 1970. Intracellular localization of enzymes in yeast. *Arch. Biochem. Biophys.* **136:** 245.

Peterson, C.L. and I. Herskowitz. 1992. Characterization of the yeast *SWI1, SWI2,* and *SWI3* genes, which encode a global activator of transcription. *Cell* **68:** 573.

Poole, B. 1971. The kinetics of disappearance of labeled leucine from the free leucine pool of rat liver and its effect on the apparent turnover of catalase and other hepatic proteins. *J. Biol. Chem.* **246:** 6587.

Poole, B., T. Higashi, and C. de Duve. 1970. The synthesis and turnover of rat liver peroxisomes. III. The size distribution of peroxisomes and the incorporation of new catalase. *J. Cell Biol.* **45:** 408.

Purdue, P.E. and P.B. Lazarow. 1994. Peroxisomal biogenesis: Multiple pathways of protein import. *J. Biol. Chem.* **269:** 30065.

———. 1995. Identification of peroxisomal membrane ghosts with an epitope-tagged integral membrane protein in yeast mutants lacking peroxisomes. *Yeast* **11:** 1045.

Rachubinski, R.A. and S. Subramani. 1995. How proteins penetrate peroxisomes. *Cell* **83:** 525.

Rattray, J.B.M., A. Schibeci, and D.K. Kidby. 1975. Lipids of yeast. *Bacteriol. Rev.* **39:** 197.

Reddy, J.K. and G.P. Mannaerts. 1994. Peroxisomal lipid metabolism. *Annu. Rev. Nutr.* **14:** 343.

Reddy, J.K., D.L. Azarnoff, and C.E. Hignite. 1980. Hypolipidaemic hepatic peroxisome proliferators form a novel class of chemical carcinogens. *Nature* **283:** 397.

Reddy, J.K., T. Suga, G.P. Mannaerts, P.B. Lazarow, and S. Subramani, Eds. 1996. Peroxisomes: Biology and role in toxicology and disease. *Ann. N.Y. Acad. Sci.* (in press).

Rehling, P., M. Albertini, and W.-H. Kunau. 1996a. New developments of protein import into peroxisomes. *Ann. N.Y. Acad. Sci.* (in press).

Rehling, P., M. Marzioch, F. Niesen, E. Wittke, M. Veenhuis, and W.-H. Kunau. 1996b. The import receptor for the peroxisomal targeting signal 2 (PTS2) in *Saccharomyces cerevisiae* is encoded by the *PAS7* gene. *EMBO J.* **15:** 2901.

Rogers, P.J. and P.R. Stewart. 1973. Mitochondrial and peroxisomal contributions to the energy metabolism of *Saccharomyces cerevisiae* in continuous culture. *J. Gen. Microbiol.* **79:** 205.

Santos, M.J., S. Hoefler, A.B. Moser, H.W. Moser, and P.B. Lazarow. 1992. Peroxisome assembly mutations in humans: Structural heterogeneity in Zellweger syndrome. *J. Cell. Physiol.* **151:** 103.

Santos, M.J., T. Imanaka, H. Shio, G.M. Small, and P.B. Lazarow. 1988. Peroxisomal membrane ghosts in Zellweger syndrome-aberrant organelle assembly. *Science* **239:** 1536.

Schutgens, R.B.H., H.S.A. Heymans, R.J.A. Wanders, H. van den Bosch, and J.M. Tager. 1986. Peroxisomal disorders: A newly recognised group of genetic diseases. *Eur. J. Pediatr.* **144:** 430.

Scott, P.J., L.P. Visentin, and J.M. Allen. 1969. The enzymatic characteristics of peroxisomes of amphibian and avian liver and kidney. *Ann. N.Y. Acad. Sci.* **168:** 244.

Shani, N., P.A. Watkins, and D. Valle. 1995. *PXA1*, a possible *Saccharomyces cerevisiae* ortholog of the human adrenoleukodystrophy gene. *Proc. Natl. Acad. Sci.* **92:** 6012.

Shimozawa, N., Y. Suzuki, T. Orii, A. Moser, H.W. Moser, and R.J.A. Wanders. 1993. Standardization of complementation grouping of peroxisome-deficient disorders and the second Zellweger patient with peroxisomal assembly factor-1 (PAF-1) defect. *Am. J. Hum. Genet.* **52:** 843.

Shimozawa, N., T. Tsukamoto, Y. Suzuki, T. Orii, Y. Shirayoshi, T. Mori, and Y. Fujiki. 1992. A human gene responsible for Zellweger syndrome that affects peroxisome assembly. *Science* **255:** 1132.

Simon, M.M., G. Adam, W. Rapatz, W. Spevak, and H. Ruis. 1991. The *Saccharomyces cerevisiae ADR1* gene is a positive regulator of transcription of genes encoding peroxisomal proteins. *Mol. Cell Biol.* **11:** 699.

Simon, M.M., M. Binder, G. Adam, A. Hartig, and H. Ruis. 1992. Control of peroxisome proliferation in *Saccharomyces cerevisiae* by *ADR1, SNF1* (*CAT1, CCR1*) and *SNF4* (*CAT3*). *Yeast* **8:** 303.

Simon, M.M., P. Pavlik, A. Hartig, M. Binder, H. Ruis, W.J. Cook, C.L. Denis, and B. Schanz. 1995. A C-terminal region of the *Saccharomyces cerevisiae* transcription factor ADR1 plays an important role in the regulation of peroxisome proliferation by fatty acids. *Mol. Gen. Genet.* **249:** 289.

Skoneczny, M. and J. Rytka. 1996. Maintenance of peroxisomal compartment in glucose repressed and anaerobically grown *Saccharomyces cerevisiae* cells. *Biochimie* **78:** 95.

Skoneczny, M., A. Chelstowska, and J. Rytka. 1988. Study of the coinduction by fatty acids of catalase A and acyl-CoA oxidase in standard and mutant *Saccharomyces cerevisiae* strains. *Eur. J. Biochem.* **174:** 297.

Slawecki, M.L., G. Dodt, S. Steinberg, A.B. Moser, H.W. Moser, and S.J. Gould. 1995. Identification of three distinct peroxisomal protein import defects in patients with peroxisome biogenesis disorders. *J. Cell Sci.* **108:** 1817.

Sloots, J.A., J.D. Aitchison, and R.A. Rachubinski. 1991. Glucose-responsive and oleic acid-responsive elements in the gene encoding the peroxisomal trifunctional enzyme of *Candida tropicalis*. *Gene* **105:** 129.

Small, G.M., L.J. Szabo, and P.B. Lazarow. 1988. Acyl-CoA oxidase contains two targeting sequences each of which can mediate protein import into peroxisomes. *EMBO J.* **7:** 1167.

Sommer, J.M. and C.C. Wang. 1994. Targeting proteins to the glycosomes of African trypanosomes. *Annu. Rev. Microbiol.* **48:** 105.

Stanway, C.A., J.M. Gibbs, and E. Berardi. 1995. Expression of the *FOX1* gene of *Saccharomyces cerevisiae* is regulated by carbon source, but not by the known glucose repression genes. *Curr. Genet.* **27:** 404.

Steffan, J.S. and L. McAlister-Henn. 1992. Isolation and characterization of the yeast gene encoding the MDH3 isozyme of malate dehydrogenase. *J. Biol. Chem.* **267:** 24708.

Subramani, S. 1993. Protein import into peroxisomes and biogenesis of the organelle. *Annu. Rev. Cell Biol.* **9:** 445.

Sulter, G.J., K. Verheyden, G. Mannaerts, W. Harder, and M. Veenhuis. 1993. The *in vitro* permeability of yeast peroxisomal membranes is caused by a 31 kDa integral membrane protein. *Yeast* **9:** 733.

Susani, M., P. Zimniak, F. Fessl, and H. Ruis. 1976. Localization of catalase A in vacuoles of *Saccharomyces cerevisiae:* Evidence for the vacuolar nature of isolated "yeast peroxisomes." *Hoppe-Seyler's Z. Physiol. Chem.* **357:** 961.

Suzuki, Y., N. Shimozawa, S. Yajima, T. Orii, S. Yokota, Y. Tashiro, T. Osumi, and T. Hashimoto. 1992. Different intracellular localization of peroxisomal proteins in fibroblasts from patients with aberrant peroxisome assembly. *Cell Struct. Funct.* **17:** 1.

Swartzman, E., M. Viswanathan, and J. Thorner. 1996. The *PAL1* gene product is a peroxisomal ATP-binding cassette transporter in the yeast *Saccharomyces cerevisiae. J. Cell Biol.* **132:** 549.

Szabo, A.S. and C.J. Avers. 1969. Some aspects of regulation of peroxisomes and mitochondria in yeast. *Ann. N.Y. Acad. Sci.* **168:** 302.

Szilard, R.K., V.I. Titorenko, M. Veenhuis, and R.A. Rachubinski. 1995. Pay32p of the yeast *Yarrowia lipolytica* is an intraperoxisomal component of the matrix protein translocation machinery. *J. Cell Biol.* **131:** 1453.

Tahotna, D., I. Hapala, E. Zinser, W. Flekl, F. Paltauf, and G. Daum. 1993. Two yeast peroxisomal proteins crossreact with an antiserum against human sterol carrier protein 2 (SCP-2). *Biochim. Biophys. Acta* **1148:** 173.

Takeshige, K., M. Baba, S. Tsuboi, T. Noda, and Y. Ohsumi. 1992. Autophagy in yeast demonstrated with proteinase-deficient mutants and conditions for its induction. *J. Cell Biol.* **119:** 301.

Tanaka, A., M. Osumi, and S. Fukui. 1982. Peroxisomes of alkane-grown yeast: Fundamental and practical aspects. *Ann. N.Y. Acad. Sci.* **386:** 183.

Terlecky, S.R., W.M. Nuttley, D. McCollum, E. Sock, and S. Subramani. 1995. The *Pichia pastoris* peroxisomal protein PAS8p is the receptor for the C-terminal tripeptide peroxisomal targeting signal. *EMBO J.* **14:** 3627.

Thieringer, R., H. Shio, Y. Han, G. Cohen, and P.B. Lazarow. 1991. Peroxisomes in *Saccharomyces cerevisiae:* Immunofluorescence analysis and import of catalase A into isolated peroxisomes. *Mol. Cell. Biol.* **11:** 510.

Thumm, M., R. Egner, B. Koch, M. Schlumpberger, M. Straub, M. Veenhuis, and D.H. Wolf. 1994. Isolation of autophagocytosis mutants of *Saccharomyces cerevisiae. FEBS Lett.* **349:** 275.

Titorenko, V.I., I. Keizer, W. Harder, and M. Veenhuis. 1995. Isolation and characterization of mutants impaired in the selective degradation of peroxisomes in the yeast

Hansenula polymorpha. *J. Bacteriol.* **177:** 357.

Tolbert, N.E. 1981. Metabolic pathways in peroxisomes and glyoxysomes. *Annu. Rev. Biochem.* **50:** 133.

Tsukada, M. and Y. Ohsumi. 1993. Isolation and characterization of autophagy-defective mutants of *Saccharomyces cerevisiae*. *FEBS Lett.* **333:** 169.

Tuttle, D.L. and W.A. Dunn, Jr. 1995. Divergent modes of autophagy in the methylotrophic yeast *Pichia pastoris*. *J. Cell Sci.* **108:** 25.

Tuttle, D.L., A.S. Lewin, and W.A. Dunn, Jr. 1993. Selective autophagy of peroxisomes in methylotrophic yeasts. *Eur. J. Cell Biol.* **60:** 283.

van den Bosch, H., R.B.H. Schutgens, R.J.A. Wanders, and J.M. Tager. 1992. Biochemistry of peroxisomes. *Annu. Rev. Biochem.* **61:** 157.

van der Klei, I.J., J. Rytka, W.-H. Kunau, and M. Veenhuis. 1990. Growth of catalase A and catalase T deficient mutant strains of *Saccharomyces cerevisiae* on ethanol and oleic acid. Growth profiles and catalase activities in relation to microbody proliferation. *Arch. Microbiol.* **153:** 513.

van der Klei, I.J., R.E. Hilbrands, G.J. Swaving, H.R. Waterham, E.G. Vrieling, V.I. Titorenko, J.M. Cregg, W. Harder, and M. Veenhuis. 1995. The *Hansenula polymorpha PER3* gene is essential for the import of PTS1 proteins into the peroxisomal matrix. *J. Biol. Chem.* **270:** 17229-17236.

van der Leij, I., M.M. Franse, Y. Elgersma, B. Distel, and H.F. Tabak. 1993. PAS10 is a tetratricopeptide-repeat protein that is essential for the import of most matrix proteins into peroxisomes of *Saccharomyces cerevisiae*. *Proc. Natl. Acad. Sci.* **90:** 11782.

van der Leij, I., M. Van Den Berg, R. Boot, M. Franse, B. Distel, and H.F. Tabak. 1992. Isolation of peroxisome assembly mutants from *Saccharomyces cerevisiae* with different morphologies using a novel positive selection procedure. *J. Cell Biol.* **119:** 153.

van Roermund, C.W.T., Y. Elgersma, N. Singh, R.J.A. Wanders, and H.F. Tabak. 1995. The membrane of peroxisomes in *Saccharomyces cerevisiae* is impermeable to NAD(H) and acetyl-CoA under *in vivo* conditions. *EMBO J.* **14:** 3480.

Van Veldhoven, P.P., W.W. Just, and G.P. Mannaerts. 1987. Permeability of the peroxisomal membrane to cofactors of β-oxidation. Evidence for the presence of a pore-forming protein. *J. Biol. Chem.* **262:** 4310.

Veenhuis, M. 1992. Peroxisome biogenesis and function in *Hansenula polymorpha*. *Cell Biochem. Funct.* **10:** 175.

Veenhuis, M. and W. Harder. 1991. Microbodies. In *The yeasts* (ed. A.H. Rose and S.J. Harrison), vol. 4, p. 601. Academic Press, London.

Veenhuis, M., A. Douma, W. Harder, and M. Osumi. 1983. Degradation and turnover of peroxisomes in the yeast *Hansenula polymorpha* induced by selective inactivation of peroxisomal enzymes. *Arch. Microbiol.* **134:** 193.

Veenhuis, M., M. Mateblowski, W.-H. Kunau, and W. Harder. 1987. Proliferation of microbodies in *Saccharomyces cerevisiae*. *Yeast* **3:** 77.

Voorn-Brouwer, T., I. van der Leij, W. Hemrika, B. Distel, and H.F. Tabak. 1993. Sequence of the *PAS8* gene, the product of which is essential for biogenesis of peroxisomes in *Saccharomyces cerevisiae*. *Biochim. Biophys. Acta* **1216:** 325.

Walton, P.A., P.E. Hill, and S. Subramani. 1995. Import of stably folded proteins into peroxisomes. *Mol. Biol. Cell* **6:** 675.

Wanders, R.J.A., M. Kos, B. Roest, A.J. Meijer, G. Schrakamp, H.S.A. Heymans, W.H.H. Tegelaers, H. van den Bosch, R.B.H. Schutgens, and J.M. Tager. 1984. Activity of peroxisomal enzymes and intracellular distribution of catalase in Zellweger syndrome. *Biochem. Biophys. Res. Commun.* **123:** 1054.

Wang, T.W., A.S. Lewin, and G.M. Small. 1992. A negative regulating element controlling transcription of the gene encoding acyl-CoA oxidase in *Saccharomyces cerevisiae. Nucleic Acids Res.* **20:** 3495.

Wang, T., Y. Luo, and G.M. Small. 1994. The *POX1* gene encoding peroxisomal acyl-CoA oxidase in *Saccharomyces cerevisiae* is under the control of multiple regulatory elements. *J. Biol. Chem.* **269:** 24480.

Wanner, G. and R.R. Theimer. 1982. Two types of microbodies in *Neurospora crassa. Ann. N.Y. Acad. Sci.* **386:** 269.

Waterham, H.R., I. Keizer-Gunnink, J.M. Goodman, W. Harder, and M. Veenhuis. 1990. Immunocytochemical evidence for the acidic nature of peroxisomes in methylotrophic yeasts. *FEBS Lett.* **262:** 17.

Wendland, M. and S. Subramani. 1993a. Cytosol-dependent peroxisomal protein import in a permeabilized cell system. *J. Cell Biol.* **120:** 675.

———. 1993b. Presence of cytoplasmic factors functional in peroxisomal protein import implicates organelle-associated defects in several human peroxisomal disorders. *J. Clin. Invest.* **92:** 2462.

Wiebel, F.F. and W.-H. Kunau. 1992. The Pas2 protein essential for peroxisome biogenesis is related to ubiquitin-conjugating enzymes. *Nature* **359:** 73.

Wiemer, E.A.C., W.M. Nuttley, B.L. Bertolaet, X. Li, U. Francke, M.J. Wheelock, U.K. Anné, K.R. Johnson, and S. Subramani. 1995. Human peroxisomal targeting signal-1 receptor restores peroxisomal protein import in cells from patients with fatal peroxisomal disorders. *J. Cell Biol.* **130:** 51.

Zhang, J.W. and P.B. Lazarow. 1995. *PEB1* (*PAS7*) in *Saccharomyces cerevisiae* encodes a hydrophilic, intraperoxisomal protein that is a member of the WD repeat family and is essential for the import of thiolase into peroxisomes. *J. Cell Biol.* **129:** 65.

———. 1996. Peb1p (Pas7p) is an intraperoxisomal receptor for the NH_2-terminal, type 2, peroxisomal targeting sequence of thiolase: Peb1p itself is targeted to peroxisomes by an NH_2-terminal peptide. *J. Cell Biol.* **132:** 325.

Zhang, J.W., Y. Han, and P.B. Lazarow. 1993a. Novel peroxisome clustering mutants and peroxisome biogenesis mutants of *Saccharomyces cerevisiae. J. Cell Biol.* **123:** 1133.

Zhang, J.W., C. Luckey, and P.B. Lazarow. 1993b. Three peroxisome protein packaging pathways suggested by selective permeabilization of yeast mutants defective in peroxisome biogenesis. *Mol. Biol. Cell 4:* 1351.

Zimniak, P., E. Hartter, W. Woloszczuk, and H. Ruis. 1976. Catalase biosynthesis in yeast: Formation of catalase A and catalase T during oxygen adaptation of *Saccharomyces cerevisiae. Eur. J. Biochem.* **71:** 393.

Zinser, E. and G. Daum. 1995. Isolation and biochemical characterization of organelles from the yeast, *Saccharomyces cerevisiae. Yeast* **11:** 493.

Zinser, E., C.D.M. Sperka-Gottlieb, E.-V. Fasch, S.D. Kohlwein, F. Paltauf, and G. Daum. 1991. Phospholipid synthesis and lipid composition of subcellular membranes in the unicellular eukaryote *Saccharomyces cerevisiae. J. Bacteriol.* **173:** 2026.

7

Cell Cycle Control in *Saccharomyces cerevisiae*

Daniel J. Lew
Department of Pharmacology and Cancer Biology
Duke University Medical Center
Durham, North Carolina 27710

Ted Weinert
Department of Molecular and Cellular Biology
University of Arizona
Tucson, Arizona 85721

John R. Pringle
Department of Biology
University of North Carolina
Chapel Hill, North Carolina 27599-3280

I. INTRODUCTION

The cell cycle comprises the set of processes that result in duplication of all cell constituents, segregation of the duplicated constituents to different parts of the cell, and division into two daughter cells. To ensure that each division produces two viable progeny, these disparate processes must be coordinated both temporally and spatially. Our understanding of the molecular basis for this coordination has grown dramatically since this topic was reviewed in the previous edition of this monograph (Pringle and Hartwell 1981), leading to a unified theory of cell cycle control that, in its essential elements, is believed to apply to all eukaryotic cells. In this chapter, we do not attempt to trace the historical development of this theory, which involved crucial insights from several experimental systems and many investigators. Instead, we attempt to provide a coherent picture of what is known about the temporal control of the *Saccharomyces cerevisiae* cell cycle in the context of the unified theory. In so doing, we have attempted to cite the papers most pertinent to our present understanding, papers of particular historical importance, and papers that provide key points of entry into the wider literature. The vastness of this wider literature precluded any attempt to cite all relevant papers, and we have doubtless inadvertently failed to cite some papers whose historical importance or current utility is equal to or greater than that of the papers that we have cited. For this, we apologize to the investigators concerned.

II. CELL CYCLE PHASES, TRANSITIONS, AND EVENTS

In many cells, the cell cycle can be rather neatly subdivided into a longer period (interphase) during which cell contents are duplicated and a shorter period (mitosis or M phase) during which the contents are segregated (Alberts et al. 1994). The transition from interphase to mitosis is dramatic, involving concerted large-scale changes in many cell structures: Chromosome condensation, nuclear envelope breakdown, formation of a microtubule spindle, reorganization of cortical actin and associated changes in cell shape, and vesiculation of membrane compartments all take place within a short time interval. A similarly dramatic reversal of these changes takes place at the end of mitosis, concomitant with cytokinesis, as the cells return to interphase. Interphase can be further subdivided into S phase, during which DNA replication takes place, and the "Gap" phases G_1 and G_2 that precede and follow S phase. This division of the cell cycle into four major phases (G_1, S, G_2, and M, with cytokinesis viewed either as marking the end of M phase or as occurring during G_1) has been adopted universally as a useful if somewhat oversimplified description of the eukaryotic cell cycle (see, e.g., Alberts et al. 1994). Both proliferating cells and cells that have ceased proliferating can generally be described unambiguously as being in a particular phase: Most nonproliferating cells are arrested in G_1, whereas some specialized cells (notably oocytes and eggs) are arrested in either G_2 or M phase. These observations highlight the possible importance of transitions between the major phases as key targets for regulation by developmental signals.

The budding mode of growth employed by *S. cerevisiae* is unusual in that the segregation of many cell constituents into the bud is concurrent with their duplication, so that there is no clean distinction between duplication and segregation phases. Nevertheless, it is possible to define G_1, S, G_2, and M phases for *S. cerevisiae*, based on the discrete intervals occupied by DNA replication and the events associated with chromosome segregation (see below). Although such a partitioning of the cell cycle provides a convenient way to refer to its several stages, it is not at all clear whether the transitions between the major phases are necessarily the most important regulatory points in the yeast (or, indeed, any) cell cycle.

Early work defined a key regulatory transition (termed "Start") in late

G_1 of the *S. cerevisiae* cell cycle (Pringle and Hartwell 1981). Cells in early G_1 (pre-Start) are competent to embark upon a number of developmental options depending on the external environment: Well-fed cells proceed with the cell cycle, nutrient-limited cells can enter a quiescent state or begin sporulation, and cells exposed to mating pheromones (and expressing the corresponding receptor) arrest in G_1 and initiate the mating program. In contrast, cells in late G_1 or subsequent phases (post-Start) typically proceed to complete the cell cycle even if deprived of nutrients or exposed to pheromone. These observations suggested the existence of a decision point (or period) in late G_1 (i.e., Start) at which cells integrate inputs from their environment and become committed to a particular developmental program (Pringle and Hartwell 1981). Studies in other organisms are consistent with the existence of a similar transition in late G_1 that governs cell cycle commitment and precedes the G_1/S boundary (for review, see Reed 1992). Progression past Start in *S. cerevisiae* triggers multiple cell cycle events, including the nearly simultaneous initiation of DNA replication, bud emergence, and duplication of the microtubule-organizing center. Thus, a "G_1/S transition" is conspicuous and easily scored in this organism, although it should be stressed that the most prominent regulatory transition (Start) appears to precede somewhat the initiation of DNA replication per se (see also Sections V.C.7 and VII.B and recent discussions by Cross [1995], Stuart and Wittenberg [1995], and Dirick et al. [1995]).

More confusing and controversial has been definition of the G_2/M transition in *S. cerevisiae*. Anaphase (spindle elongation and chromosome segregation) clearly occurs well after the end of S phase, but many hallmarks of mitosis are absent or difficult to monitor in yeast. Nuclear envelope breakdown does not occur; chromosome condensation occurs, but the small size of the yeast chromosomes makes this hard to detect; and most membrane compartments are transported to the bud throughout the budded phase, so that a mitosis-associated vesiculation of membrane compartments is not necessary for their segregation. As a result, the assembly of a bipolar microtubule spindle has been proposed as the sole available cytological marker for the entry into M phase (see, e.g., Forsburg and Nurse 1991). This view and the fact that spindle assembly typically occurs well before the beginning of anaphase (Byers and Goetsch 1975; Kilmartin and Adams 1984) have led to the common misconception (see, e.g., Russell and Nurse 1986; Forsburg and Nurse 1991; Alberts et al. 1994) that the S and M phases are overlapping (with no G_2 phase) in *S. cerevisiae* and thus that the organization of the cell cycle is fundamentally different in this organism from that seen in other eukaryotic cells. In fact, although measurements of the duration of S phase and of its timing relative to bud emergence and bud growth have given somewhat different results (Williamson 1965; Slater et al. 1977; Rivin and Fangman 1980; Surana et al. 1991; Paulovich and Hartwell 1995), presumably depending on the strains, growth conditions, and methods used, it seems clear that DNA replication is normally completed significantly before the stage (bud diameter ~0.4 x mother cell diameter: Byers and Goetsch 1975; Kilmartin and Adams 1984) at which the assembly of the short bipolar spindle occurs. In addition, several studies in which both DNA replication and spindle formation were monitored directly also indicate that the latter event normally follows the completion of the former (Surana et al. 1991; Lim et al. 1996a; Sundberg et al. 1996). Thus, the *S. cerevisiae* cell cycle does appear to have a distinct G_2 phase (and hence both S/G_2 and G_2/M transitions) even if spindle assembly is accepted as marking the beginning of M phase. However, it is also far from clear whether spindle assembly should really be regarded as marking the beginning of mitosis in *S. cerevisiae*. On the basis of the way the cell cycle is regulated (see below), some investigators have used anaphase onset to define the beginning of mitosis (see, e.g., Weinert and Hartwell 1988; Reed and Wittenberg 1990). Other investigators have preferred simply to avoid the uncertainty in defining G_2 versus early M phase in *S. cerevisiae* by referring to the interval between the end of S phase and the beginning of anaphase as the "G_2/M phase." In any case, however mitosis is defined, it is now quite clear that the fundamental principles of cell cycle control in *S. cerevisiae* do not differ markedly from those in other eukaryotes.

As more is learned about cell cycle control, the concepts of cell cycle phases and transitions have become somewhat blurred. The initial discoveries that led to current models of cell cycle control appeared to support the idea that the role of the conserved cell cycle regulators (CDKs and cyclins: see Sections IV and V) was to trigger the major cell cycle phase transitions (G_1/S and G_2/M). However, subsequent more detailed analysis has suggested instead that these cell cycle regulators may be continuously active (in different configurations) through most of the cell cycle, to initiate and sustain a wide variety of processes. For this reason, the discussion that follows treats the cell cycle primarily in terms of cell cycle "events," rather than phases or transitions. This terminology allows greater precision by focusing attention on the manifold particular processes controlled by the cell cycle regulators. At the end of the chapter, we return to a consideration of whether cell cycle phases and transitions represent meaningful units of the cell cycle.

III. LANDMARK EVENTS IN THE CELL CYCLE

Cell cycle events are discrete, stage-specific processes that occur during cell proliferation but not in nonproliferating cells; examples include DNA replication, duplication of the microtubule-organizing center, spindle assembly, and chromosome segregation. Following an earlier and still serviceable practice (Pringle and Hartwell 1981), we designate as "landmark events" those cell cycle events that can be readily observed either cytologically or biochemically. The rapidly increasing knowledge of the details of cell cycle progression has led to this term's having two somewhat different applications. In the broader application, the landmark events include all those aspects of cell cycle progression that can be monitored by an available assay, providing information about the position of cells within the cell cycle. In particular, a large number of transcripts and proteins are now known whose abundance, localization, or activity state varies reproducibly during the cell cycle. This provides an abundance of assays for monitoring cell cycle progression and the effects

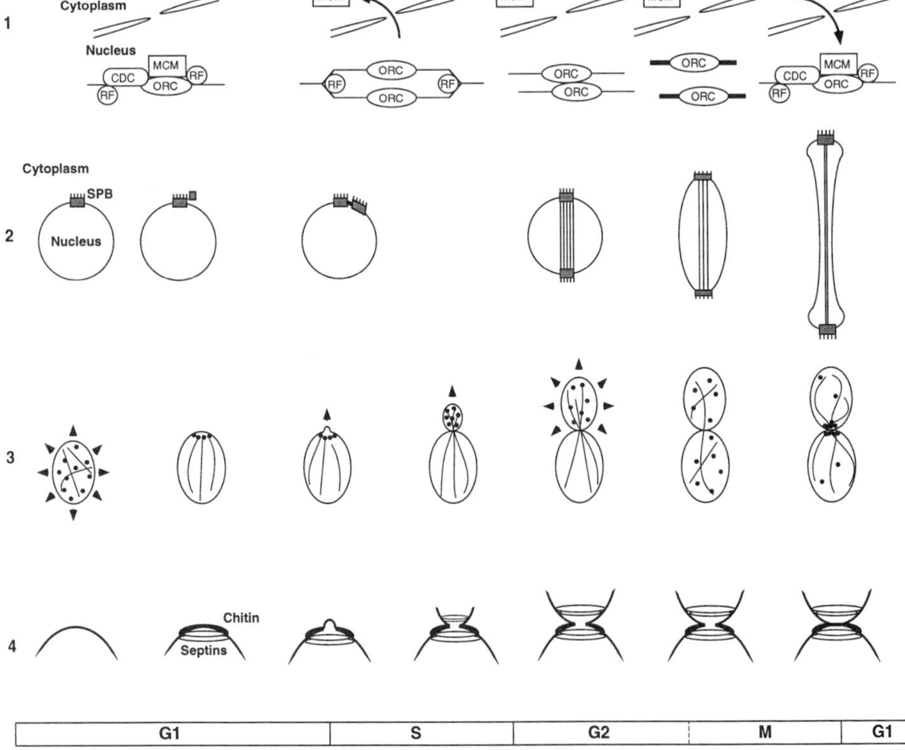

Figure 1 (See facing page for legend.)

of cell cycle perturbations that are useful in various specific applications, but we will not attempt a comprehensive listing here. In the narrower application, the landmark events include the major processes (summarized below and in Fig. 1) that actually duplicate and segregate cell constituents and produce daughter cells. These major processes involve the cell-cycle-regulated behavior of many proteins simultaneously and must be coordinated with each other during cell cycle progression.

Details of the major landmark events can be found elsewhere; what follows here is a set of brief sketches, referring to Figure 1, that are intended primarily to provide background for the discussion of cell cycle control that follows. Each row in the figure illustrates a set of events that always occurs in the order shown, although events in different rows can vary in relative timing between strains or as a result of perturbations. In general, the molecular machinery responsible for executing events in different rows is distinct.

Figure 1 Major landmark events of the *S. cerevisiae* cell cycle. See text for details. (*1*) DNA replication and the chromosome cycle. ORC represents the Origin Replication Complex (a complex of six proteins); CDC represents both Cdc6p and Cdc7p/Dbf4p; MCM represents the complex of related MCM proteins; RF represents replication factors, such as DNA polymerases and single-strand DNA-binding proteins, which mediate elongation of replication forks during S phase. The depiction of RFs at replication origins in G_1 is based on findings in metazoan nuclei; there are no relevant data yet for *S. cerevisiae.* It is unclear whether the exclusion of MCM proteins from the nucleus after G_1 reflects directed export of these proteins from the nucleus or degradation within the nucleus and accumulation of newly synthesized proteins in the cytoplasm. Thickened lines represent the chromosome condensation that occurs during mitosis. (*2*) The spindle pole body (SPB) and microtubule cycle. The nuclear envelope, SPB, SPB satellite (in G_1), and cytoplasmic and nuclear microtubules are depicted. The cytoplasmic microtubules are in fact longer and undergo more complicated behavior than depicted here. (3) Rearrangements of the actin cytoskeleton and secretory machinery. (*Closed circles*) Cortical actin patches; (*lines*) actin cables; (*arrowheads*) pattern of secretion and cell surface growth. The patterns of secretion at the stages (both very brief) indicated by the second and sixth cells are unknown. In the seventh cell, undergoing cytokinesis, secretion is directed toward the neck. (*4*) Septin assembly and septum formation. The plasma membrane is depicted along with the septin filaments (*thin lines*) and the chitin (*heavy lines*) that is deposited in the overlying cell wall and in the primary septum at the time of division. Various other proteins also accumulate at the neck in a septin-dependent manner.

A. DNA Replication and the Chromosome Cycle

S. cerevisiae cells have 16 chromosomes (per haploid set) that are replicated in a discrete interval of the cell cycle. DNA replication is initiated at well-defined origins of replication (for review, see Newlon and Theis 1993; Diffley 1995). In vivo footprints at replication origins suggest that a "preinitiation complex" of proteins is bound to the origins in G_1 and is converted to a smaller "postreplication complex" in S phase, which then persists until the next G_1 (Diffley et al. 1994; Santocanale and Diffley 1996; for review, see Diffley 1996; Stillman 1996). A complex of six proteins termed the "origin recognition complex" (ORC) binds specifically to origins of replication and is required for the initiation of replication (Bell and Stillman 1992; Bell et al. 1993, 1995; Foss et al. 1993; Loo et al. 1995; Rao and Stillman 1995; Rowley et al. 1995; Hardy 1996; Hori et al. 1996). ORC generates an in vitro footprint similar to the postreplication in vivo footprint (Diffley et al. 1994; Santocanale and Diffley 1996). These data suggest that ORC is bound to origins throughout the cell cycle and that additional proteins bind the origin-ORC complex in G_1 to generate the preinitiation complex (Fig. 1, line 1).

Several proteins have been implicated in the formation and/or regulation of preinitiation complexes. Cdc6p, an unstable protein related to the largest ORC subunit, is synthesized in late M/G_1 and is required for origin firing (Bueno and Russell 1992; Zwerschke et al. 1994; Liang et al. 1995; Piatti et al. 1995). Cdc6p is required for the establishment and maintenance of the larger preinitiation footprint (Cocker et al. 1996; Santocanale and Diffley 1996; Detweiler and Li 1997). Several related "MCM proteins," including Cdc46p, Cdc47p, Cdc54p, Mcm2p, and Mcm3p, are also required for origin firing. These proteins form a complex with each other and exhibit cell-cycle-regulated nuclear localization suggestive of a role in establishing the preinitiation complex (Hennessy et al. 1990; Yan et al. 1993; Tye 1994; Dalton and Whitbread 1995; Lei et al. 1996): In G_1, the MCM proteins are nuclear, but once replication begins, they are excluded from the nucleus and accumulate in the cytoplasm until the end of mitosis, when they reenter the nucleus (Fig. 1, line 1). Cdc45p appears to interact both with MCM proteins and with at least one ORC subunit, suggesting that it may be involved in the assembly of the preinitiation complex (Hopwood and Dalton 1996; Zou et al. 1997). Cdc7p and Dbf4p are the catalytic and regulatory subunits, respectively, of a protein kinase required for initiation of replication (Sclafani and Jackson 1994). Dbf4p binds to origins in vivo, suggesting that the Dbf4p/Cdc7p complex may regulate the activity of the preinitiation complex (Dowell et al. 1994), and genetic data suggest that the

MCM protein complex may be one target of Cdc7p/Dbf4p (Hardy et al. 1997). Cdc5p, a protein kinase implicated also in control of the end of mitosis (see Section V.C.6), may also be recruited to origin complexes by Dbf4p and play a part in their function (Hardy and Pautz 1996).

Many of the genes encoding proteins involved in DNA replication are transcribed in a cell-cycle-dependent manner (for review, see Koch and Nasmyth 1994). However, deregulated transcription of these genes does not appear to be deleterious (Falconi et al. 1993; Koch et al. 1993; Piatti et al. 1995; Detweiler and Li 1997), and the role of the cell-cycle-regulated transcription remains unclear.

Following DNA replication, the chromosomes appear to undergo a modest degree of condensation (see Wente et al., this volume). Using a fluorescence in situ hybridization (FISH) assay, Guacci et al. (1994) estimated that cells arrested in G_2/M by treatment with nocodazole condense their chromosomes about twofold. (For comparison, this assay estimates the condensation of mammalian chromosomes in mitosis to be about fivefold that of interphase chromosomes.) Upon return to G_1, the chromosomes are once more decondensed. A caveat is that so far, condensation has not been measured in cycling cells, but only in arrested cells; the true degree and timing of condensation in the normal cell cycle therefore remain uncertain.

A final chromosomal landmark event is the maturation of centromere/kinetochore complexes and their binding to microtubules (for review, see Hyman and Sorger 1995). In vivo footprinting of centromeres suggests that a centromere-binding complex is bound throughout the cell cycle (Bloom et al. 1984; Wilmen and Hegemann 1996), but in vitro assays for centromere-microtubule interactions show large differences between cells arrested in G_1 with mating pheromone and cells arrested in G_2/M with nocodazole (Kingsbury and Koshland 1991). Although this confirms the expectation that kinetochore function is cell-cycle-regulated, these assays have not yet been applied to synchronized cycling cells, and the number, timing, and extent of kinetochore alterations during the cell cycle remain unknown.

B. The Spindle Pole Body and Microtubule Cycles

The principal (or exclusive) microtubule-organizing center in the *S. cerevisiae* cell is the spindle pole body (SPB), which is embedded in the nuclear envelope and nucleates growth of both cytoplasmic and nuclear microtubules (for review, see Byers 1981; Winey and Byers 1993; Snyder 1994; Botstein et al., this volume). In G_1, a "satellite" structure, thought to represent a first step in SPB duplication, is generated (Fig. 1, line 2). Subsequently, SPB duplication occurs in a conservative process in which the new SPB is constructed adjacent to the old SPB in the nuclear envelope (Vallen et al. 1992; Donaldson and Kilmartin 1996); in most strains, duplication occurs at about the time of bud emergence. At about the same time, the SPB moves (either by migration within the nuclear envelope or by rotation of the entire nucleus) in a process that depends on the cytoplasmic microtubules (Jacobs et al. 1988), so that it is oriented toward the incipient bud site. The orientation of the SPB toward the bud site appears normally to precede the appearance of a detectable bud (Byers and Goetsch 1975; Snyder et al. 1991), although recent observations on living cells have cast some doubt on this conclusion (K. Bloom, pers. comm.). In addition, the precise timing of the orientation process relative to SPB duplication is not known. The duplicated SPBs stay side by side for a considerable period (typically until after DNA replication is completed and the bud is about 0.4 times the diameter of the mother cell; see Section II), and then one SPB migrates away to the other side of the nucleus. This SPB separation requires microtubules (Jacobs et al. 1988) and results in the generation of a short intranuclear spindle connecting the two SPBs. A variety of microtubule motors have been identified that are important for SPB separation: The kinesin-related Cin8p and Kip1p motors apparently promote sliding of nuclear microtubules to push SPBs apart (Hoyt et al. 1992; Roof et al. 1992), whereas dynein acts on cytoplasmic microtubules to pull the SPBs apart (Eshel et al. 1993; Li et al. 1993). Kar3p, another kinesin-related motor, antagonizes these forces (Saunders and Hoyt 1992; Hoyt et al. 1993).

Following SPB separation, there ensues a considerable period during which the short spindle remains relatively constant in length while the bud continues to grow (Byers and Goetsch 1975; Kilmartin and Adams 1984). During this period, the nucleus and spindle migrate to (and into) the mother-bud neck and orient along the mother-bud axis in a process that requires cytoplasmic microtubules, dynein, and the dynactin complex (Huffaker et al. 1988; Jacobs et al. 1988; Palmer et al. 1992; Sullivan and Huffaker 1992; Eshel et al. 1993; Li et al. 1993; Clark and Meyer 1994; McMillan and Tatchell 1994; Muhua et al. 1994). The spindle then elongates (marking anaphase); an initial period of rapid elongation is followed by a period of slower elongation during which the nucleus acquires an "hourglass" morphology (Yeh et al. 1995). Kinesin-related proteins and dynein collaborate in spindle elongation (Saunders et al. 1995), and tubulin polymerization must occur on interpolar microtubules to maintain spindle structure. Detailed structural analyses of the

S. cerevisiae spindle suggest that each kinetochore binds a single microtubule and that the number of interpolar microtubules decreases during spindle elongation (Winey et al. 1995). These analyses also suggest that the chromosomes do not line up at a metaphase plate before segregation. Finally, the spindle breaks down either shortly before or coincident with cytokinesis (Kilmartin and Adams 1984; Copeland and Snyder 1993).

C. Rearrangements of the Actin Cytoskeleton and Secretory Machinery

Formation of a bud requires polarization of the actin cytoskeleton and of the secretory system, localized secretion of cell wall components, and local remodeling of the cell wall (Harold 1990; Drubin 1991; Preuss et al. 1992; TerBush and Novick 1995; Drubin and Nelson 1996). The distribution of F-actin (and associated proteins) and the pattern of surface growth undergo several reorganizations that accompany and are apparently responsible for the formation of a bud of normal shape (Adams and Pringle 1984; Kilmartin and Adams 1984; Novick and Botstein 1985; Lew and Reed 1993, 1995b; Welch et al. 1994). F-actin is found in two types of structures within the cell: Cortical patches are small discs rich in F-actin that lie beneath the plasma membrane, and cables are bundles of long actin filaments (Mulholland et al. 1994; Welch et al. 1994). The distribution of these structures during the cell cycle is highly dynamic (Doyle and Botstein 1996; Waddle et al. 1996) and exhibits a characteristic series of patterns as the cell cycle proceeds (Fig. 1, line 3), as follows.

1. In late G_1 (following Start), cortical actin patches congregate in a ring at the "pre-bud" site, where the bud will emerge, and cables are oriented toward that site. Prior to this, many cells display a seemingly random (unpolarized) distribution of cables and patches, as shown in Figure 1. However, under conditions allowing rapid growth, early G_1 cells may retain a residual "postcytokinesis" actin configuration (see below) until formation of the new pre-bud site in late G_1. In addition to the cortical actin patches, a large number of other proteins involved in morphogenesis and polarized secretion accumulate at the pre-bud site, many of them in an actin-dependent manner (Ayscough et al. 1997).
2. About 15 minutes after actin polarization, a visible bud is formed. At this time, secretory vesicles cluster just beneath the bud, and most secretion and cell wall expansion are directed toward the tip of the bud. As the bud enlarges, the cortical actin patches become concentrated in the bud.

3. In small and medium-sized buds, most cell wall expansion occurs at the bud tip ("apical" growth), whereas in larger buds, cell wall expansion occurs all over the bud ("isotropic" growth). This implies that there is a change in the pattern of bud growth, which has been called the "apical/isotropic switch." The switch is quite abrupt and occurs after DNA replication but before chromosome segregation.
4. In cells with large buds, the mother/bud asymmetry in actin distribution breaks down, and actin structures are randomly distributed over the entire cell. In the unperturbed cell cycle, this stage is very brief and occurs after chromosome segregation. The pattern of secretion during this time is unknown.
5. At the end of the cell cycle, cortical actin patches become concentrated on both sides of the neck, cables orient toward the neck, and secretory vesicles cluster around the neck. At this time, secretion and cell wall synthesis occur primarily at the neck to promote cytokinesis and septation. Following septation, the cortical patches remain clustered for some time at the division site (the "postcytokinesis" configuration) and then disperse randomly or redistribute to the next pre-bud site.

D. Septin Assembly, Cell Wall Synthesis, and Septum Formation

In addition to the polarization of cell surface growth resulting from the polarization of the actin cytoskeleton and secretory system, shaping of the bud involves formation of a neck or constriction at the base of the bud and eventually of a septum to separate the mature bud from its mother. These processes require a group of related proteins called septins (Cdc3p, Cdc10p, Cdc11p, and Cdc12p) that are constituents of a set of approximately 10-nm-diameter filaments that encircle the neck just inside the plasma membrane (Fig. 1, line 4; for review, see Longtine et al. 1996). The septins assemble in a ring at the pre-bud site in late G_1 at about the same time as do the cortical actin patches. However, the assembly of the septins (and associated proteins) and the assembly of actin (and associated proteins) appear to be mutually independent (Adams and Pringle 1984; Ford and Pringle 1991; Ayscough et al. 1997; H.B. Kim et al., unpubl.). As the bud emerges, more septin filaments assemble so that the band of filaments is approximately symmetric on the mother and bud sides of the neck. The septin filaments remain in place at least until cytokinesis, and a ring of septin protein (which may or may not still be assembled into filaments) is present at the division site on both mother and daughter cells. These septin rings then disassemble during G_1.

Following assembly of the septins at the pre-bud site, a ring of chitin is deposited in the overlying cell wall; the chitin is deposited asymmetrically on the mother side of the neck and eventually becomes the bud scar that marks the division site on the mother cell (Cabib et al. 1982; Pringle et al. 1989; Cid et al. 1995; Orlean, this volume). Formation of the chitin ring requires the chitin synthase III catalytic subunit encoded by *CHS3* and several accessory proteins (Shaw et al. 1991; Bulawa 1993; Orlean, this volume). During bud growth, chitin synthesis appears to be largely or entirely shut down until the time of cytokinesis, when a chitin-containing primary septum is laid down in the neck between mother and bud. Synthesis of the primary septum is performed by a distinct chitin synthase encoded by *CHS2* (Shaw et al. 1991; Bulawa 1993; Orlean, this volume). A secondary, nonchitinous septum is then formed on either side of the primary septum, and the primary septum is degraded by chitinase (Cts1p) to allow cell separation (Kuranda and Robbins 1991). The secondary septum marking the division site on the daughter cell can be distinguished from the rest of the cell wall as a birth scar that is detectable by electron microscopy (Barton 1950; Talens et al. 1977; Chant and Pringle 1995) or by differential staining (Lew and Reed 1993; Chant and Pringle 1995), suggesting that wall composition or structure may differ in this region, although the birth scar contains little or no chitin (Pringle et al. 1989). The preceding description implies that chitin synthesis and degradation are cell-cycle-regulated, and some information on the basis for this regulation has emerged. Cts1p and Chs2p appear to be synthesized in a cell-cycle-specific manner and (at least in the case of Chs2p) degraded rapidly (Dohrmann et al. 1992; Pammer et al. 1992; Choi et al. 1994; Chuang and Schekman 1996). In contrast, although transcriptional regulation of *CHS3* (*DIT101*) has also been reported (Pammer et al. 1992), more recent evidence suggests that Chs3p activity is probably regulated posttranslationally (Choi et al. 1994; Chuang and Schekman 1996; Ziman et al. 1996). For both Chs2p and Chs3p, regulation of activity also appears to be coupled to regulation of the proteins' spatial localization (Chuang and Schekman 1996; Ziman et al. 1996; Santos and Snyder 1997).

There are also cell-cycle-dependent changes in the processes that contribute to general cell wall synthesis. Cyclic variations have been reported in exoglucanase activity, in the fraction of glucan that is alkali-insoluble (thought to reflect covalent attachment to chitin), and in cell wall permeability (thought to reflect the function of cell wall mannoproteins) (Nombela et al. 1988; De Nobel et al. 1991; Klis 1994; Orlean, this volume). It is not clear whether these changes reflect direct cell cycle regulation of the synthesis or cross-linking of major cell wall components or whether they reflect instead the changes in polarity of wall growth as described above. Several genes encoding proteins involved in cell wall synthesis are expressed in a cell-cycle-regulated manner (Igual et al. 1996). Deregulation of their transcription during the cell cycle results in sensitivity to cell-wall-perturbing agents, suggesting that the cell-cycle-regulated synthesis of these components may be important (at least in aggregate) for proper cell wall construction (Igual et al. 1996).

E. Organelle Segregation

In addition to the nucleus (whose microtubule-dependent orientation and division are discussed above), numerous other organelles must also be partitioned between mother and bud so that both progeny receive an adequate complement of organelles following cell division. In contrast to the vesiculation of many membrane compartments that occurs during mitosis in mammalian cells, most cytoplasmic organelles appear to remain intact throughout the cell cycle in yeast (for review, see Warren and Wickner 1996). The available data suggest that the endoplasmic reticulum, Golgi, mitochondria, and vacuoles all begin to populate the bud well before anaphase and that their segregation into the bud does not require microtubules (Gomes de Mesquita et al. 1991; Warren and Wickner 1996). The segregation of mitochondria and vacuoles may involve the actin cytoskeleton (Drubin et al. 1993; Simon et al. 1995; Hill et al. 1996), but mutants with specific defects in both mitochondrial and vacuolar segregation have also been isolated (Wang et al. 1996; Warren and Wickner 1996), suggesting that these organelles segregate by active processes that require a machinery(ies) distinct from that required to build a bud or carry out the nuclear cycle. In the case of vacuoles, morphologically distinct "inheritance structures" form at a specific stage of the cell cycle (when small buds are present), suggesting that the partitioning process is under cell cycle control.

F. Other Cell Cycle Events

Clearly, the list of major landmark events provided above is missing important events that must occur but cannot be readily observed with current techniques (e.g., centromere attachment to the spindle and sister chromatid disjunction). In addition, we expect that many other events will be revealed by future studies; for example, changes in nuclear envelope structure may precede or accompany nuclear division, changes in the secretion of cell-wall-modifying enzymes may accompany switches in growth polarity, and the continuing application of fluorescence techniques for protein localization is likely to reveal the existence of other protein complexes that assemble and disassemble at particular places in the cell at particular times in the cell cycle. Even the partial listing of cell cycle events provided above indicates that ordered series of changes must occur in numerous processes that are not mechanistically connected to each other (e.g., DNA replication and chitin synthesis). How these processes are coordinated with each other is the subject of the balance of this chapter.

IV. A UNIFIED THEORY OF CELL CYCLE CONTROL

A central problem in the field of cell cycle control is that of coordinating cell cycle events so that they occur in the proper order with respect to each other. For instance, chromosome segregation must follow DNA replication, because attempts to segregate unreplicated chromosomes lead to aneuploidy and attempts to segregate partially replicated chromosomes lead to chromosome breakage. Cytokinesis must, in turn, follow chromosome segregation, since premature cytokinesis leads to generation of aploid and polyploid daughter cells. In addition, each major event must occur once and only once in each cycle. Because there are many events and each major event is carried out by a distinct and awesomely complex molecular apparatus, it is by no means a trivial task to coordinate all cell cycle events so that each cell cycle successfully gives rise to two viable progeny.

The prevailing theory of cell cycle control invokes two central coordinating mechanisms (Fig. 2). The first is a biochemical oscillator, or "cell cycle clock." This clock is based on a set of serine/threonine protein kinases called CDKs (*c*yclin-*d*ependent *k*inases) that have been highly conserved in all eukaryotes studied to date. The activity of the CDKs varies cyclically as a result of covalent modifications and associations with a variety of positive and negative regulatory proteins, as discussed in detail below. It is presumed that phosphorylation of distinct sets of substrate proteins by the sequential activation of distinct CDK complexes triggers the events of the cell cycle in the proper order. What are the components of the cell cycle clock, and how does it tick? What events are controlled by the clock, and how are these events regulated? These questions are addressed in Section V.

The idea of an autonomous cell cycle clock suggests that different cell cycle events should occur with a constant timing, even if one or more events are delayed due to environmental insults or experimental manipu-

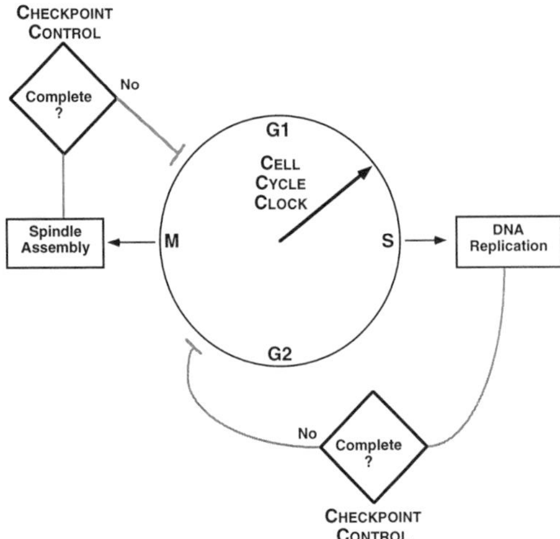

Figure 2 Cell cycle coordinating mechanisms. The two principal mechanisms for coordinating the major cell cycle events are the cell cycle clock, based on oscillations in cyclin/CDK activity, and checkpoint controls that restrain cell cycle progression until key events have been completed. Checkpoint controls are depicted as functioning through regulation of the clock, but this has yet to be established in most cases. Only some of the known checkpoint controls are indicated.

lations. This is manifestly incorrect in *S. cerevisiae,* where most cell cycle events are linked in dependent pathways (Pringle and Hartwell 1981). For instance, chromosome segregation does not occur until cells have completed DNA replication, even if replication is delayed by several hours. These observations were reconciled with the idea of a cell cycle clock by postulating a second coordinating mechanism in the form of a set of checkpoint controls. Checkpoint controls are regulatory pathways that monitor the progress of key cell cycle events and act to delay cell cycle progression if those events have not been satisfactorily completed. Thus, checkpoint controls ensure that cell cycle events occur in the proper order even if some perturbation delays performance of a particular event. What processes are monitored by checkpoint controls, and what exactly is detected by each checkpoint? How do checkpoint controls affect the ticking and/or outputs of the clock? These questions are addressed in Section VI.

To date, the questions raised above are far from being answered in detail. In what follows, we discuss the available information about the *S. cerevisiae* cell cycle in the context of these questions and examine how well the available data fit with the unified model. It should be noted that this chapter does not discuss in detail a second major problem in the field of cell cycle control, namely, the regulation of the cell cycle in response to environmental factors. *S. cerevisiae* cells have a variety of developmental options in addition to the rapid proliferation that occurs in rich media. Nutritional deprivation can lead to cell cycle arrest in G_1, to meiosis and sporulation, or to pseudohyphal growth, depending on the cell type and on the type and severity of nutrient limitation. Haploid cells respond to mating pheromones by (among other things) transiently arresting the cell cycle in G_1. Even for continuously proliferating cells, the external milieu affects the rate of mass accumulation and hence the rate of cell proliferation. These controls operate primarily at Start, which depends (among other things) on the attainment of a critical cell size (Pringle and Hartwell 1981). However, with the exception of the response to mating pheromones (for review, see Cross 1995; Wittenberg and Reed 1996), the mechanisms remain largely obscure.

V. THE CELL CYCLE CLOCK

The ticking of the cell cycle clock is manifested as cyclical changes in CDK kinase activities. The phosphorylations catalyzed by CDKs regulate a large number of processes, some of which (e.g., synthesis, activation levels, and proteolysis of CDK regulators) themselves contribute to the oscillations of CDK activity. This set of linked autoregulatory loops is thought to generate the periodicity of the clock. The importance of particular regulatory mechanisms varies among organisms and cell types, but CDK kinases are universally required for cell cycle progression, so there is a good case for discussing the clock in terms of how CDK activities are regulated and how they then orchestrate the events of the cell cycle.

A. CDKs and Cyclins

The main CDK implicated in cell cycle control in *S. cerevisiae* is Cdc28p (Reed 1992; Nasmyth 1993). Other CDKs with distinct functions have been identified (see Table 1), suggesting that the CDK mode of regulation is not restricted to cell cycle control. All CDKs are catalytically inactive as monomers and require association with positive regulatory proteins called cyclins for activity (Morgan 1995). Cyclins were first discovered in rapidly cleaving early embryonic cells and were named for

Table 1 Cyclin-dependent kinases (CDKs) in *Saccharomyces cerevisiae*

CDK	Cyclin	Function	References
Cdc28p	Cln1p-3p	cell cycle control: G_1	see text
Cdc28p	Clb1p-6p	cell cycle control: S,G_2,M	see text
Pho85p	Pho80p	phosphate metabolism[a]	1,2
Pho85p	Pcl1p (Hcs26p), Pcl2p (OrfDp)	???[b]	2
Kin28p	Ccl1p	TFIIH/mRNA transcription[c]	3,4
Ctk1p	Ctk2p	RNA polymerase II CTD kinase[d]	5
Ssn3p (Ume5p, Srb10p)	Ssn8p (Ume3p, Srb11p)	RNA polymerase II holoenzyme[e]	6,7

References: (1) Kaffman et al. 1994; (2) Lenburg and O'Shea 1996; (3) Feaver et al. 1994; (4) Cismowski et al. 1995; (5) Sterner et al. 1995; (6) Liao et al. 1995; (7) Kuchin et al. 1995.

[a]The Pho80p/Pho85p complex is involved in transcriptional regulation of *PHO5* and several other genes involved in phosphate metabolism in response to phosphate availability. Pho80p/Pho85p may also regulate one or more other genes that affect vacuole inheritance (Nicolson et al. 1996).

[b]The normal physiological role of the Pcl1p-2p/Pho85p complex is unknown, but it can function in cell cycle control under certain conditions (see also Section V.D.1). Pho85p may also have a variety of other roles involving its association with other cyclins (Huang et al. 1996; Timblin et al. 1996; Measday et al. 1997).

[c]The Ccl1p/Kin28p complex forms a part of the general transcription factor TFIIH and is required for mRNA transcription.

[d]The Ctk2p/Ctk1p complex phosphorylates the CTD (carboxy-terminal domain) of RNA polymerase II (as do several other cyclin/CDK complexes), and *ctk* mutants are cold-sensitive.

[e]The Srb11p/Srb10p complex is a component of the RNA polymerase II holoenzyme.

their strikingly cyclical accumulation during the cell cycle (Evans et al. 1983). The cyclins share a homologous domain, the "cyclin box," that is necessary and sufficient for binding and activation of CDKs (Morgan 1995) and that has now become the defining feature of cyclins: Although the majority of known cyclins do exhibit periodic accumulation during the cell cycle, several cyclins defined by cyclin box homology do not (see Table 1). The cyclins form a very diverse family that can be subdivided into different classes based on sequence homologies within the cyclin box. Different classes of cyclins have little or no homology outside of the cyclin box.

Many of the yeast cyclins were identified based on genetic suppression of mutant *cdc28* alleles, suggesting that they might interact directly with Cdc28p (Hadwiger et al. 1989b; Surana et al. 1991). Biochemical studies have confirmed that nine of the cyclins (Cln1p–3p, Clb1p–6p) indeed form complexes with Cdc28p (Table 1) (Wittenberg et al. 1990; Ghiara et al. 1991; Surana et al. 1991; Grandin and Reed 1993; Tyers et al. 1993; Schwob et al. 1994; Li and Cai 1997). The kinase activity of cyclin/Cdc28p complexes is generally measured in vitro using the test substrate histone H1, and the above nine cyclins are associated with a histone H1 kinase activity that is Cdc28p-dependent (i.e., complexes from *cdc28* mutant cells have a much reduced or thermolabile activity). Analysis of Cdc28p in extracts from synchronized cells has shown that Cdc28p levels do not fluctuate during the cell cycle and that the majority of Cdc28p is monomeric throughout the cell cycle (Wittenberg and Reed 1988). This suggests that Cdc28p itself is always in excess and that the cyclins are the limiting factor in Cdc28p activation.

Studies from many organisms have shown that cyclin/CDK complexes generally phosphorylate serine or threonine residues that are followed by a proline (Nigg 1991). Although not a requirement, the presence of nearby basic residues is a common feature of the target site, which has a consensus sequence of K/R **S/T** P X K/R (phosphorylation site in bold; X = any amino acid). Subtle differences in substrate specificity have been detected for different cyclin/CDK complexes in vitro (Peeper et al. 1993), but such studies have yet to be performed in yeast. Studies in vivo strongly suggest that different complexes phosphorylate different substrates (see below, particularly Section V.D), but it has yet to be determined to what extent these differences arise from differences in substrate specificity or in subcellular localization of the different cyclin/Cdc28p complexes. Furthermore, cyclin/Cdc28p activities in cell extracts are found in heterogeneous, slowly sedimenting complexes, suggesting that either oligomerization or association with other factors may regulate cyclin/Cdc28p activity and/or substrate specificity (Wittenberg and Reed 1988).

With one exception (Cln3p; see below), the *S. cerevisiae* cyclins are encoded by pairs of closely related genes with overlapping functions, the products of which accumulate at specific times during the cell cycle (Fig. 3). This pattern of cyclin accumulation leads to overlapping waves of activation of distinct cyclin/Cdc28p complexes, so that the profile of Cdc28p activities changes through the cell cycle in a manner consistent with the proposal that cyclin/Cdc28p acts as a clock. Each pair of cyclins can be subdivided (Fig. 3) into a "major" partner, which is expressed at a higher level and whose deletion causes a more severe phenotype, and a "minor" partner, whose deletion generally causes no detectable phenotype. This raises the question of why the minor partner has been maintained during evolution. A possible explanation is suggested by the evidence that at least one major partner during the mitotic cycle (Clb2p) is not expressed during meiosis, whereas its minor partner (Clb1p) is expressed at a much higher level than during the mitotic cycle and is critical for meiosis (Grandin and Reed 1993). Similarly, Clb4p, which has

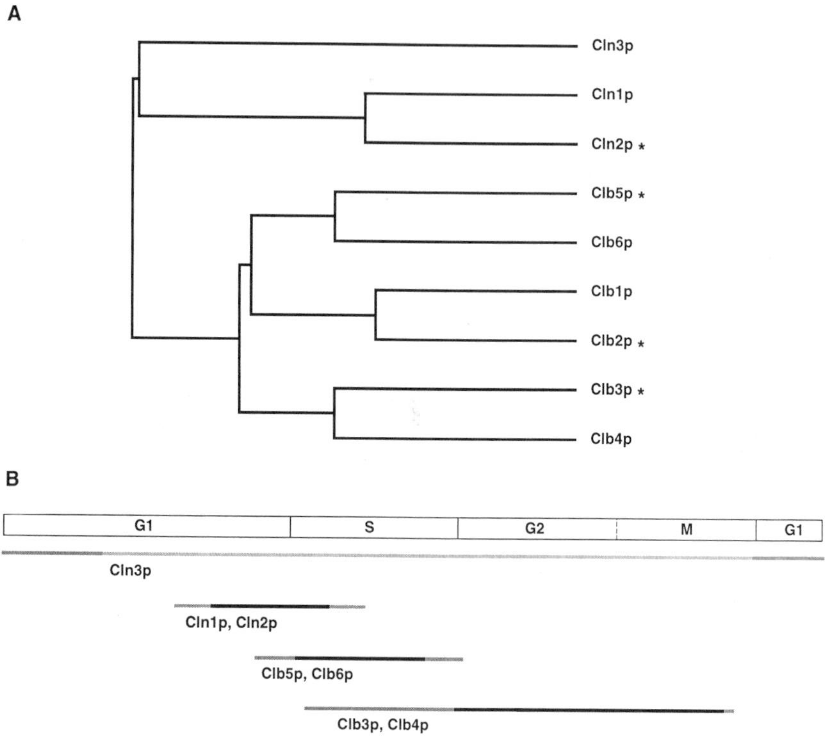

Figure 3 Cyclins that activate Cdc28p in *S. cerevisiae*. (*A*) Dendrogram of cyclins indicating degrees of sequence divergence (horizontal distances). An asterisk indicates the major partner for each cyclin pair. (*B*) Timing of Cdc28p activation by different cyclins. Darker shading indicates greater activity of Cdc28p complexes containing the indicated cyclin(s).

little impact on the mitotic cycle, becomes important during meiosis (Dahmann and Futcher 1995). Such differential expression and function may provide opportunities to optimize cyclins for subtly different tasks under different conditions.

Although the Cln and Clb cyclins can all activate the Cdc28p histone H1 kinase, the magnitude of kinase activity associated with the different cyclins is very different. A direct comparison has shown that Cln3p/Cdc28p complexes possess levels of kinase activity about two orders of magnitude lower than Cln1p/Cdc28p or Cln2p/Cdc28p complexes, which in turn have much less kinase activity than Clb/Cdc28p complexes in extracts from asynchronous cells (Tyers et al. 1993). In a separate study, it was shown that the majority (>60%) of the Cdc28p kinase activity in extracts from asynchronous cells was attributable to Clb3p/Cdc28p complexes, whereas most (>80%) of the Cdc28p kinase activity in extracts from cells arrested with nocodazole was attributable to Clb2p/Cdc28p complexes (Grandin and Reed 1993). The total Cdc28p-associated kinase activity rises from undetectable levels in early G_1 to maximal levels during nuclear division, consistent with the changes in the complexes present at different times during the cell cycle (Fig. 3B) (Reed and Wittenberg 1990). However, it should be emphasized that these studies do not distinguish between differences in kinase activity due to cyclin abundance, efficiency of cyclin/Cdc28p complex formation, or the specific activity of particular complexes. In addition, bovine histone H1 (the test substrate in these studies) may not be a good indicator of kinase activity toward the physiologically relevant substrates.

B. Posttranslational Regulation of Cyclin/Cdc28p Complexes

The kinase activity of cyclin/Cdc28p complexes can be regulated posttranslationally by phosphorylation of Cdc28p and by association with other proteins. These modes of regulation are summarized in Figure 4.

1. Phosphorylation of Cdc28p T169

Cdc28p phosphorylation has been documented at two sites, threonine 169 (T169) and tyrosine 19 (Y19). Phosphorylation at T169 appears to be required for activity (as in other CDKs), and replacement of T169 with alanine generates a mutant protein that cannot sustain cell proliferation (Fig. 4A) (Cismowski et al. 1995; Deshaies and Kirschner 1995; Lim et al. 1996b). Cdc28p T169 phosphorylation is catalyzed by Cak1p (or Civ1p), a monomeric and highly diverged protein kinase (Espinoza et al. 1996; Kaldis et al. 1996; Thuret et al. 1996). Cdc28p T169 phosphorylation does not appear to require cyclin binding (Cismowski et al. 1995), and there is no evidence as yet for any cell cycle regulation by T169 phosphorylation.

2. Phosphorylation of Cdc28p Y19

In many organisms, Cdc2p phosphorylation at Y15 inhibits the kinase activity of cyclin/Cdc2p complexes as measured in vitro (Solomon et al. 1990). Phosphorylation of the analogous Cdc28p Y19 in *S. cerevisiae* has been demonstrated, but the initial studies failed to find any inhibition of Cdc28p kinase activity as a result of Y19 phosphorylation (Amon et al. 1992). Moreover, the kinase activity associated with wild-type or

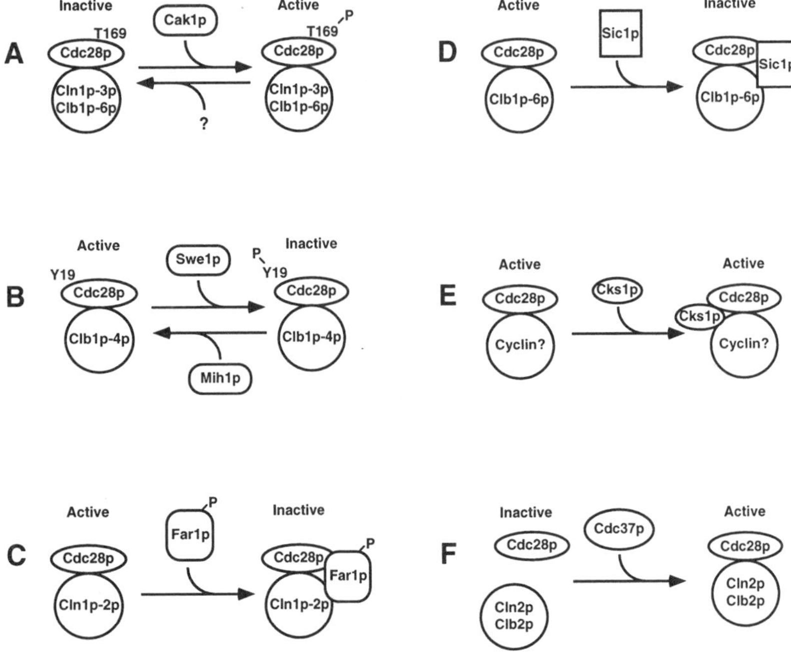

Figure 4 Posttranslational regulation of Cdc28p activity. (*A*) Cak1p (Civ1p) kinase phosphorylates threonine 169 of Cdc28p; this phosphorylation is thought to be required for the activity of all cyclin/Cdc28p complexes. The phosphatase(s) that reverses this modification is unknown. (B) Swe1p kinase phosphorylates Cdc28p (in complexes containing Clb1p–Clb4p) on tyrosine 19, leading to inhibition of Cdc28p kinase activity. The phosphatase Mih1p reverses this modification and activates Cdc28p. (*C*) Far1p binds to and inhibits the activity of Cln1p–2p/Cdc28p complexes (and perhaps Clb/Cdc28p complexes), but only when Far1p has been activated by phosphorylation in response to pheromone stimulation. (*D*) Sic1p binds to and inhibits the activity of Clb/Cdc28p complexes. (*E*) Cks1p binds to Cdc28p. It is not known whether Cks1p binds preferentially to particular cyclin/Cdc28p complexes, and the physiological role of Cks1p binding is unclear (see text). (*F*) Cdc37p is required for efficient complex formation between Cdc28p and Cln2p or Clb2p (and perhaps other cyclins). See text for more details.

Y19F mutant Clb2p/Cdc28p complexes isolated from various cell extracts appeared to be the same, further suggesting that Y19 phosphorylation was not inhibitory (Amon et al. 1992). However, the stoichiometry of Y19 phosphorylation was unknown in this study, and a more recent study (Lin and Arndt 1995) did find increased kinase activity in Y19F versus wild-type Clb2p/Cdc28p complexes from S-phase cells. Induction of high levels of Y19 phosphorylation by overexpressing Wee1p (see below) did not lead to a change in the overall Cdc28p-associated kinase activity, again suggesting that Y19 phosphorylation was not inhibitory (Sorger and Murray 1992). In this case, the stoichiometry of Y19 phosphorylation was probably high, but the amount of Clb/Cdc28p was unknown: If the Wee1p-overproducing cells also accumulated large amounts of Clb/Cdc28p in comparison to the starting population, then Cdc28p might indeed have been inhibited in those cells. A more definitive experiment assayed the activity of a fixed amount of Clb2p/Cdc28p complex following phosphorylation of Y19 in vitro and concluded that, as for Cdc2p from other organisms, Cdc28p Y19 phosphorylation indeed inhibits kinase activity (Booher et al. 1993).

The Cdc28p Y19F mutation does not appear to affect any aspect of the cell cycle or cell proliferation under standard laboratory growth conditions (Amon et al. 1992; Sorger and Murray 1992). This finding was rather surprising, because phosphorylation of the analogous Cdc2p Y15 in *Schizosaccharomyces pombe* appears to be the dominant mechanism controlling the timing of Cdc2p activation and entry into mitosis (for review, see Lew and Kornbluth 1996; MacNeill and Nurse, this volume). Studies in *S. pombe* identified kinases (Wee1p and Mik1p) and a phosphatase (Cdc25p) that control the state of Cdc2p Y15 phosphorylation. A homologous kinase (Swe1p) and phosphatase (Mih1p) have been identified in *S. cerevisiae* (Russell et al. 1989; Booher et al. 1993). Disruption of *MIH1* produces a small G_2 delay (Russell et al. 1989), whereas disruption of *SWE1* has no readily detectable effect except to eliminate the G_2 delay in *mih1* mutants (Booher et al. 1993). This suggests that Swe1p activity in the unperturbed cell cycle is low and that Mih1p activity is sufficient to eliminate any effect on cell cycle progression due to Y19 phosphorylation. Overexpression of Swe1p from the strong *GAL1* promoter arrests cells with a phenotype similar to that produced by deletion of cyclins Clb1p–4p (see below), suggesting that Clb1p–4p/Cdc28p complexes are targets for Swe1p phosphorylation and inhibition in vivo (Booher et al. 1993; Lim et al. 1996a). In contrast, Swe1p overproduction did not affect early cell cycle events dependent on Cln/Cdc28p activity in vivo, and Swe1p was unable to phosphorylate Cdc28p complexed with Cln2 in vitro (Booher et al. 1993). Thus, the cyclin partner associated with Cdc28p appears to be a critical determinant in Y19 phosphorylation by Swe1p (Fig. 4B).

Although Cdc28p Y19 phosphorylation does not appear to have a significant role in the unperturbed cell cycle, it becomes important in cells that are impaired in their ability to form a bud; in particular, a morphogenesis checkpoint control appears to employ Y19 phosphorylation to delay other aspects of cell cycle progression until a bud has been formed (see Section VI.D).

3. Binding of Sic1p and Far1p

The kinase activity associated with cyclin/Cdc28p complexes can also be inhibited through the binding of "CDK inhibitor proteins" (Peter and Herskowitz 1994b). Two such proteins have been identified to date in *S. cerevisiae*, Far1p and Sic1p (Fig. 4C,D). These have no homology to each other or to the CDK inhibitors identified to date in mammalian cells. Biochemical studies have suggested that the ability of Far1p and Sic1p to inhibit Cdc28p kinase activity depends on the cyclin moiety in the complex: Far1p specifically inhibited Cln/Cdc28p complexes, whereas Sic1p specifically inhibited Clb/Cdc28p complexes (Peter and Herskowitz 1994a; Schwob et al. 1994). The stoichiometry of binding to cyclin/Cdc28p complexes has not been determined, and the mechanism of kinase inhibition is unknown. Sic1p has been shown to lower both the K_m and V_{max} values of Cdc28p kinase for a peptide substrate, and it was suggested that binding of Sic1p reduces access even of small substrates to the active site (Mendenhall 1993). Far1p and Sic1p accumulate periodically during the cell cycle, with maximal accumulation in early G_1, when cyclins (except for Cln3p) are absent (see Fig. 3) (McKinney et al. 1993; Donovan et al. 1994; Schwob et al. 1994).

FAR1 was originally identified in a search for genes required for cell cycle arrest in response to mating pheromones (Chang and Herskowitz 1990). Deletion of *FAR1* produces no readily detectable phenotype in cells that have not been exposed to pheromone, and it is thought that Far1p is inactive as a CDK inhibitor until it is phosphorylated in response to pheromone stimulation (Peter et al. 1993). Interestingly, artificial expression of Far1p during late stages of the cell cycle allows a pheromone-induced cell cycle arrest in post-G_1 cells, suggesting that Far1p may be capable of inhibiting Clb/Cdc28p, as well as Cln/Cdc28p, complexes in vivo (McKinney and Cross 1995).

Sic1p was originally identified as a 40-kD in vitro substrate of Cdc28p that was associated with Cdc28p in cell extracts (Mendenhall et al. 1987). Deletion of *SIC1* causes rather heterogeneous defects including increased rates of chromosome loss and breakage, and, in some strain backgrounds, a *sic1* population produces many dead cells (Nugroho and Mendenhall 1994). Intriguingly, daughter cells seem much more sensitive than mother cells to loss of Sic1p (Nugroho and Mendenhall 1994).

Sic1p has been suggested to have diverse roles contributing to the oscillations of cyclin/Cdc28p activities (Donovan et al. 1994; Schwob et al. 1994; Toyn et al. 1997), but the basis for the daughter cell specificity of the *sic1* phenotype remains obscure.

4. Binding of Cks1p

Cks1p also associates with cyclin/Cdc28p complexes (Fig. 4E) (Hadwiger et al. 1989a). Binding of Cks1p occurs at a site distinct from those for cyclins and Sic1p, and Cks1p-binding does not affect activation by cyclins or inhibition by Sic1p (Hadwiger et al. 1989a; Mendenhall 1993). *CKS1* was originally identified as a multicopy suppressor of a *cdc28^{ts}* mutation (Hadwiger et al. 1989a). It is an essential gene, and Ts^{-} mutants show pleiotropic cell cycle arrest with defects in bud formation, DNA replication, and nuclear division (Tang and Reed 1993). Because cyclin/Cdc28p complexes are required for all of these events (see below), and Cks1p binds to these complexes in vitro, the simplest interpretation of the *cks1^{ts}* arrest phenotype is that Cks1p is important for the function of multiple cyclin/Cdc28p complexes. However, analysis of Cdc28p from *cks1*-arrested cells did not identify any defect in the formation of complexes with cyclins or in kinase activity in vitro, and it was therefore proposed that Cks1p is important for some more subtle aspect of Cdc28p function, such as targeting of cyclin/Cdc28p complexes to appropriate substrates (Tang and Reed 1993). Structural studies of a complex of human CDK2 with a Cks1p homolog show that binding occurs adjacent to the catalytic cleft, consistent with this proposal (Bourne et al. 1996). However, it should also be noted that the existence of genetic and physical interactions between Cks1p and Cdc28p does not prove that all *cks1* phenotypes are due to perturbations of Cdc28p function. Indeed, some *cks1* phenotypes appear to be novel and possibly unrelated to cell cycle progression; for example, *cks1* cells elaborate long bundles of cytoplasmic microtubules and show aberrant localization of mitochondrial DNA (Tang and Reed 1993).

CKS1 is homologous to *suc1* from *S. pombe*, and two human homologs have also been identified (Hayles et al. 1986; Brizuela et al. 1987; Hindley et al. 1987; Richardson et al. 1990). Although the homologs vary significantly in size (Cks1p, 18 kD; Suc1p, 13 kD; human homologs ~9 kD), the degree of sequence conservation within the central domain is high (>50% identity), and each of the homologs can rescue a *cks1* null mutant and associate with Cdc28p in vitro (Richardson et al. 1990). Studies in *S. pombe* have identified numerous genetic interactions between *suc1* and *cdc2* (Hayles et al. 1986; Brizuela et al. 1987), strengthening the suggestion that these proteins are somehow involved in regulation of CDK activity, but the mechanism remains obscure (MacNeill and Nurse, this volume).

The physical association between Cks1p or Suc1p and CDKs has been exploited to generate beads conjugated with Suc1p or Cks1p as affinity reagents for purification of CDKs from several organisms (see, e.g., Labbé et al. 1989). Suc1p beads can bind 100% of the Cdc2p present in *Xenopus* egg extracts, showing that in this case, at least, binding occurs regardless of whether or not cyclin is associated with Cdc2p (Dunphy et al. 1988). In other systems, it is not known to what extent the Cks1p (Suc1p)-Cdc28p (Cdc2p) interaction is affected by the presence of cyclins or other factors in the complex, or what population(s) of Cdc28p associates with Cks1p in vivo.

5. *Regulation of Cdc28p-Cyclin Association*

CDC37 was discovered in a search for mutants that displayed defects similar to those of *cdc28* mutants (Reed 1980). In *cdc37* mutants, Cln2p and Clb2p fail to assemble efficiently into complexes with Cdc28p (Fig. 4F) (Gerber et al. 1995). The basis for this is unclear, and genetic interactions have now been found between *CDC37* and its homologs and the genes for numerous other protein kinases (Cutforth and Rubin 1994; Valay et al. 1995; Stepanova et al. 1996; Schutz et al. 1997), suggesting that Cdc37p does not function solely to promote cyclin-Cdc28p binding. It is unclear whether the efficiency of cyclin/Cdc28p complex formation is ever regulated physiologically, but this idea received support from the finding that Cln2p association with Cdc28p was more efficient in cells lacking Clb5p or Clb6p (Basco et al. 1995). This suggests that association between Cln2p and Cdc28p may be inhibited during S phase through the action of Clb5p and Clb6p, in a process that might involve Cdc37p.

C. What Is the Basis for Periodic Cdc28p Activity?

What mechanisms control the sequential appearance and disappearance of distinct cyclin/Cdc28p activities? The genes encoding cyclins, Cdc28p inhibitors (Sic1p and Far1p), and the Cdc28p regulatory kinase Swe1p all display periodic transcription. In addition, the turnover rates (proteolysis) of some cyclins and of both known Cdc28p inhibitors have been shown to vary during the cell cycle. Cell cycle regulation of the synthesis and proteolysis of these Cdc28p regulators is in turn controlled by cyclin/Cdc28p activities. Although the detailed molecular basis for this regulation has not yet emerged, these links are beginning to provide a rough sketch of how cyclin/Cdc28p oscillations occur. In what follows, we consider the behavior of each Cdc28p regulator in turn.

1. Cln3p

Unlike the other cyclins, Cln3p is thought to be present at relatively constant levels throughout the cell cycle, although this conclusion must be tempered by the caveat that it is difficult to monitor this protein accurately because of its very low abundance (Tyers et al. 1992; Cross and Blake 1993). Recently, it was shown that *CLN3* mRNA undergoes a modest induction in early G_1 (L. Breeden, pers. comm.). Cln3p is an unstable protein because of a number of "PEST" motifs (rich in Pro, Glu, Ser, and Thr residues, as found in numerous other unstable proteins), but there is no evidence for cell cycle variations in Cln3p stability (Tyers et al. 1992, 1993; Cross and Blake 1993). Cln3p is stabilized in *cdc28* mutants, and phosphorylation of Cln3p by Cdc28p may contribute to its instability (Cross and Blake 1993; Yaglom et al. 1995; see also discussion of Cln2p, below). Interestingly, phosphorylation and subsequent degradation of Cln3p (but not Cln2p) require the molecular chaperone Ydj1p (Yaglom et al. 1996).

2. Cln1p and Cln2p

Transcription of *CLN1* and *CLN2* is periodic during the cell cycle, with maximal mRNA accumulation in late G_1 (Fig. 5A) (Wittenberg et al. 1990). This pattern of mRNA accumulation is observed for a large set of genes in yeast and appears to result from the presence of either of two promoter elements, called MCB and SCB (for review, see Breeden 1996). Sequences resembling MCB and SCB elements are present in both the *CLN1* and *CLN2* promoters and are required for efficient transcription (Cross et al. 1994; Stuart and Wittenberg 1994). MCB and SCB elements are recognized by transcription factor complexes called MBF and SBF, respectively. MBF and SBF share a common subunit, Swi6p, and contain related DNA-binding subunits, Mbp1p (MBF) and Swi4p (SBF) (Koch et al. 1993). *SWI4* is itself transcribed periodically (through MCB elements) with a periodicity similar to that of *CLN1* and *CLN2* (Foster et al. 1993). Swi6p undergoes cell-cycle-regulated localization, being nuclear from the end of mitosis through late G_1 and predominantly cytoplasmic for the rest of the cell cycle (Taba et al. 1991; Sidorova et al.

1995). However, neither the regulation of *SWI4* transcription nor the regulation of Swi6p localization is essential for the timing of MBF- or SBF-driven transcription (Breeden and Mikesell 1994; Sidorova et al. 1995). In addition, transcriptional induction upon Cln/Cdc28p activation does not require new protein synthesis (Marini and Reed 1992). In vivo footprinting suggests that SBF is bound to the *CLN2* promoter throughout G_1 and hence that activation of *CLN2* transcription in late G_1 must occur through in situ activation of SBF (Koch et al. 1996). In contrast, the *CLN2* promoter is vacant during G_2/M, suggesting that repression of *CLN2* transcription occurs by a distinct mechanism (Koch et al. 1996).

There is now good evidence that the timing of *CLN1* and *CLN2* expression is controlled by other cyclins; in particular, Cln3p turns transcription on, whereas Clb1p–4p turn transcription off (Fig. 5A). As discussed further in Section V.D.1, recent observations argue against an earlier suggestion that *CLN1* and *CLN2* transcription is induced through a positive feedback loop and suggest instead that Cln3p/Cdc28p is solely responsible for inducing *CLN1* and *CLN2* transcription at the correct time in G_1. In the absence of Clb1p–4p, *CLN1* and *CLN2* transcripts

Figure 5 Bases for periodicity of Cdc28p activity. (*A*) Transcription of *CLN1* and *CLN2* (as well as other SBF-controlled genes) is induced by Cln3p/Cdc28p and subsequently repressed by Clb1p–4p/Cdc28p. (*B*) Transcription of *CLB5* and *CLB6* (as well as other MBF-controlled genes) is induced by Cln3p/Cdc28p and subsequently repressed by unknown mechanisms. (*C*) Transcription of *CLB3* and *CLB4* is induced and repressed by unknown mechanisms. (D) Transcription of *CLB1* and *CLB2* (as well as other Mcm1p-controlled genes, including *SWI5*) is induced by Clb1p–4p/Cdc28p, but the relative contributions of the different cyclins are unclear. Repression of *CLB1* and *CLB2* may result from Cdc28p inactivation following Clb1p–4p degradation. (*E*) APC-dependent degradation of Clb1p–4p is induced during (or possibly following) anaphase, in a manner that is not fully understood but requires prior induction of Clb1p–2p/Cdc28p. The APC is inhibited by Cln1p–2p/Cdc28p in the following cycle. (*F*) Transcription of *SIC1* is induced by Swi5p, which enters the nucleus upon inactivation of Clb/Cdc28p during anaphase. Swi5p is unstable in the nucleus, so *SIC1* transcription is short-lived. (*G*) Phosphorylation of Sic1p and Far1p by cyclin/Cdc28p complexes targets them for degradation. It is thought that Cln1p–2p/Cdc28p complexes are primarily responsible for Sic1p and Far1p degradation in late G_1; Clb/Cdc28p complexes may then ensure that these proteins cannot accumulate again until the inactivation of Cdc28p upon Clb1p–4p degradation during anaphase. The timing of these various oscillations is indicated relative to the cell cycle phases by the timeline at the bottom of the figure.

remain abundant following their induction in late G_1, suggesting that Clb1p–4p (and primarily Clb1p–2p) somehow shut down SBF (Amon et al. 1993). Swi4p binds to Clb2p and can be phosphorylated by Clb2p/Cdc28p in vitro, but the functional significance of these interactions has yet to be determined (Amon et al. 1993).

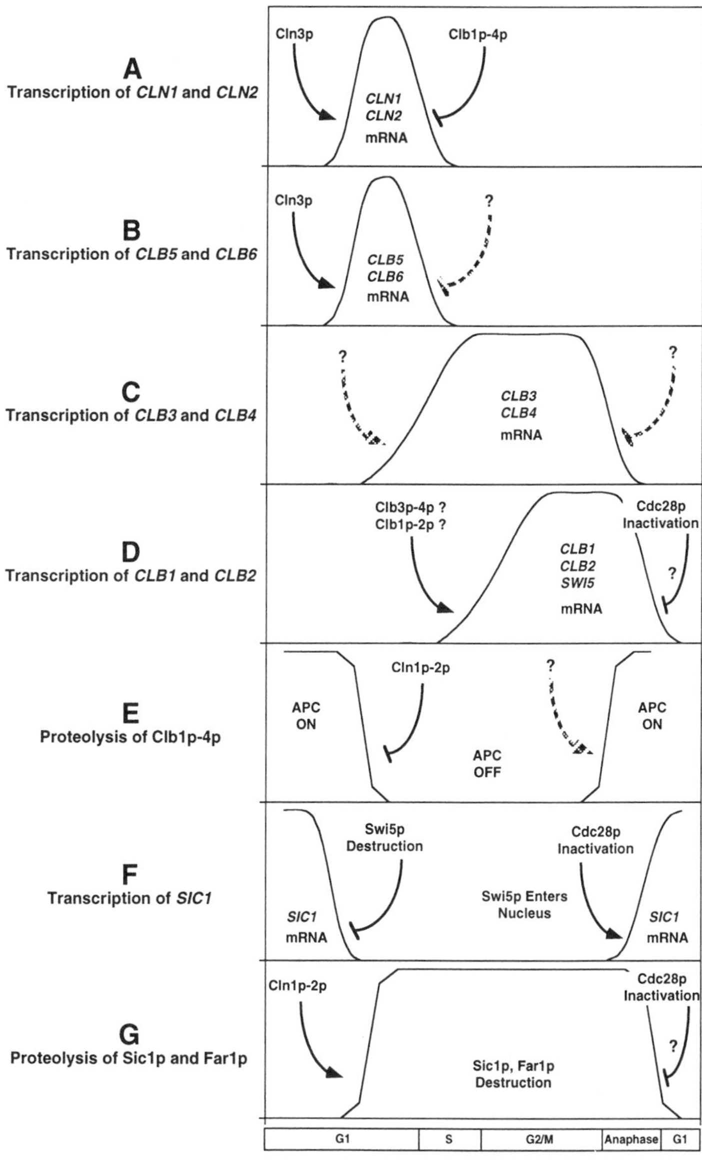

Figure 5 (*See facing page for legend.*)

Cln2p is an unstable protein (half-life ~8 minutes; Lanker et al. 1996), and Cln2p levels are closely correlated with *CLN2* mRNA levels through the cell cycle (Wittenberg et al. 1990). PEST motifs are present in Cln1p and Cln2p (as well as in Cln3p; see above) and are necessary but not sufficient for rapid Cln2p turnover (Salama et al. 1994). Recent studies suggest that the Cln2p carboxy-terminal domain (including PEST motifs) may function to promote proteolysis by providing target sites for proline-directed kinases (Lanker et al. 1996). Cln2p is phosphorylated at a number of serine and threonine residues by active Cdc28p kinase, and mutation of these sites creates a stable (half-life ~1 hour) cyclin that activates Cdc28p both in vitro and in vivo (Lanker et al. 1996). Once phosphorylated, Cln2p is recognized by a complex that appears to contain Cdc53p, Cdc34p, and Cdc4p (Mathias et al. 1996; Willems et al. 1996) and may also contain Grr1p and Skp1p (Barral et al. 1995; Bai et al. 1996; Connelly and Hieter 1996). Mysteriously, Skp1p is also found, at least in part, in the kinetochore complex (Connelly and Hieter 1996; Stemmann and Lechner 1996). Cdc34p is a ubiquitin-conjugating enzyme (Goebl et al. 1988), and binding of the Cdc34p-containing complex may lead directly to Cln2p ubiquitination and degradation (Willems et al. 1996). Cdc34p-dependent Cln2p ubiquitination has been reconstituted in vitro (Deshaies et al. 1995), strengthening the hypothesis that Cdc34p is directly responsible for Cln2p ubiquitination. However, another recent study suggested that at least part of the Cdc34p requirement for Cln2p instability might be indirect and that a separate pathway activated by Clb1p–4p/Cdc28p may contribute to Cln2p degradation (Blondel and Mann 1996). Cln2p/Cdc28p phosphorylation of Cln2p can promote its ubiquitination and then destruction, providing the opportunity for a self-limiting cycle, which may contribute to the oscillation of Cln2p (and, by extension, Cln1p) levels during the cell cycle.

3. *Clb5p and Clb6p*

Like *CLN1* and *CLN2*, *CLB5* and *CLB6* are transcribed periodically, with maximal mRNA accumulation in late G_1 (Fig. 5B) (Epstein and Cross 1992; Kühne and Linder 1993; Schwob and Nasmyth 1993). The *CLB5* and *CLB6* promoters contain MCB motifs, and it is thought that their transcription is primarily MBF-dependent (in contrast to the SBF-dependent transcription of *CLN1* and *CLN2*). The activation of transcription is apparently triggered by Cln3p/Cdc28p (Dirick et al. 1995; Stuart and Wittenberg 1995), but the molecular basis for MBF activation is not clear. Unlike transcription of *CLN1* and *CLN2,* transcription of *CLB5* and

CLB6 is repressed normally even in the absence of Clb1p–4p (Amon et al. 1993). This has led to the suggestion that the inactivation of SBF and MBF is differentially regulated. The basis for MBF inactivation (presumed to be responsible for repression of *CLB5* and *CLB6*) remains unknown.

The stability of Clb5p has been reported to be relatively constant through the cell cycle (half-life ~10 minutes in G_1 and ~20 minutes in S or G_2/M) (Seufert et al. 1995). However, it should be noted that because of the methods used for stability measurements, these data were collected in arrested cells (G_1, pheromone arrest; S, hydroxyurea arrest; G_2/M, nocodazole arrest). Thus, it is possible that the degradation rates measured reflected the arrest perturbations rather than the values obtaining during normal cell cycle progression. Clb5p is stabilized in *ubc9* mutants (defective for the ubiquitin-conjugating enzyme Ubc9p), suggesting that Clb5p degradation occurs through ubiquitination (Seufert et al. 1995). However, it is not yet clear whether Clb5p is ubiquitinated by Ubc9p or whether its stabilization in *ubc9* mutants is indirect. Clb6p stability has not yet been examined.

4. *Swe1p*

SWE1 transcripts display a periodicity similar to those of *CLN1, CLN2, CLB5,* and *CLB6,* and the *SWE1* promoter contains SCB elements (Lim et al. 1996a; Ma et al. 1996; Sia et al. 1996). Swe1p abundance has not yet been measured. The role of periodic *SWE1* transcription in late G_1 is unclear, because the Clb1p–4p/Cdc28p complexes that Swe1p has been shown to inhibit do not accumulate until later in the cell cycle. It is not yet known whether Swe1p can inhibit Clb5p–6p/Cdc28p complexes.

5. *Clb3p and Clb4p*

CLB3 and *CLB4* transcripts accumulate in S phase and G_2, later than the transcripts of *CLN1, CLN2, CLB5,* and *CLB6,* but earlier than those of *CLB1* and *CLB2* (Fig. 5C) (Surana et al. 1991; Richardson et al. 1992). The mechanisms responsible for turning these genes on and off remain entirely unknown. Stability measurements on Clb3p and Clb4p have yet to be performed, although the presence of a conserved "destruction box" motif in the amino terminus may indicate that these proteins are subject to cell-cycle-dependent degradation like that of Clb1p and Clb2p (see below).

6. Clb1p and Clb2p

CLB1 and *CLB2* transcripts accumulate in G_2 and M phases (Fig. 5D), a pattern common to many genes in yeast, including those encoding the transcription factors Swi5p and Ace2p, the protein kinases Cdc5p and Dbf2p, and the MCM proteins Cdc46p and Cdc47p (Koch and Nasmyth 1994). Transcription of several members of this family is known to require the transcription factor Mcm1p in conjunction with a poorly characterized cofactor(s) (Lydall et al. 1991; Althoefer et al. 1995; Maher et al. 1995). In vivo footprinting revealed that these factors are bound to *SWI5* and *CLB2* promoters both in G_1 (pheromone-arrested cells) and in G_2/M (nocodazole-arrested cells), suggesting that transcriptional regulation may act through modulation of the activity of a promoter-bound factor(s) (Althoefer et al. 1995).

Induction of *CLB1* and *CLB2* transcription is greatly delayed in cells lacking Clb1p–4p (Amon et al. 1993). It is not known which of these cyclins induces *CLB1* and *CLB2* transcription in the normal cell cycle; Clb3p–4p may turn on *CLB1* and *CLB2* transcription (analogous to the induction of *CLN1* and *CLN2* by Cln3p) and/or Clb1p and Clb2p may contribute to *CLB1* and *CLB2* transcription in a positive feedback loop. At the end of mitosis, Clb1p–4p are degraded (see below), and transcription of *CLB1* and *CLB2* is repressed. The basis for this repression is unknown, but a parsimonious model is that repression simply represents a reversal of the initial activation process and follows upon the proteolytic degradation of Clb1p–4p (Fig. 5D; and see below).

Studies in metazoan cells have shown that cyclins of the "B" class (related to Clb1p–6p) are stable during interphase but are degraded rapidly during mitosis (for review, see Deshaies 1995). A motif near the amino terminus of these cyclins dubbed the "destruction box" is responsible for targeting the cyclins for mitotic destruction (Glotzer et al. 1991). Destruction is due to polyubiquitination of the cyclin followed by its degradation by the proteasome. Cyclin ubiquitination is catalyzed by a large (20S) protein complex with cyclin-ubiquitin ligase activity, called the cyclosome (Sudakin et al. 1995) or the APC (anaphase promoting complex) (King et al. 1995). The APC is present throughout the cell cycle, but it is only active during anaphase (Hershko et al. 1994; King et al. 1995), probably because of regulated phosphorylation of APC components (Lahav-Baratz et al. 1995).

Like metazoan B-type cyclins, Clb1p–5p (but, interestingly, not Clb6p) contain destruction-box motifs (Ghiara et al. 1991; Surana et al. 1991; Epstein and Cross 1992; Fitch et al. 1992; Richardson et al. 1992; Schwob and Nasmyth 1993). The stability of Clb2p is dramatically lower (half-life <1 minute) in early G_1 cells than in S or G_2/M cells (half-life >60 minutes) (Amon et al. 1994). Mutation of the destruction box significantly stabilizes Clb2p in early G_1 cells (half-life ~10 minutes), suggesting that, as in metazoan cells, the destruction box confers cell-cycle-dependent instability (Amon et al. 1994). However, as noted above, Clb5p stability does not vary markedly during the cell cycle despite the presence of an apparent destruction box; the basis for this difference remains mysterious. Destabilization of Clb2p occurs during anaphase, and stabilization occurs after Start in the following G_1 (Fig. 5E) (Amon et al. 1994). An elegant screen for mutants in the mitotic Clb-destruction pathway identified novel alleles of *CDC16* and *CDC23*, implicating Cdc16p and Cdc23p in Clb2p degradation (Irniger et al. 1995). The mutant alleles produced defects in the degradation of Clb2p, but not of Clb5p or Swi5p, in anaphase and early G_1 cells. In a striking convergence of several separate lines of investigation, it was recently shown that Cdc16p, Cdc23p, and Cdc27p are conserved components of the APC (Lamb et al. 1994; Irniger et al. 1995; King et al. 1995; Tugendreich et al. 1995). Further analysis has shown that the yeast APC is a 36S particle containing at least seven proteins including Cdc26p and Apc1p in addition to Cdc16p, Cdc23p, and Cdc27p (Zachariae et al. 1996).

What turns on the pathway for Clb1p–4p destruction? Clb1p–4p degradation is not initiated in G_2 cells lacking Clb1p–2p, so presumably activation of Clb1p–2p/Cdc28p initiates a chain of events that leads to Clb1p–4p destruction via the APC. The steps between activation of Clb1p–2p/Cdc28p and APC-mediated Clb1p–4p destruction are unknown, but a number of mutations appear to interrupt the process. These include mutations in *CDC5*, *CDC15*, and *DBF2* (which encode serine/threonine protein kinases, Johnston et al. 1990; Schweitzer and Philippsen 1991; Kitada et al. 1993), *CDC14* (which encodes a protein-tyrosine phosphatase, Wan et al. 1992), *TEM1* (which encodes a Ras-related GTP-binding protein, Shirayama et al. 1994a), and *LTE1* (which encodes a guanine nucleotide exchange factor, Shirayama et al. 1994b). Conditional mutations in any of these genes arrest the cell cycle at the end of anaphase but before spindle disassembly and cytokinesis, and, in the cases that have been examined, the mutant cells retain high levels of Clb/Cdc28p kinase activity, suggesting that Clb1p–4p destruction has not been initiated (Surana et al. 1993; Toyn and Johnston 1994). Extracts from *cdc15* mutant cells arrested at the end of anaphase fail to ubiquitinate Clb2p in vitro, but if *cdc15* cells are arrested in G_1 prior to shift to the restrictive temperature, then the extracts are competent to ubiquitinate Clb2p (Zachariae and Nasmyth 1996). This suggests that

Cdc15p may have a role in activating the APC to initiate Clb1p–4p degradation but that it is not required for APC function once activation has occurred. Whether Cdc15p and the other proteins listed above function directly or indirectly in APC activation remains to be determined.

What turns off the pathway for Clb1p–4p destruction? Rapid Clb2p degradation is maintained in G_1 cells lacking Cln1p–3p, and stabilization of Clb2p follows quickly upon Cln1p–3p induction (Amon et al. 1994). This suggests that Clb2p stabilization in post-Start cells is triggered by the activation of Cln1p–3p/Cdc28p (Fig. 5E). The steps linking Cln1p–3p/Cdc28p activation to Clb2p stabilization are unknown. Although these studies were performed only for Clb2p, the fact that Clb1p, Clb3p, and Clb4p all disappear at the same time in the cell cycle as does Clb2p suggests that the stability of these cyclins may be regulated similarly. This hypothesis is supported by the finding that Clb3p, like Clb2p, is ubiquitinated in an APC-dependent manner in vitro (Zachariae and Nasmyth 1996).

7. *Sic1p and Far1p*

SIC1 transcripts accumulate in late M/early G_1 (Fig. 5F) (Donovan et al. 1994; Schwob et al. 1994), a pattern common to several genes in yeast (Koch and Nasmyth 1994). Transcription of *SIC1* depends primarily on the transcription factor Swi5p (Knapp et al. 1996; Toyn et al. 1997). *SWI5* transcripts are themselves regulated coordinately with *CLB1* and *CLB2*, accumulating in G_2/M (Koch and Nasmyth 1994). However, phosphorylation of Swi5p by Clb/Cdc28p complexes masks its nuclear localization signal, so that Swi5p remains in the cytoplasm until the inactivation of Cdc28p at the end of anaphase (Moll et al. 1991). Entry of Swi5p into the nucleus presumably allows it to induce *SIC1* and other M/G_1 transcripts, but Swi5p is unstable, and the induction is therefore short-lived (Moll et al. 1991).

FAR1 transcripts accumulate from G_2 through late G_1 in the next cell cycle (McKinney et al. 1993). This pattern requires the combined action of Mcm1p, which induces many genes in G_2/M (see above), and Ste12p, which induces many pheromone-responsive genes in G_1 (Oehlen et al. 1996).

Sic1p and Far1p both appear to be stable in late M and early G_1 phases and unstable during the rest of the cell cycle (Fig. 5G) (Schwob et al. 1994; McKinney and Cross 1995); in conjunction with the patterns of transcript accumulation, this leads to accumulation of protein only in early G_1. This pattern of protein stability is exactly the opposite of that seen with Clb2p (and presumably Clb1p, Clb3p, and Clb4p). The pattern of

Far1p accumulation helps to ensure that cells arrest only in G_1 in response to the pheromone signal (McKinney and Cross 1995). The pattern of Sic1p accumulation could allow this protein to help prevent premature induction of Clb5p–6p/Cdc28p activities at the end of G_1 (Schwob et al. 1994). Support for this idea comes from the observation that cells lacking Sic1p undergo accelerated DNA replication (Schneider et al. 1996), suggesting that Clb5p–6p/Cdc28p complexes are activated prematurely (see Section V.D.2). This is consistent with the hypothesis that Sic1p introduces a delay between activation of Cln1p–2p/Cdc28p and activation of Clb5p–6p/Cdc28p, thus ensuring that these complexes are active sequentially (see Fig. 3B). Another possible role of Sic1p may be to assist in the inactivation of Clb1p–4p/Cdc28p complexes following anaphase (Donovan et al. 1994; Toyn et al. 1997).

Both Sic1p and Far1p become phosphorylated by Cdc28p in late G_1 just before they are degraded (McKinney and Cross 1995; Schneider et al. 1996), and a *cln1 cln2* double mutant is hypersensitive to overexpression of Sic1p (Tyers 1996). This suggests that phosphorylation (primarily by Cln1p–2p/Cdc28p) targets the proteins for degradation, as seems also to be the case for Cln2p (see Section V.C.2). Indeed, mutation of Sic1p phosphorylation sites stabilizes the protein (M. Mendenhall, pers. comm.). If phosphorylation of these proteins by any cyclin/Cdc28p complex (except for Cln3p/Cdc28p) is the signal for rapid degradation, then the proteins will be unstable from late G_1, when Cln1p and Cln2p accumulate, until the end of anaphase, when Clb1p–4p are degraded; this model is consistent with the available data on Far1p and Sic1p stability. Sic1p and Far1p are also stabilized in *cdc34* mutants, suggesting that the ubiquitin-conjugating enzyme Cdc34p initiates their destruction (McKinney et al. 1993; Schneider et al. 1996). Indeed, the *cdc34* arrest phenotype appears to be due largely to the failure to degrade Sic1p, which leads to sustained inhibition of Clb/Cdc28p complexes and hence to cell cycle arrest (Schwob et al. 1994).

8. Summary

The data summarized above and in Figure 5 suggest that the orderly progression of cyclin/Cdc28p activities during the cell cycle results primarily from the regulated transcription and degradation of cyclins and Cdc28p inhibitors (for reviews, see King et al. 1996; Nasmyth 1996). Two modes of regulation appear to be involved. In the first mode, exhibited by Cln1p and Cln2p, cyclic accumulation is governed by cyclic transcription combined with a constitutive instability of the proteins. In the second mode, exhibited by Clb1p–4p, Sic1p, and Far1p, the periodicity of protein accumulation is similar even if transcription is made constitutive. In these cases, regulated degradation mechanisms prevent the proteins from accumulating at the wrong times even if the transcripts are abundant. The cell-cycle-regulated transcription of *CLB1–CLB4*, *SIC1*, and *FAR1* serves mainly to determine the rate and extent to which the encoded factors accumulate during the cell cycle interval in which the degradation mechanisms permit accumulation to occur.

Two pathways of regulated proteolysis have been uncovered, both of which are initiated by substrate ubiquitination. Ubiquitination of Clb1p–4p is catalyzed by the APC, whose activity (as measured in cell extracts or inferred from Clb accumulation patterns) is regulated during the cell cycle. In contrast, ubiquitination of Sic1p, Far1p, and Cln2p appears to be catalyzed by a Cdc34p/Cdc53p/Cdc4p complex upon phosphorylation of the substrate. Although regulation of the Cdc34p/Cdc53p/Cdc4p complex during the cell cycle has not been definitively addressed, the available data suggest that Sic1p, Far1p, and Cln2p degradation can occur at any time during the cell cycle if the proteins are appropriately phosphorylated.

Activation of Cln1p–3p/Cdc28p complexes in late G_1 promotes subsequent accumulation of Clb/Cdc28p activities by at least three mechanisms, transcriptional induction of *CLB5–CLB6*, stabilization of Clb1p–4p, and destabilization of Sic1p. In turn, the Clb/Cdc28p activities repress *CLN1* and *CLN2* transcription and later sow the seeds of their own destruction by destabilizing Clb1p–4p (and perhaps stabilizing Sic1p). Degradation of Clb1p–4p may be responsible for the decay of *CLB1* and *CLB2* transcripts, and for the induction of *SIC1* transcripts, at the end of anaphase.

In summary, the available data provide a rough but reasonably satisfying model for how cyclin/Cdc28p oscillations occur. However, there are clearly many questions still to be answered.

D. Which Events Are Controlled by Which Forms of Cdc28p?

Most temperature-sensitive (*ts*) *cdc28* alleles arrest the cell cycle at Start in G_1 (Reed 1980), as does overexpression of dominant-negative *CDC28* alleles (Mendenhall et al. 1988). DNA replication, SPB duplication, and bud initiation all fail to occur, suggesting that all of these events require Cdc28p activity. The fact that most later events can apparently be completed normally under restrictive conditions in these mutants was interpreted initially as evidence that Cdc28p function was restricted to G_1 (Reed 1980). However, an unusual temperature-sensitive allele, *cdc28-*

1N, blocked nuclear division and cytokinesis, suggesting that later cell cycle events also require Cdc28p activity (Piggott et al. 1982). Subsequent studies confirmed that other temperature-sensitive *cdc28* alleles also produce defects in late cell cycle events following prolonged shift to the restrictive temperature (Reed and Wittenberg 1990; Surana et al. 1991). Detailed examination revealed that a large number of both early and late cell cycle events could be blocked or delayed by *cdc28* mutations. Examination of *cln* and *clb* mutants confirmed this conclusion and suggested specific roles for different cyclin/Cdc28p complexes. Most of the events affected by *cln* and *clb* mutations are also affected in some *cdc28* mutants, so it seems very likely that all Cln and Clb functions are mediated through the kinase activity of cyclin/Cdc28p complexes.

No one Cln or Clb is essential for viability in *S. cerevisiae*, and even most double knockouts of *CLN* or *CLB* genes are nonlethal, indicating that there is considerable redundancy in cyclin function. Because of their timing of expression and mutant phenotypes, Cln1p–3p have been referred to as "G_1 cyclins," Clb5p–6p as "S-phase cyclins," and Clb1p–4p as "mitotic cyclins." We consider each of these sets in turn.

1. G_1 Cyclin/Cdc28p Complexes

cln1 cln2 cln3 triple knockout strains are inviable and arrest at Start with a phenotype similar to that of most *cdc28* mutants (Richardson et al. 1989; Cross 1990). Any one *CLN* gene is sufficient for viability, leading to the early hypothesis that these three cyclins have redundant roles. However, the differences in protein structure, periodicity of expression, and associated kinase activity between Cln3p, on the one hand, and Cln1p and Cln2p, on the other, made this model unattractive (for review, see Cross 1995). It was then suggested that Cln3p, as the sole cyclin present throughout G_1, acts as a "pilot light" that turns on transcription of *CLN1*, *CLN2*, and other late G_1 transcripts at the appropriate time; the genes thus induced are then responsible for triggering DNA replication, SPB duplication, and bud initiation (Fig. 6) (Cross 1988, 1995; Nash et al. 1988; Tyers et al. 1993). Disruption of *CLN3* results in delayed induction of late G_1 transcripts, whereas increased expression of *CLN3* causes premature induction (Dirick et al. 1995; Stuart and Wittenberg 1995 and pers. comm.), consistent with the hypothesis that Cln3p controls the timing of transcriptional induction. This hypothesis is also supported by the earlier observations that the presence of dominant (hyperactive) alleles of *CLN3* (originally called *WHI1-1* and *DAF1-1*), or overexpression of the wild-type gene, accelerated cells past Start and resulted in a small-

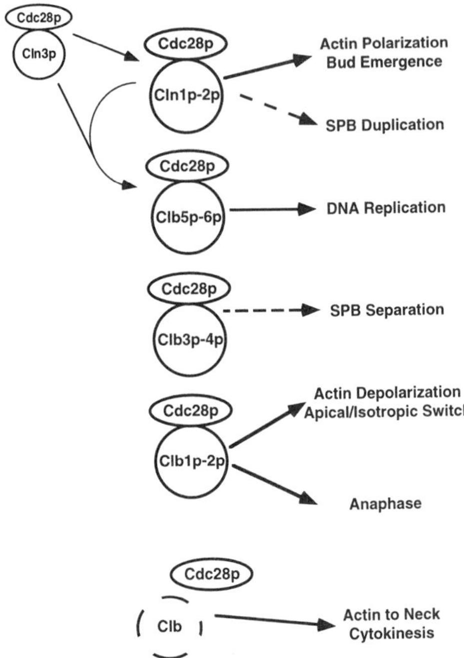

Figure 6 Landmark events controlled by cyclin/Cdc28p complexes. The complexes primarily responsible for triggering particular events are indicated. The connections between Cln1p–2p/Cdc28p and SPB duplication and between Clb3p–4p/Cdc28p and SPB separation are still uncertain, as indicated by the dashed arrows. Cln3p/Cdc28p is only implicated in controlling the landmark events indirectly, through transcriptional induction of *CLN1*, *CLN2*, *CLB5*, and *CLB6*. Cln1p–2p/Cdc28p complexes are required for timely DNA replication, but this is thought to reflect the role of these complexes in activating Clb5p–6p/Cdc28p complexes by triggering Sic1p degradation. Reorganization of actin to the neck and cytokinesis require inactivation of Clb1p–4p/Cdc28p complexes by Clb destruction, as indicated by the broken circle.

er cell size, whereas deletion of the gene caused a delay of Start and a larger cell size (Sudbery et al. 1980; Cross 1988; Nash et al. 1988). Transcriptional induction by Cln3p requires Cdc28p (Dirick and Nasmyth 1991; Marini and Reed 1992), suggesting that Cln3p/Cdc28p kinase provides the trigger for transcriptional induction. In the absence of Cln3p, transcriptional induction does occur eventually (Stuart and Wittenberg 1995), and *cln3* null mutant cells are viable.

Activation of Cln1p/Cdc28p, Cln2p/Cdc28p, or Cln3p/Cdc28p kinase in G_1-arrested *cln1 cln2 cln3* cells leads to a rapid induction of late G_1 transcripts, including *CLN1* and *CLN2* (Cross and Tinkelenberg 1991;

Dirick and Nasmyth 1991). This observation indicated the potential for positive feedback in *CLN1* and *CLN2* expression, whereby initial induction of *CLN1* or *CLN2* would lead to an increase in Cln1p–2p/Cdc28p kinase activity, which would in turn elevate *CLN1* and *CLN2* transcription in an autocatalytic positive feedback loop (Cross and Tinkelenberg 1991; Dirick and Nasmyth 1991). However, recent experiments indicate that full induction of late G_1 transcripts (including *CLN1* and *CLN2*) occurs with normal timing in *cln1 cln2 CLN3* cells, showing that Cln3p is sufficient to induce these transcripts on time and arguing against a significant contribution from the positive feedback loop (Dirick et al. 1995; Stuart and Wittenberg 1995). Nevertheless, it remains possible that the feedback loop contributes to the sharpness of transcriptional induction, even if not to the timing of its onset.

Although induction of late G_1 transcripts occurs promptly in *cln1 cln2 CLN3* cells, DNA replication and bud emergence are significantly delayed (Dirick et al. 1995; Stuart and Wittenberg 1995). (The timing of SPB duplication has not been examined in these cells.) In contrast, DNA replication and bud emergence occur prematurely in cells that express Cln2p prematurely because of a stabilizing mutation or expression from a heterologous promoter (Hadwiger et al. 1989b; Lew et al. 1992). This suggests that Cln1p–2p/Cdc28p complexes directly or indirectly trigger DNA replication and bud initiation (and probably also SPB duplication) (Fig. 6). If this is so, how do cells that lack Cln1p and Cln2p eventually succeed in replicating their DNA and forming a bud? In this regard, observations on *CLB5, CLB6, PCL1,* and *PCL2* appear to be relevant. These other cyclin genes are also transcribed in late G_1 (see above and Table 1), *cln1 cln2 clb5 clb6* cells are unable to replicate DNA and are inefficient in bud formation (Schwob and Nasmyth 1993), and *cln1 cln2 pcl1 pcl2* cells are unable to form buds (Espinoza et al. 1994; Measday et al. 1994). Thus, it appears that Clb5p, Clb6p, Pcl1p, and Pcl2p are also induced to accumulate by Cln3p and can contribute to the triggering of DNA replication and bud formation (and probably also SPB duplication), at least when Cln1p and Cln2p are absent. The Cln1p–2p requirement for timely DNA replication probably reflects the need to degrade Sic1p to allow activation of Clb5p–6p/Cdc28p complexes (see Section V.C.7), a hypothesis supported by the observation that deletion of *SIC1* rescues the inviability of a *cln1 cln2 cln3* triple mutant (Schneider et al. 1996; Tyers et al. 1996).

2. *S-phase Cyclin/Cdc28p Complexes*

Cells lacking Clb5p and Clb6p are larger and grow more slowly than wild-type cells (Epstein and Cross 1992; Kühne and Linder 1993;

Schwob and Nasmyth 1993; Basco et al. 1995). Induction of late G_1 transcripts is followed rapidly by bud emergence, but DNA replication (which is normally concurrent with bud emergence) is delayed significantly in *clb5 clb6* mutants (Schwob and Nasmyth 1993). Moreover, in the absence of Clb5p and Clb6p, DNA replication becomes absolutely dependent on both the Cln1p–2p and the Clb3p–4p cyclins (Schwob and Nasmyth 1993). Thus, Clb5p and Clb6p are probably responsible for initiating and perhaps sustaining DNA replication in the normal cell cycle (Fig. 6). In their absence, Clb3p–4p can apparently perform the same function(s), but DNA replication is delayed until sufficient Clb3p–4p accumulate.

3. *Mitotic Cyclin/Cdc28p Complexes*

Analysis of *clb3 clb4* mutants has not yet revealed any cell cycle perturbation (Fitch et al. 1992; Richardson et al. 1992). However, the kinetics of SPB separation to form a short spindle have not been examined in these cells, and, in the absence of Clb3p–4p, both the Clb5p–6p and the Clb1p–2p cyclins are required for this event (Fitch et al. 1992; Richardson et al. 1992; Schwob and Nasmyth 1993). Thus, it seems likely that Clb3p and Clb4p are normally responsible for promoting initial SPB separation (Fig. 6) but that their function can also be supplied by the other Clb cyclins.

The late-accumulating cyclins Clb1p and Clb2p are the only cyclin pair that is indispensable in *S. cerevisiae* (Surana et al. 1991; Fitch et al. 1992; Richardson et al. 1992). Cells lacking Clb2p (the major cyclin of the pair) display a marked delay in anaphase onset and in depolarization of bud growth (the apical-isotropic switch), and conditional mutants in which Clb1p and Clb2p are both inactivated are unable to perform these events (Surana et al. 1991; Fitch et al. 1992; Richardson et al. 1992; Lew and Reed 1993). Thus, Clb1p–2p control both the cytoplasmic apical-isotropic switch and nuclear spindle elongation. Premature expression of Clb1p or Clb2p induces a premature apical-isotropic switch, consistent with a rate-limiting role for these cyclins (Lew and Reed 1993). However, the timing of anaphase onset is not greatly accelerated by premature Clb1p–2p expression, and the rate-limiting control(s) for this process is unknown.

Cells lacking all Clb cyclins are able to initiate bud formation; thus, unlike DNA replication, this event only requires Cln1p–2p (Schwob et al. 1994). SPB duplication has not been examined in cells lacking all Clb cyclins, but a *clb1 clb2 clb3 clb4* mutant undergoes SPB duplication

(Fitch et al. 1992), and *cdc4* and *cdc34* mutants are able to duplicate their SPBs (Pringle and Hartwell 1981) as well as form buds, despite lacking detectable Clb/Cdc28p activity because of the failure to degrade Sic1p (Schwob et al. 1994). Thus, it seems likely that SPB duplication also does not require Clb cyclins (Fig. 6). *clb1 clb2 clb3 clb4* cells were reported to be incapable of completing DNA replication (Richardson et al. 1992). However, it now appears that the cells used in that study also lacked Clb5p, and another study found no evidence for incomplete DNA replication in *clb1 clb2 clb3 clb4* cells (Fitch et al. 1992). Thus, although these cyclins may contribute in a minor way to DNA replication (as suggested by the observations on *clb5 clb6* cells; see above), it seems likely that Clb5p–6p normally perform this role. *clb1 clb2 clb3 clb4* mutants have "activated" kinetochores as measured by an in vitro microtubule-binding assay, but they do not condense their DNA (D. Koshland, pers. comm.).

4. *Role of Mitotic Cyclin Destruction*

Overexpression of Clb1p, Clb2p, or Clb3p delays spindle disassembly and cytokinesis following anaphase, and either massive overexpression or expression of stabilized mutant forms of these cyclins blocks these events altogether (Ghiara et al. 1991; Stueland et al. 1993; Surana et al. 1993). Such cells are released and undergo cytokinesis upon inactivation of Cdc28p (through temperature shift of a $cdc28^{ts}$ mutant), showing that the arrest is due to maintenance of elevated Clb/Cdc28p activity (Ghiara et al. 1991; Lew and Reed 1993). Thus, Clb destruction and consequent inactivation of Cdc28p following anaphase are essential for cytokinesis (Fig. 6) (King et al. 1996). Interestingly, inactivation of Cdc28p prior to anaphase does not accelerate cytokinesis. Instead, there is a reversal of the apical-isotropic switch, and cells resume growth at the bud tip; anaphase and cytokinesis do eventually occur, but only after a delay (Lew and Reed 1993; D. Lew, unpubl.). Thus, the rate-limiting control(s) for cytokinesis remains unknown.

Temperature-sensitive mutants (*cdc16*, *cdc23*, *cdc26*, and *cdc27*) defective in the Clb1p–4p destruction machinery (the APC; see Section V.C.6) arrest cell cycle progression with large buds, replicated DNA, and short spindles (Pringle and Hartwell 1981). Because a failure to degrade Clb1p–4p results in arrest with long spindles and segregated chromosomes (see above), the arrest phenotype of these mutants implies that the APC has a role in promoting anaphase that is distinct from its role in targeting Clb1p–4p for rapid destruction. Indeed, the original *cdc16-1* and *cdc27-1* alleles did not produce any detectable defect in Clb2p degradation but were defective in promoting anaphase (Irniger et al. 1995). An elegant experiment showed that *cdc23* mutations did not prevent anaphase in cells that lacked sister chromatids, suggesting that the APC is required specifically for sister chromatid disjunction rather than for anaphase chromosome movement (Irniger et al. 1995).

The simplest hypothesis to explain these findings is that the APC targets other proteins, in addition to Clb1p–4p, for cell-cycle-regulated degradation (King et al. 1996). Two noncyclin APC targets have been identified to date in *S. cerevisiae*, Pds1p and Ase1p (Pellman et al. 1995; Yamamoto et al. 1996a,b). The abundances of Pds1p and Ase1p fluctuate during the cell cycle with patterns similar to that of Clb2p, and both proteins contain destruction boxes that target them for APC-dependent degradation (Pellman et al. 1995; Yamamoto et al. 1996b). Ase1p has been implicated in anaphase spindle elongation and accumulates on the spindle midzone (Pellman et al. 1995; Yamamoto et al. 1996a,b), and expression of a nondegradable mutant Ase1p delays spindle disassembly following anaphase (Juang et al. 1997). Pds1p is required for sister chromatid cohesion and plays a poorly defined part in spindle assembly. Strikingly, mutants defective in APC components require Pds1p to prevent spindle elongation (Yamamoto et al. 1996b). This suggests that Pds1p acts as an "anaphase inhibitor" that directly or indirectly prevents sister chromatid disjunction, so that APC-mediated degradation of Pds1p is necessary for anaphase onset. This model is supported by the observation that expression of a nondegradable mutant Pds1p inhibits the onset of anaphase (Cohen-Fix et al. 1996).

5. *Summary*

The picture emerging from these studies (Fig. 6) is that most (if not all) major cell cycle events require changes in Cdc28p activity and that each pair of cyclins is responsible for promoting specific events. In several cases, changes in Cdc28p activity appear to be the rate-limiting controls that trigger cell cycle events. For instance, bud initiation occurs upon Cln1p–2p/Cdc28p activation, initiation of DNA replication occurs upon Clb5p–6p/Cdc28p activation, and the apical-isotropic switch in bud growth occurs upon Clb1p–2p/Cdc28p activation, even if the timing of these Cdc28p activity changes is experimentally accelerated (Lew and Reed 1993; Schneider et al. 1996). However, in other cases, events that require particular changes in Cdc28p activity cannot be accelerated simply by accelerating the relevant Cdc28p change. For instance, anaphase onset requires Clb1p–2p/Cdc28 activation but is not accelerated by overproduction of Clb1p or Clb2p, and cytokinesis requires Clb1p–4p degradation but is not accelerated by premature inactivation of Cdc28p (Lew and Reed 1993, 1995a). In these cases, the rate-limiting controls are unknown. Anaphase onset requires one or more APC functions other than Clb1p–4p proteolysis, and the same may be true for cytokinesis; thus, APC-mediated degradation processes may be the rate-limiting controls for both events. If so, the question is merely pushed back: What is the rate-limiting control for APC activation?

Little is known about the substrates of cyclin/Cdc28p complexes that mediate the control of cell cycle events. The best understood instances pertain to the regulation of components of the cell cycle clock themselves; for instance, Sic1p is a substrate of Cdc28p complexes, and its phosphorylation probably targets it for destruction (see Section V.C.7). In many cases, we can identify putative targets that may mediate Cdc28p control of cell cycle events, but the mechanisms by which these targets are regulated remain obscure. For instance, the transcription factors Swi4p, Swi6p, Mbp1p, Mcm1p, Swi5p, and Ace2p probably mediate Cdc28p-dependent transcription of many cell-cycle-regulated genes, and APC components probably mediate Cdc28p-dependent degradation of various proteins. Orc1p–6p, Cdc6p, and the MCM proteins are likely to mediate Cdc28p control of DNA replication; the cell-cycle-regulated phosphorylation of Spc110p and Spc42p (Donaldson and Kilmartin 1996; Friedman et al. 1996) suggests that they may mediate SPB behavior; and microtubule motors (including Cin8p, Kip1p, Kar3p, and Dhc1p [Dyn1p]) probably mediate spindle assembly and elongation. Many of these proteins contain consensus Cdc28p phosphorylation sites, but analyses of whether they are phosphorylated by Cdc28p in vivo and of whether their phosphorylation has functional consequences are only in the initial stages. This is an area of intense investigation, so that rapid progress in identifying the physiologically relevant substrates of the cyclin/Cdc28p complexes is likely. We anticipate that there are many such substrates but that not all cell cycle activities are regulated directly by cyclin/Cdc28p-mediated phosphorylation (see also Section VII.B).

E. How Are Events Restricted to Once per Cell Cycle?

Cell cycle events, once triggered by changes in cyclin/Cdc28p activities, do not recur until the next cell cycle. In some cases, the singularity of cell cycle events is easily explained. For instance, sister chromatid separation, once triggered, cannot recur until the separated chromosomes have been replicated again (i.e., in the next cell cycle). In other cases, the basis for the restriction of events to once per cell cycle is not obvious.

1. DNA Replication

Once an origin of replication has fired in a particular cycle, it does not do so again, even while S-phase-promoting Clb/Cdc28p complexes persist. Several groups have investigated the basis for this restriction by isolating mutants that *do* reinitiate DNA replication without going through mitosis (Broek et al. 1991; Hayles et al. 1994; Dahmann et al. 1995; Singer et al. 1996). One important conclusion to emerge from these studies is that B-type cyclins (i.e., Clb1p–6p in *S. cerevisiae*) are required to prevent reinitiation of DNA replication within a cell cycle (Hayles et al. 1994; Dahmann et al. 1995). Given the available information on the control of origin firing in *S. cerevisiae* (see Section III.A and Fig. 1), it has been suggested that Clb/Cdc28p kinases both trigger origin firing and destroy the cell's ability to assemble new preinitiation complexes at the spent origins (Dahmann et al. 1995; Diffley 1996; Stillman 1996). Thus, once an origin has fired, it cannot assemble a new preinitiation complex until Clb cyclins are degraded following anaphase. If assembly of a preinitiation complex is indeed essential for an origin to fire, this mechanism would effectively limit origin firing to once per cell cycle. Strong support for this model has come from the observation that inhibition of Clb/Cdc28p complexes in nocodazole-arrested (pre-anaphase) cells through Sic1p overexpression leads to the appearance of preinitiation footprints at origins of replication (Dahmann et al. 1995). If Sic1p is then shut off, Clb/Cdc28p complexes are reactivated and promote origin firing, leading to rereplication.

How do Clb/Cdc28p complexes prevent the assembly of preinitiation complexes? Assembly of these complexes requires Cdc6p, an unstable protein that is synthesized in the period between the end of anaphase and the beginning of S phase (Zwerschke et al. 1994; Bell et al. 1995; Piatti et al. 1995; Cocker et al. 1996; Santocanale and Diffley 1996; Detweiler and Li 1997; Section III.A). Could the role of Clb/Cdc28p complexes be simply to ensure that Cdc6p does not accumulate once DNA replication has begun? This question was addressed by Piatti et al. (1996), who found that Clb/Cdc28p activation defined a "point of no return," following which even Cdc6p overexpression was unable to promote formation of preinitiation complexes. Thus, at least two mechanisms prevent preinitiation complex formation following Clb/Cdc28p induction: Cdc6p is not expressed, and even if some Cdc6p accumulates, it cannot be assembled into preinitiation complexes. One factor that may contribute to this regulation is the exclusion of MCM proteins from the nucleus (Hennessy et al. 1990; Yan et al. 1993; Tye 1994; Dalton and Whitbread 1995), which may also be Clb/Cdc28p-dependent.

Another finding to emerge from the search for rereplicating mutants is that mutant alleles of *CDC16* and *CDC27* permit cells to reinitiate DNA replication even in the presence of high levels of Clb/Cdc28p activity (Heichman and Roberts 1996). Cdc16p and Cdc27p are components of the APC (see Sections V.C.6 and V.D.4). An attractive explanation for these observations is that Clb/Cdc28p complexes restrict formation of preinitiation complexes by inducing the APC-dependent ubiquitination, and hence destruction, of some key preinitiation factors (Heichman and Roberts 1996).

If Clb/Cdc28p kinases both activate origin firing and prevent assembly of new preinitiation complexes, then inactivation of these kinases in G_2 should permit formation of preinitiation complexes but not rereplication of DNA (because the newly competent origins would require fresh Clb/Cdc28p kinase to fire). Thus, rereplication requires cycles of Clb/Cdc28p activation and inactivation. Deletion of the B cyclin gene *cdc13* in *S. pombe* or of *CLB1–CLB5* in *S. cerevisiae* permits rereplication of DNA, apparently in whole-genome increments (Hayles et al. 1994; S. Haase and S. Reed, pers. comm.). In the *S. cerevisiae* experiments, it appears that the level of the remaining B-type cyclin, Clb6p, oscillates even in the absence of the mitotic cyclins (S. Haase and S. Reed, pers. comm.). This observation suggests that rereplication in these strains is driven by cyclic Clb6p/Cdc28p activation.

Although the model discussed above appears to be sufficient to explain both how the cell prevents the recurrence of an entire S phase within a cell cycle and how it prevents the rereplication of portions of its genome within a given S phase, it remains possible that these two problems involve separate mechanisms. In this regard, it may be relevant that all of the rereplication mutations discussed above appear to cause repetition of entire S phases in the absence of anaphase or cytokinesis. In contrast, mutations in *DOA4* (which encodes a ubiquitin hydrolase, or ubiquitin-removing enzyme) appear to cause rereplication of only a portion of the genome (Singer et al. 1996). Although the interpretation of these observations is not yet clear, it is possible that the *doa4* mutant is defective in a mechanism specific for prevention of reinitiation from particular origins within a single S phase.

2. *Bud Formation*

Bud formation involves a polarization of secretion toward a discrete site on the cell surface. This polarization is triggered by Cln/Cdc28p activation, and secretion is subsequently depolarized in response to Clb/Cdc28p activation (Lew and Reed 1993, 1995b; see Section III.C and Fig. 1). In the absence of Clb cyclins, this depolarization does not occur, and after a period of polarized growth of the initial bud, the cell establishes a new growth site on the cell surface and forms what appears to be a second bud (Schwob et al. 1994). A similar phenotype is observed in *cdc4*, *cdc34*, and *cdc53* mutants (Pringle and Hartwell 1981; Adams and Pringle 1984; Mathias et al. 1996). These mutants fail to degrade Sic1p and hence fail to activate Clb/Cdc28p kinases (see Section V.C.7); deletion of *SIC1* in these mutants permits Clb/Cdc28p activation and prevents rebudding (Schwob et al. 1994). These provocative observations suggest that there is an inhibition of bud initiation by Clb/Cdc28p complexes. Indeed, expression of a stabilized mutant Clb1p in G_1 prevents bud formation (Lew and Reed 1993). Thus, Clb/Cdc28p kinases may contribute to the restriction of bud formation, as well as DNA replication, to once per cycle. However, it should be noted that this model does not address the equally interesting question of why a cell does not form two or more buds simultaneously during a given cell cycle.

3. *Spindle Pole Body Duplication*

Control of SPB duplication has been investigated by searching for mutants that accumulate extra SPBs (*esp* mutants). To date, no mutant has been identified that reduplicates SPBs within a cell cycle; instead, the known *esp* mutants undergo a defective mitosis and segregate extra SPBs into one of the progeny cells (McGrew et al. 1992; Pellman et al. 1995). Thus, the possible role of Clb cyclins in preventing SPB reduplication remains unresolved. Examination of cells lacking Clb1p–4p (*clb1*Δ *clb3*Δ *clb4*Δ *clb2*^{ts}; Fitch et al. 1992) or the activity of Clb1p–6p/Cdc28p complexes (*cdc34* and *cdc53* mutants; Pringle and Hartwell 1981; Mathias et al. 1996) failed to detect reduplication of SPBs, but other combinations of Clb depletions remain to be tested. Reduplication of SPBs may be prevented, at least in part, simply by restricting production of SPB structural components to around the normal time of SPB duplication (Donaldson and Kilmartin 1996).

4. *Summary*

In summary, it appears that cyclin/Cdc28p complexes are important not only for triggering cell cycle events, but also for preventing some events from occurring more than once per cell cycle. The mechanisms responsible for this are under intensive investigation.

VI. CHECKPOINT CONTROLS

Early studies on the effects of perturbing cell cycle events in *S. cerevisiae* established that certain events (e.g., DNA replication, anaphase, and cytokinesis) are linked in "dependent pathways" (Hartwell et al. 1974; Pringle and Hartwell 1981). This means that when performance of one event (e.g., DNA replication) is blocked by some experimental manipulation, performance of the later events (anaphase, cytokinesis) is also blocked or greatly delayed, indicating that execution of the later events is in some way dependent on successful completion of the earlier event. Initially, these observations appeared to argue against models in which a central clock triggers multiple cell cycle events. Instead, it appeared that the cell cycle was best viewed as a set of several dependent pathways that were mutually independent but coordinated by their common dependence on the Start event at the beginning of the cell cycle (Hartwell et al. 1974; Pringle and Hartwell 1981). Increasing appreciation of the multiple cell cycle roles and evolutionary conservation of cyclin/CDK regulators (and the support that this provided for the idea of a central cell cycle clock) thus led to an apparent paradox: How could cells maintain the observed dependency relationships if events were really triggered independently by an autonomous cell cycle clock?

This apparent paradox was resolved by the recognition that cell cycle dependencies could be established in either of two ways (Hartwell and Weinert 1989). First, the two events could be mechanistically linked; for example, completion of one event could produce a substrate necessary for execution of the second event (the "substrate-product" model, as well established for several steps in bacteriophage morphogenesis). Second, the two events could be mechanistically unlinked, but a regulatory pathway could ensure that the later event does not begin until the earlier one has been completed (the "checkpoint control" model).

To assess whether a particular dependency was the result of a mechanistic link or a checkpoint control mechanism, Hartwell and Weinert (1989) suggested an empirical criterion based on whether specific mutations or drugs could be found that uncoupled the dependent events and permitted the second event to occur even when the first event was blocked. They argued that dependencies resulting from mechanistic links would be much harder (and perhaps impossible) to uncouple than would dependencies resulting from checkpoint controls. Hence, the existence of uncoupling mutations or drugs would support the checkpoint model and identify functions in the checkpoint control pathway. For example, several mutations have now been identified that allow continued cell cycle progression even when DNA replication or nuclear division is blocked, and it is generally accepted that these, and probably other, cell cycle dependencies result from checkpoint control pathways.

Since its introduction, the term "checkpoint" has gained considerable popularity and has been used in many different contexts both within and outside of the cell cycle control field. At least three distinct but interrelated meanings have been used in relation to cell cycle control: (1) Checkpoint controls are signaling pathways that act to delay cell cycle progression when perturbations delay completion of certain cell cycle events (the original definition by Hartwell and Weinert 1989). (2) Checkpoints are critical stages (often equated with regulatory transitions) during the cell cycle at which cells pause to assess whether certain events have been completed before continuing with the cell cycle (see, e.g., Murray 1992). If the events have not been completed, the cells arrest at the checkpoint (e.g., the G_2/M boundary). (3) Checkpoint controls are molecular mechanisms for inhibiting cell cycle progression. In this usage, any negative regulator (e.g., Sic1p or Swe1p; see Section V.C) is a checkpoint regulator, whether or not regulation occurs in response to the status of cell cycle events. Thus, in current usage, checkpoints can be signaling pathways, stages in the cell cycle, or negative regulators of cell cycle progression, making the term ambiguous and sometimes confusing. In this chapter, we use exclusively the first meaning, which we believe to be the most useful definition.

Checkpoint control pathways are hypothesized to consist of a monitor or sensor that can detect the progress of one cell cycle event, a signal transducer and perhaps additional signaling components that can convey the signal to the target, and a target in the form of a regulator that can delay the initiation of one or more subsequent events. Given the information summarized in previous sections, an attractive hypothesis for how checkpoint controls can delay cell cycle events is that they act to stop the cyclin/Cdc28p cell cycle clock. Research during the past few years has focused on identifying the molecular components of checkpoint control pathways, with the goal of understanding how cell cycle events are monitored, how information is transduced, and how the cell cycle is regulated.

A. The DNA Replication Checkpoint

A delay in DNA replication caused by inhibition of dNTP synthesis (using hydroxyurea, which inhibits ribonucleotide reductase, or a *cdc8^{ts}*

mutation that encodes a thermolabile thymidylate kinase) or by inhibition of elongation factors or polymerases (e.g., using a *cdc17*^{ts} mutation that encodes a thermolabile DNA polymerase α) results in a delay of nuclear division due to the "DNA replication checkpoint" (Weinert and Hartwell 1993; Allen et al. 1994; Weinert et al. 1994). In principle, this checkpoint control could respond either to the presence of unreplicated DNA or to the presence of replication fork structures. Unlike treatments that perturb elongation of replication forks, mutations that effectively prevent *initiation* of DNA replication (e.g., depletion of Cdc6p [Piatti et al. 1995] or temperature-sensitive *cdc7* mutations in some strain backgrounds [Toyn et al. 1995]) fail to induce a subsequent delay in nuclear division. (Although these strains may be defective in a different checkpoint [see Section VI.F], the DNA replication checkpoint itself appears to be functional.) The mutant cells contain a full genome's worth of unreplicated DNA but no replication forks. Thus, unreplicated DNA per se does not appear to cause a delay in nuclear division, and it is generally assumed that the signal indicating incomplete DNA replication originates at the replication fork.

Unlike the large majority of treatments that block elongation of replication forks, certain temperature-sensitive alleles of genes encoding replication complex components permit anaphase to be initiated before completion of DNA replication at the restrictive temperature. As a consequence, cells undergo a catastrophic attempt to segregate incompletely replicated genomes, leading to DNA fragmentation, unequal segregation, and death. Such checkpoint-defective alleles have been identified for three genes: *POL2*, encoding DNA polymerase ε (Navas et al. 1995); *DPB11*, whose product interacts with DNA polymerase ε (Araki et al. 1995); and *RFC5*, which encodes a subunit of replication factor C (Sugimoto et al. 1996). The fact that mutations in replication fork components can result in checkpoint defects strengthens the hypothesis that the replication fork is the origin of the checkpoint signal and raises the exciting possibility that one or more of these components (Pol2p, Dpb11p, or Rfc5p) transmit the signal to a downstream checkpoint protein (Fig. 7).

Mutants that display defects in the DNA replication checkpoint but are fully competent to replicate DNA have also been identified in genetic screens. To date, such checkpoint mutations have been identified in two genes, *MEC1* (*ESR1, SAD3*) and *RAD53* (*SPK1, MEC2, SAD1*) (Allen et al. 1994; Weinert et al. 1994). Cells harboring checkpoint-defective *mec1* or *rad53* alleles do not display any readily detectable cell cycle defects unless DNA replication is blocked (e.g., with hydroxyurea), in which case they undergo a catastrophic mitosis (see above). This suggests that

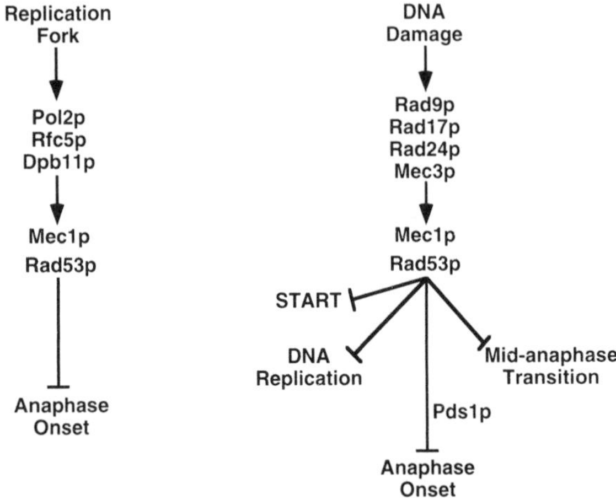

Figure 7 The DNA replication (*left*) and DNA damage (*right*) checkpoints. Both pathways involve Mec1p and Rad53p, which respond to signals emanating from DNA-bound complexes. DNA replication occurs only during S phase (by definition), and a defect delays the subsequent anaphase. DNA damage can occur at any point in the cell cycle, and the checkpoint delays different events depending on when the damage occurs. Pds1p is required for the blockage of anaphase onset by the DNA damage checkpoint but not by the DNA replication checkpoint. The figure focuses on cell cycle regulation, but many of the same checkpoint proteins are required for the transcriptional responses to stalled DNA replication or DNA damage.

the DNA replication checkpoint is not essential during the unperturbed cell cycle and only becomes critical when DNA replication is delayed. Surprisingly, however, both *MEC1* and *RAD53* are essential genes; their essential functions remain unclear.

Mec1p is a member of the phosphatidylinositol (PI) kinase family, a group of large, multidomain proteins that share greatest homology in a carboxy-terminal domain that possesses either PI 3-kinase activity or protein kinase activity in different family members (for review, see Hunter 1995). The substrates of this domain in Mec1p have not yet been determined. Interestingly, one member of this family (mammalian DNA-dependent protein kinase or DNA-PK) is recruited to DNA breaks by regulatory subunits called Ku proteins; this recruitment activates DNA-PK to phosphorylate substrate proteins (for review, see Jeggo et al. 1995). This suggests a simple model in which Mec1p is recruited to replication forks and activated to phosphorylate downstream substrates. *MEC1* relatives in fission yeast (*rad3*) and humans (*ATM*) have also been found to function in checkpoint pathways (for review, see Zakian 1995), suggesting that this role is highly conserved.

Rad53p is a protein kinase that becomes phosphorylated and activated upon treatment of cells with hydroxyurea (Sanchez et al. 1996; Sun et al. 1996). Rad53p phosphorylation in response to hydroxyurea is abolished in the checkpoint-defective mutants of *POL2* and *RFC5* and in *mec1* mutants, suggesting that Rad53p acts downstream from Pol2p and Rfc5p and downstream from or at the same step as Mec1p (Fig. 7) (Sanchez et al. 1996; Sun et al. 1996). It is possible that Rad53p is a direct substrate for Mec1p, although the relationship of these proteins to each other and the identity of their downstream targets remain to be determined.

The mechanism by which the DNA replication checkpoint delays cell cycle progression in *S. cerevisiae* is unknown. Formation of a short bipolar spindle is unaffected by blocking DNA replication (Pringle and Hartwell 1981; Weinert and Hartwell 1993), suggesting that the checkpoint acts downstream from spindle assembly to prevent anaphase. As discussed above (see Sections V.D.3 and V.D.4), anaphase onset requires both the activation of Clb1p–2p/Cdc28p kinase complexes and the APC-dependent degradation of an anaphase inhibitor, so that these processes are attractive candidates for the checkpoint target. Hydroxyurea-treated cells accumulate some Y19-phosphorylated (and thereby inhibited) Cdc28p, but mutation of *CDC28* to prevent this phosphorylation does not detectably alter the checkpoint response (Amon et al. 1992). In vitro assays suggest that Clb1p–2p/Cdc28p kinase complexes attain a respectable degree of activity in hydroxyurea-treated cells and can be activated to very high levels by overexpression of *CLB1* or *CLB2* (Stueland et al. 1993). However, this activation does not lead to anaphase in hydroxyurea-treated cells (Stueland et al. 1993). Although not conclusive, these experiments have failed to reveal a role for inhibition of Clb/Cdc28p complexes in the arrest mediated by the DNA replication checkpoint, perhaps indicating that a later step, like the destruction of an anaphase inhibitor, is regulated by the checkpoint control. Deletion of *PDS1*, which encodes the only known anaphase inhibitor, does not affect arrest in response to hydroxyurea (Yamamoto et al. 1996a,b), suggesting that other anaphase inhibitors may be involved.

If the replication forks present during a normal S phase send a checkpoint signal to inhibit anaphase, we might expect that checkpoint mutants lacking this inhibition would undergo an acceleration of nuclear division. This has not been observed in *mec1* or *rad53* mutants, suggesting that the ticking of the cell cycle clock provides ample time for completion of DNA replication in unperturbed cells. This raises the question of whether the checkpoint pathway is activated simply by the presence of normal replication forks (as assumed above) or whether abnormal replication forks (e.g., stalled forks in hydroxyurea-treated cells) are required. Stalled forks may lead to the generation of abnormal structures due to the uncoordinated action of replication fork components (e.g., continued unwinding by DNA helicases in the absence of polymerization would lead to the production of single-stranded DNA). In fact, treatments (e.g., hydroxyurea) that produce stalled forks also lead to the transcriptional induction of several genes that are not induced during a normal S phase, and this transcriptional response is defective in *mec1* or *rad53* mutants (Allen et al. 1994; Kiser and Wienert 1996; Navas et al. 1996). This is consistent with the hypothesis that the DNA replication checkpoint is only one part of a broader Mec1p- and Rad53p-dependent reaction to abnormal replication fork structures; perhaps the transcriptional response induces genes that help to repair the fork defect, whereas the checkpoint response provides time for the repair mechanisms to operate. This view of the checkpoint is particularly attractive by analogy to the cellular response to DNA damage (see below).

B. The DNA Damage Checkpoint

DNA damage caused by irradiation (with X-rays or UV light) or by a variety of chemicals (e.g., alkylating agents) evokes a damage response involving transcriptional induction of repair genes and also delays cell cycle progression. Mutations have been identified that allow continued cell cycle progression even in the presence of DNA damage, so that the normal cell cycle delays are considered to result from a checkpoint control(s). The DNA damage checkpoint is critical for the ability of cells to survive exposure to DNA damaging agents: Mutants defective in this checkpoint lose viability rapidly if their DNA is damaged, presumably because of the replication of damaged chromosomes (generating mutations) and the segregation of broken chromosomes (leading to loss of acentric chromosome fragments) (Weinert and Hartwell 1988).

DNA damage causes delays at various stages of the cell cycle (see below), but most work to date has concentrated on the delay of anaphase onset elicited by DNA damage incurred following S phase. A single double-strand DNA break appears to be sufficient to trigger a delay in anaphase onset (Sandell and Zakian 1993; S.-K. Kim and T. Weinert, unpubl.), suggesting that the checkpoint is very sensitive to damage. In addition to irradiation and chemicals, temperature-sensitive mutations in some DNA metabolism genes (e.g., *CDC9*, encoding DNA ligase [Barker et al. 1985], and *CDC13*, encoding a single-strand telomeric DNA-binding protein [Nugent et al. 1996]) have been used to induce DNA damage (Weinert and Hartwell 1993). Checkpoint mutations have been identified both through the analysis of previously isolated *rad* (radiation-sensitive) mutations and from a genetic screen for mutations that cause *cdc13* cells to die rapidly (Weinert et al. 1994). Checkpoint-defective alleles of *MEC1* and *RAD53* (see Section VI.A) and null mutations in *RAD9*, *RAD17*, *RAD24*, and *MEC3* all permit anaphase to occur inappropriately in cells harboring damage (Weinert and Hartwell 1988, 1990; Allen et al. 1994; Weinert et al. 1994).

Temperature-sensitive *cdc13* mutants complete DNA replication but accumulate single-stranded DNA at telomeres after shift to restrictive temperature, and anaphase (but not the formation of a short bipolar spindle) is blocked (Pringle and Hartwell 1981; Garvik et al. 1995). Accumulation of single-stranded DNA was reduced in *rad17*, *rad24*, and *mec3* mutants and enhanced in *rad9* mutants (Lydall and Weinert 1995). This suggests that Rad9p, Rad17p, Rad24p, and Mec3p participate in the processing of the lesion produced by the *cdc13* mutation (and presumably other lesions as well). Support for the hypothesis that these genes are involved in DNA metabolism comes from the homologies between Rad17p and a 3′-5′exonuclease (Lydall and Weinert 1995) and between Rad24p and subunits of replication factor C (Griffiths et al. 1995). Rad9p and Mec3p do not have relatives of known function. It may be that damaged templates must be processed by Rad9p, Rad17p, Rad24p, and Mec3p in order to produce structures recognized by the checkpoint pathway sensor. Alternatively, these proteins may recognize damaged templates and form a complex that both modifies the lesion and serves as a signal generator for the checkpoint pathway, analogous to the replication fork complex thought to generate the signal for the DNA replication checkpoint (see Section VI.A). In addition, the checkpoint proteins appear to be involved in the transcriptional induction of repair genes (Allen et al. 1994; Aboussekhra et al. 1996; Kiser and Weinert 1996; Navas et al. 1996). Thus, the proteins of the DNA damage checkpoint appear to have several roles in the response to DNA damage, all of which increase the cell's capacity to repair damage; these roles include cell cycle arrest at various stages of the cell cycle (see also below), transcriptional induction of repair genes, and a role (as yet not well defined) directly in DNA repair (Lydall and Weinert 1995, 1996).

The checkpoint signal(s) generated by Rad9p, Rad17p, Rad24p, and Mec3p is thought to be transmitted to Mec1p and Rad53p, as for the DNA replication checkpoint (Fig. 7). Rad53p is strongly phosphorylated and activated by DNA damage, in a manner that depends on both Mec1p and the lesion-processing components (Sanchez et al. 1996; Sun et al. 1996).

The delay of anaphase onset due to the DNA damage checkpoint involves the anaphase inhibitor Pds1p (Yamamoto et al. 1996b). The requirement for Pds1p may not be absolute, in that some cells of a *pds1* null mutant population appear to arrest following DNA damage, whereas others proceed through anaphase and cytokinesis (Yamamoto et al. 1996b). These observations suggest that the DNA damage checkpoint might delay anaphase onset by delaying the APC-mediated degradation of Pds1p (and hence sister chromatid disjunction). In *pds1* mutants, there is no Pds1p to act as the checkpoint effector, and consequently cell cycle progression occurs. It is unclear why some cells in a *pds1* population do in fact delay anaphase; perhaps there is another anaphase inhibitor in addition to Pds1p. The existence of such a second inhibitor might also explain why *pds1* cells are still responsive to the DNA replication checkpoint (i.e., still block anaphase when treated with hydroxyurea; see Section VI.A).

Depending on the stage of the cell cycle at which DNA damage occurs, the damage may cause delays in various cell cycle events (Fig. 7). In the cases that have been examined to date, the *MEC1, RAD53, RAD9, RAD17*, *RAD24*, and *MEC3* checkpoint genes are also required for damage-induced delays at stages of the cell cycle other than anaphase onset (Siede et al. 1993, 1994, 1996; Allen et al. 1994; Paulovich and Hartwell 1995; Longhese et al. 1996; Paulovich et al. 1997). Damage sustained in G_1 causes a delay in bud formation as well as in the onset of DNA replication, and at least part of the checkpoint delay appears to occur before Start (Siede et al. 1994, 1996). These observations could be explained most easily by an inhibition of Cln1p–2p/Cdc28p activation, but this has yet to be demonstrated directly, and the mechanism of arrest remains unclear. DNA damage sustained late in G_1 or during S phase causes a delay or slowing of DNA replication itself (Paulovich and Hartwell 1995). The mechanisms of this delay are unknown but may involve the Mec1p-dependent phosphorylation of replication factor A (Brush et al. 1996). DNA damage sustained following S phase causes a delay of anaphase onset, as discussed above. Finally, it appears that DNA damage can delay cell cycle progression even following anaphase onset. Video microscopy experiments on living cells have shown that cells carrying dicentric chromosomes often pause in the middle of anaphase, following sister chromatid disjunction and partial spindle elongation (Yang et al. 1997). This mid-anaphase delay is *RAD9*-dependent, suggesting that it is caused by the DNA damage checkpoint, presumably in response to breakage of the dicentric chromosomes (Yang et al. 1997), but the mechanism responsible for the delay is unknown. The fact that the DNA damage checkpoint can cause delays at so many different stages of the cell cycle suggests that it can engage a number of different cell cycle regulators. It is unclear whether different regulators are engaged depending on the stage of the cell cycle at which damage occurs or whether a panoply of regulators is mobilized whenever damage occurs, to freeze the cells where they are until the damage can be repaired.

C. The Spindle Assembly Checkpoint

Accurate chromosome segregation in mitosis requires bipolar attachment of all of the replicated chromosomes to the microtubule spindle via their centromere/kinetochore complexes. Studies in metazoan cells with large, easily visualized chromosomes have led to the view that kinetochore capture by spindle microtubules is a stochastic process, so that the time required to capture all of the kinetochores and form bipolar attachments varies significantly from cell to cell (Alberts et al. 1994). However, sister chromatid disjunction and anaphase do not occur until all of the chromosomes are bi-oriented, suggesting that anaphase onset is dependent on correct attachment of all of the chromosomes to the spindle. A single unattached kinetochore can cause a delay of anaphase in at least some metazoan cells, suggesting that proper kinetochore-spindle attachment is monitored by a sensitive checkpoint mechanism that ensures sufficient time for all of the chromosomes to form bipolar spindle attachments before chromosome segregation is allowed to proceed (for review, see Wells 1996).

Perturbation of spindle assembly in *S. cerevisiae* using a low dose of an antimicrotubule agent (nocodazole or benomyl) delays spindle elongation and subsequent cytokinesis (Li and Murray 1991; Wang and Burke 1995). Genetic screens identified six genes (*MAD1–MAD3* and *BUB1–BUB3*) required for the delay of cell cycle progression in the presence of these agents (Hoyt et al. 1991; Li and Murray 1991). Mutants harboring null alleles of these genes are viable but die rapidly, due to massive chromosome missegregation, upon exposure to drugs (or the presence of mutations) that perturb spindle structure. The loss of viability is prevented if anaphase is delayed using hydroxyurea, suggesting that the major role of the *MAD* and *BUB* gene products is to delay anaphase via a "spindle assembly checkpoint" (Hoyt et al. 1991; Li and Murray 1991). In unperturbed cell populations, *mad* and *bub* mutants display an elevated frequency of spontaneous chromosome loss, consistent with the hypothesis that this checkpoint delays anaphase in the occasional cell that takes longer than usual to complete the stochastic chromosome bi-orientation process (Hoyt et al. 1991; Li and Murray 1991).

MAD- and *BUB*-dependent delays of anaphase have been detected following perturbation of chromosome attachment through mutations in centromeres, kinetochore components, or spindle components (for review, see Rudner and Murray 1996). Cells containing a single mutant (partially functional) centromere delay anaphase in each cell cycle, suggesting that the checkpoint pathway can respond with great sensitivity to kinetochore structure or attachment problems (Pangilinan and Spencer 1996). Mutants that are blocked in SPB duplication cannot form bipolar spindles and (with one exception, discussed below) delay cytokinesis (Pringle and Hartwell 1981). Such cells contain monopolar spindles that presumably fail to attach one half of the kinetochores. The cytokinesis delay in these cells is abolished in *mad1* mutants, suggesting that it is mediated by the spindle assembly checkpoint (Weiss and Winey 1996). These findings strongly support the hypothesis that the checkpoint signal originates at unattached or improperly attached kinetochores, but it is not yet clear whether such kinetochore signals are the only means of triggering the spindle assembly checkpoint.

BUB1 encodes a protein kinase (Roberts et al. 1994), and the other spindle assembly checkpoint genes encode novel proteins whose sequences have provided few clues to their functions. Disruption of spindle assembly leads to phosphorylation of Mad1p, and this phosphorylation is dependent on Bub1p, Bub3p, and Mad2p, suggesting that these proteins act upstream of Mad1p in the checkpoint pathway (Fig. 8) (Hardwick and Murray 1995). *MAD2* homologs have been identified in vertebrates, and immunofluorescence data suggest that Mad2p is specifically localized to unattached kinetochores (Chen et al. 1996; Li and Benezra 1996). Assuming that the same holds true for yeast, this exciting observation greatly strengthens the hypothesis that improperly attached kinetochores generate the checkpoint signal. Mad2p binds directly to Mad1p, suggesting that Mad1p phosphorylation may also occur at the kinetochore (Rudner and Murray 1996). Mad3p and Bub2p are not required for Mad1p phosphorylation upon checkpoint activation (Hardwick and Murray 1995), suggesting that they act downstream from Mad1p or in a separate pathway (Fig. 8).

One gene (*MPS1*) required for SPB duplication is also known to be essential for checkpoint function (Weiss and Winey 1996). *mps1^{ts}* mutants shifted to restrictive temperature following SPB duplication undergo anaphase onset even in the presence of spindle defects (Weiss and Winey 1996), and overexpression of Mps1p blocks anaphase onset in a manner that requires all of the *MAD* and *BUB* gene products (Hardwick et al. 1996). These observations are most simply interpreted by suppos-

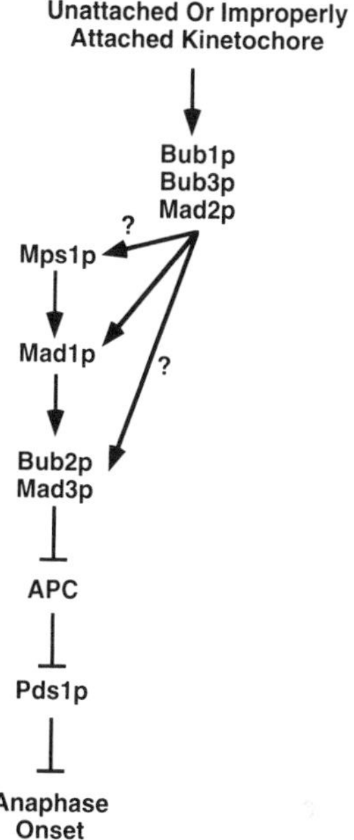

Figure 8 The spindle assembly checkpoint. Unattached or improperly attached kinetochores induce a delay of anaphase onset that requires all of the indicated proteins. The postulated functional relationships among Bub1p-3p, Mad1p-3p, and Mps1p are based largely on determinations of which proteins are and which are not required for Mad1p phosphorylation in response to spindle defects. As Pds1p is also required for the delay of anaphase onset, the delay may be obtained by blocking the APC-dependent degradation of Pds1p (and perhaps other) proteins. See text for more details.

ing that Mps1p "activation" is an early step in the checkpoint response and that Mps1p overexpression mimics this activation, generating a checkpoint-dependent block of anaphase even in the absence of spindle defects (Fig. 8). Mps1p is a protein kinase that can phosphorylate Mad1p in vitro, and overexpression of Mps1p leads to phosphorylation of Mad1p in vivo (Lauzé et al. 1995; Hardwick et al. 1996). Although Bub1p, Bub3p, and Mad2p are required for Mps1p-induced cell cycle arrest, they are not required for Mps1p-induced Mad1p phosphorylation (in contrast to the Mad1p phosphorylation induced by disruption of spindle assembly) (Hardwick et al. 1996). This suggests that Mps1p overexpression does not simply mimic a checkpoint signal and that the checkpoint genes may have more complex roles than suggested by the simple linear pathway described above.

Anaphase onset could be prevented by blocking the APC-dependent degradation of an anaphase inhibitor. Indeed, the anaphase inhibitor Pds1p is required for arrest by the spindle assembly checkpoint: *pds1* mutants precociously disjoin sister chromatids when spindle assembly is prevented (Yamamoto et al. 1996b). The hypothesis that both the DNA damage and spindle assembly checkpoints delay anaphase onset by preventing APC-dependent Pds1p destruction fits well with earlier observations that cells released from an arrest induced by the spindle assembly checkpoint could still delay anaphase in response to DNA damage, and vice versa (Weinert and Hartwell 1990). However, unlike the progression through cytokinesis in *pds1* mutants with DNA damage, *pds1* mutants with spindle defects do not proceed past anaphase and into cytokinesis (Yamamoto et al. 1996b). It has been suggested that whereas the DNA damage checkpoint acts specifically through Pds1p regulation, the spindle assembly checkpoint prevents APC-dependent degradation of multiple substrates including both Pds1p and Clb cyclins (Yamamoto et al. 1996b). Thus, *pds1* mutants with improperly assembled spindles would fail to degrade Clb cyclins and arrest in late anaphase.

D. The Cell Morphogenesis Checkpoint

Many environmental stresses, including sudden alterations in the temperature or osmolarity of the growth medium, result in a transient depolarization of the actin cytoskeleton in *S. cerevisiae*, leading to delays in bud formation (Pringle et al. 1989; Chowdhury et al. 1992; Lillie and Brown 1994). Environmentally stressed cells delay cell cycle progression to accommodate the defect in morphogenesis, thus minimizing the production of binucleate cells. Analysis of cell cycle progression in temperature-sensitive mutants (*cdc24*, *cdc42*) that are defective in bud formation revealed that these cells delay anaphase onset through regulation of the cyclin/Cdc28p cell cycle clock (Lew and Reed, 1995a,b).

G_1 *cdc24* or *cdc42* cells incubated at the restrictive temperature proceed through S phase with normal kinetics and form a short bipolar spindle, despite the absence of a bud (Adams et al. 1990; Lew and Reed 1995a). However, anaphase onset is greatly delayed (except in one atypical strain; see below), and this delay is abolished by mutations that prevent Cdc28p Y19 phosphorylation (including deletion of *SWE1*, overexpression of *MIH1*, and mutation of *CDC28* tyrosine 19 to phenylalanine; see Section V.B.2 and Fig. 4) (Lew and Reed 1995a; Sia et al. 1996). This suggests that the delay arises through a "morphogenesis checkpoint." Recognition of this checkpoint was delayed for some years because the first *cdc24* strain analyzed proceeds efficiently through nuclear division (Hartwell et al. 1974; Pringle and Hartwell 1981; Adams et al. 1990), perhaps because of a (as yet undefined) checkpoint defect. The mechanism for delaying anaphase onset through the morphogenesis checkpoint is distinct from those involved in other checkpoints and involves inhibition of Clb1p–2p/Cdc28p complexes by tyrosine phosphorylation. As with several mutations affecting other checkpoint pathways, *swe1* null mutations, $CDC28^{Y19F}$ mutations, and *MIH1* overexpression have no readily detectable effect on the unperturbed cell cycle (Sorger and Murray 1992; Sia et al. 1996; see Section V.B.2).

The defect monitored by the morphogenesis checkpoint remains obscure. In principle, the checkpoint might detect improper polarization of the actin cytoskeleton, absence of a bud, improper neck structures, or insufficient bud size. Most morphogenesis mutations perturb all of these features, but one mutation (*bed1*, also called *slc2* or *mnn10*) has been described that delays bud emergence even though the actin cytoskeleton is polarized (Mondésert and Reed 1996). *bed1* cells employ the morphogenesis checkpoint to delay anaphase onset and become multinucleate when the checkpoint is compromised (Mondésert and Reed 1996). This suggests that the checkpoint monitors defects in bud formation that lie downstream from actin polarization, although the possibility that some aspect of actin function is defective in the *bed1* mutants cannot be ruled out.

E. Checkpoint Adaptation

The morphogenesis checkpoint appears to display adaptation. Following a prolonged block of bud formation, cells undergo nuclear division and become binucleate (Lew and Reed 1995a). This adaptation requires *MIH1*, which is not required for nuclear division upon recovery from the block to bud formation (Sia et al. 1996). Interestingly, *mih1* cells were less proficient than their wild-type counterparts in surviving a prolonged (but transient) block of bud formation, suggesting that it is advantageous to permit cell cycle progression after a long delay even if the budding defect has not been repaired (Sia et al. 1996). The identification of a specific checkpoint adaptation mutant, combined with the demonstration that adaptation improves viability, suggests that this is an active process.

The DNA damage checkpoint appears to undergo a similar adaptation (Sandell and Zakian 1993). However, some studies have suggested that persistent unrepairable DNA breaks in nonessential plasmids are lethal, for reasons that are unclear but that might involve a permanent checkpoint arrest (Bennett et al. 1993, 1996). Thus, although it is likely (Weinert and Lydall 1993), it remains unproven that the DNA damage and other checkpoints display adaptation.

F. How Many Checkpoints Are There?

The empirical criterion of Hartwell and Weinert (1989) for defining a checkpoint control pathway is that a dependent series of events (e.g., performance of event B dependent on prior completion of event A) can be uncoupled by a mutation or drug (permitting B to occur without A). In its broadest usage, this definition could be used to argue that many phenomena that we have referred to as the "ticking of the cell cycle clock" constitute checkpoint pathways. For instance, activation of Clb/Cdc28p complexes ("event B") is normally dependent on prior activation of Cln/Cdc28p complexes ("event A"), in part because Cln/Cdc28p complexes trigger the degradation of the Clb/Cdc28p inhibitor Sic1p (see Section V.C.7). In *sic1* mutants, these "events" are uncoupled (Schneider et al. 1996); does this identify a checkpoint linking Clb/Cdc28p activation to Cln/Cdc28p activation? In another instance, sister chromatid disjunction ("event B") is normally dependent on prior activation of the APC ("event A"), perhaps because this complex triggers degradation of the anaphase inhibitor Pds1p (see Section V.D.4). In *pds1* mutants, these "events" are uncoupled (Yamamoto et al. 1996b); does this identify a checkpoint linking sister chromatid disjunction to activation of the APC?

Including such instances within a broad checkpoint terminology violates the spirit of the original checkpoint concept and, we think, degrades its usefulness. The original formulation embodied the idea that cells have evolved surveillance mechanisms for ensuring that error-prone cell cycle processes are allowed sufficient time for error correction before the cell proceeds with subsequent events. DNA replication and chromosome attachment to the spindle are both error-prone processes that probably take different amounts of time to complete in different individual cells (and can be perturbed easily by environmental factors). The consequences of proceeding with chromosome segregation before completing either DNA replication or chromosome attachment to the spindle are severe, and checkpoint control pathways monitoring these processes result in a significantly increased fidelity of genome inheritance, and hence evolutionary fitness. The DNA damage checkpoint, although not strictly monitoring a cell cycle "event," fits well within this checkpoint definition because DNA damage can clearly occur during cell cycle progression, and providing time for repair of the damage again improves the fidelity of cellular reproduction and hence evolutionary fitness. The morphogenesis checkpoint also seems to fit well under this rubric: Transient perturbations of actin cytoskeletal function (causing problems or delays in bud formation) are likely to occur with some frequency in the natural environment, and providing time for recovery and bud growth prior to nuclear division should improve the fidelity of cellular reproduction.

We suggest that the checkpoint terminology is best restricted to these and other instances in which the "event" monitored by the checkpoint is not simply some molecular aspect of the cell cycle clock (like activation of Cln/Cdc28p or the APC). There is no exact definition of a checkpoint control pathway, just as there is no exact definition of what constitutes a component of the cell cycle clock. However, both the clock and the checkpoint concepts have been very useful as simple organizing ideas in thinking about cell cycle control. Keeping the terms and concepts separate will help to ensure that they retain their value in clarifying, rather than obscuring, complex cell cycle control phenomena.

Given the suggested restrictive usage, is it likely that all of the checkpoint controls have been identified, or do more remain to be discovered? At present, two other plausible checkpoint pathways have been suggested. The first of these would link cytokinesis to the prior occurrence of nuclear segregation into the bud. Mutations in dynein or dynactin subunits lead to defects in nuclear migration and spindle orientation during anaphase, giving rise to cells in which both sets of segregated chromosomes are present in the mother cell and the bud is aploid (Eshel et al. 1993; Li et al. 1993; Clark and Meyer 1994; McMillan and Tatchell 1994; Muhua et al. 1994). Eventually, one set of chromosomes usually finds its way into the bud, and individual cells typically delay cytokinesis until the bud contains a daughter nucleus (Yeh et al. 1995). It has been suggested that the delay of cytokinesis in these cells is due to a checkpoint that detects defects in the movement of the daughter nucleus into the bud (Yeh et al. 1995). However, checkpoint mutants have not yet been described, and it remains possible that dynein/dynactin play a direct role in cytokinesis.

The second candidate checkpoint pathway emerged from studies of *cdc7* and *dbf4* mutations, which interfere with the initiation of DNA replication (see Section III.A). In some *cdc7* and *dbf4* strains, anaphase onset is also blocked (Pringle and Hartwell 1981; Toyn et al. 1995), whereas in others it is not (Toyn et al. 1995). Three arguments suggest that the block of anaphase onset (where it occurs) is not due to the DNA replication checkpoint. First, these mutations are expected to block formation of replication forks, the apparent source of the signal for this checkpoint (see Section VI.A). Second, in the *cdc7* and *dbf4* strains that do not display blockage of anaphase, the DNA replication checkpoint appears to be intact as judged from experiments using hydroxyurea (Toyn et al. 1995). Third, at least for a *cdc7–4*^{ts} strain, the arrest appeared to be largely unaffected by a *mec1* checkpoint mutation (Weinert et al. 1994). Thus, one possible interpretation of the available data is that the block of anaphase onset is due to a novel checkpoint that detects some feature of cells in late G_1 and that this checkpoint is not functional in the mutant strains in which anaphase occurs (Toyn et al. 1995). However, this interpretation is rendered uncertain by the possibility that particular temperature-sensitive alleles, in particular strain backgrounds, might be leaky enough that replication forks would form, leading to a signal that would trigger the DNA replication checkpoint. The *mec1* studies do not rule out this possibility because there did appear to be a small effect on the *cdc7–4*^{ts} strain (Weinert et al. 1994). Another possible interpretation of the data is that Cdc7p and Dbf4p are bifunctional, having positive roles both in the initiation of DNA replication and in the ticking of the cell cycle clock. On this interpretation, the second function would not be required for anaphase in some strains.

However these two cases are ultimately interpreted, it will not be surprising if other bona fide checkpoint pathways emerge from closer scrutiny of cell cycle kinetics in the future.

VII. PERSPECTIVES AND CONCLUSIONS

A. Interpretation of *cdc* Mutants

Much of the progress in elucidating cell cycle control mechanisms in *S. cerevisiae* has come from the analysis of *cdc* (cell division cycle) mutants, as pioneered by Hartwell and co-workers (Hartwell et al. 1970, 1973, 1974; Pringle and Hartwell 1981). *cdc* mutants are conditional mutants that arrest under restrictive conditions with a uniform terminal morphology, which has been taken to indicate arrest at a particular stage of the cell cycle. It is now clear that the uniform terminal morphologies of these mutants can reflect at least four different types of involvement of the *CDC* gene products in the cell cycle. (1) Some *CDC* genes encode proteins important for the ticking of the cell cycle clock (e.g., *CDC28, CDC4, CDC34, CDC53, CDC16, CDC23, CDC26, CDC27,* and perhaps *CDC5, CDC14,* and *CDC15*); when the gene product is absent or defective, cell cycle arrest occurs because the clock stops ticking. (2) Some *CDC* genes encode proteins important for processes that are monitored by checkpoint controls (e.g., *CDC2, CDC6, CDC7, CDC8, CDC9, CDC13, CDC17, CDC44, CDC46, CDC47, CDC54*). In these cases, the gene products are important for cell cycle events, rather than cell cycle control, and the uniform arrest morphology is due to checkpoint-mediated arrest. (3) Some *cdc* mutations that arrest cells in G_1 mimic the effects of extracellular signals that normally cause cell cycle arrest (e.g., *cdc36, cdc39 [ros1], cdc70 [gpa1, scg1]*, and *cdc72 [nmt1]* mutations induce the pheromone response [Sprague and Thorner 1992], whereas *cdc25, cdc33, cdc35 [cyr1]*, and *cdc63 [prt1]* mutations mimic a nutritional deprivation response [Hanic-Joyce et al. 1987; Brenner et al. 1988; Broach and Deschenes 1990; Thevelein 1994]). (4) Some *CDC* genes encode proteins that are involved directly in the morphogenetic steps of the cell cycle (e.g., *CDC24, CDC42, CDC3, CDC10, CDC11, CDC12*). In these cases, the uniform terminal morphology reflects the role of the gene products in bud formation and cytokinesis, rather than an actual cell cycle arrest (except as may be mediated by the morphogenesis checkpoint; see Section VI.D).

As emphasized previously (see, e.g., Pringle 1978), in none of these cases does the terminal phenotype of the *cdc* mutant correspond accurately to a normal point in the cell cycle. Whether the mutation affects the cell cycle clock directly or indirectly, the cell is not "frozen" at a particular step: Arrested cells generally contain constellations of Cdc28p regulators different from those present at any point in an unperturbed cycle, and some cell cycle events typically continue while others are blocked under restrictive conditions. For example, mutations that block DNA replication also prevent anaphase onset but do not prevent spindle assembly or the apical-isotropic switch in bud growth, both of which normally occur after DNA replication is complete. Moreover, mutations that mimic the effects of extracellular signals also typically cause the cells to acquire various properties not characteristic of cycling cells (Sprague and Thorner 1992; Werner-Washburne et al. 1993).

B. The Concept of the Cell Cycle Clock and the Nature of Start

Early discussions of how cell cycle events might be coordinated considered both models based on a cell cycle clock and models based on organization of the events into dependent pathways (perhaps with branches and points of convergence) (Mitchison 1971; Hartwell et al. 1974; Pringle 1978; Pringle and Hartwell 1981). The concept of a cell cycle clock received its most convincing early support from studies on the rapid and synchronous cleavage divisions that occur during early embryonic development in many metazoans. In particular, periodic cortical contractions (presumably reflecting attempts to undergo cleavage) continue unabated in frog eggs after DNA replication or spindle assembly is blocked and persist even in enucleated eggs. These "bags of cytoplasm" were inferred to contain an autonomous oscillator (or clock) that went on ticking in the absence of any nucleus-associated cell cycle events (for review, see Kirschner 1992). This clock was later shown to be based on cycles of cyclin/CDK activity. The data from *S. cerevisiae* also suggest that (at least under favorable growth conditions) Cdc28p is a central component of a similar cell cycle clock: Cdc28p activities oscillate in a largely self-regulated manner during the cell cycle, and changes in Cdc28p activities are required to promote many, if not all, cell cycle events.

In the most extreme version of the clock model, the detailed series of changes in Cdc28p activities promotes the detailed progression of cell cycle events, including the initiation of new cell cycles, in the correct order. Thus, each pair of cyclins is uniquely responsible for triggering a particular set of events, and each event is triggered directly by a particular cyclin/Cdc28p complex. Even when checkpoint control mechanisms are introduced to account for the dependency relationships among major cell cycle events (see Section VI), this extreme clock model faces several difficulties, which raise several centrally important questions.

First, there are hundreds of specific activities involved in cell cycle progression, and it seems unlikely that all of them are controlled directly by Cdc28p. Thus, we must ask how many (and which) cell cycle events are controlled directly by cyclin/Cdc28p-mediated phosphorylation, and which events lie downstream in pathways that begin with these directly controlled events. As noted above (see Section V.D.5), intensive efforts are under way to identify the direct, physiologically relevant substrates of the various cyclin/Cdc28p complexes. Although final judgment must clearly await more information, it seems very likely that the cell cycle control system is in fact an intermediate between the extreme clock and dependent-pathway models as considered in the early discussions; thus, the phosphorylation of particular substrates by the cyclin/Cdc28p complexes probably triggers cascades of events (in the form of linear and branched dependent pathways) that lead, in aggregate, to the execution of the major cell cycle events. At present, it does appear that each major event is triggered separately by the cyclin/Cdc28p clock, so that the major events may be connected to each other only by the clock and the checkpoint controls.

Second, with the exception of Clb1p–2p, no cyclin pair is essential, and events thought to be controlled primarily by each pair of cyclins can be accomplished (albeit generally after a delay) in the presence of sufficient quantities of other cyclins. Equally puzzling is that ectopic expression of cyclins does not (in many cases) interfere with the orderly progression of cell cycle events. These observations pose the question of how specialized the different cyclins really are. Do different cyclins target Cdc28p to different substrates in order to promote distinct events or are the different cyclins distinguished simply by their timing of expression (in which case other factors must control the detailed order of events)?

These questions have been approached by asking whether overexpression of one cyclin can rescue an otherwise lethal combination of mutations in other cyclins. For example, overexpression of Clb1p can rescue a *clb1 clb2 clb3 clb4 clb5 clb6* strain (S. Haase and S. Reed, pers. comm.), which could be taken as support for the hypothesis that the Clb cyclins differ mainly in their temporal patterns of expression. However, it also seems possible that the high level of overexpression required for this rescue could lead to phosphorylation of substrates which are very inefficient targets of the Clb1p/Cdc28p kinase or that overexpression allows the "sloppy" localization of Clb1p/Cdc28p complexes to subcellular domains where they are not normally found. It has also been observed that overexpressed Clb5p and Clb6p (Epstein and Cross 1992; Schwob and Nasmyth 1993; Basco et al. 1995), but not Clb1p or Clb2p (Lew et al. 1991), can rescue a *cln1 cln2 cln3* strain. This observation was originally interpreted as evidence that Clb1p/Cdc28p and Clb2p/Cdc28p complexes could not phosphorylate substrates important for early events of the cell cycle. However, it is now apparent that Clb1p and Clb2p are extremely unstable in G_1, and stabilized mutant Clb1p and Clb2p can in fact trigger DNA replication in *cln1 cln2 cln3* cells (Amon et al. 1994). In summary, these experiments indicate that many cyclin/Cdc28p complexes have the ability to phosphorylate substrates that are normally the targets of distinct cyclin/Cdc28p complexes, but it is unclear whether this ability has any significance beyond the obvious fact that all of the complexes share the same protein kinase catalytic subunit, Cdc28p.

Meanwhile, examination of the effects of ectopic cyclin expression on cell morphology has provided strong evidence that at least the Cln1p–2p and Clb1p–2p pairs of cyclins normally target Cdc28p to different substrates. These cyclins have antagonistic effects on polarization of the actin cytoskeleton and cell growth: Cln1p–2p trigger actin polarization and bud emergence at the end of G_1, whereas Clb1p–2p depolarize bud growth in G_2 (Lew and Reed 1993). Overexpression of Cln1p or Cln2p in G_2 causes hyperpolarization of bud growth, whereas overexpression of the stabilized mutant Clb1p or Clb2p in G_1 depolarizes the actin cytoskeleton and prevents bud emergence (Lew and Reed 1993). Thus, these cyclins exhibit the same effect on cell polarity whether they are expressed in G_1 or in G_2, arguing that the nature of the cyclin, rather than simply its time of expression, can determine the target specificity of cyclin/Cdc28p complexes. Further evidence for cyclin specificity has come from a study on the transcriptional response to mating pheromone, which revealed that Cln1p and Cln2p, but not any of the other cyclins, could shut off the pheromone response pathway (Oehlen and Cross 1994), and from comparisons of the functional specificities of Cln2p and Cln3p in a variety of assays (Levine et al. 1996).

One possibly useful perspective on this somewhat confusing situation is that the plethora of cyclins in *S. cerevisiae* may represent an evolutionary specialization of a much simpler underlying clock, and that all that is absolutely required for a (mostly) correct cell cycle progression is an oscillating level of one or two cyclin/Cdc28p kinases. In this regard, it should also be noted that viability under laboratory conditions is not a stringent test of evolutionary fitness, and it is likely that manipulations of cyclin levels that are found to be nonlethal in the laboratory are highly deleterious under natural conditions.

Finally, progression through early G_1 phase is differentially sensitive to conditions that affect the nutritional status or growth rate of the cell, and it remains unclear what circumstances during early G_1 determine whether and when a cell will progress past Start into a new cell cycle. Perhaps the appropriate metaphor for the cell cycle timer is not a clock but a "stopwatch" (whose start button must be pushed, by an adequate supply of nutrients and/or rate or amount of growth, in each new cell cycle). Understanding of this question, and thus clarification of the nature and control of Start, should come from integrating the expanding information (not reviewed here) on yeast growth and growth control with the increasingly sophisticated picture of the control of cell cycle events as described in this chapter.

In its earliest formulations (Hartwell et al. 1974; Pringle and Hartwell 1981), Start was viewed tentatively as a unitary event but with the explicit caveat that this view might be oversimplified. Indeed, it now seems clear that Start is in fact a complex of temporally and functionally separable events (Cross 1995; Dirick et al. 1995; Stuart and Wittenberg 1995). In particular, the closest approximation to the original view of Start as the event that launches the cell cycle is the activation of Cln3p/Cdc28p. As described above, this induces the various late G_1 transcripts, including those of Cln1p and Cln2p, which in turn drive the landmark events (initiation of DNA replication, bud initiation, SPB duplication) that mark the passage of Start. However, a cell that has activated Cln3p/Cdc28p is not yet fully committed to the cell cycle. It is not clear whether such a cell would stop if suddenly deprived of essential nutrients or of a part of its mass, but it clearly remains sensitive to pheromone-induced arrest until the activation of Cln1p–2p/Cdc28p has disposed of Far1p (Section V.C.7; McKinney et al. 1993; McKinney and Cross 1995). Moreover, an **a**/α cell does not become committed to mitosis vis à vis meiosis until somewhat later (Pringle and Hartwell 1981), as an arrested *cdc4[ts]* mutant can undergo meiosis directly if switched back to permissive temperature under appropriate nutritional conditions (Hirschberg and Simchen 1977). Thus, it appears that the launching of and commitment to a cell cycle (two ideas that were amalgamated in the original Start concept) are in fact separable (and that the latter even occurs in stages), much as a jetliner launches its takeoff when it begins to roll but does not become committed to takeoff until a critical speed is reached. How a cell decides when to launch a cell cycle (i.e., activate Cln3p/Cdc28p) remains essentially unknown and poses a major challenge for future research.

C. Cell Cycle Phases and Transitions Revisited

How does the clock (or stopwatch)/checkpoint picture of cell cycle control fit with earlier descriptions based on the G_1, S, G_2, and M phases and the transitions between these phases? One view that has had a powerful impact on cell cycle research is that "transitions" in CDK activity are responsible for the "transitions" between the phases of the cell cycle (see, e.g., King et al. 1994). This idea originated in part from the early observations that most *cdc28* mutants in *S. cerevisiae* arrest primarily in G_1, whereas most *cdc2* mutants in *S. pombe* arrest primarily in G_2, suggesting that CDK activity is specifically required for executing the G_1/S and G_2/M transitions. In an updated version of this model, three key changes in the clock are proposed to control the cell cycle: Activation of G_1 cyclin/CDK complexes promotes the G_1/S transition, activation of B-type cyclin/CDK complexes promotes the G_2/M transition, and activation of the APC promotes the M/G_1 transition (King et al. 1994; note the equation, in this formulation, of the beginning of anaphase with the end of mitosis).

Support for the idea that changes in the cell cycle clock are appropriately viewed as "transitions" comes from observations on the mechanisms underlying the changes in CDK activity. Positive feedback loops such as those proposed to drive accumulation of Cln1p and Cln2p in G_1

and Clb1p and Clb2p in G_2 may help to make changes in CDK activity all-or-none. Furthermore, regulated proteolysis mechanisms such as those responsible for degradation of Sic1p in late G_1 phase and Clb1p–4p in late M phase may help to make changes in CDK activity irreversible. Thus, some changes in CDK activity have properties consistent with the idea that they constitute "transitions" (i.e., decisive unidirectional changes in the state of the cell).

Another feature of this model is that the G_1/S, G_2/M, and M/G_1 transitions are key points (or "checkpoints") for regulation of the cell cycle clock by extracellular inputs or by checkpoint controls (see, e.g., Murray 1992). Indeed, extracellular signals typically arrest cell cycle progression in G_1, whereas several checkpoints arrest cell cycle progression prior to anaphase. (By analogy with checkpoint arrests in metazoan cells, investigators studying the DNA damage checkpoint in *S. cerevisiae* have referred to the stage of arrest as "G_2" [Weinert and Hartwell 1988], and investigators studying the spindle assembly checkpoint have referred to the stage of arrest as "metaphase" [Li and Murray 1991].) Cells thus appear to assess external conditions in G_1 before deciding to enter S phase, to assess whether DNA replication is complete and the DNA is undamaged before deciding to enter M phase, and to assess whether all chromosomes are properly aligned on the spindle before deciding to "exit" M phase (by beginning anaphase). Thus, the inputs from extracellular signals or checkpoint controls could be viewed simply as regulating the key changes in the state of the clock.

Despite its attractions, this view of cell cycle control has some significant problems. It is not clear that the beginning of anaphase should be equated with the end of mitosis or that cytokinesis should be ignored in describing the phases and transitions in the cell cycle. Moreover, in *S. cerevisiae*, sequential Cdc28p activation by distinct cyclins (Cln3p, Cln1p–2p, Clb5p–6p, Clb3p–4p, Clb1p–2p; see Figs. 3, 5, and 6) appears to occur at five (not two) times in the cell cycle. This level of complexity is not unique to yeast: Mammalian somatic cells also display sequential activation of at least five CDK complexes (cyclin D/Cdk4p, cyclin E/Cdk2p, cyclin A/Cdk2p, cyclin A/Cdc2p, and cyclin B/Cdc2p) (Sherr 1994), and the story may be even more complex, as indicated by the identification of several distinct cyclin B isoforms in metazoans (Gallant and Nigg 1994; Jackman et al. 1995; Kreutzer et al. 1995). Furthermore, APC-dependent degradation events may also occur at several different times: The relative timing of Pds1p, Ase1p, and Clb1p–4p degradation has not yet been accurately compared, but events dependent on these processes occur at different times (e.g., anaphase onset vs. cytokinesis). Similarly, degradation of different cyclins in metazoans occurs at different times during mitosis (Whitfield et al. 1990; Hunt et al. 1992; Sigrist et al. 1995) and may sequentially regulate different events. Complexity is also apparent in the ability of the DNA damage checkpoint to arrest cell cycle progression at at least four stages: in G_1, within S phase, in "G_2" (prior to anaphase), and within anaphase (see Section VI.B).

Given this complexity, the simple three-transition model must be updated or discarded. Does each "tick" of the cell cycle clock represent a transition? If so, then continued study of the clock may lead us to models postulating perhaps a dozen transitions in the cell cycle. Clearly, this is too many to map onto the simple four-phase cell cycle picture. Indeed, even the simple three-transition model requires some stretching to correlate CDK changes with transitions between the phases: G_1 cyclin/CDK activation clearly precedes the G_1/S boundary (by several hours in mammalian cells), and APC-dependent degradation of at least some substrates occurs prior to anaphase, well before the M/G_1 boundary. As we have suggested earlier in this chapter, a description of the cell cycle in terms of events, rather than phases and transitions, eliminates the need for such stretching and circumvents the awkward growth in the number of transitions. In this view, changes in the cell cycle clock trigger (and sometimes sustain) performance of particular cell cycle events. Regulatory inputs from extracellular signals or checkpoint controls have specific effects on the clock and may also affect cell cycle events more directly. This formulation serves to focus attention on the mechanisms by which changes in the clock affect individual events and on the effects of regulatory inputs on the ticking of the clock. The G_1, S, G_2, and M phase nomenclature remains a convenient, if inexact, way of referring to particular periods of the cell cycle, but phases, and the transitions between phases, no longer assume regulatory connotations.

D. Summary

The observations reviewed in this chapter are consistent with a model of cell cycle control in which a central cell cycle clock (or stopwatch) triggers events with a timing that is modulated by inputs from surveillance pathways called checkpoint controls. The clock is a complex oscillator centered on the Cdc28p kinase and its regulators. Direct Cdc28p regulators include cyclins (Cln1p–3p, Clb1p–6p), inhibitors (Sic1p, Far1p), kinases (Cak1p, Swe1p), and phosphatases (Mih1p). In turn, expression of these regulators is controlled by transcription factors (Swi4p, Swi6p, Mbp1p, Mcm1p, Swi5p, Ace2p) and by proteins controlling ubiquitin-mediated proteolysis (Cdc16p, Cdc23p, Cdc26p, Cdc27p, Apc1p; Cdc34p, Cdc4p, Cdc53p; Ubc9p; and perhaps Skp1p and Grr1p). This listing is undoubtedly incomplete, and further layers of regulation seem certain to be important for the "ticking of the clock." The clock is thought to regulate cell cycle events primarily through Cdc28p-mediated phosphorylation events, although, at present, the details of these links are almost entirely unknown. In addition, many of the clock components listed above clearly affect cell cycle events themselves; for instance, the transcription factors control the periodic expression of large sets of genes, and the APC ubiquitin ligase controls the periodic destruction of what may be quite a large family of proteins, including Pds1p and Ase1p in addition to cyclins. Another set of proteins that are bound to be important are the phosphatases that reverse the Cdc28p-catalyzed phosphorylations. It is not yet clear whether these are also cell-cycle-regulated or simply required constitutively.

The molecular dissection of checkpoint control pathways is just beginning. We are still largely ignorant both of the mechanisms responsible for detection of improper cell cycle events and of those responsible for arrest of cell cycle progression. It seems likely that cell cycle arrest in response to checkpoints is ultimately mediated through regulation of the cell cycle clock. However, at least some checkpoints clearly involve physiological adaptations (e.g., transcriptional induction of DNA repair genes by the DNA damage checkpoint) that are independent of the clock, and it remains possible that some checkpoint-induced delays in cell cycle processes (e.g., elongation arrest of replication forks within S phase in response to DNA damage) are also independent of the clock. Checkpoints monitoring DNA replication, DNA damage, and spindle assembly are accepted as general aspects of eukaryotic cell cycle control, but other checkpoints (such as the morphogenesis checkpoint in *S. cerevisiae*) almost certainly exist, and the number of surveillance systems contributing to the fidelity of the cell cycle is not known. This should be an area of rapid progress in the next few years.

Finally, a full understanding of cell cycle control (in both *S. cerevisiae* and other cell types) will require us to understand in more detail both how extracellular signals control passage through early G_1 and how the events of the cell cycle are arranged properly in space as well as in time.

ACKNOWLEDGMENTS

We thank Fred Cross, Curt Wittenberg, Steve Reed, Kerry Bloom, Steve Garrett, Steve Haase, Mark Watson, Dave Stuart, and Stefan Lanker for comments on the manuscript, and Linda Breeden, Doug Koshland, Steve Haase, Steve Reed, Mike Mendenhall, Dave Stuart, Curt Wittenberg, and Kerry Bloom for the communication of unpublished information. We also thank Lee Hartwell, Brian Reid, and Joe Culotti for pioneering the genetic analysis of the cell cycle, and the numerous subsequent investigators whose vigorous and penetrating analyses have made this field so exciting (if exhausting) to review. D.J.L. is supported by a grant from the National Institutes of Health (GM-53050) and by funds from the Searle Scholars Program/Chicago Community Trust. T.W. is supported by a grant from the National Institutes of Health (GM-45276). J.R.P. is supported in part by funds from the RJEG Trust.

REFERENCES

Aboussekhra, A., J.E. Vialard, D.E. Morrison, M.A. de la Torre-Ruiz, L. Cernáková, F. Fabre, and N.F. Lowndes. 1996. A novel role for the budding yeast *RAD9* checkpoint gene in DNA damage-dependent transcription. *EMBO J.* **15:** 3912.

Adams, A.E.M. and J.R. Pringle. 1984. Relationship of actin and tubulin distribution to bud growth in wild-type and morphogenetic-mutant *Saccharomyces cerevisiae. J. Cell Biol.* **98:** 934.

Adams, A.E.M., D.I. Johnson, R.M. Longnecker, B.F. Sloat, and J.R. Pringle. 1990. *CDC42* and *CDC43*, two additional genes involved in budding and the establishment of cell polarity in the yeast *Saccharomyces cerevisiae. J. Cell Biol.* **111:** 131.

Alberts, B., D. Bray, J. Lewis, M. Raff, K. Roberts, and J.D. Watson. 1994. *Molecular biology of the cell*, 3rd edition. Garland, New York.

Allen, J.B., Z. Zhou, W. Siede, E.C. Friedberg, and S.J. Elledge. 1994. The *SAD1/RAD53* protein kinase controls multiple checkpoints and DNA damage-induced transcription in yeast. *Genes Dev.* **8:** 2401.

Althoefer, H., A. Schleiffer, K. Wassmann, A. Nordheim, and G. Ammerer. 1995. Mcm1 is required to coordinate G2-specific transcription in *Saccharomyces cerevisiae. Mol. Cell. Biol.* **15:** 5917.

Amon, A., S. Irniger, and K. Nasmyth. 1994. Closing the cell cycle circle in yeast: G2 cyclin proteolysis initiated at mitosis persists until the activation of G1 cyclins in the next cycle. *Cell* **77:** 1037.

Amon, A., U. Surana, I. Muroff, and K. Nasmyth. 1992. Regulation of $p34^{CDC28}$ tyrosine phosphorylation is not required for entry into mitosis in *S. cerevisiae. Nature* **355:** 368.

Amon, A., M. Tyers, B. Futcher, and K. Nasmyth. 1993. Mechanisms that help the yeast cell cycle clock tick: G2 cyclins transcriptionally activate G2 cyclins and repress G1 cyclins. *Cell* **74:** 993.

Araki, H., S.-H. Leem, A. Phongdara, and A. Sugino. 1995. Dpb11, which interacts with DNA polymerase II(ε) in *Saccharomyces cerevisiae*, has a dual role in S-phase progression and at a cell cycle checkpoint. *Proc. Natl. Acad. Sci.* **92:** 11791.

Ayscough, K.R., J. Stryker, N. Pokala, M. Sanders, P. Crews, and D.G. Drubin. 1997. High rates of actin filament turnover in budding yeast and roles for actin in establishment and maintenance of cell polarity revealed using the actin inhibitor latrunculin-A. *J. Cell Biol.* **137:** 399.

Bai, C., P. Sen, K. Hofmann, L. Ma, M. Goebl, J.W. Harper, and S.J. Elledge. 1996. *SKP1* connects cell cycle regulators to the ubiquitin proteolysis machinery through a novel motif, the F-box. *Cell* **86:** 263.

Barker, D.G., J.H.M. White, and L.H. Johnston. 1985. The nucleotide sequence of the DNA ligase gene (*CDC9*) from *Saccharomyces cerevisiae:* A gene which is cell-cycle regulated and induced in response to DNA damage. *Nucleic Acids Res.* **13:** 8323.

Barral, Y., S. Jentsch, and C. Mann. 1995. G1 cyclin turnover and nutrient uptake are controlled by a common pathway in yeast. *Genes Dev.* **9:** 399.

Barton, A.A. 1950. Some aspects of cell division in *Saccharomyces cerevisiae. J. Gen. Microbiol.* **4:** 84.

Basco, R.D., M.D. Segal, and S.I. Reed. 1995. Negative regulation of G1 and G2 by S-phase cyclins of *Saccharomyces cerevisiae. Mol. Cell. Biol.* **15:** 5030.

Bell, S.P. and B. Stillman. 1992. ATP-dependent recognition of eukaryotic origins of DNA replication by a multiprotein complex. *Nature* **357:** 128.

Bell, S.P., R. Kobayashi, and B. Stillman. 1993. Yeast origin recognition complex functions in transcription silencing and DNA replication. *Science* **262:** 1844.

Bell, S.P., J. Mitchell, J. Leber, R. Kobayashi, and B. Stillman. 1995. The multidomain structure of Orc1p reveals similarity to regulators of DNA replication and transcriptional silencing. *Cell* **83:** 563.

Bennett, C.B., A.L. Lewis, K.K. Baldwin, and M.A. Resnick. 1993. Lethality induced by a single site-specific double-strand break in a dispensable yeast plasmid. *Proc. Natl. Acad. Sci.* **90:** 5613.

Bennett, C.B., T.J. Westmoreland, J.R. Snipe, and M.A. Resnick. 1996. A double-strand break within a yeast artificial chromosome (YAC) containing human DNA can result in YAC loss, deletion, or cell lethality. *Mol. Cell. Biol.* **16:** 4414.

Blondel, M. and C. Mann. 1996. G2 cyclins are required for the degradation of G1 cyclins in yeast. *Nature* **384:** 279.

Bloom, K.S., E. Amaya, J. Carbon, L. Clarke, A. Hill, and E. Yeh. 1984. Chromatin conformation of yeast centromeres. *J. Cell Biol.* **99:** 1559.

Booher, R.N., R.J. Deshaies, and M.W. Kirschner. 1993. Properties of *Saccharomyces cerevisiae* wee1 and its differential regulation of $p34^{CDC28}$ in response to G1 and G2 cyclins. *EMBO J.* **12:** 3417.

Bourne, Y., M.H. Watson, M.J. Hickey, W. Holmes, W. Rocque, S.I. Reed, and J.A. Tainer. 1996. Crystal structure and mutational analysis of the human CDK2 kinase complex with cell cycle-regulatory protein CksHs1. *Cell* **84:** 863.

Breeden, L. 1996. Start-specific transcription in yeast. *Curr. Top. Microbiol. Immunol.* **208:** 95.

Breeden, L. and G. Mikesell. 1994. Three independent forms of regulation affect expression of *HO, CLN1* and *CLN2* during the cell cycle of *Saccharomyces cerevisiae. Genetics* **138:** 1015.

Brenner, C., N. Nakayama, M. Goebl, K. Tanaka, A. Toh-e, and K. Matsumoto. 1988. *CDC33* encodes mRNA cap-binding protein eIF-4E of *Saccharomyces cerevisiae. Mol. Cell. Biol.* **8:** 3556.

Brizuela, L., G. Draetta, and D. Beach. 1987. $p13^{suc1}$ acts in the fission yeast cell division cycle as a component of the $p34^{cdc2}$ protein kinase. *EMBO J.* **6:** 3507.

Broach, J.R. and R.J. Deschenes. 1990. The function of *RAS* genes in *Saccharomyces cerevisiae. Adv. Cancer Res.* **54:** 79.

Broek, D., R. Bartlett, K. Crawford, and P. Nurse. 1991. Involvement of $p34^{cdc2}$ in establishing the dependency of S phase on mitosis. *Nature* **349:** 388.

Brush, G.S., D.M. Morrow, P. Hieter, and T.J. Kelly. 1996. The ATM homologue *MEC1* is required for phosphorylation of replication protein A in yeast. *Proc. Natl. Acad. Sci.* **93:** 15075.

Bueno, A. and P. Russell. 1992. Dual functions of *CDC6:* A yeast protein required for DNA replication also inhibits nuclear division. *EMBO J.* **11:** 2167.

Bulawa, C.E. 1993. Genetics and molecular biology of chitin synthesis in fungi. *Annu. Rev. Microbiol.* **47:** 505.

Byers, B. 1981. Cytology of the yeast life cycle. In *The molecular biology of the yeast* Saccharomyces: *Life cycle and inheritance* (ed. J.N. Strathern et al.), p. 59. Cold Spring Harbor Laboratory, Cold Spring Harbor, New York.

Byers, B. and L. Goetsch. 1975. Behavior of spindles and spindle plaques in the cell cycle and conjugation of *Saccharomyces cerevisiae. J. Bacteriol.* **124:** 511.

Cabib, E., R. Roberts, and B. Bowers. 1982. Synthesis of the yeast cell wall and its regulation. *Annu. Rev. Biochem.* **51:** 763.

Chang, F. and I. Herskowitz. 1990. Identification of a gene necessary for cell cycle arrest by a negative growth factor of yeast: *FAR1* is an inhibitor of a G1 cyclin, CLN2. *Cell* **63:** 999.

Chant, J. and J.R. Pringle. 1995. Patterns of bud-site selection in the yeast *Saccharomyces cerevisiae. J. Cell Biol.* **129:** 751.

Chen, R.-H., J.C. Waters, E.D. Salmon, and A.W. Murray. 1996. Association of spindle assembly checkpoint component XMAD2 with unattached kinetochores. *Science* **274:** 242.

Choi, W.-J., B. Santos, A. Durán, and E. Cabib. 1994. Are yeast chitin synthases regulated at the transcriptional or the posttranslational level? *Mol. Cell. Biol.* **14:** 7685.

Chowdhury, S., K.W. Smith, and M.C. Gustin. 1992. Osmotic stress and the yeast cytoskeleton: Phenotype-specific suppression of an actin mutation. *J. Cell Biol.* **118:** 561.

Chuang, J.S. and R.W. Schekman. 1996. Differential trafficking and timed localization of two chitin synthase proteins, Chs2p and Chs3p. *J. Cell Biol.* **135:** 597.

Cid, V.J., A. Durán, F. del Rey, M.P. Snyder, C. Nombela, and M. Sánchez. 1995. Molecular basis of cell integrity and morphogenesis in *Saccharomyces cerevisiae. Microbiol. Rev.* **59:** 345.

Cismowski, M.J., G.M. Laff, M.J. Solomon, and S.I. Reed. 1995. *KIN28* encodes a C-terminal domain kinase that controls mRNA transcription in *Saccharomyces cerevisiae* but lacks cyclin-dependent kinase-activating kinase (CAK) activity. *Mol. Cell. Biol.* **15:** 2983.

Clark, S.W. and D.I. Meyer. 1994. *ACT3:* A putative centractin homologue in *S. cerevisiae* is required for proper orientation of the mitotic spindle. *J. Cell Biol.* **127:** 129.

Cocker, J.H., S. Piatti, C. Santocanale, K. Nasmyth, and J.F.X. Diffley. 1996. An essential role for the Cdc6 protein in forming the pre-replicative complexes of budding yeast. *Nature* **379:** 180.

Cohen-Fix, O., J.-M. Peters, M.W. Kirschner, and D. Koshland. 1996. Anaphase initiation in *Saccharomyces cerevisiae* is controlled by the APC-dependent degradation of the anaphase inhibitor Pds1p. *Genes Dev.* **10:** 3081.

Connelly, C. and P. Hieter. 1996. Budding yeast *SKP1* encodes an evolutionarily conserved kinetochore protein required for cell cycle progression. *Cell* **86:** 275.

Copeland, C.S. and M. Snyder. 1993. Nuclear pore complex antigens delineate nuclear envelope dynamics in vegetative and conjugating *Saccharomyces cerevisiae. Yeast* **9:** 235.

Cross, F.R. 1988. *DAF1*, a mutant gene affecting size control, pheromone arrest, and cell cycle kinetics of *Saccharomyces cerevisiae. Mol. Cell. Biol.* **8:** 4675.

———. 1990. Cell cycle arrest caused by *CLN* gene deficiency in *Saccharomyces cerevisiae* resembles START-I arrest and is independent of the mating-pheromone signal. *Mol. Cell. Biol.* **10:** 6482.

———. 1995. Starting the cell cycle: What's the point? *Curr. Opin. Cell Biol.* **7:** 790.

Cross, F.R. and C.M. Blake. 1993. The yeast Cln3 protein is an unstable activator of Cdc28. *Mol. Cell. Biol.* **13:** 3266.

Cross, F.R. and A.H. Tinkelenberg. 1991. A potential positive feedback loop controlling *CLN1* and *CLN2* gene expression at the start of the yeast cell cycle. *Cell* **65:** 875.

Cross, F.R., M. Hoek, J.D. McKinney, and A.H. Tinkelenberg. 1994. Role of Swi4 in cell cycle regulation of *CLN2* expression. *Mol. Cell. Biol.* **14:** 4779.

Cutforth, T. and G.M. Rubin. 1994. Mutations in *Hsp83* and *cdc37* impair signaling by the sevenless receptor tyrosine kinase in *Drosophila. Cell* **77:** 1027.

Dahmann, C. and B. Futcher. 1995. Specialization of B-type cyclins for mitosis or meiosis in *S. cerevisiae. Genetics* **140:** 957.

Dahmann, C., J.F.X. Diffley, and K.A. Nasmyth. 1995. S-phase-promoting cyclin-dependent kinases prevent re-replication by inhibiting the transition of replication origins to a pre-replicative state. *Curr. Biol.* **5:** 1257.

Dalton, S. and L. Whitbread. 1995. Cell cycle-regulated nuclear import and export of Cdc47, a protein essential for initiation of DNA replication in budding yeast. *Proc. Natl. Acad. Sci.* **92:** 2514.

De Nobel, J.G., F.M. Klis, A. Ram, H. Van Unen, J. Priem, T. Munnik, and H. Van Den Ende. 1991. Cyclic variations in the permeability of the cell wall of *Saccharomyces cerevisiae. Yeast* **7:** 589.

Deshaies, R.J. 1995. The self-destructive personality of a cell cycle in transition. *Curr. Opin. Cell Biol.* **7:** 781.

Deshaies, R.J. and M. Kirschner. 1995. G1 cyclin-dependent activation of $p34^{CDC2}$ (Cdc28p) *in vitro. Proc. Natl. Acad. Sci.* **92:** 1182.

Deshaies, R.J., V. Chau, and M. Kirschner. 1995. Ubiquitination of the G1 cyclin Cln2p by a Cdc34p-dependent pathway. *EMBO J.* **14:** 303.

Detweiler, C.S. and J.J. Li. 1997. Cdc6p establishes and maintains a state of replication competence during G_1 phase. *J. Cell Sci.* **110:** 753.

Diffley, J.F.X. 1995. The initiation of DNA replication in the budding yeast cell division cycle. *Yeast* **11:** 1651.

———. 1996. Once and only once upon a time: Specifying and regulating origins of DNA replication in eukaryotic cells. *Genes Dev.* **10:** 2819.

Diffley, J.F.X., J.H. Cocker, S.J. Dowell, and A. Rowley. 1994. Two steps in the assembly of complexes at yeast replication origins in vivo. *Cell* **78:** 303.

Dirick, L. and K. Nasmyth. 1991. Positive feedback in the activation of G1 cyclins in yeast. *Nature* **351:** 754.

Dirick, L., T. Böhm, and K. Nasmyth. 1995. Roles and regulation of Cln-Cdc28 kinases at the start of the cell cycle of *Saccharomyces cerevisiae. EMBO J.* **14:** 4803.

Dohrmann, P.R., G. Butler, K. Tamai, S. Dorland, J.R. Greene, D.J. Thiele, and D.J. Stillman. 1992. Parallel pathways of gene regulation: Homologous regulators *SWI5* and *ACE2* differentially control transcription of *HO* and chitinase. *Genes Dev.* **6:** 93.

Donaldson, A.D. and J.V. Kilmartin. 1996. Spc42p: A phosphorylated component of the *S. cerevisiae* spindle pole body (SPB) with an essential function during SPB duplication. *J. Cell Biol.* **132:** 887-901.

Donovan, J.D., J.H. Toyn, A.L. Johnson, and L.H. Johnston. 1994. P40^{SDB25}, a putative CDK inhibitor, has a role in the M/G1 transition in *Saccharomyces cerevisiae. Genes Dev.* **8:** 1640.

Dowell, S.J., P. Romanowski, and J.F.X. Diffley. 1994. Interaction of Dbf4, the Cdc7 protein kinase regulatory subunit, with yeast replication origins in vivo. *Science* **265:** 1243.

Doyle, T. and D. Botstein. 1996. Movement of yeast cortical actin cytoskeleton visualized *in vivo. Proc. Natl. Acad. Sci.* **93:** 3886.

Drubin, D.G. 1991. Development of cell polarity in budding yeast. *Cell* **65:** 1093.

Drubin, D.G. and W.J. Nelson. 1996. Origins of cell polarity. *Cell* **84:** 335.

Drubin, D.G., H.D. Jones, and K.F. Wertman. 1993. Actin structure and function: Roles in mitochondrial organization and morphogenesis in budding yeast and identification of the phalloidin-binding site. *Mol. Biol. Cell* **4:** 1277.

Dunphy, W.G., L. Brizuela, D. Beach, and J. Newport. 1988. The *Xenopus cdc2* protein is a component of MPF, a cytoplasmic regulator of mitosis. *Cell* **54:** 423.

Epstein, C.B. and F.R. Cross. 1992. *CLB5:* A novel B cyclin from budding yeast with a role in S phase. *Genes Dev.* **6:** 1695.

Eshel, D., L.A. Urrestarazu, S. Vissers, J.-C. Jauniaux, J.C. van Vliet-Reedijk, R.J. Planta, and I.R. Gibbons. 1993. Cytoplasmic dynein is required for normal nuclear segregation in yeast. *Proc. Natl. Acad. Sci.* **90:** 11172.

Espinoza, F.H., J. Ogas, I. Herskowitz, and D.O. Morgan. 1994. Cell cycle control by a complex of the cyclin HCS26 (*PCL1*) and the kinase PHO85. *Science* **266:** 1388.

Espinoza, F.H., A. Farrell, H. Erdjument-Bromage, P. Tempst, and D.O. Morgan. 1996. A cyclin-dependent kinase-activating kinase (CAK) in budding yeast unrelated to vetebrate CAK. *Science* **273:** 1714.

Evans, T., E.T. Rosenthal, J. Youngblom, D. Distel, and T. Hunt. 1983. Cyclin: A protein specified by maternal mRNA in sea urchin eggs that is destroyed at each cleavage division. *Cell* **33:** 389.

Falconi, M.M., A. Piseri, M. Ferrari, G. Lucchini, P. Plevani, and M. Foiani. 1993. *De novo* synthesis of budding yeast DNA polymerase α and *POL1* transcription at the G1/S boundary are not required for entrance into S phase. *Proc. Natl. Acad. Sci.* **90:** 10519.

Feaver, W.J., J.Q. Svejstrup, N.L. Henry, and R.D. Kornberg. 1994. Relationship of CDK-activating kinase and RNA polymerase II CTD kinase TFIIH/TFIIK. *Cell* **79:** 1103.

Fitch, I., C. Dahmann, U. Surana, A. Amon, K. Nasmyth, L. Goetsch, B. Byers, and B. Futcher. 1992. Characterization of four B-type cyclin genes of the budding yeast *Saccharomyces cerevisiae. Mol. Biol. Cell* **3:** 805.

Ford, S.K. and J.R. Pringle. 1991. Cellular morphogenesis in the *Saccharomyces cerevisiae* cell cycle: Localization of the *CDC11* gene product and the timing of events at the budding site. *Dev. Genet.* **12:** 281.

Forsburg, S. and P. Nurse. 1991. Cell cycle regulation in the yeasts *Saccharomyces cerevisiae* and *Schizosaccharomyces pombe. Annu. Rev. Cell Biol.* **7:** 227.

Foss, M., F.J. McNally, P. Laurenson, and J. Rine. 1993. Origin recognition complex (ORC) in transcriptional silencing and DNA replication in *S. cerevisiae. Science* **262:**

1838.

Foster, R., G.E. Mikesell, and L. Breeden. 1993. Multiple SWI6-dependent *cis*-acting elements control *SWI4* transcription through the cell cycle. *Mol. Cell. Biol.* **13:** 3792.

Friedman, D.B., H.A. Sundberg, E.Y. Huang, and T.N. Davis. 1996. The 110-kD spindle pole body component of *Saccharomyces cerevisiae* is a phosphoprotein that is modified in a cell cycle-dependent manner. *J. Cell Biol.* **132:** 903.

Gallant, P. and E.A. Nigg. 1994. Identification of a novel vertebrate cyclin: Cyclin B3 shares properties with both A- and B-type cyclins. *EMBO J.* **13:** 595.

Garvik, B., M. Carson, and L.H. Hartwell. 1995. Single-stranded DNA arising at telomeres in *cdc13* mutants may constitute a specific signal for the *RAD9* checkpoint. *Mol. Cell. Biol.* **15:** 6128.

Gerber, M.R., A. Farrell, R.J. Deshaies, I. Herskowitz, and D.O. Morgan. 1995. Cdc37 is required for association of the protein kinase Cdc28 with G1 and mitotic cyclins. *Proc. Natl. Acad. Sci.* **92:** 4651.

Ghiara, J.B., H.E. Richardson, K. Sugimoto, M. Henze, D.J. Lew, C. Wittenberg, and S.I. Reed. 1991. A cyclin B homolog in *S. cerevisiae:* Chronic activation of the Cdc28 protein kinase by cyclin prevents exit from mitosis. *Cell* **65:** 163.

Glotzer, M., A.W. Murray, and M.W. Kirschner. 1991. Cyclin is degraded by the ubiquitin pathway. *Nature* **349:** 132.

Goebl, M.G., J. Yochem, S. Jentsch, J.P. McGrath, A. Varshavsky, and B. Byers. 1988. The yeast cell cycle gene *CDC34* encodes a ubiquitin-conjugating enzyme. *Science* **241:** 1331.

Gomes de Mesquita, D.S., R. ten Hoopen, and C.L. Woldringh. 1991. Vacuolar segregation to the bud of *Saccharomyces cerevisiae:* An analysis of morphology and timing in the cell cycle. *J. Gen. Microbiol.* **137:** 2447.

Grandin, N. and S.I. Reed. 1993. Differential function and expression of *Saccharomyces cerevisiae* B-type cyclins in mitosis and meiosis. *Mol. Cell. Biol.* **13:** 2113.

Griffiths, D.J.F., N.C. Barbet, S. McCready, A.R. Lehmann, and A.M. Carr. 1995. Fission yeast *rad17:* A homologue of budding yeast *RAD24* that shares regions of sequence similarity with DNA polymerase accessory proteins. *EMBO J.* **14:** 5812.

Guacci, V., E. Hogan, and D. Koshland. 1994. Chromosome condensation and sister chromatid pairing in budding yeast. *J. Cell Biol.* **125:** 517.

Hadwiger, J.A., C. Wittenberg, M.D. Mendenhall, and S.I. Reed. 1989a. The *Saccharomyces cerevisiae CKS1* gene, a homolog of the *Schizosaccharomyces pombe* $suc1^+$ gene, encodes a subunit of the Cdc28 protein kinase complex. *Mol. Cell. Biol.* **9:** 2034.

Hadwiger, J.A., C. Wittenberg, H.E. Richardson, M. de Barros Lopes, and S.I. Reed. 1989b. A family of cyclin homologs that control the G1 phase in yeast. *Proc. Natl. Acad. Sci.* **86:** 6255.

Hanic-Joyce, P.J., G.C. Johnston, and R.A.Singer. 1987. Regulated arrest of cell proliferation mediated by yeast *prt1* mutations. *Exp. Cell Res.* **172:** 134.

Hardwick, K.G. and A.W. Murray. 1995. Mad1p, a phosphoprotein component of the spindle assembly checkpoint in budding yeast. *J. Cell Biol.* **131:** 709.

Hardwick, K.G., E. Weiss, F.C. Luca, M. Winey, and A.W. Murray. 1996. Activation of the budding yeast spindle assembly checkpoint without mitotic spindle disruption. *Science* **273:** 953.

Hardy, C.F.J. 1996. Characterization of an essential Orc2p-associated factor that plays a role in DNA replication. *Mol. Cell. Biol.* **16:** 1832.

Hardy, C.F.J. and A. Pautz. 1996. A novel role for Cdc5p in DNA replication. *Mol. Cell Biol.* **16:** 6775.

Hardy, C.F.J., O. Dryga, S. Seematter, P.M.B. Pahl, and R.A. Sclafani. 1997. *mcm5/cdc46-bob1* bypasses the requirement for the S phase activator Cdc7p. *Proc. Natl. Acad. Sci.* **94:** 3151.

Harold, F.M. 1990. To shape a cell: An inquiry into the causes of morphogenesis of microorganisms. *Microbiol. Rev.* **54:** 381.

Hartwell, L.H. and T.A. Weinert. 1989. Checkpoints: Controls that ensure the order of cell cycle events. *Science* **246:** 629.

Hartwell, L.H., J. Culotti, and B.J. Reid. 1970. Genetic control of the cell-division cycle in yeast, I. Detection of mutants. *Proc. Natl. Acad. Sci.* **66:** 352.

Hartwell, L.H., J. Culotti, J.R. Pringle, and B.J. Reid. 1974. Genetic control of the cell division cycle in yeast. *Science* **183:** 46.

Hartwell, L.H., R.K. Mortimer, J. Culotti, and M. Culotti. 1973. Genetic control of the cell division cycle in yeast. V. Genetic analysis of *cdc* mutants. *Genetics* **74:** 267.

Hayles, J., S. Aves, and P. Nurse. 1986. *suc1* is an essential gene involved in both the cell cycle and growth in fission yeast. *EMBO J.* **5:** 3373.

Hayles, J., D. Fisher, A. Woollard, and P. Nurse. 1994. Temporal order of S phase and mitosis in fission yeast is determined by the state of the $p34^{cdc2}$-mitotic B cyclin complex. *Cell* **78:** 813.

Heichman, K.A. and J.M. Roberts. 1996. The yeast *CDC16* and *CDC27* genes restrict DNA replication to once per cell cycle. *Cell* **85:** 39.

Hennessy, K.M., C.D. Clark, and D. Botstein. 1990. Subcellular localization of yeast *CDC46* varies with the cell cycle. *Genes Dev.* **4:** 2252.

Hershko, A., D. Ganoth, V. Sudakin, A. Dahan, L.H. Cohen, F.C. Luca, J.V. Ruderman, and E. Eytan. 1994. Components of a system that ligates cyclin to ubiquitin and their regulation by the protein kinase cdc2. *J. Biol. Chem.* **269:** 4940.

Hill, K.L., N.L. Catlett, and L.S. Weisman. 1996. Actin and myosin function in directed vacuole movement during cell division in *Saccharomyces cerevisiae. J. Cell Biol.* **135:** 1535.

Hindley, J., G. Phear, M. Stein, and D. Beach. 1987. $suc1^+$ encodes a predicted 13-kilodalton protein that is essential for cell viability and is directly involved in the division cycle of *Schizosaccharomyces pombe. Mol. Cell. Biol.* **7:** 504.

Hirschberg, J. and G. Simchen. 1977. Commitment to the mitotic cell cycle in yeast in relation to meiosis. *Exp. Cell Res.* **105:** 245.

Hopwood, B. and S. Dalton. 1996. Cdc45p assembles into a complex with Cdc46p/Mcm5p, is required for minichromosome maintenance, and is essential for chromosomal DNA replication. *Proc. Natl. Acad. Sci.* **93:** 12309.

Hori, Y., K. Shirahige, C. Obuse, T. Tsurimoto, and H. Yoshikawa. 1996. Characterization of a novel *CDC* gene (*ORC1*) partly homologous to *CDC6* of *Saccharomyces cerevisiae. Mol. Biol. Cell* **7:** 409.

Hoyt, M.A., L. Totis, and B.T. Roberts. 1991. *S. cerevisiae* genes required for cell cycle arrest in response to loss of microtubule function. *Cell* **66:** 507.

Hoyt, M.A., L. He, K.K. Loo, and W.S. Saunders. 1992. Two *Saccharomyces cerevisiae* kinesin-related gene products required for mitotic spindle assembly. *J. Cell Biol.* **118:** 109.

Hoyt, M.A., L. He, L. Totis, and W.S. Saunders. 1993. Loss of function of *Saccharomyces cerevisiae* kinesin-related *CIN8* and *KIP1* is suppressed by *KAR3* motor domain mutations. *Genetics* **135:** 35.

Huang, D., I. Farkas, and P.J. Roach. 1996. Pho85p, a cyclin-dependent protein kinase, and the Snf1p protein kinase act antagonistically to control glycogen accumulation in *Saccharomyces cerevisiae. Mol. Cell. Biol.* **16:** 4357.

Huffaker, T.C., J.H. Thomas, and D. Botstein. 1988. Diverse effects of β-tubulin mutations on microtubule formation and function. *J. Cell Biol.* **106:** 1997.

Hunt, T., F.C. Luca, and J.V. Ruderman. 1992. The requirements for protein synthesis and degradation, and the control of destruction of cyclins A and B in the meiotic and mitotic cell cycles of the clam embryo. *J. Cell Biol.* **116:** 707.

Hunter, T. 1995. When is a lipid kinase not a lipid kinase? When it is a protein kinase. *Cell* **83:** 1.

Hyman, A.A. and P.K. Sorger. 1995. Structure and function of kinetochores in budding yeast. *Annu. Rev. Cell Dev. Biol.* **11:** 471.

Igual, J.C., A.L. Johnson, and L.H. Johnston. 1996. Coordinated regulation of gene expression by the transcription factor SWI4 and the protein kinase C MAP kinase pathway for yeast cell integrity. *EMBO J.* **15:** 5001.

Irniger, S., S. Piatti, C. Michaelis, and K. Nasmyth. 1995. Genes involved in sister chromatid separation are needed for B-type cyclin proteolysis in budding yeast. *Cell* **81:** 269.

Jackman, M., M. Firth, and J. Pines. 1995. Human cyclins B1 and B2 are localized to strikingly different structures: B1 to microtubules, B2 primarily to the Golgi apparatus. *EMBO J.* **14:** 1646.

Jacobs, C.W., A.E.M. Adams, P.J. Szaniszlo, and J.R. Pringle. 1988. Functions of microtubules in the *Saccharomyces cerevisiae* cell cycle. *J. Cell Biol.* **107:** 1409.

Jeggo, P.A., G.E. Taccioli, and S.P. Jackson. 1995. Menage à trois: Double strand break repair, V(D)J recombination and DNA-PK. *BioEssays* **17:** 949.

Johnston, L.H., S.L. Eberly, J.W. Chapman, H. Araki, and A. Sugino. 1990. The product of the *Saccharomyces cerevisiae* cell cycle gene *DBF2* has homology with protein kinases and is periodically expressed in the cell cycle. *Mol. Cell. Biol.* **10:** 1358.

Juang, Y.-L., J. Huang, J.-M. Peters, M.E. McLaughlin, C.-Y. Tai, and D. Pellman. 1997. APC-mediated proteolysis of Ase1 and the morphogenesis of the mitotic spindle. *Science* **275:** 1311.

Kaffman, A., I. Herskowitz, R. Tjian, and E.K. O'Shea. 1994. Phosphorylation of the transcription factor PHO4 by a cyclin-CDK complex, PHO80-PHO85. *Science* **263:** 1153.

Kaldis, P., A. Sutton, and M.J. Solomon. 1996. The Cdk-activating kinase (CAK) from budding yeast. *Cell* **86:** 553.

Kilmartin, J.V. and A.E.M. Adams. 1984. Structural rearrangements of tubulin and actin during the cell cycle of the yeast *Saccharomyces. J. Cell Biol.* **98:** 922.

King, R.W., P.K. Jackson, and M.W. Kirschner. 1994. Mitosis in transition. *Cell* **79:** 563.

King, R.W., R.J. Deshaies, J.-M. Peters, and M.W. Kirschner. 1996. How proteolysis drives the cell cycle. *Science* **274:** 1652.

King, R.W., J.-M. Peters, S. Tugendreich, M. Rolfe, P. Hieter, and M.W. Kirschner. 1995. A 20S complex containing CDC27 and CDC16 catalyzes the mitosis-specific conjugation of ubiquitin to cyclin B. *Cell* **81:** 279.

Kingsbury, J. and D. Koshland. 1991. Centromere-dependent binding of yeast minichromosomes to microtubules in vitro. *Cell* **66:** 483.

Kirschner, M. 1992. The cell cycle then and now. *Trends Biochem. Sci.* **17:** 281.

Kiser, G.L. and T.A. Weinert. 1996. Distinct roles of yeast *MEC* and *RAD* checkpoint genes in transcriptional induction after DNA damage and implications for function. *Mol. Biol. Cell* **7:** 703.

Kitada, K., A.L. Johnson, L.H. Johnston, and A. Sugino. 1993. A multicopy suppressor gene of the *Saccharomyces cerevisiae* G1 cell cycle mutant gene *dbf4* encodes a protein kinase and is identified as *CDC5*. *Mol. Cell. Biol.* **13:** 4445.

Klis, F.M. 1994. Review: Cell wall assembly in yeast. *Yeast* **10:** 851.

Knapp, D., L. Bhoite, D.J. Stillman, and K. Nasmyth. 1996. The transcription factor Swi5 regulates expression of the cyclin kinase inhibitor p40^{SIC1}. *Mol. Cell. Biol.* **16:** 5701.

Koch, C. and K. Nasmyth. 1994. Cell cycle regulated transcription in yeast. *Curr. Opin. Cell Biol.* **6:** 451.

Koch, C., A. Schleiffer, G. Ammerer, and K. Nasmyth. 1996. Switching transcription on and off during the yeast cell cycle: Cln/Cdc28 kinases activate bound transcription factor SBF (Swi4/Swi6) at Start, whereas Clb/Cdc28 kinases displace it from the promoter in G2. *Genes Dev.* **10:** 129.

Koch, C., T. Moll, M. Neuberg, H. Ahorn, and K. Nasmyth. 1993. A role for the transcription factors Mbp1 and Swi4 in progression from G1 to S phase. *Science* **261:** 1551.

Kreutzer, M.A., J.P. Richards, M.N. De Silva-Udawatta, J.J. Temenak, J.A. Knoblich, C.F. Lehner, and K.L. Bennett. 1995. *Caenorhabditis elegans* cyclin A- and B-type genes: A cyclin A multigene family, an ancestral cyclin B3 and differential germline expression. *J. Cell Sci.* **108:** 2415.

Kuchin, S., P. Yeghiayan, and M. Carlson. 1995. Cyclin-dependent protein kinase and cyclin homologs SSN3 and SSN8 contribute to transcriptional control in yeast. *Proc. Natl. Acad. Sci.* **92:** 4006.

Kühne, C. and P. Linder. 1993. A new pair of B-type cyclins from *Saccharomyces cerevisiae* that function early in the cell cycle. *EMBO J.* **12:** 3437.

Kuranda, M.J. and P.W. Robbins. 1991. Chitinase is required for cell separation during growth of *Saccharomyces cerevisiae*. *J. Biol. Chem.* **266:** 19758.

Labbé, J.-C., J.-P. Capony, D. Caput, J.-C. Cavadore, J. Derancourt, M. Kaghad, J.-M. Lelias, A. Picard, and M. Dorée. 1989. MPF from starfish oocytes at first meiotic metaphase is a heterodimer containing one molecule of cdc2 and one molecule of cyclin B. *EMBO J.* **8:** 3053.

Lahav-Baratz, S., V. Sudakin, J.V. Ruderman, and A. Hershko. 1995. Reversible phosphorylation controls the activity of cyclosome-associated cyclin-ubiquitin ligase. *Proc. Natl. Acad. Sci.* **92:** 9303.

Lamb, J.R., W.A. Michaud, R.S. Sikorski, and P.A. Hieter. 1994. Cdc16p, Cdc23p and Cdc27p form a complex essential for mitosis. *EMBO J.* **13:** 4321.

Lanker, S., M.H. Valdivieso, and C. Wittenberg. 1996. Rapid degradation of the G1 cyclin Cln2 induced by CDK-dependent phosphorylation. *Science* **271:** 1597.

Lauzé, E., B. Stoelcker, F.C. Luca, E. Weiss, A.R. Schutz, and M. Winey. 1995. Yeast spindle pole body duplication gene *MPS1* encodes an essential dual specificity protein kinase. *EMBO J.* **14:** 1655.

Lei, M., Y. Kawasaki, and B.K. Tye. 1996. Physical interactions among Mcm proteins and effects of Mcm dosage on DNA replication in *Saccharomyces cerevisiae*. *Mol. Cell. Biol.* **16:** 5081.

Lenburg, M.E. and E.K. O'Shea. 1996. Signaling phosphate starvation. *Trends Biochem. Sci.* **21:** 383.

Levine, K., K. Huang, and F.R. Cross. 1996. *Saccharomyces cerevisiae* G_1 cyclins differ in their intrinsic functional specificities. *Mol. Cell. Biol.* **16:** 6794.

Lew, D.J. and S. Kornbluth. 1996. Regulatory roles of cyclin dependent kinase phosphorylation in cell cycle control. *Curr. Opin. Cell Biol.* **8:** 795.

Lew, D.J. and S.I. Reed. 1993. Morphogenesis in the yeast cell cycle: Regulation by Cdc28 and cyclins. *J. Cell Biol.* **120:** 1305.

———. 1995a. A cell cycle checkpoint monitors cell morphogenesis in budding yeast. *J. Cell Biol.* **129:** 739.

———. 1995b. Cell cycle control of morphogenesis in budding yeast. *Curr. Opin. Genet. Dev.* **5:** 17.

Lew, D.J., V. Dulic, and S.I. Reed. 1991. Isolation of three novel human cyclins by rescue of G1 cyclin (Cln) function in yeast. *Cell* **66:** 1197.

Lew, D.J., N.J. Marini, and S.I. Reed. 1992. Different G1 cyclins control the timing of cell cycle commitment in mother and daughter cells of the budding yeast *S. cerevisiae. Cell* **69:** 317.

Li, R. and A.W. Murray. 1991. Feedback control of mitosis in budding yeast. *Cell* **66:** 519.

Li, X. and M. Cai. 1997. Inactivation of the cyclin-dependent kinase Cdc28 abrogates cell cycle arrest induced by DNA damage and disassembly of mitotic spindles in *Saccharomyces cerevisiae. Mol. Cell. Biol.* **17:** 2723.

Li, Y. and R. Benezra. 1996. Identification of a human mitotic checkpoint gene: *hsMAD2. Science* **274:** 246.

Li, Y.-Y., E. Yeh, T. Hays, and K. Bloom. 1993. Disruption of mitotic spindle orientation in a yeast dynein mutant. *Proc. Natl. Acad. Sci.* **90:** 10096.

Liang, C., M. Weinreich, and B. Stillman. 1995. ORC and Cdc6p interact and determine the frequency of initiation of DNA replication in the genome. *Cell* **81:** 667.

Liao, S.-M., J. Zhang, D.A. Jeffery, A.J. Koleske, C.M. Thompson, D.M. Chao, M. Viljoen, H.J.J. van Vuuren, and R.A. Young. 1995. A kinase-cyclin pair in the RNA polymerase II holoenzyme. *Nature* **374:** 193.

Lillie, S.H. and S.S. Brown. 1994. Immunofluorescence localization of the unconventional myosin, Myo2p, and the putative kinesin-related protein, Smy1p, to the same regions of polarized growth in *Saccharomyces cerevisiae. J. Cell Biol.* **125:** 825.

Lim, H.H., P.-Y. Goh, and U. Surana. 1996a. Spindle pole body separation in *Saccharomyces cerevisiae* requires dephosphorylation of the tyrosine 19 residue of Cdc28. *Mol. Cell. Biol.* **16:** 6385.

Lim, H.H., C.J. Loy, S. Zaman, and U. Surana. 1996b. Dephosphorylation of threonine 169 of Cdc28 is not required for exit from mitosis but may be necessary for Start in *Saccharomyces cerevisiae. Mol. Cell. Biol.* **16:** 4573

Lin, F.C. and K.T. Arndt. 1995. The role of *Saccharomyces cerevisiae* type 2A phosphatase in the actin cytoskeleton and in entry into mitosis. *EMBO J.* **14:** 2745.

Longhese, M.P., R. Fraschini, P. Plevani, and G. Lucchini. 1996. Yeast *pip3/mec3* mutants fail to delay entry into S phase and to slow DNA replication in response to DNA damage, and they define a functional link between Mec3 and DNA primase. *Mol. Cell. Biol.* **16:** 3235.

Longtine, M.S., D.J. DeMarini, M.L. Valencik, O.S. Al-Awar, H. Fares, C. De Virgilio, and J.R. Pringle. 1996. The septins: Roles in cytokinesis and other processes. *Curr. Opin. Cell Biol.* **8:** 106.

Loo, S., C.A. Fox, J. Rine, R. Kobayashi, B. Stillman, and S. Bell. 1995. The origin recognition complex in silencing, cell cycle progression, and DNA replication. *Mol. Biol. Cell* **6:** 741.

Lydall, D. and T. Weinert. 1995. Yeast checkpoint genes in DNA damage processing: implications for repair and arrest. *Science* **270:** 1488.

———. 1996. From DNA damage to cell cycle arrest and suicide: A budding yeast perspective. *Curr. Opin. Genet. Dev.* **6:** 4.

Lydall, D., G. Ammerer, and K. Nasmyth. 1991. A new role for MCM1 in yeast: Cell cycle regulation of *SWI5* transcription. *Genes Dev.* **5:** 2405.

Ma, X.-J., Q. Lu, and M. Grunstein. 1996. A search for proteins that interact genetically with histone H3 and H4 amino termini uncovers novel regulators of the Swe1 kinase in *Saccharomyces cerevisiae. Genes Dev.* **10:** 1327.

Maher, M., F. Cong, D. Kindelberger, K. Nasmyth, and S. Dalton. 1995. Cell cycle-regulated transcription of the *CLB2* gene is dependent on Mcm1 and a ternary complex factor. *Mol. Cell. Biol.* **15:** 3129.

Marini, N.J. and S.I. Reed. 1992. Direct induction of G1-specific transcripts following reactivation of the Cdc28 kinase in the absence of de novo protein synthesis. *Genes Dev.* **6:** 557.

Mathias, N., S.L. Johnson, M. Winey, A.E.M. Adams, L. Goetsch, J.R. Pringle, B. Byers, and M.G. Goebl. 1996. Cdc53p acts in concert with Cdc4p and Cdc34p to control the G1-to-S-phase transition and identifies a conserved family of proteins. *Mol. Cell. Biol.* **16:** 6634.

McGrew, J.T., L. Goetsch, B. Byers, and P. Baum. 1992. Requirement for *ESP1* in the nuclear division of *Saccharomyces cerevisiae. Mol. Biol. Cell* **3:** 1443.

McKinney, J.D. and F.R. Cross. 1995. *FAR1* and the G1 phase specificity of cell cycle arrest by mating factor in *Saccharomyces cerevisiae. Mol. Cell. Biol.* **15:** 2509.

McKinney, J.D., F. Chang, N. Heintz, and F.R. Cross. 1993. Negative regulation of *FAR1* at the Start of the yeast cell cycle. *Genes Dev.* **7:** 833.

McMillan, J.N. and K. Tatchell. 1994. The *JNM1* gene in the yeast *Saccharomyces cerevisiae* is required for nuclear migration and spindle orientation during the mitotic cell cycle. *J. Cell Biol.* **125:** 143.

Measday, V., L. Moore, J. Ogas, M. Tyers, and B. Andrews. 1994. The PCL2 (ORFD)-PHO85 cyclin-dependent kinase complex: A cell cycle regulator in yeast. *Science* **266:** 1391.

Measday, V., L. Moore, R. Retnakaran, J. Lee, M. Donoviel, A.M. Neiman, and B. Andrews. 1997. A family of cyclin-like proteins that interact with the Pho85 cyclin-dependent kinase. *Mol. Cell. Biol.* **17:** 1212.

Mendenhall, M.D. 1993. An inhibitor of $p34^{CDC28}$ protein kinase activity from *Saccharomyces cerevisiae. Science* **259:** 216.

Mendenhall, M.D., C.A. Jones, and S.I. Reed. 1987. Dual regulation of the yeast CDC28-p40 protein kinase complex: Cell cycle, pheromone, and nutrient limitation effects. *Cell* **50:** 927.

Mendenhall, M.D., H.E. Richardson, and S.I. Reed. 1988. Dominant negative protein kinase mutations that confer a G1 arrest phenotype. *Proc. Natl. Acad. Sci.* **85:** 4426.

Mitchison, J.M. 1971. *The biology of the cell cycle.* Cambridge University Press, Cambridge, United Kingdom.

Moll, T., G. Tebb, U. Surana, H. Robitsch, and K. Nasmyth. 1991. The role of phosphorylation and the CDC28 protein kinase in cell cycle-regulated nuclear import of the *S. cerevisiae* transcription factor SWI5. *Cell* **66:** 743.

Mondésert, G. and S.I. Reed. 1996. *BED1,* a gene encoding a galactosyltransferase homologue, is required for polarized growth and efficient bud emergence in *Saccharomyces cerevisiae. J. Cell Biol.* **132:** 137.

Morgan, D.O. 1995. Principles of CDK regulation. *Nature* **374:** 131.

Muhua, L., T.S. Karpova, and J.A. Cooper. 1994. A yeast actin-related protein homologous to that in vertebrate dynactin complex is important for spindle orientation and nuclear migration. *Cell* **78:** 669.

Mulholland, J., D. Preuss, A. Moon, A. Wong, D. Drubin, and D. Botstein. 1994. Ultrastructure of the yeast actin cytoskeleton and its association with the plasma membrane. *J. Cell Biol.* **125:** 381.

Murray, A.W. 1992. Creative blocks: cell-cycle checkpoints and feedback controls. *Nature* **359:** 599.

Nash, R., G. Tokiwa, S. Anand, K. Erickson, and A.B. Futcher. 1988. The *WHI1*$^+$ gene of *Saccharomyces cerevisiae* tethers cell division to cell size and is a cyclin homolog. *EMBO J.* **7:** 4335.

Nasmyth, K. 1993. Control of the yeast cell cycle by the Cdc28 protein kinase. *Curr. Opin. Cell Biol.* **5:** 166.

———. 1996. At the heart of the budding yeast cell cycle. *Trends Genet.* **12:** 405.

Navas, T.A., Y. Sanchez, and S.J. Elledge. 1996. *RAD9* and DNA polymerase ε form parallel sensory branches for transducing the DNA damage checkpoint signal in *Saccharomyces cerevisiae. Genes Dev.* **10:** 2632-2643.

Navas, T.A., Z. Zhou, and S.J. Elledge. 1995. DNA polymerase ε links the DNA replication machinery to the S phase checkpoint. *Cell* **80:** 29.

Newlon, C.S. and J.F. Theis. 1993. The structure and function of yeast ARS elements. *Curr. Opin. Genet. Dev.* **3:** 752.

Nicolson, T., B. Conradt, and W. Wickner. 1996. A truncated form of the Pho80 cyclin of *Saccharomyces cerevisiae* induces expression of a small cytosolic factor which inhibits vacuole inheritance. *J. Bacteriol.* **178:** 4047.

Nigg, E.A. 1991. The substrates of the cdc2 kinase. *Semin. Cell Biol.* **2:** 261.

Nombela, C., M. Molina, R. Cenamor, and M. Sanchez. 1988. Yeast β-glucanases: A complex system of secreted enzymes. *Microbiol. Sci.* **5:** 328.

Novick, P. and D. Botstein. 1985. Phenotypic analysis of temperature-sensitive yeast actin mutants. *Cell* **40:** 405.

Nugent, C.I., T.R. Hughes, N.F. Lue, and V. Lundblad. 1996. Cdc13p: A single-strand telomeric DNA-binding protein with a dual role in yeast telomere maintenance. *Science* **274:** 249.

Nugroho, T.T. and M.D. Mendenhall. 1994. An inhibitor of yeast cyclin-dependent protein kinase plays an important role in ensuring the genomic integrity of daughter cells. *Mol. Cell. Biol.* **14:** 3320.

Oehlen, L.J.W.M. and F.R. Cross. 1994. G1 cyclins *CLN1* and *CLN2* repress the mating factor response pathway at Start in the yeast cell cycle. *Genes Dev.* **8:** 1058.

Oehlen, L.J.W.M., J.D. McKinney, and F.R. Cross. 1996. Ste12 and Mcm1 regulate cell cycle-dependent transcription of *FAR1. Mol. Cell. Biol.* **16:** 2830.

Palmer, R.E., D.S. Sullivan, T. Huffaker, and D. Koshland. 1992. Role of astral microtubules and actin in spindle orientation and migration in the budding yeast, *Saccharomyces cerevisiae. J. Cell Biol.* **119:** 583.

Pammer, M., P. Briza, A. Ellinger, T. Schuster, R. Stucka, H. Feldmann, and M. Breitenbach. 1992. *DIT101* (*CSD2, CAL1*), a cell cycle-regulated yeast gene required for synthesis of chitin in cell walls and chitosan in spore walls. *Yeast* **8:** 1089.

Pangilinan, F. and F. Spencer. 1996. Abnormal kinetochore structure activates the spindle assembly checkpoint in budding yeast. *Mol. Biol. Cell* **7:** 1195.

Paulovich, A.G. and L.H. Hartwell. 1995. A checkpoint regulates the rate of progression through S phase in *S. cerevisiae* in response to DNA damage. *Cell* **82:** 841.

Paulovich, A.G., R.U. Margulies, B.M. Garvik, and L.H. Hartwell. 1997. *RAD9, RAD17,* and *RAD24* are required for S phase regulation in *Saccharomyces cerevisiae* in response to DNA damage. *Genetics* **145:** 45.

Peeper, D.S., L.L. Parker, M.E. Ewen, M. Toebes, F.L. Hall, M. Xu, A. Zantema, A.J. van der Eb, and H. Piwnica-Worms. 1993. A- and B-type cyclins differentially modulate substrate specificity of cyclin-cdk complexes. *EMBO J.* **12:** 1947.

Pellman, D., M. Bagget, H. Tu, and G.R. Fink. 1995. Two microtubule-associated proteins required for anaphase spindle movement in *Saccharomyces cerevisiae. J. Cell Biol.* **130:** 1373.

Peter, M. and I. Herskowitz. 1994a. Direct inhibition of the yeast cyclin-dependent kinase Cdc28-Cln by Far1. *Science* **265:** 1228.

———. 1994b. Joining the complex: Cyclin-dependent kinase inhibitory proteins and the cell cycle. *Cell* **79:** 181.

Peter, M., A. Gartner, J. Horecka, G. Ammerer, and I. Herskowitz. 1993. *FAR1* links the signal transduction pathway to the cell cycle machinery in yeast. *Cell* **73:** 747.

Piatti, S., C. Lengauer, and K. Nasmyth. 1995. Cdc6 is an unstable protein whose *de novo* synthesis in G1 is important for the onset of S phase and for preventing a 'reductional' anaphase in the budding yeast *Saccharomyces cerevisiae. EMBO J.* **14:** 3788.

Piatti, S., T. Böhm, J.H. Cocker, J.F.X. Diffley, and K. Nasmyth. 1996. Activation of S-phase-promoting CDKs in late G1 defines a "point of no return" after which Cdc6 synthesis cannot promote DNA replication in yeast. *Genes Dev.* **10:** 1516.

Piggott, J.R., R. Rai, and B.L.A. Carter. 1982. A bifunctional gene product involved in two phases of the yeast cell cycle. *Nature* **298:** 391.

Preuss, D., J. Mulholland, A. Franzusoff, N. Segev, and D. Botstein. 1992. Characterization of the *Saccharomyces* Golgi complex through the cell cycle by immunoelectron microscopy. *Mol. Biol. Cell.* **3:** 789.

Pringle, J.R. 1978. The use of conditional lethal cell cycle mutants for temporal and functional sequence mapping of cell cycle events. *J. Cell. Physiol.* **95:** 393.

Pringle, J.R. and L.H. Hartwell. 1981. The *Saccharomyces cerevisiae* cell cycle. In *The molecular biology of the yeast* Saccharomyces: *Life cycle and inheritance* (ed. J.N. Strathern et al.), p. 97. Cold Spring Harbor Laboratory, Cold Spring Harbor, New York.

Pringle, J.R., R.A. Preston, A.E.M. Adams, T. Stearns, D.G. Drubin, B.K. Haarer, and E.W. Jones. 1989. Fluorescence microscopy methods for yeast. *Methods Cell Biol.* **31:** 357.

Rao, H. and B. Stillman. 1995. The origin recognition complex interacts with a bipartite DNA binding site within yeast replicators. *Proc. Natl. Acad. Sci.* **92:** 2224.

Reed, S.I. 1980. The selection of *S. cerevisiae* mutants defective in the Start event of cell division. *Genetics* **95:** 561.

———. 1992. The role of p34 kinases in the G1 to S-phase transition. *Annu. Rev. Cell Biol.* **8:** 529.

Reed, S.I. and C. Wittenberg. 1990. Mitotic role for the Cdc28 protein kinase of *Saccharomyces cerevisiae. Proc. Natl. Acad. Sci.* **87:** 5697.

Richardson, H.E., C. Wittenberg, F.R. Cross, and S.I. Reed. 1989. An essential G1 function for cyclin-like proteins in yeast. *Cell* **59:** 1127.

Richardson, H.E., D.J. Lew, M. Henze, K. Sugimoto, and S.I. Reed. 1992. Cyclin-B homologs in *Saccharomyces cerevisiae* function in S phase and in G2. *Genes Dev.* **6:** 2021.

Richardson, H.E., C.S. Stueland, J. Thomas, P. Russell, and S.I. Reed. 1990. Human cDNAs encoding homologs of the small $p34^{Cdc28/cdc2}$-associated protein of *Saccharomyces cerevisiae* and *Schizosaccharomyces pombe*. *Genes Dev.* **4:** 1332.

Rivin, C.J. and W.L. Fangman. 1980. Cell cycle phase expansion in nitrogen-limited cultures of *Saccharomyces cerevisiae. J. Cell Biol.* **85:** 96.

Roberts, B.T., K.A. Farr, and M.A. Hoyt. 1994. The *Saccharomyces cerevisiae* checkpoint gene *BUB1* encodes a novel protein kinase. *Mol. Cell. Biol.* **14:** 8282.

Roof, D.M., P.B. Meluh, and M.D. Rose. 1992. Kinesin-related proteins required for assembly of the mitotic spindle. *J. Cell Biol.* **118:** 95.

Rowley, A., J.H. Cocker, J. Harwood, and J.F.X. Diffley. 1995. Initiation complex assembly at budding yeast replication origins begins with the recognition of a bipartite sequence by limiting amounts of the initiator, ORC. *EMBO J.* **14:** 2631.

Rudner, A.D. and A.W. Murray. 1996. The spindle assembly checkpoint. *Curr. Opin. Cell Biol.* **8:** 773.

Russell, P. and P. Nurse. 1986. *Schizosaccharomyces pombe* and *Saccharomyces cerevisiae:* A look at yeasts divided. *Cell* **45:** 781.

Russell, P., S. Moreno, and S.I. Reed. 1989. Conservation of mitotic controls in fission and budding yeast. *Cell* **57:** 295.

Salama, S.R., K.B. Hendricks, and J. Thorner. 1994. G1 cyclin degradation: The PEST motif of yeast CLN2 is necessary, but not sufficient, for rapid protein turnover. *Mol. Cell. Biol.* **14:** 7953.

Sanchez, Y., B.A. Desany, W.J. Jones, Q. Liu, B. Wang, and S.J. Elledge. 1996. Regulation of *RAD53* by the *ATM*-like kinases *MEC1* and *TEL1* in yeast cell cycle checkpoint pathways. *Science* **271:** 357.

Sandell, L.L. and V.A. Zakian. 1993. Loss of a yeast telomere: Arrest, recovery, and chromosome loss. *Cell* **75:** 729.

Santocanale, C. and J.F.X. Diffley. 1996. ORC- and Cdc6-dependent complexes at active and inactive chromosomal replication origins in *Saccharomyces cerevisiae. EMBO J.* **15:** 6671.

Santos, B. and M. Snyder. 1997. Targeting of chitin synthase 3 to polarized growth sites in yeast requires Chs5p and Myo2p. *J. Cell Biol.* **136:** 95.

Saunders, W.S. and M.A. Hoyt. 1992. Kinesin-related proteins required for structural integrity of the mitotic spindle. *Cell* **70:** 451.

Saunders, W.S., D. Koshland, D. Eshel, I.R. Gibbons, and M.A. Hoyt. 1995. *Saccharomyces cerevisiae* kinesin- and dynein-related proteins required for anaphase chromosome segregation. *J. Cell Biol.* **128:** 617.

Schneider, B.L., Q.-H. Yang, and A.B. Futcher. 1996. Linkage of replication to Start by the Cdk inhibitor Sic1. *Science* **272:** 560.

Schutz, A.R., T.H. Giddings, Jr., E. Steiner, and M. Winey. 1997. The yeast *CDC37* gene interacts with *MPS1* and is required for proper execution of spindle pole body duplication. *J. Cell Biol.* **136:** 969.

Schweitzer, B. and P. Philippsen. 1991. *CDC15,* an essential cell cycle gene in *Saccharomyces cerevisiae,* encodes a protein kinase domain. *Yeast* **7:** 265.

Schwob, E. and K. Nasmyth. 1993. *CLB5* and *CLB6,* a new pair of B cyclins involved in DNA replication in *Saccharomyces cerevisiae. Genes Dev.* **7:** 1160.

Schwob, E., T. Böhm, M.D. Mendenhall, and K. Nasmyth. 1994. The B-type cyclin kinase inhibitor $p40^{SIC1}$ controls the G1 to S transition in *S. cerevisiae* (erratum in *Cell* **84** [1996]). *Cell* **79:** 233.

Sclafani, R.A. and A.L. Jackson. 1994. Cdc7 protein kinase for DNA metabolism comes of age. *Mol. Microbiol.* **11:** 805.

Seufert, W., B. Futcher, and S. Jentsch. 1995. Role of a ubiquitin-conjugating enzyme in degradation of S- and M-phase cyclins. *Nature* **373:** 78.

Shaw, J.A., P.C. Mol, B. Bowers, S.J. Silverman, M.H. Valdivieso, A. Durán, and E. Cabib. 1991. The function of chitin synthetases 2 and 3 in the *Saccharomyces cerevisiae* cell cycle. *J. Cell Biol.* **114:** 111.

Sherr, C.J. 1994. G1 phase progression: Cycling on cue. *Cell* **79:** 551.

Shirayama, M., Y. Matsui, and A. Toh-e. 1994a. The yeast *TEM1* gene, which encodes a GTP-binding protein, is involved in termination of M phase. *Mol. Cell. Biol.* **14:** 7476.

Shirayama, M., Y. Matsui, K. Tanaka, and A. Toh-e. 1994b. Isolation of a *CDC25* family gene, *MSI2/LTE1,* as a multicopy suppressor of *ira1. Yeast* **10:** 451.

Sia, R.A.L., H.A. Herald, and D.J. Lew. 1996. Cdc28 tyrosine phosphorylation and the morphogenesis checkpoint in budding yeast. *Mol. Biol. Cell* **7:** 1657.

Sidorova, J.M., G.E. Mikesell, and L.L. Breeden. 1995. Cell cycle-regulated phosphorylation of Swi6 controls its nuclear localization. *Mol. Biol. Cell* **6:** 1641.

Siede, W., A.S. Friedberg, and E.C. Friedberg. 1993. *RAD9*-dependent G1 arrest defines a second checkpoint for damaged DNA in the cell cycle of *Saccharomyces cerevisiae. Proc. Natl. Acad. Sci.* **90:** 7985.

Siede, W., J.B. Allen, S.J. Elledge, and E.C. Friedberg. 1996. The *Saccharomyces cerevisiae MEC1* gene, which encodes a homolog of the human *ATM* gene product, is required for G1 arrest following radiation treatment. *J. Bacteriol.* **178:** 5841.

Siede, W., A.S. Friedberg, I. Dianova, and E.C. Friedberg. 1994. Characterization of G1 checkpoint control in the yeast *Saccharomyces cerevisiae* following exposure to DNA-damaging agents. *Genetics* **138:** 271.

Sigrist, S., H. Jacobs, R. Stratmann, and C.F. Lehner. 1995. Exit from mitosis is regulated by *Drosophila fizzy* and the sequential destruction of cyclins A, B and B3. *EMBO J.* **14:** 4827.

Simon, V.R., T.C. Swayne, and L.A. Pon. 1995. Actin-dependent mitochondrial motility in mitotic yeast and cell-free systems: Identification of a motor activity on the mitochondrial surface. *J. Cell Biol.* **130:** 345.

Singer, J.D., B.M. Manning, and T. Formosa. 1996. Coordinating DNA replication to produce one copy of the genome requires genes that act in ubiquitin metabolism. *Mol. Cell. Biol.* **16:** 1356.

Slater, M.L., S.O. Sharrow, and J.J. Gart. 1977. Cell cycle of *Saccharomyces cerevisiae* in populations growing at different rates. *Proc. Natl. Acad. Sci.* **74:** 3850.

Snyder, M. 1994. The spindle pole body of yeast. *Chromosoma* **103:** 369.

Snyder, M., S. Gehrung, and B.D. Page. 1991. Studies concerning the temporal and genetic control of cell polarity in *Saccharomyces cerevisiae. J. Cell Biol.* **114:** 515.

Solomon, M.J., M. Glotzer, T.H. Lee, M. Philippe, and M.W. Kirschner. 1990. Cyclin activation of $p34^{cdc2}$. *Cell* **63:** 1013.

Sorger, P.K. and A.W. Murray. 1992. S-phase feedback control in budding yeast independent of tyrosine phosphorylation of $p34^{cdc28}$. *Nature* **355:** 365.

Sprague, G.F., Jr. and J.W. Thorner. 1992. Pheromone response and signal transduction during the mating process of *Saccharomyces cerevisiae.* In *The molecular and cellular biology of the yeast* Saccharomyces: *Gene expression* (ed. E.W. Jones et al.), p. 657. Cold Spring Harbor Laboratory Press, Cold Spring Harbor, New York.

Stemmann, O. and J. Lechner. 1996. The *Saccharomyces cerevisiae* kinetochore contains a cyclin-CDK complexing homologue, as identified by *in vitro* reconstitution. *EMBO J.* **15:** 3611.

Stepanova, L., X. Leng, S.B. Parker, and J.W. Harper. 1996. Mammalian $p50^{Cdc37}$ is a protein kinase-targeting subunit of Hsp90 that binds and stabilizes Cdk4. *Genes Dev.* **10:** 1491.

Sterner, D.E., J.M. Lee, S.E. Hardin, and A.L. Greenleaf. 1995. The yeast carboxyl-terminal repeat domain kinase CTDK-I is a divergent cyclin-cyclin-dependent kinase complex. *Mol. Cell. Biol.* **15:** 5716.

Stillman, B. 1996. Cell cycle control of DNA replication. *Science* **274:** 1659.

Stuart, D. and C. Wittenberg. 1994. Cell cycle-dependent transcription of *CLN2* is conferred by multiple distinct *cis*-acting regulatory elements. *Mol. Cell. Biol.* **14:** 4788.

———. 1995. *CLN3*, not positive feedback, determines the timing of *CLN2* transcription in cycling cells. *Genes Dev.* **9:** 2780.

Stueland, C.S., D.J. Lew, M.J. Cismowski, and S.I. Reed. 1993. Full activation of $p34^{CDC28}$ histone H1 kinase activity is unable to promote entry into mitosis in checkpoint-arrested cells of the yeast *Saccharomyces cerevisiae. Mol. Cell. Biol.* **13:** 3744.

Sudakin, V., D. Ganoth, A. Dahan, H. Heller, J. Hershko, F.C. Luca, J.V. Ruderman, and A. Hershko. 1995. The cyclosome, a large complex containing cyclin-selective ubiquitin ligase activity, targets cyclins for destruction at the end of mitosis. *Mol. Biol. Cell* **6:** 185.

Sudbery, P.E., A.R. Goodey, and B.L.A. Carter. 1980. Genes which control cell proliferation in the yeast *Saccharomyces cerevisiae. Nature* **288:** 401.

Sugimoto, K., T. Shimomura, K. Hashimoto, H. Araki, A. Sugino, and K. Matsumoto. 1996. Rfc5, a small subunit of replication factor C complex, couples DNA replication and mitosis in budding yeast. *Proc. Natl. Acad. Sci.* **93:** 7048.

Sullivan, D.S. and T.C. Huffaker. 1992. Astral microtubules are not required for anaphase B in *Saccharomyces cerevisiae. J. Cell Biol.* **119:** 379.

Sun, Z., D.S. Fay, F. Marini, M. Foiani, and D.F. Stern. 1996. Spk1/Rad53 is regulated by Mec1-dependent protein phosphorylation in DNA replication and damage checkpoint pathways. *Genes Dev.* **10:** 395.

Sundberg, H.A., L. Goetsch, B. Byers, and T.N. Davis. 1996. Role of calmodulin and Spc110 interaction in the proper assembly of spindle pole body components. *J. Cell Biol.* **133:** 111.

Surana, U., A. Amon, C. Dowzer, J. McGrew, B. Byers, and K. Nasmyth. 1993. Destruction of the CDC28/CLB mitotic kinase is not required for the metaphase to anaphase transition in budding yeast. *EMBO J.* **12:** 1969.

Surana, U., H. Robitsch, C. Price, T. Schuster, I. Fitch, A.B. Futcher, and K. Nasmyth. 1991. The role of CDC28 and cyclins during mitosis in the budding yeast *S. cerevisiae. Cell* **65:** 145.

Taba, M.R.M., I. Muroff, D. Lydall, G. Tebb, and K. Nasmyth. 1991. Changes in a SWI4,6-DNA-binding complex occur at the time of *HO* gene activation in yeast. *Genes Dev.* **5:** 2000.

Talens, L.T., M. Miranda, and M.W. Miller. 1977. Electron micrography of bud formation in *Metschnikowia krissii. J. Bacteriol.* **114:** 413.

Tang, Y. and S.I. Reed. 1993. The Cdk-associated protein Cks1 functions both in G1 and G2 in *Saccharomyces cerevisiae. Genes Dev.* **7:** 822.

TerBush, D.R. and P. Novick. 1995. Sec6, Sec8, and Sec15 are components of a multisubunit complex which localizes to small bud tips in *Saccharomyces cerevisiae. J. Cell Biol.* **130:** 299.

Thevelein, J.M. 1994. Signal transduction in yeast. *Yeast* **10:** 1753.

Thuret, J.-Y., J.-G. Valay, G. Faye, and C. Mann. 1996. Civ1 (CAK in vivo), a novel Cdk-activating kinase. *Cell* **86:** 565.

Timblin, B.K., K. Tatchell, and L.W. Bergman. 1996. Deletion of the gene encoding the cyclin-dependent protein kinase Pho85 alters glycogen metabolism in *Saccharomyces cerevisiae. Genetics* **143:** 57.

Toyn, J.H. and L.H. Johnston. 1994. The Dbf2 and Dbf20 protein kinases of budding yeast are activated after the metaphase to anaphase cell cycle transition. *EMBO J.* **13:** 1103.

Toyn, J.H., A.L. Johnson, and L.H. Johnston. 1995. Segregation of unreplicated chromosomes in *Saccharomyces cerevisiae* reveals a novel G1/M-phase checkpoint. *Mol. Cell. Biol.* **15:** 5312.

Toyn, J.H., A.L. Johnson, J.D. Donovan, W.M. Toone, and L.H. Johnston. 1997. The Swi5 transcription factor of *Saccharomyces cerevisiae* has a role in exit from mitosis through induction of the cdk-inhibitor Sic1 in telophase. *Genetics* **145:** 85.

Tugendreich, S., J. Tomkiel, W. Earnshaw, and P. Hieter. 1995. CDC27Hs colocalizes with CDC16Hs to the centrosome and mitotic spindle and is essential for the metaphase to anaphase transition. *Cell* **81:** 261.

Tye, B.-K. 1994. The MCM2-3-5 proteins: Are they replication licensing factors? *Trends Cell Biol.* **4:** 160.

———. 1996. The cyclin-dependent kinase inhibitor $p40^{SIC1}$ imposes the requirement for Cln G1 cyclin function at Start. *Proc. Natl. Acad. Sci.* **93:** 7772.

Tyers, M., G. Tokiwa, and B. Futcher. 1993. Comparison of the *Saccharomyces cerevisiae* G1 cyclins: Cln3 may be an upstream activator of Cln1, Cln2 and other cyclins. *EMBO J.* **12:** 1955.

Tyers, M., G. Tokiwa, R. Nash, and B. Futcher. 1992. The Cln3-Cdc28 kinase complex of *S. cerevisiae* is regulated by proteolysis and phosphorylation. *EMBO J.* **11:** 1773.

Valay, J.-G., M. Simon, M.-F. Dubois, O. Bensaude, C. Facca, and G. Faye. 1995. The *KIN28* gene is required for both RNA polymerase II mediated transcription and phosphorylation of the Rpb1p CTD. *J. Mol. Biol.* **249:** 535.

Vallen, E.A., T.Y. Scherson, T. Roberts, K. van Zee, and M.D. Rose. 1992. Asymmetric mitotic segregation of the yeast spindle pole body. *Cell* **69:** 505.

Waddle, J.A., T.S. Karpova, R.H. Waterston, and J.A. Cooper. 1996. Movement of cortical actin patches in yeast. *J. Cell Biol.* **132:** 861.

Wan, J., H. Xu, and M. Grunstein. 1992. *CDC14* of *Saccharomyces cerevisiae*. Cloning, sequence analysis, and transcription during the cell cycle. *J. Biol. Chem.* **267:** 11274.

Wang, Y. and D.J. Burke. 1995. Checkpoint genes required to delay cell division in response to nocodazole respond to impaired kinetochore function in the yeast *Saccharomyces cerevisiae. Mol. Cell. Biol.* **15:** 6838.

Wang, Y.-X., H. Zhao, T.M. Harding, D.S. Gomes de Mesquita, C.L. Woldringh, D.J. Klionsky, A.L. Munn, and L.S. Weisman. 1996. Multiple classes of yeast mutants are defective in vacuole partitioning yet target vacuole proteins correctly. *Mol. Biol. Cell* **7:** 1375.

Warren, G. and W. Wickner. 1996. Organelle inheritance. *Cell* **84:** 395.

Weinert, T.A. and L.H. Hartwell. 1988. The *RAD9* gene controls the cell cycle response to DNA damage in *Saccharomyces cerevisiae. Science* **241:** 317.

———. 1990. Characterization of *RAD9* of *Saccharomyces cerevisiae* and evidence that its function acts posttranslationally in cell cycle arrest after DNA damage. *Mol. Cell. Biol.* **10:** 6554.

———. 1993. Cell cycle arrest of *cdc* mutants and specificity of the *RAD9* checkpoint. *Genetics* **134:** 63.

Weinert, T.A. and D. Lydall. 1993. Cell cycle checkpoints, genetic instability, and cancer. *Semin. Cancer Biol.* **4:** 129.

Weinert, T.A., G.L. Kiser, and L.H. Hartwell. 1994. Mitotic checkpoint genes in budding yeast and the dependence of mitosis on DNA replication and repair. *Genes Dev.* **8:** 652.

Weiss, E. and M. Winey. 1996. The *Saccharomyces cerevisiae* spindle pole body duplication gene *MPS1* is part of a mitotic checkpoint. *J. Cell Biol.* **132:** 111.

Welch, M.D., D.A. Holtzman, and D.G. Drubin. 1994. The yeast actin cytoskeleton. *Curr. Opin. Cell Biol.* **6:** 110.

Wells, W.A.E. 1996. The spindle-assembly checkpoint: Aiming for a perfect mitosis, every time. *Trends Cell Biol.* **6:** 228.

Werner-Washburne, M., E. Braun, G.C. Johnston, and R.A. Singer. 1993. Stationary phase in the yeast *Saccharomyces cerevisiae. Microbiol. Rev.* **57:** 383.

Whitfield, W.G.F., C. Gonzalez, G. Maldonado-Codina, and D.M. Glover. 1990. The A- and B-type cyclins of *Drosophila* are accumulated and destroyed in temporally distinct events that define separable phases of the G2-M transition. *EMBO J.* **9:** 2563.

Willems, A.R., S. Lanker, E.E. Patton, K.L. Craig, T.F. Nason, N. Mathias, R. Kobayashi, C. Wittenberg, and M. Tyers. 1996. Cdc53 targets phosphorylated G1 cyclins for degradation by the ubiquitin proteolytic pathway. *Cell* **86:** 453.

Williamson, D.H. 1965. The timing of deoxyribonucleic acid synthesis in the cell cycle of *Saccharomyces cerevisiae. J. Cell Biol.* **25:** 517.

Wilmen, A. and J.H. Hegemann. 1996. The chromatin of the *Saccharomyces cerevisiae* centromere shows cell-type specific changes. *Chromosoma* **104:** 489.

Winey, M. and B. Byers. 1993. Assembly and functions of the spindle pole body in budding yeast. *Trends Genet.* **9:** 300.

Winey, M., C.L. Mamay, E.T. O'Toole, D.N. Mastronarde, T.H. Giddings, Jr., K.L. McDonald, and J.R. McIntosh. 1995. Three-dimensional ultrastructural analysis of the *Saccharomyces cerevisiae* mitotic spindle. *J. Cell Biol.* **129:** 1601.

Wittenberg, C. and S.I. Reed. 1988. Control of the yeast cell cycle is associated with assembly/disassembly of the Cdc28 protein kinase complex. *Cell* **54:** 1061.

———. 1996. Plugging it in: Signaling circuits and the yeast cell cycle. *Curr. Opin. Cell Biol.* **8:** 223.

Wittenberg, C., K. Sugimoto, and S.I. Reed. 1990. G1-specific cyclins of *S. cerevisiae*: Cell cycle periodicity, regulation by mating pheromone, and association with the $p34^{CDC28}$ protein kinase. *Cell* **62:** 225.

Yaglom, J.A., A.L. Goldberg, D. Finley, and M.Y. Sherman. 1996. The molecular chaperone Ydj1 is required for the $p34^{CDC28}$-dependent phosphorylation of the cyclin Cln3 that signals its degradation. *Mol. Cell. Biol.* **16:** 3679.

Yaglom, J., M.H.K. Linskens, S. Sadis, D.M. Rubin, B. Futcher, and D. Finley. 1995.

p34^{Cdc28}-mediated control of Cln3 cyclin degradation. *Mol. Cell. Biol.* **15:** 731.

Yamamoto, A., V. Guacci, and D. Koshland. 1996a. Pds1p is required for faithful execution of anaphase in the yeast, *Saccharomyces cerevisiae. J. Cell Biol.* **133:** 85.

———. 1996b. Pds1p, an inhibitor of anaphase in budding yeast, plays a critical role in the APC and checkpoint pathway(s). *J. Cell Biol.* **133:** 99.

Yan, H., A.M. Merchant, and B.-K. Tye. 1993. Cell cycle-regulated nuclear localization of MCM2 and MCM3, which are required for the initiation of DNA synthesis at chromosomal replication origins in yeast. *Genes Dev.* **7:** 2149.

Yang, S.S., E. Yeh, E.D. Salmon, and K. Bloom. 1997. Identification of a mid-anaphase checkpoint in budding yeast. *J. Cell Biol.* **136:** 345.

Yeh, E., R.V. Skibbens, J.W. Cheng, E.D. Salmon, and K. Bloom. 1995. Spindle dynamics and cell cycle regulation of dynein in the budding yeast, *Saccharomyces cerevisiae. J. Cell Biol.* **130:** 687.

Zachariae, W. and K. Nasmyth. 1996. TPR proteins required for anaphase progression mediate ubiquitination of mitotic B-type cyclins in yeast. *Mol. Biol. Cell* **7:** 791.

Zachariae, W., T.H. Shin, M. Galova, B. Obermaier, and K. Nasmyth. 1996. Identification of subunits of the anaphase-promoting complex of *Saccharomyces cerevisiae. Nature* **274:** 1201.

Zakian, V.A. 1995. *ATM*-related genes: What do they tell us about functions of the human gene? *Cell* **82:** 685.

Ziman, M., J.S. Chuang, and R.W. Schekman. 1996. Chs1p and Chs3p, two proteins involved in chitin synthesis, populate a compartment of the *Saccharomyces cerevisiae* endocytic pathway. *Mol. Biol. Cell* **7:** 1909.

Zou, L., J. Mitchell, and B. Stillman. 1997. *CDC45,* a novel yeast gene that functions with the origin recognition complex and Mcm proteins in initiation of DNA replication. *Mol. Cell. Biol.* **17:** 553.

Zwerschke, W., H.-W. Rottjakob, and H. Küntzel. 1994. The *Saccharomyces cerevisiae CDC6* gene is transcribed at late mitosis and encodes a ATP/GTPase controlling S phase initiation. *J. Biol. Chem.* **269:** 23351.

8

Cell Cycle Control in Fission Yeast

Stuart A. MacNeill
Institute of Cell And Molecular Biology
University of Edinburgh, King's Buildings
Edinburgh EH9 3JR, United Kingdom

Paul Nurse
Imperial Cancer Research Fund, Lincoln's Inn Fields
London WC2A 3PX, United Kingdom

I. INTRODUCTION

A. Overview

The fission yeast *Schizosaccharomyces pombe* has proved to be a good model system for analysis of the eukaryotic cell cycle. A number of the key cell cycle regulatory molecules now being extensively studied in higher eukaryotes were first identified and characterized in fission yeast, and it has become clear that many of the regulatory networks elucidated through genetic analysis in *S. pombe* are conserved in higher eukaryotes as well (Nurse 1990). In this chapter, we consider the regulatory mechanisms that control progression through the mitotic cycle of fission yeast with particular emphasis on those mechanisms that ensure the correct timing and temporal ordering of key cell cycle transitions. However, we do not deal with the control of cytokinesis, which is discussed elsewhere in this volume (Su and Yanagida, this volume). We also consider what is known of the enzymology of DNA replication in fission yeast, whereas the accompanying chapters focus on the processes of mitosis and cytokinesis (Su and Yanagida, this volume) and of conjugation, meiosis, and sporulation (Yamamoto et al., this volume). We begin with a brief introduction to *S. pombe* and its life cycle.

B. Life Cycle and Genetics

S. pombe is a unicellular ascomycete fungus only distantly related to the budding yeast *Saccharomyces cerevisiae* (for discussion of phylogenetic relationships, see Sipiczki 1995). Fission yeast cells are rod-shaped, grow by apical extension, and divide by medial fission. The cells are usually haploid, with a haploid genome size similar to that of budding yeast at approximately 14 Mb, but they contain only three relatively long chromosomes of 5.7, 4.7, and 3.5 Mb (for review, see Su and Yanagida, this volume). Ordered cosmid and YAC (yeast artificial chromosome) banks of genomic DNA covering the entire genome have been generated (Maier et al. 1992; Hoheisel et al. 1993; Mizukami et al. 1993), and sequencing of the genome has recently begun (for information, see http://www.sanger.ac.uk). *S. pombe* cells are typically eukaryotic, sharing similarities with multicellular eukaryotes in processes such as transcription initiation (Fikes et al. 1990), intron splicing (Käufer et al. 1985), protein modification (Chappell and Warren 1989), and the mitotic cell cycle.

When provided with a supply of nutrients, fission yeast cells grow and enter the mitotic cell cycle (Fig. 1). Upon exhaustion of the nutrients, several alternative developmental fates are possible. If the culture con-

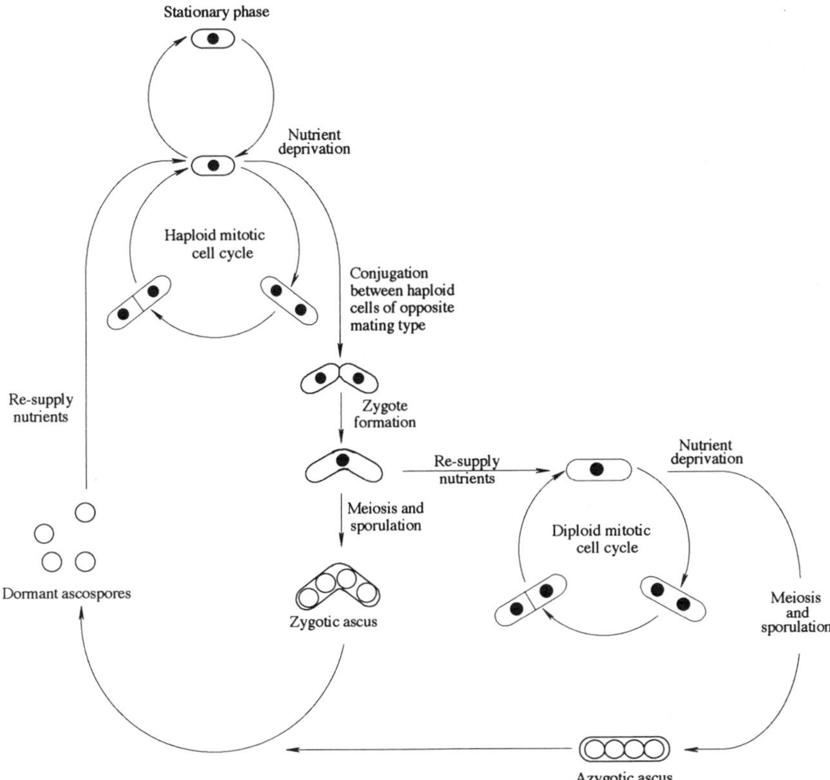

Figure 1 Life cycle of fission yeast. (*Top left*) Haploid mitotic cell cycle; (*center and lower left*) haploid cells mating to form a diploid zygote, followed by meiosis and sporulation leading to zygotic ascus formation; (*lower right*) reentry of diploid zygotes into mitotic cycle. See text for details.

tains cells of only a single mating type, then these cells exit the mitotic cell cycle and accumulate in stationary phase. In contrast, if both mating types are present (for review, see Yamamoto et al., this volume), then the cells are able to conjugate and form a diploid zygote, which can then undergo meiosis and sporulation to generate an ascus containing four haploid spores. However, if diploid zygotes are transferred into fresh media prior to the initiation of meiosis, they are then able to reenter the mitotic cycle. Subsequently, if once again deprived of nutrients, these cells can undergo meiosis and sporulation without intervening conjugation. These features of the life cycle make the fission yeast particularly amenable to genetic analysis: Genetic mapping can be carried out by tetrad dissection or by random spore analysis, and complementation analysis can be performed using zygotically derived diploids (see Leupold 1970; Gutz et al. 1974; Hayles and Nurse 1992).

II. THE MITOTIC CELL CYCLE

A. Events of the Mitotic Cell Cycle

The mitotic cell cycle of the fission yeast is similar to that of higher eukaryotes and can be divided into discrete G_1, S, G_2, and M phases (Mitchison 1970). The G_1 and S phases are relatively short. After a lengthy G_2 period, entry into mitosis (M) is marked by chromosome condensation (Robinow 1977) and by a rapid cytoskeletal reorganization, with the network of cytoplasmic microtubules typical of interphase being replaced by an intranuclear mitotic spindle (Hagan and Hyams 1988). The nuclear membrane does not break down during mitosis (McCully and Robinow 1971). The condensed chromosomes just become visible at metaphase, and the sister chromatids are pulled apart. Mitosis is followed by cytokinesis and cell division: A septum is formed across the center of the cell, and cell separation occurs when this septum is cleaved. The duration of each phase of the cycle at rapid growth rates has been calculated as approximately 0.1 of a cell cycle for both G_1 and S phases, 0.7 of a cycle for G_2, and 0.1 of a cycle for mitosis (Nasmyth et al. 1979). In rapidly growing cells, the G_1 and S phases follow immediately after the completion of nuclear division, such that at the time of cell separation, the daughter cells are already in G_2. Note that the organization of the phases of the cell cycle in the fission yeast differs from that of the budding yeast in which a premitotic spindle persists for much of the cell cycle. It has been argued that S phase and mitosis overlap in the budding yeast cell cycle and that there is therefore no discrete G_2 phase (Nurse 1985; Lew et al., this volume).

B. Cell Cycle Control Points

There are two major control points in the fission yeast mitotic cell cycle: in late G_1 before the initiation of S phase and in late G_2 before the initiation of mitosis. Start is the name given to the point or period in late G_1 when a cell becomes committed to the mitotic cell cycle (Hartwell et al. 1974; Nurse and Bissett 1981). Prior to passing Start, cells can undergo one of two alternative developmental programs. Under appropriate nutritional conditions, cells become committed to the mitotic cell cycle, whereas if the cell receives a different set of environmental stimuli, the pathway leading to the sexual cycle is activated. Passage of Start and entry into S phase are dependent on prior completion of mitosis in the previous cycle and on the cell reaching a critical minimal size (Nurse 1975; Nurse et al. 1976; Nurse and Thuriaux 1977). Under conditions where cells are unusually small (e.g., following spore germination) or under conditions of nutrient limitation, the period before Start is lengthened and passage through Start requires growth to the critical size. Normally, this size control is cryptic since in a rapidly growing population, newly born cells have already reached this critical size (Nurse and Thuriaux 1977; Nasmyth 1979). Under these circumstances, S phase occurs at a fixed time after completion of mitosis (approximately 0.1 of a cell cycle). Under poor growth conditions, the G_1 period before Start is expanded. As in the budding yeast, mating pheromones cause a transient cell cycle arrest before Start during conjugation (for review, see Yamamoto et al., this volume). The Start control is discussed at length below (see Section III).

The second major control point in the fission yeast cell cycle is located in late G_2 and determines the timing of entry into mitosis (Nurse 1975). Several functions have been identified that can be altered to bring about advancement of the G_2-M transition, with the result that the G_2 period is shortened and the cell enters mitosis at a reduced cell size compared to wild type. This in turn leads to cell division at a reduced size (called the wee phenotype; see Nurse 1975; Thuriaux et al. 1978; Nurse and Thuriaux 1980; Fantes 1981). Genetic analysis of *wee* mutants has uncovered an intricate mechanism regulating the timing of entry into mitosis that has subsequently been elaborated biochemically (Fantes 1979; Russell and Nurse 1986, 1987a,b; Gould and Nurse 1989). This is discussed further below (see Section V).

Passage of the G_2-M control point and entry into mitosis are dependent on the cell reaching a certain size (Nurse 1975) as well as on prior completion of S phase and of any repair of DNA damage that may be necessary. The size requirement for the initiation of mitosis is reflected in the cell size at cell division, as mitosis is completed and cell division follows rapidly once the size requirement has been met. As a result of their reduced cell size at division, *wee* mutant cells are not large enough to pass the G_1-S control point at birth and must grow until they reach the appropriate size. Thus, the shortening of G_2 is balanced by a lengthening of G_1, and the generation time of these cells is kept constant. Under poor growth conditions, cell size at mitosis is also reduced, leading to the expansion of G_1 noted earlier, as once again the cells must grow for a longer time to attain the critical size required for passage of Start (Fantes and Nurse 1977). This organization of the cell cycle results in a greater proportion of cells accumulating in the uncommitted pre-Start stage of the cycle when growth conditions begin to deteriorate.

C. Temporal Ordering of Cell Cycle Events

Correct temporal coordination of the major cell cycle events of DNA replication and mitosis is essential for maintaining genomic integrity: Mitosis must not begin until DNA replication is complete (and any damaged DNA present is repaired), and DNA replication in the following cell cycle must wait until mitosis is complete. In addition to the controls that exist over entry into S phase and mitosis, the initiation of cytokinesis is also tightly controlled and is dependent on the completion of certain early events of mitosis (Nurse et al. 1976; Minet et al. 1979). This control, which is necessary to ensure coordination of the nuclear division cycle (DNA replication and mitosis) with the cell division cycle (growth and cytokinesis), is discussed in Su and Yanagida (this volume).

D. Cell Cycle Mutants

A large number of conditional-lethal (cold-sensitive or temperature-sensitive) mutants defective in cell cycle progression have been isolated in a range of genetic screens. Because fission yeast cells grow by apical extension, the identification of interphase-arrested cell cycle mutants is relatively straightforward, as the cells simply continue to grow but do not divide, thus becoming highly elongated (the cdc phenotype; Nurse et al. 1976). This feature was what first attracted attention to fission yeast as a potentially useful model system, as the position of a cell in the cell cycle could be determined simply by measuring its length (for a historical perspective, see Mitchison 1957, 1990). In contrast to the *cdc* mutants, mutants blocked during mitosis do not elongate to anything like the same extent, and they typically require more detailed cytological analysis before the precise mutant phenotype can be determined (see Su and Yanagida, this volume).

In addition to these genes, a large number of new genes have been identified on the basis of their interactions with preexisting cell cycle genes or by physical methods. To date, more than 80 gene functions involved in the mitotic cell cycle have been identified. Table 1 lists these gene functions along with a brief summary of their in vivo roles. For more complete descriptions of those functions involved in mitosis and cytokinesis (detailed consideration of which lies outside the scope of this chapter), see Su and Yanagida (this volume). Gene functions implicated exclusively in the sexual cycle are described elsewhere (see Table 1 in Yamamoto et al., this volume).

A full range of genetic and molecular biological techniques comparable to those in use in studies of the budding yeast cell cycle are now available for the analysis of the cell cycle of *S. pombe*. Particularly useful is the ability to create a null allele of any cloned gene using the one-step gene-replacement technique (Rothstein 1983), to overproduce cell cycle proteins in vivo, and to mutagenize cloned genes to identify a new mutant with novel functions (for reviews, see MacNeill and Fantes 1993; Reed et al. 1995).

E. A Brief Note Concerning Nomenclature

An agreed system for genetic nomenclature in *S. pombe* has been published (Kohli 1987) and is used throughout this chapter. We have chosen to indicate *S. pombe* proteins simply in nonitalicized lowercase letters (cdc2, orc1), in part to distinguish them from budding yeast proteins (Cdc28p, Orc1p) but also to avoid confusion between proteins of similar molecular weights (e.g., $p34^{cdc16}$ and $p34^{cdc2}$). Where the same gene and/or protein has been identified in several laboratories, alternative names are indicated as, for instance, *nda1⁺/cdc19⁺* or res1/sct1. Generally, the first name given is the more widely used or is that which we expect, for various reasons, to be most widely used in the future. For instance, the gene encoding DNA polymerase α has been identified in four separate laboratories, as *pol1⁺*, *cdc29⁺*, and *swi7⁺* (Damagnez et al. 1991; Park et al. 1993; Singh and Klar 1993; Murakami and Okayama 1995). We shall use *pol1⁺* for the gene, and pol1 for the protein.

III. G_1-S CONTROL

As described in the Introduction, Start is considered to be a point (or interval) during G_1 at which the cell becomes committed to the mitotic cell cycle (Hartwell et al. 1974; Nurse and Bissett 1981). Before passing Start, two alternative developmental fates are available to the cell. In the absence of sufficient nutrients, haploid cells cease progression through the mitotic cycle and either enter stationary phase or, in the presence of cells of opposite mating type, enter the sexual cycle of conjugation, meiosis, and sporulation (diagrammed in Fig. 1; see also Yamamoto et al., this volume). If adequate nutrients are present, however, cells become committed to the mitotic cycle. Once commitment has occurred, the cell is unable to initiate sexual development until the mitotic cycle is completed and the cell has once again reached pre-Start G_1. The following sections describe what is known of the molecular mechanisms of Start control in *S. pombe*, beginning with discussion of the role of transcriptional regulation.

Table 1 Cell cycle gene functions

Gene name	Alternative names	Cloned?	Protein function	Key references
(A) G_1-S functions				
cdc2+		Y	essential cyclin-dependent protein kinase, required at G_1-S and G_2-M; at G_1-S, cdc2 forms a complex with cig2 that is regulated by rum1	1–8
cdc10+		Y	essential component of the transcription factor (MBF) that controls periodic gene expression at G_1-S	1,9
cig2+	*cyc17+*	Y	nonessential G_1 cyclin that complexes with cdc2 at G_1-S; expression of *cig2+* is periodic and under MBF control (see above)	10–13
cyc17+	*cig2+*	Y	see entry for *cig2+*	
pct1+	*res2+*	Y	see entry for *res2+*	
rep1+		Y	zinc finger protein; primary role, perhaps akin to that of rep2 (see below), appears to be in the sexual cycle	14
rep2+		Y	zinc finger protein, transcriptional activator of cdc10-res2, and possibly also cdc10-res1; expression of *rep2+* is nutritionally regulated, being reduced under starvation conditions	15
res1+	*sct1+*	Y	component of MBF transcription factor required at G_1-S; res1 binds to cdc10 and shares overlapping function with res2; nonessential	16,17
res2+	*pct1+*	Y	component of MBF transcription factor required for G_1-S; primary function is in sexual cycle but shares overlapping function with res1 in mitotic cycle; nonessential	18,19
rum1+		Y	inhibitor of cdc2-cig2 activity at G_1-S and also of cdc2-cdc13; has an important role in maintaining cell cycle dependencies; nonessential	20–22
sct1+	*res1+*	Y	see entry for *res1+*	

(B) S-phase functions				
cdc1$^+$		Y	putative small subunit of DNA polymerase δ; binds to pol3 and cdc27	1,23
cdc6$^+$	*pol3*$^+$	Y	see entry for *pol3*$^+$	
cdc17$^+$		Y	DNA ligase	24–26
cdc18$^+$		Y	regulator of S-phase onset, probably ORC-associated; periodically expressed with peak *cdc18*$^+$ mRNA levels at G_1-S; *cdc18*Δ cells are checkpoint-defective	25,27,28
cdc19$^+$	*nda1*$^+$	Y	see entry for *nda1*$^+$	
cdc20$^+$	*pol2*$^+$	Y	see entry for *pol2*$^+$	
cdc21$^+$		Y	protein of the MCM family; required for DNA replication	25,29
cdc22$^+$		Y	regulatory (large) subunit of ribonucleotide reductase; periodically expressed with peak *cdc22*$^+$ mRNA levels at G_1-S	25,30,31
cdc23$^+$		Y	required for DNA replication but function unknown; sequencing in progress	25,32
cdc24$^+$		N	required for completion of DNA replication; function unknown	25
cdc27$^+$		Y	binds to cdc1, possible role in DNA polymerase δ complex	23,25,33
cdc29$^+$	*pol1*$^+$/*swi7*$^+$	Y	see entry for *pol1*$^+$	
cdc30$^+$	*orc1*$^+$/*orp1*$^+$	Y	see entry for *orp1*$^+$	
cdt1$^+$		Y	required for DNA replication but function unknown; periodically expressed with peak mRNA levels at G_1-S; *cdt1*Δ cells are checkpoint-defective	34
cut5$^+$	*rad4*$^+$	Y	required for the initiation of DNA replication; temperature-sensitive *cut5* cells are checkpoint-defective	35,36
hsk1$^+$		Y	protein kinase, possibly ORC-associated, required for DNA replication; *hsk1*Δ cells are checkpoint-defective	37
mcm3$^+$		Y	protein of the MCM family; required for DNA replication	29
mis5$^+$		Y	protein of the MCM family; required for DNA replication	38
nda1$^+$	*cdc19*$^+$	Y	protein of the MCM family; required for DNA replication	25,39–41
nda4$^+$		Y	protein of the MCM family; required for DNA replication	39,40

(*Continued on following pages.*)

Table 1 (Continued.)

Gene name	Alternative names	Cloned?	Protein function	Key references
orc1$^+$	*cdc30*$^+$/*orp1*$^+$	Y	see entry for *orp1*$^+$	
orp1$^+$	*orc1*$^+$/*cdc30*$^+$	Y	putative ORC (origin-binding complex) component required for replication initiation; *orp1*Δ cells are checkpoint-defective	42–44
orp2$^+$		Y	putative ORC (origin-binding complex) component required for replication initiation; orp2 interacts with cdc2 and cdc18; *orp2*Δ cells are checkpoint-defective	45
pcn1$^+$		Y	PCNA, processivity factor for DNA polymerases δ and ε	46
pol1$^+$	*swi7*$^+$/*cdc29*$^+$	Y	catalytic subunit of DNA polymerase α; *pol1*Δ cells are checkpoint-defective	47–50
pol2$^+$	*cdc20*$^+$	Y	catalytic subunit of DNA polymerase ε	25,51,52
pol3$^+$	*cdc6*$^+$	Y	catalytic subunit of DNA polymerase δ	25,48,53,54
rad4$^+$	*cut5*$^+$	Y	see entry for *cut5*$^+$	
rfc2$^+$		Y	probable replication factor C (RF-C) subunit; homolog of budding yeast Rfc2p	55
suc22$^+$		Y	catalytic (small) subunit of ribonucleotide reductase	30,31
swi7$^+$	*cdc29*$^+$/*pol1*$^+$	Y	see entry for *pol1*$^+$	
(C) G_2-M functions				
cdc2$^+$		Y	essential protein kinase; required at G_1-S and G_2-M, at G_2-M, cdc2 forms a complex with cdc13 that is activated by Tyr-15 dephosphorylation catalyzed by cdc25; dominant *wee* alleles of *cdc2* advance entry into mitosis	1–8
cdc5$^+$		Y	essential DNA-binding protein that has homology with c-Myb; required in G_2 but precise function unknown	1,56
cdc13$^+$		Y	essential B-type cyclin; forms complex with cdc2; cdc13 is degraded, at least in part, in anaphase	6,7,25, 57–59
cdc25$^+$		Y	essential protein tyrosine-phosphatase; activates cdc2 at G_2-M by Tyr-15 dephosphorylation; overproduction advances entry into mitosis	5,25,60,61

cdc28⁺		N	required at G_2-M; function unknown	25
cdr1⁺	*nim1⁺*	Y	see entry for *nim1⁺*	
cig1⁺		Y	nonessential B-type cyclin; forms complex with cdc2; probably functions at G_2-M	62,63
crk1⁺	*mop1⁺/mcs6⁺*	Y	see entry for *mcs6⁺*	
mcs1⁺		N	mutants isolated as suppressors of mitotic catastrophe phenotype; function unknown	64
mcs2⁺		Y	cyclin-like protein component of cak; binds mcs6/mop1/crk1 kinase that phosphorylates cdc2 on Thr-167	64–67
mcs3⁺		N	mutants isolated as suppressors of mitotic catastrophe phenotype; function unknown	64
mcs4⁺		N	mutants isolated as suppressors of mitotic catastrophe phenotype; function unknown	64
mcs5⁺		N	mutants isolated as suppressors of mitotic catastrophe phenotype; function unknown	64
mcs6⁺	*mop1⁺/crk1⁺*	Y	protein kinase component of cak; binds to mcs2 cyclin-like protein; phosphorylates cdc2 on Thr-167	66,67
mik1⁺		Y	protein kinase that shares overlapping function with wee1, phosphorylating cdc2 on Tyr-15	68,69
mop1⁺	*mcs6⁺/crk1⁺*	Y	see entry for *mcs6⁺*	
nim1⁺	*cdr1⁺*	Y	protein kinase that inhibits wee1 by direct phosphorylation; implicated in nutritional sensing; overproduction advances entry into mitosis	70–74
ppa1⁺		Y	minor PP-2A phosphatase; shares overlapping functions with ppa2 and ppe1	75,76
ppa2⁺		Y	major PP-2A phosphatase; *ppa2*Δ cells are partially advanced into mitosis	75,76
ppe1⁺		Y	minor PP-2A phosphatase; shares overlapping functions with ppa1 and ppa2	77,78
ptc1⁺		Y	PP-2C phosphatase implicated in stress response pathway; *ptc1*Δ cells are sensitive to heat shock	79

(*Continued on following pages.*)

Table 1 (Continued.)

Gene name	Alternative names	Cloned?	Protein function	Key references
$ptc2^{+}$		Y	PP-2C phosphatase implicated in stress response pathway	80
$ptc3^{+}$		Y	PP-2C phosphatase implicated in stress response pathway; *ptc3*Δ cells are partially advanced into mitosis	80
$pyp1^{+}$		Y	protein tyrosine phosphatase; *pyp1*Δ cells are partially advanced into mitosis; pyp1 acts, with pyp2, on sty1/spc1 in stress response pathway	81–84
$pyp2^{+}$		Y	protein tyrosine phosphatase; pyp2 acts on sty1/spc1 in stress response pathway; $pyp2^{+}$ mRNA is induced under stress conditions in sty1-dependent manner	81–84
$pyp3^{+}$		Y	protein tyrosine phosphatase; acts with cdc25 to dephosphorylate cdc2 at G_2-M	85
$spc1^{+}$	$sty1^{+}$	Y	see entry for $sty1^{+}$	
$spc2^{+}$	$wis1^{+}$	Y	see entry for $wis1^{+}$	
$spt1^{+}$		Y	protein tyrosine phosphatase; may dephosphorylate cdc2 but role at G_2-M in vivo (if any) is uncertain	86
$sty1^{+}$	$spc1^{+}$	Y	MAP kinase; substrate of wis1 in stress response pathway	80,84,87
$suc1^{+}$		Y	essential function; binds to cdc2	6,88–92
$swo1^{+}$		Y	essential Hsp90 homolog; binds to wee1	93
$wee1^{+}$		Y	protein tyrosine kinase, inhibits cdc2 by phosphorylation on Tyr-15; *wee1* cells are advanced into mitosis	94–98
$win1^{+}$		Y	not yet sequenced but a possible component of wis1-sty1/spc1 pathway	99
$wis1^{+}$	$spc2^{+}$	Y	MAP kinase kinase; phosphorylates sty1/spc1 is stress response pathway	84,87,100
$wis2^{+}$		Y	heat-shock-inducible cyclophilin-like protein; $wis2^{+}$ mRNA is induced by heat shock	101,102
$wis3^{+}$		Y	isolated at multicopy suppressor of G_2-M defect; not yet sequenced	101
$wis4^{+}$		Y	MAP kinase kinase kinase; may phosphorylate wis1/spc2	101,103
$wis5^{+}$		Y	isolated at multicopy suppressor of G_2-M defect; not yet sequenced	101

(D) Mitotic functions

atb2⁺		Y	nonessential α2-tubulin	104,105
ben1⁺	*nda3*⁺	Y	see entry for *nda3*⁺	
ben4⁺		N	required for spindle function in mitosis; *ben4* mutants display genetic interactions with tubulin and actin genes	106,107
bws1⁺	*dis2*⁺	Y	see entry for *dis2*⁺	
cek1⁺		Y	nonessential protein kinase; suppressor of *cut8* mutants	108
cut1⁺		Y	required for chromosome segregation; interacts with cut2	109,110
cut2⁺		Y	required for chromosome segregation; interacts with cut1, probably regulated by cdc2	109,110
cut3⁺		Y	SMC protein; required in mitosis for chromosome condensation and segregation	109,111
cut4⁺		N	mutants display cut phenotype	109
cut6⁺		N	mutants display cut phenotype	109
cut7⁺		Y	required for SPB separation and spindle formation; a kinesin homolog	109,112,113
cut8⁺		Y	mutants display spindle elongation defect in metaphase; interacts genetically with *cut1*	108,109
cut9⁺		Y	TPR repeat protein required for spindle elongation in mitosis	109,114
cut11⁺		N	mutants display cut phenotype	109
cut12⁺		N	mutants display cut phenotype	115
cut13⁺		N	mutants display cut phenotype	115
cut14⁺		Y	SMC protein; required in mitosis for chromosome condensation and segregation	111,115
cut15⁺		N	mutants display cut phenotype	115
cut16⁺		N	mutants display cut phenotype	115
cut17⁺		N	mutants display cut phenotype	115
cut18⁺		N	mutants display cut phenotype	115
cut19⁺		N	mutants display cut phenotype	115

(*Continued on following pages.*)

Table 1 (*Continued.*)

Gene name	Alternative names	Cloned?	Protein function	Key references
dcd1$^+$	*pim1*$^+$	Y	see entry for *pim1*$^+$	
dis1$^+$		Y	microtubule-associated protein required for sister chromatid separation	116,117
dis2$^+$	*bws1*$^+$	Y	major PP-1 phosphatase; required for mitotic exit; negatively regulated by cdc2 in mitosis	116,118,119
dis3$^+$		Y	essential nuclear protein; required in mitosis but function unclear	116,120
dsk1$^+$		Y	nonessential protein kinase; interacts genetically with *dis1*	121
fyt1$^+$	*spi1*$^+$	Y	see entry for *spi1*$^+$	
gtb1$^+$		Y	essential γ-tubulin involved in microtubule nucleation	122
mis6$^+$		N	mutants display chromosome segregation defects	40
mis12$^+$		N	mutants display chromosome segregation defects	40
mts1$^+$		N	mutants are resistant to microtubule inhibitor MBC	123
mts2$^+$		Y	component of the 26S proteosome complex; essential for completion of mitosis	123
mts3$^+$		N	mutants are resistant to microtubule inhibitor MBC	123
mts4$^+$		N	mutants are resistant to microtubule inhibitor MBC	123
mts5$^+$		N	mutants are resistant to microtubule inhibitor MBC	123
nda2$^+$		Y	α1-tubulin; essential for spindle formation and elongation	39,105,124
nda3$^+$	*ben1*$^+$	Y	β-tubulin; essential for spindle formation and elongation	39,104, 125,126
nuc2$^+$		Y	TPR repeat protein required for spindle elongation in mitosis	127–129
pim1$^+$	*dcd1*$^+$	Y	RCC1 homolog; exchange factor for spi1 nuclear GTPase	78,130–132
plo1$^+$		Y	essential protein kinase; member of polo family; required for spindle function in mitosis; pol1 also promotes septum formation (see below)	133
sad1$^+$		Y	spindle pole body associated protein; *sad1* mutants form aberrant spindles	134

scn1$^+$		Y	mutations cause spindle elongation defect; interacts genetically with *cut9* and *nuc2*	114
scn2$^+$		Y	mutations cause spindle elongation defect; interacts genetically with *cut9* and *nuc2*	114
sds21$^+$		Y	minor PP-1 phosphatase; related to dis2 protein but not regulated by cdc2	75,118,119
sds22$^+$		Y	essential regulatory subunit for dis2 and sds21 PP-1 phosphatases; required mid-mitosis	135,136
spi1$^+$	*fyt1*$^+$	Y	T4/Ran nuclear GTPase homolog	78,130–132
swi6$^+$		Y	essential component of centromeres; a chromodomain protein	137
top1$^+$		Y	type I topoisomerase	138–142
top2$^+$		Y	type II topoisomerase	136–142
(E) Cytokinesis functions				
cdc3$^+$		Y	essential actin-binding protein profilin; may function as general actin organizing factor	1,143,144
cdc4$^+$		Y	essential EF-hand protein that localizes to the actin ring; a possible myosin light chain	1,143,145
cdc7$^+$		Y	protein kinase	1,143,146
cdc8$^+$		Y	essential actin-binding protein tropomyosin; may function as general actin organizing factor	1,143, 147
cdc11$^+$		N	essential for septum formation but precise function unknown	1,143
cdc12$^+$		Y	large proline-rich protein; required for actin ring formation	1,143,148
cdc14$^+$		Y	essential for septum formation but precise function unknown	1,143,149
cdc15$^+$	*rng1*$^+$	Y	essential regulator of actin ring formation	1,143,150
cdc16$^+$		Y	mutants exhibit multiple rounds of septum formation without cleavage	151,152
mid1$^+$		Y	required for placement of actin ring	148
plo1$^+$		Y	protein kinase of polo family; a regulator of septum formation; also required in mitosis for spindle function (see above)	133

(*Continued on the following pages.*)

Table 1 (Continued.)

Gene name	Alternative names	Cloned?	Protein function	Key references
rng1⁺	*cdc15*⁺	Y	see entry for *cdc15*⁺	
rng2⁺		Y	required for actin ring formation	148,153
sep1⁺		N	mutants display cell separation defect after cytokinesis but precise function unknown	154
skp1⁺		Y	glycogen synthase kinase homolog; *skp1*⁺ interacts genetically with *cdc14*	155
spl1⁺		N	mutants display aberrant septum fomation but precise function unknown	154
(F) Checkpoint functions: DNA replication and repair				
cds1⁺		Y	multicopy suppressor of *cdc29/pol1* mutant; a protein kinase, implicated in DNA replication checkpoint	156
chk1⁺		Y	protein kinase; required for cell cycle arrest in response to DNA damage and most DNA replication blocks; overproduction can rescue *cdc2-r4* allele, suggesting a link with the mitotic control	157–159
hus1⁺		Y	required for rad/hus checkpoint function	160
hus2⁺		N	mutants display sensitivity to hydroxyurea; possible checkpoint defect	160
hus3⁺		N	mutants display sensitivity to hydroxyurea; possible checkpoint defect	160
hus4⁺		N	mutants display sensitivity to hydroxyurea; possible checkpoint defect	160
hus5⁺		Y	E2 class ubiquitin-conjugating enzyme with multiple cellular roles; required in mitosis	160,161
rad1⁺		Y	probable 3′-5′ exonuclease; required for rad/hus checkpoint function	162–165
rad3⁺		Y	PI3-kinase, required for rad/hus checkpoint function	162,166
rad9⁺		Y	required for rad/hus checkpoint function	162,167,168
rad17⁺		Y	required for rad/hus checkpoint function; rad17 displays limited homology with RF-C subunits	162,169
rad24⁺		Y	14-3-3 protein homolog; required DNA damage checkpoint	162,170

rad25⁺		Y	14-3-3 protein homolog; required DNA damage checkpoint	162,170
rad26⁺		Y	required for rad/hus checkpoint function	162
rad27⁺	*chk1*⁺	Y	see entry for *chk1*⁺	

(1) Nurse et al. 1976; (2) Nurse and Thuriaux 1980; (3) Nurse and Bissett 1981; (4) Beach et al. 1982; (5) Gould and Nurse 1989; (6) Moreno et al. 1989; (7) Hayles et al. 1994; (8) Hayles and Nurse 1995; (9) Lowndes et al. 1992; (10) Bueno and Russell 1993a; (11) Connolly and Beach 1994; (12) Obara-Ishihara et al. 1994; (13) Martín-Castellanos et al. 1996; (14) Sugiyama et al. 1994; (15) Nakashima et al. 1995; (16) Tanaka et al. 1992; (17) Caligiuri and Beach 1993; (18) Miyamoto et al. 1994; (19) Zhu et al. 1994a; (20) Moreno and Nurse 1994; (21) Labib et al. 1995a; (22) Correa-Bordes and Nurse 1995; (23) MacNeill et al. 1996; (24) Nasmyth 1977; (25) Nasmyth and Nurse 1981; (26) Barker et al. 1987; (27) Kelly et al. 1993; (28) Nishitani and Nurse 1995; (29) Coxon et al. 1992; (30) Gordon and Fantes 1986; (31) Fernandez-Sarabia et al. 1993; (32) S. Aves, pers. comm.; (33) Hughes et al. 1992; (34) Hofmann and Beach 1994; (35) Saka and Yanagida 1993; (36) Saka et al. 1994a; (37) Masai et al. 1995; (38) Takahashi et al. 1994; (39) Toda et al. 1983; (40) Miyake et al. 1993; (41) Forsburg and Nurse 1994b; (42) Gavin et al. 1995; (43) Muzi-Falconi and Kelly 1995; (44) B. Grallert and P. Nurse, unpubl.; (45) Leatherwood et al. 1996; (46) Waseem et al. 1992; (47) Damagnez et al. 1991; (48) Park et al. 1993; (49) Singh and Klar 1993; (50) D'Urso and Nurse 1995; (51) G. D'Urso and P Nurse, unpubl.; (52) A. Sugino, pers. comm.; (53) Pignède et al. 1991; (54) M. Yamamoto, pers. comm.; (55) N. Reynolds et al., unpubl.; (56) Ohi et al. 1994; (57) Booher and Beach 1987; (58) Booher and Beach 1988; (59) Hagan et al. 1988; (60) Russell and Nurse 1986; (61) Millar et al. 1991; (62) Bueno et al. 1991; (63) Basi and Draetta 1995a; (64) Molz et al. 1989; (65) Molz and Beach 1993; (66) Buck et al. 1995; (67) Damagnez et al. 1995; (68) Lundgren et al. 1991; (69) Lee et al. 1994; (70) Russell and Nurse 1987b; (71) Feilotter et al. 1991; (72) Wu and Russell 1993; (73) Coleman et al. 1993; (74) Parker et al. 1993; (75) Kinoshita et al. 1990; (76) Kinoshita et al. 1993; (77) Shimanuki et al. 1993; (78) Matsumoto and Beach 1993; (79) Shiozaki et al. 1994; (80) Shiozaki and Russell 1995a; (81) Ottilie et al. 1991; (82) Ottilie et al. 1992; (83) Millar et al. 1992b; (85) Millar et al. 1995; (86) Mondesert et al. 1994; (87) Shiozaki et al. 1995b; (88) Hayles et al. 1986b; (89) Hayles et al. 1986a; (90) Hindley et al. 1987; (91) Brizuela et al. 1987; (92) Endicott et al. 1995; (93) Aligue et al. 1994; (94) Nurse 1975; (95) Russell and Nurse 1987a; (96) Featherstone and Russell 1991; (97) McGowan and Russell 1993; (98) Parker et al. 1993; (99) Ogden and Fantes 1986; (100) Warbrick and Fantes 1991; (101) Warbrick and Fantes 1992; (102) Weisman et al. 1996; (103) I. Samejima and P. Fantes, pers. comm.; (104) Hiraoka et al. 1984; (105) Adachi et al. 1986; (106) Roy and Fantes 1982; (107) Fantes 1989; (108) Samejima and Yanagida 1994b; (109) Hirano et al. 1986; (110) Uzawa et al. 1990; (111) Saka et al. 1994b; (112) Hagan and Yanagida 1990; (113) Hagan and Yanagida 1992; (114) Samejima and Yanagida 1994a; (115) Samejima et al. 1993; (116) Ohkura et al. 1988; (117) Nabeshima et al. 1995; (118) Ohkura et al. 1989; (119) Yamano et al. 1994; (120) Kinoshita et al. 1991(121) Takeuchi and Yanagida 1993; (122) Horio et al. 1991; (123) Gordon et al. 1993; (124) Toda et al. 1984; (125) Umesono et al. 1983; (126) Yamamoto 1980; (127) Hirano et al. 1988; (128) Hirano et al. 1990; (129) Kumada et al. 1995; (130) Matsumoto and Beach 1991; (131) Sazer and Nurse 1994; (132) Demeter et al. 1995; (133) Ohkura et al. 1995; (134) Hagan and Yanagida 1995; (135) Ohkura and Yanagida 1991; (136) Stone et al. 1993; (137) Ekwall et al. 1995; (138) Uemura and Yanagida 1984; (139) Uemura and Yanagida 1986; (140) Uemura et al. 1986; (141) Uemura et al. 1987a; (142) Uemura et al. 1987b; (143) Marks et al. 1992a; (144) Balasubramanian et al. 1994; (145) McCollum et al. 1995; (146) Fankhauser and Simanis 1994; (147) Balasubramanian et al. 1992; (148) Chang et al. 1996; (149) Fankhauser and Simanis 1993; (150) Fankhauser et al. 1995; (151) Minet et al. 1979; (152) Fankhauser et al. 1993; (153) Chang and Nurse 1996; (154) Sipiczki et al. 1993; (155) Plyte et al. 1996; (156) Murakami and Okayama 1995; (157) Walworth et al. 1993; (158) Al-Khodairy et al. 1994; (159) Walworth and Bernards 1996; (160) Enoch et al. 1992; (161) Al-Khodairy et al. 1995; (162) Al-Khodairy and Carr 1992; (163) Rowley et al. 1992b; (164) Sunnerhagen et al. 1990; (165) Lydall and Weinert 1995; (166) Jimenez et al. 1992; (167) Lieberman et al. 1992; (168) Murray et al. 1991; (169) Griffiths et al. 1995; (170) Ford et al. 1994.

A. Transcriptional Control at Start

Cell-cycle-regulated periodic transcription is central to the control of the early part of the cell cycle in both fission and budding yeasts. In *S. cerevisiae*, a large number of genes have been identified that are periodically expressed in late G_1. These genes, which encode a wide range of factors required for DNA synthesis, have a common *cis*-acting element (called an MCB) in their promoter sequences (for review, see Lew et al., this volume). MCB elements are necessary for the coordinated periodic expression and are recognized by a DNA-binding activity called MBF (MCB-binding factor). MBF contains Swi6p and Mbp1p and is thought to bind to MCB elements throughout the cell cycle. A second set of genes that are periodically expressed in late G_1 are regulated by a different DNA-binding activity that recognizes a distinct target sequence (SCB). This activity also contains Swi6p, in this case partnered by Swi4p (see Lew et al., this volume and references therein).

Experiments have demonstrated the existence of an MBF-like activity in fission yeast (Lowndes et al. 1992) that controls expression of at least four genes: *cdc22*$^+$, which encodes the large subunit of ribonucleotide reductase; *cdc18*$^+$, which encodes a protein required for replication initiation; *cig2*$^+$/*cyc17*$^+$, which encodes a B-type cyclin; and *cdt1*$^+$, whose precise function is unknown (Gordon and Fantes 1986; Bueno and Russell 1993a, 1994; Caligiuri and Beach 1993; Fernandez-Sarabia et al. 1993; Kelly et al. 1993; Connelly and Beach 1994; Hofmann and Beach 1994; Obara-Ishihara and Okayama 1994). In contrast to the situation in budding yeast, the vast majority of genes encoding proteins required for DNA synthesis in fission yeast, such as the catalytic subunits of the DNA polymerases or DNA ligase, are not periodically expressed, although the reason for this is unclear (White et al. 1986).

Each of the four periodically expressed genes has one or more MCB elements in its promoter region. Binding of MBF has been reported to be cell-cycle-regulated, increasing during G_1, remaining high throughout S phase and G_2, before declining during mitosis (Reymond et al. 1993). Whether this periodic fluctuation is responsible for in vivo transcriptional activation in late G_1, rather than, for example, posttranslational modification of the protein components of MBF, remains to be seen. That MBF is bound throughout S phase and G_2 suggests that periodic changes in binding cannot account for the rapid decline in transcript levels seen in vivo during this period, so some other mechanism must operate.

It seems likely that fission yeast MBF consists of two distinct DNA-binding complexes, as two complexes have been identified that have a role at Start, although their relative contributions appear to differ. Both complexes contain the essential cdc10 protein ($p85^{cdc10}$) as does the MBF activity in fission yeast extracts detected by gel-retardation assays (Aves et al. 1985; Simanis and Nurse 1989; Lowndes et al. 1992; Reymond et al. 1992, 1993; Reymond and Simanis 1993). One complex contains cdc10 together with res1/sct1 ($p72^{res1}$; Marks et al. 1992b; Tanaka et al. 1992; Caligiuri and Beach 1993), whereas in the other complex, cdc10 binds to res2/pct1 ($p73^{res2}$; Miyamoto et al. 1994; Zhu et al. 1994a).

On the basis of genetic evidence, the res1-cdc10 complex appears to function mainly at Start in the mitotic cell cycle, but it also has a minor role prior to premeiotic DNA replication in the sexual cycle. Gel-retardation assays indicate that res1 is part of the MBF activity identified in fission yeast extracts (Caligiuri and Beach 1993). In contrast, the res2-cdc10 complex acts predominantly in the sexual cycle and has only a minor role in the mitotic cycle. Although MBF activity can be reconstituted using in-vitro-synthesized res2 and cdc10 proteins, the presence of res2 in the MBF activity detectable in *S. pombe* extracts is assumed and has not been directly demonstrated (Zhu et al. 1994a).

All three proteins cdc10, res1/sct1, and res2/pct1 share significant sequence similarity to one another and to the Swi4p, Swi6p, and Mbp1p proteins that make up SBF and MBF in *S. cerevisiae* (Fig. 2) (see Lew et al., this volume). All three contain two copies of a conserved 33-amino-acid repeated motif (called the cdc10/Swi6 motif, or ankyrin motif) found in a variety of proteins, including ankyrin, the transcription factors NF-κB and GABP-β, and developmental regulators such as Dorsal, Notch, Glp-1, and lin-12 (for review, see Michaely and Bennett 1992). The function of this motif in cdc10, res1, and res2 has not been thoroughly investigated, although in the case of cdc10, it is known that mutations within the repeated motifs can abolish function (Reymond et al. 1992; see also Ayté et al. 1995). Analysis of the GABP-β protein suggests that the repeats may mediate protein-protein interactions (see Thompson et al. 1991). To date, only res1 has been shown to bind to DNA directly, through its amino terminus (Ayté et al. 1995). This part of the res1 protein includes a short region (called the res1/Swi4 motif) that is also well conserved in res2, Swi4p, and Mbp1p and which is thought likely to be directly involved in DNA binding (Fig. 2) (Miyamoto et al. 1994; Zhu et al. 1994a). Binding of res1 to cdc10 is mediated through a carboxy-terminal region in res1, overproduction of which leads to a decrease in MBF activity and cell cycle arrest in G_1, perhaps by titrating out cdc10 from the endogenous res1 protein (Ayté et al. 1995). A similar

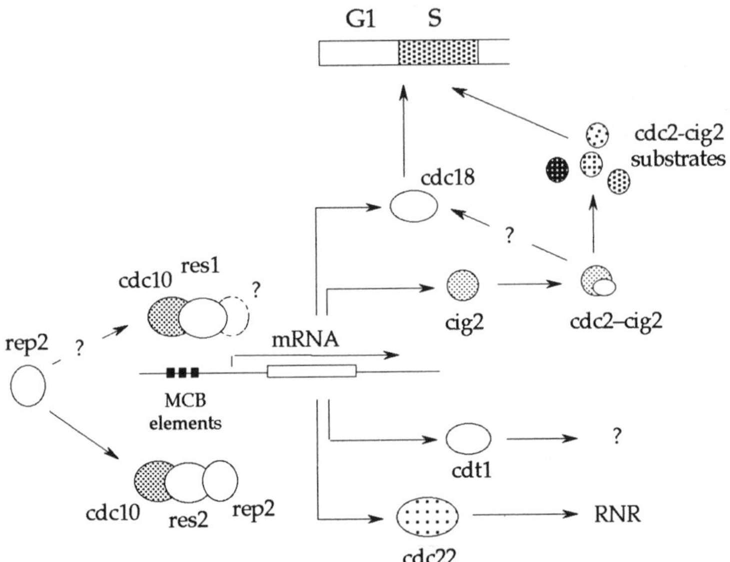

Figure 2 G_1-S regulation in fission yeast. Both transcriptional and posttranslational controls have important roles in regulating the G_1-S transition in *S. pombe*. Two cdc10-containing complexes activate transcription of at least four MCB-controlled genes required for entry into and/or progression through S phase. In the mitotic cell cycle, the major part in transcriptional activation is played by the cdc10-res1/sct1 complex and a minor part is played by the cdc10-res2/pct1 complex. The rep2 protein binds to res2 and is required to activate transcription, and it may also interact with res1 in the cdc10-res1 complex. Expression of the *rep2*⁺ gene is nutritionally regulated, thereby providing a link between the nutritional status of the cell and its propensity to pass Start. The four MCB-controlled genes identified to date are *cig2*⁺, *cdc18*⁺, *cdc22*⁺, and *cdt1*⁺. The cig2 cyclin complexes with cdc2 to promote entry into S phase, the cdc18 protein is a key regulator of entry into S phase and may be one of several substrates of cdc2-cig2 at this point in the cycle, and cdc22 is the regulatory subunit of ribonucleotide reductase (RNR). The function of cdt1 is unknown. Note that cdc2-cig2 activity is also transiently inhibited by rum1. See text for further details and references.

phenotype is seen when the carboxyl terminus of cdc10 is overproduced. Deletion of this region from the cdc10 protein results in constitutive high-level expression of MCB-controlled genes, suggesting that it may be the target of an inhibitory signal (McInerny et al. 1995).

At growth temperatures close to optimal for wild-type cells, *res1*⁺ is a nonessential gene, although *res1*Δ cells grow more slowly than wild type and are elongated, indicative of cell cycle delay (Tanaka et al. 1992;

Caligiuri and Beach 1993). At higher or lower temperatures, *res1Δ* cells are unable to grow, showing that res1 function is essential. Under these conditions, *res1Δ* cells arrest in G_1 but are still able to conjugate, i.e., they have not yet passed Start. Intriguingly, *res1Δ* cells bypass the usual nutritional requirements for entry into the sexual cycle by being able to conjugate on rich medium, so that in addition to promoting passage of Start in the mitotic cell cycle, res1 can be thought of as a repressor of sexual development. Despite being able to conjugate, *res1Δ/res1Δ* diploids sporulate only poorly, indicating that they are defective for some aspect of meiosis, presumably passage of premeiotic Start and entry into premeiotic DNA replication (Tanaka et al. 1992; Caligiuri and Beach 1993).

In contrast to *res1⁺*, *res2⁺/pct1⁺* function is not essential for progression through the mitotic cell cycle, although *res2Δ* cells grow more slowly than wild type and are elongated, particularly at lower temperatures (Miyamoto et al. 1994; Zhu et al. 1994a). Flow cytometric analysis demonstrates that elongated *res2Δ* cells are partially arrested in G_1. The *res2⁺* function is required for the sexual cycle, however. Although *res2Δ* cells are able to conjugate, the majority of asci from *res2Δ/res2Δ* diploids contain two or fewer spores and the individual nuclei within each ascus often are irregularly distributed, indicating that the sexual cycle is severely impaired. Transcription of the *res2⁺* gene is highly induced during conjugation (Obara-Ishihara and Okayama 1994), again pointing to the function of this gene as being particularly crucial during sexual development.

A model for the action of the res1-cdc10 and res2-cdc10 complexes is shown in Figure 2. In this model, the res1-cdc10 complex has the major role in promoting passage of Start and entry into S phase, whereas cdc10-res2 has only a minor role. In the absence of res1, the res2-cdc10 complex can substitute to promote passage of mitotic Start.

Recently, two further interacting functions have been identified, *rep1⁺* and *rep2⁺*. The activity of res2-cdc10 at mitotic Start is dependent on the function of rep2, a protein with a putative zinc finger domain (Nakashima et al. 1995). rep2 binds to res2 both in vivo and in vitro and is thought to be a transcriptional activator subunit of res2-cdc10 (and perhaps also of res1-cdc10; see Nakashima et al. 1995). *rep2⁺* function is essential at low temperatures where *rep2Δ* cells arrest in G_1. Interestingly, *rep2⁺* function may provide a link between the nutritional status of a cell and its propensity to pass Start. The level of *rep2⁺* mRNA falls sharply during nitrogen starvation and is restored on refeeding. Nitrogen-starved *rep2Δ* cells arrest in G_1 more rapidly than wild type, and conversely, overproduction of rep2 inhibits G_1 arrest in a *res2*⁺-dependent fashion. Taken together, these results suggest that rep2 may mediate nutritional signals to res2-cdc10. The rep1 protein may have a role similar to that of rep2 prior to premeiotic DNA replication (Sugiyama et al. 1994). Although more than twice the size of rep2, rep1 shares a degree of protein sequence similarity with rep2 and also contains a single zinc finger motif.

B. Passing Start Requires cdc2 Activity

The fission yeast cdc2 protein kinase ($p34^{cdc2}$) has a key role in the control of the mitotic cell cycle, being required in G_1 for passage of Start and entry into S phase (Nurse and Bissett 1981) and also in late G_2 for entry into mitosis (Nurse et al. 1976). Cells carrying temperature-sensitive alleles of cdc2 become arrested at one or the other of these points, or both, when shifted to the restrictive temperature (Nurse et al. 1976; Nasmyth and Nurse 1981; Beach et al. 1982; Hindley and Phear 1984; Durkacz et al. 1985, 1986; Booher and Beach 1986, 1987; Simanis and Nurse 1986; Carr et al. 1989; Fleig and Nurse 1991; MacNeill et al. 1991; Ayscough et al. 1992; Fleig et al. 1992; MacNeill and Nurse 1993). cdc2 appears also to be required for sister chromatid separation during mitosis (Labib et al. 1995b) and for the second meiotic nuclear division (Nakaseko et al. 1984; Niwa and Yanagida 1988; Grallert and Sipiczki 1989, 1990, 1991; see also Yamamoto et al., this volume).

Fission yeast cdc2 and budding yeast Cdc28p were the prototype cyclin-dependent kinases (for review, see Pines and Hunter 1995), and at least three, and probably four, cyclin proteins have been identified that are thought to interact with cdc2 in *S. pombe* (for review, see Fisher and Nurse 1995). At the G_2-M transition, cdc2 forms a complex with the essential B-type cyclin cdc13 (Booher and Beach 1987, 1988; Hagan et al. 1988; Booher et al. 1989; Moreno et al. 1989; Alfa et al. 1990; Gallagher et al. 1993). The activity of this complex is essential for entry into mitosis, and the timing of its activation determines the timing of entry into mitosis. This aspect of cdc2 function is considered at length later in this chapter (see Section V). cdc2 also complexes with the nonessential B-type cyclin cig1 whose precise function is uncertain (Bueno et al. 1991; Bueno and Russell 1993b; Connolly and Beach 1994; Basi and Draetta 1995a; Martín-Castellanos et al. 1996) and with the suc1 protein required for exit from mitosis (Hayles et al. 1986a,b; Brizuela et al. 1987; Hindley et al. 1987; Moreno et al. 1989; Basi and Draetta 1995a; Endicott et al. 1995).

At Start, cdc2 is complexed with the second nonessential B-type cyclin, cig2/cyc17 (Bueno and Russell 1993a, 1994; Connolly and Beach 1994; Obara-Ishihara and Okayama 1994; Martín-Castellanos et al. 1996). Levels of both *cig2*$^+$ mRNA and cig2 protein peak at G_1-S, and periodic transcription of the *cig2*$^+$ gene is dependent on MBF function. cig2-associated cdc2 kinase activity is also dependent on MBF and is periodic through the cell cycle (Martín-Castellanos et al. 1996). Cells lacking cig2 are viable, indicating that one or more cyclins can compensate for loss of *cig2*$^+$ function (Bueno and Russell 1993a, 1994; Connolly and Beach 1994; Obara-Ishihara and Okayama 1994; for discussion, see Martín-Castellanos et al. 1996). Indeed, it has recently been shown that the cig1 and cdc13 cyclins, both of which are thought to function primarily at G_2-M in wild-type cells, are capable of promoting S phase in the absence of cig2 (Fisher and Nurse 1996). However, deleting the *cig2*$^+$ gene delays entry into S phase and reduces cdc2 activity in vivo showing that neither cig1 nor cdc13 can fully compensate for loss of cig2 activity (Martín-Castellanos et al. 1996).

Despite being required for passage of Start and entry into S phase in fission yeast, specific G_1-S substrates of cdc2 have yet to be unambiguously identified, and thus it is not clear exactly how activation of cdc2 promotes entry into S phase. Several candidate substrates for cdc2 at G_1-S have been suggested, specifically the orp2 and cdc18 proteins (Nishitani and Nurse 1995; Leatherwood et al. 1996; discussed below) and rum1 (Moreno and Nurse 1994; Correa-Bordes and Nurse 1995; Labib et al. 1995a). cdc2 activity at G_1-S is transiently inhibited by rum1, a protein whose function is discussed in greater detail below in Section VI. Since the rum1 protein has two potential cdc2 phosphorylation sites in its sequence (Moreno and Nurse 1994), it has been suggested that rum1 inhibition of cdc2 activity may be overcome by accumulation of cig2-cdc2 (Labib et al. 1995a; Martín-Castellanos et al. 1996). A similar regulatory system exists in budding yeast. Sic1p in budding yeast regulates entry into S phase via inhibition of Cdc28p–B-type cyclin complexes (Schwob et al. 1994; see Lew et al., this volume). Recently, it has been shown that expression of Sic1p in fission yeast complements *rum1*Δ cells (unpublished results cited in Martín-Castellanos et al. 1996).

IV. DNA REPLICATION

On completion of Start, cells proceed rapidly through late G_1 and into S phase. This takes about 0.1 of a cell cycle in rapidly growing cells and is probably not extended even at slow growth rates. Several essential replication factors have already been identified in *S. pombe*, mostly through the analysis of existing cell cycle mutants, although recent years have seen physical methods such as low-stringency hybridization and polymerase chain reaction (PCR) amplification applied to the problem of cloning fission yeast homologs of replication factors first identified in budding yeast or vertebrate cells. In this section, we summarize what is known of the DNA replication machinery in fission yeast. Subsequently, (see Section VI), we deal with the controls linking the completion of DNA replication to the onset of mitosis and the initiation of S phase to the completion of mitosis in the previous cell cycle.

A. Replication Origins

Cis-acting replication origin sequences in budding yeast were first identified on the basis of their ability to promote autonomous replication of plasmids, which led to them being called autonomously replicating sequences, or ARS elements (see Lew et al., this volume). ARS elements function as origins of replication on plasmids and in the chromosome. Budding yeast ARS elements are generally 100–200 bp in size and A/T-rich. Within this region is found an exact or very close match to an 11-bp sequence called the ARS consensus sequence (ACS), mutations in which can abolish ARS function altogether. Numerous ARS elements have been isolated from fission yeast and an ACS has been identified, although unlike the situation in budding yeast, the ACS is not absolutely required for ARS function and some ARS elements seem to lack an ACS altogether (Maundrell et al. 1988; Zhu et al. 1994b; Clyne and Kelly 1995). ARS elements function as replication origins in the chromosome and in some cases have been shown to be essential for replication initiation in situ (Caddle and Calos 1994; Dubey et al. 1994; Wohlgemuth et al. 1994). It is estimated that there is one ARS element per 20–50 kb of genomic DNA.

B. S-phase Mutants

A total of 11 of the original collection of 26 cell division cycle mutants (see Table 1) (Nurse et al. 1976; Nasmyth and Nurse 1981) identified genes involved in the processes of DNA replication, although in certain cases this has become apparent only recently (see MacNeill et al. 1996). Cells carrying temperature-sensitive mutants in any of these genes undergo cell cycle arrest and cell elongation when shifted to the restrictive temperature. Generally, the cell number of an asynchronously growing culture is seen to double following shift-up, indicative of an early block to cell cycle progression. The DNA content of the arrested cells (determined by chemical or cytometric analysis) varies from 1C to 2C (Nurse et al. 1976; Nasmyth and Nurse 1981). Additional DNA replication gene functions have been identified in screens for cold-sensitive cell cycle mutants (Miyake et al. 1993), for mutants that are unable to stably maintain an artificial minichromosome (Takahashi et al. 1994), and also by physical methods (Masai et al. 1995). The following section discusses what is known about the functions of the proteins encoded by these genes.

C. Replication Factors

1. Origin-binding Complexes

A protein complex called ORC (origin recognition complex) has been identified in budding yeast that binds to origins of replication throughout the cell cycle (Bell and Stillman 1992; Diffley and Cocker 1992). ORC comprises six proteins: Each of the six is essential for DNA replication (see Lew et al., this volume). Although the existence of a similar complex in fission yeast has yet to be demonstrated biochemically, two putative ORC subunits have been identified, the Orc1p homolog orp1/orc1/cdc30 and the Orc2p homolog orp2 (Muzi-Falconi and Kelly 1995; Gavin et al. 1995; Leatherwood et al. 1996; B. Grallert and P. Nurse, unpubl.; H. Okayama, pers. comm.). The orp1/orc1/cdc30 and orp2 proteins, both of which are essential, are approximately 30% and 22% identical to budding yeast Orc1p and Orc2p, respectively. orp2 was identified on the basis of its interaction with cdc2, but it also interacts with cdc18 (discussed below). *orp1*Δ and *orp2*Δ cells are defective in DNA replication, and overproduction of orp2 results in G_2 delay (Carpenter et al. 1996; Leatherwood et al. 1996). It is possible that cdc2 might regulate ORC function via phosphorylation of orp2, which has multiple cdc2 consensus phosphorylation sites in its amino terminus, but this has yet to be tested.

The fission yeast cdc18 and hsk1 proteins are also implicated in ORC function, with cdc18 playing a particularly important part (Kelly et al. 1993; Masai et al. 1995; Nishitani and Nurse 1995). cdc18 ($p65^{cdc18}$) is required for the initiation of DNA replication and shares sequence similarity with orp1/orc1/cdc30 (Bell et al. 1995; Gavin et al. 1995; Muzi-Falconi and Kelly 1995). The *cdc18⁺* gene is periodically expressed through the cell cycle in a cdc10-dependent fashion, with *cdc18⁺*

mRNA levels peaking at the G_1-S boundary (Kelly et al. 1993; see also Section III). cdc18 protein levels also peak at this time. Significantly, gross overproduction of cdc18 (to levels ~10–20 times the peak levels observed in wild-type cells) brings about repeated rounds of DNA replication in the absence of mitosis and of continuing protein synthesis (Nishitani and Nurse 1995). The requirement for full cdc10 function is also bypassed under these conditions. However, DNA synthesized when cdc18 is overproduced is defective, suggesting that other events (e.g., activation of the cdc2-cig2 complex) may be required to bring about the orderly onset of S phase. In addition, since the level of cdc18 required to bring about entry into S phase under these conditions is considerably higher than that observed during the normal cell cycle, it is unlikely that simple accumulation of cdc18 is the sole determinant of the timing of S-phase onset.

Aside from its sequence similarity to orc1, evidence for the involvement of cdc18 with ORC comes from the analysis of the budding yeast cdc18 homolog Cdc6p, which interacts both functionally and physically with ORC (Liang et al. 1995). The precise function of Cdc6p is not known, but it has been shown to possess a nucleotide-binding site and to have ATPase activity (Zwerschke et al. 1994). Although cdc18 is approximately 28% identical to Cdc6p, the two proteins do not cross-complement. This could be due to inadequate or inappropriate expression in the heterologous yeast. Alternatively, the structures of the two proteins may simply have diverged too much to allow cross-complementation (Kelly et al. 1993). The cdc18 protein also contains several cdc2 phosphorylation sites, making it a potential target for regulation by this enzyme.

The hsk1 protein kinase is approximately 46% identical to budding yeast Cdc7p and shares many of the unusual structural features of this protein (for details, see Masai et al. 1995). Cdc7p forms a complex with Dbf4p, which acts to target Cdc7p to ORC (Jackson et al. 1993; Dowell et al. 1994), suggesting that the Cdc7p kinase, and by implication hsk1 in *S. pombe* as well, may have a role in regulating ORC activity. Further aspects of cdc18 and hsk1 function revealed by analysis of *cdc18*Δ and *hsk1*Δ cells are discussed below in Section VI.

2. *DNA Polymerase Complexes*

Eukaryotic chromosomal DNA replication requires three DNA polymerase activities. DNA polymerase α is required for replication initiation, whereas DNA polymerases δ and ε are involved in strand elongation (for review, see Kelman and O'Donnell 1994; Stillman 1994). Genes encoding the catalytic subunits of each of these three enzymes, together with a number of their accessory factors, have now been identified in fission yeast (see Table 1) and are considered here.

The catalytic subunit of DNA polymerase α in fission yeast is the pol1/swi7/cdc29 protein (Damagnez et al. 1991; Francesconi et al. 1993a; Park et al. 1993, 1995; Singh and Klar 1993; Bouvier and Baldacci 1995; D'Urso et al. 1995). DNA polymerase α activity is required for DNA replication and also for mating-type switching (see Singh and Klar 1993). In human cells, DNA polymerase α has been shown to act in concert with a heterodimeric DNA primase complex to synthesize short RNA-DNA primers at the replication fork (see Kelman and O'Donnell 1994; Stillman 1994). The single-stranded DNA-binding complex RPA, a heterotrimer of the Rpa1, Rpa2, and Rpa3 subunits, also acts at this point. The genes encoding the two primase subunits and three RPA subunits have yet to be identified in fission yeast.

An important property of *pol1*Δ cells is that they do not undergo normal cell cycle arrest in response to blocked DNA replication (D'Urso et al. 1995). Instead, the cells proceed through G_2 and mitosis with normal kinetics but rapidly lose viability, presumably as a result of attempting to segregate the unreplicated genome. This property, which is shared by *orp2*Δ, *cdc18*Δ, *cdt1*Δ, *hsk1*Δ, and *cut5*^{ts} cells (Kelly et al. 1993; Saka and Yanagida 1993; Hofmann and Beach 1994; Saka et al. 1994a; Leatherwood et al. 1996), reveals a key aspect of how the cell cycle is regulated and is considered at length below in Section VI.

The fission yeast pol3/cdc6 protein is the catalytic subunit of DNA polymerase δ (Pignède et al. 1991; Francesconi et al. 1993b; Park et al. 1993; Moussy et al. 1995; M. Yamamoto, pers. comm.), which in mammalian cells is a heterodimer of large (catalytic) and small subunits (Ng et al. 1991). cdc1 is most likely the small subunit on the basis of its sequence similarity to the small subunits of the human and bovine holoenzymes and in vitro interaction with pol3 (Zhang et al. 1995; MacNeill et al. 1996). The cdc1 protein has been shown to bind to cdc27, but the role, if any, of this latter protein in DNA polymerase δ function is unknown, as the cdc27 protein does not appear to share sequence similarity with previously identified replication factors (Hughes et al. 1992; MacNeill et al. 1996). Processivity of DNA polymerase δ requires the accessory protein PCNA (proliferating cell nuclear antigen), whose fission yeast homolog is called pcn1 (Waseem et al. 1992). In contrast to the group of genes discussed in the previous section (*orp2*, *pol1*, *cdc18*, etc.), cells carrying deletions of the *cdc6*, *cdc1*, *cdc27*, or *pcn1* genes undergo normal cell cycle arrest and cell elongation (see Section VI).

Assembly of the DNA polymerase δ–PCNA complex onto nascent DNA strand ends requires RF-C, a complex of five related proteins (Kelman and O'Donnell 1994; Stillman 1994). To our knowledge, only one of the genes encoding the anticipated five subunits of the PCNA loading complex RF-C has been identified in fission yeast (all five have been found in budding yeast; for details, see Lew et al., this volume). *rfc2*[+] was uncovered during the wholesale sequencing of chromosome I. The predicted rfc2 protein is 53% identical to budding yeast Rfc2p and 45% to human RFC2. Further analysis of rfc2 function is under way (N. Reynolds et al., unpubl.).

The cdc20/pol2 protein is the catalytic subunit of DNA polymerase ε (G. D'Urso and P. Nurse, unpubl.; A. Sugino, pers. comm.), an enzyme whose precise biochemical function in chromosomal DNA replication is still not clear, as DNA polymerase ε function is not required in mammalian in vitro replication systems (Kelman and O'Donnell 1994; Stillman 1994). However, temperature-sensitive *cdc20* mutants are unable to undergo S phase, suggesting a role either at, or just after, the initiation of DNA replication (G. D'Urso and P. Nurse, unpubl.). The fission yeast cdc17 protein is DNA ligase; cells carrying temperature-sensitive *cdc17* mutants accumulate Okazaki fragments (Nasmyth 1977; Barker et al. 1987). A similar property is displayed by temperature-sensitive *cdc24* mutants; however, the function of this gene is unknown (see Table 1) (Nasmyth and Nurse 1981).

3. *Replication Licensing Factor*

Replication licensing factor (RLF) is the name given to an activity identified in DNA replication-competent *Xenopus* egg extracts which ensures that exogenously added DNA is replicated once and only once in each in vitro cell cycle (for review, see Blow 1995; Huberman 1995). On exit from mitosis, activated RLF is thought to bind to or modify chromatin to allow DNA replication in the subsequent cell cycle. RLF is inactivated as replication occurs, and it has been proposed that since it cannot be imported into the closed nucleus, further rounds of replication are prevented until the nuclear membrane breaks down at mitosis to allow RLF access.

Purification of RLF has demonstrated that it contains *Xenopus* homologs of *S. cerevisiae* Mcm2p, Mcm3p, and Cdc46p/Mcm5p (Chong et al. 1995; Kubota et al. 1995; Madine et al. 1995), members of a family of six related proteins required for DNA replication in yeast that had earlier been shown to have localization properties similar to those expected of RLF (see Hennessy et al. 1990; Blow 1995). In fission yeast, cdc19/nda1, cdc21, nda4, and mis5 are members of this family (see Table 1) (Coxon et al. 1992; Miyake et al. 1993; Forsburg and Nurse 1994b; Takahashi et al. 1994). Each of these proteins has been shown to be required for DNA replication in *S. pombe*, but none have been subjected to immunohistochemical analysis, so it remains to be seen whether they display localization similar to their budding yeast counterparts.

D. Precursor Biosynthesis

The cdc22 and suc22 proteins in fission yeast are the regulatory and catalytic subunits of ribonucleotide reductase (Fernandez-Sarabia et al. 1993). cdc22 is the target of the widely used DNA replication inhibitor hydroxyurea (HU). Expression of *cdc22*$^{+}$ is periodic through the cell cycle, peaking at G_1-S, and is under MBF/cdc10 control (Gordon and Fantes 1986; Lowndes et al. 1992). The *suc22*$^{+}$ gene is represented by two mRNA transcripts of different sizes, the larger being extended at the 5′ end. This transcript is hardly detectable under normal growth conditions, but it is strongly induced by DNA replication blocks or by treatment of cells with DNA-damaging agents (P. Harris and P. Fantes, pers. comm.). Both genes are essential (Fernandez-Sarabia et al. 1993).

E. Histone Expression

The genes encoding the four core histones (H2A, H2B, H3, and H4) together with a single histone variant gene (H2A.F/Z) have been identified in fission yeast (Choe et al. 1985; Matsumoto and Yanagida 1985; Carr et al. 1994). In addition to a single H2B gene, there are two copies of H2A and three copies each of H3 and H4. One of the H2A genes is unlinked to any of the other core histone genes: The remaining genes are found as an H2A-H2B pair and three copies of an H3-H4 pair. Within each pair, the two genes are transcribed in opposite directions. The H2A.F/Z gene is unlinked to any of the above. Accumulation of the histone mRNAs (including that encoding H2A.F/Z) is sharply periodic, with the peak in transcript levels coincident with S phase (Aves et al. 1985; Matsumoto et al. 1987; Carr et al. 1994). The mechanism of histone mRNA accumulation is not known, but it may involve both transcriptional and posttranscriptional regulation, as is the case with the budding yeast (see Lew et al., this volume). A highly conserved 17-bp sequence found upstream of each of the core histone genes has been suggested to play a part in the transcriptional regulation of these genes (Matsumoto and Yanagida 1985).

V. G_2-M CONTROL

A. Entry into Mitosis

After a lengthy G_2 period, entry into mitosis is marked by partial chromosome condensation and by a rapid microtubular reorganization, with the network of interphase cytoplasmic microtubules disappearing, to be replaced by an intranuclear mitotic spindle (Robinow 1977; Hagan and Hyams 1988; for detailed description, see Su and Yanagida, this volume). The nuclear membrane does not break down during mitosis (McCully and Robinow 1971).

Analysis of the G_2-M transition has been greatly facilitated by the isolation of mutants that result in the timing of entry into mitosis being advanced, with the results that the mutant cells undergo cell division at a reduced cell size (called the wee phenotype; see Nurse 1975; Nurse and Thuriaux 1980). At least nine gene functions have been identified that can be mutated by either inactivation or overexpression to cause mitotic advancement (listed in Table 2). The effects on the cell cycle of advancing mitosis in this way are shown in Figure 3. As G_2 is shortened and cell size at division is reduced, so the cryptic size control at G_1-S is revealed and the G_1 period is lengthened. In this way, the generation time remains constant (Nurse 1975).

Activation of the cdc2 protein kinase is essential for entry into mitosis and represents the rate-limiting step in the process, such that all of the mutational events that bring about an advancement of mitosis are believed to do so by either directly or indirectly affecting cdc2 activity. Once past Start, cdc2 becomes associated with the B-type cyclin cdc13 (Booher and Beach 1987, 1988; Hagan et al. 1988; Booher et al. 1989; Moreno et al. 1989; Alfa et al. 1990; Gallagher et al. 1993; Hayles and Nurse 1995) but is kept inactive by phosphorylation on Tyr-15 in the ATP-binding site of the cdc2 kinase (Gould and Nurse 1989). cdc2 does not become phosphorylated on Tyr-15 until cells have passed Start (no Tyr-15 phosphorylation is seen in cells arrested at Start by inactivation of cdc10), but it is phosphorylated in S phase (Hayles and Nurse 1995). Only a very low level of cdc2-cdc13 complex can be detected before Start, which may be due to the fact that little cdc13 protein is present in this part of the cycle. The absence of significant levels of cdc2-cdc13 complex may provide a partial explanation for why cdc2 does not become tyrosine-phosphorylated at this point in the cycle, as Tyr-15 phosphorylation appears to take place only when cdc2 is complexed with cdc13 (Hayles and Nurse 1995).

The following paragraphs deal with the importance of Tyr-15 dephosphorylation of cdc2 and describe the ways in which this is regulated.

Table 2 Advancing entry into mitosis by mutation or overexpression

Recessive loss-of-function mutations	Dominant gain-of-function mutations	By overexpression only
wee1	*cdc2*[a]	cdc25
swo1[b,c]		pyp3
ppa2[c]		nim1
pyp1[c]		wis1[c]
ptc3 [c]		

All the elements listed, with the sole exception of swo1, are either protein kinases (cdc2, wee1, nim1, wis1) or phosphoprotein phosphatases (cdc25, ppa2, pyp1, ptc3, pyp3). swo1 protein binds to the wee1 protein kinase.

[a]Gross overexpression of cdc2 can also advance entry into mitosis.

[b]Advancement seen with partial loss of function mutants only, as complete loss of function is lethal.

[c]Advancement to semi-wee size only (9–12 μm vs. 7–8 μm fully wee, and 14 μm wild type). See text and Table 1 for details and references.

However, it should also be noted that the absence of Tyr-15 phosphorylation during G_1 has implications for how this stage of the cell cycle is regulated. High-level coexpression of cdc2 and cdc13 in pre-Start results in premature entry into S phase (Hayles et al. 1994). These issues are discussed further below.

B. Activation of cdc2 through Tyr-15 Dephosphorylation

The activity of the cdc2-cdc13 complex is periodic through the cell cycle, reaching a peak in late G_2 at the initiation of mitosis (Booher et al. 1989; Moreno et al. 1989; Creanor and Mitchison 1994). Activity is maintained during metaphase but falls off sharply in anaphase, when much cdc13 is proteolytically degraded (see Su and Yanagida, this volume). Experiments with dominant-negative mutant cdc2 proteins have demonstrated the importance of cdc2 activity during mitosis (Labib et al. 1995b), and the cdc2 interacting protein suc1 ($p13^{suc1}$) also has a role at this point in the cycle (Hayles et al. 1986a,b; Brizuela et al. 1987; Hindley et al. 1987; Moreno et al. 1989; Basi and Draetta 1995a; Endicott et al. 1995).

Activation of cdc2 at G_2-M is dependent on dephosphorylation of Tyr-15 primarily catalyzed by the protein tyrosine phosphatase cdc25 (Millar et al. 1991; for review, see Millar and Russell 1992). Inactivation of cdc25 results in failure to activate cdc2 and leads to cell cycle arrest in late G_2 (Russell and Nurse 1986; Gould and Nurse 1989), whereas overproduction of the enzyme leads to mitotic advancement (Russell and

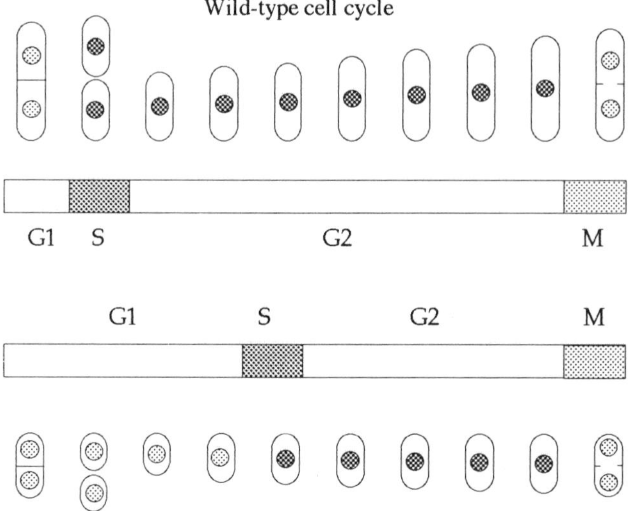

Figure 3 Cell cycle of fission yeast. (*Top*) A schematic representation of the cell cycle of wild-type fission yeast cells. There are four discrete phases, with G_1, S, and M each occupying about 0.1 (10%) of a cycle, and G_2 occupying about 0.7 (70%) of a cycle. In rich medium at optimal growth temperatures, the entire cycle is completed in approximately 200 min. Under such conditions, G_1 and S phase are completed prior to the completion of cell division, with the result that newborn cells are already in G_2. The lightly and darkly shaded nuclei represent cells with unreplicated (1C) and replicated (2C) DNA. (*Bottom*) Cell cycle of *wee* mutant cells. Advancement into mitosis results in cells undergoing cell division at about 50% the size of wild type (7 μm vs. 14 μm). To compensate for their small size at birth, newborn cells have an extended G_1 period, during which time cell division is completed. Thus, the overall length of the cycle remains constant, as the extended time spent in G_1 compensates for the shortened G_2. See text for further discussion.

Nurse 1986). cdc25 activity appears to be regulated at multiple levels. The *cdc25*$^+$ mRNA and the cdc25 protein vary in level through the cell cycle, in both cases reaching a peak at G_2-M (Ducommun et al. 1990; Moreno et al. 1990). Additionally, there is an abundance of evidence from higher eukaryotes that the activity of the cdc25 protein is regulated posttranslationally, by phosphorylation, possibly catalyzed by cdc2 itself in a positive feedback loop (for review, see Basi and Draetta 1995b). Whether this is the case for fission yeast remains to be seen.

In addition to cdc25, two other protein tyrosine phosphatases, both nonessential, have been identified that appear to be capable of dephosphorylating cdc2 at G_2-M, at least under certain conditions (see Table 1) (Millar et al. 1992a; Mondesert et al. 1994; Modesti et al. 1995). Overproduction of either pyp3 or spt1 is sufficient to rescue cells defective in cdc25 activity, even though neither protein is particularly closely related to cdc25 at the primary sequence level. pyp3 can also dephosphorylate and activate tyrosine-phosphorylated cdc2 in vitro, and *pyp3*Δ cells are slightly longer than wild type at division, indicative of a delay to the initiation into mitosis. Overproduction of pyp3 leads to mitotic advancement and cell division at a reduced cell size (Table 2) (Millar et al. 1992a). These results suggest that pyp3 may have a role in regulating cdc25 activity in vivo. In contrast, the case for a role for spt1 at G_2-M is much weaker, as *spt1*Δ cells are in all respects indistinguishable from wild type (Mondesert et al. 1994).

C. Phosphorylation of Tyr-15

Phosphorylation of Tyr-15 is carried out by the protein kinases wee1 and mik1 (Russell and Nurse 1987a; Featherstone and Russell 1991; Lundgren et al. 1991; Parker et al. 1992; McGowan and Russell 1993; Lee et al. 1994), and the activity of wee1 is regulated in turn by a third kinase, nim1/cdr1 (Russell and Nurse 1987b; Feilotter et al. 1991; Coleman et al. 1993; Parker et al. 1993; Wu and Russell 1993).

Inactivation of wee1 brings about mitotic advancement with the result that the cells undergo cell division at approximately half the size of wild type (Nurse 1975), whereas overproduction of wee1 leads to G_2 delay and cell elongation (Russell and Nurse 1987a). Cells containing two extra copies of the *wee1*$^+$ gene divide at approximately 21 μm (50% larger than wild type); those with four extra copies divide at 28 μm (100% larger) and those with six extra copies divide at 35 μm (150% larger). Higher levels of *wee1*$^+$ expression result in even greater delays (Russell and Nurse 1987a). Inactivation of wee1 rescues cells defective for cdc25 function (Fantes 1979). The wee1 protein kinase is a dual specificity enzyme, i.e., it has the potential to phosphorylate its substrates on both tyrosine and threonine/serine residues (Featherstone and Russell 1991), yet it appears to phosphorylate cdc2 on Tyr-15 only, at least under normal circumstances (Parker et al. 1992; McGowan and Russell 1993; see below).

The mik1 protein kinase is related to wee1 at the primary sequence level and has a partially overlapping function (Lundgren et al. 1991; Lee et al. 1994). *mik1*$^+$ is nonessential and *mik1*Δ cells divide at the same size as wild type. Overexpression of *mik1*$^+$ delays entry into mitosis, even in the absence of wee1 function. Simultaneous inactivation of wee1 and mik1 results in lethal premature entry into mitosis from S phase or G_2 (Lundgren et al. 1991) but not from G_1 (Hayles and Nurse 1995). The cdc2 protein is rapidly dephosphorylated on Tyr-15 under these conditions and, not surprisingly, cell viability falls. As might be expected, this dephosphorylation is delayed in cdc25 mutant cells but nevertheless goes to completion, presumably catalyzed by pyp3 and perhaps spt1 as well (Lundgren et al. 1991; Millar et al. 1992a; Mondesert et al. 1994; Modesti et al. 1995).

D. Regulation of wee1 Activity by nim1

wee1 is phosphorylated in vivo (Featherstone and Russell 1991). This phosphorylation plays a part in regulating wee1 activity and also provides a significant clue to understanding how nutritional signals are communicated to the mitotic control, as it has been shown that the nonessential nim1 protein kinase is partly responsible for phosphorylation of wee1 (Russell and Nurse 1987b, Wu and Russell 1993). Overproduction of nim1 leads to mitotic advancement (Table 2), whereas deletion of the gene causes cell division at an increased cell size compared to wild type. nim1 has been shown to phosphorylate wee1 directly in vitro and in doing so reduces the ability of wee1 to phosphorylate and inactivate cdc2 (Coleman et al. 1993; Parker et al. 1993; Wu and Russell 1993). In vivo phosphorylation of wee1 is reduced by approximately 50% in *nim1Δ* cells and is increased twofold in a nim1 overproducing strain (Wu and Russell 1993). Like wee1, nim1 is a dual specificity protein kinase.

A role for nim1 in communicating nutritional signals to the mitotic control is suggested by the observation that *nim1*$^+$ is allelic with *cdr1*$^+$ (Feilotter et al. 1991), mutants in which had previously been isolated on the basis of their altered response to changing nutritional conditions (Young and Fantes 1984, 1987). When wild-type fission yeast cells are shifted from rich medium to poor, there is a transient acceleration of entry into mitosis, with the result that the cells divide at a reduced size. Conversely, when cells are shifted from poor medium to rich, entry into mitosis is transiently delayed. This behavior is lost in *wee1* mutants (Fantes et al. 1991) and also in *cdr1* mutants (Young and Fantes 1984, 1987), indicating that these functions are required for the observed effects. Intriguingly, there is evidence indicating that wee1 is able to phosphorylate nim1, at least in vitro (Coleman et al. 1993). This raises the possibility of either positive or negative regulation of nim1 activity by wee1 via a feedback loop mechanism. Whether this will prove to be the case remains to be seen.

The wee1 kinase also appears to interact with the essential Hsp90 homolog swo1, mutant forms of which can suppress the cell cycle arrest seen when wee1 is grossly overproduced (Aligue et al. 1994). Partial loss of swo1 activity also results in cells being advanced into mitosis at a reduced cell size when compared to wild type, indicating that swo1 has a role in regulating the timing of mitosis in wild-type cells (Table 2). The wee1 and swo1 proteins are able to bind each other in vivo, and the available evidence suggests that swo1 probably acts to stabilize wee1, as wee1 protein levels are markedly reduced in *swo1* mutant cells (Aligue et al. 1994).

E. Phosphorylation of Thr-14 and Thr-167

In higher eukaryotic cells, cdc2 is phosphorylated on four residues, the equivalents of Thr-14, Tyr-15, Thr-167, and Ser-277 (for review, see Basi and Draetta 1995b). Initially, fission yeast cdc2 was thought to be phosphorylated only on Tyr-15 and Thr-167 (Gould and Nurse 1989; Gould et al. 1991; see below), and there is still no evidence of phosphorylation on Ser-277 (see Krek et al. 1992). Recently, however, low-level phosphorylation of cdc2 on Thr-14 has been detected, specifically in cells overproducing wee1 (Den Haese et al. 1995). Overproduction of mik1 does not have this effect, nor is mik1 required for Thr-14 phosphorylation in wee1 overproducing cells. Cells expressing the mutant cdc2 Ala-14 protein are partially advanced into mitosis (semi-wee) compared to wild type, an intriguing result but one that does not necessarily require cdc2 to be phosphorylated on Thr-14 in normal cells, as mutants that map some distance from the phosphorylation sites can cause mitotic advancement (Carr et al. 1989; MacNeill and Nurse 1993). Curiously, when Thr-14 is replaced by serine, this residue becomes phosphorylated even in wild-type cells and in a wee1-dependent fashion (i.e., Ser-14 phosphorylation is abolished in *wee1Δ* cells), indicating that wee1 is required either directly or indirectly for Ser-14 phosphorylation. Although the significance of these observations remains unclear, particularly with regard to whether Thr-14 (rather than the mutant Ser-14) is actually phosphorylated in wild-type cells under any circumstances, experiments with various S-phase mutants point to Ser-14 phosphorylation occurring in late S phase and G_2. No Ser-14 phosphorylation was detected in hydroxyurea-blocked early S-phase cells.

Phosphorylation of Thr-167 is essential for cdc2 function. Replacing Thr-167 with nonphosphorylatable alanine abolishes cdc2 activity altogether (Gould et al. 1991). In vertebrate cells, Thr-167 is phosphorylated by an activity known as CAK (cdk-activating kinase), which is a complex of the cdk7/MO15 and cyclin H proteins. Human cdk7 and cyclin H are also components of the essential general transcription factor TFIIH (for review, see Basi and Draetta 1995b). Fission yeast cdc2 is in all likelihood phosphorylated on Thr-167 by the cdk-cyclin complex mcs6/crk1/mop1-mcs2, where mcs6/crk1/mop1 is closely related to the mammalian cdk7 and *Xenopus* MO15 proteins, and mcs2 protein is related to cyclin H (Molz et al. 1989; Molz and Beach 1993; Buck et al. 1995; Damagnez et al. 1995). In common with its vertebrate homologs, the mcs6-mcs2 complex can also phosphorylate, at least in vitro, a peptide corresponding to the carboxy-terminal repeat domain (CTD) of RNA polymerase II, suggesting it may have a role in TFIIH function in vivo (Buck et al. 1995; Damagnez et al. 1995). Both the mcs6 kinase and its cyclin partner mcs2 are essential proteins, and the *mcs6*Δ and *mcs2*Δ phenotypes are similar, with the majority of cells arresting with a septum and what appears to be condensed chromatin, suggesting that activity of the mcs6-mcs2 complex is required in mitosis (Molz and Beach 1993; Buck et al. 1995). However, it is as yet unclear whether this phenotype results specifically from failure to phosphorylate cdc2 on Thr-167 or whether the primary defect in *mcs6*Δ or *mcs2*Δ cells might be transcriptional, i.e., failure to phosphorylate RNA polymerase II CTD repeats. Clearly, further work is required to resolve these possibilities.

F. Role of the cdc2-cig1 Complex

In considering the role of cdc2 at the G_2-M transitions, the preceding sections have focused specifically on the function of the cdc2-cdc13 complex. It is likely, however, that a second cdc2-cyclin complex, cdc2-cig1, also has a role at this point in the cell cycle (Basi and Draetta 1995a). cig1 is a B-type cyclin that is most closely related at the primary sequence level to budding yeast S-phase cyclins Clb3p and Clb4p and to mammalian cyclin G whose function is not known (see Fisher and Nurse 1995 and references therein; Lew et al., this volume). *cig1*Δ cells are viable, indicating that cig1 is nonessential, and have a normal cell cycle; early reports of an extended G_1 period having proved erroneous (Bueno et al. 1991; see Forsburg and Nurse 1994a; for correction, see Bueno and Russell 1993b).

G. Stress Responses and Mitotic Regulation

MAP kinase signal transduction pathways have been implicated in a wide range of cellular processes such as response to environmental stresses or stimuli. The architecture of these pathways is conserved across evolution: A MAP kinase kinase kinase (MAPKKK) phosphorylates and activates a MAP kinase kinase (MAPKK) which in turn phosphorylates and activates a MAP kinase (MAPK). The MAPKK is a dual specificity enzyme, capable of phosphorylating the MAPK on both tyrosine and threonine (for review, see Blumer and Johnson 1994; Johnson and Vaillancourt 1994; Herskowitz 1995).

To date, two MAP kinase pathways have been identified in fission yeast. One of these, involving the byr2, byr1/ste1, and spk1 protein kinases (MAPKKK, MAPKK, and MAPK, respectively), is required for cellular responses to mating pheromone and is considered in detail elsewhere in this volume (see Yamamoto et al., this volume). Two, perhaps three, of the corresponding proteins in the second pathway have been identified: the MAPKK homolog wis1/sty2/spc2 and its substrate, the MAPK homolog sty1/spc1 (see Fig. 4) (Warbrick and Fantes 1991; Shiozaki et al. 1994; Millar et al. 1995; Shiozaki and Russell 1995a,b). The wis4 protein, a MAPKKK homolog, most likely acts upstream of wis1 (Warbrick and Fantes 1992; I. Samejima and P. Fantes, pers. comm.). This pathway has an important role in regulating entry into mitosis under conditions of environmental stress (see below), although it is not yet clear how these effects are mediated to the cdc2 kinase.

Although neither gene is essential under normal growth conditions, inactivation of either wis1 or sty1 leads to expression of multiple phenotypes, such as sensitivity to osmotic stress, inability to grow at high temperatures or on glycerol, and, most importantly within the context of this chapter, a delay to mitotic initiation (Warbrick and Fantes 1991; Millar et al. 1995; Shiozaki and Russell 1995b). In addition, moderate overproduction of wis1 leads to mitotic advancement and cell division at a reduced cell size, indicating that wis1 is a true mitotic inducer (Warbrick and Fantes 1991).

The osmotic sensitivity displayed by cells in which the wis1-sty1/spc1 pathway is disrupted probably results from a failure to induce transcription of *gpd1*$^+$, one of two genes encoding glycerol-3-phosphate dehydrogenase in fission yeast (Aiba et al. 1995), as overexpression of *gpd1*$^+$, or the related *gpd2*$^+$ gene, is sufficient to rescue the osmosensitivity of *wis1*Δ cells. *wis1*Δ cells are also unable to induce the meiotic regulator *mei2*$^+$ (and display mating defects) and *fbp1*$^+$, which encodes fructose-1,6-bis-phosphatase, a key enzyme of the gluconeogenic pathway (see Stettler et al. 1996 and references therein). Failure to induce *fbp1*$^+$ prevents *wis1*Δ cells from utilizing glycerol as sole carbon source. The *mei2*$^+$ and *fbp1*$^+$ genes are also subject to negative regulation

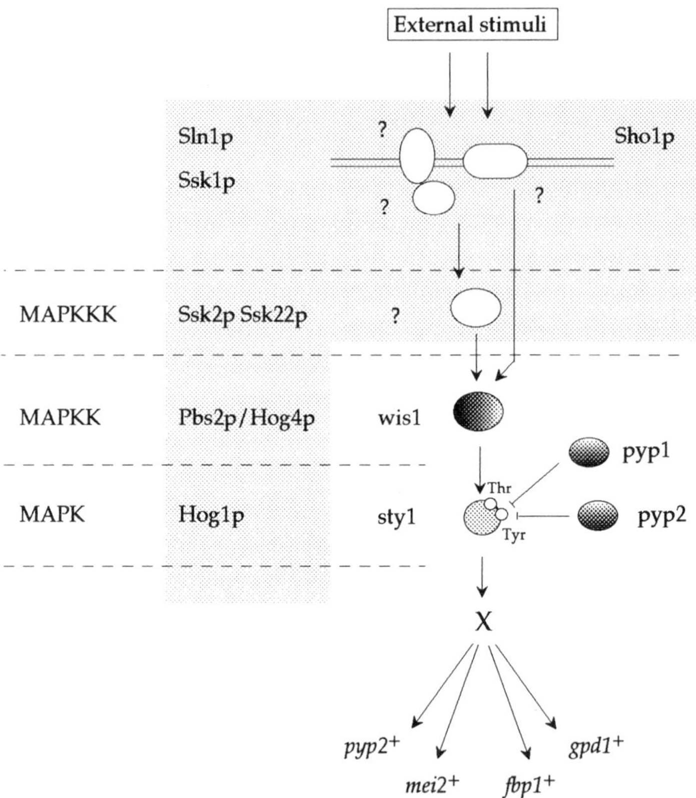

Figure 4 A stress response pathway controlling mitotic onset. A schematic representation of the wis1-sty1/spc1 MAP kinase pathway in fission yeast alongside the related Pbs2p-Hog1p pathway in budding yeast. In *S. pombe*, the MAPKK homolog wis1 phosphorylates the MAPK homolog sty1/spc1 and is probably phosphorylated by wis4, an MAPKKK homolog. Activity of the pathway is required for transcriptional induction of at least three genes regulating diverse aspects of cell metabolism—*mei2*+, *fbp1*+, and *gpd1*+—as well as for expression of *pyp2*+, whose product, the pyp2 tyrosine phosphatase, acts in a negative feedback loop to down-regulate the pathway. Directly downstream from sty1/spc1 is presumed to be a transcription factor (perhaps a heterodimer containing atf1 and pcr1), indicated as X, but direct evidence for this is lacking and it is equally possible that sty1/spc1 may exert its effects on such a factor indirectly, through phosphorylation of some other (unknown) protein. Similarly, although activity of the pathway is required for transcription of the four genes described above, it is not known whether their transcription is directly activated by X or by a transcription factor whose expression is itself controlled by X. *mei2*+ expression is probably mediated through the ste11, for example. In *S. cerevisiae*, the Pbs2p-Hog1p pathway is activated in two ways, either via the two-component sensor-receptor Sln1p-Ssk1p or by Sho1p which binds directly to Pbs2p. Homologs of Sln1p, Ssk1p, and Sho1p, should they exist, have yet to be identified in fission yeast. See text for details and references.

by cAMP, suggesting a link between the cAMP system and the wis1-sty1/spc1 pathway (Stettler et al. 1996).

The protein tyrosine phosphatases pyp1 and pyp2 (Table 1) (Ottilie et al. 1991, 1992; Millar et al. 1992b) have been shown to regulate the wis1-sty1/spc1 pathway negatively by dephosphorylating sty1/spc1 (Millar et al. 1995; Shiozaki and Russell 1995b). In wild-type cells, sty1/spc1 is rapidly tyrosine-phosphorylated following osmotic shock, but this phosphorylation is only transient and soon returns to normal levels. In *pyp2Δ* cells, however, the level of tyrosine-phosphorylated sty1 remains high, indicating that pyp2 activity is required for the decline observed in wild type. *pyp2⁺* mRNA is sharply induced following osmotic shock, but strikingly, this transcriptional induction does not occur in either *wis1Δ* or *sty1* mutant cells, implying that pyp2 inhibits sty1 as part of a negative feedback loop (diagrammed in Fig. 4).

As well as the tyrosine phosphatases pyp1 and pyp2, the type-2C protein serine-threonine phosphatases ptc1, ptc2, and ptc3 have also been implicated in the wis1-sty1 pathway (Shiozaki et al. 1994; Shiozaki and Russell 1995a,b). The PP-2C enzymes are distinct from other protein serine-threonine phosphatases in that they require magnesium for activity and are insensitive to certain inhibitors, such as okadaic acid. In addition, they have no apparent sequence homology with the PP-1, PP-2A, and PP-2B serine-threonine phosphatases described in the following section. None of the three PP-2C enzymes identified to date in fission yeast have an essential function under normal growth conditions, but several lines of evidence point to these proteins functioning in cellular stress response pathways. Expression of *ptc1⁺*, for instance, which was first identified as a suppressor of a defect in the swo1 Hsp90 homolog (Table 1), is induced by heat shock, and *ptc1Δ* cells are sensitive to such treatments (Shiozaki et al. 1994). *ptc1Δ ptc3Δ* and *ptc1Δ ptc2Δ ptc3Δ* cells also display various growth defects that can be suppressed either by the addition of osmotic stabilizers or by inactivation of either wis1 or sty1/spc1 (Shiozaki and Russell 1995a,b), and *ptc3Δ* cells undergo cell division at a somewhat reduced cell size when compared to wild type (Table 2). At present, the in vivo substrates of the PP-2C enzymes have yet to be identified, but it is tempting to speculate whether either wis1 or sty1/spc1 may be dephosphorylated by either ptc1, ptc2, or ptc3.

Although those elements upstream of wis1 and sty1/spc1 in the pathway have yet to be identified in fission yeast, with the probable exception of wis4 (see above), examination of the related osmoregulatory pathway in budding yeast may give an indication of the overall structure of the pathway (see Fig. 4) (Maeda et al. 1994, 1995). In this organism,

Hog1p and Pbs2p/Hog4p are MAPK and MAPKK proteins related to sty1 and wis1, respectively (see Warbrick and Fantes 1991; Millar et al. 1995). Two related nonessential MAPKKK proteins, Ssk2p and Ssk22p, have been implicated as acting upstream of Pbs2p/Hog4p. Further upstream, Sln1p and Ssk1p comprise a two-component sensor-receptor system that has been shown to regulate the pathway through Ssk2p and Ssk22p, whereas Sho1p interacts directly with Pbs2p through its SH3 domain (Maeda et al. 1994, 1995). It is possible, perhaps even likely, given the extent to which many diverse cellular processes are conserved between the two yeasts, that similar systems regulate the wis1-sty1 pathway in fission yeast, but confirmation of this will clearly require the identification of the corresponding fission yeast proteins. In addition to those elements upstream of wis1 in the pathway, it should be noted that the identity of the transcription factor (or factors) that presumably acts downstream from the sty1/spc1 pathway has yet to be established, although the recently characterized atf1 and pcr1 proteins (described by Takeda et al. 1995; Watanabe and Yamamoto 1996) are strong candidates.

H. PP-2A Function at G_2-M

In addition to the tyrosine phosphatases cdc25, pyp1, pyp2, and pyp3, and the type 2C (PP-2C) serine-threonine phosphatases ptc1, ptc2, and ptc2, at least five other phosphoprotein phosphatases have been implicated in various aspects of cell cycle control in *S. pombe*. The type 2A (PP-2A) enzymes ppa1 and ppa2 are implicated in mitotic control (Kinoshita et al. 1990). Although neither ppa1 nor ppa2 is essential for cell cycle progression on its own, simultaneous inactivation of both is lethal, indicating that they share at least one essential function. ppa2 is responsible for the bulk of detectable PP-2A activity and *ppa2Δ* cells are advanced into mitosis at a small cell size (see Table 2), an effect that may be mediated via cdc25, although this is by no means certain (Kinoshita et al. 1993). A third PP-2A enzyme, ppe1, also appears to have a role in mitosis and like ppa1 shares at least one essential function with ppa2 (Matsumoto and Beach 1993; Shimanuki et al. 1993).

The type-1 (PP-1) enzymes dis2 and sds21 have been clearly shown to have a key role during mitosis in fission yeast, discussion of which falls outside the scope of this chapter (see Su and Yanagida, this volume), but dis2 has also been implicated in the mitotic control (Booher and Beach 1989) and is most likely regulated by phosphorylation, by cdc2 (Yamano et al. 1994). However, a specific molecular role for dis2 in the mitotic control has yet to be found.

I. A Model for the Control of G_2-M Transition

Figure 5 shows our current understanding of how the timing of entry into mitosis is controlled in fission yeast. Central to this control is the timing of activation of the cdc2-cdc13 mitotic kinase complex via dephosphorylation of Tyr-15 in cdc2, catalyzed primarily by the cdc25 tyrosine phosphatase, assisted in all likelihood by pyp3 and perhaps also by spt1. Counteracting the activity of these three tyrosine phosphatases are the wee1 and mik1 kinases responsible for phosphorylation of Tyr-15 and perhaps also Thr-14, although this remains uncertain. Whether activation of cdc2 and entry into mitosis are triggered by an increase in tyrosine phosphatase activity or by a decrease in tyrosine kinase activity, or both, remains to be seen. As discussed above, cdc25 protein levels rise throughout G_2 to reach a peak at G_2-M. wee1 is inhibited by nim1/cdr1, the latter being first identified on the basis of its role in nutritional sensing.

VI. DEPENDENCY RELATIONSHIPS

A. Restraining Entry into Mitosis during S Phase and G_2

Entry into mitosis is prevented until DNA replication is complete, and any damaged DNA present is repaired by cellular DNA repair mechanisms (Fig. 6) (for review, see Carr 1995; D'Urso and Nurse 1995). Six gene functions have been identified in fission yeast that are required for this checkpoint control: *rad1⁺*, *rad3⁺*, *rad9⁺*, *rad17⁺*, *rad26⁺*, and *hus1⁺* (Al-Khodairy and Carr 1992; Enoch et al. 1992; Rowley et al. 1992b; Al-Khodairy et al. 1994; Kanter-Smoler et al. 1995; Lieberman 1995). Cells carrying loss-of-function mutations in any of these six genes are unable to arrest the cell cycle when DNA replication is blocked (either by mutation or by chemical inhibition) or when DNA is damaged (by UV or γ-rays), with the result that they proceed into mitosis and lose viability. In addition, the *rad/hus* genes are required for recovery from DNA replication arrest. Thus, *rad/hus* mutants treated with the DNA synthesis inhibitor hydroxyurea lose viability in S phase, considerably before entering mitosis with unreplicated DNA (Enoch et al. 1992; Al-Khodairy et al. 1994).

What are the functions of the rad/hus proteins? All six genes have been cloned, but only in three cases, specifically *rad1*, *rad3*, and *rad17*, does sequence analysis suggest a function for the encoded proteins (Sunnerhagen et al. 1990; Murray et al. 1991; Jimenez et al. 1992; Lieberman et al. 1992; Al-Khodairy et al. 1994; Griffiths et al. 1995; T. Enoch, pers. comm.). On the basis of its sequence similarity (the proteins are 23%

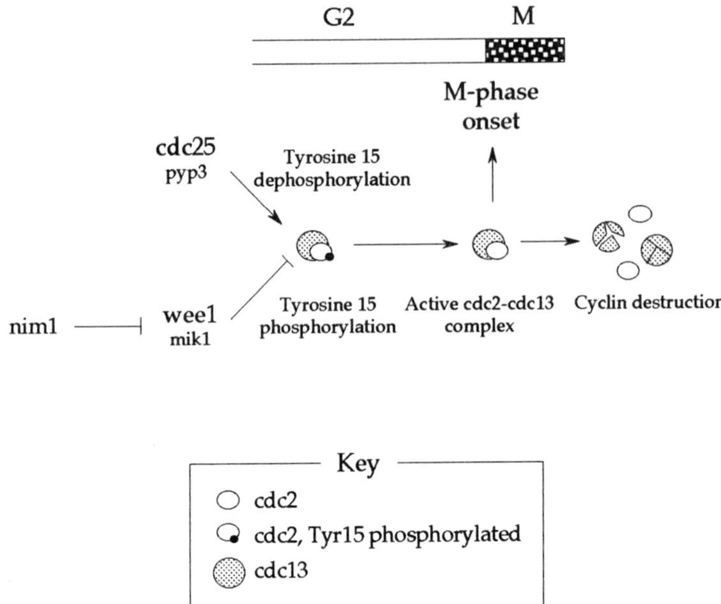

Figure 5 G_2-M regulation through tyrosine phosphorylation of cdc2. cdc2 activation at the G_2-M transition requires dephosphorylation of Tyr-15 catalyzed largely by the essential protein tyrosine phosphatase cdc25. A minor part is played by the unrelated nonessential tyrosine phosphatase pyp3, and a third such enzyme (spt1, not shown) could conceivably also play a part. Phosphorylation of cdc2 on Tyr-15 occurs post-Start and is catalyzed by the wee1 and mik1 kinases, with wee1 playing the major part. The nim1/cdr1 kinase directly phosphorylates wee1 to inhibit its activity, and it appears to have a role in nutritional sensing. Activation of the cdc2-cdc13 complex is probably achieved by changes in the proteolytic destruction of cdc13. See text for details and references.

identical) to budding yeast Rad17p, rad1 is most likely a 3′-5′exonuclease (Lydall and Weinert 1995), suggesting a direct role in the processing of DNA damage as well as in checkpoint function. The rad3 protein shares sequence similarity with members of the PI3-kinase family, including the budding yeast checkpoint protein Mec1p and the mammalian ataxia telangiectasia (ATM) protein implicated in cellular responses to DNA damage (for review, see Zakian 1995). It is unclear as yet whether rad3 is a bone fide phosphotidylinositol kinase, however, as some members of the PI3-kinase family have protein kinase activity (for discussion, see Hunter 1995).

The rad17 protein displays limited sequence similarity to the components of RF-C, a five-subunit protein complex required for loading the

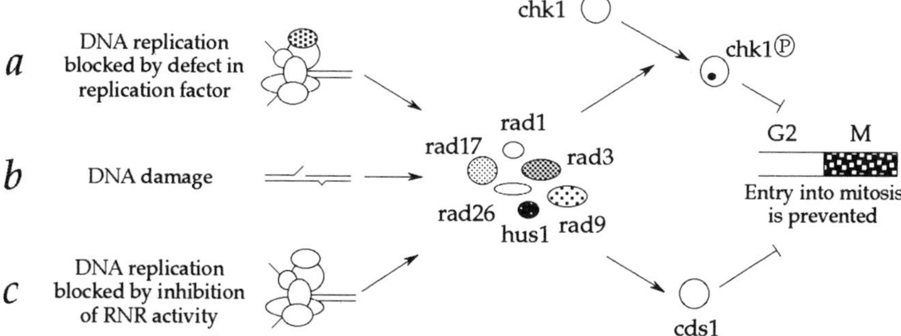

Figure 6 Checkpoints controlling the onset of mitosis. Entry into mitosis in fission yeast is prevented when damaged DNA is present (*b*) or when DNA replication is blocked either (*a*) by mutational inactivation of any one of a number of replication factors, such as DNA polymerases δ and ε or DNA ligase, or (*c*) by inhibition of ribonucleotide reductase (RNR) activity. In each case, the block to mitosis requires the functions of six rad/hus proteins, rad1 (a putative 3′-5′exonuclease), rad3 (a phosphotidylinositol kinase), rad9, rad17 (a protein that has some homology with the subunits of replication factor C), rad26, and hus1. *a* and *b*, but not *c*, also require the function of the chk1/rad27 protein kinase, which appears to lie downstream from the rad/hus proteins, whereas *c* alone requires cds1 function. See text for details and references.

DNA polymerase δ accessory factor PCNA onto nascent DNA strand ends (for review, see Stillman 1994; Griffiths et al. 1995). Although rad17 is unlikely to be a subunit of the well-characterized RF-C complex involved in DNA replication, it is possible that rad17 could be a part of a specialized RF-C-like complex with a specialized checkpoint function. Alternatively, the similarity to the RF-C proteins may reflect a shared interaction with some other component of the DNA replication and DNA repair machinery, or simply a common biochemical function.

As described above, blocking S phase generally leads to checkpoint activation and cell cycle arrest prior to entry into mitosis. However, this is not always the case. Cells carrying temperature-sensitive *cut5* mutants arrest in the early stages of S phase yet do not activate the checkpoint, suggesting that the checkpoint is inoperative in *cut5* mutant cells (Saka and Yanagida 1993; Saka et al. 1994a). In the absence of a functional checkpoint, cut5 cells proceed through mitosis with unreplicated DNA, with fatal consequences. A further demonstration that the checkpoint is inoperative under these circumstances is provided by the fact that unrestrained entry into mitosis cannot be prevented by addition of hydroxyurea (Saka et al. 1994a). Similar properties are displayed by *cdc18Δ* and *pol1Δ* cells (but not temperature-sensitive *cdc18* or *pol1* mutants, suggesting that these mutants still have sufficient residual function to activate the checkpoint) as well as by *orp1*Δ, *orp2*Δ, *cdt1*Δ, and *hsk1*Δ cells (Kelly et al. 1993; Hofmann and Beach 1994; D'Urso et al. 1995; Masai et al. 1995; Muzi-Falconi and Kelly 1995; Leatherwood et al. 1996). Thus, each of these five functions is required in order for the rad/hus checkpoint to operate.

The orp1, orp2, cdc18, hsk1, and pol1 proteins are each implicated in the very earliest stages of DNA replication: orp1 and orp2 are most likely subunits of fission yeast ORC; cdc18 and hsk1 are, by analogy with budding yeast, likely to be ORC-associated, if only transiently in G_1 (Dowell et al. 1994; Liang et al. 1995); and pol1 is required at the replication fork for synthesis of the DNA portion of the RNA-DNA primers utilized for strand elongation (see Section IV). The biochemical roles of the cut5/rad4 and cdt1 proteins are not known, but it is tempting to speculate that these proteins will also function at the replication fork (Fenech et al. 1991; Saka and Yanagida 1993; Hofmann and Beach 1994). Indeed, in *S. cerevisiae*, a gene encoding a homolog of cut5, *DBP11*, has been isolated by virtue of its ability to rescue mutants in DNA polymerase ε (Araki et al. 1995).

The fact that checkpoint control is unaffected in temperature-sensitive *cdc18* and *pol1* mutants but lost in *cdc18*Δ and *pol1*Δ cells suggests a model for checkpoint function, in which successful set up of the checkpoint requires the presence of a functional multiprotein complex at the replication fork. If orp1, orp2, cdc18, cut5, cdt1, and pol1 are required to generate such a complex, then, in the absence of any of these proteins, the complex will not form and the checkpoint will be inoperative. In the presence of each of these proteins, even in inactivate thermolabile forms, complex formation is assumed to be unaffected (even though replication cannot take place) and the checkpoint can function as normal and block entry into mitosis.

Once replication is initiated, after the pol1 step has been completed, the rad/hus checkpoint is able to prevent entry into mitosis if replication is blocked by mutation or chemical inhibition. In sharp contrast to *pol1*Δ cells defective in DNA polymerase α function, *pol3*Δ cells defective in DNA polymerase δ function undergo normal cell cycle arrest and cell elongation presumably mediated via the rad/hus checkpoint (Francesconi et al. 1993b). The same is true of the DNA polymerase-δ-related functions cdc1 and cdc27 (MacNeill et al. 1996), the catalytic subunit of DNA polymerase ε cdc20 (G. D'Urso and P. Nurse, unpubl.), and pcn1 (Waseem et al. 1992). The checkpoint could function by monitoring the activity of this complex as it elongates nascent DNA strands.

Activation of the checkpoint prevents activation of the cdc2 kinase and entry into mitosis by maintaining the inhibitory phosphorylation of the enzyme on Tyr-15 (Enoch and Nurse 1990; Enoch et al. 1991; see also Kovelman and Russell 1996). In situations where cells that have a reduced capacity to maintain Tyr-15 phosphorylation, for instance, when the cdc25 phosphatase is overproduced or when Tyr-15 phosphorylation is blocked altogether by substitution of Tyr-15 with nonphosphorylatable phenylalanine, inhibiting DNA replication does not prevent entry into mitosis (Enoch and Nurse 1990; Enoch et al. 1991). Quite how activation of the checkpoint brings about inhibition of Tyr-15 dephosphorylation remains an open question, but it is clear that neither wee1 nor cdc25 is essential for checkpoint function, as *cdc25Δ* and *wee1Δ* cells each still undergoes cell cycle arrest in response to blocked DNA replication or DNA damage (Enoch et al. 1991; Barbet and Carr 1993). An early report (Rowley et al. 1992a) suggesting that wee1 was required for radiation-induced cell cycle delay appears to have been erroneous, although *wee1Δ* cells are more sensitive to irradiation than are wild type (Barbet and Carr 1993), perhaps because loss of wee1 function directly or indirectly affects the DNA replication and repair apparatus (for discussion, see Fantes 1983; MacNeill et al. 1996).

The *chk1⁺/rad27⁺* function, which shares several similarities with those of the *rad/hus* checkpoint genes discussed previously, may provide a direct link between checkpoint activation and cdc2. *chk1⁺* is required for cell cycle arrest in response to the DNA damage caused by UV- or γ-irradiation or by treatment of cells with the DNA-alkylating agent MMS and also for cell cycle arrest when DNA replication is blocked by mutations in any one of a large range of DNA replication gene functions, such as *cdc17⁺* or *pol3⁺* (Walworth et al. 1993; Al-Khodairy et al. 1994; Carr et al. 1995; Francesconi et al. 1995; Walworth and Bernards 1996). However, *chk1Δ* cells do not arrest the cell cycle when ribonucleotide reductase activity is blocked, either by mutation in *cdc22⁺* or by hydroxyurea treatment (Walworth et al. 1993; Al-Khodairy et al. 1994). Clearly, the signals generated by these treatments must in some way differ. In addition, the kinetics of loss of viability displayed by radiation-treated *chk1Δ* cells suggest that unlike *rad1* mutants for instance, *chk1Δ* cells are not impaired in their ability to recover from blocked DNA replication or DNA damage.

The chk1/rad27 protein is a 56-kD protein kinase ($p56^{chk1}$) that is phosphorylated in response to DNA damage (UV, γ, MMS) or blocked DNA replication (except when as a result of inhibition of RNR). Strikingly, this modification is dependent on the functions of the *rad/hus* checkpoint genes *rad1*$^+$, *rad3*$^+$, *rad9*$^+$, *rad17*$^+$, and *rad26*$^+$, indicating that the chk1 protein kinase is likely to function downstream from these genes (Walworth and Bernards 1996). Strong overproduction of chk1 leads to cell cycle arrest (Walworth et al. 1993; Ford et al. 1994), and the link to cdc2 is provided by the observation that overproduction of chk1 rescues cells carrying a particular cold-sensitive mutant *cdc2* allele, *cdc2-r4* (Walworth et al. 1993).

Two further functions implicated in the DNA damage checkpoint are *rad24*$^+$ and *rad25*$^+$, which encode a functionally overlapping pair of proteins of the 14-3-3 family (Al-Khodairy et al. 1994; Ford et al. 1994; for review of 14-3-3 proteins, see Burbelo and Hall 1995). Like *chk1Δ/rad27Δ* cells, *rad24Δ* cells are highly sensitive to UV- and γ-irradiation and fail to prevent entry into mitosis when replication is blocked by mutation (of DNA ligase, *cdc17*, for instance) but do arrest in the presence of hydroxyurea. Similar phenotypes are displayed by *rad25Δ* cells, although to a much reduced degree, and *rad24Δ rad25Δ* double mutants are inviable. There is also some evidence pointing to rad24 and rad24 functioning downstream from chk1/rad27, as overproduction of the latter does not prevent colony formation in *rad24Δ* cells as it does in wild type (Ford et al. 1994).

Finally, the protein kinase encoded by *cds1*$^+$ also appears to play a part in the DNA replication checkpoint (Murakami and Okayama 1995). Indeed, it has been suggested (Walworth and Bernards 1996) that cds1 may perform a chk1-like function specific to the DNA replication checkpoint as *cds1Δ* cells are able to block mitosis following irradiation-induced DNA damage but are unable to do so when DNA replication is inhibited by hydroxyurea (or in *cdc22* mutant cells). Additionally, overproduction of cds1 can rescue the hydroxyurea sensitivity of *rad1*, *rad3*, and *rad9* mutant cells (Murakami and Okayama 1995). However, the timing of entry into mitosis in *cds1Δ* cells treated with hydroxyurea, or in *cds1Δ cdc22* double mutants, points to the cds1 protein as being required to delay entry into mitosis only once the cells have begun to leak through the initial cell cycle block; thus, prior to this, some other checkpoint system must operate. Interestingly, *cds1*$^+$ was isolated as a multicopy suppressor of a temperature-sensitive *pol1/swi7/cdc29* mutant, suggesting a link with DNA polymerase α function (Murakami and Okayama 1995).

B. Restraining Entry into Mitosis from G_1

If the function of the DNA replication checkpoint relies on the presence of active replication complexes, then how are cells in G_1 prevented from entering mitosis, when no such complexes are present? The function of the rum1 protein is important here, for when *rum1*Δ cells are arrested in G_1 prior to Start, they enter mitosis directly without replicating their DNA (Fig. 7) (Moreno and Nurse 1994). rum1 is a potent inhibitor of cdc2-cdc13 activity in vitro and acts under normal circumstances to prevent an increase in cdc2-cdc13 kinase activity during G_1 (Correa-Bordes and Nurse 1995). This is important because not all the cdc13 in a cell is destroyed during mitosis (Hayles and Nurse 1995) and what remains is able to form active mitotic kinase complexes with the tyrosine-dephosphorylated G_1 form of cdc2. As discussed earlier, the presence of active mitotic kinase in G_1 leads to lethal entry into mitosis (Hayles et al. 1994). rum1 not only binds to cdc2-cdc13 complexes, but also targets cdc13 for degradation, probably via the fission yeast equivalent of the proteolytic mechanism that has recently been described in budding yeast (Irniger et al. 1995; for further discussion, see Su and Yanagida, this volume).

The rum1 protein also has a role in determining the length of G_1 and timing of S-phase onset, possibly via the transient inhibition of cdc2-cig2 complexes or by inhibiting cdc2-cdc13 complexes (Martín-Castellanos et al. 1996). This role is confined to G_1, however, because rum1 is only present during this period (Correa-Bordes and Nurse 1995). Note that deregulated overexpression of rum1 in G_2 cells has dramatic consequences (Moreno and Nurse 1994); these will be discussed shortly.

C. Sequential Operation of Pre-Start and Post-Start Checkpoints

Taken together, the observations described in the preceding paragraphs lend themselves to a simple model is which two checkpoints, one pre-Start and the other post-Start, act sequentially to prevent entry into mitosis from G_1 or S phase, or when DNA is damaged (Nurse 1994). Cells in G_1 are prevented from entry into mitosis by the action of rum1 in inhibiting cdc2-cdc13 complexes, whereas post-Start cells rely on the DNA replication checkpoint and its effect on the tyrosine phosphorylation state of cdc2 (Enoch and Nurse 1990; Enoch et al. 1991; Moreno and Nurse 1994; Correa-Bordes and Nurse 1995). In wild-type cells, the handover from pre-Start to post-Start checkpoint control may be almost instantaneous, so that few cells ever fall between the safety nets provided by the two different checkpoint systems. When the handover is delayed, however, in *cdc18*Δ or *pol1*Δ cells, for example, the consequences are severe.

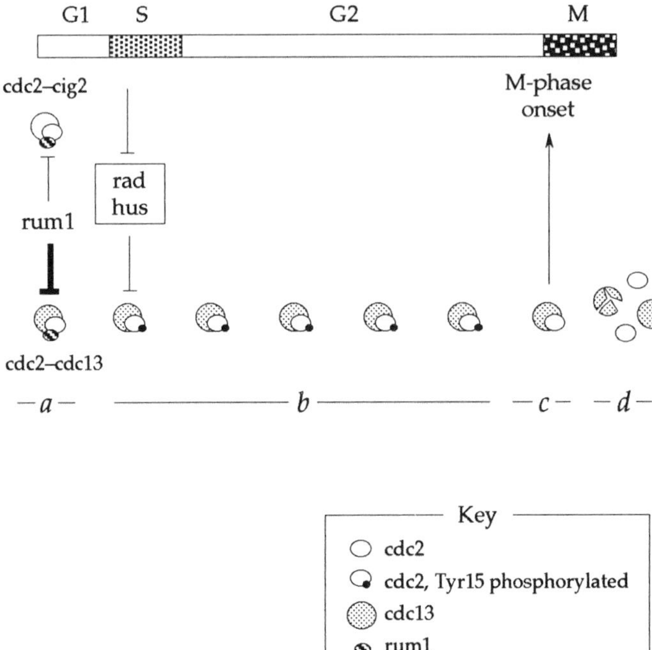

Figure 7 Summary model for restraining entry into mitosis in G_1, S phase, and G_2 cells. The rum1 protein has a key role in G_1, probably by transiently inhibiting cdc2-cig2 complex activity to regulate the timing of Start, and potently inhibiting cdc2-cdc13 complexes to prevent entry into mitosis (part *a*). Once past Start, cdc2 becomes phosphorylated on Tyr-15 by wee1 and mik1. Entry into mitosis in prevented while DNA replication is incomplete, or if damaged DNA is present, by the inhibitory action of the rad/hus checkpoint system (part *b*). Once replication is complete, and other preconditions for entry into mitosis (cell size, nutritional status) are satisfied, the cdc2-cdc13 complex is activated by Tyr-15 dephosphorylation catalyzed by cdc25 (part *c*). Later, cdc13 degradation allows exit from mitosis (part *d*), but is incomplete, so that there is potential for active (Tyr-15 dephosphorylated) cdc2-cdc13 to form in G_1. This is prevented by the action of rum1.

D. Restraining Entry into S phase

As entry into mitosis is dependent on completion of DNA replication, so the initiation of S phase is normally dependent on completion of the early events of mitosis in the previous cell cycle. Thus, S phase is restricted to occurring once per cycle. In a number of circumstances, this dependency is broken, however. These are considered below.

As discussed in the preceding pages, activity of the cdc2-cdc13 complex is essential for entry into mitosis (Moreno et al. 1989), and simultaneous overproduction of cdc2 and cdc13 can drive cells into mitosis from G_1 (Hayles et al. 1994). Subsequent analysis of *cdc13Δ* cells has shown that while they are unable to enter mitosis, the arrested cells are able to repeatedly rereplicate their DNA and become highly elongated with greatly enlarged nuclei (Hayles et al. 1994). This phenotype is not confined to *cdc13Δ* cells, as cells carrying certain mutant *cdc2* and *cdc13* alleles can be forced to undergo rereplication by nitrogen starvation followed by heat shock treatment (see Broek et al. 1991; Hayles et al. 1994). A single round of treatment results in diploidization; further rounds induce higher ploidies. It is thought that the effect of these treatments is to bring about denaturation of the mutant proteins, thereby disrupting the cdc2-cdc13 complex (Hayles et al. 1994). Not surprisingly, overproduction of rum1 in G_2 also results in rereplication, through inhibition of cdc2-cdc13 activity (Moreno and Nurse 1994; Correa-Bordes and Nurse 1995).

How does the cdc2-cdc13 complex prevent rereplication during normal cell cycles? The simplest explanation is that the cdc2-cdc13 complex phosphorylates and inhibits a key protein, or proteins, required for replication initiation (indicated as protein X in Fig. 8). cdc2-cdc13 is activated to a low level as soon as cells pass Start (Creanor and Mitchison 1994; Hayles and Nurse 1995). The rapid phosphorylation of specific substrates at this point would ensure that further initiation of DNA replication was prevented during S phase and G_2. Candidate substrates of the cdc2-cdc13 complex would include cdc18, which has been shown to have an important role in controlling replication initiation, or members of the Cdc46p/Mcm family of proteins likely to make up RLF activity. The direct inhibition of ORC-associated proteins, such as cdc18, might prevent individual replicons firing more than once per cycle, whereas phosphorylation of RLF components might have more global effects. As discussed earlier in this chapter (see Section IV), gross overproduction of cdc18 leads to repeated rounds of DNA replication in the absence of mitosis, bypassing cdc10 function at Start (Nishitani and Nurse 1995). Analysis of the rereplicated DNA by pulse-field gel electrophoresis (PFGE) assay shows that replication is incomplete, however, suggesting that additional factors other than cdc18 are required for successful DNA replication. Activation of the rad/hus checkpoint as a result of synthesis of defective DNA provides a simple explanation for the G_2 arrest seen when cdc18 is overproduced (Kelly et al. 1993).

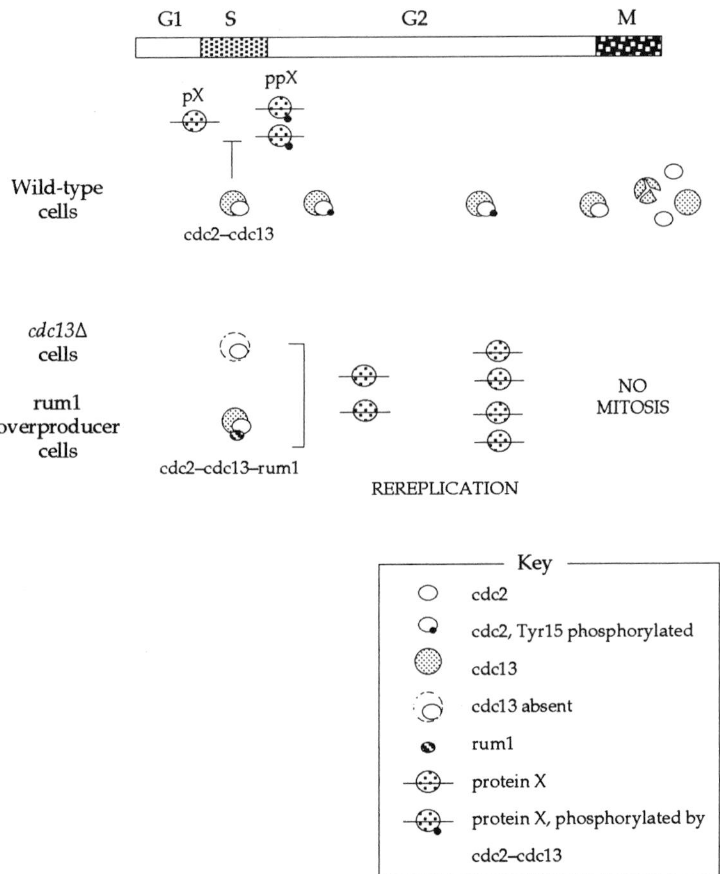

Figure 8 A model for preventing replication in G_2 cells. Activity of the cdc2-cdc13 kinase complex is required to prevent rereplication of DNA in G_2. In *cdc13Δ* cells, and in cells overproducing the rum1 protein, a potent inhibitor of cdc2-cdc13 activity, the absence of cdc2-cdc13 activity leads to rereplication in the absence of mitosis. The block to rereplication may be exerted via phosphorylation of a replication factor (protein X, indicated as pX) by cdc2-cdc13 immediately following S-phase initiation. At this time, rum1 is inactive but cdc2-cdc13 activity is insufficient to drive cells into mitosis. The identity of protein X is unknown, but candidates include cdc18 and/or any of the Cdc46p/Mcm family of proteins (see text).

E. Linking the S-M and Rereplication Checkpoints

One feature of the model discussed in the previous paragraph is that cdc2-cdc13 phosphorylation of the target replication factor (protein X in Fig. 8) ought to occur early in S phase, just after replication is initiated.

At this point in the cycle, rum1 activity must be low enough to allow cdc2-cdc13 to phosphorylate protein X to prevent rereplication but insufficient to drive cells into mitosis. In those mutants that block the initiation of DNA replication but fail to activate the rad/hus checkpoint system (such as in *pol1Δ* or *cdc18Δ* cells), it is possible that cdc2-cdc13 activity may continue to increase unchecked, with the eventual result that mitosis begins before replication is complete. If this is the case, then analysis of the phosphorylation state of cdc2 in *pol1Δ* or *cdc18Δ* cells under such circumstances may be revealing.

VII. CONCLUDING REMARKS

Investigation of the control of the fission yeast cell cycle has revealed a central role for the cdc2 cyclin-dependent protein kinase in controlling the onset of both S phase and mitosis and also in ensuring that these two events occur in the correct temporal sequence. Important challenges for the future will be to understand more fully how the cdc2 protein kinase is regulated throughout the cell cycle, in particular at the G_1-S transition, to establish how S phase and mitosis are actually brought about, and to work out how the cell monitors whether either S phase or mitosis is in progress and communicates this information to the more global cell cycle controls centered on cdc2. Given the effective tools that can be brought to bear on these problems in fission yeast, rapid progress can be expected in the future.

ACKNOWLEDGMENTS

We thank our friends and colleagues for their advice and encouragement during the writing of this chapter in Oxford (1989–1990) and London and Edinburgh (1995–1996). In particular, we thank Gennaro D'Urso, Fred Chang, Jaime Correa-Bordes, Karim Labib, Dan Fisher, Alison Woollard, Paul Kersey, Pamela Harris, Sergio Moreno, Itaru Samejima, Sophie Stettler, Peter Fantes, Masayuki Yamamoto, Hiroto Okayama, and Akio Sugino for communicating unpublished results, and Jacky Hayles, Bodo Stern, and Buzz Baum for reading and commenting on the manuscript. S.M. is supported by the Wellcome Trust. Research in P.N.'s laboratory is supported by the Imperial Cancer Research Fund.

REFERENCES

Adachi, Y., T. Toda, O. Niwa, and M. Yanagida. 1986. Differential expressions of essential and non-essential α-tubulin genes in *Schizosaccharomyces pombe. Mol. Cell. Biol.* **6:** 2168.

Aiba, H., H. Yamada, R. Ohmiya, and T. Mizuno. 1995. The osmo-inducible *gpd1*$^{+}$ gene is a target of the signaling pathway involving Wis1 MAP-kinase kinase in fission yeast. *FEBS Lett.* **376:** 199.

Alfa, C.E., B. Ducommun, D. Beach, and J.S. Hyams. 1990. Distinct nuclear and spindle pole body populations of cyclin-cdc2 in fission yeast. *Nature* **347:** 680.

Aligue, R., H. Akhavan-Niak, and P. Russell. 1994. A role for Hsp90 in cell cycle control: Wee1 tyrosine kinase activity requires interaction with Hsp90. *EMBO J.* **13:** 6099.

Al-Khodairy, F. and A.M. Carr. 1992. DNA repair mutants defining G_2 checkpoint pathways in *Schizosaccharomyces pombe. EMBO J.* **11:** 1343.

Al-Khodairy, F., T. Enoch, I.M. Hagan, and A.M. Carr. 1995. The *Schizosaccharomyces pombe hus5* gene encodes a ubiquitin conjugating enzyme required for normal mitosis. *J. Cell Sci.* **108:** 475.

Al-Khodairy, F., E. Fotou, K.S. Sheldrick, D.J.F. Griffiths, A.R. Lehmann, and A.M. Carr. 1994. Identification and characterization of new elements involved in checkpoint and feedback controls in fission yeast. *Mol. Biol. Cell* **5:** 147.

Araki, H., S.-H. Leem, A. Phongdara, and A. Sugino. 1995. Dpb11, which interacts with DNA polymerase II (ε) in *Saccharomyces cerevisiae*, has a dual role in S-phase progression and at a cell cycle checkpoint. *Proc. Natl. Acad. Sci.* **92:** 11791.

Aves, S.J., B.W. Durkacz, A. Carr, and P. Nurse. 1985. Cloning, sequencing and transcriptional control of the *Schizosaccharomyces pombe cdc10* "start" gene. *EMBO J.* **4:** 457.

Ayscough, K., J. Hayles, S.A. MacNeill, and P. Nurse. 1992. Cold-sensitive mutants of $p34^{cdc2}$ that suppress a mitotic catastrophe phenotype in fission yeast. *Mol. Gen. Genet.* **232:** 344.

Ayté, J., J.F. Leis, A. Herrera, E. Tang, H. Yang, and J.A. DeCaprio. 1995. The *Schizosaccharomyces pombe* MBF complex requires heterodimerization for entry into S phase. *Mol. Cell. Biol.* **15:** 2589.

Balasubramanian, M.K., D.M. Helfman, and S.M. Hemmingsen. 1992. A new tropomyosin essential for cytokinesis in the fission yeast *S. pombe. Nature* **360:** 84.

Balasubramanian, M.K., B.R. Hirani, J.D. Burke, and K.L. Gould. 1994. The *Schizosaccharomyces pombe cdc3*$^{+}$ gene encodes a profilin essential for cytokinesis. *J. Cell Biol.* **125:** 1289.

Barbet, N.C. and A.M. Carr. 1993. Fission yeast *wee1* protein kinase is not required for DNA damage-dependent mitotic arrest. *Nature* **364:** 824.

Barker, D.G., J.M. White, and L.H. Johnston. 1987. Molecular characterisation of the DNA ligase gene, *CDC17*, from the fission yeast *Schizosaccharomyces pombe. Eur. J. Biochem.* **162:** 659.

Basi, G. and G. Draetta. 1995a. $p13^{suc1}$ of *Schizosaccharomyces pombe* regulates two distinct forms of the mitotic cdc2 kinase. *Mol. Cell. Biol.* **15:** 2028.

———. 1995b. The cdc2 kinase: Structure, activation, and its role at mitosis in vertebrate cells. In *Cell cycle control* (ed. C. Hutchison and D.M. Glover), p. 106. IRL Press, Oxford, United Kingdom.

Beach, D., B. Durkacz, and P. Nurse. 1982. Functionally homologous cell cycle control genes in budding and fission yeast. *Nature* **300:** 706.

Bell, S.P. and B. Stillman. 1992. ATP-dependent recognition of eukaryotic origins of DNA replication by a multiprotein complex. *Nature* **357:** 128.

Bell, S.P., J. Mitchell, J. Leber, R. Kobayashi, and B. Stillman. 1995. The multidomain structure of Orc1p reveals similarity to regulators of DNA replication and transcriptional silencing. *Cell* **83:** 563.

Blow, J.J. 1995. S-phase and its regulation. In *Cell cycle control* (ed. C. Hutchison and D.M. Glover), p. 177. IRL Press, Oxford, Uited Kingdom.

Blumer, K.J. and G.L. Johnson. 1994. Diversity in function and regulation of MAP kinase pathways. *Trends Biochem. Sci.* **19:** 236.

Booher, R. and D. Beach. 1986. Site-specific mutagenesis of *cdc2*+, a cell cycle control gene of the fission yeast *Schizosaccharomyces pombe*. *Mol. Cell. Biol.* **6:** 3523.

———. 1987. Interaction between *cdc13*+ and *cdc2*+ in the control of mitosis in fission yeast: Dissociation of the G_1 and G_2 roles of the *cdc2*+ protein kinase. *EMBO J.* **6:** 3441.

———. 1988. Involvement of *cdc13*+ in mitotic control in *Schizosaccharomyces pombe*: possible interaction of the gene product with microtubules. *EMBO J.* **7:** 2321.

———. 1989. Involvement of a type 1 protein phosphatase encoded by *bws1*+ in fission yeast mitotic control. *Cell* **57:** 1009.

Booher, R.N, C.E. Alfa, J.S. Hyams, and D.H. Beach. 1989. The fission yeast cdc2/cdc13/suc1 protein kinase: Regulation of catalytic activity and nuclear localization. *Cell* **58:** 485.

Bouvier, D. and G. Baldacci. 1995. The N-terminus of fission yeast DNA polymerase α contains a basic pentapeptide that acts in vivo as a nuclear localization signal. *Mol. Biol. Cell* **6:** 1697.

Brizuela, L., G. Draetta, and D. Beach. 1987. $p13^{suc1}$ acts in the fission yeast cell division cycle as a component of the $p34^{cdc2}$ protein kinase. *EMBO J.* **6:** 3507.

Broek, D., R. Bartlett, K. Crawford, and P. Nurse. 1991. Involvement of $p34^{cdc2}$ in establishing the dependency of S phase on mitosis. *Nature* **349:** 388.

Buck, V., R. Russell, and J.B.A. Millar. 1995. Identification of a cdk-activating kinase in fission yeast. *EMBO J.* **14:** 6173.

Bueno, A. and P. Russell. 1993a. Two fission yeast B-type cyclins, Cig2 and Cdc13, have different functions in mitosis. *Mol. Cell. Biol.* **13:** 2286.

———. 1993b. A fission yeast B-type cyclin functioning early in the cell cycle (erratum). *Cell* **73:** 1050.

———. 1994. Two fission yeast B-type cyclins, Cig2 and Cdc13, have different functions in mitosis (author's correction). *Mol. Cell. Biol.* **14:** 869.

Bueno, A., H. Richardson, S.I. Reed, and P. Russell. 1991. A fission yeast B-type cyclin functioning early in the cell cycle. *Cell* **66:** 149.

Burbelo, P.D. and A. Hall. 1995. Hot numbers in signal transduction. *Curr. Biol.* **5:** 95.

Caddle, M.S. and M.P. Calos. 1994. Specific initiation at an origin of replication from *Schizosaccharomyces pombe*. *Mol. Cell. Biol.* **14:** 1796.

Caligiuri, M. and D. Beach. 1993. sct1 functions in partnership with cdc10 in a transcription complex that activates cell cycle START and inhibits differentiation. *Cell* **72:** 607.

Carpenter, P.B., P.R. Mueller, and W.G. Dunphy. 1996. Role for a *Xenopus* Orc2-related protein in controlling DNA replication. *Nature* **379:** 357.

Carr, A.M. 1995. DNA structure checkpoints in fission yeast. *Semin. Cell Biol.* **6:** 65.

Carr, A.M., S.A. MacNeill, J. Hayles, and P. Nurse. 1989. Molecular cloning and sequence analysis of mutant alleles of the fission yeast *cdc2* protein kinase gene: Implications for *cdc2*+ protein structure and function. *Mol. Gen. Genet.* **218:** 41.

Carr, A.M., M. Moudjou, N.J. Bentley, and I.M. Hagan. 1995. The *chk1* pathway is required to prevent mitosis following cell cycle arrest at start. *Curr. Biol.* **5:** 1179.

Carr, A.M., S.M. Dorrington, J. Hindley, G.A. Phear, S.J. Aves, and P. Nurse. 1994.

Analysis of a histone H2A variant from fission yeast: Evidence for a role in chromosome stability. *Mol. Gen. Genet.* **245:** 628.

Chang, F. and P. Nurse. 1996. How fission yeast fission in the middle. *Cell* **84:** 191.

Chang, F., A. Woollard, and P. Nurse. 1996. Isolation and characterization of fission yeast mutants defective in the assembly and placement of the contractile actin ring. *J. Cell Sci.* **109:** 131.

Chappell, T.G. and G. Warren. 1989. A galactosyltransferase from the fission yeast *Schizosaccharomyces pombe. J. Cell Biol.* **109:** 2693.

Choe, J., T. Schuster, and M. Grunstein. 1985. Organization, primary structure, and evolution of histone H2A and H2B genes of the fission yeast *Schizosaccharomyces pombe. Mol. Cell. Biol.* **5:** 3261.

Chong, J.P.J., H.M. Mahbubani, C.-Y. Khoo, and J.J. Blow. 1995. Purification of an MCM-containing complex as a component of the DNA replication licensing system. *Nature* **375:** 418.

Clyne, R.K. and T.J. Kelly. 1995. Genetic analysis of an ARS element from the fission yeast *Schizosaccharomyces pombe. EMBO J.* **14:** 6348.

Coleman, T.R., Z. Tang, and W.G. Dunphy. 1993. Negative regulation of the Wee1 protein kinase by direct action of the Nim1/Cdr1 mitotic inducer. *Cell* **72:** 919.

Connolly, T. and D. Beach. 1994. Interaction between the Cig1 and Cig2 B-type cyclins in the fission yeast cell cycle. *Mol. Cell. Biol.* **14:** 768.

Correa-Bordes, J. and P. Nurse. 1995. $p25^{rum1}$ orders S phase and mitosis by acting as an inhibitor of the $p34^{cdc2}$ mitotic kinase. *Cell* **83:** 1001.

Coxon, A., K. Maundrell, and S.E. Kearsey. 1992. Fission yeast *cdc21⁺* belongs to a family of proteins involved in an early step of chromosome replication. *Nucleic Acids Res.* **20:** 5571.

Creanor, J. and J.M. Mitchison. 1994. The kinetics of H1 histone kinase activation during the cell cycle of wild-type and *wee* mutants of the fission yeast *Schizosaccharomyces pombe. J. Cell Sci.* **107:** 1197.

Damagnez, V., T.P. Mäkelä, and G. Cottarel. 1995. *Schizosaccharomyces pombe* Mop1-Mcs2 is related to mammalian CAK. *EMBO J.* **14:** 6164.

Damagnez, V., J. Tillit, A.-M. de Recondo, and G. Baldacci. 1991. The *POL1* gene from the fission yeast *Schizosaccharomyces pombe* shows conserved amino acid blocks specific for eukaryotic DNA polymerases alpha. *Mol. Gen. Genet.* **226:** 182.

Demeter, J., M. Morphew, and S.Sazer. 1995. A mutation in the RCC1-related protein pim1 results in nuclear envelope fragmentation in fission yeast. *Proc. Natl. Acad. Sci.* **92:** 1436.

Den Haese, G.J., N. Walworth, A.M. Carr, and K.L. Gould. 1995. The Wee1 protein kinase regulates T14 phosphorylation of fission yeast Cdc2. *Mol. Biol. Cell* **6:** 371.

Diffley, J.F.X. and J.H. Cocker. 1992. Protein-DNA interactions at a yeast replication origin. *Nature* **357:** 169.

Dowell, S.J., P. Romanowski, and J.F.X. Diffley. 1994. Interaction of Dbf4, the Cdc7 protein kinase regulatory subunit, with yeast replication origins in vivo. *Science* **265:** 1243.

Dubey, D.D., J. Zhu, D.L. Carlson, K. Sharma, and J.A. Huberman. 1994. Three ARS elements contribute to the *ura4* replication origin region in the fission yeast *Schizosaccharomyces pombe. EMBO J.* **13:** 3638.

Ducommun, B., G. Draetta, P. Young, and D. Beach. 1990. Fission yeast cdc25 is a cell cycle regulated protein. *Biochem. Biophys. Res. Commun.* **167:** 301.

Durkacz, B., A. Carr, and P. Nurse. 1986. Transcription of the *cdc2* cell cycle control gene of the fission yeast *Schizosaccharomyces pombe*. *EMBO J.* **5:** 369.

Durkacz, B., D. Beach, J. Hayles, and P. Nurse. 1985. The fission yeast cell cycle control gene *cdc2:* Structure of the *cdc2* region. *Mol. Gen. Genet.* **201:** 543.

D'Urso, G. and P. Nurse. 1995. Checkpoints in the cell cycle of fission yeast. *Curr. Opin. Genet. Dev.* **5:** 12.

D'Urso, G., B. Grallert and P. Nurse. 1995. DNA polymerase alpha, a component of the replication initiation complex, is essential for the checkpoint coupling S phase to mitosis in fission yeast. *J. Cell Sci.* **108:** 3109.

Ekwall, K., J.-P. Javerzat, A. Lorentz, H. Schmidt, G. Cranston, and R. Allshire. 1995. The chromodomain protein Swi6: A key component at fission yeast centromeres. *Science* **269:** 1429.

Endicott, J.A., M.E. Noble, E.F. Garman, N. Brown, B. Rasmussen, P. Nurse, and L.N. Johnson. 1995. The crystal structure of p13^{suc1}, a p34^{cdc2}-interacting cell cycle control protein. *EMBO J.* **14:** 1004.

Enoch, T. and P. Nurse. 1990. Mutation of fission yeast cell cycle control genes abolishes dependence of mitosis on DNA replication. *Cell* **60:** 665.

Enoch, T., A.M. Carr, and P. Nurse. 1992. Fission yeast genes involved in coupling mitosis to the completion of DNA replication. *Genes Dev.* **6:** 2035.

Enoch, T., K. Gould, and P. Nurse. 1991. Mitotic checkpoint control in fission yeast. *Cold Spring Harbor Symp. Quant. Biol.* **56:** 406.

Fankhauser, C. and V. Simanis. 1993. The *Schizosaccharomyces pombe cdc14* gene is required for septum formation and can also inhibit nuclear division. *Mol. Biol. Cell* **4:** 531.

———. 1994. The *cdc7* protein kinase is a dosage dependent regulator of septum formation in fission yeast. *EMBO J.* **13:** 3011.

Fankhauser, C., J. Marks, A. Reymond, and V. Simanis. 1993. The *S. pombe cdc16* gene is required both for maintenance of p34^{cdc2} kinase activity and regulation of septum formation: A link between mitosis and cytokinesis? *EMBO J.* **12:** 2697.

Fankhauser, C., A. Reymond, L. Cerutti, S. Utzig, K. Hofmann, and V. Simanis. 1995. The *S. pombe cdc15* gene is a key element in the reorganization of F-actin at mitosis. *Cell* **82:** 435.

Fantes, P. 1979. Epistatic gene interactions in the control of division in fission yeast. *Nature* **279:** 428.

———. 1981. Isolation of cell size mutants of a fission yeast by a new selective method: Characterization of mutants and implications for division control mechanisms. *J. Bacteriol.* **146:** 746.

———. 1983. Control of timing of cell cycle events in fission yeast by the *wee1*$^{+}$ gene. *Nature* **302:** 153.

———. 1989. Cell cycle controls. In *Molecular biology of the fission yeast* (ed. A. Nasim et al.), p. 127. Academic Press, San Diego.

Fantes, P.A. and P. Nurse. 1977. Control of cell size at division in fission yeast by a growth-modulated size control over nuclear division. *Exp. Cell Res.* **107:** 377.

Fantes, P.A., E. Warbrick, D.A. Hughes, and S.A. MacNeill. 1991. New elements in the mitotic control of the fission yeast *Schizosaccharomyces pombe*. *Cold Spring Harbor Symp. Quant. Biol.* **56:** 605.

Featherstone, C. and P. Russell. 1991. Fission yeast p107^{wee1} mitotic inhibitor is a tyrosine/serine kinase. *Nature* **349:** 808.

Feilotter, H., P. Nurse, and P.G. Young. 1991. Genetic and molecular analysis of *cdr1/nim1* in *Schizosaccharomyces pombe*. *Genetics* **127:** 309.

Fenech, M., A.M. Carr, J. Murray, F.Z. Watts, and A.R. Lehmann. 1991. Cloning and characterization of the *rad4* gene of *Schizosaccharomyces pombe:* A gene showing short regions of sequence similarity to the human *XRCC1* gene. *Nucleic Acids Res.* **19:** 6737.

Fernandez-Sarabia, M.-J., C. McInerny, P. Harris, C. Gordon, and P.A. Fantes. 1993. The cell cycle genes *cdc22*$^{+}$ and *suc22*$^{+}$ of the fission yeast *Schizosaccharomyces pombe* encode the large and small subunits of ribonucleotide reductase. *Mol. Gen. Genet.* **238:** 241.

Fikes, J.D., D.M. Becker, F. Winston, and L. Guarente. 1990. Striking conservation of TFIID in *Schizosaccharomyces pombe* and *Saccharomyces cerevisiae*. *Nature* **346:** 291.

Fisher, D.L. and P. Nurse. 1995. Cyclins of the fission yeast *Schizosaccharomyces pombe*. *Semin. Cell Biol.* **6:** 73.

———. 1996. A single fission yeast mitotic cyclin B-p34^{cdc2} kinase promotes both S-phase and mitosis in the absence of G1 cyclins. *EMBO J.* **15:** 850.

Fleig, U.N. and P. Nurse. 1991. Expression of a dominant negative allele of *cdc2* prevents activation of the endogenous p34^{cdc2} kinase. *Mol. Gen. Genet.* **226:** 432.

Fleig, U.N., K.L. Gould, and P. Nurse. 1992. A dominant negative allele of p34^{cdc2} shows altered phosphoamino acid content and sequesters p56^{cdc13} cyclin. *Mol. Cell. Biol.* **12:** 2295.

Ford, J.C., F. Al-Khodairy, E. Fotou, K.S. Sheldrick, D.J.F. Griffiths, and A.M. Carr. 1994. 14-3-3 protein homologs required for the DNA damage checkpoint in fission yeast. *Science* **265:** 533.

Forsburg, S.L. and P. Nurse. 1994a. Analysis of the *Schizosaccharomyces pombe* cyclin puc1: evidence for a role in cell cycle exit. *J. Cell Sci.* **107:** 601.

———. 1994b. The fission yeast *cdc19*$^{+}$ gene encodes a member of the MCM family of replication proteins. *J. Cell Sci.* **107:** 2779.

Francesconi, S., W.C. Copeland, and T.S.-F. Wang. 1993a. In vivo species specificity of DNA polymerase α. *Mol. Gen. Genet.* **241:** 457.

Francesconi, S., A.-M. de Recondo, and G. Baldacci. 1995. DNA polymerase δ is required for the replication feedback control of cell cycle progression in *Schizosaccharomyces pombe*. *Mol. Gen. Genet.* **246:** 561.

Francesconi, S., H. Park, and T.S.-F. Wang. 1993b. Fission yeast with DNA polymerase δ temperature-sensitive alleles exhibits cell division cycle phenotype. *Nucleic Acids Res.* **21:** 3821.

Gallagher, I.M., C.E. Alfa, and J.S. Hyams. 1993. p63^{cdc13}, a B-type cyclin, is associated with both the nucleolar and chromatin domains of the fission yeast nucleus. *Mol. Biol. Cell* **4:** 1087.

Gavin, K.A., M. Hidaka, and B. Stillman. 1995. Conserved initiator proteins in eukaryotes. *Science* **270:** 1667.

Gordon, C.B. and P.A. Fantes. 1986. The *cdc22* gene of *Schizosaccharomyces pombe* encodes a cell-cycle regulated transcript. *EMBO J.* **5:** 2981.

Gordon, C., G. McGurk, P. Dillon, C. Rosen, and N.D. Hastie. 1993. Defective mitosis due to a mutation in the gene for a fission yeast 26S protease subunit. *Nature* **366:** 355.

Gould, K.L. and P. Nurse. 1989. Tyrosine phosphorylation of the fission yeast *cdc2*$^{+}$ protein kinase regulates entry into mitosis. *Nature* **342:** 39.

Gould, K.L., S. Moreno, D.J. Owen, S. Sazer, and P. Nurse. 1991. Phosphorylation of Thr167 is required for *Schizosaccharomyces pombe* p34^{cdc2} function. *EMBO J.* **10:** 3297.

Grallert, B. and M. Sipiczki. 1989. Initiation of the second meiotic division in *Schizosaccharomyces pombe* shares common functions with that of mitosis. *Curr. Genet.* **15:** 231.

———. 1990. Dissociation of meiotic and mitotic roles of the fission yeast *cdc2* gene. *Mol. Gen. Genet.* **222:** 473.

———. 1991. Common genes and pathways in the regulation of the mitotic and meiotic cell cycles of *Schizosaccharomyces pombe. Curr. Genet.* **20:** 199.

Griffiths, D.J.F., N.C. Barbet, S. McCready, A.R. Lehmann, and A.M. Carr. 1995. Fission yeast *rad17:* A homologue of budding yeast *RAD24* that shares regions of sequence similarity with DNA polymerase accessory proteins. *EMBO J.* **14:** 5812.

Gutz, H., H. Heslot, U. Leupold, and N. Loprieno. 1974. *Schizosaccharomyces pombe.* In *Handbook of genetics* (ed. R.C. King), pp. 395. Plenum Press, New York.

Hagan, I.M. and J.S. Hyams. 1988. The use of cell division cycle mutants to investigate the control of microtubule distribution in the fission yeast *Schizosaccharomyces pombe. J. Cell Sci.* **89:** 343.

Hagan, I. and M. Yanagida. 1990. Novel potential mitotic motor protein encoded by the fission yeast *cut7*$^+$ gene. *Nature* **347:** 563.

———. 1992. Kinesin-related cut7 protein associates with mitotic and meiotic spindles in fission yeast. *Nature* **356:** 74.

———. 1995. The product of the spindle formation gene *sad1*$^+$ associates with the fission yeast spindle pole body and is essential for viability. *J. Cell Biol.* **129:** 1033.

Hagan, I., J. Hayles, and P. Nurse. 1988. Cloning and sequencing of the cyclin-related *cdc13*$^+$ gene and a cytological study of its role in fission yeast mitosis. *J. Cell Sci.* **91:** 587.

Hartwell, L.H., J. Culotti, J.R. Pringle, and B.J. Reid. 1974. Genetic control of the cell division cycle in yeast. *Science* **183:** 46.

Hayles, J. and P. Nurse. 1992. Genetics of the fission yeast *Schizosaccharomyces pombe. Annu. Rev. Genet.* **26:** 373.

———. 1995. A pre-Start checkpoint preventing mitosis in fission yeast acts independently of p34^{cdc2} tyrosine phosphorylation. *EMBO J.* **14:** 2760.

Hayles, J., S. Aves, and P. Nurse. 1986a. *suc1* is an essential gene involved in both the cell cycle and growth in fission yeast. *EMBO J.* **5:** 3373.

Hayles, J., D. Beach, B. Durkacz, and P. Nurse. 1986b. The fission yeast cell cycle control gene *cdc2:* Isolation of a sequence *suc1* that suppresses *cdc2* mutant function. *Mol. Gen. Genet.* **202:** 291.

Hayles, J., D. Fisher, A. Woollard, and P. Nurse. 1994. Temporal order of S phase and mitosis in fission yeast is determined by the state of the p34^{cdc2}-mitotic B cyclin complex. *Cell* **78:** 813.

Hennessy, K.M., C.D. Clark, and D. Botstein. 1990. Subcellular localization of yeast CDC46 varies with the cell cycle. *Genes Dev.* **4:** 2252.

Herskowitz, I. 1995. MAP kinase pathways in yeast: For mating and more. *Cell* **80:** 187.

Hindley, J. and G.A. Phear. 1984. Sequence of the cell division gene *CDC2* from *Schizosaccharomyces pombe:* Patterns of splicing and homology to protein kinases. *Gene* **31:** 129.

Hindley, J., G. Phear, M. Stein, and D. Beach. 1987. *suc1*$^+$ encodes a predicted 13 kilodalton protein that is essential for cell viability and is directly involved in the division cycle of *Schizosaccharomyces pombe*. *Mol. Cell. Biol.* **7:** 504.

Hirano, T., Y. Hiraoka, and M. Yanagida. 1988. A temperature-sensitive mutation of the *Schizosaccharomyces pombe* gene *nuc2*$^{+}$ that encodes a nuclear scaffold-like protein blocks spindle elongation in mitotic anaphase. *J. Cell Biol.* **106:** 1171.

Hirano, T., S. Funahashi, T. Uemura, and M. Yanagida. 1986. Isolation and characterization of *Schizosaccharomyces pombe cut* mutants that block nuclear division but not cytokinesis. *EMBO J.* **5:** 2973.

Hirano, T., N. Kinoshita, K. Morikawa, and M. Yanagida. 1990. Snap helix with knob and hole: Essential repeats in *S. pombe* nuclear protein *nuc2*$^{+}$. *Cell* **60:** 319.

Hiraoka, Y., T. Toda, and M. Yanagida. 1984. The *NDA3* gene of fission yeast encodes β-tubulin: A cold-sensitive *nda3* mutation reversibly blocks spindle formation and chromosome movement in mitosis. *Cell* **39:** 349.

Hofmann, J.F.X. and D. Beach. 1994. *cdt1* is an essential target of the Cdc10/Sct1 transcription factor: Requirement for DNA replication and inhibition of mitosis. *EMBO J.* **13:** 425.

Hoheisel, J.D., E. Maier, R. Mott, L. McCarthy, A.V. Grigoriev, L.C. Schalkwyk, D. Nizetic, F. Francis, and H. Lehrach. 1993. High resolution cosmid and P1 maps spanning the 14 Mb genome of the fission yeast *S. pombe*. *Cell* **73:** 109.

Horio, T., S. Uzawa, M.K. Jung, B.R. Oakley, K. Tanaka, and M. Yanagida. 1991. The fission yeast γ-tubulin is essential for mitosis and is localized at microtubule organizing centers. *J. Cell Sci.* **99:** 693.

Huberman, J.A. 1995. A licence to replicate. *Nature* **375:** 360.

Hughes, D.A., S.A. MacNeill, and P.A. Fantes. 1992. Molecular cloning and sequence analysis of *cdc27*$^{+}$ required for the G2-M transition in the fission yeast *Schizosaccharomyces pombe*. *Mol. Gen. Genet.* **231:** 401.

Hunter, T. 1995. When is a lipid kinase not a lipid kinase? When it is a protein kinase. *Cell* **83:** 1.

Irniger, S., S. Piatti, C. Michaelis, and K. Nasmyth. 1995. Genes involved in sister chromatid separation are needed for B-type cyclin proteolysis in budding yeast. *Cell* **81:** 269.

Jackson, A.L., P.M.B. Pahl, K. Harrison, J. Rosamond, and R.A. Sclafani. 1993. Cell cycle regulation of the yeast Cdc7 protein kinase by association with the Dbf4 protein. *Mol. Cell. Biol.* **13:** 2899.

Jimenez, G., J. Yucel, R. Rowley, and S. Subramani. 1992. The *rad3*$^{+}$ gene of *Schizosaccharomyces pombe* is involved in multiple checkpoint functions and in DNA repair. *Proc. Natl. Acad. Sci.* **89:** 4952.

Johnson, G.L. and R.R. Vaillancourt. 1994. Sequential protein kinase reactions controlling cell growth and differentiation. *Curr. Opin. Cell Biol.* **6:** 230.

Kanter-Smoler, G., K.E. Knudsen, G. Jimenez, P. Sunnerhagen, and S. Subramani. 1995. Separation of phenotypes in mutant alleles of the *Schizosaccharomyces pombe* cell cycle checkpoint gene *rad1*$^{+}$. *Mol. Biol. Cell* **6:** 1793.

Kaüfer, N.F., V. Simanis, and P. Nurse. 1985. Fission yeast *Schizosaccharomyces pombe* correctly excises a mammalian RNA transcript intervening sequence. *Nature* **318:** 78.

Kelly, T.J., G.S. Martin, S.L. Forsburg, R.J. Stephen, A. Russo, and P. Nurse. 1993. The fission yeast *cdc18*$^{+}$ gene product couples S phase to START and mitosis. *Cell* **74:** 371.

Kelman, Z. and M. O'Donnell. 1994. DNA replication: Enzymology and mechanisms. *Curr. Opin. Genet. Dev.* **4:** 185.

Kinoshita, N., M. Goebl, and M. Yanagida. 1991. The fission yeast *dis3*$^{+}$ gene encodes a 110-kDa essential protein implicated in mitotic control. *Mol. Cell. Biol.* **11:** 5839.

Kinoshita, N., H. Ohkura, and M. Yanagida. 1990. Distinct, essential roles of type 1 and 2A protein phosphatases in the control of the fission yeast cell division cycle. *Cell* **63:** 405.

Kinoshita, N., H. Yamano, H. Niwa, T. Yoshida, and M. Yanagida. 1993. Negative regulation of mitosis by the fission yeast protein phosphatase ppa2. *Genes Dev.* **7:** 1059.

Kohli, J. 1987. Genetic nomenclature and gene list of the fission yeast *Schizosaccharomyces pombe. Curr. Genet.* **11:** 575.

Kovelman, R. and P. Russell. 1996. Stockpiling of Cdc25 during a DNA replication checkpoint arrest in *Schizosaccharomyces pombe. Mol. Cell. Biol.* **16:** 86.

Krek, W., J. Marks, N. Schmitz, E.A. Nigg, and V. Simanis. 1992. Vertebrate p34^{cdc2} phosphorylation site mutants: Effects upon cell cycle progression in the fission yeast *Schizosaccharomyces pombe. J. Cell Sci.* **102:** 43.

Kubota, Y., S. Mimura, S. Nishimoto, H. Takisawa, and H. Nojima. 1995. Identification of the yeast MCM3-related protein as a component of *Xenopus* DNA replication licensing factor. *Cell* **81:** 601.

Kumada, K., S. Su, M. Yanagida, and T. Toda. 1995. Fission yeast TPR-family protein nuc2 is required for G1-arrest upon nitrogen starvation and is an inhibitor of septum formation. *J. Cell Sci.* **108:** 895.

Labib, K., S. Moreno, and P. Nurse. 1995a. Interaction of *cdc2* and *rum1* regulates start and S-phase in fission yeast. *J. Cell Sci.* **108:** 3285.

Labib, K., R.A. Craven, K. Crawford, and P. Nurse. 1995b. Dominant mutants identify new roles for p34^{cdc2} in mitosis. *EMBO J.* **14:** 2155.

Leatherwood, J., A. Lopez-Girona, and P. Russell. 1996. Interaction of Cdc2 and Cdc18 with a fission yeast ORC2-like protein. *Nature* **379:** 360.

Lee, M.S., T. Enoch, and H. Piwnica-Worms. 1994. *mik1*$^{+}$ encodes a tyrosine kinase that phosphorylates p34^{cdc2} on tyrosine 15. *J. Biol. Chem.* **269:** 30530.

Leupold, U. 1970. Genetical methods for *Schizosaccharomyces pombe. Methods Cell Physiol.* **4:** 169.

Liang, C., M. Weinreich, and B. Stillman. 1995. ORC and Cdc6p interact and determine the frequency of initiation of DNA replication in the genome. *Cell* **81:** 667.

Lieberman, H.B. 1995. Extragenic suppressors of *Schizosaccharomyces pombe rad9* mutations uncouple radioresistance and hydroxyurea sensitivity from cell cycle checkpoint control. *Genetics* **141:** 107.

Lieberman, H.B., K.M. Hopkins, M. Laverty, and H.M. Chu. 1992. Molecular cloning and analysis of *Schizosaccharomyces pombe rad9*, a gene involved in DNA repair and mutagenesis. *Mol. Gen. Genet.* **232:** 367.

Lowndes, N.F., C.J. McInerny, A.L. Johnson, P.A. Fantes, and L.H. Johnston. 1992. Control of DNA synthesis genes in fission yeast by the cell cycle gene *cdc10*$^{+}$. *Nature* **355:** 449.

Lundgren, K., N, Walworth, R. Booher, M. Dembski, M. Kirschner, and D. Beach. 1991. mik1 and wee1 cooperate in the inhibitory tyrosine phosphorylation of cdc2. *Cell* **64:** 1111.

Lydall, D. and T. Weinert. 1995. Yeast checkpoint genes in DNA damage processing: Implications for repair and arrest. *Science* **270:** 1488.

MacNeill, S.A. and P.A. Fantes. 1993. Methods for analysis of the fission yeast cell cycle. In *The cell cycle—A practical approach* (ed. P. Fantes and R. Brooks), p. 93. IRL Press, Oxford, United Kingdom.

MacNeill, S.A. and P. Nurse. 1993. Mutational analysis of the fission yeast p34[cdc2] protein kinase gene. *Mol. Gen. Genet.* **236:** 415.

MacNeill, S.A., J. Creanor, and P. Nurse. 1991. Isolation, characterisation and molecular cloning of new mutant alleles of the fission yeast p34[cdc2] protein kinase gene: Identification of temperature-sensitive G2-arresting alleles. *Mol. Gen. Genet.* **229:** 109.

MacNeill, S.A., S. Moreno, N. Reynolds, P. Nurse, and P.A. Fantes. 1996. The fission yeast Cdc1 protein, a homologue of the small subunit of DNA polymerase δ, binds to Pol3 and Cdc27. *EMBO J.* **15:** 4613.

Madine, M.A., C.-Y. Khoo, A.D. Mills, and R.A. Laskey. 1995. MCM3 complex required for cell cycle regulation of DNA replication in vertebrate cells. *Nature* **375:** 421.

Maeda, T., S.M. Wurgler-Murphy, and H. Saito. 1994. A two-component system that regulates an osmosensing MAP kinase cascade in yeast. *Nature* **369:** 242.

Maeda, T., M. Takekawa, and H. Saito. 1995. Activation of yeast PBS2 MAPKK by MAPKKKs or by binding of an SH3-containing osmosensor. *Science* **269:** 554.

Maier, E., J.D. Hoheisel, L. McCarthy, R. Mott, A.V. Grigoriev, A.P. Monaco, Z. Larin, and H. Lehrach. 1992. Complete coverage of the *Schizosaccharomyces pombe* genome in yeast artificial chromosomes. *Nat. Genet.* **1:** 273.

Marks, J., C. Fankhauser, and V. Simanis. 1992a. Genetic interactions in the control of septation in *Schizosaccharomyces pombe*. *J. Cell Sci.* **101:** 801.

Marks, J., C. Fankhauser, A. Reymond, and V. Simanis. 1992b. Cytoskeletal and DNA structure abnormalities result from bypass of requirement for the *cdc10* start gene in the fission yeast *Schizosaccharomyces pombe*. *J. Cell Sci.* **101:** 517.

Martín-Castellanos, C., K. Labib, and S. Moreno. 1996. B-type cyclins regulate G1 progression in fission yeast in opposition to the p25[rum1] CDK inhibitor. *EMBO J.* **15:** 839.

Masai, H., T. Miyake, and K. Arai. 1995. *hsk1*$^+$, a *Schizosaccharomyces pombe* gene related to *Saccharomyces cerevisiae CDC7*, is required for chromosomal replication. *EMBO J.* **14:** 3094.

Matsumoto, S. and M. Yanagida. 1985. Histone gene organization of fission yeast: A common upstream sequence. *EMBO J.* **4:** 3531.

Matsumoto, S., M. Yanagida, and P. Nurse. 1987. Histone transcription in cell cycle mutants of fission yeast. *EMBO J.* **6:** 1093.

Matsumoto, T. and D. Beach. 1991. Premature initiation of mitosis in yeast lacking RCC1 or an interacting GTPase. *Cell* **66:** 347.

———. 1993. Interaction of the *pim1/spi1* mitotic checkpoint with a protein phosphatase. *Mol. Biol. Cell* **4:** 337.

Maundrell, K., A. Hutchison, and S. Shall. 1988. Sequence analysis of ARS elements in fission yeast. *EMBO J.* **7:** 2203.

McCollum, D., M.K. Balasubramanian, L.E. Pelcher, S.M. Hemmingsen, and K.L. Gould. 1995. *Schizosaccharomyces pombe cdc4*$^+$ gene encodes a novel EF-hand protein essential for cytokinesis. *J. Cell Biol.* **130:** 651.

McCully, E.K. and C.F. Robinow. 1971. Mitosis in the fission yeast *Schizosaccharomyces pombe:* A comparative study with light and electron microscopy. *J. Cell Sci.* **9:** 475.

McGowan, C.H. and P. Russell. 1993. Human wee1 kinase inhibits cell division by phosphorylating p34[cdc2] exclusively on Tyr15. *EMBO J.* **12:** 75.

McInerny, C.J., P.J. Kersey, J. Creanor, and P.A. Fantes. 1995. Positive and negative roles for cdc10 in cell cycle gene expression. *Nucleic Acids Res.* **23:** 4761.

Michaely, P. and V. Bennett. 1992. The ANK repeat: A ubiquitous motif involved in macromolecular recognition. *Trends Cell Biol.* **2:** 127.

Millar, J.B.A. and P. Russell. 1992. The cdc25 M-phase inducer: An unconventional protein phosphatase. *Cell* **68:** 407.

Millar, J.B.A., V. Buck, and M.G. Wilkinson. 1995. Pyp1 and Pyp2 PTPases dephosphorylate an osmosensing MAP kinase controlling cell size at division in fission yeast. *Genes Dev.* **9:** 2117.

Millar, J.B.A., G. Lenaers, and P. Russell. 1992a. *Pyp3* PTPase acts as a mitotic inducer in fission yeast. *EMBO J.* **11:** 4933.

Millar, J.B.A., P. Russell, J.E. Dixon, and K.L. Guan. 1992b. Negative regulation of mitosis by two functionally overlapping PTPases in fission yeast. *EMBO J.* **11:** 4943.

Millar, J.B.A., C.H. McGowan, G. Lenaers, R. Jones, and P. Russell. 1991. $p80^{cdc25}$ mitotic inducer is the tyrosine phosphatase that activates $p34^{cdc2}$ kinase in fission yeast. *EMBO J.* **10:** 4301.

Minet, M., P. Nurse, P. Thuriaux, and J.M. Mitchison. 1979. Uncontrolled septation in a cell division cycle mutant of the fission yeast *Schizosaccharomyces pombe. J. Bacteriol.* **137:** 440.

Mitchison, J.M. 1957. The growth of single cells. I. *Schizosaccharomyces pombe. Exp. Cell Res.* **13:** 244.

———. 1970. Physiological and cytological methods for *Schizosaccharomyces pombe. Methods Cell Physiol.* **4:** 131.

———. 1990. The fission yeast, *Schizosaccharomyces pombe. BioEssays* **12:** 189.

Miyake, S., N. Okishio, I. Samejima, Y. Hiraoka, T. Toda, I. Saitoh, and M. Yanagida. 1993. Fission yeast genes $nda1^+$ and $nda4^+$, mutations of which lead to S-phase block, chromatin alteration and Ca^+ suppression, are members of the *CDC46/MCM2* family. *Mol. Biol. Cell* **4:** 1003.

Miyamoto, M., K. Tanaka, and H. Okayama. 1994. $res2^+$, a new member of the $cdc10^+$/*SWI4* family, controls the "start" of mitotic and meiotic cycles in fission yeast. *EMBO J.* **13:** 1873.

Mizukami, T., W.I. Chang, I. Garkavtsev, N. Kaplan, D. Lombardi, T. Matsumoto, O. Niwa, A. Kounoso, M. Yanagida, T.G. Marr, and D. Beach. 1993. A 13 kb resolution cosmid map of the 14 Mb fission yeast genome by nonrandom sequence- tagged site mapping. *Cell* **73:** 121.

Modesti, A., P. Cirri, G. Raugei, L. Carraresi, F. Magherini, G. Manao, G. Camici, and G. Ramponi. 1995. Expression, purification and kinetic behaviour of fission yeast low M_r protein-tyrosine-phosphatase. *FEBS Lett.* **375:** 235.

Molz, L. and D. Beach. 1993. Characterization of the fission yeast *mcs2* cyclin and its associated protein kinase activity. *EMBO J.* **12:** 1723.

Molz, L., R. Booher, P. Young, and D. Beach. 1989. *cdc2* and the regulation of mitosis: Six interacting *mcs* genes. *Genetics* **122:** 773.

Mondesert, O., S. Moreno, and P. Russell. 1994. Low molecular weight protein-tyrosine phosphatases are highly conserved between fission yeast and man. *J. Biol. Chem.* **269:** 27996.

Moreno, S. and P. Nurse. 1994. Regulation of progression through the G1 phase of the cell cycle by the $rum1^+$ gene. *Nature* **367:** 236.

Moreno, S., J. Hayles, and P. Nurse. 1989. Regulation of $p34^{cdc2}$ protein kinase during mitosis. *Cell* **58:** 361.

Moreno, S., P. Nurse, and P. Russell. 1990. Regulation of mitosis by cyclic accumulation of p80^{cdc25} mitotic inducer in fission yeast. *Nature* **344:** 549.

Moussy, G., A.M. de Recondo, and G. Baldacci. 1995. Inter-species DNA polymerase δ chimeras are functional in *Saccharomyces cerevisiae*. *Eur. J. Biochem.* **231:** 45.

Murakami, H. and H. Okayama. 1995. A kinase from fission yeast responsible for blocking mitosis in S phase. *Nature* **374:** 817.

Murray, J.M., A.M. Carr, A.R. Lehmann, and F.Z. Watts. 1991. Cloning and characterisation of the *rad9* DNA repair gene from *Schizosaccharomyces pombe*. *Nucleic Acids Res.* **19:** 3525.

Muzi-Falconi, M. and T.J. Kelly. 1995. Orp1, a member of the Cdc18/Cdc6 family of S-phase regulators, is homologous to a component of the origin recognition complex. *Proc. Natl. Acad. Sci.* **92:** 12475.

Nabeshima, K., H. Kurooka, M. Takeuchi, K. Kinoshita, Y. Nakeseko, and M. Yanagida. 1995. p93^{dis1}, which is required for sister chromatid separation, is a novel microtubule and spindle pole body-associating protein phosphorylated at the Cdc2 target sites. *Genes Dev.* **9:** 1572.

Nakaseko, Y., O. Niwa, and M. Yanagida. 1984. A meiotic mutant of the fission yeast *Schizosaccharomyces pombe* that produces mature asci containing two diploid spores. *J. Bacteriol.* **157:** 334.

Nakashima, N., K. Tanaka, S. Sturm, and H. Okayama. 1995. Fission yeast Rep2 is a putative transcriptional activator subunit for the cell cycle "start" function of Res2-Cdc10. *EMBO J.* **14:** 4794.

Nasmyth, K.A. 1977. Temperature-sensitive lethal mutants in the structural gene for DNA ligase in the yeast *Schizosaccharomyces pombe*. *Cell* **12:** 1109.

———. 1979. A control acting over the initiation of DNA replication in the yeast *Schizosaccharomyces pombe*. *J. Cell Sci.* **36:** 155.

Nasmyth, K. and P. Nurse. 1981. Cell division cycle mutants altered in DNA replication and mitosis in the fission yeast *Schizosaccharomyces pombe*. *Mol. Gen. Genet.* **182:** 119.

Nasmyth, K., P. Nurse, and R.S.S. Fraser. 1979. The effect of cell mass on the cell cycle timing and duration of S-phase in fission yeast. *J. Cell Sci.* **39:** 215.

Ng, L., C.-K. Tan, K.M. Downey, and P.A. Fisher. 1991. Enzymologic mechanism of calf thymus DNA polymerase δ. *J. Biol. Chem.* **266:** 11699.

Nishitani, H. and P. Nurse. 1995. p65^{cdc18} plays a major role controlling the initiation of DNA replication in fission yeast. *Cell* **83:** 397.

Niwa, O. and M. Yanagida. 1988. Universal and essential role of MPF/*cdc2*$^+$. *Nature* **336:** 430.

Nurse, P. 1975. Genetic control of cell size at cell division in yeast. *Nature* **256:** 547.

———. 1985. Cell cycle control genes in yeast. *Trends Genet.* **1:** 51.

———. 1990. Universal control mechanism regulating onset of M-phase. *Nature* **344:** 503.

———. 1994. Ordering S phase and M phase in the cell cycle. *Cell* **79:** 547.

Nurse, P. and Y. Bissett. 1981. Gene required in G1 for commitment to cell cycle and in G2 for control of mitosis in fission yeast. *Nature* **292:** 558.

Nurse, P. and P. Thuriaux. 1977. Controls over the timing of DNA replication during the cell cycle of fission yeast. *Exp. Cell Res.* **107:** 365.

———. 1980. Regulatory genes controlling mitosis in the fission yeast *Schizosaccharomyces pombe*. *Genetics* **96:** 627.

Nurse, P., P. Thuriaux, and K. Nasmyth. 1976. Genetic control of the cell division cycle in the fission yeast *Schizosaccharomyces pombe*. *Mol. Gen. Genet.* **146:** 167.

Obara-Ishihara, T. and H. Okayama. 1994. A B-type cyclin negatively regulates conjugation via interacting with cell cycle start genes in fission yeast. *EMBO J.* **13:** 1863.

Ogden, J.E. and P.A. Fantes. 1986. Isolation of a novel type of mutation in the mitotic control of *Schizosaccharomyces pombe* whose phenotypic expression is dependent on the genetic background and nutritional environment. *Curr. Genet.* **10:** 509.

Ohi, R., D. McCollum, B. Hirani, G.J. Den Haese, X. Zhang, J.D. Burke, K. Turner, and K.L. Gould. 1994. The *Schizosaccharomyces pombe* $cdc5^+$ gene encodes an essential protein with homology to c-Myb. *EMBO J.* **13:** 471.

Ohkura, H. and M. Yanagida. 1991. *S. pombe* gene $sds22^+$ essential for a midmitotic transition encodes a leucine-rich repeat protein that positively modulates protein phosphatase-1. *Cell* **64:** 149.

Ohkura, H., I.M. Hagan, and D.M. Glover. 1995. The conserved *Schizosaccharomyces pombe* kinase plo1, required to form a bipolar spindle, the actin ring, and septum, can drive septum formation in G1 and G2 cells. *Genes Dev.* **9:** 1059.

Ohkura, H. N. Kinoshita, S. Miyatani, T. Toda, and M. Yanagida. 1989. The fission yeast $dis2^+$ gene required for chromosome disjoining encodes one of two putative type 1 protein phosphatases. *Cell* **57:** 997.

Ohkura, H., Y. Adachi, N. Kinoshita, O. Niwa, T. Toda, and M. Yanagida. 1988. Cold-sensitive and caffeine supersensitive mutants of the *Schizosaccharomyces pombe dis* genes implicated in sister chromatid separation during mitosis. *EMBO J.* **7:** 1465.

Ottilie, S., J. Chernoff, G. Hannig, C.S. Hoffman, and R.L. Erikson. 1991. A fission yeast gene encoding a protein with features of protein-tyrosine-phosphatases. *Proc. Natl. Acad. Sci.* **88:** 3455.

———. 1992. The fission yeast genes $pyp1^+$ and $pyp2^+$ encode protein tyrosine phosphatases that negatively regulate mitosis. *Mol. Cell. Biol.* **12:** 5571.

Park, H., R. Davis, and T.S.-F. Wang. 1995. Studies of *Schizosaccharomyces pombe* DNA polymerase α at different stages of the cell cycle. *Nucleic Acids Res.* **23:** 4337.

Park, H., S. Francesconi, and T. S.-F. Wang. 1993. Cell cycle expression of two replicative DNA polymerases α and δ from *Schizosaccharomyces pombe*. *Mol. Biol. Cell* **4:** 145.

Parker, L.L., S. Atherton-Fessler, and H. Piwnica-Worms. 1992. $p107^{wee1}$ is a dual-specificity kinase that phosphorylates $p34^{cdc2}$ on tyrosine 15. *Proc. Natl. Acad. Sci.* **89:** 2917.

Parker, L.L., S.A. Walter, P.G. Young, and H. Piwnica-Worms. 1993. Phosphorylation and inactivation of the mitotic inhibitor Wee1 by the *nim1/cdr1* kinase. *Nature* **363:** 736.

Pignède, G., D. Bouvier, A.-M. de Recondo, and G. Baldacci. 1991. Characterization of the *POL3* gene product from *Schizosaccharomyces pombe* indicates inter-species conservation of the catalytic subunit of DNA polymerase δ. *J. Mol. Biol.* **222:** 209.

Pines, J. and T. Hunter. 1995. Cyclin-dependent kinases: An embarrassment of riches? In *Cell cycle control* (ed. C. Hutchison and D.M. Glover), p. 144. IRL Press, Oxford, United Kingdom.

Plyte, S.E., A. Feoktistova, J.D. Burke, J.R. Woodgett, and K.L. Gould. 1996. *Schizosaccharomyces pombe* $skp1^+$ encodes a protein kinase related to mammalian glycogen synthase kinase 3 and complements a *cdc14* cytokinesis mutant. *Mol. Cell. Biol.* **16:** 179.

Reed, S.I., C.J. Hutchison, and S. MacNeill. 1995. A brief history of the cell cycle. In *Cell cycle control* (ed. C. Hutchison and D.M. Glover), p. 16. IRL Press, Oxford, United Kingdom.

Reymond, A. and V. Simanis. 1993. Domains of p85^{cdc10} required for function of the fission yeast DSC-1 factor. *Nucleic Acids Res.* **21:** 3615.

Reymond, A., J. Marks, and V. Simanis. 1993. The activity of *S. pombe* DSC-1-like factor is cell cycle regulated and dependent on the activity of p34^{cdc2}. *EMBO J.* **12:** 4325.

Reymond, A., S. Schmidt, and V. Simanis. 1992. Mutations in the *cdc10* start gene of *Schizosaccharomyces pombe* implicate the region of homology between *cdc10* and *SWI6* as important for p85^{cdc10} function. *Mol. Gen. Genet.* **234:** 449.

Robinow, C.F. 1977. The number of chromosomes in *Schizosaccharomyces pombe:* Light microscopy of stained preparations. *Genetics* **87:** 491.

Rothstein, R.J. 1983. One-step gene disruption in yeast. *Methods Enzymol.* **101:** 202.

Rowley, R., J. Hudson, and P.G. Young. 1992a. The *wee1* protein kinase is required for radiation-induced mitotic delay. *Nature* **356:** 353.

Rowley, R., S. Subramani, and P.G. Young. 1992b. Checkpoint controls in *Schizosaccharomyces pombe: rad1*. *EMBO J.* **11:** 1335.

Roy, D. and P.A. Fantes. 1982. Benomyl resistant mutants of *Schizosaccharomyces pombe* cold-sensitive for mitosis. *Curr. Genet.* **6:** 195.

Russell, P. and P. Nurse. 1986. *cdc25*$^+$ functions as an inducer in the mitotic control of fission yeast. *Cell* **45:** 145.

———. 1987a. Negative regulation of mitosis by *wee1*$^+$, a gene encoding a protein kinase homolog. *Cell* **49:** 559.

———. 1987b. The mitotic inducer *nim1*$^+$ functions in a regulatory network of protein kinase homologs controlling the initiation of mitosis. *Cell* **49:** 569.

Saka, Y. and M. Yanagida. 1993. Fission yeast *cut5*$^+$, required for S phase onset and M phase restraint, is identical to the radiation-damage repair gene *rad4*$^+$. *Cell* **74:** 383.

Saka, Y., P. Fantes, T. Sutani, C. McInerny, J. Creanor, and M. Yanagida. 1994a. Fission yeast cut5 links nuclear chromatin and M phase regulator in the replication checkpoint control. *EMBO J.* **13:** 5319.

Saka, Y., T. Sutani, Y. Yamashita, S. Saitoh, M. Takeuchi, Y. Nakaseko, and M. Yanagida. 1994b. Fission yeast cut3 and cut14, members of a ubiquitous protein family, are required for chromosome condensation and segregation in mitosis. *EMBO J.* **13:** 4938.

Samejima, I. and M. Yanagida. 1994a. Bypassing anaphase by fission yeast *cut9* mutation: Requirement of *cut9*$^+$ to initiate anaphase. *J. Cell Biol.* **127:** 1655.

———. 1994b. Identification of *cut8*$^+$ and *cek1*$^+$, a novel protein kinase gene, which complement a fission yeast mutation that blocks anaphase. *Mol. Cell. Biol.* **14:** 6361.

Samejima, I., T. Matsumoto, Y. Nakaseko, D. Beach, and M. Yanagida. 1993. Identification of seven new *cut* genes involved in *Schizosaccharomyces pombe* mitosis. *J. Cell Sci.* **105:** 135.

Sazer, S. and P. Nurse. 1994. A fission yeast RCC1-related protein is required for the mitosis to interphase transition. *EMBO J.* **13:** 606.

Schwob, E., T. Böhm, M. Mendenhall, and K. Nasmyth. 1994. The B-type cyclin kinase inhibitor p40^{Sic1} controls the G1 to S transition in *S. cerevisiae*. *Cell* **79:** 233.

Shimanuki, M., N. Kinoshita, H. Ohkura, T. Yoshida, T. Toda, and M. Yanagida. 1993. Isolation and characterization of the fission yeast protein phosphatase gene *ppe1*$^+$ involved in cell shape and mitosis. *Mol. Biol. Cell.* **4:** 303.

Shiozaki, K. and P. Russell 1995a. Counteractive roles of protein phosphatase 2C (PP2C) and a MAP kinase kinase homolog in the osmoregulation of fission yeast. *EMBO J.* **14:** 492.

———. 1995b. Cell-cycle control linked to extracellular environment by MAP kinase pathway in fission yeast. *Nature* **378:** 739.

Shiozaki, K., H. Akhavan-Niaki, C.H. McGowan, and P. Russell. 1994. Protein phosphatase 2C, encoded by *ptc1*$^+$, is important in the heat shock response of *Schizosaccharomyces pombe*. *Mol. Cell. Biol.* **14:** 3742.

Simanis, V. and P. Nurse. 1986. The cell cycle control gene *cdc2*$^+$ of fission yeast encodes a protein kinase potentially regulated by phosphorylation. *Cell* **45:** 261-268.

———. 1989. Characterisation of the fission yeast *cdc10*$^+$ protein that is required for commitment to the cell cycle. *J. Cell Sci.* **92:** 51.

Singh, J. and A.J.S. Klar. 1993. DNA polymerase-α is essential for mating-type switching in fission yeast. *Nature* **361:** 271.

Sipiczki, M. 1995. Phylogenesis of fission yeasts. Contradictions surrounding the origin of a century old genus. *Antonie Leeuwenhoek* **68:** 119.

Sipiczki, M., B. Grallert, and I. Miklos. 1993. Mycelial and syncytial growth in *Schizosaccharomyces pombe* induced by novel septation mutations. *J. Cell Sci.* **104:** 485.

Stettler, S., E. Warbrick, S. Prochnik, S. Mackie, and P. Fantes. 1996. The wis1 signal transduction pathway is required for expression of cAMP-repressed genes in fission yeast. *J. Cell Sci.* **109:** 1927.

Stillman, B. 1994. Smart machines at the DNA replication fork. *Cell* **78:** 725.

Stone, E.M., H. Yamano, N. Kinoshita, and M. Yanagida. 1993. Mitotic regulation of protein phosphatases by the fission yeast sds22 protein. *Curr. Biol.* **3:** 13.

Sugiyama, A., K. Tanaka, K. Okazaki, H. Nojima, and H. Okayama. 1994. A zinc finger protein controls the onset of premeiotic DNA synthesis of fission yeast in a Mei2- independent cascade. *EMBO J.* **13:** 1881.

Sunnerhagen, P., B.L. Seaton, A. Nasim, and S. Subramani. 1990. Cloning and analysis of a gene involved in DNA repair and recombination, the *rad1* gene of *Schizosaccharomyces pombe*. *Mol. Cell. Biol.* **10:** 3750.

Takahashi, K., H. Yamada, and M. Yanagida. 1994. Fission yeast minichromosome loss mutants *mis* cause lethal aneuploidy and replication abnormality. *Mol. Biol. Cell* **5:** 1145.

Takeda, T., T. Toda, K. Kominami, A. Kohnosu, M. Yanagida, and N. Jones. 1995. *Schizosaccharomyces pombe atf1*$^+$ encodes a transcription factor required for sexual development and entry into stationary phase. *EMBO J.* **14:** 6193.

Takeuchi, M. and M. Yanagida. 1993. A mitotic role for a novel fission yeast protein kinase dsk1 with cell cycle stage dependent phosphorylation and localization. *Mol. Biol. Cell* **4:** 247.

Tanaka, K., K. Okazaki, N. Okazaki, T. Ueda, A. Sugiyama, H. Nojima, and H. Okayama. 1992. A new *cdc* gene required for S-phase entry of *Schizosaccharomyces pombe* encodes a protein similar to the *cdc10*$^+$ and *SWI4* gene products. *EMBO J.* **11:** 4923.

Thompson, C.C., T.A. Brown, and S.L. McKnight. 1991. Convergence of Ets- and Notch-related structural motifs in heteromeric DNA binding complex. *Science* **253:** 762.

Thuriaux, P., P. Nurse, and B. Carter. 1978. Mutants altered in the control coordinating cell division with cell growth in the fission yeast *Schizosaccharomyces pombe*. *Mol. Gen. Genet.* **161:** 215.

Toda, T., K. Umesono, A. Hirata, and M. Yanagida. 1983. Cold-sensitive nuclear division arrest mutants of the fission yeast *Schizosaccharomyces pombe*. *J. Mol. Biol.* **168:** 251.

Toda, T., Y. Adachi, Y. Hiraoka, and M. Yanagida. 1984. Identification of the pleiotropic cell division cycle gene *NDA2* as one of two different α-tubulin genes in *Schizosaccharomyces pombe*. *Cell* **37:** 233.

Uemura, T. and M. Yanagida. 1984. Isolation of type I and II DNA topoisomerase mutants from fission yeast: Single and double mutants show different phenotypes in cell growth and chromatin organization. *EMBO J.* **3:** 1737.

———. 1986. Mitotic spindle pulls but fails to separate chromosomes in type II DNA topoisomerase mutants: Uncoordinated mitosis. *EMBO J.* **5:** 1003.

Uemura, T., K. Morikawa, and M. Yanagida. 1986. The nucleotide sequence of the fission yeast DNA topoisomerase II gene: Structural and functional relationships to other DNA topoisomerases. *EMBO J.* **5:** 2355.

Uemura, T., K. Morino, S. Uzawa, K. Shiozaki, and M. Yanagida. 1987a. Cloning and sequencing of *Schizosaccharomyces pombe* DNA topoisomerase I gene, and effect of gene disruption. *Nucleic Acids Res.* **15:** 9727.

Uemura, T., H. Ohkura, Y. Adachi, K. Morino, K. Shiozaki, and M. Yangida. 1987b. DNA topoisomerase II is required for condensation and separation of mitotic chromosomes in *S. pombe*. *Cell* **50:** 917.

Umesono, K., T. Toda, S. Hayashi, and M. Yanagida. 1983. Two cell division cycle genes *NDA2* and *NDA3* of the fission yeast *Schizosaccharomyces pombe* control microtubular organization and sensitivity to anti-mitotic benzimidazole compounds. *J. Mol. Biol.* **168:** 271.

Uzawa, S., I. Samejima, T. Hirano, K. Tanaka, and M. Yanagida. 1990. The fission yeast *cut1*$^+$ gene regulates spindle pole body duplication and has homology to the budding yeast *ESP1* gene. *Cell* **62:** 913.

Walworth, N.C. and R. Bernards. 1996. *rad*-dependent response of the *chk1*-encoded protein kinase at the DNA damage checkpoint. *Science* **271:** 353.

Walworth, N., S. Davey, and D. Beach. 1993. Fission yeast *chk1* protein kinase links the rad checkpoint pathway to *cdc2*. *Nature* **363:** 368.

Warbrick, E. and P.A. Fantes. 1991. The wis1 protein kinase is a dosage-dependent regulator of mitosis in *Schizosaccharomyces pombe*. *EMBO J.* **10:** 4291.

———. 1992. Five novel elements involved in the regulation of mitosis in fission yeast. *Mol. Gen. Genet.* **232:** 440.

Waseem, N.H., K. Labib, P. Nurse, and D.P. Lane. 1992. Isolation and analysis of the fission yeast gene encoding polymerase δ accessory protein PCNA. *EMBO J.* **11:** 5111.

Watanabe, Y. and M. Yamamoto. 1996. *Schizosaccharomyces pombe pcr1*$^+$ encodes a CREB/ATF protein involved in regulation of gene expression for sexual development. *Mol. Cell. Biol.* **16:** 704.

Weisman, R., J. Creanor, and P. Fantes. 1996. A multicopy suppressor of a cell cycle defect in *S. pombe* encodes a heat shock-inducible 40 kDa cyclophilin-like protein. *EMBO J.* **15:** 447.

White, J.H.M., D.G. Barker, P. Nurse, and L.H. Johnston. 1986. Periodic transcription as a means of regulating gene expression during the cell cycle: Contrasting modes of expression of DNA ligase genes in budding and fission yeast. *EMBO J.* **5:** 1705.

Wohlgemuth, J.G., G.H. Bulboaca, M. Moghadam, M.S. Caddle, and M.P. Calos. 1994. Physical mapping of origins of replication in the fission yeast *Schizosaccharomyces pombe*. *Mol. Biol. Cell* **5:** 839.

Wu, L. and P. Russell. 1993. Nim1 kinase promotes mitosis by inactivating Wee1 tyrosine kinase. *Nature* **363:** 738.

Yamamoto, M. 1980. Genetic analysis of resistant mutants to antimitotic benzimidazole compounds in *Schizosaccharomyces pombe*. *Mol. Gen. Genet.* **180:** 231.

Yamano, H., K. Ishii, and M. Yanagida. 1994. Phosphorylation of dis2 protein phosphatase at the C-terminal cdc2 consensus and its potential role in cell cycle regulation. *EMBO J.* **13:** 5310.

Young, P.G. and P.A. Fantes. 1984. Changed division response mutants function as allosuppressors. In *Growth, cancer and the cell cycle* (ed. P. Skehan and S.J. Friedman), p. 279. Humana Press, New York.

———. 1987. *Schizosaccharomyces pombe* mutants affected in their division response to starvation. *J. Cell Sci.* **88:** 295.

Zakian, V.A. 1995. *ATM*-related genes: What do they tell us about functions of the human gene? *Cell* **82:** 685.

Zhang, J., C.-K. Tan, B. McMullen, K.M. Downey, and A.G. So. 1995. Cloning of the cDNAs for the small subunits of bovine and human DNA polymerase δ and chromosomal location of the human gene (POLD2). *Genomics* **29:** 179.

Zhu, Y., T. Takeda, K. Nasmyth, and N. Jones. 1994a. *pct1*$^{+}$, which encodes a new DNA binding partner of p85^{cdc10}, is required for meiosis in the fission yeast *Schizosaccharomyces pombe*. *Genes Dev.* **8:** 885.

Zhu, J., D.L. Carlson, D.D. Dubey, K. Sharma, and J.A. Huberman. 1994b. Comparison of the two major ARS elements of the *ura4* replication origin region with the other ARS elements in the fission yeast, *Schizosaccharomyces pombe*. *Chromosoma* **103:** 414.

Zwerschke, W., H.-W. Rottjakob, and H. Küntzel. 1994. The *Saccharomyces cerevisiae CDC6* gene is transcribed at late mitosis and encodes a ATP/GTPase controlling S-phase initiation. *J. Biol. Chem.* **269:** 23351.

9

Mitosis and Cytokinesis in the Fission Yeast, *Schizosaccharomyces pombe*

Sophia S.Y. Su and Mitsuhiro Yanagida
Department of Biophysics
Faculty of Science
Kyoto University
Kyoto 606-01, Japan

I. INTRODUCTION

Schizosaccharomyces pombe cells are rod-shaped with round capped ends (see Fig. 2). They have a uniform diameter of about 3.5 μm and grow by extension at both ends with a length ranging from 7 μm to 16 μm (Mitchison 1970; Miyata et al. 1978). Whereas the yeast *Saccharomyces cerevisiae* cells divide by budding, *S. pombe* cells divide by fission. *S. pombe* cells are more stable as a haploid than as a diploid. Although diploid zygotes can arise through conjugation of two haploid cells, diploid cells tend to undergo an immediate meiosis to form haploid spores. Under favorable conditions, fission yeast reproduce asexually through the mitotic cell cycle. Typical of all eukaryotic cells, the cell cycle is composed of discrete G_1, S, G_2, and M phases (pre-DNA synthesis, DNA synthesis, post-DNA synthesis, and mitotic phases, respectively). Among the various phases, the events in the mitotic phase are the most dramatic and most striking. Replicated DNAs become precisely separated into two daughter nuclei during mitosis. Cytokinesis then divides the cell equally into two daughter cells each with identical genetic materials. The aim of this chapter is to describe the critical events that occur in mitosis and the gene products that are known to affect these events in the yeast *S. pombe*. Gene products that control the final septation and cell separation events following mitosis are also discussed (for review, see Fankhauser and Simanis 1994b).

Although both *S. pombe* and *S. cerevisiae* are ideal for laboratory investigations and serve as excellent model organisms for the study of eukaryotic processes, *S. pombe* has certain advantages over *S. cerevisiae* when studying the processes in mitosis. *S. pombe* has three observable chromosomes that clearly condense during mitosis, whereas *S. cerevisiae* has 16 chromosomes whose condensation is cytologically unclear. Thus, individual chromosome behavior in fission yeast mitosis can be followed by cytological methods. *S. pombe* mitosis is also closely similar to mitosis in higher eukaryotes. The mitotic spindle is present only for a short time during mitosis and the spindle pole body remains undivided until mitosis, which is more typical of higher eukaryotes (McCully and Robinow 1971; Hiraoka et al. 1984; Hagan and Hyams 1988; Robinow and Hyams 1989).

The initiation of mitosis must be regulated and coordinated with other cell cycle events to ensure proper nuclear and cell division. First of all, cells must be undergoing mitotic cell division and not meiotic differentiation (Fantes 1989; Hayles and Nurse 1992). Second, cell growth is required to attain a critical cell mass necessary for nuclear division (Fantes 1977, 1989; Fantes and Nurse 1978). Third, DNA synthesis must be completed to provide the available substrates (duplicated chromosomes) for subsequent nuclear division. Finally, a number of regulatory controls must act in concert to prevent mitosis under inappropriate situations, such as incomplete replication or unrepaired DNA damage (for review, see Hartwell and Weinert 1989; Sheldrick and Carr 1993; Nurse 1994b; D'Urso and Nurse 1995). These regulatory controls are largely responsible for the proper order of events in the cell division cycle. The genetic control of cell cycle (MacNeill and Nurse, this volume) and the processes of mating and sporulation (Yamamoto et al., this volume) in *S. pombe* are covered in other chapters in this volume. For this chapter, we briefly describe the importance of these topics with respect to mitosis and cytokinesis.

A. Commitment to Mitotic Cell Division or Meiotic Differentiation

In the fission yeast life cycle, the decision to reproduce asexually or to undergo sexual differentiation is determined by the presence of mating partners and the nutritional environment (for review, see Nasim et al. 1989; Egel et al. 1990; Hayles and Nurse 1992). Meiosis in the fission yeast is tightly linked to the mating process. Nitrogen starvation triggers conjugation in strains of opposite mating types and immediately leads to sporulation. Therefore, the divergence from mitotic cell division to meiosis involves the participation of mating-type genes (*mat1-P, mat1-M*) as well as numerous gene products that transduce the pheromone response through the MAP kinase pathway (such as *ras1*$^+$, *gpa1*$^+$, *byr1*$^+$, and *byr2*$^+$) (Nadin-Davis and Nasim 1990; Xu et al. 1994). Gene products that detect and transduce the nutritional status of the cell (such as *cgs1*$^+$, *pka1*$^+$, *cyr1*$^+$, and *cgs2*$^+$/*pde1*$^+$) are also important determinants for sexual differentiation (Maeda et al. 1990, 1994; DeVoti et al. 1991; Mochi-zuki and Yamamoto 1992). Controls governing the initiation of meiosis are described by Yamamoto et al. (this volume).

The commitment to the mitotic cell division cycle occurs in a narrow window in the G_1 phase of the cell cycle prior to Start (Nurse and Bissett 1981). After Start, cells can no longer conjugate and must proceed through the cell cycle. Among the numerous cell cycle genes, *cdc2*$^+$ and *cdc10*$^+$ have been shown to be critical for this control point. The finer points in the control of Start are discussed further by MacNeill and Nurse (this volume).

B. The Mitotic Cell Division Cycle of Fission Yeast

One of the major features of the *S. pombe* cell cycle is that the duration of G_1 is very short and G_2 is long. The G_1 phase is negligible in wild-type cells and is normally completed at the time of cell division; it only becomes apparent in poor media or in certain mutant strains (Fantes 1977; Fantes and Nurse 1977; Nurse and Thuriaux 1980). A number of genes are known to function during the G_1/S transition and are involved in regulating Start. These genes are reviewed in the *S. pombe* cell cycle chapter in this volume (MacNeill and Nurse). The S phase is also brief, and the exact timing of its initiation and completion is unknown. At cytokinesis, the S phase has already been completed.

The G_2 phase of *S. pombe* corresponds to the long growing stage and spans three quarters of the cell cycle. Since the region of growth is associated with the localization of actin granules, growth extension can be monitored by observing the actin distribution in the cell (see Fig. 2) (Marks and Hyams 1985; Marks et al. 1986). In early G_2 phase immediately after cell division, actin localization is restricted to the old end of the cell (which existed in the previous cell cycle), so growth is monopolar. After reaching a certain cell length of about 9.0–9.5 μm, actin granules redistribute to both ends, and the new end starts to grow as well. This point in the G_2 phase is called *n*ew *e*nd *t*ake *o*ff, or NETO (Mitchison and Nurse 1985; May and Mitchison 1986). As cells reach their maximum length (14–16 μm), the G_2 phase ends and is followed by nuclear and cell division.

Mitchison (1970) defined the progress of the cell cycle from the birth of a new cell at 0 to the end of cell division at 1. By this definition, rapidly growing wild-type cells at 0 are in early G_2 and mitosis occurs at 0.75 of the cell cycle. Therefore, a large fraction of exponentially growing *S. pombe* cells is in the G_2 phase and contain a 2C quantity of DNA. Cells containing a 1C quantity of DNA are not usually observed for wild-type strains growing in rich media.

C. Cell Morphogenesis

A number of studies have recently begun to examine the basis for polarized growth in *S. pombe* (Snell and Nurse 1993; Nurse 1994a). The organization of the cytoskeleton may direct cell wall deposition which in turn determines the cell shape. This hypothesis is reenforced by results showing that inhibitors of cytoskeleton assembly and mutations disrupting cytoskeletal organization cause abnormal cell shapes (Umesono et al. 1983b; Hiraoka et al. 1984; Kobori et al. 1989).

Several protein kinases and phosphatases are involved in directing polarized growth, possibly by regulating actin distribution within the cell. Strains with mutations in protein kinase C (*pck1*$^+$ and *pck2*$^+$/*sts6*$^+$), casein kinase II (*orb5*$^+$/*cka1*$^+$ and *ckb1*$^+$), and a novel protein kinase (*kin1*$^+$) display delocalized actin distribution and have abnormal pear-like or round cells (Levin and Bishop 1990; Toda et al. 1993; Roussou and Draetta 1994; Snell and Nurse 1994). Mutants of the regulatory subunits of protein phosphatase type 2A (PP2A) (*paa1*$^+$ and *pab1*$^+$), of a PP2A homolog (*ppe1*$^+$), or of calcineurin (*ppb1*$^+$) are also round or form abnormal branching patterns with deregulated cytoskeletal distribution (Shimanuki et al. 1993; Yoshida et al. 1994; Kinoshita et al. 1996). In contrast, mutations in the *ssp1*$^+$ protein kinases cause actin localization to one end only, corresponding to monopolar growth. The loss of *ssp1*$^+$ can restore the rod-like cell shape in a *ppe1* or *sts5* mutant, suggesting that *ssp1*$^+$, *ppe1*$^+$, and *sts5*$^+$ gene products participate in regulating growth polarity associated with NETO (Matsusaka et al. 1995).

The *ras*-like and *rho*-like gene products and their regulators (such as *ras1*$^+$, *ral1*$^+$/*scd1*$^+$, *ral3*$^+$/*scd2*$^+$, *cdc42*$^+$, *ste10*$^+$, and *cwg2*$^+$) are also intimately involved in directing polarized growth. In addition to the cell morphology defect, these mutants fail to project polarized growth toward a mating partner (Fukui et al. 1986; Nadin-Davis et al. 1986; Fukui and Yamamoto 1988; Leupold 1991; Fawell et al. 1992; Díaz et al. 1993; Miller and Johnson 1994; Xu et al. 1994). The *cwg2* mutant displays reduced cell wall polysaccharide synthesis activity (1-3 β-D-glucanase) as well, suggesting a link between the establishment of cell polarity and cell wall synthesis (Díaz et al. 1993).

II. GENOME AND CHROMOSOMAL FEATURES

A. Genome Size

The genomic DNA size of an *S. pombe* haploid cell is approximately 14 Mb, and it is organized into three relatively long chromosomes. Genetic analyses of more than 230 markers established three linkage groups (Kohli et al. 1977; Gygax and Thuriaux 1984; Kohli 1987). Cytological observations and pulse-field gel electrophoresis (PFGE) confirmed the genetic linkage data (Robinow 1977; Umesono et al. 1983a; Niwa and Yanagida 1985; Smith et al. 1987; Robinow and Hyams 1989). Detailed PFGE analyses using the *Not*I restriction sites indicated that chromosomes I, II, and III are 5700, 4700, and 3500 kb long, respectively (Fan et al. 1988). Genomic DNAs of *S. pombe* have been covered by ordered cosmid banks (Hoheisel et al. 1993; Mizukami et al. 1993). Nucleotide sequencing of the entire genomic DNAs has currently been undertaken by several laboratories.

B. Structural Features of the Chromosomes

1. ARS

ARS (autonomously replicating sequences) are chromosomal DNA fragments capable of promoting autonomous replication and extrachromosomal maintenance of transforming circular DNAs. *S. pombe* ARS elements were isolated on the basis of their high-frequency transformation and unstable transformant phenotypes (Sakaguchi and Yamamoto 1982; Toda et al. 1984b; Maundrell et al. 1985, 1988; Heyer et al. 1986; Wright et al. 1986; Zhu et al. 1992; Wohlgemuth et al. 1994). Two-dimensional gel electrophoresis mapping of ARS has demonstrated that an individual ARS element can initiate DNA replication at its native chromosomal locus. ARS elements in clusters may interfere with each others' activities in a hierarchical fashion (Dubey et al. 1994; Wohlgemuth et al. 1994).

Nucleotide sequence determination of a number of *S. pombe* chromosomal fragments with ARS properties revealed that they are AT-rich and contain an 11-bp consensus sequence, 5′-(A/T)PuTTTATTTA(A/T)-3′, similar to that of *S. cerevisiae* (Maundrell et al. 1988; Zhu et al. 1994). Surprisingly, some functional ARS elements lack the *S. pombe* ARS consensus sequence (PACS), and deletion of this consensus sequence bears no noticeable effect on ARS activity (Johnston and Barker 1987; Maundrell et al. 1988; Zhu et al. 1994). Therefore, PACS is not the determinant for ARS function. Although there is speculation that multiple asymmetric A+T-rich sequences in ARS elements may be significant in fostering ARS activity, it is unclear what sequence defines the ARS function in *S. pombe*.

2. *Telomeres*

Linear DNA molecules in higher eukaryotic cells are either degraded or integrated into a chromosome through recombination unless the ends are adequately protected with telomeric repeats. Telomeres have been cloned from *S. pombe*, and they have been demonstrated to be functional in fission yeast as well (Sugawara 1989). The terminal repeat array of fission yeast telomeres is about 300 bp in length with a consensus sequence $TTAC_{0\text{-}1}A_{0\text{-}1}C_{0\text{-}1}G_{1\text{-}8}$ running toward the end of the chromosome. Similar sequences were also found at the ends of a rescued minichromosome generated by γ-irradiation, suggesting that *S. pombe* telomerase adds de novo telomeric repeats to the broken ends (Matsumoto et al. 1987). The four telomeres of chromosomes I and II bear additional repeats of similar sequences (designated telomere-associated sequences, or TAS) adjacent to the terminal repeats (Sugawara 1989). The two telomeres of chromosome III lack TAS but have the terminal repeats placed directly adjacent to ribosomal DNA (Sugawara 1989; Hoheisel et al. 1993; Mizukami et al. 1993).

Placement of telomeric sequences next to genes in *S. pombe* also results in reversible repression of telomere-adjacent genes as observed for *S. cerevisiae* and *Drosophila* (Gottschling et al. 1990; Tartof and Bremer 1990; Gottschling 1992; Nimmo et al. 1994). This kind of varied gene expression associated with adjacent telomeric sequences is termed position effect variegation (PEV). The degree of repression correlates not only with the distance from the telomeric repeats, but also with the orientation of the gene with respect to the telomere (Nimmo et al. 1994). Telomere-associated PEV observed with adjacent genes suggests that the telomere may assemble specialized chromatin structures leading to intermediate expression or repression. Report of the formation of nonnucleosomal structure at the fission yeast telomere supports this hypothesis (Chikashige et al. 1989).

3. *Centromere Structure and Function*

Centromeres are specialized regions of the eukaryotic chromosome that are required for the faithful transmission of chromosomes during cell division. Aneuploidy in *S. pombe* is mostly lethal. Only a haploid disome for chromosome III is viable, but it is highly unstable (Niwa and Yanagida 1985). In higher eukaryotes, kinetochore proteins assemble on the centromere DNA to form a structure that acts as a microtubule attachment point. The centromere is thought to be the site where the two sister chromatids are held together after replication. In *S. cerevisiae*, the centromere is localized to a short segment of less than 200 bp, which is nonfunctional in *S. pombe* (Panzeri et al. 1985; Clarke 1990). The centromeres of *S. pombe* are large (40–120 kb) and more complex (Fig. 1) (Clarke and Carbon 1985; Nakaseko et al. 1986, 1987; Fishel et al. 1988; Chikashige et al. 1989; Hahnenberger et al. 1989; Niwa et al. 1989; Murakami et al. 1991). These regions rarely undergo meiotic recombination and do not encode protein sequences (see Nakaseko et al. 1986, 1987; Niwa et al. 1989; Takahashi et al. 1991). A large number of transfer RNA genes are found symmetrically arranged in the centromere regions of all three chromosomes; however, it is not clear whether or not these tRNA genes are transcribed (see Kuhn et al. 1991; Takahashi et al. 1991).

Studies have shown that the centromere regions span 40 kb for *cen1*, 70 kb for *cen2*, and 110 kb for *cen3* (Nakaseko et al. 1986; Fishel et al. 1988; Chikashige et al. 1989; Murakami et al. 1991). The basic structural motif that is common to all three *S. pombe* centromeres is composed of a central core (cnt or CC) immediately flanked by inverted repeats. This latter portion is made up of the outer repetitive sequence (otr) and the inner most repeat (imr or CAR) (Fig. 2) (Clarke and Baum 1990; Hahnenberger et al. 1991; Murakami et al. 1991; Takahashi et al. 1992; Steiner and Clarke 1994). The otr domain consists of repetitive elements called dg and dh (Nakaseko et al. 1986, 1987) (also called K and L, respectively; Clarke and Carbon 1985; Fishel et al. 1988). The imr domains are large inverted repeats that are characteristic of each individual chromosome, and they directly flank the central core (Murakami et al. 1991; Takahashi et al. 1992). The sequences of *cnt1* and *cnt3* are asymmetric, and they contain a highly homologous segment (designated tm) that is absent in *cnt2*, suggesting that the tm sequences may not be essential for centromere function (see Murakami et al. 1991; Takahashi et al. 1992). Chromosome III has the largest repetitive region among the three centromere regions (see Chikashige et al. 1989; Murakami et al. 1991).

Artificial minichromosomes have been useful for studying centromere function. Linear or circular minichromosomes containing the minimum centromeric motifs (including a pair each of otrs and imrs, and one cnt) are functional (Niwa et al. 1986, 1989; Hahnenberger et al. 1989; Matsumoto et al. 1990). Inclusion of selectable markers in the minichromosome allows for a convenient assay for chromosome stability in

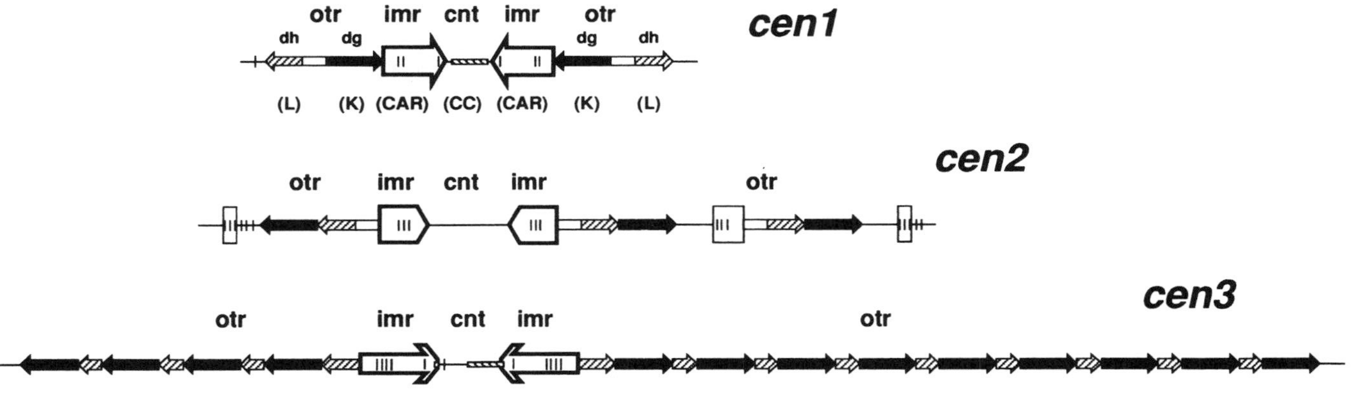

Figure 1 Organization of the fission yeast centromere DNAs. Schematic representation of each of the three centromeres (*cen1*, *cen2*, and *cen3*) is shown. The nomenclature of various regions is as presented by Takahashi et al. (1992). The K/L/CAR/CC nomenclature of Hahnenberger et al. (1991) is also indicated in parenthesis. The outer repetitive region (otr) is composed of repetitive elements, dg and dh (indicated by the *closed arrows* and *hatched arrows*, respectively). The *large open arrows* indicate the inner most repeat regions (imr). The imr sequences are largely different for individual *cen* genes. The central core (cnt) is also distinct for each *cen*, except for the tm regions (*hatched box*) in *cen1* and *cen3*. The distal parts of imr2 are duplicated within otr2 (indicated in *open boxes*). The vertical lines indicate tRNA-like genes present within the centromere regions (Takahashi et al. 1991). Bar, 10 kb.

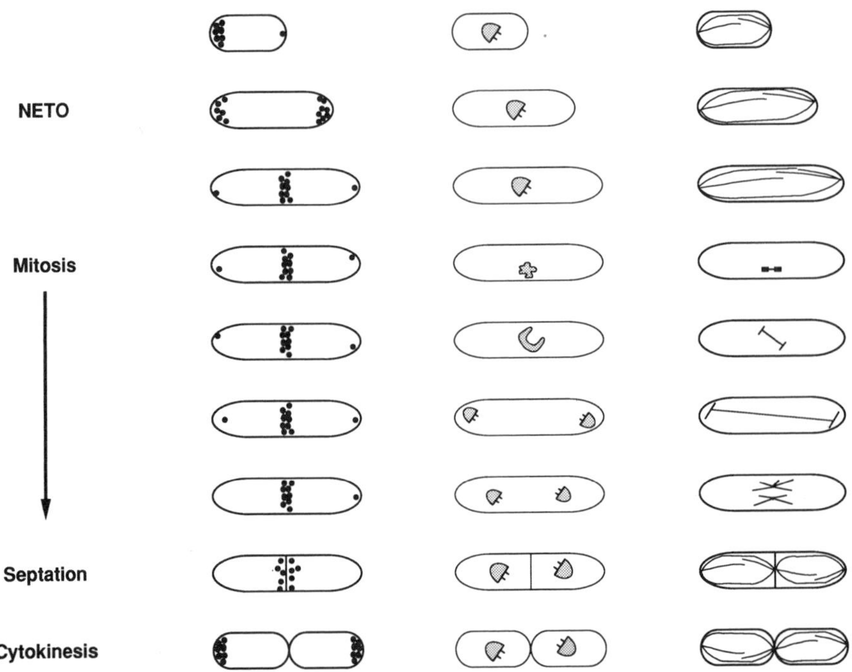

Figure 2 Nuclear and cytoskeletal reorganization in mitosis. The schematic representation of the nuclear and cytoskeletal alterations in fission yeast mitosis is based on Robinow and Hyams (1989), Hagan and Hyams (1988), Marks and Hyams (1985), Marks et al. (1986), and Toda et al. (1981). (*Center panel*) Changes in nuclear morphology as observed using the fluorescent probe, DAPI. The nucleus appears hemispherical in interphase. As the cell enters mitosis, the condensed nucleus transforms into a U-shape intermediate and then separates into two daughter nuclei. Finally, the daughter nuclei migrate toward the centers of the bisected cell. (*Left panel*) Actin localization using anti-actin antibodies or rhodamine-conjugated phalloidin. At early interphase, actin granules (*closed circles*) concentrate at the old end of the cell, correlating with the growing end. When the cell reaches NETO, actin granules are distributed to both ends, corresponding to bidirectional growth. At late G_2, they disappear from the ends and redistribute to the medial plane, which marks the site of later septum formation. After the completion of mitosis and septum formation, actin granules relocate to the old (preexisting) end of the cell, from where growth resumes. (*Right panel*) Restructuring of microtubules as seen using anti-tubulin antibodies. The interphase cell contains a well-organized cytoplasmic microtubule array. Concomitant with the disappearance of cytoplasmic microtubules at mitosis, the SPB duplicates (appearing as two intense foci) and generates a short mitotic spindle. The spindle then elongates as the daughter nuclei divide. Aster microtubules also appear tangential to the spindle pole at this stage. Following nuclear division, the mitotic spindle breaks down. Two MTOCs appear at the cell equator and regenerate the cytoplasmic microtubule array as they converge. Septation and cytokinesis complete the separation of the two daughter cells.

various centromeric constructs. Both cnt and imr provide essential centromere functions because these regions are required for maintaining the stability of minichromosomes (Hahnenberger et al. 1991; Takahashi et al. 1992; Baum et al. 1994). Although the loss of the otr element does not affect mitotic minichromosome stability per se, it drastically impairs the maintenance of sister chromatid attachment in meiosis I, indicating that otr sequences are indispensable for full centromere function (Niwa et al. 1989; Clarke and Baum 1990; Hahnenberger et al. 1991). The minimal sequence sufficient for a functional centromere requires only the centromeric dg (K-type) repeat in conjunction with the central core of *cen2* (Baum et al. 1994). Increasing copies of dg repeats stabilize the minichromosomes, suggesting that this region is a major contributor of centromere function (Baum et al. 1994).

Several studies have shown that the chromatin structure of centromere regions does not exhibit typical nucleosomal packaging (Polizzi and Clarke 1991; Takahashi et al. 1992). The integrated genes within or adjacent to the centromeric repeat units not only display nonnucleosomal structure, but also demonstrate variegated expression similar to that observed in telomeres (Allshire et al. 1994). These results imply that the centromere may constitute a specialized site for the assembly of distinct structures.

An epigenetic control seems to act at the centromere loci as well. When a partially functional minichromosome (missing an imr) is transformed into cells, transformants harboring either a functional (mitotically stable) or a nonfunctional (mitotically destabilized) minichromosome are retrieved (Steiner and Clarke 1994). Since these two classes of transformants did not arise from changes in centromere content, structural arrangement, or chemical modification of the minichromosome, it seems possible that the formation of functional and nonfunctional kinetochore complexes may dictate their difference. A higher-order structure resulting from the folding of the centromere region with the assembly of kinetochore proteins may be necessary to specify a functional centromere. The formation of centromere-specific protein complexes may spread to genes integrated into the centromere, thus resulting in various degrees of epigenetic gene repression (Allshire et al. 1994). Minimal centromere DNA sequences may at times be able to acquire the proper centromere chromatin structure and remain functional. Otherwise, the centromere becomes nonfunctional. Since many centromere-binding proteins have been identified in other eukaryotes and a number of protein-binding sites have been identified within a dg (K) repeat by gel mobility shift assay, fission yeast most certainly form centromere-specific complexes that will specify centromere functions (Willard 1990; Baum et al. 1994).

III. CYTOLOGY OF MITOTIC EVENTS

Mitosis involves a complex series of sequential events including dramatic cytoskeletal reorganization and nuclear events such as chromosome condensation, segregation, and separation (Fig. 3) (McCully and Robinow 1971; Toda et al. 1981). Mitotic events can be visualized utilizing diamidinophenyl-indole (DAPI), rhodamine-conjugated phalloidin or

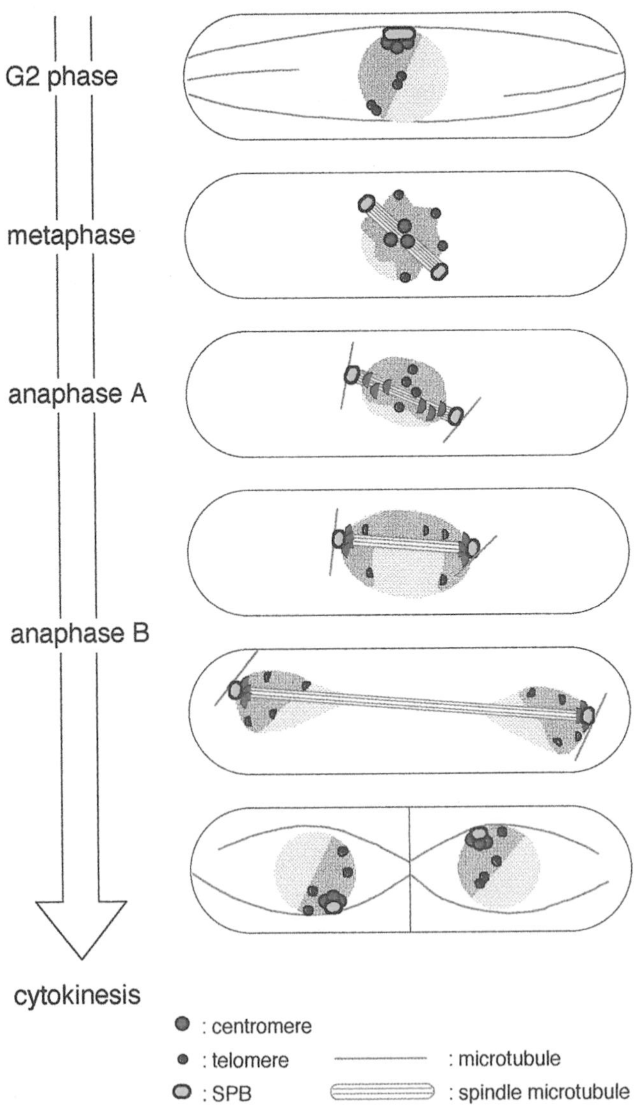

Figure 3 (*See facing page for legend.*)

indirect immunofluorescence microscopy using antibodies against components of the mitotic apparatus (such as tubulin and spindle pole body protein), and conventional electron microscopy. DAPI staining reveals stepwise structural alterations of the chromosomes (Toda et al. 1981). Rhodamine-conjugated phalloidin, anti-actin, anti-tubulin, and anti-Sad1 antibodies detect cytoskeletal changes that accompany chromosome segregation (Marks and Hyams 1985; Hagan and Hyams 1988; Hagan and Yanagida 1995). Freeze-substitution electron microscopy enables detailed visualization of the structure of mitotic components (McCully and Robinow 1971; Tanaka and Kanbe 1986; Ding et al. 1993). Recent application of fluorescence in situ hybridization (FISH) technology allows the visualization of specific chromosomal DNA sequences, such as centromeres and telomeres, during mitosis, which further enhances our knowledge of the mode of chromosome segregation (Uzawa and Yanagida 1992; Funabiki et al. 1993; Saka et al. 1994). From the results obtained through these various techniques, the dramatic nuclear and cytoskeletal events occurring during mitosis can be deduced.

A. Actin Reorganization

Actin structures can be visualized by indirect immunofluorescence microscopy using anti-actin antibodies or rhodamine-conjugated phalloidin. Actin granules and fibers undergo striking structural alteration in the cell cycle, and their distribution parallels the growing ends of the cells (Fig. 2, left) (Marks and Hyams 1985; Marks et al. 1986; Alfa and Hyams 1990). In newly divided cells, actin granules localize first to the growing old end until NETO occurs. Then actin granules localize to both ends of the cell, coincident to bidirectional growth extension. In late G_2 and M phases, actin granules disappear from the ends as cell elongation ceases and gather at the cell equator in a ring which marks the site of later septum formation. Upon the completion of septation, actin granules depart

Figure 3 Nuclear events leading to sister chromatid separation. In interphase cells, centromeres cluster close to the SPB, whereas telomeres associate loosely opposite the SPB at the nuclear envelope. As the cell enter mitosis, the SPB duplicates and forms a mitotic spindle. The centromeres align at the center of the spindle, whereas the telomeres dissociate from the nuclear envelope at metaphase. Sister centromeres split in anaphase A and move toward the opposite poles of the cell. It is then followed by separation of sister telomeres. In anaphase B, spindle elongation fully separates the chromosomes to opposite poles of the cell while maintaining close association with their respective SPBs. Finally, the divided nuclei, which have resumed their interphase configuration, are separated by the septum to generate two daughter cells. (Figure is based on the review by Yanagida [1995].)

from the cell equator and move to the opposite ends of each daughter cell to initiate new cell growth.

B. Mitotic Spindle Formation and Elongation

Chromosome segregation is mediated by the mitotic spindle apparatus, a complex structure containing spindle pole bodies (SPBs), microtubules (MTs), and chromosomes (for review, see Page and Snyder 1993). SPBs are composed of electron-dense materials assembled at the nuclear envelope (McCully and Robinow 1971; Tanaka and Kanbe 1986). Immunofluorescence microscopy using anti-tubulin antibodies as well as electron microscopy readily reveals the spindle dynamics in mitosis (Hiraoka et al. 1984; Tanaka and Kanbe 1986; Hagan and Hyams 1988; Alfa and Hyams 1990; Ding et al. 1993).

Prior to mitosis, a well-developed cytoplasmic array of microtubules is observed in interphase cells (Hagan and Hyams 1988; Alfa and Hyams 1990). Cytoplasmic microtubules are aligned in an array extending between the tips of the cell, correlating with cell length (Fig. 2 right; Fig. 3). At the onset of mitosis, the interphase microtubule array disappears, and two intense foci of anti-tubulin staining appear corresponding to the separation of duplicated SPBs (Hagan and Hyams 1988). Separation of the SPBs precedes the formation of the short spindle (~2 μm). Electron microscopy studies establish that this spindle is composed of kinetochore microtubules (KMTs) and nonkinetochore microtubules (NKMTs, including continuous pole-to-pole and discontinuous MTs). KMTs link the spindle pole to the chromosome through interaction with an electron-dense kinetochore structure. NKMTs may participate in the establishment and elongation of the spindle apparatus. At the midzone of the short spindle, KMTs may have captured the chromosomes at their centromeres. Discontinuous MTs originating from either spindle pole also appear to overlap and interdigitate with each other at the midzone of the short spindle (Ding et al. 1993). Together, the interaction of KMTs, pole-to-pole MTs, and discontinuous MTs may facilitate the segregation of sister centromeres toward their respective SPBs. This motion comprises anaphase A (see below). Once the chromosomes have arrived at the poles, the KMTs are no longer evident, possibly through the shortening of KMTs.

The spindle elongates in anaphase B at a rate of about 1 μm per minute to a final length of 12–15 μm, spanning the entire length of the cell (Fig. 3) (Hiraoka et al. 1984; Tanaka and Kanbe 1986; Hagan and Hyams 1988; Hagan et al. 1990; Ding et al. 1993). The elongated spindle lacks KMTs and continuous MTs but is composed of discontinuous MTs emanating from each pole and interdigitating at the midzone region in a highly organized, square-packed arrangement (Ding et al. 1993). An explanation for the loss of continuous MTs in the elongated spindle is that continuous MTs may actually be interdigitating MTs that are long enough to reach the opposite poles, and the sliding and extension of antiparallel MTs result in the presence of interdigitating MTs only at the end of anaphase B. During this stage of spindle elongation, astral MTs also appear tangential to the spindle poles (Hagan and Hyams 1988). Although it is unclear if astral MTs participate in chromosome segregation, there are forces that act in or on the mitotic spindle during anaphase and telophase to pull the sister chromatids to opposite ends of the cell.

The mechanism of spindle elongation in *S. pombe* has been analyzed in vitro using *nuc2* mutant cells, which arrest with a metaphase spindle at the restrictive temperature (Masuda et al. 1990). Spindle elongation can occur in *nuc2* cells at the permissive temperature, suggesting that the machinery for spindle elongation remains intact despite its defect in chromosome segregation. Spindle elongation requires the addition of exogenous tubulin and ATP, indicating that microtubule polymerization and motor proteins (such as dynein and kinesin) participate in this process.

C. Regeneration of Interphase Microtubule Array

Following nuclear division, the mitotic spindle breaks down and cytoplasmic microtubule arrays are regenerated. Coincident with the arrival of the nuclei at the ends of the cell, two dots of anti-tubulin fluorescence appear at the cell equator (Fig. 2, right). These two dots reflect the reformation of the interphase microtubule organizing centers (MTOCs), which are distinct from the SPBs (Fig. 3) (Hagan and Hyams 1988; Alfa and Hyams 1990). The MTOCs radiate microtubules for the regeneration of the cytoplasmic microtubule network. As septation progresses, the MTOCs move toward the center of the cell and reestablish the interphase microtubule array. It is probable that MTOCs act as a marker for reestablishing the rod-like cell shape in newly divided cells. In cells treated with antimicrotubule drugs or in tubulin mutants *nda2* and *nda3/ben1*, the nucleus cannot be maintained at a fixed position and the cells become swelled or branched (Walker 1982; Toda et al. 1983, 1984a; Umesono et al. 1983a,b; Hiraoka et al. 1984). Cytoplasmic microtubules may be important in positioning the nucleus at the equator of the cell and in establishing the cell polarity.

D. Chromosome Behavior

The nucleus of *S. pombe* in the interphase cell is spherical with a diameter of 2–3 μm and is enclosed by the nuclear membrane. The nuclear structure of *S. pombe* has been investigated by electron microscopy (Robinow 1977; Tanaka and Kanbe 1986) and by fluorescence microscopy using the fluorescent probe DAPI (Toda et al. 1981; Yanagida et al. 1986). The interphase nucleus consists of an electron-dense hemisphere containing the nonchromosomal RNA-rich (nucleolar) region and a DAPI-staining chromosomal hemisphere (Fig. 2, center; Fig. 3). Short DAPI-stained rods protruding into the nonchromosomal region are tandemly arranged rDNA clusters (Hirano et al. 1989; Uzawa and Yanagida 1992).

In early mitosis, fission yeast chromosomes condense approximately three- to fourfold in volume (Toda et al. 1981). Individual mitotic chromosomes are usually difficult to distinguish in wild-type cells because they are clustered. However, in thiabendazole-treated cells or in certain mitotic mutants (such as *nda2*, *nda3*, *nuc2*, *cut7*, and *dis* mutants), individual DAPI-stained chromosomes appear to be hypercondensed and can be observed in fluorescence photomicrographs (Toda et al. 1981; Umesono et al. 1983a; Hiraoka et al. 1984; Hirano et al. 1988; Ohkura et al. 1988; Hagan and Yanagida 1990). As the chromosomes continue through mitosis, they appear as a U-shaped intermediate with the arms of the U facing opposite ends of the cell (Fig. 2, center) (Toda et al. 1981). The center of the U constricts and the arms then separate into two smaller hemispheres of daughter nuclei moving toward the ends of the cell. Following septum formation, each of the daughter nuclei migrates back to the center of each cell. Electron microscopy and indirect immunofluorescence light microscopy demonstrate that the nuclear envelope remains intact throughout mitosis and elongates into a dumbbell structure around the separated nuclei until the connection is broken in late telophase (Tanaka and Kanbe 1986; Pidoux and Armstrong 1993). The entire sequence of mitotic events occurs within the intact nuclear envelope.

Chromosome segregation in mitosis can be examined in detail by applying the FISH technique (Uzawa and Yanagida 1992; Funabiki et al. 1993). Using probes unique to the centromere region, the precise location of the centromeres during the course of mitosis can be observed (Fig. 3). In G_2 interphase cells, centromeres seem to cluster in the vicinity of the SPB at the outer membrane of the nuclear envelope (Funabiki et al. 1993). Upon entry into mitosis, the SPB duplicates, and the centromere cluster dissociates from the SPB to become distributed along the short mitotic spindle. Individual centromeres can be seen positioning themselves at irregular points along the spindle, suggesting that they may undergo a series of oscillations before being captured by the mitotic spindle. A metaphase plate seems to form as centromeres gather at the center of the spindle, resembling the alignment of chromosomes at metaphase. During the metaphase/anaphase transition, sister centromeres split and move toward opposite SPBs prior to full spindle extension. This motion defines anaphase A movement. Finally, spindle elongation in anaphase B fully separates the chromosomes to opposite poles of the cell while maintaining the close association of centromeres to their respective SPBs. Early mitotic events such as chromosome/spindle interaction, metaphase alignment, and sister centromere separation can be observed using FISH, whereas DAPI-staining does not reveal the details of chromosome division in early mitosis (Toda et al. 1981; Hiraoka et al. 1984; Tanaka and Kanbe 1986; Uzawa and Yanagida 1992; Funabiki et al. 1993). This demonstrates the power of the FISH method in dissecting minute and yet important events in mitosis.

The movement of chromosome ends in mitosis can also be visualized by applying FISH to unique sequences in the telomere regions (Fig. 3) (Funabiki et al. 1993). In G_2 interphase cells, the telomeres cluster loosely and associate with the nuclear envelope. At mitosis, telomere clustering is disrupted. Sister telomeres separate after centromere separation, implying that the force which separates sister chromatids may drag and pull sister telomeres apart.

IV. CONTROLS LEADING TO THE ONSET OF MITOSIS

A. Cdc2p–Cyclin B Complex

The Ccdc2p kinase is a major element controlling the G_1/S transition as well as the entry into mitosis (Fig. 4) (for review, see Hindley and Phear 1984; Gould and Nurse 1989; Nurse 1990; Ducommun et al. 1991; Gould et al. 1991; King et al. 1994). Cdc2p alone does not have protein kinase activity but acts in concert with Cdc13p mitotic cyclin for full kinase activity (Hagan et al. 1988). The loss of *cdc13*$^+$ function prevents Cdc2p kinase activation and subsequent mitotic entry (Booher and Beach 1987, 1988; Solomon et al. 1988; Moreno et al. 1989). Genetic and biochemical experiments have shown that Cdc2p phosphorylation is required for its interaction with Cdc13p (Booher et al. 1989; Gould et al. 1991; Fleig et al. 1992). Similar to cyclins of other organisms, accumulation of Cdc13p product contributes to the activation of Cdc2p kinase, which then induces entry into mitosis (Alfa et al. 1989; Booher et al. 1989; Moreno et al. 1989). Degradation of Cdc13p cyclin leads to Cdc2p inactivation and mitotic exit.

The Cdc2p-Cdc13p complex establishes the temporal order of S phase and M phase in the cell cycle (for review, see Nurse 1994b;

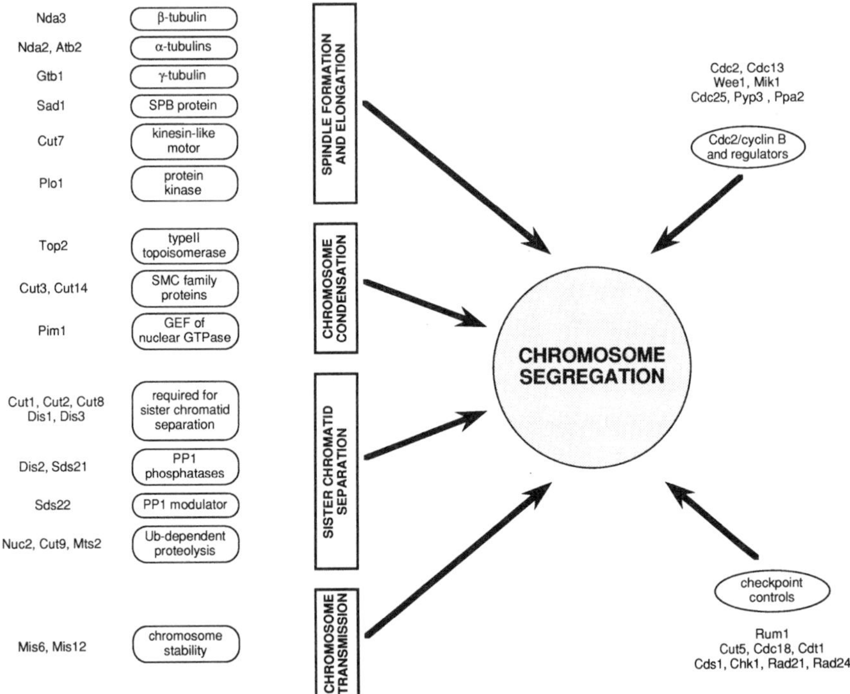

Figure 4 Roles of mitotic gene products in chromosome segregation. The major events that facilitate chromosome segregation can be classified into spindle formation and elongation, chromosome condensation, sister chromatid separation, and chromosome transmission. The gene products that participate in each of these processes are shown. See text for a further description.

O'Connell and Nurse 1994). The loss of mitotic B-type cyclin causes cells to bypass nuclear division and cytokinesis and undergo repeated rounds of DNA synthesis (Hayles et al. 1994). Specific *cdc2* alleles also result in DNA rereplication without mitosis (Broek et al. 1991; Enoch and Nurse 1991). Conversely, overexpression of Cdc2p and Cdc13p in G_1 causes cells to enter mitosis without undergoing replication (Hayles et al. 1994). These results suggest that the formation of the Cdc2p-Cdc13p complex is required for the G_2/M progression.

B. Regulators of Cdc2p Kinase

Regulators that alter Cdc2p kinase activity have profound effects on the onset of mitosis (Fig. 4). In particular, the dephosphorylation of Tyr-15 of Cdc2p kinase is required for its activation (Gould and Nurse 1989).

Wee1p and Mik1p tyrosine kinases are mitotic inhibitors that phosphorylate Tyr-15, resulting in the inactivation of Cdc2p kinase (Featherstone and Russell 1991; Lundgren et al. 1991; Parker et al. 1992; Lee et al. 1994). Nim1p/Cdr1p kinase is a negative regulator of *wee1*$^+$, thus activating mitosis by relieving the mitotic inhibition of *wee1*$^+$ (Russell and Nurse 1987; Feilotter et al. 1991; Parker et al. 1993; Wu and Russell 1993). Since *wee1*$^+$ and *nim1*$^+$ gene products are required to give normal cell cycle timing for M phase and response to the nutritional status of the cell, the cell mass limitation that controls mitosis most likely acts through the Wee1p pathway (Nurse 1975; Fantes and Nurse 1977, 1978; Thuriaux et al. 1978; Russell and Nurse 1987; Feilotter et al. 1991). Conversely, Cdc25p and Pyp3p tyrosine phosphatases are mitotic inducers that antagonize the Wee1p and Mik1p action by dephosphorylating Tyr-15 of Cdc2p, thus activating Cdc2p kinase (Russell and Nurse 1986, 1987; Kumagai and Dunphy 1991; Millar et al. 1991, 1992a; Millar and Russell 1992). Coordination of these mitotic inhibitors and activators leads to elevated Cdc2p kinase activity which facilitates the subsequent mitotic events (Moreno et al. 1989, 1990).

C. Involvement of Other Protein Phosphatases in G_2/M Transition

Other gene products have also been identified that affect G_2/M transition by regulating Cdc2p kinase indirectly through the Wee1p or Cdc25p pathways. PP2A, encoded by *ppa1*$^+$ and *ppa2*$^+$, displays overlapping okadaic-acid-sensitive PP2A activities. The *ppa2*$^+$ gene codes for the predominate PP2A activity in wild-type cells (Kinoshita et al. 1990, 1993). Residual PP2A activity in Δ*ppa2* causes cell size reduction (semi-wee phenotype). This effect is exacerbated to produce a *wee1*-like phenotype when PP2A activity is further suppressed by okadaic acid, indicating that PP2A catalytic activity is involved in the negative regulation of mitosis (Kinoshita et al. 1990, 1993). Since Δ*ppa2* can partially suppress the mitotic block in *cdc25* mutants, and the Δ*ppa2 wee1* double mutant is lethal at 36°C but does not exhibit mitotic catastrophe as in the *wee1 mik1* double mutant, these genetic data support the hypothesis of a PP2A-dependent regulation of Cdc2p acting either through Cdc25p or through Wee1p (Kinoshita et al. 1991b, 1993; Lundgren et al. 1991).

Similarly to *ppa1*$^+$ and *ppa2*$^+$, *pyp1*$^+$ and *pyp2*$^+$ encode functionally overlapping protein tyrosine phosphatases (in which *pyp1*$^+$ has the dominate role) that also negatively regulate mitosis (Millar et al. 1992b; Ottilie et al. 1992). The *pyp1* phenotype and its genetic interaction with other mitotic mutants resemble that of *wee1* in every way (Millar et al.

1992b). However, mitotic delay observed in strains overexpresssing *pyp1*$^+$ or *pyp2*$^+$ requires an active *wee1*$^+$ function. These results establish that protein tyrosine dephosphorylation by *pyp1*$^+$ and *pyp2*$^+$ contributes to the regulation of mitosis by promoting the inhibitory Wee1p pathway.

V. ISOLATION OF MITOTIC MUTANTS

Many *S. pombe* mitotic mutants have been isolated by a variety of methods. The most conventional approach involves cytological screens of conditional lethal mutants that display cell cycle defects. Analyses of temperature-sensitive *cdc* (*c*ell *d*ivision *c*ycle) and cold-sensitive *nda* (*n*uclear *d*ivision *a*rrest) mutants have defined a number of unlinked genes involved in various cell cycle events (Nurse et al. 1976; Nasmyth and Nurse 1981; Toda et al. 1983). Cytological screens for aberrant chromatin structures and uncoordinated mitosis phenotypes result in the identification of novel classes of mutants affecting nuclear division, including *nuc* (*nuc*lear structure alteration), *cut* (*c*ell *u*ntimely *t*orn), and *dis* (*d*efect *i*n *s*ister chromatid disjunction) mutants (Hirano et al. 1986, 1988; Ohkura et al. 1988; Samejima et al. 1993). Utilization of minichromosomes also allows for the identification of *mis* (*m*inichromosome *i*n*s*tability) mutants displaying increased rates of chromosome instability (Takahashi et al. 1994). Other mitotic mutants can be isolated by drug resistance or hypersensitivity (*ben* and *mts*) or screened through enzymatic assays (*top*) (Yamamoto 1980; Umesono et al. 1983b; Uemura and Yanagida 1984). Mutants displaying specific defects in the process of mitosis can be classified by further genetic analyses and cytological examinations.

Further understanding of the gene products participating in mitosis can be gained by cloning and characterizing the genes responsible for the defects. Because those genes with conditional lethal mutations can be cloned by complementation, their products can be analyzed using standard molecular techniques developed for *S. pombe* (for review, see Beach and Nurse 1981; MacNeill and Fantes 1993). The deduced amino acid sequence from the cloned gene may then provide information about its gene function. With the gene available, recombinant protein can be produced and antisera prepared and used to investigate the biochemical function and cellular localization of the gene product.

Gene products of other organisms that are known to be involved in mitosis can be inferred to have similar functions in fission yeast. Assuming that essential mitotic genes are conserved, isolation of the homologous gene in fission yeast is possible using standard cloning techniques. A few examples of this method are the isolations of histones, γ-tubulin (*gtb1*$^{+}$), PP2A (*ppa1*$^{+}$ and *ppa2*$^{+}$), and PP2A-related phosphatase (*ppe1*$^{+}$) (Matsumoto and Yanagida 1985; Kinoshita et al. 1990; Horio et al. 1991; Shimanuki et al. 1993).

VI. MAJOR EVENTS IN MITOSIS

Analyses of mutant phenotypes have demonstrated several potentially independent mitotic pathways, namely, spindle formation and elongation, chromosome condensation, chromosome separation, septation, and cytokinesis (Uemura and Yanagida 1986; Yanagida 1989). Mitotic events in these pathways are temporally and spatially coordinated, but they become uncoupled in certain mitotic mutants. In mutants that block the G_2 to M transition (such as *cdc2* and *cdc25*), further mitotic events do not take place, consistent with the notion that the execution of these mitotic events is dependent on the controls that initiate the entry into mitosis. In *cut* mutants, however, septation and cytokinesis proceed despite blockage of earlier cell cycle events, indicating that mitotic events can become uncoupled in mutants that cannot adequately coordinate the distinct mitotic pathways. We describe here the elements responsible for controlling these individual mitotic events through the characterization of mutant phenotypes and of their gene products (Figs. 4 and 5).

A. Components for Spindle Formation and Elongation

Genes encoding components of the mitotic spindle or regulators of spindle formation and elongation are required for chromosome disjunction (for review, see Snyder 1994; Masuda 1995). Mutants with defects in these gene products may display an abnormal or no mitotic spindle, whereas other mitotic events, such as chromosome condensation and septation, can still occur. The identification of these types of mutants defines spindle formation as a separate pathway that can be uncoupled from the other mitotic events.

1. The Tubulin Genes: α, β, and γ Tubulins

The major component of the spindle apparatus is the microtubule. The α and β tubulins, organized in heterodimers, form the structural subunits of microtubules (Hiraoka et al. 1984; Toda et al. 1984b; Adachi et al. 1986). The γ tubulins are localized to the SPB and MTOC at which micro-

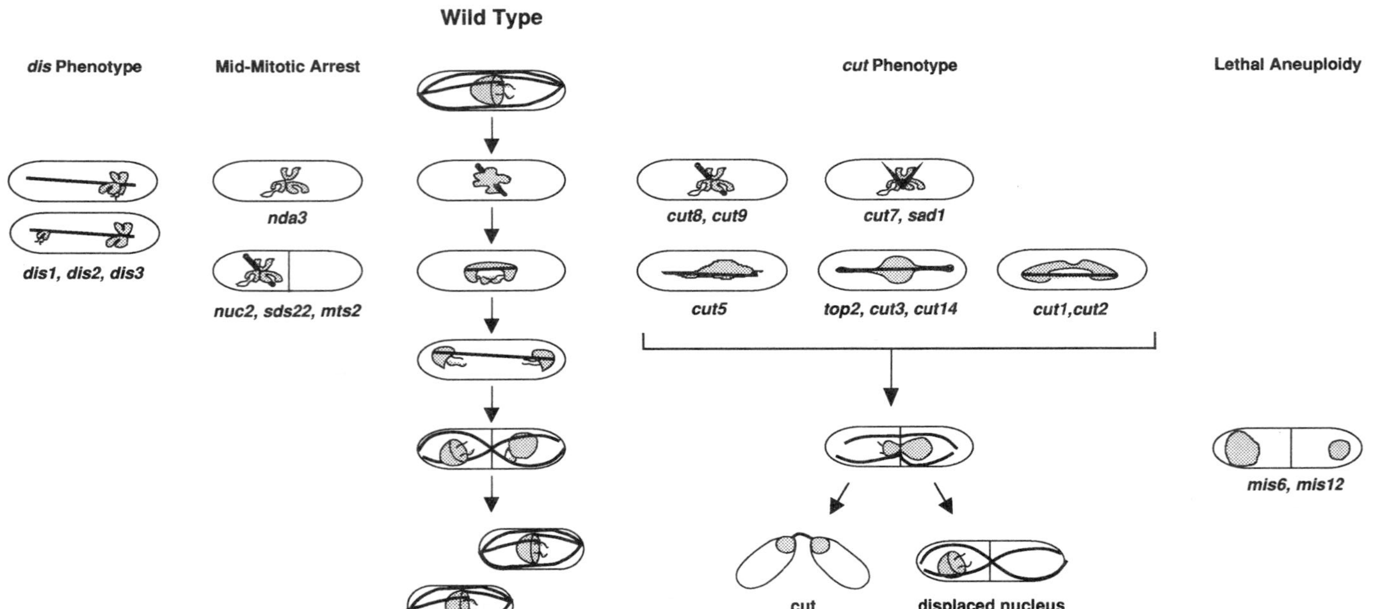

Figure 5 Phenotypes of mitotic mutants. Many mitotic mutants are classified into different categories on the basis of their phenotypes. They exhibit metaphase arrest, *dis*, *cut*, or lethal aneuploid phenotypes. The major defects displayed by these mutants are illustrated. See text for a further description.

tubules nucleate (Horio et al. 1991). Detailed analyses of the defective phenotypes in tubulin mutants and of their gene products revealed different cellular roles for these tubulin genes.

There are two α-tubulin genes and one β-tubulin gene in *S. pombe*. The α1-tubulin (*nda2*$^+$) and β-tubulin (*nda3*$^+$/*ben1*$^+$) genes are essential for growth, whereas the α2-tubulin gene (*atb2*$^+$) is dispensable (Hiraoka et al. 1984; Adachi et al. 1986). Cold-sensitive *nda2* and *nda3* mutants display highly condensed chromosomes that fail to separate in mitosis (Fig. 5). No mitotic spindle can be observed in *nda3-311* at the restrictive temperature, but reversible shift to the permissive temperature enables chromosome segregation concomitant with spindle formation and elongation (Hiraoka et al. 1984; Kanbe et al. 1990). Thus, β tubulins comprise the major components required for spindle formation and elongation. Cells with the *nda2-52* mutation arrest with an abnormal mitotic spindle and a displaced nucleus or partially separated chromosomes, suggesting that α2 tubulins are also required for nuclear division (Toda et al. 1984a). Since α2-tubulin can compensate for the loss of α1-tubulin, and α1- and α2-tubulin expressions are differentially regulated, *nda2*$^+$ and *atb2*$^+$ may perform different functions in vivo (Toda et al. 1984a; Adachi et al. 1986).

At the onset of mitosis, the mitotic spindle nucleates from the SPB. An in vitro study has shown that the mitotic, but not interphase, SPB can assemble spindle microtubules, and this activity is dependent on an activated Cdc2p kinase (Masuda et al. 1990, 1992). The association of Cdc2p–cyclin B complexes to SPBs at early stages of mitosis further supports a link between mitotic spindle nucleation and Cdc2p kinase activation (Alfa and Hyams 1990). γ-tubulin, an essential product encoded by *gtb1*$^+$, colocalizes with the SPB and MTOC and is a likely component involved in nucleating mitotic as well as cytoplasmic microtubule assembly (Horio et al. 1991; Masuda et al. 1992).

2. *Spindle-pole-body-associated Protein:* sad1$^+$

The *sad1*$^+$ gene encodes an essential product that resides with the SPB throughout the mitotic cell cycle (Funabiki et al. 1993; Hagan and Yanagida 1995). It differs from the localization of γ-tubulin in that Sad1p-specific antibodies do not associate with the central MTOCs which appear at the end of mitosis. Mutants lacking *sad1*$^+$ function form aberrant V- or X-shaped spindles and consequently fail to segregate their chromosomes (Fig. 5) (Hagan and Yanagida 1990, 1995). Since the *sad1*$^+$ gene product contains a putative membrane-spanning region and its overexpression leads to accumulation at the nuclear periphery, Sad1p may be a structural component of the SPB that integrates into the nuclear membrane during SPB separation and spindle formation.

3. *Kinesin-related Motor:* cut7$^+$

cut7$^+$ encodes a product with structural similarity to the *bimC*-related kinesin family, and like *sad1*, the loss of *cut7*$^+$ function causes defective V-shaped spindle structures instead of normal linear mitotic spindles (Fig. 5) (Hagan and Yanagida 1990, 1995). Indirect immunofluorescence microscopy with anti-Cut7p antibodies detects intense localization at the SPB throughout mitosis, but not with MTOCs or cytoplasmic or astral microtubules (Hagan and Yanagida 1990, 1992). These images suggest that Cut7p may participate specifically in intranuclear spindle function. It is not a general component of MTOCs, and unlike kinesin, it is not involved in other microtubule-based events, such as vesicle transport (Horio et al. 1991; Hagan and Yanagida 1992). Anti-Cut7p antibodies also detect its antigen along the center of the spindle in a stage-dependent manner. Central spindle staining is weak until spindle elongation begins, at which time Cut7p appears to be distributed at defined points along the spindle. It then gathers at the midzone on further elongation. Cut7p may function as a motor that directs microtubule movement, as suggested by its localization and sequence homology. The unusual V-shaped spindle formed in *cut7* mutants also suggests that *cut7*$^+$ performs a second function. The *cut7*$^+$ gene product is required to reorientate the microtubules from a parallel to an antiparallel configuration in which polar microtubules interdigitate. Thus, the coordinate activities of these two functions enable SPB separation and spindle formation.

4. *Protein Kinase:* plo1$^+$

The gene *plo1*$^+$ has been isolated as a homolog of *Drosophila polo* and budding yeast *CDC5* (Llamazares et al. 1991; Kitada et al. 1993; Ohkura et al. 1995). One defect caused by a disruption of *plo1*$^+$ is a block in the progression through mitosis. The *plo1* mutant displays a condensed nucleus and monopolar spindle, similar to those seen in the *Drosophila polo* mutant (Llamazares et al. 1991; Ohkura et al. 1995). The fact that either loss of function or overexpression of *plo1*$^+$ yields cells with monopolar spindles suggests that proper level or timing of *plo1*$^+$ activation may be required for bipolar spindle formation. The similar monopolar spindle defect in *plo1* and *cut7* suggests that the Cut7p motor protein and Plo1p kinase may have a close functional relationship (Hagan and Yanagida 1990, 1992).

B. Chromosome Condensation and Decondensation

When spindle formation is disrupted by chemical treatment or in mutants, chromosome condensation can still occur. This suggests that chromosome condensation and spindle formation are processes that can proceed independently up to a certain point. Mutants that affect one process without affecting the other have been isolated. For example, *nda3* and *nuc2* represent mutations that allow chromosome hypercondensation but are unable to form or elongate mitotic spindles, respectively. In contrast, *top2*, *cut3*, and *cut14* represent mutations that fail to compact chromosomes but still allow abnormal chromosome segregation and subsequent cytokinesis to proceed. Since the chromosomes in these mutants fail to separate fully, the chromatin becomes severed by the proceeding septation and cytokinesis. This phenotype is formally defined as "*cut*" (Hirano et al. 1986; Samejima et al. 1993). In this section, we focus on genes that regulate appropriate chromosome condensation in mitosis.

1. Topoisomerases: top1$^+$ *and* top2$^+$

Topoisomerase activities of decatenation/catenation, knotting/unknotting, and relaxation of DNA molecules have been shown to affect DNA topology and have a number of critical roles in replication, recombination, transcription, condensation, and segregation of chromosomes (for review, see Yanagida and Wang 1987; Yanagida 1990; Yanagida and Sternglanz 1990). Type I topoisomerases create single-strand nicks, whereas type II topoisomerases make double-strand cuts to relax and to decatenate interlinked DNA molecules. Type I (*top1*$^+$) and type II (*top2*$^+$) topoisomerases have been cloned, and mutant alleles have been isolated from an in vitro screen for low topoisomerase activities (Uemura and Yanagida 1984; Uemura et al. 1986, 1987a). The *top1* mutants remain viable, since Top2p function can compensate for the loss of Top1p activity (see below).

In contrast, *top2* mutants are inviable and display a severe cell division defect. Compacted chromosomes cannot form at the time of mitosis, but the chromatin is pulled by elongating mitotic spindles (Uemura and Yanagida 1984, 1986). This pulling force ultimately yields cells exhibiting the typical "*cut*" phenotype, which was first observed in this mutant (Fig. 5). Taking advantage of the reversible block in spindle formation of *nda3-311*, the requirement of Top2p topoisomerase at different stages in mitosis has been investigated (Uemura et al. 1987b). Temperature-shift experiments using combinations of *nda3-311* and both temperature-sensitive (*ts*) and cold-sensitive (*cs*) alleles of *top2* demonstrated that Top2p activity is required for chromosome condensation early in mitosis as well as for sister chromosome separation in anaphase. Chromosome painting, an application of FISH which allows the visualization of a particular chromosome, confirms that one of the defects of *top2* is an inability to condense chromosomes (Saka et al. 1994). The cold-sensitive *top2* allele has a mutation in the ATP-binding domain, whereas the temperature-sensitive *top2* allele has a mutation in the DNA-binding domain (Shiozaki and Yanagida 1991). Thus, DNA association and double-stranded DNA cleavage activity are necessary for *top2*$^+$ function.

The phenotype of the *top1 top2* double mutant is strikingly different from that of the *top2* mutant alone. Instead of displaying a "*cut*" phenotype like *top2*, the *top1 top2* double mutant exhibits an aberrant ring-shaped nuclear structure and a collapsed nucleolus immediately upon shift to the nonpermissive temperature (Uemura and Yanagida 1984; Hirano et al. 1989). This suggests that DNA relaxation activity is required throughout the cell cycle to maintain normal chromatin structure, and this activity is provided by Top1p in the *top2* mutant. Top2p activity can partially substitute for Top1p function to maintain chromatin integrity, but it may be absolutely required in mitosis. This may explain why *top1* mutants are viable and *top2* mutants only remain viable until entering mitosis (Uemura and Yanagida 1984, 1986). Nevertheless, both *top1*$^+$ and *top2*$^+$ in wild-type cells act to preserve the chromatin structure in interphase and to facilitate proper chromosome condensation and separation in mitosis.

2. *SMC Family Genes:* cut3$^+$ *and* cut14$^+$

Cytological screens for mutants that exhibit a "*cut*" phenotype similar to *top2* identified a number of new complementation groups, some designated *cut1-cut9* and *cut11-cut19* (Hirano et al. 1986; Samejima et al. 1993; I. Hagan and J.R. McIntosh, unpubl.). Among these mutants, *cut3* and *cut14* in particular resemble *top2*. At the restrictive temperature during mitosis, chromosomes in *cut3* and *cut14* mutants cannot condense and portions of the chromatin are pulled toward the opposing poles along the extended mitotic spindle, at times displaying ϕ-shaped or streaked nuclear structures (Fig. 5) (Saka et al. 1994). Chromosome painting has illustrated that chromosomes in *cut3* and *cut14* mutants cannot compact sufficiently in mitosis. In late mitosis, centromeres migrate to opposite poles, but chromosome arms fail to separate and appear streaked along the elongated spindle. Since chromosome condensation may provide the rigidity required for normal separation under the pulling force generated by the mitotic spindle, the inability to compact DNA may cause the chromatin to become tangled and unable to separate.

Null alleles of *cut3* also display fibrous or extended nuclear chromatin in interphase, suggesting that Cut3p may be required to maintain the chromatin architecture in interphase as well. Therefore, Cut3p and Cut14p may be important for organizing the higher-order chromatin structures throughout the cell cycle.

Genetic interactions of *cut3*, *cut14*, *top1*, and *top2* further suggest that they control similar processes (Saka et al. 1994). In addition to the fact that the "*cut*" phenotypes of *cut3*, *cut14*, and *top2* are similar and these mutants display a defect in chromosome condensation, cold-sensitive *top2* is synthetic lethal in combination with temperature-sensitive *cut3*, a common phenomenon for mutations affecting interacting proteins. The *top1 cut3* double mutant also displays an abnormal ring-like nuclear structure similar to that of *top1 top2* (Uemura and Yanagida 1984; Saka et al. 1994). Overexpression of *cut14*$^+$, on the other hand, can partially relieve the temperature-dependent growth defect in *cut3*, suggesting some overlapping function or complex formation. Overexpression of *cut3*$^+$ cannot bypass the *cut14* phenotype, so they are not functionally interchangeable. Cut3p and Cut14p may together maintain and regulate the changes in chromatin structure, but each performs their particular roles in facilitating chromosome condensation in mitosis.

Sequence analyses of *cut3*$^+$ and *cut14*$^+$ reveal that they belong to a novel protein family termed SMC, reported to be essential for chromosome condensation and required for mitotic chromosome disjunction (for review, see Strunnikov et al. 1993; Chuang et al. 1994; Hirano and Mitchson 1994; Peterson 1994; Saitoh et al. 1994; Saka et al. 1994; Gasser 1995; Hirano 1995; Hirano et al. 1995). These proteins are large and share a common head-rod-tail configuration consisting of a putative amino-terminal NTP-binding domain, a central coiled-coil region, and a putative carboxy-terminal helix-loop-helix DNA-binding region. The mechanism of how these proteins interact with the chromosomal DNA and mediate condensation is unknown. A DNA-dependent ATP-driven motor activity has been proposed to bring distant segments of DNA together and facilitate condensation (Gasser 1995; Hirano et al. 1995; Yanagida 1995). Condensed chromosome will then provide adequate rigidity to prevent tangled chromatin and allow for the proper segregation of sister chromatids.

3. *Nuclear GTPase Components:* pim1$^+$ *and* spi1$^+$

In contrast to the phenotypes of *top2*, *cut3*, and *cut14* mutants, the loss of *pim1*$^+$ leads to inappropriate chromosome condensation. The *pim1/dcd1*^{ts}

mutant exhibits condensed chromosomes at all stages of the cell cycle, even without the completion of S phase or the *cdc25*$^+$ mitotic inducer (Matsumoto and Beach 1991). However, it is not clear whether inappropriate chromosome condensation is the result of premature initiation of mitosis or a defect in decondensation after exit from mitosis (Sazer and Nurse 1994). Recent analysis of the *pim1* mutant demonstrates that the nuclear envelope undergoes extensive fragmentation following mitosis (Demeter et al. 1995).

The *pim1*$^+$ gene encodes a homolog of the human RCC1 nuclear protein, which acts as an guanine nucleotide exchange factor (GEF) for the nuclear GTPases, Ran/TC4 (for review, see Bischoff and Ponstingl 1991; Matsumoto and Beach 1991; Tartakoff and Schneiter 1995). Suppressors of the *pim1* defect have identified the Pim1p-interacting GTPase, *spi1*$^+$. This gene sequence shares homology with the human Ran/TC4 GTPase, and the *spi1* mutant displays a *pim1*-like phenotype as well (Matsumoto and Beach 1991; Sazer and Nurse 1994). Since products of the nuclear GTPase cycle (NGC) in other organisms have been demonstrated to affect chromatin structure as well as other biological events, Pim1p and Spi1p may participate in the maintenance of nuclear and chromosome architecture in the decondensed interphase state through a Ras-like GTPase activity (Dasso 1993; Demeter et al. 1995). The loss of a PP2A-related phosphatase, *ppe1*$^+$/*esp1*$^+$, is able to overcome the *pim1* defect, suggesting that the Pim1p-dependent NGC regulating chromosome condensation and decondensation is controlled in part by protein phosphorylation (Matsumoto and Beach 1993; Shimanuki et al. 1993).

C. Regulation of the Metaphase/Anaphase Transition

1. *Ubiquitin-dependent Proteolysis:* mts2$^+$

A screen for mutants resistant to the microtubule poison methylbenzylcarbamylate (MBC) has identified *mts2*, a metaphase-defective mutation (Gordon et al. 1993). Mutants display condensed chromosomes with short spindles typical of metaphase cells (Fig. 5). However, chromosomes fail to segregate, leading to chromosome nondisjunction. The *mts2*$^+$ gene codes for an essential functional homolog of the human S4 subunit of the 26S proteosome. Another subunit of the proteosome complex MSS1 (human S7 cDNA) can rescue the temperature-sensitive but not the null allele of *mts2*$^+$, suggesting that Mts2p is likely to interact with other components of the 26S proteosome in fission yeast as well. Consistent with sequence homology data, the *mts2* mutant accumulates polyubiquitinated proteins, indicating that it is defective in ubiquitin-dependent protein degradation. Thus, ubiquitin-dependent proteolysis seems to be essential for the progression of mitosis.

2. *TPR Repeat Proteins:* nuc2$^+$ *and* cut9$^+$

Both *nuc2*$^+$ and *cut9*$^+$ encode products containing repeated blocks of 34 amino acid repeats (termed tetratricopeptide repeat, TPR) (Hirano et al. 1990; Samejima and Yanagida 1994a). Several members of the TPR gene family in budding yeast, filamentous fungi, humans, and fission yeast have essential roles in mitosis (for review, see Goebl and Yanagida 1991). The *nuc2*$^+$ gene product shares homology with *A. nidulans bimA* and *S. cerevisiae* and human *CDC27* (Hirano et al. 1990; O'Donnell et al. 1991; Tugendreich et al. 1993). The *cut9*$^+$ gene product is homologous to *S. cerevisiae CDC16* (Sikorski et al. 1990; Samejima and Yanagida 1994a). *nuc2* and *cut9* mutants share a common defect at the metaphase stage; anaphase is blocked by the failure to elongate the mitotic spindle in both mutants (Fig. 5) (Hirano et al. 1988; Samejima and Yanagida 1994a). These two fission yeast TPR products may regulate a common process in the onset of anaphase.

The phenotypes of *nuc2* and *cut9* are different despite their common defect in spindle elongation. In *nuc2* mutants, short spindles run through the center of the highly condensed chromosomes aligned at the metaphase plate, but the spindles cannot elongate (Hirano et al. 1988). Centromeres remain clustered at the midmitotic spindle and sister chromatids cannot separate (Funabiki et al. 1993). These mutants arrest with condensed chromosomes displaced from the center and with a septum at the cell equator (Fig. 5). Therefore, the *nuc2* mutant is defective in the progression through mitosis and subsequent cytokinesis, although septation does occur.

The *cut9* mutant leads to the accumulation of highly condensed chromosomes with short spindles at the same mid-mitosis stage as in *nuc2* (Fig. 5) (Samejima and Yanagida 1994a). Unlike *nuc2*, however, post-anaphase events continue in the absence of spindle elongation. Subsequent to metaphase, the short spindle degrades as post-anaphase cytoplasmic microtubules reform and then septation and cytokinesis follow. The chromosomes are either displaced or positioned at the center of the cell, resulting in a cell with a displaced or "*cut*" nucleus. The *cut9* mutation seems to abolish the dependency of post-anaphase events on anaphase. Anaphase is bypassed, which leads to cells that cannot segregate sister chromatids but still allows post-anaphase interphase microtubule arrays to form and cell division to occur. The *cut9⁺* gene product may be a component of a negative feedback control in which a block or delay in the metaphase/anaphase transition activates a restraint system for subsequent post-anaphase events until the completion of chromosome segregation. In the case of other metaphase-arrest mutants such as *nuc2* and *nda3*, the *cut9* feedback control is operational so they become arrested at metaphase and are prevented from proceeding through mitosis. Alternatively, the *cut9* allele may retain partial function. Residual activity in the *cut9* allele may be sufficient for post-anaphase events but not enough for spindle elongation at the metaphase/anaphase transition.

Genetic evidence suggests that Nuc2p and Cut9p may interact with each other or with other products to regulate anaphase initiation and progression (Hirano et al. 1988; Samejima and Yanagida 1994a). The *nuc2 cut9* double mutant cannot form colonies even at the permissive temperature, so the combination of these two mutations cause synthetic lethality. Two temperature-sensitive *cut9* suppressors (cold-sensitive *scn1* and *scn2*), which enable *cut9* mutants to form colonies at the restrictive temperature, are synthetic lethal with *nuc2* as well. Both *scn1* and *scn2* mutations cause a late anaphase defect. These mutants are unable to elongate their spindles fully, so chromosomes become partially segregated. Because these mutants affect the initiation or progression of anaphase, it seems likely that *nuc2*$^+$, *cut9*$^+$, *scn1*$^+$, and *scn2*$^+$ may be involved in controlling the same processes. Genetic interactions among these mutants are also consistent with the notion that products of *nuc2*$^+$, *cut9*$^+$, *scn1*$^+$, and *scn2*$^+$ interact and form complexes that have pivotal roles in governing the onset of anaphase and subsequent events. All mutation sites in *nuc2* and *cut9* alleles lie in the TPR repeat regions, which may disrupt inter- or intramolecular interactions that are dependent on a postulated snap helix motif (Hirano et al. 1990; Goebl and Yanagida 1991; Samejima and Yanagida 1994a).

Immunoprecipitation experiments have indicated that Nuc2p and Cut9p do indeed form a complex, consistent with findings in *S. cerevisiae* homologs, Cdc27p and Cdc16p, respectively (Lamb et al. 1994; Y. Yamada and K. Kumada, unpubl.). Cdc27p, Cdc16p, and a third TPR protein, Cdc23p, form a complex in which interaction between Cdc27p and Cdc23p occurs through a TPR unit (Lamb et al. 1994). Nuc2p homologs, BimA and Cdc27Hs, as well as the Cut9p homolog, Cdc16Hs, colocalize to the SPB and mitotic spindle of HeLa cells (Mirabito and Morris 1993; Tugendreich et al. 1995). Furthermore, a 20S particle (the anaphase-promoting complex), which contains Cdc27p, Cdc16p, and possibly Cdc23p, mediates the ubiquitin-dependent proteolysis of cyclin B

(Irniger et al. 1995; King et al. 1995). In fact, Nuc2p and Cut9p are components of a similar 20S complex in fission yeast (H. Yamada and K. Kumada, unpubl.). The implication from these results in *S. cerevisiae* and humans is that the *S. pombe* Nuc2p and Cut9p products are involved in the degradation of essential mitotic gene products, possibly Cdc13p cyclin B.

3. cut8$^+$ *and* cek1$^+$

The mutant *cut8* also displays a metaphase defect similar to those described above. In these mutants, chromosomes become highly condensed and short spindles accumulate. Some mutant cells exhibit partially extended spindles but chromosomes fail to separate, resulting in the "*cut*" phenotype (Fig. 5) (Samejima and Yanagida 1994b). The *cut8*$^+$ sequence is homologous to the budding yeast *DBF8*, and the phenotype of *dbf8* resembles that of *cut8*, suggesting that these two products may perform similar functions (Houman and Holm 1994; Samejima and Yanagida 1994b).

In addition to *cut8*$^+$, the *cut1*$^+$ gene can also overcome the *cut8* defect (Samejima and Yanagida 1994b). The *cut8* phenotype is similar to *cut1* with respect to spindle formation and elongation, suggesting that they may have overlapping roles (see below) (Uzawa et al. 1990). Furthermore, *cut1* and *cut8* display an accumulation of diploidized cells after a brief heat treatment at the nonpermissive temperature (Creanor and Mitchison 1990; Samejima et al. 1993; Samejima and Yanagida 1994b). However, the continued DNA synthesis observed in *cut1 cdc11* does not occur in the *cut8 cdc11* double mutant; septation and cytokinesis took place in the *cut8 cdc11* mutant (Creanor and Mitchison 1990; Samejima and Yanagida 1994b).

A high dosage of *cek1*$^+$ has been found to overcome the defect in *cut8* (Samejima and Yanagida 1994b). The *cek1*$^+$ gene encodes a protein kinase that is nonessential, and the mutant does not exhibit any noticeable phenotype. Although the function of *cek1*$^+$ is unknown, Cek1p kinase may modify Cut8p to activate its gene function or may simply bypass the *cut8*$^+$ function.

4. Type I Protein Phosphatases and Related Genes: dis2$^+$, sds21$^+$, *and* sds22$^+$

dis2-11 is one of three cold-sensitive *dis* mutants that undergo chromosome condensation, spindle formation, and elongation but are defective in sister chromatid separation (Fig. 5) (Ohkura et al. 1988). The non-disjoined chromosomes migrate along the elongated spindle, resulting in unequal distribution of condensed chromosomes to both ends of the cell. All *dis* mutants retain high H1 kinase activity at the restrictive temperature, indicative of an arrest in mid-mitosis despite spindle elongation (Stone et al. 1993; Kinoshita et al. 1991b). The *dis2*⁺ gene encodes a protein highly homologous to mammalian type-1 protein phosphatase (PP1), and diminished PP1 activity accounts for the mitotic defects in *dis2* (Ohkura et al. 1989; Kinoshita et al. 1990). Thus, the dephosphorylation activity of the *dis2*⁺ gene product is critical for the transition from metaphase to anaphase.

In the course of cloning the *dis2*⁺ gene, another *S. pombe* PP1 gene (*sds21*⁺) capable of suppressing the *dis2-11* mutation had been identified (Ohkura et al. 1989). The sequences of *dis2*⁺ and *sds21*⁺ are 79% identical to each other. Although the *dis2-11* allele is lethal at the restrictive temperature, the null alleles of either *dis2*⁺ or *sds21*⁺ are viable. The *dis2 sds21* double disruptant, however, is lethal, indicating that *dis2*⁺ and *sds21*⁺ share overlapping essential functions. Assays of phosphorylase phosphatase activity in Δ*dis2* and Δ*sds21* have confirmed this result and suggest that the Dis2p phosphatase accounts for the major PP1 activity required for mitotic exit (Kinoshita et al. 1990).

A clone that bypasses the *wee1* suppression of *cdc25* (*bws1*⁺) has been shown to be identical to *dis2*⁺ (Booher and Beach 1989). Since *wee1*⁺ and *cdc25*⁺ activities are tightly regulated in the control of Cdc2p kinase, the isolation of *bws1*⁺ implies that Dis2p phosphatase is also involved in this highly regulated system. The presence of a carboxy-terminal Cdc2p target consensus sequence (TPPR) further suggests that Dis2p is likely to be regulated by Cdc2p kinase through cell-cycle-dependent phosphorylation. Indeed, Dis2p is intensely phosphorylated in mitotically arrested *nuc2* and weakly phosphorylated in G_2-arrested *cdc2* cells (Yamano et al. 1994). Phosphorylation occurs predominately at residue Thr-316 at the Cdc2p target site. In vitro experiments confirm that mitotically activated Cdc2p kinase directly phosphorylates Dis2p at Thr-316 and inhibits Dis2p phosphatase activity. For Sds21p, which lacks the TPPR consensus sequence, or for Dis2p-APPR, in which TPPR is altered to APPR, Thr-316-specific phosphorylation does not occur in vivo or in vitro and their respective phosphatase activities remain unchanged (Yamano et al. 1994). Sequence comparison of Dis2p with other PP1s reveals that the carboxy-terminal TPPR sequence is conserved among PP1s of numerous organisms (particularly in human, rabbit, and rat), but not all PP1s retain this sequence (Cohen 1989; Yamano et al. 1994). The *sds21*⁺ or *dis2* truncation alleles lacking the carboxyl TPPR sequence can rescue the complete loss of PP1 (Δ*dis2* Δ*sds21* double mutant), which further suggests that this region is not essential for phosphatase activity (Yamano et al. 1994). Instead, the TPPR sequence may serve as a motif for negative regulation by Cdc2p kinase.

The effect of potential negative regulation of Dis2p PP1 by Cdc2p kinase is striking when overexpression phenotypes of *dis2*+ and *sds21*+ are examined (Yamano et al. 1994; Ishii et al. 1996). Elevated *sds21*+ expression blocks cell cycle progression and causes extreme G_2 delay with elongated cells, whereas *dis2*+ overexpression exerts no visible effect. The differential effects are attributed to the carboxyl TPPR sequence. Alteration of the *sds21*+ carboxy-terminal sequence to the TPPR sequence abolishes the G_2 delay phenotype with overexpression, and conversely, the *dis2*-APPR allele causes cell elongation as seen for overexpression of *sds21*+. PP1 lacking the TPPR sequence may not be able to respond to Cdc2p-dependent negative regulation, and thus elevated levels of PP1 expression causes cell division delay. With this motif, Cdc2p kinase can adequately manipulate Dis2p phosphatase activity in a cell-cycle-dependent manner through phosphorylation at the carboxy-terminal Cdc2p kinase consensus site.

The major role of Dis2p phosphatase in the cell cycle may be to facilitate exit from mitosis and reentry into interphase (Yanagida et al. 1992; Yamano et al. 1994; Ishii et al. 1996). As Cdc2p kinase is activated early in mitosis, Dis2p phosphatase becomes inactive through phosphorylation at Thr-316, and mitosis proceeds. Inactivation of Cdc2p kinase in anaphase allows for the restoration of Dis2p phosphatase activity through autodephosphorylation or through another phosphatase and signals the end of metaphase concomitant with the onset of anaphase. This view of the role of Dis2p may partially explain the phenotype of *dis2-11*. The *dis2-11* mutant can enter but not exit mitosis because PP1 activity is nearly absent in the cell (Ohkura et al. 1988; Kinoshita et al. 1991b). In cells overexpressing unregulated *sds21*+ or *dis2*-APPR, high levels of PP1 activities may persist and dephosphorylate substrates that require phosphorylation for the advancement through mitosis. Although Cdc2p kinase activity is high in these cells, mitotic progression is severely hampered (K. Ishii, unpubl.).

In addition to *dis2*+ and *sds21*+, *sds22*+ can also suppress the *dis2-11* defect (Ohkura et al. 1989; Ohkura and Yanagida 1991). Expression of *sds22*+ cannot substitute for the loss of both *dis2*+ and *sds21*+, which strongly suggests that it is not simply another PP1 (Ohkura and Yanagida 1991; Stone et al. 1993). Sequence analysis verifies this conclusion. The *sds22*+ gene encodes a product that contains a tandem array of leucine-rich repeats (LRRs), each 22 amino acids in length (Ohkura and Yanagida 1991; Stone et al. 1993). *sds22*⁺ is also an essential gene that is highly conserved among various eukaryotes, which may reflect its functional importance (Hisamoto et al. 1995; MacKelvie et al. 1995).

The 11 LRR units have been predicted to form a distinct β-α structure and have been shown to be important for physically complexing with either Dis2p or Sds21p (Stone et al. 1993; Kobe and Deisenhofer 1994). Association of Dis2p or Sds21p with Sds22p seems to alter the substrate specificities of the Dis2p and Sds21p phosphatases (Stone et al. 1993). Whereas Dis2p and Sds21p immunoprecipitable complexes display both H1 kinase and phosphorylase phosphatase activities, Dis2p-Sds22p and Sds21p-Sds22p complexes cannot recognize phosphorylase as a substrate. Sds21p may be able to suppress the *dis2* defect by associating with Sds22p and augment Sds21p PP1 activity by targeting specific proteins whose dephosphorylation is necessary for mitotic exit.

Temperature-sensitive and null alleles of *sds22* mutants are arrested at mid-mitosis with condensed chromosomes and a short spindle (Fig. 5). This resembles strains that lack both *dis2*⁺ and *sds21*⁺ PP1 activities (Ohkura and Yanagida 1991; Stone et al. 1993; Ishii et al. 1996). Mutants that do not have Sds22p may not be able to target proteins adequately for dephosphorylation, and thus they behave as if they completely lack PP1 activity and exhibit a mid-mitotic arrest phenotype. Interestingly, a high dosage of *sds21*⁺ can rescue Δ*sds22*, suggesting that the defect in Δ*sds22* involves inefficient PP1 activity (Ohkura and Yanagida 1991).

5. dis3⁺

dis3-54 is another allele-specific mutation that causes a similar chromosome nondisjunction defect as in *dis2-11* (Ohkura et al. 1988). The *dis3*⁺ gene encodes an essential nuclear protein that oligomerizes in its native state (Kinoshita et al. 1991a). In addition to the similar phenotype and cellular localization to *dis2*, *dis3-54* is synthetic lethal with *dis2-11*, and elevated dosage of *dis3*⁺ is able to reverse the phenotype of a *wee1 cdc25* double mutant, suggesting that these gene products may have overlapping functions. Furthermore, the predicted *dis3*⁺ sequence is similar to the budding yeast gene *SSD1*/*SRK1*, a suppressor of protein phosphatase *SIT4* and of the regulatory subunit of cAMP-dependent protein kinase *BCY1* (Kinoshita et al. 1991a; Sutton et al. 1991; Wilson et al. 1991). These data favor the hypothesis that *dis2*⁺ and *dis3*⁺ are related. A multicopy plasmid carrying *dis3*⁺ can compensate for the loss of protein phosphatase *ppe1*$^+$, suggesting that *dis3*$^+$ may participate in protein dephosphorylation during mitosis (Shimanuki et al. 1993). However, the level of PP1 activity monitored using the phosphorylase substrate is not reduced in *dis3-54*, arguing against this hypothesis (Kinoshita et al. 1991a). The role of *dis3*$^+$ remains obscure. Very recently, Dis3p was shown to interact directly with Ran, a nuclear G-protein, and enhance the guanine-nucleotide-exchanging activity of RCC1, a regulator for Ran (Noguchi et al. 1996).

D. Sister Chromatid Separation

1. dis1$^+$

The *dis1*$^+$ gene encodes a 93-kD product required for sister chromatid separation. Unlike *dis2*$^+$ and *dis3*$^+$, it does not resemble any known protein with enzymatic activity (Nabeshima et al. 1995). It contains six Cdc2p target sites that are phosphorylated in vivo and a central region rich in proline, serine, threonine, and basic amino acids. This latter region is similar to regions found in certain microtubule-associating proteins (MAPs), such as MAP4 and tau (Aizawa et al. 1991; Kanai et al. 1992). Bacterially produced Dis1p binds to microtubules directly in an in-vitro-binding assay that is dependent on the central region (Nakaseko et al. 1996). Furthermore, it associates with microtubules in vivo, as shown by fluorescence microscopy using anti-Dis1p antibody and Dis1p-GFP (green fluorescent protein) fusion (Chalfie et al. 1994; Nabeshima et al. 1995). These observations suggest that Dis1p indeed possesses microtubule-binding activity. In vitro phosphorylation of Dis1p with purified Cdc2p kinase reduces the microtubule-binding ability (Nakaseko et al. 1996). Thus, Dis1p displays an affinity for microtubules that is dependent on the phosphorylation state of Dis1p.

The GFP-tagged Dis1p is a valuable tool for visualizing protein localization as well as spindle dynamics in living fission yeast cells (Nabeshima et al. 1995). In interphase cells, Dis1p-GFP colocalizes with cytoplasmic microtubules. In early mitosis, Dis1p-GFP reduces its association with the microtubules while its association with the mitotically active SPBs intensifies. Phosphorylation of Dis1p by Cdc2p kinase in mitosis may facilitate the migration from microtubules to SPBs. When Cdc2p kinase activity diminishes during the metaphase/anaphase transition, a portion of Dis1p-GFP regains its binding to microtubules, first associating with the elongating spindle and then with the cytoplasmic microtubules as they regenerate. The reassociation of Dis1p to microtubules may result from the concomitant decrease in phosphorylated Dis1p and accumulation of dephosphorylated Dis1p. The SPB-binding domain of Dis1p, which resides in the carboxyl terminus, is sufficient to rescue the *dis1* mutation, although it is still not clear how Dis1p functions. The mitotic form of Dis1p may be essential to mediate sister chromatid separation through its interaction with other components of the SPB, such as cyclin B, the mitotic motor Cut7p, and the spindle formation protein Sad1p (Alfa et al. 1990; Hagan and Yanagida 1992, 1995). The interphase form of Dis1p may maintain the cytoplasmic microtubule network or associate with cytoplasmic microtubules as a dock until the next mitosis.

2. dis1-*suppressing Kinase:* dsk1$^+$

Isolation of *dis1*-complementing genes results in the identification of *dsk1$^+$*, which is able to rescue the *dis1* defect fully (Takeuchi and Yanagida 1993). The *dsk1$^+$* gene product is a serine/threonine protein kinase with a hydrophilic spacer region within the kinase domain. Loss of *dsk1$^+$* shows no significant defect. The kinase activity of Dsk1p demonstrates narrow substrate preference and is cell-cycle-dependent, peaking in M phase. Dsk1p autophosphorylation in mitosis may enhance its own kinase activity. Interestingly, Dsk1p is normally excluded from the nucleus, but it accumulates in the nucleus during M phase. Dsk1p phosphorylation of specific targets in the nucleus may promote mitotic exit such that overexpression of *dsk1$^+$* is able to overcome the mitotic arrest defect in *dis1*. Since elevated expression of *dsk1$^+$* in wild-type cells delays progression from G_2 to M, cytoplasmic Dsk1p may block mitotic entry in G_2. The spacer region in Dsk1p may regulate its cell cycle localization because *dsk1*-sp (*dsk1* allele missing the spacer) is present in the nucleus even in interphase. It is possible that phosphorylation of Dsk1p alters the spacer domain in a way that blocks nuclear transport. Since *dsk1$^+$* suppresses the loss of *dis1*, Dsk1p kinase may activate a bypass pathway for Dis1p function. The identification of in vivo substrates of Dsk1p kinase is important to assess its biological role in mitosis.

3. cut1$^+$ *and* cut2$^+$

The *cut1$^+$* and *cut2$^+$* gene products seem to be functionally coordinated (Hirano et al. 1986; Uzawa et al. 1990). Cytologically, mutations in either gene allow all mitotic events to take place normally except chromosome segregation. FISH analysis of *cut1* verifies that its primary defect lies in its failure to complete chromosome segregation (Fig. 5).

Centromeres and most of the chromatin DNA separate normally with the elongating spindle, but the telomeres remain in the undivided region (Uzawa et al. 1990; Funabiki et al. 1993). When septum formation and cytokinesis are blocked by a *cdc11* mutation in either *cut1* and *cut2*, DNA synthesis continues through several rounds despite the absence of nuclear division (Creanor and Mitchison 1990; Uzawa et al. 1990; Funabiki et al. 1996b). Both *cut1 cdc11* and *cut2 cdc11* double mutants produce large polyploid nuclei and multiple aberrant SPBs. The dependency of DNA replication and SPB duplication on the completion of mitosis is lost in the absence of functional Cut1p or Cut2p products. Since *cut1* and *cut2* are synthetic-lethal and *cut1*$^+$ expression can overcome the *cut2* mutation, these genetic data suggest that Cut1p and Cut2p interact and have related functions (Uzawa et al. 1990). A recent experiment has confirmed that Cut1p and Cut2p associate physically (Funabiki et al. 1996a).

The *cut1*$^+$ gene product consists of a putative ATP-binding domain, a helical coil region, a Ca^{++}-binding pocket, and a carboxy-terminal region homologous to the *S. cerevisiae ESP1* and *A. nidulans BimB* gene products (Uzawa et al. 1990; May et al. 1992). The *esp1* and *bimB* mutants also accumulate multiple spindles, similar to the *cut1 cdc11* mutant (Baum et al. 1988; May et al. 1992; McGrew et al. 1992). Although *ESP1* and *cut1*$^+$ are not functional homologs, the *ESP1* homologous region is required for *cut1* self-complementation, indicating that this region is essential for *cut1*$^+$ function (Uzawa et al. 1990). The behavior of Cut1p in terms of cell-cycle-specific localization and protein levels parallels that of Cut2p, which further emphasizes their related roles in chromosome segregation (see below).

The *cut2*$^+$ gene product does not show significant homology with known proteins (Uzawa et al. 1990). During the cell cycle, it undergoes dramatic changes in phosphorylation and in cellular localization (Funabiki et al. 1996b). Cut2p contains a Cdc2p target site at Ser-109 that is phosphorylated in vivo (H. Funabiki, unpubl.). Phosphorylated Cut2p accumulates in M phase in parallel with elevated H1 kinase activity. The phosphorylation of Cut2p may trigger the changes in localization of the protein in mitosis. Prior to the onset of mitosis, Cut2p is nuclear. In mitosis, it accumulates around the metaphase spindle and finally disappears in anaphase (Funabiki et al. 1996b).

Not only does Cut2p alter its localization in mitosis, but Cut2p levels also vary with the cell cycle. Cut2p level is highest in metaphase and is reduced by half after entering anaphase or in G_1-arrested cells (Funabiki et al. 1996b). The amino terminus of *cut2*$^+$ contains a sequence weakly resembling the cyclin destruction box, which may account for the increased stability of Cut2Δ81 (Cut2p lacking its amino terminus) in G_1. Cut2p degradation appears to be dependent on a ubiquitin-mediated proteolytic system because Cut2p levels do not diminish in G_1 in a *cdc10 cut9* double mutant. Cut2p may be a direct or indirect target for Cut9p-dependent proteolysis, particularly during the metaphase/anaphase transition.

The timely destruction of Cut2p may be essential for the metaphase/anaphase progression. Moderate expression of the amino truncated Cut2Δ80 allele blocks sister chromatid separation due to the presence of the undegradable Cut2p (Funabiki et al. 1996b). Cut2p has been proposed to be required for holding sister chromatids together. Alternatively, it may inhibit the machinery that separates sister chromatids. Therefore, the degradation of Cut2p will be necessary to facilitate sister separation. The presence of undegradable Cut2p would block sister chromatid segregation by maintaining the attachment between sister chromatids.

E. Chromosome Transmission

High-fidelity chromosome transmission is necessary to ensure correct segregation of DNA to each daughter cell. In budding yeast, mutations that cause a high rate of minichromosome loss identify genes involved in a number of general cellular functions (e.g., DNA replication and transcription) as well as those specifically affecting mitosis, such as genes affecting spindle formation or kinetochore functions. In fission yeast, mutations that cause high-frequency loss of minichromosomes have been isolated and define 12 genetic loci, designated *mis* (*mi*nichromosome *ins*tability) (Takahashi et al. 1994).

Among these mutants, some display an interphase arrest phenotype, whereas others clearly display missegregation of chromosomes. Both *mis5* and *mis11* exhibit a *cdc* phenotype, although *mis5* is defective in DNA replication and *mis11* is not (Takahashi et al. 1994). *mis5*$^+$ encodes a novel MCM family gene product essential for the onset of DNA replication similar to *cdc21*$^+$, *nda1*$^+$, and *nda4*$^+$ (Coxon et al. 1992; Miyake et al. 1993; Takahashi et al. 1994). *mis11*$^+$, on the other hand, is identical to *prp2*$^+$, which is involved in RNA splicing (Potashkin et al. 1993; Takahashi et al. 1994). The *mis4* mutant displays a high frequency of "*cut*" cells with condensed fiber-like chromosomes and hypersensitivity to hydroxyurea and UV, suggesting that chromosomes may have missegregated due to defects in DNA replication or repair. The *mis6* and *mis12* mutants undergo unequal chromosome segregation, producing large and small daughter nuclei and frequent production of lethal aneuploids (Fig. 5). Using rDNA as a probe for FISH analysis in these mutants, it is clear that the chromosomes are not equally segregated to both daughters. This is a strong indication that *mis6*$^+$ and *mis12*$^+$ gene products are particularly critical for controlling faithful chromosome transmission.

VII. SEPTATION AND CYTOKINESIS

The onset of cytokinesis is tightly coordinated with mitosis to bring about correct nuclear and cell division (for review, see Chang and Nurse 1993; Fankhauser and Simanis 1994b; D'Urso and Nurse 1995). Identification of several septation mutants has revealed that nuclear division can occur in the absence of septum formation (for review, see Fankhauser and Simanis 1994b). In addition, the initiation of septation is not necessarily dependent on the successful completion of nuclear division (Nurse et al. 1976; Minet et al. 1979; Hirano et al. 1986; Samejima et al. 1993). Mutants defective in septation and cytokinesis fall into different groups. Mutants producing multinucleated cells with no detectable septum are defined as early septation mutants. Mutants producing cells with a disorganized septum dividing the fully separated daughter nuclei are defined as late septation mutants. Mutants producing multiseptated or "*cut*" cells fail to coordinate septum formation properly with other cell cycle events. Most of the *cut* mutants are defective in mitosis and have been discussed above. In this section, genes are discussed that are specifically involved in the biosynthesis and regulation of septum formation and cytokinesis (for review, see Fankhauser and Simanis 1994b).

A. Cytological Events Leading to Cytokinesis

The final stage of the fission yeast cell cycle is marked by septum formation and subsequent cytokinesis which completes cell division (Johnson et al. 1982; Robinow and Hyams 1989; Fankhauser and Simanis 1994b). Upon completion of mitosis, actin particles that were concentrated at both growing ends of the cell are relocated to the cell equator (Fig. 6) (Marks and Hyams 1985; Marks et al. 1986; Kanbe et al. 1989). The position of the actin contractile ring anticipates the site of septum formation. As mentioned previously, the MTOC regenerates cytoplasmic microtubules from the cell equator as well. The plasma membrane invaginates at this site to form the primary septum, where septum wall materials (detectable with the fluorescent dye Calcofluor) are deposited.

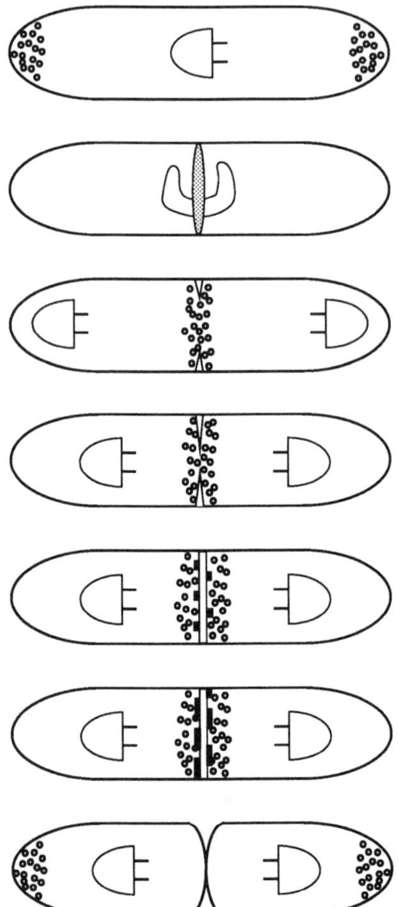

Figure 6 Stages of septum formation and cytokinesis in fission yeast. In interphase, actin granules (*open circles*) are localized at both growing ends of the cell. At the onset of mitosis, actin gathers at the medial plane to form a hollow ring overlying the nucleus (*shaded oval*). At the end of mitosis, the plasma membrane invaginates at this site to form the primary septum (*open boxes*). Secondary septum (*closed boxes*) forms on either side of the primary septa, which is then dissolved away to facilitate cell separation. Following the completion of the cell plate, actin granules relocate to the preexisting old ends of the cell. (Adapted from Fankhauser and Simanis 1994b.)

Secondary septa form on either side of the primary one, which is then "dissolved" away to bring about cell separation. When cells have completed the cell plate and have undergone cytokinesis, actin relocates to the preexisting old end of the cell and marks the growing phase of the cell cycle.

B. Actin-associated Elements

cdc3, *cdc4*, *cdc8*, and *cdc12* mutants are late septation mutants. These mutants undergo normal DNA replication and mitosis but have ill-formed septa, producing elongated, dumbbell-shaped cells with multiple nuclei (Nurse et al. 1976). Fine structural studies have suggested that the products of these genes may be required for normal F-actin function (Streiblová et al. 1984). This conclusion is supported by sequence and immunolocalization analyses of *cdc8*$^+$ and *cdc3*$^+$ (Balasubramanian et al. 1992, 1994). The *cdc8*$^+$ gene encodes an essential tropomyosin, which is known to be an actin-binding protein (Balasubramanian et al. 1992). Anti-Cdc8p immunoreative signals are found in well-separated binucleated cells as a medial band coincident to the actin ring. The essential *cdc3*$^+$ gene product is an *S. pombe* profilin, a known monomeric actin (G-actin)-binding protein. Cdc3p colocalizes with F-actin in the growing ends of the cell as well as in the medial plane of binucleated cells (Balasubramanian et al. 1994). The loss of the *cdc3*$^+$ gene product partially preserves the actin patches associated with interphase cells but eliminates the formation of F-actin contractile rings after nuclear division. The septum materials are delocalized, possibly because the *cdc3* mutant lacks the medial actin ring to guide their placement. Therefore, Cdc8p and Cdc3p are likely to interact with actin molecules and form essential components of the F-actin contractile ring during cytokinesis.

The importance of the actin cytoskeleton in cytokinesis is further reflected by the effects of overproducing actin and profilin. Excess actin in the cell causes multiple septation events and abnormal cytokinesis (Fantes 1989). Overproduction of profilin from either fission (Cdc3p) or budding yeast (Pfy1p) also leads to a *cdc3*-like septation defect (Balasubramanian et al. 1994). This phenotype results from the presence of excess profilin which sequesters actin monomers and prevents them from polymerizing into F-actin filaments. Together, these results suggest that gene products involved in the synthesis and maintenance of an F-actin contractile ring are essential for the cell division event.

C. Regulation of Septum Formation

1. Genetic Interaction of Septation Genes

Mutants *cdc7*, *cdc11*, *cdc14*, *cdc15*, and *cdc16* represent two classes of genes regulating septum formation (Nurse et al. 1976; Minet et al. 1979). *cdc7*, *cdc11*, *cdc14*, and *cdc15* display the phenotype of early septation mutants, producing highly elongated cells with multiple nuclei (Nurse et al. 1976). In these mutants, replication and mitosis occur normally despite the absence of septum formation (Nurse et al. 1976; Fankhauser and Simanis 1993, 1994a). Conversely, the *cdc16* mutant allows the periodic formation of septa in the absence of cell cycle progression, thus producing cells with multiple septa (Fankhauser et al. 1993). The *cdc16* multiseptation phenotype is blocked in the presence of a *cdc2* mutation, further emphasizing the dependency of cytokinesis events on prior initiation of mitosis. Double-mutant analyses have shown that *cdc15*+ activity is required for expression of the *cdc16* mutant phenotype, but it functions downstream from *cdc7*+, *cdc11*+, and *cdc14*+ (Fig. 7) (Marks et al. 1992). Several lines of genetic evidence indicate a strong interaction among *cdc7*+, *cdc14*+, *cdc11*+, and *cdc16*+ gene products (Marks et al. 1992). *cdc7*, *cdc14*, and *cdc16* mutants are synthetic-lethal in all double-mutant combinations. The *cdc11* mutant displays allele-specific lethality with *cdc7* and *cdc14* and allele-specific suppression with *cdc16*. Thus, the genetic control of septation may involve the activation of a complex containing Cdc7p, Cdc11p, Cdc14p, and Cdc16p that targets Cdc15p (directly or indirectly) for septum formation. The completion of a single septum may then generate a signal which acts through Cdc16p to inactivate the complex and allow cells to proceed to cell separation (Fig. 7).

2. cdc7+ *Protein Kinase*

Molecular analysis of *cdc7*+ has demonstrated that it is an indispensable serine/threonine protein kinase whose phosphorylation activity is essential for cytokinesis (Fankhauser and Simanis 1994a). Overexpression of *cdc7*+, in a dose-dependent manner, opposes the effect caused by the loss of *cdc7*+ and produces multiple septa without cleavage similar to *cdc16* mutants (Minet et al. 1979; Fankhauser and Simanis 1994a). This similarity suggests that Cdc7p kinase may either modify Cdc16p directly or regulate Cdc16p activity indirectly. The dependency of cytokinesis on the previous mitosis remains intact in G_2-arrested *cdc2* or *cdc25* mutants overexpressing *cdc7*+, because elevated levels of Cdc7p cannot overcome the mitotic entry requirement for cytokinesis (Fankhauser and Simanis 1994a). Overexpression of the catalytically inactive *cdc7*-K38R allele (with a lysine to arginine mutation in the kinase domain) in a wild-type strain surprisingly yields elongated multinucleated cells equivalent to cells lacking *cdc7*+ function. It is possible that massive quantities of inactive Cdc7p may titrate essential components interacting with Cdc7p to initiate septum formation. The *cdc11* defects can be suppressed by

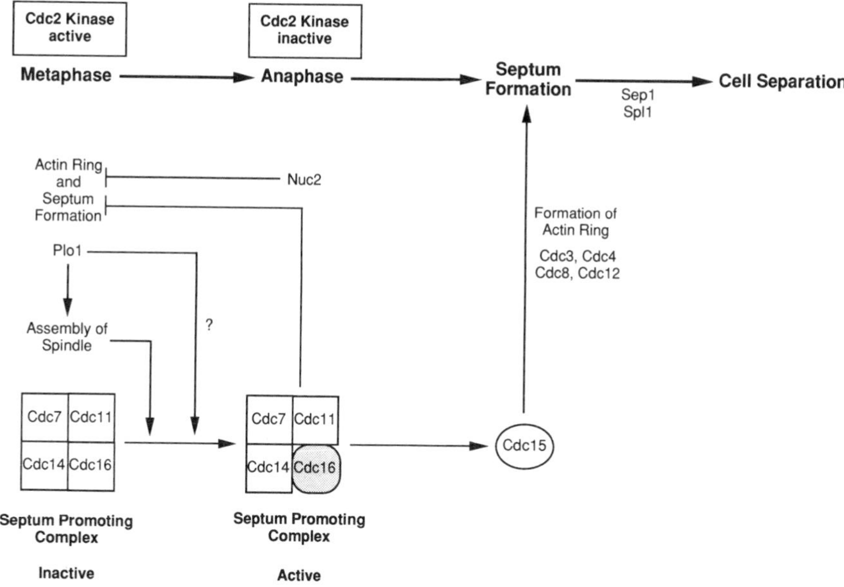

Figure 7 Model for the regulation of septum formation in fission yeast. During the G_2/M transition, Cdc2p kinase is activated and cells assemble a mitotic spindle that requires Plo1p kinase activity. The septum-promoting complex includes Cdc16p in the A form, which is required to maintain high levels of Cdc2p kinase activity in mitosis and inhibit the activities of Cdc7p, Cdc11p, and Cdc14p for actin ring formation and subsequent septation. After the completion of spindle formation at the metaphase/anaphase transition, Cdc16p is transformed to the B form, which allows the decay of Cdc2p kinase in a mechanism independent of *cdc16*+ and renders the septum-promoting complex active. Plo1p kinase also participates in initiating septum formation, possibly by activating the septum-promoting complex. The active complex then acts in part through Cdc15p to assemble the components properly for the contractile ring (Cdc3p, Cdc4p, Cdc8p, and Cdc12p) and form a septum. Finally, Sep1p and Spl1p are required to separate the two daughter cells. After the formation of a single septum, the B form of Cdc16p may be retransformed into the A form to block further rounds of septation until the next anaphase. In addition to mediating the progression through the metaphase/anaphase transition, Nuc2p has a critical role in coordinating mitosis with septation by restraining septum formation until the completion of mitosis. Thereby, the temporal order of the cell cycle is preserved. (Model adapted primarily from D'Urso and Nurse [1995] and Marks et al. [1992] with various modifications].) See text for a further description.

high Cdc7p levels, but functional *cdc14*+ and *cdc15*+ are necessary for the manifestation of the Cdc7p overproduction phenotype. These results provide additional support for the interaction between Cdc7p and

Cdc11p and the activities of Cdc14p and Cdc15p to activate a septum formation pathway (Fig. 7).

3. cdc14$^+$

The *cdc14*$^+$ gene product is an essential 28-kD polypeptide containing a potential amphipathic helix and a leucine zipper that may mediate interactions with other septation proteins, such as Cdc7p, Cdc11p, and Cdc15p (Fankhauser and Simanis 1993). Unlike *cdc7*$^+$, excess production of Cdc14p causes G_2 arrest instead of multiple septation. The inhibition of nuclear division seems to act through the Wee1p inhibitory pathway since this effect can be relieved in *wee1* or *cdc2.1w* (a *cdc2* allele that is insensitive to Wee1p activity) strains. Thus, *cdc14*$^+$ seems to have two roles. It is required for the initiation of septum formation and can also delay the onset of mitosis. Although the physiological relevance of mitotic inhibition is unknown, the dual functionality of Cdc14p may ensure the temporal order of the cell cycle by coordinating the events in mitosis and cytokinesis.

4. Septum-promoting Factor: Protein Kinase plo1$^+$

In addition to the defect in mitotic progression, mutation in *plo1*$^+$ also causes a block in septation leading to multinucleated cells (Ohkura et al. 1995). Whereas *cdc11* and *cdc14* still form actin rings despite their inability to form septa, *plo1* mutants fail either to form actin rings or to deposit septum materials, indicating that *plo1*$^+$ is likely to function early in the pathway for cytokinesis (Marks et al. 1986; Fankhauser and Simanis 1994b; Ohkura et al. 1995). Surprisingly, overexpression of *plo1*$^+$ drives septation in G_1- or G_2-arrested cells without apparent mitotic events, such as activated Cdc2p activity, spindle formation, and nuclear division. Since activation of Cdc2p–cyclin B is not necessary for the induction of septum formation by *plo1*$^+$, Plo1p may be a key inducer for septum formation and cytokinesis. Cdc2p kinase may directly or indirectly activate a set of mediators which induces a variety of mitotic events. Plo1p may be one such mediator which at some time is required for the progression through mitosis via formation of a functional bipolar spindle and at a later time functions as a major septum-promoting factor (Fig. 7).

D. Cell Separation

Cell separation marks the final step of the cell division cycle and is achieved by the action of β-glucanase which dissolves the primary septum (Horisberger and Rouvet-Vauthey 1985). Two mutants have been identified that participate in the cell separation process (Sipiczki et al. 1993). *sep1* mutants accumulate branching mycelium-like cells but express normal β-glucanase activity, indicating that *sep1*$^+$ is not involved in the regulation of β-glucanase. *spl1* mutants form a syncytium of one to five uncleaved cells with aberrant septa due to irregular deposition of cell wall material. *sep1*$^+$ and *spl1*$^+$ may perform closely related functions in that both display similar phenotypes and interact with the *cdc4* mutation to produce a more severe septation and cell separation defect. The *sep1* mutation may be associated with a cytoskeletal defect since it causes increased resistance to benomyl as well. The actual roles of Sep1p and Spl1p are unknown, but it is possible that they may be involved in the organization of cytoskeletal components associated with cytokinesis.

VIII. COORDINATION OF MITOSIS WITH CYTOKINESIS

The cytological and biochemical changes that the cell goes through during mitosis are very dramatic and in many ways can be detrimental to the cell if not properly controlled. Mitotic events need to be tightly regulated and restrained until cells are ready for mitosis. For example, cells must be restrained from entering mitosis if they have not completed DNA synthesis. Otherwise, the resulting daughters will lack a complete complement of DNA information and eventually die. This coordination invokes a number of control mechanisms, including the pre-Start G_1 checkpoint, replication checkpoint, and DNA damage checkpoint. These topics will be covered extensively in the chapter on the genetic control of the cell cycle in *S. pombe* (MacNeill and Nurse). Similarly, cytokinesis also needs to be regulated to prevent cell separation prior to the completion of mitosis. If such coordination is lost, chromosomes may become severed by premature cytokinesis, for example, leading to lethality. The gene products that participate in the coordination between mitosis and the subsequent cytokinesis as well as entry into the next cell cycle are discussed in this section.

A. *cdc16*$^+$ in Spindle Checkpoint

The *cdc16*$^+$ gene product maintains the orderly progression of the cell cycle by preventing untimely septum formation until the end of mitosis. The *cdc16*$^+$ gene product is homologous to budding yeast *BUB2*, which displays a multibudded mutant phenotype analogous to *cdc16* and can rescue the *cdc16* temperature-sensitive defect (Hoyt et al. 1991; Fankhauser et al. 1993). Similar to *bub2* mutants, *cdc16* cells fail to maintain high Cdc2p kinase activity in response to the lack of a mitotic spindle when cells are treated with the microtubule polymerization inhibitor thiabendazole. *cdc16* continues to undergo multiseptation even when mitosis has been blocked by thiabendazole treatment. The implication of these results are consistent with that of *BUB2* (Hoyt et al. 1991); i.e., Cdc16p may be a component of a feedback mechanism that monitors spindle integrity and maintains elevated Cdc2p kinase activity to prevent premature mitotic exit (Fig. 7). Simultaneously, Cdc16p restrains the septum activation machinery from initiating septation, possibly by interfering with Cdc7p, Cdc11p, or Cdc14p, until the onset of anaphase (Fig. 7).

B. *nuc2$^+$*

The *nuc2$^+$* gene product may be an important link coordinating mitotic events with cytokinesis and reentry into the next cell cycle. In addition to its role in mitosis, Nuc2p has other roles that become evident when the phenotypes of the temperature-sensitive *nuc2-663* allele at different temperatures and of *nuc2$^+$* overexpression are examined. Whereas loss of function of the *nuc2-663* allele at 36°C leads to metaphase arrest with a displaced nucleus, this mutant at lower restrictive temperatures (30°C and 33°C) attempts cell separation in the absence of nuclear division, resulting in "*cut*" cells (Hirano et al. 1988; Kumada et al. 1995). At 26°C, instead of blocking S-phase entry in response to nitrogen starvation, *nuc2-663* becomes arrested in G_2. The inability of *nuc2* to arrest in G_1 may be attributable to this mating defect at the permissive temperature (Hirano et al. 1988).

Overexpression of *nuc2$^+$*, in contrast, leads to multiple nuclear division cycles without septation, thus resulting in elongated multinucleated cells similar to the early septation mutants. If the effects of various temperatures on the *nuc2-663* allele reflect various levels of Nuc2p activities, the function of Nuc2p may vary dramatically dependent on the level of Nuc2p activity. Its primary function may be to traverse through the metaphase/anaphase transition, but Nuc2p may also be required to restrain septation until the end of mitosis and to promote progression into the next cell cycle. How Nuc2p accomplishes all these roles is unknown. However, it seems possible that Nuc2p interaction with different proteins through the TPR motif may trigger different Nuc2p functions accordingly (Hirano et al. 1990; Goebl and Yanagida 1991).

IX. SUMMARY

The study of mitosis and cytokinesis has been particularly revealing in the fission yeast *S. pombe*. Fission yeast is amenable to classical genetic analysis, and thus identification of genes required for cellular function can be accomplished through analysis of mutants defective in mitosis or cytokinesis. Recent advances in molecular, biochemical, and cytological techniques have also aided the investigation of the biological functions of these gene products. Since mitosis in fission yeast resembles that of higher eukaryotes, *S. pombe* in particular is an ideal model for the study of eukaryotic mitosis.

Mitosis and cytokinesis clearly comprise the most important events in the cell division cycle of *S. pombe*. As described in this chapter, this involves chromosome condensation, spindle formation and elongation, sister chromatid separation, septation, and cytokinesis. These different pathways seem to act independently to some extent, but these cellular events must coordinate with each other for a successful mitosis. Phosphorylation, dephosphorylation, and proteolysis of mitotic gene products for the large part may facilitate such coordination and orchestrate mitotic events. The molecular mechanisms that execute mitosis are very complex. From the identification of many gene products participating in these processes, our understanding of the regulation and mechanism of mitosis and cytokinesis has greatly advanced. Certain enzymatic activities act on the chromosomes themselves to condense the chromosomes, so that chromosomes can separate without tangling. Components of the mitotic spindle assemble and interact with regulators to generate the force for chromosome segregation. A number of gene products act in concert to trigger chromosome segregation; among these are dephosphorylation and proteolytic activities. Finally, the activation and proper placement of the septum enable the separation of two daughters with identical genetic complements. Although much information has been gained in the last several years, the molecular details of nuclear and cell division remain a mystery. We await eagerly for future biochemical and molecular studies to unravel this mystery.

ACKNOWLEDGMENTS

We are extremely grateful to all members of our laboratory for their help in the preparation of this chapter. We extend special thanks to Y. Nakaseko, D. Michaelson, Y. Adachi, H. Funabiki, Y. Saka, and K. Ishii for their discussion and critical comments. Research in this laboratory is supported by grants from the Ministry of Education, Science and Culture of Japan. S.S.Y.S. is supported by the Japan Society for the Promotion of Science.

REFERENCES

Adachi, Y., T. Toda, O. Niwa, and M. Yanagida. 1986. Differential expressions of essential and nonessential α-tubulin genes in *Schizosaccharomyces pombe*. *Mol. Cell. Biol.* **6:** 2168.

Aizawa, H., Y. Emori, A. Mori, H. Murofushi, H. Sakai, and K. Suzuki. 1991. Functional analyses of the domain structure of microtubule-associated protein-4 (MAP-U). *J. Biol. Chem.* **266:** 9841.

Alfa, C.E. and J.S. Hyams. 1990. Distribution of tubulin and actin through the cell division cycle of the fission yeast *Schizosaccharomyces japonicus* var. *versatilis:* A comparison with *Schizosaccharomyces pombe*. *J. Cell Sci.* **96:** 71.

Alfa, C.E., R. Booher, D. Beach, and J.S. Hyams. 1989. Fission yeast cyclin: Subcellular localization and cell cycle regulation. *J. Cell Sci. Suppl.* **12:** 9.

Alfa, C.E., B. Ducommun, D. Beach, and J.S. Hyams. 1990. Distinct nuclear and spindle pole body populations of cyclin-cdc2 in fission yeast. *Nature* **347:** 680.

Allshire, R.C., J.-P. Javerzat, N.J. Redhead, and G. Cranston. 1994. Position effect variegation at fission yeast centromeres. *Cell* **76:** 157.

Balasubramanian, M.K., D.M. Helfman, and S.M. Hemmingsen. 1992. A new tropomyosin essential for cytokinesis in the fission yeast *S. pombe*. *Nature* **360:** 84.

Balasubramanian, M.K., B.R. Hirani, J.D. Burke, and K.L. Gould. 1994. The *Schizosaccharomyces pombe cdc3*$^{+}$ gene encodes a profilin essential for cytokinesis. *J. Cell Biol.* **125:** 1289.

Baum, M., V.K. Ngan, and L. Clarke. 1994. The centromeric K-type repeat and the central core are together sufficient to establish a functional *Schizosaccharomyces pombe* centromere. *Mol. Biol. Cell* **5:** 747.

Baum, P., C. Yip, L. Goetsch, and B. Byers. 1988. A yeast essential for regulation of spindle pole duplication. *Mol. Cell. Biol.* **8:** 5386.

Beach, D. and P. Nurse. 1981. High-frequency transformation of the fission yeast *Schizosaccharomyces pombe*. *Nature* **290:** 140.

Bischoff, F.R. and H. Ponstingl. 1991. Catalysis of guanine nucleotide exchange on Ran by the mitotic regulator RCC1. *Nature* **354:** 80.

Booher, R. and D. Beach. 1987. Interaction between *cdc13*$^{+}$ and *cdc2*$^{+}$ in the control of mitosis in fission yeast: Dissociation of the G1 and G2 roles of the *cdc2*$^{+}$ protein kinase. *EMBO J.* **6:** 3441.

———. 1988. Involvement of *cdc13*$^{+}$ in mitotic control in *Schizosaccharomyces pombe:* Possible interaction of the gene product with microtubules. *EMBO J.* **7:** 2321.

———. 1989. Involvement of a type 1 protein phosphatase encoded by *bws1*$^{+}$ in fission yeast mitotic control. *Cell* **57:** 1009.

Booher, R.N., C.E. Alfa, J.S. Hyams, and D.H. Beach. 1989. The fission yeast cdc2/cdc13/suc1 protein kinase: Regulation of catalytic activity and nuclear localization. *Cell* **58:** 485.

Broek, D., R. Bartlett, K. Crawford, and P. Nurse. 1991. Involvement of p34^{cdc2} in establishing the dependency of S phase on mitosis. *Nature* **349:** 388.

Chalfie, M., Y. Tu, G. Euskirchen, W.W. Ward, and D.C. Prasher. 1994. Green fluorescent protein as a marker for gene expression. *Science* **263:** 802.

Chang, F. and P. Nurse. 1993. Finishing the cell cycle: Control of mitosis and cytokinesis in fission yeast. *Trends Genet.* **9:** 333.

Chikashige, Y., N. Kinoshita, Y. Nakaseko, T. Matsumoto, S. Murakami, O. Niwa, and M. Yanagida. 1989. Composite motifs and repeat symmetry in *S. pombe* centromeres: Direct analysis by integration of *Not*I restriction sites. *Cell* **57:** 739.

Chuang, P.-T., D.G. Albertson, and B.J. Meyer. 1994. DPY-27: A chromosome condensation protein homolog that regulates *C. elegans* dosage compensation through association with the X chromosome. *Cell* **79:** 459.

Clarke, L. 1990. Centromeres of budding and fission yeasts. *Trends Genet.* **6:** 150.

Clarke, L. and M.P. Baum. 1990. Functional analysis of a centromere from fission yeast: A role for centromere-specific repeated DNA sequences. *Mol. Cell. Biol.* **10:** 1863.

Clarke, L. and J. Carbon. 1985. The structure and function of yeast centromeres. *Annu. Rev. Genet.* **19:** 29.

Cohen, P. 1989. The structure and regulation of protein phosphatases. *Annu. Rev. Biochem.* **58:** 453.

Coxon, A., K. Maundrell, and S.E. Kearsey. 1992. Fission yeast *cdc21$^+$* belongs to a family of proteins involved in an early step of chromosome replication. *Nucleic Acids Res.* **20:** 5571.

Creanor, J. and J.M. Mitchison. 1990. Continued DNA synthesis after a mitotic block in the double mutant *cut1 cdc11* of the fission yeast *Schizosaccharomyces pombe*. *J. Cell Sci.* **96:** 435.

Dasso, M. 1993. RCC1 in the cell cycle: The regulator of chromosome condensation takes on new roles. *Trends Biochem. Sci.* **18:** 96.

Demeter, J., M. Morphew, and S. Sazer. 1995. A mutation in the RCC1-related protein *pim1* results in nuclear envelope fragmentation in fission yeast. *Proc. Natl. Acad. Sci.* **92:** 1436.

DeVoti, J., G. Seydoux, D. Beach, and M. McLeod. 1991. Interaction between *ran1$^+$* protein kinase and cAMP dependent protein kinase as negative regulators of fission yeast meiosis. *EMBO J.* **10:** 3759.

Díaz, M., Y. Sanchez, T. Bennett, C.R. Sun, C. Godoy, F. Tamanoi, A. Duran, and P. Perez. 1993. The *Schizosaccharomyces pombe cwg2$^+$* gene codes for the β-subunit of a geranylgeranyltransferase type I required for β-glucan synthesis. *EMBO J.* **12:** 5245.

Ding, R., K.L. McDonald, and J.R. McIntosh. 1993. Three-dimensional reconstruction and analysis of mitotic spindles from the yeast, *Schizosaccharomyces pombe*. *J. Cell Biol.* **120:** 141.

Dubey, D.D., J. Zhu, D.L. Carlson, K. Sharma, and J.A. Huberman. 1994. Three ARS elements contribute to the *ura4* replication origin region in the fission yeast, *Schizosaccharomyces pombe*. *EMBO J.* **13:** 3638.

Ducommun, B., P. Brambilla, M.-A. Félix, B.R. Franza, Jr., E. Karsenti, and G. Draetta. 1991. cdc2 phosphorylation is required for its interaction with cyclin. *EMBO J.* **10:** 3311.

D'Urso, G. and P. Nurse. 1995. Checkpoints in the cell cycle of fission yeast. *Curr. Opin. Genet. Dev.* **5:** 12.

Egel, R., O. Nielsen, and D. Weilguny. 1990. Sexual differentiation in fission yeast. *Trends Genet.* **6:** 369.

Enoch, T. and P. Nurse. 1991. Coupling M phase and S phase: Controls maintaining the dependence of mitosis on chromosome replication. *Cell* **65:** 921.

Fan, J.-B., Y. Chikashige, C.L. Smith, O. Niwa, M. Yanagida, and C.R. Cantor. 1988. Construction of a *Not*I restriction map of the fission yeast *Schizosaccharomyces pombe* genome. *Nucleic Acids Res.* **17:** 2801.

Fankhauser, C. and V. Simanis. 1993. The *Schizosaccharomyces pombe cdc14* gene is required for septum formation and can also inhibit nuclear division. *Mol. Biol. Cell* **4:** 531.

———. 1994a. The *cdc7* protein kinase is a dosage dependent regulator of septum formation in fission yeast. *EMBO J.* **13:** 3011.

———. 1994b. Cold fission: Splitting the *pombe* cell at room temperature. *Trends Cell Biol.* **4:** 96.

Fankhauser, C., J. Marks, A. Reymond, and V. Simanis. 1993. The *S. pombe cdc16* gene is required both for maintenance of $p34^{cdc2}$ kinase activity and regulation of septum formation: A link between mitosis and cytokinesis? *EMBO J.* **12:** 2697.

Fantes, P.A. 1977. Control of cell size and cycle time in *Schizosaccharomyces pombe. J. Cell Sci.* **24:** 51.

———. 1989. Cell cycle controls. In *Molecular biology of the fission yeast* (ed. A. Nasim et al.), p. 127. Academic Press, San Diego.

Fantes, P. and P. Nurse. 1977. Control of cell size at division in fission yeast by a growth-modulated size control over nuclear division. *Exp. Cell Res.* **107:** 377.

———. 1978. Control of the timing of cell division in fission yeast: Cell size mutants reveal a second control pathway. *Exp. Cell Res.* **115:** 317.

Fawell, E., S. Bowden, and J. Armstrong. 1992. A homologue of the *ras*-related *CDC42* gene from *Schizosaccharomyces pombe. Gene* **114:** 153.

Featherstone, C. and P. Russell. 1991. Fission yeast $p107^{wee1}$ mitotic inhibitor is a tyrosine/serine kinase. *Nature* **349:** 808.

Feilotter, H., P. Nurse, and P.G. Young. 1991. Genetic and molecular analysis of *cdr1/nim1* in *Schizosaccharomyces pombe. Genetics* **127:** 309.

Fishel, B., H. Amstutz, M. Baum, J. Carbon, and L. Clarke. 1988. Structural organization and functional analysis of centromeric DNA in the fission yeast *Schizosaccharomyces pombe. Mol. Cell. Biol.* **8:** 754.

Fleig, U.N., K.L. Gould, and P. Nurse. 1992. A dominant negative allele of $p34^{cdc2}$ shows altered phosphoamino acid content and sequesters $p56^{cdc13}$ cyclin. *Mol. Cell. Biol.* **12:** 2295.

Fukui, Y. and M. Yamamoto. 1988. Isolation and characterization of *Schizosaccharomyces pombe* mutants phenotypically similar to $ras1^-$. *Mol. Gen. Genet.* **215:** 26.

Fukui, Y., T. Kozasa, Y. Kaziro, T. Takeda, and M. Yamamoto. 1986. Role of a *ras* homolog in the life cycle of *Schizosaccharomyces pombe. Cell* **44:** 329.

Funabiki, H., K. Kumada, and M. Yanagida. 1996a. Fission yeast Cut1 and Cut2 essential for sister chromatid separation concentrate along the metaphase spindle and form a large complex. *EMBO J.* (in press).

Funabiki, H., I. Hagan, S. Uzawa, and M. Yanagida. 1993. Cell cycle-dependent specific positioning and clustering of centromeres and telomeres in fission yeast. *J. Cell Biol.* **121:** 961.

Funabiki, H., H. Yamano, K. Kumada, K. Nagao, T. Hunt, and M. Yanagida. 1996b. Cut2 proteolysis required for sister-chromatid separation in fission yeast. *Nature* **381:** 438.

Gasser, S.M. 1995. Coiling up chromosomes. *Curr. Biol.* **5:** 357.

Goebl, M. and M. Yanagida. 1991. The TPR snap helix: A novel protein repeat motif from mitosis to transcription. *Trends Biochem. Sci.* **16:** 173.

Gordon, C., G. McGurk, P. Dillon, C. Rosen, and N.D. Hastie. 1993. Defective mitosis due to a mutation in the gene for a fission yeast 26S protease subunit. *Nature* **366:** 355.

Gottschling, D.E. 1992. Telomere-proximal DNA in *Saccharomyces cerevisiae* is refractory to methyltransferase activity *in vivo. Proc. Natl. Acad. Sci.* **89:** 4062.

Gottschling, D.E., O.M. Aparicio, B.L. Billington, and V.A. Zakian. 1990. Position effect at *S. cerevisiae* telomeres: Reversible repression of Pol II transcription. *Cell* **63:** 751.

Gould, K.L. and P. Nurse. 1989. Tyrosine phosphorylation of the fission yeast *cdc2*$^{+}$ protein kinase regulates entry into mitosis. *Nature* **342:** 39.

Gould, K.L., S. Moreno, D.J. Owen, S. Sazer, and P. Nurse. 1991. Phosphorylation at Thr 167 is required for *Schizosaccharomyces pombe* p34^{cdc2} function. *EMBO J.* **10:** 3297.

Gygax, A. and P. Thuriaux. 1984. A revised chromosome map of the fission yeast *Schizosaccharomyces pombe. Curr. Genet.* **8:** 85.

Hagan, I.M. and J.S. Hyams. 1988. The use of cell division cycle mutants to investigate the control of microtubule distribution in the fission yeast *Schizosaccharomyces pombe. J. Cell Sci.* **89:** 343.

Hagan, I.M. and M. Yanagida. 1990. Novel potential mitotic motor protein encoded by the fission yeast *cut7*$^{+}$ gene. *Nature* **347:** 563.

———. 1992. Kinesin-related cut7 protein associates with mitotic and meiotic spindles in fission yeast. *Nature* **356:** 74.

———. 1995. The product of the spindle formation gene *sad1*$^{+}$ associates with the fission yeast spindle pole body and is essential for viability. *J. Cell Biol.* **129:** 1033.

Hagan, I., J. Hayles, and P. Nurse. 1988. Cloning and sequencing of the cyclin-related *cdc13*$^{+}$ gene and a cytological study of its role in fission yeast mitosis. *J. Cell Sci.* **91:** 587.

Hagan, I.M., P.N. Riddle, and J.S. Hyams. 1990. Intramitotic controls in the fission yeast *Schizosaccharomyces pombe:* The effect of cell size on spindle length and the timing of mitotic events. *J. Cell Biol.* **110:** 1617.

Hahnenberger, K.M., J. Carbon, and L. Clarke. 1991. Identification of DNA regions required for mitotic and meiotic functions within the centromere of *Schizosaccharomyces pombe* chromosome I. *Mol. Cell. Biol.* **11:** 2206.

Hahnenberger, K.M., M.P. Baum, C.M. Polizzi, J. Carbon, and L. Clarke. 1989. Construction of functional artificial minichromosomes in the fission yeast *Schizosaccharomyces pombe. Proc. Natl. Acad. Sci.* **86:** 577.

Hartwell, L.H. and T.A. Weinert. 1989. Checkpoints: Controls that ensure the order of cell cycle events. *Science* **246:** 629.

Hayles, J. and P. Nurse. 1992. Genetics of the fission yeast *Schizosaccharomyces pombe. Annu. Rev. Genet.* **26:** 373.

Hayles, J., D. Fisher, A. Woollard, and P. Nurse. 1994. Temporal order of S phase and mitosis in fission yeast is determined by the state of the p34^{cdc2}-mitotic B cyclin complex. *Cell* **78:** 813.

Heyer, W.D., M. Sipiczki, and J. Kohli. 1986. Replicating plasmids in *Schizosaccharomyces pombe:* Improvement of symmetric segregation by a new genetic element. *Mol. Cell Biol.* **6:** 80.

Hindley, J. and G.A. Phear. 1984. Sequence of the cell division gene *CDC2* from *Schizosaccharomyces pombe*; patterns of splicing and homology to protein kinases. *Gene* **31:** 129.

Hirano, T. 1995. Biochemical and genetic dissection of mitotic chromosome condensation. *Trends Biochem. Sci.* **20:** 357.

Hirano, T. and T. Mitchison. 1994. A heterodimeric coiled-coil protein required for mitotic chromosome condensation in vitro. *Cell* **79:** 449.

Hirano, T., Y. Hiraoka, and M. Yanagida. 1988. A temperature-sensitive mutation of the *Schizosaccharomyces pombe* gene *nuc2*$^{+}$ that encodes a nuclear scaffold-like protein blocks spindle elongation in mitotic anaphase. *J. Cell Biol.* **106:** 1171.

Hirano, T., T.J. Mitchison, and J.R. Swedlow. 1995. The SMC family: From chromosome condensation to dosage compensation. *Curr. Opin. Cell Biol.* **7:** 329.

Hirano, T., S. Funahashi, T. Uemura, and M. Yanagida. 1986. Isolation and characterization of *Schizosaccharomyces pombe cut* mutants that block nuclear division but not cytokinesis. *EMBO J.* **5:** 2973.

Hirano, T., N. Kinoshita, K. Morikawa, and M. Yanagida. 1990. Snap helix with knob and hole: Essential repeats in *S. pombe* nuclear protein *nuc2*$^{+}$. *Cell* **60:** 319.

Hirano, T., G. Konoha, T. Toda, and M. Yanagida. 1989. Essential roles of the RNA polymerase I largest subunit and DNA topoisomerases in the formation of fission yeast nucleolus. *J. Cell Biol.* **108:** 243.

Hiraoka, Y., T. Toda, and M. Yanagida. 1984. The *NDA3* gene of fission yeast encodes β-tubulin: A cold-sensitive *nda3* mutation reversibly blocks spindle formation and chromosome movement in mitosis. *Cell* **39:** 349.

Hisamoto, N., D.L. Frederick, K. Sugimoto, K. Tatchell, and K. Matsumoto. 1995. The *EGP1* gene may be a positive regulator of protein phosphatase type 1 in the growth control of *Saccharomyces cerevisiae. Mol. Cell. Biol.* **15:** 3767.

Hoheisel, J.D., E. Maier, R. Mott, L. McCarthy, A.V. Grigoriev, L.C. Schalkwyk, D. Nizetic, F. Francis, and H. Lehrach. 1993. High resolution cosmid and P1 maps spanning the 14 Mb genome of the fission yeast *S. pombe. Cell* **73:** 109.

Horio, T., S. Uzawa, M.K. Jung, B.R. Oakley, K. Tanaka, and M. Yanagida. 1991. The fission yeast γ-tubulin is essential for mitosis and is localized at microtubule organizing centers. *J. Cell Sci.* **99:** 693.

Horisberger, M. and M. Rouvet-Vauthey. 1985. Cell wall architecture of the fission yeast *Schizosaccharomyces pombe. Experientia* **41:** 748.

Houman, F. and C. Holm. 1994. *DBF8*, an essential gene required for efficient chromosome segregation in *Saccharomyces cerevisiae. Mol. Cell. Biol.* **14:** 6350.

Hoyt, M.A., L. Totis, and B.T. Roberts. 1991. *S. cerevisiae* genes required for cell cycle arrest in response to loss of microtubule function. *Cell* **66:** 507.

Irniger, S., S. Piatti, C. Michaelis, and K. Nasmyth. 1995. Genes involved in sister chromatid separation are needed for B-type cyclin proteolysis in budding yeast. *Cell* **81:** 269.

Ishii, T., K. Kumada, T. Toda, and M. Yanagida. 1996. Requirement of PP1 phosphatase and 20S cyclosome/APC for the onset of anaphase is lessened by the dosage increase of a novel gene *sds23*$^{+}$. *EMBO J.* (in press).

Johnson, B.F., G.B. Calleja, B.Y. Yoo, M. Zuker, and I.J. McDonald. 1982. Cell division: Key to cellular morphogenesis in the fission yeast, *Schizosaccharomyces. Int. Rev. Cytol.* **75:** 167.

Johnston, L.H. and D.G. Barker. 1987. Characterisation of an autonomously replicating sequence from the fission yeast *Schizosaccharomyces pombe. Mol. Gen. Genet.* **207:** 161.

Kanai, Y., J. Chen, and N. Hirokawa. 1992. Microtubule bundling by tau proteins *in vivo:*

Analysis of functional domains. *EMBO J.* **11:** 3953.

Kanbe, T., I. Kobayashi, and K. Tanaka. 1989. Dynamics of cytoplasmic organelles in the cell cycle of the fission yeast *Schizosaccharomyces pombe:* Three-dimensional reconstruction from serial sections. *J. Cell Sci.* **94:** 647.

Kanbe, T., Y. Hiraoka, K. Tanaka, and M. Yanagida. 1990. The transition of cells of the fission yeast β-tubulin mutant *nda3-311* as seen by freeze-substitution electron microscopy: Requirement of functional tubulin for spindle pole body duplication. *J. Cell Sci.* **96:** 275.

King, R.W., P.K. Jackson, and M.W. Kirschner. 1994. Mitosis in transition. *Cell* **79:** 563.

King, R.W., J.M. Peters, S. Tugendreich, M. Rolfe, P. Hieter, and M.W. Kirschner. 1995. A 20S complex containing CDC27 and CDC16 catalyzes the mitosis-specific conjugation of ubiquitin to cyclin B. *Cell* **81:** 279.

Kinoshita, K., T. Nemoto, K. Nabeshima, H. Kondoh, H. Niwa, and M. Yanagida. 1996. The regulatory subunits of fission yeast protein phosphatase 2A (PP2A) affect cell morphogenesis, cell wall synthesis, osmoregulation and cytokinesis. *Genes Cells* **1:** 29.

Kinoshita, N., M. Goebl, and M. Yanagida. 1991a. The fission yeast *dis3*$^+$ gene encodes a 110-kDa essential protein implicated in mitotic control. *Mol. Cell. Biol.* **11:** 5839.

Kinoshita, N., H. Ohkura, and M. Yanagida. 1990. Distinct, essential roles of type 1 and 2A protein phosphatases in the control of the fission yeast cell division cycle. *Cell* **63:** 405.

Kinoshita, N., H. Yamano, H. Niwa, T. Yoshida, and M. Yanagida. 1993. Negative regulation of mitosis by the fission yeast protein phosphatase *ppa2*$^+$. *Genes Dev.* **7:** 1059.

Kinoshita, N., H. Yamano, F. Le Bouffant-Sladeczek, H. Kurooka, H. Ohkura, E.M. Stone, M. Takeuchi, T. Toda, T. Yoshida, and M. Yanagida. 1991b. Sister-chromatid separation and protein dephosphorylation in mitosis. *Cold Spring Harbor Symp. Quant. Biol.* **56:** 621.

Kitada, K., A.L. Johnson, L.H. Johnston, and A. Sugino. 1993. A multicopy suppressor gene of the *Saccharomyces cerevisiae* G1 cell cycle mutant gene *dbf4* encodes a protein kinase and is identified as *CDC5*. *Mol. Cell. Biol.* **13:** 4445.

Kobe, B. and J. Deisenhofer. 1994. The leucine-rich repeat: A versatile binding motif. *Trends Biochem. Sci.* **19:** 415.

Kobori, H., N. Yamada, A. Taki, and M. Osumi. 1989. Actin is associated with the formation of the cell wall in reverting protoplasts of the fission yeast *Schizosaccharomyces pombe*. *J. Cell Sci.* **94:** 635.

Kohli, J. 1987. Genetic nomenclature and gene list of the fission yeast *Schizosaccharomyces pombe*. *Curr. Genet.* **11:** 575.

Kohli, J., H. Hottinger, P. Munz, A. Strauss, and P. Thuriaux. 1977. Genetic mapping in *Schizosaccharomyces pombe* by mitotic and meiotic analysis and induced haploidization. *Genetics* **87:** 471.

Kuhn, R.M., L. Clarke, and J. Carbon. 1991. Clustered tRNA genes in *Schizosaccharomyces pombe* centromeric DNA sequence repeats. *Proc. Natl. Acad. Sci.* **88:** 1306.

Kumada, K., S. Su, M. Yanagida, and T. Toda. 1995. Fission yeast TPR-family protein nuc2 is required for G1-arrest upon nitrogen starvation and is an inhibitor of septum formation. *J. Cell Sci.* **108:** 895.

Kumagai, A. and W.G. Dunphy. 1991. The cdc25 protein controls tyrosine dephosphorylation of the cdc2 protein in a cell-free system. *Cell* **64:** 903.

Lamb, J.R., W.A. Michaud, R.S. Sikorski, and P.A. Hieter. 1994. Cdc16p, Cdc23p and

Cdc27p form a complex essential for mitosis. *EMBO J.* **13:** 4321.

Lee, M.S., T. Enoch, and H. Piwnica-Worms. 1994. *mik1*$^{+}$ encodes a tyrosine kinase that phosphorylates p34^{cdc2} on tyrosine 15. *J. Biol. Chem.* **269:** 30530.

Leupold, U. 1991. Characterization of a partially fertile ras1-like *ste10*-UGA nonsense mutant of fission yeast. *Curr. Genet.* **20:** 75.

Levin, D.E. and J.M. Bishop. 1990. A putative protein kinase gene (*kin1*$^{+}$) is important for growth polarity in *Schizosaccharomyces pombe. Proc. Natl. Acad. Sci.* **87:** 8272.

Llamazares, S., A. Moreira, A. Tavares, C. Girdham, B.A. Spruce, C. Gonzalez, R.E. Karess, D.M. Glover, and C.E. Sunkel. 1991. *polo* encodes a protein kinase homolog required for mitosis in *Drosophila. Genes Dev.* **5:** 2153.

Lundgren, K., N. Walworth, R. Booher, M. Dembski, M. Kirschner, and D. Beach. 1991. mik1 and wee1 cooperate in the inhibitory tyrosine phosphorylation of cdc2. *Cell* **64:** 1111.

MacKelvie, S.H., P.D. Andrews, and M.J.R. Stark. 1995. The *Saccharomyces cerevisiae* gene *SDS22* encodes a potential regulator of the mitotic function of yeast type 1 protein phosphatase. *Mol. Cell. Biol.* **15:** 3777.

MacNeill, S.A. and P.A. Fantes. 1993. Methods for analysis of the fission yeast cycle. In *The cell cycle* (ed. P. Fantes and R. Brooks.), p. 93. IRL Press, Oxford, United Kingdom.

Maeda, T., N. Mochizuki, and M. Yamamoto. 1990. Adenylyl cyclase is dispensable for vegetative cell growth in the fission yeast *Schizosaccharomyces pombe. Proc. Natl. Acad. Sci.* **87:** 7814.

Maeda, T., Y. Watanabe, H. Kunitomo, and M. Yamamoto. 1994. Cloning of the *pka1* gene encoding the catalytic subunit of the cAMP-dependent protein kinase in *Schizosaccharomyces pombe. J. Biol. Chem.* **269:** 9632.

Marks, J. and J.S. Hyams. 1985. Localization of F-actin through the cell division cycle of *Schizosaccharomyces pombe. Eur. J. Cell Biol.* **39:** 27.

Marks, J., C. Fankhauser, and V. Simanis. 1992. Genetic interactions in the control of septation in *Schizoaccharomyces pombe. J. Cell Sci.* **101:** 801.

Marks, J., I.M. Hagan, and J.S. Hyams. 1986. Growth polarity and cytokinesis in fission yeast: The role of the cytoskeleton. *J. Cell Sci. Suppl.* **5:** 229.

Masuda, H. 1995. The formation and functioning of yeast mitotic spindles. *BioEssays* **17:** 45.

Masuda, H., M. Sevik, and W.Z. Cande. 1992. In vitro microtubule-nucleating activity of spindle pole bodies in fission yeast *Schizosaccharomyces pombe:* Cell cycle-dependent activation in *Xenopus* cell-free extracts. *J. Cell Biol.* **117:** 1055.

Masuda, H., T. Hirano, M. Yanagida, and W.Z. Cande. 1990. In vitro reactivation of spindle elongation in fission yeast *nuc2* mutant cells. *J. Cell Biol.* **110:** 417.

Matsumoto, S. and M. Yanagida. 1985. Histone gene organization of fission yeast: A common upstream sequence. *EMBO J.* **4:** 3531.

Matsumoto, T. and D. Beach. 1991. Premature initiation of mitosis in yeast lacking RCC1 or an interacting GTPase. *Cell* **66:** 347.

———. 1993. Interaction of the pim1/spi1 mitotic checkpoint with a protein phosphatase. *Mol. Biol. Cell* **4:** 337.

Matsumoto, T., S. Murakami, O. Niwa, and M. Yanagida. 1990. Construction and characterization of centric circular and acentric linear chromosomes in fission yeast. *Curr. Genet.* **18:** 323.

Matsumoto, T., K. Fukui, O. Niwa, N. Sugawara, J.W. Szostak, and M. Yanagida. 1987.

Identification of healed terminal DNA fragments in linear minichromosomes of *Schizosaccharomyces pombe*. *Mol. Cell. Biol.* **7:** 4424.

Matsusaka, T., D. Hirata, M. Yanagida, and T. Toda. 1995. A novel protein kinase gene *ssp1$^+$* is required for alteration of growth polarity and actin localization in fission yeast. *EMBO J.* **14:** 3325.

Maundrell, K., A. Hutchison, and S. Shall. 1988. Sequence analysis of ARS elements in fission yeast. *EMBO J.* **7:** 2203.

Maundrell, K., A.P.H. Wright, M. Piper, and S. Shall. 1985. Evaluation of heterologous ARS activity in *S. cerevisiae* using cloned DNA from *S. pombe*. *Nucleic Acids Res.* **13:** 3711.

May, G.S., C.A. McGoldrick, C.L. Holt, and S.H. Denison. 1992. The *bimB3* mutation of *Aspergillus nidulans* uncouples DNA replication from the completion of mitosis. *J. Biol. Chem.* **267:** 15737.

May, J.W. and J.M. Mitchison. 1986. Length growth in fission yeast cells measured by two novel techniques. *Nature* **322:** 752.

McCully, E.K. and C.F. Robinow. 1971. Mitosis in the fission yeast *Schizosaccharomyces pombe:* A comparative study with light and electron microscopy. *J. Cell Sci.* **9:** 475.

McGrew, J.T., L. Goetsch, B. Byers, and P. Baum. 1992. Requirement for *ESP1* in the nuclear division of *Saccharomyces cerevisiae*. *Mol. Biol. Cell* **3:** 1443.

Millar, J.B.A. and P. Russell. 1992. The cdc25 M-phase inducer: An unconventional protein phosphatase. *Cell* **68:** 407.

Millar, J.B.A., G. Lenaers, and P. Russell. 1992a. *Pyp3* PTPase acts as a mitotic inducer in fission yeast. *EMBO J.* **11:** 4933.

Millar, J.B.A., P. Russell, J.E. Dixon, and K.L. Guan. 1992b. Negative regulation of mitosis by two functionally overlapping PTPases in fission yeast. *EMBO J.* **11:** 4943.

Millar, J.B.A., C.H. McGowan, G. Lenaers, R. Jones, and P. Russell. 1991. $p80^{cdc25}$ mitotic inducer is the tyrosine phosphatase that activates $p34^{cdc2}$ kinase in fission yeast. *EMBO J.* **10:** 4301.

Miller, P.J. and D.I. Johnson. 1994. Cdc42p GTPase is involved in controlling polarized cell growth in *Schizosaccharomyces pombe*. *Mol. Cell. Biol.* **14:** 1075.

Minet, M., P. Nurse, P. Thuriaux, and J.M. Mitchison. 1979. Uncontrolled septation in a cell division cycle mutant of the fission yeast *Schizosaccharomyces pombe*. *J. Bacteriol.* **137:** 440.

Mirabito, P.M. and N.R. Morris. 1993. BIMA, a TPR-containing protein required for mitosis, localizes to the spindle pole body in *Aspergillus nidulans*. *J. Cell Biol.* **120:** 959.

Mitchison, J.M. 1970. Physiological and cytological methods for *Schizosaccharomyces pombe*. *Methods Cell Physiol.* **4:** 131.

Mitchison, J.M. and P. Nurse. 1985. Growth in cell length in the fission yeast *Schizosaccharomyces pombe*. *J. Cell Sci.* **75:** 357.

Miyake, S., N. Okishio, I. Samejima, Y. Hiraoka, T. Toda, I. Saitoh, and M. Yanagida. 1993. Fission yeast genes *nda1$^+$* and *nda4$^+$*, mutations of which lead to S-phase block, chromatin alteration and Ca^+ suppression, are members of the *CDC46/MCM2* family. *Mol. Biol. Cell* **4:** 1003.

Miyata, H., M. Miyata, and M. Ito. 1978. The cell cycle in the fission yeast, *Schizosaccharomyces pombe*. I. Relationship between cell size and cycle time. *Cell Struct. Funct.* **3:** 39.

Mizukami, T., W.I. Chang, I. Garkavtsev, N. Kaplan, D. Lombardi, T. Matsumoto, O. Niwa, A. Kounosu, M. Yanagida, T.G. Marr, and D. Beach. 1993. A 13 kb resolution cosmid map of the 14 Mb fission yeast genome by nonrandom sequence-tagged site mapping. *Cell* **73:** 121.

Mochizuki, N. and M. Yamamoto. 1992. Reduction in the intracellular cAMP level triggers initiation of sexual development in fission yeast. *Mol. Gen. Genet.* **233:** 17.

Moreno, S., J. Hayles, and P. Nurse. 1989. Regulation of p34^{cdc2} protein kinase during mitosis. *Cell* **58:** 361.

Moreno, S., P. Nurse, and P. Russell. 1990. Regulation of mitosis by cyclic accumulation of p80^{cdc25} mitotic inducer in fission yeast. *Nature* **344:** 549.

Murakami, S., T. Matsumoto, O. Niwa, and M. Yanagida. 1991. Structure of the fission yeast centromere *cen3:* Direct analysis of the reiterated inverted region. *Chromosoma* **101:** 214.

Nabeshima, K., H. Kurooka, M. Takeuchi, K. Kinoshita, Y. Nakaseko, and M. Yanagida. 1995. p93^{dis1}, which is required for sister chromatid separation, is a novel microtubule and spindle pole body-associating protein phosphorylated at the Cdc2 target sites. *Genes Dev.* **9:** 1572.

Nadin-Davis, S.A. and A. Nasim. 1990. *Schizosaccharomyces pombe ras1* and *byr1* are functionally related genes of the *ste* family that affect starvation-induced transcription of mating-type genes. *Mol. Cell. Biol.* **10:** 549.

Nadin-Davis, S.A., A. Nasim, and D. Beach. 1986. Involvement of *ras* in sexual differentiation but not in growth control in fission yeast. *EMBO J.* **5:** 2963.

Nakaseko, Y., N. Kinoshita, and M. Yanagida. 1987. A novel sequence common to the centromere regions of *Schizosaccharomyces pombe* chromosomes. *Nucleic Acids Res.* **15:** 4705.

Nakaseko, Y., Nabeshima, K. Kinoshita, and M. Yanagida. 1996. Dissection of fission yeast microtubule associating protein p93Dis1: Regions implicated in regulated localization and microtubule interaction. *Genes Cells* **1:** 633.

Nakaseko, Y., Y. Adachi, S. Funahashi, O. Niwa, and M. Yanagida. 1986. Chromosome walking shows a highly homologous repetitive sequence present in all the centromere regions of fission yeast. *EMBO J.* **5:** 1011.

Nasim, A., P. Young, and B.F. Johnson, eds. 1989. *Molecular biology of the fission yeast.* Academic Press, San Diego.

Nasmyth, K. and P. Nurse. 1981. Cell division cycle mutants altered in DNA replication and mitosis in the fission yeast *Schizosaccharomyces pombe. Mol. Gen. Genet.* **182:** 119.

Nimmo, E.R., G. Cranston, and R.C. Allshire. 1994. Telomere-associated chromosome breakage in fission yeast results in variegated expression of adjacent genes. *EMBO J.* **13:** 3801.

Niwa, O. and M. Yanagida. 1985. Triploid meiosis and aneuploidy in *Schizosaccharomyces pombe:* An unstable aneuploid disomic for chromosome III. *Curr. Genet.* **9:** 463.

Niwa, O., T. Matsumoto, and M. Yanagida. 1986. Construction of a minichromosome by deletion and its mitotic and meiotic behaviour in fission yeast. *Mol. Gen. Genet.* **203:** 397.

Niwa, O., T. Matsumoto, Y. Chikashige, and M. Yanagida. 1989. Characterization of *Schizosaccharomyces pombe* minichromosome deletion derivatives and a functional allocation of their centromere. *EMBO J.* **8:** 3045.

Noguchi, E., N. Hayashi, Y. Azuma, T. Seki, M. Nakamura, N. Nakashima, M.

Yanagida, X. He, U. Mueller, S. Sazer, and T. Nishimoto. 1996. Dis3, implicated in meiotic control, directly binds to Ran and enhances the GEF activity of RCC1. *EMBO J.* (in press).

Nurse, P. 1975. Genetic control of cell size at cell division in yeast. *Nature* **256:** 547.

———. 1990. Universal control mechanism regulating onset of M-phase. *Nature* **344:** 503.

———. 1994a. Fission yeast morphogenesis-posing the problems. *Mol. Biol. Cell* **5:** 613.

———. 1994b. Ordering S phase and M phase in the cell cycle. *Cell* **79:** 547.

Nurse, P. and Y. Bissett. 1981. Gene required in G1 for commitment to cell cycle and in G2 for control of mitosis in fission yeast. *Nature* **292:** 558.

Nurse, P. and P. Thuriaux. 1980. Regulatory genes controlling mitosis in the fission yeast *Schizosaccharomyces pombe*. *Genetics* **96:** 627.

Nurse, P., P. Thuriaux, and K. Nasmyth. 1976. Genetic control of the cell division cycle in the fission yeast *Schizosaccharomyces pombe*. *Mol. Gen. Genet.* **146:** 167.

O'Connell, M.J. and P. Nurse. 1994. How cells know they are in G1 or G2. *Curr. Opin. Cell Biol.* **6:** 867.

O'Donnell, K.L., A.H. Osmani, S.A. Osmani, and N.R. Morris. 1991. *bimA* encodes a member of the tetratricopeptide repeat family of proteins and is required for the completion of mitosis in *Aspergillus nidulans*. *J. Cell Sci* **99:** 711.

Ohkura, H. and M. Yanagida. 1991. *S. pombe* gene *sds22*$^+$ essential for a midmitotic transition encodes a leucine-rich repeat protein that positively modulates protein phosphatase-1. *Cell* **64:** 149.

Ohkura, H., I.M. Hagan, and D.M. Glover. 1995. The conserved *Schizosaccharomyces pombe* kinase plo1, required to form a bipolar spindle, the actin ring, and septum, can drive septum formation in G1 and G2 cells. *Genes Dev.* **9:** 1059.

Ohkura, H., N. Kinoshita, S. Miyatani, T. Toda, and M. Yanagida. 1989. The fission yeast *dis2*$^+$ gene required for chromosome disjoining encodes one of two putative type 1 protein phosphatases. *Cell* **57:** 997.

Ohkura, H., Y. Adachi, N. Kinoshita, O. Niwa, T. Toda, and M. Yanagida. 1988. Cold-sensitive and caffeine supersensitive mutants of the *Schizosaccharomyces pombe dis* genes implicated in sister chromatid separation during mitosis. *EMBO J.* **7:** 1465.

Ottilie, S., J. Chernoff, G. Hannig, C.S. Hoffman, and R.L. Erikson. 1992. The fission yeast genes *pyp1*$^+$ and *pyp2*$^+$ encode protein tyrosine phosphatases that negatively regulate mitosis. *Mol. Cell. Biol.* **12:** 5571.

Page, B.D. and M. Snyder. 1993. Chromosome segregation in yeast. *Annu. Rev. Microbiol.* **47:** 231.

Panzeri, L., L. Landonio, A. Stotz, and P. Philippsen. 1985. Role of conserved sequence elements in yeast centromere DNA. *EMBO J.* **4:** 1867.

Parker, L.L., S. Atherton-Fessler, and H. Piwnica-Worms. 1992. p107^{wee1} is a dual-specificity kinase that phosphorylates p34^{cdc2} on tyrosine 15. *Proc. Natl. Acad. Sci.* **89:** 2917.

Parker, L.L., S.A. Walter, P.G. Young, and H. Piwnica-Worms. 1993. Phosphorylation and inactivation of the mitotic inhibitor Wee1 by the *nim1/cdr1* kinase. *Nature* **363:** 736.

Peterson, C.L. 1994. The SMC family: Novel motor proteins for chromosome condensation? *Cell* **79:** 389.

Pidoux, A.L. and J. Armstrong. 1993. The BiP protein and the endoplasmic reticulum of

Schizosaccharomyces pombe: Fate of the nuclear envelope during cell division. *J. Cell Sci.* **105:** 1115.

Polizzi, C. and L. Clarke. 1991. The chromatin structure of centromeres from fission yeast: Differentiation of the central core that correlates with function. *J. Cell Biol.* **112:** 191.

Potashkin, J., K. Naik, and K. Wintz-Hunter. 1993. U2AF homolog required for splicing in vivo. *Science* **262:** 573.

Robinow, C.F. 1977. The number of chromosomes in *Schizosaccharomyces pombe:* Light microscopy of stained preparations. *Genetics* **87:** 491.

Robinow, C.F. and J.S. Hyams. 1989. General cytology of fission yeasts. In *Molecular biology of the fission yeast* (ed. A. Nasim et al.), p. 273. Academic Press, San Diego.

Roussou, I. and G. Draetta. 1994. The *Schizosaccharomyces pombe* casein kinase II α and β subunits: Evolutionary conservation and positive role of the β subunit. *Mol. Cell. Biol.* **14:** 576.

Russell, P. and P. Nurse. 1986. *cdc25*$^{+}$ functions as an inducer in the mitotic control of fission yeast. *Cell* **45:** 145.

———. 1987. The mitotic inducer *nim1*$^{+}$ functions in a regulatory network of protein kinase homologs controlling the initiation of mitosis. *Cell* **49:** 569.

Saitoh, N., I.G. Goldberg, E.R. Wood, and W.C. Earnshaw. 1994. ScII: An abundant chromosome scaffold protein is a member of a family of putative ATPases with an unusual predicted tertiary structure. *J. Cell Biol.* **127:** 303.

Saka, Y., T. Sutani, Y. Yamashita, S. Saitoh, M. Takeuchi, Y. Nakaseko, and M. Yanagida. 1994. Fission yeast *cut3* and *cut14*, members of a ubiquitous protein family, are required for chromosome condensation and segregation in mitosis. *EMBO J.* **13:** 4938.

Sakaguchi, J. and M. Yamamoto. 1982. Cloned *ura1* locus of *Schizosaccharomyces pombe* propagates autonomously in this yeast assuming a polymeric form. *Proc. Natl. Acad. Sci.* **79:** 7819.

Samejima, I. and M. Yanagida. 1994a. Bypassing anaphase by fission yeast *cut9* mutation: Requirement of *cut9*$^{+}$ to initiate anaphase. *J. Cell Biol.* **127:** 1655.

———. 1994b. Identification of *cut8*$^{+}$ and *cek1*$^{+}$, a novel protein kinase gene, which complement a fission yeast mutation that blocks anaphase. *Mol. Cell. Biol.* **14:** 6361.

Samejima, I., T. Matsumoto, Y. Nakaseko, D. Beach, and M. Yanagida. 1993. Identification of seven new *cut* genes involved in *Schizosaccharomyces pombe* mitosis. *J. Cell Sci.* **105:** 135.

Sazer, S. and P. Nurse. 1994. A fission yeast RCC1-related protein is required for the mitosis to interphase transition. *EMBO J.* **13:** 606.

Sheldrick, K.S. and A.M. Carr. 1993. Feedback controls and G2 checkpoints: Fission yeast as a model system. *BioEssays* **15:** 775.

Shimanuki, M., N. Kinoshita, H. Ohkura, T. Yoshida, T. Toda, and M. Yanagida. 1993. Isolation and characterization of the fission yeast protein phosphatase gene *ppe1*$^{+}$ involved in cell shape control and mitosis. *Mol. Biol. Cell* **4:** 303.

Shiozaki, K. and M. Yanagida. 1991. A functional 125-kDa core polypeptide of fission yeast DNA topoisomerase II. *Mol. Cell. Biol.* **11:** 6093.

Sikorski, R.S., M.S. Boguski, M. Goebl, and P. Hieter. 1990. A repeating amino acid motif in *CDC23* defines a family of proteins and a new relationship among genes required for mitosis and RNA synthesis. *Cell* **60:** 307.

Sipiczki, M., B. Grallert, and I. Miklos. 1993. Mycelial and syncytial growth in

Schizosaccharomyces pombe induced by novel septation mutations. *J. Cell Sci.* **104:** 485.

Smith, C.L., T. Matsumoto, O. Niwa, S. Kico, J.-B. Fan, M. Yanagida, and C.R. Cantor. 1987. An electrophoretic karyotype for *Schizosaccharomyces pombe* by pulsed field gel electrophoresis. *Nucleic Acids Res.* **15:** 4481.

Snell, V. and P. Nurse. 1993. Investigations into the control of cell form and polarity: The use of morphological mutants in fission yeast. *Development* (suppl.), p. 289.

———. 1994. Genetic analysis of cell morphogenesis in fission yeast—A role for casein kinase II in the establishment of polarized growth. *EMBO J.* **13:** 2066.

Snyder, M. 1994. The spindle pole body of yeast. *Chromosoma* **103:** 369.

Solomon, M., R. Booher, M. Kirschner, and D. Beach. 1988. Cyclin in fission yeast. *Cell* **54:** 738.

Steiner, N.C. and L. Clarke. 1994. A novel epigenetic effect can alter centromere function in fission yeast. *Cell* **79:** 865.

Stone, E.M., H. Yamano, N. Kinoshita, and M. Yanagida. 1993. Mitotic regulation of protein phosphatases by the fission yeast sds22 protein. *Curr. Biol.* **3:** 13.

Streiblová, E., J. Hašek, and E. Jelke. 1984. Septum pattern in ts mutants of *Schizosaccharomyces pombe* in defective genes *cdc3*, *cdc4*, *cdc8* and *cdc12*. *J. Cell Sci.* **69:** 47.

Strunnikov, A.V., V.L. Larionov, and D. Koshland. 1993. *SMC1:* An essential yeast gene encoding a putative head-rod-tail protein is required for nuclear division and defines a new ubiquitous protein family. *J. Cell Biol.* **123:** 1635.

Sugawara, N. 1989. "DNA sequences at the telomeres of the fission yeast *S. pombe.*" Ph.D. thesis, Harvard University, Cambridge, Massachusetts.

Sutton, A., D. Immanuel, and K.T. Arndt. 1991. The *SIT4* protein phosphatase functions in late G1 for progression into S phase. *Mol. Cell. Biol.* **11:** 2133.

Takahashi, K., H. Yamada, and M. Yanagida. 1994. Fission yeast minichromosome loss mutants *mis* cause lethal aneuploidy and replication abnormality. *Mol. Biol. Cell* **5:** 1145.

Takahashi, K., S. Murakami, Y. Chikashige, O. Niwa, and M. Yanagida. 1991. A large number of tRNA genes are symmetrically located in fission yeast centromeres. *J. Mol. Biol.* **218:** 13.

Takahashi, K., S. Murakami, Y. Chikashige, H. Funabiki, O. Niwa, and M. Yanagida. 1992. A low copy number central sequence with strict symmetry and unusual chromatin structure in fission yeast centromere. *Mol. Biol. Cell* **3:** 819.

Takeuchi, M. and M. Yanagida. 1993. A mitotic role for a novel fission yeast protein kinase dsk1 with cell cycle stage dependent phosphorylation and localization. *Mol. Biol. Cell* **4:** 247.

Tanaka, K. and T. Kanbe. 1986. Mitosis in the fission yeast *Schizosaccharomyces pombe* as revealed by freeze-substitution electron microscopy. *J. Cell Sci.* **80:** 253.

Tartakoff, A.M. and R. Schneiter. 1995. The nuclear GTPase cycle: Promoting peripheralization? *Trends Cell Biol.* **5:** 5.

Tartof, K.D. and M. Bremer. 1990. Mechanisms for the construction and development control of heterochromatin formation and imprinted chromosome domains. *Development* (suppl.), p. 35.

Thuriaux, P., P. Nurse, and B. Carter. 1978. Mutants altered in the control coordinating cell division with cell growth in the fission yeast *Schizosaccharomyces pombe*. *Mol. Gen. Genet.* **161:** 215.

Toda, T., M. Shimanuki, and M. Yanagida. 1993. Two novel protein kinase C-related genes of fission yeast are essential for cell viability and implicated in cell shape control. *EMBO J.* **12:** 1987.

Toda, T., M. Yamamoto, and M. Yanagida. 1981. Sequential alterations in the nuclear chromatin region during mitosis of the fission yeast *Schizosaccharomyces pombe:* Video fluorescence microscopy of synchronously growing wild-type and cold-sensitive *cdc* mutants by using a DNA-binding fluorescent probe. *J. Cell Sci.* **52:** 271.

Toda, T., Y. Adachi, Y. Hiraoka, and M. Yanagida. 1984a. Identification of the pleiotropic cell division cycle gene *NDA2* as one of two different α-tubulin genes in *Schizosaccharomyces pombe. Cell* **37:** 233.

Toda, T., Y. Nakaseko, O. Niwa, and M. Yanagida. 1984b. Mapping of rRNA genes by integration of hybrid plasmids in *Schizosaccharomyces pombe. Curr. Genet.* **8:** 93.

Toda, T., K. Umesono, A. Hirata, and M. Yanagida. 1983. Cold-sensitive nuclear division arrest mutants of the fission yeast *Schizosaccharomyces pombe. J. Mol. Biol.* **168:** 251.

Tugendreich, S., J. Tomkiel, W. Earnshaw, and P. Hieter. 1995. CDC27Hs colocalizes with CDC16Hs to the centrosome and mitotic spindle and is essential for the metaphase to anaphase transition. *Cell* **81:** 261.

Tugendreich, S., M.S. Boguski, M.S. Seldin, and P. Hieter. 1993. Linking yeast genetics to mammalian genomes: Identification and mapping of the human homolog of *CDC27* via the expressed sequence tag (EST) data base. *Proc. Natl. Acad. Sci.* **90:** 10031.

Uemura, T. and M. Yanagida. 1984. Isolation of type I and II DNA topoisomerase mutants from fission yeast: Single and double mutants show different phenotypes in cell growth and chromatin organization. *EMBO J.* **3:** 1737.

———. 1986. Mitotic spindle pulls but fails to separate chromosomes in type II DNA topoisomerase mutants: Uncoordinated mitosis. *EMBO J.* **5:** 1003.

Uemura, T., K. Morikawa, and M. Yanagida. 1986. The nucleotide sequence of the fission yeast DNA topoisomerase II gene: Structural and functional relationships to other DNA topoisomerases. *EMBO J.* **5:** 2355.

Uemura, T., K. Morino, S. Uzawa, K. Shiozaki, and M. Yanagida. 1987a. Cloning and Sequencing of *Schizosaccharomyces pombe* DNA topoisomerase I gene, and effect of gene disruption. *Nucleic Acids Res.* **15:** 9727.

Uemura, T., H. Ohkura, Y. Adachi, K. Morino, K. Shiozaki, and M. Yanagida. 1987b. DNA topoisomerase II is required for condensation and separation of mitotic chromosomes in *S. pombe. Cell* **50:** 917.

Umesono, K., Y. Hiraoka, T. Toda, and M. Yanagida. 1983a. Visualization of chromosomes in mitotically arrested cells of the fission yeast *Schizosaccharomyces pombe. Curr. Genet.* **7:** 123.

Umesono, K., T. Toda, S. Hayashi, and M. Yanagida. 1983b. Two cell division cycle genes *NDA2* and *NDA3* of the fission yeast *Schizosaccharomyces pombe* control microtubular organization and sensitivity to anti-mitotic benzimidazole compounds. *J. Mol. Biol.* **168:** 271.

Uzawa, S. and M. Yanagida. 1992. Visualization of centromeric and nucleolar DNA in fission yeast by fluorescence in situ hybridization. *J. Cell Sci.* **101:** 267.

Uzawa, S., I. Samejima, T. Hirano, K. Tanaka, and M. Yanagida. 1990. The fission yeast *cut1*$^{+}$ gene regulates spindle pole body duplication and has homology to the budding yeast *ESP1* gene. *Cell* **62:** 913.

Walker, G.M. 1982. Cell cycle specificity of certain antimicrotubular drugs in *Schizosaccharomyces pombe. J. Gen. Microbiol.* **128:** 61.

Willard, H.F. 1990. Centromeres of mammalian chromosomes. *Trends Genet.* **6:** 410.

Wilson, R.B., A.A. Brenner, T.B. White, M.J. Engler, J.P. Gaughran, and K. Tatchell. 1991. The *Saccharomyces cerevisiae SRK1* gene, a suppressor of *bcy1* and *ins1*, may be involved in protein phosphatase function. *Mol. Cell. Biol.* **11:** 3369.

Wohlgemuth, J.G., G.H. Bulboaca, M. Moghadam, M.S. Caddle, and M.P. Calos. 1994. Physical mapping of origins of replication in the fission yeast *Schizosaccharomyces pombe*. *Mol. Biol. Cell* **5:** 839.

Wright, A.P.H., K. Maundrell, and S. Shall. 1986. Transformation of *Schizosaccharomyces pombe* by non-homologous unstable integration of plasmids in the genome. *Curr. Genet.* **10:** 503.

Wu, L. and P. Russell. 1993. Nim1 kinase promotes mitosis by inactivating Wee1 tyrosine kinase. *Nature* **363:** 738.

Xu, H.-P., M. White, S. Marcus, and M. Wigler. 1994. Concerted action of RAS and G proteins in the sexual response pathways of *Schizosaccharomyces pombe*. *Mol. Cell. Biol.* **14:** 50.

Yamamoto, M. 1980. Genetic analysis of resistant mutants to antimitotic benzimidazole compounds in *Schizosaccharomyces pombe*. *Mol. Gen. Genet.* **180:** 231.

Yamano, H., K. Ishii, and M. Yanagida. 1994. Phosphorylation of dis2 protein phosphatase at the C-terminal cdc2 consensus and its potential role in cell cycle regulation. *EMBO J.* **13:** 5310.

Yanagida, M. 1989. Gene products required for chromosome separation. *J. Cell Sci. Suppl.* **12:** 213.

———. 1990. Higher order chromosome structure in yeast. *J. Cell. Sci.* **96:** 1.

———. 1995. Frontier questions about sister chromatid separation in anaphase. *BioEssays* **17:** 519.

Yanagida, M. and R. Sternglanz. 1990. Genetics of DNA topoisomerases. In *DNA topology and its biological effects* (ed. J.C. Wang and N.R. Cozzarelli), p. 299. Cold Spring Harbor Laboratory Press, Cold Spring Harbor, New York.

Yanagida, M. and J.C. Wang. 1987. Yeast DNA topoisomerases and their structural genes. In *Nucleic acids and molecular biology* (ed. F. Eckstein and D.M.J. Lilley), p. 196. Springer-Verlag, Berlin.

Yanagida, M., N. Kinoshita, E.M. Stone, and H. Yamano. 1992. Protein phosphatases and cell division cycle control. In *Ciba Found. Symp.* **170:** 130.

Yanagida, M., Y. Hiraoka, T. Uemura, S. Miyake, and T. Hirano. 1986. Control mechanisms of chromosome movement in mitosis of fission yeast. In *Yeast cell biology* (ed. J. Hicks), p. 279. A.R. Liss, New York.

Yoshida, T., T. Toda, and M. Yanagida. 1994. A calcineurin-like gene *ppb1*$^+$ in fission yeast: Mutant defects in cytokinesis, cell polarity, mating and spindle pole body positioning. *J. Cell Sci.* **107:** 1725.

Zhu, J., C. Brun, H. Kurooka, M. Yanagida, and J.A. Huberman. 1992. Identification and characterization of a complex chromosomal replication origin in *Schizosaccharomyces pombe*. *Chromosoma* **102:** S7-S16.

Zhu, J., D.L. Carlson, D.D. Dubey, K. Sharma, and J.A. Huberman. 1994. Comparison of the two major ARS elements of the *ura4* replication origin region with other ARS elements in the fission yeast, *Schizosaccharomyces pombe*. *Chromosoma* **103:** 414.

10

The Pathway of Cell and Nuclear Fusion during Mating in *S. cerevisiae*

Lorraine Marsh
Department of Cell Biology
Albert Einstein College of Medicine of Yeshiva University
Bronx, New York 10461

Mark D. Rose
Department of Molecular Biology
Princeton University
Princeton, New Jersey 08544-1014

I. INTRODUCTION

In the yeast *Saccharomyces cerevisiae*, the process of conjugation begins with two haploid cells of opposite mating type (**a** and α) which ultimately fuse to form a single diploid zygote (for recent reviews, see Cross et al. 1988; Sprague and Thorner 1994). Along the way, the cells must recognize the presence and spatial position of the partner cell via the mutual exchange of sexual pheromones. Cells of opposite mating type adhere and begin the orderly removal of intervening cell wall and plasma membrane to complete cell fusion. As the cells fuse, the nuclei move together and fuse to produce a single diploid nucleus. The diploid zygote then reenters the cell cycle and mitosis gives rise to diploid buds. The diploid cells then may divide indefinitely or, under appropriate conditions, enter the meiotic pathway to produce haploid spores. The haploid spore progeny can then recapitulate the conjugation pathway. The conjugation pathway includes elements of signal transduction, cell polarity, membrane fusion, and microtubule function. As such, it is a microcosm whose elements are of particular interest to cell biologists. Because of the unique power of the genetics of the pathway, considerable progress has been made in the last decade. In this chapter, we review the pathway of conjugation in yeast, beginning with nuclear fusion, about which the most is known, and then finishing with the pathway of cell fusion. The material on nuclear fusion is reprinted with permission, with minor modification, from a recent review on the pathway of nuclear fusion in yeast (Rose 1996).

II. OVERVIEW OF NUCLEAR FUSION

In *S. cerevisiae*, as in other fungi, the nuclear envelope remains intact throughout all stages of mitosis, meiosis, and conjugation (Byers 1981a).

The permanence of the nuclear envelope makes necessary several adaptations that influence the architecture and functions of the nucleus. The first concerns the localization of proteins required to act on intranuclear structures during mitosis. For example, the presence of microtubules in both the nucleus and the cytoplasm implies that yeast tubulin must contain signals for nuclear localization. Clearly, such signals would not be required in organisms that break down the nuclear envelope during mitosis.

Second, the sole microtubule-organizing center, called the spindle pole body (SPB) or spindle plaque, must nucleate both cytoplasmic and nuclear microtubules (Byers et al. 1978; Hyams and Borisy 1978; Kilmartin 1994; Snyder 1994). This is accomplished by virtue of the special architecture of the spindle plaque. Consisting of at least three distinct layers, the spindle plaque is a disk embedded within the plane of the nuclear envelope, much in the fashion of a nuclear pore. The outer layer of the spindle plaque nucleates cytoplasmic microtubules whose functions are to mediate nuclear movement and orientation (for a recent review of the role of cytoplasmic microtubules in fungi, see Morris et al. 1995). The inner layer of the spindle plaque nucleates the microtubules that mediate spindle assembly, spindle elongation, and chromosome segregation (Page and Snyder 1993; Rose et al. 1993; Kilmartin 1994; Snyder 1994). One benefit of the laminar organization of the spindle plaque is that it allows independent regulation of the nuclear and cytoplasmic microtubules.

The third major adaptation of the closed mitosis in yeast concerns conjugation and meiosis. A diploid nucleus is produced by direct fusion of the two parental haploid nuclei. The yeast nuclear fusion pathway has been the subject of particular study because of several distinctive attributes (Rose 1991, 1996). Most importantly, the formation of the final diploid nucleus is exquisitely sensitive to the functions of both microtubules and the nuclear envelope, two areas of intensive interest in their own right. However, since mating and nuclear fusion are not essential for viability, mutations were readily isolated that block nuclear fusion without leading simultaneously to a serious block in cell division. Some of these mutations cause defects in a nonessential function of otherwise essential cell structures (e.g., spindle plaque, microtubules, and nuclear envelope). Nevertheless, the specificity of the mutant defects suggested that many of the gene products acted relatively directly in the nuclear fusion pathway. The promise of this approach was that additional genetic and molecular analyses would expand the set of genes and proteins known to be components of the essential structures. Similarly, homologs of interesting genes might be found that act in related functions in the mitotic cycle.

A highly detailed description of the yeast mating pathway would be beyond the scope of this review and has been excellently and thoroughly covered elsewhere (Cross et al. 1988; Sprague and Thorner 1994). For current purposes, we touch on some of the salient features of nuclear fusion from earlier studies. The detailed cytology of conjugation and yeast nuclear fusion have been extensively described (Byers and Goetsch 1975; Byers 1981b; Hašek et al. 1987; Rose and Fink 1987; Read et al. 1992) and reviewed (Cross et al. 1988; Rose 1991; Sprague and Thorner 1994). A schematic view of the pathway is shown in Figure 1. In response to mating pheromone, cells develop a characteristic pointed mating "projection." The projection causes the cell to become pear-shaped, sometimes called a "shmoo" in reference to a defunct cartoon strip. As discussed below, the cell walls of the mating cells adhere and then break down in the region of adhesion. Plasma membrane fusion occurs very rapidly thereafter, since unfused cells are not normally found as a distinct intermediate. Prior to cell fusion, the nuclei are positioned close to the tip of the mating projection. As observed by immunofluorescence microscopy, a microtubule bundle emanates from the SPB and terminates close to the tip of the projection. The use of tubulin mutants (Read et al. 1992) and microtubule depolymerizing drugs (Delgado and Conde 1984; Hasek et al. 1987) indicates that the position of the nucleus at this point is dependent on the microtubules, and it is likely that prepositioning of the nucleus is important for efficient nuclear fusion. Coincident with plasma membrane fusion, microtubules connect the two nuclei, producing a small spindle-like array. The nuclei move together and nuclear envelope fusion ensues along one edge of the spindle plaque. The resultant single fused spindle plaque has a residual crease along the suture line. As the SPBs fuse, the nuclear membranes fuse such that the lumens of the two nuclear envelopes and the two nucleoplasms become continuous.

A. Membrane Fusion

The detailed events during nuclear membrane fusion have not been carefully described largely because of the dense staining of the spindle plaque by electron microscopy techniques. Therefore, the exact number of membrane fusion events and their topology are not known. Theoretical considerations indicate that either one or three membrane fusion events must occur to give rise to a single nucleus with a single spindle

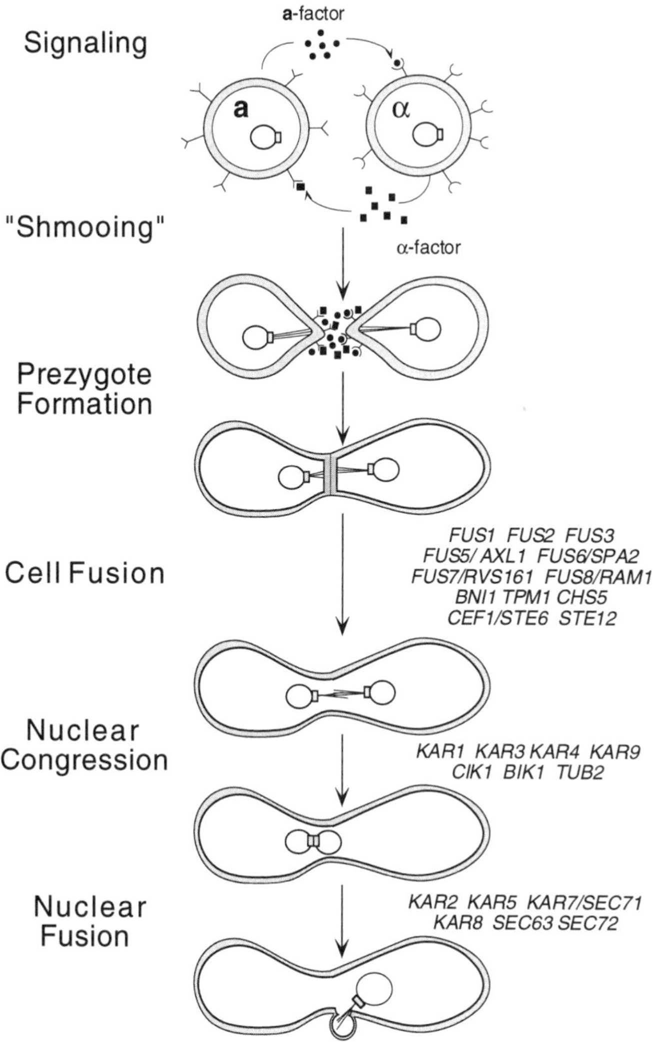

Figure 1 The pathway of nuclear fusion in yeast (see text for description). (Modified, with permission, from Rose 1996.)

plaque (Fig. 2). Assuming that the spindle plaque is embedded in the nuclear envelope with a topology similar to that of a nuclear pore, then the SPB is a place where the inner and outer nuclear envelopes are continuous. As such, a single membrane fusion event along one edge of the pore, followed by membrane flow, is theoretically sufficient to fuse the two nuclei. By this "one-step" model, the two spindle plaques would likewise fuse along the same edges. Thus, along that edge, the former

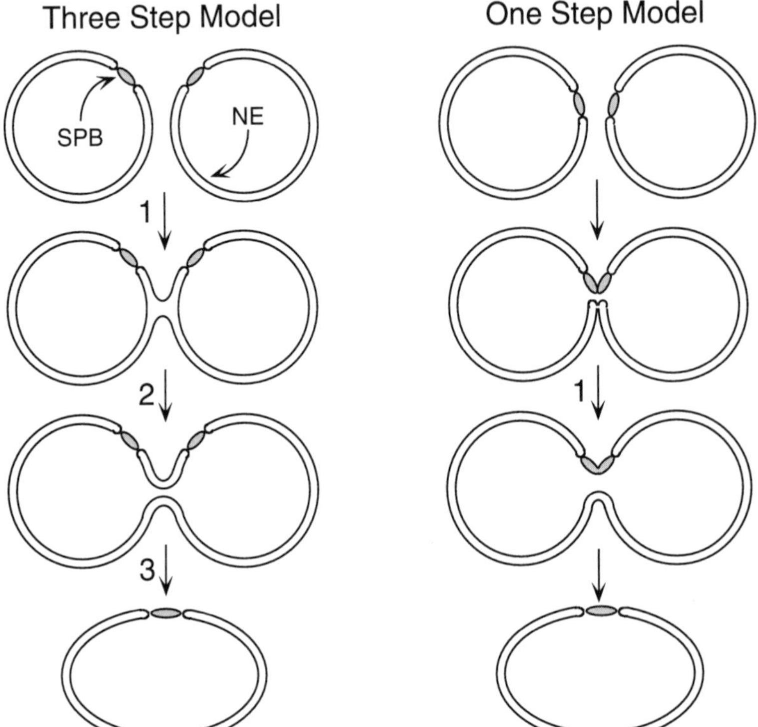

Figure 2 Two models for nuclear envelope fusion (detailed in the text). The numbers refer to the specific membrane fusion events. (SPB) Spindle pole body; (NE) nuclear envelope. (Reprinted, with permission, from Rose 1996.)

boundaries between the spindle plaques and membranes become sites of membrane fusion and spindle plaque fusion. Topological considerations show that this overall fusion event opens a pore between the two nucleoplasms. Subsequent dilation of the pore leads to completion of nuclear fusion.

The alternate model is somewhat easier to visualize. In the "three-step" model, complete nuclear fusion arises from a series of three fusion events, first between the outer nuclear envelope membranes, then between the inner nuclear envelopes, and finally between the two SPBs. The first fusion event results in continuity between the lumens of the two nuclear envelopes, whereas the second fusion event results in the continuity of the two nucleoplasms. Finally, to produce the single spindle plaque observed in the zygote, the two SPBs must fuse. Given the topology of the embedded spindle plaques, this last event must also involve membrane fusion. Fusion would be a "pinching" down between the inner and outer membranes, such that the two pores become one. There is no strong evidence to support either model, but the first one appears to be more consistent with the available data, although there is a strong appeal to the apparently simpler second model.

B. Regulation of Nuclear Fusion

Several earlier observations indicated that some components of the nuclear fusion pathway must be regulated by response to mating pheromone. The most important were experiments to examine the fate of nuclei during spheroplast fusion experiments (Curran and Carter 1986; Rose et al. 1986). These showed that mitotic nuclei were generally incompetent for nuclear fusion. However, if cells were pretreated with mating pheromone, the nuclei fused with high efficiency, approaching that of normal matings. Other experiments suggested that the two nuclei have only a narrow window of opportunity for fusion. For example, for recessive mutations that block nuclear fusion, subsequent buds are no more likely to be diploid than the initial bud (Lancashire and Mattoon 1979). The simplest interpretation of this observation is that the zygote becomes equivalent to a nonmating **a**/α diploid and therefore refractory to further response to mating pheromones. Taking these data together, it is clear that the nuclear fusion pathway is likely to depend on two classes of genes, some induced by mating pheromone and others constitutively expressed, such as tubulin. These points are returned to below.

III. APPROACHES TO THE STUDY OF NUCLEAR FUSION

Until recently, nuclear fusion has been studied exclusively by either genetic or cytological methods. The development of cell-free in vitro assays for the fusion of nuclear and endoplasmic reticulum (ER)-derived membranes promises to elucidate the detailed mechanism of the membrane fusion steps (Latterich and Schekman 1994; Latterich et al. 1995). Similarly, the development of assays for the functions of individual gene products has helped clarify their roles in nuclear congression.

A. Karyogamy Mutations

The mutations that block nuclear fusion can be broadly classified into two groups: unilateral mutations, which block fusion when mutants are mated to wild-type cells, and bilateral mutations, which only manifest their defects when two mutant cells are mated together. Of necessity, the first set of karyogamy mutants were either unilateral (e.g., *kar1-1*, *kar2-1*, and *kar3-1*) (Conde and Fink 1976; Polaina and Conde 1982) or discovered by serendipity (e.g., *bik1-1*) (Berlin et al. 1990). The basis of the bilateral mutants can be readily understood as recessive mutations affecting proteins that are diffusible in the zygote. An alternate possibility is that they affect proteins which are not required to act on both nuclei. However, no clear examples of this class exist in nuclear fusion. Likewise, there are several mechanisms by which unilateral mutants can arise. One type of unilateral mutation produces a dominant-negative protein (e.g., *kar3-1*). A second type is composed of recessive mutations in proteins that are not diffusible between the nuclei or that cannot be functionally assembled into the nuclear structure during the brief "window of opportunity."

The initial isolation of the unilateral mutants depended on the properties of the unique "cytoductant" cells that arise after a failure in nuclear fusion (Conde and Fink 1976; Lancashire and Mattoon 1979). Cytoductants are defined as the haploid cells that contain the nucleus from one parent and the cytoplasm from the other parent. In practice, the cytoductant is selected to contain the *kar*$^-$ mutant nucleus by using a preexisting recessive drug resistance mutation to select against diploid buds. The cytoplasm of the other parent is selected on the basis of its mitochondrial genome. These methods are reviewed in Berlin et al. (1991) and Rose (1991).

Because of the biases in the selection for the unilateral mutants, several screens have been developed specifically to isolate bilateral mutants. In principle, these all require that after mutagenesis, the mating types of the candidate mutants can be switched, the candidates are then allowed to mate with themselves, and the mutants are recognized by the consequences of a "self-sterile" mating defect. One attempt used the *HO* gene to switch mating type (Berlin et al. 1991). The mating defect was recognized by the presence of residual fertile cells of both mating types in colonies that grew up from mutagenized *HO* cells. However, no new *kar* mutants were reported from that screen.

A novel screen (Kurihara et al. 1994) used plasmid segregation to generate cells capable of mating as either α or **a** cells. This system takes advantage of the fact that cells deleted at the mating-type locus mate like **a** cells, but the deletion is recessive to the wild-type *MAT*α gene on a plasmid. The initial strain mates as an α; at each cell division, a few percent of the cells missegregate the plasmid, generating **a**-like cells that mate with their sisters. Such matings generate cells of diploid (or higher) ploidy in the population. The number of diploid cells is measured in a semiquantitative way using a "plasmid loop-out" test (Chan and Botstein 1993). In this system, one gene (e.g., *HIS3*) is disrupted by an integrated plasmid bearing the sole wild-type copy of a second gene (e.g., *TRP1*). The plasmid can "loop out" by virtue of homologous recombination between repeated sequences from the disrupted gene. The looped-out plasmid lacks an origin of replication and is rapidly lost. Thus, in this example, the starting strain is His^- and Trp^+. The loop out is His^+ and Trp^-. Only diploids that carry two copies of the disruption can be both His^+ and Trp^+ simultaneously. Wild-type strains bearing the construct frequently papillate to His^+, Trp^+. Mutants with defects anywhere in the mating pathway are detected by their reduced frequency of papillation. Microscopic examination then allows the different types of mutants to be sorted into classes. By this route, mutations in five additional genes, *KAR4*, *KAR5*, *KAR7*, *KAR8*, and *KAR9*, were identified (Kurihara et al. 1994).

The new karyogamy mutants fell into two distinct classes, depending on the final position of the nuclei in unbudded zygotes. The class I mutants contained nuclei separated by approximately 3.3 μm, determined by the position of the fluorescently stained DNA. In contrast, the class II mutants contained nuclei separated by 0.3 μm or less. The class I genes consist of *KAR1*, *KAR3*, *TUB2*, *BIK1*, *CIK1*, *KAR4*, and *KAR9*. This class was determined to be specifically defective in nuclear congression (Fig. 1) because of the failure of the nuclei to become closely apposed, the identification of several microtubule-associated proteins in the group, and the observation of aberrant cytoplasmic microtubules in shmoos and zygotes for many of the mutants.

The class II genes were composed of *KAR2*, *KAR5*, *KAR7*, and *KAR8*. Immunofluorescence and electron microscopy studies demonstrated that the nuclei become closely apposed in the zygotes, but the nuclear envelopes remain separate and unfused over a large region of close contact. Cytoplasmic microtubules appeared to be normal in both shmoos and zygotes. For these reasons and the results described below, this set of mutants has been recognized as being defective specifically in nuclear envelope fusion (Fig. 2). As expected, strains containing both class I and class II mutations show a class I defect, indicating that the putative nuclear congression step does occur prior to fusion. Intriguingly, the nuclear envelopes in the class II zygotes appeared to be connected by thin membranous connections. Their appearance in mutants blocked in nuclear envelope fusion raises the interesting possibility that the bridges may represent the products of aborted membrane fusion events. The morphology of the bridges varies somewhat between the different mutants, lending credence to the hypothesis that they arise from different membrane fusion intermediates.

B. Biochemical Assays of Nuclear/Endoplasmic Reticulum Fusion

Recently, Latterich and Schekman (1994) developed a novel cell-free assay for the fusion of membranes derived from the nuclear envelope and/or endoplasmic reticulum. Their assay utilizes a mutation (*gls1*) in the glucosidase responsible for the removal of a terminal glucosyl residue initially present on core oligosaccharides added to proteins in the ER. A precursor protein, prepro-α-factor, is translocated into the ER lumen of the *gls1* mutant strain. An ER-enriched membrane fraction is then incubated in the presence of a similar membrane fraction from a wild-type (*GLS1*$^+$) strain. When membrane fusion occurs, the luminal contents of the two sets of membranes mix, leading to trimming of the core oligosaccharides and a 3-kD reduction in the molecular mass of the prepro-α-factor. Using microsomes, as much as 20% of the precursor protein can be trimmed in the reaction. Isolated nuclei were up to 40% efficient. It is unclear whether these numbers correspond to the true efficiency of membrane fusion or whether a higher efficiency is masked by unproductive fusion reactions between membranes derived from the same strain. Membrane fusion required ATP, intact membranes, and membrane-associated proteins. Unlike other membrane fusion systems, it was independent of added cytosol and Ca^{++}.

Examination of various secretion-defective mutants revealed the surprising fact that the ER fusion assay was independent of genes required for membrane fusion in the secretory pathway (e.g., *SEC17* and *SEC18*, encoding the yeast SNAP and NSF, respectively). Furthermore, antibodies directed against components required for ER vesicle trafficking, Sec17p, Sar1p, Sec23p, and Sec13p, were ineffective at inhibiting the fusion reaction in vitro. In contrast, mutations in Kar2p, the yeast BiP homolog, produced profound defects in fusion in vitro. Different temperature-sensitive mutations in *kar2* produced allele-specific temperature-sensitive defects in vitro. Depletion of the wild-type protein resulted in a nonconditional defect. Taken together, these results imply that Kar2p has a direct, albeit unspecified, role in ER membrane fusion.

Several aspects of the in vitro ER fusion assay suggest that it is a good model for the events that occur during nuclear fusion. Most important, there is a relatively good correlation between the genetic requirements for nuclear fusion and in vitro ER fusion. All of the remaining class II karyogamy mutants (*kar5*, *kar7*, and *kar8*) exhibited an ER fusion defect in vitro (Kurihara et al. 1994). Furthermore, the in vitro ER fusion was not affected by mutations in several genes that had not been observed to cause a nuclear fusion defect in vivo (i.e., *sec17* and *sec18*). However, in other potentially revealing ways, the reaction is not like nuclear fusion. First, it does not require the activities of the class I genes, *KAR1* and *KAR3*. It is reasonable that large nuclei may require microtubule-dependent movement for congression in the dense cytoplasm. In vitro, it is likely that the requirement for active movement can be overcome by the smaller size of the microsomes and the reduced viscosity of the diluted cytosol. Second, when bilateral mutants were examined (*kar5*, *kar7*, and *kar8*), they were found to exhibit unilateral defects in vitro; i.e., mutations in either the donor or the acceptor membranes alone were sufficient to block membrane fusion. This discrepancy can be explained by the observation that all of the class II genes encode ER membrane proteins (Ng and Walter 1996; C. Beh et al., in prep.), and it is likely that the proteins must be properly localized to function. The conditions of the in vitro fusion assay were not compatible with translocation of proteins into the ER. Therefore, even if wild-type precursor proteins were available for translocation, they could not be transported to their site of function.

In vivo, of course, translocation can occur and the relevant nascent protein precursors should be available for import into the mutant nuclear envelope. Given that, we must still explain the unilateral nature of recessive *kar2* mutants. Kar2p is thought to be present at very high concentrations within the lumen of the ER/nuclear envelope. Therefore, the "window of opportunity" is likely to be too short for sufficient wild-type Kar2p to accumulate in the mutant nuclear envelope.

Other discrepancies between the two systems are potentially more disquieting. First, the in vitro assay occurs efficiently between membranes from mitotic extracts and was not stimulated by addition of cytosol from pheromone-induced cells (Latterich and Schekman 1994). This raises three distinct possibilities. First, the assay may be measuring a distinct reaction, some of whose components are shared with nuclear fusion. Second, any putative rate-limiting pheromone-induced components may be luminal or integral membrane proteins and therefore were not present in the cytosol. This possibility is also consistent with the observation that, so far, all of the class II genes encoded ER membrane proteins (Ng and Walter 1996; C. Beh et al., in prep.). Third, any pheromone-induced components may be restricted to the nuclear congression part of the process. The third possibility is ruled out by the observation that at least one of the class II proteins, Kar5p, is strongly induced by mating pheromone (C. Beh and M. Rose, in prep.).

Direct evidence in support of the possibility that the two reactions are distinct comes from the discovery that the in vitro reaction is strongly dependent on a homolog of Sec18p, encoded by *CDC48* (Latterich et al. 1995). Mutations in *cdc48* result in a block in mitosis during anaphase that was originally interpreted as a spindle elongation defect (Frohlich et al. 1991). Reexamination of the data suggests that it can also be interpreted as resulting from a defect in "nuclear fission." In either case, it is clear that the lack of a requirement for Sec18p in the in vitro reaction is because of the existence of an alternate Sec18p-like protein. However, of crucial significance to nuclear fusion, Cdc48p does not seem to be required for nuclear fusion in vivo (C. Beh and M. Rose, unpubl.). Thus, the two reactions may not be equivalent but would still share components. This view is consistent with general observations that most, if not all, of the proteins required for nuclear fusion are recruited from other pathways in the cell.

What then are the normal mitotic functions of the class II genes? Of particular significance to this question was the subsequent observation that mammalian homologs of Cdc48p are also required for the fusion reactions that reform the Golgi after mitosis (Acharya et al. 1995; Mellman 1995; Rabouille et al. 1995). Possibly, they are required for fusion reactions that are part of the mechanism of nuclear fission or other types of nuclear envelope/ER remodeling. These results also suggest that there may yet be another Sec18p homolog that is induced by mating pheromone and is required for nuclear membrane fusion during mating.

IV. NUCLEAR CONGRESSION

As described above, karyogamy can be viewed as having two major steps: nuclear congression and nuclear fusion. Nuclear congression is highly dependent on normal microtubule function, and the in vivo assays for nuclear fusion are exquisitely sensitive to perturbations in their functions. As described below, molecular and biochemical characterizations of some of the required proteins have allowed the construction of specific models for nuclear congression.

A. *KAR3*

On the basis of sequence identity, Kar3p was identified as an unusual kinesin-related protein (Meluh and Rose 1990). Like kinesin (Yang et al. 1989), Kar3p is predicted to have three domains: a motor domain sharing 39% identity with kinesin, a central coiled-coil domain similar to kinesin, and a small globular "nonmotor" domain unrelated to kinesin. The order of the domains is opposite between the two proteins; the motor domain is amino-terminal for kinesin and carboxy-terminal for Kar3p. *KAR3* is most closely related to a handful of other carboxy-terminal kinesins that have been subsequently discovered (Endow 1991). Two of these are notable, ncd from *Drosophila* (McDonald and Goldstein 1990; McDonald et al. 1990; Walker et al. 1990) and KlpA from *Aspergillus nidulans* (O'Connell et al. 1993). At 46% identity, the ncd motor domain is more closely related to Kar3p than it is to other members of the family. The ncd protein was the first example of a "minus-end-oriented" kinesin-related protein. Whereas authentic kinesin and certain other members of the kinesin superfamily move preferentially toward the "plus end" of the microtubule, ncd moves toward the minus end. KlpA appears to be most closely related to Kar3p. Although it appears to have no essential function in mating and meiosis in *Aspergillus*, KlpA expressed in *S. cerevisiae* can partially suppress the mating defect of a *kar3* mutation (O'Connell et al. 1993).

Two different types of mutations in Kar3p have been reported to affect nuclear fusion. Several dominant-negative mutations (e.g., *kar3-1*) give rise to a unilateral defect. These all map within the conserved ATP-binding site (Meluh and Rose 1990; Roof et al. 1991; Meluh 1992), and based on the analogy to kinesin, these should cause the protein to bind tightly to microtubules but be incapable of movement. In contrast, null alleles constructed in vitro are fully recessive and bilateral, showing that wild-type Kar3p is diffusible in the zygote and that the dominant-negative mutants block nuclear fusion. In zygotes made from the null mutants, both nuclei show properly oriented but highly elongated cytoplasmic microtubule bundles that do not interact with each other. In zygotes made from the *kar3-1* mutant, the two elongated microtubule bundles are continuous with each other. These phenotypic observations suggest that Kar3p is required to "cross-bridge" the two bundles of microtubules and thereby provide the force to move the two nuclei together.

Hybrids of Kar3p fused to β-galactosidase localize to two sites in shmoos: the distal ends of the cytoplasmic microtubules and the SPB (Meluh and Rose 1990; Vallen et al. 1992a; Page et al. 1994). The hybrids demonstrated that the amino-terminal half of Kar3p contains sequences for microtubule association which are independent of the motor domain. Using antibodies to authentic Kar3p, mutant Kar3-1p was found all along the cytoplasmic microtubules and, in zygotes, was particularly prominent in the region of overlapping microtubule bundles. The localization data are consistent with the model of Kar3p cross-bridging the cytoplasmic microtubules via its two different domains for microtubule association and thereby providing the force for nuclear congression by sliding them one against another. Localization of Kar3p to the SPB in shmoos is dependent on one specific domain of the SPB protein Kar1p (Vallen et al. 1992a) and is not affected by treatment with nocodazole (Meluh and Rose 1990). These results suggest that the SPB localization of Kar3p reflects a second function for Kar3p in nuclear fusion and that the role of Kar1p in nuclear fusion is mediated by Kar3p.

The directionality of Kar3p movement was investigated using a hybrid protein containing the motor domain fused to glutathione-*S*-transferase (GST) (Endow et al. 1994). The hybrid protein, expressed and purified from *Escherichia coli*, exhibited all of the properties of an authentic kinesin-like motor protein; it is a microtubule-stimulated ATPase, it binds to microtubules in a nucleotide-sensitive fashion, and it can move microtubules at 1–2 μm/min in a minus-end-directed fashion. Concurrently, authentic Kar3p was identified as a minus-end-oriented motor activity associated with kinetochore protein fractions purified from *S. cerevisiae* (Middleton and Carbon 1994). Thus, there can be little doubt that Kar3p is a minus-end-oriented motor, like ncd. This finding is gratifying in that it is consistent with the simplest model for nuclear congression. It is generally thought that the cytoplasmic microtubules are oriented with their "plus ends" extending away from the SPBs. A motor moving the nuclei together by acting on the overlapping microtubules between them must be minus-end-oriented.

A second activity of the GST-Kar3p hybrid was discovered in the in vitro motor assays (Endow et al. 1994). It was noticed that the taxol-stabilized microtubules were shrinking at a rate of 0.4 μm/min as they moved during the assay. Microtubules incubated in the absence of GST-Kar3p were stable. Microtubules nucleated from flagellar axonemes were also relatively stable in the presence of GST-Kar3p. Since the axoneme-grown microtubules have free plus ends but not minus ends, the clear implication is that Kar3p destabilizes microtubules from their minus ends. This observation is consistent with two, otherwise puzzling, in vivo phenotypes associated with *kar3* and *kar1* mutants. First, the mitotic defects of *kar3* null mutants can be partially suppressed by the presence of low concentrations of benomyl, a microtubule depolymerizing drug (Roof et al. 1991; Meluh 1992). Although there are other interpretations for this fact, it suggests that the absence of Kar3p stabilizes microtubules and that there is a requirement for Kar3p-mediated depolymerization at some stage of the cell cycle. Second, both *kar1* and *kar3* mutants are notable for the excessive length of the cytoplasmic microtubules in shmoos and zygotes (Rose and Fink 1987; Vallen et al. 1992a). That the absence of wild-type Kar1p function acts locally on the cytoplasmic microtubules can be seen in matings between a *kar1* mutant and wild type (Rose and Fink 1987). In this case, a long microtubule bundle is seen emanating from only one nucleus. The simplest interpretation of these results is that some Kar1p-dependent microtubule length-determining function is present at the SPB. Given that Kar3p-β-galactosidase localization is dependent on Kar1p and that GST-Kar3p destabilizes microtubules from their minus end, the implication is that one of the roles of Kar3p in congression is the depolymerization of the cytoplasmic microtubule bundle.

Depolymerization may seem to be a superfluous function in congression, given the much faster rate of directed movement. However, further considerations suggest that such a function might be essential to nuclear fusion. Without some mechanism for removing the microtubules, nuclei can come no closer than the length of the longest cytoplasmic microtubules in the bundle. Indeed, the appearance of zygotes produced by crosses between *KAR1*$^+$ and *kar1*$^-$ strains suggests that the nuclei are kept separate by the elongated microtubule bundle. Although this discussion has assumed that depolymerization occurs during congression, it is likely that depolymerization occurs prior to cell fusion. As noted above, *kar1* and *kar3* shmoos already have elongated cytoplasmic microtubules prior to cell fusion (Rose and Fink 1987; Vallen et al. 1992a). Therefore, a negative regulator of cytoplasmic microtubule length must be active in shmoos and defective in the mutants.

In wild-type shmoos, the nucleus moves into the shmoo projection with the SPB oriented toward the tip of the projection. A variety of studies have clearly shown that the position of the nucleus at this time depends on the integrity of the cytoplasmic microtubules. The *kar1* and *kar3* mutations do not appear to affect nuclear orientation per se, but they do affect whether the nucleus migrates into the shmoo projection. Mutations affecting a membrane protein, Fus2p, affect both nuclear orientation and nuclear fusion (Trueheart et al. 1987; Elion et al. 1995). As discussed below, Fus2p is required for cell fusion and is localized to the tip of the shmoo projection. In contrast, mutations in cytoplasmic dynein (Eshel et al. 1993; Li et al. 1993), required for mitotic nuclear migration, do not affect nuclear fusion. One model that is consistent with these data is that Fus2p helps anchor and stabilize the plus ends of the cytoplasmic microtubules at a cortical site in the shmoo tip. Acting at the SPB, Kar3p-dependent depolymerization of the cytoplasmic microtubules would draw the nucleus into the shmoo projection.

B. *CIK1*

CIK1 was identified in a genetic screen designed to find mutations that affect both nuclear fusion and chromosome stability (hence the name *CIK1* for *c*hromosome *i*nstability and *k*aryogamy) (Page and Snyder 1992). The *cik1* deletion is viable but temperature-sensitive for growth and exhibits an elevated rate of chromosome loss during mitotic growth. During mating, the deletion strain shows a severe bilateral nuclear fusion defect, similar to that of a *kar3* deletion. Like *kar3* mutants, *cik1* mutants exhibit a congression defect with elongated cytoplasmic microtubules which are oriented properly. *CIK1* mRNA is strongly induced during mating, indicating that it also has a specific role during nuclear fusion. Cik1p is 594 amino acids in length and is predicted to possess a central 300-amino-acid coiled-coil domain. Authentic Cik1p can be detected at the SPB in both mitotic and mating cells. In shmoos, Cik1p-β-galactosidase hybrids show a localization pattern identical to that of Kar3p-β-galactosidase hybrids (Page et al. 1994). The localization of each hybrid is dependent on the integrity of the other protein (i.e., Cik1p-β-galactosidase is dependent on Kar3p and Kar3p-β-galactosidase is dependent on Cik1p). Furthermore, both hybrids are dependent on Kar1p for localization to the SPB but not to the cytoplasmic microtubules. In mitotic cells, the two proteins can be shown to interact in "two-hybrid" experiments. In mating cells, the two proteins coimmunoprecipitate. Taken together, these data strongly suggest that Cik1p is a "light chain" for Kar3p during mating.

C. Mitotic Roles for *KAR3* and *CIK1*

CIK1 and *KAR3* are induced by pheromone, but they both have defects in mitosis. Although both exhibit an elevated frequency of chromosome loss, additional data suggest that Cik1p and Kar3p may have different functions in mitotic cells. First, the two deletion mutants exhibit somewhat different mitotic phenotypes: *kar3* mutants are extremely slow growing at all temperatures (Meluh and Rose 1990) and *cik1* mutants are only temperature-sensitive (Page and Snyder 1992). Up to 30–40% of *kar3* deletion cells are inviable and arrested in the G_2/M of the cell cycle (large bud, short spindle, undivided nucleus) (Meluh and Rose 1990). The *cik1* deletion does not result in such a severe defect. Second, mutations in *kar3* suppress mutations in *cin8* and *kip1* (Saunders and Hoyt 1992; Hoyt et al. 1993), whereas *cik1* mutations do not (Page et al. 1994). Cin8p and Kip1p (Roof et al. 1991, 1992; Hoyt et al. 1992) belong to a subfamily of kinesin-related proteins, some of whose members have been shown to be plus-end-oriented microtubule motors (Endow 1991; Barton and Goldstein 1996). Cin8p and Kip1p (as well as other members of the family) have been implicated as having essential roles in spindle assembly (Roof et al. 1991, 1992; Hoyt et al. 1992) and spindle elongation (Saunders et al. 1995). This observation implies that Cik1p is not required for this aspect of Kar3p function in mitosis. Third, both Cik1p-β-galactosidase and Kar3p-β-galactosidase hybrids localize in the nucleus, near the spindle poles, in mitotic cells, but the localization of Kar3p-β-galactosidase is independent of Cik1p (although the converse does not appear to be true) (Page et al. 1994). Finally, although the two proteins are clearly in a complex in mating cells, they do not readily coimmunoprecipitate in mitotic cells (Page et al. 1994). Taken together, these data strongly suggest that Kar3p and Cik1p are not physically associated in mitotic cells and have somewhat different mitotic roles. It seems likely that the two proteins physically associate as a regulated response to mating pheromone.

In mitosis, Kar3p is in the nucleus where it is engaged in spindle dynamics (Meluh 1992; Page et al. 1994). Biochemical data implicate Kar3p as playing a part as the kinetochore motor (Middleton and Carbon 1994). Its minus-end-oriented motor activity is consistent with a role in anaphase A, the movement of the chromosomes toward the spindle poles. Because yeast chromosomes do not strongly condense in mitosis, clear evidence for anaphase A in *S. cerevisiae* has not yet been published. However, detailed experiments with fluorescent in situ hybridization to detect the behavior of individual chromosomes promise to provide strong evidence in this regard (D. Koshland, pers. comm.). Given the fact that *KAR3* is not an essential gene, it is initially surprising to suggest that Kar3p might be required for anaphase A. However, anaphase B in yeast results in at least an eightfold increase in the length of the mitotic spindle, and this should be more than sufficient to separate the chromosomes even without anaphase A. The specific role for Cik1p in mitosis is less clear, but its localization near the SPB probably reflects a role in spindle function (Page and Snyder 1992; Page et al. 1994).

One clear function of the regulated interaction between Kar3p and Cik1p is to effect a change in the specific localization of Kar3p (Page et al. 1994). Upon treatment with mating pheromone, both Kar3p-β-galactosidase and Cik1p-β-galactosidase move into the cytoplasm where they localize to the cytoplasmic microtubules and SPB. In *cik1* mutants, Kar3p-β-galactosidase stays in the nucleus, demonstrating the requirement for Cik1p in the relocalization. The specific mechanism by which the two proteins change their interaction and localization is not known, although Cik1p is reported to be modified in manner that suggests phosphorylation. It is not known whether either protein is associated with other specific proteins in mitosis.

D. *KAR1*

As discussed above, it is clear that Kar1p is required to localize Kar3p and Cik1p to the SPB. *KAR1* is essential for mitotic growth, and temperature-sensitive mutations result in a block in the initial stage of SPB duplication (Rose and Fink 1987). The Kar1p-β-galactosidase hybrids were localized to the newly replicated SPB (Vallen et al. 1992b). This remarkable result indicated that Kar1p was likely to be an SPB component and that the interaction of Kar1p with SPB components might be cell-cycle-regulated. Furthermore, the finding that the new SPB is preferentially segregated into the bud implied that the nucleus has an inherent polarity that is aligned with the cell axis. After SPB duplication, the new SPB remains oriented toward the bud, and the old SPB moves around to the opposite side of the nucleus during spindle assembly. These results imply that the new SPB must specifically be tethered to a cortical site during the critical movement. Conclusive evidence that Kar1p is an SPB component came from immunoelectron microscopy studies which showed that the protein is present in the "half-bridge" structure associated with the spindle plaque (Spang et al. 1995).

Kar1p is organized into at least three functional domains (Vallen et al. 1992a). A central domain is required for SPB duplication and is both necessary and sufficient to target Kar1p-β-galactosidase hybrid proteins to the SPB. Deletions in this interval do not affect nuclear fusion. An amino-terminal domain is required only for nuclear fusion and is necessary for the mating-specific localization of Kar3p-β-galactosidase to the SPB. A third domain of Kar1p, important for both mitosis and karyogamy, includes the hydrophobic carboxyl terminus and is sufficient to target a β-galactosidase-Kar1p hybrid to the nuclear envelope. The essential mitotic regions of *KAR1* comprise only a small portion (~20%) of the coding sequence of this 432-residue protein. Overall, these data suggest that Kar1p is composed of two or more protein-binding domains tethered to the nuclear envelope via the hydrophobic tail.

The mitotic function of Kar1p was initially revealed by genetic studies of a unique temperature-sensitive deletion in the essential SPB domain of *KAR1* (Vallen et al. 1994). Both dominant and high-copy suppressors were found to be allelic to *CDC31*. *CDC31* is required for SPB duplication, and it encodes a calmodulin-like protein most closely related to centrin, a highly conserved protein found in the centrosome of many eukaryotes (Baum et al. 1986; Baron and Salisbury 1988; Huang et al. 1988). Temperature-sensitive *cdc31* mutations result in a defect in SPB duplication identical to that of the *kar1* mutation (Baum et al. 1986), and Cdc31p is also localized to cytoplasmic side of the SPB half-bridge (Spang et al. 1993; Biggins and Rose 1994). One *CDC31* suppressor allele conferred a temperature-sensitive defect in SPB duplication, which was countersuppressed by recessive mutations in *KAR1* (Vallen et al. 1994). The strongest *CDC31* alleles suppressed a complete deletion of *KAR1*. These "bypass" suppressors indicate that Kar1p acts through Cdc31p to initiate SPB duplication and that in the suppressor alleles, Cdc31p must act independently of Kar1p, possibly via a downstream effector.

Proof of a direct interaction between Kar1p and Cdc31p came from in vitro binding studies (Biggins and Rose 1994; Spang et al. 1995). Cdc31p, expressed and purified from *E. coli*, binds to Kar1p immobilized on filters. As predicted by the suppressor studies, deletion mapping showed that Cdc31p interacts specifically with a short sequence within the essential SPB duplication domain of Kar1p (Biggins and Rose 1994; Spang et al. 1995). The Cdc31p-binding sequence contains three acidic amino acids not found in calmodulin-binding peptides, which may explain why Kar1p binds Cdc31p but not calmodulin (Spang et al. 1995). Cdc31p was also able to bind to the carboxyl terminus of Nuf1p/Spc110p, a major calmodulin-binding component of the SPB (Spang et al. 1995). However, the significance of this interaction in vivo is unclear.

Aside from the suppressor studies, immunofluorescence microscopy demonstrated that the interaction between Cdc31p and Kar1p is functionally significant in vivo. In a *kar1* mutant, Cdc31p fails to localize to the SPB (Biggins and Rose 1994; Spang et al. 1995). In *kar1* mutants containing the dominant *CDC31* suppressor mutations, Cdc31p again localized to the SPB (Biggins and Rose 1994). Furthermore, the localization of Cdc31p to the SPB was affected by the overexpression of Kar1p and Kar1p-β-galactosidase such that Cdc31p accumulates in aggregates formed by the overexpressed proteins (Biggins and Rose 1994). On the other hand, localization of Kar1p to the SPB is independent of Cdc31p (Spang et al. 1995). Taking these data together, we proposed that the essential function of Kar1p is to localize Cdc31p to the SPB and that this interaction is normally required for SPB duplication. Given that the suppressor mutant forms of Cdc31p can localize to the SPB in the absence of Kar1p, there must be additional Cdc31p-interacting proteins in the SPB, and the suppressors may enhance their interaction in some way.

Several studies are currently under way to identify the additional Cdc31p-interacting proteins as these hold the promise of being critical functional components in the initiation of SPB assembly.

E. Other Genes Affecting Nuclear Congression

Given the nature of their mutant defects and the functions of the identified gene products, all of the nuclear congression mutants seem likely to affect microtubule function. In the following cases, specific functions have not yet been identified for the gene products.

1. BIK1

Identified serendipitously, *BIK1* is required for the complete array of microtubule-associated functions in yeast (Berlin et al. 1990). Although the gene is not essential for viability, deletion causes a severe bilateral defect in nuclear fusion, elevated rates of chromosome loss, and frequent missegregation of the nucleus during mitosis. The *bik1* null mutants contain very short or undetectable cytoplasmic microtubules in both mating and mitosis. Nuclear orientation/migration is defective, and mitosis often occurs within the mother cell, leading to formation of both binucleate and anucleate cells. Cells overexpressing Bik1p initially have long cytoplasmic microtubules and abnormally short spindle microtubules. After longer periods of overexpression, cells lose all microtubules and arrest at G_2/M. In addition, synthetic lethality was observed in double-mutant strains containing mutations in both *BIK1* and either of the genes encoding α and β tubulins (*TUB1* and *TUB2*). Consistent with these phenotypes, Bik1p colocalizes with both cytoplasmic and nuclear microtubules. From these results, it has been proposed that Bik1p is required stoichiometrically for the formation or stabilization of microtubules during mitosis and conjugation. That Bik1p may have a direct role in microtubule stability is supported by the finding that it contains a microtubule-binding domain that is conserved in CLIP-170, a mammalian protein implicated in binding endocytic carrier vesicles to microtubules (Pierre et al. 1992, 1994), and the *Drosophila* Glued protein, a component of the dynein complex (Waterman-Storer and Holzbaur 1996).

Recent genetic evidence suggests that *BIK1* is not essential because of the presence of a second gene, *ASE1* (Pellman et al. 1995). *ASE1* is a nonessential gene that encodes a novel protein which localizes specifically to the mitotic spindle midzone. Strains lacking both *ASE1* and *BIK1* are inviable, and a conditional *bik1 ase1* double mutant appears to be defective for anaphase spindle elongation. These results suggest that

Bik1p and Ase1p may have partially overlapping functions during spindle elongation. However, *ase1* mutants do not exhibit defects in nuclear migration and congression. Therefore, it is likely that their activities do not overlap for functions associated with the cytoplasmic microtubules.

2. KEM1

One of the more puzzling genes involved in nuclear fusion is *KEM1* (also known as *XRN1*, *DST2*, *SEP1*, and *RAR5*) (Kim et al. 1990; Larimer et al. 1992; Bahler et al. 1994; Interthal et al. 1995; Johnson and Kolodner 1995), originally identified on the basis that *kem1* mutations enhanced the nuclear fusion defect of a *kar1* mutation (Kim et al. 1990). The *kem1* mutants exhibit a bilateral nuclear fusion defect on their own and exhibit a number of defects that implicate a role in microtubule-associated functions (Interthal et al. 1995). These include benomyl sensitivity, elevated rates of chromosome loss, impaired SPB separation, and defective nuclear migration to the bud neck. Purified Kem1p promotes polymerization of tubulin from both porcine brain and *S. cerevisiae*. Furthermore, Kem1p cosediments with these microtubules in sucrose gradient centrifugation. Genetic analysis of double-mutant strains containing a mutation in *KEM1* and either *TUB1* or *TUB2* further suggests an interaction between Kem1p and microtubules. Taken together, these results are all highly suggestive of a direct role for Kem1p in microtubule function. Despite these results, several studies have independently identified Kem1p as being a 5′-3′ exoribonuclease (Larimer et al. 1992), a DNA strand transferase (Bahler et al. 1994; Heyer et al. 1995), and a G4 DNA nuclease (Liu and Gilbert 1994; Liu et al. 1995). Genetic studies have also identified it as being required for plasmid stability, and *kem1* mutations are synthetically lethal with mutations in the *SKI2* and *SKI3* genes implicated in the regulation of killer toxin expression (Johnson and Kolodner 1995). *kem1* mutants clearly have a number of pleiotropic defects, but it is unclear which if any of the known mutant defects are the direct result of loss of Kem1p function. An intriguing possibility is that the disparate functions of Kem1p belie a functionally relevant coupling of microtubule binding to DNA strand exchange, possibly related to its role in chromosome pairing and exchange.

3. KAR4 *and* KAR9

Two genes recently identified by bilateral nuclear congression mutations are *KAR4* and *KAR9* (Kurihara et al. 1994). The phenotype of *kar4* mutants in mating is identical to that of *kar3* and *cik1* mutants, resulting in elongated cytoplasmic microtubules that are properly oriented along the zygote axis, toward the opposite nucleus. These results suggest that Kar4p may function with or be required for Kar3p/Cik1p function during mating. Recent data indicate that Kar4p is a transcription factor that is specifically required for the pheromone-dependent induction of *KAR3* and *CIK1* (Kurihara et al. 1996).

The *kar9* mutation causes a defect in nuclear orientation in zygotes. Unlike the *bik1* mutation, cytoplasmic microtubules are present, although they are not properly oriented. Other mitotic phenotypes indicate that *KAR9* has a general role in cytoplasmic microtubule function in mitosis (R. Miller and M. Rose, in prep.). One attractive model is that Kar9p helps orient the cytoplasmic microtubules in mating and mitosis.

4. FUS2

Originally identified in a screen for mutants that show a partial defect in cell fusion during mating, *fus2* mutants also show a fairly strong defect in nuclear fusion (Elion et al. 1995). *FUS2* is highly induced only in haploid cells treated with mating pheromone. Fus2p localizes to novel vesicular or cytoskeletal structures at the tip of the shmoo projection and at the junction between cells in the course of fusion. The 617-residue protein is predicted to be largely coiled-coil, and, consistent with a cytoskeletal function, it fractionates in a high-speed pellet and is not easily extracted by detergents, high salt, or chaotropic agents. The *fus2* mutant defect in cell fusion is partially suppressed by expression (and overexpression) of the unrelated membrane protein, Fus1p. The *fus1* mutants exhibit a cell fusion defect similar to that of *fus2*, and the double mutant shows a much stronger cell fusion defect than either mutant alone (McCaffrey et al. 1987; Trueheart et al. 1987). However, the *fus1* mutant does not show a nuclear fusion defect. Therefore, it is likely that the nuclear fusion defect of *fus2* is not simply a secondary consequence of the cell fusion defect. In *fus2* mutant zygotes in which cell fusion but not nuclear fusion has occurred, the defect appears to be in nuclear congression. In these zygotes, the cytoplasmic microtubules are found to be misoriented with respect to the zygote axis. One attractive model is that Fus2p serves as part of the cortical site for directing the attachment and orientation of the cytoplasmic microtubules in the shmoo. Less clear is the role in cell fusion, since neither Fus1p nor Fus2p is required for the localization of the other protein.

5. CIN *Genes*

The *CIN* genes were identified in a genetic screen for mutants exhibiting an elevated rate of chromosome loss (Hoyt et al. 1990; Stearns et al.

1990). Mutations in *CIN1*, *CIN2*, and *CIN4* cause cells to be supersensitive to benomyl, cold-sensitive, and defective in both nuclear migration and nuclear fusion. In the cold, the mutants lose assembled microtubules and arrest in the cell cycle at G_2/M. The mutations show a wide variety of genetic interactions with mutations in the tubulin genes. These results suggest that Cin1p, Cin2p, and Cin4p contribute generally to microtubule stability. The finding that *CIN4* encodes a small GTP-binding protein suggests that Cin4p may be acting in a regulatory fashion.

6. CDC *Genes*

Several cell division cycle genes have been implicated in the nuclear fusion pathway, including *CDC4*, *CDC28*, *CDC34*, and *CDC37* (Dutcher and Hartwell 1982, 1983a,b). Matings between these mutants and wild type at the nonpermissive temperature result in significant increases in the frequency of cytoductant formation, signaling the failure in nuclear fusion (albeit, except for the *cdc4* mutant, most of the defects are much weaker than those in the *kar* mutants). Cdc28p is the *S. cerevisiae* homolog of the mitosis promoting factor (MPF) cell division cycle kinase (Beach et al. 1982; Ferguson et al. 1986; Dunphy et al. 1988).

Cdc4p contains limited homology with β-transducin (Peterson et al. 1984; Fong et al. 1986) and is required for spindle assembly (separation of the duplicated SPBs). The *cdc4* mutants produce multiple rounds of aberrant buds at the nonpermissive temperature (Adams and Pringle 1984). The specific function of Cdc4p in this context is unknown. Recently, a transducin repeat containing microtubule-associated protein has been identified, raising the possibility that Cdc4p acts through the microtubules (Li and Suprenant 1994). However, the fact that multiple rounds of budding occur suggests that Cdc28p kinase activity remains high at the nonpermissive temperature. If so, the defect may be due to prolonged activation of the kinase at the nonpermissive temperature.

Mutations in *CDC34* also lead to unseparated SPBs and a multibudded phenotype (Goebl et al. 1988). Cdc34p is a ubiquitin-conjugating protein (Goebl et al. 1988, 1994) that is required of the degradation of the G_1 cyclin Cln3p (Tyers et al. 1992; Yaglom et al. 1995) as well as a potent inhibitor of the B cyclins (Sic40p) (Schwob et al. 1994). This mutant is prevented from entering S phase because the activity of the G_1 cyclins remains high while the B cyclins stay low. Thus, this mutant may also be defective for nuclear fusion because Cdc28p remains active for its G_1 kinase activity. A specific role for Cdc37p has not yet been clearly identified; however, a recent report suggests that it is a positive regulator of Cdc28p kinase activity required for the association between the kinase and multiple cyclins (Gerber et al. 1995).

These four mutants suggest that there are two types of Cdc28p-associated defects in nuclear fusion. Mutations that lead to reduced levels of activity and mutations that lead to prolonged activation in G_1 lead to nuclear fusion defects. It seems likely that the mechanism of the defects must be distinct. A role for Cdc28p in nuclear fusion is somewhat puzzling given that the Cdc28p kinase is thought to be inactive during mating (Elion et al. 1990, 1991; Peter et al. 1993; Tyers and Futcher 1993). Possibly, the kinase is required prior to cell fusion or in a previous cell cycle to activate a nondiffusible component of the fusion pathway. On the other hand, Cdc28p kinase activity must be inhibited during pheromone arrest by Fus3p and Far1p for normal mating to proceed (Elion et al. 1990, 1991; Peter et al. 1993; Tyers and Futcher 1993). Since several components of the nuclear fusion pathway are also utilized for different functions during mitosis, continued activation of the cell cycle may inhibit nuclear fusion.

F. A Model for Nuclear Congression

Taken together, these various observations suggest a coherent model for the mechanism of nuclear congression and the roles played by specific proteins (see Figs. 1 and 3). As the cell responds to pheromone, the axis of cell polarity changes such that it becomes oriented toward the mating partner. A shmoo projection forms and grows toward the partner. In the tip of the shmoo projection are cortical proteins that mark the site at which cell fusion and ultimately nuclear fusion will occur. One component of this site is likely to be Fus2p. A second component may be Kar9p. Cytoplasmic microtubules are captured by the cortical site, and the nucleus moves toward the tip of the shmoo projection. Several proteins are required for the stability of the cytoplasmic microtubules (e.g., Tub2p, Bik1p, Cin1p, Cin2p, and Cin4p), and mutations in these genes affect nuclear fusion at this initial stage of nuclear movement. The force for nuclear movement at this point is not clear, but their mutant phenotypes suggest that it may come from depolymerization of the cytoplasmic microtubules at the SPB, mediated by Kar3p in association with Cik1p and Kar1p. As the cell walls between the two cells and plasma membranes break down and fuse, the cytoplasmic microtubules from the two nuclei interdigitate. The two sets of microtubules are cross-bridged by Kar3p and Cik1p. Kar4p is also required for this step, but it

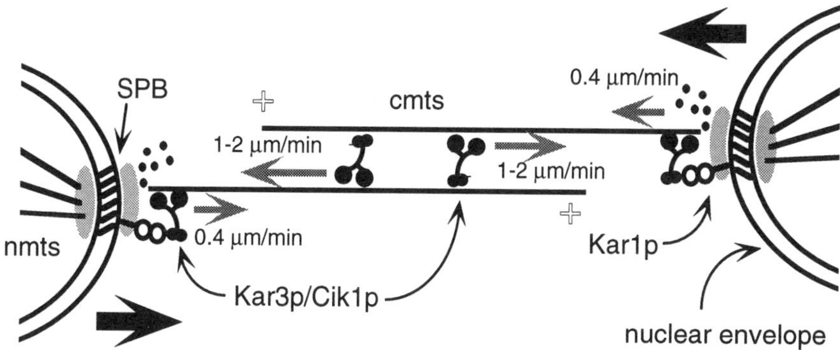

Figure 3 A model for Kar3p/Cik1p in nuclear congression (see text for details). (SPB) Spindle pole body; (nmts) nuclear microtubules; (cmts) cytoplasmic microtubules; (+) the plus ends of the microtubules. Small dots at the minus ends of the microtubules depict free tubulin released from the end of the depolymerizing microtubules. Small arrows depict the movement of Kar3p/Cik1p relative to the microtubules, and large arrows depict the overall movement of the nuclei. (Reprinted, with permission, from Rose 1996.)

does not have a direct role. The microtubules slide together due to the minus-end-oriented motor activity of Kar3p. As a result of this movement and the continued shrinking of the microtubules, the two nuclei become closely apposed. Nuclear envelope fusion then occurs in the vicinity of the SPB.

V. NUCLEAR MEMBRANE FUSION

In the last few years, we have witnessed the exciting coalescence of several different lines of research into the mechanism of intracellular membrane fusion. Research in three different systems, transport between Golgi compartments, fusion of synaptic vesicles with the presynaptic membranes, and identification of genes required for transport from the ER to the Golgi in yeast, demonstrated that one basic mechanism is operative in each type of membrane fusion (Schekman 1992; Bennett and Scheller 1993, 1994; Ferro-Novick and Jahn 1994; Rothman 1994; Rothman and Warren 1994). In each case, an ATPase, called NSF for *N*-ethylmaleimide-sensitive factor (encoded by *SEC18* in yeast), interacts with a protein(s) called SNAP for soluble NSF attachment protein (encoded by *SEC17*), and the NSF-SNAP complex then interacts with any one of a series of membrane-bound proteins called SNAREs, for SNAP receptors. The NSF-SNAP complex is a general factor that acts at each stage of the secretory pathway, whereas the SNAREs are specific to the stage of secretory or endocytic pathway. Specific v-SNAREs are present on the transport vesicles, and the t-SNAREs are present on the target membranes. The formation and function of the complex are then further regulated by interaction with a host of additional proteins, including a host of small GTPases of the Rab protein family. In yeast, many different SNAREs and Rabs have been identified that act at each stage of the secretory pathway (Ferro-Novick and Jahn 1994).

Integral to the SNARE hypothesis model is the idea that the SNAREs and Rabs generate the specificity in the membrane fusion reactions. This seems particularly important given that fusion is occurring between membranes of different compartments (so-called heterotypic fusion reactions), and a mechanism for ensuring targeting to the appropriate membranes must exist. In the case of nuclear fusion, two membranes of identical origin are involved. In principle, it is possible that the nuclei from the two different mating types could have different SNAREs and thereby conform to the standard model. However, two lines of evidence argue that the nuclei of the same mating type can fuse with near wild-type efficiency during mating. The first came from spheroplast fusion experiments performed with mating-type **a** cells treated with α-factor (Curran and Carter 1986; Rose et al. 1986). The second came from the demonstration that the mating-type-specific pheromones and pheromone receptors are the major, if not the only, determinants of sexual identity in yeast (Bender and Sprague 1986, 1989). The implication is that, aside from the transcription factors that specify mating-type-specific genes, internal structures are identical between the two mating types. Thus, for nuclear fusion, membranes of identical origin and therefore presumably bearing the identical set of SNAREs must be capable of fusion.

The second difference between the heterotypic membrane fusion events in the secretory pathway and nuclear fusion is the lack of dependence on *SEC18*. As mentioned above, a second NSF-like molecule, Cdc48p, has been found to be responsible for the in vitro fusion of nuclear membranes from mitotic cells (see Latterich et al. 1995). It is possible that the in vitro fusion reaction might also represent a heterotypic reaction in that it might be occurring between the peripheral ER and nuclear envelope. This would be consistent with the lack of a requirement for Cdc48p in nuclear fusion in vivo. Alternatively, it is possible that an additional NSF-like protein is induced during the mating response. In any event, it will be essential in the coming years to identify the proteins that are directly responsible for nuclear membrane fusion. Candidates for these are among the set of class II nuclear fusion mutants.

A. *KAR2*

KAR2 encodes the yeast homolog of the ER-resident Hsp70 protein, also known as BiP (Normington et al. 1989; Rose et al. 1989). Its association with folding and misfolded proteins has clearly demonstrated the role for BiP and yeast Kar2p as a chaperone in the folding of proteins in the secretory pathway (Simons et al. 1995). In addition, several in vivo and in vitro experiments demonstrate a critical role for Kar2p in the translocation of proteins into the ER (Vogel et al. 1990; Sanders et al. 1992; Brodsky and Schekman 1993; Brodsky et al. 1995). Evidence from in vitro experiments suggests that Kar2p may have two functions during translocation (Sanders et al. 1992): One is to bind the translocating peptide as it appears in the ER and the second is to "reset" components of the translocation apparatus to allow initial binding to the nascent chain.

The role for Kar2p in nuclear fusion is less clear. The in vitro ER/nuclear envelope fusion data (Latterich and Schekman 1994) suggest that it has a direct role in membrane fusion. It is very unlikely that Kar2p is required for karyogamy because of its role in translocation per se. First, there is no correlation between the severity of different *kar2* mutations in karyogamy versus translocation (Vogel et al. 1990; Vogel 1993). Second, *kar2* mutations affect the in vitro assay under conditions in which translocation does not occur (Latterich and Schekman 1994). Given the current knowledge of the roles of peripheral and integral membrane proteins in membrane fusion, it is initially somewhat surprising that a luminal protein would be required (Stegmann et al. 1989; White 1992). A possible role for Kar2p comes from the recognition that complete fusion of membranes catalyzed by the influenza hemagglutinin requires its transmembrane domain; with only the ectodomains, mixing of lipids but not luminal contents occurs (Kemble et al. 1994; Melikyan et al. 1995). One interpretation of these results is that the ectodomain is required for docking and initiation of membrane fusion, whereas the transmembrane domain has an essential role in pore dilation. It seems reasonable to suggest that a chaperone activity might be required during pore dilation in some membrane fusion events. That such luminal proteins have not yet appeared as a requirement for fusion in the secretory pathway suggests either that they are not essential in this pathway or that the existing mutant screens are biased against their isolation.

B. Other Proteins Interacting with Kar2p

Hsp70 proteins generally act in concert with activating proteins sharing homology with the DnaJ protein from *E. coli* (Cyr et al. 1994). Kar2p interacts with at least one integral membrane protein that contains a luminal DnaJ domain called Sec63p (Scidmore et al. 1993). *SEC63* was identified in two very different genetic screens. The first screen looked for mutations that block the translocation of secreted proteins (Deshaies and Schekman 1987), and the second screen looked for mutations that block the import of proteins into the nucleus (Sadler et al. 1989). It seems clear that Sec63p has a direct role in translocation, possibly by localizing and/or activating Kar2p (Brodsky and Schekman 1993; Scidmore et al. 1993). Sec63p has been isolated in complexes with several other proteins required for translocation, including Sec61p, Sec62p, Sec71p, Sec72p, and Kar2p (Deshaies et al. 1991; Brodsky and Schekman 1993; Feldheim et al. 1993; Kurihara and Silver 1993). The function of Sec63p in nuclear import is less clear, but it is of interest that the mutations that block this process largely map to a cytoplasmic domain of the protein (Feldheim et al. 1992; Nelson et al. 1993).

Recently, a role for Sec63p in nuclear fusion has appeared from the isolation of an unusual allele of Sec63p bearing a carboxy-terminal truncation of 27 amino acids (Ng and Walter 1996). The *sec63-201* allele exhibits a strong bilateral class II nuclear fusion defect. This allele also causes a strong block in translocation, but other alleles of *sec63* that also cause strong blocks in translocation result in a much weaker defect or no defects in karyogamy. The allele specificity argues that the nuclear fusion defect does not arise from a simple block in translocation.

Given the identification of both Kar2p and Sec63p as proteins required for nuclear fusion, it was natural to examine the requirement for other proteins known to interact with Kar2p and Sec63p. Accordingly, deletions of both *sec71* and *sec72* were found to result in bilateral nuclear fusion defects (Ng and Walter 1996). Sec71p is an integral ER membrane protein and Sec72p is a peripheral ER membrane protein (Green et al. 1992; Feldheim et al. 1993; Kurihara and Silver 1993). Deletions in both genes result in viable cells that are temperature-sensitive for viability and translocation. At the same time, *KAR7* was cloned and identified as the being the same gene as *SEC71* (V. Brizzio et al., in prep.). Deletions of *SEC71* and *SEC72* also result in a temperature-sensitive defect in the in vitro ER fusion assay (V. Brizzio et al., in prep.). Thus, Sec71p and Sec72p also have roles in nuclear fusion. However, the temperature sensitivity of the deletion strains in vitro suggests that they may have only an indirect role, such as stabilizing the fusion complex. No or very mild defects were found for mutations in *sec61* and *sec62* (Ng and Walter 1996; V. Brizzio et al., in prep.). However, it is possible that stronger defects might arise from different mutant alleles.

Taken together, these results demonstrate that at least four of the six proteins identified as components of the translocation complex are also required for nuclear envelope fusion. The allele specificity of the mutations strongly suggests that their role in nuclear fusion is independent of their role in translocation. Possible functions might include membrane recognition, docking, or catalyzing the membrane fusion reaction. However, it will be difficult to assign specific functions to these proteins in nuclear fusion without detailed biochemical studies of the reaction in vitro.

C. *KAR5* and *KAR8*

Two other genes identified as being required for nuclear membrane fusion are *KAR5* and *KAR8* (Kurihara et al. 1994). Preliminary evidence indicates that Kar5p is a novel luminal ER membrane protein (C. Beh and M. Rose, in prep.). Kar5p is induced during mating and presumably has a critical and specific role during mating. Interestingly, in shmoos, Kar5p is found to cluster near the SPB, the site at which nuclear membrane fusion normally initiates. It is tempting to speculate that Kar5p may have a role in recruiting the proteins required for membrane fusion to the region of the SPB. Although Kar5p may have a mating-specific role in membrane fusion, it is required in the in vitro fusion assay made from mitotic extracts. Therefore, it seems likely that Kar5p is also required for a mitotic function involved with ER/nuclear envelope fusion. However, the deletion is viable and the defects are more subtle than those of *cdc48* mutants. The normal mitotic function of this protein remains to be determined.

KAR8 appears to encode an additional DnaJ-like protein (D. Huddler and M. Rose, unpubl.). Sequence analysis suggests that the protein is an integral membrane protein, and the DnaJ domain is predicted to be luminal. Kar8p appears to be relatively specific to nuclear fusion and cannot compensate for Sec63p in translocation and vice versa. From the appearance of the putative nuclear fusion intermediates in *kar8* mutants, it is tempting to speculate that Kar8p is required to chaperone a somewhat downstream step in the membrane fusion reaction.

VI. PERSPECTIVES FOR NUCLEAR FUSION

In summary, the karyogamy pathway has been separated into at least two major steps: congression and nuclear envelope fusion. Congression is the major function of the cytoplasmic microtubules during conjugation.

Several mitotic microtubule-dependent functions are recruited to effect the nuclear movements that constitute congression. Nuclear fusion similarly recruits proteins that have additional and distinct primary roles during mitotic growth. The identification of the roles of *KAR* genes in mitotic growth and the identification of a role in karyogamy for genes known from mitosis have led to a great enrichment of our understanding. However, much work needs to be done before our understanding is complete.

One major question concerns the mechanism of the nuclear membrane fusion pathway. How many membranes actually fuse? If two membranes fuse, are there separate components that mediate the fusion of the internal membranes? Which proteins provide membrane recognition and which catalyze the fusion reaction? Is there an equivalent normal mitotic pathway of nuclear envelope and ER fusion in which the *KAR* genes participate? How is this pathway related to the pathway of nuclear envelope reassembly that occurs during mitosis in animals and plants?

Although better understood, several issues also remain in the description of the pathway of nuclear congression. One of these concerns the regulation of the pathway. How are Kar3p and Cik1p recruited from nuclear locations to the cytoplasm? Is there an additional regulation that prevents their inappropriate induction and relocalization when the cells first respond to pheromone? What resets the functions of these proteins such that they are competent for reentry into the cell cycle? What is the exact role of Kar1p, and does it act solely through Kar3p and Cik1p? Is there a strong requirement for nuclear migration and orientation prior to cell fusion? What are the mitotic functions of Kar3p and Cik1p? Does Kar3p have different associated light chains at different stages of the cell cycle?

Given the complexity of the process and the fact that the mutant screens have not been saturated, it seems likely that more genes will be discovered that have specific roles in the nuclear fusion pathway. Our understanding of the biochemical steps in the pathway is in its infancy. We look forward to future progress and the continued development of an integrated model of the pathway incorporating both genetic and biochemical breakthroughs.

VII. OVERVIEW OF CELL FUSION DURING MATING

Much less is known about the mechanism of the cell fusion and cytoplasmic mixing event, plasmogamy, that precedes nuclear fusion than karyogamy itself. Our best picture of the stages of mating come from

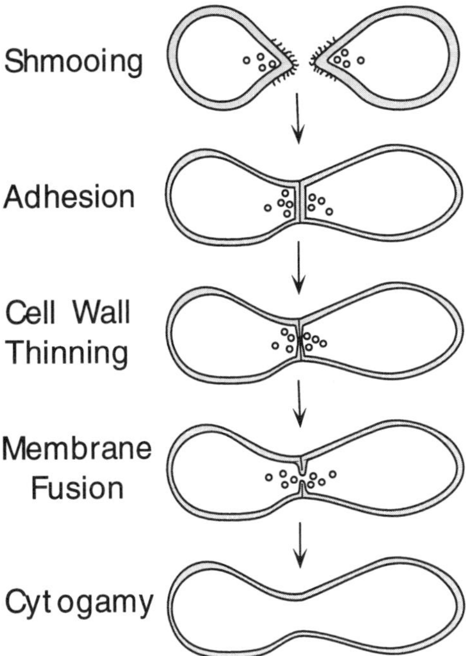

Figure 4 The steps of cell fusion (see text for details). Cells expressing the agglutinins (indicated by the fimbriae on the shmoo projection) first bind or clump with one another. At this stage, they may exhibit nearly normal mitotic ovoid/spherical morphology. As mating progresses, cells become deformed in the region of contact and the cell wall thins. A conjugation tube develops in the region of fusion. Dissolution of the cell wall septum in a localized region of cell-cell contact permits membrane fusion and cytoplasmic mixing. Remodeling of the cell wall continues after cytoplasmic mixing. The final product of mating is a zygote with a smooth, bilobed appearance and a well-knitted wall that shows little evidence that it originated from more than one cell.

direct observation of mating cells. Electron microscopy of mixtures of cells of **a** and α mating type reveals a sequence of events diagrammed in Figure 4. Cells first bind or clump with one another, probably via agglutinins described below. At this stage, they may exhibit nearly normal mitotic ovoid/spherical morphology. As mating progresses, cells become deformed in the region of contact and the cell wall thins. A conjugation tube develops in the region of fusion. Dissolution of the cell wall septum in a localized region of cell-cell contact permits membrane fusion and cytoplasmic mixing. The plasma membranes of the two cells presumably fuse at this point as well. Nuclear fusion proceeds as described earlier, often within the conjugation tube. Electron microscopy suggests that remodeling of the cell wall continues after cytoplasmic mixing. The final product of mating is a zygote with a smooth, bilobed appearance and a well-knitted wall which shows little evidence that it originated from more than one cell. Budding from the zygote proceeds as with a normal diploid cell. In rich medium at 30°C, the mating process occurs in 2–4 hours (Osumi et al. 1974).

A number of features of the mating process deserve comment because of their unique biological requirements. Cells of the two mating types must orient their cytoskeletons toward their partners since the fusion reaction involves highly asymmetric changes in the cell wall (Jackson and Hartwell 1990; Segall 1993; Chenevert 1994). Mobile organisms have the option of finding a preexisting site on the cell surface of their partners, but such an approach would be inefficient for nonmotile organisms like budding yeast. Pheromones appear to direct cell asymmetry, but cell wall factors have not been ruled out as accessories. Cell wall fusion requires specialized remodeling with tight spatial control. A cell wall seal must be built up around the periphery of the region of cell-cell contact to firmly attach the mating cells to one another. In the center of the contact region, cell wall material is dissolved to permit plasma membrane contact and cell fusion. It is not known if specific factors are needed for plasma membrane fusion. Tight regulation of cell wall synthesis and degradation might be coordinated by both partners (Cross et al. 1988). The cell wall remodeling problem is akin to digging a tunnel from both sides of a mountain and hoping to meet in the middle. Most striking is that the entire mating process occurs efficiently in hypotonic medium, indicating that cell wall integrity is maintained throughout.

A. Morphology of Mating

Electron microscopy has been used extensively to study fusion. These studies have provided descriptions of the steps of fusion but have been limited by the relative nonspecificity of structures in the cell wall. Comparisons of morphology and ultrastructure of fusion intermediates of various yeast species have provided some sense of the essence of the process (Crandall et al. 1977). Unfortunately, such studies have not contributed the same level of molecular insight as the analogous work on nuclear fusion. The disperse, heterogeneous, and complex nature of the cell wall contrasts with the discrete SPB, microtubules, and nuclear membrane.

Treatment of haploid cells with pheromone mimics early stages of mating. Pheromone treatment leads to an accumulation of vesicles at the shmoo tip (Byers 1981a). Especially interesting are vesicles with the characteristics of filosomes, implicated in rapid cell wall remodeling at the growing tips of filamentous fungi (Byers 1981a; Baba et al. 1989). Specialized exocytosis is characteristic of polarized cell remodeling from lower to higher eukaryotes. The cell wall at the tip of projections of pheromone-treated cells looks "fuzzy" with fimbriae that may bear agglutinins (Lipke et al. 1976).

The cell wall of mating yeast changes in the region of fusion. The outer layer of the vegetative cell wall is principally mannoprotein and appears under usual electron microscopy staining conditions as a thin electron-dense layer. The mannoprotein layer is involved in controlling the permeability of the cell wall but not its mechanical strength (de Nobel et al. 1990; Klis 1994). The inner layer of the cell wall is glucan-rich and provides strength (Osumi et al. 1974). It appears as a thick electron-transparent layer. Closest to the membrane is another layer of integral, or membrane-associated, proteins not always resolved by electron microscopy. Pheromone treatment leads to thinning of the glucan layer.

The septum separating conjugating *S. cerevisiae* cells initially seems to consist of a tripartite glucan-mannoprotein-glucan structure. The glucan layers of the cell wall thin in the region of the septum separating cells (Osumi et al. 1974; Baba et al. 1989). Additional insight comes from studies of the budding yeast *Hansenula wingei* whose intermediate stages of mating seem to be more accessible than those of *S. cerevisiae* (Crandall et al. 1977). In electron micrographs of mating in *H. wingei*, loss of the mannoprotein layer can be observed at a stage just prior to cytoplasmic mixing (Conti and Brock 1965). In *H. wingei*, the septum is devoid of layers at this late stage in fusion. A progression thus seems to occur during mating. The first contact of cells leads to a glucan-mannoprotein-mannoprotein-glucan structure. Partial cell wall fusion in the septum region creates a glucan-mannoprotein-glucan structure, in which each cell may retain control of its "own" glucan layer. The final stage of cell wall fusion may create a glucan-only septum which both cells potentially can remodel. The prezygotes accumulated by fusion mutants provide greater access to intermediate forms in mating (McCaffrey et al. 1987; Trueheart et al. 1987; Elion et al. 1990, 1995; Elia and Marsh 1996; Brizzio et al. 1996). Certain fusion mutants of *S. cerevisiae* may block mating at the glucan-mannoprotein-glucan stage of cell fusion, whereas others may block at the glucan-only stage (Trueheart et al. 1987; Elion et al. 1990; Elia and Marsh 1996). It is not certain whether the "prezygote" forms that accumulate in mating mixtures of fusion-defective mutants represent true intermediates in the mating process, although they do have a superficial similarity.

Cell fusion appears in most cases to start at a single spot and expand (Osumi et al. 1974). Cytoplasmic mixing may occur in the absence of complete septum breakdown, since in *fus1, fus2* mutants, nuclear fusion can occur in the presence of an apparently intact septum (Trueheart et al. 1987). Few electron photomicrographs of the critical step in which the septum breaks down and membrane fusion occurs are available for *S. cerevisiae*. Rarely, it appears that large segments of cell wall and membrane detach at the fusion septum, leaving a substantial channel between cells (Osumi et al. 1974; Elia and Marsh 1996; Brizzio et al. 1996). It is difficult to rule out artifacts, however, since these structures are observed rarely. It has not been possible to separate membrane fusion from cell wall fusion during mating in *S. cerevisiae*. Cytoplasmic mixing appears to be accompanied by plasma membrane fusion without loss of membrane integrity.

B. Isolation of Fusion-defective Mutants

Mutants defective in the fusion step of mating have been valuable in the study of cell fusion. The genes defined by such mutants provide the best source of information on molecular aspects of fusion. Still, it must be confessed that fusion genes have not yet really provided specific molecular models for fusion.

Mutations affecting pheromone production or response (*ste* mutations) cause sterility. The sterile mutants can reduce mating frequency by six orders of magnitude or more. Sterile strains do not initiate mating and lack both prezygotes and zygotes in their mating mixtures. Fusion-defective (*fus*) mutants are defective only in cell fusion and have other mating responses more or less intact. Mutants defective only in the cell fusion stage of mating rarely exhibit strong mating defects when mated to a normal partner (~0–50-fold mating reduction) (McCaffrey et al. 1987; Trueheart et al. 1987; Elion et al. 1990; Liu and Bretscher 1992; Kurihara et al. 1994). The weak effects on mating of defects in genes involved in cell fusion may indicate that the products can be provided by either partner or may increase efficiency of mating without being essential. The basic phenotype of fusion-defective (*fus*) mutants is that they accumulate intermediates in the mating pathway (prezygotes). The prezygotes usually resemble zygotes, but they retain a septum separating the mating cells that prevents cytoplasmic mixing. Observation of prezygotes requires microscopic examination of mating mixtures. Normal strains accumulate fewer than 1 prezygote per 100 zygotes in a typical mating reaction (Trueheart et al. 1987; Sprague 1991; Kurihara et al. 1994; Elia and Marsh 1996; Brizzio et al. 1996). The key in development of screens for fusion mutants has been to reduce the number of candidate strains that must be subjected to the final, tedious process of microscopic screening.

Identification of fusion-defective mutants has often been dependent on serendipity. Both *FUS1* and *FUS2* were first detected more by astute observation than by genetic strategy. In both cases, a bilateral mating defect, observed in crosses performed for other reasons, was traced to a defect in a *FUS* gene. Mutations in *FUS3* were identified by the presence of a strong mating defect only when mated to a *fus1* or *fus2* strain (Berlin et al. 1991). The *FUS3* example highlights strategic application of the concept that mutations with defects in fusion show only weak overall mating defects because the process is controlled at multiple, redundant levels (Berlin et al. 1991). Thus, a search for mutants with a strong mating defect to a *fus* partner but not to a wild-type partner is essentially a synthetic sterile screen, even though the two mutations lie in different cells. It is hard to predict in advance whether all genes involved in fusion will exhibit synthetic interactions with known *FUS* genes.

A screen designed to identify bilateral sterile mutants of any type is described above in the karyogamy section. In addition to the karyogamy mutants, this screen yielded mutations in known *FUS* genes (e.g., *FUS2*) as well as several new genes (e.g., *RAM1*, *AXL1*, and *SPA2*) (Kurihara et al. 1994; Brizzio et al. 1996). Several of the mutants identified contained mutations acting in a unilateral fashion.

Other approaches, based on serendipity and observation, yield *fus* mutants for reasons that are not yet entirely clear. One screen involves mating to an *sst2* partner. Mutations in *sst2* increase the response to pheromone and reduce mating slightly (Chan and Otte 1982). A screen for mutants defective in mating to an *sst2* partner revealed that several had mutations in known *FUS* genes, *FUS1*, *FUS2*, and *FUS3* (C. Boone, pers. comm.). Expanding the screen yielded several new genes required for fusion, including *SPA2*, *CHS5*, and *BNI1*. A screen for mutants defective in mating to a *far1* strain also yielded fusion-defective mutants including defects in *SPA2*, *AXL1*, *STE6*, and *FPS1* (Davis et al. 1993; J. Phillips and I. Herskowitz, pers. comm.). *FAR1* has a role both in pheromone-induced control of the cell cycle and in orientation toward pheromone gradients (Davis et al. 1993; Chenevert 1994).

Many of the screens above rely on features of fusion mutants that may not be universal. For example, several fusion mutants have a bilateral mating defect and are especially defective in mating to another *fus* mutant, but this does not mean that all *fus* mutants have that phenotype. To take another approach, a large screen for strains with weak mating defects was performed. This screen required mutations to act unilaterally. The defective strains were then screened microscopically for fusion defects during mating to a wild-type partner (Elia and Marsh 1996). Mutants with partial defects in *STE6* and *STE12* were identified. New *fus* mutations appear to be represented among mutants with very weak unilateral mating defects.

Pheromone-inducible genes provide a group of genes highly enriched for mating functions. *FUS1* was cloned by one group as a gene highly induced by pheromone. Several other known fusion genes (e.g., *FUS2*) are highly induced by pheromone.

The mutant hunts described above partially overlap because some of the same genes were identified by a variety of screens. Nevertheless, it seems likely that many genes involved in fusion have not been identified. For example, essential or duplicated genes may not have been detected for genetic reasons.

C. Cell Polarity and Fusion

The molecular basis of cell fusion would be more evident if we knew more of the proteins involved. As things stand, we have isolated pieces of a jigsaw puzzle that hint at the cell systems involved. Many of the genes that seem to have a fairly specific role in fusion are localized in a highly polarized fashion to the site of fusion or affect the ability of cells to polarize in pheromone. In most cases, however, we have only indirect evidence for a specific interaction with cytoskeleton or intracellular trafficking. From the hints so far, it seems that actin-based systems are involved in fusion (Liu and Bretscher 1992). Nearly all of the group of fusion proteins contain motifs that might indicate interaction with actin, e.g., homology with myosin or SH3 domains (McCaffrey et al. 1987; Trueheart et al. 1987; Gehrung and Snyder 1990; Liu and Bretscher 1992). This excitement must be tempered by the lack of biochemistry to demonstrate such interaction. It is also possible that specialized vesicular trafficking is important, and this may account for the role of actin. The genes identified so far as having a role in fusion do not seem to have a specific role in determining the site of polarization of the cell, but rather affect the general ability to become polarized or represent highly localized products that might be involved in fusion or as docking molecules for vesicles.

It seems likely that microtubules are not required, since tubulin mutations or nocodazole inhibits nuclear fusion but not shmoo or zygote formation (Delgado and Conde 1984; Hašek et al. 1987; M. Rose, unpubl.). Furthermore, the zygotes formed in such experiments do not show accumulation of prezygotes. However, mutations in one gene required for cell fusion do have associated microtubule and nuclear congression defects (*fus2*), suggesting that these two pathways may intersect at this level (Elion et al. 1995).

VIII. GENES REQUIRED FOR CELL FUSION

A. *FUS1*

FUS1 has an apparently specific role in cell fusion during mating. Mutants defective in *FUS1* exhibit no vegetative defects and initiate mating normally, but they fail to complete the cell wall digestion required for cells to fuse. A certain redundancy of function is observed since *fus1* strains mate poorly to another *fus1* strain, but mate well to a *FUS1* partner (Trueheart et al. 1987). Although wild-type x *fus1* crosses do not exhibit an obvious mating defect, prezygotes still accumulate in mating mixtures, suggesting that fusion is slowed but not blocked under these circumstances. Mating of a *fus1* strain has a greater than normal dependence on Ca^{++}. On low-calcium medium, or upon addition of EGTA to medium, the mating frequency of a *fus1* strain is greatly reduced (Elion et al. 1995). The prezygotes formed in *fus1* crosses include a wide variety of forms and do not point to a specific step within the fusion process that might be defective. Prezygotes are resistant to sonication, indicating that they have progressed beyond mere agglutination to partial cell wall fusion (McCaffrey et al. 1987; Trueheart et al. 1987). Treatment with Zymolyase separates the partners into two shmoo-like cells with flattened ends, suggesting that the cell wall and plasma membrane of each cell remain separate and distinct.

FUS1 was discovered independently by two approaches. Strains carrying a large *HIS4* to *LEU2* deletion exhibited a bilateral mating defect, suggesting that a gene in this region functioned in mating. Subsequent genetic and complementation studies identified *FUS1* as the relevant gene. An independent study identified control regions in the yeast genome that cause transcriptional induction by α-factor when placed upstream of the bacterial *lacZ* gene. The *FUS1* promoter was isolated by virtue of its strong α-factor induction (see McCaffrey et al. 1987). In both instances, additional analysis revealed the specific role of *FUS1* in fusion.

FUS1 encodes a 512-residue serine/threonine-rich membrane protein with a single transmembrane spanning domain (McCaffrey et al. 1987; Trueheart et al. 1987). The amino-terminal region lies outside the cell and is heavily *O*-glycosylated but not *N*-glycosylated (Trueheart and Fink 1989). The absence of *N*-glycosylation is significant since consensus *N*-glycosylation sites are predicted to lie within the cell. Protease sensitivity studies confirm the orientation in the membrane (Trueheart and Fink 1989). The carboxy-terminal, intracellular, portion of Fus1p contains three regions of weak homology with the myosin heavy chain and a potential SH3 domain that might interact with the cytoskeleton. Although a direct role of extracellular regions in cell fusion was originally hypothesized, subsequent experiments showed that the transmembrane and extracellular portion of Fus1p were partially dispensable for its fusion function (J. Trueheart, pers. comm.). A truncated protein remains membrane-associated, suggesting interaction with other membrane proteins, or the cytoskeleton. Fus1p is localized by immunofluorescence to the tips of shmoos (McCaffrey et al. 1987; Trueheart et al. 1987). In zygotes, Fus1p is localized to the neck region where fusion has occurred. It is presumed that Fus1p would localize to the septum of the fusion region.

Since *FUS1* was identified as an α-factor-regulated gene, it is not surprising that it is among the most strongly pheromone-controlled genes. The induction of *FUS1* by **a**-factor in *MAT*α strains or α-factor in *MAT***a** strains ranges from 20-fold to 120-fold depending on strains and conditions (McCaffrey et al. 1987; Trueheart et al. 1987). The upstream region of *FUS1* contains three copies of the pheromone response element (PRE) hexamer TGAAACA (Trueheart et al. 1987; Hagen et al. 1991). Although *FUS1* basal expression of mRNA in a haploid cell is low, the expression is at least 40-fold lower in an **a**/α diploid. The tightly regulated expression of *FUS1* is not unexpected for a product that functions specifically in cell fusion. The potent regulation of *FUS1* by pheromones has led to its treatment as a generic pheromone-induced gene. However, there are reasons to believe that not all pheromone-responsive genes are controlled identically (see, e.g., the karyogamy genes *KAR3* and *CIK1* above). From a teleological perspective, differential regulation of genes required at different stages of mating might make sense.

B. *FUS2*

Mutations in *FUS2* lead to a phenotype very similar to that of *FUS1* mutations. Again, the mating fusion defect is most striking when both mating cells are defective (bilateral defect). *FUS2* was identified as a naturally occurring mutation in a laboratory strain (Trueheart et al. 1987). What might have been treated as an unpleasant "messiness" in strains turned out to be highly informative (Trueheart et al. 1987; Elion et al. 1995). Early results suggested that *FUS1* and *FUS2* might have overlapping functions. Mutations of one gene strengthen the phenotype of mutations in the other. For example, a *fus1, fus2* strain mates more poorly than a *fus1* strain. A *fus1* strain mates poorly to a *fus2* strain. A strong mating defect is seen in a *fus1, fus2* x *fus1, fus2* cross. All combinations of *fus1* and *fus2* matings accumulate prezygotes of similar appearance (Trueheart et al. 1987). A second piece of data suggesting redundant functions for *FUS1* and *FUS2* was the partial cross-suppression of mutants by high-copy plasmids bearing clones of the other gene. Despite these genetic hints of overlapping function, subsequent studies suggest that the roles of *FUS1* and *FUS2* may be quite different.

Like Fus1p, Fus2p is localized to the shmoo tip and the junction of mating cells. However, Fus1p appears at the plasma membrane, whereas Fus2p localizes at vesicles or vesicle-like structures just below the cell surface (Elion et al. 1995). An important question is whether the vesicles are coming or going. Are they delivering degradative enzymes and fusogens or removing cell wall and debris that block membrane apposition and fusion? The 617-residue Fus2p has overall weak similarity to dystrophin and myosin-like protein (Mlp1) which could indicate cytoskeletal interactions. Fus2p pellets with particulate fractions that contain vesicles and cytoskeletal components (Elion et al. 1995).

FUS2 is regulated in a manner that ensures its expression only in mating cells (Elion et al. 1990). Like *FUS1*, expression is quite low in vegetative cells but is highly induced by pheromone. Two copies of the PRE hexamer are found upstream of *FUS2* as well as a copy of a 14-nucleotide motif found upstream of *FUS1*, of unknown significance. Fus2p is undetectable soon after cell fusion, suggesting that it is unstable in the newly formed diploid cell (Elion et al. 1990).

The issue of whether *FUS1* and *FUS2* act on related processes during fusion cannot be resolved by genetics alone. One creative approach to this problem has been to "expand the phenotype" of each mutant to see if differences can be found. As mentioned above, the mating of *fus1* mutants is sensitive to reduced levels of calcium in the mating medium. Mating of *fus2* mutants is resistant to this treatment. The reverse pattern holds for sensitivity to the drug polymyxin B. Strains with mutations in *FUS2* are more sensitive to effects of polymyxin B on mating than strains with mutations in *FUS1*. Most significantly, *fus2* mutations lead to defects in nuclear alignment and karyogamy that are absent in *fus1* strains (see karyogamy section above). Taken together, these results hint that *FUS1* and *FUS2* have different functions. Thus, defects in distinct facets of the cell fusion program may lead to prezygotes with similar morphologies.

It is interesting to speculate that *FUS1* and *FUS2* might be involved in targeting or controlling specialized exocytotic or endocytotic processes required for cell fusion. In this regard, it is interesting that *rvs161* (*end6*) mutants exhibit fusion defects during mating (V. Brizzio and M. Rose, in prep.). *RVS161* encodes a homolog of the synaptic vesicle protein amphiphysin (Munn et al. 1995; Sivadon et al. 1995). *RVS161* also affects endocytosis and bipolar budding and has genetic interactions with actin (Munn et al. 1995; Sivadon et al. 1995). The potential roles for regulated trafficking events would seem plentiful. Gfp-Rvs161p localizes to the shmoo tip where it might have a specific role in trafficking required for fusion (V. Brizzio and M. Rose, in prep.).

C. *SPA2*

The highly polarized expression of Fus1p and Fus2p raises issues of the role of cell polarization and the cytoskeleton in fusion. *SPA2* was identified in several screens for genes required for fusion in mating. Spa2p is a large protein (1466 residues) with repeat elements potentially forming a coiled-coil structure found in some cytoskeletal proteins (Gehrung and Snyder 1990). Spa2p was originally identified using human autoimmune antibodies that recognize the centrosome. It may act at sites of polarized growth in vegetative and mating yeast (Snyder 1989). Spa2p localizes to the tip of mating projections in a more focused pattern than actin and is found in zygotes at the site of cell fusion. Mutants defective in *SPA2* have slight defects in bud site selection but are largely normal for vegetative growth (Snyder 1989). In response to pheromone, the mutant cells swell, become slightly oblong, and undergo cell cycle arrest, but they fail to make pointed projections (Chenevert 1994). *SPA2*-defective strains exhibit a bilateral mating defect (Gehrung and Snyder 1990; Kurihara et al. 1994). It is not clear if the fusion defect is secondary to a failure to polarize (e.g., a failure to deliver specific cell wall modifying enzymes to a concentrated spot on the cell surface) or if Spa2p has a fusion-specific role. The complex genetic interactions of *SPA2* with genes involved in morphogenesis, cytokinesis, and control of polarity provide a cornucopia of possible roles in fusion.

D. *BNI1*

Another gene required for fusion in *S. cerevisiae* is *BNI1* (C. Boone, pers. comm.). *BNI1* is an *S. cerevisiae* homolog of the *Schizosaccharomyces pombe fus1, fus1(Sp)*. The *fus1(Sp)* gene is unrelated to the *S. cerevisiae FUS1* gene but is similarly required for cell fusion during mating (Peterson et al. 1995). The Fus1(*Sp*) protein is pheromone-inducible in *S. pombe* and localizes to the site of cell fusion, suggesting a specific role in fusion (Peterson et al. 1995). Higher-organism homologs of *BNI1* include the *Drosophila* genes *cappuccino* and *diaphanous* and the vertebrate *limb deformity* locus. These genes have roles in morphogenesis and cell polarity (Emmons et al. 1995). *BNI1* has roles in the daughter-specific accumulation of the Ash1p regulator and in bipolar bud site selection (Jansen et al. 1996). The higher-organism homologs are specifically implicated in cell polarity processes that use microtubules to move products to one side of the cell. Defects in *BNI1* block polarization following exposure to pheromone (C. Boone, pers. comm.).

E. *TPM1*

The tropomyosin gene, *TPM1*, is required for efficient fusion (Liu and Bretscher 1992). Tropomyosin interacts with actin, itself localized to the projection tip of shmoos and the zone of cell fusion during conjugation (Hašek et al. 1987; Liu and Bretscher 1992). Mutations in *TPM1* lead to defects, like *spa2* and *bni1*, in polarization of pheromone-treated cells (Gehrung and Snyder 1990; Liu and Bretscher 1992). Vesicles accumulate in the mutants, and transit of agglutinin, but not invertase secretion, is slowed. It is tempting to imagine that a pathway for polarized export of cell wall proteins is being affected.

F. The Two Classes of Cell Fusion Mutants

A crucial difference between two groups of mutants that block fusion is evident. The first group, *spa2*, *bni1*, and *tpm1*, grossly disrupts the morphology and reduces the polarity of the pheromone-responding cell, whereas the second group, *fus1* and *fus2*, does not. The possibility that the fusion blockade induced by *spa2*, *bni1*, or *tpm1* mutations might be secondary to nonspecific defects in cell polarization is difficult to rule out at the moment. Certainly, many models for cell fusion rely on highly spatially regulated changes in the cell wall that superficially would seem to depend on a highly polarized cell cytoskeleton.

Actin cortical patches are observed at the tip of projections of pheromone-treated cells (Hašek et al. 1987). A number of other cytoskeletal proteins, especially those that interact with actin, also localize to the tip of the shmoo, or the fusion region of the zygote (Welch et al. 1994). Examples include Myo2p and Smy1p (Lillie and Brown 1994). Their role, if any, in cell fusion is uncertain, however. During vegetative growth, they localize to the site of bud initiation and growth; this site and the shmoo tip are essentially equivalent with respect to the localization of many actin-based elements of the cytoskeleton. Although actin-based processes are essential for cell fusion, it is unclear if the requirement is based on the generic role of actin in cell polarization or a more specialized role.

G. Role of Pheromones and Pheromone Response in Fusion

Pheromone synthesis and response are required for all stages of mating. A question remains concerning the specific role of pheromone signaling during fusion. Is signaling required merely to maintain levels of mating proteins or does it contribute more specifically to fusion? One model for the control of the progress through the various stages of mating is that at each subsequent step, higher levels of pheromone signaling are needed than the previous step. For example, genes required for cell fusion might require a stronger signal for expression than genes involved early in the process, such as agglutination. This model predicts that an intermediate level of pheromone, or reduced signaling through the pathway, might cause cells to arrest during mating as prezygotes. Several fusion-defective mutants appear to potentially fit this model.

FUS3-defective strains exhibit a bilateral fusion defect similar to those of *FUS1*- and *FUS2*-defective strains. However, the *fus3* strains, unlike *fus1, fus2*, often initiate mating and then stop and resume budding and vegetative growth. *FUS3* encodes one of the two partially redundant MAP kinases in the pheromone response kinase cascade. Transcription of *FUS3* itself is induced by pheromone (Elion et al. 1990). Mutations in *FUS3* were initially detected because of their defect in mating to a *fus1, fus2* strain relative to mating to wild type (Berlin et al. 1991). The fusion-defective *fus3* alleles (e.g., *fus3-1* and *fus3-2*), unlike null alleles, conferred little or no reduction of induction of *FUS1* or *FUS2*, indicating that they did not act by reducing expression of these genes. The *fus3-2* strain exhibited a reduction in morphological changes in response to α-factor, suggesting that this effect might underlie the fusion defect. However, the *fus3-1* strain actually elaborated more elongate projections in response to pheromone than the parental strain, indicating that morphological changes did not bear a simple relationship to cell fusion. One interesting aspect of the partially defective *fus3* mutants was a failure to undergo G_1 arrest when exposed to pheromone (Elion et al. 1990; Fujimura 1990). However, the cell cycle arrest defect is not required for the fusion defect of *fus3* mutants (Fujimura 1994).

Special alleles of *STE12*, the gene encoding the transcriptional regulator phosphorylated by Fus3p (Elion et al. 1993), have also been isolated as cell fusion mutations (L. Elia and L. Marsh, unpubl.). The PRE elements upstream of *FUS1* and *FUS2* bind Ste12p, which is required for induction (Hagen et al. 1991; Elion et al. 1993). The fusion-defective *ste12* strains exhibit normal growth arrest in response to pheromone. Perhaps fusion requires more complete activation of Ste12p than other aspects of pheromone response. Such a model would be consistent with the observation that full pheromone levels are required for efficient fusion (Brizzio et al. 1996). However, the induction of *FUS1* and *FUS2* is not reduced by the fusion-defective alleles of *FUS3* or *STE12* (Elion et al. 1990; Fujimura 1994; L. Elia and L. Marsh, unpubl.), ruling out action through these fusion genes. Ste12p is a DNA-binding molecule with many partners and might require a special coregulator for some fusion genes much as Kar4p combines with Ste12p to mediate expression of some karyogamy genes.

Although it seems that *FUS3* and *STE12* may have a special role in fusion, the remaining question is whether a reduction in pheromone signaling alone can affect fusion. Two unilateral fusion mutants, *fus5* and *fus8*, identified in a screen for unilateral or bilateral sterile mutants (see karyogamy section above), proved to affect **a**-factor production (Kurihara et al. 1994; Brizzio et al. 1996). *FUS5* is allelic with *AXL1* and *FUS8* is allelic with *RAM1*, both of which are involved in **a**-factor processing (Goodman et al. 1990; Fujita et al. 1994; Adames et al. 1995; Brizzio et al. 1996). Mutations in these genes reduce **a**-factor expression, but they also affect a number of other cellular functions. *AXL1* is a key determinant of the axial budding pattern of haploid cells, since mutations in it lead to a bipolar budding pattern (Fujita et al. 1994; Adames et al. 1995). Axl1p endoprotease activity is involved in processing of **a**-factor but is not required for bud site control (Adames et al. 1995). Ram1p is required for farnesyl transferase activity (Goodman et al. 1990). Mutants defective in *RAM1* exhibit pleiotropic effects on **a**-factor, Ras, and other isoprenylated molecules.

Although both *AXL1* and *RAM1* are complex and interact with morphogenic and **a**-factor processing pathways, it seemed that the simplest explanation for the fusion defect was that reduced **a**-factor led to a reduced response in the *MAT*α mating partner. To test if it were reduced

a-factor or some other feature of the mutants that created the fusion defect, strains were mated under conditions that compensate for their pheromone defects (Chan et al. 1983). Overexpression of *MFA1*, coding for **a**-factor, or matings to a *MAT*α pheromone-supersensitive (*sst2*) partner resulted in partial suppression of the fusion defect of the *axl1* and *ram1* strains. These experiment show that the reduction of **a**-factor expression conferred by the *ram1* and *axl1* mutations was the phenotype relevant to the fusion defect. The fusion defect of **a**-factor processing mutants could also be reproduced by underexpression of **a**-factor in a *MAT***a** strain or α-factor in a *MAT*α strain by appropriate regulation of their structural genes (Brizzio et al. 1996). Taken together, these results strongly suggest that completion of cell fusion requires more pheromone than the initiation of conjugation.

STE6 encodes the **a**-factor transporter and is required for mating by *MAT***a** cells (Kuchler et al. 1989; Michaelis 1993). Ste6p is polarized in mating cells, principally in vesicles (Kuchler et al. 1993; Berkower et al. 1994). Strains with defects in *STE6* have been identified by several groups in screens for fusion-defective strains. The best characterized fusion-defective alleles were isolated in a large-scale nonselective screen for unilateral fusion mutants (Elia and Marsh 1996). Null mutations in *STE6* cause an extreme mating defect in which cells do not even begin to initiate the mating process (Michaelis 1993). The obvious possibility is that these partial-function *ste6* alleles act in a fashion similar to that of *ram1* or *axl1*. Surprisingly, the fusion defects of *ste6*(*cef*) (cell fusion) alleles, unlike the *ram1* or *axl1* mutations, are not suppressed by conditions designed to overcome the **a**-factor defect. Overexpression of the **a**-factor structural gene *MFA2* in *ste6*(*cef*) strains increases **a**-factor export but does not restore fusion. Prezygotes accumulate to the same extent in matings to a wild-type partner or mating to an *sst2* partner, which is hypersensitive to **a**-factor (Elia and Marsh 1996). Perhaps the role of Ste6p in fusion is more complex than that of other components of **a**-factor metabolism.

H. The Cell Membrane and Cell Wall in Cell Fusion

The cell wall and membrane are the two components of the cell most directly involved in fusion and cytoplasmic exchange. The cell wall literature has been reviewed recently (Klis 1994; Cid et al. 1995). The following discussion focuses on issues of greatest potential relevance to fusion. Although many changes are observed in the cell wall during fusion, the significance of these changes is not always clear. Genetic evidence for roles of membrane and cell wall proteins in the fusion process is notable by its absence. Cell wall remodeling must occur to permit mixing of the contents of the mating cells. On the basis of electron microscopy studies, the remodeling involves localized loss of cell wall material. Distinct steps of remodeling seem to be involved, but it is not clear if the *fus* mutants block a specific step or if they simply slow the many steps of fusion. Although mating-specific cell wall remodeling enzymes might be required, it is also possible that vegetative enzymes are redeployed during mating. The nature of the membrane fusion event is even less certain. Membrane fusion could involve an uncatalyzed fusion of closely opposed membranes. However, in all known biological systems that involve membrane fusion, the event is facilitated (White 1992). The *fus* mutants accumulate prezygotes with a visible cell wall septum between the mating pairs. It is not certain what the phenotype of a mutant defective specifically in the membrane fusion stage of mating would be. Overall, one is left with the impression that, on a molecular level, important aspects of fusion remain a black box.

An intriguing set of experiments implicate sterols in cell fusion during mating. A role for sterols in the membrane fusion step of fusion might be easily interpretable, but the sterol effect seems to affect the same cell wall remodeling step as the *fus* mutants. Ergosterol is required for *S. cerevisiae* viability. To replace sterols, *hem1, erg1* mutants unable to synthesize ergosterol were fed exogenous sterols. Either ergosterol, the natural sterol, or the related compound stigmasterol permitted vegetative growth (Tomeo et al. 1992). Only ergosterol functioned to permit efficient mating. The stigmasterol-treated cells accumulated prezygotes during mating. Stigmasterol does not interfere with mating since a mixture of stigmasterol and ergosterol (even when stigmasterol is in excess) permits efficient mating. Thus, it seems that some part of the fusion process requires a specific sterol structure (Tomeo et al. 1992).

Like the *fus* mutants, sterol-altered yeast mutants initiate mating and undergo shape changes normally, but they fail to complete degradation of the septum separating the mating partners. The possible roles of sterols in this process are numerous. Although pheromone responses such as morphological changes are superficially intact in sterol-substituted cells, subtle differences might exist. Trafficking within the cell, endocytosis, and exocytosis might be affected. The activity of the enzymes modifying the cell wall might be directly or indirectly changed. One last possibility that should be seriously considered is that a membrane fusion blockade might lead to the classic *fus* phenotype of prezygote accumulation. We do not know at what stage membrane fusion normally occurs in mating. If fusion normally occurs before complete wall degradation has occurred (e.g., through an inconspicuous pore in the cell walls), then a membrane fusion blockade might be present as a prezygote with an intact septum between the cells.

Possibly related to the effects of sterol substitution are the effects of the lipid-binding antibiotic polymyxin B. Polymyxin B can inhibit mating at concentrations that do not affect viability at least in part through an ability to block agglutination (Boguslawski 1986). Polymyxin B also increases the fusion defect of *fus2* strains (Elion et al. 1995) and can promote an accumulation of prezygotes in wild-type mating mixtures (L. Elia and L. Marsh, unpubl.). The effects of polymyxin B might be quite complex, since in other organisms it can activate protein kinase C (PKC) and other signaling pathways as well as bind liposaccharides. As with the sterol-substitution experiments, it seems possible that the effects of polymyxin B on fusion might be indirect.

Integral membrane proteins might have roles in membrane fusion or signaling to coordinate changes in the cell wall (White 1992). Currently, there are no good candidates for integral membrane proteins involved in fusion. Fus1p spans the membrane, but the membrane-spanning and extracellular portions of the protein, as described above, appear at least partially dispensable. Ste6p and Fps1p are polytopic proteins genetically implicated in fusion which might conceivably have roles in cell communication.

It is perhaps surprising that no cell wall synthesis enzymes have been directly implicated in the cell wall remodeling steps of cell fusion. Chitin is deposited in the subapical region of the shmoo tip and is found at the site of cell fusion in zygotes (Lipke et al. 1976). *CHS5* is required, directly or indirectly, for chitin synthetase III activity (Cid et al. 1995). Disruption of *CHS5* leads to cell fusion defects (Spellig et al. 1994; C. Boone, pers. comm.). Mutations in *CHS5* arose in several screens for strains with altered cell walls (for review, see Cid et al. 1995). Strains defective in *CHS5* have altered chitin deposition, are resistant to the dye Calcofluor White, and are defective in chitin synthetase III activity (CSIII). Since CSIII activity is required for the deposition of chitin on the shmoo tip (see below), it is tempting to infer a requirement for chitin deposition in fusion. However, mutations in other genes required for CSIII activity (*CHS3* and *CHS4*) apparently do not lead to a mating fusion defect, although they reduce or eliminate chitin deposition in the shmoo and lead to morphological changes (Cid et al. 1995). Chs5p is a heavy neurofilament protein (NF-H) homolog containing heptad repeats followed by a lysine-rich carboxy-terminal domain and is unlikely to be a catalytic subunit of chitin synthetase (Cid et al. 1995; C. Boone, pers. comm.). If Chs5p has a cytoskeletal role like NF-H, it is possible that it participates in the transport or localization of activities in addition to CSIII. Although *CHS5* is included in this section because of its connection with the cell wall, it could have as easily been grouped with cytoskeletal components such as tropomyosin. *CHS5* mRNA is not induced by α-factor, and Chs5p is known to be active in vegetative cells. Thus, this protein may be another example of a nonmating-specific protein that nevertheless has a role in mating.

Other cell wall synthesis enzymes may well have a role in fusion, but their roles are less clear. Mutations in *GNS1* lead to a reduction in 1,3-β-glucan in the cell wall. *GNS1* encodes an integral membrane protein that may be a structural component of 1,3-β-glucan synthetase. The *gns1* mutant strains are sickly and exhibit reduced mating that might indicate a role for 1,3-β-glucan synthetase in mating or indicate a nonspecific requirement for an intact cell wall in mating (el-Sherbeini and Clemas 1995).

The process of cell fusion is hazardous. Yeast cells maintain a positive turgor pressure and protect themselves from lysis with a strong cross-linked cell wall. Mating requires the exposure of a portion of the cell membrane in the fervent hope that there is another yeast doing the same thing on the other side. Although the details of how overall osmotic control is integrated into the mating program are unknown, it seems logical that such control exists (Brewster et al. 1993). It was previously mentioned that treatment of haploid cells with the appropriate pheromone does not lead to cell lysis (Lipke et al. 1976; L. Elia and L. Marsh, unpubl.). Strains defective in *PKC1* or other genes on the PKC signaling pathway do lyse on exposure to pheromone, indicating the importance of normal homeostasis (Paravicini et al. 1992; Irie et al. 1993; Shimizu et al. 1994). The Pkc1p-controlled MAP kinase Slt2p/Mpk1p is activated in pheromone-treated cells in a process that requires the mating pheromone gene *STE20* (Zarzov et al. 1996). An interesting question is whether the PKC pathway is activated directly (e.g., by cross-talk with the pheromone MAP kinase cascade) or as a consequence of morphological changes induced by pheromone. Strains defective in *PKC1* mate successfully, but the zygotes are distorted, fragile, and frequently lyse. The lysis occurs in a pattern which suggests that mating has left the region of cell joining especially vulnerable, although lysis occurs in other regions as well (Levin and Errede 1995; L. Elia, pers. comm.). In general, *PKC1* seems to be more important for repairing damage to the cell wall during mating than in controlling the actual fusion process. Consistent with a role for osmotic regulation in cell fusion is the discovery that a mutation in *FPS1* leads to a fusion defect (Luyten et al. 1995; J. Phillips and I. Herskowitz, pers. comm.). *FPS1* encodes a multiple-span integral membrane protein of the MIP family implicated in osmotic regulation via control of intracellular glycerol. This family of proteins is involved in a range of activities including signaling, morphogenesis, and transport. Fps1p may have functions in addition to regulating glycerol transport (Luyten et al. 1995).

I. Exposure of Haploid Cells to Pheromone as a Model for Mating Cells

Haploid cells exposed to pheromones in the absence of mating partners have served as a model to study changes associated with mating (Cross et al. 1988). Much of what we know about changes in the cell wall during mating is dependent on such studies. Haploid cells treated with pheromone undergo morphological changes that make them reminiscent of half of a zygote. Pheromone responses can be detected in these cells, making them valuable for signal transduction studies (for review, see Cross et al. 1988; Marsh et al. 1991; Herskowitz 1995). Thus, one approach to studying mating is to examine pheromone-induced changes in haploid cells. Changes in the cell wall after pheromone treatment could reflect the remodeling changes occurring during mating. A limitation of this approach is that the biological significance of pheromone-induced changes has not usually been tested.

Cells exposed to pheromone in the absence of a partner may not be a complete model for mating cells. Although mating cells completely lose their cell wall in the region of contact to permit fusion, shmoos remain osmotically stable with a cell wall covering the tip of the cell (Lipke et al. 1976). It is possible that some cell contact signal is required for fusion in addition to pheromone. Tight cell contact may serve to provide pressure on the fusion zone during mating, and this osmotic stabilization may serve as a contact signal in the region of cell wall degradation, for example, by relieving a *PKC*-mediated inhibition of cell wall degradation. In the absence of such a signal, compensatory changes in the cell wall may occur to prevent lysis. It may be difficult to distinguish primary pheromone-induced changes in cells from secondary responses designed to maintain cell integrity.

Another concern is that shmoos are generated by treating cells with pheromone in solution. This isotropic pheromone treatment does not duplicate the pheromone gradients that cells experience in normal mating (Segall 1993; Chenevert 1994). These pheromone gradients are presumed to have an important role in the coordination of joint cell wall remodeling during cell fusion. Shmoos, themselves, are not normal intermediates in mating and are not competent to mate. It appears that pheromone exposure in the absence of a partner renders cells temporarily unable to conjugate with a genuine partner (Crandall et al. 1977). The value of studies of cells exposed to synthetic or purified pheromones should not blind us to their limits. Overall, the aspects of yeast mating dependent on both partners remain understudied.

J. Changes in the Cell Wall

The most striking and relevant changes in the cell wall during mating involve induction of agglutinin molecules. Mating of yeasts in a liquid environment is potentially a problem, since cells need to stay stuck together long enough to initiate fusion. The cell surface agglutinins solve this problem by providing tight cell-cell adhesion. Whether they might have additional roles in mating is unclear.

The α-agglutinin (*AGα1*) decorates fimbriae on the cell wall. It originates as a glycosylphosphatidylinositol (GPI)-anchored protein attached to the plasma membrane and is apparently transferred to glucan molecules, permitting movement to the outer cell wall (Doi et al. 1989; Hauser and Tanner 1989; Lipke et al. 1989; Lipke and Kurjan 1992). One of the **a**-agglutinin subunits, Aga2p, is bound to the cell wall via disulfide bonds to the anchorage subunit Aga1p (Roy et al. 1991; Lipke and Kurjan 1992; Cappellaro et al. 1994; Yamaguchi et al. 1994). The α- and **a**-agglutinins bind tightly to each other in a highly species-specific manner (Crandall et al. 1977; Lipke and Kurjan 1992). Although these molecules serve as paradigms for mating-specific changes in the cell wall and seem to be ideal molecules to serve as signals for cell-cell contact, there is no evidence to support any role except mediating cell attachment. The agglutinins are required for mating in liquid medium where cell-cell interactions are transient and unstable, but not for mating on solid medium where cell-cell contacts are stable (Lipke et al. 1989).

The content of cell wall polymers is altered, and there is a reduction in the mannosylation of mannan deposited at the shmoo tip following pheromone treatment (Lipke et al. 1976; Klis 1994; Shaw et al. 1994). Glucan increases, but the cell becomes more sensitive to glucanases, suggesting a change in structure or reduction in cross-linking (Lipke et al. 1976; Crandall et al. 1977). Morphologically, the cell wall layer that contains 1,3-β-glucan appears to thin at the shmoo tip and at the conjugation septum. Interestingly, *FKS2*, which encodes a 1,3-β-glucan synthetase subunit, is up-regulated in cells treated with pheromone (Mazur et al. 1995). Induction of *FKS2* is not immediate and requires calcium influx and calcineurin, in addition to the usual pheromone pathway which might indicate a secondary response. The level of the GPI-linked protein Gas1p/Ggp1p is reduced during mating (Popolo et al. 1993). Although Gas1p is a major cell wall protein, it is not known whether it has a specific role in mating.

Chitin increases and is deposited on the neck, but not the very tip, of the shmoo. The neck of the shmoo may correspond to the region sealing cells together in a zygote, since a ring of chitin is observed at the zone of fusion of zygotes (Crandall et al. 1977; Cid et al. 1995). Chitin is probably deposited as part of the inner section of the glucan-rich layer of the cell wall (Klis 1994). CSI and CSIII activities are induced by pheromone, but CSII is down-regulated (Shaw et al. 1994). CSII is principally involved in synthesis of the bud septum and its down-regulation by pheromone seems to be sensible. *GFA1/GCN1*, which is required for synthesis of the hexosamine precursor of chitin, is up-regulated by α-factor. However, chitin deposition is not required for fusion, since *chs3* mutants fail to lay down chitin during mating but undergo fairly normal mating (Cid et al. 1995). Chitin induction may be part of a redundant set of cell wall changes or a repair process (Cid et al. 1995).

A priori, it seems highly likely that there is a role in cell fusion for pheromone-induced changes in cell wall remodeling. However, it has proven difficult to determine the biological significance of specific changes in the yeast cell wall observed during mating. It is possible that changes in the properties of the cell wall are the result of many incremental changes in the various polymers and that mutations with effects on a single polymer have more dramatic effects on vegetative growth than mating. This model might explain the absence of cell wall biosynthesis genes (with the exceptions described) among the *FUS* genes.

K. Other Gamete Fusion Systems

A look at gamete fusion systems in lower and higher eukaryotes suggests dramatic differences embedded in common themes. These other systems have been reviewed and only a glimpse of their complexity is offered here. Strains of the smut fungus *Ustilago maydis* of the same mating type can fuse if exposed to a pheromone type to which they respond (Spellig et al. 1994). This phenomenon is observed in various forms in a number of filamentous fungi. The basic observation that cells of a single mating type can undergo homotypic cell wall, plasma membrane, and sometimes nuclear fusion has important implications. In essence, it suggests that conjugating cells do not depend on their mating-type differences for most of the program of fusion. Such an observation is consistent with findings in *S. cerevisiae* that (1) most cell and nuclear fusion genes do not act in a cell-type-specific manner, (2) defects in *FUS* genes are enhanced dramatically when both partners are defective, and (3) the only mating-type-specific components of conjugation are the pheromones and receptors (Bender and Sprague 1986, 1989). The existence of homotypic cell fusion in fungi raises important questions about the mechanism by which cell wall remodeling is spatially coordinated by two cells to allow membrane contact and formation of a seal surrounding the fusion site.

It is useful to compare the fusion of haploid yeast to metazoan sperm-egg fusion. Although motile sperm attraction to egg would seem to have little similarity to the sessile yeast gametes, the orientation of yeast cells toward a pheromone gradient suggests at least a potentially related effect on the cytoskeleton. Sperm penetration of the egg zona pellucida requires a regulated exocytosis, the acrosome reaction (Fraser 1995). Tantalizing evidence, described above, supports the concept that a regulated exocytosis might be involved in yeast fusion. Sperm-egg contact seems to result in signaling and adhesion via integrin/disintegrin-related molecules (Myles et al. 1994). This contact may also trigger membrane fusion (White 1992; Myles 1993). A specific cell-cell contact signal is not known in yeast. These last stages of interaction between cells just preceding membrane fusion are the most poorly understood in yeast.

IX. PERSPECTIVES FOR CELL FUSION

The process of cell fusion remains a mystery at a molecular level. The current evidence suggests that most of the control of the process lies in the cytoplasm. If we allow guilt by association (or homology), the proteins involved will be involved with the cytoskeleton and in vesicle trafficking. One of the most exciting prospects is that yeast mating may turn out to be a model for regulated exocytosis. The membrane fusion step should also provide interest if approaches can be developed for its study.

ACKNOWLEDGMENTS

We thank the many past and current members of our laboratories that have worked on the cell and nuclear fusion pathway and contributed their parts of the puzzle. We also thank our many collaborators and colleagues including Charlie Boone, Sharyn Endow, Ira Herskowitz, Martin Latterich, Jennifer Philips, Randy Schekman, Mike Snyder, Mark Winey, and others. Their many contributions and discussions of unpublished data have been critical to this work and to continued progress. M.D.R. is supported by National Institutes of Health grants 37739 and 52526. L.M. is supported in part by American Cancer Society grant VM-143 and Cancer Core support grant P30 CA-13330 from the National Cancer Institute.

REFERENCES

Acharya, U., R. Jacobs, J.-M. Peters, N. Watson, M.G. Farquhar, and V. Malhotra. 1995. The formation of Golgi stacks from vesiculated Golgi membranes requires two distinct fusion events. *Cell* **82:** 895.

Adames, N., K. Blundell, M.N. Ashby, and C. Boone. 1995. Role of yeast insulin-degrading enzyme homologs in propheromone processing and bud site selection. *Science* **270:** 464.

Adams, A.E. and J.R. Pringle. 1984. Relationship of actin and tubulin distribution to bud growth in wild-type and morphogenetic-mutant *Saccharomyces cerevisiae. J. Cell Biol.* **98:** 934.

Baba, M., N. Baba, Y. Ohsumi, K. Kanaya, and M. Osumi. 1989. Three-dimensional analysis of morphogenesis induced by mating pheromone α factor in *Saccharomyces cerevisiae. J. Cell Sci.* **94:** 207.

Bahler, J., G. Hagens, G. Holzinger, H. Scherthan, and W.D. Heyer. 1994. *Saccharomyces cerevisiae* cells lacking the homologous pairing protein p175SEP1 arrest at pachytene during meiotic prophase. *Chromosoma* **103:** 129.

Baron, A.T. and J.L. Salisbury. 1988. Identification and localization of a novel, cytoskeletal, centrosome-associated protein in PtK2 cells. *J. Cell Biol.* **107:** 2669.

Barton, N.R. and L.S.B. Goldstein. 1996. Going mobile: Microtubules and chromosome segregation. *Proc. Natl. Acad. Sci.* **93:** 1735.

Baum, P., C. Furlong, and B. Byers. 1986. Yeast gene required for spindle pole body duplication: Homology of its product with Ca^{2+}-binding proteins. *Proc. Natl. Acad. Sci.* **83:** 5512.

Beach, D., B. Durkacz, and P. Nurse. 1982. Functionally homologous cell cycle control genes in budding and fission yeast. *Nature* **300:** 706.

Bender, A. and G.F. Sprague, Jr. 1986. Yeast peptide pheromones **a**-factor and α-factor activate a common response mechanism in their target cells. *Cell* **47:** 929.

———. 1989. Pheromones and pheromone receptors are the primary determinants of mating specificity in the yeast *Saccharomyces cerevisiae. Genetics* **121:** 463.

Bennett, M.K. and R.H. Scheller. 1993. The molecular machinery for secretion is conserved from yeast to neurons. *Proc. Natl. Acad. Sci.* **90:** 2559.

———. 1994. A molecular description of synaptic vesicle membrane trafficking. *Annu. Rev. Biochem.* **63:** 63.

Berkower, C., D. Loayza, and S. Michaelis. 1994. Metabolic instability and constitutive endocytosis of *STE6*, the α-factor transporter of *Saccharomyces cerevisiae. Mol. Biol. Cell* **5:** 1185.

Berlin, V., C.A. Styles, and G.R. Fink. 1990. BIK1, a protein required for microtubule function during mating and mitosis in *Saccharomyces cerevisiae*, colocalizes with tubulin. *J. Cell Biol.* **111:** 2573.

Berlin, V., J.A. Brill, J. Trueheart, J.D. Boeke, and G.R. Fink. 1991. Genetic screens and selections for cell and nuclear fusion mutants. *Methods Enzymol.* **194:** 774.

Biggins, S. and M.D. Rose. 1994. Direct interaction between yeast spindle pole body components: Kar1p is required for Cdc31p localization to the spindle pole body. *J. Cell Biol.* **125:** 843.

Boguslawski, G. 1986. Polymyxin B nonapeptide inhibits mating in *Saccharomyces cerevisiae. Antimicrob. Agents Chemother.* **29:** 330.

Brewster, J.L., T. De Valoir, N.D. Dwyer, E. Winter, and M.C. Gustin. 1993. An osmosensing signal transduction pathway in yeast. *Science* **259:** 1760.

Brizzio, V., A.E. Gammie, G. Nijbroek, S. Michaelis, and M.D. Rose. 1996. Cell fusion during yeast mating requires high levels of **a**-factor mating pheromone. *J. Cell Biol.* **135:** (in press).

Brodsky, J.L. and R. Schekman. 1993. A Sec63p-BiP complex from yeast is required for protein translocation in a reconstituted proteoliposome. *J. Cell Biol.* **123:** 1355.

Brodsky, J.L., J. Goeckeler, and R. Schekman. 1995. BiP and Sec63p are required for both co- and posttranslational protein translocation into the yeast endoplasmic reticulum. *Proc. Natl. Acad. Sci.* **92:** 9643.

Byers, B. 1981a. Cytology of the yeast life cycle. In *The molecular biology of the yeast* Saccharomyces: *Life cycle and inheritance* (ed. J.N. Strathern et al.), p. 59. Cold Spring Harbor Laboratory, Cold Spring Harbor, New York.

———. 1981b. Multiple roles of the spindle pole bodies in the life cycle of *Saccharomyces cerevisiae. Alfred Benzon Symp.* **16:** 119.

Byers, B. and L. Goetsch. 1975. Behavior of spindles and spindle plaques in the cell cycle and conjugation in *Saccharomyces cerevisiae. J. Bacteriol.* **124:** 511.

Byers, B., K. Shriver, and L. Goetsch. 1978. The role of spindle pole bodies and modified microtubule ends in the initiation of microtubule assembly in *Saccharomyces cerevisiae. J. Cell Sci.* **30:** 331.

Cappellaro, C., C. Baldermann, R. Rachel, and W. Tanner. 1994. Mating type-specific cell-cell recognition of *Saccharomyces cerevisiae:* Cell wall attachment and active sites of **a**- and α-agglutinin. *EMBO J.* **13:** 4737.

Chan, C.S. and D. Botstein. 1993. Isolation and characterization of chromosome-gain and increase-in-ploidy mutants in yeast. *Genetics* **135:** 677.

Chan, R.K. and C.A. Otte. 1982. Physiological characterization of *Saccharomyces cerevisiae* mutants supersensitive to G_1 arrest by **a** factor and α factor pheromones. *Mol. Cell. Biol.* **2:** 21.

Chan, R.K., L.M. Melnick, L.C. Blair, and J. Thorner. 1983. Extracellular suppression allows mating by pheromone-deficient sterile mutants of *Saccharomyces cerevisiae. J. Bacteriol.* **155:** 903.

Chenevert, J. 1994. Cell polarization directed by extracellular cues in yeast. *Mol. Biol. Cell* **5:** 1169.

Cid, V.J., A. Duran, F. Del Rey, M.P. Snyder, C. Nombela, and M. Sanchez. 1995. Molecular basis of cell integrity and morphogenesis in *Saccharomyces cerevisiae. Microbiol. Rev.* **59:** 345.

Conde, J. and G.R. Fink. 1976. A mutant of *Saccharomyces cerevisiae* defective for nuclear fusion. *Proc. Natl. Acad. Sci.* **73:** 3651.

Conti, S.F. and T.D. Brock. 1965. Electron microscopy of cell fusion in conjugating *Hansenula wingei. J. Bacteriol.* **90:** 524.

Crandall, M., R. Egel, and V.L. Mackay. 1977. Physiology of mating in three yeasts. *Adv. Microb. Physiol.* **15:** 307.

Cross, F., L.H. Hartwell, C. Jackson, and J.B. Konopka. 1988. Conjugation in *Saccharomyces cerevisiae. Annu. Rev. Cell Biol.* **44:** 29.

Curran, B.P. and B.L. Carter. 1986. α-Factor enhancement of hybrid formation by protoplast fusion in *Saccharomyces cerevisiae* II. *Curr. Genet.* **10:** 943.

Cyr, D.M., T. Langer, and M.G. Douglas. 1994. DnaJ-like proteins: Molecular chaperones and regulators of Hsp70. *Trends Biochem. Sci.* **19:** 176.

Davis, N.G., J.L. Horecka, and G.F. Sprague, Jr. 1993. *cis-* and *trans-*acting functions required for endocytosis of the yeast pheromone receptors. *J. Cell Biol.* **122:** 53.

Delgado, M.A. and J. Conde. 1984. Benomyl prevents nuclear fusion in *Saccharomyces cerevisiae. Mol. Gen. Genet.* **193:** 188.

de Nobel, J.G., F.M. Klis, J. Priem, T. Munnik, and H. van den Ende. 1990. The glucanase-soluble mannoproteins limit cell wall porosity in *Saccharomyces cerevisiae. Yeast* **6:** 491.

Deshaies, R.J. and R. Schekman. 1987. A yeast mutant defective at an early stage in import of secretory precursors into the endoplasmic reticulum. *J. Cell Biol.* **105:** 633.

Deshaies, R.J., S.L. Sanders, D.A. Feldheim, and R. Schekman. 1991. Assembly of yeast Sec proteins involved in translocation into the endoplasmic reticulum into a membrane-bound multisubunit complex. *Nature* **349:** 806.

Doi, S., K. Tanabe, M. Watanabe, M. Yamaguchi, and M. Yoshimura. 1989. An α-specific gene, *SAG1*, is required for sexual agglutination in *Saccharomyces cerevisiae. Curr. Genet.* **15:** 393.

Dunphy, W.G., L. Brizuela, D. Beach, and J. Newport. 1988. The *Xenopus* cdc2 protein is a component of MPF, a cytoplasmic regulator of mitosis. *Cell* **54:** 423.

Dutcher, S.K. and L.H. Hartwell. 1982. The role of *S. cerevisiae* cell division cycle genes in nuclear fusion. *Genetics* **100:** 175.

———. 1983a. Genes that act before conjugation to prepare the *Saccharomyces cerevisiae* nucleus for caryogamy. *Cell* **33:** 203.

———. 1983b. Test for temporal or spatial restrictions in gene product function during the cell division cycle. *Mol. Cell. Biol.* **3:** 1255.

Elia, L. and L. Marsh. 1996. Role of the ABC transporter Ste6 in cell fusion during yeast conjugation. *J. Cell Biol.* **135:** 741.

Elion, E.A., J.A. Brill, and G.R. Fink. 1991. FUS3 represses CLN1 and CLN2 and in concert with KSS1 promotes signal transduction. *Proc. Natl. Acad. Sci.* **88:** 9392.

Elion, E.A., P.L. Grisafi, and G.R. Fink. 1990. *FUS3* encodes a cdc^{2+}/CDC28-related kinase required for the transition from mitosis into conjugation. *Cell* **60:** 649.

Elion, E.A., B. Satterberg, and J.E. Kranz. 1993. *FUS3* phosphorylates multiple components of the mating signal transduction cascade: Evidence for *STE12* and *FAR1*. *Mol. Biol. Cell* **4:** 495.

Elion, E.A., J. Trueheart, and G.R. Fink. 1995. Fus2 localizes near the site of cell fusion and is required for both cell fusion and nuclear alignment during zygote formation. *J. Cell Biol.* **130:** 1283.

el-Sherbeini, M. and J.A. Clemas. 1995. Cloning and characterization of *GNS1:* A *Saccharomyces cerevisiae* gene involved in synthesis of 1,3-β-glucan in vitro. *J. Bacteriol.* **177:** 3227.

Emmons, S., H. Phan, J. Calley, W. Chen, B. James, and L. Manseau. 1995. *cappuccino*, a *Drosophila* maternal effect gene required for polarity of the egg and embryo, is related to the vertebrate limb deformity locus. *Genes Dev.* **9:** 2482.

Endow, S.A. 1991. The emerging kinesin family of microtubule motor proteins. *Trends Biochem. Sci.* **16:** 221.

Endow, S.A., S.J. Kang, L.L. Satterwhite, M.D. Rose, V.P. Skeen, and E.D. Salmon. 1994. Yeast Kar3 is a minus-end microtubule motor protein that destabilizes micro tubules preferentially at the minus ends. *EMBO J.* **13:** 2708.

Eshel, D.E., L.A. Urrestarazu, S. Vissers, J.C. Jauniaux, J.C. van Vliet-Reedijk, R.J. Planta, and I.R. Gibbons. 1993. Cytoplasmic dynein is required for normal nuclear segregation in yeast. *Proc. Natl. Acad. Sci.* **90:** 11172.

Feldheim, D., J. Rothblatt, and R. Schekman. 1992. Topology and functional domains of Sec63p, an endoplasmic reticulum membrane protein required for secretory protein translocation. *Mol. Cell. Biol.* **12:** 3288.

Feldheim, D., K. Yoshimura, A. Admon, and R. Schekman. 1993. Structural and functional characterization of Sec66p, a new subunit of the polypeptide translocation apparatus in the yeast endoplasmic reticulum. *Mol. Biol. Cell* **4:** 931.

Ferguson, J., J.Y. Ho, T.A. Peterson, and S.I. Reed. 1986. Nucleotide sequence of the yeast cell division cycle start genes *CDC28, CDC36, CDC37,* and *CDC39*, and a structural analysis of the predicted products. *Nucleic Acids Res.* **14:** 6681.

Ferro-Novick, S. and R. Jahn. 1994. Vesicle fusion from yeast to man. *Nature* **370:** 191.

Fong, H.K.W., J.B. Hurley, R.S. Hopkins, R. Miake-Lye, M.S. Johnson, R.F. Doolittle, and M.I. Simon. 1986. Repetitive segmental structure of the transducin β subunit homology with the *CDC4* gene and identification of related mRNAs. *Proc. Natl. Acad. Sci.* **83:** 2162.

Fraser, L.R. 1995. Cellular biology of capacitation and the acrosome reaction. *Hum. Reprod.* (suppl.1) **10:** 22.

Frohlich, K.-U., H.W. Fries, M. Rudiger, R. Erdmann, D. Botstein, and D. Mecke. 1991. Yeast cell cycle protein CDC48p shows full-length homology to the mammalian protein VCP and is a member of a protein family involved in secretion, peroxisome formation, and gene expression. *J. Cell Biol.* **114:** 443.

Fujimura, H. 1990. Molecular cloning of the *DAC2/FUS3* gene essential for pheromone-induced G_1-arrest of the cell cycle in *Saccharomyces cerevisiae. Curr. Genet.* **18:** 395.

———. 1994. Yeast homolog of mammalian mitogen-activated protein kinase, FUS3/DAC2 kinase, is required both for cell fusion and for G_1 arrest of the cell cycle and morphological changes by the *cdc37* mutation. *J. Cell Sci.* **107:** 2617.

Fujita, A., C. Oka, Y. Arikawa, T. Katagai, A. Tonouchi, S. Kuhara, and Y. Misumi. 1994. A yeast gene necessary for bud-site selection encodes a protein similar to insulin-degrading enzyme. *Nature* **372:** 567.

Gehrung, S. and M. Snyder. 1990. The *SPA2* gene of *Saccharomyces cerevisiae* is important for pheromone-induced morphogenesis and efficient mating. *J. Cell Biol.* **111:** 1451.

Gerber, M.R., A. Farrell, R.J. Deshaies, I. Herskowitz, and D.O. Morgan. 1995. Cdc37 is required for association of the protein kinase Cdc28 with G_1 and mitotic cyclins. *Proc. Natl. Acad. Sci.* **92:** 4651.

Goebl, M.G., L. Goetsch, and B. Byers. 1994. The Ubc3 (Cdc34) ubiquitin-conjugating enzyme is ubiquitinated and phosphorylated in vivo. *Mol. Cell. Biol.* **14:** 3022.

Goebl, M.G., J. Yochem, S. Jentsch, J.P. McGrath, A. Varshavsky, and B. Byers. 1988. The yeast cell cycle gene *CDC34* encodes a ubiquitin-conjugating enzyme. *Science* **241:** 1331.

Goodman, L.E., S.R. Judd, C.C. Farnsworth, S. Powers, M.H. Gelb, J.A. Glomset, and F. Tamanoi. 1990. Mutants of *Saccharomyces cerevisiae* defective in the farnesylation of Ras proteins. *Proc. Natl. Acad. Sci.* **87:** 9665.

Green, N., H. Fang, and P. Walter. 1992. Mutants in three novel complementation groups inhibit membrane protein insertion into and soluble protein translocation across the endoplasmic reticulum membrane of *Saccharomyces cerevisiae. J. Cell Biol.* **116:** 597.

Hagen, D.C., G. McCaffrey, and G.F. Sprague, Jr. 1991. Pheromone response elements are necessary and sufficient for basal and pheromone-induced transcription of the *FUS1* gene of *Saccharomyces cerevisiae. Mol. Cell. Biol.* **11:** 2952.

Hasek, J., I. Rupes, J. Svobodova, and E. Streiblova. 1987. Tubulin and actin topology during zygote formation of *Saccharomyces cerevisiae. J. Gen. Microbiol.* **133:** 3355.

Hauser, K. and W. Tanner. 1989. Purification of the inducible α-agglutinin and molecular cloning of the gene. *FEBS Lett.* **255:** 290.

Herskowitz, I. 1995. MAP kinase pathways in yeast: For mating and more. *Cell* **80:** 187.

Heyer, W.D., A.W. Johnson, U. Reinhart, and R.D. Kolodner. 1995. Regulation and intracellular localization of *Saccharomyces cerevisiae* strand exchange protein 1 (Sep1/Xrn1/Kem1), a multifunctional exonuclease. *Mol. Cell. Biol.* **15:** 2728.

Hoyt, M.A., T. Stearns, and D. Botstein. 1990. Chromosome instability mutants of *Saccharomyces cerevisiae* that are defective in microtubule-mediated processes. *Mol. Cell. Biol.* **10:** 223.

Hoyt, M.A., L. He, K.K. Loo, and W.S. Saunders. 1992. Two *Saccharomyces cerevisiae* kinesin-related gene products required for mitotic spindle assembly. *J. Cell Biol.* **118:** 109.

Hoyt, M.A., L. He, L. Totis, and W.S. Saunders. 1993. Loss of function of *Saccharomyces cerevisiae* kinesin-related *CIN8* and *KIP1* is suppressed by *KAR3* motor domain mutations. *Genetics* **135:** 35.

Huang, B., A. Mengersen, and V.D. Lee. 1988. Molecular cloning of cDNA for caltractin, a basal body-associated Ca^{2+}-binding protein: Homology in its protein sequence with calmodulin and the yeast *CDC31* gene product. *J. Cell Biol.* **107:** 133.

Hyams, J.S. and G.G. Borisy. 1978. Nucleation of microtubules in vitro by isolated spindle pole bodies of the yeast *Saccharomyces cerevisiae. J. Cell Biol.* **78:** 401.

Interthal, H., C. Bellocq, J. Bahler, V.I. Bashkirov, S. Edelstein, and W.D. Heyer. 1995. A role of Sep1 (= Kem1, Xrn1) as a microtubule-associated protein in *Saccharomyces cerevisiae. EMBO J.* **14:** 1057.

Irie, K., M. Takase, K.S. Lee, D.E. Levin, H. Araki, K. Matsumoto, and Y. Oshima. 1993. *MKK1* and *MKK2*, which encode *Saccharomyces cerevisiae* mitogen-activated protein kinase-kinase homologs, function in the pathway mediated by protein kinase C. *Mol. Cell. Biol.* **13:** 3076.

Jackson, C.L. and L.H. Hartwell. 1990. Courtship in *S. cerevisiae:* Both cell types choose mating partners by responding to the strongest pheromone signal. *Cell* **63:** 1039.

Jansen, R.P., C. Dowzer, C. Michaelis, M. Galova, and K. Nasmyth. 1996. Mother cell-specific *HO* expression in budding yeast depends on the unconventional myosin Myo4p and other cytoplasmic proteins. *Cell* **84:** 687.

Johnson, A.W. and R.D. Kolodner. 1995. Synthetic lethality of *sep1* (*xrn1*) *ski2* and *sep1* (*xrn1*) *ski3* mutants of *Saccharomyces cerevisiae* is independent of killer virus and suggests a general role for these genes in translation control. *Mol. Cell. Biol.* **15:** 2719.

Kemble, G.W., T. Danieli, and J.M. White. 1994. Lipid-anchored influenza hemagglutinin promotes hemifusion, not complete fusion. *Cell* **76:** 383.

Kilmartin, J.V. 1994. Genetic and biochemical approaches to spindle function and chromosome segregation in eukaryotic microorganisms. *Curr. Opin. Cell Biol.* **6:** 50.

Kim, J., P.O. Ljungdahl, and G.R. Fink. 1990. *kem* mutations affect nuclear fusion in *Saccharomyces cerevisiae. Genetics* **126:** 799.

Klis, F.M. 1994. Cell wall assembly in yeast. *Yeast* **10:** 851.

Kuchler, K., H.G. Dohlman, and J. Thorner. 1993. The **a**-factor transporter (*STE6* gene product) and cell polarity in the yeast *Saccharomyces cerevisiae. J. Cell Biol.* **120:** 1203.

Kuchler, K., R.E. Sterne, and J. Thorner. 1989. *Saccharomyces cerevisiae STE6* gene product: A novel pathway for protein export in eukaryotic cells. *EMBO J.* **13:** 3973.

Kurihara, L.J., B.G. Stewart, A.E. Gammie, and M.D. Rose. 1996. Kar4p, a karyogamy-specific component of the yeast pheromone response pathway. *Mol. Cell. Biol.* **16:** 3990.

Kurihara, L.J., C.T. Beh, M. Latterich, R. Schekman, and M.D. Rose. 1994. Nuclear congression and membrane fusion: Two distinct events in the yeast karyogamy pathway. *J. Cell Biol.* **126:** 911.

Kurihara, T. and P. Silver. 1993. Suppression of a *sec63* mutations identifies a novel component of the yeast endoplasmic reticulum translocation apparatus. *Mol. Biol. Cell* **4:** 919.

Lancashire, W.E. and J.R. Mattoon. 1979. Cytoduction: A tool for mitochondrial genetic studies in yeast. Utilization of the nuclear-fusion mutation kar 1-1 for transfer of drug r and mit genomes in *Saccharomyces cerevisiae. Mol. Gen. Genet.* **170:** 333.

Larimer, F.W., C.L. Hsu, M.K. Maupin, and A. Stevens. 1992. Characterization of the *XRN1* gene encoding a 5′→3′exoribonuclease: Sequence data and analysis of disparate protein and mRNA levels of gene-disrupted yeast cells. *Gene* **120:** 51.

Latterich, M. and R. Schekman. 1994. The karyogamy gene *KAR2* and novel proteins are required for ER-membrane fusion. *Cell* **78:** 87.

Latterich, M., K.-U. Frohlich, and R. Schekman. 1995. Membrane fusion and the cell cycle: Cdc48 participates in the fusion of ER membranes. *Cell* **82:** 885.

Levin, D.E. and B. Errede. 1995. The proliferation of MAP kinase signaling pathways in yeast. *Curr. Opin. Cell Biol.* **7:** 197.

Li, Q. and K.A. Suprenant. 1994. Molecular characterization of the 77-kDa echinoderm microtubule-associated protein. Homology to the β-transducin family. *J. Biol. Chem.* **269:** 31777.

Li, Y.Y., E. Yeh, T. Hays, and K. Bloom. 1993. Disruption of mitotic spindle orientation in a yeast dynein mutant. *Proc. Natl. Acad. Sci.* **90:** 10096.

Lillie, S.H. and S.S. Brown. 1994. Immunofluorescence localization of the unconventional myosin, Myo2p, and the putative kinesin-related protein, Smy1p, to the same regions of polarized growth in *Saccharomyces cerevisiae. J. Cell Biol.* **125:** 825.

Lipke, P.N. and J. Kurjan. 1992. Sexual agglutination in budding yeasts: Structure, function, and regulation of adhesion glycoproteins. *Microbiol. Rev.* **56:** 180.

Lipke, P.N., A. Taylor, and C.E. Ballou. 1976. Morphogenic effects of α-factor on *Saccharomyces cerevisiae* **a** cells. *J. Bacteriol.* **127:** 610.

Lipke, P.N., D. Wojciechowicz, and J. Kurjan. 1989. *AGa1* is the structural gene for the *Saccharomyces cerevisiae* α-agglutinin, a cell surface glycoprotein involved in cell-cell interactions during mating. *Mol. Cell. Biol.* **9:** 3155.

Liu, H. and A. Bretscher. 1992. Characterization of *TPM1* disrupted yeast cells indicates an involvement of tropomyosin in directed vesicular transport. *J. Cell Biol.* **118:** 285.

Liu, Z. and W. Gilbert. 1994. The yeast *KEM1* gene encodes a nuclease specific for G4 tetraplex DNA: Implication of in vivo functions for this novel DNA structure. *Cell* **77:** 1083.

Liu, Z., A. Lee, and W. Gilbert. 1995. Gene disruption of a G4-DNA-dependent nuclease in yeast leads to cellular senescence and telomere shortening. *Proc. Natl. Acad. Sci.* **92:** 6002.

Luyten, R., J. Albertyn, W.F. Skibbe, B.A. Prior, J. Ramos, J.M. Thevelein, and S. Hohmann. 1995. Fps1, a yeast member of the MIP family of channel proteins, is a facilitator for glycerol uptake and efflux and is inactive under osmotic stress. *EMBO J.* **14:** 1360.

Marsh, L., A.M. Neiman, and I. Herskowitz. 1991. Signal transduction during pheromone response in yeast. *Annu. Rev. Cell Biol.* **7:** 699.

Mazur, P., N. Morin, W. Baginsky, M. el-Sherbeini, J.A. Clemas, J.B. Nielsen, and F. Foor. 1995. Differential expression and function of two homologous subunits of yeast 1,3-β-D-glucan synthase. *Mol. Cell. Biol.* **15:** 5671.

McCaffrey, G., F.J. Clay, K. Kelsay, and G.F. Sprague, Jr. 1987. Identification and regulation of a gene required for cell fusion during mating of the yeast *Saccharomyces cerevisiae. Mol. Cell. Biol.* **7:** 2680.

McDonald, H.B. and L.S.B. Goldstein. 1990. Identification and characterization of a gene encoding a kinesin-like protein in *Drosophila. Cell* **61:** 991.

McDonald, H.B., R.J. Stewart, and L.S.B. Goldstein. 1990. The kinesin-like ncd protein of *Drosophila* is a minus end-directed microtubule motor. *Cell* **63:** 1159.

Melikyan, G.B., J.M. White, and F.S. Cohen. 1995. GPI-anchored influenza hemagglutinin induces hemifusion to both red blood cell and planar bilayer membranes. *J. Cell Biol.* **131:** 679.

Mellman, I. 1995. Enigma variations: Protein mediators of membrane fusion. *Cell* **82:** 869.

Meluh, P.B. 1992. "*KAR3*, a kinesin-related gene required for nuclear fusion, mitosis and meiosis in *Saccharomyces cerevisiae*." Ph.D. thesis, Princeton University, Princeton, New Jersey.

Meluh, P.B. and M.D. Rose. 1990. *KAR3*, a kinesin-related gene required for yeast nuclear fusion. *Cell* **60:** 1029.

Michaelis, S. 1993. *STE6*, the yeast **a**-factor transporter. *Semin. Cell Biol.* **4:** 17.

Middleton, K. and J. Carbon. 1994. *KAR3*-encoded kinesin is a minus-end-directed motor that functions with centromere binding proteins (CBF3) on an in vitro yeast kinetochore. *Proc. Natl. Acad. Sci.* **91:** 7212.

Morris, N.R., X. Xiang, and S.M. Beckwith. 1995. Nuclear migration advances in fungi. *Trends Cell Biol.* **5:** 278.

Munn, A.L., B.J. Stevenson, M.I. Geli, and H. Riezman. 1995. *end5, end6, end7:* Mutations that cause actin delocalization and block the internalization step of endocytosis in *Saccharomyces cerevisiae. Mol. Biol. Cell* **6:** 1721.

Myles, D.G. 1993. Molecular mechanisms of sperm-egg membrane binding and fusion in mammals. *Dev. Biol.* **158:** 35.

Myles, D.G., L.H. Kimmel, C.P. Blobel, J.M. White, and P. Primakoff. 1994. Identification of a binding site in the disintegrin domain of fertilin required for sperm-egg fusion. *Proc. Natl. Acad. Sci.* **91:** 4195.

Nelson, M.K., T. Kurihara, and P.A. Silver. 1993. Extragenic suppressors of mutations in the cytoplasmic C terminus of *SEC63* define five new genes in *Saccharomyces cerevisiae. Genetics* **134:** 159.

Ng, D.T.W. and P. Walter. 1996. ER membrane protein complex required for nuclear fusion. *J. Cell Biol.* **132:** 499.

Normington, K., K. Kohno, Y. Kozutsumi, M.J. Gething, and J. Sambrook. 1989. *S. cerevisiae* encodes an essential protein homologous in sequence and function to mammalian BiP. *Cell* **57:** 1223.

O'Connell, M.J., P.B. Meluh, M.D. Rose, and N.R. Morris. 1993. Suppression of the *bimC4* mitotic spindle defect by deletion of *klpA*, a gene encoding a *KAR3*-related kinesin-like protein in *Aspergillus nidulans. J. Cell Biol.* **120:** 153.

Osumi, M., C. Shimoda, and N. Yanagishima. 1974. Mating reaction in *Saccharomyces cerevisiae*. Changes in the fine structure during the mating reaction. *Arch. Microbiol.* **97:** 27.

Page, B.D. and M. Snyder. 1992. *CIK1:* A developmentally regulated spindle pole body-associated protein important for microtubule functions in *Saccharomyces cerevisiae. Genes Dev.* **6:** 1414.

———. 1993. Chromosome segregation in yeast. *Annu. Rev. Microbiol.* **47:** 231.

Page, B.D., L.L. Satterwhite, M.D. Rose, and M. Snyder. 1994. Localization of the Kar3 kinesin heavy chain-related protein requires the Cik1 interacting protein. *J. Cell Biol.* **124:** 507.

Paravicini, G., M. Coooper, L. Friedli, D.J. Smith, J.L. Carpentier, L.S. Klig, and M.A. Payton. 1992. The osmotic integrity of the yeast cell wall requires a functional *PKC1* gene product. *Mol. Cell. Biol.* **12:** 4896.

Pellman, D., M. Bagget, H. Tu, and G.R. Fink. 1995. Two microtubule-associated proteins required for anaphase spindle movement in *Saccharomyces cerevisiae. J. Cell Biol.* **130:** 1373.

Peter, M., A. Gartner, J. Horecka, G. Ammerer, and I. Herskowitz. 1993. *FAR1* links the signal transduction pathway to the cell cycle machinery in yeast. *Cell* **73:** 747.

Peterson, J., D. Weilguny, R. Egel, and O. Nielsen. 1995. Characterization of *fus1* of *Schizosaccharomyces pombe:* A developmentally controlled function needed for conjugation. *Mol. Cell. Biol.* **15:** 3697.

Peterson, T.A., J. Yochem, B. Byers, M.F. Nunn, P.H. Duesberg, R.F. Doolittle, and S.I. Reed. 1984. A relationship between the yeast cell cycle genes *CDC4* and *CDC36* and the ets sequence of oncogenic virus E26. *Nature* **309:** 556.

Pierre, P., R. Pepperkok, and T.E. Kreis. 1994. Molecular characterization of two functional domains of CLIP-170 in vivo. *J. Cell Sci.* **107:** 1909.

Pierre, P., J. Scheel, J.E. Rickard, and T.E. Kreis. 1992. CLIP-170 links endocytic vesicles to microtubules. *Cell* **70:** 887.

Polaina, J. and J. Conde. 1982. Genes involved in the control of nuclear fusion during the sexual cycle of *Saccharomyces cerevisiae. Mol. Gen. Genet.* **186:** 253.

Popolo, L., P. Cavadini, M. Vai, and L. Alberghina. 1993. Transcript accumulation of the the *GGP1* gene, encoding a yeast GPI-anchored glycoprotein, is inhibited during arrest in the G_1 phase and during sporulation. *Curr. Genet.* **24:** 382.

Rabouille, C., T.P. Levine, J.-M. Peters, and G. Warren. 1995. An NSF-like ATPase, p97, and NSF mediate cisternal regrowth from mitotic Golgi fragments. *Cell* **82:** 905.

Read, E.B., H.H. Okamura, and D.G. Drubin. 1992. Actin- and tubulin-dependent functions during *Saccharomyces cerevisiae* mating projection formation. *Mol. Biol. Cell* **3:** 429.

Roof, D.M., P.B. Meluh, and M.D. Rose. 1991. Multiple kinesin-related proteins in yeast mitosis. *Cold Spring Harbor Symp. Quant. Biol.* **56:** 693.

———. 1992. Kinesin-related proteins required for assembly of the mitotic spindle. *J. Cell Biol.* **118:** 95.

Rose, M.D. 1991. Nuclear fusion in yeast. *Annu. Rev. Microbiol.* **45:** 539.

———. 1996. Nuclear fusion in the yeast, *Saccharomyces cerevisiae. Annu. Rev. Cell Dev. Biol.* **12:** 663.

Rose, M.D. and G.R. Fink. 1987. *KAR1*, a gene required for function of both intranuclear and extranuclear microtubules in yeast. *Cell* **48:** 1047.

Rose, M.D., S. Biggins, and L.L. Satterwhite. 1993. Unravelling the tangled web at the microtubule-organizing center. *Curr. Opin. Cell Biol.* **5:** 105.

Rose, M.D., L.M. Misra, and J.P. Vogel. 1989. *KAR2*, a karyogamy gene, is the yeast homolog of the mammalian BiP/GRP78 gene. *Cell* **57:** 1211.

Rose, M.D., B.R. Price, and G.R. Fink. 1986. *Saccharomyces cerevisiae* nuclear fusion requires prior activation by alpha factor. *Mol. Cell. Biol.* **6:** 3490.

Rothman, J.E. 1994. Mechanisms of intracellular protein transport. *Nature* **372:** 55.

Rothman, J.E. and G. Warren. 1994. Implications of the SNARE hypothesis for intracellular membrane topology and dynamics. *Curr. Biol.* **4:** 220.

Roy, A., C.F. Lu, D. Marykwas, P.N. Lipke, and J. Kurjan. 1991. The *AGA1* gene is involved in cell surface attachment of the *Saccharomyces cerevisiae* cell adhesion glycoprotein **a**-agglutinin. *Mol. Cell. Biol.* **11:** 4196.

Sadler, I., A. Chiang, T. Kurihara, J. Rothblatt, J. Way, and P. Silver. 1989. A yeast gene important for protein assembly into the endoplasmic reticulum and the nucleus has homology to DnaJ, an *Escherichia coli* heat shock protein. *J. Cell Biol.* **109:** 2665.

Sanders, S.L., K.M. Whitfield, J.P. Vogel, M.D. Rose, and R.W. Schekman. 1992. Sec61p and BiP directly facilitate polypeptide translocation into the ER. *Cell* **69:** 353.

Saunders, W.S. and M.A. Hoyt. 1992. Kinesin-related proteins required for structural integrity of the mitotic spindle. *Cell* **70:** 451.

Saunders, W.S., D. Koshland, D. Eshel, I.R. Gibbons, and M.A. Hoyt. 1995. *Saccharomyces cerevisiae* kinesin- and dynein-related proteins required for anaphase chromosome segregation. *J. Cell Biol.* **128:** 617.

Schekman, R. 1992. Genetic and biochemical analysis of vesicular traffic in yeast. *Curr. Opin. Cell Biol.* **4:** 587.

Schwob, E., T. Bohm, M.D. Mendenhall, and K. Nasmyth. 1994. The B-type cyclin kinase inhibitor p40SIC1 controls the G_1 to S transition in *S. cerevisiae. Cell* **79:** 233.

Scidmore, M.A., H.H. Okamura, and M.D. Rose. 1993. Genetic interactions between *KAR2* and *SEC63*, encoding eukaryotic homologues of DnaK and DnaJ in the endoplasmic reticulum. *Mol. Biol. Cell* **4:** 1145.

Segall, J.E. 1993. Polarization of yeast cells in spatial gradients of α mating factor. *Proc. Natl. Acad. Sci.* **90:** 8332.

Shaw, J.A., P.C. Mol, B. Bowers, S.J. Silverman, M.H. Valdivieso, A. Duran, and E. Cabib. 1994. The function of chitin synthetases 2 and 3 in the *Saccharomyces cerevisiae* cell cycle. *J. Cell Biol.* **114:** 111.

Shimizu, J., K. Yoda, and M. Yamasaki. 1994. The hypo-osmolarity-sensitive phenotype of the *Saccharomyces cerevisiae hpo2* mutant is due to a mutation in *PKC1*, which regulates expression of β-glucanase. *Mol. Gen. Genet.* **242:** 641.

Simons, J.F., S. Ferro-Novick, M.D. Rose, and A. Helenius. 1995. BiP/Kar2p serves as a molecular chaperone during carboxypeptidase Y folding in yeast. *J. Cell Biol.* **130:** 41.

Sivadon, P., F. Bauer, M. Aigle, and M. Crouzet. 1995. Actin cytoskeleton and budding pattern are altered in the yeast *rvs161* mutant: The Rvs161 protein shares common domains with the brain protein amphiphysin. *Mol. Gen. Genet.* **246:** 485.

Snyder, M. 1989. The *SPA2* protein of yeast localizes to sites of cell growth. *J. Cell Biol.* **108:** 1419.

———. 1994. The spindle pole body of yeast. *Chromosoma* **103:** 369.

Spang, A., I. Courtney, U. Fackler, M. Matzner, and E. Schiebel. 1993. The calcium-binding protein cell division cycle 31 of *Saccharomyces cerevisiae* is a component of the half bridge of the spindle pole body. *J. Cell Biol.* **123:** 405.

Spang, A., I. Courtney, K. Grein, M. Matzner, and E. Schiebel. 1995. The Cdc31p-binding protein Kar1p is a component of the half bridge of the yeast spindle pole body. *J. Cell Biol.* **128:** 863.

Spellig, T., M. Bolker, F. Lottspeich, F.W. Frank, and R. Kahmann. 1994. Pheromones trigger filamentous growth in *Ustilago maydis. EMBO J.* **13:** 1620.

Sprague, G.F., Jr. 1991. Assay of yeast mating reaction. *Methods Enzymol.* **194:** 77.

Sprague, G.F., Jr. and J.W. Thorner. 1994. Pheromone response and signal transduction during the mating process of *Saccharomyces cerevisiae*. In *The molecular and cellular biology of the yeast* Saccharomyces: *Gene expression* (ed. E.W. Jones et al.), vol. 2, p. 657. Cold Spring Harbor Laboratory Press, Cold Spring Harbor, New York.

Stearns, T., M.A. Hoyt, and D. Botstein. 1990. Yeast mutants sensitive to antimicrotubule drugs define three genes that affect microtubule function. *Genetics* **124:** 251.

Stegmann, T., R.W. Doms, and A. Helenius. 1989. Protein-mediated membrane fusion. *Annu. Rev. Biophys. Biophys. Chem.* **18:** 187.

Tomeo, M.E., G. Fenner, S.R. Tove, and L.W. Parks. 1992. Effect of sterol alterations on conjugation in *Saccharomyces cerevisiae. Yeast* **8:** 1015.

Trueheart, J. and G.R. Fink. 1989. The yeast cell fusion protein *FUS1* is *O*-glycosylated and spans the plasma membrane. *Proc. Natl. Acad. Sci.* **86:** 9916.

Trueheart, J., J.D. Boeke, and G.R. Fink. 1987. Two genes required for cell fusion during yeast conjugation: Evidence for a pheromone-induced surface protein. *Mol. Cell. Biol.* **7:** 2316.

Tyers, M. and B. Futcher. 1993. Far1 and Fus3 link the mating pheromone signal transduction pathway to three G_1-phase Cdc28 kinase complexes. *Mol. Cell. Biol.* **13:** 5659.

Tyers, M., G. Tokiwa, R. Nash, and B. Futcher. 1992. The Cln3-Cdc28 kinase complex of *S. cerevisiae* is regulated by proteolysis and phosphorylation. *EMBO J.* **11:** 1773.

Vallen, E.A., M.A. Hiller, T.Y. Scherson, and M.D. Rose. 1992a. Separate domains of *KAR1* mediate distinct functions in mitosis and nuclear fusion. *J. Cell Biol.* **117:** 1277.

Vallen, E.A., W. Ho, M. Winey, and M.D. Rose. 1994. Genetic interactions between *CDC31* and *KAR1*, two genes required for duplication of the microtubule organizing center in *Saccharomyces cerevisiae. Genetics* **137:** 407.

Vallen, E.A., T.Y. Scherson, T. Roberts, K. van Zee, and M.D. Rose. 1992b. Asymmetric mitotic segregation of the yeast spindle pole body. *Cell* **69:** 505.

Vogel J. 1993. "*KAR2*, the yeast homologue of mammalian BiP/GRP78." Ph.D. thesis, Princeton University, Princeton, New Jersey.

Vogel, J.P., L.M. Misra, and M.D. Rose. 1990. Loss of BiP/GRP78 function blocks translocation of secretory proteins in yeast. *J. Cell Biol.* **110:** 1885.

Walker, R.A., E.D. Salmon, and S.A. Endow. 1990. The *Drosophila claret* segregation protein is a minus-end directed motor molecule. *Nature.* **347:** 780.

Waterman-Storer, C.M. and E.L. Holzbaur. 1996. The product of the *Drosophila* gene, *Glued*, is the functional homologue of the $p150^{Glued}$ component of the vertebrate dynactin complex. *J. Biol. Chem.* **271:** 1153.

Welch, M.D., D.A. Holtzman, and D.G. Drubin. 1994. The yeast actin cytoskeleton. *Curr. Opin. Cell Biol.* **6:** 110.

White, J.M. 1992. Membrane fusion. *Science* **258:** 917.

Yaglom, J., M.H. Linskens, S. Sadis, D.M. Rubin, B. Futcher, and D. Finley. 1995. $p34^{cdc28}$-mediated control of Cln3 cyclin degradation. *Mol. Cell. Biol.* **15:** 731.

Yamaguchi, M., K. Yoshida, and N. Yanagishima. 1994. Sexual agglutination substances require a "carrier" glycoprotein for integration into the cell wall of *Saccharomyces cerevisiae. Microbiology* **140:** 2217.

Yang, J.T., R.A. Laymon, and L.S.B. Goldstein. 1989. A three-domain structure of kinesin heavy chain revealed by DNA sequence and microtubule binding analyses. *Cell* **56:** 879.

Zarzov, P., C. Mazzoni, and C. Mann. 1996. The *SLT2* (*MPK1*) MAP kinase is activated during periods of polarized cell growth in yeast. *EMBO J.* **15:** 83.

11

Meiosis and Sporulation in *Saccharomyces cerevisiae*

Martin Kupiec,[1,2] Breck Byers,[1]
Rochelle E. Esposito,[3] and Aaron P. Mitchell[4*]

[1]Department of Genetics, University of Washington
Seattle, Washington 98198-7360

[2]Department of Molecular Microbiology and Biotechnology
Tel Aviv University
Tel Aviv, Israel

[3]Department of Molecular Genetics and Cell Biology
University of Chicago
Chicago, Illinois 60637

[4]Department of Microbiology
Columbia University
New York, New York 10032

*All four authors contributed equally to the preparation of this review.

I. INTRODUCTION

A. The Importance of Meiosis

Meiosis has a central role in the sexual reproduction of nearly all eukaryotes. The major genetic events that occur during this process are critical for generating genetic diversity and producing offspring with normal chromosome numbers. The stages and transition periods in the two successive divisions that comprise the process are exquisitely regulated and differ significantly in timing and chromosome behavior from the corresponding stages of mitosis. The first division deviates most dramatically from mitotic cell division in its prolonged DNA S phase, greatly extended prophase in which homologous pairing and recombination occur, anaphase cohesion of sister chromatids, and disjunction of homologs (for review, see Baker et al. 1976). In contrast, the second meiotic division has a shortened G_1, lacks its own S and G_2 phases, and is otherwise similar to mitosis, utilizing many of the same gene products to disjoin the previously replicated sister chromatids. These dramatic changes in chromosome behavior offer a fascinating look at the regulatory mechanisms that control both cell division and cell differentiation. For example, what triggers meiotic development? Is there a single irreversible commitment point similar to mitotic Start for all meiotic events? How does the synaptonemal complex develop and what specific role(s) does it have in meiosis? What factors regulate the assembly of meiosis I and meiosis II spindles and the pattern of chromosome segregation in each division? How is the progression of events coordinated through two meiotic divisions? Do they function on a single sequential dependent pathway? How do meiotic controls interface with those of the mitotic cell division cycle?

B. Advantages of Yeast for Studies of Meiosis

Baker's yeast, a relatively simple eukaryote, has unique advantages for detailed analysis of meiosis: (1) Sporulation can easily be triggered by a simple change of nutritional conditions, and landmark events are conveniently monitored in populations of relatively homogeneous single cells. (2) The meiotic products remain in association, allowing precise reconstruction of the recombination and segregation events taking place in the two divisions. (3) Features of the yeast life cycle facilitate the isolation of both dominant and recessive meiotic mutations. (4) The ability to interrupt meiosis and recover viable cells by the return-to-growth pathway permits one to monitor the timing and progression of events as they occur. (5) Well-developed genetics and sophisticated recombinant DNA technology offer powerful tools to examine specific gene functions by gene replacement procedures. On the other hand, some aspects of yeast meiosis differ from the predominant pattern among eukaryotes, including (1) the absence of nuclear membrane breakdown and cytokinesis associated with each meiotic chromosome division, (2) the duration of specific stages and the overall level of genetic recombination per physical length of DNA, (3) the manner in which the meiotic products are packaged and maturated, and (4) the ability of the resulting haploid cells to resume mitotic cell division directly. These differences must be borne in mind as we attempt to utilize information about yeast meiosis to extend our understanding of meiosis in yeast to other organisms.

C. Rationale and Approaches Used in the Yeast System

Most current studies of meiotic mechanisms in yeast entail the identification and isolation of genes that are required to carry out the process. Together with a detailed description of cytological, genetic, and molecular events that take place in the wild type, the analysis of how mutants progress through meiotic landmarks has identified points of genetic control for more detailed analysis. These approaches have already yielded significant insight into the dependency relationships and coordinating functions occurring during meiosis and spore formation. In conjunction with modern technologies in genetic and cell biology, a thorough understanding of the structure, function, and regulation of all of the critical genes required for meiosis and sporulation is expected to emerge.

Three complementary approaches have been been employed to identify genes that are essential for meiosis and spore development: (1) classical genetics to recover mutants specifically defective in sporulation; (2) analysis of previously known gene mutations, particularly those affecting the cell cycle and DNA repair processes, for their effects on meiosis; and (3) the assay of transcription products to identify genes that are specifically expressed during meiosis, enabling use of reverse genetics to isolate the wild-type genes and construct deletion alleles for phenotypic analysis. The criteria used to define defective essential genes rely primarily on selections or screens for the reduction or alteration of the end product of meiotic development. These include changes in the level of total asci (Esposito and Esposito 1969; R.E. Esposito and Esposito 1974), the number and ploidy of spores per ascus (Klapholz and Esposito 1980a), spore viability (Rockmill and Roeder 1988; Engebrecht and Roeder 1989), and the presence of specific spore wall components (Briza et al. 1990). In addition to these general criteria, others include assays for specific landmarks such as genetic recombination and homolog disjunction, as cited later.

We will not review specific detection systems used to recover sporulation-defective mutants since this has been done recently (Esposito et al. 1991). However, the bypass rationale, which has shown increasing value as a means of identifying sporulation mutations in regulatory functions that might not otherwise be uncovered, deserves specific mention. This approach, pioneered by Dawes (1975), involves detection of mutants that allow cells to bypass the requirement for specific conditions (such as starvation) or essential genes (such as mating-type) to initiate or proceed through the meiotic divisions (Hopper et al. 1975; Kassir and Simchen 1976; Shilo et al. 1978; Matsumoto et al. 1983; Tatchell et al. 1985). This method does not depend on failure to complete the process properly but rather on acquisition of the capacity to do meiosis and spore formation (or specific subprocesses) under normally restrictive conditions. The method has been adapted, in conjunction with cloning procedures, to recover wild-type genes that function (often in high copy number) to potentiate one or more landmarks events or that can suppress specific mutant defects to permit meiotic progression in the absence of typical controls (Kassir et al. 1988; Smith and Mitchell 1989; Yoshida et al. 1990).

Once essential genes are identified, the key determinants that specifically promote meiotic differentiation are expected to be found among those genes that act uniquely in meiosis and spore formation as opposed to those that are also required for other processes. The criterion that has evolved to define experimentally such sporulation-specific essential genes relies upon their expression and function only in **a**/α cells during the starvation conditions required for sporulation. Thus, genes that are expressed and function in **a**/α cells under nonstarvation conditions or in **a**/**a** or α/α cells under starvation conditions are not considered unique to the meiotic process, which requires both appropriate cell type and nutritional signals to proceed as detailed in the other reviews cited above.

D. Sporulation Landmarks

In baker's yeast, meiosis is controlled by the interaction of two genetic systems: one that responds to cell type and one that senses nutritional conditions. It typically occurs in *MAT***a**/*MAT*α diploids deprived of glucose and nitrogen, in the presence of a nonfermentable carbon source. In contrast to higher eukaryotes, the entire process of meiosis takes place within the confines of a single cell. Following initiation and premeiotic

DNA synthesis, three major changes in cellular morphogenesis ensue: (1) Formation of synaptonemal complexes (SCs) during prophase of meiosis I and their subsequent degradation prior to anaphase I. (2) Spindle pole body (SPB) duplication, SPB separation, and assembly of intranuclear spindles at meiosis I and II, respectively. Apparent "binucleate" and "tetranucleate" cells can be seen by DAPI staining, following the segregation first of homologous centromeres (meiosis I) and then of sister centromeres (meiosis II). (3) SPB modification (enlargement of the outer plaque) at meiosis II, initiation of prospore wall growth at the site of each modified SPB, and enclosure into spores of the four nuclear lobes containing the haploid genomes. Additional spore wall material is deposited and mature asci form, containing all of the products of a single meiosis within the original mother cell.

E. Recent Advances

When meiosis and sporulation were first reviewed in this series of volumes (Esposito and Klapholz 1981), no meiosis or sporulation-specific genes had yet been cloned, few genes had been assigned to specific morphogenic events, and little was known about their gene products or regulation. Since then, more than 30 meiosis and sporulation-specific genes have been isolated and evidence has been obtained for developmental transcriptional regulation of many of them. The basic architecture of the regulatory cascade involved in both cell type and nutritional control of meiotic initiation has begun to emerge (Malone 1990; Mitchell 1994). New strategies to identify specific classes of meiotic mutants have been developed based on the unique ability of *spo13* mutants to bypass meiosis I and to undergo haploid meiosis. Finally, the visualization of synaptonemal complexes by spreading techniques (Dresser and Giroux 1988), rapid screens for the presence of spore wall components (Briza et al. 1990), and localization of specific gene products by antibody tagging (see, e.g., Hollingsworth et al. 1990; Sym et al. 1993; Bishop 1994) have significantly advanced the analysis of mutant phenotypes.

In the previous volumes of this series, the genetic control of meiosis and spore formation and the cytology of *Saccharomyces cerevisiae* were independently reviewed, in part emphasizing the different goals of studies conducted to that time (Byers 1981; Esposito and Klapholz 1981). The initial priorities were to describe the precise sequence of morphological changes and to identify specific genes that were required during sporulation. Progress in these areas over the ensuing 15 years has now made it possible to synthesize a more comprehensive view of meiotic development. We focus our attention on the cell biological aspects of meiosis and sporulation and direct the reader's attention to numerous excellent reviews, some of which are listed here (Dickinson 1988; Mitchell 1988, 1994; Simchen and Kassir 1989; McLeod 1989; Miller 1989; Malone 1990; Broach 1991; Esposito et al. 1991; Petes et al. 1991).

II. INITIATION OF MEIOSIS AND MEIOTIC DNA REPLICATION

A. Overview of the Initiation of Meiosis

Entry into meiosis occurs only in **a**/α cells, and only in response to starvation. All known properties of **a**/α cells, including sporulation ability, depend on activity of the repressor, **a**1-α2. **a**1-α2 is a heterodimer of the *MAT***a***1* and *MAT*α*2* gene products (see Herskowitz et al. 1992). **a**1-α2 promotes meiosis through effects on two genes: It represses expression of *RME1* and it stimulates expression of *IME4* (also known as *SPO8*; Fig. 1). Rme1p is an inhibitor of meiosis that is expressed in **a** and α cells. Ime4p is a positive regulator of meiosis that is expressed only in **a**/α cells and only in response to starvation.

The effects of Rme1p and Ime4p are exerted on expression of *IME1*, which specifies a positive regulator of meiosis. *IME1* is expressed at low levels in growing **a**/α cells and in growing or starved **a** and α cells; its expression under these circumstances depends on two additional gene products: Mck1p, which is a protein kinase, and Rim1p (see below). *IME1* is expressed at high levels after starvation of **a**/α cells, reflecting activation by Ime4p and, possibly, increased activity of Mck1p- and Rim1p-dependent pathways. Ime1p is phosphorylated by the protein kinase Rim11p. One role of Ime1p is to stimulate expression of many early meiotic genes, products of which have more direct roles in meiotic prophase events. These genes are repressed in the absence of Ime1p by action of Sin3p, Rpd3p, and the DNA-binding protein Ume6p. In the presence of Ime1p and active Rim11p, Ume6p is transformed from a negative to a positive regulator that induces early meiotic gene expression. These genes and others that are relevant to later meiotic processes are grouped by function in Table 1. Mutant phenotypes are described in Table 2 at the end of this chapter.

B. Nutritional Control of Entry into Meiosis

Sporulation occurs only in the presence of a nonfermentable carbon source and upon limitation for some nutrient (for review, see Esposito and Klapholz 1981; Miller 1989). Laboratory sporulation media general-

a or α cell, growing or starved

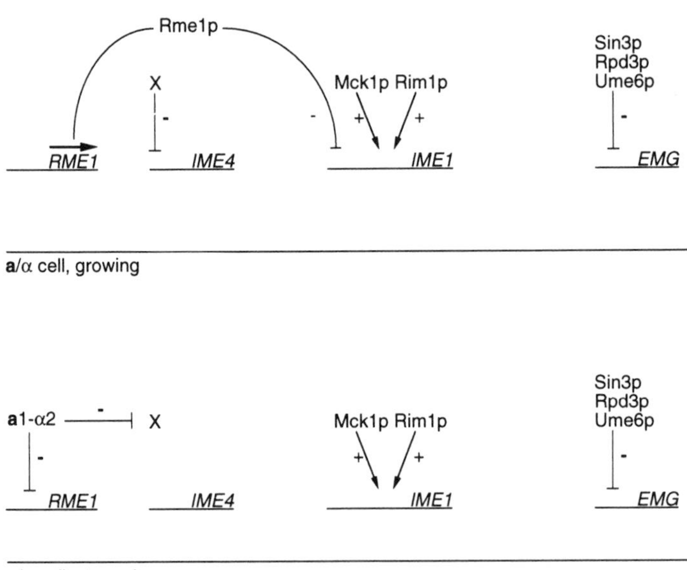

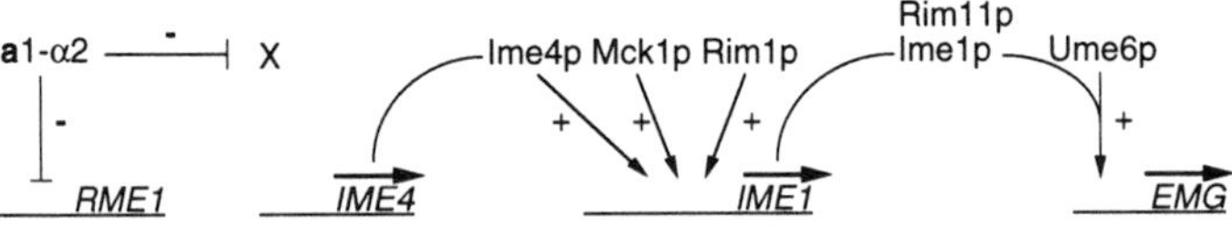

Figure 1 Overview of early meiotic gene regulation. In **a** or α cells, Rme1p represses *IME1* expression. In addition, *IME4* is not expressed, perhaps due to repression by an unidentified protein X. Mck1p and Rim1p stimulate *IME1* expression at low levels. Early meiotic genes (*EMG*) are repressed by activities of Sin3p, Rpd3p, and Ume6p. In growing **a**/α cells, **a**1-α2 represses *RME1* and may relieve *IME4* repression. In starved **a**/α cells, *IME4* is expressed, permitting elevation of *IME1* expression. In conjunction with protein kinase Rim11p, Ime1p converts Ume6p into an activator, stimulating *EMG* expression.

ly contain acetate as a carbon source, but ethanol, pyruvate, and glycerol also permit sporulation to occur (Freese et al. 1982). Nutrient deprivation is usually accomplished by nitrogen limitation, but limitation for carbon, phosphate, sulfate, and some nucleotides or amino acids has been used (Freese et al. 1982; Varma et al. 1985). Growth limitation in the absence of a fermentable carbon can occur in stationary phase, after glucose has been fermented, and some strains sporulate after reaching stationary phase in a rich growth medium (see Esposito and Klapholz 1981).

Deprivation for glucose and nitrogen is required not only to initiate sporulation, but also to complete the process. In wild-type strains, cells that initiate sporulation will resume mitotic growth if nutrients become available up until a point called "commitment to meiosis," "commitment to sporulation," or "commitment to haploidization" (R.E. Esposito and Esposito 1974; see Esposito and Klapholz 1981), which corresponds approximately to the pachytene phase. Thus, a sporulating diploid population will yield diploid cells if nutrients are provided prior to commitment; they will yield spores that germinate into haploid cells if nutrients are provided after commitment. Experiments in which precommitment populations are provided with nutrients are often called "return-to-growth" or "pullback" experiments. In sporulation-defective mutant strains, cells arrested at stages later than pachytene, even after meiosis I or meiosis II, can resume mitotic growth without completing spore formation if nutrients are provided, suggesting that the divisions themselves do not commit cells to proceed through spore formation (Honigberg et al. 1992; Honigberg and Esposito 1994).

Why might sporulation occur only upon growth limitation? One idea is based on the observation that spores are more resistant than unsporulated cells to a variety of harsh environments (for review, see Miller 1989); starvation may signal onset of such unfavorable conditions. A second idea is that the main selective advantage of sporulation is generation of new genotypes through meiosis; the process would occur only in poor environments because the time required to sporulate—equivalent to five to ten mitotic generations—would cause a severe selective disadvantage to sporulating cells if growth were the alternative.

1. cAMP

cAMP acts as a negative regulator of sporulation through effects on cAMP-dependent protein kinase activity (for review, see Broach 1991). One set of observations supporting this role comes from conditional-lethal mutants that lack cAMP or cAMP-dependent protein kinase activity. Under restrictive conditions, such mutants sporulate and express *IME1* in the absence of nitrogen limitation (Matsumoto et al. 1983; Smith and Mitchell 1989; Matsuura et al. 1990). Sporulation also occurs, but to a much lesser extent, in the presence of glucose (Matsumoto et al. 1983). Considering that starvation is required for sporulation at both initiation and commitment steps, cAMP depletion must permit these two

Table 1 Roles of genes in meiosis and sporulation

(A) *Genes involved in nutritional control and initiation of meiosis:*			
ARD1	*NAT1*	*RPD3/SDI2*	*SSN6/CYC8*
BCY1	*RAS1*	*SGA1*	*STA1*
DSM1	*RAS2*	*SIN4/TSF3*	*SWI1/ADR6/GAM3*
GAL11/SPT13	*RES1*	*SME2*	*TPK1, 2, 3*
GCD1	*RGR1*	*SME3*	*TUP1*
GUA1	*RIM1*	*SNF1*	*UME1*
IDS2	*RIM4*	*SNF4*	*UME2*
IME1	*RIM8*	*SPD1*	*UME3/SSN8/SRB11*
IME2	*RIM9*	*SPD3*	*UME4/SIN3/RPD1*
IRA1, IRA2	*RIM11/MDS1*	*SPD4*	*UME5/SSN3/SRB10*
MAT	*RIM131*	*SPE2*	*UME6/CAR80*
MCK1/YPK1/IME3	*RIM15*	*SPO8/IME4*	*VAN1, 2*
MSI1	*RME1*	*SPT15*	*YPT1*
(B) *Genes involved in DNA replication, checkpoint control, and progression through meiosis:*			
CDC2/POL3	*CDC20*	*CDC40*	*PRI1*
CDC4	*CDC21/TMP1*	*HFD1*	*RAD9*
CDC5	*CDC23*	*HID1*	*RAD53/MEC2/SAD1*
CDC7	*CDC25*	*KEX2*	*RFA1*
CDC8	*CDC28*	*MEC1/ESR1/SAD3*	*SHP1*
CDC9	*CDC30/PGI1*	*MPS1*	*SPO1*
CDC13	*CDC31*	*MPS2*	*SPO12*
CDC14	*CDC33*	*NDC1*	*SPO13*
CDC16	*CDC35/CYR1*	*NDT80*	*SPO14*
CDC17/POL1	*CDC36*	*POL30*	
CDC19	*CDC39/NOT1*	*POL4*	

(C) *Genes involved in recombination, synapsis, and segregation of chromosomes:*			
BDF1	*MRE2*	*RAD51*	*SAE1*
CEP1	*MRE3*	*RAD52*	*SAE2/COM1*
CHL1	*MRE11/RAD58*	*RAD54*	*SAE3*
DIS1	*MSH2*	*RAD55*	*SEP1*
DMC1/MED1	*MSH3*	*RAD57*	*SIR2*
HFD1	*MSH4*	*RAP1*	*SPO11*
HOP1	*MSH5*	*REC1*	*SRS2*
ISC2	*NDT80*	*REC3*	*TID1*
MEI4	*PMS1*	*REC4*	*TOP2*
MEI5/SPO12T	*PMS2/MLH1*	*REC102*	*XRS2*
MEK1/MRE4	*PMS3/MSH6*	*REC103*	*ZIP1*
MER1	*PSO4*	*REC104*	*ZIP2*
MER2/REC107	*RAD6*	*REC114*	
MER3	*RAD50*	*RED1*	
(D) *Genes involved in ascus formation and spore wall morphogenesis:*			
CAK1	*PRB1*	*SPO5*	*SPR6*
CAL1/SCD2	*SMK1*	*SPO20*	*SPS1*
DIT1	*SPO2*	*SPR1*	*SPS18*
DIT2	*SPO3*	*SPR2*	*SPS100*
GCN1	*SPO4*	*SPR3*	*SSG1*
(E) *Genes that either affect meiosis and sporulation by unknown mechanisms or have no apparent role despite meiosis-specific expression:*			
BDF1	*PRE1*	*SPO9*	*SPO51*
BLM1	*RAD2*	*SPO10*	*SPO53*
GSG1	*RAD6*	*SPO15/VPS1*	*SPO-T1-23*
ISC10	*RPK1*	*SPO16*	*SPS19*
LYT1	*SOD1*	*SPO17*	*SSA3*
PEP4	*SOD2*	*SPO18*	*UBI4*
PRC1	*SPO7*	*SPO50*	*UBR1*

temporally distinct events to occur. A second set of observations comes from mutants with high, constitutive protein kinase activity: These mutants fail to sporulate or express *IME1* in response to starvation (Matsumoto et al. 1983; Matsuura et al. 1990). They also have pleiotropic defects in the starvation response, such as failure to accumulate glycogen or arrest as unbudded cells. These findings, along with more extensive physiological studies of the mutants, led to the simple hypothesis that nitrogen might inhibit sporulation simply by stimulating cAMP accumulation.

Two observations have called this view into question. First, mutants with low constitutive protein kinase A activity (and lacking adenylate cyclase) grow if provided with nitrogen and sporulate properly in response to nitrogen limitation (Cameron et al. 1988). Thus, a cAMP-independent system must be able to stimulate sporulation in response to nitrogen limitation. Second, measurement of cAMP levels indicates that the presence of glucose, rather than the presence of nitrogen, stimulates cAMP accumulation: Addition of glucose to glucose-limited cultures causes a rapid increase in cAMP levels (Eraso and Gancedo 1985), and intracellular cAMP levels drop precipitously as glucose is depleted from the growth medium (Russell et al. 1993). One simple model is that two nutritional signaling pathways govern sporulation, for example, the carbon-responsive cAMP pathway and a non-cAMP pathway which responds to nitrogen. Hyperactivity or inactivity of the cAMP pathway may override the other one. A second model is that the carbon-responsive cAMP pathway governs sensitivity of a nitrogen-sensing pathway. Both of these models account for the finding that carbon limitation can promote sporulation in the presence of nitrogen (Freese et al. 1982); presumably this circumstance would be mimicked by cAMP limitation in the presence of nitrogen.

2. *GTP*

GTP has been proposed to act as a signal molecule in nutritional control of meiosis (Varma et al. 1985; Olempska-Beer and Freese 1987). This idea stems from analysis of cells sporulating in response to guanine limitation in medium containing acetate and nitrogen. GTP pools decrease tenfold within 2 hours after guanine starvation. The decline is not due to general loss of energy charge, because ATP pools remain high. Under these circumstances, there is essentially no decrease in cAMP pools (Olempska-Beer and Freese 1987). Thus, GTP or a metabolic derivative is a candidate for a signal molecule that may act downstream or independently of cAMP.

The response of GTP pools is not specific for guanine limitation: GTP pools drop during methionine starvation (again in acetate- and nitrogen-containing medium), and these conditions also lead to sporulation (Varma et al. 1985). A decline in GTP pools occurs upon limitation for cAMP, thus suggesting that some or all effects of cAMP limitation may result from reduced GTP levels (Pall 1988). Unfortunately, many questions concerning GTP pools remain unanswered, such as their response to nitrogen limitation, where in the sporulation program (initiation, commitment) they may exert an effect, and whether artificial maintenance of high GTP pools might block sporulation.

3. The MCK1, SME3, *and* RIM1/8/9/13 *Pathways*

Mck1p is a serine-threonine-tyrosine protein kinase (Dailey et al. 1990) with three known functions. First, it is as a positive regulator of meiotic gene expression: Increased *MCK1* dosage causes elevated *IME1* expression, and *mck1* null mutations cause reduced *IME1* expression and slow, inefficient sporulation (Neigeborn and Mitchell 1991; Su and Mitchell 1993a). Second, Mck1p is required for ascus maturation: *mck1* mutants accumulate immature asci even when *IME1* is overexpressed, so Mck1p has two distinct roles in the sporulation program. Third, Mck1p has a positive role in mitotic centromere behavior: Increased *MCK1* dosage suppresses partially defective centromere mutations (Shero and Hieter 1991) and temperature-sensitive defects in the centromere-binding proteins Cbf2p (also called Ndc10p or Ctf14p) and Cbf5p (W. Jiang et al. 1995). Conversely, *mck1* null mutations cause benomyl hypersensitivity, chromosome missegregation, and cold-sensitive lethality (Shero and Hieter 1991). Benomyl hypersensitivity and cold-sensitive lethality are not suppressed by *IME1* overexpression. The observation that Mck1p can phosphorylate Cbf2p in vitro makes Cbf2p a good candidate for an in vivo substrate (W. Jiang et al. 1995). Thus, Mck1p may have three substrates, each with a distinct function, or Mck1p (and, perhaps, Cbf2p) may lie in a pathway that has these three diverse functions.

The fact that *mck1* mutants express *IME1*, albeit at reduced levels, indicates that another pathway or gene product must also stimulate *IME1* expression. *RIM1*, *RIM8*, *RIM9*, and *RIM13* define a second *IME1* activation pathway. Recessive mutations in these *RIM* genes cause a number of phenotypes, including cold-sensitive growth, reduced *IME1* expression, and slow inefficient sporulation (Su and Mitchell 1993a). The *rim1/8/9/13* sporulation defect arises from a defect in *IME1* expression, because expression of *IME1* from a heterologous promoter permits *rim1/8/9/13* mutants to sporulate efficiently. These *RIM* gene products were initially thought to act in a single pathway or complex for two reasons. First, the mutants have common phenotypes. Second, *rim rim* double mutants and *rim* single mutants have equally severe phenotypes. These inferences have been supported by more detailed studies of Rim1p. Rim1p is a zinc finger protein (Su and Mitchell 1993b) that exists in two forms that differ by 8 kD (Li and Mitchell 1997). In *rim8, rim9*, or *rim13* mutants, only the larger Rim1p form is detectable. It is thought that Rim8p, Rim9p, and Rim13p are required for carboxy-terminal proteolytic cleavage of Rim1p because carboxy-terminal Rim1p deletion derivatives suppress all *rim8, rim9,* and *rim13* mutant defects (Li and Mitchell 1997).

Mck1p acts independently of Rim1/8/9/13p to stimulate *IME1* expression, because *mck1 rim* double mutants have a more severe sporulation defect than *rim* or *mck1* single mutants (Su and Mitchell 1993a). Analogous observations indicate that Ime4p stimulates *IME1* expression independently of either Mck1p or Rim1/8/9/13p. (Note: These experiments were conducted in strains in which *ime4* mutations do not abolish sporulation; Su and Mitchell 1993b.) The signals that Mck1p and Rim1/8/9/13p transmit are unknown. Present evidence argues that they act independently of Rme1p, in that Rme1p acts through a region of the *IME1* promoter different from that of Mck1p or Rim1/8/9/13p (W. Li, unpubl.), and because an *rme1*Δ mutation does not suppress *mck1* or *rim1/8/9/13* defects. They also appear not to respond to cAMP, because overexpression of cAMP phophodiesterases permits increased *IME1* expression in *mck1* or *rim1* mutants (L. Neigeborn, pers. comm.). Although this observation formally indicates that a cAMP-dependent signal can be transmitted independently of either gene product, it does not rule out the possibility that multiple gene products, including Mck1p and Rim1p, respond to cAMP levels. Similarities in the structure and regulation of Rim1p and PacC, an *Aspergillus* acid-alkaline response regulator (Orejas et al. 1995; Tilburn et al. 1995), suggest that these proteins may represent a conserved signal transduction pathway. Indeed, external pH does affect Rim1p processing, but it is not clear whether pH governs sporulation through these effects on Rim1p (Li and Mitchell 1997).

SME3 is another candidate for a nutritional control gene, based on its regulation and dosage effects (Kawaguchi et al. 1992). *SME3* RNA accumulates to much higher levels in stationary phase cells than in exponential phase cells, regardless of cell type, as might be expected if carbon, nitrogen, growth, or cAMP levels governed *SME3* expression. A high-copy *SME3* plasmid causes increased *IME1* expression and sporulation in the presence of glucose or ammonia. However, an *sme3* mutation does not affect sporulation efficiency. Thus, Sme3p may stimulate sporulation only when it is overexpressed, or perhaps other gene products have overlapping functions. The relationships between Sme3p and other meiotic regulators (such as Mck1p and Rim1/8/9/13p) are unclear.

C. Cell-type Requirement for Entry into Meiosis

Roman et al. (1955) first reported that **a**/α diploids were able to sporulate, whereas diploids of cell-type **a** or α (of genotype **a**/**a** and α/α, respectively) were not. More detailed genetic analysis of the mating-type locus revealed that *MAT***a***1* and *MAT*α*2* were the critical *MAT* genes required for sporulation (Kassir and Simchen 1976; Strathern et al. 1981). *MAT***a***1* and *MAT*α*2* specify subunits of a transcriptional repressor, **a**1-α2, that is required for all known properties of **a**/α cells (**a**1-α2 is described in detail by Herskowitz et al. 1992). In some genetic backgrounds, sporulation can occur despite defects in *MAT***a***1* and *MAT*α*2*. Some such strains have a recessive *rme1* mutation (also called *csp1;* Hopper et al. 1975; Kassir and Simchen 1976), which does not affect mating ability in any obvious way. Others have mutations (e.g., *sir1* through *sir4*) that derepress *HML* and/or *HMR*, the normally silent copies of *MAT* information (Laurenson and Rine 1992); mild or infrequent *HML/R* expression permits sporulation without causing an obvious mating defect. In addition, the mutations *RES1-1*, *rgr1-100*, and *ume1* through *ume6* permit some or all of the sporulation program to occur in haploids or diploids expressing only one mating-type gene (i.e., non-**a**/α cells; Table 1).

Which specific meiotic events require *MAT***a***1* and *MAT*α*2*? Studies comparing diploids or disomes of different cell types indicate that meiotic DNA synthesis, elevated meiotic gene conversion, spore formation, and the sporulation gene expression program depend on **a**1-α2 activity (Roth and Lusnak 1970; Roth and Fogel 1971; Hopper and Hall 1975a; Kassir and Simchen 1976; Wagstaff et al. 1982; Clancy et al. 1983; Percival-Smith and Segall 1984; Kaback and Feldberg 1985; Wang et al. 1987; de Massy et al. 1994). Starved non-**a**/α cells become arrested prior to Start in G_1 phase of the cell cycle. Why is the presence of both *MAT***a** and *MAT*α a useful signal for initiation of meiosis? Typically diploid cells express both *MAT* alleles, whereas haploid cells do not. Diploidy is a prerequisite for the reductional (meiosis I) division, in which pairs of homologous chromosomes segregate; haploid cells lack the genetic material required to complete this division. Thus, the requirement for **a**/α cell-type signals prevents haploid cells from initiating a developmental program that they cannot successfully complete.

1. MAT**a**1 *and* MATα2

a1-α2 governs expression of two genes, *RME1* and *IME4*, to permit meiosis to occur. *RME1* specifies an inhibitor of meiosis that is repressed by **a**1-α2 (Kassir and Simchen 1976; Rine et al. 1981; Mitchell and Herskowitz 1986). *IME4* specifies a positive regulator of meiosis that is ultimately activated by **a**1-α2 (Shah and Clancy 1992). Repression of *RME1* follows a simple paradigm: **a**1-α2 binds to an operator site at –400 (relative to the *RME1*-coding region); this operator is necessary and sufficient for repression (Covitz et al. 1991). The stimulation of *IME4* expression was first proposed to be indirect, through repression by **a**1-α2 of a hypothetical repressor of *IME4*, to be consistent with the known function of **a**1-α2 as a repressor (Shah and Clancy 1992). Two recent observations suggest that the idea of indirect induction may be correct but that it may occur through a slightly more unusual route. First, there is an antisense transcript adjacent to *IME4*, present only in **a** and α haploid cells and not in **a**/α cells. Second, insertions in the *IME4* 3′ region that block accumulation of the antisense transcript permit elevated *IME4* expression (M. Clancy, pers. comm.). Thus, Clancy has suggested that **a**1-α2 may indirectly induce *IME4*, through repression of the antisense mRNA, whose transcription may prevent *IME4* expression in haploid cells.

2. RME1

Three observations indicate that *RME1* transmits a cell-type signal. First, *RME1* expression is repressed by **a**1-α2 (Mitchell and Herskowitz 1986). Second, *rme1* loss-of-function mutations permit sporulation in **a**/α and non-**a**/α cells, implying that the wild-type *RME1* gene negatively regulates sporulation in non-**a**/α cells (Kassir and Simchen 1976; Rine et al. 1981; Mitchell and Herskowitz 1986). Third, expression of *RME1* in **a**/α cells from a multicopy plasmid or constitutive promoter blocks sporulation, so repression of *RME1* in **a**/α cells is required to execute sporulation (Mitchell and Herskowitz 1986; Granot et al. 1989).

Rme1p is a C_2H_2 zinc finger protein that directly represses *IME1*.

Rme1p binds in vitro to a site derived from a region 2 kbp upstream of *IME1* (Covitz and Mitchell 1993). A mutation in this site permits greater *IME1* expression in the presence of Rme1p (e.g., as evidenced by sporulation ability), so binding is relevant in vivo. The fact that repression is not fully relieved by mutation of the site may indicate that there are other Rme1p-binding sites through which repression of *IME1* is exerted (Covitz and Mitchell 1993). Indeed, apparent titration of Rme1p by a high-copy plasmid carrying far upstream *IME1* sequences has been taken to mean that an Rme1p-binding site may lie 4 kbp upstream of *IME1* (Granot et al. 1989). Repression of *IME1* seems to be the only mechanism by which Rme1p inhibits meiosis, because expression of *IME1* from Rme1p-independent promoters permits efficient sporulation in the presence of Rme1p (Kassir et al. 1988; Smith and Mitchell 1989; Covitz et al. 1991).

Rme1p does not act alone to repress *IME1;* in fact, an isolated Rme1p site functions as an Rme1p-dependent activation site, not a repression site. Repression depends on both an Rme1p site and the neighboring 300-bp segment from the *IME1* upstream region. Thus, repression may depend on multiple functional sites, on precise spacing between an Rme1p site and a second site, or on the spacing between an Rme1p site and the target promoter (Covitz and Mitchell 1993). Repression by Rme1p is thought to depend on a local change in chromatin structure, because repression depends on Sin4p and Rgr1p (Covitz et al. 1994), two proteins implicated by other studies in maintenance of chromatin structure (Table 1) (Jiang and Stillman 1992; Y.W. Jiang et al. 1995).

Current observations suggest that repression by Rme1p may be exerted only in starved, nongrowing, non-**a**/α cells, since both *RME1* and *rme1* strains express *IME1* at similar levels during vegetative growth (Kassir et al. 1988; Neigeborn and Mitchell 1991). However, whereas *RME1* strains maintain the same low level of *IME1* RNA after nitrogen starvation, *rme1* strains accumulate 5–30-fold higher *IME1* mRNA levels after starvation (Kassir et al. 1988; Smith and Mitchell 1989). Interestingly, starvation of **a** and α haploid cells results in a 10-fold increase of *RME1* mRNA (Covitz and Mitchell 1993), with greatest accumulation in the G_1 phase of the cell cycle (Toone et al. 1995). Thus, it is possible that elevated levels of Rme1p present after starvation are necessary for the protein to function efficiently in repression. Finally, elevated Rme1p may well have other roles in permitting non-**a**/α cells to respond appropriately to starvation or G_1 arrest, as indicated by the finding that Rme1p is an ancillary activator of transcription of *CLN2*, which specifies a G_1 cyclin (Toone et al. 1995).

3. IME4

Three observations indicate that *IME4* transmits a cell-type-specific signal (Shah and Clancy 1992). First, *IME4* RNA levels are higher in **a**/α cells than in non-**a**/α cells, particularly after starvation. Second, *ime4/spo8* loss-of-function mutations block sporulation or, in some strains, reduce sporulation efficiency (Esposito et al. 1972; Su and Mitchell 1993b), indicating that *IME4* positively regulates sporulation. Detailed analyses of meiotic landmarks in *spo8* mutants demonstrate that *IME4/SPO8* is required for premeiotic DNA synthesis, one of the earliest meiotic events dependent on **a**/α expression (M.A. Esposito and Esposito 1974). Third, expression of *IME4* from a multicopy plasmid permits sporulation in non-**a**/α cells. Ime4p is formally considered to be a positive regulator of *IME1* expression because an *ime4* insertion mutation reduces *IME1* RNA levels and sporulation, but the mechanism by which Ime4p stimulates *IME1* expression is not yet clear.

IME4 may transmit a starvation signal as well as a cell-type signal since *IME4* expression is much greater in starved **a**/α cells than in growing **a**/α cells (Shah and Clancy 1992). However, *IME4* expression is clearly not the only pathway that exerts nutritional control of *IME1* expression. For example, the *RES1-1* mutation permits *IME1* expression in *ime4* null mutants; in such *RES1-1 ime4* strains, *IME1* expression is still stimulated by nitrogen starvation. Therefore, stimulation of *IME1* expression by starvation probably occurs through both *IME4*-dependent and *IME4*-independent pathways. As discussed above, Mck1p and Rim1/8/9/13p may also transmit *IME4*-independent starvation signals.

Ime4p may have a second role in sporulation in addition to activation of *IME1* expression. This inference is based on the behavior of two independently isolated suppressors, *DSM1*, which suppresses *spo8-1,* and *RES1*, which suppresses an *ime4::LEU2* disruption, respectively. The semidominant *DSM1-1* temperature-sensitive allele, located on chromosome VII, permits a partial bypass of the *spo8/spo8* block: At a restrictive temperature, DNA synthesis and intragenic recombination are restored, but not intergenic exchange or spore formation (Fast 1978). Similarly, in the presence of the semidominant *RES1-1* allele, *ime4::LEU2* homozygotes express *IME1* efficiently but sporulate very poorly (Shah and Clancy 1992). In the latter case, introduction of an *IME4* plasmid complements the sporulation defect, so the phenotype is not a consequence of an independent secondary mutation. The conclusion is complicated by the finding that a high-copy *IME1* plasmid suppresses the *ime4::LEU2* sporulation defect efficiently (Shah and Clancy 1992). These observations have been reconciled with the suggestion that overexpression of *IME1* might permit Ime1p to substitute for Ime4p in its second role, whereas normal levels of Ime1p (in the *RES1-1 ime4::LEU2* strain) cannot substitute for Ime4p (Shah and Clancy 1992).

4. RES1

RES1-1 is a partially dominant mutation that permits sporulation of non-**a**/α cells by allowing expression of *IME1*, independently of cell type, after starvation (Kao et al. 1990). *RES-1* does not act, indirectly, through an effect on *IME4* expression because *RES1-1* permits *IME1* expression in an *ime4* null mutant (Shah and Clancy 1992). *RES1-1* has no effect on *rme1-lacZ* expression and is thought to act independently of *RME1* activity because *RES1-1 rme1::LEU2* double mutants have a more severe phenotype (i.e., they sporulate twofold more efficiently in a non-**a**/α background) than *rme1::LEU2* mutants (Kao et al. 1990). If one discounts this twofold difference in sporulation, then the *RES1* gene product may disrupt Rme1p-dependent repression. Otherwise, *RES1-1* may affect either a step downstream from Ime4p or a gene product that influences *IME1* expression independently of Rme1p and Ime4p, such as Mck1p and Rim1/8/9/13p. Because the nature of the *RES1-1* mutation is unclear, it is possible that the *RES1* gene product is either a positive or negative regulator of *IME1* expression. It should also be noted that the relationship between *RES1-1* and *DSM1-1*, an allele that exhibits similar behavior and maps to the general region of the *RME1* locus, is unknown, but it raises the intriguing possibility that both of these semidominant mutations may actually be alleles of *RME1*.

D. Meiotic DNA Synthesis

The first landmark event of the sporulation program is initiation of DNA synthesis. This round of DNA synthesis is generally called "premeiotic" S phase because it precedes the meiotic nuclear divisions. It is defined as the DNA synthesis phase that occurs in starved **a**/α cells and not in non-**a**/α cells (for a review of earlier studies, see Esposito and Klapholz 1981). Although many common genes are utilized in both mitotic and premeiotic DNA synthesis (see below and Table 1), premeiotic DNA synthesis also depends on a number of genes that are dispensable for mitotic DNA synthesis, including *IME1* (Kassir et al. 1988), *SPO7*, *SPO8*, and *SPO9* (M.S. Esposito and Esposito 1974), *MEI1* and *MEI2,3* (Roth 1976), and *SPO17* and *SPO18* (Smith et al. 1988). Conversely, several gene products (e.g., *CDC7*, *CDC28*, *CDC36*, *CDC39*), act prior to mitotic DNA synthesis in the cell division cycle, but after meiotic DNA synthesis in the sporulation program (Schild and Byers 1978; Shuster and Byers 1989), indicating that regulation of DNA synthesis during meiosis is at least, in part, unique.

Meiotic DNA synthesis extends for at least 65 minutes, based on pulse-labeling experiments (Williamson et al. 1983), and is thus three times longer than the mitotic DNA synthesis phase. Despite this apparent difference, the two types of S phases employ the same basic machinery, including polymerase α catalytic subunits (Budd et al. 1989) and primase subunits (Longhese et al. 1993). One simple explanation for S-phase length differences is that different replication origins may be activated. That this explanation is unlikely is based on analysis of replication bubbles in a 200-kb region of chromosome III, demonstrating that the same three origins are active in both mitotic and meiotic S phases (Collins and Newlon 1994a). In addition, several potential origins (cloned as autonomous replication sequence [ARS] elements) remain inactive during both types of S phases. Overall replication fork movement occurs at comparable rates in the two S phases (Johnston et al. 1982b), and there is no indication of local impediments to fork movement in meiosis (Collins and Newlon 1994a). Collins and Newlon (1994a) pointed out that interpretation of pulse-labeling experiments rests on several untested assumptions and that reexamination of meiotic S-phase length may be warranted. At present, however, the basis for lengthening of meiotic DNA synthesis remains unclear.

The meiotic S phase appears to have a broad role in organizing the sporulation program. Experiments with temperature-sensitive DNA synthesis mutants indicate that meiotic DNA synthesis is generally required for genetic recombination, meiosis I, meiosis II, and spore formation (for review, see Esposito and Klapholz 1981; Budd et al. 1989), and the transcription of at least one sporulation-specific gene (Kao et al. 1989). The mechanism by which meiotic DNA synthesis influences other events is not known.

E. Meiotic Gene Expression

Many genes are expressed at much higher levels in starved **a**/α cells than in growing **a**/α cells or in starved non-**a**/α cells. These genes have been called meiotic genes, meiosis-specific genes, or sporulation-specific genes. As indicated in Figure 2, they lie in classes defined by their times of expression: early genes are expressed during prophase, middle genes are expressed during late prophase and the meiotic divisions, and mid-

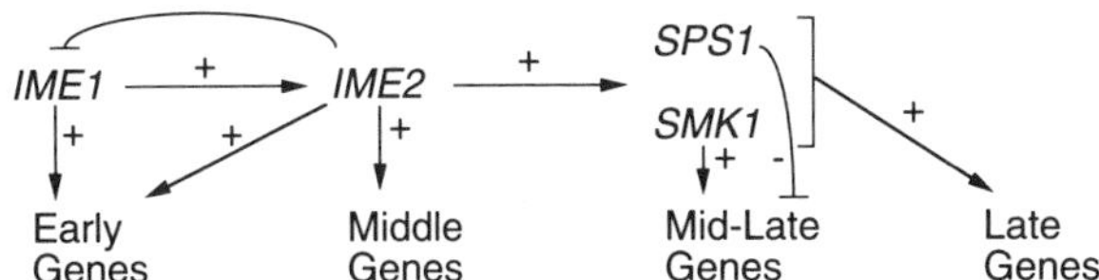

Figure 2 Transcriptional dependency relationships. Ime1p stimulates expression of *IME2* and early genes. Ime2p permits down-regulation of *IME1* and stimulates expression of early and middle genes. Among the middle genes are *SPS1* and *SMK1*. Smk1p stimulates expression of mid-late and late genes; Sps1p inhibits expression of mid-late genes and stimulates expression of late genes.

late genes and late genes are expressed during the meiotic divisions and spore packaging, respectively (for review, see Magee 1987; Mitchell 1994). In addition, many genes expressed during meiosis are also expressed under other circumstances, including genes involved in DNA replication and repair, nuclear division, heat shock, and metabolism (Kaback and Feldberg 1985; Johnston et al. 1986; Madura and Prakash 1990; Raymond and Kleckner 1993a; Szent-Gyorgyi 1995).

Meiotic genes have been identified through strategies based on differential function and differential expression. The first strategy has involved identifying mutations that cause specific defects in sporulation; many such mutations turned out to lie in genes expressed only during sporulation (for review, see Esposito et al. 1991; see, e.g., Wang et al. 1987; Briza et al. 1990; Engebrecht and Roeder 1990; Hollingsworth et al. 1990; Atcheson 1991). The second strategy has involved identifying genes expressed only in sporulating cells, for example, through differential hybridization (Clancy et al. 1983; Percival-Smith and Segall 1984; Gottlin-Ninfa and Kaback 1986), by screening *lacZ* fusion libraries or transposon insertions (Coe et al. 1992; Burns et al. 1994), or by cloning genes encoding sporulation-specific enzyme activities (Yamashita and Fukui 1985; San Segundo et al. 1993)

1. Control of Early Meiotic Genes

The molecular mechanisms responsible for early meiotic gene expression are detailed below. In overview, their regulation occurs as follows: The early meiotic genes are repressed in nonmeiotic cells by a mechanism that depends on a nearby upstream repression sequence (URS1) site and on Ume6p, an URS1-binding protein. This repression also requires Sin3p and Rpd3p, which may act indirectly by binding to (or otherwise modifying) Ume6p. Starvation of **a**/α cells stimulates *IME4* expression; Ime4p (in conjunction with Mck1p- and Rim1p-dependent pathways) stimulates

IME1 expression. Ime1p then binds to the protein kinase Rim11p and becomes phosphorylated. Ume6p is converted into an activator by Ime1p and Rim11p, perhaps by formation of an Ime1p-Ume6p complex. Activation depends on Ume6p and the URS1 site as well as proteins acting through a neighboring UAS_H or T_4C site. One early meiotic gene product, Ime2p, further stimulates early meiotic gene expression independently of Ime1p.

a. URS1 and Neighboring Sites. The URS1 site was first identified as a repression site in the *CAR1* 5′ region through an analysis of *CAR1* expression in nonmeiotic cells (Sumrada and Cooper 1987). Related sites have been identified in regulatory regions of many genes with diverse functions (Buckingham et al. 1990; Luche et al. 1990; Park and Craig 1991; Lopes et al. 1993; Szent-Gyorgyi 1995). Included among these are almost all early meiotic genes (Buckingham et al. 1990; for review, see Mitchell 1994). The element is typically found in the 5′ upstream regions of meiosis-specific genes, but in at least one case (*SPO11*), it exerts it regulatory effect from an internal position within the coding region of the gene (Atcheson 1991). Mutations of URS1 sites at meiotic genes cause low-level constitutivity (Buckingham et al. 1990; Vershon et al. 1992; Bowdish and Mitchell 1993), which has two consequences. In nonmeiotic (vegetative) cells, URS1 mutations cause increased expression relative to the wild-type gene. In meiotic cells, URS1 mutations cause decreased expression relative to the wild-type gene. Therefore, in early meiotic promoters, URS1 behaves as a negative site in nonmeiotic cells and as a positive site in meiotic cells.

At many early meiotic genes, URS1 sites are accompanied by either a UAS_H site or a T_4C site, which function as transcriptional enhancers (Vershon et al. 1992; Bowdish and Mitchell 1993; for review, see Mitchell 1994). In isolation, these sites act as weak upstream activation sequences (UASs) in nonmeiotic cells. For the *HOP1* gene, it has been shown that UAS_H UAS activity is repressed by URS1 in nonmeiotic cells (Vershon et al. 1992). There is a synergistic interaction between URS1 and either accompanying site in meiotic cells, because T_4C and URS1, or UAS_H and URS1, stimulate greater expression together than the sum of their individual contributions (Vershon et al. 1992; Bowdish and Mitchell 1993; Bowdish et al. 1995). It is presently unclear whether most early meiotic promoter regions fit a simple paradigm in which a meiotic UAS governs expression from a typical TATA region. Although the UAS_H site has UAS activity (the intact *HOP1* promoter can confer meiosis-specific expression to a *hop1-lacZ* reporter gene), no promoter segment from *HOP1* has meiosis-specific UAS activity, as assayed with a heterologous TATA and initiation region (Vershon et al. 1992). In contrast, the *IME2* URS1 upstream region does have meiosis-specific UAS activity (Bowdish and Mitchell 1993; Bowdish et al. 1995).

Recent studies have shown that Abf1p is the UAS_H-binding protein responsible for meiotic gene activation (Gailus-Durner et al. 1996). A protein-DNA complex containing UAS_H was first detected by gel-shift assays (Prinz et al. 1995). Such an assay was used to monitor UAS_H-binding activity during purification. Partial protein sequence determination indicated that Abf1p was the binding protein, as verified by anti-Abf1p supershift results (Gailus-Durner et al. 1996). Abf1p is essential for viability, so its functional role in activation at UAS_H could not be confirmed by a null mutation. However, extensive mutagenesis of UAS_H reveals that its functional sequence requirements match the Abf1p-binding site consensus (Gailus-Durner et al. 1996). These findings are of particular significance because Abf1p sites are also required for activation of later classes of meiotic genes (see below).

In addition to the two enhancer elements described above, a 7-bp repression element, called ARE (auxillary repression element), has also been identified. The ARE element is required for full repression by URS1 in the *HSP82* promoter (Szent-Gyorgyi 1995) and in the *SPO13* promoter (Buckingham 1992). Not all early promoters contain URS1 and its associated enhancer and repression sites. For example, the promoter region of *IME1*, one of the earliest meiosis-specific genes to be expressed, lacks URS1 (Shefer-Vaida et al. 1995). Significantly, this promoter has detectable activity in growing (nonmeiotic) cells. Different types of early meiotic promoters may be useful for fine-tuning relative expression levels or kinetics.

b. UME6 *and* RFA1/2/3. *UME6* was identified as a negative regulator of early meiotic genes in nonmeiotic cells, because *ume6* point and null mutations lead to elevated expression of these genes (Bowdish and Mitchell 1993; Strich et al. 1994; Bowdish et al. 1995). The deduced structure of Ume6p suggested that the protein binds DNA and thus likely functions directly as a repressor. For example, it has characteristic residues of C_6Zn_2 zinc-cluster DNA-binding proteins, and the relevant cysteine residues are required for function (Strich et al. 1994). Two lines of evidence indicate that Ume6p in fact binds to URS1, a *cis*-acting element in early meiotic promoters also required for repression: (1) a recombinant Ume6p zinc-cluster domain specifically gel shifts a DNA fragment containing the *SPO13* URS1 element (Strich et al. 1994) and

(2) purified Ume6p directly footprints on URS1 (Anderson et al. 1995). During vegetative growth, Ume6p specifically represses meiotic genes in the early expression class, including *SPO11*, *SPO13*, *SPO16*, and *IME2*; *ume6* mutations do not permit vegetative expression of middle and mid-late genes, such as *SPO12*, *SPS2*, *SPS4*, and *DIT1* (Strich et al. 1994; Hepworth et al. 1995; Steber and Esposito 1995). On the other hand, Ume6p repression is not confined to meiotic genes; it also represses several nonmeiotic, URS1-containing genes, including *CAR1* and *INO1* (Park et al. 1992; Lopes et al. 1993; Strich et al. 1994). These findings argue strongly that Ume6p is either a repressor or the DNA-binding subunit of a repression complex that acts at URS1 sites in nonmeiotic cells.

In light of these observations, it is surprising that a *ume6* null mutation does not affect one of the major URS1-protein complexes detectable in crude extracts (Park et al. 1992). This complex, first called BUF (for *b*inding *U*RS1 *f*actor), has been purified and characterized in detail (Luche et al. 1992, 1993). Cloning of genes for the major BUF subunits has revealed that they are Rfa1p and Rfa2p, two subunits of replication factor A (RF-A, also known as RP-A; Luche et al. 1993). The third RF-A subunit is also present in BUF preparations (Luche et al. 1993). RF-A is best known as a single-stranded DNA-binding protein required for DNA replication (Brill and Stillman 1991). In principle, RF-A might bind only to single-stranded URS1 oligonucleotides that contaminate the double-stranded URS1 probes. However, Luche et al. (1993) have demonstrated that single-stranded oligonucleotides do not compete for RF-A binding to URS1 double-stranded probes. Because RF-A is essential for viability (Brill and Stillman 1991), it is not possible to test RF-A null mutants for defects in URS1 site activities. Thus, the role of RF-A at URS1 sites in vivo is unclear at present.

UME6 is transcribed during both mitosis and meiosis, suggesting that its repression function is modified during meiosis (Strich et al. 1994). Indeed, Ume6p has been proposed to also function as a positive regulator of early meiotic genes in meiotic cells (Bowdish and Mitchell 1993; Strich et al. 1994; Bowdish et al. 1995; Steber and Esposito 1995). This idea is supported by a number of diverse observations. First, Ime1p is unable to stimulate *IME2* UAS activity in a *ume6* mutant (Bowdish and Mitchell 1993). *ume6* mutants exhibit a low level of expression of other early meiotic transcripts, for example, *SPO11* and *SPO13*, which also do not increase in response to *IME1* expression (Bowdish et al. 1995; Steber and Esposito 1995). Second, most *ume6* cells arrest during prophase early in meiosis; they fail to form recombinants or undergo haploidization and display significant delays in the induction of mid and mid/late transcripts, for example, *SPO12*, *SPS2*, and *DIT1* (Steber and Esposito 1995). Third, the Ume6p-binding site (URS1) functions as a weak, Ime1p-dependent activation sequence in starved cells (Bowdish et al. 1995). Fourth, the two functions of Ume6p, repression in nonmeiotic cells and activation in meiotic cells, have been separated by a *ume6* missense mutation first called *rim16-12* (Mitchell and Bowdish 1992; Bowdish et al. 1995). Repression is unaffected in the mutant, because early meiotic genes are not expressed during vegetative growth and the URS1 site functions as a repression site. In contrast, activation by Ime1p is severely defective and sporulation levels are reduced. Finally, Ume6p fusions to the DNA-binding domains of lexA or Gal4p can activate transcription of reporter genes fused to the lexA- or Gal4p-binding sites, dependent on the presence of Ime1p (Bowdish et al. 1995; Rubin-Bejerano et al. 1996). In the case of lexA-Ume6p, it has been shown that activation is abolished if the *rim16-12* substitution is introduced into the fusion. Thus, activation by the fusions and activation of early meiotic genes have similar genetic requirements. These results taken together strongly argue that Ume6p functions as a positive regulator in the presence of Ime1p. Activation by lexA-Ume6p occurs through a lexA-binding site, so it is unlikely that URS1-bound RF-A is absolutely required for meiotic gene activation.

It should be noted that some variation has been observed in the level of expression of early meiotic transcripts in different *ume6* mutants during vegetative growth. For example, the *ume6-2* point mutation expresses several early meiotic transcripts during vegetative growth at levels comparable to sporulating wild-type cells (Strich et al. 1994), whereas null alleles in different strain backgrounds exhibit derepressed vegetative levels significantly lower than sporulating wild-type cells (Bowdish et al. 1995; Steber and Esposito 1995). These differences may be due to genetic background, the nature of the *ume6* alleles, or the synchrony of the sporulating wild-type populations used as positive controls.

c. IME1 *and* RIM11. Ime1p is a positive regulator of most or all meiotic genes, because *ime1* null mutations prevent detectable expression of all meiotic genes that have been examined (Engebrecht and Roeder 1990; Mitchell et al. 1990; Yoshida et al. 1990; Galbraith and Malone 1992; Pittman et al. 1993), with the exception of *IME4* (Shah and Clancy 1992), and block premeiotic DNA synthesis, gene conversion, and spore formation (Kassir et al. 1988). *IME1* is expressed at very low levels in growing cells and at 5–30-fold higher levels after starvation in **a**/α cells (Kassir et al. 1988; Smith and Mitchell 1989). *IME1*, like most meiotic genes, is expressed only transiently during meiosis, so the magnitude estimated for its expression during meiosis reflects both the increase per cell and the synchrony of the sporulating population.

The mechanism of Ime1p action is not known, but suggestive experiments support the simple idea that Ime1p functions as a positive transcription factor. Ime1p is a tyrosine-rich protein that does not resemble any known transcription factors (Smith et al. 1990; Sherman et al. 1993). However, fusion of Ime1p to either the lexA or Gal4p DNA-binding domain yields a potent transcriptional activator (Smith et al. 1993; Mandel et al. 1994). There are four correlations between natural Ime1p function and transcriptional activation by these fusion proteins. First, *ime1* missense mutations that prevent sporulation impair transcriptional activation of lexA-Ime1p fusions. Second, deletion of the tyrosine-rich Ime1p activation domain impairs natural Ime1p function. Third, replacement of the Ime1p tyrosine-rich activation domain with conventional acidic activation domains preserves natural Ime1p function. Fourth, transcriptional activation by lexA-Ime1p fusions depends on Rim11p, a protein kinase that phosphorylates Ime1p and is required for meiotic gene expression and sporulation (see below). The observation that an Ime1p-β-galactosidase fusion protein is concentrated in the nucleus (Smith et al. 1993) is also consistent with a direct role for Ime1p in transcription. Does Ime1p act only to stimulate early meiotic gene expression? The requirement for Ime1p in sporulation extends until commitment to haploidization, as assessed with an *ime1* temperature-sensitive mutation (reported by Mandel et al. 1994). By that point, early genes have been expressed for some time and may even be shut off. Thus, Ime1p may stimulate expression of several classes of meiotic genes or have some other role in sporulation that is unrelated to gene expression.

Rim11p, a constitutively expressed protein kinase (Bowdish et al. 1994), has a close functional relationship with Ime1p. *ime1* and *rim11* mutations block expression of early meiotic genes (Bowdish et al. 1994), and both mutations can be partially suppressed through expression of *IME2* (Smith and Mitchell 1989; Yoshida et al. 1990; Mitchell and Bowdish 1992). The fact that Rim11p is required for transcriptional activation by lexA-Ime1p suggested that Rim11p might modify Ime1p (Smith et al. 1993). Indeed, Ime1p is detectable as a phosphoprotein in Rim11p immune complexes. Phosphorylation of Ime1p in vitro depends on Rim11p protein kinase activity and is diminished by an *ime1* missense mutation that reduces biological activity (Bowdish et al. 1994). Thus, there is a good correlation between the ability of Rim11p to phosphorylate Ime1p and the ability of Ime1p to activate meiotic genes.

Rim11p is a close structural relative to Mck1p kinase. Indeed, *RIM11* was isolated as a multicopy clone that suppresses an *mck1* mutation (Puziss et al. 1994). Suppression is broad: *RIM11* restores benomyl resistance, low-temperature growth, and *IME1* promoter activity (Puziss et al. 1994; K.S. Bowdish and A.P. Mitchell, unpubl.). There is no evidence that Rim11p normally has a role in centromere function similar to Mck1p, because *rim11* mutants have none of the growth defects of *mck1* mutants, and a *rim11 mck1* double mutant has equally severe phenotypes as an *mck1* single mutant (Puziss et al. 1994). On the other hand, *rim11* and *mck1* mutations both cause reduced *IME1* mRNA levels, so it is possible that Rim11p and Mck1p have a common role in stimulating *IME1* expression. Puziss et al. (1994) have shown that Rim11p and Mck1p will phosphorylate an artificial substrate for glycogen synthase kinase-3, their mammalian homolog, so suppression probably results from overlapping substrate specificity.

Rim11p is not the only gene product that appears to have roles in both *IME1* expression and activity: Ume6p does, as well. A *ume6* null mutation causes elevated *IME1* mRNA levels (Bowdish et al. 1995; Steber and Esposito 1995), and the nonactivatable *rim16-12* allele of *UME6* causes reduced *IME1* mRNA levels (Bowdish et al. 1995). Ume6p may act through an unusual or indirect mechanism at the *IME1* promoter, because there is no obvious URS1 site flanking *IME1*. The argument that Rim11p and Ume6p are required for Ime1p activity (in addition to expression) comes from strains that express *IME1* from the *GAL1* promoter: Ime1p polypeptide levels are the same in wild-type, *rim11*, and *rim16-12* strains, but Ime1p is unable to activate a meiotic promoter or stimulate sporulation in the *rim11* and *rim16-12* strains (Mitchell and Bowdish 1992; Bowdish and Mitchell 1993). These observations are consistent with the recent report that expression of Ime1p in growing cells stimulates *IME1* promoter activity (Shefer-Vaida et al. 1995). Thus, Ime1p, Rim11p, and Ume6p apparently stimulate most early meiotic promoters through the Ume6p-binding site (URS1) and the *IME1* promoter through an as yet undefined site.

What is the mechanism by which Ime1p, Rim11p, and Ume6p promote induction of meiotic transcription? The simplest view at present is that Ime1p activates early meiotic genes by binding to Ume6p and providing a transcriptional activation domain, thereby converting Ume6p from a repressor to an activator (Smith et al. 1993; Bowdish et al. 1994, 1995; Steber and Esposito 1995). Phosphorylation of Ime1p by Rim11p may be required for exposure of the activation domain or for interaction with Ume6p. Support for this view has come from a recent two-hybrid analysis demonstrating a Rim11p-dependent interaction between Ime1p and Ume6p (Rubin-Bejerano et al. 1996). An alternative possibility is that Ime1p acts as a regulatory subunit for Rim11p kinase, permitting phosphorylation of Ume6p and conversion of Ume6p to an activator. Further studies are required to determine more precisely the roles of Ime1p and Rime11 in this process.

d. SIN3 *and* RPD3. *sin3* (also known as *rpd1* and *ume4*) and *rpd3* null mutations permit expression of early meiotic genes in vegetative cells, independently of **a**1-α2 or Ime1p (Strich et al. 1989; Vidal and Gaber 1991; Vidal et al. 1991; Bowdish and Mitchell 1993). Thus, Sin3p and Rpd3p are formally negative regulators of early genes. They act in the same functional pathway, because *sin3 rpd3* double mutants have the same phenotypes as either single mutant (Vidal and Gaber 1991). In principle, they may act downstream or independently of Ime1p. The finding that Ime1p can still stimulate *IME2* promoter activity in a *sin3* null mutant argues that Ime1p and Sin3p act independently (Bowdish and Mitchell 1993). Moreover, if Ime1p acts through Sin3p and Rpd3p, then one might expect that *sin3* and *rpd3* mutations would permit sporulation of *ime1* mutants, but *sin3* and *rpd3* mutations cause a sporulation defect even in *IME1*-containing strains (Strich et al. 1989; Vidal and Gaber 1991; Vidal et al. 1991). The sporulation defect may result to a slight extent from derepressed *RME1* expression in the mutants (Vidal and Gaber 1991). However, a more compelling explanation is that *sin3* mutations cause delayed expression or failure to express later meiotic genes (Yoshimoto et al. 1992; A. Helms and R.E. Esposito, unpubl.) and block sporulation at late stages of meiosis after genetic recombination (A. Helms and R.E. Esposito, unpubl.). *sin3* and *rpd3* null mutations also cause elevated repressed levels and diminished derepressed levels of expression of several nonmeiotic genes (Breeden and Nasmyth 1987; Sternberg et al. 1987; Vidal and Gaber 1991; Vidal et al. 1991; Hudak et al. 1994). Therefore, Sin3p and Rpd3p have both positive and negative regulatory roles in nonmeiotic gene expression.

In meiosis, Sin3p may function more directly as a repressor and contribute to nutritional regulation of early meiotic genes. *SIN3* mRNA levels are higher in growing cells than in nutrient-limited stationary-phase cells (Wang et al. 1990), suggesting that nutrient limitation may create a permissive state for meiotic gene expression by down-regulation of Sin3p-dependent repression. The observation that Sin3p is concentrated in the nucleus initially suggested that it has a direct role in transcription (Wang et al. 1990). More recent studies demonstrate that a lexA-Sin3p fusion protein acts as a repressor of a reporter gene containing lexA-binding sites (Wang and Stillman 1993). Whether repression by lexA-Sin3p requires Rpd3p is not known. Neither Sin3p nor Rpd3p resembles DNA-binding proteins (Wang et al. 1990; Vidal and Gaber 1991; Vidal et al. 1991). What might be the DNA-binding subunit of a hypothetical Sin3p repression complex? From the standpoint of early meiotic genes, Ume6p seems to be the best candidate. Indeed, recent studies have shown that Ume6 repression is dependent on Sin3p (K. Struhl; I. Rubin-Bejerano and Y. Kassir; both unpubl.). This idea is also consistent with the observation that URS1 mutations relieve repression by Sin3p of the *IME2* UAS (Bowdish and Mitchell 1993). However, the spectrum of nonmeiotic genes affected by *ume6* mutations differs from those affected by *sin3* and *rpd3* mutations (Strich et al. 1994). Thus, Sin3p and Rpd3 might act in conjunction Ume6p and other DNA-binding proteins or with a different URS1-binding protein, such as RF-A.

e. UME2, UME3, *and* UME5. The abundance of early meiotic transcripts is exquisitely regulated by control of transcription and RNA stability. Several early transcripts have among the shortest half-lives of yeast messages of approximately 3 minutes (Surosky and Esposito 1992). Thus, when transcriptional induction ceases, they rapidly disappear. For *SPO13*, translation is required for mRNA instability, so *spo13* nonsense alleles produce more stable transcripts (Surosky and Esposito 1992). In addition, early transcripts are more unstable in glucose than in acetate media; this phenomenon may be one mechanism by which glucose inhibits sporulation (Surosky and Esposito 1992).

Ume2p and Ume5p are two negative regulators of early meiotic genes that act at the level of RNA stability (Strich et al. 1994). In *ume2* and *ume5* mutants, several early, middle, and mid-late meiotic transcripts are twofold more stable, whereas a variety of nonmeiotic transcripts are unaffected (Strich et al. 1994). A *ume2 ume5* double mutant has no more severe a phenotype than either single mutant, indicating that Ume2p and Ume5p act in the same pathway (Surosky et al. 1994). Ume2p and Ume5p apparently act independently of the translation-destabilization pathway, because a *spo13* nonsense allele produces a more stable transcript than wild-type *SPO13* in a *ume5* mutant. Instead, Ume5p and Ume2p appear to act in the glucose-regulated RNA turnover pathway, since *SPO13* mRNA is equally stable in glucose and acetate media in a *ume5* mutant (Surosky et al. 1994).

Ume5p, also known as Ssn3p, is a cyclin-dependent protein kinase homolog (Surosky et al. 1994; Kuchin et al. 1995). It is associated with a cyclin C homolog, Ume3p (or Ssn8p), which is required for Ume5p kinase activity in immune complexes (Kuchin et al. 1995; R. Strich, unpubl.). Whether Ume2p is also required for Ume5p kinase activity is not known. Ume5p and Ume3p are part of the SRB complex, which copurifies with RNA polymerase II (reported by Kuchin et al. 1995). The finding that Ume5p and Ume3p are required for transcriptional expression of many diverse genes is consistent with their having a role in RNA polymerase II activity (Kuchin et al. 1995). This close biochemical and functional relationship between Ume5p and Ume3p raises the question of how only Ume5p (and not Ume3p) affects RNA stability (Surosky and Esposito 1992). The nature of the *ume3* and *ume5* mutants in which RNA stabilty was measured has not been established, so this issue may be resolved by measurement of RNA stability in bona fide null mutants. It is also possible that Ume5p associates with Ume3p to govern transcription and with a different cyclin homolog to govern RNA stability.

f. IME2. *IME2* is one of the early meiotic genes activated by Ime1p. Ime2p can activate most of early meiotic genes that Ime1p activates, as indicated by two observations (Mitchell et al. 1990). First, *ime2* null mutants accumulate early meiotic transcripts slowly after starvation (despite their overexpression of *IME1*; see below). Second, expression of *IME2* (from an *IME1*-independent *GAL1-IME2* fusion gene) in an *ime1* null mutant permits expression of many other early meiotic genes. One exception, *REC114*, is activated by Ime1p but not by Ime2p (Pittman et al. 1993). Therefore, activation by Ime2p is independent of Ime1p.

Ime2p is a nuclear serine-threonine protein kinase (Yoshida et al. 1990; Kominami et al. 1993). It likely acts to stimulate early meiotic gene transcription, rather than to stabilize transcripts, because it can stimulate its own promoter through 5′-nontranscribed sequences (Bowdish and Mitchell 1993). Although the mechanism through which Ime2p activates early meiotic genes is unclear, it stands to reason that the target site is common to early meiotic genes. A very simple idea is that Ime2p may phosphorylate a URS1-binding protein, such as Ume6p or RF-A, to effect activation. Ime2p has two additional roles: It is a negative regulator of *IME1* RNA accumulation, because *ime2* null mutants overexpress *IME1* (Smith and Mitchell 1989), and it is a positive regulator of middle and late meiotic genes, because *ime2* null mutants are unable to express these genes (Mitchell et al. 1990; Hepworth et al. 1995; Sia and Mitchell 1995). Thus, Ime2p first replaces Ime1p as an activator of early genes and then promotes a transition in which early genes are shut off and middle genes are expressed. It has been unclear how Ime2p carries out different functions (i.e., activation of early and middle genes) at different times. Recently, the early and later functions of Ime2p have been separated genetically through *ids2* mutations (Sia and Mitchell 1995). *ids2* null mutations interefere with middle gene expression but have no effect on activation of early genes by Ime2p. Ids2p levels are higher in growing cells than in meiotic cells, so Ids2p may act indirectly to influence Ime2p-dependent functions. Nevertheless, *IDS2* provides a genetic basis for the temporal change in Ime2p-dependent functions.

g. MER1 *and meiosis-specific splicing.* Although most meiosis-specific gene products are regulated at the transcriptional level, the *MER2* gene product is restricted to meiotic cells through a meiosis-specific splicing reaction. *MER2* (also known as *REC107*; Cool and Malone 1992) is essential for meiotic recombination and production of viable spores. It is transcribed in both vegetative and meiotic cells but is spliced only in meiotic cells (Engebrecht et al. 1991). Splicing depends on *MER1*, an early meiotic gene (Engebrecht and Roeder 1990; Engebrecht et al. 1991). Expression of *MER1* in vegetative cells permits efficient *MER2* splicing, suggesting that Mer1p might be the only meiosis-specific product required for this splicing reaction. On the other hand, it recently has been shown that *MRE2*, which encodes a protein with two RNA-binding motifs, is also required for the splicing of the *MER2* message. *mre2* mutants are phenotypically similar to *mer1* mutants (Engebrecht and Roeder 1990; Ajimura et al. 1993), and introduction of an intronless version of *MER2* into *mre2* mutants similarly restores synapsis and gene conversion without correcting the defect in reciprocal recombination (H. Ogawa, pers. comm.).

The *MER2* 5′ splice junction differs from the consensus yeast donor sequence (Engebrecht et al. 1991), and a mutation to consensus bypasses the Mer1p requirement for *MER2* splicing (Nandabalan et al. 1993). *MER2* is also unusual in having a fairly long 5′ exon, deletion of which also bypasses the Mer1p requirement (Nandabalan and Roeder 1995). Mer1p binds in vitro to sequences within the 5′ exon and intron and may either recruit the general splicing machinery or alter *MER2* RNA secondary structure to facilitate interaction with splicing machinery (Nandabalan and Roeder 1995). Finally, genetic studies argue that Mer1p may have other substrates, in addition to *MER2* RNA. *mer1* mutations cause recombination defects and spore inviability, as do *mer2* mutations (Engebrecht and Roeder 1989). Expression of a *MER2* cDNA only partially suppresses *mer1* defects (Engebrecht et al. 1991). Thus, it seems likely that Mer1p has an additional function in meiosis, such as in splicing another precursor RNA.

2. *Control of Middle and Late Meiotic Genes*

Toward the end of meiotic prophase, early meiotic genes are shut off and a new set of genes, the middle genes, are expressed. Included among the middle genes are *SPS1* and *SMK1*, which specify protein kinases required for spore wall formation (Friesen et al. 1994; Krisak et al. 1994), and *SPO12*, which is required for reductional (meiosis I) division (Malavasic and Elder 1990; S. Frackman and R.E. Esposito, unpubl.). Shortly afterward mid-late genes are expressed, including *DIT1* and *DIT2*, which are divergently transcribed genes required for synthesis of spore wall dityrosine (Briza et al. 1990). Finally, late genes are expressed, including *SPS100*, which is required for spore wall maturation (Law and Segall 1988), *SPR1*, which specifies an exoglucanase (Muthukumar et al. 1993), *SGA1*, which specifies a glucoamylase (Yamashita and Fukui 1985), and *SPR3*, which specifies a homolog of bud neck filament proteins (Ozsarac et al. 1995). Because expression of middle, mid-late, and late genes from different laboratoaries has not been extensively compared, there is some uncertainty about these expression-class assignments.

Regulation of these sets of genes is not understood in detail, although analysis of regulatory sequences, described below, provides a starting point. Current understanding provides at least a framework to explain how successive sets of genes are activated (Fig. 2). Ime2p, an early gene product, stimulates middle gene expression (Mitchell et al. 1990; Hepworth et al. 1995; Sia and Mitchell 1995). Sps1p and Smk1p, two middle gene products, stimulate late gene expression (Friesen et al. 1994; Krisak et al. 1994). (Regulation of mid-late genes does not fit in as neatly and is discussed in the section on Sps1p and Smk1p.) The three regulators are all protein kinases, so they presumably act indirectly to stimulate expression of target genes. They may phosphorylate transcription factors or act more indirectly, such as by stimulating an event to which gene expression is coupled.

a. Middle/Late Promoter Sequences. The promoters of *SPS4, SGA1, SPR3,* and *SPS18-SPS19* have been analyzed in some detail (Kihara et al. 1991; Coe et al. 1994a,b; Hepworth et al. 1995; Ozsarac et al. 1995). Two observations distinguish these genes from early meiotic genes: They are expressed later than early genes, and (for *SPS4* and *SGA1*) their expression is much more dependent on *IME2* than is expression of early genes. Two functional sequence elements have been identified in *SGA1* 5′sequences (Kihara et al. 1991). One functions as a UAS when inserted into a heterologous promoter. The reporter gene containing the *SGA1* UAS is repressed by either glucose or ammonia. The second element (called the *SGA1* NRE) functions as a negative site in a heterologous promoter. The reporter gene containing the *SGA1* NRE is regulated similarly to a middle meiotic gene: Expression depends on starvation, **a**1-α2, and *IME1* and *IME2*. Thus, it appears that redundant controls, operating through the UAS and NRE, restrict *SGA1* expression to starved cells. A site related to the *SGA1* NRE lies far upstream of *SPS4* (Kihara et al. 1991), but deletion of NRE homology does not affect *SPS4* expression (Hepworth et al. 1995). Thus, the site at *SPS4* may be either redundant or nonfunctional.

SPS4 has a single 15-bp region that functions as a sporulation-specific UAS (Hepworth et al. 1995). The 15-bp element is centered 125 bp upstream of the *SPS4* RNA start site. An adjacent upstream sequence improves UAS activity considerably but does not provide UAS activity on its own. Both the *SPS4* UAS and the flanking stimulatory sequence are highly homologous to sites flanking the mid-late *DIT1* gene. Thus, the *SPS4* regulatory sites may be employed for sporulation-specific activation of several genes. Like the URS1 site of early genes, the *SPS4* UAS may bind a vegetative cell protein that becomes an activator during meiosis (Hepworth et al. 1995). Two observations indicate the presence of a binding protein in vegetative cells. First, *SPS4* UAS-binding activity is detectable in extracts of nonmeiotic cells; no binding occurs to a mutant UAS derivative. Second, the *SPS4* UAS causes weak repression in nonmeiotic cells when inserted into the nonmeiotic *CYC1* promoter; no repression is exerted by the mutant UAS derivative. Repression may result simply from protein binding and may not mean that the UAS acts as a repression site in nonmeiotic cells (Hepworth et al. 1995).

The *SPR3* 5′ region includes two sites found in other later genes (Ozsarac et al. 1995): There is a 15/19-bp match to a site from the *SPR6* 5′ region and a palindrome, PAL, that is present in the *SPS1*, *SPS19*, and *SPS100* 5′ regions. There is also a binding site for the transcription factor ABF1. Mutational analysis indicates that the ABF1 site is required for *SPR3* expression during sporulation. Ozsarac et al. (1995) have identified ABF1-binding sites flanking meiotic genes of early, middle, and late expression classes, so ABF1 may have a broad role in activation of meiotic genes. For example, ABF1 may act in conjunction with proteins at the PAL site to stimulate meiosis-specific transcription (Ozsarac et al. 1995).

The promoter of the divergently transcribed *SPS18* and *SPS19* genes has also been analyzed in some detail (Coe et al. 1994a,b). A 25-bp intragenic region is required for expression of both genes. The region does not resemble known protein-binding sites. Whether this region is sufficient to confer sporulation-specific expression is unknown.

b. SPS1 and SMK1. *SPS1* and *SMK1* specify protein kinases that have a regulatory role in spore packaging. Both genes are expressed only during meiosis, as members of the middle gene class (Percival-Smith and Segall 1984; Friesen et al. 1994; Krisak et al. 1994). Sps1p is a homolog of the protein kinase Ste20p, which acts upstream of the pheromone-responsive MAP kinase cascade (Leberer et al. 1992; Ramer and Davis 1993). Smk1p is a homolog of MAP kinases such as Fus3p and Kss1p (see Sprague and Thorner 1992), which are downstream components of MAP kinase cascades. Both *sps1* and *smk1* null mutations cause defects in spore wall formation. These defects may arise, in part, from the reduced expression of late genes observed in the mutants (Friesen et al. 1994; Krisak et al. 1994). On the basis of their common functions and structural features, it has been proposed that these gene products may act in a sporulation-specific MAP kinase cascade (Friesen et al. 1994; Krisak et al. 1994).

Sps1p and Smk1p have opposite roles in expression of the mid-late gene *DIT1:* An *sps1* mutant overexpresses *DIT1*, whereas an *smk1* mutant underexpresses *DIT1*. This apparent phenotypic difference may indicate either that Sps1p has a second function in addition to the presumed stimulation of Smk1p activity or that Smk1p has an Sps1p-independent function. Recent studies have shown that Ste20p may have a function in cytokinesis of vegetative cells that is independent of the pheromone-response MAP kinase cascade (Cvrčková et al. 1995). Such a second role for Sps1p may explain the different roles of Sps1p and Smk1p in *DIT1* expression.

III. PROPHASE I AND MEIOTIC RECOMBINATION

A. Classical View of Meiotic Pairing and Exchange in Eukaryotes

A striking feature of meiosis is the high level of recombination that occurs during the prolonged meiotic prophase. Crossovers between homologs are crucial for appropriate segregation of the bivalents in meiosis I, for the chiasmata that are created by crossing over act as sites of attachment between the homologous chromosomes during their alignment on the metaphase plate (for review, see von Wettstein et al. 1984). When the meiosis I spindle begins to exert tension on kinetochores toward opposite poles, the bonding provided by the chiasmata provides the resistance needed to stably align the chromosomes equidistant between the poles (Nicklas 1977). At anaphase I, chiasmata are resolved, permitting the chromosomes to be pulled to their respective poles. In the absence of crossing over, no chiasmata are formed, alignment fails, and homologs undergo nondisjunction. Therefore, mechanisms are required to ensure the formation of at least one chiasma between each pair of homologs. During the prolonged meiotic prophase in most eukaryotes, homologs assume a side-by-side alignment as the proteinaceous ribbon-like structure called the synaptonemal complex (SC) forms between them. This tripartite structure consists of a medial element flanked by two lateral elements that are intimately associated with the chromatin. It is during this stage of meiosis that meiotic recombination takes place. The importance of synapsis in recombination is further emphasized by the finding that, for organisms that have SC in only a limited portion of the genome, recombination is similarly restricted to that region; this may include cases where no meiotic recombination takes place in one of the sexes, such as in male *Drosophila* (for review, see von Wettstein et al. 1984).

What is the functional relationship between SC formation and recombination? Is SC assembly a prerequisite for meiotic recombination or might SC formation and recombination proceed in parallel as two independent processes? It is widely accepted on the basis of the correlation between SC and recombination proficiency in numerous organisms that SC has an active role in recombination, and yeast is no exception. In yeast, 16 segments of SC can be observed, one for each chromosome (Byers and Goetsch 1975; Dresser and Giroux 1988). Most mutations in yeast that disrupt SC formation lead in parallel to reductions in the levels of meiotic recombination (as detailed below). Furthermore, no SC is detected in the nucleolus, correlating with the relative absence of recombination between the rDNA repeats present here (Petes and Botstein 1977). Conversely, an increased extent of time spent in pachytene (when SCs are present) may account for observed increases in recombination. For example, Davidow and Byers (1984) found that many laboratory strains (designated as homozygous for *pac1*) become arrested at pachytene if the meiotic cultures are transferred to elevated temperature (36.5°C). The arrested cells exhibit progressive increases in gene conversion and postmeiotic segregation with increasing times of arrest. Mutations of several *CDC* functions (Horesh et al. 1979; Shuster and Byers

1989) similarly cause pachytene arrest and result in elevated levels of recombination, suggesting that prolonged presence of SC results in increased levels of recombination.

There are, on the other hand, many findings indicating that SC formation and recombination are independent of one another: (1) In *Schizosaccharomyces pombe* (Egel-Mitani et al. 1982; Bahler et al. 1993), *Ustilago maydis* (Fletcher 1981), *Aspergillus* (Egel-Mitani et al. 1982), and *Tetrahymena* (Orias 1986), genetic recombination occurs without any visible SC. (2) In *S. cerevisiae*, homologous sequences located at nonhomologous positions in the genome efficiently recombine during meiosis despite the fact that SCs are not evident except between homologs (Jinks-Robertson and Petes 1986). However, temporary SC formation between nonhomologous sequences cannot be ruled out in these cases of "ectopic recombination." (3) Certain mutants that are strongly defective in SC formation (e.g., *red1*, *hop1*) are not completely devoid of recombination. Again, no definitive conclusion can be drawn, since components of the SC may persist and continue to function independently of the fully assembled structure.

B. Recombination Mechanisms in Yeast

Classical genetic studies have established general principles about the nature of meiotic recombination, thereby leading to the development of useful models for genetic exchange (for review, see Petes et al. 1991). Analysis of meiotic end products has shown that nonreciprocal recombination (gene conversion) and reciprocal recombination (crossing over) take place after DNA replication. Crossing over involves the reciprocal breakage and rejoining of chromatids, whereas gene conversion entails the transfer of sequence information from one chromatid to another. These two types of recombination appear to be interdependent since approximately half of the gene conversion events detected in unselected tetrads show an associated crossing over (Fogel and Hurst 1967), and selected crossovers for closely flanking markers frequently display gene conversion (Borts and Haber 1987; Willis and Klein 1987; Symington and Petes 1988). A reasonable interpretation of these phenomena is that gene conversion and crossing over are mechanistically related, and many recombination models postulate an intermediate that can be resolved to yield conversion either with or without associated crossing over. This intermediate is believed to contain heteroduplex DNA, either strand of the DNA duplex being derived from each parental chromatid. At sites where the sequences differ, the resulting mismatches may stimulate repair processes, leading either to gene conversion or to restoration of the parental genotype.

Many specific pathways of molecular events have been postulated to explain how meiotic recombination takes place. Although the Aviemore model (Meselson and Radding 1975) was long believed to account for the association of gene conversions with sites of crossing over, the important role of double-strand breaks (DSBs) in meiotic recombination has come to dominate thinking about meiosis in recent years (Szostak et al. 1983; Sun et al. 1989; Kleckner et al. 1991). For the present review, which focuses on the cytological aspects of meiosis, we will simplify our considerations by limiting our attention to the double-strand break model (Fig. 3). Briefly, although recombination might be initiated by interruptions either in a single strand (a nick) or in both strands (a DSB), the present model posits that a DSB is the predominant initiating event in meiosis. This is followed by resection of the 5′ends to give exposed 3′ends, which then can invade the homologous chromosome to create an intermediate containing two Holliday junctions. Some heteroduplex would be created by the displacement of one of the parental strands by the invading single strand, whereas further heteroduplex formation could arise from branch migration of the junctions. Finally, isomerization and resolution lead to either of two possible outcomes: (1) The two DNA duplexes may become rejoined in an exchanged configuration or (2) they may retain their original configuration of flanking markers. In either case, the region where the recombination occurred may contain a short segment of heteroduplex DNA. Repair of this heteroduplex DNA may lead either to conversion or to restoration, whereas an absence of repair permits the mismatch to persist through both meiotic divisions. DNA replication in the first postmeiotic cell cycle then results in formation of a sectored colony, with either half carrying genetic information from each parent. This latter pathway, termed postmeiotic segregation (PMS), provides compelling genetic evidence for the formation of heteroduplex DNA in meiosis.

An alternative viewpoint holds that there are mechanistically distinct pathways of events leading either to gene conversion or to crossing over, and the two modes of recombination often coincide simply because they both derive from a common initiating event. Findings that have been taken to support this alternative view include the following: (1) The degree of association between conversion and crossover varies between loci (Fogel et al. 1981). (2) Some substrates, such as direct repeats, rarely show any crossing over in association with gene conversion (Klein and Petes 1981; Klar and Strathern 1984; Klein 1984; Jackson and Fink

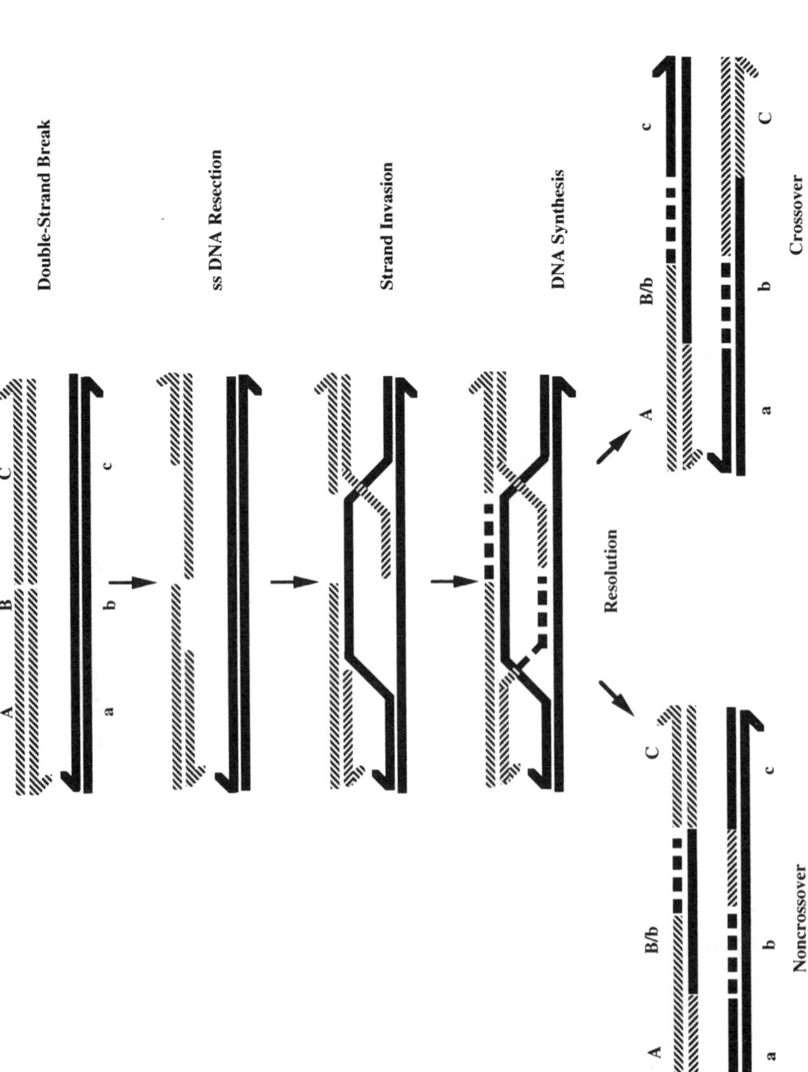

Figure 3 Stages in the double-strand break model for meiotic recombination, as described in the text.

1985). (3) Certain mutations permit normal levels of gene conversion but cause reductions in the level of crossing over (such as *mer1+MER2*, *zip1*, *msh4*, as described later). (4) Crossovers display chiasma interference, whereas gene conversion events fail to interfere either with each other or with crossovers (for review, see Carpenter 1987; Sym and Roeder 1994). (5) Although crossovers usually are associated with conversion tracts, about 10% of gene conversion tracts analyzed in one study were found to be separated from the point of crossing over by a marker segregating 2:2 (Symington and Petes 1988).

Certain models (Smithies and Powers 1986; Carpenter 1987; Roeder 1990) have sought to explain this variation in gene conversion frequencies by invoking two stages of recombination. In the first stage, the search for homology is carried out at the DNA level by strand invasion. If extensive homology is not found, the chromosomes dissociate, leaving a tract of heteroduplex that might be resolved as a gene conversion as the search for homology is repeated. If homology is found, SC assembly ensues, bringing the homologs together throughout their length as the cell progresses into the phase of crossing over. It would then be only during this second round of strand invasion, leading to SC assembly, that the resulting gene conversions would be accompanied by crossovers. These models may provide a reasonable explanation for ectopic recombination, wherein recombination occurs between homologous sequences located in different positions of the genome. Strand exchange reactions would take place on a genome-wide basis, rather than being limited to homologs juxtaposed to one another within fully formed SC. Short regions of heteroduplex may be unstable, permitting gene conversion but undergoing dissociation before crossing over can be completed. When recombination is measured between naturally occurring repeats, such as Ty elements, gene conversion is not associated with crossing over (Kupiec and Petes 1988), a fact that is consistent with this model. On the other hand, ectopic conversion between artificial repeats shows normal levels of association with crossing over (Jinks-Robertson and Petes 1986; Lichten et al. 1987), so the model outlined here fails to account fully for some of the phenomena under consideration.

C. Evidence for Double-strand Break-initiated Recombination

In recent years, evidence has accumulated indicating that meiotic recombination is initiated by DSBs. Meiotic DSBs were first detected at certain loci that showed elevated levels of meiotic recombination ("hot spots"), such as *ARG4* (Nicolas et al. 1989), a *HIS4:LEU2* insertion allele on chromosome III (Cao et al. 1990), *HIS4* (Nag and Petes 1993), YCR47-48 (Zenvirth et al. 1992; Goldway et al. 1993b), and *CYS3* (Ohta et al. 1994). These DSBs are transient in nature, disappearing before the appearance of heteroduplex DNA (Cao et al. 1990; Padmore et al. 1991; Goyon and Lichten 1993). More recent studies (Game 1992; Zenvirth et al. 1992; Wu and Lichten 1994) have shown that DSBs can be detected not only at specific hot spots, but also at other sites situated at intervals of 10–50 kb along the chromosomes. Not all of these sites suffer DSB formation at the same frequency, and a smaller chromosome may experience as few as one DSB per meiosis.

Several DSB sites have been shown to reside at or near potential transcription promoters (Wu and Lichten 1994, 1995; Ohta et al. 1994), coinciding with sites that are hypersensitive to DNase I. However, whereas DSB formation is meiosis-specific, the hypersensitivity to DNase I does not differ between mitotic and meiotic growth. These findings may indicate that features of chromatin structure established before meiosis play a part in determining the sites of initiation of meiotic recombination. On the other hand, studies by Ohta et al. (1994) have shown that sensitivity to micrococcal nuclease (MNase) increases during meiosis at sites that display increased DSB formation. Constructs that increase or decrease recombination in the *ARG4* locus show corresponding changes both in MNase sensitivity and in the extent of DSB formation. It remains to be resolved why chromatin changes that take place at DSB sites upon entry into meiosis potentiate attack by MNase I while susceptibility to DNase I remains unchanged. Recently, the exact locations of DSBs have been mapped to the nucleotide level at the *HIS4:LEU2* hot spot (Keeney and Kleckner 1995) as well as at the *ARG4* and *YCR47-48* loci (J. Liu et al. 1995) and near the *CYS3* locus (de Massy et al. 1995). In all of these cases, DSBs occurred at multiple, irregularly spaced sites within an approximately 150-bp region encompassed by the segment of nuclease-hypersensitive chromatin. No consensus sequence could be discerned within or near these DSB sites. One study (J. Liu et al. 1995) concluded that breaks occur with a two-nucleotide 5′overhang, whereas the other (de Massy et al. 1995) indicated that breaks occur at the same position in both strands, generating blunt ends. It is unclear whether these differences reflect different types of breaks at different loci or are due to technical differences between the studies. Following breakage, there is rapid 5′-strand resection of the broken ends to create 3′single-stranded DNA overhangs about 600 bp in length (Sun et al. 1991; Bishop et al. 1992). In strains carrying the *rad50S* mutation (see below), DSBs accumulate rather than being processed in this manner, so *rad50S* provides a valuable tool for mapping and quantifying DSBs. Interestingly, in these strains, a protein can be detected bound to the 5′end of the breaks. A possible explanation for the absence of resection is that the enzyme involved in DSB formation becomes covalently bound to the 5′ends upon cleavage; in a *rad50S* background, the enzyme may be incapable of releasing its substrate, precluding further processing (de Massy et al. 1995; Keeney and Kleckner 1995; J. Liu et al. 1995). (See Note Added in Proof.)

What is the evidence that DSBs initiate recombination? Numerous studies have shown a correlation between the level of DSBs and the level of recombination at various loci (see, e.g., Sun et al. 1991; White et al. 1991; Wu and Lichten 1994). Additionally, artificially created DSBs can increase the level of meiotic recombination in wild-type strains (Kolodkin et al. 1986; Malkova et al. 1996); breaks created by X-rays can partially alleviate the recombination defect of the *spo11* mutant, which ordinarily fails to contain any DSBs (Thorne and Byers 1993). Furthermore, several mutations that decrease or increase meiotic recombination have parallel effects on the local level of DSB formation (Nicolas et al. 1989; Rocco et al. 1992; de Massy and Nicolas 1993; Ohta et al. 1994; Fan et al. 1995; Wu and Lichten 1995). Careful measurements at several loci have shown that the number of DSBs, as quantified from the stable forms created in a *rad50S* strain, could account for most or all of the recombination events seen (Rocco et al. 1992; de Massy and Nicolas 1993; Wu and Lichten 1994, 1995).

Additional evidence for the role of DSBs in meiotic recombination was provided by the use of artificial or nonessential chromosomes. Ross et al. (1992) showed that the addition of a 12.5-kb fragment carrying the *ARG4* hot spot/DSB site allowed crossing over and normal disjunction of two λ-based artificial chromosomes that otherwise would show nondisjunction in about one quarter of all asci. Similarly, Sears et al. (1994) showed that the insertion of a 6-kb yeast fragment carrying the *HIS4::LEU2* hot spot/DSB site into two heterologous yeast artificial chromosomes (YACs) led to a strong increase in recombination and prevented misdivision. Using chromosomal fragments, Goldway et al. (1993a) showed that specific fragments of chromosome III that were present in addition to the normal diploid complement caused an increase in nondisjunction of the intact chromosomes III. Assuming that the chromosomal fragments contain pairing sites that competed with those on the endogenous chromosomes, a genetic screen was devised to search for plasmid-borne sequences that would increase nondisjunction when present in high copy number. One plasmid identified in this manner contained a 7.5-kb insert from the YCR47-48 region. Notably, this same region was present on a chromosomal fragment that caused elevated nondisjunction. Furthermore, this region contains a very strong DSB site that is associated with a hot spot for recombination (Zenvirth et al. 1992; Goldway et al. 1993a). Thus, a genetic screen designed to identify pairing sites produced a site for DSB formation, further strengthening the evidence for an association among DSBs, pairing, and recombination. Finally, DSBs have also been detected in haploid strains undergoing meiosis (de Massy et al. 1994; Gilbertson and Stahl 1994), demonstrating that DSB formation can occur independently of homologous pairing. On the other hand, when both homologs are present (in a diploid), the level of DSBs detected on one chromosome is slightly reduced by sequence differences in the corresponding region of the homolog (A. Nicolas, pers. comm.), suggesting that homologous interactions that are allowed to occur do, in fact, influence DSB formation.

D. The Synaptonemal Complex in Yeast

Our knowledge of cytological events occurring in meiotic prophase derives largely from study of meiocytes that are more amenable to microscopic analysis than yeast. Since the pioneering of thin-section electron microscopy of meiosis by Moses (1956), innumerable studies have revealed a striking structural homology among eukaryotes: Early in meiotic prophase (at leptotene), the chromosomes undergo condensation about an elongate *axial element*, which then progressively becomes aligned in parallel with its homolog throughout zygotene. At pachytene (when crossing over is believed to occur), the paired chromosomes are joined along their length by the synaptonemal complex, consisting of two darkly staining *lateral elements* (formerly axial elements) spaced about 100 nm apart on either side of an electron transparent *central region*. Chromatin coats the outer surfaces of the lateral elements. It is presumed that the underlying 250-nm chromatin strand is arrayed in a series of loops of approximately equivalent contour length attached at intervals along the length of the lateral element (Moens 1994). Dissolution of the SC upon entry into diplotene is nearly complete but short segments may be retained and undergo modification within the chiasmata (von Wettstein et al. 1984). These structures, in concert with the mutual adhesion of sister chromatids to one another, are thought to confer the cohesive force between chromosomes that is crucial for proper alignment of homologs on the metaphase plate for meiosis I (Nicklas 1977).

Early electron microscopy of meiotic yeast by Moens and Rapport (1971a,b) revealed that the SC in this organism is relatively thin and indistinct, so features of SC assembly are poorly defined. Appearance of the SC proper was seen to be preceded by the appearance of a globular *dense body*, which stains darkly like the lateral elements of the SC and/or the *polycomplex body*, which appears to represent an assemblage of SC-like material similar to that seen in other organisms. Polycomplexes may also accumulate in yeast nuclei at the conclusion of pachytene (Zickler and Olson 1975). No compelling evidence has been presented for any functional role for either structure, and it seems likely that both dense bodies and polycomplex bodies represent nonfunctional aggregation of SC components before or after their utilization in synapsis. Reconstruction of pachytene nuclei from serial sections has allowed resolution of yeast chromosomes to a degree not previously attained and has improved definition of the yeast karyotype (Byers and Goetsch 1975). The contour length of the visualizable segments of SC extending between their insertions on the nuclear envelope can be taken to represent the length of individual bivalents extending from telomere to telomere. The aggregate length of SC per cell computed on this basis, considered with respect to the known DNA content of the yeast genome, indicates that the DNA molecule within each pachytene bivalent must be condensed about 100-fold relative to its extended length as B-form DNA. This so-called "packing ratio" of 100:1 contrasts with a packing ratio of nearly 10,000:1 for human bivalents, which must be considerably more tightly condensed but nevertheless display the same approximately 100-nm spacing between their lateral elements.

Thin-section electron microscopy of meiotic yeast (Fig. 4) also reveals small globular dense structures situated at intervals along the central element of the SC (Byers and Goetsch 1975). The spacing of these so-called *recombination nodules* is consistent with their marking the sites of meiotic recombination. Elegant characterization of the analogous structures in *Drosophila* oocytes by Carpenter (1975) demonstrated the occurrence of these nodules in two distinct sizes. The smaller ones, which are often more ellipsoidal in form, arise at an early stage and are nearly randomly arrayed along the bivalent; these are believed to represent incipient sites of recombination or gene conversion. In contrast, the larger nodules, which are more nearly spherical, appear only later in pachytene and are spaced appropriately to represent sites of crossing over. This distinction in sizes of recombination nodules has not been shown for yeast. Although these structures have not been demonstrated reliably in the more recently developed methods for spreading the

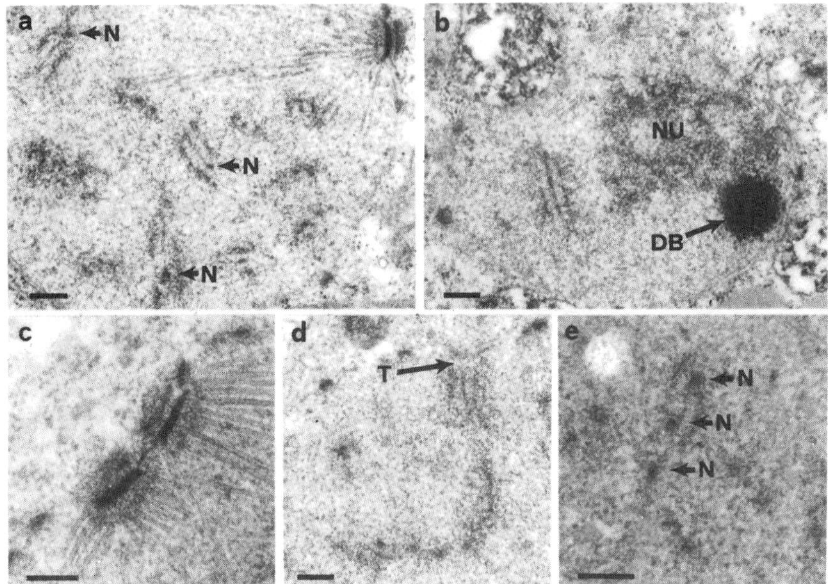

Figure 4 Features of pachytene yeast nuclei seen by thin-section electron microscopy. (*a*) Segments of synaptonemal complex bear recombination nodules (N) adjacent to the central element. (*b*) Situated next to the nucleolus (NU) in this cells is the dense body (DB), which appears to contain constituents of the synaptonemal complex. (*c*) The duplicated spindle pole bodies remain in this side-by-side configuration throughout meiotic prophase. (*d*) The telomere (T) of this meiotic bivalent abuts the nuclear envelope as is typical. (*e*) These recombination nodules (N) are closely spaced along the synaptonemal complex, reflecting the high frequency of crossing over in meiotic yeast. (Reprinted, with permission, from Byers and Goetsch 1975.)

nuclear contents (described below), certain gene products, such as Dmc1p and Rad51p, have been inferred to reside within them on the basis of a punctate distribution for antibody staining by fluorescence microscopy (Bishop 1994), as shown in Figure 7.

The restricted ability for quantitative characterization of synapsis from serial sections eventually encouraged exploration of the chromosomal spreading methods that had proven to be so instructive about meiotic prophase in other eukaryotes (Dresser and Moses 1980), and it was clear that yeast bivalents, although relatively more fragile and indistinct, could be visualized in lysates (Goetsch and Byers 1982). The coincident observation of spindle microtubules in these images confirmed the finding from serial section microscopy that the chromosomes are fully detached from the spindle microtubules at pachytene but may regain attachment immediately thereafter. Significant technical improvements by Dresser and Giroux (1988) have made characterization of yeast pachytene bivalents routine and have enabled workers to describe the kinetics of synapsis and the meiotic divisions accurately in both wild-type and mutant strains (see Fig. 7). Application of these methods has generally shown *S. cerevisiae* to be unremarkable in its synaptic behavior among eukaryotes, thus establishing this process as yet another target for effective use of yeast as a model organism.

One notable feature that may differ, however, is in the manner of entry into synapsis: Whereas many organisms display distinct axial elements at leptotene and their progressive alignment to form SC throughout zygotene, fully formed unpaired axial elements have not been detected in wild-type yeast. Although there is apparent pairing of the homologs early in meiotic prophase as evidenced by the juxtaposition of homologous DNA sequences revealed by chromosomal "painting" (Scherthan et al. 1992), no axial elements are seen. Later, short segments of SC begin to appear and gradually coalesce with one another in the space intervening between homologs until a continuous SC extending from telomere to telomere is present. It seems likely that the failure to detect axial elements at leptotene realistically reflects unique chromosomal behavior in yeast, rather than an inadequacy of the methods used to preserve such structures, for unpaired axial elements are clearly seen in certain mutant strains, as discussed later.

The apparent absence of any condensation into a discrete axial element before synapsis prevents resolution of relevant issues concerning ectopic recombination, in which homologous sequences artificially added to nonhomologous chromosomes undergo both gene conversion and reciprocal recombination (Jinks-Robertson and Petes 1985). Namely, it is of interest to learn whether or not SC is actually required for meiotic crossing over, as might be concluded if one could establish conclusively that SC is never formed between nonhomologous chromosomes. In the absence of continuous axial elements, it is difficult to establish whether or not there is chromosomal continuity between the isolated segments of SC present early in synapsis. Accordingly, some segments might join nonhomologs, leaving open the possibility that ectopic recombination occurs within such segments rather than independently of the SC.

A third feature of synapsis evident from the spreading method concerns the situation within the nucleolus. Serial sectioning of SC had failed to resolve whether SC is present here (Byers and Goetsch 1975), but spreading establishes that the chromosomal axes of the two homologs traverse the nucleolus independently (Dresser and Giroux 1988). Furthermore, staining of spread pachytene nuclei with antibodies against two components of the SC—*ZIP1* (Sym et al. 1993) and *HOP1* (F. Klein, pers. comm.)—have shown a specific absence of these proteins in the nucleolar segments of the chromosomes. An intriguing correlation with the absence of SC here is seen in the striking paucity of meiotic crossing over within these nuclear DNA sequences (Petes and Botstein 1977). Perhaps these findings indicate that the absence of SC, or of some protein that is similarly required for its assembly, accounts for the deficit of reciprocal recombination within the cluster of ribosomal DNA sequences.

E. Relationship between Double-strand Breaks and Synaptonemal Complex Formation

Is synapsis normally required for DSB formation? If not, do DSBs have a role in facilitating synapsis? Although definitive answers to these questions are not yet available, we note three lines of evidence supporting the idea that at least the formation of DSBs is not dependent on the synapsis of homologous chromosomes: (1) Several meiotic mutants fail to form mature SC but show DSB formation (Cao et al. 1990; Bishop et al. 1992). Conversely, no deletion mutant has been described that shows SC formation in the absence of DSBs. (2) DSBs occur in haploid meiosis, where no homolog is available for normal synapsis (de Massy et al. 1994; Gilbertson and Stahl 1994). However, SCs are formed between nonhomologs under these conditions (Loidl et al. 1994), so the formation of DSBs in haploids does not prove that DSB formation is independent of SC formation. (3) Studies carried out with a strain that undergoes highly synchronous meiosis (SK1) have allowed the establishment of the temporal relationship between DSB formation and synapsis (Padmore et al. 1991). It was shown that DSBs arise prior to the formation of SC and disappear as mature full-length segments of SC accumulate. Together, these results rule out the possibility that SC formation is a prerequisite for DSB formation, but it remains possible that initiation of recombination (DSB formation) is required for the formation of stable SCs or that the two processes are independent of each other. The components of the SC appear to self-assemble, probably by protein-protein interactions (for review, see Moens 1994). Even in the absence of chromosomes, SC components form structures of similar morphology, such as the so-called polycomplex. In some organisms, such as *Ascaris suum*, polycomplexes assemble in the cytoplasm prior to meiotic prophase and subsequently are transported to the nucleus (Fiil and Moens 1973). SCs have also been shown to form between nonhomologous chromosomes, especially in haploids and in cells carrying chromosomal aberrations, again indicating that formation of SC may occur independently of the synapsis of homologs (Loidl et al. 1991; Dresser et al. 1994).

F. Search for Homology and Presynaptic Alignment (Pairing)

If the SCs do not mediate homologous alignment of the chromosomes, what are the alternative mechanisms? Some models propose that homologous pairing is mediated by interactions between proteins that bind to specific sites on the chromosome (Comings and Riggs 1971; Holliday 1977; Stern and Hotta 1987), but little direct evidence has been garnered in support of this idea. Other models posit that the search for homology principally involves DNA-DNA interactions, assisted perhaps by strand transfer enzymes similar to the *Escherichia coli* RecA protein, which can detect homology between single-stranded DNA and double-stranded DNA molecules and promote strand exchange in such regions (see Cox and Lehman 1987). RecA homologs in yeast (the products of *RAD51* and *DMC1*) have been shown to be active in meiosis (see below). Following DSB formation and 5′-end resection, single-stranded DNA is created at the broken ends. These single-stranded DNA molecules may then interact with these RecA-like proteins to mediate a search for homology by means of strand invasion, creating heteroduplex DNA. If sufficient homology is found, then SC formation may proceed, bringing the homologs together throughout their length (Hotta et al. 1984; Smithies and Powers 1986; Carpenter 1987). However, although DSBs arise early in the meiotic process, heteroduplex DNA is undetectable until much later, concomitant with the appearence of mature recombinant molecules (Nag and Petes 1993; Goyon and Lichten 1993; Schwacha and Kleckner 1994, 1995). Since strand invasion alone should generate heteroduplex DNA, one must entertain the possibility that heteroduplex DNA formed at early stages is too short or too unstable to be detected.

An alternative mode of sequence-specific homology searching might utilize other types of unstable intermediates. Notably, the two homologous duplexes might become associated with each other by paranemic interactions, in which there is a side-by-side alignment without any intertwining of the strands. Kleckner et al. (1991) have proposed that a search of the entire genome for homology may depend first on an early search based on unstable pairing interactions of this sort. Once provisionally paired sites are established, these intermediates might proceed into DSB formation, leading to synapsis. Thus, in this view of homology search, the mitotic mechanisms for the repair of double-stranded DNA breaks, which utilize the undamaged homolog as a template for repair, may have evolved into a mechanism for pairing homologs at meiotic prophase. Although the search for homology occurs throughout the genome, some sites may be preferred due to some aspect of chromatin structure or chromosome topology. These are the hot spots for recombination, which also show high levels of DSBs (Zenvirth et al. 1992).

Meiotic chromosomal pairing has in recent years been successfully explored by means of fluorescence in situ hybridization (FISH) techniques, in which probes for sequences broadly represented on an entire chromosome serve to "paint" the nuclear location of both homologs during meiosis (Scherthan et al. 1992; Loidl et al. 1994; Weiner and Kleckner 1994; Nag et al. 1995). The use of differently colored probes for each of two chromosomes allows the distance between hybridizing signals to be measured. A spatial association between homologs (same color) can be compared with that between nonhomologous chromosomes (different colors), thereby providing a basis for distinguishing significant interactions between homologs from the random colocalization of the nonhomolog controls. Strikingly, the kinetics of pairing appear to be nearly indistinguishible from those of SC appearance. On the other hand, in rapidly sporulating strain SK1, some nuclei showed paired chromosomes even at the time of transfer to sporulation medium (t = 0). The frequency of these early pairings declined during premeiotic DNA synthesis and later increased again with the onset of SC formation. It appears uncertain whether the earlier phase of pairing represents specific homologous interactions in mitosis or cells that have entered into meiosis during the period of culture intended for presporulation growth. Interestingly, FISH-detected homologous pairing can occur independently of both DSB formation and SC formation. *hop1* and *mer1* mutants fail to form any SC, but they have high levels of pairing (Loidl et al. 1994; Weiner and Kleckner 1994; Nag et al. 1995). Even mutants that completely lack DSBs (*spo11*, *rad50del*) or display DSBs that are not resected (*rad50S*) or do not lead to strand invasion (*rad51*, *dmc1*) still show some degree of pairing, albeit at a lower level than the wild type. Accordingly, these mutants clearly affect pairing as well as DSB and SC formation (see below).

G. Detection of Intermediates

Recently, branched DNA molecular complexes containing sequences derived from both parental chromosomes have been detected (Collins and Newlon 1994b; Schwacha and Kleckner 1994, 1995). These "joint molecules" may well represent recombination intermediates since they first appear after DSB formation, accumulate to a level proportional to the levels of DSB formation and genetic recombination, and disappear at the time when heteroduplex DNA and recombinant molecules are detected (Schwacha and Kleckner 1994). Furthermore, they fail to be formed between homeologous chromosomes or in mutants that are deficient in DSB formation or processing, such as *spo11*, *rad50del*, and *rad50S* (Collins and Newlon 1994a; Schwacha and Kleckner 1994). The molecules described by the two groups behave somewhat differently on two-dimensional gels, and it is not clear whether they represent the same intermediate.

Melting of these joint DNA molecules and electrophoretic display of the products have shown them to be composed exclusively of intact full-length single strands, all of which are nonrecombinant for flanking markers (Schwacha and Kleckner 1994). An intermediate of this sort must not contain a simple Holliday junction, which necessarily would include two recombinant single strands together with two parental strands, but this sort of intermediate might contain two (or a multiple of two) Holliday junctions. The original application of two-dimensional gel methods to meiotic yeast DNA by Bell and Byers (1983) had yielded a population of branched molecules that were shown by electron microscopy to lack any molecules with the open-centered appearance characteristic of single Holliday junctions. However, the molecules seen included many that were interpretable as hemicatenations or complexes joined by double Holliday junctions separated by 100–1000 bp. Interestingly, the proportion of intersister joint molecules is substantially lower than that of interhomolog joint molecules, implying that the mechanisms which discriminate in favor of interhomolog interactions in meiotic recombination already have been exerted by this stage of the pathway. In light of observations that the frequency of sister chromatid exchange (SCE) in haploid meiosis is higher than that of a normal diploid (Wagstaff et al. 1985), it seems that SCE is prevented in some manner by the homolog, perhaps by its outcompeting the sister strand. The basis for such competition remains obscure, however, since there was no increase in SCE in a strain carrying a homologous chromosome that is so diverged in DNA sequence as to preclude formation of any interhomolog joint molecules (Collins and Newlon 1994b).

More recently, further analysis of these joint molecules has shown that single DNA strands isolated from them have undergone transfer of genetic information locally (in the manner of a gene conversion near the

DSB site) and thus are true recombination intermediates (Schwacha and Kleckner 1995). When treated with Holliday-junction-resolving enzymes, such as RuvC or T4Endo VII, the joint molecules are resolved into both parental and recombinant duplexes, as expected for molecular complexes that are joined by two Holliday junctions (as predicted by the DSB-initiated model for recombination [see Fig. 3]). Unligated single-stranded DNA ends, or heteroduplex DNA, which also are predicted by the model, could not be detected at this stage. This may reflect their short lifetimes, a technical inability to detect them, or possibly both. It is now proposed that two rounds of heteroduplex formation and repair may take place, the first one being carried out efficiently during the formation of joint molecules and the second occurring less efficiently after their resolution (Schwacha and Kleckner 1995).

H. Assays for the Analysis of Meiotic Parameters

1. Gene Conversion

Typically, meiotic recombination is followed in a diploid organism by observing among the progeny the generation of new combinations of alleles at linked loci. Tetrad analysis in yeast readily reveals that most meiotic recombination of this sort consists of classical crossing over, where reciprocal sets of recombined products are generated on the participating pair of chromosomes. However, tetrad analysis also reveals the occurrence of gene conversions (1:3 and 3:1 segregations) that involve the replacement of a short segment of sequence information on on duplex from the other without altering the sequence on the donor chromatid. Most alleles show meiotic levels of conversion in the range of 0.5% to 20% (Fogel et al. 1981), although higher levels of conversion have been detected for certain alleles and in certain genetic backgrounds (Nag et al. 1989). Because gene conversion is a local phenomenon that may or may not be accompanied by crossing over, it is most readily detected as an intragenic recombination between two mutant alleles at the same locus. For convenience, one simply selects recombinants by plating for prototrophic spores arising from a diploid that is heteroallelic for auxotrophic alleles of the gene (such as one bearing two different alleles at *ade2*). Although the diploid parent is auxotrophic for adenine, Ade^+ recombinants can be quantitated by selection on plates lacking adenine. Gene conversion can also be observed to take place between homologous sequences that are transplaced to heterologous positions in the genome (ectopic recombination; Jinks-Robertson and Petes 1986). Ectopic recombination of this sort generally is found to be regulated by the same genes that control allelic recombination, including *spo11* and *rad50del* (Steele et al. 1991), as well as *red1* (Rockmill and Roeder 1990), *rec102* (Bhargava et al. 1992), and *mek1* (Rockmill and Roeder 1991).

2. *Reduction to Haploidy*

Although reciprocal exchange is most definitively monitored by tetrad analysis, searches for recombinational defects generally must rely on simple plating assays. A convenient and powerful approach for this purpose is the use of strains heterozygous for genes involved in drug resistance, such as the *CAN1* gene (which confers sensitivity to the arginine analog canavanine) and the *CYH* genes (which confer sensitivity to cyclohexamide). Resistance to either drug is recessive, and so the heterozygous diploid strains are unable to grow on plates containing the drug. Half of all haploid spores, however, will be resistant to that drug and can be selected on appropriate plates. Linkage of other genes can then be evaluated.

3. *Direct Repeat Recombination*

Intrachromosomal recombination (also called direct repeat recombination, or DRR) can occur between sequences that share homology and are located at different sites in the same chromosome. Such configurations are relatively easy to construct; integration of a plasmid carrying a mutated or truncated gene and a selectable marker will cause a duplication of the gene, separated by plasmid sequences and the marker. Recombination may reconstruct a functional gene that can be selected on dropout plates. The recombinants are usually divided into those that retain the marker (such as a gene conversion event in which one copy of the gene acts as a donor and the other as a recipient) and those in which the marker is lost. This latter category is sometimes assumed to occur by reciprocal exchange, but can in fact arise by different mechanisms, including single-strand annealing and unequal sister chromatid exchange (for review, see Liefshitz et al. 1995).

4. *Return-to-growth Protocols*

Yeast cells that are induced to undergo meiosis can return to vegetative growth even after they have been committed to undergo meiotic levels of recombination (Sherman and Roman 1963; R.E. Esposito and Esposito 1974). This protocol, called "return-to-growth," allows one to monitor commitment to recombination in strains that are unable to produce viable spores.

5. *Physical Monitoring of Crossing Over*

Using strains heterozygous for restriction enzyme site polymorphisms makes it possible to monitor, by Southern blot analysis, the appearence of recombinant molecules whose sizes differ from those of the parental DNA. This method has been successfully used to characterize the point at which the recombination pathway is blocked for several meiotic mutants (Borts et al. 1986; Schwacha and Kleckner 1994). Pulse-field gel electophoresis has also been used to monitor recombination events physically (Game et al. 1989). Both methods have the advantage that one measures recombination levels irrespective of viability.

6. *DSB Monitoring*

In Southern blot analysis of meiotic cultures, using as probes genes that show a high level of recombination, the creation of DSBs leads to the appearance of faint diffuse bands. Strand-specific probes have shown in wild-type strains that the 5′end is rapidly resected, whereas the 3′end remains intact (Cao et al. 1990; Sun et al. 1989, 1991). *rad50S* mutants fail to undergo 5′resection, thereby allowing unprocessed DSBs to accumulate (Alani et al. 1990), and thus DSB formation can be monitored quantitatively in the usual Southern blot analysis or after pulse-field gel electrophoresis (see, e.g., Padmore et al. 1991; Zenvirth et al. 1992).

7. *Fluorescence In Situ Hybridization*

In situ hybridization of spread nuclei, using large chromosome-specific probes (FISH), has permitted definition of the locations and condensation states of chromosomes throughout meiosis (Loidl et al. 1994; Weiner and Kleckner 1994; Nag et al. 1995). If differently colored dyes are used for different chromosomes, it is possible to monitor the distance between homologs (two signals of the same color) with precision. As the homologs become paired, these signals coalesce into a single area of signal. The distances between differently colored signals (representing nonhomologous chromosomes) serves as a control for cases in which the signals are juxtaposed by chance.

8. *Heteroduplex DNA Formation*

Two different methods have been developed to monitor heteroduplex formation. In one of them (Goyon and Lichten 1993), denaturing gel electrophoresis was used to follow mismatches in the DNA at the *ARG4* gene, and it could be shown that the mismatches chosen are poorly recognized by the mismatch repair machinery. The second method takes advantage of the fact that foldbacks (hairpins) generated by small palindromic insertions consisting of different restriction site insertions are not attacked by the mismatch correction system of yeast (Nag and Petes 1993). Strains were constructed that were heteroallelic for two different palindromic insertions (different restriction enzyme sites) at the same site in the *HIS4* gene. Heteroduplex DNA that includes the mutant alleles contains a cruciform structure in which the restriction sites reside in single-stranded DNA loops and consequently are resistant to digestion.

9. Joint DNA Molecules

The use of two-dimensional gels has allowed the detection of intermediates in recombination that carry information from both parental DNAs (Bell and Byers 1983; Brewer and Fangman 1991). These gels separate DNA molecules in one dimension according to size, and in the other according to shape, generating an arc of locations for linear molecules. Branched molecules can be detected because they respond differently to these electrophoretic conditions and are displaced from the main arc. In this way, two groups have isolated molecules that are good candidates for recombination intermediates (Collins and Newlon 1994b; Schwacha and Kleckner 1994, 1995).

I. Isolation of Recombination-defective Mutants

1. Meiosis-specific Genes

Since reciprocal recombination is a prerequisite for faithful chromosome segregation, mutants defective for crossing over suffer extensive nondisjunction, producing aneuploid spores that mostly are inviable. Because meiosis in *S. cerevisiae* represents a distinct pattern of cell division that is restricted to cultures transferred to sporulation media, potential meiosis-specific mutants can be isolated and propagated on growth media with good viability. Mutants can then be identified by assessing the ability of an aliquot of these cells to produce viable spores. This approach led to identification of *spo11* (Esposito and Esposito 1969), *red1* (Rockmill and Roeder 1988), *mer1* (Engebrecht and Roeder 1989), and *mek1* (Rockmill and Roeder 1991). A search for sporulation-specific genes that are expressed early in meiosis identified a yeast homolog of the bacterial RecA gene, *DMC1* (Bishop et al. 1992). Recently, Burns et al. (1994) developed a large-scale screen to identify meiosis-specific genes. Diploid strains containing random *lacZ* insertions were screened for meiosis-specific expression. Fifty-five strains showing meiotic induction of the fused *lacZ* gene were identified. They were found to be mutant either in known genes, including *SPO11, HOP1, ZIP1*, and *MER1*, or at some novel loci. From these studies, it was estimated that there are about 100 (93–135) meiotically induced genes, of which 30–50 may be required for sporulation and spore viability.

2. Screening of Known Mitotic Mutants

Many mutations that affect the mitotic cell cycle also have effects on meiosis. For example, *CDC* genes that have a central role in the mitotic cell cycle, especially those required for DNA replication and nuclear division, have also been found to be essential for the meiotic process (Simchen 1974; Schild and Byers 1978). In addition, many mutations originally isolated on the basis of their sensitivity to DNA-damaging agents, such as the *rad50* series (Game 1983), are defective in both mitotic and meiotic recombination and therefore fail in meiosis (Petes et al. 1991).

3. Return-to-growth Studies

Yeast cells in the earlier stages of meiosis can return to vegetative growth even after they have been committed to undergo meiotic levels of recombination (Sherman and Roman 1963; R.E. Esposito and Esposito 1974). Haploid strains disomic for chromosome III, when transferred to sporulation medium, will perform meiotic DNA synthesis and recombination, but they will die in an attempt to perform two meiotic divisions. But if these cells are returned to vegetative growth before entry into the meiotic divisions, viable cells that show meiotic levels of recombination are obtained. This method has been successfully used by Fogel and Roth (1974) and by Ogawa and co-workers (Leem and Ogawa 1992; Ajimura et al. 1993) to identify recessive mutants that reduce the levels of conversion (*con* and *mre* mutants). Similarly, Fogel and co-workers exploited this system to isolate mutants exhibiting increased levels of conversion, including *pms1* and *pms2* (Fogel et al. 1981; Williamson et al. 1985; Jeyaprakash et al. 1994).

4. Use of the spo13 *Mutation*

The *spo13* mutation (Klapholz and Esposito 1980a,b) allows yeast cells to undergo a single meiotic division. Diploid *spo13* strains initiate meiosis normally, undergo DNA synthesis and meiotic recombination, and then enter into a single division, producing two diploid spores instead of four haploid ones (Fig. 5). Marker segregation in these diploid spores showed that the single division usually was equational (mitosis-

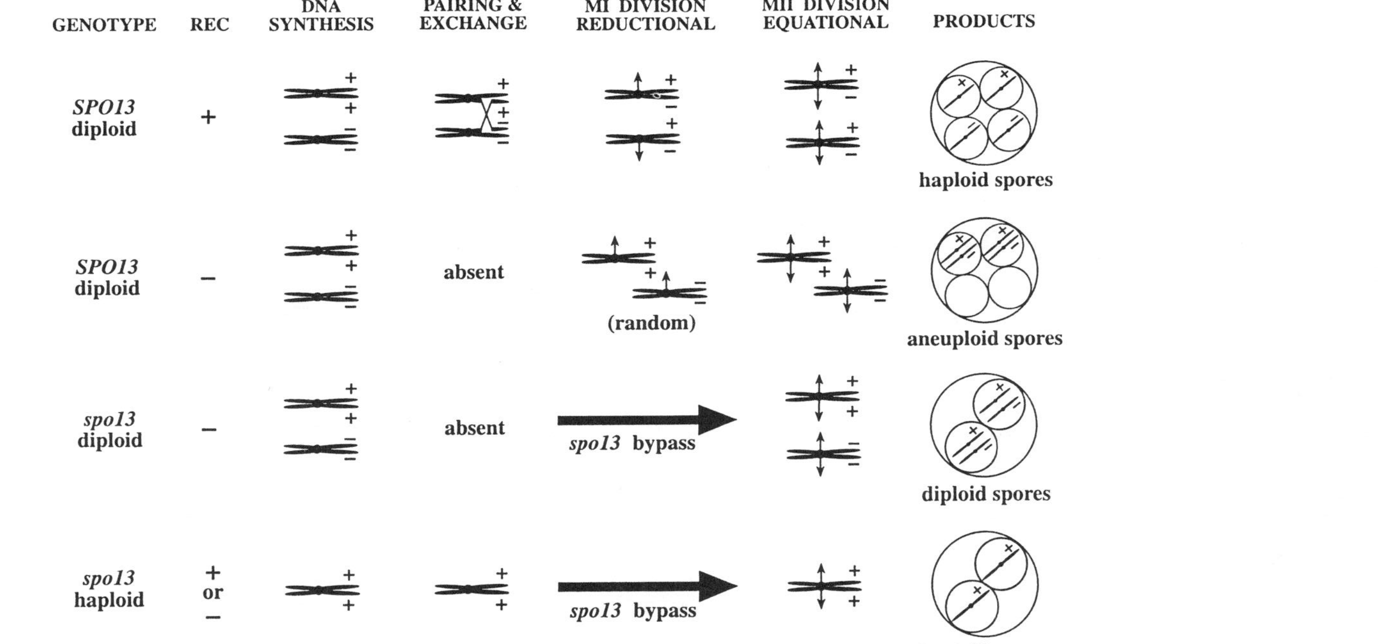

Figure 5 Representation of the manner in which mutants defective in pairing and exchange (*second line*) cause nondisjunction at meiosis I (MI DIVISION, REDUCTIONAL), whereas bypass of the reductional division in *spo13* strains (*third line*) prevents aneuploidy. Similarly, *spo13* eliminates the reductional division in the sporulation of haploid cells, leading to the production of haploid meiotic progeny (*fourth line*).

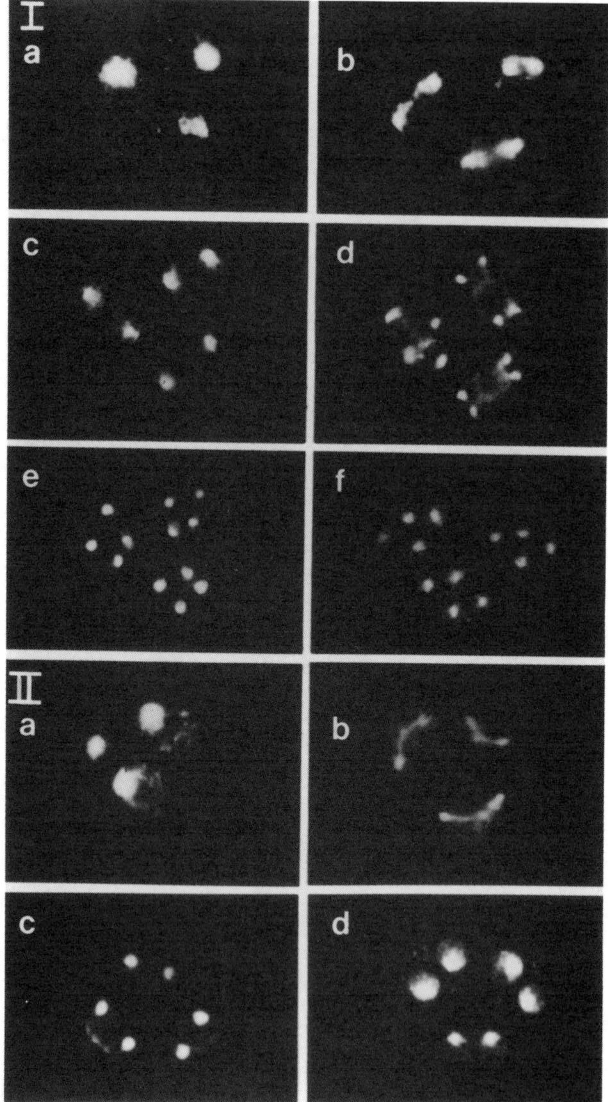

Figure 6 DNA staining with 4′,6-diamidino-2-phenylindole (DAPI) reveals the synchronous progression through the meiotic divisions, as seen in representative sets of three incipient asci per field. (I) Wild-type cells are shown in meiotic prophase (*a*), in meiosis I (*b*), at telophase I (*c*), at late anaphase II (*d*), and after meiosis II (*e,f*). (II) *spo13* cells are seen at prophase (*a*), in anaphase (*b*), and after completion of the sole meiotic division that occurs in this mutant (*c,d*). (Courtesy of S. Klapholz and R.E. Esposito.)

like) (Klapholz and Esposito 1980b). Since there is no necessity for homologs to disjoin from one another in an equational division, recombination defects do not prevent viable sporulation in a *spo13* strain (Fig. 6) (Malone and Esposito 1981). In addition, *spo13* haploid strains expressing both the **a** and α information undergo a single-division meiosis to produce two viable spores (Wagstaff et al. 1982), allowing the isolation of recessive mutations affecting recombination (*hop1*, Hollingsworth and Byers 1989; *mei4*, Menees and Roeder 1989). Not all recombination-deficient mutations can be rescued by *spo13*. *rad52* cells, for example, die in meiosis even if also made homozygous for *spo13*, probably because the wild-type *RAD52* function is needed for a relatively late stage of recombination. Once recombination is initiated in the absence of *RAD52* function, lethal lesions may arise in the DNA that lead to inviability (Malone and Esposito 1981; Malone 1983). *rad52* strains can, however, be rescued from this inviability by additional mutations that block the recombination pathway at an early stage, thereby preventing formation of the lethal intermediate. In a screen for mutants that would rescue a *spo13 rad52* double mutant, Malone et al. (1991) isolated *rec102*, *rec104*, and *rec114*, as well as new alleles of *SPO11*, *MER1*, *MER2*, *MEI4*, and *RAD50*.

5. *Classification of Recombination-defective Mutations*

As explained above, homologs do not usually segregate from each other in the single meiosis of *spo13* mutants. Therefore, failure to recombine per se does not lead to aneuploidy and spore death. On the basis of their phenotype in the presence of *spo13*, mutants can be divided into two broad classes: Class-1 mutants are rescued by *spo13* and are therefore presumed to be defective in an early function that serves to initiate recombination. Class-2 mutants are not rescued, presumably because they generate defective intermediates that cause spore death even when (as in *spo13* strains) recombination is no longer needed for proper segregation. Certain class-2 mutants can, however, be rescued by *spo13* together with a class-1 mutation that prevents initiation of recombination (see Malone 1983; Petes et al. 1991).

J. Ordering Recombination Functions

Our current understanding of meiotic prophase derives largely from the effects of various mutations on the level of recombination, its distribution, and associated perturbations of synapsis. In recent years, considerable progress has been made in isolating mutations as well as in developing more informative assays for specifying their effects on the chromosomes. In analyzing the phenotypes of such mutants, several key issues should be kept in mind: (1) Different yeast strains show different phenotypes for the same mutation. Two striking examples are *zip1* and *dmc1*. Whereas *zip1* null mutants do not complete meiosis in a standard yeast background (Sym et al. 1993), strains homozygous for the same allele are able to complete meiosis and give 60% spore viability in an SK1 background (Sym and Roeder 1995). There seems to be an exception of the opposite type for *dmc1:* These mutants fail to produce spores in SK1 (Bishop et al. 1992), but they are able to sporulate normally in other backgrounds, where they show subtle effects in the level of crossing over (*med1*, Bishop et al. 1992). (2) Different laboratories use different techniques for analyzing SC formation. For example, although gentle spreading or sectioning methods may reveal intact SC for a particular genotype, other (perhaps harsher) methods may cause the dissociation of weakened structures and show only individual axial elements. For example, in the analysis of the *rad50S* mutation in some strains, Alani et al. (1990) observed only occasional stretches of SC, whereas Loidl et al. (1994) reported nearly complete formation of SC in 7% of the cells after 7 hours. As shown in Table 1, most mutants can be classified into groups of genes that act early (class 1) or late (class 2) in the recombination-synapsis pathway. Further subgrouping is based on the severity of their effects and on the stage at which the meiotic defect is evident.

1. Class 1, Group 1

A large subgroup of class-1 mutants seem to be affected in very early functions that are required for DSB formation. This category includes *mei4, mer1, mer2, mre2, mre11, rad50del, rec102, rec104, rec114, spo11del,* and *xrs2*. Although not all available assays have been performed on all members of this class, many common phenotypes are known. It has been demonstrated that no DSBs are formed in *spo11, rad50del, xrs2, mer2,* and *mre11* strains (Cao et al. 1990; Ivanov et al. 1992; Johzuka and Ogawa 1995; Rockmill et al. 1995b) and that all subsequent events are also absent. Only short axial elements (AEs) and no full-length SCs are formed (Alani et al. 1990; Cao et al. 1990), there is no heteroduplex DNA (Nag and Petes 1993; Nag et al. 1995), and neither joint molecules (Collins and Newlon 1994b; Schwacha and Kleckner 1994) nor recombinant molecules can be detected (Borts et al. 1986; Nag et al. 1995; Schwacha and Kleckner 1994). In addition, no commitment to gene conversion (GC) or crossing over (CO) takes place (Engebrecht et al. 1990; Cool and Malone 1992; Ivanov et al. 1992; Ajimura et al. 1993; Rockmill et al. 1995b). Rad50p, Xrs2p, and Mre11p probably interact to form a complex (Johzuka and Ogawa 1995).

No meiotic recombination is seen for *rec102* (Bhargava et al. 1992; Cool and Malone 1992), *mer2* (Cool and Malone 1992), *mei4* (Menees and Roeder 1989), *rec104* (Galbraith and Malone 1992), or *rec114* (Pittman et al. 1993). Heteroduplex DNA is undetectable in *mei4* and *rec102* mutants, and FISH analysis has shown that chromatin is condensed, but pairing is reduced, in these two mutants (Nag et al. 1995). A *mer1* mutant displays about 10% of wild-type GC and CO and a clear defect in chromosomal synapsis. Most cells show unpaired AEs of various lengths, although about 10% of the cells have full-length SCs (Engebrecht and Roeder 1990). Heteroduplex DNA levels are also about 10% of wild type. *MER1* encodes a meiosis-specific RNA splicing factor that is required for processing the transcript of *MER2* and probably that of at least one additional gene (Engebrecht et al. 1991; Nandabalan et al. 1993; Nandabalan and Roeder 1995). In the absence of Mer1p, the *MER2* transcript is spliced at 10% of the wild-type level. The *mer1* defects in GC and SC formation can be attributed to the defect in *MER2* RNA splicing, since introduction of an intronless version of *MER2* into *mer1* fully restores conversion and synapsis (see below). Mutations at the *MRE2* gene display similar phenotypes (Ajimura et al. 1993), and it recently has been shown that *MRE2* is also required for the splicing of *MER2* (H. Ogawa, pers. comm.).

In summary, phenotypes of class 1/group 1 mutants indicate that the complete formation of tripartite SC (but not AE formation alone) is dependent on successful completion of early functions, such as DSB formation, which are probably also required for recombination. Accordingly, if these mutants are defective in some aspect of homolog recognition, or in the actual initiation of recombination, then AE nucleation must be independent of these functions, although progression into the formation of mature SC appears to depend on these functions. Further support for a role in early recombination functions comes from findings that mutations of this group eliminate the reductional segregation of centromere-proximal markers that sometimes are observed in *spo13* meiosis, presumably because recombination interferes in some manner with the pathway to an equational division (Malone 1983 [*rad50*]; Klapholz et al. 1985 [*spo11*]; Engebrecht and Roeder 1989 [*mer1*]; Menees and Roeder 1989 [*mei4*]; Galbraith and Malone 1992 [*rec104*]; Pittman et al. 1993 [*rec114*]).

During haploid meiosis, where no homologs are present, AEs form, but some SCs are clearly seen; this must result from the synapsis of nonhomologous chromosomes (Loidl et al. 1991). Some class-1 mutants have been tested as haploids and were found to remain unsynapsed. Although this seems to imply that the wild-type gene products are required for nonhomologous synapsis, this conclusion must be approached with caution, since the spreading conditions used in these experiments differed from those used in the analysis of diploids. Conceivably, nonhomologous synapsis may also take place in the relevant mutant diploids but lead to a type of SC that is more easily disrupted.

Recent experiments using an artificially induced DSB during meiosis have shown that *rad50del* and *xrs2* mutants retained the capacity to complete gene conversion at the site of the break. However, many of the observed events were aberrant, indicating that the gene products of *RAD50* and *XRS2* must perform at least one more function in meiotic recombination besides the creation of DSBs (Malkova et al. 1996).

2. *Class 1, Group 2*

A second subgroup of class-1 mutants consists of *mer1+MER2*, *mre2+MER2*, *msh4*, *msh5*, and *dmc1* (*med1*). In these mutants, SC formation appears to be complete and there are normal (or nearly normal) levels of GC, but reduced levels of CO. As explained above, *MER1* and *MRE2* appear to be required for the proper meiosis-specific splicing of the *MER2* transcripts and possibly those of other meiotic genes (Engebrecht et al. 1991; H. Ogawa, pers. comm.). *mer1* cells make unpaired AEs with occasional short stretches of tripartite SC (Engebrecht and Roeder 1990). Overexpression of a spliced *MER2* cDNA restores wild-type SC and GC levels but does not completely suppress the CO defect. The defect in CO seen in *mer1* and *mre2* strains presumably is caused by the lack of splicing of an unknown additional gene(s). When COs do occur, they fail to ensure proper disjunction at meiosis I. Overexpression of *MER2* cDNA not only restores SC formation and GC, but also increases the production of COs that ensure proper segregation (Engebrecht et al. 1990). These authors therefore suggest that only those crossovers that occur in the context of a SC are capable of promoting proper chromosomal segregation. Recent studies (Storlazzi et al. 1995) have shown that the overexpression of *MER2* restores the level of DSB formation, which precedes SC formation. It is likely, then, that the restoration of GC by *MER2* is a consequence of the increased levels of DSBs seen. On the other hand, CO does not occur at wild-type levels in these strains.

MSH4 was isolated in a hunt for genes whose expression is induced during meiosis (Burns et al. 1994). Although the *MSH4* product shows homology with proteins involved in mismatch repair, such as that of the bacterial *MutS* gene, *msh4* strains are proficient in mismatch repair. They show normal levels of GC and a twofold reduced level of CO, and immunofluorescence experiments reveal that Msh4p localizes to discrete sites on synapsed chromosomes (Ross-Macdonald and Roeder 1994). Another MutS homolog, *MSH5*, was identified in a hunt for mutants defective in interhomolog recombination. *msh5* mutants are phenotypically similar to *msh4* mutants, and it is suggested that the two gene products may act together, perhaps as a heterodimer (Hollingsworth et al. 1995).

DMC1 was first isolated from a meiosis-specific cDNA library and identified as a RecA homolog (Bishop et al. 1992). The *med1* mutation, later shown to be allelic to *dmc1*, was isolated in a screen for meiotic-lethal mutants and shows normal levels of GC and an approximate twofold reduction in the level of CO relative to the wild type (Rockmill and Roeder 1994). Tests of doubly mutant strains have placed *DMC1* in different epistasis groups both from *MER1* (group 1) and from the *RED1/HOP1/MEK1* group (group 3). Electron microscopy has revealed that synapsis is delayed but that some apparently normal SCs eventually form (Bishop et al. 1992; Rockmill et al. 1995a) In SK1 strains, a late prophase arrest occurs in the absence of *DMC1* that is not rescued by *spo13* (Bishop et al. 1992). A delay in prophase rather than an arrest is seen in the BR strain background used in the Roeder lab (Rockmill et al. 1995a). Thus, the BR strains may be defective in a checkpoint system that senses the presence of some abnormal intermediate and causes arrest in the SK1 background. Alternatively, the type of arrest shown by SK1 may depend on the rapid pace at which this strain progresses through meiosis, resulting in the absence of coordination between different pathways and leading to failure of the overall process. In the slower BR background, a greater time allotment may permit coordinate completion of all essential functions even in the absence of the *DMC1* gene product.

In summary, the phenotypes of class 1/group 2 mutants demonstrate again that GC can be separated from CO. Moreover, there appears to be a strong association between proficiency in GC and the presence of SC, as is best shown by the joint restoration of these two features (but not CO) in the high-copy *MER2* suppression of the *mer1* or *mre2* defects.

3. Class 1, Group 3

A third subgroup of class-1 mutants is composed of *hop1*, *red1*, and *mek1*. Neither *hop1* nor *red1* shows any formation of SC components, and both mutants display reduced (but not abolished) levels of GC and CO in meiosis. Interestingly, there are locus-specific defects in interchromosomal GC in *red1* strains (Rockmill and Roeder 1990), although the level of meiotic intrachromosomal recombination remains normal in both *hop1* and *red1* (Hollingsworth and Byers 1989; Rockmill and Roeder 1990). Thus, both GC and some reciprocal recombination can take place in the total absence of SC structures. The *HOP1* product contains an essential zinc-binding domain and localizes to the meiotic chromosomes (Hollingsworth et al. 1990). Antibodies to Red1p also localize to the SC as well as to unsynapsed AEs (Nag et al. 1995). Epistasis analysis indicates that *RED1* and *HOP1* function in the same pathway (Rockmill and Roeder 1990) together with *MEK1* (see below). In addition, high-copy-number plasmids carrying *RED1* are able to suppress the defects of certain *hop1* mutations, suggesting that the two proteins interact, presumably within the SC (Friedman et al. 1994), and biochemical characterization of Hop1p provides additional evidence for protein-protein association (K. Muniyappa, pers. comm.). Interestingly, overexpression of *REC104*, an early-acting gene, partially suppresses the phenotype of a *hop1* deletion allele, indicating that some aspects of the *HOP1* function can be bypassed if some early function is provided in excess. A simple interpretation of the phenotypes of these mutants is that Hop1p and Red1p are essential components of the SC. The fact that interchromosomal recombination is largely unaffected and considerable interchromosomal recombination persists in *red1* mutants contributes to the viewpoint that SC formation is not an absolute requirement for the production of recombinants. Increased levels of recombination intermediates throughout the genome resulting from overexpression of *REC104* somehow leads to improved segregation in the absence of proper synapsis. It has been suggested (Weiner and Kleckner 1994) that the formation of recombination intermediates may stabilize pairing interactions. Mutants for *mei4*, *rec102*, and *hop1* all are deficient in the induction in meiotic recombination, whereas *red1* mutants still show about 25% of wild-type levels of CO. Nevertheless, *red1* strains share with mutants for these other loci a similarly strong depression in their levels of homolog pairing, as assessed by FISH (Nag et al. 1995), so it remains doubtful whether recombination necessarily is accompanied by pairing of the homologs.

The level of DSBs in a *hop1* mutant is reduced tenfold (Schwacha and Kleckner 1994). Since DSBs normally appear well before any elements of the SC are detectable (Padmore et al. 1991), *HOP1* may be controlling chromosomal behavior in some manner that is independent of any later role it may have within the SC. Additional evidence for early function by *HOP1*, as well as *RED1*, is found in the fact that *hop1* and *red1* mutations share with the early recombination genes (*rec102*, *rec104*, *rec114*) the ability to rescue the arrest caused by *rad50S* (Mao-Draayer et al. 1996). This indicates that *HOP1* and *RED1* provide functions that are required for entry into the *rad50S* block. Therefore, these functions must be executed before, or at the time of, DSB formation, well before SC usually is formed. However, in contrast to class 1/group 1 mutants, neither *red1* nor *hop1* prevents the inviability of spores produced by a *rad52 spo13* strain, confirming that these represent different functions. Accordingly, *HOP1* and *RED1* were classified as "early synapsis" genes, as opposed to "early exchange" (class 1, group 1) genes (Mao-Draayer et al. 1996). Joint intermediates appear later in *hop1* strains than in wild-type strains and at a reduced level; they are composed mainly of sister chromatids rather than homologs (Schwacha and Kleckner 1995). This defect may, of course, reflect the deficiency for synapsis, since the resolution of DSBs that have failed to become aligned with the homolog may well require an interaction with the intact (and presumably neighboring) sister strand.

MEK1 shows homology with serine-threonine protein kinases. The *mek1* mutant shows apparently normal SCs but has reduced levels of GC and CO. Although this phenotype is quite different from that of *hop1* and *red1*, the phenotypes of double mutants show that *mek1* is epistatic to *red1* and *hop1* (Rockmill and Roeder 1990, 1991). The level of recombination in *mek1*, *red1*, and *hop1* strains is similar, but *mek1* mutants show higher spore viability in a *SPO13* background (30% for *mek1* vs. ~1% for the others; Rockmill and Roeder 1991). This difference could be explained by the fact that *mek1* mutants still produce SCs, and crossovers that occur in the context of an SC are more efficient in providing for appropriate disjunction at meiosis I. The activity of the *MEK1* kinase must also be required for some aspect of recombination or SC complex modification, since *mek1* strains do not show significant reductions in levels of recombination. Additional evidence for a role of these three genes in recombination comes from the fact that they eliminate the aberrant dyads observed in an *spo13* background (Hollingsworth and Byers 1989 [*hop1*]; Rockmill and Roeder 1990 [*red1*]; Rockmill and Roeder 1991 [*mek1*]), presumably because the recombination intermediates that ordinarily would be formed interfere with entry into the *spo13*-dependent equational division.

4. Class-2 (Late) Mutants

Class-2 mutants are defined by their inability to be rescued from inviability by the *spo13*-mediated division that rescues class-1 mutants.

Class-2 genes include not only those that are required in vegetative growth for the recombinational repair of DSBs (*RAD51*, *RAD52*, *RAD54*, *RAD55*, and *RAD57*), but also two meiosis-specific genes (*DMC1* and *ZIP1*). In addition, although a null allele of *RAD50* belongs to class 1, the *rad50S* allele should be assigned to class 2 on the basis of its failure to be rescued by *spo13*. Class-2 mutants are presumed to produce dead spores in *spo13* meiosis because entry into the recombinational process generates chromosomal lesions that cannot be repaired due to deficiency for specific late-acting functions.

One type of class-2 mutant displays defects in the conversion of DSBs into recombinational intermediates which join the homologs together. *rad50S* simply accumulates unprocessed DSBs (Alani et al. 1990; Cao et al. 1990), whereas *rad51* and *dmc1* mutants show 5′ resection of DSBs but do so more extensively than wild-type strains (Bishop et al. 1992; Shinohara et al. 1992). Southern blot analysis shows that both *rad51* and *dcm1* cells produce about 10% the normal level of recombinant DNA molecules, whereas *rad50S* cells form none. In *rad50S* cells, DSBs seem to arise at normal frequencies but are not processed further (Alani et al. 1990; Cao et al. 1990). Analysis of the broken ends accumulating in these strains has shown the presence of a protein bound to the 5′ strand (de Massy et al. 1995; Keeney and Kleckner 1995; J. Liu et al. 1995). Although *RAD50* appears unlikely to encode the meiotic endonuclease itself, the product may act rather directly at the site of breakage, for DSB formation proceeds to a great extent. But the mutant cells are highly deficient in subsequent steps of the pathway, including release of the enzyme from the broken ends. Interestingly, when *rad50S* mutants are returned to mitotic growth after accumulation of DSBs, the breaks are repaired efficiently (with little loss of viability) by a nonrecombinogenic pathway. This has led to the suggestion that the action of the endonuclease may be reversible in a manner similar to that shown by topoisomerases (de Massy et al. 1995; J. Liu et al. 1995).

The *RAD51* and *DMC1* gene products show a high level of sequence identity to the bacterial strand exchange protein RecA and have also been shown to form filaments on double-stranded DNA (Ogawa et al. 1993a; Story et al. 1993). Furthermore, yeast Rad51p can act as a strand-exchange protein in vitro, as can the product of the human counterpart (Benson et al. 1994; Sung and Robberson 1995). The products of *DMC1* and *RAD51* both colocalize to discrete "nucleoids" (Fig. 7) in meiotic chromosomes (Bishop 1994), but the two proteins seem to have rather different roles: The Dmc1p complexes are undetectable in a *rad51* strain, whereas the Rad51p complexes are retained indefinitely in a *dmc1*

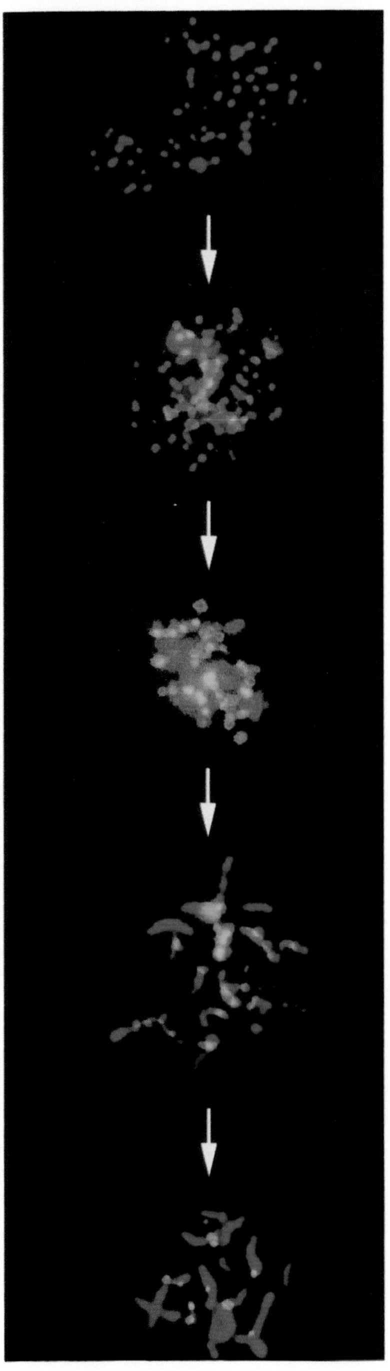

Figure 7 Formation of Dmc1p-dependent subnuclear complexes prior to assembly of the central region of the synaptonemal complex. Immunolocalization images representing successive stages in synapsis reveal that sites of assembly by the recA homolog, Dmc1p (*green*), appear prior to formation of linear structures revealed by antibody to Zip1p (*red*). Sites of colocalization (*yellow*) suggest that the successive replacement of Dmc1p by Zip1p involves a brief stage at which both proteins are present in the same region. Such colocalization does not appear to occur simultaneously in all chromosomes (see Bishop 1994).

mutant. It has been suggested that these "nucleoids" represent recombination nodules (Bishop 1994).

Examples of potential late recombinational functions are those conferred by *RAD52* and *RAD57*, which seem to be necessary for the resolution of recombination intermediates. *rad52* mutants are defective in almost all aspects of mitotic recombination (for review, see Petes et al. 1991). In meiosis, they show normal levels of DSBs (L. Xu and N. Kleckner, unpubl.) and substantial levels of heteroduplex DNA and physical recombinants (Borts et al. 1986; Nag and Petes 1993), but they fail to produce viable spores. A few recombinants can be obtained in a return-to-growth protocol (Resnick et al. 1986), suggesting that under certain conditions, the potentially lethal intermediates can be resolved mitotically. A *rad57* mutant shows normal levels of DNA recombinant molecules (Borts et al. 1986), but it also produces inviable spores, presumably because of a deficiency for a late-acting function.

Although *RAD54* and *RAD55* had previously been reported not to be required for meiotic recombination, new alleles of these genes were obtained in a hunt for mutants with reduced recombination (Ajimura et al. 1993). At least in some backgrounds, deletion of *RAD54* results in spore inviability, as is also the case for *rad51* strains (B. Liefshitz and M. Kupiec, unpubl.). Rad55p and Rad57p also share some homology with the RecA gene product (Kans and Mortimer 1991; Lovett 1994), and mutants have similar mitotic phenotypes (Liefshitz et al. 1995). *rad51* and *rad54* mutants also share common mitotic phenotypes (Liefshitz et al. 1995). The predicted *RAD54* gene product has weak homology with helicases (Emery et al. 1991). Another helicase gene, *SRS2*, interacts genetically with *RAD51* (Aboussekhra et al. 1992). Although recombination is normal in *srs2* mutants, both meiotic divisions are delayed (Palladino and Klein 1992).

The *SEP1* gene product is capable of catalyzing strand transfer reactions in vitro, but mutations in this gene have highly pleiotropic phenotypes, as indicated by their having been isolated as defective in RNA metabolism, microtubule function, regulation of autonomous replication, and telomere stability (for review, see Heyer et al. 1995). In meiosis, *sep1* mutants arrest at pachytene, but the SCs that are formed show aberrant morphology (Tishkoff et al. 1991, 1995; Bahler et al. 1994). Although levels of DSB formation seem normal and recombined DNA molecules can be detected (Bahler et al. 1994), there is substantial variation in the level of meiotic recombination between different loci, strains, and protocols, possibly reflecting the pleiotropic nature of the mutant phenotypes (Dykstra et al. 1991; Tishkoff et al. 1991, 1995;

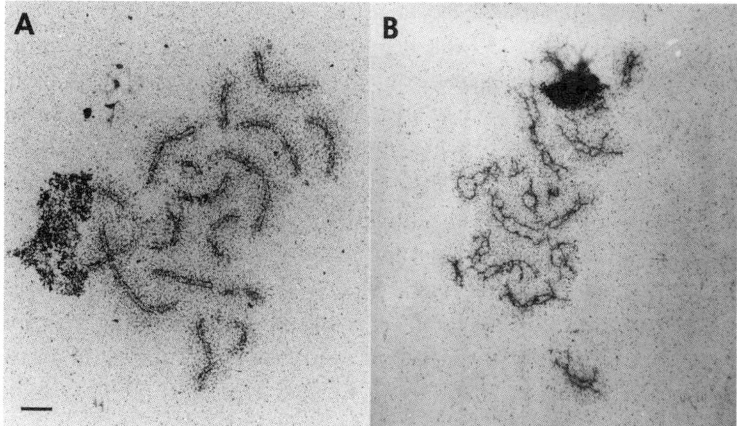

Figure 8 Electron microscopy of yeast meiotic chromosomes prepared by surface spreading and silver nitrate staining. (*A*) Fully synapsed chromosomes from a pachytene nucleus. (Micrograph by K.-S. Tung, courtesy of G.S. Roeder.) Bar, 1 μm. (*B*) Spreading of a *zip1* nucleus reveals full-length axial elements that are homologously paired only at axial associations. (Reprinted, with permission, from Sym et al. 1993 [copyright Cell Press].)

Bahler et al. 1994). *sep1* arrest at pachytene resembles the effects of *dmc1* and *zip1* (see below), but *sep1* is unique among these three in the failure of its arrest to be alleviated by blocking the initiation of recombination. The meiotic arrest of *sep1* mutants therefore cannot be attributed solely to the failure to process DSBs (Tishkoff et al. 1995).

Finally, *ZIP1* encodes a structural component of the central region of the SC (Sym et al. 1993; Sym and Roeder 1994, 1995). In the absence of the Zip1p, AEs (axial elements) become aligned in parallel, but the distance between them (as seen in spread preparations) is variable and greater than the distance between lateral elements in mature SCs (Fig. 8). Anti-Zip1p antibodies localize to the SC but not to unsynapsed AEs. Strikingly, insertional mutations that should increase the length of the protein in its presumed coiled-coil configuration also increase the overall distance between the lateral elements (Sym and Roeder 1995). *zip1* mutants are only mildly reduced for recombination or spore viability in the SK1 background, but they show a striking reduction in chiasma interference (Sym and Roeder 1994), wherein adjacent crossovers are spaced farther apart than would be dictated by chance. Chiasma interference had previously been noted to be widespread among eukaryotes, including *S. cerevisiae*, but restricted to species that display SC during meiosis. The absence of this phenomenon in species lacking SC, such as *S. pombe* and

A. nidulans, clearly implicated SC in the interference mechanism. Consistent with this viewpoint, it has been shown that late recombination nodules, which are intimately associated with the SC during the latter stages of pachytene, display a similarly nonrandom distribution along the SC (Carpenter 1975; Laurie and Jones 1981). Tetrad analysis has shown that chiasma interference is abolished not only in *zip1* mutants (Sym and Roeder 1994), but also in strains mutant for *msh4* (G.S. Roeder, pers. comm.) which encodes another component of the SC. These findings thereby confirm and extend the evidence that the SC assists in some manner to confer chiasma interference on the distribution of crossovers. Interestingly, COs that form in *zip1* mutants are able to ensure chromosome disjunction, whereas those formed in *mer1* or *red1* strains, which also show 10–20% of wild-type levels of exchange but lack complete SCs, do not ensure proper disjunction (Engebrecht et al. 1990; Rockmill and Roeder 1991). These results are consistent with an additional role for the SC in the creation of functional chiasmata. *zip1* mutants show normal (or elevated) levels of hDNA, indicating that Zip1p acts late in the recombination process (Nag et al. 1995).

K. Distribution of Crossovers

Besides the crucial insight into the mechanism of chiasma interference (described above) as provided by analysis of *ZIP1* (Sym and Roeder 1994), additional relevant information is provided by experiments in which the sizes of yeast chromosomes were altered experimentally. Yeast chromosomes normally range in size from 250 to 1250 kb. On the basis of theoretical calculations, it was concluded that the smallest chromosomes must undergo more recombination per unit physical length than the bigger chromosomes in order to ensure proper segregation in all meioses (Kaback et al. 1989). When chromosome I, the smallest chromosome, was bisected to create two functional half-chromosomes, the recombination rates for the bisected halves (when made homozygous) were increased. Conversely, when chromosomes I and II were fused by a reciprocal translocation, thereby locating chromosome I sequences within the context of a larger chromosome, recombination was decreased (Kaback et al. 1992). These experiments demonstrate a size-dependent control on recombination levels in yeast, such that smaller chromosomes will more likely undergo crossing over in any one meiosis than expected simply from their lengths. Although the relevant mechanisms remain obscure, the authors note that the observed phenomena could be explained by an increased strength of chiasma interference for the larger chromosomes (Kaback et al. 1992).

In addition to the effects of chiasma interference, the distribution of crossover events is also controlled in many eukaryotes by proximity to specific chromosomal components. Notably, *Drosophila melanogaster* and *S. pombe* have been shown to be strikingly repressed for crossing over in the regions surrounding the centromeres (Dobzhansky 1931; Nakaseko et al. 1987), but conflicting results have been obtained for *S. cerevisiae* (Lambie and Roeder 1986; Symington and Petes 1988). Comparison of the genetic map with the DNA sequence data now available for most yeast chromosomes (Mortimer et al. 1992) fails to reveal any evidence for significant repression of recombination in centromere-proximal regions of the yeast map.

Telomeres similarly may repress adjacent meiotic recombination in some species, but probably not significantly in yeast. The ends of yeast chromosomes are usually heterogeneous in length and are occupied by various families of repeated sequences (Louis and Haber 1990; Louis et al. 1994). Single-copy genes present in this milieu therefore vary in their distance from the physical chromosomal end, complicating the analysis of whether proximity to the telomeres affects the level of recombination. It is nevertheless clear that repeated subtelomeric sequences are able to undergo meiotic recombination with similarly located sequences on other chromosomes (Louis and Haber 1990; see below).

In addition to possible centromere and telomere effects, the distribution of meiotic crossing over on the remainder of the chromosome is clearly nonrandom. Early experiments showed that the mitotic and meiotic genetic maps differ (Malone et al. 1980). "Cold spots" and "hot spots" for recombination have been described and confirmed by sequencing. For example, the *MAT-CRY1* (Larkin and Woolford 1983) and *TRP1-CEN4* (Meddle et al. 1985) regions show quite low recombination per physical length in comparison with the rest of the genome. Examples of hot spots are the *CDC24-PYK1* interval on chromosome I (Coleman et al. 1986) and the *THR4* region on chromosome III (Goldway et al. 1993a,b). There also are regions that show strikingly increased rates of gene conversion, including *ARG4* (Schultes and Szostak 1990), *HIS2* (Malone et al. 1992), and *CYS3* (Ohta et al. 1994). A complication in identifying relevant sequences is that these frequencies often are influenced by genetic background, such as the situation for *HIS4* (Detloff et al. 1992: Nag and Petes 1993) and by the nearby integration of foreign sequences (Cao et al. 1990; Stapleton and Petes 1991; de Massy and Nicolas 1993; White et al. 1993; Wu and Lichten 1995). In some manner, these differences in recombination levels seem to result from changes in chromatin configuration (Wu and Lichten 1994, 1995; Ohta et al. 1994) and reflect the differential susceptibility to DSB formation in different backgrounds (Zenvirth et al. 1992; Goldway et al. 1993a,b; Wu and Lichten 1994).

Another region of the genome that behaves uniquely in meiotic recombination is the rDNA—a cluster of about 200 tandemly arranged 9-kb repeats located on chromosome XII. The genetic map of this region shows a 100-fold lower frequency of reciprocal exchanges per DNA length than other regions (Petes and Botstein 1977), whereas unequal sister-strand recombination occurs frequently (Petes 1980). It may be relevant that electron microscopy (Byers and Goetsch 1975; Dresser and Giroux 1988) has failed to reveal the presence of any SC in the nucleolus during meiosis. Consistently, antibody staining for Zip1p (Sym et al. 1993) and Hop1p (F. Klein and B. Byers, unpubl.) shows these SC components to be distinctly absent from the nucleolar regions of spread meiotic preparations. These findings are consistent with an important role for SC in crossing over between homologs and suggest that it may also aid in distinguishing between the sister chromatid and the homolog. However, the observed differential effects on intersister versus interhomolog recombination may simply reflect the fact that no other part of the genome is so rich in repeated sequences. In contrast to rDNA sequences, the subtelomeric Y′ repeats frequently undergo recombination with homologous sequences in nonhomologous positions in subtelomeric regions of other chromosomes (Louis and Haber 1990; Louis et al. 1994). Ectopic recombination by subtelomeric Y′ repeats with homologous internal sequences also occurs, but much less frequently than between two subtelomeric regions (Huckle and Louis 1995).

Dispersed repeated sequences in yeast also show peculiar behavior with respect to recombination. Ectopic recombination between Ty elements, a family of approximately 6-kb-long repeated sequences present in about 80 copies per diploid genome, occurs at a low level relative to certain artificially created repeats (Kupiec and Petes 1988). Furthermore, those gene conversions that do occur between Ty elements are characterized by a relatively low level of reciprocal exchange. One notes that a high level of exchange between dispersed repeated sequences should cause extreme genomic instability, since this would generate chromosomal aberrations, such as translocations, duplications, and deletions in almost every meiosis (Kupiec and Petes 1988). The relevant mechanism for inhibition of Ty recombination remains unknown. Although Ty elements have also been shown to be defective in damage-induced mitotic recombination (Parket and Kupiec 1992), mitotic recombination can be induced either by a site-directed DSB (Parket et al. 1995) or by increased transcription (Nevo-Caspi and Kupiec 1994). Perhaps a global regulation of recombination between repeated sequences is conferred by topological or conformational changes in chromatin, such as by the action of topoisomerases. Notably, mutations in the three topoisomerases of yeast (*TOP1*, *TOP2*, and *TOP3*) cause an increase in rDNA recombination (Christman et al. 1988; Gangloff et al. 1994); *top3* mutants also show increased recombination between the long terminal repeats (LTRs) of Ty elements (Wallis et al. 1989).

Additionally, transcription appears to affect recombination significantly, perhaps by virtue of its effects on topology or chromatin configuration. When inserted in a novel genomic position, a fragment of the rDNA repeat that contains promoter sequences increases both intrachromosal and interchromosomal recombination (Keil and Roeder 1984). Furthermore, a transcriptional terminator abolishes this induction, showing that the inducing effect is exerted through transcription per se (Voelkel-Meiman et al. 1987). Similarly, Ty recombination is increased when the Ty promoter is replaced by the highly inducible *GAL1* promoter (Nevo-Caspi and Kupiec 1994), and this induction is also prevented by a transcriptional terminator (Y. Nevo-Caspi and M. Kupiec, in prep.). Intrachromosomal mitotic recombination at the *GAL10* locus is also increased by transcription (Thomas and Rothstein 1989). Although these examples are mitotic, transcription has also been shown to interfere with recombination in meiotic cells (Rocco et al. 1992). Moreover, *sir2*-dependent defects in chromatin structure and transcription are accompanied by effects on both mitotic and meiotic recombination at the rDNA locus (Gottlieb and Esposito 1989). Recent studies at the *HIS4* locus suggest that transcription factors may interact with the meiosis-specific endonuclease that catalyzes DSB formation (Fan and Petes 1996). There clearly must be important relationships between transcription factors, chromatin accessability, and recombination hot-spot activity that remain to be elucidated (Wu and Lichten 1994, 1995; Ohta et al. 1994; Fan and Petes 1996).

L. Overview of Meiotic Synapsis and Recombination

The analysis of mutants has led to several conclusions about synapsis and recombination: (1) Synapsis is not absolutely required for meiotic recombination, for some mutants with no detectable SC (*red1*, *hop1*) still show some meiotic recombination (Hollingsworth and Byers 1989; Rockmill and Roeder 1990). Conversely, certain mutants that lack meiotic recombination can form either AEs (*rad50S*, Alani et al. 1990) or complete SCs (*spo11-1*, Klapholz et al. 1985). Several mutants with SC of normal appearance are reduced for recombination (*med1/dmc1*, Rockmill and Roeder 1994; *mek1*, Rockmill and Roeder 1991; *mer1+MER2*, Engebrecht et al. 1990; *msh4*, Ross-Macdonald and Roeder 1994). (2) COs and GCs appear at the same time but are genetically separable, as exemplified by mutants of the class-1/group-2 category (*mer1+MER2*, *mre2+MER2*, *msh4*, *msh5*, and *med1*). These mutants are defective in CO but show nearly normal levels of GC. Other mutants, such as *hop1* and *red1*, are more severely affected in their ability to produce COs than GC events. (3) GC and SC formation appear to be functionally related. This is exemplified by the restoration of both SC formation and high GC levels when *MER2* is overexpressed in a *mer1* strain (Engebrecht and Roeder 1990; Storlazzi et al. 1995). (4) COs alone are insufficient to ensure proper disjunction; they appear competent only if formed in the context of the SC (Engebrecht et al. 1990). Consistent with this viewpoint, an asynaptic *red1* mutant forms COs but nondisjoins (Rockmill and Roeder 1990). We note in this regard not only that addition of a hot spot for DSB formation to an artificial chromosome is sufficient to enhance proper meiotic disjunction (Ross et al. 1992; Sears et al. 1994), but also that random DSBs created by X-rays in a recombination-deficient *spo11* strain increase spore viability, presumably by enhancing proper disjunction, but the enhancement is marginal (Thorne and Byers 1993). It seems likely that only naturally formed DSBs, which are processed within the SC, are capable of forming functional chiasmata efficiently. (5) The formation of AEs is independent of many early recombinational functions but is dependent on *RED1* (Rockmill and Roeder 1990) and *HOP1* (Hollingsworth et al. 1990). AE formation can occur in the absence of DSBs but seems often to be reduced in parallel with DSB formation, as evidenced by the different levels seen for *rad50del*, *rad50S*, and *dmc1* strains (Alani et al. 1990; Cao et al. 1990; Bishop et al. 1992). On the other hand, SC formation is strongly correlated with the execution of early functions.

How does meiotic recombination proceed? When cells enter meiosis, certain regions of the genome are particularly susceptible to DSB formation. These regions often coincide with the promoters of genes and are contained within chromatin of unique properties (Wu and Lichten 1994). Such regions of DSB formation are nonrandomly distributed in the genome and numerous enough that the chiasmata eventually formed are sufficient to ensure the appropriate disjunction of most, if not all, chromosomes at meiosis I. Whether or not chromosomal pairing (as defined by FISH) is an important precondition for DSB formation remains to be established. The molecular nature of the meiotic nuclease(s) responsible for DSB formation is undefined, but the activity clearly is dependent in some manner on the activities of several genes, including *SPO11, RAD50, XRS2, MRE11, REC102, REC104, REC114, MER2, MEI4*, and *HOP1*. Once the DSB is formed, resection of the 5′ ends ensues. This resection appears to be essential for recombination since mutants that lack DSBs or end processing do not show meiotic induction of CO or GC (see, e.g., Cao et al. 1990; Ivanov et al. 1992). Mutants that extend processing beyond the *rad50S*-blocked step and show some degree of end processing (such as *rad51* and *dmc1*) display at least limited recombination (Bishop et al. 1992; Shinohara et al. 1992).

Coincident with synapsis, DSBs are processed to form branched intermediates that have the characteristics expected from the model for DSB-initiated recombination (see Fig. 3). They consist of paternal and maternal DNA strands that are intertwined by two Holliday junctions and are fully ligated (Schwacha and Kleckner 1994, 1995). By this point, then, homologous sequences have come into intimate contact, presumably by the action of recA-like strand-transfer proteins, including Dmc1p and Rad51p. The relationship between mature SC appearance and joint molecule formation remains to be clarified. From mutational analysis, we know that AE formation is independent of DSB creation and processing, whereas complete SC formation and DSB processing require many of the same functions. This suggests an essential role for the formation of joint molecules in SC assembly, but the alternative possibility that SC formation facilitates joint molecule formation cannot be excluded. It is not yet known how homologs are distinguished from sister chromatids in joint molecule formation. Cohesion of sister chromatids is probably essential for the stability of chiasma, as suggested by the segregational defects of *DIS1* mutants. Physical analysis of the joint molecules shows that although the majority join homologs (Collins and Newlon 1994b; Schwacha and Kleckner 1994), some join sister chromatids, especially when formed in *hop1*, *red1*, and *mek1* strains. Presumably, the abundance of these latter forms reflects the related defects in normal synapsis of homologs, whereas the presence of such forms in normal meiosis might suggest a role for them in sister:sister adhesion.

Elucidating the ensuing steps of meiotic recombination will require further study. Theoretical considerations demand the resolution of joint intermediates and the repair of mismatches. Heteroduplex DNA is not detected until the time when joint molecules are resolved into recombinant DNA sequences. Resolution of joint intermediates might lead to either reciprocal (CO) or nonreciprocal (GC) products (Storlazzi et al. 1995), but it is unknown whether these two types of products derive from distinct pathways or are simply two modes of resolution of the same intermediate. The parts played in such processing here or at other stages by genes of the recombinational DNA-repair group, such as *RAD52* and *RAD57*, remain to be defined. Heteroduplex DNA, which clearly is created near sites of crossing over, is not readily detected until the later stages of meiotic prophase. Additionally, the manner in which the products of *ZIP1* and *MSH4*/*MSH5* establish the distribution of COs, thus helping to ensure at least one chiasma per chromosome (Sym and Roeder 1994; S. Roeder, pers. comm.), demands further analysis.

IV. CHROMOSOME SEGREGATION IN THE MEIOTIC DIVISIONS

A. SPB/Spindle Cycle Landmarks

The reductional segregation of centromeres is a key feature of meiosis I, distinguishing this mode of nuclear division from mitosis or meiosis II. Thin-section electron microscopy in yeast has failed, however, to reveal any distinctive structural difference in formation of the spindle for meiosis I. As in mitosis, a cell entering meiosis bears a single unduplicated spindle pole body (SPB) embedded in the nuclear envelope. A bundle of about 20–30 microtubules emanating from the nucleoplasmic surface constitutes a half-spindle, whereas fewer microtubules extend outward from the cytoplasmic surface. One edge of the SPB abuts the half-bridge, which appears as a darker-staining portion of the nuclear envelope (Byers and Goetsch 1974). The half-bridge has a role in duplication of the SPB in preparation for both mitosis and meiosis: Creation of the new SPB is initiated by the formation of the satellite—a small patch of similar-appearing material on the cytoplasmic surface of the half-bridge distal to the SPB proper. SPB duplication is subsequently completed by expansion of the satellite and its insertion into the envelope, whereupon it gains an array of nuclear microtubules and becomes a second SPB. This transition from satellite-bearing single SPB to a completely duplicated pair of SPBs normally coincides with Start in the mitotic cell cycle and closely precedes entry into S phase. In meiosis, SPB duplication similarly occurs at about the time when DNA synthesis begins.

Once initiation of the meiotic pathway has taken place, the SPBs remain in the side-by-side configuration throughout S phase and prophase of meiosis I. Later, when synaptonemal complexes begin to dissociate (marking the end of pachytene), the bridge joining the two SPBs is cleaved and the SPBs move apart to form the spindle for meiosis I. Although this spindle is morphologically similar to one formed for mitosis (Fig. 9), anaphase spindle elongation in meiosis I (as well as in meiosis

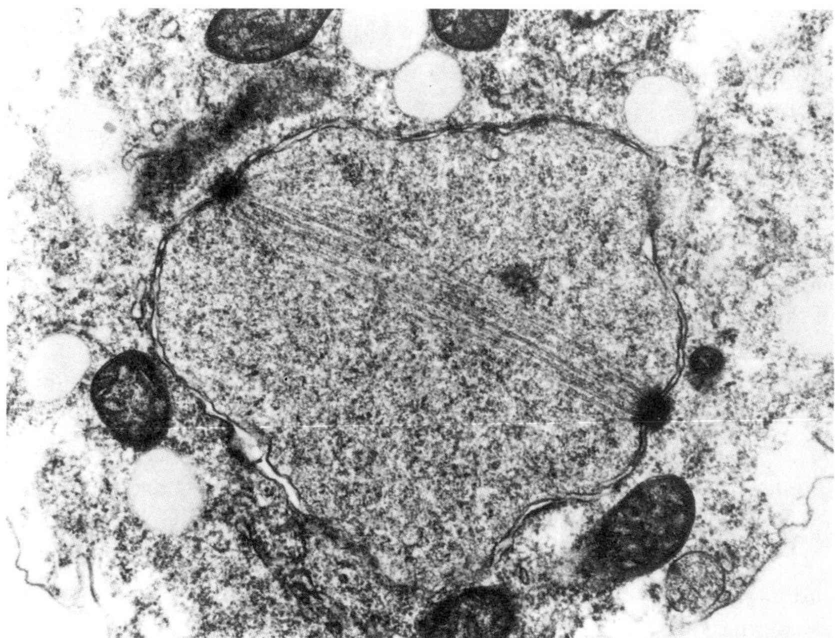

Figure 9 Electron microscopy of thin section through a meiosis I nucleus shows the typical spindle structure, which is similar to that seen during mitosis. (Courtesy of S. Klapholz and R.E. Esposito.)

II) must differ somewhat from mitotic anaphase because of the more restricted size of the meiotic cell. Since no bud forms in meiosis, meiotic spindles can elongate only to the diameter of the ascal wall. Meiosis I also differs in the events that follow spindle elongation: In mitosis, a thin remnant of the spindle persists in the neck of the bud until the cleavage furrow closes on it, effecting nuclear division and cytokinesis simultaneously (Byers and Goetsch 1976). In meiosis I, there is a partial constriction of the nuclear envelope in the equatorial region, but the spindle remains intact and nuclear division is incomplete (Moens and Rapport 1971a), despite the apparent formation of two nuclei when viewed by staining with DAPI or propidium iodide for light microscopy.

Following meiosis I, both SPBs undergo duplication in preparation for meiosis II, leading to the presence of two side-by-side pairs of SPBs within the single nucleus. These pairs later undergo separation from one another as the meiosis I spindle disappears and two spindles for meiosis II are formed. The two meiosis II spindles frequently assume a crossed configuration within the single nucleus (Moens and Rapport 1971a). Appearing at this stage is a striking modification of each SPB that later plays a critical part in spore wall formation (as described later). The cytoplasmic component layer of each SPB (the *outer plaque*) becomes expanded both in its width (tangential to the nuclear surface) and in its thickness. This modification arises while the duplicated SPBs are still in the side-by-side configuration and persists throughout meiosis II. Elongation of these two spindles for anaphase II forces the spindle poles outward, thereby creating four lobes extending outward from the surface of the formerly spherical nucleus.

Electron microscopy has revealed that yeast meiosis is incomplete not only in the first division, but also in this final segregation of nuclear lobes at meiosis II. One sees that a developing spore wall envelops each pole of the meiosis spindle: A portion of the nucleus that is more darkly staining in electron microscopy than the rest generally is excluded from any of the four meiotic products, remaining in the ascal space. Since all chromosomes must have been segregated to the spore nuclei, only non-chromosomal elements of the original nucleus could be omitted from packaging into the nascent spores. Brewer et al. (1980) have shown in fact that all chromosomal elements, including rDNA, are transmitted fully through meiosis, whereas there is partial loss of the yeast 2-micron plasmid, perhaps among the nuclear components that remain in the ascus space.

B. Chromosomal Behavior in the First Meiotic Division

As noted above, there is a prolonged delay in meiosis I between the times of SPB duplication and spindle formation. The duplicated SPBs remain in the side-by-side configuration throughout S phase and the prolonged meiotic prophase. Chromosomal behavior, which is difficult to discern in most phases of the yeast life cycle, becomes evident at this stage due to formation of synaptonemal complexes, which are readily detectable by electron microscopy. It is only at pachytene, when SCs extend from telomere to telemere, that we can be certain of the positions of the chromosomes within the nucleus. Two features are evident from visualization by thin sectioning: First, in yeast, as in many other organisms (see von Wettstein et al. 1984), both telomeres of a chromosome are associated with the nuclear envelope, so that the SC in favorable views appears to extend perpendicularly from the inner face of the envelope into the nucleoplasmic space. It has been suggested for some species that these attachments to the envelope aid synapsis by limiting the locus of homologous regions to the nuclear cortex, thereby facilitating the search for homology between homologs. This tantalizing idea has been cast in doubt, however, on the grounds both that synapsis often can be seen to initiate in proximal regions of the chromatids and that synapsis proceeds readily in organisms where envelope attachment is lacking (Goldstein and Moens 1976). Futhermore, it has been shown in yeast that ring chromosomes, which have no telomeres, suffer no deficit in crossing over relative to the linear control (Haber and Thorburn 1984), so telomeric localization on the envelope may well be insignificant to synapsis.

Another notable feature of the pachytene bivalents evident from electron microscopy is the absence of any apparent association with the half-spindles extending into the nuclear interior from the side-by-side SPBs (Byers and Goetsch 1975). Although the centromeres have not been visualized microscopically, they may be assumed to lie somewhere along the contours of the SC or even to specify a modification in the structure of SC (as seen in other species) immediately adjacent to where they lie (Moses 1977). During pachytene, spindle microtubules form a compact bundle with one end attached to the SPB. The other end of the microtubule bundle lies within the nuclear interior, where it often lies distant from any profiles of SCs, excluding the possibility that these microtubules are attached to the chromosomes. Spindle microtubules were seen to regain their association with the still-condensed chromosomes as the spindle for meiosis I was formed (Goetsch and Byers 1982).

The molecular analysis of the centromere in *S. cerevisiae* is facilitated by its extraordinary small size, spanning less than 200 bp along the DNA helix (see Newlon 1988; Clarke 1990; Hegemann and Fleig 1993). Sequence similarities and extensive experimental manipulation have defined the segments required for various centromere functions, which include the stabilization of plasmids in single-copy number and the regular segregation of plasmids to ascospores (Panzeri et al. 1985; Futcher and Carbon 1986; Cumberledge and Carbon 1987; Cottarel et al. 1989; Gaudet and Fitzgerald 1989). Centromere DNA (CEN DNA) sequences are seen to consist of three centromeric DNA elements in linear array: CDE1, CDEII, and CDEIII. CDEI is a highly conserved 8-bp sequence with bilateral symmetry. CDEII is 78–86 bp in length and shows little conservation except that it is more than 90% A and T residues. This is followed by the 26-bp-long CDEIII sequence, which also is bilaterally symmetric and contains a highly conserved central core. Saturation mutagenesis has revealed important roles for CDEI and CDEIII in mitotic segregation, whereas small changes in CDEII do not significantly affect centromere function (Hegemann and Fleig 1993). In meiosis, mutations in CDEI have only minor effects, whereas alterations in CDEII substantially disrupt centromere function in both meiotic divisions (Panzeri et al. 1985; Cumberledge and Carbon 1987; Gaudet and Fitzgerald 1989; Sears et al. 1995). It has been suggested (Gaudet and Fitzgerald 1989; Murphy et al. 1991) that the AT-rich sequence may be responsible for inducing a bend in the DNA that is required for meiotic centromere function.

CEN DNA adopts a specific chromatin structure that is essential for centromere function (Bloom et al. 1989). Several proteins have been shown to bind CEN DNA with specificity; these include the *CBF1* product, which binds to CDEI. The Cbf1 protein is not essential, but in its absence, chromosome losses are elevated tenfold in vegetative growth (Cai and Davis 1990; Mellor et al. 1991). These mutants have also shown a general increase in plasmid nondisjunction during both meiotic divisions (Masison and Baker 1992). A complex of three proteins (CBF3) binds to CDEIII in a phosphorylation-dependent manner (Lechner and Carbon 1991). Interestingly, a screen designed to isolate extragenic high-dosage suppressors of mutations in CDEIII yielded *MCK1*, which encodes a protein kinase known to be required for entry into meiosis (Neigeborn and Mitchell 1991; Shero and Hieter 1991).

It remains to be established whether any meiosis-specific proteins interact with centromere sequences (perhaps binding to CDEII) to provide for the cosegregation of sister centromeres that is unique to meiosis I. A long-standing model for the reductional division has held that sister centromere duplication was specifically delayed in some manner until after meiosis I. The possibility that this behavior might be exerted not by associated proteins but by delaying centromeric DNA replication has now been excluded by using two-dimensional gel techniques to analyze replication forks in premeiotic S phase (Collins and Newlon 1994a). These workers noted a minor delay in replication fork progression at the centromere of chromosome III, but this delay was no longer than one that also occurs in mitotic S phase, and it clearly was too brief to account for a postponement of centromere replication until after meiosis I. Accordingly, the basis in sister cosegregation seems likely to lie in the behavior of other centromere-associated components. Cytologically, it has been found in male *Drosophila* that the kinetochore (the cytological equivalent of the centromere) first appears in meiosis I as a single spherical body encompassing the centric regions of both sister chromatids. This single structure then undergoes a striking morphological rearrangement to form two individual kinetochores (one on each chromatid) just as the chromosomes attach to the spindle in a manner such that both appropriately become directed toward the same pole (Goldstein 1981). Whether analogous structures exist on the much smaller centromeres of yeast remains to be determined.

C. Homolog Association for the Reductional Division

Besides the requirement for sister-centromere association, successful formation of chiasmata is crucial for proper reductional segregation in meiosis I. This has been found to be true in every known situation where crossing over ordinarily occurs. In those meiotic pathways where crossing over is absent, special mechanisms have evolved to ensure adhesion of the homologs to one another as they align on the metaphase plate for meiosis I (von Wettstein et al. 1984). In male *Drosophila*, the so-called collochores perform this function (Cooper 1964), whereas the synaptonemal complex is retained and modified to serve this role in female *Bombyx* (Rasmussen 1977). On the other hand, reductional division generally depends on the retention of chiasmata at the sites of crossing over to hold the homologs together. This permits the bivalents to interact with the spindle in a stable manner dependent on the apparent balance of tension as force is exerted toward either spindle pole by spindle fibers attached to the centromeres of either homolog (Nicklas 1977).

In species where chromosomes can readily be seen, the univalents produced when chiasmata are lacking due to recombinational defects are unable to congress to the metaphase plate or to remain stably associated with the spindle and nondisjunction occurs. Analogous observations have been obviated in yeast by the small size of the chromosomes and their apparent lack of much, if any, condensation during division (Moens and Rapport 1971a,b; Byers and Goetsch 1975), but there is compelling evidence for an analogous dependency of proper disjunction on synapsis and crossing over. Several mutations described above (notably *spo11, mer1, mer2, red1, rad50*, and *hop1*) that reduce or eliminate recombination during meiotic prophase ultimately result in spore lethality. Evidence that this lethality arises because of attempted reductional division in the absence of chiasmata is provided by the restoration of spore viability when recombination-deficient mutants are also made homozygous for *spo13*. As first shown for *rad50* and *spo11*, a single division gives rise to two viable nonrecombinant spores of the same ploidy as the parental strain (Malone and Esposito 1981; Klapholz et al. 1985). As described above, this has been exploited to distinguish mutations that fail to undergo recombination from mutations, such as *rad52* (Malone 1983), that are deleterious independently of disjunctional defects.

Further evidence that crossing over guides disjunction is provided by definitive experiments of Surosky and Tye (1988), who created chromosomes that had homologous arms but had their centromeres replaced by those from other chromosomes. The resulting faithful disjunction clearly showed that proper segregation depends on homology in the arms, where pairing and crossing over could occur, rather than in the centromeres themselves. However, recent experiments indicate that exchanges differ in their ability to promote faithful reductional segregation of the homologs, with crossovers that occur near the telomere being less effective than those that arise more proximally (Ross et al. 1996).

D. Global Control of Meiotic Division by Cyclin-dependent Kinase

The global control of nuclear division cycles by CDKs (cyclin-dependent kinases) in yeast has been more thoroughly explored in mitotic cells than in meiosis. The formation and action of the mitotic spindle depend on the activity of a CDK, which consists of a serine-threonine protein kinase encoded by *CDC28* complexed with a cyclin that acts as an activating regulatory subunit (for review, see Draetta 1994). B-type cyclins accumulate in G_2 and M phases, serving to potentiate *CDC28* activity that is necessary, but not sufficient, for mitosis (Bartlett and Nurse 1990). Following anaphase, the cyclins are ubiquitinated and destroyed, leading to dissolution of the spindle and exit from mitosis (Murray et al. 1989; Surana et al. 1993). Six B-type cyclins have been described in yeast (*CLB1-6*; for review, see Nasmyth 1993). Among them, *CLB5* and *CLB6* are most directly required for S phase, whereas the other four act later to control the mitotic division. The *CLB2* gene product acts as the main mitotic cyclin; although a *clb2* mutant shows only a slight phenotype in isolation, *clb2* deletions in combination with mutants for any of the other late-acting cyclin genes (*clb1*, *clb3*, or *clb4*) have profoundly deleterious effects (Fitch et al. 1992; Richardson et al. 1992).

In meiosis, *CLB2* is expressed at very low levels (Grandin and Reed 1993), and its deletion has no meiotic phenotype (Grandin and Reed 1993; Dahmann and Futcher 1995). The main role for cyclins in meiosis is played instead by the products of *CLB1* and *CLB4*. Strains lacking *CLB1* have low sporulation efficiencies, and *clb1 clb4* double mutants produce asci containing two diploid spores (Dahmann and Futcher 1995). These defects appear to reflect roles for these cyclins in the activation of the *CDC28* kinase function for sporulation, since *cdc28* cells share with mutants for other Start genes (*cdc36* and *cdc39*) the ability to cause meiotic arrest at pachytene, when the SPBs remain in the side-by-side configuration (Shuster and Byers 1989). Many of the *cdc28* cells that are permitted to execute this earlier function at the permissive temperature suffer a defect in meiosis II, leading to the formation of diploid dyads. In this case, electron microscopy has shown that the two pairs of SPBs appearing after meiosis I fail to undergo separation for meiosis II spindle formation. Nevertheless, outer plaque and prospore wall formation proceed and two diploid spores arise. Although the meiosis II defect is partially penetrant (affecting ~20% of cells) and not temperature-sensitive, it seems to reflect loss of the *CDC28* kinase activity, as indicated by the fact that the analogous *S. pombe* mutation (*cdc2*) also causes such single division meiosis (Niwa and Yanagida 1988). It seems likely that defects in Cdc28-cyclin activity in meiosis cause a failure in separation of SPBs and in meiosis II spindle formation, causing an exit from meiosis after the first meiotic division. This aberrant pathway of spore formation may share certain key features with that caused by *spo13*, as described below.

E. Role of *SPO13* in the Meiotic Divisions

Yeast strain ATCC4117, as originally described by Grewal and Miller (1972), undergoes a single meiotic division to yield dyads of diploid spores. The cytological basis was clarified by work of Moens (1974), who discovered that the enlarged outer plaque on the SPBs, which makes its appearance during meiosis II in the wild type, arises instead at the time of the single meiotic division shown by this strain. As in the wild type, outer plaque formation is quickly followed by the initiation of prospore wall formation, such that each spindle pole gains a spore wall and just two, rather than four, spore walls are induced to form in the mutant ascus. Detailed characterization of the genetic basis for this behavior by Klapholz and Esposito (1980a,b) showed that this strain contains relevant genetic elements at two loci. Strains homozygous for either *spo12* or *spo13* undergo single-division meiosis. Recombination occurred normally in either *spo12* or *spo13* meiosis, but most centromere-linked markers remain heterozygous, failing to undergo reductional segregation in the manner typical of wild-type meiosis I. This variation on the normal meiotic pathway can therefore be viewed as representing a bypass of meiosis I followed by the normal execution of meiosis II. It was also reported that individual dyads may show some reductional segregation for some centromeres and equational segregation for others. Such mixed patterns of segregation have been established more widely by analyzing the meiosis of *spo13* strains bearing heterozygous markers at several centromeres (Hugerat and Simchen 1993). Furthermore, it is now clear that the relative tendency for either pattern of centromere segregation is influenced by genetic background as well as by the extent of synapsis and/or recombination on the chromosomes in question (Hollingsworth and Byers 1989).

Notwithstanding the low level of reductional devision, the predominant equational pattern provides a valuable tool for dissection of the meiotic pathway, as it enables one to recover viable progeny following a meiotic prophase that is deficient for synapsis and/or recombination. Although either *spo12* or *spo13* individually causes single-division meiosis, the functions of the two loci are distinguishable. Spore inviability resulting from recombination-free meiosis (such as occurs in strains homozygous for *rad50* or *spo11*) is corrected when the strains also are made homozygous for *spo13*, consistent with a bypass of the first meiotic division. But homozygosity for *spo12* fails to restore viability to these recombinationally defective mutants (Malone and Esposito 1981). The basis for this difference remains obscure. The *spo13*-dependent single equational division also allows haploid cells that are provided with appropriate heterozygosity at the mating-type locus to sporulate and form dyads containing two viable haploid spores (Wagstaff et al. 1982). In haploid cells carrying two copies of chromosome III, all the monosomic chromosomes segregate equationally, whereas the two copies of chromosome III frequently show reductional segregation (Wagstaff et al. 1982; Engebrecht and Roeder 1989; Gottlieb et al. 1989; Hollingsworth and Byers 1989).

It has been proposed (McCarroll and Esposito 1994) that the *SPO13* gene product has its central role in the regulation of the meiotic divisions by controlling their timing. Specifically, *SPO13* may execute a function that serves to delay the first meiotic division until the chromosomes or segregation apparatus has been appropriately modified to perform a reductional segregation. Furthermore, *SPO13* may also be involved in preventing a return to interphase and DNA synthesis following the first division. Several observations support this model: (1) *spo13* strains often produce three- and four-spored asci if meiosis takes place at very low temperatures or in the presence of low levels of hydroxyurea. Under these conditions, the rate of progression through the early stages of meiosis is markedly reduced and any additional delay that normally would depend on *SPO13* may no longer be required. Conversely, overexpression of *SPO13* during sporulation imposes an unusually long delay at meiosis I but does not affect meiosis II. (2) Overexpression of *SPO13* in vegetative cells causes arrest in G_2 or M phase. This arrest depends on the activity of *CDC28*, which encodes the catalytic subunit of the CDK.

(3) In meiosis, *cdc28* mutants arrest in pachytene at elevated temperatures. At lower temperatures, many cells complete the first meiotic division and bypass the second one, forming dyads of diploid spores (Shuster and Byers 1989). Overexpression of *SPO13* during meiosis causes an approximate 50-fold increase in the fraction of asci with three and four spores; overexpression of *SPO13*, however, cannot replace or bypass the *CDC28* function, but seems instead to act by stabilizing or enhancing the kinase activity.

These findings provide the basis for the following hypothesis (McCarroll and Esposito 1994): Each chromosome must be modified in some manner for it to enter into a reductional segregation. For example, some provision must be made to keep the sister centromeres together during meiosis I. The *SPO13* gene product would act by delaying entry into the division until such modification was complete. Accordingly, in the absence of *SPO13* function, the cell would enter division prematurely; those chromosomes that had been modified would undergo reductional segregation, whereas those that had not would segregate equationally by "default," thus leading to a mixed division. Conditions that slowed the onset of meiosis (such as low temperature or hydroxyurea) would result in increased reductional segregation, as well as in the fraction of cells that undergo two divisions. Conversely, if the onset of division were accelerated, then less reductional segregation would occur. It is perhaps relevant that some recombinational defects accelerate progression through prophase, for this might result in the observed tendency for increased equational division among such mutants (Hollingsworth and Byers 1989).

Why then do *spo13* cells undergo a single division? This remains obscure, but it seems likely that *SPO13* enhances the activity of the *CDC28* kinase, thus preventing full exit from M phase following the first round of chromosome segregation. This would allow entry to meiosis II without a return to interphase and a subsequent round of DNA synthesis. This proposal is consistent not only with the single-division phenotype of *spo13* cells, but also with the suppression of *cdc28* meiotic phenotype by *SPO13* overexpression. At moderately elevated temperatures, *cdc28* cells may contain enough kinase to carry out meiosis I, but not meiosis II. Increasing the *SPO13* gene product might produce enough kinase activity to proceed through the second division. The similar but nonidentical behavior of *spo12* mutants remains to be explored more fully. It has been found that the product of this gene can interact in vegetative cells with the products of the *DBF2* and *DBF20* genes, which encodes other cell cycle protein kinases (Toyn and Johnston 1994). The part that these kinases may play in meiosis is unknown. Unlike *spo13*, deletion of *SPO12* does not restore spore viability to *rec*⁻ mutants (Malone and Esposito 1981).

F. Mutations Affecting the Meiotic Divisions

One additional pathway leading to the formation of dyads of diploid spores came to light from studies in meiosis of the mutations isolated for defects in the mitotic cell cycle. Two mutations that cause arrest late in mitosis (*cdc5* and *cdc14*) affect meiosis in different ways depending on the time of transfer to nonpermissive conditions (Schild and Byers 1980). Early transfer to the restrictive temperature results in arrest at meiosis I with a disrupted spindle; later transfer yields two diploid spores that differ for centromere-linked markers, indicating that reductional segregation has occurred. Cytological examination revealed that two spindles for meiosis II are formed but they fail to elongate. Outer plaques then arise on schedule and a prospore wall is induced to form adjacent to each of them, but insufficient pole separation in each meiosis II spindle permits the incipient prospore walls to coalesce, forming a single spore wall that surrounds the entire spindle, thereby creating a diploid spore. It remains unknown how such spores, which contain two SPBs and the DNA content appropriate for a diploid cell in G_1, recover to resume normal growth upon germination.

A similar but distinct pattern of diploid dyad formation has been analyzed by Sharon and Simchen (1990a,b), who challenged *cdc5* cells early in meiosis with the restrictive temperature and then returned them to permissive conditions. The dyads of diploid spores showed a mixed pattern of marker segregation, with some chromosomes predominantly undergoing reductional segregation and others in the same dyad undergoing equational segregation. The specificity of this chromosome-specific segregational bias was explored by swapping centromeric segments of chromosomes favoring either type of segregation. The bias was shown to depend on the centromere-proximal sequences, rather than on more distal segments of the chromosomes, suggesting that chromosome-specific responses reflected differences near the sites of spindle attachment. The cytological basis was not explored, but one may surmise that a prolonged arrest in meiosis I was accompanied by progressive detachment of the centromeres from the spindle fibers in a chromosome-specific manner, followed by reattachment to the spindle in an equational orientation.

Diploid dyads resulting from failure of the equational division (meiosis II) also result from temperature sensitivity of mutations that cause monopolar mitosis after failure of SPB duplication. The prototypic monopolar mutation, *cdc31*, is known to cause failure of SPB duplication in the second cycle after challenge by elevated temperature (Winey et al. 1993). In meiosis, *cdc31* strains are similarly competent for SPB duplication in the first division (meiosis I) but then fail in meiosis II. Despite this failure, outer plaque formation ensues and two diploid spores are formed. Another diploidizing mutation, *ndc1* (Thomas and Botstein 1986), which forms only one functional SPB in mitosis, also leads to the production of diploid dyads in meiosis. In summary, defects in SPB duplication and spindle formation prevent spindle formation when meiosis I is arrested, but certain defects in meiosis II fail to inhibit progression into outer plaque and prospore wall formation, so viable diploid spores are recovered from the defective meiosis.

An additional mode of single-division meiosis was reported by Klar et al. (1979), who exploited the ability of dikaryons derived from karyogamy-defective matings to undergo sporulation in an unusual manner. Although a simple dikaryon formed from two haploid parents failed to sporulate, one generated between a *MAT***a**/*MAT***a** diploid and a *MAT*α haploid sporulated and two of the spores were haploid, bearing markers derived exclusively from the haploid parent. This shows not only that sporulation does not require that *MAT***a** and *MAT*α reside in the same nucleus, but also that one nucleus must be diploid. Furthermore, it seemed possible that the haploid nucleus had completely failed to participate in the earlier phases of the meiotic process, only undergoing an equational division upon reentry into the spore formation pathway. But making this nucleus disomic for chromosome III (and homoallelic for *MAT*) revealed that recombination took place between markers borne on the disomic chromatids before the single equational division. The mechanisms controlling this single division, which necessarily was equational in the manner of *spo13* meiosis, remain to be identified.

G. Other Segregational Defects

TOP2 encodes the sole topoisomerase II in yeast (Goto and Wang 1984). *top2* mutants are unable to separate intertwined sister chromatids during vegetative growth and therefore suffer considerable chromosome breakage and nondisjunction, leading to cell death (Holm et al. 1985). On the other hand, *top2* mutants arrest in meiosis at pachytene, before spindle formation, and remain viable, as shown by their ability to form colonies upon plating on growth medium (Rose and Holm 1993). *rad50 top2* double mutants proceed beyond meiosis I, suggesting that the meiotic arrest caused by *top2* alone is triggered by the presence on a recombination intermediate, perhaps one that interlocks the homologs in a manner that requires topoisomerase activity for its resolution (Rose et al. 1990; Rose and Holm 1993).

Several mutants have been isolated in *Drosophila* that appear to be defective in sister chromatid cohesion (for review, see Sekelsky and Hawley 1995). A screen for yeast mutants that generate aneuploid spores yielded two dominant alleles at the *DIS1* locus (Rockmill and Fogel 1988). *DIS1-1* and *DIS1-2* strains generate disomes for several chromosomes at a rate that is 100-fold higher than that of the wild type. Recombination frequencies are normal, indicating that a general recombination deficiency is not the cause of the nondisjunction. Since in most cases the disomes are generated during the first meiotic division, it was proposed that they result from precocious sister chromatid separation. Because mitotic nondisjunction is also strongly elevated in these mutants, the affected mechanisms clearly are not functions that are restricted to the meiotic pathway.

H. Meiotic Checkpoints

In vegetative cells, the efficacy of stage-specific processes often is ensured by mechanisms that confer a dependency of later functions on the successful completion of earlier events. Genes responsible for this type of coordination, termed *checkpoint* control (Weinert and Hartwell 1989), have been identified by mutants unable to maintain such control. Mutations in *RAD9, RAD17, RAD24, MEC1, MEC2*, and *MEC3* cause increased sensitivity to radiation or to mutational defects in chromosome processing, as well as showing enhanced lethality upon inhibition of DNA replication by treatment with hydroxyurea. *RAD9* has been shown to have a similar role in meiosis, where a temperature-sensitive *cdc13* mutation causes an arrest in meiotic prophase that is relieved by deletion of *RAD9* (Weber and Byers 1992). Furthermore, although X-ray treatments in the initial stages of meiosis arrest progression independent of *RAD9* function, a *RAD9*-dependent arrest by X-rays can be demonstrated later in meiotic prophase (Thorne and Byers 1993) and methylmethane sulfonate treatments have similar effects (M. Kupiec and B. Byers, unpubl.).

These findings raise the possibility that other mutational defects preventing completion of the sporulation pathway also represent the responses of checkpoint control monitoring the efficacy of earlier events. The prevalence of arrests at pachytene by various other mutations, including *cdc28* (Shuster and Byers 1989), thereby leaving the cells capable of reverting to mitotic growth, suggests that this may be a point of meiotic arrest under checkpoint control. Notably, several mutants defective in chromosomal synapsis (*zip1*, Sym et al. 1993) and in chromosomal metabolism (*dmc1*, Bishop et al. 1992; *sep1*, Tishkoff et al. 1995; *top2*, Rose and Holm 1993) become arrested at this stage and retain viability for prolonged periods, as though being arrested in response to the presence of unprocessed recombination intermediates. Consistent with this viewpoint, mutations in early-acting genes, such as *rad50*, prevent these prophase arrests (Sym et al. 1993; Rose and Holm 1993).

Which checkpoint functions may be relevant to meiotic progression? Although artificial chromosomal damage is being subjected to checkpoint control by a system involving *RAD9*, a *rad9* deletion strain formed normal levels of fully viable spores that displayed the same level of meiotic recombination as the wild-type control (Weber and Byers 1992). Furthermore, *RAD9* itself seems not to be important for the prophase arrests resulting from the *dmc1*, *sep1*, or *zip1* mutations (Bishop et al. 1992; Roeder 1995; Tishkoff et al. 1995). On the other hand, recent results suggest that the checkpoint genes *RAD17*, *RAD24*, and *MEC1* are required for the meiotic arrests caused by *dmc1* (Lydall et al. 1996; G.S. Roeder, unpubl.). *RAD24* is also required for *zip1* arrest, suggesting that the checkpoint defined by *zip1* arrest may be the same as that defined by *dmc1* arrest. The roles of *RAD17* and *MEC1* in *zip1* arrest have yet to be tested. Checkpoint control of meiosis clearly demands further study, and workers in this area will have to bear in mind that differences in genetic background may bear heavily on the results. We note, for example, that *zip1* causes meiotic arrest in some backgounds (such as RB) but not in fast-sporulating SK1 strains (Sym and Roeder 1994). Conversely, *dmc1* causes arrest in SK1 but not in RB strains. In crosses between strains of different backgounds, the arrest phenotype segregates as a multilocus trait (G.S. Roeder, pers. comm.).

I. Distributive Segregation and Size-dependent Effects

In *Drosophila*, a specialized pathway has been shown to provide for the reductional segregation of nonrecombinant chromosomes in meiosis I (Grell 1970). This mechanism, termed distributive disjunction, is independent of homology between the participating chromosomes, and experiments with deletion and translocation chromosomes have demonstrated that the mechanism discriminates on the basis of chromosome size (Hawley 1989; Carpenter 1991). It has been proposed that the distributive segregation system regularly serves as the basis for proper meiotic disjunction in *Drosophila* oocytes of the small fourth chromosome, which ordinarily does not recombine. The viewpoint that the relevant mechanism involves a second round of pairing once recombination had failed has now been cast in doubt by the cytological demonstration that nonexchange fourth chromosomes are situated near either pole of the spindle at metaphase I rather than lying on the metaphase plate (Theurkauf and Hawley 1992). These authors suggest that their independent orientation derives from a balance between the well-characterized poleward forces exerted by the spindle on centromeres and opposing forces that interact with the chromosome arms, thereby accounting for variation in spindle position with the length of the arms.

A distributive system capable of ensuring the appropriate segregation of small chromosomes and plasmids has also been found in *Saccharomyces*. Early experiments involving circular minichromosomes showed that they disjoined in most of the cases in meiosis I (Clarke and Carbon 1980; Fitzgerald-Hayes et al. 1982). Subsequent experiments carried out with linear artificial chromosomes showed the same effect (Dawson et al. 1986; Mann and Davis 1986; Kaback 1989), yielding proper disjunction of nonrecombinant artificial chromosomes in 60–90% of the cases. Using strains monosomic for both chromosome I and chromosome III, Guacci and Kaback (1991) showed that these two nonhomologous chromosomes disjoined from each other in about 90% of the meioses. These frequencies suggest that the yeast distributive disjunctional system functions less efficiently than the corresponding system in *Drosophila* (Hawley 1989).

Loidl et al. (1994) recently used both electron microscopy and FISH to show that monosomic chromosomes actually pair during pachytene before disjoining from one another. Thus, distributive segregation in yeast appears to involve physical synapsis. The SC formed between the nonhomologous chromosomes adopted conformations suggest that this pairing occurs primarily in centromeric or telomeric regions. However, telomeres seem unlikely to be essential to the mechanism, since distributive disjunction also acts effectively on circular minichromosomes in yeast (Mann and Davis 1986; Kaback 1989; Guacci and Kaback 1991).

V. SPORE WALL FORMATION

The ascospore wall, rather than arising by budding from the mother cell as in vegetative growth, is formed de novo within the interior of the sporulating cell. The complex morphogenetic program for initiation of ascospore wall formation is intimately associated with the SPBs at the

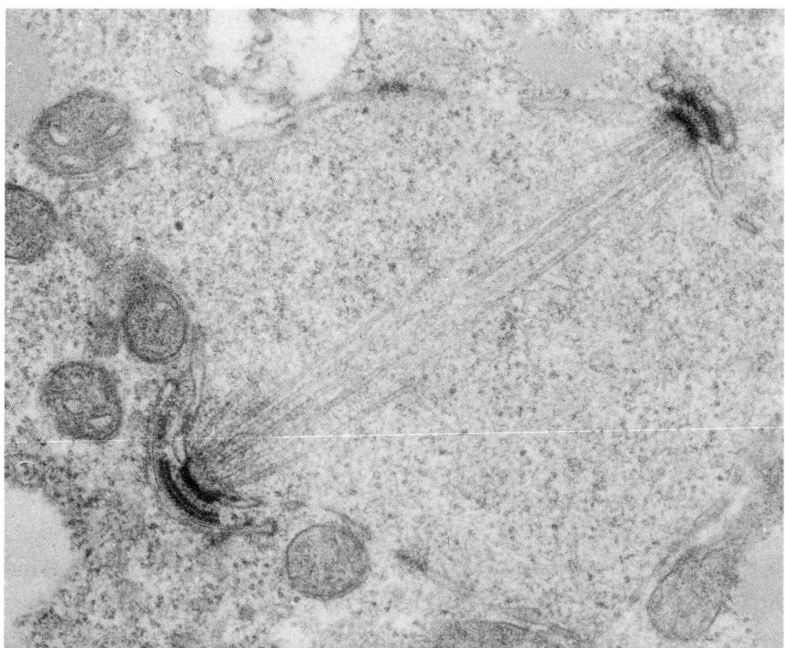

Figure 10 Meiosis II spindle displaying the typical enlargement of the outer plaque of each spindle pole body. The flattened sac being assembled adjacent to each outer plaque eventually will engulf the underlying nuclear lobe to form the prospore wall. (Reprinted, with permission, from Byers 1981.)

completion of meiosis II (Moens and Rapport 1971a). Since neither meiotic division in yeast is complete, a meiotic cell at anaphase II has a single nucleus bearing four lobes with a SPB situated at the tip of each lobe. At this stage, each SPB undergoes a structural transformation, the outer plaque becoming thicker and also expanding at its margins to overlie a larger area of the nuclear surface than that of the central plaque that underlies it (Fig. 10). Then, cytoplasmic vesicles (~50–100 nm in diameter) begin to fuse against the cytoplasmic surface of the outer plaque, forming a flattened sac that remains closely applied to the outer surface as the materials of prospore wall begin to be deposited within its flattened lumen (Lynn and Magee 1970). The margin of the prospore wall progressively expands along the surface of the nuclear envelope, extending beyond the extent of the outer plaque and eventually encompassing most of the nuclear lobe. Subsequently, the prospore wall dissociates from the nuclear surface as cytoplasmic components invade the intervening space, and continued growth at the free edge leads to closure of the curved surface to create a complete spherical shell. Concomitantly, the nuclear lobe is cut free from the rest of the nuclear mass to form an individual haploid nucleus, and the surrounding cytoplasm is segregated from other cytoplasmic components that will be discarded in the ascus space. Once completely formed, other layers of spore wall materials are then added to its outer surface.

The enlarged outer plaque is implicated in the initiation of prospore wall formation not only by the structural apposition of these two structures, but also by altered pathways of spore formation in mutants. Notably, delaying the completion of meiotic prophase by incubating *pac1* cells at the restrictive temperature for several hours before allowing sporulation to resume results in the generation of dyads of haploid spores (Davidow et al. 1980). This effect reflects a starvation response to gradual exhaustion of the sporulation medium, for formation of full tetrads was restored by transferring the arrested culture to fresh medium at the time of temperature downshift. Strikingly, the two spores of each dyad formed in exhausted medium differ for most centromere-proximal markers, thus identifying them as nonsister progeny. Consistent with this genetic evidence, electron microscopy showed that only one of the two daughter nuclei derived from each meiosis II spindle becomes encapsulated by a prospore wall. This appears to result from only the SPB at this pole (perhaps the parental one) gaining an expanded outer plaque, suggesting that cells respond to starvation by limiting the number of SPBs that acquire this modification. It is speculated that this mode of structural regulation effectively limits the number of spore walls that the ascus will attempt to complete, thus conserving limited resources to ensure survival of at least two haploid genomes. Uno et al. (1985) demonstrated in *cyr1* homozygotes that a similar limitation in spore wall initiation at the level of SPB modification was not limited to the formation of nonsister spores, so SPB age must not be a crucial determinant for this response.

The encapsulation of each spore nucleus by the prospore wall is followed by further steps of spore wall formation and maturation. The completed wall is a multilayered affair with its thick inner layer being derived largely from the former prospore wall. This part of the wall, which appears transparent by standard methods of thin-section electron microscopy, is believed to consist largely of glucans and mannans and thereby to be quite similar to vegetative cell walls (Briza et al. 1988). Lying outside these polysaccharides are two distinct thin layers, the inner one being mostly composed of *chitosan*, a type of chitin (an *N*-acetylglucosamine polymer) in which the action of chitin deacetylation has removed the acetyl residues to create a polymer that resists digestion by

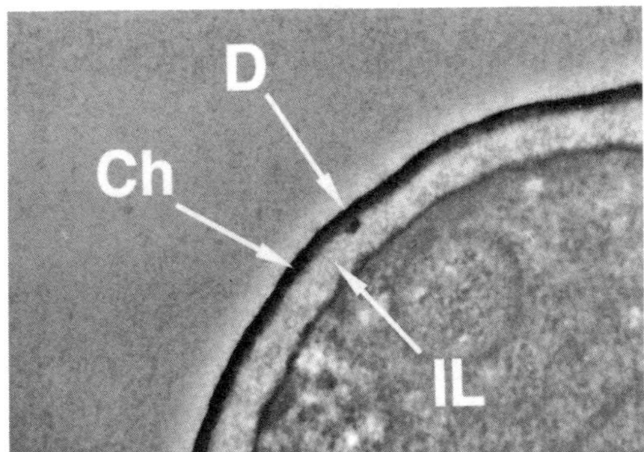

Figure 11 Layers of the spore wall. (D) Dityrosine; (Ch) chitosan; (IL) inner layer. (Micrograph courtesy of M. Breitenbach.)

chitinase, glusulase, and zymolyase (Fig. 11). Studies of the chitin synthetic pathway have revealed three chitin synthesis genes with partially overlapping functions; among the three chitin synthetase gene known in yeast, the one encoded by *DIT101/CAL1/CDS2/KTI1* (see Choi et al. 1994) is most clearly implicated in chitosan synthesis (for review, see Shaw et al. 1991). Spores formed by cells mutant for this gene are viable but defective in maturation of the walls, lacking both outer layers and being sensitive to enzymatic digestion. The outermost wall of the spore is composed primarily of highly insoluble protein that is stabilized by tyrosine cross-links joining tyrosine residues in both the D and L configurations (Briza et al. 1986). A systematic search for genes that are subject to transcriptional activation during sporulation identified several *DIT* loci that have key roles in outer wall formation (Briza et al. 1990). Mutants lacking these activities are viable but can be identified by their failure to display the dityrosine-dependent fluorescence seen when wild-type spores are exposed to ultraviolet irradiation. Key steps in the formation of the dityrosine layer involve a modification of tyrosine residues under control of *DIT1* and the dimerization per se by the *DIT1* product (Briza et al. 1994).

Other genes affecting spore wall formation have been detected in screens to identify loci subject to enhanced transcription late in sporulation. One gene thus identified is *SPS1*, which encodes a serine/threonine protein kinase that is essential for normal maturation of the outer layers of the spore wall (Friesen et al. 1994). Another protein kinase, encoded by *SMK1* which was isolated in a PCR screen for homology with MAP kinases, has a similar phenotype (Krisak et al. 1994). This suggests that a signal transduction pathway (perhaps one including *SPS1*) may serve to activate spore maturation at the appropriate stage. The downstream targets of such regulation might well include *DIT1* (see above) and/or *SPR1*, which is transcriptionally activated late in meiosis and which encodes a specific glucanase that could function in wall maturation by processing the ends of glucan polymers in the wall (Muthukumar et al. 1993). Deletion of this gene prevents the development of the thermoresistance that normally characterizes ascospores.

Germination of ascospores, which is initiated by exposure to fresh growth medium, must entail extensive modification of the spore wall to permit its outgrowth and formation of the initial bud. Relevant mutations may therefore be identified by defects in spore germination. Such mutations include *ubc1*, which encodes a ubiquitin-conjugating enzyme and thereby plays a key part in the targeting of specific proteins for degradation (Seufert and Jentsch 1991), and *alg7*, which is defective in initiating the *N*-glycosylation of proteins via the dolichol-mediated pathway (Kukuruzinska and Lennon 1995). Other mutants of potential relevance are in *GDR1* and *GDR2*, which were identified by mutations that are deregulated for control of germination (Dickinson and Cole 1991).

DENOUEMENT: It was not so many years ago that yeast was suspected by many of simply masquerading among "true" eukaryotes, therefore casting doubt that its meiotic mechanisms would serve as an adequate model for those in plants and animals. Our increasing body of knowledge about yeast over the past couple of decades has confirmed the highly conserved nature of many crucial meiotic processes in yeast, opening up the investigation of these processes to the extraordinarily powerful molecular genetics that yeast studies have come to enjoy. The explosive growth of the literature on yeast meiosis and sporulation has precluded our offering a truly comprehensive treatment of all that is now known, and we apologize to those whose contributions were neglected. We have attempted in this review to bring the exciting recent advances of this field into focus in order that we might perceive what to expect in the future. That future is particularly promising in light of the accessibility to DNA sequence information for the entire yeast genome, making *Saccharomyces cerevisiae* the first fully sequenced organism that performs meiosis. This resource will enable workers to identify relevant genes more readily and to focus their research on unraveling the molecular mechanisms that are crucial for synapsis, recombination, and chromatid segregation during meiosis in yeast and other eukaryotes.

Note Added in Proof

Recent work by two groups has now identified the protein bound to the cleaved DNA ends following DSB formation in *rad50S* meiosis as the product of the *SPO11* gene (Bergerat et al. 1997; Keeney et al. 1997). Intriguingly, this work also demonstrates striking sequence similarity between Spo11p to a novel class of type II topoisomerases recently found in the Archaea. These findings suggest that Spo11p serves enzymatically to create the DSB by action of a conserved tyrosine that attacks the phosphodiester backbones of both DNA strands and remains covalently bonded to the resulting terminal phosphate residues.

ACKNOWLEDGMENTS

We are grateful for advice and materials provided by numerous colleagues in our own labs and elsewhere, especially including D.K. Bishop, M. Breitenbach, J. Englebrecht, C.N. Giroux, N.M. Hollingsworth, Y. Kassir, S. Keeney, N. Kleckner, R.E. Malone, B. McKee, C.W. Moore, A.M. Neiman, H. Ogawa, T. Ogawa, G.S. Roeder, B. Rockmill, R. Strich, and E. Winter. And we thank Brian Washburn for invaluable technical assistance.

Table 2 Phenotypes of mutants affecting sporulation

Locus	Function	Reference
ARD1	Complexes with the catalytic subunit of *N*-α-acetyltransferase (Nat1). Required for full repression of *HML*, entry into G_0, and sporulation.	Park and Szostak (1992)
BCY1	Regulatory subunit of cAPK. Required for stationary phase, heat shock resistance, glycogen accumulation, and sporulation. *bcy1* is constitutive for cAPK activity and causes meiotic arrest prior to premeiotic DS.	Matsumoto et al. (1983); Cameron et al. (1988)
BDF1	Protein with two bromodomain motifs. Required for wild-type growth rate and sporulation. Localizes to the nucleus and to chromosomes in spread meiotic nuclei but is excluded from the nucleolus.	Chua and Roeder (1995)
BLM5	Required for resistance to bleomycin, ionizing radiation, and oxidative damage by hydrogen peroxide. Required for sporulation. Mutant arrests in meiosis at the mononucleate stage.	Moore (1991)
CAK1	Cyclin-dependent kinase activating kinase. Phosphorylates and activates Cdc28 and is required for vegetative growth. Required for proper spore wall morphogenesis. High-copy *CAK1* suppresses spore wall assembly defects of *smk1* missense mutants. mRNA increases in mid sporulation.	Kaldis et al. (1996); Thuret et al. (1996); Espinoza et al. (1996)
CAL1/SCD2	Chitin synthase. Required for chitosan synthesis during spore wall maturation.	Valdivieso et al. (1991); Shaw et al. (1991); Pammer et al. (1992); Choi et al. (1994)
CDC2/POL3	Catalytic subunit of DNA polymerase δ. Essential for mitotic and meiotic DS. Dispensable for meiotic SPB duplication but required for SCs and full intragenic Rec, SPB separation, and spindle formation. *cdc2* arrests at the mononucleate stage with duplicated SPBs and no complete spindles.	Simchen (1974); Schild and Byers (1978); Sitney et al. (1989)
CDC4	β-subunit of large G proteins with similarity to β-transducin. Dispensable for both mitotic and meiotic SPB duplication. Essential for mitotic but not premeiotic DS, wild-type levels of SCs, intragenic recombination, SPB separation, and spindles. *cdc4* arrests in meiosis at the mononucleate stage with duplicated SPBs.	Simchen (1974); Byers and Goetsch (1975); Zamb and Roth (1977); Horesh et al. (1979)
CDC5	Protein kinase. Required for the G_2/M transition in mitosis. Dispensable for premeiotic DS and Rec but required for tripartite SCs, haploidization and spores. *cdc5-ts* forms SCs lacking central elements and arrests either at MI with broken spindles or at MII with short spindles. Late shifts to a restrictive temperature result in reductional dyads; each spore con-	Simchen (1974); Horesh et al. (1979); Schild and Byers (1980); Simchen et al. (1981); Sharon and Simchen

Gene	Description	References
	tains an entire MII short spindle with unseparated chromatids. In some strains at a semi-permissive temperature, chromosomes segregate reductionally or equationally depending on the centromere.	(1990a,b)
CDC7	Ser/Thr protein kinase. Required for mitotic DS but dispensable for premeiotic DS and SPB duplication. Required for SCs, meiotic Rec, SPB separation, and spindles. Mutant arrests with duplicated SPBs and no spindles; the SPBs eventually enlarge, invaginate from the nuclear envelope into the center of the nucleus, sometimes fragmenting into three or four smaller SPBs. In heterozygotes, *cdc7* spores fail to germinate. mRNA is induced in meiosis and peaks during recombination.	Simchen (1974); Schild and Byers (1978); Horesh et al. (1979); Patterson et al. (1986); Sclafani et al. (1988); Hollingsworth and Sclafani (1990, 1993); Buck et al. (1991)
CDC8	Thymidylate kinase. Essential for mitotic DS. Required for premeiotic DS, SCs, Rec, MI, MII, and spores. *cdc8* arrests at the mononucleate stage with duplicated SPBs and no spindles.	Simchen (1974); Zamb and Roth (1977); Schild and Byers (1978)
CDC9	DNA ligase. Essential for mitosis and meiosis. Dispensable for intragenic Rec, but required for haploidization and spores.	Simchen (1974); Johnston and Nasmyth (1978); Johnston et al. (1982a)
CDC13	Binds to single-stranded TG_{1-3} telomere G-tails and is required for the G_2/M transition in mitosis. Dispensable for premeiotic DS but required for SCs, Rec, MI, MII, and spores. The *cdc13* G_2 meiotic arrest is dependent on the *RAD9* checkpoint function.	Simchen (1974); Weber and Byers (1992); Thorne and Byers (1993); Garvik et al. (1995); V. Zakain (unpubl.); V. Lundblad (unpubl.)
CDC14	Soluble tyrosine-specific protein phosphatase. Required for mitosis and sporulation. Mutants have the same phenotype as *cdc5*.	Simchen (1974); Schild and Byers (1980); Sharon and Simchen (1990a,b)
CDC16	Putative metal-binding nucleic-acid-binding protein. A component of anaphase-promoting complex required for the G_2/M transition in mitosis and degradation of mitotic cyclins. Interacts with Cdc23p and Cdc27p to catalyze the conjugation of ubiquitin to cyclin B. Required for sporulation.	Simchen (1974); Icho and Wickner (1987); Irniger et al. (1995); King et al. (1995); Tugendreich et al. (1995)
CDC17/POL1	DNA polymerase α. Required for mitotic DS, premeiotic DS, Rec, and full sporulation.	Simchen (1974); Budd et al. (1989)
CDC19	Pyruvate kinase. Required for Start A in the cell cycle and sporulation.	Verdier et al. (1989)

Continued on following pages.

Table 2 (*Continued.*)

Locus	Function	Reference
CDC20	Homolog of β-transducin. Required for mitosis and sporulation.	Simchen (1974)
CDC21/TMP1	Thymidylate synthetase. Required for continued DS in mitosis and meiosis, SCs, and intragenic Rec. *cdc21* arrests at the mononucleate stage with duplicated SPBs.	Brendel and Fath (1974); Game (1976); Bisson and Thorner (1977); Schild and Byers (1978)
CDC23	A component of anaphase-promoting complex required for the G_2/M transition in mitosis and degradation of mitotic cyclins. Interacts with Cdc16p and Cdc27p to catalyze the conjugation of ubiquitin to cyclin B. Required for sporulation.	Simchen (1974); Irniger et al. (1995); King et al. (1995); Tugendreich et al. (1995)
CDC25	GDP/GTP exchange protein. Required for RAS activation, Start B in the cell cycle, and Glu and N repression of sporulation in rich medium; *cdc25-ts* transiently arrests in G_1 and sporulates precociously like *cyr1* or *ras2*. Carboxy-terminal 38 residues are critical for Spo^+ during starvation; when altered cAMP increases and cells are Spo^-.	Shilo et al. (1978); Tripp and Pinon (1986); Munder et al. (1988); Munder and Kuntzel (1989)
CDC28	Catalytic subunit of cyclin-dependent protein kinase. Required for Start B in mitosis. Required for SPB separation at MI and MII. At a restrictive temperature, *cdc28* arrests at pachytene at the mononucleate stage with duplicated SPBs and no spindles; at a semi-permissive temperature, mutants complete MI, forming reductional dyads due to failure of MII SPB separation and enclosure of each MI product into a spore.	Reed et al. (1985); Shuster and Byers (1989)
CDC30, PGI1	Isozymes for phosphoglucose isomerase. Required for gluconeogenesis and sporulation.	Dickinson et al. (1988)
CDC31	Calcium-binding protein component of SPBs. Required for SPB duplication in mitosis and MII. Localizes to half-bridges and interacts with *KAR1*. *cdc31* forms reductional dyads with unduplicated SPBs.	Spang et al. (1993); Vallen et al. (1994); Biggins and Rose (1994)
CDC33	elF-4E mRNA Cap-binding protein. Required for Start A of cell cycle and sporulation. *cdc33* has normal cAMP pools, G_1 arrest, and is not suppressed by cAPK mutants, suggesting that sporulation is independent of the cAMP pathway.	Verdier et al. (1989)
CDC35/CYR1	Adenylate cyclase. Required for Start A of cell cycle, and Glu and N repression of sporulation; *cyr1* transiently arrests in G_1 and sporulates precociously. Amino-terminal domain is dispensable for mitotic G_1 arrest after N-starvation, but required for Spo^+; when altered, cAMP levels remain high and cells continue to bud with abnormal spindles.	Shilo et al. (1978); Matsumoto et al. (1983); Uno et al. (1990)
CDC36	Nuclear protein negatively regulating basal transcription. Required for Start B in mitosis. Re-	Shuster and Byers (1989)

Gene	Description	References
	quired for MI SPB separation. *cdc36* at a restrictive temperature arrests at pachytene at the mononucleate stage with duplicated SPBs and no spindles.	
CDC39/NOT1/ROS1	Encodes nuclear protein negatively regulating basal transcription. Required for Start B in mitosis, and SPB separation at MI. *cdc39* at a restrictive temperature arrests at pachytene at the mononucleate stage with duplicated, unseparated SPBs and no spindles.	Shuster and Byers (1989)
CDC40	β-transducin homolog. Essential for mitosis only at 36°C (null is viable at 25°C). Required for proper timing of DS at all temperatures and completion of DS at a restrictive temperature in both mitosis and meiosis, and for completion of meiotic Rec, MI, MII, and spores. Null arrests at the mononucleate stage at a restrictive temperature.	Kassir and Simchen (1978); Vaisman et al. (1995)
CDEI, CDEII, CDEIII	Conserved centromere elements sufficient for mitotic and meiotic segregation. CDEI (8 bp) is required for mitotic segregation, although most assays indicate that it is dispensable for meiotic chromosome segregation; it is needed for MI plasmid disjunction. CDEII (78–86 bp A-T-rich region) is specifically required for MI disjunction. CDEIII (25 bp) is essential for both mitotic and meiotic segregation.	Cottarel et al. (1989); Gaudet and Fitzgerald (1989); Cumberledge and Carbon (1987); Panzeri et al. (1985); Sears et al. (1995)
CEP1	CDEI centromere-binding protein required for mitotic segregation and normal growth rate. Null requires Met, suggesting that Cep1p is a transcription factor. Null causes precocious sister segregation at MI and reduced spore viability.	Cumberledge and Carbon (1987); Cai and Davis (1990); Baker and Masison (1990); Mellor et al. (1991); Masison and Baker (1992)
CHL1	Required for mitotic chromosome segregation. Needed for wild-type levels of meiotic Rec and spore viability.	Liras et al. (1978)
DIS1	Required for mitotic segregation, MI segregation, and spore viability.	Rockmill and Fogel (1988)
DIT1	First enzyme in dityrosine synthesis pathway converting L-tyrosine to *N*-formyl-L-tyrosine. Dispensable for spores and spore viability but required for dityrosine accumulation in the outer spore wall, spore wall maturation, and resistance to ether and lytic enzymes. Spore-autonomous function in heterozygotes. mRNA is transcribed mid/late during spore wall formation.	Briza et al. (1990, 1994, 1996)
DIT2	Cytochrome P450. Dispensable for spores and spore viability. Required for dityrosine biosynthesis and dityrosine accumulation in the outer spore wall, spore wall maturation, and resistance to ether and lytic enzymes. Dit2p catalyzes oxidation of *N*-formyl tyrosine to	Briza et al. (1990, 1994); P. Briza and M. Breitenbach (unpubl.)

Continued on following pages.

Table 2 (*Continued.*)

Locus	Function	Reference
	N,N-bisformyl dityrosine in vitro. Spore-autonomous function in heterozygotes. mRNA is transcribed mid/late during spore wall formation .	
DMC1/MED1	Meiosis-specific RecA homolog. Dispensable for DSBs, SCs, and gene conversion in return to growth assay. Required for full pairing by in situ hybridization assay, wild-type time of appearance of SCs, axial associations in *zip1* mutant, and reciprocal Rec by return to growth assay. Dmc1p colocalizes with Rad51p to discrete subnuclear sites in nuclear spreads during mid prophase, briefly colocalizes with Zip1p, and then disappears by pachytene. *dmc1* accumulates SC-related dense body, has processed DSBs with longer 3′ ends, and either delays prophase or arrests in prophase at the mononucleate stage with duplicated SPBs and no spindles (dependent on *MEC1*⁻, *RAD17*⁻, *RAD24*, and independent of *RAD9*). *spo13* partially suppresses *dmc1* meiotic lethality. *spo11* and *rad50* are epistatic to *dmc1*, suggesting a mid/late Rec function. mRNA is induced early in meiosis.	Bishop et al. (1992); Bishop (1994); Rockmill and Roeder (1994); Rockmill et al. (1995a); Lydall et al. (1996); Weiner and Kleckner (1994)
DSM1	Semidominant ts allele on chromosome VII that partially suppresses *spo8-1-ts*. Restores premeiotic DS, intragenic Rec, but not reciprocal Rec or ascus formation to *spo8-1* at a restrictive temperature. *DSM1* alone is dispensable for sporulation.	Fast (1978)
GAL11/SPT13	Component of the RNA polymerase II holoenzyme complex. Regulates transcription of a diverse array of genes. Required for mating and sporulation.	Fassler and Winston (1988); Kim et al. (1994)
GCD1/TRA3	Regulator of amino acid biosynthesis. Required for sporulation.	Shilo et al. (1978); Hinnebusch and Fink (1983)
GCN1	Dispensable for meiosis and spore formation. Required for the outer spore wall layer, spore refractility, resistance to β-glucanases, and ascospore maturation.	Whelan and Ballou (1975); Ballou et al. (1977)
GSG1	Dispensable for mitosis, premeiotic DS, and meiotic Rec but required for spores. Mutants enter premeiotic DNA synthesis later than wild type; block in spore formation is not bypassed by *spo11*. mRNA increases slightly during sporulation.	Kaytor and Livingston (1995); J. Engebrecht (unpubl.)
GUA1	GMP synthetase. *gua1* has low GTP and is Spo$^+$ in synthetic acetate even when intracellular cAMP is high; thus, low GTP (not cAMP), may be critical for sporulation. Since *gua1* is Spo$^-$ when starved for guanine in synthetic glucose, Glu repression of sporulation is independent of GTP effect.	Olempska-Beer and Freese (1987); Dujardin et al. (1994)
HFD1	Dispensable for meiosis but required for proper outer plaque modification of MII SPBs and formation of tetrad asci. In *hft1-1*, only one member of each the duplicated MII SPBs be-	Okamoto and Iino (1982)

	comes modified and nucleates spore wall formation producing dyads containing nonsister haploid spores.	
HID1	Required for disome delay of sporulation in some strains. Required for *zip1* mutant arrest.	B. Rockmill and S. Roeder (unpubl.)
HOP1	Meiosis-specific zinc finger protein which localizes to lateral elements and unsynapsed axial elements. Dispensable for basal level of intrachromosomal Rec. Required for full chromosome condensation, DSBs, heteroduplex DNA, tripartite SCs, gene conversion, reciprocal Rec, increased intrachramosomal Rec in haploid meiosis, and viable spores. Interacts with Red1p and Rec104p; Red1p required for SC localization. *hop1* exhibits variable chromosome pairing (~30–90% of wild type) depending on the strain. *hop1* is rescued by *spo13* and is classified as an early Rec function. mRNA is induced early in meiosis.	Hollingsworth and Byers (1989); Hollingsworth et al. (1990); Vershon et al. (1992); Hollingsworth and Johnson (1993); Friedman et al. (1994); Weiner and Kleckner (1994); Nag et al. (1995)
IDS2	Encodes an *IME2*-dependent signal required for down-regulation of *IME1*, autoregulation of the *IME2* UAS, and for expression of middle and late (but not early) sporulation genes.	Sia and Mitchell (1995)
IME1	Dispensable for mitosis. Required for transcription of meiotic genes. Ime1p is recruited to the promotors of early meiotic genes by a Rim11p-regulated interaction with Ume6p. *IME1* is regulated by cell type and nutritional controls and is one of the earliest sporulation-specific genes induced in meiosis. It is directly repressed by Rme1. Overexpression in rich medium causes induction of early meiotic genes in SK1, and early, middle, and late genes during N starvation in all strains.	Kassir et al. (1988); Granot et al. (1989); Matsuura et al. (1990); Smith et al. (1990, 1993); Mandel et al. (1994); Steber and Esposito (1995); Rubin-Bejerano et al. (1996)
IME2	Ser/Thr protein kinase. Dispensable for mitosis. Positively regulated by *IME1*. Stimulates early, middle, and late gene expression and negatively regulates *IME1*. High-copy *IME2* stimulates meiotic Rec without starvation and permits meiosis in an *IME1* deletion.	Smith and Mitchell (1989); Mitchell et al. (1990); Yoshida et al. (1990)
IRA1, IRA2	Encodes GTPase-activating protein. Negatively regulates cAPK by antagonizing *CDC25*. Mutants are N-starvation-sensitive, heat-shock-sensitive, and suppress lethality of *cdc25*, but not *cyr1* or *ras1, ras2*. Required for sporulation.	Tanaka et al. (1989, 1990); Matsuura et al. (1990)
ISC2	Encodes meiosis-specific protein required for sporulation. Homologous to meiosis-specific lily cDNA clone *LIM15*, with limited similarity to *RAD57*.	Kobayashi et al. (1993)
ISC10	Encodes meiosis-specific protein required for sporulation.	Kobayashi et al. (1993)
KEX2	Encodes a processing protease. Dispensable for premeiotic DS and meiotic Rec but partially required for MI and/or MII.	Leibowitz and Wickner (1976)

Continued on following pages.

Table 2 (*Continued.*)

Locus	Function	Reference
LYT1	Required for mitosis and sporulation. *lyt1* at a restrictive temperature buds at a distal instead of axial position and undergoes *cdc*-like autolysis when buds reach the size of mother cells. The mitotic but not the Spo$^-$ phenotype is suppressible by high-copy *SPO12*.	Molero et al. (1993)
MATa and *MATα*	Required for premeiotic DS, Rec, MI, MII, and spores. The **a**1-α2 negative regulator promotes *IME1* transcription through at least two parallel pathways, one dependent on *RME1* and the other on *SPO8/IME4*.	Hopper and Hall (1975b); Kassir and Simchen (1976); Mitchell and Herskowitz (1986); Shah and Clancy (1992)
MCK1/YPK1/IME3	Phosphotyrosyl Ser/Thr protein kinase. Dispensable for mitosis, but required for chromosome segregation, benomyl resistance, basal *IME1* transcription in mitosis, and wild-type growth rate. Required for *IME1* induction in meiosis and for ascus maturation independently of *IME1*. High copy accelerates early gene expression while mutations delay and decrease sporulation. mRNA is transcribed constitutively.	Shero and Hieter (1991); Neigeborn and Mitchell (1991); W. Jiang et al. (1995)
MEC1/ESR1/SAD3	Protein is similar to phosphatidylinositol(PI)3-kinases required for DNA damage-induced checkpoint response in G_1, S/M, intra S, and G_2/M in mitosis. Regulates phosphorylation of Rad53p; overproduction of Rad53p rescues some *mec1* alleles. Required for *dmc1* arrest and meiotic Rec.	Kato and Ogawa (1994); Weinert et al. (1994); Sanchez et al. (1996)
MEI4	Required for full sporulation. Dispensable for axial cores but required for full chromosome pairing seen by in situ hybridization, heteroduplex DNA, SCs, meiotic intra- and interchromosomal gene conversion, reciprocal Rec and viable spores. *mei4* has near wild-type chromosome condensation, executes both divisions with a delay in MII, is rescued by *spo13* and is epistatic to *rad52*, and is classified as an early Rec function. mRNA is meiosis-specific and has an 88-bp intron at 5′ end spliced independently of *MER1*.	Menees and Roeder (1989); Menees et al. (1992); Nag et al. (1995)
MEI5/SPO12T	Required for meiotic reciprocal Rec, SC disassembly, and MI progression from prophase to M phase. Meiotic arrest is suppressed by *spo11* and *rad50*. RNA is induced in meiosis.	Giroux et al. (1993)
MEK1/MRE4	Meiosis-specific Ser/Thr protein kinase. Dispensable for chromosome pairing and chromosome condensation seen by in situ hybridization. Required for full DSBs, normal-length SCs, meiotic Rec, and spore viability. *mek1* is rescued by *spo13* and classified an early Rec function. Interacts with Hop1 and thought to phosphorylate it. mRNA is induced in meiosis.	Rockmill and Roeder (1991); Leem and Ogawa (1992); Nag et al. (1995); N. Hollingsworth (unpubl.)

MER1	Encodes protein with RNA-binding motifs required for *MRE2*-dependent mRNA splicing. Dispensable for axial elements in meiosis but required for full chromosome pairing and chromosome condensation seen by in situ hybridization, wild-type level of SCs, heteroduplex DNA, gene conversion and reciprocal Rec, and spore viability. Partially suppressed by multicopy *MER2* (see *MER2* below).	Engebrecht and Roeder (1989, 1990); Engebrecht et al. (1990); Weiner and Kleckner (1994); Nag et al. (1995); Ogawa et al. (1995)
MER2/REC107	Dispensable for mitotic Rec. Dispensable for axial elements, but required for chromosome pairing seen by in situ hybridization, DSBs, gene conversion, and tripartite SCs. *MER2* mRNA binds to and is specifically spliced by Mer1p in meiosis. Multicopy *MER2* restores gene conversion and SCs to *mer1* mutant, but not reciprocal Rec or viability. mRNA is transcribed constitutively.	Engebrecht et al. (1991); Malone et al. (1991); Cool and Malone (1992); Nandabalan and Roeder (1995); Rockmill et al. (1995b)
MER3	Required for a late step in meiotic Rec and sporulation. preRNA is spliced dependent on *MRE2*.	H. Ogawa (unpubl.)
MPS1	Essential Ser/Thr/Tyr kinase. Required for SPB duplication and a mitotic checkpoint function. *mps1*, in contrast to *mps2*, does not arrest in the absence of SPB duplication and monopolar spindle formation, or nocodazole treatment. Required for sporulation.	Winey et al. (1991); Lauze et al. (1995); Weiss and Winey (1996); M. Winey (unpubl.)
MPS2	Dispensable for SPB duplication, but required for function of newly formed SPBs in mitosis and bipolar spindles. Required for proper SPB modification in MII. *mps2* modifies only one half of each duplicated MII SPB; spore walls subsequently enclose each MI genome producing reductional dyads.	Winey et al. (1991)
MRE2	RNA-binding protein. Dispensable for mitosis and premeiotic DS. Required for meiosis-specific splicing of *MER2* and *MER3*, DSBs, SCs, Rec, and viable spores. Rescued by *spo13*, suggesting an early Rec function.	Ajimura et al. (1993); Ogawa et al. (1995)
MRE3	Dispensable for mitosis and for meiotic DSBs. Required for meiotic DSB repair, reciprocal Rec associated with conversion, and viable spores.	T. Ogawa (unpubl.)
MRE11	MMS-sensitive and hyper-Rec in mitosis. Dispensable for premeiotic DS, but required for both DSB formation and resection, SCs, meiotic Rec, and viable spores. *mre11* is rescued by *spo13* and epistatic to *rad50S*, suggesting it is an early Rec function. Localizes to discrete sites in *rad50S*. Mre11p, Rad50p, Mer2p, and Xrs2p function in a complex by immunoprecipitation and two-hybrid analyses; mutations in these genes have similar phenotypes. mRNA is induced in meiosis.	Ajimura et al. (1993); Johzuka and Ogawa (1995); Ogawa et al. (1995); F. Klein (unpubl.)

Continued on following pages.

Table 2 (*Continued.*)

Locus	Function	Reference
MSH2	MutS homolog encoding major mismatch repair activity in mitosis and meiosis; suppresses postmeiotic segregation and homologous recombination and is required for gene conversion polarity gradients (possibly by restricting heteroduplex tract length), and high spore viability. Functions with Pms1p and Pms2/Mlh1p in a complex which interacts with either Pms3/Msh6p to repair single-base and insertion-deletion mispairs or Msh3p to repair only insertion-deletion mispairs.	Reenan and Kolodner (1992); Alani et al. (1994); Prolla et al. (1994b); Selva et al. (1995); Marsischky et al. (1996)
MSH3	MutS homolog. Acts in mismatch repair in mitosis and meiosis but to a lesser extent than *MSH2*. Required for microsatellite stability. Forms a complex with Msh2p to repair insertion-deletion mispairs; redundant with Pms3/Msh6p in repair of insertion-deletion mispairs. *msh3* has normal SCs and slightly elevated PMS.	New et al. (1993); Strand et al. (1995); Marsischky et al. (1996)
MSH4	MutS homolog. Dispensable for DNA repair. Required for full levels of reciprocal exchange and spore viability. Localizes to discrete sites on meiotic chromosomes.	Ross-Macdonald and Roeder (1994)
MSH5	MutS homolog. No mitotic phenotype. Dispensable for DNA repair and meiotic intrachromosomal reciprocal Rec. Required for full reciprocal Rec between homologs, and spore viability. *msh5* is epistatic to *msh4*, suggesting they act in the same pathway.	Hollingsworth et al. (1995)
MSI1	Negative regulator of cAMP induction; homologous to β subunit of guanine nucleotide-binding proteins. Multicopy suppressor of high cAMP in *ira1* and RAS^{val19}. *msl1* causes Spo⁻ in *ira1* and RAS^{val19} but not *bcy1* mutants.	Ruggieri et al. (1989)
NAT1	*N*-α-acetyltransferase. Required for entry into stationary phase, heat shock-resistance, **a**-mating-type functions, and sporulation.	Lee et al. (1989); Mullen et al. (1989); Park and Szostak (1992)
NDC1	Encodes protein with multiple putative transmembrane domains. Dispensable for mitotic SPB duplication, but required for insertion of nascent SPBs into the nuclear envelope. *ndc1* parental SPBs form monopolar spindles in mitosis. Required for MII.	Thomas and Botstein (1986); Winey et al. (1993)
NDT80	No mitotic phenotype detected. Dispensable for DSBs, SCs, and SPB duplication. Required for full meiotic Rec. Null arrests at the mononucleate stage with duplicated SPBs and no spindles and is not rescued by *spo11* or *rad50*. mRNA is sporulation-specific, appearing after *DMC* (probably in the middle expression class).	Xu et al. (1995)
PEP4	Required for haploidization and sporulation. *pep4* exhibits a 60–70% reduction in total protein degradation during sporulation.	Zubenko and Jones (1981)

PMS1	MutL homolog. Required for mismatch repair in mitosis and meiosis, low levels of post-meiotic segregation, and high spore viability; dispensable for homeologous recombination. Acts in a complex with Pms2/Mlh1p and Msh2p to repair mismatched DNA.	Williamson et al. (1985); Kramer et al. (1989); Bishop et al. (1989); Borts et al. (1990); Bailis and Rothstein (1990); Alani et al. (1994)
PMS2/MLH1	MutL homolog. Required for mismatch repair in mitosis and meiosis, low levels of post-meiotic segregation, and high spore viability. Forms a complex with Pms1p and Msh2p to repair mismatched DNA.	Prolla et al. (1994a); Jeyaprakash et al. (1996)
PMS3/MSH6	MutS homolog. Required for DNA repair in mitosis and meiosis, low levels of postmeiotic segregation, and high spore viability. Forms a complex with Msh2p to repair both single-base and insertion-deletion mispairs. Redundant with Msh3p in repair of insertion-deletion mispairs.	Williamson et al. (1985); Jeyaprakash et al. (1994); Marsischky et al. (1996)
POL1	DNA polymerase α. mRNA cell-cycle-regulated (increases in G_1 and peaks in S). Essential for mitotic and meiotic DNA replication, meiotic Rec, and sporulation. Dispensable for repair of X-ray damage.	Budd et al. (1989)
POL30	Proliferating cell nuclear antigen (PCNA); accessory factor for DNA polymerase δ. Required for DNA replication and repair. mRNA increases in G_1, peaks in S in mitosis, and increases prior to DS in meiosis. Required for viability in *cdc44*, *rad50*, *rad52*, or *rad57* backgrounds.	Bauer and Burgers (1990); C. Holm (unpubl.)
POL4	Homologous to human DBA polymerase β. Deletion shows increased gene conversion. mRNA induced in meiosis.	Shimizu et al. (1993); Leem et al. (1994); Budd and Campbell (1995)
PRB1	Proteinase B. Dispensable for haploidization and sporulation, but needed for full protein degradation during sporulation, and proper spore morphology. Mutant has smaller spores than wild-type embedded in a thick matrix.	Zubenko et al. (1979); Zubenko and Jones (1981)
PRC1	Proteinase C. Dispensable for haploidization and sporulation, but required for full protein degradation during sporulation.	Wolf and Fink (1975); Zubenko and Jones (1981)
PRE1	Proteinase E subunit. Required for mitotic division and sporulation. *pre1* accumulates ubiquitin-protein conjugates.	Heinemeyer et al. (1991)
PRI1	DNA primase I. mRNA is cell-cycle-regulated like *POL1* and peaks in meiosis, during premeiotic DS.	Johnston et al. (1990)

Continued on following pages.

Table 2 (*Continued.*)

Locus	Function	Reference
PSO4	Wild-type confers resistance to 8-methoxypsoralen-sensitive, UV, MMS, and X-rays. Required for error-prone repair, induced mitotic Rec, and sporulation.	Henriques et al. (1989)
RAD2	Required for UV-damage repair. mRNA cell-cycle-regulated, and induced by DNA damage and by meiosis (different *cis*-sites utilized in damage and meiotic induction).	Siede et al. (1989)
RAD6	Encodes a ubiquitin-conjugating (E2) enzyme. Required for UV-induced mutagenesis, postreplication repair, repression of retrotransposition, and sporulation. Dispensable for premeiotic DS, but required for meiotic gene conversion and haploidization. Separate domains of Rad6p interact with Ubr1 (an E3 ubiquitin ligase needed for multiubiquitination), and Rad18p (a single-stranded DNA-binding protein). The carboxy-terminal 23 residues are critical for sporulation and histone polyubiquitinating activity, but not UV repair or induced mutagenesis. Mutations in *srs2* suppress *rad6* radiation sensitivity but not Spo$^-$ defect. *rad6* forms Rec intermediates, is not rescued by either *spo13* or *rad50 spo13*, and is thought to be Spo$^-$ due to lesions independent of meiotic Rec pathways. mRNA is induced by UV and starvation.	Game and Mortimer (1974); Game et al. (1980) Malone and Esposito (1981); Borts et al. (1986) Kupiec and Simchen (1986); Sung et al. (1988); Sung et al. (1990); Schiestl et al. (1990); Watkins et al. (1993)
RAD9	Required for DNA damaged-induced G_2 arrest in mitosis. Required for ionizing radiation-induced G_1 arrest, and *cdc13*-induced G_2 arrest in meiosis.	Weber and Byers (1992); Thorne and Byers (1993)
RAD50	Protein contains a purine-binding domain, two heptad repeats, and a hydrophobic tail. Dispensable for mitotic Rec, but required for X-ray damage repair. Dispensable for premeiotic DS, axial cores, and meiotic rDNA intrachromosomal Rec. Required for full chromosome pairing seen by in situ hybridization, formation and processing of DSBs, gene conversion and reciprocal Rec in non-rDNA, tripartite SCs, heteroduplex DNA, and viable spores. *rad50-1* or null is rescued by *spo13*, rescues *rad52 spo13*, and is classified as early Rec function. *rad50S* arrests at prophase and accumulates unprocessed DSBs breaks; Rad50p interacts with Mre11p and Xrs2p in two-hybrid and immunoprecipitation analysis; it colocalizes to spots with Mre11p and Xrs2p in a *rad50S* background.	Game and Mortimer (1974); Malone and Esposito (1981); Malone (1983); Borts et al. (1986); Gottlieb et al. (1989); Alani et al. (1989, 1990); Cao et al. (1990); Nag and Petes (1993); Raymond and Kleckner (1993a,b); Weiner and Kleckner (1994); Ogawa et al. (1995)
RAD51	RecA homolog. Required for X-ray damage repair and gene conversion but not intrachromosomal reciprocal (pop-out) Rec in mitosis. Dispensable for premeiotic DS but needed for proper processing of meiotic DSBs, proper time of appearance of SCs, axial associations in *zip1*, meiotic gene conversion and reciprocal Rec, and viable spores. Null accumu-	Game and Mortimer (1974); Morrison and Hastings (1979); Aboussekhra et al. (1992); Shinohara et al.

	lates processed DSBs with long heterogeneous tails. Rad51p colocalizes to ~65 spots with Dmc1p prior to synapsis (independently of *ZIP1* and *DMC1*) and interacts with Rad52p and Rad55p by two-hybrid analysis. mRNA increases in meiosis.	(1992); Ogawa et al. (1993a); Milne and Weaver (1993); Donovan et al. (1994); Bishop (1994); Rockmill et al. (1995a); Hays et al. (1995); Johnson and Symington (1995)
RAD52	Required for X-ray damage repair and various types of intra- and interchromosomal mitotic Rec, including HO switching and plasmid exchange. Dispensable for premeiotic DS, DSBs, SCs, and heteroduplex formation, but generally required for completion of meiotic Rec, although recombinants appear transiently in *rad52* in return-to-growth assay. DSB ends are excessively recessed in mutant. *rad52* is rescued by *rad50 spo13*, but not *spo13*, and is classified as late Rec gene. *RAD52* controls the level of a 72-kD endo-exonuclease in log phase and sporulation. Some alleles are partially meiotic Rec^+ and Spo^+, e.g., *rad52-20*, which is suppressed by *srs2* null (see *SRS2* below) and by *RAD51* overexpression. Interacts with Rad51p by two-hybrid analysis. mRNA is induced in meiosis during Rec.	Game and Mortimer (1974); Prakash et al. (1980); Malone and Esposito (1981); Malone (1983); Borts et al. (1986); Resnick et al. (1986); Chow and Resnick (1988); Cole et al. (1989); Giroux et al. (1989); Sugawara and Haber (1992); Nag and Petes (1993); Ogawa et al (1993b); Kaytor and Livingston (1994); Schild (1995)
RAD53/MEC2 SAD1	Protein kinase. Required for DNA damage-induced checkpoint arrest in G_1, S/M, and G_2/M in mitosis. Mec1p and Tel1p regulate Rad53p phosphorylation and Rad53 overproduction rescues some *mec1* alleles. Dispensable for sporulation but required for wild-type spore viability.	Game and Mortimer (1974); Allen et al. (1994); Weinert et al. (1994)
RAD54	Required for X-ray damage repair, mitotic Rec, and full meiotic Rec. mRNA increases in meiosis.	Game and Mortimer (1974); Cole and Mortimer (1989); Cole et al. (1989)
RAD55	RecA homolog (related to *DMC1*, *RAD51*, *RAD57*). Required for X-ray damage repair, full sporulation and viable spores. Interacts with Rad51p and Rad57p by two-hybrid analysis.	Game and Mortimer (1974); Lovett and Mortimer (1987); Lovett (1994); Hays et al. (1995); Johnson and Symington (1995)

Continued on following pages.

Table 2 (Continued.)

Locus	Function	Reference
RAD57	RecA homolog (similar to *DMC1*, *RAD51*, and *RAD55*). Required for X-ray damage repair, meiotic Rec, wild-type levels of sporulation, and viable spores. Interacts with Rad55p by two-hybrid analysis.	Game and Mortimer (1974); Game et al. (1980); Borts et al. (1986); Hays et al. (1995); Johnson and Symington (1995)
RAD58/XRS4	Dispensable for mitotic Rec. Required for meiosis. *rad58* is rescued by *rad50 spo13*, but not *spo13*, suggesting it is a late Rec function.	Chepurnaya et al. (1995)
RAP1	DNA-binding protein involved in gene silencing. Rap1p-binding site is required for the gene conversion polarity gradients and high meiotic Rec at the 5′end of *HIS4*. Localized cytologically to telomeres.	White et al. (1991); Klein et al. (1992)
RAS1, RAS2	GTP-binding protein which in GTP-bound form activates adenylate cyclase and cAPK. Activated by *CDC25* and negatively regulated by *IRA1* and *IRA2*. Prevents sporulation in rich medium.	Tatchell et al. (1985); Toda et al. (1985)
REC1	Required for mitotic intra- and intergenic Rec and for sporulation.	Esposito and Brown (1990)
REC3	Required for mitotic intra- and intergenic Rec and for sporulation.	Esposito and Brown (1990)
REC4	Required for mitotic intra- and intergenic Rec and for sporulation.	Esposito and Brown (1990)
REC102	23-kD protein containing a putative leucine zipper. Dispensable for mitotic Rec, DNA damage repair, axial elements, and meiotic chromosome condensation. Required for wild-type level of chromosome pairing seen by in situ hybridization, tripartite SCs, heteroduplex DNA, gene conversion and reciprocal Rec, proper timing of MI, and viable spores. *rec102* executes MI early; mutant is rescued by *spo13* and is epistatic to *rad52*, and is classified as early Rec function. mRNA is induced early in sporulation.	Malone et al. (1991); Bhargava et al. (1992); Cool and Malone (1992); Nag et al. (1995); R. Malone (unpubl.)
REC103	Dispensable for mitotic but required for meiotic Rec, and spore viability. *rec103* is rescued by *spo13* and is epistatic to *rad52 spo13*, suggesting it is an early Rec gene. mRNA is induced early in meiosis.	Cool and Malone (1992)
REC104	Dispensable for mitotic Rec and axial elements in meiosis. Required for tripartite SCs, meiotic Rec, and spore viability. *rec104* executes MI early; it is rescued by *spo13* and is epistatic to *rad52 spo13* and is classified as an early Rec gene. mRNA is induced in meiosis.	Malone et al. (1991); Galbraith and Malone (1992); R. Malone (unpubl.)

REC114	Dispensable for mitotic Rec and axial elements in meiosis but required for SCs, meiotic Rec, and spore viability. *rec114* executes MI early, is rescued by *spo13* and is epistatic to *rad52 spo13*, and is classified as an early Rec gene. mRNA induced early in sporulation.	Cool and Malone (1992); R. Malone (unpubl.)
RED1	Required for full chromosome pairing and chromosome condensation seen by in situ hybridization, axial elements, stable localization of Hop1p and SCs. At *HIS2* required for normal levels of DSBs in *rad50S*, heteroduplex DNA, full gene conversion at specific loci, full reciprocal Rec, and spore viability. Localizes to chromosome cores independently of Mei4p and Spo11p. *red1* is rescued by *spo13* and epistatic to *rad52* and is classified as early Rec function. *mek1* and *hop1* are on the same pathway as *red1*, which is independent of the *mer1* pathway by epistasis analysis. *RED1* suppresses DSB repair in *dmc1* mutants. mRNA is induced in meiosis.	Rockmill and Roeder (1988); Thompson and Roeder (1989); Rockmill and Roeder (1990); Nag et al. (1995); D. Bishop (unpubl.); A.V. Smith and G.S. Roeder (unpubl.)
RES1	*RES1* is a dominant allele allowing *IME1* induction and sporulation of **a/a** or α/α in the presence of high-copy *RME1*.	Kao et al. (1990)
RFA1	69-kD subunit of the heterotrimeric RPA ssDNA-binding protein. Required for DNA-damage repair, full levels of gene conversion, and sporulation. *rfa1* repair defects are suppressed by high-copy *RAD52*.	Firmenich et al. (1995); Smith and Rothstein (1995)
RGR1	Essential gene affecting chromatin structure, transcriptional regulation of diverse genes, and sporulation. Required for glucose repression, *HO* repression, *RME1* repression, and sporulation. *rgr1-ts* allows sporulation of ***a***/α diploids overexpressing *RME1*. Interacts with Sin4p, Gal11p, and an unknown 50-kD polypeptide.	Sakai et al. (1988); Stillman et al. (1994); Covitz et al. (1994); Y.W. Jiang et al. (1995); Li et al. (1995)
RIM1	Required for wild-type growth (at 17°C), *IME1* expression, and proper time of appearance of mature spores. *RIM1,8,9*, and *13* act in a single pathway (the *RIM1* pathway), functioning in parallel with *MCK1* by epistasis analysis. *Note:* it not known if *RIM1* acts in the same pathway as *IME4/SPO8*.	Su and Mitchell (1993a,b)
RIM4	Required for *IME2* expression. *rim4*, *rim11*, and *rim15* mutants cannot be suppressed by overexpression of *IME1*.	Su and Mitchell (1993a)
RIM8	Required for wild-type growth (at 17°C), *IME1* expression, and time of appearance of mature spores. *RIM1,8,9*, and *13* act in a single pathway (the *RIM1* pathway), functioning in parallel with *MCK1* by epistasis analysis.	Su and Mitchell (1993a,b)
RIM9	Required for wild-type growth (at 17°C), *IME1* expression, and time of appearance of mature spores. *RIM1,8,9*, and *13* act in a single pathway (the *RIM1* pathway), functioning in parallel with *MCK1* by epistasis analysis.	Su and Mitchell (1993a,b)

Continued on following pages.

Table 2 (*Continued.*)

Locus	Function	Reference
RIM13	Required for wild-type growth (at 17ºC), *IME1* expression, and time of appearance of mature spores. *RIM1,8,9*, and *13* act in a single pathway (the *RIM1* pathway), functioning in parallel to *MCK1* by epistasis analysis.	Su and Mitchell (1993a,b)
RIM11/MDS1	Ser/Thr protein kinase. Required for Ime1p phosphorylation, association of the Ime1p-Ume6p meiotic activatior, early meiotic gene expression, and sporulation. *rim11* is epistatic to lethality of *IME1* overexpression in haploids and permits Ime1p accumulation. High-copy suppressor of *mck1-cs*. Some alleles are Spo^+ and sporulate slowly.	Mitchell and Bowdish (1992); Bowdish et al. (1994); Rubin-Bejerano et al. (1996)
RIM15	Required for *IME2* expression. *rim4*, *rim11*, and *rim15* mutants cannot be suppressed by overexpression of *IME1*.	Bowdish and Mitchell (1993)
RME1	Zinc finger protein. Mediates cell-type control of sporulation. Directly repressed by **a**1-α2 regulator. Negatively regulates *IME1* and sporulation. *rme1* allows **a**/**a** and α/α diploids to sporulate, and ***a*** and α haploids to form viable spores in the presence of *spo13*.	Hopper and Hall (1975a); Kassir and Simchen (1976); Mitchell and Herskowitz (1986); Margolskee (1988); Covitz and Mitchell (1993)
RPD3/SDI2	Histone deacetylase. Required for vegetative repression of early meiosis-specific as well as nonmeiotic genes. Required for sporulation. Mutant epistasis analysis indicates that *RPD3* acts in the same pathway as *UME4/SIN3*.	Vidal and Gaber (1991); Bowdish and Mitchell (1993); A. Carmen et al. (unpubl.)
RPK1	Encodes essential Ser/Thr protein kinase. Expression increases late in meiosis.	Poch et al. (1994)
SAE1	Positively regulates entry into and exit from pachytene; required for full Rec and full spore formation. *sae1* is delayed in entry into and exit from pachytene and completion of MI; transiently arrests in prophase when Rec is initiated (e.g., in $SPO11^+$) but not in the absence of Rec (e.g., in a *spo11* null).	A. Mckee and N. Kleckner (unpubl.)
SAE2/COM1	Dispensable for mitosis (slightly MMS-sensitive) and meiotic DSBs. Required for pairing, processing of DSBs, full gene conversion and tripartite SCs, proper timing of MI, wild-type spore morphology, spore viability, and wild-type level of asci. *sae2* causes prophase arrest in wild-type *SPO11* which allows initiation of Rec, but not in a *spo11* null.	F. Klein (unpubl.); A. Mckee and N. Kleckner (unpubl.)
SAE3	Dispensable for meiotic DSBs and low-level Rec. Required for full Rec, MI, MII, and spores. Positively regulates entry into and exit from pachytene. *sae3* has hyper-resected DSBs (like *dmc1*), reduced Rec, and is delayed in entry into pachytene. *sae3* causes prophase arrest in	A. McKee and N. Kleckner (unpubl.)

	wild-type *SPO11* which allows initiation of Rec, but not in a *spo11* null. mRNA is meiosis-specific.	
SEP1/DST2/KEM1/RAR5/SKI1/STPb/XRM	DNA strand-transfer protein β/exoribonuclease I. Catalyzes the formation of hybrid DNA in vitro, has 5′ to 3′ exonuclease activity on DNA and RNA, binds to G_4 tetraplex DNA and cuts in a single-stranded region 5′ to the G_4 structure. Dispensable for mitotic Rec but is required to maintain telomere length in mitosis. In meiosis, dispensable for DSBs, gene conversion, and SCs but required for proper time of appearance of DSBs and gene conversions, full reciprocal Rec, SPB separation, spindles, and spores. *sep1* arrests with duplicated SPBs, fully synapsed chromosomes, and no spindles, independently of *RAD9*. Not rescued by *spo13* or *rad50spo13*. Double-mutant analysis suggests Sep1 acts in a separate pathway from Dmc1p and Rad51p. Protein increases severalfold in meiotic cells.	Kim et al. (1990); Dykstra et al. (1991); Tishkoff et al. (1991); Muhlrad and Parker (1992); Bahler et al. (1994); Z. Liu et al. (1995); Tishkoff et al. (1995)
SGA1	Intracellular sporulation-specific glucoamylase involved in glycogen degradation. Induced during starvation of **a**/α late in sporulation, but dispensable for sporulation. Upstream region contains a UAS subject to nutritional control and an NRE regulated by *IME2*.	Colonna and Magee (1978); Yamashita and Fukui (1985); Dranginis (1989); Pugh et al. (1989); Pugh and Clancy (1990); Kihara et al. (1991)
SHP1	Suppressor of *GLC7* high-copy lethality (catalytic subunit of protein phosphatase 1). Required for wild-type growth rate, glycogen accumulation, and sporulation.	Zhang et al. (1995)
SIN4/TSF3	Required for full repression of *RME1* and a variety of other genes. Interacts with Rgr1p as part of the RNA polymerase II holoenzyme. Involved in regulating chromatin structure.	Covitz et al. (1994); Y.W. Jiang et al. (1995); Li et al. (1995)
SIR2	Suppresses mitotic and meiotic intra- and interchromosomal rDNA Rec (10–15-fold). *RAD52* and *RAD50* are dispensable for basal level rDNA exchange in *SIR2* but are required for increase exchange in *sir2*.	Gottlieb et al. (1989)
SME2	High-copy suppressor of N, but not Glu, repression of late *Spo* gene expression and sporulation. Thought to be an activator of late meiosis-specific genes. Dispensable for sporulation.	Kawaguchi et al. (1992)
SME3	High-copy suppressor of N and Glu repression, and partially of *MAT* control of sporulation. High-copy induces early (*IME1* and *IME2*) and late (*SGA1*) genes during N^+, Glu^+ sporulation. mRNA is induced in stationary phase. Dispensable for sporulation.	Kawaguchi et al. (1992)
SMK1	Sporulation-specific MAP kinase. Required for proper spore wall morphogenesis.	Krisak et al. (1994)
SNF1	Protein kinase required for release from Glu repression, invertase expression, and sporulation. High-copy *MSI1* and *PDE2* partially suppress *snf1* Spo^- defect.	Hubbard et al. (1992)

Continued on following pages.

Table 2 (*Continued.*)

Locus	Function	Reference
SNF4	Encodes a protein that associates with Snf1p. Required for the release from Glu repression, invertase expression, and sporulation. High-copy *MSI1* and *PDE2* partially suppresses *snf4* Spo⁻.	Hubbard et al. (1992)
SOD1	Superoxide dismutase (*SOD1* = Cu-Zn-SOD). Required for sporulation.	Liu et al. (1992)
SOD2	Superoxide dismutase (*SOD2* = Mn-SOD). Required for sporulation	Liu et al. (1992)
SPD1	Mediates N repression of sporulation in rich medium; *spd1* sporulates in stationary phase.	Dawes (1975); Vezinhet et al. (1979); Dawes and Calvert (1984)
SPD3	Mediates N repression of sporulation in rich medium; *spd3* sporulates in stationary phase.	Dawes (1975); Vezinhet et al. (1979); Dawes and Calvert (1984)
SPD4	Mediates N repression of sporulation in rich medium; *spd4* sporulates in stationary phase.	Dawes (1975); Vezinhet et al. (1979); Dawes and Calvert (1984)
SPE2	*S*-adenosylmethionine decarboxylase. Required for haploidization and sporulation in the absence of exogenous spermine and spermidine.	Cohn et al. (1978)
SPO1	Encodes a protein with high similarity to phospholipase B. Dispensable for mitosis, premeiotic DS, SCs, and Rec, but required for meiotic SPB duplication, MI, MII, and spore formation. *spo1-1* and null arrests as mononucleate cells with unduplicated SPBs.	Esposito et al. (1970); R.E. Esposito and Esposito (1974); Moens et al. (1974); Tevzadze et al (1996)
SPO2	Dispensable for mitosis, premeiotic DS, Rec, MI, MII. Required for nuclear membrane integrity at MI and MII, and localized prospore wall growth at the nuclear envelope. In a *spo2* mutant, the nucleus divides into two separate bodies at MI, and again at MII; prospore walls initiate at SPBs normally but instead of elongating adjacent to the nuclear membrane, they grow independently and curl backward, forming small anucleate spores and free unenclosed nuclei.	Esposito et al. (1970); R.E. Esposito and Esposito (1974); Moens et al. (1974)
SPO3	Dispensable for mitosis, premeiotic DS, Rec, M, MII, and initiation of prospore walls. Required for spore wall elongation, coordination of spore wall closure with MII segregation, and spore maturation. *spo3-1* at a semipermisive temperature produces asci with one or two randomly packaged haploid spores; at a restrictive temperature, prospore walls grow and	Esposito et al. (1970); R.E. Esposito and Esposito (1974); M.S. Esposito and Esposito (1974); Honigberg

	close prior to the completion of MII, resulting in immature aneuploid and/or anucleate spores. Multinucleate cells completing MII, but blocked in spore development, bud and resume division in return to growth assay.	and Esposito (1994); Moens et al. (1974)
SPO4	Dispensable for mitosis, premeiotic DS, Rec, M, MII, and prospore wall initiation. Required for spore wall elongation, the coordination of spore wall closure with MII segregation, and spore wall maturation. *spo4-1* has similar phenotype to *spo3-1*.	Esposito et al. (1972); M.S. Esposito and Esposito (1974); Esposito and Esposito (1978); P. Moens (unpubl.)
SPO5	Dispensable for mitosis, premeiotic DS, Rec, MI, MII, and prospore wall initiation. Required for spore wall elongation, the coordination of spore wall closure with MII segregation, and spore wall maturation. *spo5-1* has similar phenotype to *spo3-1*.	Esposito et al. (1972); M.S. Esposito and Esposito (1974); Esposito and Esposito (1978); P. Moens (unpubl.)
SPO7	Encodes a constitutively expressed transcript and protein of unknown function. Dispensable for mitosis but required normal mutation rate. Required for premeiotic DS, Rec, MI, MII, glycogen degradation, and spores. Most *spo7* cells arrest at the mononucleate stage with unduplicated SPBs and numerous cytoplasmic membrane whorls.	Esposito et al. (1972); Esposito and Esposito (1975); M.S. Esposito and Esposito (1974); Whyte et al. (1990); P. Moens (unpubl.)
SPO8/IME4	Required for induction of *IME1* and sporulation. Regulated by cell type and nutritional controls. Cell-type control is independent of *RME1*.	Esposito et al. (1972); M.S. Esposito and Esposito (1974); Shah and Clancy (1992); I. Yamashita and M. Clancy (unpubl.)
SPO9	Dispensable for mitosis. Required for premeiotic DS, SPB duplication, MI, MII, and spores.	Esposito et al. (1972); M.S. Esposito and Esposito (1974)
SPO10	Dispensable for mitosis and premeiotic DS. Required for SPB duplication, MI, MII, and spores. *spo10* arrests at the mononucleate stage with unduplicated SPBs and accumulates aggregates of SCs.	Esposito et al. (1972); M.S. Esposito and Esposito (1974); Esposito and Esposito (1978); Moens and Kundu (1982)
SPO11	Encodes one of the earliest meiosis-specific Rec functions. Dispensable for mitosis, premeiotic DS, SPB duplication, MI, MII, and spores. Required for chromosome pairing seen	Esposito et al. (1972); Klapholz et al. (1985); Wag-

Continued on following pages.

Table 2 (*Continued.*)

Locus	Function	Reference
	by in situ hybridization, DSBs, SCs, Rec (in non-rDNA and rDNA), full level of asci, and spore viability. *spo11* executes MI early, is rescued by *spo13* and is epistatic to *rad52 spo13*, and is classified as an early Rec function. It fails to form any Rec intermediates and is partially suppressed by X-irradiation (by induced DSBs), in meiosis. The *spo11-1-ts* mutant is DSB$^-$, Rec$^-$, and exhibits tripartite SC structures at a restrictive temperature, suggesting SC formation may be partially independent of DSBs and Rec. mRNA is induced early in meiosis; meiosis-specific regulation is dependent on a URS1 element within the coding region. Protein is covalently bound to DSBs (for references, see Note Added in Proof).	staff et al. (1985); Borts et al. (1986); Atcheson et al. (1987); Giroux et al. (1989); Cao et al. (1990); Thorne and Byers (1993); Collins and Newlon (1994b); Weiner and Kleckner (1994); C. Atcheson and R. Esposito (unpubl.)
SPO12	Encodes a 20-kD protein with negatively charged carboxyl terminus required for function. Thought to be a positive regulator of exit from M phase in mitosis and meiosis. Dispensable for mitosis but required for normal growth rate. Dispensable for premeiotic DS, Rec, MII, and spores. Required for two division meiosis, specifically for MI. Loss of function in mitosis results in delay in G_2. Loss of function in meiosis results in a prolonged pachytene stage and presence of SCs, a single MII-like equational division at the time of MII, and dyad asci containing two diploid spores. Gain of function in mitosis suppresses M-phase anaphase arrest caused by overexpression of *CLB2 deg*$^-$ and mutants (e.g., *dbf2-ts*). mRNA is cell-cycle-regulated (with *DBF2*) in mitosis and increases 5–10x in meiosis. Protein interacts with Dbf2p and Dbf20p kinases.	Klapholz and Esposito (1980a,b); Malavasic and Elder (1990); Parkes and Johnston (1992); Toyn and Johnston (1993); C. Atcheson and R. Esposito (unpubl.)
SPO13	Encodes a 33.4-kD basic protein that localizes to the nucleus. Thought to be a meiosis-specific negative regulator of M phase during MI, coordinating sister-chromatid/centromere cohesion with nuclear division. Dispensable for mitosis, premeiotic DS, Rec, and spores. Required for two-division meiosis, generally for MI. Loss of function typically results in a single equational MII-like division (with some chromosomes segregating reductionally or aberrantly depending on strain background), occurring slightly earlier or at the time of wild-type MI, and dyad asci containing two diploid or near diploid spores. *spo13* rescues the meiotic lethality of early Rec$^-$ mutants and Rec$^+$ haploids via bypass of the MI division. Gain of function causes a *CDC28*-dependent arrest at M phase in mitosis and a delay in MI nuclear division. mRNA is induced early in meiosis peaking at MI. Protein has transcriptional activation activity in one-hybrid analysis.	Klapholz and Esposito (1980a,b); Wang et al. (1987); Buckingham et al. (1990); Hugerat and Simchen (1993); McCarroll and Esposito (1994); L. Henninger et al. (unpubl.)

SPO14	Encodes a phospholipase D homolog. Dispensable for mitosis, premeiotic DS, Rec, MI. Required for commitment to meiosis, MII, and spores. Most *spo14* mutant cells arrest at the binucleate stage; a small fraction proceed to the tetranucleate stage. Unlike the wild type, *spo14* cells can return to growth after either MI, or MII. mRNA is induced in mid-meiosis.	Honigberg et al. (1992); Honigberg and Esposito (1994); Rose et al. (1995); G. Tevzadze and R. Esposito (unpubl.)
SPO15/VPS1	Encodes a protein with significant homology to dynamin, a microtubule bundling protein. Dispensable fo premeiotic DS and Rec. Required for SPB separation, MI, and MII.	Yeh et al. (1991)
SPO16	Dispensable for mitosis but required for full sporulation, e.g., full Rec, haploidization, and spores. Contains a long untranslated 3' end which overlaps *SPO12*, an adjacent gene encoded on the complementary strand. Translation of the complementary region is dispensable for meiosis. mRNA is specifically induced early in meiosis.	Malavasic and Elder (1990); S. Frackman and R. Esposito (unpubl.)
SPO17	Dispensable for mitosis. Required for premeiotic DS and subsequent meiotic landmarks, and sporulation-specific amyloglucosidase activity.	Smith et al. (1988)
SPO18	Dispensable for mitosis. Required for premeiotic DS and subsequent meiotic landmarks, and sporulation-specific amyloglucosidase activity.	Smith et al. (1988)
SPO20	Dispensable for mitosis and meiosis but required to package nuclei into spores. Homologous to Sec9 and mammalian SNAP-25 family proteins involved in vesicle fusion with plasma membrane.	A. Neiman (unpubl.)
SPO50	Dispensable for mitosis. Required for premeiotic DS, Rec, and spores. *spo50* prevents precocious sporulation of *spd1* mutant in rich medium.	Calvert and Dawes (1984); Dickinson and Hewlins (1991)
SPO51	Dispensable for mitosis. Required for sporulation.	Calvert and Dawes (1984)
SPO53	Dispensable for mitosis. Required for sporulation.	Calvert and Dawes (1984)
SPOT1-11	Dispensable for mitosis. Required for premeiotic DS, MI, MII, and spores.	Tsuboi (1983); Tanaka and Tsuboi (1985)
SPOT12-15	Dispensable for mitosis and premeiotic S. Required for MI, MII, and spores.	Tsuboi (1983)
SPOT16-18	Dispensable for mitosis, premeiotic DS, and MI. Required for MII and spores.	Tsuboi (1983)
SPOT19-23	Dispensable for mitosis, premeiotic DS, MI, MII. Required for spores.	Tsuboi (1983)
SPR1	Homolog of *EXG1* vegetative exo-1,3-β-glucanase. Required for full heat resistance of spores. mRNA is sporulation-specific.	Primerano et al. (1988); Muthukumar et al. (1993)
SPR2	Encodes sporulation-regulated transcript and 10-nm filament family protein.	Primerano et al. (1988); J. Fares and J. Pringle (unpubl.)

Continued on following pages.

Table 2 (*Continued.*)

Locus	Function	Reference
SPR3	Encodes homolog of the *CDC3/10/11/12* bud neck microfilament genes. Dispensable for mitosis and sporulation. mRNA expressed at time of MI.	Kao et al. (1989)
SPR6	Dispensable for mitosis and sporulation; increases efficiency of sporulation in some SPR-defective strains	Kallal et al. (1990); Ozsarac et al. (1995)
SPS1	Ser/Thr protein kinase homologous to *STE20*. Dispensable for mitosis. Expressed mid/late in meiosis and required for spore wall formation.	Friesen et al. (1994)
SPS18	Encodes a 34-kD protein containing a putative zinc-binding domain and a basic region. Dispensable for mitosis. Required for full sporulation and normal spore walls. mRNA induced late in sporulation; transcribed divergently from *SPS19*.	Coe et al. (1994a)
SPS19	Encodes a 31-kD protein that has homology with alcohol dehydrogenases. mRNA induced during sporulation; transcribed divergently from *SPS18*.	Coe et al. (1994a)
SPS100	ORF encodes 34.2-kD protein containing potential signal sequence and numerous glycosylation sites. Dispensable for mitosis, sporulation, and spore viability, but required for proper timing of spore wall maturation. mRNA induced late in sporulation.	Law and Segall (1988)
SPT15	TFIID TATA-binding factor. Mutants grow slowly and are defective in mating and sporulation.	Eisenmann et al. (1989)
SRS2	DNA helicase. Wild-type suppresses mitotic recombination. Loss of function results in *RAD52*-dependent hyper-Rec, suggesting Rec suppression occurs by antagonizing the Rad52 recombinational repair pathway. Required for proper timing of commitment to meiotic recombination and the transition from MI to MII. Some mutant alleles have lower spore viability which is not rescued by *spo13*, suggesting they affect a late Rec function. mRNA increases in meiosis during recombination.	Lawrence and Christensen (1979) Aboussekhra et al. (1989); Schiestl et al. (1990); Aboussekhra et al. (1992); Palladino and Klein (1992); Milne et al. (1995); Heude et al. (1995)
SSA3	Hsp70 cognate gene. mRNA repressed by Glu, increases in stationary phase and sporulation. Dispensable for sporulation.	Werner et al. (1989)
SSG1	Encodes sporulation-specific exoglucanase. Mutant sporulates but has delayed ascus maturation.	San Segundo et al. (1993)
SSN6/CYC8	Negative regulator (acts in a complex with Tup1p) of a variety of genes in mitosis and meiosis. Required for sporulation.	Schultz et al. (1990); Keleher et al. (1992); J. Segall, un-

		publ.)
STA1	Secreted glucoamylase. Repressed by **a**/α during growth in Glu and is present in haploids. mRNA is induced during sporulation.	Dranginis (1989)
SUP3	Ochre-suppressing allele reduces premeiotic DS, Rec, MI, and MII and inhibits spore formation. Intragenic second-site mutations which eliminate ochre suppression, but do not restore wild type tRNA function, suppress Spo$^-$, suggesting readthrough causes Spo$^-$ phenotype.	Rothstein et al. (1977)
SUP5, SUP11	Ochre suppressing alleles increases meiosis I nondisjunction 16–160-fold. Disomes are nonrecombinant suggesting readthough results in aberrant pairing, Rec, or Seg function (hygromycin also causes readthrough and meiotic nondisjunction)	Louis and Haber (1989)
SWI1/ADR6/ GAM3	Part of the SNF-SWI complex required for transcriptional activation of diverse genes (including alcohol dehydrogenase). Required for sporulation. mRNA is constitutive.	Taguchi and Young (1987); Cairns et al. (1994); Peterson and Herskowitz (1992)
TID1	Encodes putative helicase similar to *RAD54*. Dispensable for premeiotic DS, SPB, duplication, and separation at MI and MII. Dispensable for gene conversion but required for proper processing of reciprocal crossovers. Interacts with *Dmc1*. Mutant arrests at the mononucleate stage.	M. Dresser (unpubl.)
TOP2	Dispensable for premeiotic DS and Rec but required for MI. Mutant arrests at the mononucleate stage; Rec$^-$ mutants suppress the MI block, suggesting that *TOP2* resolves recombinant chromosomes. *TOP2*-encoded protein localizes to axial cores in meiosis.	Rose et al. (1990); Klein et al. (1992)
TOP3	Homologous to bacterial type I topoisomerases. Suppresses recombination; *top3* is *RAD1*-dependent hyper-Rec in mitosis, suggesting that *top3* damage is channeled to Rec repair pathway by *RAD1*. Required for sporulation.	Bailis et al. (1992)
TPK1,TPK2, TPK3	Encodes subunits of cAPK. Mutants in each of these suppress Spo$^-$ of *bcy1*.	Toda et al. (1987); Cameron et al. (1988)
TUP1	Similarity to β subunits of G proteins. Participates in general negative regulation of multiple pathways (in a complex with Ssn6) and is required for sporulation. Needed for **a**1-α2 repression of *RME1* and α2 repression of **a**-specific genes.	Wickner (1974); Williams and Trumbly (1990); Mukai et al. (1991); Keleher et al. (1992)
UBI4	Encodes polyubiquitin function. mRNA induced after DNA damage and sporulation, peaking at the time of the nuclear divisions.	Treger et al. (1988)
UBR1	Encodes 225-kD protein required for ubiquitin-dependent proteolysis and full sporulation.	Bartel et al. (1990)
UME1	Negative regulator of early meiotic gene expression. Dispensable for mitosis and instability of early meiotic mRNAs. Required for repression of early meiotic transcription. *ume1* does	Strich et al. (1989); R. Surosky et al. (unpubl.)

Continued on following pages.

Table 2 (*Continued.*)

Locus	Function	Reference
	not require *MAT***a**/*MAT*α, N and Glu starvation, *IME1* or *IME2* for derepressed expression in mitosis.	
UME2/SRB9	Negative regulator of early meiotic gene expression. Dispensable for mitosis. Required for repression of early meiotic transcription and instability of early meiotic mRNAs. *ume2* does not require *MAT***a**/*MAT*α, N and Glu starvation, *IME1* or *IME2* for derepressed expression in mitosis.	Strich et al. (1989); Surosky et al. (1994); R. Strich (unpubl.)
UME3/SSN8/SRB11	Encodes cyclin C homolog. Negative regulator of early meiotic gene expression. Dispensable for mitosis and instability of early meiotic mRNAs. Required for repression of early meiotic transcription. *ume3* does not require *MAT***a**/*MAT*α, N and Glu starvation, *IME1* or *IME2* for derepressed expression in mitosis. Activity is regulated by sporulation-specific ubiquitin-mediated degradation.	Strich et al. (1989); Kuchin et al. (1995); Liao et al. (1995); R. Surosky et al. (unpubl.); K. Cooper et al. (unpubl.)
UME4/RPD1/SIN3/SID1	Encodes 175-kD protein containing four PAH domains. Negative regulator of early meiotic gene expression. Dispensable for mitosis and instability of early meiotic mRNAs. Required for Ume6p repression of early meiotic transcription (and many other genes not involved in meiosis and spore formation) during mitosis. Required for re-repression or early meiotic genes at late stages of meiosis and for proper timing and full induction of meiosis-specific transcription. *ume4* does not require *MAT***a**/*MAT*α, N and Glu starvation, *IME1* or *IME2* for derepressed expression in mitosis.	Strich et al. (1989); Wang et al. (1990); Vidal et al. (1991); R. Surosky et al. (unpubl.); K. Struhl (unpubl.); A. Helms and R. Esposito (unpubl.)
UME5/SSN3/SRB10/ARE1	Encodes protein kinase. Negative regulator of early meiotic gene expression. Dispensable for mitosis. Required for repression of early meiotic gene transcription and instability of early meiotic mRNAs. *ume5* does not require *MAT***a**/*MAT*α, N and Glu starvation, *IME1* or *IME2* for derepressed expression in mitosis. *UME5*-mediated mRNA instability is glucose-dependent.	Strich et al. (1989); Surosky et al. (1994); Kuchin et al. (1995); Liao et al. (1995)
UME6/CAR80	Encodes C6 zinc finger URS1-binding protein that is a key regulator required for both repression and induction of early meiotic genes, and for sporulation. Dispensable for mitosis and instability of early meiotic mRNAs. Required for repression of early meiotic genes in mitosis, their activation in meiosis, completion of premeiotic DS, Rec, MI, MII, and spores. *ume6* does not require *MAT***a**/*MAT*α, starvation, *IME1* or *IME2* for derepressed expression in mitosis, and is epistatic to lethality of *IME1* overexpression in haploids. Ume6p requires Ume4p for mitotic repression and interacts with and requires Ime1p and Rim11p for induction of meiosis-specific transcription.	Strich et al. (1989); Park et al. (1992); Strich et al. (1994); Bowdish et al. (1995); Anderson et al. (1995); Steber and Esposito (1995); Rubin-Bejerano et al. (1996)

VAN1, VAN2	Required for vanadate sensitivity, entry into stationary phase, heat shock resistance, outer chain glycosylation of secreated proteins, and Glu, N repression of sporulation in rich medium. Mutants sporulate in rich medium and form inviable spores.	Kanik and Neff (1990); Kanik et al. (1995)
VMA1	Encodes a protein with three regions (ABC) which are spliced to yield AC(69K) and B (50K); AC is a vacuolar (H+)-Atpase and B is a site-specific endonuclease homologous to HO. Cleavage is meiosis-specific and induces conversion.	Gimble and Thorner (1992)
XRS2	Required for DNA repair but dispensable for mitotic Rec (*xrs2* is hyper-Rec during vegetative growth). Required for DSBs, meiotic Rec, and spore viability. *xrs2* is rescued by *spo13* and is epistatic to *rad52*, and is classified as early Rec function.	Ivanov et al. (1992); Chepurnaia et al. (1993)
YPT1	Encodes GDP/GTP-binding protein required for secretion. Essential for mitosis and required for sporulation. *ypt1* causes lethality during N starvation.	Segev and Botstein (1987)
ZIP1	Encodes 100-kD protein that is thought to form the transverse filaments of SCs. Dispensable for mitosis, axial cores, pairing, and chromosome condensation seen by in situ hybridization, and axial associations between lateral elements of homologous chromosomes. Required for intimately pairing of lateral elements in SCs and crossover interference. *zip1* exhibits meiotic conversion and reciprocal Rec in return to growth assay, but no recombined molecules during sporulation, suggesting complete synapsis is not required for Rec initiation. Cells arrest in prophase I at the mononucleate stage with duplicated but unseparated SPBs; Rec^- (e.g., *mei4*) bypasses *zip1* arrest and allows asci to form but spores are inviable, suggesting the arrest mechanism fails to occur in the absence of recombination. Zip1p localizes to SC central region.	Sym et al. (1993); Sym and Roeder (1994); Nag et al. (1995); Sym and Roeder (1995)
ZIP2	Dispensable for chromosome pairing and axial element formation, but required for synapsis and wild-type levels of reciprocal Rec. Zip2p localizes to discrete sites on wild-type chromosomes (i.e., to the axial associations observed in *zip1* strains which may mark sites of synaptic initiation). Zip2p colocalizes with Mre11p in a *rad50S* strain.	P. Chua and G.S. Roeder (unpubl.)

Abbreviations: (DS) DNA synthesis; (MI) meiosis I; (MII) meiosis II; (Rec) recombination; (SC) synaptonemal complex; (SPB) spindle pole body. For additional references on the map locations and identification of gene functions refer to the *Saccharomyces* Genome Data Base.

REFERENCES

Aboussekhra, A., R. Chanet, A. Adjiri, and F. Fabre. 1992. Semidominant suppressors of Srs2 helicase mutations of *Saccharomyces cerevisiae* map in the *RAD51* gene, whose sequence predicts a protein with similarities to procaryotic RecA proteins. *Mol. Cell. Biol.* **12:** 3224.

Aboussekhra, A., R. Chanet, Z. Zgaga, C.C. Cassier, M. Heude, and F. Fabre. 1989. *RADH*, a gene of *Saccharomyces cerevisiae* encoding a putative DNA helicase involved in DNA repair. Characteristics of *radH* mutants and sequence of the gene. *Nucleic Acids Res.* **17:** 7211.

Ajimura, M., S.H. Leem, and H. Ogawa. 1993. Identification of new genes required for meiotic recombination in *Saccharomyces cerevisiae. Genetics* **133:** 51.

Alani, E., R. Padmore, and N. Kleckner. 1990. Analysis of wild-type and *rad50* mutants of yeast suggests an intimate relationship between meiotic chromosome synapsis and recombination. *Cell* **61:** 419.

Alani, E., R.A. Reenan, and R.D. Kolodner. 1994. Interaction between mismatch repair and genetic recombination in *Saccharomyces cerevisiae. Genetics* **137:** 19.

Alani, E., S. Subbiah, and N. Kleckner. 1989. The yeast *RAD50* gene encodes a predicted 153-kD protein containing a purine nucleotide-binding domain and two large heptad-repeat regions. *Genetics* **122:** 47.

Allen, J. B., Z. Zhou, W. Siede, E.C. Friedberg, and S.J. Elledge. 1994. The SAD1/RAD53 protein kinase controls multiple checkpoints and DNA damage-induced transcription in yeast. *Genes Dev.* **8:** 2401.

Anderson, S.F., C.M. Steber, R.E. Esposito, and J.E. Coleman. 1995. *UME6*, a negative regulator of meiosis in *Saccharomyces cerevisiae*, contains a C-terminal Zn_2Cys_6 binuclear cluster that binds the URS1 DNA sequence in a zinc-dependent manner. *Protein Sci.* **4:** 1832.

Atcheson, C.L. 1991. "Meiosis-specific regulation of the *SPO11* gene of the yeast *Saccharomyces cerevisiae.*" Ph.D. thesis, University of Chicago, Illinois.

Atcheson, C.L., B. DiDomenico, S. Frackman, R.E. Esposito, and R.T. Elder. 1987. Isolation, DNA sequence, and regulation of a meiosis-specific eukaryotic recombination gene. *Proc. Natl. Acad. Sci.* **84:** 8035.

Bahler, J., T. Wyler, J. Loidl, and J. Kohli. 1993. Unusual nuclear structures in meiotic prophase of fission yeast: A cytological analysis. *J. Cell Biol.* **121:** 241.

Bahler, J., G. Hagens, G. Holzinger, H. Scherthan, and W.D. Heyer. 1994. *Saccharomyces cerevisiae* cells lacking the homologous pairing protein p175SEP1 arrest at pachytene during meiotic prophase. *Chromosoma* **103:** 129.

Bailis, A.M. and R. Rothstein. 1990. A defect in mismatch repair in *Saccharomyces cerevisiae* stimulates ectopic recombination between homeologous genes by an excision repair dependent process. *Genetics* **126:** 535.

Bailis, A.M., L. Arthur, and R. Rothstein. 1992. Genome rearrangement in *top3* mutants of *Saccharomyces cerevisiae* requires a functional *RAD1* excision repair gene. *Mol. Cell. Biol.* **12:** 4988.

Baker, B.S., A.T. Carpenter, M.S. Esposito, R.E. Esposito, and L. Sandler. 1976. The genetic control of meiosis. *Annu. Rev. Genet.* **10:** 53.

Baker, R.E. and D.C. Masison. 1990. Isolation of the gene encoding the *Saccharomyces cerevisiae* centromere-binding protein CP1. *Mol. Cell. Biol.* **10:** 2458.

Ballou, C., S. Maitra, J. Walker, and W. Whelan. 1977. Developmental defects associated with glucosamine auxotrophy in *Saccharomyces cerevisiae. Proc. Natl. Acad. Sci.* **74:**

4351.

Bartel, B., I. Wunning, and A. Varshavsky. 1990. The recognition component of the N-end rule pathway. *EMBO J.* **9:** 3179.

Bartlett, R. and P. Nurse. 1990. Yeast as a model system for understanding the control of DNA replication in eukaryotes. *BioEssays* **12:** 457.

Bauer, G.A. and P.M. Burgers. 1990. Molecular cloning, structure and expression of the yeast proliferating cell nuclear antigen gene. *Nucleic Acids Res.* **18:** 261.

Bell, L.R. and B. Byers. 1983. Homologous association of chromosomal DNA during yeast meiosis. *Cold Spring Harbor Symp. Quant. Biol.* **47:** 829.

Benson, F.E., A. Stasiak, and S.C. West. 1994. Purification and characterization of the human Rad51 protein, an analogue of *E. coli* RecA. *EMBO J.* **13:** 5764.

Bergerat, A., B. de Massy, D. Gadelle, P.-C. Varoutes, A. Nicolas, and P. Forterre. 1997. An atypical topoisomerase II from Archea with implications for meiotic recombination (comments). *Nature* **386:** 414.

Bhargava, J., J. Engebrecht, and G.S. Roeder. 1992. The *rec102* mutant of yeast is defective in meiotic recombination and chromosome synapsis. *Genetics* **130:** 59.

Biggins, S. and M.D. Rose. 1994. Direct interaction between yeast spindle pole body components: Kar1p is required for Cdc31p localization to the spindle pole body. *J. Cell Biol.* **125:** 843.

Bishop, D.K. 1994. RecA homologs Dmc1 and Rad51 interact to form multiple nuclear complexes prior to meiotic chromosome synapsis. *Cell* **79:** 1081.

Bishop, D.K., J. Andersen, and R.D. Kolodner. 1989. Specificity of mismatch repair following transformation of *Saccharomyces cerevisiae* with heteroduplex plasmid DNA. *Proc. Natl. Acad. Sci.* **86:** 3713.

Bishop, D.K., D. Park, L. Xu, and N. Kleckner. 1992. *DMC1:* A meiosis-specific yeast homolog of *E. coli* recA required for recombination, synaptonemal complex formation, and cell cycle progression. *Cell* **69:** 439.

Bisson, L. and J. Thorner. 1977. Thymidine 5′-monophosphate-requiring mutants of *Saccharomyces cerevisiae* are deficient in thymidylate synthetase. *J. Bacteriol.* **132:** 44.

Bloom, K., M. Kenna, and M. Saunders. 1989. *cis*- and *trans*-acting factors affecting the structure of yeast centromeres. *J. Cell Sci. Suppl.* **12:** 231.

Borts, R.H. and J.E. Haber. 1987. Meiotic recombination in yeast: Alteration by multiple heterozygosities. *Science* **237:** 1459.

Borts, R. H., M. Lichten, and J.E. Haber. 1986. Analysis of meiosis-defective mutations in yeast by physical monitoring of recombination. *Genetics* **113:** 551.

Borts, R. H., W.Y. Leung, W. Kramer, B. Kramer, M. Williamson, S. Fogel, and J.E. Haber. 1990. Mismatch repair-induced meiotic recombination requires the *pms1* gene product. *Genetics* **124:** 573.

Bowdish, K.S. and A.P. Mitchell. 1993. Bipartite structure of an early meiotic upstream activation sequence from *Saccharomyces cerevisiae*. *Mol. Cell. Biol.* **13:** 2172.

Bowdish, K.S., H.E. Yuan, and A.P. Mitchell. 1994. Analysis of RIM11, a yeast protein kinase that phosphorylates the meiotic activator *IME1*. *Mol. Cell. Biol.* **14:** 7909.

———. 1995. Positive control of yeast meiotic genes by the negative regulator *UME6*. *Mol. Cell. Biol.* **15:** 2955.

Breeden, L. and K. Nasmyth. 1987. Cell cycle control of the yeast *HO* gene: *cis*- and *trans*-acting regulators. *Cell* **48:** 389.

Brendel, M. and W.W. Fath. 1974. Isolation and characterization of mutants of *Sac-*

charomyces cerevisiae auxotrophic and conditionally auxotrophic for 5′-dTMP. *Z. Naturforsch. C Biosci.* **29:** 733.

Brewer, B.J. and W.L. Fangman. 1991. Mapping replication origins in yeast chromosomes. *BioEssays* **13:** 317.

Brewer, B.J., V.A. Zakian, and W.L. Fangman. 1980. Replication and meiotic transmission of yeast ribosomal RNA genes. *Proc. Natl. Acad. Sci.* **77:** 6739.

Brill, S.J. and B. Stillman. 1991. Replication factor-A from *Saccharomyces cerevisiae* is encoded by three essential genes coordinately expressed at S phase. *Genes Dev.* **5:** 1589.

Briza, P., M. Eckerstorfer, and M. Breitenbach. 1994. The sporulation-specific enzymes encoded by the *DIT1* and *DIT2* genes catalyze a two-step reaction leading to a soluble LL- dityrosine-containing precursor of the yeast spore wall. *Proc. Natl. Acad. Sci.* **91:** 4524.

Briza, P., M. Breitenbach, A. Ellinger, and J. Segall. 1990. Isolation of two developmentally regulated genes involved in spore wall maturation in *Saccharomyces cerevisiae. Genes Dev.* **4:** 1775.

Briza, P., A. Ellinger, G. Winkler, and M. Breitenbach. 1988. Chemical composition of the yeast ascospore wall. The second outer layer consists of chitosan. *J. Biol. Chem.* **263:** 11569.

Briza, P., G. Winkler, H. Kalchhauser, and M. Breitenbach. 1986. Dityrosine is a prominent component of the yeast ascospore wall. A proof of its structure. *J. Biol. Chem.* **261:** 4288.

Briza, P., H. Kalchhauser, E. Pittenauer, G. Allmaier, and M. Breitenbach. 1996. N,N′-Bisformyl dityrosine is an in vivo precursor to the yeast ascospore wall. *Eur. J. Biochem.* **239:** 124.

Broach, J.R. 1991. *RAS* genes in *Saccharomyces cerevisiae:* Signal transduction in search of a pathway. *Trends Genet.* **7:** 28.

Buck, V., A. White, and J. Rosamond. 1991. *CDC7* protein kinase activity is required for mitosis and meiosis in *Saccharomyces cerevisiae. Mol. Gen. Genet.* **227:** 452.

Buckingham, L.E. 1992. "Regulation of *SPO13*, a meiosis-specific gene of the yeast *Saccharomyces cerevisiae*." Ph.D. thesis, University of Chicago, Illinois.

Buckingham, L.E., H.T. Wang, R.T. Elder, R.M. McCarroll, M.R. Slater, and R.E. Esposito. 1990. Nucleotide sequence and promoter analysis of *SPO13*, a meiosis-specific gene of *Saccharomyces cerevisiae. Proc. Natl. Acad. Sci.* **87:** 9406.

Budd, M.E. and J.L. Campbell. 1995. Purification and enzymatic and functional characterization of DNA polymerase β-like enzyme, POL4, expressed during yeast meiosis. *Methods Enzymol.* **262:** 108.

Budd, M.E., K.D. Wittrup, J.E. Bailey, and J.L. Campbell. 1989. DNA polymerase I is required for premeiotic DNA replication and sporulation but not for X-ray repair in *Saccharomyces cerevisiae. Mol. Cell. Biol.* **9:** 365.

Burns, N., B. Grimwade, M.P.B. Ross, E.Y. Choi, K. Finberg, G.S. Roeder, and M. Snyder. 1994. Large-scale analysis of gene expression, protein localization, and gene disruption in *Saccharomyces cerevisiae. Genes Dev.* **8:** 1087.

Byers, B. 1981. Cytology of the yeast life cycle. In *The molecular biology of the yeast* Saccharomyces: *Life cycle and inheritance* (ed. J.N. Strathern et al.), p. 59. Cold Spring Harbor Laboratory, Cold Spring Harbor, New York.

Byers, B. and L. Goetsch. 1974. Duplication of spindle plaques and integration of the yeast cell cycle. *Cold Spring Harbor Symp. Quant. Biol.* **38:** 123.

———. 1975. Electron microscopic observations on the meiotic karyotype of diploid and tetraploid *Saccharomyces cerevisiae. Proc. Natl. Acad. Sci.* **72:** 5056.

———. 1976. A highly ordered ring of membrane-associated filaments in budding yeast. *J. Cell Biol.* **69:** 717.

Cai, M. and R.W. Davis. 1990. Yeast centromere binding protein CBF1, of the helix-loop-helix protein family, is required for chromosome stability and methionine prototrophy. *Cell* **61:** 437.

Cairns, B.R., Y.J. Kim, M.H. Sayre, B.C. Laurent, and R.D. Kornberg. 1994. A multi-subunit complex containing the *SWI1/ADR6, SWI2/SNF2, SWI3, SNF5*, and *SNF6* gene products isolated from yeast. *Proc. Natl. Acad. Sci.* **91:** 1950.

Calvert, G.R. and I.W. Dawes. 1984. Initiation of sporulation in *Saccharomyces cerevisiae.* Mutations preventing initiation. *J. Gen. Microbiol.* **130:** 615.

Cameron, S., L. Levin, M. Zoller, and M. Wigler. 1988. cAMP-independent control of sporulation, glycogen metabolism, and heat shock resistance in *S. cerevisiae. Cell* **53:** 555.

Cao, L., E. Alani, and N. Kleckner. 1990. A pathway for generation and processing of double-strand breaks during meiotic recombination in *S. cerevisiae. Cell* **61:** 1089.

Carpenter, A.T.C. 1975. Electron microscopy of meiosis in *Drosophila melanogaster* females: The recombination nodule—A recombination-associated structure at pachytene? *Proc. Natl. Acad. Sci.* **72:** 3186.

———. 1987. Gene conversion, recombination nodules, and the initiation of meiotic synapsis. *BioEssays* **6:** 232.

———. 1991. Distributive segregation: Motors in the polar wind (review)? *Cell* **64:** 885.

Chepurnaya, O.V., S.A. Kozhin, V.T. Peshekhonov, and V.G. Korolev. 1995. *RAD58 (XRS4)*—A new gene in the *RAD52* epistasis group. *Curr. Genet.* **28:** 274.

Chepurnaia, O.V., V.T. Peshekhonov, T.N. Kozhina, and V.G. Korolev. 1993. The *xrs2* gene controls recombination repair in yeast. *Genetika* **29:** 571.

Choi, W.J., B. Santos, A. Durán, and E. Cabib. 1994. Are yeast chitin synthases regulated at the transcriptional or the posttranslational level? *Mol. Cell. Biol.* **14:** 7685.

Chow, T.Y. and M.A. Resnick. 1988. An endo-exonuclease activity of yeast that requires a functional *RAD52* gene. *Mol. Gen. Genet.* **211:** 41.

Christman, M.F., F.S. Dietrich, and G.R. Fink. 1988. Mitotic recombination in the rDNA of *S. cerevisiae* is suppressed by the combined action of DNA topoisomerases I and II. *Cell* **55:** 413.

Chua, P. and G.S. Roeder. 1995. Bdf1, a yeast chromosomal protein required for sporulation. *Mol. Cell. Biol.* **15:** 3685.

Clancy, M.J., M.B. Buten, D.J. Straight, A.L. Kennedy, R.M. Partridge, and P.T. Magee. 1983. Isolation of genes expressed preferentially during sporulation in the yeast *Saccharomyces cerevisiae. Proc. Natl. Acad. Sci.* **80:** 3000.

Clarke, L. 1990. Centromeres of budding and fission yeasts (review)? *Trends Genet.* **6:** 150.

Clarke, L. and J. Carbon. 1980. Isolation of a yeast centromere and construction of functional small circular chromosomes. *Nature* **287:** 504.

Coe, J.G., K. Hamberg, and I.W. Dawes. 1994a. Mapping of the divergently transcribed sporulation-specific genes *SPS18* and *SPS19* to the left arm of chromosome XIV of *Saccharomyces cerevisiae. Yeast* **10:** 833.

Coe, J.G., L.E. Murray, and I.W. Dawes. 1994b. Identification of a sporulation-specific promoter regulating divergent transcription of two novel sporulation genes in *Sac-*

charomyces cerevisiae. Mol. Gen. Genet. **244:** 661.

Coe, J.G., L.E. Murray, C.J. Kennedy, and I.W. Dawes. 1992. Isolation and characterization of sporulation-specific promoters in the yeast *Saccharomyces cerevisiae. Mol. Microbiol.* **6:** 75.

Cohn, M.S., C.W. Tabor, and H. Tabor. 1978. Isolation and characterization of *Saccharomyces cerevisiae* mutants deficient in S-adenosylmethionine decarboxylase, spermidine, and spermine. *J. Bacteriol.* **134:** 208.

Cole, G.M. and R.K. Mortimer. 1989. Failure to induce a DNA repair gene, *RAD54*, in *Saccharomyces cerevisiae* does not affect DNA repair or recombination phenotypes. *Mol. Cell. Biol.* **9:** 3314.

Cole, G.M., D. Schild, and R.K. Mortimer. 1989. Two DNA repair and recombination genes in *Saccharomyces cerevisiae, RAD52* and *RAD54*, are induced during meiosis. *Mol. Cell. Biol.* **9:** 3101.

Coleman, K.G., H.Y. Steensma, D.B. Kaback, and J.R. Pringle. 1986. Molecular cloning of chromosome I DNA from *Saccharomyces cerevisiae:* Isolation and characterization of the *CDC24* gene and adjacent regions of the chromosome. *Mol. Cell. Biol.* **6:** 4516.

Collins, I. and C.S. Newlon. 1994a. Chromosomal DNA replication initiates at the same origins in meiosis and mitosis. *Mol. Cell. Biol.* **14:** 3524.

———. 1994b. Meiosis-specific formation of joint DNA molecules containing sequences from homologous chromosomes. *Cell* **76:** 65.

Colonna, W.J. and P.T. Magee. 1978. Glycogenolytic enzymes in sporulating yeast. *J. Bacteriol.* **134:** 844.

Comings, D.E. and A.D. Riggs. 1971. Molecular mechanisms of chromosome pairing, folding, and function. *Nature* **233:** 48.

Cool, M. and R.E. Malone. 1992. Molecular and genetic analysis of the yeast early meiotic recombination genes *REC102* and *REC107/MER2. Mol. Cell. Biol.* **12:** 1248.

Cooper, K.W. 1964. Meiotic conjunctive elements not involving chiasmata. *Proc. Natl. Acad. Sci.* **52:** 1248.

Cottarel, G., J.H. Shero, P. Hieter, and J.H. Hegemann. 1989. A 125-base-pair *CEN6* DNA fragment is sufficient for complete meiotic and mitotic centromere functions in *Saccharomyces cerevisiae. Mol. Cell. Biol.* **9:** 3342.

Covitz, P.A. and A.P. Mitchell. 1993. Repression by the yeast meiotic inhibitor *RME1*. *Genes Dev.* **7:** 1598.

Covitz, P.A., I. Herskowitz, and A.P. Mitchell. 1991. The yeast RME1 gene encodes a putative zinc finger protein that is directly repressed by a1-α2. *Genes Dev.* **5:** 1982.

Covitz, P.A., W. Song, and A.P. Mitchell. 1994. Requirement for *RGR1* and *SIN4* in *RME1*-dependent repression in *Saccharomyces cerevisiae. Genetics* **138:** 577.

Cox, M.M. and I.R. Lehman. 1987. Enzymes of general recombination. *Annu. Rev. Biochem.* **56:** 229.

Cumberledge, S. and J. Carbon. 1987. Mutational analysis of meiotic and mitotic centromere function in *Saccharomyces cerevisiae. Genetics* **117:** 203.

Cvrčková, F., C. De Vergilio, E. Manser, J.R. Pringle, and K. Nasmyth. 1995. Ste20-like protein kinases are required for normal localization of cell growth and for cytokinesis in budding yeast. *Genes Dev.* **9:** 1817.

Dahmann, C. and B. Futcher. 1995. Specialization of B-type cyclins for mitosis or meiosis in *S. cerevisiae. Genetics* **140:** 957.

Dailey, D., G.L. Schieven, M.Y. Lim, and H. Marquardt. 1990. Novel yeast protein kinase (*YPK1* gene product) is a 40- kilodalton phosphotyrosyl protein associated with protein-tyrosine kinase activity. *Mol. Cell. Biol.* **10:** 6244.

Davidow, L.S. and B. Byers. 1984. Enhanced gene conversion and postmeiotic segregation in pachytene-arrested *Saccharomyces cerevisiae. Genetics* **106:** 165.

Davidow, L.S., L. Goetsch, and B. Byers. 1980. Preferential occurrence of nonsister spores in two-spored asci of *Saccharomyces cerevisiae:* Evidence for regulation of spore-wall formation by the spindle pole body. *Genetics* **94:** 581.

Dawes, I.W. 1975. Study of cell development using depressed mutations. *Nature* **255:** 707.

Dawes, I.W. and G.R. Calvert. 1984. Initiation of sporulation in *Saccharomyces cerevisiae.* Mutations causing derepressed sporulation and G_1 arrest in the cell division cycle. *J. Gen. Microbiol.* **130:** 605.

Dawson, D.S., A.W. Murray, and J.W. Szostak. 1986. An alternative pathway for meiotic chromosome segregation in yeast. *Science* **234:** 713.

de Massy, B. and A. Nicolas. 1993. The control in *cis* of the position and the amount of the *ARG4* meiotic double-strand break of *Saccharomyces cerevisiae. EMBO J.* **12:** 1459.

de Massy, B., F. Baudat, and A. Nicolas. 1994. Initiation of recombination in *Saccharomyces cerevisiae* haploid meiosis. *Proc. Natl. Acad. Sci.* **91:** 11929.

de Massy, B., V. Rocco, and A. Nicolas. 1995. The nucleotide mapping of DNA double-strand breaks at the *CYS3* initiation site of meiotic recombination in *Saccharomyces cerevisiae. EMBO J.* **14:** 4589.

Detloff, P., M.A. White, and T.D. Petes. 1992. Analysis of a gene conversion gradient at the *HIS4* locus in *Saccharomyces cerevisiae. Genetics* **132:** 113.

Dickinson, J.R. 1988. The metabolism of sporulation in yeast. *Microbiol. Sci.* **5:** 121.

Dickinson, J.R. and D. Cole. 1991. A rapid and convenient screening technique for developmental pathway mutants of *Saccharomyces cerevisiae. FEMS Microbiol. Lett.* **68:** 245.

Dickinson, J.R. and M.J. Hewlins. 1991. 13C NMR analysis of a developmental pathway mutation in *Saccharomyces cerevisiae* reveals a cell derepressed for succinate dehydrogenase. *J. Gen. Microbiol.* **137:** 1033.

Dickinson, J.R., M.E. Smith, T.R. Swanson, A.S. Williams, and J.M. Wingfield. 1988. The *cdc30* mutation in *Saccharomyces cerevisiae* affects phosphoglucose isomerase, the cell cycle and sporulation. *J. Gen. Microbiol.* **134:** 2475.

Dobzhansky, T. 1931. Translocations involving the second and fourth chromosomes of *Drosophila melanogaster. Genetics* **16:** 629.

Donovan, J.W., G.T. Milne, and D.T. Weaver. 1994. Homotypic and heterotypic protein associations control Rad51 function in double-strand break repair. *Genes Dev.* **8:** 2552.

Draetta, G.F. 1994. Mammalian G1 cyclins. *Curr. Opin. Cell Biol.* **6:** 842.

Dranginis, A.M. 1989. Regulation of *STA1* gene expression by *MAT* during the life cycle of *Saccharomyces cerevisiae. Mol. Cell. Biol.* **9:** 3992.

Dresser, M.E. and C.N. Giroux. 1988. Meiotic chromosome behavior in spread preparations of yeast. *J. Cell Biol.* **106:** 567.

Dresser, M.E. and M.J. Moses. 1980. Synaptonemal complex karyotyping in spermatocytes of the Chinese hamster (*Cricetulus griseus*). IV. Light and electron microscopy of synapsis and nucleolar development by silver staining. *Chromosoma* **76:** 1.

Dresser, M.E., D.J. Ewing, S.N. Harwell, D. Coody, and M.N. Conrad. 1994. Nonhomologous synapsis and reduced crossing over in a heterozygous paracentric inversion in *Saccharomyces cerevisiae. Genetics* **138:** 633.

Dujardin, G., M. Kermorgant, P. Slonimski, and H. Boucherie. 1994. Cloning and sequencing of the GMP synthetase-encoding gene of *Saccharomyces cerevisiae. Gene* **139:** 127.

Dykstra, C.C., K. Kitada, A.B. Clark, R.K. Hamatake, and A. Sugino. 1991. Cloning and characterization of *DST2*, the gene for DNA strand transfer protein β from *Saccharomyces cerevisiae. Mol. Cell. Biol.* **11:** 2583.

Egel-Mitani, M., L.W. Olson, and R. Egel. 1982. Meiosis in *Aspergillus nidulans:* Another example for lacking synaptonemal complexes in the absence of crossover interference. *Hereditas* **97:** 179.

Eisenmann, D.M., C. Dollard, and F. Winston. 1989. *SPT15*, the gene encoding the yeast TATA binding factor TFIID, is required for normal transcription initiation *in vivo. Cell* **58:** 1183.

Emery, H.S., D. Schild, D.E. Kellogg, and R.K. Mortimer. 1991. Sequence of *RAD54*, a *Saccharomyces cerevisiae* gene involved in recombination and repair. *Gene* **104:** 103.

Engebrecht, J. and G.S. Roeder. 1989. Yeast *mer1* mutants display reduced levels of meiotic recombination. *Genetics* **121:** 237.

———. 1990. *MER1*, a yeast gene required for chromosome pairing and genetic recombination, is induced in meiosis. *Mol. Cell. Biol.* **10:** 2379.

Engebrecht, J., J. Hirsch, and G.S. Roeder. 1990. Meiotic gene conversion and crossing over: Their relationship to each other and to chromosome synapsis and segregation. *Cell* **62:** 927.

Engebrecht, J.A., M.K. Voelkel, and G.S. Roeder. 1991. Meiosis-specific RNA splicing in yeast. *Cell* **66:** 1257.

Eraso, P. and J.P. Gancedo. 1985. Use of glucose analogs to study the mechanism of glucose-mediated cAMP increase in yeast. *FEBS Lett.* **191:** 51.

Espinoza, F.H., A. Farrell, H. Erdjument-Bromage, P. Tempst, and D.O. Morgan. 1996. A cyclin-dependent kinase-activating kinase (CAK) in budding yeast unrelated to vertebrate CAK. *Science* **273:** 1714.

Esposito, M.S. and J.T. Brown. 1990. Conditional hyporecombination mutants of three *REC* genes of *Saccharomyces cerevisiae. Curr. Genet.* **17:** 7.

Esposito, M.S. and R.E. Esposito. 1969. The genetic control of sporulation in *Saccharomyces.* I. The isolation of temperature-sensitive sporulation-deficient mutants. *Genetics* **61:** 79.

———. 1974. Genes controlling meiosis and spore formation in yeast. *Genetics* **78:** 215.

———. 1978. Aspects of the general control of meiosis and ascospore development inferred from the study of *spo* (sporulation-deficient) mutants of *Saccharomyces cerevisia. Biol. Cell* **33:** 93.

Esposito, M.S., F.M. Bolotin, and R.E. Esposito. 1975. Antimutator activity during mitosis by a meiotic mutant of yeast. *Mol. Gen. Genet.* **139:** 9.

Esposito, M.S., R.E. Esposito, M. Arnaud, and H.O. Halvorson. 1970. Conditional mutants of meiosis in yeast. *J. Bacteriol.* **104:** 202.

Esposito, R.E. and M.S. Esposito. 1974. Genetic recombination and commitment to meiosis in *Saccharomyces. Proc. Natl. Acad. Sci.* **71:** 3172.

Esposito, R.E. and S. Klapholz. 1981. Meiosis and ascospore development. In *The molecular biology of the yeast* Saccharomyces: *Life cycle and inheritance* (ed. J.N. Strathern et al.), p. 211. Cold Spring Harbor Laboratory, Cold Spring Harbor, New York.

Esposito, R.E., M. Dresser, and M. Breitenbach. 1991. Identifying sporulation genes, visualizing synaptonemal complexes, and large-scale spore and spore wall purification. *Methods Enzymol.* **194:** 110.

Esposito, R.E., N. Frink, P. Bernstein, and M.S. Esposito. 1972. The genetic control of sporulation in *Saccharomyces.* II. Dominance and complementation of mutants of meiosis and spore formation. *Mol. Gen. Genet.* **114:** 241.

Fan, Q.-Q. and T.D. Petes. 1996. Relationship between nuclease-hypersensitive sites and meiotic recombination hot spot activity at the *HIS4* locus of *Saccharomyces cerevisiae. Mol. Cell. Biol.* **16:** 2037.

Fan, Q., F. Xu, and T.D. Petes. 1995. Meiosis-specific double-strand DNA breaks at the *HIS4* recombination hot spot in the yeast *Saccharomyces cerevisiae:* Control in *cis* and *trans. Mol. Cell. Biol.* **15:** 1679.

Fassler, J.S. and F. Winston. 1988. Isolation and analysis of a novel class of suppressor of Ty insertion mutations in *Saccharomyces cerevisiae. Genetics* **118:** 203.

Fast, D.E. 1978. "Premeiotic DNA synthesis and recombination in the yeast *Saccharomyces.*" Ph.D. thesis, University of Chicago, Illinois.

Fiil, A. and P.B. Moens. 1973. The development, structure and function of modified synaptonemal complexes in mosquito oocytes. *Chromosoma* **41:** 37.

Firmenich, A.A., A.M. Elias, and P. Berg. 1995. A novel allele of *Saccharomyces cerevisiae RFA1* that is deficient in recombination and repair and suppressible by *RAD52. Mol. Cell. Biol.* **15:** 1620.

Fitch, I., C. Dahmann, U. Surana, A. Amon, K. Nasmyth, L. Goetsch, B. Byers, and B. Futcher. 1992. Characterization of four B-type cyclin genes of the budding yeast *Saccharomyces cerevisiae. Mol. Biol. Cell* **3:** 805.

Fitzgerald-Hayes, M., J.M. Buhler, T.G. Cooper, and J. Carbon. 1982. Isolation and subcloning analysis of functional centromere DNA (*CEN11*) from *Saccharomyces cerevisiae* chromosome XI. *Mol. Cell. Biol.* **2:** 82.

Fletcher, H.L. 1981. A search for synaptonemal complexes in *Ustilago maydis. J. Cell Sci.* **50:** 171.

Fogel, S. and D.D. Hurst. 1967. Meiotic gene conversion in yeast tetrads and the theory of recombination. *Genetics* **57:** 455.

Fogel, S. and R. Roth. 1974. Mutations affecting meiotic gene conversion in yeast. *Mol. Gen. Genet.* **130:** 189.

Fogel, S., R.K. Mortimer, and K. Lusak. 1981. Mechanisms of meiotic gene conversion, or "Wanderings on a foreign strand." In *The molecular biology of the yeast* Saccharomyces: *Life cycle and inheritance* (ed. J.N. Strathern et al.), p. 289. Cold Spring Harbor, Laboratory, Cold Spring Harbor, New York.

Freese, E.B., M.I. Chu, and E. Freese. 1982. Initiation of yeast sporulation of partial carbon, nitrogen, or phosphate deprivation. *J. Bacteriol.* **149:** 840.

Friedman, D.B., N.M. Hollingsworth, and B. Byers. 1994. Insertional mutations in the yeast *HOP1* gene: Evidence for multimeric assembly in meiosis. *Genetics* **136:** 449.

Friesen, H., R. Lunz, S. Doyle, and J. Segall. 1994. Mutation of the *SPS1*-encoded protein kinase of *Saccharomyces cerevisiae* leads to defects in transcription and morphology during spore formation. *Genes Dev.* **8:** 2162.

Futcher, B. and J. Carbon. 1986. Toxic effects of excess cloned centromeres. *Mol. Cell. Biol.* **6:** 2213.

Gailus-Durner, V., J. Xie, C. Chintamaneni, and A. Vershon. 1996 Participation of the yeast activator Abf1 in meiosis-specific expression of the *HOP1* gene. *Mol. Cell. Biol.* **16:** 2777.

Galbraith, A.M. and R.E. Malone. 1992. Characterization of *REC104*, a gene required for early meiotic recombination in the yeast *Saccharomyces cerevisiae*. *Dev. Genet.* **13:** 392.

Game, J.C. 1976. Yeast cell-cycle mutant *cdc21* is a temperature-sensitive thymidylate auxotroph. *Mol. Gen. Genet.* **146:** 313.

———. 1983. Radiation-sensitive mutants and repair in yeast. In *Yeast genetics: Fundamental and applied aspects* (ed. J.F.T. Spencer et al.), p. 109. Springer-Verlag, Berlin.

———. 1992. Pulsed-field gel analysis of the pattern of DNA double-strand breaks in the *Saccharomyces* genome during meiosis. *Dev. Genet.* **13:** 485.

Game, J.C. and R.K. Mortimer. 1974. A genetic study of x-ray sensitive mutants in yeast. *Mutat. Res.* **24:** 281.

Game, J.C., K.C. Sitney, V.E. Cook, and R.K. Mortimer. 1989. Use of a ring chromosome and pulsed-field gels to study interhomolog recombination, double-strand DNA breaks and sister-chromatid exchange in yeast. *Genetics* **123:** 695.

Game, J.C., T.J. Zamb, R.J. Braun, M. Resnick, and R.M. Roth. 1980. The role of radiation (*rad*) genes in meiotic recombination in yeast. *Genetics* **94:** 51.

Gangloff, S., J.P. McDonald, C. Bendixen, L. Arthur, and R. Rothstein. 1994. The yeast type I topoisomerase Top3 interacts with Sgs1, a DNA helicase homolog: A potential eukaryotic reverse gyrase. *Mol. Cell. Biol.* **14:** 8391.

Garvik, B., M. Carson, and L. Hartwell. 1995. Single-stranded DNA arising at telomeres in *cdc13* mutants may constitute a specific signal for the *RAD9* checkpoint. *Mol. Cell. Biol.* **15:** 6128.

Gaudet, A. and H.M. Fitzgerald. 1989. Mutations in *CEN3* cause aberrant chromosome segregation during meiosis in *Saccharomyces cerevisiae*. *Genetics* **121:** 477.

Gilbertson, L.A. and F.W. Stahl. 1994. Initiation of meiotic recombination is independent of interhomologue interactions. *Proc. Natl. Acad. Sci.* **91:** 11934.

Gimble, F.S. and J. Thorner. 1992. Homing of a DNA endonuclease gene by meiotic gene conversion in *Saccharomyces cerevisiae*. *Nature* **357:** 301.

Giroux, C.N., M.E. Dresser, and H.F. Tiano. 1989. Genetic control of chromosome synapsis in yeast meiosis. *Genome* **31:** 88.

Giroux, C.N., S.L. Misra, M. Modesti, and J. Guerreiro. 1993 Decision points and the genetic control of meiotic progression. In *Meiosis II: Contemporary approaches to the study of meiosis* (ed.F.P. Haseltine and S. Heyner), p. 1. AAAS Press, Washington, D.C.

Goetsch, L. and B. Byers. 1982. Meiotic cytology of *Saccharomyces cerevisiae* in protoplast lysates. *Mol. Gen. Genet.* **187:** 54.

Goldstein, L.S. 1981. Kinetochore structure and its role in the first meiotic division in male *Drosophila*. *Cell* **25:** 591.

Goldstein, P. and P.B. Moens. 1976. Karyotype analysis of *Ascaris lumbricoides var. suum*. Male and female pachytene nuclei by 3-D reconstruction from electron microscopy of serial sections. *Chromosoma* **58:** 101.

Goldway, M., T. Arbel, and G. Simchen. 1993a. Meiotic nondisjunction and recombination of chromosome III and homologous fragments in *Saccharomyces cerevisiae*. *Genetics* **133:** 149.

Goldway, M., A. Sherman, D. Zenvirth, T. Arbel, and G. Simchen. 1993b. A short chromosomal region with major roles in yeast chromosome III meiotic disjunction, recombination and double strand breaks. *Genetics* **133:** 159.

Goto, T. and J.C. Wang. 1984. Yeast DNA topoisomerase II is encoded by a single-copy, essential gene. *Cell* **36:** 1073.

Gottlieb, S. and R.E. Esposito. 1989. A new role for a yeast transcriptional silencer gene, *SIR2*, in regulation of recombination in ribosomal DNA. *Cell* **56:** 771.

Gottlieb, S., J. Wagstaff, and R.E. Esposito. 1989. Evidence for two pathways of meiotic intrachromosomal recombination in yeast. *Proc. Natl. Acad. Sci.* **86:** 7072.

Gottlin-Ninfa, E. and D.B. Kaback. 1986. Isolation and functional analysis of sporulation-induced transcribed sequences from *Saccharomyces cerevisiae. Mol. Cell. Biol.* **6:** 2185.

Goyon, C. and M. Lichten. 1993. Timing of molecular events in meiosis in *Saccharomyces cerevisiae:* Stable heteroduplex DNA is formed late in meiotic prophase. *Mol. Cell. Biol.* **13:** 373.

Grandin, N. and S.I. Reed. 1993. Differential function and expression of *Saccharomyces cerevisiae* B-type cyclins in mitosis and meiosis. *Mol. Cell. Biol.* **13:** 2113.

Granot, D., J.P. Margolskee, and G. Simchen. 1989. A long region upstream of the *IME1* gene regulates meiosis in yeast. *Mol. Gen. Genet.* **218:** 308.

Grell, E.H. 1970. Distributive pairing: Mechanism for segregation of compound autosomal chromosomes in oocytes of *Drosophila melanogaster. Genetics* **65:** 65.

Grewal, N.S. and J.J. Miller. 1972. Formation of asci with two diploid spores by diploid cells of *Saccharomyces. Can. J. Microbiol.* **18:** 1897.

Guacci, V. and D.B. Kaback. 1991. Distributive disjunction of authentic chromosomes in *Saccharomyces cerevisiae. Genetics* **127:** 475.

Haber, J.E. and P.C. Thorburn. 1984. Healing of broken linear dicentric chromosomes in yeast. *Genetics* **106:** 207.

Hawley, R.S. 1989. Genetic and molecular analysis of a simple disjunctional system in *Drosophila melanogaster. Prog. Clin. Biol. Res.* **311:** 277.

Hays, S.L., A.A. Firmenich, and P. Berg. 1995. Complex formation in yeast double-strand break repair: Participation of Rad51, Rad52, Rad55, and Rad57 proteins. *Proc. Natl. Acad. Sci.* **92:** 6925.

Hegemann, J.H. and U.N. Fleig. 1993. The centromere of budding yeast. *BioEssays* **15:** 451.

Heinemeyer, W., J.A. Kleinschmidt, J. Saidowsky, C. Escher, and D.H. Wolf. 1991. Proteinase yscE, the yeast proteasome/multicatalytic-multifunctional proteinase: Mutants unravel its function in stress induced proteolysis and uncover its necessity for cell survival. *EMBO J.* **10:** 555.

Henriques, J.A., E.J. Vicente, K.V. Leandro da Silva, and A.C. Schenberg. 1989. *PSO4:* A novel gene involved in error-prone repair in *Saccharomyces cerevisiae. Mutat. Res.* **218:** 111.

Hepworth, S.R., L.K. Ebisuzaki, and J. Segall. 1995. A 15-base-pair element activates the *SPS4* gene midway through sporulation in *Saccharomyces cerevisiae. Mol. Cell. Biol.* **15:** 3934.

Herskowitz, I., J. Rine, and J.N. Strathern. 1992. Mating-type determination and mating-type interconversion in *Saccharomyces cerevisiae.* In *The molecular and cellular biology of the yeast* Saccharomyces: *Gene expression* (ed. E.W. Jones et al.), p. 583. Cold Spring Harbor Laboratory Press, Cold Spring Harbor, New York.

Heude, M., R. Chanet, and F. Fabre. 1995. Regulation of the *Saccharomyces cerevisiae* Srs2 helicase during the mitotic cell cycle, meiosis and after irradiation. *Mol. Gen. Genet.* **248:** 59.

Heyer, W.D., A.W. Johnson, U. Reinhart, and R.D. Kolodner. 1995. Regulation and intracellular localization of *Saccharomyces cerevisiae* strand exchange protein 1 (Sep1/Xrn1/Kem1), a multifunctional exonuclease. *Mol. Cell. Biol.* **15:** 2728.

Hinnebusch, A.G. and G.R. Fink. 1983. Positive regulation in the general amino acid control of *Saccharomyces cerevisiae*. *Proc. Natl. Acad. Sci.* **80:** 5374.

Holliday, R. 1977. Recombination and meiosis. *Philos. Trans. R. Soc. Lond. B Biol. Sci.* **277:** 359.

Hollingsworth, N.M. and B. Byers. 1989. *HOP1:* A yeast meiotic pairing gene. *Genetics* **121:** 445.

Hollingsworth, N.M. and A.D. Johnson. 1993. A conditional allele of the *Saccharomyces cerevisiae HOP1* gene is suppressed by overexpression of two other meiosis-specific genes: *RED1* and *REC104*. *Genetics* **133:** 785.

Hollingsworth, N.M., L. Goetsch, and B. Byers. 1990. The *HOP1* gene encodes a meiosis-specific component of yeast chromosomes. *Cell* **61:** 73.

Hollingsworth, N.M., L. Ponte, and C. Halsey. 1995. *MSH5*, a novel *MutS* homolog, facilitates meiotic reciprocal recombination between homologs in *Saccharomyces cerevisiae* but not mismatch repair. *Genes Dev.* **9:** 1728.

Hollingsworth, R.E., Jr. and R.A. Sclafani. 1990. DNA metabolism gene *CDC7* from yeast encodes a serine (threonine) protein kinase. *Proc. Natl. Acad. Sci.* **87:** 6272.

———. 1993. Yeast pre-meiotic DNA replication utilizes mitotic origin ARS1 independently of *CDC7* function. *Chromosoma* **102:** 415.

Holm, C., T. Goto, J.C. Wang, and D. Botstein. 1985. DNA topoisomerase II is required at the time of mitosis in yeast. *Cell* **41:** 553.

Honigberg, S.M. and R.E. Esposito. 1994. Reversal of cell determination in yeast meiosis: Postcommitment arrest allows return to mitotic growth. *Proc. Natl. Acad. Sci.* **91:** 6559.

Honigberg, S.M., C. Conicella, and R.E. Espositio. 1992. Commitment to meiosis in *Saccharomyces cerevisiae:* Involvement of the *SPO14* gene. *Genetics* **130:** 703.

Hopper, A.K. and B.D. Hall. 1975a. Mating type and sporulation in yeast. I. Mutations which alter mating-type control over sporulation. *Genetics* **80:** 41.

———. 1975b. Mutation of a heterothallic strain to homothallism. *Genetics* **80:** 77.

Hopper, A.K., J. Kirsch, and B.D. Hall. 1975. Mating type and sporulation in yeast. II. Meiosis, recombination, and radiation sensitivity in an α-α diploid with altered sporulation control. *Genetics* **80:** 61.

Horesh, O., G. Simchen, and A. Friedmann. 1979. Morphogenesis of the synapton during yeast meiosis. *Chromosoma* **75:** 101.

Hotta, Y., S. Tabata, and H. Stern. 1984. Replication and nicking of zygotene DNA sequences. Control by a meiosis-specific protein. *Chromosoma* **90:** 243.

Hubbard, E.J., X.L. Yang, and M. Carlson. 1992. Relationship of the cAMP-dependent protein kinase pathway to the SNF1 protein kinase and invertase expression in *Saccharomyces cerevisiae*. *Genetics* **130:** 71.

Huckle, T.C. and E.J. Louis. 1995. Yeast telomeres are sequestered from the rest of the genome during recombination. *Yeast* **11:** S131.

Hudak, K.A., J.M. Lopes, and S.A. Henry. 1994. A pleiotropic phospholipid biosynthetic regulatory mutation in *Saccharomyces cerevisiae* is allelic to *sin3* (*sdi1, ume4, rpd1*). *Genetics* **136:** 475.

Hugerat, Y. and G. Simchen. 1993. Mixed segregation and recombination of chromosomes and YACs during single-division meiosis in *spo13* strains of *Saccharomyces cerevisiae*. *Genetics* **135:** 297.

Icho, T. and R.B. Wickner. 1987. Metal-binding, nucleic acid-binding finger sequences in the *CDC16* gene of *Saccharomyces cerevisiae. Nucleic Acids Res.* **15:** 8439.

Irniger, S., S. Piatti, C. Michaelis, and K. Nasmyth. 1995. Genes involved in sister chromatid separation are needed for B-type cyclin proteolysis in budding yeast. *Cell* **81:** 269.

Ivanov, E.L., V.G. Korolev, and F. Fabre. 1992. *XRS2*, a DNA repair gene of *Saccharomyces cerevisiae*, is needed for meiotic recombination. *Genetics* **132:** 651.

Jackson, J.A. and G.R. Fink. 1985. Meiotic recombination between duplicated genetic elements in *Saccharomyces cerevisiae. Genetics* **109:** 303.

Jeyaprakash, A., R. Das Gupta, and R. Kolodner. 1996. *Saccharomyces cerevisiae pms2* mutations are alleles of *MLH1* and *pms2-2* corresponds to a hereditary nonpolyposis colorectal carcinoma-causing missense mutations. *Mol. Cell. Biol.* **16:** 3008.

Jeyaprakash, A., J.W. Welch, and S. Fogel. 1994. Mutagenesis of yeast MW104-1B strain has identified the uncharacterized *PMS6* DNA mismatch repair gene locus and additional alleles of existing *PMS1, PMS2* and *MSH2* genes. *Mutat Res.* **325:** 21.

Jiang, W., M.Y. Lim, H.J. Yoon, J. Thorner, G.S. Martin, and J. Carbon. 1995. Overexpression of the yeast MCK1 protein kinase suppresses conditional mutations in centromere-binding protein genes *CBF2* and *CBF5. Mol. Gen. Genet.* **246:** 360.

Jiang, Y.W. and D.J. Stillman. 1992. Involvement of the SIN4 global transcriptional regulator in the chromatin structure of *Saccharomyces cerevisiae. Mol. Cell. Biol.* **12:** 4503.

Jiang, Y.W., P.R. Dohrmann, and D.J. Stillman. 1995. Genetic and physical interactions between yeast *RGR1* and *SIN4* in chromatin organization and transcriptional regulation. *Genetics* **140:** 47.

Jinks-Robertson, S. and T.D. Petes. 1985. High-frequency meiotic gene conversion between repeated genes on nonhomologous chromosomes in yeast. *Proc. Natl. Acad. Sci.* **82:** 3350.

———. 1986. Chromosomal translocations generated by high-frequency meiotic recombination between repeated yeast genes. *Genetics* **114:** 731.

Johnson, R.D. and L.S. Symington. 1995. Functional differences and interactions among the putative RecA homologs Rad51, Rad55, and Rad57. *Mol. Cell. Biol.* **15:** 4843.

Johnston, L.H. and K. Nasmyth. 1978. *Saccharomyces cerevisiae* cell cycle mutant *cdc9* is defective in DNA ligase. *Nature* **274:** 891.

Johnston, L.H., A.L. Johnson, and D.G. Barker. 1986. The expression in meiosis of genes which are transcribed periodically in the mitotic cell cycle of budding yeast. *Exp. Cell Res.* **165:** 541.

Johnston, L.H., A.L. Johnson, and J.C. Game. 1982a. The effect of the *cdc9* mutation on premeiotic DNA synthesis in the yeast *Saccharomyces cerevisiae. Exp. Cell Res.* **141:** 63.

Johnston, L.H., D.H. Williamson, A.L. Johnson, and D.J. Fennell. 1982b. On the mechanism of premeiotic DNA synthesis in the yeast*Saccharomyces cerevisiae. Exp. Cell Res.* **141:** 53.

Johnston, L.H., J.H. White, A.L. Johnson, G. Lucchini, and P. Plevani. 1990. Expression of the yeast DNA primase gene, *PRI1*, is regulated within the mitotic cell cycle and in meiosis. *Mol. Gen. Genet.* **221:** 44.

Johzuka, K. and H. Ogawa. 1995. Interaction of Mre11 and Rad50: Two proteins required for DNA repair and meiosis-specific double-strand break formation in *Saccharomyces cerevisiae. Genetics* **139:** 1521.

Kaback, D.B. 1989. Meiotic segregation of circular plasmid-minichromosomes from intact chromosomes in *Saccharomyces cerevisiae*. *Curr. Genet.* **15:** 385.

Kaback, D.B. and L.R. Feldberg. 1985. *Saccharomyces cerevisiae* exhibits a sporulation-specific temporal pattern of transcript accumulation. *Mol. Cell. Biol.* **5:** 751.

Kaback, D.B., H.Y. Steensma, and P. de Jonge. 1989. Enhanced meiotic recombination on the smallest chromosome of *Saccharomyces cerevisiae*. *Proc. Natl. Acad. Sci.* **86:** 3694.

Kaback, D.B., V. Guacci, D. Barber, and J.W. Mahon. 1992. Chromosome size-dependent control of meiotic recombination. *Science* **256:** 228.

Kaldis, P., A. Sutton, and M.J. Solomon. 1996. The Cdk-activating kinase (CAK) from budding yeast. *Cell* **86:** 553.

Kallal, L.A., M. Bhattacharyya, S.N. Grove, R.F. Iannacone, T.A. Pugh, D.A. Primerano, and M.J. Clancy. 1990. Functional analysis of the sporulation-specific *SPR6* gene of *Saccharomyces cerevisiae*. *Curr. Genet.* **18:** 293.

Kanik, E.C. and N. Neff. 1990. Vanadate-resistant mutants of *Saccharomyces cerevisiae* show alterations in protein phosphorylation and growth control. *Mol. Cell. Biol.* **10:** 898.

Kanik, E. C., E. Montalvo, and N. Neff. 1995. Sodium orthovanadate-resistant mutants of *Saccharomyces cerevisiae* show defects in Golgi-mediated protein glycosylation, sporulation and detergent resistance. *Genetics* **140:** 933.

Kans, J.A. and R.K. Mortimer. 1991. Nucleotide sequence of the *RAD57* gene of *Saccharomyces cerevisiae*. *Gene* **105:** 139.

Kao, G., J.C. Shah, and M.J. Clancy. 1990. An *RME1*-independent pathway for sporulation control in *Saccharomyces cerevisiae* acts through *IME1* transcript accumulation. *Genetics* **126:** 823.

Kao, G., D.G. Mannix, B.L. Holaway, M.C. Finn, A.E. Bonny, and M.J. Clancy. 1989. Dependence of inessential late gene expression on early meiotic events in *Saccharomyces cerevisiae*. *Mol. Gen. Genet.* **215:** 490.

Kassir, Y. and G. Simchen. 1976. Regulation of mating and meiosis in yeast by the mating- type region. *Genetics* **82:** 187.

———. 1978. Meiotic recombination and DNA synthesis in a new cell cycle mutant of *Saccharomyces cerevisiae*. *Genetics* **90:** 49.

Kassir, Y., D. Granot, and G. Simchen. 1988. *IME1*, a positive regulator gene of meiosis in *S. cerevisiae*. *Cell* **52:** 853.

Kato, R. and H. Ogawa. 1994. An essential gene, *ESR1*, is required for mitotic cell growth, DNA repair and meiotic recombination in *Saccharomyces cerevisiae*. *Nucleic Acids Res.* **22:** 3104.

Kawaguchi, H., M. Yoshida, and I. Yamashita. 1992. Nutritional regulation of meiosis-specific gene expression in *Saccharomyces cerevisiae*. *Biosci. Biotechnol. Biochem.* **56:** 289.

Kaytor, M.D. and D.M. Livingston. 1994. *Saccharomyces cerevisiae RAD52* alleles temperature- sensitive for the repair of DNA double-strand breaks. *Genetics* **137:** 933.

———. 1995. *GSG1*, a yeast gene required for sporulation. *Yeast* **11:** 1147.

Keeney, S. and N. Kleckner. 1995. Covalent protein-DNA complexes at the 5′ strand termini of meiosis-specific double-strand breaks in yeast. *Proc. Natl. Acad. Sci.* **92:** 11274.

Keeney, S., C.N. Giroux, and N. Kleckner. 1997. Meiosis-specific DNA double-strand breaks are catalyzed by Spo11, a member of a widely conserved protein family. *Cell*

88: 375.

Keil, R.L. and G.S. Roeder. 1984. *cis*-Acting, recombination-stimulating activity in a fragment of the ribosomal DNA of *S. cerevisiae*. *Cell* **39:** 377.

Keleher, C.A., M.J. Redd, J. Schultz, M. Carlson, and A.D. Johnson. 1992. Ssn6-Tup1 is a general repressor of transcription in yeast. *Cell* **68:** 709.

Kihara, K., M. Nakamura, R. Akada, and I. Yamashita. 1991. Positive and negative elements upstream of the meiosis-specific glucoamylase gene in *Saccharomyces cerevisiae*. *Mol. Gen. Genet.* **226:** 383.

Kim, J., P.O. Ljungdahl, and G.R. Fink. 1990. *kem* mutations affect nuclear fusion in *Saccharomyces cerevisiae*. *Genetics* **126:** 799.

Kim, Y. J., S. Bjorklund, Y. Li, M. H. Sayre, and R.D. Kornberg. 1994. A multiprotein mediator of transcriptional activation and its interaction with the C-terminal repeat domain of RNA polymerase II. *Cell* **77:** 599.

King, R.W., J.M. Peters, S. Tugendreich, M. Rolfe, P. Hieter, and M.W. Kirschner. 1995. A 20S complex containing CDC27 and CDC16 catalyzes the mitosis-specific conjugation of ubiquitin to cyclin B. *Cell* **81:** 279.

Klapholz, S. and R.E. Esposito. 1980a. Isolation of *SPO12-1* and *SPO13-1* from a natural variant of yeast that undergoes a single meiotic division. *Genetics* **96:** 567.

———. 1980b. Recombination and chromosome segregation during the single division meiosis in *SPO12-1* and *SPO13-1* diploids. *Genetics* **96:** 589.

Klapholz, S., C.S. Waddell,, and R.E. Esposito. 1985. The role of the *SPO11* gene in meiotic recombination in yeast. *Genetics* **110:** 187.

Klar, A.J. and J.N. Strathern. 1984. Resolution of recombination intermediates generated during yeast mating type switching. *Nature* **310:** 744.

Klar, A.J., S. Fogel, and D.N. Radin. 1979. Switching of a mating-type **a** mutant allele in budding yeast *Saccharomyces cerevisiae*. *Genetics* **92:** 759.

Kleckner, N., R. Padmore, and D.K. Bishop. 1991. Meiotic chromosome metabolism: One view. *Cold Spring Harbor Symp. Quant. Biol.* **56:** 729.

Klein, F., T. Laroche, M.E. Cardenas, J.F. Hofmann, and D. Schweizer. 1992. Localization of RAP1 and topoisomerase II in nuclei and meiotic chromosomes of yeast. *J. Cell Biol.* **117:** 935.

Klein, H.L. 1984. Lack of association between intrachromosomal gene conversion and reciprocal exchange. *Nature* **310:** 748.

Klein, H.L. and T.D. Petes. 1981. Intrachromosomal gene conversion in yeast. *Nature* **289:** 144.

Kobayashi, T., Y. Hotta, and S. Tabata. 1993. Isolation and characterization of a yeast gene that is homologous with a meiosis-specific cDNA from a plant. *Mol. Gen. Genet.* **237:** 225.

Kolodkin, A.L., A.J. Klar, and F.W. Stahl. 1986. Double-strand breaks can initiate meiotic recombination in *S. cerevisiae*. *Cell* **46:** 733.

Kominami, K., Y. Sakata, M. Sakai, and I. Yamashita. 1993. Protein kinase activity associated with the *IME2* gene product, a meiotic inducer in the yeast *Saccharomyces cerevisiae*. *Biosci. Biotechnol. Biochem.* **57:** 1731.

Kramer, W., B. Kramer, M.S. Williamson, and S. Fogel. 1989. Cloning and nucleotide sequence of DNA mismatch repair gene *PMS1* from *Saccharomyces cerevisiae:* Homology of *PMS1* to procaryotic *MutL* and *HexB*. *J. Bacteriol.* **171:** 5339.

Krisak, L., R. Strich, R.S. Winters, J.P. Hall, M.J. Mallory, D. Kreitzer, R.S. Tuan, and E. Winter. 1994. *SMK1*, a developmentally regulated MAP kinase, is required for spore wall assembly in *Saccharomyces cerevisiae. Genes Dev.* **8:** 2151.

Kuchin, S., P. Yeghiayan, and M. Carlson. 1995. Cyclin-dependent protein kinase and cyclin homologs *SSN3* and *SSN8* contribute to transcriptional control in yeast. *Proc. Natl. Acad. Sci.* **92:** 4006.

Kukuruzinska, M.A. and K. Lennon. 1995. Diminished activity of the first N-glycosylation enzyme, dolichol-P-dependent N-acetylglucosamine-1-P transferase (GPT), gives rise to mutant phenotypes in yeast. *Biochim. Biophys. Acta* **1247:** 51.

Kupiec, M. and T.D. Petes. 1988. Meiotic recombination between repeated transposable elements in *Saccharomyces cerevisiae. Mol. Cell. Biol.* **8:** 2942.

Kupiec, M. and G. Simchen. 1986. Regulation of the *RAD6* gene of *Saccharomyces cerevisiae* in the mitotic cell cycle and in meiosis. *Mol. Gen. Genet.* **203:** 538.

Lambie, E.J. and G.S. Roeder. 1986. Repression of meiotic crossing over by a centromere (*CEN3*) in *Saccharomyces cerevisiae. Genetics* **114:** 769.

Larkin, J.C. and J.L.J. Woolford. 1983. Molecular cloning and analysis of the *CRY1* gene: A yeast ribosomal protein gene. *Nucleic Acids Res.* **11:** 403.

Laurenson, P. and J. Rine. 1992. Silencers, silencing, and heritable transcriptional states. *Microbiol. Rev.* **56:** 543.

Laurie, D.A. and G.H. Jones. 1981. Inter-individual variation in chiasma distribution in *Chorthippus brunneus. Heredity* **47:** 409.

Lauze, E., B. Stoelcker, F.C. Luca, E. Weiss, A.R. Schutz, and M. Winey. 1995. Yeast spindle pole body duplication gene *MPS1* encodes an essential dual specificity protein kinase. *EMBO J.* **14:** 1655.

Law, D.T. and J. Segall. 1988. The *SPS100* gene of *Saccharomyces cerevisiae* is activated late in the sporulation process and contributes to spore wall maturation. *Mol. Cell. Biol.* **8:** 912.

Lawrence, C.W. and R.B. Christensen. 1979. Metabolic suppressors of trimethoprim and ultraviolet light sensitivities of *Saccharomyces cerevisiae rad6* mutants. *J. Bacteriol.* **139:** 866.

Leberer, E., D. Dignard, D. Harcus, and D.Y. Thomas. 1992. The protein kinase homologue Ste20p is required to link the yeast pheromone response G-protein βγ subunits to downstream signalling components. *EMBO J.* **11:** 4815.

Lechner, J. and J. Carbon. 1991. A 240 kd multisubunit protein complex, CBF3, is a major component of the budding yeast centromere. *Cell* **64:** 717.

Lee, F.J., L.W. Lin, and J.A. Smith. 1989. N^{α} acetylation is required for normal growth and mating of *Saccharomyces cerevisiae. J. Bacteriol.* **171:** 5795.

Leem, S.H. and H. Ogawa. 1992. The *MRE4* gene encodes a novel protein kinase homologue required for meiotic recombination in *Saccharomyces cerevisiae. Nucleic Acids Res.* **20:** 449.

Leem, S.H., P.A. Ropp, and A. Sugino. 1994. The yeast *Saccharomyces cerevisiae* DNA polymerase IV: Possible involvement in double strand break DNA repair. *Nucleic Acids Res.* **22:** 3011.

Leibowitz, M.J. and R.B. Wickner. 1976. A chromosomal gene required for killer plasmid expression, mating, and spore maturation in *Saccharomyces cerevisiae. Proc. Natl. Acad. Sci.* **73:** 2061.

Li, W. and A.P. Mitchell. 1997. Proteolytic activation of Rim1p, a positive regulator of yeast sporulation and invasive growth. *Genetics* **145:** 63.

Li, Y., S. Bjorklund, Y.W. Jiang, Y.J. Kim, W.S. Lane, D.J. Stillman, and R.D. Kornberg. 1995. Yeast global transcriptional regulators Sin4 and Rgr1 are components of mediator complex/RNA polymerase II holoenzyme. *Proc. Natl. Acad. Sci.* **92:** 10864.

Liao, S.M., J. Zhang, D.A. Jeffery, A.J. Koleske, C.M. Thompson, D.M. Chao, M. Viljoen, H.J. van Vuuren, and R.A. Young. 1995. A kinase-cyclin pair in the RNA polymerase II holoenzyme. *Nature* **374:** 193.

Lichten, M., R.H. Borts, and J.E. Haber. 1987. Meiotic gene conversion and crossing over between dispersed homologous sequences occurs frequently in *Saccharomyces cerevisiae*. *Genetics* **115:** 233.

Liefshitz, B., A. Parket, R. Maya, and M. Kupiec. 1995. The role of DNA repair genes in recombination between repeated sequences in yeast. *Genetics* **140:** 1199.

Liras, P., J. McCuster, S. Mascioli, and J.E. Haber. 1978. Characterization of a mutation in yeast causing nonrandom chromosome loss during mitosis. *Genetics* **88:** 651.

Liu, J., T.C. Wu, and M. Lichten. 1995. The location and structure of double-strand DNA breaks induced during yeast meiosis: Evidence for a covalently linked DNA-protein intermediate. *EMBO J.* **14:** 4599.

Liu, X.F., I. Elashvili, E.B. Gralla, J.S. Valentine, P. Lapinskas, and V.C. Culotta. 1992. Yeast lacking superoxide dismutase. Isolation of genetic suppressors. *J. Biol. Chem.* **267:** 18298.

Liu, Z., A. Lee, and W. Gilbert. 1995. Gene disruption of a G4-DNA-dependent nuclease in yeast leads to cellular senescence and telomere shortening. *Proc. Natl. Acad. Sci.* **92:** 6002.

Loidl, J., K. Nairz, and F. Klein. 1991. Meiotic chromosome synapsis in a haploid yeast. *Chromosoma* **100:** 221.

Loidl, J., H. Scherthan, and D.B. Kaback. 1994. Physical association between nonhomologous chromosomes precedes distributive disjunction in yeast. *Proc. Natl. Acad. Sci.* **91:** 331.

Longhese, M.P., L. Jovine, P. Plevani, and G. Lucchini. 1993. Conditional mutations in the yeast DNA primase genes affect different aspects of DNA metabolism and interactions in the DNA polymerase α-primase complex. *Genetics* **133:** 183.

Lopes, J.M., K.L. Schulze, J.W. Yates, and J.P. Hirsch. 1993. The *INO1* promoter of *Saccharomyces cerevisiae* includes an upstream repressor sequence (URS1) common to a diverse set of yeast genes. *J. Bacteriol.* **175:** 4235.

Louis, E.J. and J.E. Haber. 1989. Nonrecombinant meiosis I nondisjunction in *Saccharomyces cerevisiae* induced by tRNA ochre suppressors. *Genetics* **123:** 81.

———. 1990. The subtelomeric Y' repeat family in *Saccharomyces cerevisiae:* An experimental system for repeated sequence evolution. *Genetics* **124:** 533.

Louis, E.J., E.S. Naumova, A. Lee, G. Naumov, and J.E. Haber. 1994. The chromosome end in yeast: Its mosaic nature and influence on recombinational dynamics. *Genetics* **136:** 789.

Lovett, S.T. 1994. Sequence of the *RAD55* gene of *Saccharomyces cerevisiae:* Similarity of RAD55 to prokaryotic RecA and other RecA-like proteins. *Gene* **142:** 103.

Lovett, S.T. and R.K. Mortimer. 1987. Characterization of null mutants of the *RAD55* gene of *Saccharomyces cerevisiae:* Effects of temperature, osmotic strength and mating type. *Genetics* **116:** 547.

Luche, R.M., W.C. Smart, and T.G. Cooper. 1992. Purification of the heteromeric protein binding to the URS1 transcriptional repression site in *Saccharomyces cerevisiae*. *Proc. Natl. Acad. Sci.* **89:** 7412.

Luche, R.M., R. Sumrada, and T.G. Cooper. 1990. A *cis*-acting element present in multiple genes serves as a repressor protein binding site for the yeast *CAR1* gene. *Mol. Cell.*

Biol. **10:** 3884.

Luche, R.M., W.C. Smart, T. Marion, M. Tillman, R.A. Sumrada, and T.G. Cooper. 1993. *Saccharomyces cerevisiae* BUF protein binds to sequences participating in DNA replication in addition to those mediating transcriptional repression (URS1) and activation. *Mol. Cell. Biol.* **13:** 5749.

Lydall, D., Y. Nikolsky, D. K. Bishop, and T. Weinert. 1996. A meiotic recombination checkpoint controlled by mitotic checkpoint genes. *Nature* **383:** 840.

Lynn, R.R. and P.T. Magee. 1970. Development of the spore wall during ascospore formation in *Saccharomyces cerevisiae*. *J. Cell Biol.* **44:** 688.

Madura, K. and S. Prakash. 1990. The *Saccharomyces cerevisiae* DNA repair gene *RAD2* is regulated in meiosis but not during the mitotic cell cycle. *Mol. Cell. Biol.* **10:** 3256.

Magee, P.T. 1987. Transcription during meiosis. In *Meiosis* (ed. P. Moens), p. 355. Academic Press, Orlando, Florida.

Malavasic, M.J. and R.T. Elder. 1990. Complementary transcripts from two genes necessary for normal meiosis in the yeast *Saccharomyces cerevisiae*. *Mol. Cell. Biol.* **10:** 2809.

Malkova, A., L. Ross, D. Dawson, M.F. Hoekstra, and J.E. Haber. 1996. Meiotic recombination initiated by a double-strand break in *rad50Δ* yeast cells otherwise unable to initiate meiotic recombination. *Genetics* **143:** 741.

Malone, R.E. 1983. Multiple mutant analysis of recombination in yeast. *Mol. Gen. Genet.* **189:** 405.

———. 1990. Dual regulation of meiosis in yeast. *Cell* **61:** 375.

Malone, R.E. and R.E. Esposito. 1981. Recombinationless meiosis in *Saccharomyces cerevisiae*. *Mol. Cell. Biol.* **1:** 891.

Malone, R.E., J.E. Golin, and M.S. Esposito. 1980. Mitotic versus meiotic recombination in *Saccharomyces cerevisiae*. *Curr. Genet.* **1:** 241.

Malone, R.E., S. Bullard, S. Lundquist, S. Kim, and T. Tarkowski. 1992. A meiotic gene conversion gradient opposite to the direction of transcription. *Nature* **359:** 154.

Malone, R.E., S. Bullard, M. Hermiston, R. Rieger, M. Cool, and A. Galbraith. 1991. Isolation of mutants defective in early steps of meiotic recombination in the yeast *Saccharomyces cerevisiae*. *Genetics* **128:** 79.

Mandel, S., K. Robzyk, and Y. Kassir. 1994. *IME1* gene encodes a transcription factor which is required to induce meiosis in *Saccharomyces cerevisiae*. *Dev. Genet.* **15:** 139.

Mann, C. and R.W. Davis. 1986. Meiotic disjunction of circular minichromosomes in yeast does not require DNA homology. *Proc. Natl. Acad. Sci.* **83:** 6017.

Mao-Draayer, Y., A.M. Galbraith, D.L. Pittman, M. Cool, and R.E. Malone. 1996. Analysis of meiotic recombination pathways in the yeast *Saccharomyces cerevisiae*. *Genetics* **144:**71.

Margolskee, J.P. 1988. The sporulation capable (*sca*) mutation of *Saccharomyces cerevisiae* is an allele of the *SIR2* gene. *Mol. Gen. Genet.* **211:** 430.

Marsischky, G.T., N. Filosi, M.F. Kane, and R. Kolodner. 1996. Redundancy of *Saccharomyces cerevisiae MSH3* and *MSH6* in *MSH2*-dependent mismatch repair. *Genes Dev.* **10:** 407.

Masison, D.C. and R.E. Baker. 1992. Meiosis in *Saccharomyces cerevisiae* mutants lacking the centromere-binding protein CP1. *Genetics* **131:** 43.

Matsumoto, K., I. Uno, and T. Ishikawa. 1983. Initiation of meiosis in yeast mutants defective in adenylate cyclase and cyclic AMP-dependent protein kinase. *Cell* **32:** 417.

Matsuura, A., M. Treinin, H. Mitsuzawa, Y. Kassir, I. Uno, and G. Simchen. 1990. The adenylate cyclase/protein kinase cascade regulates entry into meiosis in *Saccharomyces cerevisiae* through the gene *IME1*. *EMBO J.* **9:** 3225.

McCarroll, R.M. and R.E. Esposito. 1994. *SPO13* negatively regulates the progression of mitotic and meiotic nuclear division in *Saccharomyces cerevisiae*. *Genetics* **138:** 47.

McLeod, M. 1989. Regulation of meiosis: From DNA binding protein to protein kinase. *BioEssays* **11:** 9.

Meddle, C.C., P. Kumar, J. Ham, D.A. Hughes, and I.R. Johnston. 1985. Cloning of the *CDC7* gene of *Saccharomyces cerevisiae* in association with centromeric DNA. *Gene* **34:** 179.

Mellor, J., J. Rathjen, W. Jiang, C.A. Barnes, and S.J. Dowell. 1991. DNA binding of CPF1 is required for optimal centromere function but not for maintaining methionine prototrophy in yeast. *Nucleic Acids Res.* **19:** 2961.

Menees, T.M. and G.S. Roeder. 1989. *MEI4*, a yeast gene required for meiotic recombination. *Genetics* **123:** 675.

Menees, T.M., M.P.B. Ross, and G.S. Roeder. 1992. *MEI4*, a meiosis-specific yeast gene required for chromosome synapsis. *Mol. Cell. Biol.* **12:** 1340.

Meselson, M.S. and C.M. Radding. 1975. A general model for genetic recombination. *Proc. Natl. Acad. Sci.* **72:** 358.

Miller, J.J. 1989. Sporulation in *Saccharomyces cerevisiae*. In *The yeasts* (ed. A.H. Rose and J.S. Harrison), p. 491. Academic Press, San Diego.

Milne, G.T. and D.T. Weaver. 1993. Dominant negative alleles of *RAD52* reveal a DNA repair/recombination complex including Rad51 and Rad52. *Genes Dev.* **7:** 1755.

Milne, G.T., T. Ho, and D.T. Weaver. 1995. Modulation of *Saccharomyces cerevisiae* DNA double-strand break repair by *SRS2* and *RAD51*. *Genetics* **139:** 1189.

Mitchell, A.P. 1988. Two switches govern entry into meiosis in yeast. *Prog. Clin. Biol. Res.* **267:** 47.

———. 1994. Control of meiotic gene expression in *Saccharomyces cerevisiae*. *Microbiol. Rev.* **58:** 56.

Mitchell, A.P. and K.S. Bowdish. 1992. Selection for early meiotic mutants in yeast. *Genetics* **131:** 65.

Mitchell, A.P. and I. Herskowitz. 1986. Activation of meiosis and sporulation by repression of the *RME1* product in yeast. *Nature* **319:** 738.

Mitchell, A.P., S.E. Driscoll, and H.E. Smith. 1990. Positive control of sporulation-specific genes by the *IME1* and *IME2* products in *Saccharomyces cerevisiae*. *Mol. Cell. Biol.* **10:** 2104.

Moens, P.B. 1974. Modification of sporulation in yeast strains with two- spored asci (*Saccharomyces, Ascomycetes*). *J. Cell Sci.* **16:** 519.

———. 1994. Molecular perspectives of chromosome pairing at meiosis. *BioEssays* **16:** 101.

Moens, P.B. and S.C. Kundu. 1982. Meiotic arrest and synaptonemal complexes in yeast *ts spo10 (Saccharomyces cerevisiae)*. *Can. J. Biochem.* **60:** 284.

Moens, P.B. and E. Rapport. 1971a. Spindles, spindle plaques, and meiosis in the yeast *Saccharomyces cerevisiae* (Hansen). *J. Cell Biol.* **50:** 344.

———. 1971b. Synaptic structures in the nuclei of sporulating yeast, *Saccharomyces cerevisiae* (Hansen). *J. Cell Sci.* **9:** 665.

Moens, P.B., R.E. Esposito, and M.S. Esposito. 1974. Aberrant nuclear behavior at meiosis and anucleate spore formation by sporulation-deficient (*SPO*) mutants of *Saccharomyces cerevisiae*. *Exp. Cell Res.* **83:** 166.

Molero, G., R.M. Yuste, A. Montesi, A. Vázquez, C. Nombela, and M. Sanchez. 1993. A *cdc*-like autolytic *Saccharomyces cerevisiae* mutant altered in budding site selection is complemented by *SPO12*, a sporulation gene. *J. Bacteriol.* **175:** 6562.

Moore, C.W. 1991. Further characterizations of bleomycin-sensitive (*blm*) mutants of *Saccharomyces cerevisiae* with implications for a radiomimetic model. *J. Bacteriol.* **173:** 3605.

Morrison, D.P. and P.J. Hastings. 1979. Characterization of the mutator mutation *mut5-1*. *Mol. Gen. Genet.* **175:** 57.

Mortimer, R.K., C.R. Contopoulou, and J.S. King. 1992. Genetic and physical maps of *Saccharomyces cerevisiae* (edition 11). *Yeast* **8:** 817.

Moses, M.J. 1956. *J. Biophys. Biochem. Cytol.* **2:** 215.

———. 1977. The synaptonemal complex and meiosis. In *Molecular human cytogenetics* (ed. R.S. Sparkes et al.), p. 101. Academic Press, New York.

Muhlrad, D. and R. Parker. 1992. Mutations affecting stability and deadenylation of the yeast *MFA2* transcript. *Genes Dev.* **6:** 2100.

Mukai, Y., S. Harashima, and Y. Oshima. 1991. AAR1/TUP1 protein, with a structure similar to that of the β subunit of G proteins, is required for **a**1-α2 and α2 repression in cell type control of *Saccharomyces cerevisiae. Mol. Cell. Biol.* **11:** 3773.

Mullen, J.R., P.S. Kayne, R.P. Moerschell, S. Tsunasawa, M. Gribskov, M. Colavito-Shepanski, M. Grunstein, F. Sherman, and R. Sternglanz. 1989. Identification and characterization of genes and mutants for an N-terminal acetyltransferase from yeast. *EMBO J.* **8:** 2067.

Munder, T. and H. Kuntzel. 1989. Glucose-induced cAMP signaling in *Saccharomyces cerevisiae* is mediated by the CDC25 protein. *FEBS Lett.* **242:** 341.

Munder, T., M. Mink, and H. Kuntzel. 1988. Domains of the *Saccharomyces cerevisiae CDC25* gene controlling mitosis and meiosis. *Mol. Gen. Genet.* **214:** 271.

Murphy, M.R., D.M. Fowlkes, and H.M. Fitzgerald. 1991. Analysis of centromere function in *Saccharomyces cerevisiae* using synthetic centromere mutants. *Chromosoma* **101:** 189.

Murray, A.W., M.J. Solomon, and M.W. Kirschner. 1989. The role of cyclin synthesis and degradation in the control of maturation promoting factor activity. *Nature* **339:** 280.

Muthukumar, G., S.H. Suhng, P.T. Magee, R.D. Jewell, and D.A. Primerano. 1993. The *Saccharomyces cerevisiae SPR1* gene encodes a sporulation-specific exo-1,3-β-glucanase which contributes to ascospore thermoresistance. *J. Bacteriol.* **175:** 386.

Nag, D.K. and T.D. Petes. 1993. Physical detection of heteroduplexes during meiotic recombination in the yeast *Saccharomyces cerevisiae. Mol. Cell. Biol.* **13:** 2324.

Nag, D.K., M.A. White, and T.D. Petes. 1989. Palindromic sequences in heteroduplex DNA inhibit mismatch repair in yeast. *Nature* **340:** 318.

Nag, D.K., H. Scherthan, B. Rockmill, J. Bhargava, and G.S. Roeder. 1995. Heteroduplex DNA formation and homolog airing in yeast meiotic mutants. *Genetics* **141:** 75.

Nakaseko, Y., N. Kinoshita, and M. Yanagida. 1987. A novel sequence common to the centromere regions of *Schizosaccharomyces pombe* chromosomes. *Nucleic Acids Res.* **15:** 4705.

Nandabalan, K. and G.S. Roeder. 1995. Binding of a cell-type-specific RNA splicing factor to its target regulatory sequence. *Mol. Cell. Biol.* **15:** 1953.

Nandabalan, K., L. Price, and G.S. Roeder. 1993. Mutations in U1 snRNA bypass the requirement for a cell type-specific RNA splicing factor. *Cell* **73:** 407.

Nasmyth, K. 1993. Control of the yeast cell cycle by the Cdc28 protein kinase. *Curr. Opin. Cell Biol.* **5:** 166.

Neigeborn, L. and A.P. Mitchell. 1991. The yeast *MCK1* gene encodes a protein kinase homolog that activates early meiotic gene expression. *Genes Dev.* **5:** 533.

Nevo-Caspi, Y. and M. Kupiec. 1994. Transcriptional induction of Ty recombination in yeast. *Proc. Natl. Acad. Sci.* **91:** 12711.

New, L., K. Liu, and G.F. Crouse. 1993. The yeast gene *MSH3* defines a new class of eukaryotic *MutS* homologues. *Mol. Gen. Genet.* **239:** 97.

Newlon, C.S. 1988. Yeast chromosome replication and segregation. *Microbiol. Rev.* **52:** 568.

Nicklas, R.B. 1977. Chromosome distribution: Experiments on cell hybrids and *in vitro*. *Philos. Trans. R. Soc. Lond. B Biol. Sci.* **277:** 267.

Nicolas, A., D. Treco, N.P. Schultes, and J.W. Szostak. 1989. An initiation site for meiotic gene conversion in the yeast *Saccharomyces cerevisiae*. *Nature* **338:** 35.

Niwa, O. and M. Yanagida. 1988. Universal and essential role of MPF/*cdc2*$^{+}$. *Nature* **336:** 430.

Ogawa, H., K. Johzuka, T. Nakagawa, S.H. Leem, and A.H. Hagihara. 1995. Functions of the yeast meiotic recombination genes, *MRE11* and *MRE2*. *Adv. Biophys.* **31:** 67.

Ogawa, T., X. Yu, A. Shinohara, and E.H. Egelman. 1993a. Similarity of the yeast RAD51 filament to the bacterial RecA filament. *Science* **259:** 1896.

Ogawa, T., A. Shinohara, A. Nabetani, T. Ikeya, X. Yu, E.H. Egelman, and H. Ogawa. 1993b. RecA-like recombination proteins in eukaryotes: Functions and structures of *RAD51* genes. *Cold Spring Harbor Symp. Quant. Biol.* **58:** 567.

Ohta, K., T. Shibata, and A. Nicolas. 1994. Changes in chromatin structure at recombination initiation sites during yeast meiosis. *EMBO J.* **13:** 5754.

Okamoto, S. and T. Iino. 1982. Genetic block of outer plaque morphogenesis at the second meiotic division in an *hfd1-1* mutant of *Saccharomyces cerevisiae*. *J. Gen. Microbiol.* **128:** 1309.

Olempska-Beer, Z. and E. Freese. 1987. Initiation of meiosis and sporulation in *Saccharomyces cerevisiae* does not require a decrease in cyclic AMP. *Mol. Cell. Biol.* **7:** 2141.

Orejas, M., E.A. Espeso, J. Tilburn, S. Sarkar, H.N. Arst, Jr., and M.A. Penalva. 1995. Activation of the *Aspergillus* PacC transcription factor in response to alkaline ambient pH requires proteolysis of the carboxy-terminal moiety. *Genes Dev.* **9:** 1622.

Orias, E. 1986. Ciliate conjugation. In *The molecular biology of ciliated protozoa* (ed. J.G. Gall), p. 45. Academic Press, New York.

Ozsarac, N., M. Bhattacharyya, I.W. Dawes, and M.J. Clancy. 1995. The *SPR3* gene encodes a sporulation-specific homologue of the yeast CDC3/10/11/12 family of bud neck microfilaments and is regulated by ABFI. *Gene* **164:** 157.

Padmore, R., L. Cao, and N. Kleckner. 1991. Temporal comparison of recombination and synaptonemal complex formation during meiosis in *S. cerevisiae*. *Cell* **66:** 1239.

Pall, M.L. 1988. Cyclic AMP control of GTP pools in *Saccharomyces cerevisiae*. *Biochem. Biophys. Res. Commun.* **150:** 1144.

Palladino, F. and H.L. Klein. 1992. Analysis of mitotic and meiotic defects in *Saccharomyces cerevisiae SRS2* DNA helicase mutants. *Genetics* **132:** 23.

Pammer, M., P. Briza, A. Ellinger, T. Schuster, R. Stucka, H. Feldmann, and M. Breitenbach. 1992. *DIT101 (CSD2, CAL1)*, a cell cycle-regulated yeast gene required for synthesis of chitin in cell walls and chitosan in spore walls. *Yeast* **8:** 1089.

Panzeri, L., L. Landonio, A. Stotz, and P. Philippsen. 1985. Role of conserved sequence elements in yeast centromere DNA. *EMBO J.* **4:** 1867.

Park, E.C. and J.W. Szostak. 1992. *ARD1* and *NAT1* proteins form a complex that has N-terminal acetyltransferase activity. *EMBO J.* **11:** 2087.

Park, H.D., R.M. Luche, and T.G. Cooper. 1992. The yeast *UME6* gene product is required for transcriptional repression mediated by the CAR1 URS1 repressor binding site. *Nucleic Acids Res.* **20:** 1909.

Park, H.O. and E.A. Craig. 1991. Transcriptional regulation of a yeast *HSP70* gene by heat shock factor and an upstream repression site-binding factor. *Genes Dev.* **5:** 1299.

Parkes, V. and L. H. Johnston. 1992. *SPO12* and *SIT4* suppress mutations in *DBF2*, which encodes a cell cycle protein kinase that is periodically expressed. *Nucleic Acids Res.* **20:** 5617.

Parket, A. and M. Kupiec. 1992. Ectopic recombination between Ty elements in *Saccharomyces cerevisiae* is not induced by DNA damage. *Mol. Cell. Biol.* **12:** 4441.

Parket, A., O. Inbar, and M. Kupiec. 1995. Recombination of Ty elements in yeast can be induced by a double-strand break. *Genetics* **140:** 67.

Patterson, M., R.A. Sclafani, W.L. Fangman, and J. Rosamond. 1986. Molecular characterization of cell cycle gene *CDC7* from *Saccharomyces cerevisiae. Mol. Cell. Biol.* **6:** 1590.

Percival-Smith, A. and J. Segall. 1984. Isolation of DNA sequences preferentially expressed during sporulation in *Saccharomyces cerevisiae. Mol. Cell. Biol.* **4:** 142.

Peterson, C.L. and I. Herskowitz. 1992. Characterization of the yeast *SWI1*, *SWI2*, and *SWI3* genes, which encode a global activator of transcription. *Cell* **68:** 573.

Petes, T.D. 1980. Unequal meiotic recombination within tandem arrays of yeast ribosomal DNA genes. *Cell* **19:** 765.

Petes, T.D. and D. Botstein. 1977. Simple Mendelian inheritance of the reiterated ribosomal DNA of yeast. *Proc. Natl. Acad. Sci.* **74:** 5091.

Petes, T.D., R.E. Malone, and L.S. Symington. 1991. Recombination in yeast. In *The molecular and cellular biology of the yeast* Saccharomyces: *Genome dynamics, protein synthesis, and energetics* (ed. J.R. Broach et al.), p. 407. Cold Spring Harbor Laboratory Press, Cold Sping Harbor, New York.

Pittman, D., W. Lu, and R.E. Malone. 1993. Genetic and molecular analysis of *REC114*, an early meiotic recombination gene in yeast. *Curr. Genet.* **23:** 295.

Poch, O., E. Schwob, F. de Fraipont, A. Camasses, R. Bordonn'e, and R.P. Martin. 1994. *RPK1*, an essential yeast protein kinase involved in the regulation of the onset of mitosis, shows homology to mammalian dual-specificity kinases. *Mol. Gen. Genet.* **243:** 641.

Prakash, S., L. Prakash, W. Burke, and B.A. Montelone. 1980. Effects of the *RAD52* gene on recombination in *Saccharomyces cerevisiae. Genetics* **94:** 31.

Primerano, D., G. Muthukumar, S.H. Suhng, and P.T. Magee. 1988. Molecular characterization of two sporulation regulated (*SPR*) genes, one of which is involved in spore development. *Yeast* **4:** S54.

Prinz, S., F. Klein, H. Auer, D. Schweizer, and M. Primig. 1995. A DNA binding factor (UBF) interacts with a positive regulatory element in the promoters of genes expressed during meiosis and vegetative growth in yeast. *Nucleic Acids Res.* **23:** 3449-56.

Prolla, T.A., D.M. Christie, and R.M. Liskay. 1994a. Dual requirement in yeast DNA mismatch repair for *MLH1* and *PMS1*, two homologs of the bacterial *mutL* gene. *Mol. Cell. Biol.* **14:** 407.

Prolla, T.A., Q. Pang, E. Alani, R.D. Kolodner, and R.M. Liskay. 1994b. *MLH1, PMS1,* and *MSH2* interactions during the initiation of DNA mismatch repair in yeast. *Science* **265:** 1091.

Pugh, T.A. and M.J. Clancy. 1990. Differential regulation of *STA* genes of *Saccharomyces cerevisiae. Mol. Gen. Genet.* **222:** 87.

Pugh, T.A., J.C. Shah, P.T. Magee, and M.J. Clancy. 1989. Characterization and localization of the sporulation glucoamylase of *Saccharomyces cerevisiae. Biochim. Biophys. Acta* **994:** 200.

Puziss, J.W., T.A. Hardy, R.B. Johnson, P.J. Roach, and P. Hieter. 1994. *MDS1*, a dosage suppressor of an *mck1* mutant, encodes a putative yeast homolog of glycogen synthase kinase 3. *Mol. Cell. Biol.* **14:** 831.

Ramer, S.W. and R.W. Davis. 1993. A dominant truncation allele identifies a gene, *STE20*, that encodes a putative protein kinase necessary for mating in *Saccharomyces cerevisiae. Proc. Natl. Acad. Sci.* **90:** 452.

Rasmussen, S.W. 1977. The transformation of the synaptonemal complex into the "elimination chromatin" of *Bombyx mori* oocytes. *Chromosoma* **60:** 205.

Raymond, W.E. and N. Kleckner. 1993a. Expression of the *Saccharomyces cerevisiae RAD50* gene during meiosis: Steady-state transcript levels rise and fall while steady-state protein levels remain constant. *Mol. Gen. Genet.* **238:** 390.

———. 1993b. RAD50 protein of *S. cerevisiae* exhibits ATP-dependent DNA binding. *Nucleic Acids Res.* **21:** 3851.

Reed, S.I., J.A. Hadwiger, and A.T. Lorincz. 1985. Protein kinase activity associated with the product of the yeast cell division cycle gene *CDC28. Proc. Natl. Acad. Sci.* **82:** 4055.

Reenan, R.A. and R.D. Kolodner. 1992. Characterization of insertion mutations in the *Saccharomyces cerevisiae MSH1* and *MSH2* genes: Evidence for separate mitochondrial and nuclear functions. *Genetics* **132:** 975.

Resnick, M.A., J. Nitiss, C. Edwards, and R.E. Malone. 1986. Meiosis can induce recombination in *rad52* mutants of *Saccharomyces cerevisiae. Genetics* **113:** 531.

Richardson, H., D.J. Lew, M. Henze, K. Sugimoto, and S.I. Reed. 1992. Cyclin-B homologs in *Saccharomyces cerevisiae* function in S phase and in G_2. *Genes Dev.* **6:** 2021.

Rine, J., G.F. Sprague, Jr., and I. Herskowitz. 1981. *rme1* mutation of *Saccharomyces cerevisiae:* Map position and bypass of mating type locus control of sporulation. *Mol. Cell. Biol.* **1:** 958.

Rocco, V., B. de Massy, and A. Nicolas. 1992. The *Saccharomyces cerevisiae ARG4* initiator of meiotic gene conversion and its associated double-strand DNA breaks can be inhibited by transcriptional interference. *Proc. Natl. Acad. Sci.* **89:** 12068.

Rockmill, B. and S. Fogel. 1988. *DIS1:* A yeast gene required for proper meiotic chromosome disjunction. *Genetics* **119:** 261.

Rockmill, B. and G.S. Roeder. 1988. *RED1:* A yeast gene required for the segregation of chromosomes during the reductional division of meiosis. *Proc. Natl. Acad. Sci.* **85:** 6057.

———. 1990. Meiosis in asynaptic yeast. *Genetics* **126:** 563.

———. Roeder. 1991. A meiosis-specific protein kinase homolog required for chromosome synapsis and recombination. *Genes Dev.* **5:** 2392.

———. 1994. The yeast *med1* mutant undergoes both meiotic homolog nondisjunction and precocious separation of sister chromatids. *Genetics* **136:** 65.

Rockmill, B., M. Sym, H. Scherthan, and G.S. Roeder. 1995a. Roles for two RecA homologs in promoting meiotic chromosome synapsis. *Genes Dev.* **9:** 2684.

Rockmill, B., J.A. Engebrecht, H. Scherthan, J. Loidl, and G.S. Roeder. 1995b. The yeast *MER2* gene is required for chromosome synapsis and the initiation of meiotic recombination. *Genetics* **141:** 49.

Roeder, G.S. 1990. Chromosome synapsis and genetic recombination: Their roles in meiotic chromosome segregation. *Trends Genet.* **6:** 385.

———. 1995. Sex and the single cell: Meiosis in yeast. *Proc. Natl. Acad. Sci.* **92:** 10450.

Roman, H., M.M. Phillips, and S.M. Sands. 1955. Studies of polyploid *Saccharomyces cerevisiae*. I. Tetraploid segregation. *Genetics* **40:** 546.

Rose, D. and C. Holm. 1993. Meiosis-specific arrest revealed in DNA topoisomerase II mutants. *Mol. Cell. Biol.* **13:** 3445.

Rose, D., W. Thomas, and C. Holm. 1990. Segregation of recombined chromosomes in meiosis I requires DNA topoisomerase II. *Cell* **60:** 1009.

Rose, K., S.A. Rudge, M.A. Frohman, A.J. Morris, and J. Engebrecht. 1995. Phospholipase D signaling is essential for meiosis. *Proc. Natl. Acad. Sci.* **92:** 12151.

Ross, L.O., R. Maxfield, and D. Dawson. 1996. Exchanges are not equally able to enhance meiotic chromosome segregation in yeast. *Proc. Natl. Acad. Sci.* **93:** 4979.

Ross, L.O., D. Treco, A. Nicolas, J.W. Szostak, and D. Dawson. 1992. Meiotic recombination on artificial chromosomes in yeast. *Genetics* **131:** 541.

Ross-Macdonald, P. and G.S. Roeder. 1994. Mutation of a meiosis-specific *MutS* homolog decreases crossing over but not mismatch correction. *Cell* **79:** 1069.

Roth, R. 1976. Temperature-sensitive yeast mutants defective in meiotic recombination and replication. *Genetics* **83:** 675.

Roth, R. and S. Fogel. 1971. A selective system for yeast mutants defective in meiotic recombination. *Mol. Gen. Genet.* **112:** 295.

Roth, R. and K. Lusnak. 1970. DNA synthesis during yeast sporulation: Genetic control of an early developmental event. *Science* **168:** 493.

Rothstein, R.J., R.E. Esposito, and M.S. Esposito. 1977. The effect of ochre suppression on meiosis and ascospore formation in *Saccharomyces*. *Genetics* **85:** 35.

Rubin-Bejerano, I., S. Mandel, K. Robzyk, and Y. Kassir. 1996. Induction of meiosis in *Saccharomyces cerevisiae* depends on conversion of the transcriptional repressor Ume6 to a positive regulator by its regulated association with the transcriptional activator Ime1. *Mol. Cell. Biol.* **16:** 2518.

Ruggieri, R., K. Tanaka, M. Nakafuku, Y. Kaziro, A. Toh-e, and K. Matsumoto. 1989. *MSI1*, a negative regulator of the RAS-cAMP pathway in *Saccharomyces cerevisiae*. *Proc. Natl. Acad. Sci.* **86:** 8778.

Russell, M., R.J. Bradshaw, D. Markwardt, and W. Heideman. 1993. Changes in gene expression in the Ras/adenylate cyclase system of *Saccharomyces cerevisiae:* Correlation with cAMP levels and growth arrest. *Mol. Biol. Cell* **4:** 757.

Sakai, A., Y. Shimizu, and F. Hishinuma. 1988. Isolation and characterization of mutants which show an oversecretion phenotype in *Saccharomyces cerevisiae*. *Genetics* **119:** 499.

San Segundo, P., J. Correa, C.R. Vazquez de Aldana, and F. del Rey. 1993. *SSG1*, a gene encoding a sporulation-specific 1,3-β-glucanase in *Saccharomyces cerevisiae*. *J. Bacteriol.* **175:** 3823.

Sanchez, Y., B.A. Desany, W.J. Jones, Q. Liu, B. Wang, and S.J. Elledge. 1996. Regulation of *RAD53* by the ATM-like kinases *MEC1* and *TEL1* in yeast cell cycle checkpoint pathways. *Science* **271:** 357.

Scherthan, H., J. Loidl, T. Schuster, and D. Schweizer. 1992. Meiotic chromosome condensation and pairing in *Saccharomyces cerevisiae* studied by chromosome painting. *Chromosoma* **101:** 590.

Schiestl, R.H., S. Prakash, and L. Prakash. 1990. The *SRS2* suppressor of *rad6* mutations of *Saccharomyces cerevisiae* acts by channeling DNA lesions into the *RAD52* DNA repair pathway. *Genetics* **124:** 817.

Schild, D. 1995. Suppression of a new allele of the yeast *RAD52* gene by overexpression of *RAD51*, mutations in *srs2* and *ccr4*, or mating- type heterozygosity. *Genetics* **140:** 115.

Schild, D. and B. Byers. 1978. Meiotic effects of DNA-defective cell division cycle mutations of *Saccharomyces cerevisiae. Chromosoma* **70:** 109.

———. 1980. Diploid spore formation and other meiotic effects of two cell-division-cycle mutations of *Saccharomyces cerevisiae. Genetics* **96:** 859.

Schultes, N.P. and J.W. Szostak. 1990. Decreasing gradients of gene conversion on both sides of the initiation site for meiotic recombination at the *ARG4* locus in yeast. *Genetics* **126:** 813.

Schultz, J., C.L. Marshall, and M. Carlson. 1990. The N-terminal TPR region is the functional domain of *SSN6*, a nuclear phosphoprotein of *Saccharomyces cerevisiae. Mol. Cell. Biol.* **10:** 4744.

Schwacha, A. and N. Kleckner. 1994. Identification of joint molecules that form frequently between homologs but rarely between sister chromatids during yeast meiosis. *Cell* **76:** 51.

———. 1995. Identification of double Holliday junctions as intermediates in meiotic recombination. *Cell* **83:** 783.

Sclafani, R.A., M. Patterson, J. Rosamond, and W.L. Fangman. 1988. Differential regulation of the yeast *CDC7* gene during mitosis and meiosis. *Mol. Cell. Biol.* **8:** 293.

Sears, D.D., P. Hieter, and G. Simchen. 1994. An implanted recombination hot spot stimulates recombination and enhances sister chromatid cohesion of heterologous YACs during yeast meiosis. *Genetics* **138:** 1055.

Sears, D.D., J.H. Hegemann, J.H. Shero, and P. Hieter. 1995. *cis*-Acting determinants affecting centromere function, sister- chromatid cohesion and reciprocal recombination during meiosis in *Saccharomyces cerevisiae. Genetics* **139:** 1159.

Segev, N. and D. Botstein. 1987. The ras-like yeast *YPT1* gene is itself essential for growth, sporulation, and starvation response. *Mol. Cell. Biol.* **7:** 2367.

Sekelsky, J.J. and R.S. Hawley. 1995. The bond between sisters (comment). *Cell* **83:** 157.

Selva, E.M., L. New, G.F. Crouse, and R.S. Lahue. 1995. Mismatch correction acts as a barrier to homeologous recombination in *Saccharomyces cerevisiae. Genetics* **139:** 1175.

Seufert, W. and S. Jentsch. 1991. Yeast ubiquitin-conjugating enzymes involved in selective protein degradation are essential for cell viability. *Acta Biol. Hung.* **42:** 27.

Shah, J.C. and M.J. Clancy. 1992. *IME4*, a gene that mediates *MAT* and nutritional control of meiosis in *Saccharomyces cerevisiae. Mol. Cell. Biol.* **12:** 1078.

Sharon, G. and G. Simchen. 1990a. Centromeric regions control autonomous segregation tendencies in single-division meiosis of *Saccharomyces cerevisiae. Genetics* **125:** 487.

———. 1990b. Mixed segregation of chromosomes during single-division meiosis of *Saccharomyces cerevisiae. Genetics* **125:** 475.

Shaw, J.A., P.C. Mol, B. Bowers, S.J. Silverman, and M.H. Valdivieso. 1991. The function of chitin synthases 2 and 3 in the *Saccharomyces cerevisiae* cell cycle. *J. Cell Biol.* **114:** 111.

Shefer-Vaida, M., A. Sherman, T. Ashkenazi, K. Robzyk, and Y. Kassir. 1995. Positive and negative feedback loops affect the transcription of *IME1*, a positive regulator of meiosis in *Saccharomyces cerevisiae*. *Dev. Genet.* **16:** 219.

Sherman, A., M. Shefer, S. Sagee, and Y. Kassir. 1993. Post-transcriptional regulation of *IME1* determines initiation of meiosis in *Saccharomyces cerevisiae*. *Mol. Gen. Genet.* **237:** 375.

Sherman, F. and H. Roman. 1963. Evidence for two types of allelic recombination in yeast. *Genetics* **48:** 255.

Shero, J.H. and P. Hieter. 1991. A suppressor of a centromere DNA mutation encodes a putative protein kinase (*MCK1*). *Genes Dev.* **5:** 549.

Shilo, V., G. Simchen, and B. Shilo. 1978. Initiation of meiosis in cell cycle initiation mutants of *Saccharomyces cerevisiae*. *Exp. Cell Res.* **112:** 241.

Shimizu, K., C. Santocanale, P.A. Ropp, M.P. Longhese, and P. Plevani. 1993. Purification and characterization of a new DNA polymerase from budding yeast *Saccharomyces cerevisiae*. A probable homolog of mammalian DNA polymerase β. *J. Biol. Chem.* **268:** 27148.

Shinohara, A., H. Ogawa, and T. Ogawa. 1992. Rad51 protein involved in repair and recombination in *S. cerevisiae* is a RecA-like protein. *Cell* **69:** 457.

Shuster, E.O. and B. Byers. 1989. Pachytene arrest and other meiotic effects of the start mutations in *Saccharomyces cerevisiae*. *Genetics* **123:** 29.

Sia, R.A. and A.P. Mitchell. 1995. Stimulation of later functions of the yeast meiotic protein kinase Ime2p by the *IDS2* gene product. *Mol. Cell. Biol.* **15:** 5279.

Siede, W., G.W. Robinson, D. Kalainov, T. Malley, and E.C. Friedberg. 1989. Regulation of the *RAD2* gene of *Saccharomyces cerevisiae*. *Mol. Microbiol.* **3:** 1697.

Simchen, G. 1974. Are mitotic functions required in meiosis? *Genetics* **76:** 745.

Simchen, G. and Y. Kassir. 1989. Genetic regulation of differentiation towards meiosis in the yeast *Saccharomyces cerevisiae*. *Genome* **31:** 95.

Simchen, G., Y. Kassir, C.O. Horesh, and A. Friedmann. 1981. Elevated recombination and pairing structures during meiotic arrest in yeast of the nuclear division mutant *cdc5*. *Mol. Gen. Genet.* **184:** 46.

Sitney, K.C., M.E. Budd, and J.L. Campbell. 1989. DNA polymerase III, a second essential DNA polymerase, is encoded by the *S. cerevisiae CDC2* gene. *Cell* **56:** 599.

Smith, H.E. and A.P. Mitchell. 1989. A transcriptional cascade governs entry into meiosis in *Saccharomyces cerevisiae*. *Mol. Cell. Biol.* **9:** 2142.

Smith, H.E., S.E. Driscoll, R.A. Sia, H.E. Yuan, and A.P. Mitchell. 1993. Genetic evidence for transcriptional activation by the yeast *IME1* gene product. *Genetics* **133:** 775.

Smith, H.E., S.S. Su, L. Neigeborn, S.E. Driscoll, and A.P. Mitchell. 1990. Role of *IME1* expression in regulation of meiosis in *Saccharomyces cerevisiae*. *Mol. Cell. Biol.* **10:** 6103.

Smith, J. and R. Rothstein. 1995. A mutation in the gene encoding the *Saccharomyces cerevisiae* single-stranded DNA-binding protein Rfa1 stimulates a *RAD52*-independent pathway for direct-repeat recombination. *Mol. Cell. Biol.* **15:** 1632.

Smith, L.M., L.G. Robbins, A. Kennedy, and P.T. Magee. 1988. Identification and characterization of mutations affecting sporulation in *Saccharomyces cerevisiae*. *Genetics* **120:** 899.

Smithies, O. and P.A. Powers. 1986. Gene conversions and their relation to homologous chromosome pairing. *Philos. Trans. R. Soc. Lond. B Biol. Sci.* **312:** 291.

Spang, A., I. Courtney, U. Fackler, M. Matzner, and E. Schiebel. 1993. The calcium-binding protein cell division cycle 31 of *Saccharomyces cerevisiae* is a component of the half bridge of the spindle pole body. *J. Cell Biol.* **123:** 405.

Sprague, G.F., Jr. and J.W. Thorner. 1992. Pheromone response and signal transduction during the mating process of *Saccharomyces cerevisiae.* In *The molecular and cellular biology of the yeast* Saccharomyces: *Gene expression* (ed. E.W. Jones et al.), p. 657. Cold Spring Harbor Laboratory Press, Cold Spring Harbor, New York.

Stapleton, A. and T.D. Petes. 1991. The Tn3 β-lactamase gene acts as a hotspot for meiotic recombination in yeast. *Genetics* **127:** 39.

Steber, C.M. and R.E. Esposito. 1995. *UME6* is a central component of a developmental regulatory switch controlling meiosis-specific gene expression. *Proc. Natl. Acad. Sci.* **92:** 12490.

Steele, D.F., M.E. Morris, and R.S. Jinks. 1991. Allelic and ectopic interactions in recombination-defective yeast strains. *Genetics* **127:** 53.

Stern, H. and Y. Hotta. 1987. The biochemistry of meiosis. In *Meiosis* (ed. P.B. Moens), p. 303. Academic Press, Orlando, Florida.

Sternberg, P.W., M.J. Stern, I. Clark, and I. Herskowitz. 1987. Activation of the yeast *HO* gene by release from multiple negative controls. *Cell* **48:** 567.

Stillman, D.J., S. Dorland, and Y. Yu. 1994. Epistasis analysis of suppressor mutations that allow *HO* expression in the absence of the yeast SW15 transcriptional activator. *Genetics* **136:** 781.

Storlazzi, A., L. Xu, L. Cao, and N. Kleckner. 1995. Crossover and noncrossover recombination during meiosis: Timing and pathway relationships. *Proc. Natl. Acad. Sci.* **92:** 8512.

Story, R.M., D.K. Bishop, N. Kleckner, and T.A. Steitz. 1993. Structural relationship of bacterial RecA proteins to recombination proteins from bacteriophage T4 and yeast. *Science* **259:** 1892.

Strand, M., M.C. Earley, G.F. Crouse, and T.D. Petes. 1995. Mutations in the *MSH3* gene preferentially lead to deletions within tracts of simple repetitive DNA in *Saccharomyces cerevisiae*. *Proc. Natl. Acad. Sci.* **92:** 10418.

Strathern, J., J. Hicks, and I. Herskowitz. 1981. Control of cell type in yeast by the mating type locus. The α1-α2 hypothesis. *J. Mol. Biol.* **147:** 357.

Strich, R., M.R. Slater, and R.E. Esposito. 1989. Identification of negative regulatory genes that govern the expression of early meiotic genes in yeast. *Proc. Natl. Acad. Sci.* **86:** 10018.

Strich, R., R.T. Surosky, C. Steber, E. Dubois, F. Messenguy, and R.E. Esposito. 1994. *UME6* is a key regulator of nitrogen repression and meiotic development. *Genes Dev.* **8:** 796.

Su, S.S. and A.P. Mitchell. 1993a. Identification of functionally related genes that stimulate early meiotic gene expression in yeast. *Genetics* **133:** 67.

———. 1993b. Molecular characterization of the yeast meiotic regulatory gene *RIM1*. *Nucleic Acids Res.* **21:** 3789.

Sugawara, N. and J.E. Haber. 1992. Characterization of double-strand break-induced recombination: Homology requirements and single-stranded DNA formation. *Mol. Cell. Biol.* **12:** 563.

Sumrada, R.A. and T.G. Cooper. 1987. Ubiquitous upstream repression sequences control activation of the inducible arginase gene in yeast. *Proc. Natl. Acad. Sci.* **84:** 3997.

Sun, H., D. Treco, and J.W. Szostak. 1991. Extensive 3′-overhanging, single-stranded DNA associated with the meiosis-specific double-strand breaks at the *ARG4* recombination initiation site. *Cell* **64:** 1155.

Sun, H., D. Treco, N.P. Schultes, and J.W. Szostak. 1989. Double-strand breaks at an initiation site for meiotic gene conversion. *Nature* **338:** 87.

Sung, P. and D.L. Robberson. 1995. DNA strand exchange mediated by a RAD51-ssDNA nucleoprotein filament with polarity opposite to that of RecA. *Cell* **82:** 453.

Sung, P., S. Prakash, and L. Prakash. 1988. The *RAD6* protein of *Saccharomyces cerevisiae* polyubiquitinates histones, and its acidic domain mediates this activity. *Genes Dev.* **2:** 1476.

———. 1990. Mutation of cysteine-88 in the *Saccharomyces cerevisiae RAD6* protein abolishes its ubiquitin-conjugating activity and its various biological functions. *Proc. Natl. Acad. Sci.* **87:** 2695.

Surana, U., A. Amon, C. Dowzer, J. McGrew, B. Byers, and K. Nasmyth. 1993. Destruction of the CDC28/CLB mitotic kinase is not required for the metaphase to anaphase transition in budding yeast. *EMBO J.* **12:** 1969.

Surosky, R.T. and R.E. Esposito. 1992. Early meiotic transcripts are highly unstable in *Saccharomyces cerevisiae. Mol. Cell. Biol.* **12:** 3948.

Surosky, R.T. and B.K. Tye. 1988. Meiotic disjunction of homologs in *Saccharomyces cerevisiae* is directed by pairing and recombination of the chromosome arms but not by pairing of the centromeres. *Genetics* **119:** 273.

Surosky, R.T., R. Strich, and R.E. Esposito. 1994. The yeast *UME5* gene regulates the stability of meiotic mRNAs in response to glucose. *Mol. Cell. Biol.* **14:** 3446.

Sym, M. and G.S. Roeder. 1994. Crossover interference is abolished in the absence of a synaptonemal complex protein. *Cell* **79:** 283.

———. 1995. Zip1-induced changes in synaptonemal complex structure and polycomplex assembly. *J. Cell Biol.* **128:** 455.

Sym, M., J.A. Engebrecht, and G.S. Roeder. 1993. *ZIP1* is a synaptonemal complex protein required for meiotic chromosome synapsis. *Cell* **72:** 365.

Symington, L.S. and T.D. Petes. 1988. Expansions and contractions of the genetic map relative to the physical map of yeast chromosome III. *Mol. Cell. Biol.* **8:** 595.

Szent-Gyorgyi, C. 1995. A bipartite operator interacts with a heat shock element to mediate early meiotic induction of *Saccharomyces cerevisiae HSP82. Mol. Cell. Biol.* **15:** 6754.

Szostak, J.W., W.T.L. Orr, R.J. Rothstein, and F.W. Stahl. 1983. The double-strand-break repair model for recombination. *Cell* **33:** 25.

Taguchi, A.K. and E.T. Young. 1987. The cloning and mapping of *ADR6*, a gene required for sporulation and for expression of the alcohol dehydrogenase II isozyme from *Saccharomyces cerevisiae. Genetics* **116:** 531.

Tanaka, H. and M. Tsuboi. 1985. Cloning and mapping of the sporulation gene, *spoT7*, in *Saccharomyces cerevisiae. Mol. Gen. Genet.* **199:** 21.

Tanaka, K., K. Matsumoto, and E.A. Toh. 1989. *IRA1*, an inhibitory regulator of the RAS-cyclic AMP pathway in *Saccharomyces cerevisiae. Mol. Cell. Biol.* **9:** 757.

Tanaka, K., M. Nakafuku, F. Tamanoi, Y. Kaziro, K. Matsumoto, and E.A. Toh. 1990. *IRA2*, a second gene of *Saccharomyces cerevisiae* that encodes a protein with a domain homologous to mammalian ras GTPase-activating protein. *Mol. Cell. Biol.* **10:** 4303.

Tatchell, K., L. C. Robinson, and M. Breitenbach. 1985. *RAS2* of *Saccharomyces cerevisiae* is required for gluconeogenic growth and proper response to nutrient limitation. *Proc. Natl. Acad. Sci.* **82:** 3785.

Tevzadze, G.G., A.R. Mushegian, and R.E. Esposito. 1996. The *SPO1* gene product required for meiosis in yeast has a high similarity to phospholipase B enzymes. *Gene* **177:**253.

Theurkauf, W.E. and R.S. Hawley. 1992. Meiotic spindle assembly in *Drosophila* females: Behavior of nonexchange chromosomes and the effects of mutations in the *nod* kinesin-like protein. *J. Cell Biol.* **116:** 1167.

Thomas, B.J. and R. Rothstein. 1989. Elevated recombination rates in transcriptionally active DNA. *Cell* **56:** 619.

Thomas, J.H. and D. Botstein. 1986. A gene required for the separation of chromosomes on the spindle apparatus in yeast. *Cell* **44:** 65.

Thompson, E.A. and G.S. Roeder. 1989. Expression and DNA sequence of *RED1*, a gene required for meiosis I chromosome segregation in yeast. *Mol. Gen. Genet.* **218:** 293.

Thorne, L.W. and B. Byers. 1993. Stage-specific effects of X-irradiation on yeast meiosis. *Genetics* **134:** 29.

Thuret, J., J. Valay, G. Faye, and C. Mann. 1996. Civ1 (CAK in vivo), a novel Cdk-activating kinase. *Cell* **86:** 565.

Tilburn, J., S. Sarkar, D.A. Widdick, E.A. Espeso, M. Orejas, M. Mungroo, M.A. Penalva, and H.N. Arst, Jr. 1995. The *Aspergillus* PacC zinc-finger transcription factor mediates regulation of both acid- and alkaline-expressed genes by ambient pH. *EMBO J.* **14:** 779.

Tishkoff, D.X., A.W. Johnson, and R.D. Kolodner. 1991. Molecular and genetic analysis of the gene encoding the *Saccharomyces cerevisiae* strand exchange protein Sep1. *Mol. Cell. Biol.* **11:** 2593.

Tishkoff, D.X., B. Rockmill, G.S. Roeder, and R.D. Kolodner. 1995. The *sep1* mutant of *Saccharomyces cerevisiae* arrests in pachytene and is deficient in meiotic recombination. *Genetics* **139:** 495.

Toda, T., S. Cameron, P. Sass, M. Zoller, and M. Wigler. 1987. Three different genes in *S. cerevisiae* encode the catalytic subunits of the cAMP-dependent protein kinase. *Cell* **50:** 277.

Toda, T., I. Uno, T. Ishikawa, S. Powers, T. Kataoka, D. Broek, S. Cameron, J. Broach, K. Matsumoto, and M. Wigler. 1985. In yeast, RAS proteins are controlling elements of adenylate cyclase. *Cell* **40:** 27.

Toone, W.M., A.L. Johnson, G.R. Banks, J.H. Toyn, D. Stuart, C. Wittenberg, and L.H. Johnston. 1995. Rme1, a negative regulator of meiosis, is also a positive activator of G_1 cyclin gene expression. *EMBO J.* **14:** 5824.

Toyn, J.H. and L.H. Johnston. 1993. Spo12 is a limiting factor that interacts with the cell cycle protein kinases Dbf2 and Dbf20, which are involved in mitotic chromatid disjunction. *Genetics* **135:** 963.

———. 1994. The Dbf2 and Dbf20 protein kinases of budding yeast are activated after the metaphase to anaphase cell cycle transition. *EMBO J.* **13:** 1103.

Treger, J.M., K.A. Heichman, and K. McEntee. 1988. Expression of the yeast *UBI4* gene increases in response to DNA-damaging agents and in meiosis. *Mol. Cell. Biol.* **8:** 1132.

Tripp, M.L. and R. Pinon. 1986. Control of the cAMP pathway by the cell cycle start function, *CDC25*, in *Saccharomyces cerevisiae*. *J. Gen. Microbiol.* **132:** 1143.

Tsuboi, M. 1983. The isolation and genetic analysis of sporulation-deficient mutants in *Saccharomyces cerevisiae*. *Mol. Gen. Genet.* **191:** 17.

Tugendreich, S., J. Tomkiel, W. Earnshaw, and P. Hieter. 1995. CDC27Hs colocalizes with CDC16Hs to the centrosome and mitotic spindle and is essential for the metaphase to anaphase transition. *Cell* **81:** 261.

Uno, I., K. Matsumoto, A. Hirata, and T. Ishikawa. 1985. Outer plaque assembly and spore encapsulation are defective during sporulation of adenylate cyclase-deficient mutants of *Saccharomyces cerevisiae. J. Cell Biol.* **100:** 1854.

Uno, I., T. Oshima, A. Hirata, and T. Ishikawa. 1990. The functional domain of adenylate cyclase associated with entry into meiosis in *Saccharomyces cerevisiae. J. Bacteriol.* **172:** 102.

Vaisman, N., A. Tsouladze, K. Robzyk, S. Ben Yehuda, M. Kupiec, and Y. Kassir. 1995. The role of *Saccharomyces cerevisiae* Cdc40p in DNA replication and mitotic spindle formation and/or maintenance. *Mol. Gen. Genet.* **247:** 123.

Valdivieso, M.H., P.C. Mol, J.A. Shaw, E. Cabib, and A. Durán. 1991. *CAL1*, a gene required for activity of chitin synthase 3 in *Saccharomyces cerevisiae. J. Cell Biol.* **114:** 101.

Vallen, E.A., W. Ho, M. Winey, and M.D. Rose. 1994. Genetic interactions between *CDC31* and *KAR1*, two genes required for duplication of the microtubule organizing center in *Saccharomyces cerevisiae. Genetics* **137:** 407.

Varma, A., E.B. Freese, and E. Freese. 1985. Partial deprivation of GTP initiates meiosis and sporulation in *Saccharomyces cerevisiae. Mol. Gen. Genet.* **201:** 1.

Verdier, J.M., J.H. Camonis, and M. Jacquet. 1989. Cloning of *CDC33:* A gene essential for growth and sporulation which does not interfere with cAMP production in *Saccharomyces cerevisiae. Yeast* **5:** 79.

Vershon, A.K., N.M. Hollingsworth, and A.D. Johnson. 1992. Meiotic induction of the yeast *HOP1* gene is controlled by positive and negative regulatory sites. *Mol. Cell. Biol.* **12:** 3706.

Vezinhet, F., J.H. Kinnaird, and I.W. Dawes. 1979. The physiology of mutants derepressed for sporulation in *Saccharomyces cerevisiae. J. Gen. Microbiol.* **115:** 391.

Vidal, M. and R.F. Gaber. 1991. *RPD3* encodes a second factor required to achieve maximum positive and negative transcriptional states in *Saccharomyces cerevisiae. Mol. Cell. Biol.* **11:** 6317.

Vidal, M., R. Strich, R. E. Esposito, and R. F. Gaber. 1991. *RPD1* (*SIN3/UME4*) is required for maximal activation and repression of diverse yeast genes. *Mol. Cell. Biol.* **11:** 6306.

Voelkel-Meiman, K., R.L. Keil, and G.S. Roeder. 1987. Recombination-stimulating sequences in yeast ribosomal DNA correspond to sequences regulating transcription by RNA polymerase I. *Cell* **48:** 1071.

von Wettstein, D., S.W. Rasmussen, and P.B. Holm. 1984. The synaptonemal complex in genetic segregation. *Annu. Rev. Genet.* **18:** 331.

Wagstaff, J.E., S. Klapholz, and R.E. Esposito. 1982. Meiosis in haploid yeast. *Proc. Natl. Acad. Sci.* **79:** 2986.

Wagstaff, J.E., S. Klapholz, C.S. Waddell, L. Jensen, and R.E. Esposito. 1985. Meiotic exchange within and between chromosomes requires a common Rec function in *Saccharomyces cerevisiae. Mol. Cell. Biol.* **5:** 3532.

Wallis, J.W., G. Chrebet, G. Brodsky, and M. Rolfe. 1989. A hyper-recombination mutation in *S. cerevisiae* identifies a novel eukaryotic topoisomerase. *Cell* **58:** 409.

Wang, H. and D.J. Stillman. 1993. Transcriptional repression in *Saccharomyces cerevisiae* by a SIN3-LexA fusion protein. *Mol. Cell. Biol.* **13:** 1805.

Wang, H., I. Clark, P.R. Nicholson, I. Herskowitz, and D.J. Stillman. 1990. The *Saccharomyces cerevisiae SIN3* gene, a negative regulator of *HO*, contains four paired amphipathic helix motifs. *Mol. Cell. Biol.* **10:** 5927.

Wang, H.T., S. Frackman, J. Kowalisyn, R.E. Esposito, and R. Elder. 1987. Developmental regulation of *SPO13*, a gene required for separation of homologous chromosomes at meiosis I. *Mol. Cell. Biol.* **7:** 1425.

Watkins, J.F., P. Sung, S. Prakash, and L. Prakash. 1993. The extremely conserved amino terminus of *RAD6* ubiquitin-conjugating enzyme is essential for amino-end rule-dependent protein degradation. *Genes Dev.* **7:** 250.

Weber, L. and B. Byers. 1992. A *RAD9*-dependent checkpoint blocks meiosis of *cdc13* yeast cells. *Genetics* **131:** 55.

Weiner, B.M. and N. Kleckner. 1994. Chromosome pairing via multiple interstitial interactions before and during meiosis in yeast. *Cell* **77:** 977.

Weinert, T. and L. Hartwell. 1989. Control of G2 delay by the *rad9* gene of *Saccharomyces cerevisiae*. *J. Cell Sci. Suppl.* **12:** 145.

Weinert, T.A., G.L. Kiser, and L.H. Hartwell. 1994. Mitotic checkpoint genes in budding yeast and the dependence of mitosis on DNA replication and repair. *Genes Dev.* **8:** 652.

Weiss, E. and M. Winey. 1996. The *Saccharomyces cerevisiae* spindle pole body duplication gene *MPS1* is part of a mitotic checkpoint. *J. Cell Biol.* **132:** 111.

Werner, W.M., J. Becker, S.J. Kosic, and E.A. Craig. 1989. Yeast Hsp70 RNA levels vary in response to the physiological status of the cell. *J. Bacteriol.* **171:** 2680.

Whelan, W.L. and C.E. Ballou. 1975. Sporulation in D-glucosamine auxotrophs of *Saccharomyces cerevisiae:* Meiosis with defective ascospore wall formation. *J. Bacteriol.* **124:** 1545.

White, M.A., M. Dominska, and T.D. Petes. 1993. Transcription factors are required for the meiotic recombination hotspot at the *HIS4* locus in *Saccharomyces cerevisiae*. *Proc. Natl. Acad. Sci.* **90:** 6621.

White, M.A., M. Wierdl, P. Detloff, and T.D. Petes. 1991. DNA-binding protein RAP1 stimulates meiotic recombination at the *HIS4* locus in yeast. *Proc. Natl. Acad. Sci.* **88:** 9755.

Whyte, W., L.H. Keopp, J. Lamb, J.C. Crowley, and D.B. Kaback. 1990. Molecular cloning of chromosome I DNA from *Saccharomyces cerevisiae:* Isolation, characterization and regulation of the *SPO7* sporulation gene. *Gene* **95:** 65.

Wickner, R.B. 1974. Mutants of *Saccharomyces cerevisiae* that incorporate deoxythymidine-5′-monophosphate into deoxyribonucleic acid in vivo. *J. Bacteriol.* **117:** 252.

Williams, F.E. and R.J. Trumbly. 1990. Characterization of *TUP1*, a mediator of glucose repression in *Saccharomyces cerevisiae*. *Mol. Cell. Biol.* **10:** 6500.

Williamson, D.H., L.H. Johnston, D.J. Fennell, and G. Simchen. 1983. The timing of the S phase and other nuclear events in yeast meiosis. *Exp Cell Res.* **145:** 209.

Williamson, M.S., J.C. Game, and S. Fogel. 1985. Meiotic gene conversion mutants in *Saccharomyces cerevisiae*. I. Isolation and characterization of *pms1-1* and *pms1-2*. *Genetics* **110:** 609.

Willis, K.K. and H.L. Klein. 1987. Intrachromosomal recombination in *Saccharomyces cerevisiae:* Reciprocal exchange in an inverted repeat and associated gene conversion. *Genetics* **117:** 633.

Winey, M., L. Goetsch, P. Baum, and B. Byers. 1991. *MPS1* and *MPS2:* Novel yeast genes defining distinct steps of spindle pole body duplication. *J. Cell Biol.* **114:** 745.

Winey, M., M.A. Hoyt, C. Chan, L. Goetsch, D. Botstein, and B. Byers. 1993. *NDC1:* A nuclear periphery component required for yeast spindle pole body duplication. *J. Cell Biol.* **122:** 743.

Wolf, D. and G. Fink. 1975. Proteinase C (carboxypeptidase Y) mutant of yeast. *J. Bacteriol.* **123:** 1150.

Wu, T.C. and M. Lichten. 1994. Meiosis-induced double-strand break sites determined by yeast chromatin structure. *Science* **263:** 515.

———. 1995. Factors that affect the location and frequency of meiosis- induced double-strand breaks in *Saccharomyces cerevisiae. Genetics* **140:** 55.

Xu, L., M. Ajimura, R. Padmore, C. Klein, and N. Kleckner. 1995. *NDT80*, a meiosis-specific gene required for exit from pachytene in *Saccharomyces cerevisiae. Mol. Cell. Biol.* **15:** 6572.

Yamashita, I. and S. Fukui. 1985. Transcriptional control of the sporulation-specific glucoamylase gene in the yeast *Saccharomyces cerevisiae. Mol. Cell. Biol.* **5:** 3069.

Yeh, E., R. Driscoll, M. Coltrera, A. Olins, and K. Bloom. 1991. A dynamin-like protein encoded by the yeast sporulation gene *SPO15. Nature* **349:** 713.

Yoshida, M., H. Kawaguchi, Y. Sakata, K. Kominami, M. Hirano, H. Shima, R. Akada, and I. Yamashita. 1990. Initiation of meiosis and sporulation in *Saccharomyces cerevisiae* requires a novel protein kinase homologue. *Mol. Gen. Genet.* **221:** 176.

Yoshimoto, H., M. Ohmae, and I. Yamashita. 1992. The *Saccharomyces cerevisiae* GAM2/SIN3 protein plays a role in both activation and repression of transcription. *Mol. Gen. Genet.* **233:** 327.

Zamb, T.J. and R. Roth. 1977. Role of mitotic replication genes in chromosome duplication during meiosis. *Proc. Natl. Acad. Sci.* **74:** 3951.

Zenvirth, D., T. Arbel, A. Sherman, M. Goldway, S. Klein, and G. Simchen. 1992. Multiple sites for double-strand breaks in whole meiotic chromosomes of *Saccharomyces cerevisiae. EMBO J.* **11:** 3441.

Zhang, S., S. Guha, and F. C. Volkert. 1995. The *Saccharomyces SHP1* gene, which encodes a regulator of phosphoprotein phosphatase 1 with differential effects on glycogen metabolism, meiotic differentiation, and mitotic cell cycle progression. *Mol. Cell. Biol.* **15:** 2037.

Zickler, D. and L.W. Olson. 1975. The synaptonemal complex and the spindle plaque during meiosis in yeast. *Chromosoma* **50:** 1.

Zubenko, G.S. and E.W. Jones. 1981. Protein degradation, meiosis and sporulation in proteinase-deficient mutants of *Saccharomyces cerevisiae. Genetics* **97:** 45.

Zubenko, G.S., A.P. Mitchell, and E.W. Jones. 1979. Septum formation, cell division, and sporulation in mutants of yeast deficient in proteinase B. *Proc. Natl. Acad. Sci.* **76:** 2395.

12

Mating and Sporulation in *Schizosaccharomyces pombe*

Masayuki Yamamoto, Yoshiyuki Imai, and Yoshinori Watanabe
Department of Biophysics and Biochemistry
Graduate School of Science, University of Tokyo
Hongo, Tokyo 113, Japan

I. INTRODUCTION

Haploid cells of the fission yeast *Schizosaccharomyces pombe* display one of the two mating types denoted as h^+ (or P) and h^- (or M). Homothallic strains, denoted as h^{90}, change their mating type during proliferation (for review, see Klar, these volumes), whereas heterothallic strains are fixed as either h^+ or h^-. *S. pombe* cells are most stable in the haploid state and are essentially asexual as long as they are well fed. They initiate sexual development only under starvation conditions. Thus, unlike *Saccharomyces cerevisiae*, homothallic *S. pombe* cells keep growing as haploids and do not undergo conjugation in rich medium. When starved for nutrients, especially for nitrogen, haploid cells mate to form zygotes, which then undergo meiosis and generate haploid spores (zygotic ascospores). Zygotes can grow as diploids if they are transferred to rich medium immediately after conjugation. However, these diploid cells undergo meiosis and form spores (azygotic ascospores) at the end of vegetative growth. The absence of glucose is not mandatory for mating and meiosis in *S. pombe*, but these events proceed more efficiently if the concentration of glucose in the medium is lowered.

Nitrogen starvation induces expression of genes essential for sexual development, which results in the enhanced production of mating pheromones and their receptors. Cells are then able to recognize the pheromone secreted by cells of the opposite mating type, and h^+ and h^- cells elongate conjugation tubes toward each other. The cells agglutinate in pairs at the tips of these tubes, fuse, and then undergo karyogamy. Although the cell shape of *S. pombe* is rod-like and that of *S. cerevisiae* is ovoid, the morphological changes observed during conjugation in these two yeast species are not very different.

Genetic analysis of sexual development in fission yeast stems from an early study by Bresch et al. (1968), who characterized meiosis-defective mutants. The first report of a sterile *S. pombe* mutant (i.e., a mutant defective in mating) appeared somewhat later (Thuriaux et al. 1980), followed by a more extensive study of sterile mutants (Girgsdies 1982). Analyses of mutants isolated in these and subsequent studies, including molecular cloning of the responsible genes, have led to the view that the sexual development of *S. pombe* is regulated by two major signal transduction pathways, one responding to nutritional conditions and the other to the mating pheromones. This chapter summarizes current knowledge of the molecular mechanisms that regulate mating, meiosis, and sporulation in *S. pombe* and contrasts these mechanisms with those regulating the same processes in *S. cerevisiae* (for detailed review, see Kupiec et al., this volume).

II. NUTRITIONAL SIGNALING

A. Negative Regulation of Sexual Development by cAMP

An important clue to the mechanisms that regulate the entry into sexual development in *S. pombe* came with the isolation of the *pat1* (also called *ran1*) mutants (Iino and Yamamoto 1985a; Nurse 1985). Cells carrying the *pat1-114* mutation show temperature-sensitive growth and initiate ectopic meiosis and sporulation at the restrictive temperature, irrespective of nutritional conditions and the ploidy of the cells (i.e., even in the haploid state). These results suggest that Pat1p is a factor that prevents cells from entering meiosis when they should be in the mitotic cell cycle (Beach et al. 1985; Iino and Yamamoto 1985a,b; Nurse 1985). In addition, *pat1-114* cells mate efficiently on rich medium at semirestrictive temperatures, suggesting that Pat1p may also be involved in the control of mating (Beach et al. 1985; Nurse 1985). As discussed further below (Section IV.B), Pat1p is a protein kinase that inactivates Mei2p, a critical factor for the entry to meiosis (Watanabe et al. 1997). It possibly affects gene expression required for mating as well (Nielsen and Egel 1990; Li and McLeod 1996). Inactivation of Pat1p is indeed a physiological process for the initiation of meiosis, which results from nutritional starvation and mating pheromone signaling (McLeod et al. 1987; McLeod and Beach 1988; for review, see Yamamoto 1996).

Meanwhile, the addition of cAMP to the medium was found to inhibit sexual development in *S. pombe* (Calleja et al. 1980). The addition of cAMP could also suppress the ectopic meiosis in the *pat1-114* mutant (Beach et al. 1985), apparently by blocking the expression of the *mei2* gene, encoding Mei2p (Watanabe et al. 1988). It has also been shown that the level of cAMP decreases when cells are starved for nitrogen (Fukui et al. 1986b; Maeda et al. 1990; Mochizuki and Yamamoto 1992). These observations indicate that cAMP may have a critical role in the regulation of entry into sexual development in *S. pombe*. Analyses of cAMP-related mutants, as summarized below, have demonstrated unambiguously that this is the case. The relevant genes and their products are listed in Table 1, A and D.

1. Adenylyl Cyclase and CAP

The gene encoding fission yeast adenylyl cyclase, *cyr1*, was cloned by using the *S. cerevisiae* counterpart, *CYR1*, as the probe (Yamawaki-Kataoka et al. 1989; Young et al. 1989; Maeda et al. 1990). Disruption of *cyr1* is not lethal, although the disruptant (*cyr1*Δ) grows more slowly than the wild type. The level of cAMP is virtually zero, and the activity of adenylyl cyclase is not detectable in *cyr1*Δ cells. These cells prove to be highly derepressed for sexual development: If *cyr1*Δ cells of the two mating types are mixed, they readily mate and sporulate even in rich medium (Maeda et al. 1990; Kawamukai et al. 1991).

CAP (or Srv2p) is a bifunctional protein known to complex with adenylyl cyclase in *S. cerevisiae* (Fedor-Chaiken et al. 1990; Field et al. 1990). Its amino-terminal region interacts with adenylyl cyclase, whereas its carboxy-terminal region apparently regulates cell morphology and responses to nutritional starvation (Field et al. 1990; Gerst et al. 1991). The *S. pombe* homolog, Cap, is similarly bifunctional (Kawamukai et al. 1992). Disruption of the carboxy-terminal region of this protein results in temperature-sensitive growth and anomalies in cell morphology and ascus formation. Deletion of the entire gene causes ectopic mating and sporulation in rich medium, in addition to the above phenotypes. Simultaneous overexpression of the *cyr1* and *cap1* genes produces a higher level of adenylyl cyclase activity than overexpression of *cyr1* alone (Kawamukai et al. 1992)[1] From these observations, it has been concluded that *S. pombe* Cap is an activator of adenylyl cyclase, like *S. cerevisiae* CAP.

2. *cAMP Phosphodiesterase and the Regulatory Subunit of PKA*

The *cgs1* and *cgs2* genes were identified in a screen for mutations suppressing the growth deficiency of the *pat1-114* mutant at the restrictive temperature (DeVoti et al. 1991). Cloning of these genes and subsequent analysis of their products indicated that *cgs1* encodes the regulatory subunit of cAMP-dependent protein kinase (PKA) and that *cgs2* encodes cAMP phosphodiesterase (DeVoti et al. 1991). The *pde1* gene was isolated as a high-copy suppressor of the sterility of an *S. pombe* strain expressing *S. cerevisiae* adenylyl cyclase and was found to be identical to *cgs2* (Mochizuki and Yamamoto 1992). The activity of PKA is no longer dependent on cAMP in *cgs1*-defective cells (DeVoti et al. 1991), and the level of intracellular cAMP is about four times higher in *cgs2/pde1*Δ cells than in wild type (Mochizuki and Yamamoto 1992). Cells defective in either of the two genes are sterile and meiosis-deficient. In addition, whereas wild-type cells become shorter at the end of vegetative growth, the mutant cells do not. They attain higher cell numbers than the wild type at stationary phase, but many dead cells are present when growth is saturated. These observations have been thought to indicate that these cells are unable to enter G_0 phase under nutritional starvation (DeVoti et al. 1991).

3. *The Catalytic Subunit of PKA and a Homolog of Budding Yeast Sch9p Kinase*

The *pka1* gene, which encodes the catalytic subunit of PKA, was isolated as a clone that inhibits mating and sporulation if carried in multiple copies in *S. pombe* cells (Maeda et al. 1994). This gene is most similar to

[1]To unify the nomenclature of *S. pombe* genes, we call genes originally named *cap* (Kawamukai et al. 1992), *krp* (Davey et al. 1994), and *cdc42sp* (Miller and Johnson 1994) as *cap1*, *krp1*, and *cdc42*, respectively, in this chapter.

Table 1 Genes involved in mating, meiosis, and sporulation in *S. pombe*

Gene	Product	Size (amino acids)	References
(A) *Genes related to nutritional signaling and the cAMP cascade*			
gpa2 (*git8*)	G-protein α-subunit	354	Isshiki et al. (1992); Nocero et al. (1994)
cyr1 (*git2*)	adenylyl cyclase	1692	Yamawaki-Kataoka et al. (1989); Young et al. (1989); Maeda et al. (1990); Kawamukai et al. (1991); Hoffman and Winston (1991)
cgs2 (*pde1*)	cAMP phosphodiesterase	346	DeVoti et al. (1991); Mochizuki and Yamamoto (1992)
pka1 (*git6*)	catalytic subunit of PKA	512	Maeda et al. (1994); Jin et al. (1995)
cgs1	regulatory subunit of PKA	411	DeVoti et al. (1991)
sck1	homolog of *S. cerevisiae* Sch9p	696	Jin et al. (1995)
cap1	homolog of *S. cerevisiae* Cap/Srv2p	551	Kawamukai et al. (1992)
ste11 (*aff1*)	HMG family transcription factor	468	Watanabe et al. (1988); Sipiczki (1988); Kitamura et al. (1990); Sugimoto et al. (1991)
sak1	RXF family DNA-binding protein	734	Wu and McLeod (1995)
pac1	double-stranded RNase	363	Iino et al. (1991)
pac2	regulation of *ste11* expression	235	Kunitomo et al. (1995)
(B) *Mating-type genes*			
mat1-Pc	required for conjugation in h^+ cells	118	Kelly et al. (1988)
mat1-Pi (*Pm*)	homeobox protein, required for meiosis	159	Kelly et al. (1988); Nielsen et al. (1992)
mat1-Mc	HMG family transcription factor, required for conjugation in h^- cells	181	Kelly et al. (1988); Sinclair et al. (1990); Dooijes et al. (1993)
mat1-Mi (*Mm*)	required for meiosis	42	Kelly et al. (1988)

(C) *Mating-type-specific sterility genes*			
h^+-specific sterility genes			
map1	putative transcription factor; homolog of *S. cerevisiae MCM1*	–	Egel (1973b); Nielsen et al. (1996); Yabana and Yamamoto (1996)
map2	P-factor precursor	201	Egel (1973b); Imai and Yamamoto (1994)
map3	M-factor receptor	365	Tanaka et al. (1993)
map4	putative agglutinin	–	Y. Imai and M. Yamamoto (unpubl.)
sxa1	aspartyl protease	533	Imai and Yamamoto (1992)
h^--specific sterility genes			
mam1	unknown	–	Egel (1973b)
mam2	P-factor receptor	348	Egel (1973b); Kitamura and Shimoda (1991)
mam3	putative agglutinin	–	Y. Imai and M. Yamamoto (unpubl.)
mam4	farnesylcysteine carboxyl methyl-transferase	–	Imai et al. (1997)
mfm1	M-factor precursor	42	Davey (1992)
mfm2	M-factor precursor	44	Davey (1992)
mfm3	M-factor precursor	41	Kjaerulff et al. (1994)
sxa2	serine carboxyl peptidase	507	Imai and Yamamoto (1992); Ladds et al. (1996)
(D) *Sterility genes and other genes relevant to mating*			
gpa1	G-protein α-subunit coupled to pheromone receptors	407	Obara et al. (1991)
ras1 (*ste5*)	Ras homolog	219	Girgsdies (1982); Fukui and Kaziro (1985); Fukui et al. (1986b); Nadin-Davis et al. (1986); Lund et al. (1987)

(*Continued on following pages.*)

Table 1 (Continued)

Gene	Product	Size (amino acids)	References
gap1 (*sar1*)	GAP for Ras1	766	Imai et al. (1991); Wang et al. (1991a)
ste6	GEF for Ras1	911	Girgsdies (1982); Hughes et al. (1990)
ral2	putative Ras1 activator	611	Fukui and Yamamoto (1988); Fukui et al. (1989)
byr1(*ste1*)	protein kinase (MAPKK)	340	Thuriaux et al. (1980); Nadin-Davis and Nasim (1988)
byr2 (*ste8*)	protein kinase (MAPKKK)	659	Michael and Gutz (1987); Wang et al. (1991b)
spk1	protein kinase (MAPK)	372	Toda et al. (1991)
scd1 (*ral1*)	putative GEF for Cdc42p	834	Fukui and Yamamoto (1988); Chang et al. (1994)
scd2 (*ral3*)	homolog of *S. cerevisiae* Bem1p	536	Fukui and Yamamoto (1988); Chang et al. (1994)
cdc42	Rho-type GTPase	192	Miller and Johnson (1994)
shk1/*pak1*	Pak-type protein kinase	658	Marcus et al. (1995); Ottilie et al. (1995)
byr3	seven zinc finger motifs	179	Xu et al. (1992)
zfs1	two zinc finger motifs	404	Kanoh et al. (1995)
fus1	required for cell fusion; homolog of the vertebrate formins	1372	Petersen et al. (1995)
ste2	unknown	–	Girgsdies (1982)
ste3	unknown	–	Girgsdies (1982)
ste4	leucine zipper motif	264	Girgsdies (1982); Okazaki et al. (1991)
ste7	unknown	–	Michael and Gutz (1987)
ste9	unknown	–	Leupold and Sipiczki (1991)
ste10	unknown	–	Leupold and Sipiczki (1991)
ste12	unknown	–	Kitamura et al. (1990)
ste13	putative RNA helicase	485	Kitamura et al. (1990); Maekawa et al. (1994)

(E) *Genes related to meiosis and sporulation*			
pat1 (*ran1*)	protein kinase	470	Iino and Yamamoto (1985a, 1985b); Nurse (1985); Beach et al. (1985); McLeod and Beach (1986, 1988)
mei2	RNA-binding protein required for premeiotic DNA synthesis and meiosis I	750	Watanabe et al. (1988); Watanabe and Yamamoto (1994)
mei3	inhibitor of Pat1p kinase	148	McLeod et al. (1987); McLeod and Beach (1988); Li and McLeod (1996)
mei4	putative transcription factor with *fkh* motif required for meiosis I	–	S. Horie et al. (unpubl.)
mes1	required for meiosis II	101	Kishida et al. (1994)
sme2	meiRNA required for meisois I	~0.5 kb	Watanabe and Yamamoto (1994)
spo5 (*mes2*)	RNA-binding protein	–	Y. Watanabe and M. Yamamoto (unpubl.)
spo1-4,spo6-20	not known	–	Bresch et al. (1968); Kishida and Shimoda (1986)
cam1	calmodulin	150	Takeda and Yamamoto (1987); Takeda et al. (1989)
(F) *Genes involved both in meiosis and in the mitotic cell cycle*			
cdc2 (*tws1*)	protein kinase	297	Nurse and Bissett (1981); Nakaseko et al. (1984); Beach et al. (1985); Niwa and Yanagida (1988); Grallert and Sipiczki (1990); Iino et al. (1995)
puc1	cyclin	359	Forsburg and Nurse (1991, 1994)
cig2 (*cyc17*)	cyclin	411	Bueno and Russell (1993); Connolly and Beach (1994); Obara-Ishihara and Okayama (1994)
rum1	regulator of G_1 progression	230	Moreno and Nurse (1994)

(*Continued on following page.*)

Table 1 (Continued)

Gene	Product	Size (amino acids)	References
cdc10	MCB-binding transcription factor	767	Nurse and Bissett (1981); Beach et al. (1985); Aves et al. (1985); Lowndes et al. (1992)
cdc22	ribonucleotide reductase large subunit	811	Fernandez-Sarabia et al. (1993)
res1 (*sct1*)	MCB-binding transcription factor	637	Tanaka et al. (1992); Caligiuri and Beach (1993)
res2 (*pct1*)	MCB-binding transcription factor	657	Miyamoto et al. (1994); Zhu et al. (1994)
rep1	zinc finger protein	472	Sugiyama et al. (1994)
cdc13	B-type cyclin	482	Booher and Beach (1988); Hagan et al. (1988); Grallert and Sipiczki (1991); Iino et al. (1995)
cdc25	protein tyrosine phosphatase	580	Grallert and Sipiczki (1990); Iino et al. (1995)

S. cerevisiae TPK2, one of the three genes that encode the catalytic subunit of PKA in budding yeast (Toda et al. 1987). Disruption of *pka1* causes phenotypes almost identical to those of the *cyr1*Δ strain; i.e., the disruptant grows more slowly than the wild type and mates and sporulates even in rich medium (Maeda et al. 1994). Overexpression of *pka1* suppresses the ectopic meiosis produced by the *pat1-114* mutation, as does a high level of intracellular cAMP (see above). These observations confirm that cAMP inhibits sexual development through activation of PKA in *S. pombe* cells.

The *S. cerevisiae SCH9* gene encodes a protein kinase whose overproduction can suppress loss of function of the three *TPK* genes (Toda et al. 1988). Disruption of *SCH9* causes no obvious phenotype, and the function of Sch9p is thought to be redundant with that of PKA. The *sck1* gene, which encodes a fission yeast homolog of *SCH9*, was isolated as a high-copy suppressor of *git3* (Jin et al. 1995), one of the *git* mutations described by Hoffman and Winston (1990) (see also below). Disruption of *sck1* produces no obvious phenotype, whereas overexpression of *sck1* can suppress the growth retardation and inhibit ectopic mating in the *pka1*Δ strain (Jin et al. 1995). Thus, the status of *sck1* in *S. pombe* is very similar to that of *SCH9* in *S. cerevisiae*, in that both are dispensable for normal cellular activities but can replace the function of PKA if overexpressed. It is interesting that these genes have been conserved in the two distantly related yeast species, although they are not absolutely necessary for cell growth. Only when combined with disruption of *pka1* has disruption of *sck1* shown a phenotype: The recovery of growth from stationary phase is retarded in a *pka1*Δ strain, and it is more severely retarded in a *pka1*Δ *sck1*Δ double mutant (Jin et al. 1995).

4. *Heterotrimeric G-protein as an Apparent Activator of Adenylyl Cyclase*

The *S. pombe gpa2* gene was isolated as a clone that hybridizes to a *Dictyostelium discoideum* cDNA encoding a heterotrimeric G-protein α-subunit (Isshiki et al. 1992). The deduced gene product of *gpa2* is approximately 40% identical to a number of Gα subunits identified in higher organisms. Subsequent analysis revealed the relevance of *gpa2* to the cAMP cascade. A *gpa2*Δ strain shows phenotypes quite reminiscent of those of a *cyr1*Δ strain: It grows slowly, is smaller in cell size, and undergoes ectopic sexual development in rich medium. In addition, the level of intracellular cAMP in *gpa2*Δ is only one third of the wild-type level (Isshiki et al. 1992).

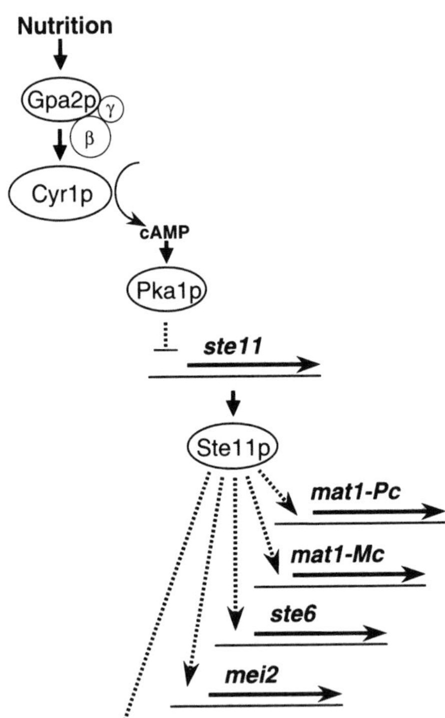

Figure 1 Nutritional signaling pathway leading to the induction of expression of *ste11*, which encodes an HMG family transcription factor that regulates a number of genes required for mating and/or meiosis. The cAMP cascade appears to mediate this signaling. Activation of PKA is inhibitory to *ste11* expression. Regulation through transcription is indicated by dotted lines.

Mutations are known that block the intrinsic GTPase activity of mammalian Gα and hence fix the protein in the activated state (Landis et al. 1989). Analogous mutations, namely, replacements of Arg-176 by histidine and Glu-202 by leucine, were introduced into *gpa2*. Cells carrying either of these mutant *gpa2* alleles show a slightly elevated cAMP level and are moderately inhibited in mating. Although *gpa2-R176H* or *gpa2-Q202L* alone causes these mild effects, combination of either mutation with disruption of *cgs2/pde1* results in a remarkable consequence. The level of cAMP in the double-mutant cells is about 20 times higher than that in wild-type cells (Isshiki et al. 1992). In contrast, disruption of *pde1* alone elevates the level of cAMP only fourfold (Mochizuki and

Yamamoto 1992). Thus, Gpa2p is likely to affect the level of cAMP by modulating the activity of adenylyl cyclase but not that of phosphodiesterase (Fig. 1). In addition, we can speculate on the basis of these observations that the level of cAMP in *S. pombe* cells is dually regulated, i.e., by modulating both its synthesis and its degradation. Although it appears unequivocal that Gpa2p stimulates adenylyl cyclase in response to the abundance of nutrients in the environment, it remains totally unknown how Gpa2p is controlled by nutritional conditions.

B. Gene Expression Under the Regulation of PKA

1. Ste11p, a Key Transcription Factor for Mating and Meiosis

Many genes that promote sexual development are transcriptionally activated under nitrogen starvation in *S. pombe*. Most of these genes carry one to five copies of a common motif named the TR-box (TTCTTTGTTY; Sugimoto et al. 1991) in their 5′ upstream regions. Such genes include the mating-type genes *mat1-P* and *mat1-M* (Sugimoto et al. 1991; Aono et al. 1994); genes involved in meiosis, such as *mei2* (Sugimoto et al. 1991) and *rep1* (Sugiyama et al. 1994); and genes involved in mating, such as *ste4* (Okazaki et al. 1991), *ste6* (Hughes et al. 1994), and *fus1* (Petersen et al. 1995). Expression of all these genes depends on the function of the *ste11* gene, which encodes a protein of the high-mobility-group (HMG) family (Sugimoto et al. 1991). Ste11p binds to the TR-box, which has been demonstrated to be an essential *cis* element for the expression of *mei2* (Sugimoto et al. 1991). Interestingly, *mat1-Pi* and *fus1*, transcription of which occurs in response to mating pheromone signal under nitrogen starvation, also carry the TR-box as an essential *cis* element for their expression and are controlled by Ste11p (Aono et al. 1994; Petersen et al. 1995). This suggests that the role of Ste11p in gene regulation may not be simple.

The *ste11* (or *aff1*) gene was originally defined by a mutation suppressing *pat1*-driven ectopic meiosis (Sipiczki 1988; Watanabe et al. 1988), and a *ste11* mutant was also found in a recent screen for sterile mutants (Kitamura et al. 1990). Disruption of *ste11* was found to render the cells completely sterile and sporulation-deficient, without affecting cell growth (Sugimoto et al. 1991). Artificial overexpression of *ste11* results in ectopic mating and sporulation, regardless of the nutritional conditions (Sugimoto et al. 1991). These observations and the characteristics of Ste11p discussed above indicate that this protein is a key transcription factor regulating sexual development in *S. pombe* (Fig. 1).

2. *Regulation of* ste11 *Expression*

Expression of *ste11* is induced in response to nitrogen starvation and is under the control of PKA (Sugimoto et al. 1991), which appears to account for the inducibility of expression of other genes under nitrogen starvation (Fig. 1). Because *ste11* itself carries TR-box motifs in its 5′ upstream region that appear to be essential for its expression, Ste11p is likely to stimulate its own production (Sugimoto et al. 1991; H. Kunitomo and M. Yamamoto, unpubl.). Besides Ste11p itself, two putative transcription factors are implicated in the regulation of *ste11* expression. One is a putative heterodimer of basic-zipper proteins encoded by *atf1* (also called *gad7*) (Takeda et al. 1995; Kanoh et al. 1996) and *pcr1* (Watanabe and Yamamoto 1996). This transcription factor is apparently regulated by a MAP kinase cascade responsive to stress, which includes Wis1p as an MAPKK (Warbrick and Fantes 1991) and Spc1p (also called Sty1p or Phh1p) as an MAPK (Millar et al. 1995; Shiozaki and Russell 1995; Kato et al. 1996). Spc1p phosphorylates Atf1p (Shiozaki and Russel 1996; Wilkinson et al. 1996). Atf1p and Pcr1p are required for the full expression of *ste11*, but whether or not they directly regulate *ste11* remains unknown. Furthermore, their activity does not appear to be regulated critically by PKA, despite the fact that these factors can bind to the consensus cAMP-response element (CRE) sequence (Takeda et al. 1995; Kanoh et al. 1996). The other is a zinc finger protein, which directly regulates *ste11* (H. Kunitomo and M. Yamamoto, unpubl.). It remains to be tested whether this transcription factor is phosphorylated and negatively regulated by PKA.

C. Other Factors That May Be Involved in Nutritional Signaling

The *pac1* and *pac2* genes also function as high-copy suppressors of *pat1-114*. Cells overexpressing *pac1* or *pac2* do not express *ste11* under starvation conditions and are sterile and sporulation-deficient. The *pac1* gene is essential for cell growth and encodes a double-stranded RNase (Iino et al. 1991). This suggests that overproduction of Pac1p RNase may cause degradation of a specific RNA molecule(s), either mRNA or something else, that is essential for the induction of *ste11* expression. However, such a target of Pac1p RNase has not yet been identified.

The *pac2* gene encodes a novel protein of 235 amino acids that is not essential for cell growth (Kunitomo et al. 1995). Cells defective in *pac2* initiate sexual development under less stringent starvation conditions than do wild-type cells. Consistent with this observation, expression of *ste11* is induced in *pac2*Δ cells only when they are partially starved. Cells overexpressing *pac2* mate and sporule poorly, and expression of *ste11* is not induced by nitrogen starvation in such cells. Genetic analysis has shown that the level of PKA activity and the magnitude of *pac2* expression can complement each other to some extent in the repression of sexual development, but neither of them is epistatic over the other (Kunitomo et al. 1995). This suggests that Pac2p may regulate *ste11* expression via a signaling pathway independent of the cAMP cascade.

The *sak1* gene was isolated as a high-copy suppressor of the meiotic defect in the *cgs1* mutant (Wu and McLeod 1995). It encodes a DNA-binding protein of 87 kD that belongs to the RFX family and is essential for cell growth. The *mei2* gene, which encodes a key regulator of meiosis (see below), is not expressed in the *cgs1* mutant. Overexpression of *sak1* restores the ability to induce *mei2* expression and perform meiosis in the *cgs1* mutant strain. It is unknown whether or not Sak1p regulates transcription of *mei2* directly.

The *esc1* mutation was identified as a suppressor of *pat1-114* (Benton et al. 1993). *esc1* encodes a putative transcription factor with a basic helix-loop-helix motif similar to those of human myogenic differentiation inducers MyoD and Myf-5. Transcription of *esc1* is induced in response to nitrogen starvation but is largely independent of Ste11p. Disruption of *esc1* results in a reduced efficiency of sexual differentiation. Esc1p may promote sexual differentiation by modulating responses to a decrease in the intracellular cAMP level (Benton et al. 1993).

D. cAMP and the Regulation of a Gluconeogenic Gene

The *S. pombe fbp1* gene, which encodes fructose-1,6-bisphosphatase, a key enzyme for gluconeogenesis, is not expressed in the presence of glucose. Mutants have been isolated that express *fbp1* constitutively (Hoffman and Winston 1990). These mutants define ten genes, named *git1–git10*. Seven of them appeared to act to produce a glucose-induced cAMP signal (Byrne and Hoffman 1993). Subsequent analyses have revealed that *git2* is allelic to *cyr1* (Hoffman and Winston 1991), *git6* to *pka1* (Jin et al. 1995), and *git8* to *gpa2* (Nocero et al. 1994). Thus, a reduction of PKA activity clearly leads to the derepression of *fbp1* expression. However, Ste11p is not involved in the control of *fbp1* expression (T. Maeda and M. Yamamoto, unpubl.), suggesting that two regulatory pathways branch downstream from PKA, one leading to *ste11* and the other to *fbp1*.

Depletion of glucose from the medium normally results in an immediate and sharp decrease in the level of cAMP in *S. pombe* cells but does not cause sexual development. In contrast, depletion of nitrogen results in a gradual and moderate decrease in the cAMP level and induces sexual development (Mochizuki and Yamamoto 1992). It is unclear whether this difference simply reflects the pool size of each nutrient or whether there may be an intricate regulatory network for nutrient sensing and sexual development. The latter possibility appears more likely, and Gpa2p is apparently involved in monitoring both glucose and a nitrogen source (Isshiki et al. 1992; Nocero et al. 1994). Thus, it will be particularly interesting to see how the activity of Gpa2p is controlled by nutrients, and whether *ste11* and *fbp1* are regulated by common transcription factors.

III. MATING PHEROMONE SIGNALING

A. Genes Controlling Mating Responses

1. Mating-type Genes

The mating type of an *S. pombe* cell is determined by the DNA sequence at the *mat1* locus. If it is P (*mat1-P*), the cell assumes the h^+ mating type, and if it is M (*mat1-M*), the cell assumes the h^- mating type. Both *mat1-P* and *mat1-M* consist of two divergently transcribed genes, termed *mat1-Pc* and *mat1-Pi* (also called *mat1-Pm*), and *mat1-Mc* and *mat1-Mi* (also called *mat1-Mm*), respectively (Table 1B) (Kelly et al. 1988). The function of *mat1-Pc* and that of *mat1-Mc* are essential for both mating and sporulation, whereas *mat1-Pi* and *mat1-Mi* are dispensable for mating but not for sporulation (Kelly et al. 1988; see below). Expression of all four genes is enhanced under starvation conditions. *mat1-Pc* and *mat1-Mc* show certain levels of basal expression in rich medium (Kelly et al. 1988). Expression of *mat1-Pi* and *mat1-Mi* requires mating pheromone signaling (Nielsen et al. 1992; Willer et al. 1995), which normally operates only under starvation conditions (see below). The product of *mat1-Mc* is an HMG-box protein, and the product of *mat1-Pi* carries a homeobox-like domain (Kelly et al. 1988; Sinclair et al. 1990), suggesting that they probably function as transcription factors. It has been shown that the HMG box in Mat1-Mcp can bind specifically to the AACAAAG heptamer in vitro (Dooijes et al. 1993). This sequence overlaps with that of the TR-box, the recognition site for Ste11p. How *S. pombe* cells use Ste11p and Mat1-Mcp differentially is still an unanswered question. Mat1-Pcp and Mat1-Mip are also likely to be transcription factors, but they show no homology with known proteins.

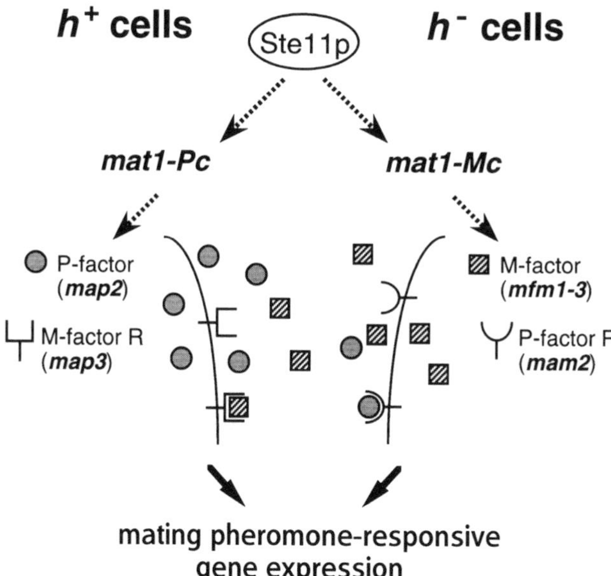

Figure 2 Communication by the mating pheromones. Upon activation by starvation (see Fig. 1), Ste11p activates *mat1-Pc* in h^+ cells and *mat1-Mc* in h^- cells. These genes promote the production of mating pheromones and their receptors in a mating-type-specific manner. The pheromone signal is mediated by Gpa1p and a MAP kinase cascade and induces expression of *mat1-Pi* in h^+ cells, *mat1-Mi* in h^- cells, and both of them in h^+/h^- diploid cells (for details, see Fig. 5).

2. *Mating-type-specific Genes*

The products of *mat1-Pc* and *mat1-Mc* apparently elicit expression of other genes that endow mating-type-specific features to each cell type (Fig. 2). Some of these genes and their products have been characterized through analyses of mating-type-specific sterile mutants. To date, four h^+-specific sterile mutants, called *map1–map4*, and four h^--specific sterile mutants, called *mam1–mam4*, have been isolated (Table 1C) (Egel 1973b; Tanaka et al. 1993; Imai et al. 1997). Mating-type-specific gene expression has been demonstrated for *map2*, which encodes the mating pheromone P-factor; for *map3*, which encodes the M-factor receptor; and for *mam2*, which encodes the P-factor receptor (Kitamura and Shimoda 1991; Tanaka et al. 1993; Imai and Yamamoto 1994). *S. pombe* has three genes encoding M-factor, named *mfm1–mfm3*. These genes were not identified by mating-type-specific sterile mutations because of their redundancy, but their transcription is h^--specific (Davey 1992; Kjaerulff et al. 1994). Another gene transcribed in an h^--specific manner, *sxa2*, encodes a protease that degrades P-factor (Imai and Yamamoto 1992; see below).

map4 and *mam3* encode putative agglutinins (Y. Imai and M. Yamamoto, unpubl.), and they are also expressed in a mating-type-specific manner. *map1* encodes a putative transcription factor related to mammalian serum responsive factor (SRF) and *S. cerevisiae* Mcm1p (Nielsen et al. 1996; Yabana and Yamamoto 1996). It turned out that the function of Map1p is not mating-type-specific in a strict sense (Yabana and Yamamoto 1996): *map1* is transcribed in both h^+ and h^- cells. Whereas Map1p is absolutely indispensable for the expression of h^+-specific genes, it is also required for the full activation of h^--specific genes. It has been suggested that Map1p forms a complex with Mat1-Pcp when it induces h^+-specific gene expression (Nielsen et al. 1996; Yabana and Yamamoto 1996). *mam4* is a homolog of *S. cerevisiae STE14*, which encodes farnesylcysteine carboxyl-methyltransferase, an enzyme involved in the modification of the **a**-factor mating pheromone (Hrycyna and Clarke 1990; Marr et al. 1990; Hrycyna et al. 1991; Sapperstein et al. 1994); Mam4p possesses this enzymatic activity (Imai et al. 1997; see also below). *mam4* is transcribed in both h^+ and h^- cells, despite the mating-type-specific phenotype of the mutant (Imai et al. 1997).

mam2, *mam3*, and *mfm1–mfm3* all carry the sequence AACAAAG in their 5′ upstream region (Kitamura and Shimoda 1991; Davey 1992; Kjaerulff et al. 1994; Y. Imai and M. Yamamoto, unpubl.), and *sxa2* carries this sequence in the opposite orientation (Imai and Yamamoto 1992). This may mean that Mat1-Mcp directly regulates transcription of these genes.

B. The *S. pombe* Mating Pheromones

Because the mating pheromones of *S. pombe* are produced only under starvation conditions and their activities are relatively weak, it was only recently that *S. pombe* cells were shown to secrete substances that can induce elongation of conjugation tubes in cells of the opposite mating type (Fukui et al. 1986a; Leupold 1987).

1. *Structural Features of the Pheromones*

a. M-factor. M-factor, secreted by h^- cells, was purified from culture medium by assaying the ability to induce an increase in the cell volume of h^+ cells. It was then identified as a nonapeptide with its carboxy-terminal cysteine carboxy-methylated and S-farnesylated (Fig. 3, top) (Davey

P-factor (23 amino acids)

Thr-Tyr-Ala-Asp-Phe-Leu-Arg-Ala-Tyr-Gln-Ser-Trp-
Asn-Thr-Phe-Val-Asn-Pro-Asp-Arg-Pro-Asn-Leu

M-factor (9 amino acids)

⌐ S-farnesyl
Tyr-Thr-Pro-Lys-Val-Pro-Tyr-Met-Cys-OCH_3

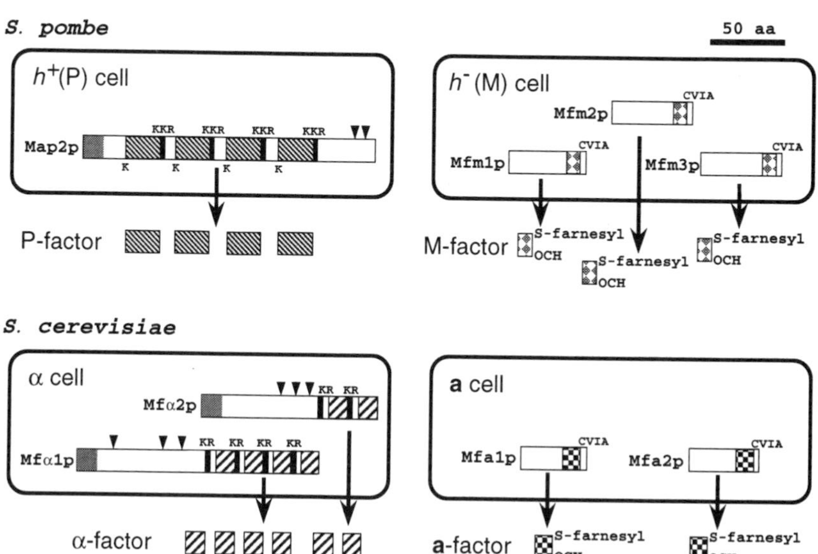

Figure 3 (*Top*) Structures of the *S. pombe* mating pheromones P-factor and M-factor. (*Bottom*) Schematic illustration of the production of mature mating pheromones from their precursors in *S. pombe* and (for comparision) in *S. cerevisiae*. The arrowheads indicate possible N-linked glycosylation sites.

1991, 1992). Carboxy methylation of M-factor is apparently catalyzed by Mam4p (Imai et al. 1997). Each of the three *mfm* genes encodes a precursor containing a single copy of the mature M-factor sequence within it (Fig. 3, bottom) (Davey 1992; Kjaerulff et al. 1994). The *mfm* genes are functionally redundant, and h^- cells become sterile only when all of them are knocked out (Kjaerulff et al. 1994).

b. P-Factor. The structure of P-factor was first deduced from the DNA sequence of the gene thought to encode it. Cloning and sequencing of an h^+-specific sterility gene, *map2*, revealed that the deduced gene product contains four repeats of 23 amino acids flanked by basic residues (Imai and Yamamoto 1994). Two of these repeats have identical sequences, but the other two differ from them in two and three amino acids, respectively. To verify that *map2* encodes P-factor, a substance that has the ability to induce formation of conjugation tubes in h^- *sxa2*Δ cells was purified from the culture medium of h^+ cells overexpressing *map2*. (The *sxa2*Δ strain was used as a tester because it is hypersensitive to P-factor.) Analysis of this substance indicated that it is a peptide of 23 amino acids whose sequence coincides with that of two of the four repeats (Fig. 3, top). Thus, *map2* is clearly the structural gene for P-factor. The purified P-factor and a chemically synthesized peptide of the same sequence gave indistinguishable specific activities in the induction of conjugation tube formation (Fig. 4), suggesting that P-factor is unmodified (Imai and Yamamoto 1994). Although the structure of the *map2* gene suggests that P-factor is heterogeneous, this has not been confirmed biochemically.

The deduced P-factor precursor bears a signal sequence, two putative N-linked glycosylation sites, and four repeats of mature P-factor (Fig. 3, bottom). Each repeat is flanked by Lys and Lys-Lys-Arg residues, suggesting that these basic amino acids provide sites for cleavage. Indeed, a protease Krp1p (originally called Krp; see Footnote 1) has been identified as a processing enzyme for P-factor (Davey et al. 1994). Krp1p is a member of the subtilisin-like family and is homologous to *S. cerevisiae* Kex2p. Krp1p produced in *Xenopus* oocytes cleaves a peptide after dibasic residues, most preferentially -Lys-Arg-, and has been demonstrated to be able to process the P-factor precursor (Davey et al. 1994). Except for Krp1p, enzymes involved in processing P-factor are not known.

c. Homology between the S. pombe *and* S. cerevisiae *Pheromone Systems. S. cerevisiae* **a**-factor is a dodecapeptide with its carboxy-terminal cysteine carboxy-methylated and *S*-farnesylated, like M-factor, whereas *S. cerevisiae* α-factor is a tridecapeptide without modification, like P-factor. Both **a**-factor and M-factor are encoded in a single copy in each gene, whereas both α-factor and P-factor are encoded in multiple copies in each gene, and their precursors share several structural features (Fig. 3, bottom). Thus, although there is no significant sequence homology between the *S. cerevisiae* and *S. pombe* mating pheromones, it appears highly likely that the mating pheromone systems in these distantly related yeast species have stemmed from a common origin. The similarity of the mating-pheromone receptors between the two yeast species (see below) supports this view, and the similarity of the mating pheromone system in *Ustilago maydis* (Bölker et al. 1992) provides additional support.

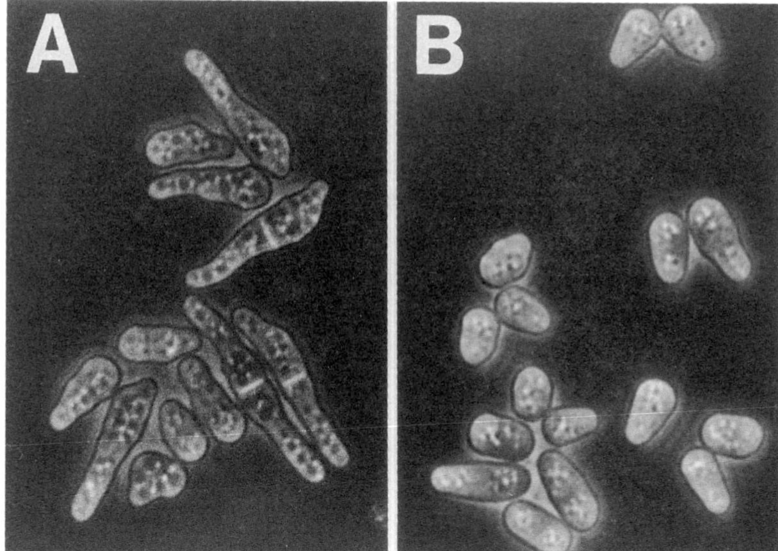

Figure 4 Projection of conjugation tubes in response to P-factor. Cells (h^- *sxa2*Δ) were mixed with chemically synthesized P-factor at 3 units/ml (*A*) or with no P-factor (*B*) under starvation conditions. For experimental details, see Imai and Yamamoto (1994).

2. *Biological Activities of the Pheromones*

a. Cytostatic Activity. The mating pheromones of *S. cerevisiae* arrest the cell cycle of the mating partner in G_1 (Bücking-Throm et al. 1973; for review, see Cross et al. 1988; Marsh et al. 1991). This activity appears to be indispensable for the mating of *S. cerevisiae* cells, because they mate in the presence of rich nutrition and need to ensure that the two conjugants contribute equal amounts of DNA to the zygote. It was unclear, however, whether the *S. pombe* pheromones would have a similar activity, because *S. pombe* cells mate under nitrogen starvation, which by itself causes cells to arrest in G_1. Use of synthetic mating pheromones enabled us and others to answer this question critically. For these experiments, either *cyr1*Δ or *pat1-114* cells were used as the recipient, because they are potentiated toward sexual development and express the mating-pheromone receptors and other essential factors even when they are growing. The results indicated that both P-factor (Imai and Yamamoto 1994) and M-factor (Davey and Nielsen 1994) can arrest the cell cycle of the recipient in G_1. The arrest caused by the pheromone is transient, and the recipient cells resume vegetative growth in several hours, indicating that *S. pombe* has an adaptation mechanism. The cytostatic activity of the pheromones may be important to ensure that the cells arrest properly in G_1, because *S. pombe* can enter a quiescent state from G_2 phase under certain nutritional conditions (Bostock 1970; Costello et al. 1986). Recent analysis has suggested that the pheromones block the onset of S phase by inhibiting the *cdc2* kinase associated with a G_1-specific cyclin and that increasing cell size is likely to provide a mechanism for pheromone adaptation (Stern and Nurse 1997). Unlike *S. cerevisiae*, *S. pombe* pheromones do not appear to inhibit transcription factors for S-phase gene expression (Stern and Nurse 1997).

b. Induction of Gene Expression. In addition to causing G_1 arrest, the mating pheromones induce a new program of gene expression in the recipient cells. *mat1-Pi* and *map4* are expressed only when h^+ cells are exposed to M-factor under nitrogen-starvation conditions (Nielsen et al. 1992; Yabana and Yamamoto 1996), and *sxa2* and *mam3* are expressed only in h^- cells exposed to P-factor under nitrogen starvation (Imai and Yamamoto 1994; Yabana and Yamamoto 1996; see Fig. 5 as discussed in more detail below). Expression of *map2* (Imai and Yamamoto 1994), *mam2* (Xu et al. 1994), *ste6* (Hughes et al. 1994), and *mat1-Mi* (Willer et al. 1995) has been shown to be initially induced by nitrogen starvation and further enhanced by mating-pheromone signaling. Unlike *S. cerevisiae*, *S. pombe* does not turn off the pheromone recognition system when diploid, and pheromone signaling results in the induction of both *mat1-Pi* and *mat1-Mi* expression in h^+/h^- diploid cells. This coexpression in turn induces expression of *mei3*, a gene essential for the initiation of meiosis (McLeod et al. 1987; Willer et al. 1995).

c. Requirement for Pheromone Signaling for Meiosis. Diploid cells that are either homozygous for a *map1* defect or defective in *mat1-Pc* at the heterozygous mating-type locus are unable to sporulate. However, these cells form spores if they have wild-type h^+ cells in their vicinity (Egel 1973b; Leupold et al. 1989). This suggested that P-factor may have the ability to induce sporulation in these mutant cells. The requirement for mating-pheromone signaling for meiosis has been demonstrated unambiguously by using mutants deficient in the signaling pathway. For example, diploid cells defective in *map1* do not sporulate, even with the aid of wild-type h^+ cells, if they are simultaneously defective in *mam2*, the gene for the P-factor receptor (Kitamura and Shimoda 1991). In addition, diploid cells defective in *gpa1*, which encodes a Gα protein coupled to the pheromone receptors (see below), do not initiate meiosis (Obara et al. 1991). Finally, diploid cells defective in either *mam2* or *map3* (encoding the M-factor receptor) are able to sporulate, but diploid cells defective in both of these genes do not enter meiosis (Tanaka et al. 1993). These results confirm that pheromone signaling is essential for meiosis and indiate that either P-factor signaling or M-factor signaling alone is sufficient to induce meiosis. This has been confirmed more directly with the use of synthetic pheromones (Willer et al. 1995). The sporulation deficiency of the *map1* and *mat1-Pc* mutants presumably reflects a lack of both the P-factor and the M-factor receptor in such cells.

3. Degradation of the Mating Pheromones

Analysis of *S. pombe* mutants that are hypersensitive to the mating pheromones has defined three genes: *sxa1*, *sxa2*, and *gap1* (Imai et al. 1991; Imai and Yamamoto 1992). Cells defective in *sxa1* exhibit reduced mating efficiency in an h^+-specific manner, whereas cells defective in *sxa2* do so in an h^--specific manner. Cells defective in *gap1*, which encodes the GTPase-activating protein for Ras1 (see below), mate poorly irrespective of the mating type. The deduced products of *sxa1* and *sxa2* appear to be an aspartyl protease and a serine protease, respectively (Imai and Yamamoto 1992). Cells defective in *sxa2* are two orders of magnitude more sensitive to P-factor than wild-type cells and cannot mate properly, but they mate efficiently if they are placed in the vicinity of wild-type h^- cells. Hence, Sxa2p was suggested to be an extracellular protease that degrades P-factor (Imai and Yamamoto 1992). Subsequent biochemical analysis has confirmed this view and demonstrated that Sxa2p removes the carboxy-terminal leucine residue from P-factor (Ladds et al. 1996).

In *S. cerevisiae*, *BAR1* (*SST1*) encodes an extracellular protease that is secreted by **a**-cells and degrades α-factor (MacKay et al. 1988). As discussed above, α-factor and P-factor may be evolutionarily homologous. This might suggest that Bar1p and Sxa2p would also be homologous. However, this does not appear to be the case, because Bar1p is not a serine protease, like Sxa2p, but an aspartyl protease, like Sxa1.

It is likely that Sxa1p is a protease that degrades M-factor, but there is as yet no biochemical evidence supporting this hypothesis. Expression of *sxa1* is enhanced by nitrogen starvation, but only two- to threefold. It is not enhanced by the M-factor pheromone signal, in contrast to the fact that expression of *sxa2* is induced only when cells are exposed to P-factor (Imai and Yamamoto 1992).

Whereas wild-type h^+ cells extend visible conjugation tubes readily in response to M-factor, wild-type h^- cells extend conjugation tubes very poorly in response to P-factor (Fukui et al. 1986a; Leupold 1987; Imai and Yamamoto 1992; Davey and Nielsen 1994). The effective concentration of P-factor around h^- cells appears to be considerably reduced by the strong activity of Sxa2p protease (Imai and Yamamoto 1994; Ladds et al. 1996).

C. Mating Pheromone Receptors and Receptor-coupled G-Protein

The genes for the P-factor and M-factor receptors were found among mating-type-specific sterility genes. Cells defective in *mam2* are unable to respond to P-factor and hence are sterile in an h^- background, whereas cells defective in *map3* are unable to respond to M-factor and hence are sterile in an h^+ background. The predicted products of *mam2* and *map3* are putative serpentine membrane-spanning proteins (Kitamura and Shimoda 1991; Tanaka et al. 1993). Mam2p shows similarity to *S. cerevisiae* Ste2p, the α-factor receptor, whereas Map3p shows similarity to *S. cerevisiae* Ste3p, the **a**-factor receptor. These and additional observations strongly suggest that Mam2p and Map3p are the P-factor and M-factor receptors, respectively (Kitamura and Shimoda 1991; Tanaka et al. 1993). However, the binding of the pheromones to their receptors has not yet been demonstrated biochemically. The similarity between the *S. pombe* and *S. cerevisiae* receptors, although confined largely to the membrane-spanning domains, again supports the view that α-factor is akin to P-factor and **a**-factor is akin to M-factor.

Because Mam2p and Map3p are apparent seven-transmembrane-domain receptors, it seemed likely that they would transmit a signal to a heterotrimeric G-protein(s). The *S. pombe gpa1* gene was isolated in a hybridization screen using rat Gi1α and Gxα cDNAs as the probes, and subsequent analysis suggested strongly that it encodes the α-subunit of the G-protein coupled to Mam2p and Map3p (Obara et al. 1991); thus, its function seems to be fully distinct from that of the *gpa2*-encoded Gα involved in transmitting the nutritional signal (see Section II.A.4). Disruption of *gpa1* does not affect cell growth but renders cells completely sterile, regardless of mating type. Moreover, cells carrying an activated allele of *gpa1*, *gpa1-Q244L*, display a morphological alteration reminiscent of conjugation-tube formation if they are placed under nitrogen starvation. Notably, this morphological response can be seen in both h^+ and h^- heterothallic cells with no nearby mating partner, suggesting that activation of Gpa1p can substitute for stimulation by mating pheromone and, furthermore, that Gpa1p is coupled to both the P-factor and M-factor receptors (Fig. 5). It is intriguing that Gα appears to be responsible

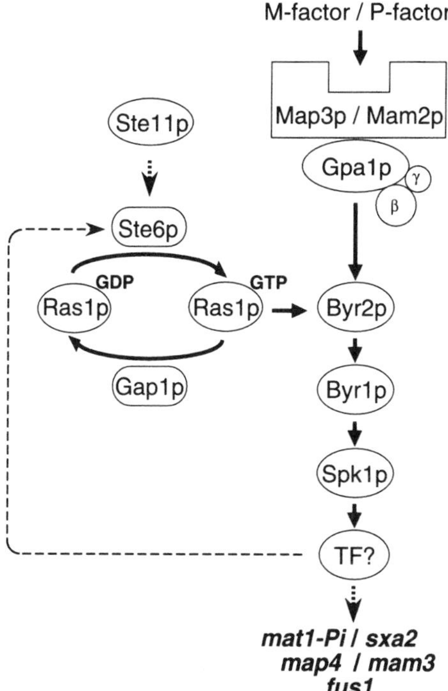

Figure 5 Mating-pheromone signaling pathway in *S. pombe*. The MAP kinase cascade constituted by Byr2p, Byr1p, and Spk1p is under the regulation both of the G-protein coupled to the pheromone receptors and of the Ras protein. Both pathways are promoted by starvation, which leads to an activation of Ste11p (see Fig. 1), which then induces expression both of the Ras-activator Ste6p (as shown here) and of the pheromones and their receptors (through its induction of *mat1-Pc* and *mat1-Mc*; see Figs. 1 and 2). Pheromone signaling can also boost itself by enhancing expression of *ste6*. The putative transcription factor downstream from Spk1 has not been identified. Expression of the genes listed in the panel is inducible almost exclusively in response to the mating pheromone signal under starved conditions. In contrast, expression of *mat1-Mi, map2,* and *mam2*, although not illustrated here, can be induced initially by nutritional starvation and further enhanced by the pheromone signal, like that of *ste6*.

for transmission of the pheromone signal in *S. pombe*, because a complex of Gβ and Gγ (Ste4p and Ste18p) has been shown to transmit the signal in *S. cerevisiae* (Dietzel and Kurjan 1987; Miyajima et al. 1987; Whiteway et al. 1989).

The immediate effector of Gpa1p has not yet been identified, although a MAP-kinase cascade is known to function downstream (see below). The β and γ subunits that presumably cooperate with Gpa1p have also not yet been identified. Recently, an *S. pombe* gene encoding a putative Gβ has been isolated in two independent screens, one using polymerase chain reaction (PCR) (Kim et al. 1996) and the other a rechecking of the *git* genes discussed above (C. Hoffman, pers. comm.). It is not yet conclusive whether this Gβ cooperates with Gpa1p, Gpa2p, or an as yet unidentified Gα.

D. The Ras Pathway in Fission Yeast

1. *Ras1p, the* S. pombe *Ras Homolog*

S. pombe has a single homolog of the mammalian *ras* proto-oncogene, named *ras1* (Fukui and Kaziro 1985; Nadin-Davis et al. 1986). *ras1* is not essential for cell growth, and *ras1*-disrupted cells grow mitotically at the same rate as wild-type cells. However, *ras1*Δ cells are shorter and rounder than wild-type cells. Haploid *ras1*Δ cells do not mate at all, and diploid *ras1*Δ cells constructed by cell fusion sporulate very inefficiently (Fukui et al. 1986b; Nadin-Davis et al. 1986). Cells defective in *ras1* can secrete mating pheromones but are unable to respond to them, whereas cells carrying $ras1^{Val17}$, an activated allele, are hypersensitive to the pheromones and readily extend conjugation tubes (Fig. 6) (Fukui et al. 1986b; Nadin-Davis et al. 1986). Expression of mating pheromone-responsive genes, such as *mat1-Pi*, is hardly seen in *ras1*Δ cells but is induced excessively in $ras1^{Val17}$ cells (Nadin-Davis and Nasim 1990; Nielsen et al. 1992). These observations have led to the proposal that Ras1p modulates sensitivity to the mating pheromones (Fig. 5) (for review, see Hughes and Yamamoto 1993). Although $ras1^{Val17}$ cells are hypersensitive to the pheromones, most of them fail to conjugate properly, eventually giving a very low mating frequency. *ras1* has been shown to be allelic to *ste5*, a sterility gene defined previously (Girgsdies 1982; Lund et al. 1987).

2. *Regulatory Factors for Ras1p*

Biochemical analyses of mammalian Ras proteins have shown that they are activated when GTP is bound. Bound GTP is hydrolyzed to GDP by the intrinsic GTPase activity of Ras, and the resulting GDP-bound Ras is inactive. The GTPase activity is strengthened by the function of proteins such as GAP (GTPase-activating protein) or NF1. On the other hand, the activation of Ras (i.e., conversion of the GDP-bound to the GTP-bound

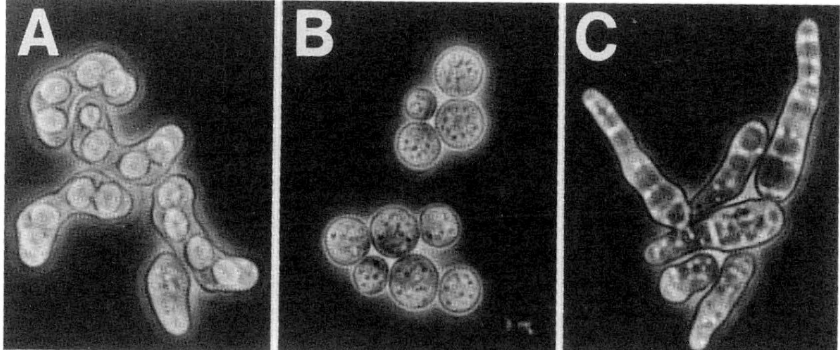

Figure 6 Phenotypes of *ras1* mutants. (*A*) Asci formed by wild-type cells placed under nutritional starvation. (*B*) Cells defective in *ras1*, placed under the same conditions. They are completely inert in mating and are deformed in cell morphology. (*C*) Cells carrying the *ras1*^{val17} mutation, which constitutively activates Ras1p. They form exaggerated conjugation tubes and ultimately mate only poorly.

form) is catalyzed by GEFs (guanine nucleotide exchange factors) (for review, see Barbacid 1987; Bourne et al. 1990, 1991).

The putative GAP for Ras1 (the product of *gap1;* Table 1D) emerged from the analysis of mutants hypersensitive to the mating pheromones, as described above (Imai et al. 1991). The same gene was isolated independently as a high-copy suppressor of the mating deficiency of the *ras1*^{Val17} mutant and was named *sar1* (Wang et al. 1991a). The product of *gap1/sar1* is 25% identical to *S. cerevisiae* Ira1p, 20% identical to human NF1, and 20% identical to mammalian GAP in the core domain. These related proteins are known to enhance the GTPase activity of Ras proteins. A *gap1/sar1*-disrupted strain is hypersensitive to mating pheromone and exhibits phenotypes quite similar to those of *ras1*^{Val17}. However, the *gap1*Δ *ras1*Δ strain is completely sterile like *ras1*Δ. These results indicate that Gap1p is an upstream factor that negatively regulates Ras1p activity and is unlikely to serve as an effector of Ras1p, which is a role postulated for GAP in the mammalian Ras system (McCormick 1989).

The putative GEF for Ras1p was identified in a study of *S. pombe ste* genes (Hughes et al. 1990). The *ste6* gene product was found to be homologous to the *S. cerevisiae CDC25* and *SDC25* gene products, which were the first Ras activators characterized. Genetic analysis suggested that Ste6p functions upstream of Ras1p (Hughes et al. 1990). Expression of *ste6* is regulated by Ste11p and is further enhanced by mating-pheromone signaling (Hughes et al. 1994). This indicates that the pheromone signal can be boosted by a positive feedback loop (Fig. 5).

The *ste6*Δ strain is defective in mating but is scarcely impaired in sporulation, and its cell morphology is not deformed (Hughes et al. 1990). This contrasts with complete loss of *ras1* function, which causes deformed cell shape and poor sporulation, and suggests that the *ste6*Δ strain still has sufficient Ras1p activity to support normal cell morphology and competence to sporulate, but not to allow mating. This observation can be interpreted in two ways: Ras1p may have a basal activity even in the complete absence of its GEF or an *S. pombe* may have another GEF in addition to Ste6p. An interesting observation in this regard is that *ral2* defects in which cause sterility, poor sporulation, and deformed cell morphology like those of *ras1*Δ apparently function upstream of *ras1* (Fukui et al. 1989; Imai et al. 1991). This suggests that Ral2p may be a second GEF for Ras1p, but no biochemical data supporting this are yet available, and no homolog of Ral2p has been reported to date in other species.

E. The MAP-Kinase Cascade in Pheromone Signaling

1. Genes Encoding the Kinases of the Cascade

byr1 and *byr2* (Table 1D) were isolated as high-copy suppressors of the low sporulation frequency of *ras1*Δ/*ras1*Δ diploid cells (Nadin-Davis and Nasim 1988; Wang et al. 1991b). *byr1* is allelic to *ste1* and *byr2* is allelic to *ste8*, genes identified previously by sterile mutations (Thuriaux et al. 1980; Michael and Gutz 1987). Cells defective in *byr1* or *byr2* are sterile and do not sporulate. The *byr1* gene product shows 43% identity to vertebrate MAP kinase kinase (MAPKK) in the catalytic domain (Crews et al. 1992; Seger et al. 1992; Kosako et al. 1993; Wu et al. 1993), whereas the *byr2* gene product shows 35% identity to vertebrate MAP kinase kinase kinase (MAPKKK) (Lange-Carter et al. 1993; Nishida and Gotoh 1993). In addition, *spk1*, which was isolated in a screening for genes whose overexpression confers staurosporine resistance (Toda et al. 1991), but later turned out not to have such an ability (Shimanuki et al. 1995), has been shown to encode a homolog of MAP kinase (Toda et al. 1991). An *spk1*Δ strain is also sterile and sporulation-deficient. Spk1p is tyrosine-phosphorylated during mating-pheromone signaling, and this phosphorylation depends on Byr1p (Gotoh et al. 1993). These findings suggest that a MAP kinase cascade comprising Byr2p, Bry1p, and Spk1p operates downstream from Ras in fission yeast (Fig. 5), as is the case in mammals (for review, see Marshall 1995). Physical interactions of Byr2p with mammalian Ras (Van Aelst et al. 1993) and *S. pombe* Ras1 (Masuda et al. 1995) have been demonstrated.

The MAP-kinase cascade of *S. pombe* appears to be functionally homologous to those in other species in many respects. In *S. cerevisiae*, a MAP-kinase cascade functions downstream from the mating-pheromone receptors and the G-protein coupled to them (for review, see Herskowitz 1995). Ste11p(MAPKKK), Ste7p(MAPKK), and Fus3p/Kss1p(MAPKs) constitute this cascade (Gartner et al. 1992; Errede et al. 1993; Neiman and Herskowitz 1994). Some of these factors are exchangeable with their counterparts in *S. pombe* (Styrkársdóttir et al. 1992; Neiman et al. 1993). Moreover, ERK2, a member of the mammalian MAP kinase family, has been shown to complement partially an *spk1* mutation (Neiman et al. 1993). *Xenopus* MAP kinase can also partially complement *spk1*, and its overexpression disturbs mating and sporulation in wild-type *S. pombe* cells, as does overexpression of *spk1* (Gotoh et al. 1993). Although expression of mammalian MAPKK alone has no effect, this protein can complement a *byr1* mutation if mammalian Raf1 is provided simultaneously as a MAPKKK in the mutant cells (Hughes et al. 1993).

2. *Regulators of the MAP-Kinase Cascade*

The results described above indicate that Ras1p is a regulator of the Byr2-Byr1-Spk1 MAP-kinase cascade. Furthermore, genetic evidence indicates that Gpa1p (i.e., the Gα coupled to the mating-pheromone receptors) also functions upstream of this MAP-kinase cascade (Neiman et al. 1993; Xu et al. 1994). The *mam2* gene (encoding the P-factor receptor) is poorly expressed in cells defective in *ras1* or *gpa1*, but its expression can be rescued by overexpression of either *byr1* or *byr2*. Curiously, however, expression of *mam2* does not depend on the function of *spk1*, suggesting that Byr2p-Byr1p regulates *mam2* expression via a factor other than Spk1p (Xu et al. 1994).

All in all, *S. pombe* appears to provide an interesting experimental system in which a MAP-kinase cascade is dually controlled by Ras and by a heterotrimeric G-protein (Fig. 5). It seems clear that the mating pheromones are the ligands that activate the G-protein in this system, but it is unclear whether Ras1p is mediating an external signal, as mammalian Ras does (Marshall 1995), except that it in part mediates the response to nutritional starvation through expression of *ste6*. It also remains to be answered whether the control of Byr2p by Gpa1p is direct or indirect.

F. Cell Morphology and Mating Ability

It is clear that Ras1p and the MAP-kinase cascade constitute a pathway, so it seems paradoxical that cells defective in *byr2*, *byr1*, or *spk1* are morphologically normal, whereas *ras1Δ* cells are deformed. This suggests that Ras1p controls another pathway, which affects cell morphology. To characterize this pathway, mutants that have phenotypes similar to those of *ras1Δ* have been isolated. The *ral1–ral4* mutants (Fukui and Yamamoto 1988) and the *scd1* and *scd2* mutants (Chang et al. 1994) have round cell morphology and are sterile (Table 1D). The possible function of Ral2p as a GEF for Ras1p was discussed above.

Sequence analysis of *scd1* and *scd2* revealed that they are homologous to *S. cerevisiae CDC24* and *BEM1*, respectively (Chang et al. 1994), and sequence analysis of the *ral* genes showed that *ral1* and *ral3* are identical to *scd1* and *scd2*, respectively (C. Kitayama and M. Yamamoto, unpubl.). *S. cerevisiae* Cdc24p is a GEF for the small GTP-binding protein Cdc42p, a member of the Rho/Rac family (Miyamoto et al. 1987, 1991; Johnson and Pringle 1990; Zheng et al. 1994). *S. cerevisiae* Bem1p is a protein with two SH3 (*src* homology 3) domains that is required for the establishment of cell polarity both during bud formation and during mating (Chenevert et al. 1992). A homolog of *S. cerevisiae CDC42*, *cdc42* (see Footnote 1), has been isolated from *S. pombe*. Cells defective in *cdc42* are inviable and exhibit a round cell morphology as a terminal phenotype (Miller and Johnson 1994). Two-hybrid analyses suggest that Ras1p, Scd1p, Scd2p, and Cdc42p interact physically, leading to the proposal that Scd1p is a second effector of Ras1p (Chang et al. 1994). In *S. cerevisiae*, Rsr1p (Bender and Pringle 1989; Chant and Herskowitz 1991), a protein of the Ras superfamily that is most closely related to a human *ras* suppressor Rap/K-Rev (Kawata et al. 1988; Pizon et al. 1988; Kitayama et al. 1989), has been shown to interact directly with Cdc24p (Ruggieri et al. 1992; Zheng et al. 1995). Thus, it is likely that *S. pombe* Ras1p performs two types of functions that are carried out separately by Ras and Rap/K-Rev in other species.

S. pombe shk1/pak1, isolated by PCR (Marcus et al. 1995; Ottilie et al. 1995), encodes a homolog of the *S. cerevisiae STE20* and *CLA4* gene products and the mammalian p65^{PAK} protein kinase, which are believed to be effectors of Cdc42 in their respective systems (Manser et al. 1994; Cvrcková et al. 1995; Simon et al. 1995; Zhao et al. 1995). Spores carrying an *shk1Δ* allele germinate, divide one to several times, and cease proliferation with roundish cell shape like that of *cdc42Δ* cells. Complex formation between Shk1p/Pak1p and Cdc42p was demonstrated, and overexpression of *shk1/pak1* alleviated the mating disability of cells expressing a dominant-negative form of Cdc42p. These results suggest that Shk1p functions downstream from Cdc42p in *S. pombe*.

Because overexpression of *byr1* or *byr2* does not restore the ability to mate in either *ras1*Δ cells or any of the morphological mutants discussed above, it is likely that the second pathway downstream from Ras1p has a function essential for mating, in addition to its function in the maintenance of cell morphology. It is an attractive hypothesis that loss of polarity disturbs both the normal cylindrical cell shape of vegetative cells and the proper projection of conjugation tubes in *S. pombe*, the latter being essential for mating.

The *fus1* mutant achieves cell contact and agglutination during mating, but the cell walls separating the mating partners are not degraded and cytoplasmic fusion does not take place (Bresch et al. 1968; Egel 1973a). Fus1p carries a proline-rich potential SH3-binding site (Petersen et al. 1995) and an FH2 (formin homology 2) domain, the latter of which has been found in cytokinesis/cell polarity proteins such as *S. pombe* Cdc12p, *S. cerevisiae* Bni1p, *Drosophila Diaphanous* and *Cappuccino*, and vertebrate formin (Kohno et al. 1996). Expression of *fus1* is induced by nitrogen starvation and requires a pheromone signal in both h^+ and h^- cells and is regulated by Ste11p (Petersen et al. 1995). Although the *fus1* mutant has a phenotype very similar to that of the *S. cerevisiae fus1* mutant, the two gene products share no homology and appear to have different roles in the process of cell fusion. Interestingly, however, *S. cerevisiae* Fus1p is a typical membrane protein carrying an SH3 domain, whereas *S. pombe* Fus1p is a potential SH3-binding protein. A Fus1p-fusion protein has been shown to localize to the projection tip during conjugation in *S. pombe* (Petersen et al. 1995).

G. Other Genes Affecting the Pheromone Signaling Pathway

Like *byr1* and *byr2*, *byr3* was identified as a high-copy suppressor of the low sporulation frequency of *ras1*Δ/*ras1*Δ diploid cells (Xu et al. 1992). The *byr3* gene product carries seven repeats of the $CX_2CX_4HX_4C$ zinc finger motif and shows structural similarity to human CNBP (cellular nucleic-acid-binding protein). Byr3p binds to DNA in vitro, and a *byr3*Δ strain has a reduced ability to mate. How Byr3p interacts with the pheromone signaling pathway remains unclear.

Overexpression of *byr4* can also partially suppress the sporulation defect of the *ras1* mutant (Song et al. 1996). Byr4p shows homology only with the product of *S. cerevisiae* YJR053W identified in the genome project. Unlike *byr1*, *byr2*, and *byr3*, *byr4* is essential for vegetative growth. Cells with no Byr4p arrest growth with multiple septa and frequently show nuclei separated unequally, suggesting that *byr4* is required for proper karyokinesis. Strong overexpression of *byr4* inhibits late steps of cytokinesis, including contraction of the actin ring, septation, and actin relocation. Because overexpression of *byr4* causes more severe effects in the *ras1* and *scd1* mutants than in the wild type, a link between the function of these genes and cytokinesis has been implicated (Song et al. 1996).

zfs1 was isolated as a high-copy suppressor of the sterility caused by overexpression of Pac1 double-stranded RNase. The *zfs1* gene product is similar to a mouse growth factor-inducible nuclear protein Nup475. It carries two $CX_8CX_5CX_3H$ zinc fingers and is apparently located in nuclei. The *zfs1*Δ strain has a reduced ability to mate and sporulate, which can be suppressed partially by overexpression of *gpa1*, *ras1*, *byr1*, or *byr2* (Kanoh et al. 1995). The function of *zfs1* appears to be required for the full activation of the pheromone signaling pathway, but it is unknown how Zfs1p achieves this.

The *MRA1* gene was isolated as a high-copy-number suppressor of the mating deficiency caused by a specific mutation in the effector region of Ras1p (Hakuno and Yamamoto 1996). Mra1p shows no specific feature in structure, but its apparent homologs exist in *S. cerevisiae* and rice. Mra1p is essential for cell growth, is apparently required for the promotion of mating, but is not involved in the maintenance of cell morphology. The two functions of Mra1p could be separated by mutations, suggesting that it may be a bifunctional protein. Overexpression of *mra1* can suppress the mating inefficiency caused by either overexpression of *gap1* or loss of function of *zfs1*. Thus, it is suspected that Mra1p may be a novel downstream factor of Ras1p, although further evidence is required to estabish this conclusion (Hakuno and Yamamoto 1996).

IV. MEIOSIS AND SPORULATION

A. Entry to Meiosis: An Overview

Zygotes of *S. pombe* generated by conjugation proceed directly to the meiotic pathway under starvation conditions (Egel 1973a). They can proliferate as diploids if shifted to rich medium immediately after conjugation, and these diploid cells undergo meiosis when they exhaust the available nutrients. In either case, meiotic cells follow essentially the same pathway. They arrest transiently in G_1 and then initiate one round of DNA replication called premeiotic DNA synthesis. This doubles the DNA content of each cell to the 4C level. The cells then undergo high-frequency meiotic recombination between homologous chromosomes and perform two consecutive nuclear divisions, called the first (meiosis I) and second (meiosis II) meiotic division. Meiosis I is reductional (i.e., homologous chromosomes are segregated), whereas meiosis II is equational (i.e., sister chromatids generated by premeiotic DNA synthesis are separated). Meiosis II results in the production of four nuclei, each containing a 1C (haploid) content of DNA, which are subsequently encapsulated and form mature spores.

Mating and sporulation are separate, differentially regulated events in *S. cerevisiae*. However, these two events are regulated largely by common signaling pathways in *S. pombe*, and we can even consider them to be the former and latter halves of a single process. We have discussed above the signaling pathways controlling conjugation, and we briefly summarize here the regulation common to the two events.

1. Nutritional Signaling

Diploid cells initiate meiosis only under poor nutritional conditions, where the level of intracellular cAMP is reduced (Mochizuki and Yamamoto 1992). Expression of *ste11*, induced in response to the decrease in cAMP level, is essential for the initiation of meiosis. Essentially the same signaling pathway appears to be used to recognize and respond to nutrient limitation in haploid and diploid cells (see Fig. 1).

2. Mating Pheromone Signaling

Pheromone signaling stimulated by either P-factor or M-factor is essential for the initiation of meiosis, as discussed above. Unlike the situation in *S. cerevisiae*, the pheromone-recognition pathway is not shut off in *S. pombe* diploid cells. Although it is unclear whether diploid cells stimulate themselves by an autocrine or a paracrine mechanism, it seems clear that the same intracellular signaling pathway is activated in haploid and diploid cells. This activation eventually induces expression of *mat1-Pi* in h^+ cells, of *mat1-Mi* in h^- cells, and of both in diploid cells (see Fig. 5) (Kelly et al. 1988; Willer et al. 1995). Coexpression of *mat1-Pi* and *mat1-Mi* enables cells to express *mei3* and to enter the meiotic pathway (see below).

B. Molecules Regulating the Initiation of Meiosis

1. Genetic Analysis

Mutants of *S. pombe* that are able to mate but cannot initiate or complete meiosis have been isolated and classified genetically into five groups, *mei1–mei4* and *mes1* (Table 1E) (Bresch et al. 1968). The mutations are recessive. *mei1*, *mei2*, and *mei3* mutants arrest before premeiotic DNA synthesis, the *mei4* mutant arrests before meiosis I, and the *mes1* mutant arrests before meiosis II. We discuss the *mei1–mei3* mutants here and the other mutants later in this review.

As described in Section II.A, the *pat1-114* mutant initiates ectopic meiosis at the restrictive temperature. Although the *mei1*, *mei2*, and *mei3* mutants all arrest before premeiotic DNA synthesis and are indistinguishable in this respect, differences became evident when these mutations were combined with *pat1-114*. The *pat1-114 mei1* and *pat1-114 mei3* double mutants could perform ectopic meiosis, like a *pat1-114* single mutant, but the *pat1-114 mei2* double mutant was unable to do so. Furthermore, the *pat1-114 mei2* double mutant could even grow at temperatures at which the *pat1-114* single mutant halts growth and initiates meiosis. This indicates that loss of *mei2* function can neutralize the conspicuous effects of the *pat1-114* mutation (Beach et al. 1985; Iino and Yamamoto 1985a,b). These observations led to the hypothesis that the final decision to enter meiosis is made by Pat1p as a negative factor and Mei2p as a positive factor and that the functions of Mei1p and Mei3p are required at an earlier stage (Iino and Yamamoto 1985b). Indeed, *mei1* has been found to be allelic to *mat1-Pi* (Egel 1984), and Mei3p has been shown to be an inhibitor of Pat1p function (McLeod and Beach 1988; Li and McLeod 1996). Loss of function of either *mat1-Pi* or *mat1-Mi* blocks induction of *mei3* expression and hence results in the *mei* phenotype (see below).

2. *The Pat1p Kinase and Its Inhibitor, Mei3p*

pat1 (denoted as *ran1* in some publications) was cloned by chromosome walking and complementation (Beach et al. 1985). The deduced gene product was found to have homology with serine/threonine protein kinases (McLeod and Beach 1986), and the protein kinase activity of Pat1p was demonstrated by autophosphorylation of the bacterially produced protein in vitro (McLeod and Beach 1988), and later by using Ste11p (Li and McLeod 1996) and Mei2p (Watanabe et al. 1997) as substrates (the biological significance of this phosphorylation is discussed below). *mei3* is expressed only when *mat1-Pi* and *mat1-Mi* are coexpressed, and artificial expression of *mei3* has been shown to lead to the initiation of ectopic meiosis as seen in the *pat1-114* mutant (McLeod et al. 1987; Willer et al. 1995). Pat1p and Mei3p appear to form a stable complex in vivo (as judged from their coimmunoprecipitation from cell lysates), and bacterially produced Mei3p can inhibit the kinase activity of Pat1p in vitro (McLeod and Beach 1988). Furthermore, Mei3p carries a sequence that is similar to the sequences targeted by Pat1p but definitely distinct from them in that it has arginine instead of serine or threonine at the deduced phosphorylation site (Li and McLeod 1996). This sequence became phosphorylatable when the arginine residue was substituted by serine, indicating that Mei3p is likely to employ a pseudosubstrate mechanism for its inhibition of Pat1p (Li and McLeod 1996). These findings suggest strongly that Mei3p functions as an inducer of meiosis by inhibiting Pat1p kinase activity. In addition, they account well for the two major physiological prerequisites for meiosis, namely, nutritional starvation and diploidy: Expression of *mat1-Pi* and *mat1-Mi* requires nutritional starvation, and coexpression of the two genes is possible only in h^+/h^- diploid cells. *mei3* is expressed in response to this coexpression, and Mei3p inactivates Pat1p, which is constitutively expressed during the mitotic cell cycle (McLeod and Beach 1986). This scheme indicates that inactivation of Pat1p kinase is a key step for the initiation of meiosis under physiological conditions, although the *pat1* mutations were originally identified as inducers of uncontrolled meiosis (Fig. 7).

3. *Mei2p, a Critical Inducer of the Meiotic Cell Cycle*

mei2 is essential for entry into meiosis (Fig. 8) (Bresch et al. 1968) and is transcriptionally activated by Ste11p under nitrogen starvation (Shimoda et al. 1987; Watanabe et al. 1988; Sugimoto et al. 1991). Loss of *mei2* function completely suppresses *pat1*-driven ectopic meiosis (Beach et al. 1985; Iino and Yamamoto 1985a,b; Watanabe et al. 1988). This ectopic meiosis can also be suppressed by loss of *ste11* function, by any mutation leading to the constitutive activation of PKA, or by overexpression of *pac1* or *pac2*, as discussed in Section II.A–C. Significantly, *mei2* expression is completely blocked in the *ste11* mutant and is severely impaired under the other conditions mentioned.

Pat1p kinase phosphorylates Mei2p on two residues, namely, Ser-438 and Thr-527, both in vivo and in vitro (Watanabe et al. 1997). Notably, expression of mutant Mei2p carrying alanine in these two positions can provoke meiosis without inactivation of Pat1p, indicating that Mei2p is the critical target of Pat1p in the induction of meiosis (Watanabe et al. 1997). These results lead to the view that the function of Mei2p is a key determinant for commitment of cells to premeiotic DNA synthesis and subsequent meiotic events, whereas Pat1p kinase has the role of inactivating Mei2p expressed either inappropriately during the mitotic cell

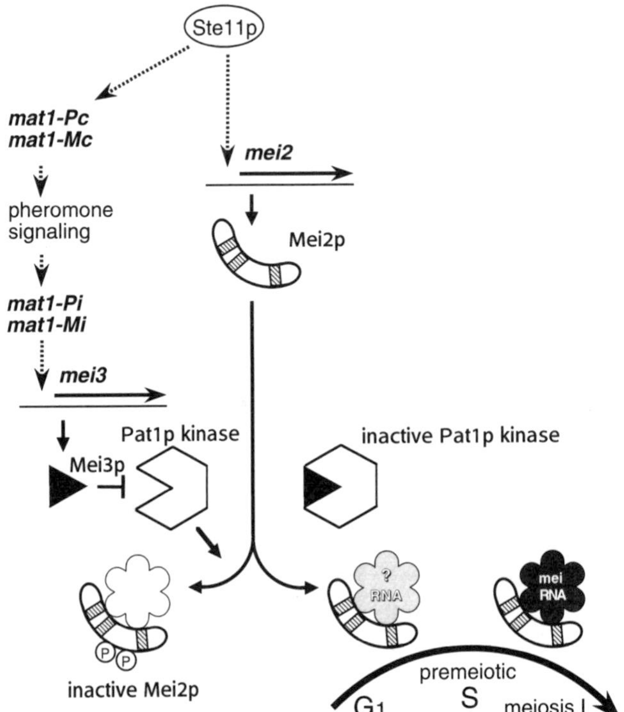

Figure 7 Regulation of the initiation and progression of meiosis. Nutritional limitation induces the expression of *ste11* (see Fig. 1), which in turn induces the expression of both *mei2* and (via *mat1-Pc*, *mat1-Mc*, and the pheromone-signaling pathway) *mat1-Pi* and *mat1-Mi*. Coexpression of *mat1-Pi* and *mat1-Mi* in turn results in the induction of *mei3* expression. Mei3p inhibits the protein kinase activity of Pat1p, which allows Mei2p to express its function. Mei2p has three RNA-recognition motifs and functions in the promotion of both premeiotic DNA synthesis and meiosis I. It cooperates with meiRNA (the *sme2* gene product) to promote meiosis I and may cooperate with a distinct RNA species to promote premeiotic DNA synthesis.

cycle or in starved haploid cells (Fig. 7). The activity of Mei2p thus appears to be doubly up-regulated by nutritional starvation, i.e., both by activation of gene expression and by inactivation of a kinase that blocks its function.

Mei2p appears to function at two points in the induction of meiosis. A *mei2*Δ strain apparently always stays in the mitotic cell cycle and never initiates premeiotic DNA synthesis. However, an h^+/h^- diploid strain homozygous for *mei2-33*, a temperature-sensitive allele of *mei2*, undergoes premeiotic DNA synthesis and then arrests before meiosis I if shifted to nitrogen-free medium at semirestrictive temperature (Watana-

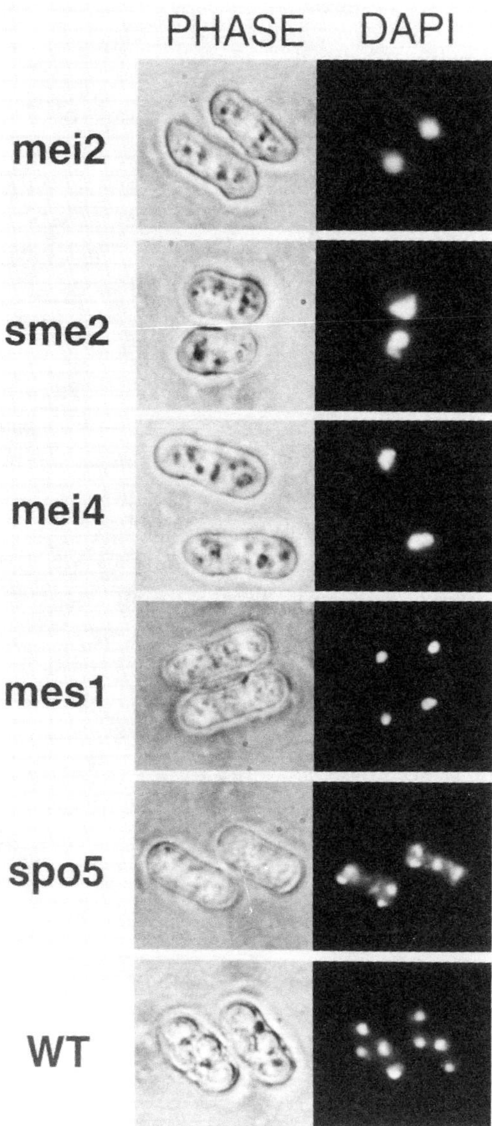

Figure 8 Phenotypes of meiosis-defective mutants. Diploid cells lacking the indicated gene function were incubated under starvation conditions, and the stage of meiosis in which they arrested was determined by phase-contrast and fluorescence microscopy. Cells were stained with the DNA-binding dye DAPI (4′,6-diamidino-2-phenylindole) to visualize their nuclei.

be and Yamamoto 1994). The same strain does not initiate premeiotic DNA synthesis at fully restrictive temperature. Thus, Mei2p appears to have an essential function in the promotion of meiosis I, in addition to its function in inducing premeiotic DNA synthesis.

4. Characterization of Mei2p as an RNA-binding Protein

a. RNA-binding Ability of Mei2p. mei2 has a coding capacity of 750 amino acids (Watanabe et al. 1988). Careful inspection suggested that Mei2p carries three RNA-recognition motifs (RRMs) (Birney et al. 1993; Watanabe and Yamamoto 1994; the two groups assigned different sets of two RRMs). An RNA molecule that can bind specifically to Mei2p was identified in screening an *S. pombe* genomic library for high-copy suppressors of *mei2-33*. A suppressor was isolated that can efficiently restore progression of meiosis I in the mutant strain, and the gene responsible for the suppression was delimited and named *sme2*. Surprisingly, *sme2* carries only small open reading frames (ORFs) that can be mutated without affecting gene function, suggesting that this gene may function via an RNA product (Watanabe and Yamamoto 1994). *sme2* is transcribed into two types of polyadenylated transcripts, which are 508/507 and 440/439 bases in size (exclusive of the poly[A] tail) and are collectively denoted as meiRNA. The large and small meiRNA species are identical except that the large one has an additional 3′ sequence. Labeled meiRNA molecules synthesized in a cell-free transcription system were shown to bind to Mei2p in vitro (Watanabe and Yamamoto 1994). They bound to Mei2p specifically in the presence of carrier RNA, but it was also noted that Mei2p has the ability to bind RNA nonspecifically. Anti-Mei2p antibodies precipitated meiRNA with Mei2p from an extract of meiotic *S. pombe* cells, indicating that meiRNA indeed forms a complex with Mei2p in vivo (Watanabe and Yamamoto 1994).

b. Requirement of meiRNA for Meiosis I. Various single-base replacements in meiRNA scarcely affected its function (Y. Watanabe and M. Yamamoto, unpubl.), but substitution of the middle region of *sme2* with a marker gene effectively inactivated it. The resulting *sme2Δ* strain, which lacks meiRNA, was not impaired in vegetative growth or premeiotic DNA synthesis but was unable to perform meiosis I (Fig. 8) (Watanabe and Yamamoto 1994). This indicates that meiRNA performs a function required specifically for meiosis I, in cooperation with Mei2p. Consistent with this interpretation, overexpression of *sme2* could not restore premeiotic DNA synthesis in a *mei2Δ* strain.

c. Possible Involvement of Other RNAs in Meiosis. The above results show clearly that an RNA molecule is essential for the progression of meiosis, an unanticipated result. However, it is also clear that the function of meiRNA is confined to promotion of meiosis I and is not relevant to premeiotic DNA synthesis. Thus, it is interesting that cells arrest prior to premeiotic DNA synthesis, not prior to meiosis I, if a single mutation is introduced into the carboxy-terminal RRM of Mei2p so that it is no longer able to bind meiRNA (Watanabe and Yamamoto 1994). This suggests that the RNA-binding ability of Mei2p is also indispensable for its function in induction of premeiotic DNA synthesis and that an RNA molecule other than meiRNA is involved in this function (see Fig. 7). This intriguing possibility remains to be substantiated. In addition, it is not yet clear what kind(s) of molecular function(s) Mei2p will execute in cooperation with RNA molecules in the course of meiosis.

5. *Targets of Pat1p Kinase Other Than Mei2p*

Pat1p phosphorylates Ste11p in vitro (Li and McLeod 1996). Although the physiological significance of this phosphorylation is yet to be demonstrated, it is conceivable that the phosphorylated form of Ste11p is less active than the dephosphorylated form, because overexpression of *pat1* is inhibitory for both mating and meiosis (McLeod and Beach 1988). Artificial inactivation of Pat1p results in activation of transcription of the mating-type genes (Nielsen and Egel 1990), which is also consistent with the above idea. Thus, possible inhibition of Ste11p activity by Pat1p appears to explain why *pat1* temperature-sensitive mutants initiate mating at semirestrictive temperatures and to support the proposal that partial inactivation of Pat1p provokes mating and its full inactivation leads to meiosis (Nielsen and Egel 1990). So far, however, Mei3p is the only known inhibitor of Pat1p kinase, and no physiological means has been demonstrated that inactivates Pat1p in haploid cells. As shown in Figure 7, inactivation of Pat1p by Mei3p depends on the function of Ste11p. Therefore, we like to infer that the major significance of dephosphorylation of Ste11p in the absence of Pat1p kinase activity will be to reinforce the commitment to meiosis through a feedback mechanism, as long as its physiological relevance to the regulation of mating remains rigorously unproved.

Pat1p kinase phosphorylates Mei2p both in vitro and in vivo. Weak expression of the dephosphorylated form of Mei2p could switch the cell cycle from mitotic to meiotic (Watanabe et al. 1997), even in *ste11*-deleted cells (Y. Watanabe and M. Yamamoto, unpubl.). This indicates that the most critical control exerted by Pat1p kinase over entry to meiosis is phosphorylation of Mei2p. However, meiosis induced by the mutant Mei2p was not as efficient as physiologically induced meiosis (Watanabe et al. 1997). This raises the possibility that Pat1p may also regulate an as yet unidentified factor(s) that is not absolutely necessary for meiosis but can facilitate its efficient progression.

C. Cytology of Meiotic Prophase

1. Nuclear Movements in Prophase

Early morphological studies showed that *S. pombe* nuclei assume a characteristic elongated "horsetail" morphology during premeiotic S phase and meiotic prophase (Robinow 1977). The recent development of video-assisted microscopic techniques has allowed observation of the dynamic movements of living nuclei during meiosis. Combining these techniques with FISH (fluorescence in situ hybridization) and immunofluorescence allowed Chikashige et al. (1994) and Svoboda et al. (1995) to clarify the nature of the nuclear movements. The horsetail nucleus moves back and forth in the cell, in movements apparently mediated by a bundle of cytoplasmic microtubles. At early prophase, the chromosome telomeres cluster near the spindle pole body (SPB), and this complex appears to lead the movement of the nucleus. This is an interesting observation that contrasts with the fact that centromeres lead the movement of chromosomes during mitosis (Funabiki et al. 1993). Subsequent analysis of nuclear reorganization that switches the position of centromeres and telomeres has revealed a two-step mechanism; first, association of telomeres with the SPB and, second, dissociation of centromeres from the SPB (Chikashige et al. 1997). The first step can take place in the haploid state in response to mating pheromone, but the second step appears to require conjugation-related events.

The telomeres are always observed as a single cluster in horsetail nuclei, whereas the centromeres of a pair of homologous chromosomes appear either as one spot or as two separate spots (Chikashige et al. 1994). These results suggest that the pairing of homologous chromosomes may start at their telomeres rather than at their centromeres. The movement of horsetail nuclei may aid pairing, recombination, and segregation of homologous chromosomes by mechanically shuffling them with their ends tied together. The clustering of telomeres in *S. pombe* is reminiscent of the "bouquet" structures reported previously in meiotic prophase nuclei of insects and plants (Loidl 1990).

As discussed above, Mei2p plays an essential part for meiosis I. Curiously, Mei2p has been found in a single "dot," closely opposed to but clearly separated from the SPB, in meiotic prophase nuclei (Watanabe et al. 1997). Formation of this dot appears to be a prerequisite for the promotion of the first division by Mei2p (Y. Watanabe and M. Yamamoto, unpubl.), but its precise formation remains totally unknown.

2. *Lack of SC Formation*

The lack of synaptonemal complex (SC) formation in meiotic prophase has been reported as a peculiar feature of in *S. pombe* meiosis. Instead of SCs, 25–30 linear elements, which can be stained with silver nitrate, are observed in this phase (Olson et al. 1978; Hirata and Tanaka 1982; Bähler et al. 1993). These linear elements first appear as short filaments when premeiotic DNA synthesis is completed, then elongate and align parallel to each other, and finally disintegrate and disappear. The SPB duplicates and separates following these events, and a meiosis I spindle is assembled. Because the association of chromatin loops with the linear elements has been suggested by electron microscopic observations of spread nuclei (Bähler et al. 1993), these elements are believed to correspond to the axial elements seen as SC precursors in other organisms such as *S. cerevisiae*. Because the number of linear elements in one nucleus clearly exceeds the number of chromosomes (six in a diploid), it is likely that several elements are formed discontinuously on a given chromosome. These linear elements may be necessary for chromatin condensation or may provide a scaffold facilitating the pairing process as well as the resolution of chromosomal interlocks (Bähler et al. 1993). They also appear to have a role in holding sister chromatids together (Molnar et al. 1995).

Chromosome pairing, meiotic recombination, and chromosome segregation thus proceed normally without formation of SCs in *S. pombe* (Kohli and Bähler 1994; Scherthan et al. 1994). Although this is somewhat unusual, certain *S. cerevisiae* mutants provide parallels. For example, the *zip1* mutant produces axial elements normally but fails to form SC structure; nevertheless, it is able to perform chromosome pairing, recombination, and chromosome segregation without serious difficulties (Sym and Roeder 1994). The only anomaly observed with this mutant was an absence of crossover interference during meiotic recombination. A lack of crossover interference is also apparent in *S. pombe* (Snow 1979; Egel et al. 1980; Munz et al. 1989; Munz 1994), consistent with the proposal (Egel 1978; Egel et al. 1980; King and Mortimer 1990) that a function of the SC is to generate crossover interference (and thus promote the distribution of crossover events). *Aspergillus nidulans* joins *S. pombe* as another organism that has abandoned the formation of SC, and hence crossover interference, during the course of evolution (Egel-Mitani et al. 1982).

D. Meiotic Recombination

Meiotic recombination is important for the maintenance of chromosome integrity, chromosome segregation, and generation of genomic diversity. Meiotic recombination is several hundred- to one thousand-fold more frequent than mitotic recombination. The overall frequency of crossing-over in *S. pombe* has been estimated to be approximately 45 events per meiosis, which is a sum of 19 events for chromosome I, 15 for chromosome II, and 11 for chromosome III (Egel 1989; Munz et al. 1989).

1. The ade6-M26 *Recombination Hot Spot*

The *ade6* gene has been a major target for the study of meiotic recombination in *S. pombe*, and marker effects have been examined extensively using nearly 400 *ade6* mutant alleles (Kohli and Bähler 1994). Among these, the *ade6-M26* allele (G at position 1010 to T, creating a UGA stop codon in the *ade6* ORF) is of special interest. The *M26* mutation increases the frequency of gene conversion at *ade6* by nearly 10-fold, suggesting that a recombination hot spot was generated by a single-base substitution (Gutz 1971). *M26* does not stimulate mitotic recombination (Schuchert and Kohli 1988; Ponticelli et al. 1988), but it increases the frequency of reciprocal exchange events during meiosis by about 15-fold (Schuchert and Kohli 1988; Grimm et al. 1994). By systematic mutational analysis, a heptameric DNA sequence (ATGACGT) surrounding the *M26* mutation site was shown to be responsible for the hot spot activity (Schuchert et al. 1991). On the supposition that this sequence provides a binding site for a recombination-promoting protein, purification of *M26*-binding proteins was pursued using *M26* DNA-affinity column chromatography. A heterodimer consisting of proteins named Mts1p and Mts2p was identified in this analysis (Wahls and Smith 1994). For a number of single-base substitutions introduced in the *M26* region, the ability to bind the Mts1p/Mts2p heterodimer in vitro and the activity as a hot spot in vivo were strictly correlated, suggesting that the binding of Mts1p/Mts2p indeed promotes meiotic recombination (Wahls and Smith 1994). These data suggest that homologous recombination may be regulated primarily by discrete sites on the DNA and by proteins that interact with these sites. The identities and precise functions of Mts1p and Mts2p remain to be elucidated.

2. *Recombination Mutants*

Mutants defective in meiotic recombination have been obtained easily in *S. cerevisiae* by taking advantage of the fact that they often show poor spore viability because of abnormal chromosome segregation following the failure of recombination. This isolation strategy has not been applicable to *S. pombe:* Because there are only three chromosomes, abnormal chromosome segregation does not reduce spore viability below about 10%. In addition, many radiation-sensitive (*rad*) mutants have been characterized in *S. cerevisiae* and shown to be relevant to meiosis or meiotic recombination, but this approach is also not well advanced in *S. pombe*. Although several *S. pombe rad* mutants have been examined, they were found not to be relevant to meiotic recombination (Grossenbacher-Grunder and Thuriaux 1981). In addition, *S. pombe* mutants *rec1* through *rec5* have been isolated (Gysler-Junker et al. 1991), but they are only relevant to mitotic, and not to meiotic, recombination. Thus, for some time, the *swi5* mutant (affected in mating-type switching) was the only *S. pombe* mutant that showed a defect in meiotic recombination (Schmidt et al. 1987).

Recently, however, a large collection of *S. pombe* mutants defective in meiotic recombination was obtained by exploiting the observation that the chromosomal *ade6-M26* allele converts to *ade6*$^+$ by recombination at a high frequency during meiosis if the cell carries another defective *ade6* allele on a plasmid (Ponticelli and Smith 1989; DeVeaux et al. 1992). The mutants isolated in this screen were subsequently shown to be defective also in recombination between homologous chromosomes. Thirty-nine mutations, all recessive, have defined 16 complementation groups, named *rec6* through *rec21*, which are spread throughout the genome. Five *rec* genes (*rec6,7,8,10,12*) have been cloned and sequenced. All of these genes are transcriptionally activated in meiotic cells (Lin et al. 1992; Lin and Smith 1994, 1995). The *rec12* gene shares homology with *S. cerevisiae SPO11*. Mutations in *rec8*, *rec10*, and *rec11* reduce recombination frequencies in a 2-Mb region on chromosome III but in no other region (DeVeaux and Smith 1994). This region-specific deficiency observed in a limited set of mutants suggests that there may be multiple activators of meiotic recombination, each specific for a limited set of loci or a chromosome region. Further analysis of the *rec* genes may clarify the validity of this hypothesis (see, e.g., Molnar et al. 1995).

The *S. cerevisiae RAD51* gene product is a homolog of the *E. coli* RecA protein and has crucial roles in both mitotic and meiotic recombination. An *S. pombe RAD51* homolog, *rad51*, has also been cloned (Shinohara et al. 1993). This gene failed to complement an *S. cerevisiae rad51* mutant, and, although likely, its involvement in meiotic recombination has yet to be proven in *S. pombe*.

E. Control of Meiotic M phase

1. Meiosis I

The first meiotic division takes place following meiotic recombination. In this division, homologous chromosomes segregate to the opposite poles of the spindle, and the two sister chromatids of each duplicated chromosome do not separate. The *mei4* mutant, isolated in an early study by Bresch et al. (1968), executes premeiotic DNA synthesis but does not undergo the first meiotic division and arrests with nuclei of the horsetail morphology (Egel and Egel-Mitani 1974; Olson et al. 1978; Shimoda et al. 1985; Y. Hiraoka, pers. comm.). The SPB is not duplicated in the *mei4* mutant, and electron microscopic observations suggest that the linear elements are fully formed but arrested in the elongated state. These elements apparently do not follow the disintegration program that occurs late in prophase in the wild type (Bähler et al. 1993). These observations suggest that the *mei4* mutant arrests in meiotic prophase I.

mei4 has been cloned by complementation, and its deduced product appears to be a member of the Fork Head family (S. Horie et al., unpubl.). The *Drosophila fork head* (*fkh*) gene encodes a transcription factor that is required for embryonic differentiation. Evidence that *S. pombe* Mei4p also functions as a transcription factor is accumulating (S. Horie et al., unpubl.).

2. Commitment to Meiosis

When cells of the *mei4* mutant are arrested after premeiotic DNA synthesis, they are already committed to meiosis; i.e., the cells do not resume the mitotic cell cycle even if they are shifted to rich nutrition (Egel 1973a). From the results of physiological experiments, it has been believed that *S. pombe* cells commit themselves to meiosis once they initiate premeiotic DNA synthesis (Beach et al. 1985; Bähler et al. 1991;

Sugiyama et al. 1994). However, analysis of cells defective in *sme2*, the gene for meiRNA, suggests otherwise. These cells arrest after premeiotic DNA synthesis and before the first meiotic division (Watanabe and Yamamoto 1994), and interestingly, they can return to the mitotic cell cycle if transferred to rich medium (Y. Watanabe and M. Yamamoto, unpubl.). This suggests that premeiotic DNA synthesis is not a determinant of the commitment and that the critical point of no return to mitosis occurs between the stage governed by Mei2p/meiRNA and that governed by Mei4p. *S. cerevisiae* has been shown to establish the commitment to meiosis after premeiotic DNA synthesis (see Kupiec et al., this volume). Thus, the two yeast species may be more similar than previously thought with regard to the commitment to meiosis. However, an alternative interpretation of the *sme2* results may be that the lack of meiRNA itself interferes with the establishment of commitment to meiosis. Further studies will be necessary to resolve these issues.

3. Meiosis II

Meiosis I and II proceed within a single nuclear envelope in *S. cerevisiae*. In contrast, meiosis I produces two separate daughter nuclei in *S. pombe*, and meiosis II then divides these two nuclei almost simultaneously into four (Olson et al. 1978; Hirata and Tanaka 1982). Segregation of sister chromatids in meiosis II mimics mitotic chromosome segregation, and it is thought to be regulated mostly by the same cell cycle regulators. However, the *mes1* mutation causes no apparent defect in mitosis but blocks meiosis II (Fig. 8) (Bresch et al. 1968). Thus, it is evident that meiosis II requires at least one specific function and is not entirely the same as mitosis. Electron microscopic observations have shown that meiosis I proceeds normally in *mes1*-defective cells but that duplication of the SPB in each daughter nucleus, which is the initial step of meiosis II, is completely blocked (Shimoda et al. 1985). Furthermore, spore wall formation does not occur in this mutant. This last observation is notable because certain other mutants such as *tws1* (an allele of *cdc2*) do not carry out meiosis II but permit spore wall formation, producing two diploid spores (see below).

Analysis of genomic and cDNA clones of *mes1* indicates that it carries a short intron of 75 nucleotides, and the mature mRNA appears to encode a small protein of 101 amino acids that has no significant homology with known proteins (Kishida et al. 1994). Interestingly, the spliced transcript is strictly specific to meiotic cells, and only unspliced mRNA of 0.9 kb in length is detectable in vegetative or meiosis-deficient cells.

Furthermore, only the spliced transcript can produce Mes1p that is functional in meiosis II (Kishida et al. 1994). Similar meiosis-specific splicing has been observed previously with the *S. cerevisiae MER2* gene, which provides a function required for meiotic recombination (Engebrecht et al. 1991). To date, *S. cerevisiae MER2* and *S. pombe mes1* are the two clear examples that demonstrate a critical role for specific splicing in the control of meiosis, but it appears highly likely that similar mechanisms are employed in higher organisms.

Because the spliceosome is a complex of proteins and RNA molecules, one interesting speculation is that the splicing of *mes1* mRNA may depend on the Mei2p-meiRNA complex. However, *mes1* mRNA can be spliced in the absence of meiRNA at least under some conditions (Y. Watanabe and M. Yamamoto, unpubl.), suggesting that this speculation is not perfectly correct.

F. Spore Formation

Ascospore development in *S. pombe* appears to be very similar to that in *S. cerevisiae*. Briefly, the SPBs differentiate into multilayered plaques at meiosis II, and the forespore membrane (a flattened, closed sac) develops on the cytoplasmic side of the differentiated plaques. The forespore membrane eventually encloses each divided nucleus together with cytoplasm-containing organelles (Yoo et al. 1973; Tanaka and Hirata 1982). Apparent fusion of cytoplasmic vesicles to the growing forespore membrane has been observed. The forespore membrane in its early stage of development appeared to be connected to the endoplasmic reticulum (ER), leading to the proposal that this membrane originates directly from the ER (Tanaka and Hirata 1982). However, the possible role of the Golgi in generating the forespore membrane requires further analysis. When the forespore membrane has surrounded the nucleus, the spore wall begins to form in the space between the inner and outer membranes. The mature spore wall of *S. pombe* is rich in amylose-like materials and is stained dark brown by iodine vapor, which thus serves as a convenient marker to monitor sporulation.

By screening for mutants that fail to stain with iodine, 20 complementation groups (*spo1–spo20*) have been shown to be required for the normal encapsulation of spore nuclei (Table 1E) (Bresch et al. 1968; Kishida and Shimoda 1986). However, most of these genes have not yet been cloned, and little information is available about their functions. An exception is *spo5*. Cells defective in this gene appear to arrest in meiosis II with the chromosome region divided but not fully separated (Fig. 8) (Hirata and Shimoda 1992). The deduced *spo5* gene product is a putative RNA-binding protein, and *spo5* expression is induced already during meiosis I (Y. Watanabe and M. Yamamoto, unpubl.). Although the *spo* mutants were originally named for their defects in spore wall formation, their functions are not necessarily related directly to the synthesis of spore wall materials. Instead, some of them, like *spo5*, appear to be defective primarily in the completion of meiosis (Kishida and Shimoda 1986).

Detailed electron microscopic analysis of the meiosis-II-specific modification of the SPB and formation of the forespore membrane was carried out in *spo4*, *spo5*, *spo14*, and *spo18* mutants (Hirata and Shimoda 1994). SPB modification and initiation of forespore-membrane formation were apparently normal in the *spo5* and *spo14* mutants, but further development of the forespore membrane was not properly organized. In contrast, the SPB was not modified and no indication of forespore-membrane formation was seen in the *spo4* and *spo18* mutants, although SPB duplication and nuclear division were apparently normal. These results suggest that modification of the SPB at meiosis II is a prerequisite for formation of the forespore membrane.

Because the *mes1* mutant is defective in both duplication and modification of the SPB, it is not surprising that no spore encapsulation occurs in this mutant (Shimoda et al. 1985). In contrast, the SPB is presumably modified appropriately without nuclear division in cells that produce two diploid spores while skipping meiosis II, such as *cdc2*(*tws1*) or *cdc25*, under certain conditions. Consistent with this interpretation, genetic analysis has shown that these cells have a defect in a later stage of meiosis than *mes1* (Grallert and Sipiczki 1989). Mutant cells carrying the *cdc2-33* allele, which allows production of two-spored asci at a high frequency (Iino et al. 1995), arrest in the mitotic cell cycle with duplicated SPBs at the restrictive temperature (King and Hyams 1982). Hence, it is possible that the SPB in the diploid spores has undergone duplication, at least partially or incompletely. An observation consistent with this possibility has been reported (Grallert and Sipiczki 1991; see below).

The full activity of calmodulin also appears to be required for the encapsulation of spores in *S. pombe*. A point mutation in the calmodulin-encoding *cam1* gene (Takeda and Yamamoto 1987) was found to make the protein unstable during nitrogen starvation and resulted in a failure of spore-wall formation, although the mutants were apparently able to perform meiosis I and II (Takeda et al. 1989). How calmodulin participates in spore-wall formation is not known.

G. Function of Cell Cycle Regulators in Meiosis

1. Regulation of Meiotic Start

It is clear that mitosis and meiosis share many events, including DNA replication and chromosome segregation. However, it is also evident that these homologous events are not regulated in exactly the same manner. Thus, interesting questions are whether a regulatory factor essential for the mitotic cell cycle is also essential for the meiotic cell cycle and, if so, whether it is used in the same or different ways in the two cycles. Comparison of meiosis with mitosis in these respects will be essential for a full understanding of meiotic regulation.

Cells decide whether to start a new round of the cell cycle while they are in the G_1 phase. If they find that all the prerequisites are met, they initiate DNA replication and normally undergo all the cell cycle events until they return to G_1. The point of the cell cycle at which cells commit themselves to the entry into S phase is called "Start" in yeasts (Hartwell 1974) and the "Restriction Point" in mammals (Pardee et al. 1978). G_1 phase also serves as the point of exit for cell differentiation, which includes entry into the meiotic pathway. Early genetic analysis suggested that two cell cycle genes, *cdc2* and *cdc10*, control mitotic Start in *S. pombe* (Nurse and Bissett 1981).

The meiotic cell cycle of *S. pombe* diploid cells starts from the G_1 arrest induced by nutritional starvation. As discussed above, the function of Mei2p is required specifically for the meiotic G_1/S transition (i.e., Start of the meiotic cell cycle). Current knowledge of other factors required for meiotic Start is summarized below (see also Table 1F).

a. cdc2 *and G_1 Cyclin.* The *cdc2* gene of *S. pombe* encodes a serine/threonine protein kinase that has key roles in the progression of both G_1 and G_2 phases (Nurse 1990). Studies of cell cycle control in budding yeast, fission yeast, and mammalian cells have established that the mitotic Start or Restriction Point is regulated by cooperation between protein kinases of the Cdc2 family (CDKs, for "cyclin-dependent kinases") and G_1 cyclins (Reed 1992; Sherr 1994). A requirement for *cdc2* function for meiotic Start has also recently been demonstrated in *S. pombe* (Iino et al. 1995). This contrasts with the report that *CDC28*, the *S. cerevisiae* homolog of *cdc2*, is dispensable for premeiotic DNA synthesis (Shuster and Byers 1989). The reason for this apparent discrepancy requires further investigation, but one possible explanation could be that a factor other than Cdc28p can provide the CDK function required for premeiotic DNA synthesis in *S. cerevisiae*. The recent finding that the *S. cerevisiae PHO85* gene encodes a CDK that also functions in the mitotic G_1/S transition (Espinoza et al. 1994; Measday et al. 1994) may support this view.

In addition to Cdc13p, which is a B-type cyclin that functions in the G_2/M transition (Booher and Beach 1988; Hagan et al. 1988), four putative cyclins have been identified in *S. pombe* (for review, see Fisher and Nurse 1995). Some of these have been suggested to be G_1 cyclins required for mitotic Start, but no critical evidence for this is yet available, and the situation appears to be rather confusing (Fisher and Nurse 1995). *cig2/cyc17*, which encodes one of these putative cyclins, can be a high-copy suppressor of *pat1-114* (Bueno and Russell 1993; Connolly and Beach 1994; Obara-Ishihara and Okayama 1994). *puc1*, which can complement loss of function of *S. cerevisiae* G_1 cyclins, encodes another putative cyclin and also shows high-copy suppression of *pat1-114* (Forsburg and Nurse 1991, 1994). Although disruptions of these genes cause no obvious defects in the mitotic cell cycle, the disruptants have been found to be accelerated in their entry into the meiotic pathway (Forsburg and Nurse 1994; Obara-Ishihara and Okayama 1994), leading to a speculation that these cyclins may function as meiotic inhibitors.

Overexpression of *rum1* causes repeats of S phase without M phase, and the *rum1* disruptant is incapable of G_1 arrest and defective in mating and meiosis (Moreno and Nurse 1994). This suggests that proper G_1 arrest ensured by the function of *rum1* is essential for the onset of sexual development.

b. cdc10 *and Functionally Related Genes. cdc10* of *S. pombe* shows homology with *S. cerevisiae SWI6*, and its product is a transcription factor required for the initiation of S phase. Cdc10p activates genes that carry MCBs (*Mlu*I cell cycle boxes) in their 5′ upstream regions (Aves et al. 1985; Lowndes et al. 1992). The target genes include *cdc22*, which encodes the large subunit of deoxyribonucleotide reductase, a key enzyme for the synthesis of substrates for DNA replication. The transcription of *cdc22* is specifically induced near the G_1/S transition (Lowndes et al. 1992; Fernandez-Sarabia et al. 1993). A requirement for the functions of both *cdc10* and *cdc22* for premeiotic DNA synthesis has been demonstrated (Beach et al. 1985; Grallert and Sipiczki 1991; Iino et al. 1995). Transcription of *cdc22* peaks prior to the initiation of premeiotic DNA synthesis, mimicking the situation in the mitotic cell cycle (Iino et al. 1995).

Two additional genes, *res1/sct1* and *res2/pct1*, were isolated as high-copy suppressors of the mitotic G_1 arrest caused by *cdc10* mutations. The products of these genes associate with Cdc10p, as *S. cerevisiae*

Swi4p and Mbp1p do with Swi6p, and the resulting complexes regulate expression of genes having MCBs (Tanaka et al. 1992; Caligiuri and Beach 1993; Miyamoto et al. 1994; Zhu et al. 1994). Cdc10p-Res1p and Cdc10p-Res2p complexes are functionally redundant, although the former appears to be used predominantly in mitosis and the latter in meiosis (Miyamoto et al. 1994; Zhu et al. 1994).

rep1 is another high-copy suppressor of *cdc10*. Expression of *rep1* is regulated by Ste11p, and its function is required for the initiation of premeiotic DNA synthesis but not for mitotic DNA synthesis. Rep1p is a zinc finger protein and has been shown to be involved either directly or indirectly in the induction of *res2* expression (Sugiyama et al. 1994).

2. *Regulation of the Meiotic Divisions*

The G_2/M transition in the mitotic cell cycle of *S. pombe* is regulated by MPF (M-phase promoting factor), which is a complex of Cdc2p and a B-type cyclin encoded by *cdc13*. A tyrosine protein phosphatase encoded by *cdc25* activates Cdc2p by dephosphorylation. These three *cdc* genes are absolutely required for mitosis. Meiosis II is regarded as analogous to mitosis, and it appears that these three genes are indeed also involved in meiosis II. Cells carrying a temperature-sensitive mutation in *cdc2*, *cdc13*, or *cdc25* often produce two-spored asci, apparently without undergoing meiosis II, if sporulation is induced at semi-restrictive temperature (Beach et al. 1985; Grallert and Sipiczki 1991). Moreover, a non-temperature-sensitive allele of *cdc2*, originally called *tws1*, causes frequent production of two-spored asci, although it does not affect the mitotic cell cycle (Nakaseko et al. 1984; Niwa and Yanagida 1988; Grallert and Sipiczki 1990). It has been demonstrated genetically that the diploid spores produced by the *tws1* mutant have not undergone meiosis II (Nakaseko et al. 1984). Furthermore, a spore in which a dividing nucleus was encapsulated within the spore wall was observed in a two-spored ascus produced by a mutant partially defective in *cdc13* (Grallert and Sipiczki 1991). These results strongly suggest that the full activity of MPF is essential for meiosis II. In mitosis, the *cdc2-3w* mutation suppresses *cdc25-22*, and the *cdc13-c1* mutation suppresses *cdc2-59*, both in an allele-specific manner. Conservation of this allele-specific suppression has been observed in meiosis, suggesting that these gene products interact similarly in mitosis and meiosis (Grallert and Sipiczki 1991).

MPF was originally defined as a factor promoting meiosis I in *Xenopus* oocytes (Masui and Markert 1971). Ironically, the above observations demonstrated a requirement for MPF for meiosis II in *S. pombe* but left open the question of whether it is required also for meiosis I. The observations suggest that meiosis I requires a lower MPF activity than does meiosis II, if, indeed, it is required at all. Two kinds of experimental difficulties have hampered attempts to answer this question. First, because meiosis is intrinsically temperature-sensitive and does not proceed properly above 34°C, it is difficult to inactivate completely the products of temperature-sensitive *cdc* alleles during meiosis. Second, the temperature shift may cause temperature-sensitive *cdc* mutants to arrest in the preceding mitotic cell cycle rather than in the meiotic cell cycle. Careful experiments to avoid these difficulties were carried out with and without the use of the *pat1-114* mutant (Iino et al. 1995). The results showed that the function of *cdc2* is essential for premeiotic DNA synthesis and that *cdc13* and *cdc25* are essential for meiosis I, suggesting that MPF function is required for meiosis I in *S. pombe.* However, although it appears likely, it is still uncertain whether *cdc2* contributes to this MPF activity, because the possible requirement for *cdc2* for meiosis I could not be differentiated experimentally from that for premeiotic DNA synthesis (Iino et al. 1995).

3. *Transcriptional Regulation of* cdc *Genes during Meiosis*

Expression of *cdc22*, which has MCB elements, peaks prior to premeiotic DNA synthesis, as discussed above. In contrast, *cdc13* and *cdc25* are not expressed significantly during the meiotic G_1 phase, but their expression (especially that of *cdc25*) is induced dramatically between premeiotic DNA synthesis and the first meiotic division. This induction depends both on the function of *mei4* and on the completion of DNA synthesis (Iino et al. 1995). The dependence of *cdc25* expression on the completion of DNA synthesis may be important as a checkpoint for meiosis I, because the *rad1* gene, which is responsible for the checkpoint control coordinating DNA replication and chromosome segregation in the mitotic cell cycle (Rowley et al. 1992), does not appear to operate as such in the meiotic cell cycle (Iino et al. 1995). The checkpoint mechanisms that operate in meiosis represent an intriguing problem barely investigated to date.

V. CONCLUDING REMARKS

Most of the *S. pombe* genes discussed in this chapter are summarized in Table 1, and Figure 9 illustrates the stages of meiosis and sporulation and the genes required for each stage. We summarize below the regulation of mating and meiosis in *S. pombe* and then compare it with the situation in *S. cerevisiae.*

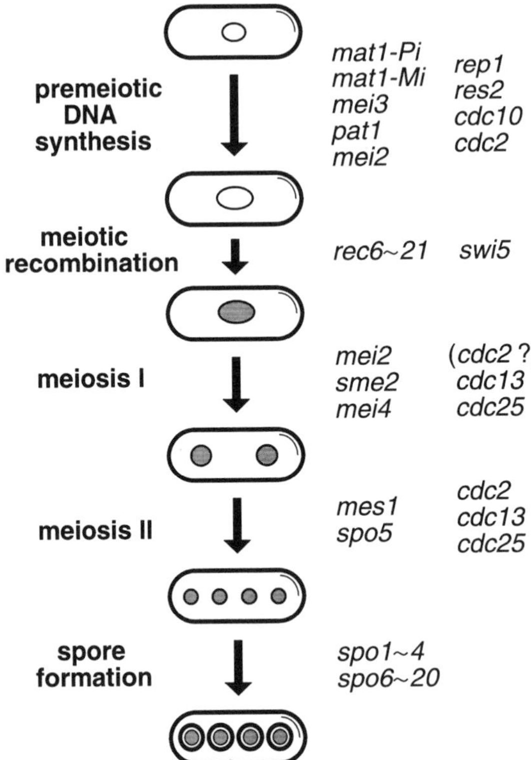

Figure 9 Summary of genes relevant to the regulation of meiosis and spore formation in *S. pombe*. Genes listed are required for the promotion of each event, except for *pat1*, which negatively regulates the initiation of premeiotic DNA synthesis. The requirement for *cdc2* for meiosis I is speculative.

A. Outline of the Regulation

1. Mating

S. pombe initiates mating and meiosis in response to nutritional starvation. Starvation is recognized by an as yet unknown mechanism, and the resulting signal is transmitted via the *gpa2*-encoded Gα protein to adenylyl cyclase, thus down-regulating its activity. The level of intracellular cAMP thus decreases in response to starvation, which lowers the activity of cAMP-dependent protein kinase (PKA). The low PKA activity induces expression of *ste11*, which encodes an HMG family transcription factor. This induction may result from dephosphorylation of a transcription factor controlling *ste11*, as discussed above. Ste11p controls many genes that are required for mating and/or meiosis (see Fig. 1).

Ste11p induces expression of *mat1-Pc* in h^+ cells and of *mat1-Mc* in h^- cells. Expression of these genes enables h^+ cells to express the genes encoding P-factor (*map2*) and the M-factor receptor (*map3*) and h^- cells to express the genes encoding M-factor (*mfm1-3*) and the P-factor receptor (*mam2*). Upon expression of these mating-type-specific genes, h^+ and h^- cells begin communicating with each other by means of mating pheromones (Fig. 2). Stimulation by mating pheromones from cells of the opposite mating type induces transient G_1 arrest and a new program of gene expression in the recipient cells, including the induction of expression of *mat1-Pi* in h^+ cells and of *mat1-Mi* and *sxa2* (encoding a protease to degrade P-factor) in h^- cells. Other genes are also likely to be regulated similarly to facilitate the formation of conjugation tubes and subsequent complex mating processes. Cell fusion is mediated by the *fus1* gene product and presumably also by additional factors not yet identified.

The binding of each mating pheromone to its receptor activates a G-protein, the α-subunit of which is encoded by *gpa1*. Activated Gpa1p transmits a signal to the downstream MAP-kinase cascade consisting of Byr2p (MAPKKK), Byr1p (MAPKK), and Spk1p (MAPK) (see Fig. 5). Spk1p is likely to regulate some transcription factor(s), although no candidates have yet been identified. Ras1p, the single *S. pombe* homolog of mammalian Ras oncoprotein, is also essential for mating. Ras1p activity is required to achieve full activation of the MAP-kinase cascade by the pheromone signal (Fig. 5). Ras1p appears to be activated by nutritional starvation through induction of *ste6* expression. *ste6* encodes a GEF for Ras1p, and its expression is regulated by Ste11p. Expression of *ste6* is further enhanced by stimulation by mating pheromone, producing a positive feedback loop for pheromone signaling (Fig. 5). It is not known whether the activity of Ras1p is regulated more directly by some factor other than the expression of *ste6*. Ras1p also controls a pathway governing cell morphology, in addition to the MAP-kinase cascade. Pheromone signaling via the MAP-kinase cascade is not completely shut off in the absence of Ras1p. Hence, the strict inability of *ras1*Δ cells to mate may be attributable, at least in part, to their morphological defects (i.e., their inability to form conjugation tubes properly).

2. *Meiosis*

The switch from the mitotic to the meiotic cell cycle is regulated by two determinants. One is the induction of expression of *mei2*, which encodes a critical positive factor for promotion of premeiotic DNA synthesis. The other is the inactivation of the Pat1p kinase, which otherwise phosphorylates and inactivates Mei2p. Both effects result from nutritional starvation. Expression of *mei2* is regulated directly by Ste11p. Inactivation of the Pat1p kinase occurs much more indirectly. *S. pombe* diploid cells apparently keep the mating-pheromone signaling pathway operational under nutritional starvation, and, in response to this signaling, they express *mat1-Pi* and *mat1-Mi*. Coexpression of these genes induces expression of *mei3*, which encodes a protein that binds to Pat1p and inactivates its kinase activity. Then, free from inhibition, Mei2p executes a function critical to initiate premeiotic DNA synthesis (see Fig. 7).

Mei2p is an RNA-binding protein whose function is required not only for premeiotic DNA synthesis, but also for meiosis I. Interestingly, an RNA molecule of about 0.5 kb, named meiRNA, complexes with Mei2p to perform the function necessary for meiosis I. Cells lacking meiRNA arrest before the first meiotic division, clearly demonstrating the involvement of this RNA in meiotic regulation. Circumstantial evidence suggests that Mei2p may cooperate with another RNA species to promote premeiotic DNA synthesis. Despite this progress, the critical molecular function(s) of Mei2p remains to be elucidated.

Nuclei at prophase of meiosis I assume a typical shape called "horsetail." They move back and forth in the cell, led by the SPB associated with the chromosomal telomeres. A *mei4* mutant arrests at this stage. This gene encodes a putative transcription factor, but not much is known about its function.

S. pombe cells do not form synaptonemal complexes (SC) during meiotic prophase. Instead, several linear elements are seen in association with the chromosomes at this stage. The lack of SC is thought to correlate with the lack of crossover interference in this organism. Collection of *S. pombe* mutants defective in meiotic recombination started relatively recently. To date, 16 *rec* genes relevant to meiotic recombination have been defined, but their molecular functions remain largely unknown. The core sequence of a hot spot for meiotic recombination has been shown to be ATGACGT, and a heterodimeric protein binding to it (Mts1p-Mts2p) has been identified.

A *mes1* mutant arrests between meiosis I and meiosis II. The transcript of this gene undergoes meiosis-specific splicing, without which functional Mes1p is not produced. The *spo5* mutant has a defect in a late stage of meiosis II and the development of the spore wall. The *spo5* gene product is a putative RNA-binding protein. The precise molecular functions of these gene products also remain to be elucidated, as do those of numerous other *spo* genes that may be involved in completion of meiosis or in spore formation itself.

cdc2, which encodes a key protein kinase regulating the G_1/S and G_2/M transitions in the mitotic cell cycle, is also essential for premeiotic DNA synthesis and for meiosis II. It is probably also required for meiosis I, but this has not yet been established. *cdc13*, which encodes a B-type cyclin that cooperates with the Cdc2p kinase in the G_2/M transition, and *cdc25*, which encodes a tyrosine protein phosphatase that activates the Cdc2p kinase, are both also required for both meiosis I and meiosis II. Expression of *cdc25* is induced upon completion of premeiotic DNA synthesis in the meiotic cell cycle; this may function as a checkpoint connecting premeiotic DNA synthesis and the first meiotic division.

B. *S. pombe* vs. *S. cerevisiae:* Evolutionary Twists

The general scheme of mating, meiosis, and sporulation appears to be well conserved in the distantly related yeast species *S. cerevisiae* and *S. pombe*. However, if we consider the underlying molecular mechanisms, we see a number of interesting contrasts. We conclude this review by highlighting examples of such contrasts in order to stimulate thinking about the basic strategy and evolution of sexual reproduction systems.

S. cerevisiae cells conjugate in the presence of rich nutrition, which means that their mating-type genes are functional during vegetative growth. The mating-type genes of *S. cerevisiae* are so-called master genes, occupying the most upstream position in a cascade and themselves controlled by nothing. In contrast, the mating-type genes of *S. pombe* are expressed and functional only when cells are starved. These differences are likely to be reflections of the fact that the wild *S. cerevisiae* has a predominant diplophase, whereas wild *S. pombe* has a predominant haplophase.

It has been shown that the level of cAMP decreases in sporulating *S. cerevisiae* cells, responding mainly to the glucose starvation that is a prerequisite for sporulation in this species. Ras1p and Ras2p are responsible for the activation of adenylyl cyclase in this yeast. In *S. pombe*, nutritional starvation, including glucose starvation, also leads to a decrease in the intracellular cAMP level, but adenylyl cyclase appears to be regulated by a heterotrimeric G-protein in this case. The situation in *S. pombe* is reminiscent of that in mammalian cells, where adenylyl cyclase is activated by a heterotrimeric G-protein in response to hormonal stimulation. However, the overall structures of *S. pombe* and *S. cerevisiae* adenylyl cyclases are very similar and quite different from the ion-channel-like structure of mammalian adenylyl cyclase.

Other contrasts in signal transduction can be seen in the pathways of mating-pheromone response. In both yeast species, pheromones, pheromone receptors, a heterotrimeric G-protein, and a MAP-kinase cascade constitute the pathway. As discussed above, the pheromones and their receptors appear to be evolutionarily homologous, as do the MAP-kinase cascades. However, the G-proteins function quite differently. Gα transmits the mating signal to the downstream elements in *S. pombe*, whereas in *S. cerevisiae*, Gβ•γ transmits the signal and Gα is involved in adaptation. Moreover, *S. pombe* Ras1p is involved in regulation of the MAP-kinase cascade, a role for Ras that is not seen in *S. cerevisiae*. This role is reminiscent of the mammalian Ras system, where Ras protein regulates Raf1, which functions as a MAPKKK. Mammalian Ras receives a signal from a receptor tyrosine kinase, but it is not known whether *S. pombe* Ras1p ever responds to an external signal by a mechanism other than changes in gene expression of its activator, Ste6p.

Once diploidized, *S. cerevisiae* cells shut off the mating-pheromone signaling pathway. In contrast, pheromone signaling is essential for meiosis in *S. pombe*. The initiation of meiosis is regulated by the negative factor Rme1p and the positive factors Ime1p and Ime2p in *S. cerevisiae* (see Kupiec et al., this volume), whereas the initiation of meiosis in *S. pombe* is regulated by the negative factor Pat1p and the positive factor Mei2p. Remarkably, there is no detectable similarity among these factors. Furthermore, Rme1p represses the function of Ime1p by blocking its gene expression, whereas Pat1p inactivates Mei2p by phosphorylation.

S. cerevisiae rad mutants were informative as to mechanisms of meiotic recombination when they were combined with *spo13*, a mutation that allows meiotic cells to skip meiosis I (Klapholz and Esposito 1980; see Kupiec et al., this volume). No mutation analogous to *spo13* has been found in *S. pombe*. It is unclear whether the apparent lack of such a mutant in *S. pombe* reflects any fundamental feature of the meiotic mechanisms in this organism.

Presumably, different organisms have different strategies for initiating the meiotic cell cycle because they have adapted their reproduction to different conditions. However, the scheme for premeiotic DNA synthesis and the two meiotic divisions appears to be very widely conserved. Thus, it will be very important to identify the most upstream factor in the meiotic pathway that is common to various species. Because Mei2p is required both for the initiation of the meiotic cell cycle and for the promotion of meiosis I, it is a candidate to be such a factor. Our recent analyses have shown that the flowering plant *Arabidopsis thaliana* has a protein structurally similar to Mei2p (T. Kuromori and M. Yamamoto, unpubl.). Whether RNA-protein complexes such as Mei2p-meiRNA have critical roles in the regulation of sexual reproduction in organisms other than *S. pombe* is undoubtedly one of the key questions to be answered.

ACKNOWLEDGMENTS

We thank Yuichi Iino, Jürg Bähler, and John Pringle for helpful comments on the manuscript. M.Y. thanks all his collaborators whose work has been cited in this review.

REFERENCES

Aono, T., H. Yanai, F. Miki, J. Davey, and C. Shimoda. 1994. Mating pheromone-induced expression of the *mat1-Pm* gene of *Schizosaccharomyces pombe:* Identification of signalling components and characterization of upstream controlling elements. *Yeast* **10:** 757.

Aves, S.J., B.W. Durkacz, A. Carr, and P. Nurse. 1985. Cloning, sequencing and transcriptional control of the *Schizosaccharomyces pombe cdc10* "start" gene. *EMBO J.* **4:** 457.

Bähler, J., P. Schuchert, C. Grimm, and J. Kohli. 1991. Synchronized meiosis and recombination in fission yeast: Observations with *pat1-114* diploid cells. *Curr. Genet.* **19:** 445.

Bähler, J., T. Wyler, J. Loidl, and J. Kohli. 1993. Unusual nuclear structures in meiotic prophase of fission yeast: A cytological analysis. *J. Cell Biol.* **121:** 241.

Barbacid, M. 1987. *ras* genes. *Annu. Rev. Biochem.* **56:** 779.

Beach, D., L. Rodgers, and J. Gould. 1985. *RAN1*$^+$ controls the transition from mitotic division to meiosis in fission yeast. *Curr. Genet.* **10:** 297.

Bender, A. and J.R. Pringle. 1989. Multicopy suppression of the *cdc24* budding defect in yeast by *CDC42* and three newly identified genes including the *ras*-related gene *RSR1*. *Proc. Natl. Acad. Sci.* **86:** 9976.

Benton, B.K., M.S. Reid, and H. Okayama. 1993. A *Schizosaccharomyces pombe* gene that promotes sexual differentiation encodes a helix-loop-helix protein with homology to MyoD. *EMBO J.* **12:** 135.

Birney, E., S. Kumar, and A.R. Krainer. 1993. Analysis of the RNA-recognition motif and RS and RGG domains: Conservation in metazoan pre-mRNA splicing factors. *Nucleic Acids Res.* **21:** 5803.

Bölker, M., M. Urban, and R. Kahmann. 1992. The *a* mating type locus of *U. maydis* specifies cell signaling components. *Cell* **68:** 441.

Booher, R. and D. Beach. 1988. Involvement of *cdc13*$^+$ in mitotic control in *Schizosaccharomyces pombe:* Possible interaction of the gene product with microtubules. *EMBO J.* **7:** 2321.

Bostock, C.J. 1970. DNA synthesis in the fission yeast *Schizosaccharomyces pombe*. *Exp. Cell Res.* **60:** 16.

Bourne, H.R., D.A. Sanders, and F. McCormick. 1990. The GTPase superfamily: A conserved switch for diverse cell functions. *Nature* **348:** 125.

———. 1991. The GTPase superfamily: Conserved structure and molecular mechanism. *Nature* **349:** 117.

Bresch, C., G. Müller, and R. Egel. 1968. Genes involved in meiosis and sporulation of a yeast. *Mol. Gen. Genet.* **102:** 301.

Bücking-Throm, E., W. Duntze, L.H. Hartwell, and T.R. Manney. 1973. Reversible arrest of haploid yeast cells at the initiation of DNA synthesis by a diffusible sex factor. *Exp. Cell Res.* **76:** 99.

Bueno, A. and P. Russell. 1993. Two fission yeast B-type cyclins, Cig2 and Cdc13, have different functions in mitosis (Correction [1994] *Mol. Cell. Biol.* **14:** 869). *Mol. Cell. Biol.* **13:** 2286.

Byrne, S.M. and C.S. Hoffman. 1993. Six *git* genes encode a glucose-induced adenylate cyclase activation pathway in the fission yeast *Schizosaccharomyces pombe. J. Cell Sci.* **105:** 1095.

Caligiuri, M. and D. Beach. 1993. Sct1 functions in partnership with Cdc10 in a transcription complex that activates cell cycle START and inhibits differentiation. *Cell* **72:** 607.

Calleja, G.B., B.F. Johnson, and B.Y. Yoo. 1980. Macromolecular changes and commitment to sporulation in the fission yeast *Schizosaccharomyces pombe. Plant Cell Physiol.* **21:** 613.

Chang, E.C., M. Barr, Y. Wang, V. Jung, H.-P. Xu, and M.H. Wigler. 1994. Cooperative interaction of *S. pombe* proteins required for mating and morphogenesis. *Cell* **79:** 131.

Chant, J. and I. Herskowitz. 1991. Genetic control of bud site selection in yeast by a set of gene products that constitute a morphogenic pathway. *Cell* **65:** 1203.

Chenevert, J., K. Corrado, A. Bender, J. Pringle, and I. Herskowitz. 1992. A yeast gene (*BEM1*) necessary for cell polarization whose product contains two SH3 domains. *Nature* **356:** 77.

Chikashige, Y., D.-Q. Ding, Y. Imai, M. Yamamoto, T. Haraguchi, and Y. Hiraoka. 1997. Meiotic nuclear reorganization: Switching the position of centromeres and telomeres in the fission yeast *Schizosaccharomyces pombe. EMBO J.* **16:** 193.

Chikashige, Y., D.-Q. Ding, H. Funabiki, T. Haraguchi, S. Mashiko, M. Yanagida, and Y. Hiraoka. 1994. Telomere-led premeiotic chromosome movement in fission yeast. *Science* **264:** 270.

Connolly, T. and D. Beach. 1994. Interaction between the Cig1 and Cig2 B-type cyclins in the fission yeast cell cycle. *Mol. Cell. Biol.* **14:** 768.

Costello, G., L. Rogers, and D. Beach. 1986. Fission yeast enters the stationary phase G_0 state from either mitotic G_1 or G_2. *Curr. Genet.* **11:** 119.

Crews, C.M., A. Alessandrini, and R.L. Erikson. 1992. The primary structure of MEK, a protein kinase that phosphorylates the *ERK* gene product. *Science* **258:** 478.

Cross, F., L.H. Hartwell, C. Jackson, and J.B. Konopka. 1988. Conjugation in *Saccharomyces cerevisiae. Annu. Rev. Cell Biol.* **4:** 429.

Cvrčková, F., C. De Virgilio, E. Manser, J.R. Pringle, and K. Nasmyth. 1995. Ste20-like protein kinases are required for normal localization of cell growth and for cytokinesis in budding yeast. *Genes Dev.* **9:** 1817.

Davey, J. 1991. Isolation and quantitation of M-factor, a diffusible mating factor from the fission yeast *Schizosaccharomyces pombe. Yeast* **7:** 357.

———. 1992. Mating pheromones of the fission yeast *Schizosaccharomyces pombe:* Purification and structural characterization of M-factor and isolation and analysis of two genes encoding the pheromone. *EMBO J.* **11:** 951.

Davey, J. and O. Nielsen. 1994. Mutations in *cyr1* and *pat1* reveal pheromone-induced G1 arrest in the fission yeast *Schizosaccharomyces pombe. Curr. Genet.* **26:** 105.

Davey, J., K. Davis, Y. Imai, M. Yamamoto, and G. Matthews. 1994. Isolation and characterization of krp, a dibasic endopeptidase required for cell viability in the fission yeast *Schizosaccharomyces pombe. EMBO J.* **13:** 5910.

DeVeaux, L.C. and G.R. Smith. 1994. Region-specific activators of meiotic recombination in *Schizosaccharomyces pombe. Genes Dev.* **8:** 203.

DeVeaux, L.C., N.A. Hoagland, and G.R. Smith. 1992. Seventeen complementation groups of mutations decreasing meiotic recombination in *Schizosaccharomyces pombe. Genetics* **130:** 251.

DeVoti, J., G. Seydoux, D. Beach, and M. McLeod. 1991. Interaction between *ran1*$^+$ protein kinase and cAMP dependent protein kinase as negative regulators of fission yeast meiosis. *EMBO J.* **10:** 3759.

Dietzel, C. and J. Kurjan. 1987. The yeast *SCG1* gene: A Gα-like protein implicated in the **a**- and α-factor response pathway. *Cell* **50:** 1001.

Dooijes, D., M. van de Wetering, L. Knippels, and H. Clevers. 1993. The *Schizosaccharomyces pombe* mating-type gene *mat*-Mc encodes a sequence-specific DNA-binding high mobility group box protein. *J. Biol. Chem.* **268:** 24813.

Egel, R. 1973a. Commitment to meiosis in fission yeast. *Mol. Gen. Genet.* **121:** 277.

———. 1973b. Genes involved in mating type expression of fission yeast. *Mol. Gen. Genet.* **122:** 339.

———. 1978. Synaptonemal complex and crossing-over: Structural support or interference? *Heredity* **41:** 233.

———. 1984. Two tightly linked silent cassettes in the mating-type region of *Schizosaccharomyces pombe. Curr. Genet.* **8:** 199.

———. 1989. Mating-type genes, meiosis, and sporulation. In *Molecular biology of the fission yeast* (ed. A. Nasim et al.), p. 31. Academic Press, San Diego.

Egel, R. and M. Egel-Mitani. 1974. Premeiotic DNA synthesis in fission yeast. *Exp. Cell Res.* **88:** 127.

Egel, R., J. Kohli, P. Thuriaux, and K. Wolf. 1980. Genetics of the fission yeast *Schizosaccharomyces pombe. Annu. Rev. Genet.* **14:** 77.

Egel-Mitani, M., L.W. Olson, and R. Egel. 1982. Meiosis in *Aspergillus nidulans:* Another example for lacking synaptonemal complexes in the absence of crossover interference. *Hereditas* **97:** 179.

Engebrecht, J., K. Voelkel-Meiman, and G.S. Roeder. 1991. Meiosis-specific RNA splicing in yeast. *Cell* **66:** 1257.

Errede, B., A. Gartner, Z. Zhou, K. Nasmyth, and G. Ammerer. 1993. MAP kinase-related FUS3 from *S. cerevisiae* is activated by STE7 *in vitro. Nature* **362:** 261.

Espinoza, F.H., J. Ogas, I. Herskowitz, and D.O. Morgan. 1994. Cell cycle control by a complex of the cyclin HCS26 (PCL1) and the kinase PHO85. *Science* **266:** 1388.

Fedor-Chaiken, M., R.J. Deschenes, and J.R. Broach. 1990. *SRV2*, a gene required for *RAS* activation of adenylate cyclase in yeast. *Cell* **61:** 329.

Fernandez-Sarabia, M.-J., C. McInerny, P. Harris, C. Gordon, and P. Fantes. 1993. The cell cycle genes *cdc22*$^+$ and *suc22*$^+$ of the fission yeast *Schizosaccharomyces pombe* encode the large and small subunits of ribonucleotide reductase. *Mol. Gen. Genet.* **238:** 241.

Field, J., A. Vojtek, R. Ballester, G. Bolger, J. Colicelli, K. Ferguson, J. Gerst, T. Kataoka, T. Michaeli, S. Powers, M. Riggs, L. Rodgers, I. Wieland, B. Wheland, and M. Wigler. 1990. Cloning and characterization of *CAP*, the *S. cerevisiae* gene encoding the 70 kd adenyly cyclase-associated protein. *Cell* **61:** 319.

Fisher, D. and P. Nurse. 1995. Cyclins of the fission yeast *Schizosaccharomyces pombe*. *Semin. Cell Biol.* **6:** 73.

Forsburg, S.L. and P. Nurse. 1991. Identification of a G1-type cyclin *pucl*$^+$ in the fission yeast *Schizosaccharomyces pombe*. *Nature* **351:** 245.

———. 1994. Analysis of the *Schizosaccharomyces pombe* cyclin puc1: Evidence for a role in cell cycle exit. *J. Cell Sci.* **107:** 601.

Fukui, Y. and Y. Kaziro. 1985. Molecular cloning and sequence analysis of a *ras* gene from *Schizosaccharomyces pombe*. *EMBO J.* **4:** 687.

Fukui, Y. and M. Yamamoto. 1988. Isolation and characterization of *Schizosaccharomyces pombe* mutants phenotypically similar to *ras1*$^-$. *Mol. Gen. Genet.* **215:** 26.

Fukui, Y., Y. Kaziro, and M. Yamamoto. 1986a. Mating pheromone-like diffusible factor released by *Schizosaccharomyces pombe*. *EMBO J.* **5:** 1991.

Fukui, Y., S. Miyake, M. Satoh, and M. Yamamoto. 1989. Characterization of the *Schizosaccharomyces pombe ral2* gene implicated in activation of the *ras1* gene product. *Mol. Cell. Biol.* **9:** 5617.

Fukui, Y., T. Kozasa, Y. Kaziro, T. Takeda, and M. Yamamoto. 1986b. Role of a *ras* homolog in the life cycle of *Schizosaccharomyces pombe*. *Cell* **44:** 329.

Funabiki, H., I.M. Hagan, S. Uzawa, and M. Yanagida. 1993. Cell cycle-dependent specific positioning and clustering of centromeres and telomeres in fission yeast. *J. Cell Biol.* **121:** 961.

Gartner , A., K. Nasmyth, and G. Ammerer. 1992. Signal transduction in *Saccharomyces cerevisiae* requires tyrosine and threonine phosphorylation of FUS3 and KSS1. *Genes Dev.* **6:** 1280.

Gerst, J.E., K. Ferguson, A. Vojtek, M. Wigler, and J. Field. 1991. CAP is a bifunctional component of the *Saccharomyces cerevisiae* adenylyl cyclase complex. *Mol. Cell. Biol.* **11:** 1248.

Girgsdies, O. 1982. Sterile mutants of *Schizosaccharomyces pombe:* Analysis by somatic hybridization. *Curr. Genet.* **6:** 223.

Gotoh, Y., E. Nishida, M. Shimanuki, T. Toda, Y. Imai, and M. Yamamoto. 1993. *Schizosaccharomyces pombe* Spk1 is a tyrosine-phosphorylated protein functionally related to *Xenopus* mitogen-activated protein kinase. *Mol. Cell. Biol.* **13:** 6427.

Grallert, B. and M. Sipiczki. 1989. Initiation of the second meiotic division in *Schizosaccharomyces pombe* shares common functions with that of mitosis. *Curr. Genet.* **15:** 231.

———. 1990. Dissociation of meiotic and mitotic roles of the fission yeast *cdc2* gene. *Mol. Gen. Genet.* **222:** 473.

———. 1991. Common genes and pathways in the regulation of the mitotic and meiotic cell cycles of *Schizosaccharomyces pombe*. *Curr. Genet.* **20:** 199.

Grimm, C., J. Bähler, and J. Kohli. 1994. *M26* recombinational hotspot and physical conversion tract analysis in the *ade6* gene of *Schizosaccharomyces pombe*. *Genetics* **135:** 41.

Grossenbacher-Grunder, A.M. and P. Thuriaux. 1981. Spontaneous and UV-induced recombination in radiation-sensitive mutants of *Schizosaccharomyces pombe*. *Mutat. Res.* **81:** 37.

Gutz, H. 1971. Site specific induction of gene conversion in *Schizosaccharomyces pombe*. *Genetics* **69:** 317.

Gysler-Junker, A., Z. Bodi, and J. Kohli. 1991. Isolation and characterization of *Schizosaccharomyces pombe* mutants affected in mitotic recombination. *Genetics* **128:**

495.

Hagan, I., J. Hayles, and P. Nurse. 1988. Cloning and sequencing of the cyclin-related *cdc13*$^+$ gene and a cytological study of its role in fission yeast mitosis. *J. Cell Sci.* **91:** 587.

Hakuno, F. and M. Yamamoto. 1996. The *Schizosaccharomyces pombe mra1* gene, which is required for cell growth and mating, can suppress the mating inefficiency caused by a deficit in the *ras1* activity. *Genes Cells* **1:** 303.

Hartwell, L.H. 1974. *Saccharomyces cerevisiae* cell cycle. *Bacteriol. Rev.* **38:** 164.

Herskowitz, I. 1995. MAP kinase pathways in yeast: For mating and more. *Cell* **80:** 187.

Hirata, A. and C. Shimoda. 1992. Electron microscopic examination of sporulation-deficient mutants of the fission yeast *Schizosaccharomyces pombe. Arch. Microbiol.* **158:** 249.

———. 1994. Structural modification of spindle pole bodies during meiosis II is essential for the normal formation of ascospores in *Schizosaccharomyces pombe*: Ultrastructural analysis of *spo* mutants. *Yeast* **10:** 173.

Hirata, A. and K. Tanaka. 1982. Nuclear behavior during conjugation and meiosis in the fission yeast *Schizosaccharomyces pombe. J. Gen. Appl. Microbiol.* **28:** 263.

Hoffman, C.S. and F. Winston. 1990. Isolation and characterization of mutants constitutive for expression of the *fbp1* gene of *Schizosaccharomyces pombe. Genetics* **124:** 807.

———. 1991. Glucose repression of transcription of the *Schizosaccharomyces pombe fbp1* gene occurs by a cAMP signaling pathway. *Genes Dev.* **5:** 561.

Hrycyna, C.A. and S. Clarke. 1990. Farnesyl cysteine C-terminal methyltransferase activity is dependent upon the *STE14* gene product in *Saccharomyces cerevisiae. Mol. Cell. Biol.* **10:** 5071.

Hrycyna, C.A., S.K. Sapperstein, S. Clarke, and S. Michaelis. 1991. The *Saccharomyces cerevisiae STE14* gene encodes a methyltransferase that mediates C-terminal methylation of **a**-factor and RAS proteins. *EMBO J.* **10:** 1699.

Hughes, D.A. and M. Yamamoto. 1993. *ras* and signal transduction during sexual differentiation in the fission yeast *Schizosaccharomyces pombe.* In *Signal transduction: Prokaryotic and simple eukaryotic systems* (ed. J. Kurjan and B.L. Taylor), p. 123. Academic Press, New York.

Hughes, D.A., A. Ashworth, and C.J. Marshall. 1993. Complementation of *byr1* in fission yeast by mammalian MAP kinase kinase requires coexpression of Raf kinase. *Nature* **364:** 349.

Hughes, D.A., Y. Fukui, and M. Yamamoto. 1990. Homologous activators of *ras* in fission and budding yeast. *Nature* **344:** 355.

Hughes, D.A., N. Yabana, and M. Yamamoto. 1994. Transcriptional regulation of a Ras nucleotide-exchange factor gene by extracellular signals in fission yeast. *J. Cell Sci.* **107:** 3635.

Imai, Y., J. Davey, M. Kawagishi-Kobayashi, and M. Yamamoto. 1997. Genes encoding farnesyl cysteine carboxyl methyltransferase in *Schizosaccharomyces pombe* and *Xenopus laevis. Mol. Cell. Biol.* **17:** 1543.

Iino, Y. and M. Yamamoto. 1985a. Mutants of *Schizosaccharomyces pombe* which sporulate in the haploid state. *Mol. Gen. Genet.* **198:** 416.

———. 1985b. Negative control for the initiation of meiosis in *Schizosaccharomyces pombe. Proc. Natl. Acad. Sci.* **82:** 2447.

Iino, Y., Y. Hiramine, and M. Yamamoto. 1995. The role of *cdc2* and other genes in meiosis in *Schizosaccharomyces pombe*. *Genetics* **140:** 1235.

Iino, Y., A. Sugimoto, and M. Yamamoto. 1991. *S. pombe pac1*$^+$, whose overexpression inhibits sexual development, encodes a ribonuclease III-like RNase. *EMBO J.* **10:** 221.

Imai, Y. and M. Yamamoto. 1992. *Schizosaccharomyces pombe sxa1*$^+$ and *sxa2*$^+$ encode putative proteases involved in the mating response. *Mol. Cell. Biol.* **12:** 1827.

———. 1994. The fission yeast mating pheromone P-factor: Its molecular structure, gene structure, and ability to induce gene expression and G_1 arrest in the mating partner. *Genes Dev.* **8:** 328.

Imai, Y., S. Miyake, D.A. Hughes, and M. Yamamoto. 1991. Identification of a GTPase-activating protein homolog in *Schizosaccharomyces pombe*. *Mol. Cell. Biol.* **11:** 3088.

Isshiki, T., N. Mochizuki, T. Maeda, and M. Yamamoto. 1992. Characterization of a fission yeast gene, *gpa2*, that encodes a Gα subunit involved in the monitoring of nutrition. *Genes Dev.* **6:** 2455.

Jin, M., M. Fujita, B.M. Culley, E. Apolinario, M. Yamamoto, K. Maundrell, and C.S. Hoffman. 1995. *sck1*, a high copy number suppressor of defects in the cAMP-dependent protein kinase pathway in fission yeast, encodes a protein homologous to the *Saccharomyces cerevisiae* SCH9 kinase. *Genetics* **140:** 457.

Johnson, D.I. and J.R. Pringle. 1990. Molecular characterization of *CDC42*, a *Saccharomyces cerevisiae* gene involved in the development of cell polarity. *J. Cell Biol.* **111:** 143.

Kanoh, J., A. Sugimoto, and M. Yamamoto. 1995. *Schizosaccharomyces pombe zfs1*$^+$ encoding a zinc-finger protein functions in the mating pheromone recognition pathway. *Mol. Biol. Cell* **6:** 1185.

Kanoh, J.,Y. Watanabe, M. Ohsugi, Y. Iino, and M. Yamamoto. 1996. *Schizosaccharomyces pombe gad7*$^+$ encodes a phosphoprotein with a bZIP domain, which is required for proper G_1 arrest and gene expression under nitrogen starvation. *Genes Cells* **1:** 391.

Kato, T., K. Okazaki, H. Murakami, S. Stetter, P.A. Fantes, and H. Okayama. 1996. Stress signal, mediated by Hog1-like MAP kinase, controls sexual development in fission yeast. *FEBS Lett.* **378:** 207.

Kawamukai, M., K. Ferguson, M. Wigler, and D. Young. 1991. Genetic and biochemical analysis of the adenylyl cyclase of *Schizosaccharomyces pombe*. *Cell Regul.* **2:** 155.

Kawamukai, M., J. Gerst, J. Field, M. Riggs, L. Rodgers, M. Wigler, and D. Young. 1992. Genetic and biochemical analysis of the adenylyl cyclase-associated protein, cap, in *Schizosaccharomyces pombe*. *Mol. Biol. Cell* **3:** 167.

Kawata, M., Y. Matsui, J. Kondo, T. Hishida, Y. Teranishi, and Y. Takai. 1988. A novel small molecular weight GTP-binding protein with the same putative effector domain as the *ras* proteins in bovine brain membranes. Purification, determination of primary structure, and characterization. *J. Biol. Chem.* **263:** 18965.

Kelly, M., J. Burke, M. Smith, A. Klar, and D. Beach. 1988. Four mating-type genes control sexual differentiation in the fission yeast. *EMBO J.* **7:** 1537.

Kim, D.-U., S.-K. Park, K.-S. Chung, M.-U. Choi, and H.-S. Soo. 1996. The G protein β subunit Gpb1 of *Schizosaccharomyces pombe* is a negative regulator of sexual development. *Mol. Gen. Genet.* **252:** 20.

King, J.S. and R.K. Mortimer. 1990. A polymerization model of chiasma interference and corresponding computer simulation. *Genetics* **126:** 1127.

King, S.M. and J.S. Hyams. 1982. Interdependence of cell cycle events in *Schizosaccharomyces pombe*. Terminal phenotypes of cell division cycle mutants arrested during DNA synthesis and nuclear division. *Protoplasma* **110:** 54.

Kishida, M. and C. Shimoda. 1986. Genetic mapping of eleven *spo* genes essential for ascospore formation in the fission yeast *Schizosaccharomyces pombe*. *Curr. Genet.* **10:** 443.

Kishida, M., T. Nagai, Y. Nakaseko, and C. Shimoda. 1994. Meiosis-dependent mRNA splicing of the fission yeast *Schizosaccharomyces pombe mes1*$^{+}$ gene. *Curr. Genet.* **25:** 497.

Kitamura, K. and C. Shimoda. 1991. The *Schizosaccharomyces pombe mam2* gene encodes a putative pheromone receptor which has a significant homology with the *Saccharomyces cerevisiae* Ste2 protein. *EMBO J.* **12:** 3743.

Kitamura, K., T. Nakagawa, and C. Shimoda. 1990. Novel sterile mutants of the fission yeast *Schizosaccharomyces pombe* which are defective in their response to starvation. *Curr. Genet.* **18:** 315.

Kitayama, H., Y. Sugimoto, T. Matsuzaki, Y. Ikawa, and M. Noda. 1989. A *ras*-related gene with transformation suppressor activity. *Cell* **56:** 77.

Kjaerulff, S., J. Davey, and O. Nielsen. 1994. Analysis of the structural genes encoding M-factor in the fission yeast *Schizosaccharomyces pombe*: Identification of a third gene, *mfm3*. *Mol. Cell. Biol.* **14:** 3895.

Klapholz, S. and R.E. Esposito. 1980. Recombination and chromosome segregation during the single division meiosis in *spo12-1* and *spo13-1* diploids. *Genetics* **96:** 589.

Kohli, J. and J. Bähler. 1994. Homologous recombination in fission yeast: Absence of crossover interference and synaptonemal complex. *Experientia* **50:** 295.

Kohno, H., K. Tanaka, A. Mino, M. Umikawa, H. Imamura, T. Fujiwara, Y. Fujita, K. Hotta, H. Qadota, T. Watanabe, Y. Ohya, and Y. Takai. 1996. Bni1p implicated in cytoskeletal control is a putative target of Rho1p small GTP binding protein in *Saccharomyces cerevisiae*. *EMBO J.* **15:** 6060.

Kosako, H., E. Nishida, and Y. Gotoh. 1993. cDNA cloning of MAP kinase kinase reveals kinase cascade pathways in yeasts to vertebrates. *EMBO J.* **12:** 787.

Kunitomo, H., A. Sugimoto, C.R.M. Wilkinson, and M. Yamamoto. 1995. *Schizosaccharomyces pombe pac2*$^{+}$ controls the onset of sexual development via a pathway independent of the cAMP cascade. *Curr. Genet.* **28:** 32.

Ladds, G., E.M. Rasmussen, T. Young, O. Nielsen, and J. Davey. 1996. The *sxa2*-dependent inactivation of the P-factor mating pheromone in the fission yeast *Schizosaccharomyces pombe*. *Mol. Microbiol.* **20:** 35.

Landis, C.A., S.B. Masters, A. Spada, A.M. Pace, H.R. Bourne, and L. Vallar. 1989. GTPase inhibiting mutations activate the α chain of G_s and stimulate adenylyl cyclase in human pituitary tumours. *Nature* **340:** 692.

Lange-Carter, C.A., C.M. Pleiman, A.M. Gardner, K.J. Blumer, and G.L. Johnson. 1993. A divergence in the MAP kinase regulatory network defined by MEK kinase and Raf. *Science* **260:** 315.

Leupold, U. 1987. Sex appeal in fission yeast. *Curr. Genet.* **12:** 543.

———. 1991. Characterization of a partially fertile *ras1*-like *ste10-UGA* nonsense mutant of fission yeast. *Curr. Genet.* **20:** 75.

Leupold, U. and M. Sipiczki. 1991. Sterile UGA nonsense mutants of fission yeast. *Curr. Genet.* **20:** 67.

Leupold, U., O. Nielsen, and R. Egel. 1989. Pheromone-induced meiosis in *P*-specific mutants of fission yeast. *Curr. Genet.* **15:** 403.

Li, P. and M. McLeod. 1996. Molecular mimicry in development: Identification of *ste11*$^{+}$ as a substrate and *mei3*$^{+}$ as a pseudosubstrate inhibitor of *ran1*$^{+}$ kinase. *Cell* **87:** 869.

Lin, Y. and G.R. Smith. 1994. Transient, meiosis-induced expression of the *rec6* and *rec12* genes of *Schizosaccharomyces pombe. Genetics* **136:** 769.

———. 1995. Molecular cloning of the meiosis-induced *rec10* gene of *Schizosaccharomyces pombe. Curr. Genet.* **27:** 440.

Lin, Y., K.L. Larson, R. Dorer, and G.R. Smith. 1992. Meiotically induced *rec7* and *rec8* genes of *Schizosaccharomyces pombe. Genetics* **132:** 75.

Loidl, J. 1990. The initiation of meiotic chromosome pairing: The cytological view. *Genome* **33:** 759.

Lowndes, N.F., C.J. McInerny, A.L. Johnson, P.A. Fantes, and L.H. Johnston. 1992. Control of DNA synthesis genes in fission yeast by the cell-cycle gene *cdc10*$^{+}$. *Nature* **355:** 449.

Lund, P.M., Y. Hasegawa, K. Kitamura, C. Shimoda, Y. Fukui, and M. Yamamoto. 1987. Mapping of the *ras1* gene of *Schizosaccharomyces pombe. Mol. Gen. Genet.* **209:** 627.

MacKay, V.L., S.K. Welch, M.Y. Insley, T.R. Manney, J. Holly, G.C. Saari, and M.L. Parker. 1988. The *Saccharomyces cerevisiae BAR1* gene encodes an exported protein with homology to pepsin. *Proc. Natl. Acad. Sci.* **85:** 55.

Maeda, T., N. Mochizuki, and M. Yamamoto. 1990. Adenylyl cyclase is dispensable for vegetative cell growth in the fission yeast *Schizosaccharomyces pombe. Proc. Natl. Acad. Sci.* **87:** 7814.

Maeda, T., Y. Watanabe, H. Kunitomo, and M. Yamamoto. 1994. Cloning of the *pka1* gene encoding the catalytic subunit of the cAMP-dependent protein kinase in *Schizosaccharomyces pombe. J. Biol. Chem.* **269:** 9632.

Maekawa, H., T. Nakagawa, Y. Uno, K. Kitamura, and C. Shimoda. 1994. The *ste13*$^{+}$ gene encoding a putative RNA helicase is essential for nitrogen starvation-induced G_1 arrest and initiation of sexual development in the fission yeast *Schizosaccharomyces pombe. Mol. Gen. Genet.* **244:** 456.

Manser, E., T. Leung, H. Salihuddin, Z. Zhao, and L. Lim. 1994. A brain serine/threonine protein kinase activated by Cdc42 and Rac1. *Nature* **367:** 40.

Marcus, S., A. Polverino, E. Chang, D. Robbins, M.H. Cobb, and M.H. Wigler. 1995. Shk1, a homolog of the *Saccharomyces cerevisiae* Ste20 and mammalian p65^{PAK} protein kinases, is a component of a Ras/Cdc42 signaling module in the fission yeast *Schizosaccharomyces pombe. Proc. Natl. Acad. Sci.* **92:** 6180.

Marr, R.S., L.C. Blair, and J. Thorner. 1990. *Saccharomyces cerevisiae STE14* gene is required for COOH-terminal methylation of **a**-factor mating pheromone. *J. Biol. Chem.* **265:** 20057.

Marsh, L., A.M. Neiman, and I. Herskowitz. 1991. Signal transduction during pheromone response in yeast. *Annu. Rev. Cell Biol.* **7:** 699.

Marshall, C.J. 1995. Specificity of receptor tyrosine kinase signaling: Transient versus sustained extracellular signal-regulated kinase activation. *Cell* **80:** 179.

Masuda, T., K. Kariya, M. Shinkai, T. Okada, and T. Kataoka. 1995. Protein kinase Byr2 is a target of Ras1 in the fission yeast *Schizosaccharomyces pombe. J. Biol. Chem.* **270:** 1979.

Masui, Y. and C.L. Markert. 1971. Cytoplasmic control of nuclear behavior during meiotic maturation of frog oocytes. *J. Exp. Zool.* **177:** 129.

McCormick, F. 1989. *ras* GTPase activating protein: Signal transmitter and signal terminator. *Cell* **56:** 5.

McLeod, M. and D. Beach. 1986. Homology between the *ran1*$^{+}$ gene of fission yeast and protein kinases. *EMBO J.* **5:** 3665.

———. 1988. A specific inhibitor of the *ran1*⁺ protein kinase regulates entry into meiosis in *Schizosaccharomyces pombe*. *Nature* **332:** 509.

McLeod, M., M. Stein, and D. Beach. 1987. The product of the *mei3*⁺ gene, expressed under control of the mating-type locus, induces meiosis and sporulation in fission yeast. *EMBO J.* **6:** 729.

Measday, V., L. Moore, J. Ogas, M. Tyers, and B. Andrews. 1994. The PCL2 (ORFD)-PHO85 cyclin-dependent kinase complex: A cell cycle regulator in yeast. *Science* **266:** 1391.

Michael, H. and H. Gutz. 1987. Sterility (*ste*) genes of *Schizosaccharomyces pombe*. *Yeast* **3:** 5.

Millar J.B.A., V. Buck, and M.G. Wilkinson. 1995. Pyp1 and Pyp2 PTPases dephosphorylate an osmosensing MAP kinase controlling cell size at division in fission yeast. *Genes Dev.* **9:** 2117.

Miller, P.J. and D.I. Johnson. 1994. Cdc42p GTPase is involved in controlling polarized cell growth in *Schizosaccharomyces pombe*. *Mol. Cell. Biol.* **14:** 1075.

Miyajima, I., M. Nakafuku, N. Nakayama, C. Brenner, A. Miyajima, K. Kaibuchi, K. Arai, Y. Kaziro, and K. Matsumoto. 1987. *GPA1*, a haploid-specific essential gene, encodes a yeast homolog of mammalian G protein which may be involved in mating factor signal transduction. *Cell* **50:** 1011.

Miyamoto, M., K. Tanaka, and H. Okayama. 1994. *res2*⁺, a new member of the *cdc10*⁺/*SWI4* family, controls the "start" of mitotic and meiotic cycles in fission yeast. *EMBO J.* **13:** 1873.

Miyamoto, S., Y. Ohya, Y. Ohsumi, and Y. Anraku. 1987. Nucleotide sequence of the *CLS4* (*CDC24*) gene of *Saccharomyces cerevisiae*. *Gene* **54:** 125.

Miyamoto, S., Y. Ohya, Y. Sano, S. Sakaguchi, H. Iida, and Y. Anraku. 1991. A *DBL*-homologous region of the yeast *CLS4*/*CDC24* gene product is important for Ca^{2+}-modulated bud assembly. *Biochem. Biophys. Res. Commun.* **181:** 604.

Mochizuki, N. and M. Yamamoto. 1992. Reduction in the intracellular cAMP level triggers initiation of sexual development in fission yeast. *Mol. Gen. Genet.* **233:** 17.

Molnar, M., J. Bähler, M. Sipiczki, and J. Kohli. 1995. The *rec8* gene of *Schizosaccharomyces pombe* is involved in linear element formation, chromosome pairing and sister-chromatid cohesion during meiosis. *Genetics* **141:** 61.

Moreno, S. and P. Nurse. 1994. Regulation of progression through the G_1 phase of the cell cycle by the *rum1*⁺ gene. *Nature* **367:** 236.

Munz, P. 1994. An analysis of interference in the fission yeast *Schizosaccharomyces pombe*. *Genetics* **137:** 701.

Munz, P., K. Wolf, J. Kohli, and U. Leupold. 1989. Genetic overview. In *Molecular biology of the fission yeast* (ed. A. Nasim et al.), p. 1. Academic Press, San Diego.

Nadin-Davis, S.A. and A. Nasim. 1988. A gene which encodes a predicted protein kinase can restore some functions of the *ras* gene in fission yeast. *EMBO J.* **7:** 985.

———. 1990. *Schizosaccharomyces pombe ras1* and *byr1* are functionally related genes of the *ste* family that affect starvation-induced transcription of mating-type genes. *Mol. Cell. Biol.* **10:** 549.

Nadin-Davis, S.A., A. Nasim, and D. Beach. 1986. Involvement of *ras* in sexual differentiation but not in growth control in fission yeast. *EMBO J.* **5:** 2963.

Nakaseko, Y., O. Niwa, and M. Yanagida. 1984. A meiotic mutant of the fission yeast *Schizosaccharomyces pombe* that produces mature asci containing two diploid spores. *J. Bacteriol.* **157:** 334.

Neiman, A.M. and I. Herskowitz. 1994. Reconstitution of a yeast protein kinase cascade *in vitro:* Activation of the yeast MEK homologue STE7 by STE11. *Proc. Natl. Acad. Sci.* **91:** 3398.

Neiman, A.M., B.J. Stevenson, H.-P. Xu, G.F. Sprague, Jr., I. Herskowitz, M. Wigler, and S. Marcus. 1993. Functional homology of protein kinases required for sexual differentiation in *Schizosaccharomyces pombe* and *Saccharomyces cerevisiae* suggests a conserved signal transduction module in eukaryotic organisms. *Mol. Biol. Cell* **4:** 107.

Nielson, O. and R. Egel. 1990. The *pat1* protein kinase controls transcription of the mating-type genes in fission yest. *EMBO J.* **9:** 1401.

Nielsen, O., J. Davey, and R. Egel. 1992. The *ras1* function of *Schizosaccharomyces pombe* mediates pheromone-induced transcription. *EMBO J.* **11:** 1391.

Nielsen, O., T. Friis, and S. Kjaerulff. 1996. The *Schizosaccharomyces pombe map1* gene encodes an SRF/MCM1-related protein required for *P*-cell specific gene expression. *Mol. Gen. Genet.* **253:** 387.

Nishida, E. and Y. Gotoh. 1993. The MAP kinase cascade is essential for diverse signal transduction pathways. *Trends Biochem. Sci.* **18:** 128.

Niwa, O. and M. Yanagida. 1988. Universal and essential role of MPF/*cdc2*$^+$. *Nature* **336:** 430.

Nocero, M., T. Isshiki, M. Yamamoto, and C.S. Hoffman. 1994. Glucose repression of *fbp1* transcription in *Schizosaccharomyces pombe* is partially regulated by adenylate cyclase activation by a G protein α subunit encoded by *gpa2* (*git8*). *Genetics* **138:** 39.

Nurse, P. 1985. Mutants of the fission yeast *Schizosaccharomyces pombe* which alter the shift between cell proliferation and sporulation. *Mol. Gen. Genet.* **198:** 497.

———. 1990. Universal control mechanism regulating onset of M-phase. *Nature* **344:** 503.

Nurse, P. and Y. Bissett. 1981. Gene required in G_1 for commitment to cell cycle and G_2 for control of mitosis in fission yeast. *Nature* **292:** 558.

Obara, T., M. Nakafuku, M. Yamamoto, and Y. Kaziro. 1991. Isolation and characterization of a gene encoding a G-protein α subunit from *Schizosaccharomyces pombe:* Involvement in mating and sporulation pathways. *Proc. Natl. Acad. Sci.* **88:** 5877.

Obara-Ishihara, T. and H. Okayama. 1994. A B-type cyclin negatively regulates conjugation via interacting with cell cycle "start" genes in fission yeast. *EMBO J.* **13:** 1863.

Okazaki, N., K. Okazaki, K. Tanaka, and H. Okayama. 1991. The *ste4*$^+$ gene, essential for sexual differentiation of *Schizosaccharomyces pombe*, encodes a protein with a leucine zipper motif. *Nucleic Acids Res.* **19:** 7043.

Olson, L.W., U. Edén, M. Egel-Mitani, and R. Egel. 1978. Asynaptic meiosis in fission yeast? *Hereditas* **89:** 189.

Ottilie, S., P.J. Miller, D.I. Johnson, C.L. Creasy, M.A. Sells, S. Bagrodia, S.L. Forsburg, and J. Chernoff. 1995. Fission yeast *pak1*$^+$ encodes a protein kinase that interacts with Cdc42p and is involved in the control of cell polarity and mating. *EMBO J.* **14:** 5908.

Pardee, A.B., R. Dubrow, J.L. Hamlin, and R.F. Kletzien. 1978. Animal cell cycle. *Annu. Rev. Biochem.* **47:** 715.

Petersen, J., D. Weilguny, R. Egel, and O. Nielsen. 1995. Characterization of *fus1* of *Schizosaccharomyces pombe:* A developmentally controlled function needed for conjugation. *Mol. Cell. Biol.* **15:** 3697.

Pizon, V., P. Chardin, I. Lerosey, B. Olofsson, and A. Tavitian. 1988. Human cDNAs rap1 and rap2 homologous to the *Drosophila* gene Dras 3 encode proteins closely related to *ras* in the "effector" region. *Oncogene* **3:** 201.

Ponticelli, A.S. and G.R. Smith. 1989. Meiotic recombination-deficient mutants of *Schizosaccharomyces pombe. Genetics* **123:** 45.

Ponticelli, A.S., E.P. Sena, and G.R. Smith. 1988. Genetic and physical analysis of the *M26* recombination hotspot of *Schizosaccharomyces pombe. Genetics* **119:** 491.

Reed, S.I. 1992. The role of p34 kinases in the G_1 to S-phase transition. *Annu. Rev. Cell Biol.* **8:** 529.

Robinow, C.F. 1977. The number of chromosomes in *Schizosaccharomyces pombe:* Light microscopy of stained preparations. *Genetics* **87:** 491.

Rowley, R., S. Subramani, and P.G. Young. 1992. Checkpoint controls in *Schizosaccharomyces pombe: rad1. EMBO J.* **11:** 1335.

Ruggieri, R., A. Bender, Y. Matsui, S. Powers, Y. Takai, J.R. Pringle, and K. Matsumoto. 1992. *RSR1*, a *ras*-like gene homologous to K*rev*-1 (*smg21A/rap1A):* Role in the development of cell polarity and interactions with the Ras pathway in *Saccharomyces cerevisiae. Mol. Cell. Biol.* **12:** 758.

Sapperstein, S., C. Berkower, and S. Michaelis. 1994. Nucleotide sequence of the yeast *STE14* gene, which encodes farnesylcysteine carboxyl methyltransferase, and demonstration of its essential role in **a**-factor export. *Mol. Cell. Biol.* **14:** 1438.

Scherthan, H., J. Bähler, and J. Kohli. 1994. Dynamics of chromosome organization and pairing during meiotic prophase in fission yeast. *J. Cell Biol.* **127:** 273.

Schmidt, H., P. Kapitza, and H. Gutz. 1987. Switching genes in *Schizosaccharomyces pombe:* Their influence on cell viability and recombination. *Curr. Genet.* **11:** 303.

Schuchert, P. and J. Kohli. 1988. The *ade6-M26* mutation of *Schizosaccharomyces pombe* increases the frequency of crossing over. *Genetics* **119:** 507.

Schuchert, P., M. Langsford, E. Käslin, and J. Kohli. 1991. A specific DNA sequence is required for high frequency of recombination in the *ade6* gene of fission yeast. *EMBO J.* **10:** 2157.

Seger, R., D. Seger, F.J. Lozeman, N.G. Ahn, L.M. Graves, J.S. Campbell, L. Ericsson, M. Harrylock, A.M. Jensen, and E.G. Krebs. 1992. Human T-cell mitogen-activated protein kinase kinases are related to yeast signal transduction kinases. *J. Biol. Chem.* **267:** 25628.

Sherr, C.J. 1994. G_1 phase progression: Cycling on cue. *Cell* **79:** 551.

Shimanuki, M., Y. Saka, M. Yanagida, and T. Toda. 1995. A novel essential fission yeast gene *pad1*$^+$ positively regulates *pap1*$^+$-dependent transcription and is implicated in the maintenance of chromosome structure. *J. Cell Sci.* **108:** 569.

Shimoda, C., A. Hirata, M. Kishida, T. Hashida, and K. Tanaka. 1985. Characterization of meiosis-deficient mutants by electron microscopy and mapping of four essential genes in the fission yeast *Schizosaccharomyces pombe. Mol. Gen. Genet.* **200:** 252.

Shimoda, C., M. Uehira, M. Kishida, H. Fujioka, Y. Iino, Y. Watanabe, and M. Yamamoto. 1987. Cloning and analysis of transcription of the *mei2* gene responsible for initiation of meiosis in the fission yeast *Schizosaccharomyces pombe. J. Bacteriol.* **169:** 93.

Shinohara, A., H. Ogawa, Y. Matsuda, N. Ushio, K. Ikeo, and T. Ogawa. 1993. Cloning of human, mouse and fission yeast recombination genes homologous to *RAD51* and *recA*. *Nat. Genet.* **4:** 239.

Shiozaki, K. and P. Russell. 1995. Cell-cycle control linked to extracellular environment by MAP kinase pathway in fission yeast. *Nature* **378:** 739.

———. 1996. Conjugation, meiosis and the osmotic stress response are regulated by Spc1 kinase via Atf1 transcription factor in fission yeast. *Genes Dev.* **10:** 2276.

Shuster, E.O. and B. Byers. 1989. Pachytene arrest and other meiotic effects of the Start mutations in *Saccharomyces cerevisiae*. *Genetics* **123:** 29.

Simon, M.-N., C. De Virgilio, B. Souza, J.R. Pringle, A. Abo, and S.I. Reed. 1995. Role for the Rho-family GTPase Cdc42 in yeast mating-pheromone signal pathway. *Nature* **376:** 702.

Sinclair, A.H., P. Berta, M.S. Palmer, J.R. Hawkins, B.L. Griffiths, M.J. Smith, J.W. Foster, A.-M. Frischauf, R. Lovell-Badge, and P.N. Goodfellow. 1990. A gene from the human sex-determining region encodes a protein with homology to a conserved DNA-binding motif. *Nature* **346:** 240.

Sipiczki, M. 1988. The role of sterility genes (*ste* and *aff*) in the initiation of sexual development in *Schizosaccharomyces pombe*. *Mol. Gen. Genet.* **213:** 529.

Snow, R. 1979. Maximum likelihood estimation of linkage and interference from tetrad data. *Genetics* **92:** 231.

Song, K., K.E. Mach, C.Y. Chen, T. Reynolds, and C.F. Albright. 1996. A novel suppressor of *ras1* in fission yeast, *byr4*, is a dosage-dependent inhibitor of cytokinesis. *J. Cell Biol.* **133:** 1307.

Stern, B. and P. Nurse. 1997. Fission yeast pheromone blocks S-phase by inhibiting the G_1 cyclin B-p34^{cdc2} kinase. *EMBO J.* **16:** 534.

Styrkársdóttir, U., R. Egel, and O. Nielsen. 1992. Functional conservation between *Schizosaccharomyces pombe ste8* and *Saccharomyces cerevisiae STE11* protein kinases in yeast signal transduction. *Mol. Gen. Genet.* **235:** 122.

Sugimoto, A., Y. Iino, T. Maeda, Y. Watanabe, and M. Yamamoto. 1991. *Schizosaccharomyces pombe ste11*$^+$ encodes a transcription factor with an HMG motif that is a critical regulator of sexual development. *Genes Dev.* **5:** 1990.

Sugiyama, A., K. Tanaka, K. Okazaki, H. Nojima, and H. Okayama. 1994. A zinc finger protein controls the onset of premeiotic DNA synthesis of fission yeast in a Mei2-independent cascade. *EMBO J.* **13:** 1881.

Svoboda, A., J. Bähler, and J. Kohli. 1995. Microtubule-driven nuclear movements and linear elements as meiosis-specific characteristics of the fission yeasts *Schizosaccharomyces versatilis* and *Schizosaccharomyces pombe. Chromosoma* **104:** 203.

Sym, M. and G.S. Roeder. 1994. Crossover interference is abolished in the absence of a synaptonemal complex protein. *Cell* **79:** 283.

Takeda, T. and M. Yamamoto. 1987. Analysis and *in vivo* disruption of the gene coding for calmodulin in *Schizosaccharomyces pombe*. *Proc. Natl. Acad. Sci.* **84:** 3580.

Takeda, T., Y. Imai, and M. Yamamoto. 1989. Substitution at position 116 of *Schizosaccharomyces pombe* calmodulin decreases its stability under nitrogen starvation and results in a sporulation-deficient phenotype. *Proc. Natl. Acad. Sci.* **86:** 9737.

Takeda, T., T. Toda, K.-I. Kominami, A. Kohnosu, M. Yanagida, and N. Jones. 1995. *Schizosaccharomyces pombe atf1*$^+$ encodes a transcription factor required for sexual development and entry into stationary phase. *EMBO J.* **14:** 6193.

Tanaka, K. and A. Hirata. 1982. Ascospore development in the fission yeasts *Schizosaccharomyces pombe* and *S. japonicus*. *J. Cell Sci.* **56:** 263.

Tanaka, K., J. Davey, Y. Imai, and M. Yamamoto. 1993. *Schizosaccharomyces pombe map3*$^+$ encodes the putative M-factor receptor. *Mol. Cell. Biol.* **13:** 80.

Tanaka, K., K. Okazaki, N. Okazaki, T. Ueda, A. Sugiyama, H. Nojima, and H. Okayama. 1992. A new cdc gene required for S phase entry of *Schizosaccharomyces pombe* encodes a protein similar to the *cdc10*$^+$ and *SWI4* gene products. *EMBO J.* **11:** 4923.

Thuriaux, P., M. Sipiczki, and P.A. Fantes. 1980. Genetical analysis of a sterile mutant by protoplast fusion in the fission yeast *Schizosaccharomyces pombe. J. Gen. Microbiol.* **116:** 525.

Toda, T., M. Shimanuki, and M. Yanagida. 1991. Fission yeast genes that confer resistance to staurosporine encode an AP-1-like transcription factor and a protein kinase related to the mammalian ERK1/MAP2 and budding yeast *FUS3* and *KSS1* kinases. *Genes Dev.* **5:** 60.

Toda, T., S. Cameron, P. Sass, and M. Wigler. 1988. *SCH9*, a gene of *Saccharomyces cerevisiae* that encodes a protein distinct from, but functionally and structurally related to, cAMP-dependent protein kinase catalytic subunits. *Genes Dev.* **2:** 517.

Toda, T., S. Cameron, P. Sass, M. Zoller, and M. Wigler. 1987. Three different genes in *S. cerevisiae* encode the catalytic subunits of the cAMP-dependent protein kinase. *Cell* **50:** 277.

Van Aelst, L., M. Barr, S. Marcus, A. Polverino, and M. Wigler. 1993. Complex formation between RAS and RAF and other protein kinases. *Proc. Natl. Acad. Sci.* **90:** 6213.

Wahls, W.P. and G.R. Smith. 1994. A heteromeric protein that binds to a meiotic homologous recombination hot spot: Correlation of binding and hot spot activity. *Genes Dev.* **8:** 1693.

Wang, Y., M. Boguski, M. Riggs, L. Rodgers, and M. Wigler. 1991a. *Sar1*, a gene from *Schizosaccharomyces pombe* encoding a protein that regulates *ras1*. *Cell Regul.* **2:** 453.

Wang, Y., H.-P. Xu, M. Riggs, L. Rodgers, and M. Wigler. 1991b. *byr2*, a *Schizosaccharomyces pombe* gene encoding a protein kinase capable of partial suppression of the *ras1* mutant phenotype. *Mol. Cell. Biol.* **11:** 3554.

Warbrick, E. and P.A. Fantes. 1991. The wis1 protein kinase is a dosage-dependent regulator of mitosis in *Schizosaccharomyces pombe. EMBO J.* **10:** 4291.

Watanabe, Y. and M. Yamamoto. 1994. *S. pombe mei2*$^+$ encodes an RNA-binding protein essential for premeiotic DNA synthesis and meiosis I, which cooperates with a novel RNA species meiRNA. *Cell* **78:** 487.

———. 1996. *Schizosaccharomyces pombe pcr1*$^+$ encodes an ATF/CRE-binding protein involved in the regulation of gene expression for sexual development. *Mol. Cell. Biol.* **16:** 704.

Watanabe, Y., Y. Iino, K. Furuhata, C. Shimoda, and M. Yamamoto. 1988. The *S. pombe mei2* gene encoding a crucial molecule for commitment to meiosis is under the regulation of cAMP. *EMBO J.* **7:** 761.

Watanabe, Y., S. Shinozaki-Yabana, Y. Chikashige, Y. Hiraoka, and M. Yamamoto. 1997. Phosphorylation of RNA-binding protein controls cell cycle switch from mitotic to meiotic in fission yeast. *Nature* **386:** 187.

Whiteway, M., L. Hougan, D. Dignard, D.Y. Thomas, L. Bell, G.C. Saari, F.J. Grant, P. O'Hara, and V.L. MacKay. 1989. The *STE4* and *STE18* genes of yeast encode potential β and γ subunits of the mating factor receptor-coupled G protein. *Cell* **56:** 467.

Wilkinson, M.G., M. Samuels, T. Takeda, W.M. Toone, J.C. Shieh, T. Toda, J.B.A. Millar, and N. Jones. 1996. The Atf1 transcription factor is a target for the Sty1 stress-activated MAP kinase pathway in fission yeast. *Genes Dev.* **10:** 2289.

Willer, M., L. Hoffmann, U. Styrkársdóttir, R. Egel, J. Davey, and O. Nielsen. 1995. Two-step activation of meiosis by the *mat1* locus in *Schizosaccharomyces pombe. Mol. Cell. Biol.* **15:** 4964.

Wu, J., J.K. Harrison, L.A. Vincent, C. Haystead, T.A.J. Haystead, H. Michel, D.F. Hunt, K.R. Lynch, and T.W. Sturgill. 1993. Molecular structure of a protein-tyrosine/threonine kinase activating p42 mitogen-activated protein (MAP) kinase: MAP kinase kinase. *Proc. Natl. Acad. Sci.* **90:** 173.

Wu, S.-Y. and M. McLeod. 1995. The *sak1*$^{+}$ gene of *Schizosaccharomyces pombe* encodes an RFX family DNA-binding protein that positively regulates cyclic AMP-dependent protein kinase-mediated exit from the mitotic cell cycle. *Mol. Cell. Biol.* **15:** 1479.

Xu, H.-P., M. White, S. Marcus, and M. Wigler. 1994. Concerted action of RAS and G proteins in the sexual response pathways of *Schizosaccharomyces pombe. Mol. Cell. Biol.* **14:** 50.

Xu, H.-P., T. Rajavashisth, N. Grewal, V. Jung, M. Riggs, L. Rodgers, and M. Wigler. 1992. A gene encoding a protein with seven zinc finger domains acts on the sexual differentiation pathways of *Schizosaccharomyces pombe. Mol. Biol. Cell* **3:** 721.

Yabana, N. and M. Yamamoto. 1996. *Schizosaccharomyces pombe map1*$^{+}$ encodes a MADS-box-family protein required for cell-type-specific gene expression. *Mol. Cell. Biol.* **16:** 3420.

Yamamoto, M. 1996. The molecular control mechanisms of meiosis in fission yeast. *Trends Biochem. Sci.* **21:** 18.

Yamawaki-Kataoka, Y., T. Tamaoki, H.-R. Choe, H. Tanaka, and T. Kataoka. 1989. Adenylate cyclases in yeast: A comparison of the genes from *Schizosaccharomyces pombe* and *Saccharomyces cerevisiae. Proc. Natl. Acad. Sci.* **86:** 5693.

Yoo, B.Y., G.B. Calleja, and B.F. Johnson. 1973. Ultrastructural changes of the fission yeast (*Schizosaccharomyces pombe*) during ascospore formation. *Arch. Microbiol.* **91:** 1.

Young, D., M. Riggs, J. Field, A. Vojtek, D. Broek, and M. Wigler. 1989. The adenylyl cyclase gene from *Schizosaccharomyces pombe. Proc. Natl. Acad. Sci.* **86:** 7989.

Zhao, Z.-S., T. Leung, E. Manser, and L. Lim. 1995. Pheromone signalling in *Saccharomyces cereivisiae* requires the small GTP-binding proten Cdc42p and its activator *CDC24. Mol. Cell. Biol.* **15:** 5246.

Zheng, Y., A. Bender, and R.A. Cerione. 1995. Interactions among proteins involved in bud-site selection and bud-site assembly in *Saccharomyces cerevisiae. J. Biol. Chem.* **270:** 626.

Zheng, Y., R. Cerione, and A. Bender. 1994. Control of the yeast bud-site assembly GTPase Cdc42: Catalysis of guanine nucleotide exchange by Cdc24 and stimulation of GTPase activity by Bem3. *J. Biol. Chem.* **269:** 2369.

Zhu, Y., T. Takeda, K. Nasmyth, and N. Jones. 1994. *pct1*$^{+}$, which encodes a new DNA-binding partner of p85^{cdc10}, is required for meiosis in the fission yeast *Schizosaccharomyces pombe. Genes Dev.* **8:** 885.

Index